Hormones and Reproduction of Vertebrates

Hormones and Reproduction of Vertebrates, Volume 1—Fishes
Hormones and Reproduction of Vertebrates, Volume 2—Amphibians
Hormones and Reproduction of Vertebrates, Volume 3—Reptiles
Hormones and Reproduction of Vertebrates, Volume 4—Birds
Hormones and Reproduction of Vertebrates, Volume 5—Mammals

Hormones and Reproduction of Vertebrates

Volume 5: Mammals

David O. Norris
Department of Integrative Physiology
University of Colorado
Boulder, Colorado

Kristin H. Lopez
Department of Integrative Physiology
University of Colorado
Boulder, Colorado

AMSTERDAM • BOSTON • HEIDELBERG • LONDON • NEW YORK
OXFORD • PARIS • SAN DIEGO • SAN FRANCISCO
SINGAPORE • SYDNEY • TOKYO

Academic Press is an imprint of Elsevier

Academic Press is an imprint of Elsevier
32 Jamestown Road, London NW1 7BY, UK
30 Corporate Drive, Suite 400, Burlington, MA 01803, USA
525 B Street, Suite 1800, San Diego, CA 92101-4495, USA

First edition 2011

Cover images
Front cover image: Sea lion. Photo courtesy of Russell Purdy.
Back cover image: The brown antechinus (*Antechinus stuartii*) is a marsupial mouse that exihibits a semelparous reproductive pattern where all males die after breeding. Photo courtesy of Dr. Bronwyn McCallan. Photographer Fritz Geiser.

British Library Cataloguing-in-Publication Data
A catalogue record for this book is available from the British Library

Library of Congress Cataloging-in-Publication Data
A catalog record for this book is available from the Library of Congress

ISBN: 978-0-12-374932-1 (Set)
ISBN: 978-0-12-375009-9 (Volume 1)
ISBN: 978-0-12-374931-4 (Volume 2)
ISBN: 978-0-12-374930-7 (Volume 3)
ISBN: 978-0-12-374929-1 (Volume 4)
ISBN: 978-0-12-374928-4 (Volume 5)

For information on all Academic Press publications
visit our website at elsevierdirect.com

Typeset by TNQ Books and Journals Pvt Ltd.
www.tnq.co.in

Printed and bound by CPI Group (UK) Ltd, Croydon, CR0 4YY
Transferred to Digital Printing, 2013

Dedication

Richard Evan Jones

This series of five volumes on the hormones and reproduction of vertebrates is appropriately dedicated to our friend and colleague of many years, Professor Emeritus Richard Evan Jones, who inspired us to undertake this project. Dick spent his professional life as a truly comparative reproductive endocrinologist who published many papers on hormones and reproduction in fishes, amphibians, reptiles, birds, and mammals. Additionally, he published a number of important books including *The Ovary* (Jones, 1975, Plenum Press), *Hormones and Reproduction in Fishes, Amphibians, and Reptiles* (Norris and Jones, 1987, Plenum Press), and a textbook, *Human Reproductive Biology* (Jones & Lopez, 3rd edition 2006, Academic Press). Throughout his productive career he consistently stressed the importance of an evolutionary perspective to understanding reproduction and reproductive endocrinology. His enthusiasm for these subjects inspired all with whom he interacted, especially the many graduate students he fostered, including a number of those who have contributed to these volumes.

Contents

Series Preface xv
Volume Preface xvii
Contributors xix

1. Sexual Differentiation of the Mammalian Brain 1

1. Introduction 1
2. Historical Overview 2
   2.1. Phoenix, Goy, Gerall, and Young: 1959 2
   2.2. Modern Views—Where Science and Society Collide 2
3. The Organizational/Activational Hypothesis of Sexual Differentiation 3
   3.1. Basic Principles of Hormone Action 3
   3.2. Sensitive Periods 5
   3.3. The Role of Aromatization 5
   3.4. Masculinization, Feminization, and Defeminization 6
   3.5. Brain vs. Behavior 6
   3.6. Challenges to the Dogma 8
4. Advances in Mechanisms 10
   4.1. Cell Birth and Cell Death 10
   4.2. Old Players with New Roles 11
   4.3. New Players with Old Roles 12
   4.4. The Need for and Consequences of Cell-to-cell Communication 14
5. Summary and Future Directions 16
References 17

2. Neuroendocrine Control of Gonadotropins in Mammals 25

1. Introduction to Gonadotropins 25
   1.1. The Discovery of Hypothalamic Releasing Factors 25
   1.2. The Organization of the Mammalian Pituitary Gland 26
2. Gonadotropin-releasing Hormone (GnRH) 27
   2.1. The Development of Gonadotropin-releasing Hormone (GnRH) Neurons: Cell Fate Specification 28
   2.2. Development of Gonadotropin-releasing Hormone (GnRH) Neurons: Neuronal Migration 29
   2.3. Development of Gonadotropin-releasing Hormone (GnRH) Neurons: Axonal Targeting 30
   2.4. Gonadotropin-releasing Hormone (GnRH) Neurons are Anatomically Unique 30
   2.5. Gonadotropin-releasing Hormone (GnRH) is Structurally Conserved 31
   2.6. Gonadotropin-releasing Hormone (GnRH) Release is Pulsatile 33
3. Gonadotropin-releasing Hormone (GnRH) Receptors 34
   3.1. Gonadotropin-releasing Hormone (GnRH) Receptor Regulation in Gonadotropes 35
   3.2. GnRH-I Receptor Signaling Pathways 36
   3.3. Gonadotropin-releasing Hormone (GnRH) Pulsatility Differentially Regulates Gonadotropins (GTHs) 36
4. Conclusions 37
References 38

3. Endocrine and Paracrine Regulation of Mammalian Spermatogenesis 45

1. Overview of Spermatogenesis 45
2. Regulation of Stem Spermatogonia Via Growth Factors and Paracrine Interactions 46
   2.1. Characteristics of Stem Spermatogonia 46
   2.2. Replication and Differentiation of Stem Spermatogonia in their Niche 46
   2.3. Glial-derived Neurotrophic Factor (GDNF) and Other Growth Factors in the Stem Spermatogonia Niche 47
   2.4. Regulation of Sertoli Cell Function by Differentiated Spermatogenetic Cells and the Implications of this Regulation for the Function of the Stem Spermatogonial Niche 48
3. Endocrine Regulation of Spermatogenesis 49
   3.1. Leydig Cells and Testosterone (T) Production 49

3.2. Mechanism(s) by which Testosterone (T) Acts to Drive Spermatogenesis: The Role of Androgen Receptors (ARs) 49
3.3. Androgen Receptor (AR) Expression is Stage-specific and Specific Stages of Spermatogenesis are Androgen-dependent 50
3.4. Testosterone (T) and Spermatogenesis 51
3.5. Male Hormonal Contraception 52
3.6. Male Infertility 52
4. Future Directions 52
Acknowledgments 53
References 53

4. The Endocrinology of the Mammalian Ovary 59

1. Introduction 59
2. Embryogenesis of the Ovary 59
2.1. Oogenesis 60
2.2. Folliculogenesis 60
2.2.1. Ovarian steroidogenesis 62
2.3. Puberty 63
3. The Ovarian Cycle 63
3.1. The Follicular Phase of the Ovarian Cycle 64
3.2. Ovulation 64
3.3. The Luteal Phase of the Ovarian Cycle 65
4. Ovarian Lifespan and Reproductive Aging 66
5. Ovarian Events and Reproductive Cycles of Selected Mammals 66
5.1. The Four-day-cycling Rat 67
5.2. The Ewe 67
5.3. Women 68
5.4. The Elephant 69
6. Future Research Needs 69
References 70

5. Hormones and Pregnancy in Eutherian Mammals 73

1. Introduction 73
1.1. Corpus Luteum (CL) Formation and Regression 73
1.2. Pregnancy Recognition Signals for Maintenance of the Corpus Luteum (CL) 74
1.3. Luteinizing Hormone (LH), Chorionic Gonadotropin (CG), and their Receptor (LHCGR) 74
1.4. Progesterone (P_4) and Progesterone Receptors (PRs) 75
1.5. Relaxin (RLX) 76
1.6. Chorionic Somatomammotropin Hormone (CSH1) 77
2. The Menstrual Cycle, Luteolysis, and Pregnancy in Primates (*Homo Sapiens* and *Macaca Mulatta*) 77
2.1. The Menstrual Cycle 77
2.2. Pregnancy 77
3. Estrous Cycles, Luteolysis, Pseudopregnancy, Delayed Implantation, and Pregnancy in Subprimate Mammals 81
3.1. Rodents: Rat (*Rattus norvegicus*), Mouse (*Mus musculus*), Guinea Pig (*Cavia porcellus*) 81
3.1.1. The estrous cycle and luteolysis 81
3.1.2. Pseudopregnancy and pregnancy 81
3.1.3. Rodent decidual prolactin (PRL)-like proteins and placental lactogens 82
3.2. Ruminants 82
3.2.1. The estrous cycle and luteolysis 82
3.2.2. Pregnancy 83
3.2.3. Cows 84
3.2.4. Goats 84
3.2.5. Sheep 84
3.3. Swine (*Sus domestica*) 84
3.3.1. The estrous cycle and luteolysis 84
3.3.2. Pregnancy 85
3.4. Horses (*Equus ferus caballus*) 87
3.4.1. The estrous cycle and luteolysis 87
3.4.2. Pregnancy 87
3.5. Rabbits (*Oryctolagus cuniculus*) 87
3.5.1. The ovarian cycle and estrus 87
3.5.2. Pregnancy 88
3.6. Domestic Cats (*Felis catus*) 89
3.6.1. The estrous cycle 89
3.6.2. Pregnancy 89
3.7. Domestic Dogs (*Canis lupus familiaris*) 90
3.7.1. The ovarian cycle and estrus 90
3.7.2. Pregnancy 90
4. Summary and Conclusions 90
5. Future Research 91
References 92

6. The Comparative Physiology of Parturition in Mammals: Hormones and Parturition in Mammals 95

1. Introduction 95
2. What Determines Gestation Length? 97
3. How are Fetal Maturation and Parturition Synchronized? 100
4. How Does the Fetus Signal the Initiation of Labor? 101
4.1. A More Complicated Case: The Corpus Luteum-dependent Species 101

4.2. The Role of the Fetal Hypothalamus–pituitary–adrenal (HPA) Axis in the Initiation of Parturition 102
4.3. Signals from the Fetal Lungs 104
5. How are the Uterotonic Mechanisms Activated? 104
5.1. The Sex Steroids: Progesterone (P_4) and Estrogens 104
5.2. Inflammatory Mediators: Prostaglandins (PGs) and Cytokines 108
5.3. Oxytocin (OXY) 110
5.4. Electrical Conductivity of the Myometrium 110
6. Conclusions 111
References 112

7. Stress and Reproduction in Mammals 117

1. Introduction 117
2. Effects of Food Restriction 119
3. Effects of Stress on Immunity and Reproduction 121
4. Effects of Other Stressors 124
4.1. Females 124
4.2. Males 125
5. Role of HPA Activation (Glucocorticoids and CRH) 126
6. Implications for Wildlife 128
7. Conclusions 129
Acknowledgements 130
References 130

8. Behavioral Neuroendocrinology of Reproduction in Mammals 139

1. Introduction 139
2. Description of Male and Female Sexual Behavior (MSB/FSB) 139
3. The Role of Gonadal Steroids in Sexual Behavior 141
4. Neural Circuitry Regulating Sexual Behavior 142
4.1. Chemosensory Inputs to Male Sexual Behavior (MSB) 142
4.2. Neural Circuitry Regulating Lordosis 142
5. Steroid Receptors and Sexual Behavior 144
5.1. Males 144
5.2. Females 145
6. Ligand-independent Activation of Steroid Receptors 146
7. Relevant Neurotransmitters and Neuropeptides 147
7.1. Females 147
7.1.1. Norepinephrine (NE) 148
7.1.2. Acetylcholine (ACh) 149
7.1.3. Gonadotropin-releasing hormone (GnRH) 149
7.1.4. Opiates 149
7.2. Neurotransmitters that Act in Both Sexes 149
7.2.1. Dopamine (DA) 150
7.2.2. Serotonin (5-HT) 150
7.2.3. Oxytocin (OXY) 151
7.2.4. α-Melanocyte-stimulating hormone (MSH, or melanotropin) 152
8. Gonad-independent Sexual Behavior 152
9. Genetics of Sexual Behavior 153
10. Neuroendocrinology of Human Sexual Behavior 156
10.1. Men 156
10.2. Women 157
11. Summary 158
References 159

9. Pheromones and Reproduction in Mammals 175

1. Introduction 175
1.1. Scope 175
1.2. Pheromones 175
1.3. Chemosensory Systems 176
1.4. Organization 176
2. Social Odors and Physiology 176
2.1. Responses by Females 176
2.1.1. Puberty acceleration 176
2.1.2. Puberty delay 177
2.1.3. Inhibition of ovarian cyclicity 178
2.1.4. Facilitation of ovarian cyclicity 178
2.1.5. Ovarian synchrony 179
2.1.6. Pregnancy blockage 179
2.2. Responses by Males 180
2.2.1. Reproductive development 180
2.2.2. Sperm allocation 180
2.2.3. Hormone release 180
3. Social Odors and Behavior 181
3.1. Responses by Females 181
3.1.1. Attraction and investigation 181
3.1.2. Scent marking 182
3.2. Responses by Males 182

3.2.1. Attraction and investigation 182
3.2.2. Vocalizations 183
3.3. Copulatory Behavior 184
4. Conclusions 184
References 186

10. Reproductive Endocrinology of Prototherians and Metatherians 195

1. The Prototherians and Metatherians: The 'Other' Mammals 195
1.1. Prototherians: Monotremes 195
1.2. Metatherians: Marsupials 196
2. Endocrine Control of Sexual Differentiation in Monotremes and Marsupials 196
2.1. Monotremes 196
2.2. Marsupials 196
3. The Ovarian Cycle in Monotremes and Marsupials 197
3.1. Monotremes 197
3.2. Marsupials 197
3.2.1. Type 1: Short gestation, long luteal phase 199
3.2.2. Type 2: Short gestation, prolonged luteal phase 199
3.2.3. Type 3: Long gestation, delayed luteal phase 199
4. Control of Ovarian Cycles in Monotremes and Marsupials 201
4.1. Monotremes 201
4.2. Marsupials 201
5. Hormonal Regulation of Birth in Monotremes and Marsupials 202
5.1. Monotremes 202
5.2. Marsupials 202
6. Embryonic Diapause and Gestational Developmental Arrest 203
6.1. Embryonic Diapause 203
6.2. Gestational Developmental Arrest 203
7. Regulation of the Testicular Cycle in Monotremes and Marsupials 204
7.1. Monotremes 204
7.2. Marsupials 204
7.2.1. Type 1: Seasonal reproductive cycle followed by complete spermatogenetic collapse 204
7.2.2. Type 2: Seasonal reproductive cycle with testicular involution 207
7.2.3. Type 3: Seasonal reproductive cycle with continuous spermatogenetic activity 207
8. Control of Testicular Function in Monotremes and Marsupials 208
8.1. Monotremes 208
8.2. Marsupials 208
9. Summary and Conclusions 208
References 208

11. Hormones and Reproductive Cycles in Rodents 215

1. Introduction 215
2. Food Availability 215
2.1. Neuroendocrine Mechanisms Regulating the Reproductive Response to Food Availability 216
2.1.1. Developmental effects of food availability 217
2.1.2. Hibernation 217
3. Photoperiod 217
3.1. Seasonal Breeding 217
3.2. Neural Mechanisms for Photoperiodic Information 218
3.3. Kisspeptin 219
3.4. RFamide-related Peptide (RFRP) 219
3.5. Thyroid Hormones 219
3.6. Pubertal Timing 220
3.7. Intraspecies Variation in Photoperiod Responsiveness 220
3.8. Spontaneous Recrudescence 221
3.9. Female Reproduction and Sex Differences in Response to Photoperiod 221
3.10. Unmasking Techniques and Non-photoresponsive Species 221
4. Water Availability 222
5. Temperature 223
6. Social Cues 223
6.1. Reproductive Suppression in Social Mole-rats 223
6.1.1. Naked mole-rats (*Heterocephalus glaber*) 224
6.1.2. Damaraland mole-rats (*Cryptomys damarensis*) 224
6.1.3. Mashona mole-rats (*Cryptomys darlingi*) 225
6.1.4. Ansells mole-rats (*Fukomys anselli*) 225
6.1.5. Common mole-rats (*Cryptomys hottentotus hottentotus*) 225
6.1.6. Highveld mole-rats (*Cryptomys hottentotus pretoriae*) 226
6.2. Reproductive Suppression in Solitary Mole-rats 226
6.2.1. Cape dune mole-rats (*Bathyergus suillus*) 226

6.2.2. Cape mole-rats (*Georychus capensis*) 226
6.3. Reproductive Suppression and Kin Recognition in Other Rodent Species 226
6.3.1. Cactus mice (*Peromyscus eremicus*) 226
6.3.2. Prairie voles (*Microtus ochrogaster*) 226
6.3.3. Pine voles (*Microtus pinetorum*) 227
6.3.4. Mongolian gerbils (*Meriones unguiculatus*) 227
6.3.5. Alpine marmots (*Marmota marmota*) and field studies of other species 227
6.4. Spontaneous vs. Induced (Reflex) Ovulators 227
6.5. Postpartum and Lactational Estrus 228
7. Small Rodent Population Cycles 230
8. Other Ecological Factors Affecting Reproduction 230
8.1. Personality and Individual Differences 230
8.2. Parasite Load 230
8.3. Compounds Available in Green Plants 230
9. Conclusions 230
9.1. Species Differences 230
9.2. Sex Differences 231
9.3. Individual Differences 231
9.4. Field vs. Laboratory Studies 231
References 231

12. Hormones and Reproductive Cycles in Bats 241

1. Introduction 241
1.1. The Hypothalamus–pituitary–gonad (HPG) Axis in Bats 241
2. Bat Reproductive Cycles 242
2.1. Patterns of Reproductive Cycles 242
2.1.1. Reproduction in temperate-zone bats 242
2.1.1.1. Reproductive pattern type I 242
2.1.1.2. Reproductive pattern type II 242
2.1.2. Reproduction in tropical bats 243
2.1.2.1. Seasonal monoestry 243
2.1.2.2. Bimodal polyestry 244
2.1.2.3. Seasonal polyestry 244
2.1.2.4. Aseasonal polyestry 244
3. Reproductive Delays in Bats 244
3.1. Delayed Ovulation 244
3.2. Delayed Fertilization 245
3.3. Delayed Implantation 247
3.4. Delayed Development 248
4. Hormones and Reproduction in Bats 249
4.1. Hypothalamic Regulation of Pituitary Function 250
4.2. Seasonal Dynamics of the Gonadotropin-releasing Hormone (GnRH) Neuronal System 250
4.3. Gonadotropins 251
4.4. Prolactin (PRL) 253
5. Hormones and Reproduction in Male Bats 253
5.1. Reproductive Patterns in Male Bats 254
5.1.1. Temperate-zone male bats 254
5.1.1.1. Reproductive pattern subtype I 254
5.1.1.2. Reproductive pattern subtype II 254
5.1.2. Tropical/subtropical-zone bats 254
5.1.2.1. Single restricted seasonal breeding (seasonal monoestry in females) 255
5.1.2.2. Single extended seasonal breeding 255
5.1.2.3. Single extended asynchronized breeding 255
5.1.2.4. Single seasonal breeding with sperm stored throughout the year 255
5.1.2.5. Seasonal bimodal breeding 255
5.1.2.6. Seasonal bimodal breeding with sperm stored throughout the year 255
5.1.2.7. Multiple seasonal breeding (seasonal female polyestry) 255
5.1.2.8. Continuous breeding throughout the year (aseasonal polyestry) 255
5.2. Morphological Changes in Male Reproductive Tissue 255
5.3. Spermatogenesis and its Regulation 256
5.4. Hormonal Changes in Males 257
6. Hormones and Reproduction of Female Bats 257
6.1. Anatomy and Physiology of the Female Reproductive Tract 259
6.1.1. Reproductive asymmetry 259
6.1.1.1. Alternating dominance 259
6.1.1.2. Dextral dominance 259
6.1.1.3. Sinistral dominance 260
6.1.1.4. Contralateral dominance 260
6.1.1.5. Symmetric dominance 260
6.1.2. Unilateral endometrial involvement 260

6.1.3. Menstruation 261
6.2. Ovarian Morphology 261
6.2.1. Follicular development 261
6.2.1.1. Pattern I: Prolonged survival of Graafian follicles 261
6.2.1.2. Pattern II: Unilateral follicular development and ovulation 262
6.3. Ovarian Steroidogenesis and Peripheral Hormone Concentrations 262
6.4. The Corpus Luteum (CL) 264
6.5. Induced Ovulation 265
7. Environmental Factors and Additional Hormones Affecting Reproduction in Bats 266
7.1. The Roles of Hibernation, Fat Accumulation, and Leptin 266
7.2. Environmental Factors Regulating Reproduction 268
7.3. Non-Hypothalamus–pituitary–gonad (HPG) Hormones Affecting Reproduction in Bats 271
7.3.1. The pineal gland 271
7.3.2. The thyroid 272
7.3.3. The adrenal cortex 273
7.3.4. The pancreas 275
8. Conclusions and Future Directions 275
References 277

13. Hormones and Reproductive Cycles in Primates 291

1. Overview of the Primates 291
2. Testicular Function and its Neuroendocrine Control 291
2.1. The Hypothalamic–pituitary–gonad (HPG) Axis and Androgen Secretion in Males 292
2.2. Spermatogenesis 293
3. Ovarian Function and its Neuroendocrine Control 298
3.1. Overview 298
3.2. Cyclical Changes in the Ovaries 299
3.2.1. Folliculogenesis 299
3.2.2. Ovulation 301
3.2.3. Corpus luteum formation, function, and regression 301
3.3. Cyclical Changes in the Uterus 301
3.4. Cyclical Changes in the Oviducts 303
3.5. Cyclical Changes in the Cervix 303
3.6. Cyclical Changes in the Vagina 303
3.7. Cyclical Changes in the External Genitalia and Sexual Skin 303
3.8. Neuroendocrine Control of the Ovarian Cycle 304
4. Puberty 304
4.1. Adrenarche 304
4.2. Gonadarche 305
4.3. Neural Control of Gonadarche 305
4.4. Timing of Puberty 306
5. Pregnancy 307
5.1. Overview 307
5.2. Steroids 308
5.3. Chorionic Gonadotropin (CG) 308
5.4. Chorionic Somatomammotropins (CSs) 309
5.5. Corticotropin-releasing Hormone (CRH) 309
5.6. Leptin 309
6. Lactation 310
7. Reproductive Aging 310
8. Sexual Behavior 311
8.1. Description 311
8.2. Hormonal Influences on Sexual Behavior 312
9. Environmental Influences on Reproduction 313
9.1. Reproductive Seasonality 313
9.2. Social Influences on Reproduction 314
9.2.1. Males 314
9.2.1.1. Social influences on reproductive maturation in males 314
9.2.1.2. Social influences on reproduction in adult males 315
9.2.2. Females 315
9.2.2.1. Social influences on reproductive maturation in females 315
9.2.2.2. Social influences on reproduction in adult females 316
9.3. Energetics of Reproduction 316
9.3.1. Introduction 316
9.3.2. Energetics of hypothalamic–pituitary function 317
9.3.3. Energetics of gonadal function 317
9.3.4. Energetics of pregnancy 317
9.3.5. Energetics of lactation 318
10. Conclusions and Future Directions 318
References 319

14. Endocrine-disrupting Chemicals (EDCs) in Mammals 329

1. Introduction 329
2. Endocrine-disrupting Chemicals that Affect Reproduction 331

3. Mechanisms of Endocrine-disrupting Chemical (EDC) Action 337
3.1. Binding to Members of the Nuclear Receptor Superfamily 337
3.1.1. Genomic estrogen receptors (ERs) 338
3.1.2. Nongenomic actions of estrogens 338
3.1.3. The androgen receptor (AR) 338
3.1.4. Progesterone receptors (PRs) 339
3.1.5. Glucocorticoid receptor (GR) 339
3.1.6. Peroxisome proliferator-activated receptors (PPARs) 339
3.1.7. Pregnane X receptor (PXR) 339
3.1.8. The constitutive active receptor (CAR) 340
3.1.9. Thyroid hormone receptors (TRs) 340
3.2. Other Mechanisms of Endocrine-disrupting Chemical (EDC) Action 340
3.2.1. Chemicals that alter steroid hormone synthesis 340
3.2.2. Chemicals that alter steroid hormone metabolism 340
3.2.3. Small molecule enhancers of steroid receptor activity 340
3.2.4. Combined mechanisms of action 341
4. Adult vs. Developmental Effects 341
4.1. Hormone Action in Adulthood 341
4.2. Hormone Action During Development 341
4.3. Developmental Endocrine-disrupting Chemical (EDC) Exposure Programs Adult Gene Expression 342
5. Levels of Xenoestrogens Relative to Endogenous Estrogens *in Utero* 343
6. Endpoints 344
6.1. Puberty 344
6.1.1. Effects on puberty onset in humans 345
6.2. Behavior 345
6.2.1. Effects on human behavior 346
6.3. Fertility 346
6.3.1. Decreased fertility in animal models 346
6.3.1.1. Pesticides 346
6.3.1.2. Polychlorinated biphenyls (PCBs) 347
6.3.1.3. Phthalates 347
6.3.1.4. Bisphenol A (BPA) 347
6.3.1.5. Diethylstilbestrol (DES) 347
6.3.1.6. Phytoestrogens 347
6.3.2. Effects on fertility in humans 347
6.4. Lesions of the Female Reproductive Tract 349
6.4.1. Effects on female reproductive tract lesions in humans 350
6.5. Endometriosis 350
6.6. Mammary Cancer 351
6.6.1. Effects on mammary cancer in humans 352
6.7. Testicular Dysgenesis Syndrome (TDS) 353
6.8. Testicular Cancer 353
6.8.1. Effects on testicular cancer in humans 354
6.9. Cryptorchidism 354
6.9.1. Effects on cryptorchidism in humans 356
6.10. Hypospadias 356
6.10.1. Effects on hypospadias in humans 357
6.11. Prostate Cancer 358
6.11.1. Effects on prostate cancer in humans 358
7. Conclusion 359
References 360
Species Index 373
Subject Index 375
Color plates

Preface

Hormones and Reproduction of Vertebrates Preface to the Series

Every aspect of our physiology and behavior is either regulated directly by hormones or modified by their actions, as exemplified by the essential and diverse roles of hormones in reproductive processes. Central to the evolutionary success of all vertebrates are the regulatory chemicals secreted by cells that control sexual determination, sexual differentiation, sexual maturation, reproductive physiology, and reproductive behavior. To understand these processes and their evolution in vertebrates, it is necessary to employ an integrated approach that combines our knowledge of endocrine systems, genetics and evolution, and environmental factors in a comparative manner. In addition to providing insight into the evolution and physiology of vertebrates, the study of comparative vertebrate reproduction has had a considerable impact on the biomedical sciences and has provided a useful array of model systems for investigations that are of fundamental importance to human health. The purpose of this series on the hormones and reproduction of vertebrates is to bring together our current knowledge of comparative reproductive endocrinology in one place as a resource for scientists involved in reproductive endocrinology and for students who are just becoming interested in this field.

In this series of five volumes, we have selected authors with broad perspectives on reproductive endocrinology from a dozen countries. These authors are especially knowledgeable in their specific areas of interest and are familiar with both the historical aspects of their fields and the cutting edge of today's research. We have intentionally included many younger scientists in an effort to bring in fresh viewpoints. Topics in each volume include sex determination, neuroendocrine regulation of the hypothalamus–pituitary–gonadal (HPG) axis, separate discussions of testicular and ovarian functions and control, stress and reproductive function, hormones and reproductive behaviors, and comparisons of reproductive patterns. Emphasis on the use of model species is balanced throughout the series with comparative treatments of reproductive cycles in major taxa.

Chemical pollution and climate change pose serious challenges to the conservation and reproductive health of wildlife populations and humans in the twenty-first century, and these issues must be part of our modern perspective on reproduction. Consequently, we have included chapters that specifically deal with the accumulation of endocrine-disrupting chemicals (EDCs) in the environment at very low concentrations that mimic or block the critical functions of our reproductive hormones. Many authors throughout the series also have provided information connecting reproductive endocrinology to species conservation.

The series consists of five volumes, each of which deals with a major traditional grouping of vertebrates: in volume order, fishes, amphibians, reptiles, birds, and mammals. Each volume is organized in a similar manner so that themes can be easily followed across volumes. Terminology and abbreviations have been standardized by the editors to reflect the more common usage by scientists working with this diverse assembly of organisms we identify as vertebrates. Additionally, we have provided indices that allow readers to locate terms of interest, chemicals of interest, and particular species. A glossary of abbreviations used is provided with each chapter.

Finally, we must thank the many contributors to this work for their willingness to share their expertise, for their timely and thoughtful submissions, and for their patience with our interventions and requests for revisions. Their chapters cite the work of innumerable reproductive biologists and endocrinologists whose efforts have contributed to this rich and rewarding literature. And, of course, our special thanks go to our editor, Patricia Gonzalez of Academic Press, for her help with keeping us all on track and overseeing the incorporation of these valuable contributions into the work.

David O. Norris

Kristin H. Lopez

Preface to Hormones and Reproduction, Vol. 5 Mammals

The vast majority of literature on the reproductive endocrinology of mammals is focused on laboratory mice, laboratory rats, and humans. In this volume, we have made a conscious effort to present a more comparative approach to the reproductive endocrinology of mammals. This volume therefore offers a phylogenetically broad view of mammalian reproduction while incorporating the most recent biochemical and endocrine research. We have devoted specific chapters to the treatment of wild species of a wide range of mammalian groups including monotremes and marsupials, rodents, bats, and primates. Notably missing are the marine mammals for which there are not sufficient data to provide a complete picture of reproductive endocrinology. Also not given separate treatment are the large terrestrial carnivores for which studies of reproductive endocrinology are limited mostly to captive zoo animals, nor have we chosen to feature domestic species such as sheep, cattle, dogs, and cats that are covered in the veterinary literature.

This volume is organized in the same format as the other volumes in the Series, beginning with treatments of sex determination, neuroendocrine regulation of the hypothalamus-pituitary-gonad axis, and specific functions of the ovaries and testes. Pregnancy and parturition are covered in separate chapters, whereas lactation and reproductive behaviors are discussed throughout many of the chapters. The effects of stress hormones and pheromones are featured, followed by discussions of the reproductive cycles of selected mammalian groups. The final chapter discusses the disruption of endocrine activities in humans and other mammals as a consequence of widespread environmental exposures to estrogenic chemicals, and implications for future reproduction of wild mammalian populations and of our own species.

Contributors

Karen L. Bales
University of California, Davis, CA, USA

Fuller W. Bazer
Texas A&M University, Texas, TX, USA

Joseph M. Beeman
University of Missouri, Columbia, MO, USA

Kunwar P. Bhatnagar
University of Louisville, Louisville, KY, USA

T.R. Brown
Johns Hopkins Bloomberg School of Public Health, Baltimore, MD, USA

Wilson C.J. Chung
University of Colorado, Boulder, CO, USA

Caroline M. Hostetler
University of California, Davis, CA, USA

J.P. Jarow
Johns Hopkins School of Medicine, Baltimore, MD, USA

Graham Jenkin
Monash Institute of Medical Research (MIMR), Clayton, Victoria, Australia

Desiree L. Krebs-Kraft
University of Maryland School of Medicine, Baltimore, MD, USA

Amitabh Krishna
Banaras Hindu University, Varanasi, India

Kristin H. Lopez
University of Colorado at Boulder, Boulder, CO, USA

Bronwyn M. McAllan
The University of Sydney, Sydney, NSW, Australia

Margaret M. McCarthy
University of Maryland School of Medicine, Baltimore, MD, USA

Sam Mesiano
Case Western Reserve University, Cleveland, OH, USA

Susan C. Nagel
University of Missouri, Columbia, MO, USA

Bridget A. Niebruegge
University of Missouri, Columbia, MO, USA

David O. Norris
University of Colorado at Boulder, Boulder, CO, USA

Toni R. Pak
Loyola University Stritch School of Medicine, Maywood, IL, USA

Jin Ho Park
University of Massachusetts, Boston, MA, USA

Katherine E. Pelch
University of Missouri, Columbia, MO, USA

Aras Petrulis
Georgia State University, Atlanta, GA, USA

Marilyn B. Renfree
The University of Melbourne, Victoria, Australia

Emilie F. Rissman
University of Virginia School of Medicine, Charlottesville, VA, USA

Julienne N. Rutherford
University of Illinois at Chicago, Chicago, IL, USA

Wendy Saltzman
University of California, Riverside, CA, USA

Geoff Shaw
The University of Melbourne, Victoria, Australia

Roger Smith
The University of Newcastle, NSW, Australia

Thomas E. Spencer
Texas A&M University, Texas, TX, USA

Suzette D. Tardif
University of Texas Health Science Center, San Antonio, TX, USA

Lynda Uphouse
Texas Woman's University, Denton, TX, USA

Stacey R. Winkeler
University of Missouri, Columbia, MO, USA

W.W. Wright
Johns Hopkins Bloomberg School of Public Health, Baltimore, MD, USA

I. Ross Young
Monash University, Clayton, Victoria, Australia

Barry R. Zirkin
Johns Hopkins Bloomberg School of Public Health, Baltimore, MD, USA

Chapter 1

Sexual Differentiation of the Mammalian Brain

Desiree L. Krebs-Kraft and Margaret M. McCarthy
University of Maryland School of Medicine, Baltimore, MD, USA

SUMMARY

Sexual differentiation of the mammalian brain requires the activity of gonadal hormones. The organizational/activational hypothesis of sexual differentiation postulates that the organizational effects of these gonadal hormones on brain development occur early in life, during a sensitive period, whereas activational effects in adulthood ensure appropriate and timely sex-specific behaviors. Testosterone (T), or estradiol (E_2) produced by aromatization of T, is responsible for masculinization of the brain and the expression of normal male reproductive behavior in adulthood. In the absence of early exposure to gonadal hormones, the brain is, by default, feminized and mediates normal female reproductive behavior in adulthood. In this chapter we will review the role of apoptosis, cell genesis and synaptogenesis in establishing neuroanatomical sex differences. We will also look at how these processes are regulated by γ-aminobutyric acid (GABA), glutamate, prostaglandins, focal adhesion kinase, paxillin, and kisspeptin to act as the mechanisms of hormone action on sexual differentiation of the brain.

1. INTRODUCTION

An individual's sex is among the most salient of biological cues for determining both intrinsic physiology and behavior and external responses from conspecifics. In mammals, sex is determined genetically, with the homogametic sex being female (XX) and the heterogametic sex being male (XY). The phenotype of the undifferentiated gonad undergoes sex determination as a result of the presence or absence of the *SRY* gene (sex-determining region of the Y chromosome). This gene codes for testis-determining factor (TDF), which will initiate a gene expression cascade that ultimate leads to development of a testis (Koopman, Gubbay, Vivian, Goodfellow, & Lovell-Badge, 1991). In the absence of this gene, the undifferentiated gonadal anlage becomes an ovary. The remaining sexual characteristics are a byproduct of gonadal secretions; therefore, this process is referred to as sex determination and the phenotype of the body, including the brain, is the product of sexual differentiation. This is not to say that genetics plays no additional role in physiology or behavior, but, for endpoints that are directly relevant to reproduction, the importance of gonadal hormones and signaling peptides cannot be overstated.

Successful reproduction requires more than functioning gonads and a viable reproductive tract. The animal must also display the appropriate behaviors in response to the appropriate conditions, including but not limited to the presence of a sexually receptive partner with which to mate. In the overwhelming majority of animals, expression of this behavior is constrained by internal variables, such as viable sperm in males or temporal relation to ovulation in females, as well as environmental variables including the availability of food, shelter, and nesting sites, which may be directly related to the season. The brain is critical to integrating both the internal and external variables to achieve optimal coordination of physiology and behavior. The optimality of this coordination varies between males and females because of the constraints of internal fertilization and requirements for extensive postnatal maternal care by mammalian females. Sexual differentiation in response to gonadal hormones ensures that the brain and the body are in sync; in other words, that brain sex matches gonadal sex. The two critical endpoints are the control of gonadotropin (GTH) secretion by the anterior pituitary, which is pulsatile in males but exhibits a midcycle ovulation-inducing surge in females, and sexual behavior. Male rodents exhibit mounting and thrusting toward sexually receptive females, whereas females adopt a sexually receptive posture referred to as lordosis that permits access by the male. The sex-specific control of GTH secretion and sexual behavior are both results of hormonally mediated sexual differentiation of the brain, and the mechanisms by which these occur are the topic of this review.

2. HISTORICAL OVERVIEW

2.1. Phoenix, Goy, Gerall, and Young: 1959

Every field has its icons and for the discipline of behavioral neuroendocrinology there is no question that a single paper published in *Endocrinology* in 1959 by trainees in the laboratory of William Young at the University of Kansas (Phoenix, Goy, Gerall, & Young, 1959) has determined the direction of research for the past 50 years. This seminal paper brought coherence to what had previously been a disparate collection of observations about hormonal effects on the reproductive behavior of males and females. The authors proposed a novel hypothesis, that early hormonal actions predisposed adult behavior, and then conducted a series of careful experiments in guinea pigs (*Cavia porcellus*) that tested the predictions stemming from this hypothesis. Now codified as the organizational/activational hypothesis of hormone action, the authors demonstrated that treatment of pregnant guinea pigs with testosterone (T) masculinized their female offspring such that they displayed male sexual behavior in adulthood, but only when these females were treated with T as adults. Females whose brains were organized in a masculinized fashion did not display maletypic behavior as adults if treated with estradiol (E_2) and progesterone (P_4), the hormones that normally activate female sexual behavior. This demonstrated the two-step actions of hormones to organize responses early in development that must then be acted upon in adulthood (Figure 1.1). Ironically, the notion that hormones would act on the brain to organize its development and thereby permanently determine adult behaviors was not well accepted at the time, as demonstrated by the authors' repeated reference to a hormone-responsive 'tissue,' with mention of the brain only as a potential candidate. Regardless, the organizational/activational hypothesis, as a sturdy framework, has withstood many challenges and provides a strong foundation against which contrasting or alternative interpretations can be tested. This hypothesis continues to provide the benefits of dogma by serving as a guiding principle (Becker et al., 2005) as well as a yielding doctrine for more modern synthesis (McCarthy & Konkle, 2005).

2.2. Modern Views—Where Science and Society Collide

Sexual differentiation of the brain is a fascinating biological phenomenon of tremendous interest in its own

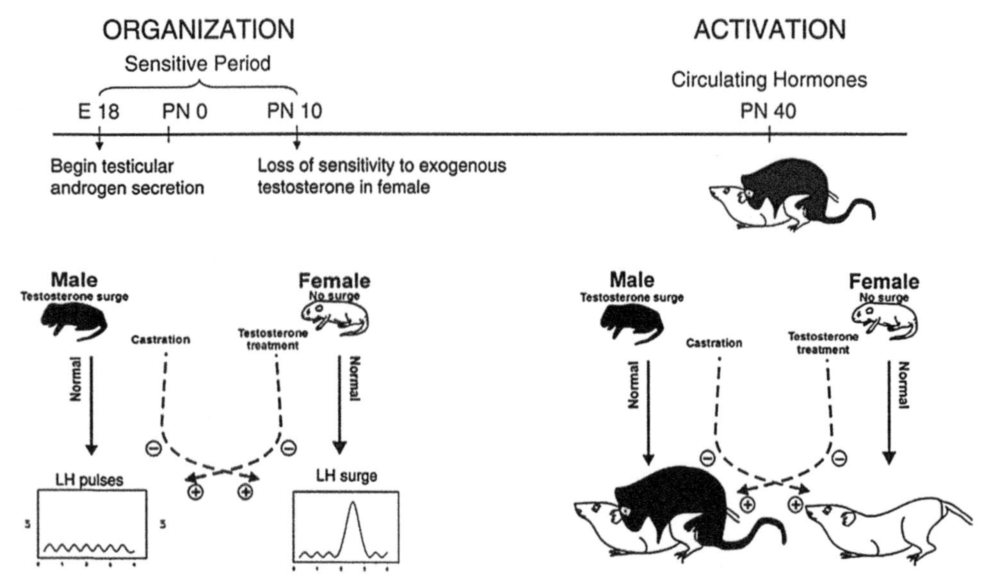

FIGURE 1.1 The organizational/activational hypothesis. This hypothesis proposes that early hormone effects organize the brain and adult hormonal effects are constrained by prior exposure (Phoenix, Goy, Gerall, & Young, 1959). Establishment of sex differences in physiology and behavior is a function of differential gonadal steroid synthesis during a perinatal sensitive period. In rats, the production of testicular androgens begins around embryonic day 18 and defines the onset of the sensitive period. The ovary remains quiescent, and a lack of exposure to androgens, and their aromatized product, estradiol (E_2), is essential for normal female brain development. Treatment of females with exogenous testosterone (T) results in its aromatization to E_2 and masculinization of adult brain and behavior. The developmental time point at which the female becomes insensitive to the masculinizing effects of exogenous T operationally defines the end of the sensitive period. As adults, males show only pulsatile release of luteinizing hormone (LH) from the pituitary, while females exhibit a large surge in LH release to induce ovulation at the midpoint of the estrous cycle. Likewise, only adult males exhibit the masculine pattern of sexual behavior of mounting a female, while only females adopt the sexually receptive posture termed lordosis. Exposure of neonatal females to T, which is aromatized to E_2, during the sensitive period will render them sterile and sexually unreceptive. Male and female adult patterns are determined by hormonal organization during development but are dependent on the activation of adult sex-specific hormones. *Reprinted with permission from McCarthy (2008).*

right and also as a valuable heuristic model for brain development. Several important discoveries about the basic principles establishing phenotypic variation in neural function have arisen directly from the study of hormonally mediated sexual differentiation. Unlike many other corners of biology, however, there is also an important social or cultural component to work on sex differences in the mammalian brain. This contributes to an exaggerated emphasis on rudimentary findings in simple organisms as being relevant for the human condition and a greater need for precision and restraint in reaching conclusions. Everyone is ready and willing to believe that men and women are different, but it is how they differ and what this means that are the sources of considerable contention. Discussion of this topic could fill volumes and consensus would still not be reached. None-the-less, it is worth asking whether the organizational/activational hypothesis applies to humans (*Homo sapiens*) and sex differences in human behavior. The very short answer is that we do not know. This is in part because we do not do experiments on humans and therefore resort to relying on Mother Nature to provide so-called 'natural experiments.' These include infants or children exposed to abnormal or heterotypic steroid levels because of medical treatments or disease states, genetic mutations, and exposures to endocrine-disrupting chemicals (EDCs) (see Chapter 14, this volume) resulting in excessive steroid production, select hormone insensitivity, or endocrine disruption. Study of these individuals is consistent with some developmental organizational effects of steroids on adult human behavior that are relevant to reproduction, but, in both this and other arenas, the magnitude and significance of the effects remains highly controversial (Berenbaum, 1999; Hines, 2002). The tendency to overinterpret results from animal studies as relevant to humans is nowhere greater than in the area of cognition. The notion that 'boys are better at math and girls are better at language' is as frequently and thoughtlessly repeated as 'an apple a day keeps the doctor away'. Animal studies are often considered to provide the unbiased-by-culture proof of sex differences in cognition, and yet we have no animals in which to examine either mathematics or language. In fact, when animal studies purporting sex differences in cognition, predominantly spatial learning, are carefully examined, there is much more smoke than fire (see McCarthy & Konkle, 2005), and the same is true for many common cognitive assessments in humans (Spelke, 2005). Does this mean there are no sex differences in the human brain? Certainly not, and indeed the growth trajectories of male and female human brains are markedly different (Lenroot et al., 2007), as are many neurological endpoints. But it is important that appropriate perspective is applied when evaluating both animal and human studies.

3. THE ORGANIZATIONAL/ ACTIVATIONAL HYPOTHESIS OF SEXUAL DIFFERENTIATION

There are several seminal events that built on the original Phoenix report that have influenced the course of the field ever since. One of the earliest was the realization that the guinea pig, the animal model used by Phoenix and colleagues, is not the best species in which to address these questions due to its small litter size, long gestation, and precocial young. The fact that the seminal organizational events occurred prenatally both confounded investigation by the hormones of pregnancy and precluded the ability to manipulate males and females independently. The laboratory rat soon replaced the guinea pig as an ideal species for study due to it large litter size, short gestation, and the observation that the sensitive period for sexual differentiation began prenatally but extended well into the postnatal period as well.

3.1. Basic Principles of Hormone Action

Steroid hormones are lipid-soluble, which allows them to cross the extracellular membrane to the inside of the cell. Classical models of steroid action suggest that steroid hormones bind to specific nuclear receptor proteins once inside the cell. In the absence of steroid hormones, the nuclear steroid receptors in various target tissues are associated with chaperone proteins, such as heat-shock proteins, in a transcriptionally inactive state (DeFranco, Ramakrishnan, & Tang, 1998). These chaperone proteins keep the receptor in a high affinity state for binding of receptor ligands. Binding of steroid hormone to its cognate nuclear receptor causes the inhibitory heat-shock proteins to dissociate and the receptor to become activated. Steroid receptor activation leads to receptor dimerization and facilitates the ability of estrogen receptors (ERs) and androgen receptors (ARs) to bind to specific DNA sequences, called hormone response elements, on the promoters of target genes (Figure 1.2). Before interacting with DNA, the steroid receptors may bind to other proteins, forming part of a transcriptional complex. This complex consists of multiple proteins called coactivators or corepressors that affect DNA transcription and are recruited to promote or inhibit the expression of target genes (Enmark & Gustafsson, 1999; He, Kemppainen, Voegel, Gronemeyer, & Wilson, 1999; Roy et al., 1999; Klinge, Kaur, & Swanson, 2000; Osborne, Zhao, & Fuqua, 2000; Nilsson et al., 2001; Heinlein & Chang, 2002). As a result of this principal mechanism of action, steroid receptors are considered nuclear transcription factors and members of a large superfamily of nuclear receptors (Beato & Klug, 2000).

Steroid hormone receptors may have multiple isoforms. For example, E_2 binds to two ER isoforms, ERα and ERβ. The existence of multiple ER isoforms allows cells to

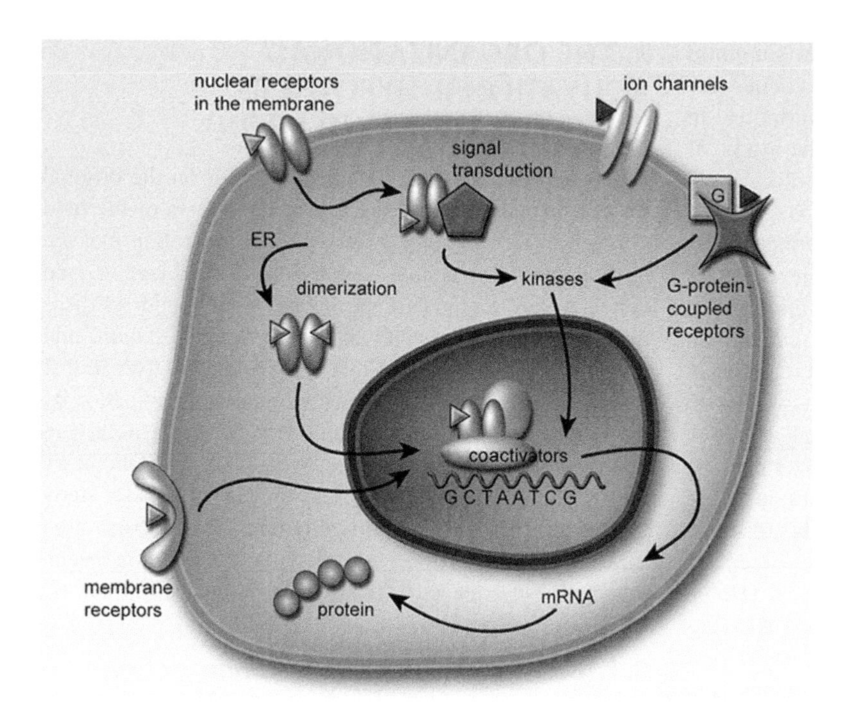

FIGURE 1.2 Steroid receptor types. Receptors of steroid hormones are nuclear transcription factors and membrane receptors that interact with kinases. For example, estradiol (E_2) binds to an intracellular receptor protein (ER) that dimerizes and translocates to the nucleus. The receptor complex, made of the receptor protein and coactivators, binds to the palindromic sequences (GCTAATCG) that constitute the estrogen response element (ERE) and affect gene transcription. In addition to this long-term genomic effect of E_2, the ER has been associated with the cell membrane of neurons where it can interact directly with signal transduction pathways such as that involving mitogen-activated protein kinase (MAPK). Estradiol also affects ion channels and G-protein-coupled receptors, independent of the ER. *Reprinted with permission from McCarthy (2008).*

selectively regulate responses to E_2. In contrast, the AR has only one form for both T and 5α-dihydrotestosterone (DHT) binding, but their effects on gene expression can be quite distinct.

In addition to the slow 'genomic' actions of steroid hormones that take hours or days for final changes in protein expression to occur, there are also rapid actions of steroids. For example, E_2 administration increases cyclic-3',5'-adenosine monophosphate (cAMP) in the uterus within 15 seconds (Szego & Davis, 1967). The 'nongenomic' actions of steroid hormones occur rapidly (milliseconds to minutes), and are initiated at the plasma membrane, resulting in the activation of kinases and signal transduction pathways, including calcium influx, within target cells (Figure 1.2). The rapid effects of steroid hormones can be mediated by direct binding to a specific binding site on the target cell, through the classical intracellular steroid receptors, and by distinct nonclassical membrane receptors, such as a G-protein-coupled receptor (Michels & Hoppe, 2008). The discovery of a novel G-protein-coupled membrane ER, GPR30 (Prossnitz, Arterburn, & Sklar, 2007), which is unrelated to nuclear ERs and localized to the endoplasmic reticulum, is another substrate for rapid estrogen signaling (Revankar, Cimino, Sklar, Arterburn, & Prossnitz, 2005; Thomas, Pang, Filardo, & Dong, 2005).

Thus, it is apparent that steroid hormones can directly or indirectly regulate gene transcription. Steroid hormones that bind to classical steroid nuclear receptors cause binding of AR and ER to DNA sequences on promoters of genes, therefore directly affecting DNA transcription to promote or inhibit the expression of target genes (Enmark & Gustafsson, 1999; He et al., 1999; Roy et al., 1999; Klinge, 2000; Osborne et al., 2000; Nilsson et al., 2001; Heinlein & Chang, 2002). The traditional nuclear steroid receptors can also indirectly regulate gene transcription by serving as signal transduction proteins that interact with various signaling pathways in the cytoplasm of a cell. In particular, ER can activate kinases, enzymes that phosphorylate and thereby activate other proteins. This sets in motion a signaling cascade that eventually ends in the nucleus with the induction of gene transcription (Singh, Setalo, Guan, Frail, & Toran-Allerand, 2000). A number of second messengers are produced, including cAMP and the inositol phospholipids. In order for membrane receptors to influence the expression of genes, the signal transduction pathway must reach the nucleus (Aronica, Kraus, & Katzenellenbogen, 1994). Ultimately, the increased calcium released by inositol triphosphate (IP_3) and cAMP influences gene expression by activating cAMP-response-element-binding protein (CREB), protein kinase A (PKA) (Aronica et al., 1994; Zhou, Watters, & Dorsa, 1996; Abraham, Todman, Korach, & Herbison, 2004), mitogen-activated protein kinase (MAPK) (Watters, 1998; Singh, Setalo, Guan, Warren, & Toran-Allerand, 1999; Kuroki, Fukushima, Kanda, Mizuno, & Watanabe, 2000), and phosphatidylinositol 3-kinase (PI3K) (Singh et al., 1999; Znamesnsky, Akama, McEwen, & Milner, 2003). The discovery of the myriad ways in which steroids can impact on cell function greatly expanded the potential mechanisms by which steroid-mediated sexual differentiation of the brain could occur, with both classic 'genomic' and non-classic 'rapid effects' identified to date.

3.2. Sensitive Periods

A common principle in brain development is the restricted time course of critical events that must unfold in a scripted order to provide the template for subsequent phases of the maturational process to be successful. For example, in the visual system there is a period of uncoordinated excitation in the retina followed by activity-dependent pruning of synapses in the visual cortex to generate patterns of neuronal input that provide for the selective response to particular stimuli. If visual input to the retina is restricted during a critical developmental window, the ability to generate the appropriate pattern of synaptic input is forever lost.

Similarly, early in development there is a sensitive or critical period for sexual differentiation of the brain (MacLusky & Naftolin, 1981; Arnold & Gorski, 1984; Nevison, Brown, & Dixson, 1997). A critical period also can be thought of as a limited window of development in which an event can occur because of enhanced sensitivity to an environmental stimulus, which may be endogenous or exogenous (Figure 1.1). During the last days of gestation in the rat, as early as embryonic day 18, and as early as the first trimester of pregnancy in primates, the testes of the developing male begin to produce significant quantities of T (Weisz & Ward, 1980; Rhoda, Corbier, & Roffi, 1984). This timed exposure to T is necessary to induce a phenotypic male brain and the formation of the secondary sex characteristics, including the epididymis, vas deferens, and male genitalia (Jost, 1947). From embryonic day 18 until around postnatal day 10, exogenous treatment of female rodents with T can override development as a female and induce a phenotypic male brain. For instance, a single injection of T within the first week of life disrupts normal cyclicity in female rodents during adulthood (Barraclough, 1961; Barraclough & Gorski, 1962). After postnatal day 10, treatment of female rodents with T has no effect on brain development because it has been permanently changed to be phenotypically female (MacLusky & Naftolin, 1981; Arnold & Gorski, 1984), signaling the end of the sensitive period.

Sensitive periods of steroid hormone effects on brain organization are species-specific. In rats (*Rattus norvegicus*), the sensitive period spans the late gestation/early postnatal period (Goldman & Gorski, 1971); however, the sensitive period in more precocious mammals is earlier in gestation. The sensitive period occurs at midgestation in guinea pigs (Goy, Bridson, & Young 1964), during the first trimester in monkeys (Herman, Jones, Mann, & Wallen, 2000), and midgestation to postnatal development in sheep (Clarke, Scaramuzzi, & Short, 1976; Wood, Mehta, Herbosa, & Foster, 1995; Foster, Padmanabhan, Wood, & Robinson, 2002). Even within one species the timing of the sensitive period may vary as a function of the endpoint. In rats and mice (*Mus musculus*), the timing for sexual differentiation of GTH secretion is earlier than that for sexual behavior (Diaz, Fleming, & Rhees, 1995), illustrating the point that different neural substrates are modified to produce distinct outcomes.

The critical period for sexual differentiation of the brain is early in development and usually perinatal, meaning around birth. However, there is also evidence for an important effect of the elevated hormones associated with puberty on the final differentiation of the brain (Arnold & Breedlove, 1985; Romeo, 2003; Sisk & Foster, 2004; Sisk & Zehr, 2005; Wallen, 2005). The absence of gonadal hormones during adolescent brain development in Syrian hamsters (*Mesocricetus auratus*) significantly reduces T-induced mounting and intromissions in adulthood (Schulz et al., 2004). The presence of pubertal gonadal hormones is also important for defeminization in Syrian hamsters. Specifically, males that are castrated prior to puberty and treated with estrogens and P_4 show equal levels of lordosis to females hormonally primed in adulthood (Schulz et al., 2003). Gonadal hormones also organize agonistic behaviors, such as flank marking and aggression, and anxiety-related behaviors, such as locomotor activity in the open field, during puberty in males and in some cases females (Sisk & Zehr, 2005). It is still unclear whether adolescence serves as a sensitive period for steroid-dependent remodeling of the brain, and further experiments are needed to elucidate this possibility.

3.3. The Role of Aromatization

As noted above, initial studies on sexual differentiation of the brain involved T treatment of pregnant guinea pigs. Unforeseen at the time was an important difference between rats and guinea pigs: the critical role of androgens vs. estrogens. That E_2 is the masculinizing hormone in rodents was codified by the aromatization hypothesis (DonCarlos, McAbee, Ramer-Quinn, & Stancik, 1995; Roselli & Klosterman, 1998; Vinader-Caerols, Collado, Segovia, & Guillamon, 2000), which reconciled the divergent observations that exogenous administration of E_2 was as or more effective than T at masculinizing behavior, that the critical brain regions expressed ERs, and that the synthetic enzyme, E_2 synthase or aromatase ($P450_{aro}$), the enzyme responsible for the conversion of T to E_2, was highly active in the neonatal brain (McCarthy, 2008). An additional critical piece of the puzzle was the determination that the steroid-binding globulin, α-fetoprotein, preferentially binds estrogens in rodents, thereby allowing T to freely enter cells where it is converted to E_2 (Soloff, Swartz, Pearlmutter, & Kithier, 1976; MacLusky, Lieberburg, & McEwen, 1979). Only decades later did it become apparent that androgens mediate sexual differentiation of the primate (*Macaca mulatta*) brain (Wallen, 2005), presumably including humans, suggesting that guinea pigs may be a far better model than rodents for understanding the biological basis of gender biases in humans. Determining whether this is true depends on whether the cellular mechanisms establishing sex differences in the brain are common in primates and rodents. Antecedent to that is determining the cellular mechanisms of sexual differentiation in rodents. Understanding the cellular mechanisms of organizing effects of hormones first requires a complete characterization of the behavioral and neuroanatomical phenomena that are the subjects of organization.

3.4. Masculinization, Feminization, and Defeminization

Sexual differentiation of the brain is a developmental process whereby gonadal steroids act during a perinatal sensitive period on the undifferentiated neural substrate to permanently alter it so that ultimately brain phenotype will match gonadal phenotype. Recent studies of potential contributions of chromosomal sex to brain phenotype have further confirmed the supremacy of steroids when it comes to reproductively related endpoints (Arnold, Rissman, & De Vries, 2003; Arnold et al., 2004). Central to the process of differentiation is the conversion of testicular-derived T to E_2 within neurons. Most sexually dimorphic areas of the brain contain substantial levels of both $P450_{aro}$ (CYP19) and high densities of ER (Pfaff & Keiner, 1973; Roselli, Ellinwood, & Resko, 1984; Roselli & Resko, 1993; MacLusky, Walters, Clark, & Toran-Allerand, 1994; DonCarlos et al., 1995; DonCarlos, 1996). Maternal estrogens are sequestered in the peripheral circulation of the fetus by α-fetoprotein, which has very little affinity for androgens, allowing the testicular T to reach and influence fetal target tissues, including the brain. Treatment with T, but not the nonaromatizable androgen DHT, mimics many of the effects of estrogens, and normal masculinization of the brain is prevented subsequent to disruption of $P450_{aro}$ during the sensitive period (McEwen, Lieberburg, Chaptal, & Krey, 1977; Vreeburg, Van der Vaart, & Van der Shoot, 1977; Mong, Glaser, & McCarthy, 1999; Amateau & McCarthy, 2002a; 2002b; Bakker, Honda, & Balthazart, 2002; Bakker, Honda, Harada, & Balthazart, 2002). This basic principle that the male brain is masculinized by local conversion of E_2 is elucidated by the aromatization hypothesis (Naftolin et al., 1975) and expanded on by others (McEwen et al., 1977; Vreeburg et al., 1977). More recently, advances have been made regarding the mechanisms by which E_2 exerts a masculinizing influence on the brain.

Studies attempting to elucidate the mechanistic basis of sex differences in the brain benefit from a robust and reliable endpoint: sexual behavior in the albino laboratory rat. Whether an adult rat responds to sexual advances from another adult rat with a male response (mounting, intromitting, and ejaculating) or a female response (lordosis) is entirely dependent on two variables: (1) the hormonal milieu during a perinatal sensitive window that must match (2) the hormonal milieu of the adult. For male behavior, the neonate must be exposed to a critical level of neuronal E_2 matched with a critical level of circulating T as an adult. For female behavior, the neonate must *not* be exposed to a critical level of neuronal E_2 matched with a threshold level of E_2 sequential with P_4 in the adult. Male sexual behavior in the laboratory rat is opportunistic and readily expressed whenever a receptive female is present. Female sexual behavior is physiologically constrained to be expressed only in proximity to ovulation. A fascinating but mechanistically unexplained distinction in male and female sexual behavior is the impact of experience. Males improve with practice and if T is removed (via castration), will continue to exhibit high levels of copulatory behavior that only gradually extinguishes over a period of months. Females, by contrast, get it right the first time and every time and will only exhibit lordosis under the proper hormonal umbrella. If steroids are eliminated, so is the behavior. In terms of ultimate causation, the adaptive basis for the distinction is obvious: there is no benefit but is possible cost to females mating outside the window of opportunity for conception, whereas males maximize fitness by never missing an opportunity to share the wealth. Proximately, however, this dichotomy in the nature of the plasticity attendant to both behaviors is suggestive of distinct and separate neuronal circuitries, be they physically or merely functionally so.

So how does the brain mediate these distinct strategies in males and females? A useful framework for investigating mechanistic questions of sexual differentiation is the operationally defined and distinct processes of masculinization, feminization, and defeminization. *Masculinization* refers to an active developmental process initiated by gonadal steroids during the perinatal sensitive period followed by expression of normal male copulatory behavior in adulthood. *Feminization* is essentially what happens in the absence of masculinization, meaning that it is the default pathway leading to expression of lordosis under the proper hormonal conditions in adulthood. *Defeminization* is distinct from but normally occurs in tandem with masculinization and refers to the process whereby the ability to express female sexual behavior is lost. Defeminization appears to be exclusively regulated by E_2, whereas masculinization involves both estrogens and androgens (Whalen & Edwards, 1967; Vreeburg et al., 1977; Auger, Tetel, & McCarthy, 2002).

3.5. Brain vs. Behavior

There are multiple sexually dimorphic brain regions as a result of the organizational effects of gonadal hormones.

Regional volumetric sex differences have been found in the preoptic area (POA), spinal nucleus of the bulbocavernosus (SNB), bed nucleus of the stria terminalis (BNST), anteroventral periventricular nucleus (AVPV), and the medial amygdala (mAMG) of the laboratory rat. The sexually dimorphic nucleus of the POA (SDN-POA) (Gorski, Gordon, Shryne, & Southam, 1978), the SNB (Nordeen, Nordeen, Sengelaub, & Arnold, 1985; Freeman, Watson, & Breedlove, 1996), the BNST (Guillamon, Segovia, & Del Abril, 1988; Hines, Allen, & Gorski, 1992), and the mAMG (Mizukami, Nishizuka, & Arai, 1983; Cooke, Tabibnia, & Breedlove, 1999) are larger in males than in females. The volume of the sex difference in the SDN-POA is so robust that it is visible to the naked eye, being approximately 3–5 times larger in males than in females (Gorski et al., 1978). In contrast, the AVPV is larger in females than males (Simerly, Swanson, & Gorski, 1985; Murakami & Arai, 1989; Sumida, Nishizuka, Kano, & Arai, 1993). Gonadal hormones establish these sex differences in volume by affecting the number of cells in each brain region. While the direction of the sex difference varies by brain region, in all cases the mechanism by which the volumetric difference is achieved is the same: differential cell death. The SNB, SDN-POA, and BNST are larger in males than females because gonadal hormones promote cell survival. Males and females begin with the same number of neurons, but many of these die in females due to the lack of trophic support from T and/or E_2 at a critical time in development (Gorski et al., 1978; Nordeen et al., 1985; Guillamon et al., 1988; Freeman et al., 1996; Forger et al., 2004). Conversely, in the AVPV, E_2 promotes cell death, resulting in a smaller volume nucleus in males (Murakami & Arai, 1989; Simerly, 1989; Sumida et al., 1993). There are also sex differences in the connectivity between brain regions. Specifically, the projection from the BNST to the AVPV contains a greater number of galanin and GABAergic fibers in males than females (Gu & Simerly, 1997; Ibanez, Gu, & Simerly, 2001; Polston & Simerly, 2003; Polston, Gu, & Simerly, 2004). In contrast, the number of substance P fibers projecting from the BNST to the POA is greater in females than males (Polston & Simerly, 2003; Polston et al., 2004). Importantly, these projections form a functional circuit controlling sexually dimorphic GTH secretion (Gu & Simerly, 1997; Herbison, 1998; De Vries & Simerly, 2002) and are organized by prenatal exposure to gonadal hormones (Simerly, 2002).

In addition to volumetric and cell number differences induced by early hormone exposure in the brain, there are also robust sex differences in synaptic patterning.

Perinatal exposure to T exerts region-specific effects on dendritic spines and branching in the developing hypothalamus. The arcuate nucleus (ARC) of males contains twice the number of axosomatic and half the number of axodendritic spine synapses as that of females (Matsumoto & Arai, 1983). Early T exposure decreases the dendritic spine density and axodendritic spine synapses in the ARC (Mong, Kurzweil, Davis, Rocca, & McCarthy, 1996; Mong & McCarthy, 1999; Mong, Roberts, Kelly, & McCarthy, 2001). Perinatal E_2 exposure has the opposite effect in the POA. Males have two to three times more dendritic spines than females, and treating neonatal females with E_2 will produce the male pattern of dendritic spines and permit the expression of T-induced male sexual behavior in adulthood (Amateau & McCarthy, 2002a; 2004). In males, the ventromedial nucleus (VMN) is only slightly larger than in females but contains more than three times as many dendritic spine and shaft synapses as that of females (Matsumoto & Arai, 1983). These sex differences in synaptic patterning are detected as early as postnatal day three and are still present at postnatal day 100 (Matsumoto & Arai, 1986; Pozzo-Miller & Aoki, 1991), which is consistent with the organizational effects of gonadal hormones. The nature of the sex difference in the synaptic pattern is fundamentally different in the VMN vs. the POA and ARC. In the latter two nuclei there are greater numbers of dendritic spine synapses per unit dendrite, meaning that the density of synapses has increased. In the VMN, males have overall more dendritic spine synapses because the dendrites are longer and branch more frequently, but the density of spine synapses per unit dendrite is the same in males and females (Mong et al., 2001; Todd, Schwarz, Mong, & McCarthy, 2007). The way in which this difference in synaptic pattern impacts on neuronal function is entirely unknown.

Neonatal exposure to gonadal hormones also affects cell morphology of astrocytes in sexually dimorphic brain regions. Testosterone or its metabolized product, E_2, increases the complexity of astrocytes in the ARC toward a more stellate morphology with more primary processes and branches (Mong & McCarthy, 1999; Mong et al., 1999). There is no evidence of ERs in astrocytes of the ARC (Mong & McCarthy, 1999). Instead, E_2 indirectly promotes astrocyte differentiation by enhancing the synthesis and release of GABA in neighboring neurons (Mong, Nunez, & McCarthy, 2002). These changes in astrocyte complexity are correlated with a reduction in the formation of dendritic spine synapses in males (Mong et al., 2001). Males also have more complex astrocytes in the POA than females (Amateau & McCarthy, 2002b); however, enhanced astrocyte differentiation in the POA is correlated with greater formation of dendritic spine synapses in males, the opposite of that seen in the ARC. The question of how hormonal modulation of astrocyte complexity modulates neuronal complexity is an important area for future study.

The GTHs provide an excellent example of neurochemical sex differences that are a result of organizational effects of gonadal hormones in the brain. Neurons expressing gonadotropin-releasing hormone (GnRH) are located in the POA, BNST, and ARC of the hypothalamus of many species (e.g., there are no GnRH neurons in the ARC of rats and mice). Gonadotropin-releasing hormone is the regulator of the pulsatile release of luteinizing hormone (LH) from the anterior pituitary in both males and females and there are no obvious sex differences in the number, distribution, or morphology of GnRH neurons. The GTHs regulate reproduction in females through a massive surge in LH at the appropriate time in the cycle to induce ovulation. Estradiol regulates the synchronized firing of GnRH neurons to induce the LH surge via a positive feedback mechanism in females but not males (Becu-Villalobos & Libertun, 1995). The masculinized brain does not respond to E_2 with positive feedback and there is therefore no LH surge (Diaz et al., 1995). A proposed mechanism for this sex difference in the LH surge is enhanced excitatory glutamatergic input onto GnRH neurons in females (Becu-Villalobos & Libertun, 1995), and this input appears to originate in the AVPV, which is larger in females than males (Gu & Simerly, 1997).

While anatomical sex differences are clearly important, it is also possible that parameters less easily measured but equally important, such as excitability, are different in males and females. In addition to electrophysiological measures, the activity of neurons can be assessed indirectly via markers of cell activity such as the immediate early gene, c-fos, or the phosphorylated form of CREB, pCREB. Males have enhanced levels of c-fos and pCREB in the immature hypothalamus during early postnatal development (Auger, Hexter, & McCarthy, 2001; Olesen & Auger, 2005) and the sex difference in pCREB is the result of neonatal exposure to T (Auger et al., 2001). Estradiol enhances the depolarizing action of GABA in the developing hypothalamus by increasing the magnitude of depolarization-induced calcium influx and extending the developmental duration of depolarizing GABA. This delays the onset of inhibitory GABA and provides the brain with critical excitation during development (Nunez, Bambrick, Krueger, & McCarthy, 2005; Perrot-Sinal, Sinal, Reader, Speert, & McCarthy, 2007). The functional significance of enhanced and extended depolarizing GABA early in development is an important topic for future research.

Aggressive behavior in adulthood is linked to concentrations of androgens during sensitive periods of development. For example, *in-utero* exposure to T induces organizational effects in the brain that are later expressed in terms of aggressive behavior in male and female hyenas (Glickman et al., 1992; Glickman, Short, & Renfree, 2005). These same females also express male-like characteristics during play-fighting and other sexually dimorphic behaviors (Glickman et al., 1992; 2005). Those females exposed to the greatest amounts of T *in utero* are likely to be the most aggressive (Glickman et al., 1992; 2005). Although T declines in females more than in males upon reaching maturity, the hormone's early impact means a lifelong tendency toward aggressive behavior (Glickman et al., 1992; 2005). The underlying neural circuits for aggressive behavior in hyenas are unclear; however, the brain regions associated with aggression in rodents include the medial POA and anterior hypothalamic areas, the medial BNST, and the mAMG (Albert, Jonik, & Walsh, 1992; Ferris et al., 1997), all regions that exhibit a high degree of sexual dimorphism.

3.6. Challenges to the Dogma

A scientific principle is only called a dogma when it is being challenged. The organizational/activational hypothesis as dogma has proved highly effective for providing form and substance to various challenges, which range from the questioning of the generalities of specific details to paradigm shifts in the way we view the origin and meaning of sex differences in behavior. Many challenges are really just modifications or qualifications to the strength of the predictions made by the dogma. For instance, while the organizational/activational hypothesis infers that early hormone effects are permanent, it has been known since prior to the articulation of the hypothesis that there is some degree of plasticity in adult sexual behavior. Treatment of females with high levels of androgens for an extended period of time will induce some degree of maletypic behavior (Sodersten, 1984), and the converse is also true in males (Olster & Blaustein, 1988), but the sensitivity to heterotypic hormones (i.e., T in females) is much less than for homotypic hormones (i.e., E_2 in females). These observations do not negate the dogma; they merely modify it to not be so absolute. Two more substantive challenges are the importance of genetics, and the notion that some organizational effects are designed to make the sexes more similar as opposed to different. These challenges also do not negate the central tenets of the organizational/activational hypothesis but they do provide a richer and more nuanced view of sex differences in behavior.

Decades of study on hormonal induction of sex differences in the brain have established the profound influences of these signaling molecules on brain development, but these studies have been confounded in that all mammalian male neurons have an XY genotype and all female neurons have an XX genotype. This simple fact had been largely ignored, or, if acknowledged, set aside as unaddressable in part because we did not know the genetic basis of testis development. The discovery of *SRY*, the sex-determining region of the Y chromosome, in the late 1990s provided the tools needed for separation of genetic sex from gonadal sex. This single gene initiates an expression cascade that leads to the production of a testis (Lovell-Badge, 1992). If *SRY* is absent or the SRY protein dysfunctional, the gonadal anlage will become an ovary. This discovery allowed for generation of mice that have the *sry* gene on an autosome, thereby allowing for genetic crosses that produce an XX animal with testes or an XY animal with ovaries. When combined with XX animals with an ovary and XY animals with a testis, these constitute what has come to be known as the four core genotypes (Arnold et al., 2004). The comparison of an XX female with an XY female allows for detection of genetic effects while comparison of an XX female with an XX male allows for detection of hormonal effects (Figure 1.3). Initial studies of this model unambiguously supported the dogma that early hormonal effects determine adult reproductive physiology and behavior (De Vries et al., 2002). However, when investigation strayed beyond those behaviors directly relevant to reproduction, a number of interesting examples of genetic sex being a pertinent variable emerged. Behavioral endpoints influenced by genetic sex include aggression (Gatewood et al., 2006) and habit formation (Quinn, Hitchcott, Umeda, Arnold, & Taylor, 2007), and neuroanatomical endpoints include vasopressin innervation of the forebrain (De Vries et al., 2002) and the number of dopamine-expressing neurons in the brainstem (Dewing et al., 2006). This is not to say that hormones are without effect on these endpoints, but rather that hormones cannot explain all the variance.

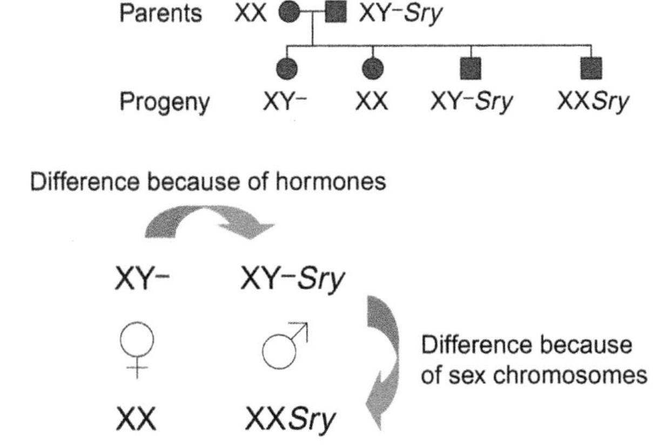

FIGURE 1.3 The four core genotypes used in studies of sexual differentiation of mice. These genotypes are developed by manipulating the mouse testis-determining gene *Sry*. The *Sry* gene is deleted from the Y chromosome, producing a Y^- chromosome, and is then added onto an autosome, to create an XY^-*Sry* genotype. Breeding these XY^-*Sry* males to XX females produces four types of offspring: XX females, XY^- females, XX*Sry* males, and XY^-*Sry* males. The XX and XY^- mice both have ovaries but differ in the genetic sex of their cells, so that differences between these females can be attributed to the complement of sex chromosomes. The XX*Sry* and XY^-*Sry* mice both have testes; therefore, these genotypes test for differences produced by sex chromosomes acting in a masculine hormonal environment. Comparing XY*Sry* and XY^- mice or XX and XX*Sry* mice allows for the observation of differences in gonadal secretions because the similar sex complements have different gonadal hormones due to the presence or absence of the *Sry* gene. *Reprinted with permission from Arnold and Burgoyne (2004).*

Several critical questions remain. What are the genes on the X or Y chromosome that contribute to sex differences in brain and behavior? How do hormones and genetics interact to produce the final phenotype? What is the relative importance of genes vs. gonads for various endpoints? With the advent of the four core genotypes, progress can be made on this interesting front.

When considering the biological basis of sex differences in the brain, the emphasis is, quite naturally, on sex differences. However, there are times when this emphasis appears misplaced or may even distort the functional interpretation of observed differences. Arginine vasopressin (AVP) has emerged as a critical regulator of complex social behaviors, in particular those that involve lasting recognition between two individuals; i.e., pair bonding and parenting (Lim & Young, 2006). The strength of a pair bond and the degree of parenting behavior varies enormously between males and females and across species. Many mammalian species show little pair bonding, or it may be essential for the female to recognize and remember former mates, whereas it is inconsequential for the male. Likewise, in many mammalian species the care of offspring is entirely in the realm of the female, with only modest input from the male, at best. Voles have proven particularly useful in comparing and contrasting the various mating systems as one species will exhibit strong pair bonding and parenting by both sexes, whereas another closely related species exhibits polygamy and unisex parenting (Carter, Williams, Witt, & Insel, 1992; Insel & Hulihan, 1995). The patterning of vasopressin innervation is highly sexually dimorphic in most rodents, and has been found to vary between species of vole as a function of reproductive strategy (De Vries & Miller, 1998; Lonstein, Rood, & De Vries, 2005; Lim & Young, 2006). Clarity was brought to this confusing collection of observations by the theoretical advance that, in monogamous species showing biparental behavior, one of the organizational hormonal effects is to induce pair bonding capacity in males (A. DeVries, M. DeVries, Taymans, & Carter, 1996). Thus, in this case, males are more like females in order to maximize their own reproductive success.

A conceptually similar interpretation has been proposed for sex differences in the hippocampus. Reports of hormonal effects on adult learning and memory extend back as far as 1926 (Ball, 1926) and remain an intensely studied topic to this day. However, the concept of hormonal modulation of learning and memory is frequently equated with sexual differentiation of learning and memory (McCarthy & Konkle, 2005). Emphasis is on the hippocampus since this is a critical brain region for spatial learning and animal models cannot be exploited to examine the two major areas of learning of interest in humans: verbal and mathematical learning. Conclusions regarding sex differences in spatial learning, which rely predominantly on the Morris water maze, have been challenged of late by the assertion that what researchers are really detecting is a sex difference in learning strategy and/or stress reactivity during the testing paradigm (Johnston & File, 1991; Perrot-Sinal, Kostenuik, Ossenkopp, & Kavaliers, 1996; Barros & Ferigolo, 1998; Kanit et al., 2000; Bielajew et al., 2003; Beiko, Lander, Hampson, Boon, & Cain, 2004; Jonasson, 2005). Similar arguments have been made for sex differences in learning in humans (Spelke, 2005). This raises the general question: is the hippocampus sexually dimorphic and, if so, is it subject to sexual differentiation along the lines of the organizational/activational hypothesis? The best answer at the moment is that it is unclear. The central tenet of the organizational/activational hypothesis is that steroid levels are elevated in one sex or the other during the critical period for sexual differentiation. This has been established clearly for reproductively relevant brain regions, such as the POA and hypothalamus, but does not hold true for the cortex or hippocampus (Amateau, Alt, Stamps, & McCarthy, 2004). When tissue levels of E_2 are quantified in these telencephalic brain regions, the levels are the same in male and female rats. Converging evidence in the adult is consistent with a view that neurons or astrocytes of some brain regions, in particular the hippocampus, are capable of *de novo* steroidogenesis of E_2 from cholesterol (Prange-Kiel, Wehrenberg, Jarry, & Rune, 2003; Hojo et al., 2004). This also may be true in the developing hippocampus, with the female brain synthesizing E_2 locally in order to equalize levels with those of the male, who generates E_2 from peripherally derived T (Nunez & McCarthy, 2008). This provocative hypothesis has not been rigorously tested but, if upheld, would be another example of one sex adapting strategies to be similar in physiology to the other.

4. ADVANCES IN MECHANISMS

4.1. Cell Birth and Cell Death

Perinatal exposure to gonadal hormones can permanently organize the volume of a brain region by influencing cell genesis, cell survival, or cell migration. To date, there is limited evidence suggesting that neonatal exposure to gonadal hormones promotes cell proliferation in the brain (Zhang, Konkle, Zup, & McCarthy, 2008). There is currently no evidence that cell migration is involved in the organizational effects of gonadal hormones on volumetric sex differences (Jacobson & Gorski, 1981; Park, Baum, Paredes, & Tobet, 1996). In fact, most of the evidence suggests that neonatal exposure to gonadal steroids likely promotes sex differences in brain and behavior by differentially affecting apoptosis in a brain region-dependent manner. For example, most of the well-characterized volumetric sex differences in the brain involve differential cell death during a perinatal sensitive period, with more cells dying in one sex than the other (Nordeen et al., 1985; Murakami & Arai, 1989; Davis, Popper, & Gorski, 1996; McCarthy, Besmer, Jacobs, Keidan, & Gibbs, 1997; Forger, 2006). During early perinatal development, the density and number of cells in the SDN-POA of males and females is similar. During the first postnatal week, however, cells in the SDN-POA of females die and establish sex differences in final volume. Neonatal exposure to T and/or E_2 decreases apoptosis or enhances cell survival, resulting in a larger SDN nucleus in males vs. females (Breedlove & Arnold, 1983; Nordeen et al., 1985; Del Abril, Segovia, & Guillamon, 1987; Davis et al., 1996). In contrast, the AVPV is larger in females than males. This is due to the ability of neonatal E_2 to enhance apoptosis of dopaminergic cells in males (Simerly, 2002). Further, mice lacking functional Bax, a proapoptotic protein, show an increased cell number in the AVPV of males (Forger et al., 2004), suggesting that apoptosis is involved in the sex difference. Gonadal steroid hormones promote apoptosis in the AVPV of males via ERα and ERβ. For example, the larger volume of the AVPV of females is precluded in female mice lacking ERα and ERβ (Bodo, Kudwa, & Rissman, 2006), and via neonatal selective activation of either receptor in females (Patchev, Gotz, & Rohde, 2004).

Sex differences in hippocampal morphology tend to be small (McCarthy & Konkle, 2005); however, males consistently have larger regional volumes than females (Allen, Hines, Shryne, & Gorski, 1989; Allen & Gorski, 1990; Madeira & Lieberman, 1995; Isgor & Sengelaub, 1998; Rabinowicz, Dean, Petetot, & De Courten-Myers, 1999; Nunez, Alt, & McCarthy, 2003a; 2003b). The larger hippocampal volume in males is associated with a greater cell number in males vs. females (Conejo et al., 2003; Hilton, Nunez, & McCarthy, 2003; Nunez et al., 2003a; 2003b). Although most volumetric sex differences occur via an effect on apoptosis, there is no evidence for a sex difference in cell death in the developing hippocampus (Nunez et al., 2003b). Further, there is a sex difference in cell genesis in the developing hippocampus, such that males have more new cells than females (Zhang et al., 2008). Interestingly, this sex difference in cell genesis is modulated by the neonatal hormonal milieu. Specifically, E_2, T, and DHT treatment enhance cell genesis in the developing hippocampus of females (Zhang et al., 2008). Consistent with this finding, both ARs and ERs are present in the hippocampus, and are highly elevated during development (Sar & Stumpf, 1973; Lieberburg, MacLusky, & McEwen, 1977; Handa, Hines, Schoonmaker, Shryne, & Gorski, 1986; Roselli, 1991; Kerr, Allore, Beck, & Handa, 1995; Ivanova & Beyer, 2000; Solum & Handa, 2001; Brannvall, Bogdanovic, Korhonen, & Lindholm, 2005; Wang, Irwin, & Brinton, 2006).

Several sex differences that are organized by gonadal steroid levels during early development have been reported in the mAMG. Males have greater nuclear volume (Mizukami et al., 1983), synaptic input (Nishizuka & Arai, 1983), and dendritic spine density (Nishizuka & Arai, 1981a; 1981b) than females. The medial amygdala contains dense ARs and ERs (Simerly, Chang, Muramatsu, & Swanson, 1990), often within the same neurons (Wood & Newman, 1995) and $P450_{aro}$ activity (Roselli et al., 1984). Previous studies have demonstrated that the sexually dimorphic mAMG is influenced by circulating gonadal hormone levels at puberty and adulthood (Cooke et al., 1999); however, this does not preclude organizational effects of gonadal hormones. Our laboratory hypothesized that, like in the hippocampus, hormone-induced sex differences in cell genesis account for sex differences in the regional volume of the AMG. Surprisingly, we found that female rat pups have a greater cell density in the developing mAMG and that this sex difference is the result of greater cell proliferation in females (Krebs-Kraft & McCarthy, unpublished findings). More experiments are needed to more accurately determine whether hormone exposure is affecting cell proliferation, differentiation, and/or survival in the developing mAMG, highlighting another important area for future research.

4.2. Old Players with New Roles

Views on GABA action have changed considerably over the past decade. Once considered the primary source of inhibition in the brain, we now know it to be a principle source of excitation via depolarization-induced calcium influx through voltage-sensitive calcium channels (VSCC). This action is most prominent developmentally and appears to be present throughout the brain (McCarthy, Auger, & Perrot-Sinal, 2002). The calcium influx induced following depolarizing GABA action places it squarely in the camp of being a trophic factor, although many of the aspects of its potential trophic action have not been well explored. The excitatory effects of GABA are mediated by the $GABA_A$ receptor, a chloride ionophore, and the relative transmembrane chloride gradient. Whether $GABA_A$ receptor activation results in chloride influx or efflux is determined by the transmembrane chloride concentration gradient, which is in turn determined by the activity and expression of chloride cotransporters (Plotkin et al., 1997a; Plotkin, Snyder, Hebert, & Delpire, 1997; Rivera, 1999; Ganguly, Schinder, Wong, & Poo, 2001). During the neonatal period, the reversal potential for chloride (E_{Cl-}) is positive relative to the resting membrane potential (Barna, Kuhnt, & Siklos, 2001), resulting in a net outward driving force upon chloride when $GABA_A$ receptors open, and membrane depolarization sufficient to open VSCCs, primarily of the L type (Leinekugel, Tseeb, Ben-Ari, & Bregestovski, 1995; Obrietan & Van den Pol, 1995; Owens, Boyce, Davis, & Kriefstein, 1996). As development progresses, E_{Cl-} becomes negative relative to the resting membrane potential, thus shifting the driving force on chloride to inward and leading to $GABA_A$ receptor-mediated hyperpolarization, the primary basis for synaptic inhibition in the mature brain.

The gradual developmental shift from GABA-mediated excitation to inhibition is controlled by changes in chloride cotransporter expression and/or activity, including NKCC1 and KCC2, along with chloride channels such as ClC2 (Schwartz-Bloom & Sah, 2001). NKCC1 promotes chloride transport into the cell along with sodium and potassium, and its expression is high in the neonatal brain but declines with advancing age. Conversely, KCC2 promotes chloride efflux and its expression is low neonatally but increases as development progresses such that, by the end of the second postnatal week, KCC2 levels are elevated and NKCC1 levels are significantly decreased (Plotkin et al., 1997a; 1997b; Delpire, 2000). Consequently, around this time, $GABA_A$ receptor activation results in chloride influx and membrane hyperpolarization (Stein, Hermans-Borgmeyer, Jentsch, & Hubner, 2004).

Our laboratory previously showed that E_2 enhances the depolarizing action of GABA in developing hypothalamic neurons by increasing the magnitude of the calcium transient with each depolarization, increasing the number of neurons that respond to $GABA_A$ receptor activation with a calcium transient, and extending the developmental duration of depolarizing GABA action (Perrot-Sinal, Davis, Gregerson, Kao, & McCarthy, 2001). The cellular mechanism(s) regulating E_2-induced increases in depolarizing GABA action are currently unknown. To date our efforts have focused on the potential for E_2 modulation of the chloride cotransporters. We hypothesize that E_2 maintains a high (relative) intracellular chloride gradient over a prolonged developmental period by increasing the expression and/or activity of NKCC1, the chloride cotransporter involved in chloride influx, and/or decreasing KCC2 and ClC2, both of which promote chloride efflux. Levels of NKCC1 are only marginally increased by E_2 treatment. However, the activity of this transporter is phosphorylation-dependent (Dowd & Forbush, 2003; Gimenez & Forbush, 2003). We recently discovered that E_2 treatment increased the amount of activated NKCC1 using a phosphorylation-specific antibody (generous gift of B. Forbush, Yale University) and found a marked increase compared to vehicle-treated control females (Perrot-Sinal et al., 2007). Increased activity of pNKCC1 should increase the intracellular concentration of chloride and enhance the depolarizing effects of $GABA_A$ receptor activation, leading to increased calcium influx via the L-type VSCCs, suggesting a plausible mechanism for how E_2 enhances muscimol-induced calcium influx.

The amino acid glutamate is a fundamental building block of proteins as well as a dominant excitatory neurotransmitter in the mammalian central nervous system (CNS). Its actions are generally rapid and mediated via two varieties of ionotropic receptors: *N*-methyl-D-aspartate (NMDA) and alpha-amino-3-hydroxy-5-methyl-4-isoxazolepropionic acid (AMPA), and a class of metabotropic G-protein-coupled receptors referred to as the mGluR. As it is ubiquitous and so essentially fundamental, glutamate has not generally been considered a reasonable candidate for mediating a process as specific and selective as sexual differentiation of the brain. Once again, however, the facts prove otherwise. In the POA, glutamate is a component of the actions of prostaglandin E_2 (PGE_2), as evidenced by the ability of AMPA receptor antagonists to block the induction of dendritic spines by the PGE_2, as reviewed in detail in Amateau & McCarthy, 2002a. Activation of the AMPA receptor only partly accounts for PGE_2 actions, however, and the cellular messengers transducing the remainder of the signal remain unknown. However, a role for PKA is emerging (Wright & McCarthy, unpublished observation).

In contrast to the POA, the role for glutamate in sexual differentiation of the hypothalamus is central. The VMN is a critical brain region for the control of female sexual behavior; thus, it is a potential substrate mediating feminization or defeminization of the brain. There is a sex difference in the neuronal architecture of the VMN but of a slightly different nature than that seen in the POA. Dendrites of male VMN neurons are longer and branch more frequently than those of female VMN neurons, and as a consequence also have more dendritic spine synapses (Mong et al., 1999; Todd et al., 2007). In VMN neurons, E_2 binds to its cogent receptor and rapidly promotes the release of glutamate from presynaptic terminals, which in turn activates postsynaptic NMDA and AMPA receptors, leading to calcium influx, activation of MAPK, and promotion of dendritic spine formation (Schwarz, Liang, Thompson, & McCarthy, 2008). The enhanced glutamate release requires E_2-induced activation of PI3K, a kinase involved in a wide range of cellular functions, and this occurs as rapidly as one hour after steroid exposure. Neither the activation of PI3K nor the enhanced glutamate release by E_2 require protein synthesis, but they do require the ER, specifically ERα. The construction and maintenance of new spines in the post-synaptic cell do not require protein synthesis, and they do not require the activation of ER (Schwarz et al., 2008). Thus, there is an essential requirement for cell-to-cell communication for the establishment of sexually dimorphic synaptic patterning in the VMN (Figure 1.4). This is similar to the POA, in which cell-to-cell communication is also critical, and yet different in that there is no apparent role for astrocytes in the VMN but there is in the POA.

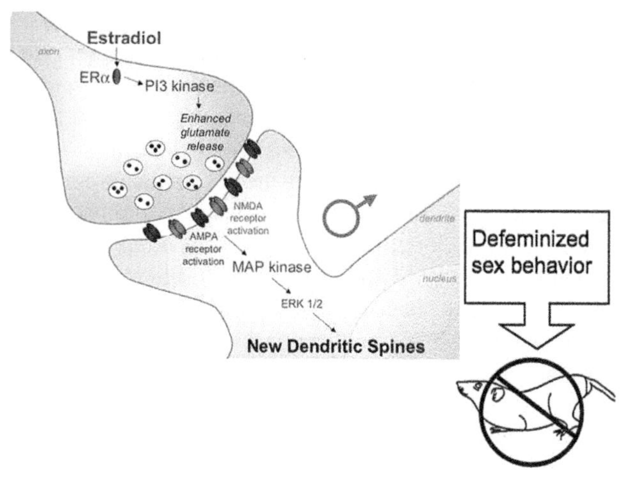

FIGURE 1.4 A model showing estradiol (E_2)-induced cell-to-cell glutamatergic communication, which can permanently organize the ventromedial nucleus (VMN) of the hypothalamus into a masculine phenotype. The ventromedial nucleus (VMN) is the critical brain region controlling expression of female sexual behavior. Dendrites on neurons in this nucleus branch more frequently in males, and also have an overall greater number of spine synapses. This sex difference is established during the perinatal sensitive period and is a function of estradiol (E_2) action in the male brain. Estradiol promotes presynaptic release of glutamate following activation of phosphatidylinositol 3-kinase (PI3K). Glutamate release enhances alpha-amino-3-hydroxy-5-methyl-4-isoxazolepropionic acid (AMPA) and *N*-methyl-D-aspartate (NMDA) glutamate receptor activity, which activate the mitogen-activated protein kinase (MAPK) pathway, leading to spine formation and possibly dendritic branching. Estradiol-induced increase in dendritic spines can be blocked by antagonism of these glutamate receptors and mimicked by the application of an NMDA receptor-specific agonist. Adult females treated with glutamate receptor agonists neonatally have defeminized sexual behavior. *Reprinted with permission from Schwarz (2008).*

The observation that E_2 promotes glutamate release to induce dendritic spines in the hypothalamus predicts that neonatal glutamate administration would be sufficient to induce defeminization. Conversely, blocking glutamate transmission neonatally should disrupt E_2-induced defeminization. Both of these predictions have proved true when assessed behaviorally or anatomically (Schwarz & McCarthy, 2008a; Schwarz et al., 2008). However, there is also a positive effect of glutamate receptor activation on the organization of male sexual behavior, suggesting a functional connection between the cellular events of masculinization and defeminization.

4.3. New Players with Old Roles

Prostaglandins are derived from arachidonic acid following cyclinization by the cyclooxygenase enzymes COX-1 and COX-2 (Kaufmann, Andreasson, Isakson, & Worley, 1997). Usually associated with inflammation, PGE_2 is emerging as an important modulator of neural activity both by indirect effects on glia (Bezzi et al., 1998; Sanzgiri,

Araque, & Haydon, 1999), and direct effects on neurons (Rage, Lee, Ma, & Ojeda, 1997; Kasai & Mizumura, 2001; McCullough et al., 2004). In the mature POA, PGE_2 is a potent stimulator of GnRH release and plays an essential role in the maturation of the pulse generator at puberty (Ma, Berg-von der Emde, Rage, Wetsel, & Ojeda, 1997); therefore, our laboratory determined whether PGE_2 might also play a role in the developing POA, and more specifically in the process of masculinization. During the sensitive period for sexual differentiation, activation of ERs in the POA induces expression of COX-2 (and COX-1), which leads to a seven-fold increase in PGE_2, but not the other prostanoids (Amateau & McCarthy, 2004). Prostaglandin E_2 induces the formation of neuronal dendritic spines via a mechanism that involves, at least in part, the release of glutamate and the activation of AMPA, but not NMDA glutamate receptors (Amateau & McCarthy, 2002a). Research by others reveals a calcium-dependent release of glutamate from astrocytes following stimulation by PGE_2 (Bezzi et al., 1998), and astrocytes of the POA are responsive to E_2 and consequently are sexually dimorphic (Amateau & McCarthy, 2002b). This has led us to propose a working model in which PGE_2 released from E_2-responsive neurons in the male promotes glutamate release from astrocytes, which in turn promotes the induction and/or maturation of dendritic spines (Figure 1.5). Treating

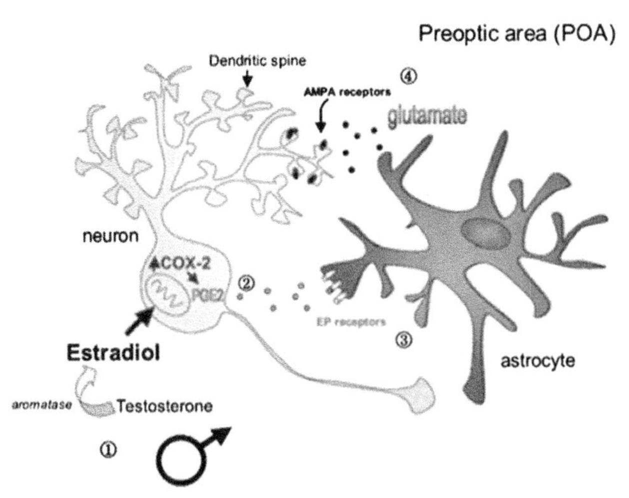

FIGURE 1.5 A model of estradiol-induced sexually dimorphic patterning. Prostaglandin E_2 (PGE_2)-induced neuronal/astrocytic crosstalk occurs in the preoptic area (POA), which is critical for controlling the expression of male sexual behavior. The POA exhibits some of the most robust sex differences in the brain. (1) Male POA neurons have around twice as many dendritic spines as those in females, and this level of spines can be induced in females by estradiol (E_2), aromatized from testosterone (T), during the perinatal sensitive period. (2) Estradiol enhances COX-2, an inducible enzyme, and induces PGE_2 synthesis. (3) Activation of E-prostanoid (EP) receptors by PGE_2 stimulates glutamate release from astrocytes in a calcium-dependent manner, and (4) glutamate induces the formation of dendritic spines, in an alpha-amino-3-hydroxy-5-methyl-4-isoxazolepropionic acid (AMPA) receptor-dependent manner. *Reproduced from Schwarz and McCarthy (2008b).*

females with either E_2 or PGE_2 results in a male-like pattern of dendritic spines, and blocking COX1/2 activation in males produces a female-like level of spines. That this is truly an organizational effect in the classic sense is confirmed by maintenance of the induced changes well into adulthood (Amateau & McCarthy, 2002a; 2004; Wright, Burks, & McCarthy, 2008).

Prostaglandins act via multiple receptors but for PGE_2 the four principals are prostaglandin E receptors 1–4 (EP1–4) (Burks, Wright, & McCarthy, 2007). Each receptor has a different affinity for PGE_2 and a unique signal transduction profile, yet the degree of crosstalk and overlap between the receptors is substantial (Regan, 2003). There are no detectable sex differences in the mRNA levels for each receptor during the sensitive period for sexual differentiation or after (Burks et al., 2007), and it appears instead that it is sex differences in the levels of COX enzymes that drive the process of masculinization (Amateau & McCarthy, 2004). All four receptors for PGE_2 are expressed in the neonatal POA, and all four have been implicated as contributing to PGE_2-induced masculinization (Burks et al., 2007), but it does not seem that all four are necessary. Convergent evidence implicates EP2 and EP4 as the dominant receptors involved (Wright et al., 2008). Activation of either EP2 or EP4 increases PKA signaling, and recent findings indicate that disrupting PKA neonatally blocks masculinization (Wright & McCarthy, unpublished observation). The activation of PKA may also explain the apparently positively reinforcing effects of PGE_2. A single intracerebroventricular injection of PGE_2 to newborn females is sufficient to induce full masculinization of dendritic spine density and adult sexual behavior (Wright et al., 2008). The activation of PKA following binding of PGE_2 to EP2 or EP4 is capable of inducing more PGE_2 production by upregulating the synthesis and activity of the COX enzymes via a glycogen synthase kinase (GSK)-mediated pathway (Regan, 2003). This positive feed-forward effect of PGE_2 likely involves both COX-1 and COX-2, which can be found in neurons, astrocytes, and microglia. Thus, distinguishing the precise cellular site of E_2 action in this system is challenging and may ultimately prove to be unimportant as the role for cell-to-cell communication emerges as a critical component of the process of masculinization.

The next frontier in understanding PGE_2-induced masculinization is to place the findings in the POA into the context of the entire circuitry controlling male and female sexual behavior. The increase in dendritic spines on POA neurons must reflect an increase in afferent input, but from whence is unknown. Does the increase in dendritic spines reflect more afferent input from the same source or the establishment of a new source of input? The relevant sources of afferent input to the POA are predominantly olfactory, including the medial amygdala, but there are also important reciprocal connections with the VMN of the hypothalamus, a critical brain region in the control of female sexual behavior.

An additional group of signaling molecules of interest is focal adhesion kinase (FAK) and its associated proteins, including paxillin (Ren et al., 2004). These molecules were first of interest in the study of metastatic cancers but have since been found to be integral to neuronal growth and branching (Yang, Ma, Chen, Wang, & Lee, 2003; Rico et al., 2004; Robles & Gomez, 2006). Focal adhesion kinase and paxillin are both significantly elevated in the neonatal female rat hypothalamus and are downregulated by E_2 (Speert et al., 2007). The decrease in FAK and paxillin induced by E_2 is correlated with an increase in neurite complexity, suggesting a plausible mechanism establishing the longer and more frequently branching dendrites of males. The observation that E_2 decreases FAK and paxillin is unusual since most, if not all, other molecules implicated in sexual differentiation involve an increase in expression in response to estrogens or androgens. Given that defeminization is the active removal of the capacity for female sexual behavior, it is plausible that suppression of the endogenously high levels of FAK and paxillin by E_2 is a mechanism to achieve that end. To our knowledge, there have been no clear examples of experimental disruption of feminization, with the exception of studies suggesting a role for E_2 in normal feminization (McCarthy, Schlenker, & Pfaff, 1993; Bakker et al., 2002b). Sex differences in neurochemistry also may offer clues as to the significance of sex differences in brain structure. An example of this concerns the AVPV. Female mice have more neurons in the AVPV than do males (Forger et al., 2004). A subset of these neurons express Kiss1-mRNA and its gene product, kisspeptin (Kp), and female mice have 10 times as many Kp neurons in the AVPV as do males (Clarkson & Herbison, 2006). In rats, the sex difference in Kp expression is close to absolute, with males expressing almost no Kp (Kauffman et al., 2007). In both species, this sex difference depends on early exposure to gonadal steroids (Kauffman et al., 2007; Gonzalez-Martinez, De Mees, Douhard, Szpirer, & Bakker, 2008) and may contribute to the sex difference in LH release. Kisspeptin-containing projections are found in close apposition to GnRH neurons (Clarkson & Herbison, 2006), and in both rats and mice Kp triggers an LH surge by stimulating GnRH neurons (Gottsch et al., 2004; Irwig et al., 2004), although indirectly. Interestingly, E_2 treatment increases Kiss1 mRNA in the AVPV but decreases it in the ARC, which contains a nondimorphic group of Kp neurons; it has been suggested that the former group contributes to the surge and the latter to the negative feedback of GTH secretion (Dungan, Clifton, & Steiner, 2006). It is tempting to speculate that the higher number of Kp neurons in the AVPV gives females the ability to respond to E_2 treatment with an LH surge, an ability that males lack. Kisspeptin is probably not the only factor, however. For example, female rats also have twice as many neurons in AVPV that express markers of both glutamatergic and GABAergic signaling (Ottem, Godwon, Krishnan, & Petersen, 2004). These dual-phenotype neurons may synapse on GnRH neurons to control the switch from negative to positive feedback of E_2 that occurs around ovulation.

4.4. The Need for and Consequences of Cell-to-cell Communication

Steroid receptors, such as ERα and ERβ, are distributed unevenly throughout the brain and therefore are differentially expressed at high levels in some cells in some brain regions but not in others. Moreover, within a particular brain region, there is no *a priori* reason to believe that every neuron or astrocyte is capable of responding to an endocrine signal to the same degree. For this reason, an early research emphasis was placed on mapping the brain according to receptor expression, with the reasonable assumption that only neurons that express a particular steroid receptor, such as ER, will respond to an E_2-initiated signal. When considering changes in dendritic morphology, it was also reasonably assumed that only those cells expressing steroid receptors would respond with an organizational change in morphology. In other words, steroid-mediated sexual differentiation of the brain was assumed to be cell-autologous, with a mixed population of neurons in which some cells were forever changed by early steroid action and others were essentially untouched. This simplistic view does appear to be true in simple systems, such as fruit flies (*Drosophila melanogaster*), but in mammalian systems the reality is much more complex. This is evident both at the level of cellular mechanisms, as well as in terms of the types of changes observed. Specifically, the cellular mechanisms reviewed above all involve a diffusible factor released from one cell to impact another. The proper sexually dimorphic innervation of the AVPV by the pBNST requires the release of a target-derived factor that attracts the growing axons. The synthesis and release of the diffusible factor is under the control of E_2, but the innervation itself is not (Polston & Simerly, 2003). An analogous situation exists in the spinal cord where a muscle-derived factor is critical to the androgen-dependent survival of the motor neurons of the SNB, and yet there is no evidence that the muscle cells express ARs (Freeman et al., 1996). On a more local level, the estrogen-induced increased synthesis and release of GABA from neurons of the ARC is critical to the sexually dimorphic morphology of the neighboring astrocytes, which neither make GABA nor express ERs (Mong et al., 1999; Mong & McCarthy, 1999; 2002; Mong et al., 2002). γ-aminobutyric acid increases the stellation of arcuate astrocytes, and this change in morphology is directly correlated with a suppression of dendritic spines, and, again, the neurons in which spinal density is being altered do not need to possess an ER to have been affected by the actions of estrogens. A similar scenario is found in the POA, where PGE_2 is released from one cell type, presumably neurons, to act on another, astrocytes (Amateau & McCarthy, 2002a). Finally, the best example of cell-to-cell communication is found in the VMN, where E_2 binds to ER in the presynaptic cell to promote glutamate release, which then acts on the postsynaptic cell and permanently organizes its dendritic morphology (Schwarz et al., 2008). The use of latrotoxin to induce the release of glutamate and mimic the actions of E_2 has definitively demonstrated that there is no need for ER in the postsynaptic neuron, regardless of whether or not it is present.

Central to observations of dual sexuality in adult animals is the nature of the neural circuits controlling male vs. female sexual behavior. As male and female sexual behaviors are so different in terms of motor patterns and hormonal dependence, it is often assumed that there are two distinct neural circuits that independently control each behavioral repertoire. Further, in an appealing analogy to the Müllerian and Wolffian duct systems, which give rise to the female vs. male reproductive tracts, respectively, masculinization could be seen as the retention of the male circuit and defeminization as the active degeneration (i.e., by Müllerian-inhibiting hormone (AMH)) of the female circuit. The shortcoming of this analogy is the ability to actually identify separate male and female neural circuits. There are clear nodes that are more critical to one sex than the other, such that an intact and functioning POA is critical for male sexual behavior and the same is true for the VMN and female sexual behavior. These two nuclei, however, are intimately connected with each other as well as being part of a larger network of nuclei integrating sensory stimuli and motivational cues to coordinate a coherent behavioral response. Careful students of anatomy such as Newman, Simerly, and Devries have long suggested that there is only one neural network controlling sexual behavior but that it is differentially weighted toward expression of either male or female motor patterns in response to specific stimuli (Newman, 1999; De Vries & Simerly, 2002). That this issue remains unresolved 50 years after the codifying of the organizational/activational hypothesis is testament to the complexity of the neural and hormonal control of reproduction.

Can studies on the mechanism of sexual differentiation provide any insight into this issue? At first blush, the observation of clearly separate processes for masculinization and feminization favors the view that there are distinct circuitries for male vs. female sexual behavior. For example, PGE_2-induced masculinization of females results in adult animals that show perfectly normal male sex behavior under conditions of high T, and perfectly normal female sex behavior following sequential E_2 and P_4 (Todd, Schwarz, & McCarthy, 2005). The converse is also true: if normal masculinization in males is blocked by inhibiting COX-2 activity, those males will show no sex behavior in adulthood, either male or female. Thus, if one uses the Müllerian/Wolffian duct analogy, PGE-masculinized females have both systems, whereas males treated neonatally with a COX inhibitor have neither system. As mentioned above, however, it is also possible to generate animals with a phenotype half way between fully masculinized and fully feminized (Schwarz & McCarthy, 2008a), and under some circumstances there is tremendous plasticity in adult sexual behavior. Does this mean that early organization of the neural substrate is not occurring? To answer this question, we must first identify what organization of the neural substrate means.

The purest examples of early organizational actions of gonadal steroids are arguably those involving differential cell death, and the two best examples of that are the larger SDN and the smaller AVPV in males. Each of these subnuclei begins as a similarly sized collection of neurons in both sexes, and then neurons die selectively in one sex due to the lack or presence of E_2, resulting in a significantly smaller nuclear volume. Cell death is permanent, and this is therefore a permanent organizational difference. However, even this seemingly irrefutable statement is challenged by the recent observation that continuing cell genesis around puberty contributes to the maintenance of the volumetric sex difference in both the SDN and AVPV (Ahmed et al., 2008). More importantly, neither of these volumetric sex differences appears to have any relevance to sexual behavior (the SDN may play a role in partner preference, whereas the AVPV is critical to the control of GTH secretion, which is sexually dimorphic). This leaves the sexual differentiation of synaptic patterning as a potential mediator of male vs. female sex behavior. As reviewed above, in the POA, males have a two- to three-fold greater density of dendritic spine synapses, whereas in the VMN males have two- to three-fold more dendritic spines because the dendrites are longer and branch more frequently. These hormonally induced differences are enduring, and levels of a marker of dendritic spines in the POA correlate with measures of masculine sexual behavior in adults. This correlation is intriguing in that it suggests a direct relationship between the number of dendritic spines and the magnitude of the expression of male sexual behavior. The more POA spine synapses, which are excitatory, the more intense is the behavior. In other words, the system is *weighted* towards male behavior by increased excitatory input onto the critical POA neurons. Females still have plenty of POA dendritic spine synapses but simply not enough to put them over a threshold necessary for expressing male sexual behavior. Steroid-induced neuronal plasticity in the adult frequently involves induction of dendritic spine synapses, and an important component of organization is likely a reduction in sensitivity of the adult brain to the sextypic hormonal profile of the opposite sex and a retention of sensitivity to one's own sextypic hormonal profile (Sodersten, 1984). The idea that steroid sensitivity is organized is not new, but the past focus has been on changes in steroid receptor levels; we are now proposing that perhaps the sensitivity is at the level of the neural plasticity, as opposed to the steroid receptors themselves. The threshold for activation of behavior is thereby determined by the previously organized sensitivity of neuronal plasticity, and the threshold can be shifted by the intensity and duration of activational hormone treatment to induce plasticity. Functionally, this must have consequences for how the brain is being organized to influence adult behavior, but precisely how remains unclear and is one of the most important future challenges that must be met in order to genuinely advance our understanding of sexual differentiation of the mammalian brain.

5. SUMMARY AND FUTURE DIRECTIONS

We have reviewed the evidence that sexual differentiation of the mammalian brain requires the activity of gonadal hormones during a sensitive period. This early exposure to gonadal hormones organizes the brain, whereas exposure to hormones in adulthood ensures appropriate and timely sex-specific behaviors; this concept is termed the organizational/activational hypothesis. Perinatal exposure to hormones such as T and E_2 is responsible for masculinization of the brain and the expression of normal male reproductive behavior in adulthood. In the absence of early exposure to gonadal hormones, the brain is, by default, feminized and mediates normal female reproductive behavior in adulthood. Testosterone and E_2 are steroids that act at receptors that are nuclear transcription factors that regulate gene expression but also have rapid effects via membrane receptors. There are multiple sexually dimorphic brain regions as a result of the organizational effects of gonadal hormones but the most pronounced are found in the hypothalamus and preoptic area. Neuroanatomical sex differences range from large volumetric differences of entire brain areas to differences in individual cell morphology and synaptic patterning. Importantly, both types of sex difference can be established by T or E_2 exposure during development and result in sex-specific physiology and behaviors.

There is a multitude of mechanisms for the organizational effects of gonadal hormones on brain and behavior; however, there is still a great deal to learn regarding the specific mechanisms of organizational actions of hormones on the developing brain and the consequences of this organization for adult brain function and behavior. The organizational effects of hormones are sufficiently enduring to be considered permanent and thereby constrain hormonal influences in adulthood. Fifty years ago, the organizational concept meshed well with the perception of the brain as a largely static organ in adulthood in which no new cells were born and neurons were connected via hard-wired circuits. Developmental hormonal effects on cell survival or synaptic connectivity would therefore *de facto* be permanent. With an increasingly sophisticated understanding of hormone action on the brain, the concept of the brain as a plastic and dynamic organ that undergoes frequent cytoarchitectural reorganization has gained increasing acceptance. Adolescent and adult neurogenesis and rapid formation and retraction of synapses are just two of the means by which a mature brain learns and responds to a changing environment. Further, we reviewed evidence that the adolescent brain and subsequent behavior is organized by gonadal steroid hormones during puberty. It is not clear whether this second organization of the brain that occurs during adolescence is an additional sensitive period that reorganizes the brain and behavior or an extension of the perinatal sensitive period (Schulz, Zehr, Salas-Ramirez, & Sisk, 2009). Finally, males can be induced to show female-typical behavior in adulthood, and, likewise, normal females will mount other females or even mount males on occasion (Beach, 1948). The degree of behavioral plasticity and in which sex the behavior predominates varies by species and strain (Goy & Goldfoot, 1975). Thus, a more liberal interpretation of the organizational/activational hypothesis allows for organizational events to be enduring but ultimately reversible.

In light of this dynamic plasticity, how can we explain the enduring organizational effects of developmental hormone exposure as the result of permanent changes in cell number or dendritic morphology? The epigenetic modification of critical genes is a reasonable candidate for organizing an enduring effect on a neural circuit that could, under specific circumstances, be at least partly undone, and this is likely to be a fruitful future area of research. Control of gene expression both across the lifespan and in some instances across generations is regulated by the degree of methylation of specific nucleotides or histones on the associated chromatin, as well as acetylation of histones. Steroid receptors affiliate closely with enzymes controlling acetylation and are potent regulators of epigenetic changes in peripheral tissues (see McCarthy et al., 2009). In the brain, histone acetylation induced by neonatal testosterone has been implicated in the differential cell death that leads to sex differences in the principle nucleus of the bed nucleus of the stria terminalis (Murray, Hien, De Vries, & Forger, 2009).

Experience is also a potent modulator of brain development and behavior. It is clear that maternal care, stress, and social interactions can all have lasting effects on brain development and behavior and can be either adaptive or maladaptive (Meaney & Szyf, 2005; Weaver, 2009). Thus, another important area of investigation is to determine the mechanisms by which environmental factors mediate the effects of gonadal steroid hormones on behavior. For example, adolescent social experiences can prevent the reproductive deficits produced by peripubertal gonadectomy (Schulz, Molenda-Figueira, & Sisk, 2009). In contrast, hormones may also indirectly influence the brain by altering social interactions and experience, leaving this an interesting avenue of exploration.

Lastly, while we are making considerable advances in the mechanistic basis of sex differences in brain areas relevant to reproduction—i.e., the preoptic area, hypothalamus, and spinal cord—we are lacking a similar level of achievement regarding other brain areas. Resolving this deficit is important for two reasons. First is the generalized assumption that the magnitude and impact of sex differences is equal in nonreproductive brain areas to that seen in the POA and hypothalamus, leading to the equally false assumption that sex differences in cognition and emotionality are robust and enduring, when in fact there is considerable evidence to the contrary. Preliminary evidence suggests that the mechanisms mediating sex differences in nonreproductive brain areas are distinct from those established in the reproductive axis, beginning with the observation of a sex difference in cell birth, not death, in the developing hippocampus (Zhang et al., 2008). The second reason is the greater probability that sex differences outside the reproductive axis may offer some explanation of the large gender bias in the frequency of neurological diseases and disorders of mental health, with boys being at substantially higher risk of autism and related spectrum disorders, attention deficit hyperactivity disorder, and early onset schizophrenia, while women suffer disproportionately from major depression, anxiety, panic, and feeding disorders (see Martel, 2009). That male-biased disorders largely have their origins in development and female-biased disorders are generally postpubertal in onset offer important clues into the nature of the gender bias and highlight the importance of a thorough in-depth understanding of how hormones impact on the developing brain.

ABBREVIATIONS

AMH Müllerian-inhibiting hormone
AMPA Alpha-amino-3-hydroxy-5-methyl-4-isoxazolepropionic acid
AR Androgen receptor
ARC Arcuate nucleus
AVP Arginine vasopressin
AVPV Anteroventral periventricular nucleus
BNST Bed nucleus of the stria terminalis
cAMP Cyclic-3',5'-adenosine monophosphate
ClC2 see Chloride channels
COX Cyclooxygenase (COX1, COX2)
CNS Central nervous system
CREB cAMP-response-element-binding protein
DHT 5α-dihydrotestosterone
E_2 Estradiol
EDC Endocrine-disrupting chemical
EP E-prostanoid
EP1–4 Prostaglandin E receptors 1–4
ER Estrogen receptor (ERα, ERβ)
ERE Estrogen response element
FAK Focal adhesion kinase
GABA γ-aminobutyric acid
GnRH Gonadotropin-releasing hormone
GPR30 G-protein-coupled estrogen receptor
GSK Glycogen synthase kinase
GTH Gonadotropin
IP_3 Inositol triphosphate
KCC2 See Chloride channels
Kp Kisspeptin
LH Luteinizing hormone
mAMG Medial amygdala
MAPK Mitogen-activated protein kinase
mGluR Metabotropic G-protein-coupled receptor
NKCC1 See Chloride channels
NMDA *N*-methyl-D-aspartate
P_4 Progesterone
$P450_{aro}$ Aromatase
pCREB Phosphorylated cAMP response element binding
PGE_2 Prostaglandin E_2
PI3K Phosphatidylinositol 3-kinase
PKA Protein kinase A
POA Preoptic area
SDN Sexually dimorphic nucleus
SNB Spinal nucleus of the bulbocavernosus
SRY Sex-determining region Y
T Testosterone
TDF Testis-determining factor
VMN Ventromedial nucleus
VGCC See VSCC
VSCC Voltage-sensitive calcium channels

REFERENCES

Abraham, I. M., Todman, M. G., Korach, K. S., & Herbison, A. E. (2004). Critical *in vivo* roles for classical estrogen receptors in rapid estrogen actions on intracellular signaling in mouse brain. *Endocrinology, 145*, 3055–3061.

Ahmed, E. I., Zehr, J. L., Schulz, K. M., Lorenz, B. H., Doncarlos, L. L., & Sisk, C. L. (2008). Pubertal hormones modulate the addition of new cells to sexually dimorphic brain regions. *Nat. Neurosci., 11*, 995–997.

Albert, D. J., Jonik, R. H., & Walsh, M. L. (1992). Hormone-dependent aggression in male and female rats: experiential, hormonal, and neural foundations. *Neurosci. Biobehav. Rev., 16*, 177–192.

Allen, L. S., & Gorski, R. A. (1990). Sex difference in the bed nucleus of the stria terminalis of the human brain. *J. Comp. Neurol., 302*, 697–706.

Allen, L. S., Hines, M., Shryne, J. E., & Gorski, R. A. (1989). Two sexually dimorphic cell groups in the human brain. *J. Neurosci., 9*, 497–506.

Amateau, S. K., & McCarthy, M. M. (2002a). A novel mechanism of dendritic spine plasticity involving estradiol induction of prostaglandin-E2. *J. Neurosci., 22*, 8586–8596.

Amateau, S. K., & McCarthy, M. M. (2002b). Sexual differentiation of astrocyte morphology in the developing rat preoptic area. *J. Neuroendocrinol., 14*, 904–910.

Amateau, S. K., & McCarthy, M. M. (2004). Induction of PGE2 by estradiol mediates developmental masculinization of sex behavior. *Nat. Neurosci., 7*, 643–650.

Amateau, S. K., Alt, J. J., Stamps, C. L., & McCarthy, M. M. (2004). Brain estradiol content in newborn rats: sex differences, regional heterogeneity, and possible *de novo* synthesis by the female telencephalon. *Endocrinology, 145*, 2906–2917.

Arnold, A. P., & Breedlove, S. M. (1985). Organizational and activational effects of sex steroids on brain and behavior: a reanalysis. *Horm. Behav., 19*, 469–498.

Arnold, A. P., & Burgoyne, P. S. (2004). Are XX and XY brain cells intrinsically different? *Trends in Endocrinology and Metabolism, 15*, 6–11.

Arnold, A. P., & Gorski, R. A. (1984). Gonadal steroid induction of structural sex differences in the central nervous system. *Ann. Rev. Neurosci., 7*, 413–442.

Arnold, A. P., Rissman, E. F., & De Vries, G. J. (2003). Two perspectives on the origin of sex differences in the brain. *Ann. NY Acad. Sci., 1007*, 176–188.

Arnold, A. P., Xu, J., Grisham, W., Chen, X., Kim, Y. H., & Itoh, Y. (2004). Minireview: Sex chromosomes and brain sexual differentiation. *Endocrinology, 145*, 1057–1062.

Aronica, S. M., Kraus, W. L., & Katzenellenbogen, B. S. (1994). Estrogen action via the cAMP signaling pathway: stimulation of adenylate cyclase and cAMP-regulated gene transcription. *Proc. Natl. Acad. Sci. USA, 91*, 8517–8521.

Auger, A. P., Hexter, D. P., & McCarthy, M. M. (2001). Sex difference in the phosphorylation of cAMP response element binding protein (CREB) in neonatal rat brain. *Brain. Res., 890*, 110–117.

Auger, A. P., Tetel, M. J., & McCarthy, M. M. (2002). Steroid receptor coactivator-1 (SRC-1) mediates the development of sex-specific brain morphology and behavior. *Proc. Natl. Acad. Sci., 97*, 7551–7555.

Bakker, J., Honda, S., Harada, N., & Balthazart, J. (2002a). Sexual partner preference requires a functional aromatase (CYP 19) gene in male mice. *Horm. Behav., 42*, 158–171.

Bakker, J., Honda, S., N., H., & Balthazart, J. (2002b). The aromatase knock-out mouse provides new evidence that estradiol is required during development in the female for the expression of sociosexual behaviors in adulthood. *J. Neurosci., 22*, 9104–9112.

Ball, J. (1926). The female sex cycle as a factor in learning in the rat. *Am. J. Physiol., 78*, 533–536.

Barna, B., Kuhnt, U., & Siklos, L. (2001). Chloride distribution in the CA1 region of newborn and adult hippocampus by light microscopic histochemistry. *Histochem. Cell. Biol., 115*, 105–116.

Barraclough, C. A. (1961). Production of anovulatory, sterile rats by single injections of testosterone propionate. *Endocrinology, 68*, 62–67.

Barraclough, C. A., & Gorski, R. A. (1962). Studies on mating behavior in the androgen-sterilized female rat in relation to the hypothalamic regulation of sexual behaviour. *J. Endocrinol., 25*, 175–182.

Barros, H. M., & Ferigolo, M. (1998). Ethopharmacology of imipramine in the forced-swimming test: gender differences. *Neurosci. Biobehav. Rev., 23*, 279–286.

Beach, F. A. (1948). *Hormones and Behavior.* New York: Paul B. Hoeber, Inc.

Beato, M., & Klug, J. (2000). Steroid hormone receptors: an update. *Hum. Reprod. Update., 6*, 225–236.

Becker, J. B., Arnold, A. P., Berkley, K. J., Blaustein, J. D., Eckel, L. A., Hampson, E., et al. (2005). Strategies and methods for research on sex differences in brain and behavior. *Endocrinology, 146*, 1650–1673.

Becu-Villalobos, D., & Libertun, C. (1995). Development of gonadotropin-releasing hormone (GnRH) neuron regulation in the female rat. *Cell. Mol. Neurobiol., 15*, 165–176.

Beiko, J., Lander, R., Hampson, E., Boon, F., & Cain, D. P. (2004). Contribution of sex differences in the acute stress response to sex differnces in water maze performance. *Behav. Brain. Res., 151*, 239–253.

Berenbaum, S. A. (1999). Effects of early androgens on sex-typed activities and interests in adolescents with congenital adrenal hyperplasia. *Horm. Behav., 35*, 102–110.

Bezzi, P., Carmignoto, G., Pasti, L., Vesce, S., Rossi, D., Rizzini, B. L., et al. (1998). Prostaglandins stimulate calcium-dependent glutamate release in astrocytes. *Nature, 391*, 281–285.

Bielajew, C., Konkle, A. T., Kentner, A. C., Baker, S. L., Stewart, A., Hutchins, A. A., et al. (2003). Strain and gender specific effects in the forced swim test: effects of previous stress exposure. *Stress, 6*, 269–280.

Bodo, C., Kudwa, A. E., & Rissman, E. F. (2006). Both estrogen receptor-alpha and -beta are required for sexual differentiation of the anteroventral periventricular area in mice. *Endocrinology, 147*, 415–420.

Brannvall, K., Bogdanovic, N., Korhonen, L., & Lindholm, D. (2005). 19-Nortestosterone influences neural stem cell proliferation and neurogenesis in the rat brain. *Eur. J. Neurosci., 21*, 871–878.

Breedlove, S. M., & Arnold, A. P. (1983). Hormonal control of a developing neuromuscular system. I. Complete demasculinization of the male rat spinal nucleus of the bulbocavernosus using the anti-androgen flutamide. *J. Neurosci., 3*, 417–423.

Burks, S. R., Wright, C. L., & McCarthy, M. M. (2007). Exploration of prostanoid receptor subtype regulating estradiol and prostaglandin E2 induction of spinophilin in developing preoptic area neurons. *Neuroscience, 146*, 1117–1127.

Carter, C. S., Williams, J. R., Witt, D. M., & Insel, T. R. (1992). Oxytocin and social bonding. *Ann. NY Acad. Sci., 652*, 204–211.

Clarke, I. J., Scaramuzzi, R. J., & Short, R. V. (1976). Sexual differentiation of the brain: endocrine and behavioural responses of androgenized ewes to oestrogen. *J. Endocrinol., 71*, 175–176.

Clarkson, J., & Herbison, A. E. (2006). Postnatal development of kisspeptin neurons in mouse hypothalamus; sexual dimorphism and projections to gonadotropin-releasing hormone neurons. *Endocrinology, 147*, 5817–5825.

Conejo, N. M., Gonzalez-Pardo, H., Pedraza, C., Navarro, F. F., Vallejo, G., & Arias, J. L. (2003). evidence for sexual difference in astrocyte of adult rat hippocampus. *Neuroscience Letters, 339*, 119–122.

Cooke, B. M., Tabibnia, G., & Breedlove, S. M. (1999). A brain sexual dimorphism controlled by adult circulating androgens. *Proc. Natl. Acad. Sci. USA, 96*, 7538–7540.

Davis, E. C., Popper, P., & Gorski, R. A. (1996). The role of apoptosis in sexual differentiation of the rat sexually dimorphic nucleus of the preoptic area. *Brain. Res., 734*, 10–18.

De Vries, G. J., & Miller, M. A. (1998). Anatomy and function of extrahypothalamic vasopressin systems in the brain. *Prog. Brain. Res., 119*, 3–20.

De Vries, G. J., & Simerly, R. B. (2002). Anatomy, development and funtion of sexually dimorphic neural circuits in the mammalian brain. In D. W. Pfaff, A. P. Arnold, A. M. Etgen, S. E. Fahrbach, & R. T. Rubin (Eds.). *Hormones, Brain and Behavior, Vol. 4* (1st ed. pp. 137–192). New York, NY: Academic Press.

DeVries, A. C., DeVries, M. B., Taymans, S. E., & Carter, C. S. (1996). The effects of stress on social preferences are sexually dimorphic in prairie voles. *Proc. Natl. Acad. Sci. USA, 93*, 11980–11984.

De Vries, G. J., Rissman, E. F., Simerly, R. B., Yang, L. Y., Scordalakes, E. M., Auger, C. J., et al. (2002). A model system for study of sex chromosome effects on sexually dimorphic neural and behavioral traits. *J. Neurosci., 22*, 9005–9014.

DeFranco, D. B., Ramakrishnan, C., & Tang, Y. (1998). Molecular chaperones and subcellular trafficking of steroid receptors. *J. Steroid. Biochem. Mol. Biol., 65*, 51–58.

Del Abril, A., Segovia, S., & Guillamon, A. (1987). The bed nucleus of the stria terminalis in the rat: regional sex differences controlled by gonadal steroids early after birth. *Dev. Brain. Res., 32*, 295–300.

Delpire, E. (2000). Cation-Chloride Cotransporters in Neuronal Communication. *News. Physiol. Sci., 15*, 309–312.

Dewing, P., Chiang, C. W., Sinchak, K., Sim, H., Fernagut, P. O., Kelly, S., et al. (2006). Direct regulation of adult brain function by the male-specific factor SRY. *Curr. Biol., 16*, 415–420.

Diaz, D. R., Fleming, D. E., & Rhees, R. W. (1995). The hormone-sensitive early postnatal periods for sexual differentiation of feminine behavior and luteinizing hormone secretion in male and female rats. *Developmental Brain Research, 86*, 227–232.

DonCarlos, L. L. (1996). Developmental profile and regulation of estrogen receptor (ER) mRNA expression in the preoptic area of prenatal rats. *Dev. Brain. Res., 20*, 224–233.

DonCarlos, L. L., McAbee, M., Ramer-Quinn, D. S., & Stancik, D. M. (1995). Estrogen receptor mRNA levels in the preoptic area of neonatal rats are responsive to hormone manipulation. *Brain. Res. Dev. Brain. Res., 84*, 253–260.

Dowd, B. F., & Forbush, B. (2003). PASK (proline-alanine-rich STE20-related kinase), a regulatory kinase of the Na–K–Cl cotransporter (NKCC1). *J. Biol. Chem., 278*, 27347–27353.

Dungan, H. M., Clifton, D. K., & Steiner, R. A. (2006). Minireview: kisspeptin neurons as central processors in the regulation of gonadotropin-releasing hormone secretion. *Endocrinology, 147*, 1154–1158.

Enmark, E., & Gustafsson, J. A. (1999). Oestrogen receptors—an overview. *J. Intern. Med., 246*, 133–138.

Ferris, C. F., Melloni, R. H., Jr., Koppel, G., Perry, K. W., Fuller, R. W., & Delville, Y. (1997). Vasopressin/serotonin interactions in the anterior hypothalamus control aggressive behavior in golden hamsters. *J. Neurosci., 17*, 4331–4340.

Forger, N. G. (2006). Cell death and sexual differentiation of the nervous system. *Neuroscience, 138*, 929–938.

Forger, N. G., Rosen, G. J., Waters, E. M., Jacob, D., Simerly, R. B., & De Vries, G. J. (2004). Deletion of Bax eliminates sex differences in the mouse forebrain. *PNAS, 101*, 13666–13671.

Foster, D. L., Padmanabhan, V., Wood, R. I., & Robinson, J. E. (2002). Sexual differentiation of the neuroendocrine control of gonadotrophin secretion: concepts derived from sheep models. *Reprod. Suppl., 59*, 83–99.

Freeman, L. M., Watson, N. V., & Breedlove, S. M. (1996). Androgen spares androgen-insensitive motoneurons from apoptosis in the spinal nucleus of the bulbocavernosus in rats. *Horm. Behav., 30*, 424–433.

Ganguly, K., Schinder, A. F., Wong, S. T., & Poo, M. (2001). GABA itself promotes the developmental switch of neuronal GABAergic responses from excitation to inhibition. *Cell, 105*, 521–532.

Gatewood, J. D., Wills, A., Shetty, S., Xu, J., Arnold, A. P., Burgoyne, P. S., et al. (2006). Sex chromosome complement and gonadal sex influence aggressive and parental behaviors in mice. *J. Neurosci., 26*, 2335–2342.

Gimenez, I., & Forbush, B. (2003). Short-term stimulation of the renal Na–K–Cl cotransporter (NKCC2) by vasopressin involves phosphorylation and membrane translocation of the protein. *J. Biol. Chem., 278*, 26946–26951.

Glickman, S. E., Frank, L. G., Licht, P., Yalcinkaya, T., Siiteri, P. K., & Davidson, J. (1992). Sexual differentiation of the female spotted hyena. One of nature's experiments. *Ann. NY Acad. Sci., 662*, 135–159.

Glickman, S. E., Short, R. V., & Renfree, M. B. (2005). Sexual differentiation in three unconventional mammals: spotted hyenas, elephants and tammar wallabies. *Horm. Behav., 48*, 403–417.

Goldman, B. D., & Gorski, R. A. (1971). Effects of gonadal steroids on the secretion of LH and FSH in neonatal rats. *Endocrinology, 89*, 112–115.

Gonzalez-Martinez, D., De Mees, C., Douhard, Q., Szpirer, C., & Bakker, J. (2008). Absence of gonadotropin-releasing hormone 1 and Kiss1 activation in alpha-fetoprotein knockout mice: prenatal estrogens defeminize the potential to show preovulatory luteinizing hormone surges. *Endocrinology, 149*, 2333–2340.

Gorski, R. A., Gordon, J. H., Shryne, J. E., & Southam, A. M. (1978). Evidence for a morphological sex difference within the medial preoptic area of the rat brain. *Brain Res, 148*, 333–346.

Gottsch, M. L., Cunningham, M. J., Smith, J. T., Popa, S. M., Acohido, B. V., Crowley, W. F., et al. (2004). A role for kisspeptins in the regulation of gonadotropin secretion in the mouse. *Endocrinology, 145*, 4073–4077.

Goy, R. W., & Goldfoot, D. A. (1975). Neuroendocrinology: animal models and problems of human sexuality. *Arch. Sex. Behav., 4*, 405–420.

Goy, R. W., Bridson, W. E., & Young, W. C. (1964). Period of maximal susceptibility of the prenatal female guinea pig to masculinizing actions of testosterone propionate. *J. Comp. Physiol. Psychol., 57*, 166–174.

Gu, G. B., & Simerly, R. B. (1997). Projections of the sexually dimorphic anteroventral periventricular nucleus in the female rat. *J. Comp. Neurol., 384*, 142–164.

Guillamon, A., Segovia, S., & Del Abril, A. (1988). Early effects of gonadal steroids on the neuron number in the medial posterior region and the lateral division of the bed nucleus of the stria terminalis in the rat. *Brain. Res. Dev. Brain. Res., 44*, 281–290.

Handa, R. J., Hines, M., Schoonmaker, J. N., Shryne, J. E., & Gorski, R. A. (1986). Evidence that serotonin is involved in the sexually dimorphic development of the preoptic area in the rat brain. *Dev. Brain. Res., 30*, 278–282.

He, B., Kemppainen, J. A., Voegel, J. J., Gronemeyer, H., & Wilson, E. M. (1999). Activation function 2 in the human androgen receptor ligand binding domain mediates interdomain communication with the NH (2)-terminal domain. *J. Biol. Chem., 274*, 37219–37225.

Heinlein, C. A., & Chang, C. (2002). Androgen receptor (AR) coregulators: an overview. *Endocr. Rev., 23*, 175–200.

Herbison, A. E. (1998). Multimodal influence of estrogen upon gonadotropin-releasing hormone neurons. *Endocr. Rev., 19*, 302–330.

Herman, R. A., Jones, B., Mann, D. R., & Wallen, K. (2000). Timing of prenatal androgen exposure: anatomical and endocrine effects on juvenile male and female rhesus monkeys. *Horm. Behav., 38*, 52–66.

Hilton, G. D., Nunez, J. L., & McCarthy, M. M. (2003). Sex differences in response to kainic acid and estradiol in the hippocampus of newborn rats. *Neuroscience, 116*, 383–391.

Hines, M. (2002). Sexual differentiation of human brain and behavior. In D. Pfaff (Ed.), *Hormones, Brain and Behavior, Vol. 5* (pp. 425–462). London, UK: Academic Press.

Hines, M., Allen, L. S., & Gorski, R. A. (1992). Sex differences in subregions of the medial nucleus of the amygdala and the bed nucleus of the stria terminalis of the rat. *Brain. Res., 579*, 321–326.

Hojo, Y., Hattori, T. A., Enami, T., Furukawa, A., Suzuki, K., Ishii, H. T., et al. (2004). Adult male rat hippocampus synthesizes estradiol from pregnenolone by cytochromes P45017alpha and P450 aromatase localized in neurons. *Proc. Natl. Acad. Sci. USA, 101*, 865–870.

Ibanez, M. A., Gu, G., & Simerly, R. B. (2001). Target-dependent sexual differentiation of a limbic–hypothalamic neural pathway. *J. Neurosci., 21*, 5652–5659.

Insel, T. R., & Hulihan, T. J. (1995). A gender-specific mechanism for pair bonding: oxytocin and partner preference formation in monogamous voles. *Behav. Neurosci., 109*, 782–789.

Irwig, M. S., Fraley, G. S., Smith, J. T., Acohido, B. V., Popa, S. M., Cunningham, M. J., et al. (2004). Kisspeptin activation of gonadotropin releasing hormone neurons and regulation of KiSS-1 mRNA in the male rat. *Neuroendocrinology, 80*, 264–272.

Isgor, C., & Sengelaub, D. R. (1998). Prenatal gonadal steroids affect adult spatial behavior, CA1 and CA3 pyramidal cell morphology in rats. *Horm. Behav., 34*, 183–198.

Ivanova, T., & Beyer, C. (2000). Ontogenetic expression and sex differences of aromatase and estrogen receptor-alpha/beta mRNA in the mouse hippocampus. *Cell. Tissue. Res., 300*, 231–237.

Jacobson, C. D., & Gorski, R. A. (1981). Neurogenesis of the sexually dimorphic nucleus of the preoptic area in the rat. *J. Comp. Neurol., 196*, 519–529.

Johnston, A. L., & File, S. E. (1991). Sex differences in animal tests of anxiety. *Physiol. Behav., 49*, 245–250.

Jonasson, Z. (2005). Meta-analysis of sex differences in rodent models of learning and memory: a review of behavioral and biological data. *Neurosci. Biobehav. Rev., 28*, 811–825.

Jost, A. (1947). Reserches sur la différenciation sexuelle de l'embryon de lapin. *Arch. Anat. Microsc. Morphol. Exp., 36*, 271–315.

Kanit, L., Taskiran, D., Yilmaz, O. A., Balkan, B., Demirgoren, S., Furedy, J. J., et al. (2000). Sexually dimorphic cognitive style in rats emerges after puberty. *Brain. Res. Bull., 52*, 243–248.

Kasai, M., & Mizumura, K. (2001). Effects of PGE_2 on neurons from rat dorsal root ganglia in intact and adjuvant-inflamed rats: role of NGF on PGE_2-induced depolarization. *Neurosci. Res., 41*, 345–353.

Kauffman, A. S., Park, J. H., McPhie-Lalmansingh, A. A., Gottsch, M. L., Bodo, C., Hohmann, J. G., et al. (2007). The kisspeptin receptor GPR54 is required for sexual differentiation of the brain and behavior. *J. Neurosci., 27*, 8826–8835.

Kaufmann, W., Andreasson, K., Isakson, P., & Worley, P. (1997). Cyclooxygenases and the central nervous system. *Prostaglandins, 54*, 601–624.

Kerr, J. E., Allore, R. J., Beck, S. G., & Handa, R. J. (1995). Distribution and hormonal regulation of androgen receptor (AR) and AR messenger ribonucleic acid in the rat hippocampus. *Endocrinology, 136*, 3213–3221.

Klinge, C. M. (2000). Estrogen receptor interaction with co-activators and co-repressors. *Steroids, 65*, 227–251.

Klinge, C. M., Kaur, K., & Swanson, H. I. (2000). The aryl hydrocarbon receptor interacts with estrogen receptor alpha and orphan receptors COUP-TFI and ERRalpha1. *Arch. Biochem. Biophys., 373*, 163–174.

Koopman, P., Gubbay, J., Vivian, N., Goodfellow, P., & Lovell-Badge, R. (1991). Male development of chromosomally female mice transgenic for *Sry*. *Nature, 351*, 117–121.

Kuroki, Y., Fukushima, K., Kanda, Y., Mizuno, K., & Watanabe, Y. (2000). Putative membrane-bound estrogen receptors possibly stimulate mitogen-activated protein kinase in the rat hippocampus. *Eur. J. Pharmacol., 400*, 205–209.

Leinekugel, X., Tseeb, V., Ben-Ari, Y., & Bregestovski, P. (1995). Synaptic $GABA_A$ activation induces Ca^{2+} rise in pyramidal cells and interneurons from rat neonatal hippocampal slices. *J. Physiol., 487*, 319–329.

Lenroot, R. K., Gogtay, N., Greenstein, D. K., Wells, E. M., Wallace, G. L., Clasen, L. S., et al. (2007). Sexual dimorphism of brain developmental trajectories during childhood and adolescence. *Neuroimage, 36*, 1065–1073.

Lieberburg, I., MacLusky, N. J., & McEwen, B. S. (1977). 5alpha-Dihydrotestosterone (DHT) receptors in rat brain and pituitary cell nuclei. *Endocrinology, 100*, 598–607.

Lim, M. M., & Young, L. J. (2006). Neuropeptidergic regulation of affiliative behavior and social bonding in animals. *Horm. Behav., 50*, 506–517.

Lonstein, J. S., Rood, B. D., & De Vries, G. J. (2005). Unexpected effects of perinatal gonadal hormone manipulations on sexual differentiation of the extrahypothalamic arginine-vasopressin system in prairie voles. *Endocrinology, 146*, 1559–1567.

Lovell-Badge, R. (1992). The role of *Sry* in mammalian sex determination. *Ciba. Found. Symp., 165*, 162–179, discussion 179–182.

Ma, Y. J., Berg-von der Emde, K., Rage, F., Wetsel, W. C., & Ojeda, S. R. (1997). Hypothalamic astrocytes respond to transforming growth factor-alpha with the secretion of neuroactive substances that stimulate the release of luteinizing hormone-releasing hormone. *Endocrinology, 138*, 19–25.

MacLusky, N. J., & Naftolin, F. (1981). Sexual differentiation of the central nervous system. *Science, 211*, 1294–1302.

MacLusky, N. J., Lieberburg, I., & McEwen, B. S. (1979). The development of estrogen receptor systems in the rat brain: perinatal development. *Brain. Res., 178*, 129–142.

MacLusky, N. J., Walters, M. J., Clark, A. S., & Toran-Allerand, C. D. (1994). Aromatase in the cerebral cortex, hippocampus, and midbrain: ontogeny and developmental implications. *Molec. Cell. Neurosci., 5*, 691–698.

Madeira, M. D., & Lieberman, A. R. (1995). Sexual dimorphism in the mammalian limbic system. *Prog. Neurobiol., 45*, 275–333.

Martel, M. M. (2009). Research review: a new perspective on attention-deficit/hyperactivity disorder: emotion dysregulation and trait models. *J. Child. Psychol. Psychiatry., 50*, 1042–1051.

Matsumoto, A., & Arai, Y. (1983). Sex difference in volume of the ventromedial nucleus of the hypothalamus in the rat. *Endocrinol. Jpn, 30*, 227–280.

Matsumoto, A., & Arai, Y. (1986). Male–female differences in synaptic organization of the ventromedial nucleus of the hypothalamus in the rat. *Neuroendocrinolgy, 42*, 232–236.

McCarthy, M. M. (2008). Estradiol and the developing brain. *Physiol. Rev., 88*, 91–124.

McCarthy, M. M., & Konkle, A. T. (2005). When is a sex difference not a sex difference? *Front. Neuroendocrinol., 26*, 85–102.

McCarthy, M. M., Auger, A. P., Bale, T. L., De Vries, G. J., Dunn, G. A., Forger, N. G., et al. (2009). The epigenetics of sex differences in the brain. *J. Neurosci., 29*, 12815–12823.

McCarthy, M. M., Auger, A. P., & Perrot-Sinal, T. S. (2002). Getting excited about GABA and sex differences in the brain. *TINS, 25*, 307–312.

McCarthy, M. M., Besmer, H. R., Jacobs, S. C., Keidan, G. M., & Gibbs, R. B. (1997). Influence of maternal grooming, sex and age on Fos immunoreactivity in the preoptic area of neonatal rats: implications for sexual differentiation. *Dev. Neurosci., 19*, 488–496.

McCarthy, M. M., Schlenker, E. H., & Pfaff, D. W. (1993). Enduring consequences of neonatal treatment with antisense oligodeoxynucleotides to estrogen receptor messenger ribonucleic acid on sexual differentiation of rat brain. *Endocrinology, 133*, 433–439.

McCullough, L., Wu, L., Haughey, N., Liang, X., Hand, T., Wang, Q., et al. (2004). Neuroprotective function of the PGE2 EP2 receptor in cerebral ischemia. *J. Neurosci., 24*, 257–268.

McEwen, B. S., Lieberburg, I., Chaptal, C., & Krey, L. C. (1977). Aromatization: important for sexual differentiation of the neonatal rat brain. *Hormones and Behavior, 9*, 249–263.

Meaney, M. J., & Szyf, M. (2005). Environmental programming of stress responses through DNA methylation: life at the interface between a dynamic environment and a fixed genome. *Dialogues. Clin. Neurosci., 7*, 103–123.

Michels, G., & Hoppe, U. C. (2008). Rapid actions of androgens. *Front. Neuroendocrinol., 29*, 182–198.

Mizukami, S., Nishizuka, M., & Arai, Y. (1983). Sexual differences in nuclear volume and its ontogeny in the rat amygdala. *Exper. Neurol., 79*, 569–575.

Mong, J. A., & McCarthy, M. M. (1999). Steroid-induced developmental plasticity in hypothalamic astrocytes: implications for synaptic patterning. *J. Neurobiol., 40*, 602–619.

Mong, J. A., & McCarthy, M. M. (2002). Ontogeny of sexually dimorphic astrocytes in the neonatal rat arcuate. *Brain. Res. Dev. Brain. Res., 139*, 151–158.

Mong, J. A., Glaser, E., & McCarthy, M. M. (1999). Gonadal steroids promote glial differentiation and alter neuronal morphology in the developing hypothalamus in a regionally specific manner. *J. Neurosci., 19*, 1464–1472.

Mong, J. A., Kurzweil, R. L., Davis, A. M., Rocca, M. S., & McCarthy, M. M. (1996). Evidence for sexual differentiation of glia in rat brain. *Horm. Behav., 30*, 553–562.

Mong, J. A., Nunez, J. L., & McCarthy, M. M. (2002). GABA mediates steroid-induced astrocyte differentiation in the neonatal rat hypothalamus. *J. Neuroendocrinol., 14*, 45–55.

Mong, J. A., Roberts, R. C., Kelly, J. J., & McCarthy, M. M. (2001). Gonadal steroids reduce the density of axospinous synapses in the developing rat arcuate nucleus: an electron microscopy analysis. *J. Comp. Neurol., 432*, 259–267.

Murakami, S., & Arai, Y. (1989). Neuronal death in the developing sexually dimorphic periventricular nucleus of the preopti area in the female rat: effect of neonatal androgen treatment. *Neurosci. Lett., 102*, 185–190.

Murray, E. K., Hien, A., De Vries, G. J., & Forger, N. G. (2009). Epigenetic control of sexual differentiation of the bed nucleus of the stria terminalis. *Endocrinology, 150*, 4241–4247.

Naftolin, F., Ryan, K. J., Davies, I. J., Reddy, V. V., Flores, F., Petro, Z., et al. (1975). The formation of estrogens by central neuroendocrine tissues. *Recent Prog. Horm. Res., 31*, 295–319.

Nevison, C. M., Brown, G. R., & Dixson, A. F. (1997). Effects of altering testosterone in early infancy on social behaviour in captive yearling rhesus monkeys. *Physiol. Behav., 62*, 1397–1403.

Newman, S. W. (1999). The medial extended amygdala in male reproductive behavior. A node in the mammalian social behavior network. *Ann. NY Acad. Sci., 877*, 242–257.

Nilsson, S., Makela, S., Treuter, E., Tujague, M., Thomsen, J., Andersson, G., et al. (2001). Mechanisms of estrogen action. *Physiol. Rev., 81*, 1535–1565.

Nishizuka, M., & Arai, Y. (1981a). Organizational action of estrogen on synaptic pattern in the amygdala: implications for sexual differentiation of the brain. *Brain. Res., 213*, 422–426.

Nishizuka, M., & Arai, Y. (1981b). Sexual dimorphism in synaptic organization in the amygdala and its dependence on neonatal hormone environment. *Brain. Res., 212*, 31–38.

Nishizuka, M., & Arai, Y. (1983). Male–female differences in the intra-amygdaloid input to the medial amygdala. *Exp. Brain. Res., 52*, 328–332.

Nordeen, E. J., Nordeen, K. W., Sengelaub, D. R., & Arnold, A. P. (1985). Androgens prevent normally occurring cell death in a sexually dimorphic spinal nucleus. *Science., 229*, 671–673.

Nunez, J. L., & McCarthy, M. M. (2008). Androgens predispose males to GABA(A)-mediated excitotoxicity in the developing hippocampus. *Exp. Neurol.*.

Nunez, J. L., Alt, J. J., & McCarthy, M. M. (2003a). A new model for prenatal brain damage. I. GABAA receptor activation induces cell death in developing rat hippocampus. *Exp. Neurol., 181*, 258–269.

Nunez, J. L., Alt, J. J., & McCarthy, M. M. (2003b). A novel model for prenatal brain damage. II. Long-term deficits in hippocampal cell number and hippocampal-dependent behavior following neonatal GABAA receptor activation. *Exp. Neurol., 181*, 270–280.

Nunez, J. L., Bambrick, L. L., Krueger, B. K., & McCarthy, M. M. (2005). Prolongation and enhancement of gamma-aminobutyric acid receptor mediated excitation by chronic treatment with estradiol in developing rat hippocampal neurons. *Eur. J. Neurosci., 21*, 3251–3261.

Obrietan, K., & Van den Pol, A. N. (1995). GABA neurotransmission in the hypothalamus: developmental reversal from Ca2+ elevating to depressing. *The Journal of Neuroscience, 15*, 5065–5077.

Olesen, K. M., & Auger, A. P. (2005). Sex differences in Fos protein expression in the neonatal rat brain. *J. Neuroendocrinol., 17*, 255–261.

Olster, D. H., & Blaustein, J. D. (1988). Progesterone facilitation of lordosis in male and female Sprague-Dawley rats following priming with estradiol pulses. *Horm. Behav., 22*, 294–304.

Osborne, C. K., Zhao, H., & Fuqua, S. A. (2000). Selective estrogen receptor modulators: structure, function, and clinical use. *J. Clin. Oncol., 18*, 3172–3186.

Ottem, E. N., Godwon, J. G., Krishnan, S., & Petersen, S. L. (2004). Dual-phenotype GABA/glutamate neurons in adult preoptic area: Sexual dimorphism and function. *J. Neurosci., 24*, 8097–8105.

Owens, D. F., Boyce, L. H., Davis, M. B. E., & Kriefstein, A. R. (1996). Excitatory GABA responses in embryonic and neonatal cortical slices demonstrated by gramicidin perforated-patch recordings and calcium imaging. *J. Neurosci., 16*, 6414–6423.

Park, J. J., Baum, M. J., Paredes, R. G., & Tobet, S. A. (1996). Neurogenesis and cell migration into the sexually dimorphic preoptic area/anterior hypothalamus of the fetal ferret. *J. Neurobiol., 30*, 315–328.

Patchev, A. V., Gotz, F., & Rohde, W. (2004). Differential role of estrogen receptor isoforms in sex-specific brain organization. *Faseb. J., 18*, 1568–1570.

Perrot-Sinal, T. S., Davis, A. M., Gregerson, K. A., Kao, J. P. Y., & McCarthy, M. M. (2001). Estradiol enhances excitatory gamma-aminobutyric acid-mediated calcium signaling in neonatal hypothalamic neurons. *Endocrinology, 143*, 2238–2243.

Perrot-Sinal, T. S., Kostenuik, M. A., Ossenkopp, K. P., & Kavaliers, M. (1996). Sex differences in performance in the Morris water maze and the effects of initial nonstationary hidden platform training. *Behav. Neurosci., 110*, 1309–1320.

Perrot-Sinal, T. S., Sinal, C. J., Reader, J. C., Speert, D. B., & McCarthy, M. M. (2007). Sex differences in the chloride cotransporters, NKCC1 and KCC2, in the developing hypothalamus. *J. Neuroendocrinol., 19*, 302–308.

Pfaff, D. W., & Keiner, M. (1973). Atlas of estradiol-concentrating cells in the central nervous system of the female rat. *J. Comp. Neurol., 151*, 121–158.

Phoenix, C. H., Goy, R. W., Gerall, A. A., & Young, W. C. (1959). Organizing action of prenatally administered testosterone propionate on the tissues mediating mating behavior in the female guinea pig. *Endocrinology, 65*, 369–382.

Plotkin, M. D., Kaplan, M. R., Peterson, L. N., Gullans, S. R., Herbert, S. C., & Delpire, E. (1997a). Expression of the Na–K–2Cl cotransporter BSC2 in the nervous system. *American Journal of Physiology, 272*, C173–C183.

Plotkin, M. D., Snyder, E. Y., Hebert, S. C., & Delpire, E. (1997b). Expression of the Na–K–2Cl- cotransporter is developmentally regulated in postnatal rat brains: a possible mechanism underlying GABA's excitatory role in immature brain. *Journal of Neurobiology, 33*, 781–795.

Polston, E. K., & Simerly, R. B. (2003). Sex-specific patterns of galanin, cholecystokinin, and substance P expression in neurons of the principal bed nucleus of the stria terminalis are differentially reflected within three efferent preoptic pathways in the juvenile rat. *J. Comp. Neurol., 465*, 551–559.

Polston, E. K., Gu, G., & Simerly, R. B. (2004). Neurons in the principle nucleus of the bed nuclei of the stria terminalis provide a sexually dimorphic GABAergic input to the anteroventral periventricular nucleus of the hypothalamus. *Neuroscience, 123*, 793–803.

Pozzo-Miller, L. D., & Aoki, A. (1991). Stereological analysis of the hypothalamic ventromedial nucleus II. Hormone induced changes in the synaptogenic pattern. *Developmental Brain Research, 61*, 189–196.

Prange-Kiel, J., Wehrenberg, U., Jarry, H., & Rune, G. M. (2003). Para/autocrine regulation of estrogen receptors in hippocampal neurons. *Hippocampus, 13*, 226–234.

Prossnitz, E. R., Arterburn, J. B., & Sklar, L. A. (2007). GPR30: A G protein-coupled receptor for estrogen. *Mol. Cell. Endocrinol., 265-266*, 138–142.

Quinn, J. J., Hitchcott, P. K., Umeda, E. A., Arnold, A. P., & Taylor, J. R. (2007). Sex chromosome complement regulates habit formation. *Nat. Neurosci., 10*, 1398–1400.

Rabinowicz, T., Dean, D. E., Petetot, J. M., & De Courten-Myers, G. M. (1999). Gender differences in the human cerebral cortex: more neurons in males; more processes in females. *J. Child. Neurol., 14*, 98–107.

Rage, F., Lee, B. J., Ma, Y. J., & Ojeda, S. R. (1997). Estradiol enhances prostaglandin E2 receptor gene expression in luteinizing hormone-releasing hormone (LHRH) neurons and facilitates the LHRH response to PGE2 by activating a glia-to-neuron signaling pathway. *J. Neurosci., 17*, 9145–9156.

Regan, J. W. (2003). EP2 and EP4 prostanoid receptor signaling. *Life Sci., 74*, 143–153.

Ren, X. R., Ming, G. L., Xie, Y., Hong, Y., Sun, D. M., Zhao, Z. Q., et al. (2004). Focal adhesion kinase in netrin-1 signaling. *Nat. Neurosci., 7*, 1204–1212.

Revankar, C. M., Cimino, D. F., Sklar, L. A., Arterburn, J. B., & Prossnitz, E. R. (2005). A transmembrane intracellular estrogen receptor mediates rapid cell signaling. *Science, 307*, 1625–1630.

Rhoda, J., Corbier, P., & Roffi, J. (1984). Gonadal steroid concentrations in serum and hypothalamus of the rat at birth: aromatization of testosterone to 17 beta-estradiol. *Endocrinology, 114*, 1754–1760.

Rico, B., Beggs, H. E., Schahin-Reed, D., Kimes, N., Schmidt, A., & Reichardt, L. F. (2004). Control of axonal branching and synapse formation by focal adhesion kinase. *Nat. Neurosci., 7*, 1059–1069.

Rivera, C., Voipio, J., Payne, J. A., Ruusuvuori, E., Lahtinen, H., Lamsa, K., et al. (1999). The K+/Cl- co-transporter KCC2 renders GABA hyperpolarizing during neuronal maturation. *Nature, 397*, 251–255.

Robles, E., & Gomez, T. M. (2006). Focal adhesion kinase signaling at sites of integrin-mediated adhesion controls axon pathfinding. *Nat. Neurosci., 9*, 1274–1283.

Romeo, R. D. (2003). Puberty: a period of both organizational and activational effects of steroid hormones on neurobehavioural development. *J. Neuroendocrinol., 15*, 1185–1192.

Roselli, C. E. (1991). Sex differences in androgen receptors and aromatase activity in microdissected regions of the rat brain. *Endocrinology, 128*, 1310–1316.

Roselli, C. E., & Klosterman, S. A. (1998). Sexual differentiation of aromatase activity in the rat brain: effects of perinatal steroid exposure. *Endocrinology, 139*, 3193–3201.

Roselli, C. E., & Resko, J. A. (1993). Aromatase activity in the rat brain: hormonal regulation and sex differences. *J. Steroid. Biochem. Mol. Biol., 44*, 499–508.

Roselli, C. E., Ellinwood, W. E., & Resko, J. A. (1984). Regulation of brain aromatase activity in rats. *Endocrinology, 114*, 192–200.

Roy, A. K., Lavrovsky, Y., Song, C. S., Chen, S., Jung, M. H., Velu, N. K., et al. (1999). Regulation of androgen action. *Vitam. Horm., 55*, 309–352.

Sanzgiri, R. P., Araque, A., & Haydon, P. G. (1999). Prostaglandin E(2) stimulates glutamate receptor-dependent astrocyte neuromodulation in cultured hippocampal cells. *J. Neurobiol., 41*, 221–229.

Sar, M., & Stumpf, W. E. (1973). Autoradiographic localization of radioactivity in the rat brain after the injection of 1,2-3H-testosterone. *Endocrinology, 92*, 251–256.

Schulz, K. M., Molenda-Figueira, H. A., & Sisk, C. L. (2009a). Back to the future: The organizational–activational hypothesis adapted to puberty and adolescence. *Horm. Behav., 55*, 597–604.

Schulz, K. M., Richardson, H. N., Romeo, R. D., Morris, J. A., Lookingland, K. J., & Sisk, C. L. (2003). Medial preoptic area dopaminergic responses to female pheromones develop during puberty in the male Syrian hamster. *Brain. Res., 988*, 139–145.

Schulz, K. M., Richardson, H. N., Zehr, J. L., Osetek, A. J., Menard, T. A., & Sisk, C. L. (2004). Gonadal hormones masculinize and defeminize reproductive behaviors during puberty in the male Syrian hamster. *Horm. Behav., 45*, 242–249.

Schulz, K. M., Zehr, J. L., Salas-Ramirez, K. Y., & Sisk, C. L. (2009b). Testosterone programs adult social behavior before and during, but not after, adolescence. *Endocrinology.*

Schwartz-Bloom, R. D., & Sah, R. (2001). gamma-Aminobutyric acid(A) neurotransmission and cerebral ischemia. *J. Neurochem., 77*, 353–371.

Schwarz, J. M. (2008). *A critical role for glutamate in the organizational effects of estradiol on the developing hypothalamus. Unpublished doctoral dissertation.* Baltimore, MD: University of Maryland, Baltimore.

Schwarz, J. M., & McCarthy, M. M. (2008a). The role of neonatal NMDA receptor activation in defeminization and masculinization of sex behavior in the rat. *Horm. Behav., 54*, 662–668.

Schwarz, J. M., & McCarthy, M. M. (2008b). Steroid-induced sexual differentiation of the developing brain: multiple pathways, one goal. *Journal of Neurochemistry, 105*, 1561–1572.

Schwarz, J. M., Liang, S.-L., Thompson, S. M., & McCarthy, M. M. (2008). Estradiol induces hypothalamic dendritic spines by enhancing glutamate release: A mechanism for organizational sex differences. *Neuron, 58*, 584–598.

Simerly, R. B. (1989). Hormonal control of the development and regulation of tyrosine hydroxylase expression within a sexually dimorphic population of dopaminergic cells in the hypothalamus. *Brain. Res. Mol. Brain. Res., 6*, 297–310.

Simerly, R. B. (2002). Wired for reproduction: organization and development of sexually dimorphic circuits in the mammalian forebrain. *Annu. Rev. Neurosci., 25*, 507–536.

Simerly, R. B., Chang, C., Muramatsu, M., & Swanson, L. W. (1990). Distribution of androgen and estrogen receptor mRNA-containing cells in the rat brain: an *in situ* hybridization study. *J. Comp. Neurol., 294*, 76–95.

Simerly, R. B., Swanson, L. W., & Gorski, R. A. (1985). The distribution of monoaminergic cells and fibers in a periventricular preoptic nucleus involved in the control of gonadotropin release: immunohistochemical evidence for a dopaminergic sexual dimorphism. *Brain. Res., 330*, 55–62.

Singh, M., Setalo, G., Jr., Guan, X., Frail, D. E., & Toran-Allerand, C. D. (2000). Estrogen-induced activation of the mitogen-activated protein kinase cascade in the cerebral cortex of estrogen receptor-alpha knock-out mice. *J. Neurosci., 20*, 1694–1700.

Singh, M., Setalo, G., Jr., Guan, X., Warren, M., & Toran-Allerand, C. D. (1999). Estrogen-induced activation of mitogen-activated protein kinase in cerebral cortical explants: convergence of estrogen and neurotrophin signaling pathways. *J. Neurosci., 19*, 1179–1188.

Sisk, C. L., & Foster, D. L. (2004). The neural basis of puberty and adolescence. *Nat. Neurosci., 7*, 1040–1047.

Sisk, C. L., & Zehr, J. L. (2005). Pubertal hormones organize the adolescent brain and behavior. *Front Neuroendocrinol, 26*, 163–174.

Sodersten, P. (1984). Sexual differentiation: Do males differ from females in behavioral sensitivity to gonadal hormones? *Prog. Brain. Res., 61*, 257–270.

Soloff, M. S., Swartz, S. K., Pearlmutter, A. F., & Kithier, K. (1976). Binding of 17beta-estradiol by variants of alpha-fetoprotein in rat amniotic fluid. *Biochim. Biophys. Acta., 427*, 644–651.

Solum, D., & Handa, R. J. (2001). Localization of estrogen receptor alpha (ER alpha) in pyramidal neurons of rat hippocampus. *Brain. res. Dev. Brain. Res., 128*, 165–175.

Speert, D. B., Konkle, A. T., Zup, S. L., Schwarz, J. M., Shiroor, C., Taylor, M. E., et al. (2007). Focal adhesion kinase and paxillin: novel regulators of brain sexual differentiation? *Endocrinology, 148*, 3391–3401.

Spelke, E. S. (2005). Sex differences in intrinsic aptitude for mathematics and science?: a critical review. *Am. Psychol., 60*, 950–958.

Stein, V., Hermans-Borgmeyer, I., Jentsch, T. J., & Hubner, C. A. (2004). Expression of the KCl cotransporter KCC2 parallels neuronal maturation and the emergence of low intracellular chloride. *J. Comp. Neurol., 468*, 57–64.

Sumida, H., Nishizuka, M., Kano, Y., & Arai, Y. (1993). Sex differences in the anteroventral periventricular nucleus of the preoptic area and in the related effects of androgen in prenatal rats. *Neurosci. Lett., 151*, 41–44.

Szego, C. M., & Davis, J. S. (1967). Adenosine 3',5'-monophosphate in rat uterus: acute elevation by estrogen. *Proc. Natl. Acad. Sci. USA, 58*, 1711–1718.

Thomas, P., Pang, Y., Filardo, E. J., & Dong, J. (2005). Identity of an estrogen membrane receptor coupled to a G protein in human breast cancer cells. *Endocrinology, 146*, 624–632.

Todd, B. J., Schwarz, J. M., & McCarthy, M. M. (2005). Prostaglandin-E2: a point of divergence in estradiol-mediated sexual differentiation. *Horm. Behav., 48*, 512–521.

Todd, B. J., Schwarz, J. M., Mong, J. A., & McCarthy, M. M. (2007). Glutamate AMPA/kainate receptors, not GABA(A) receptors, mediate estradiol-induced sex differences in the hypothalamus. *Dev. Neurobiol., 67*, 304–315.

Vinader-Caerols, C., Collado, P., Segovia, S., & Guillamon, A. (2000). Estradiol masculinizes the posteromedial cortical nucleus of the amygdala in the rat. *Brain. Res. Bull., 53*, 269–273.

Vreeburg, J. T., Van der Vaart, P. D., & Van der Shoot, P. (1977). Prevention of central defeminization but not masculinization in male rats by inhibition neonatally of oestrogen biosynthesis. *J. Endocrinol., 74*, 375–382.

Wallen, K. (2005). Hormonal influences on sexually differentiated behavior in nonhuman primates. *Front. Neuroendocrinol., 26*, 7–26.

Wang, J. M., Irwin, R. W., & Brinton, R. D. (2006). Activation of estrogen receptor alpha increases and estrogen receptor beta decreases apolipoprotein E expression in hippocampus *in vitro* and *in vivo*. *Proc. Natl. Acad. Sci. USA, 103*, 16983–16988.

Watters, J. J. a. D., D.M. (1998). Transcriptional effects of estrogen on neurotensin gene expresssion involve cAMP/protein kinase A-dependent signalling mechanisms. *J. Neurosci. 18*, 6672–6680.

Weaver, I. C. (2009). Shaping adult phenotypes through early life environments. *Birth Defects Res. C. Embryo. Today, 87*, 314–326.

Weisz, J., & Ward, I. L. (1980). Plasma testosterone and progesterone titers of pregnant rats, their male and female fetuses and neonatal offspring. *Endocrinology, 106*, 306–313.

Whalen, R., & Edwards, D. (1967). Hormonal determinants of the development of masculine and feminine behavior in male and female rats. *Anat. Rec., 157*, 173–180.

Wood, R. I., & Newman, S. W. (1995). The medial amygdaloid nucleus and medial preoptic area mediate steroidal control of sexual behavior in the male Syrian hamster. *Horm. Behav., 29*, 338–353.

Wood, R. I., Mehta, V., Herbosa, C. G., & Foster, D. L. (1995). Prenatal testosterone differentially masculinizes tonic and surge modes of luteinizing hormone secretion in the developing sheep. *Neuroendocrinology, 62*, 238–247.

Wright, C. L., Burks, S. R., & McCarthy, M. M. (2008). Identification of prostaglandin E2 receptors mediating perinatal masculinization of adult sex behavior and neuroanatomical correlates. *Dev. Neurobiol., 68*.

Yang, Y., Ma, Y., Chen, S., Wang, C., & Lee, E. (2003). Focal adhesion kinase is required, but not sufficient, for the induction of long-term potentiation in dentate gyrus neurons *in vivo*. *J. Neurosci., 15*, 4072–4080.

Zhang, J. M., Konkle, A. T., Zup, S. L., & McCarthy, M. M. (2008). Impact of sex and hormones on new cells in the developing rat hippocampus: a novel source of sex dimorphism? *Eur. J. Neurosci., 27*, 791–800.

Zhou, Y., Watters, J. J., & Dorsa, D. M. (1996). Estrogen rapidly induces the phosphorylation of the cAMP response element binding protein in rat brain. *Endocrinology, 137*, 2163–2166.

Znamesnsky, V., Akama, K. T., McEwen, B. S., & Milner, T. A. (2003). Estrogen levels regulate the subcellular distribution of phosphorylated Akt in hippocampal CA1 dendrites. *J. Neurosci., 23*, 2340–2347.

Chapter 2

Neuroendocrine Control of Gonadotropins in Mammals

Toni R. Pak* and Wilson C.J. Chung†

*Loyola University Stritch School of Medicine, Maywood, IL, USA, †University of Colorado, Boulder, CO, USA

SUMMARY

Central control of gonadotropin function is driven by gonadotropin-releasing hormone (GnRH). Numerous neurotrophic factors, such as steroid hormones, neurotransmitters, and neuroactive ions, contribute to the regulation of GnRH gene expression and pulsatile release. Moreover, there is an extensive degree of conservation and redundancy in the GnRH system across all vertebrate species, underscoring the paramount role GnRH has in conveying reproductive competency. This past decade has witnessed an exponential rise in our understanding of the molecular mechanisms responsible for transducing the GnRH signal to its downstream targets; e.g., the elucidation of specific phosphorylation sites on cellular kinases, the identification of transcription factors and scaffolding proteins, and sequence determination of gene promoter enhancer motifs. Despite these recent advances, more effort is needed towards understanding the genetic and epigenetic mechanisms that contribute to infertility and towards the development of new animal models for the study of reproductive disorders.

1. INTRODUCTION TO GONADOTROPINS

The hypothalamus–pituitary–gonad (HPG) axis comprises the anatomical framework regulating the complex physiological processes required for reproductive competence and species propagation. The coordinated actions of this three-tiered biological system begin at the level of the brain with a vast array of neuronal networks providing regulatory inputs that ultimately converge on specific neuronal subtypes in the hypothalamus. Neuroendocrine factors released from the hypothalamus target the anterior pituitary stimulating the synthesis and secretion of large glycoprotein molecules that ultimately control the development, growth, and maintenance of the gonads in both males and females. Named for their ability to stimulate gonadal growth, these glycoproteins are collectively referred to as gonadotropic hormones or gonadotropins (GTHs).

1.1. The Discovery of Hypothalamic Releasing Factors

The idea that the brain communicates with the pituitary gland to regulate reproduction was championed through the life works of Geoffrey Harris (1913–1971) (Harris, 1948a; 1948b; Green & Harris, 1949). Using electrostimulation of specific brain regions, Harris demonstrated that the brain plays an important role in regulating ovulation in rats and rabbits. Harris postulated that the neuronal terminals situated at the capillary plexus of the median eminence, first observed by the renowned neuroanatomist Santiago Ramón y Cajal (Cajal, 1911), might be a source for humoral factors originating in the brain to contact and regulate the GTH-producing cells in the anterior pituitary. Through an intricate series of experiments whereby the pituitary vessels were severed or whole pituitaries transplanted into distant anatomical locations, Harris showed that an intact vascular connection between the anterior pituitary and the hypothalamus was required for pituitary-mediated maintenance of gonadal function. However, he lacked the technical capabilities to conclusively track and identify the group of neuropeptides that we collectively refer to today as hypothalamic releasing factors. Later, Roger Guillemin and Andrew Schally investigated the possible existence of a hypothalamic releasing factor that stimulated the release of corticotropin (ACTH) from the anterior pituitary (Guillemin & Rosenberg, 1955; Guillemin, Hearn, Cheek, & Householder, 1957; Schally & Guillemin, 1959). Their search was finally realized by another laboratory led by Joachim Spiess, Jean Rivier, and Wylie Vale, when corticotropin-releasing factor (CRF) was isolated from sheep and fully sequenced in 1983 (Spiess, Rivier, Rivier, & Vale, 1981). However, it was through the simultaneous, yet independent, efforts of Guillemin and Schally (Amoss et al., 1971; Schally et al., 1971) that the sequence was determined for the hypothalamic releasing factor controlling the secretion of GTHs from the anterior pituitary: gonadotropin-releasing hormone (GnRH). The Nobel Prize in Physiology and Medicine was awarded in 1977 to Guillemin and Schally for their part in isolating GnRH and thyrotropin-releasing hormone. Due to the pioneering efforts of Harris and many others that he collaborated with, the notion of direct hypothalamic control of the pituitary gland to regulate reproductive functions is now a cornerstone of basic neuroendocrinology principles.

The pituitary GTHs, luteinizing hormone (LH) and follicle-stimulating hormone (FSH), are members of a large family of glycoprotein hormones comprised of two polypeptide subunits (α and β) connected by an oligosaccharide chain. Glycoproteins all share a common α subunit polypeptide chain that is encoded by a single gene that varies depending on the species. However, between-species comparisons have shown that the α subunit amino acid composition is highly conserved, with approximately 100, 82, and 75% sequence identity between bovine/ovine, bovine/equine, and bovine/human, respectively (Bousfield, Butnev, Gotschall, Baker, & Moore, 1996; Dalpathado et al., 2006). The functional specificity of each hormone is conferred by the presence of a unique β subunit containing 110–111 amino acids for FSHβ and 117–121 amino acids for LHβ. Both the α and β subunits have multiple N- and *O*-linked glycosylation sites that allow for the addition of the oligosaccharide chain. The specific site of glycosylation and end-terminal residues of the oligosaccharide chain, such as sulfate or sialic acid, are important for determining proper protein folding, metabolic clearance rates, and biological activity of the GTH hormones (Ulloa-Aguirre & Timossi, 2000; Ulloa-Aguirre, Maldonado, Damian-Matsumura, & Timossi, 2001; Ulloa-Aguirre, Timossi, Barrios-de-Tomasi, Maldonado, & Nayudu, 2003).

The importance of LH and FSH in mammalian reproduction has been well defined, with LH playing a major role in the regulation of steroidogenesis, and FSH primarily regulating gametogenesis, in both sexes (Catt et al., 1980; Leung & Armstrong, 1980; Means et al., 1980). Despite these broadly defined roles, it is important to note that LH and FSH often work in concert to achieve their ultimate biological outcomes. For example, in female mammals LH is primarily responsible for the formation of ovarian corpora lutea (a process termed 'luteinization'), which are a major source of progesterone (P_4) and estrogens. Ovarian estrogens are derived from the androgen precursor testosterone (T), which is converted intracellularly to 17β-estradiol (E_2) by the aromatase enzyme. Luteinizing hormone is required for the synthesis of T in the thecal cells; however, FSH is required for the initiation of aromatase activity and the subsequent conversion of T to E_2 in the granulosa cells. Similarly, in males, FSH is the primary hormone regulating the progression of spermatocyte development into mature sperm, yet LH has been shown to play a role in the later stages of spermiogenesis.

During the last decade, transgenic mouse models have been designed to more clearly identify the full spectrum of physiological functions for LH and FSH. These genetically modified mouse models include those that are lacking the β subunit of either LH or FSH, those lacking one or both GTH receptors, and those that have excessive LH production (Kumar, 2007b). As expected, animals lacking the LHβ subunit are hypogonadal and have very low circulating levels of T, estrogens, and P_4 (Ma, Dong, Matzuk, & Kumar, 2004; Kumar, 2007a). In LHβ-null male mice, the Leydig cells were sparse and of an immature phenotype, reflecting their inability to produce T. Presumably, a lack of testicular interstitial T is the main causative factor for hypogonadism in the male LHβ-null mice, as T replacement was shown to rescue the hypogonadal phenotype (Huhtaniemi, Zhang, Kero, Hamalainen, & Poutanen, 2002). Interestingly, females had normal primary and secondary ovarian follicles but lacked antral follicles, all stages of oocytes, and corpora lutea. Moreover, the thecal cell layer was well developed in all follicles, suggesting that in females neither LHβ nor ovarian steroid hormones are required for the initiation and early progression of follicular development. The LHβ-null male and female mouse phenotype essentially mimics that of mice lacking the LH receptor (LuRKO). In both transgenic models, the developmental process of masculinization was normal, such that male mice lacking LH receptors, or the LHβ subunit, did not differ from wild-type males (Huhtaniemi et al., 2002). These results indicate that embryonic masculinization is independent of LH action in the testes, which confirms previous studies showing that placental-derived chorionic gonadotropin (CG) is capable of stimulating sufficient steroidogenesis to promote sexual differentiation. However, reports of human patients with naturally occurring mutations in the LH receptor describe failed embryonic masculinization with a resulting pseudohermaphroditic phenotype, suggesting that humans lack some of the compensatory mechanisms present in mice. Collectively, data generated from the development of these genetically engineered mouse models confirm many of the early physiological experiments that originally defined the roles of LH and FSH and have expanded our knowledge of the molecular mechanisms mediating the actions of GTHs.

1.2. The Organization of the Mammalian Pituitary Gland

The pituitary gland is connected to the basal surface of the hypothalamus by the infundibular stalk and is anatomically situated in the sella turcica region of the sphenoid bone (Figure 2.1(a–b)). Pituitary organogenesis begins coincidently with neurulation in humans, at approximately 25–26 days post ovulation (Carnegie stage 12) (Solov'ev,

Bogdanov, Panteleev, & Yanin, 2008). Formation of the pituitary gland during development arises from two apparently distinct embryonic tissues, resulting in separate functional lobes. The adenohypophysis, or anterior lobe, arises from an ectodermally derived diverticulum in the developing oral cavity known as Rathke's pouch. Consequently, neuroendocrine control of the adenohypophysis is accomplished mainly through humoral connections with the hypophysial portal vasculature system. The adenohypophysis of most mammals consists of three subdivisions (Green, 1949): pars tuberalis (PT), pars intermedia (PI), and pars distalis (PD). The neurohypophysis arises from a downward extension of the diencephalon called the infundibulum. Two neurohemal areas, the median eminence (ME) and the pars nervosa (PN), comprise the neurohypophysis (Green, 1951). The PN is also referred to as the posterior lobe of the pituitary. Neuroglial cells and nerve fibers coming from hypothalamic neurons make up the cytoarchitecture of the neurohypophysis, supplying direct neural innervation to the posterior lobe. New evidence from human embryos demonstrates that the rudimentary epidermal and neural lobes of the pituitary are in close anatomical proximity throughout development, despite their apparently distinct embryological origins (Solov'ev et al., 2008). Moreover, analyses of the developing Rathke's pouch indicate that it has a potential neural crest origin, resulting in it being a 'mixed' neural/ectoderm structure (Trandafir, Sipot, & Froicu, 1990; Solov'ev et al., 2008). These data suggest that, at least in humans, the two pituitary lobes might share a more similar embryonic origin than previously considered. Similarly, in birds and fishes, the adenohypophysis originates from a thickening of neural ectoderm cells clustered along the anterior edge of the neural plate (Whitlock, 2005). The cells fated to become adenohypophysis migrate away from the developing olfactory placode coincident with the movement and subsequent establishment of cranial neural crest cells. Taken together, these data provide evidence supporting the hypothesis that a neural origin for the development of the adenohypophysis is highly conserved across all vertebrates, including mammals.

The adenohypophysis undergoes further differentiation at approximately Carnegie stage 20 and becomes anatomically divided into the PD (the anterior lobe in humans), PT, and PI. Initially, the anterior wall of the Rathke's pouch develops into the largest portion of the anterior lobe, the PD. Further anatomical differentiation occurs later in development when a small outgrowth of the anterior lobe grows along the infundibular stalk, surrounding it on either side to form the PT. A separate extension arises from the posterior wall of the Rathke's pouch, forming the PI, which is vestigial in many mammals, including humans. Each subdivision contains distinct cell types characterizing the unique functions of that region. The gonadotrope cells reside in the PD, accounting for approximately 20% of all cell types, and they are responsible for making LH and FSH.

The hypothalamo–hypophysial portal system consists of a portal vein connecting two capillary beds supplied by the superior and inferior hypophysial arteries (Figure 2.1 (c)) (Gorczyca & Hardy, 1987). The first capillary bed is in the ME, located at the proximal end of the infundibular stalk in a section of the basal hypothalamic–pituitary interface. Axons from multiple parvocellular neurons in the hypothalamus converge upon the ME and release neurohormones into the portal system at this juncture. The portal vein runs from the ME along the distal portion of the infundibular stalk and provides a connection with the epithelial cells in the anterior lobe via the second capillary bed. Gonadotropins synthesized by cells in the adenohypophysis are then released into systemic circulation via hypophysial vein drainage into the dural venous sinus.

2. GONADOTROPIN-RELEASING HORMONE (GnRH)

Gonadotropin secretion from the pituitary is regulated first and foremost by GnRH (also known as luteinizing hormone-releasing hormone (LHRH)). Gonadotropin-releasing hormone is the established upstream regulator of GTH secretion, although additional neuroendocrine factors have been proposed during the last three decades. Further, GnRH is obligatory for reproductive competency across all vertebrate species, including humans. Much of the research in the field of reproductive neuroendocrinology today is centered on understanding how GnRH is regulated at the genomic level, the nature of the molecular and post-translational events required for production of the mature peptide, the factors that impinge on GnRH neurons to modulate their activity, and how GnRH is developmentally and evolutionarily derived. The nomenclature for GnRH has evolved over the years as newly discovered forms of GnRH are sequenced and characterized from multiple species across a range of diverse phyla. Currently, there is a general consensus among researchers that there are at least three distinct subtypes of GnRH: GnRH-I, GnRH-II, and GnRH-III. It is important to note that the axons of GnRH-II- and GnRH-III-expressing neurons do not target the anterior pituitary, their anatomical distribution is distinct from that of GnRH-I, and their precise physiological function is still highly debated. The focus of this chapter is the neuroendocrine control of GTHs; therefore, only the form of GnRH that is known to be directly involved in regulating pituitary gonadotropes (GnRH-I) will be discussed and will henceforth be referred to simply as GnRH. In addition, the goal of this chapter is to provide the reader with the most current information stemming from research conducted over the last decade. See Gore (2002) for a more comprehensive and historical overview of GnRHs.

2.1. The Development of Gonadotropin-releasing Hormone (GnRH) Neurons: Cell Fate Specification

Although mammalian GnRH neurons are found in the preoptic area (POA) and anterior hypothalamus postnatally, it is accepted that GnRH neurons arise from the olfactory placodes (OP) located on the medial wall of the nasal pit. From here they migrate along the nasal cavity through the cribriform plate into the central nervous system (CNS) during the course of embryonic development (Schwanzel-Fukuda & Pfaff, 1989; Wray, Grant, & Gainer, 1989). Landmark studies showed that the vast majority (~80%) of mouse GnRH neurons undergo their last division in the OP between embryonic day (E) 9.5 and E10.5 (Wray et al., 1989). Interestingly, these postmitotic OP progenitor cells do not express detectable levels of GnRH mRNA or peptide prior to E11.5 (Schwanzel-Fukuda & Pfaff, 1989; Wray et al., 1989). In this sense, the period of GnRH neuron fate specification can be defined to occur between E9.5 and E11.5. The absence of measurable GnRH expression prior to E11.5 remains a major hurdle for studying the molecular processes that regulate GnRH neuronal fate specification, because at present the only marker for GnRH neurons is GnRH itself. Therefore, postmitotic OP progenitor cells that are destined to become GnRH neurons cannot be identified during a large portion of fate specification. Notwithstanding this major caveat, much evidence has accumulated about fate specification of the mouse GnRH neuronal system over the years.

The OPs are transient thickenings of the ectoderm on the ventral lateral aspects of the embryonic head that invaginate/ingress, ultimately forming epithelia and multiple structures of the nose. Two specific types of epithelia arise from the OP: (1) the nonsensory respiratory epithelium, which is localized proximal to the nasal pit and (2) the sensory olfactory epithelium, which is distal to the nasal pit. The sensory olfactory epithelium differentiates into the main olfactory epithelium of the chemosensory system and the vomeronasal organ (VNO) epithelium, which is part of the pheromone receptive system. Spatio—temporal mapping studies in mouse embryos were helpful in elucidating the origins of GnRH neurons and revealed that GnRH neurons first emerge from the secondary recess of the OP on E11.5. These studies support the main hypothesis that GnRH neuron progenitor cells originate from the sensory olfactory epithelium. However, there are alternative views that indicate that GnRH neurons may originate from the anterior pituitary placode and cranial neural crest (Whitlock, 2005).

Given that OP progenitor cells are the most likely candidates from which GnRH neurons develop, it makes sense that the multiple spatiotemporal cues that govern the induction and differentiation of the OP are also important for the emergence of GnRH neurons. Not surprisingly, the developmental genetic program of the OP is complex in nature, as attested by the early expression of a multitude of transcription factors, such as Dlx3, Dlx5, Mash1, and Pax6 (Wray, 2002). For instance, GnRH neurons are absent in the small-eye (Sey) mouse, which harbors a point mutation in the Pax6 gene (Dellovade, Pfaff, & Schwanzel-Fukuda, 1998). This deletion of GnRH neurons seems to be a secondary consequence of the premature cessation of OP induction, a process that starts prior to the presence of GnRH neurons in the OP (Dellovade et al., 1998).

Early GnRH neurons coexpress olfactory epithelial transcription factors, such as Olf-1 and GATA-4 (Wray, 2002), which suggests that they may be involved in the fate specification process of GnRH neurons. However, Olf-1 mutant mice did not show any defects in their GnRH neuronal system or reproductive function (Garel, Marin, Grosschedl, & Charnay, 1999), which indicates that Olf-1 may not be involved in the ontogenesis or fate specification of GnRH neurons. As for GATA-4 mutant mice, they die at E9.5 prior to the emergence of GnRH neurons (Lawson & Mellon, 1998). Thus, the precise role of GATA-4 during early GnRH neuronal development remains to be determined. Early GnRH neurons coexpress a retinoic acid-inducible basic helix-span-helix DNA-binding protein called AP-2α (Luscher, Mitchell, Williams, & Tjian, 1989; Kramer & Wray, 2000; Kramer, Krishnamurthy, Mitchell, & Wray, 2000). Similarly to the previously mentioned colocalized transcription factors, a deletion of AP-2α in mice does not affect the development of the GnRH neuronal system (Wray, 2002).

Fibroblast growth factor (FGF) signaling is important for GnRH neuron fate specification. In mice, FGF signaling is mediated by four specific types of FGF receptor (FGFRs) (Itoh & Ornitz, 2004). Immunocytochemical localization studies showed that the E10.5 mouse OP contained abundant levels of FGFR1, FGFR2, and FGFR3, but not FGFR4 protein (Gill, Moenter, & Tsai, 2004). Further, the presence of FGFR1 and FGFR3, but not FGFR2, protein has been detected in E15.5 and postnatal day 3 GnRH neurons (Gill et al., 2004). These observations are consistent with studies showing that FGFR1 and FGFR3 mRNA is highly expressed in immortalized mouse GnRH neuronal cell lines (Tsai, Werner, & Weiner, 1995). However, the most robust data indicating that FGF signaling plays a role in GnRH neuronal fate specification is derived from *in-vitro* nasal explants isolated from E10.5 embryos. In this experimental paradigm, the number of GnRH neurons in the nasal explants increased after two or three days *in vitro* (Gill et al., 2004), which could be prevented by the FGFR antagonist SU5402 (Gill et al., 2004). Further, examination of the total number of GnRH neurons in the entire heads of newborn FGFR1 hypomorphs (i.e., knock-down mice) found that the number of GnRH neurons was around 88% lower in homozygous FGFR1 hypomorphs than their wild-type littermates (Chung, Moyle, & Tsai, 2008). On the other hand, the number of GnRH neurons was unaffected in newborn FGFR3-null mice (Chung et al., 2008). These mouse transgenic studies established that the activation of FGFR1, but not FGFR3, is required for the normal number of GnRH neurons in the newborn mouse head. These whole-animal studies corroborated the earlier nasal explant studies and supported the idea that activation of FGFR function may be a key pathway required for the emergence of GnRH neurons (Gill et al., 2004).

Of the 22 identified FGF ligands, there are several lines of evidence indicating that FGF8 is the most likely FGF candidate to be involved in the early development of GnRH neurons. First, FGF8 mRNA is highly expressed in the OP from E9.5 onwards (Crossley & Martin, 1995; Kawauchi et al., 2005). This expression profile suggests that FGF8 signaling is critically important for the induction and differentiation of the mouse OP. Second, a partial loss of FGF8 function resulted in the absence of the olfactory bulbs and VNO (Meyers, Lewandoski, & Martin, 1998; Chung et al., 2008). The absence of the VNO has a special relevance given that the VNO, like the GnRH neurons, originates from the secondary recess of the OP (Wray et al., 1989). Third, FGF8 binds and readily activates FGFR signaling, including FGFR1, in cell lines (Falardeau et al., 2008). Indeed, the GnRH neuronal system in newborn homozygous FGF8 hypomorphs was completely eliminated (Chung et al., 2008; Falardeau et al., 2008). An equally striking observation is that newborn heterozygous FGF8 hypomorphs had approximately 50% fewer GnRH neurons remaining as compared to wild-type littermates (Chung et al., 2008; Falardeau et al., 2008). Further examination showed that the emergence of GnRH neurons in homozygous FGF8 hypomorphs was already deleted at E11.5 as well as at all subsequent time points examined (Chung et al., 2008). Together with the previously described nasal explant studies, these data strongly suggest that FGF8 signaling acting through FGFR1 is critical for GnRH neuron fate specification in the mouse OP.

2.2. Development of Gonadotropin-releasing Hormone (GnRH) Neurons: Neuronal Migration

The organization and functionality of the GnRH neuronal system depends on the accurate migration from the OP through the cribriform plate to the POA/hypothalamus region. Migrating GnRH neurons follow the olfacto–vomeronasal nerve axons towards the CNS. The majority of the olfacto–vomeronasal nerve axons target the main and accessory olfactory bulbs; however, a subset of the olfacto–vomeronasal nerve axons turn towards the POA and hypothalamus (Wray, Key, Qualls, & Fueshko, 1994; Tobet & Schwarting, 2006). It is this caudal branch of the olfacto–vomeronasal nerve that migrating GnRH neurons follow towards the POA and hypothalamus. The extension and targeting of the olfacto–vomeronasal nerve axons towards the main and accessory olfactory bulbs, the POA, and the hypothalamus are tightly regulated by guidance cues, such as slits and semaphorins (Wray et al., 1994; Tobet & Schwarting, 2006).

Gonadotropin-releasing hormone neurons depend on the correct targeting of the caudal olfacto–vomeronasal nerve axons in order to migrate into the POA and hypothalamus. Indeed, the caudal olfacto–vomeronasal nerve axons will aberrantly target the cortex in colorectal cancer DCC (i.e., receptor for netrin-1) mutant mice. Consequently, migrating GnRH neurons will travel to the cortex instead of the POA and hypothalamus (Schwarting, Kostek, Bless, Ahmad, & Tobet, 2001). In contrast to DCC's role in olfacto–vomeronasal nerve axon targeting, recent studies showed that FGF8 signaling is required for olfacto–vomeronasal nerve axons to be able to enter and contact the most anterior region of the brain. Currently, the mechanisms underlying the inability of olfacto–vomeronasal nerve axons to enter the embryonic FGF8 hypomorphic brain are unknown. The elongation of olfacto–vomeronasal nerve axons is accompanied by migrating olfactory ensheathing cells, which are glia that originate from the olfactory epithelium (Tennent & Chuah, 1996; Tisay & Key, 1999). These olfactory ensheathing cells express high levels of FGFR1 and are permissive for the elongation/growth of olfactory axons (Chuah & Au, 1994; Hsu et al., 2001). Together these studies indicate that the extension of olfacto–vomeronasal nerve axons into the CNS may be due to FGF8 acting on FGFR1-expressing olfactory ensheathing cells. However, it is important to emphasize that GnRH neurons were already missing in these FGF8 hypomorphic mice prior to migration. Thus, the cause of this GnRH system defect in these mice may be failure to emerge, not failure to migrate.

Many guidance cues that regulate olfacto–vomeronasal nerve axon extension and targeting do not seem to play a direct and active role in migrating GnRH neurons. However, recently studies in organotypic nasal explants and brain slices showed that the GABAergic system may be directly involved in regulating GnRH neuronal movement. For instance, the activation of $GABA_A$ receptors decreased GnRH neuronal movement, whereas live video microscopy showed that $GABA_A$ receptor inhibition reduced the ability of GnRH neurons to turn towards the POA and hypothalamus (Tobet & Schwarting, 2006). Similarly, nasal embryonic LHRH factor (NELF), which is expressed in olfacto–vomeronasal nerve axons and early GnRH neurons, may act as a directional guidance cue for migrating GnRH neurons. Indeed, NELF expression is rapidly shut down in GnRH neurons that are traveling towards the POA and hypothalamus. In contrast, GnRH neurons that migrate towards the olfactory bulbs seem to continue to express low levels of NELF (Kramer & Wray, 2000). Therefore, GnRH neuronal migration may not only depend on the correct extension and targeting of olfacto–vomeronasal nerve axons, but also on factors that regulate GnRH neuronal movement.

2.3. Development of Gonadotropin-releasing Hormone (GnRH) Neurons: Axonal Targeting

The majority of GnRH neurons target their axons to the external zone of the ME. Gonadotropin-releasing hormone is then released into the fenestrated primary capillary plexus to reach the hypophysial portal veins and ultimately the anterior pituitary (Figures 2.1 and 2.2). Gonadotropin-releasing hormone axons innervating the ME can be detected as early as E14 (Livne, Gibson, & Silverman, 1993). Organotypic culture studies using the POAs from newborn mice showed that the medial basal hypothalamus, in which the ME is localized, secretes soluble factors that attract GnRH axons (Gibson, Ingraham, & Dobrjansky, 2000). To date, many of these soluble chemoattractive factors remain elusive. However, one particular factor that has been identified is FGF2, which is highly expressed in the endothelial cells and ventricular tanycytes of the ME. Indeed, FGF2 caused stronger outgrowth and longer axon extension of preoptic GnRH neurons in organotypic cultures (Gibson et al., 2000). More recently, it was shown that the addition of FGF2 significantly stimulated neurite outgrowth in cultured E15.5 GnRH neurons (Gill & Tsai, 2006). Studies in mouse embryos expressing a dominant-negative form of the FGFR (dnFGFR), which is selectively made in GnRH neurons, further corroborate the importance of FGF signaling in GnRH axonal outgrowth. Indeed, GnRH neurons isolated from dnFGFR mouse embryos exhibit attenuated neurite outgrowth in culture (Gill & Tsai, 2006), suggesting that FGF signaling is an endogenous cue that maintains GnRH axon growth.

2.4. Gonadotropin-releasing Hormone (GnRH) Neurons are Anatomically Unique

The unique anatomical arrangement and limited number (est. 800–1500) of GnRH-expressing neurons in the mouse brain hindered progress towards gaining a greater understanding of the regulatory factors that modulate GnRH neuronal activity. The postnatal neurons are diffusely scattered throughout the basal forebrain, as opposed to being clustered into distinct nuclei. In rodents, GnRH cell bodies are situated mainly in the POA, diagonal band of Broca (DBB), medial septum, organum vasculosum of the lamina terminalis (OVLT), and anterior hypothalamic area (AHA). There are subtle anatomical differences between rodents and other mammals, such as primates, guinea pigs, and ungulates, but the general rostral to caudal distribution is similar. Despite their scattered distribution, individual GnRH neurons have exceptionally long axons that enable the entire population of GnRH neurons in the brain to converge upon the ME of the hypothalamus and target the anterior pituitary via the hypophysial portal vasculature system (Figures 2.1(b–c) and 2.2).

Early morphological analyses described GnRH neurons as having a fusiform-shaped cell body with one (unipolar) or two (bipolar) axonal projections and sparse diminutive dendritic branches (King, Tobet, Snavely, & Arimura, 1982; Merchenthaler, Gorcs, Setalo, Petrusz, &

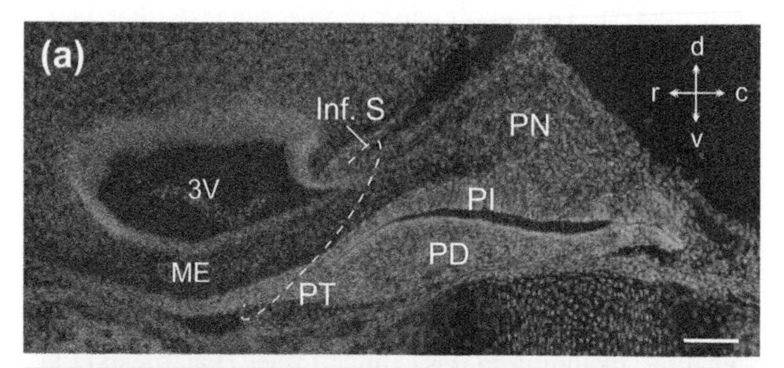

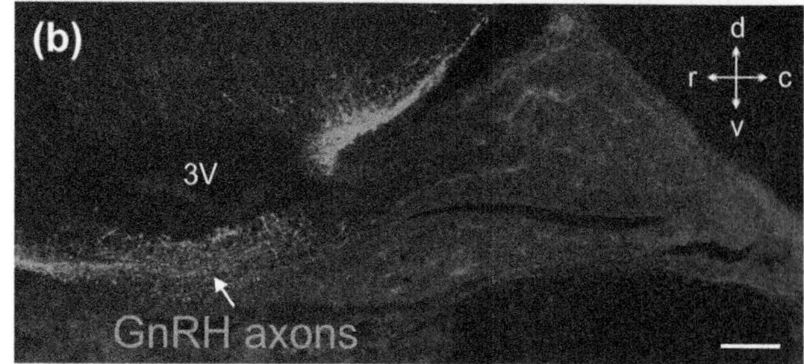

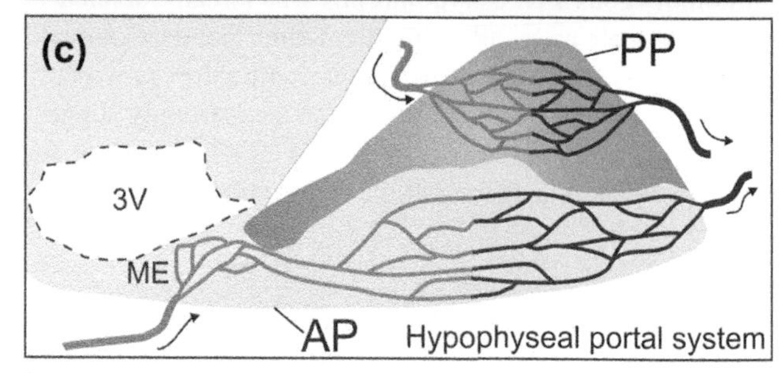

FIGURE 2.1 Parasagittal photomicrographs (a–b) and schematic drawing (c) of the postnatal day 0 mouse pituitary. (a) Anatomical organization of the pituitary as visualized by 4',6-diamidino-2-phenylindole (DAPI) nuclear stain (blue). (b) Photomicrograph depicting gonadotropin-releasing hormone (GnRH) immunofluorescent (red) axons terminating in the median eminence (arrow). (c) Schematic drawing of the hypophysial portal system in the pituitary. Note the capillary bed in the median eminence, in which GnRH peptide is released into the portal vein system in order to stimulate the release of gonadotropins from gonadotropes in the anterior pituitary (AP). Bar =100 μm. 3V, third ventricle; c, caudal; d, dorsal; Inf. S, infundibular stalk; ME, median eminence; PD, pars distalis; PI, pars intermedialis; PN, pars nervosa; PP, posterior pituitary; PT, pars tuberalis; r, rostral; v, ventral. See color plate section.

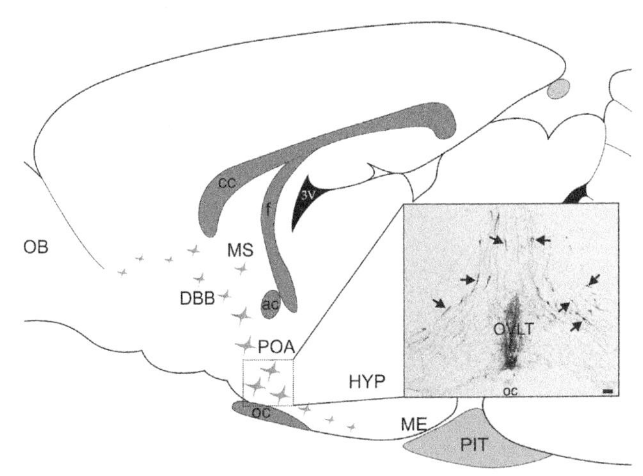

FIGURE 2.2 Schematic representation of the rostral-to-caudal distribution of gonadotropin-releasing hormone (GnRH) neurons (in red) in the postnatal rodent brain. Gonadotropin-releasing hormone neuron density is represented by the size of the red neurons. 3V, third ventricle; ac, anterior commissure; cc, corpus callosum; DBB, diagonal band of Broca; f, fornix; HYP, hypothalamus; ME, median eminence; MS, medial septum; OB, olfactory bulb; oc, optic chiasm; PIT, pituitary; POA, preoptic area. Inset is coronal photomicrograph of GnRH neurons (brown) at the level of the organum vasculosum lamina terminalis (OVLT), visualized using immunocytochemistry. Bar = 50 μm. See color plate section.

Flerko, 1984). However, these studies were limited because GnRH neurons could only be identified using immunological detection of the GnRH peptide and nonspecific neuronal tracers. Recent advances in technology have provided the tools for researchers to thoroughly map the axonal and dendritic pathways of GnRH neurons in order to gain a better understanding of the entire GnRH system. Several transgenic mouse and rat models have been generated in which enhanced green fluorescent protein (eGFP) is expressed when under the control of a GnRH gene promoter (GnRH-eGFP) (Suter et al., 2000; Spergel, Kruth, Shimshek, Sprengel, & Seeburg, 2001; Fujioka et al., 2003). The resulting animal model retains normal parameters of sexual maturation and reproductive function, yet the GnRH neurons are clearly visible when viewed under light microscopes fitted with fluorescent filters. Direct visualization of the GnRH neurons in a fresh-brain-slice-tissue preparation has allowed researchers to prepare brains by intracellularly injecting low-molecular-weight molecules, which can then be labeled with various fluorescent markers for precise morphological analyses of neuronal morphology. Using this cell-filling strategy, it was shown that, in addition to their exceptionally long axons, GnRH neurons have extraordinarily long dendrites, some up to 1000 μm in length, which can course alongside the axon and terminate in the ME (Campbell, Han, & Herbison, 2005). The dendrites also have numerous spines that run the entire length of each dendritic process, thereby providing multiple regulatory sites for presynaptic neurons.

Another important technological innovation in the study of GnRH neuronal morphology has been the implementation of tract-tracing agents, such as pseudorabies virus and barley lectin, specifically targeted to GnRH neurons in order to clearly define their afferent and efferent pathways (Boehm, Zou, & Buck, 2005; Yoon, Enquist, & Dulac, 2005; Wintermantel et al., 2006). The pseudorabies virus PRV Ba2001 is a conditionally activated retrograde transporter that will only replicate in the presence of the protein Cre recombinase (DeFalco et al., 2001). Generation of transgenic GnRH-Cre animals (Yoon et al., 2005; Wolfe et al., 2008) allows the PRV Ba2001 to be injected into the brain, but the virus will only infect, replicate, and be transported by GnRH-expressing cells. By tagging the viral protein with eGFP, multiple orders of afferent inputs to GnRH neurons can be reliably and accurately identified, depending on the duration of the infection period. Similarly, the use of barley lectin as a neuronal tracer has the added advantage that it is transported in both a retrograde and anterograde manner, allowing it to generate a complete picture of GnRH afferent and efferent projections. Cumulatively, the data obtained from using these novel methods have demonstrated a wide network of afferent signaling pathways to GnRH neurons arising from hypothalamic, cortical, limbic, and brainstem regions (Boehm et al., 2005; Yoon et al., 2005; Wintermantel et al., 2006; Campbell & Herbison, 2007). Of note, considerable afferent projections have been identified as originating in regions involved in mediating olfaction (Boehm et al., 2005; Yoon et al., 2005). Earlier studies conclusively linked the importance of olfactory and pheromonal cues with GnRH and the subsequent display of mating behaviors (Moss & McCann, 1975; Dudley, Vale, Rivier, & Moss, 1981; Shipley, Murphy, Rizvi, Ennis, & Behbehani, 1996; Wood & Coolen, 1997; Wu, Glucksman, Roberts, & Mani et al., 2006b). These new data provide an anatomical framework that corroborates the findings from earlier behavioral studies.

2.5. Gonadotropin-releasing Hormone (GnRH) is Structurally Conserved

The primary GnRH transcript encodes a large polypeptide (pro-GnRH) that is cleaved by enzymatic processes into the following four products: a signal peptide, a mature GnRH decapeptide, a proteolytic processing peptide, and a 56-amino-acid GnRH-associated peptide (GAP) (Figure 2.3(b)). The functionally mature GnRH peptide is comprised of 10 amino acids that are highly conserved across a variety of vertebrate and invertebrate animals. Notably, the first four amino acids on the carboxy terminus and the last two amino acids on the amino terminus are identical across nearly all vertebrate species studied to date

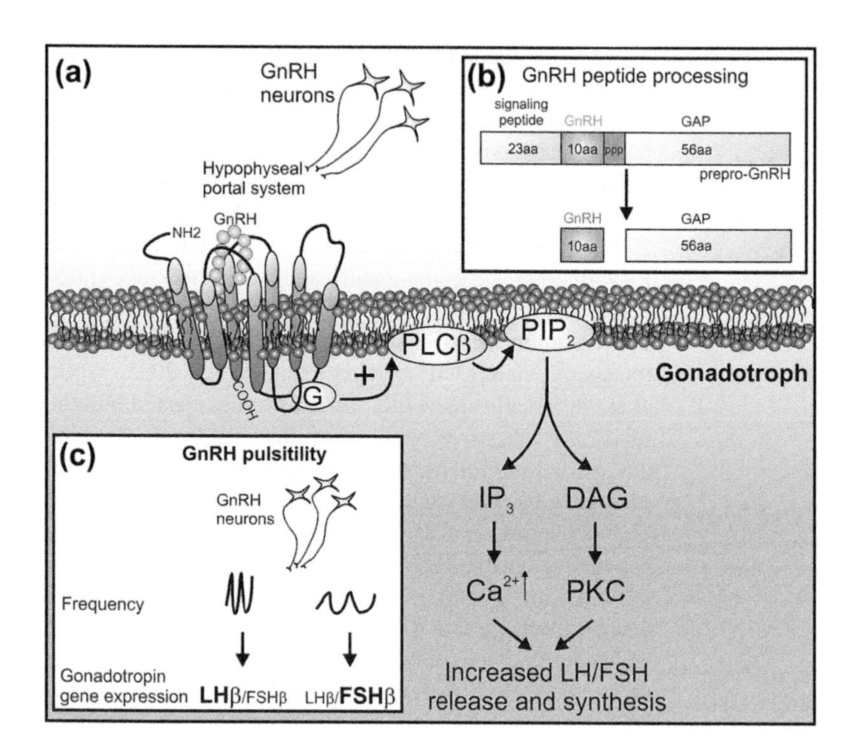

FIGURE 2.3 Schematic pathway of gonadotropin-releasing hormone (GnRH)-dependent activation of gonadotropes. (a) Upon release in the hypophysial portal system, the GnRH peptide (green) will bind the GnRH receptor (red), which is a G-protein-coupled receptor (GPCR), with seven trans-membrane alpha-helices that are connected by extracellular and intracellular loops. The activation of the GnRH receptors is conveyed by G-proteins, which signal to phospholipase C (PLCβ) to hydrolyze phosphatidylinositol 4,5 biphosphate (PIP_2) into inositol 1,4,5-triphosphate (IP_3) and diacylglycerols (DAG). This sequence of events results in the mobilization of intracellular calcium by IP_3 and the activation of PKC by DAG. (b) Schematic representation of GnRH peptide processing from the GnRH prepropeptide. aa, amino acid; GAP, GnRH-associated protein; ppp, proteolytic processing peptide. (c) Schematic representation of how GnRH pulse frequency results in the preferential expression of LHβ or FSHβ transcription. Indeed, high GnRH pulse frequency (once every 30–60 minutes) favors LHβ, whereas slow GnRH pulse frequency (once every two to three hours) results in the preferential transcription of FSHβ. See color plate section.

TABLE 2.1 Amino acid sequence of GnRH peptides identified to date (See color plate section)

DEUTEROSTOMIA	1 2 3 4 5 6 7 8 9 10
Mammal	pGlu-His-Trp-Ser-Tyr-Gly-Leu-Arg-Pro-Gly-NH2
Guinea pig	pGlu-His-Trp-Ser-Tyr-Gly-Val-Arg-Pro-Gly-NH2
Chicken-I	pGlu-His-Trp-Ser-Tyr-Gly-Leu-Gln-Pro-Gly-NH2
Chicken-II	pGlu-His-Trp-Ser-His-Gly-Trp-Tyr-Pro-Gly-NH2
Salmon	pGlu-His-Trp-Ser-Tyr-Gly-Trp-Leu-Pro-Gly-NH2
Dogfish	pGlu-His-Trp-Ser-His-Gly-Trp-Leu-Pro-Gly-NH2
Catfish	pGlu-His-Trp-Ser-His-Gly-Leu-Asn-Pro-Gly-NH2
Herring	pGlu-His-Trp-Ser-His-Gly-Leu-Ser-Pro-Gly-NH2
Medaka	pGlu-His-Trp-Ser-Phe-Gly-Leu-Ser-Pro-Gly-NH2
Lamprey-I	pGlu-His-Tyr-Ser-Leu-Glu-Trp-Lys-Pro-Gly-NH2
Lamprey-III	pGlu-His-Trp-Ser-His-Asp-Trp-Lys-Pro-Gly-NH2
Rana	pGlu-His-Trp-Ser-Tyr-Gly-Leu-Trp-Pro-Gly-NH2
Sea Bream	pGlu-His-Trp-Ser-Tyr-Gly-Leu-Ser-Pro-Gly-NH2
Whitefish	pGlu-His-Trp-Ser-Tyr-Gly-Met-Asn-Pro-Gly-NH2
Tunicate-I	pGlu-His-Trp-Ser-Asp-Tyr-Phe-Lys-Pro-Gly-NH2
Tunicate-II	pGlu-His-Trp-Ser-Leu-Cys-His-Ala-Pro-Gly-NH2
Tunicate-III	pGlu-His-Trp-Ser-Tyr-Glu-Phe-Met-Pro-Gly-NH2
Tunicate-IV	pGlu-His-Trp-Ser-Asn-Gln-Leu-Thr-Pro-Gly-NH2
Tunicate-V	pGlu-His-Trp-Ser-Tyr-Glu-Tyr-Met-Pro-Gly-NH2
Tunicate-VI	pGlu-His-Trp-Ser-Lys-Gly-Tyr-Ser-Pro-Gly-NH2
Tunicate-VII	pGlu-His-Trp-Ser-Tyr-Ala-Leu-Ser-Pro-Gly-NH2
Tunicate-VIII	pGlu-His-Trp-Ser-Leu-Ala-Leu-Ser-Pro-Gly-NH2
Tunicate-IX	pGlu-His-Trp-Ser-Asn-Lys-Leu-Ala-Pro-Gly-NH2
PROTOSTOMIA	**1 2 3 4 5 6 7 8 9 10 11 12**
Octopus	pGln-Asn-Tyr-His-Phe-Ser-Asn-Gly-Trp-His-Pro-Gly-NH2
*Aplysia	pGln-Asn-Tyr-His-Phe-Ser-Asn-Gly-Trp-Tyr-Ala-NH2
*Owl Limpet	pGln-His-Tyr-His-Phe-Ser-Asn-Gly-Trp-Lys-Ser-NH2
*Annelid	pGln-Asn-Tyr-His-Phe-Ser-His-Gly-Trp-Phe-Pro-NH2

* Sequence has not yet verified by mass spectrometry.
Conserved amino acid positions are shown in red.

(Table 2.1). As might be expected, these highly conserved sequences represent important regulatory regions for GnRH function. For example, in many species (including mammals), these exact amino acid sequences are required for proper binding and activation of its cognate receptors. However, the amino acids located in the center of the peptide (positions five, six, seven, and eight) tend to be more variable. One possible explanation is that variation in this region confers ligand : receptor specificity between different species. This is supported by studies demonstrating that, in the mammalian form of GnRH, a substitution of arginine at position eight significantly reduces ligand activity in mammals, yet enhances activity and binding affinity at nonmammalian GnRH receptors (GnRHR).

Once the mature mammalian decapeptide has been released from the axon terminals, it is rapidly degraded. During the first step of degradation, a prolyl endopeptidase cleaves the c-terminus at Gly-10, resulting in the non-apeptide (GnRH-1-9). Next, the metalloendopeptidase EP24.15 splits the peptide into two halves by cleavage at the Try-5 and Gly-6 bond. Importantly, one of the resulting products (GnRH-1-5) has recently been shown to possess distinct functional properties that serve to modulate the secretion of its parent compound through an autoregulatory mechanism (Wu, Mani, Glucksman, & Roberts, 2005; Roberts, Mani, Woller, Glucksman, & Wu, 2007). Moreover, GnRH-1-5 facilitated lordosis behavior in female rats, highlighting a mechanism that possibly enables the synchronization of critical behavioral and physiological elements of reproductive function and thereby maximizes the potential for reproductive success (Wu, Glucksman, Roberts, & Mani, 2006a).

2.6. Gonadotropin-releasing Hormone (GnRH) Release is Pulsatile

A defining feature of GnRH is that it is episodically secreted from axon terminals at the ME, with peptide release occurring at an average frequency of once per hour (Belchetz, Plant, Nakai, Keogh, & Knobil, 1978; Knobil, Plant, Wildt, Belchetz, & Marshall, 1980; Wildt et al., 1981). Individual GnRH pulses are then closely followed by a corresponding pulse of LH and/or FSH release from pituitary gonadotropes (Kawakami, Uemura, & Hayashi, 1982; Wilson et al., 1984; Silverman, Wilson, Kesner, & Knobil, 1986). The episodic oscillation of GnRH neuronal firing occurs in hypothalamic explant cultures, where the GnRH neurons are completely isolated from all afferent inputs as well as in immortalized GnRH-expressing cell lines (i.e., GT1-7) (Bourguignon & Franchimont, 1984; Kim & Ramirez, 1986; Wetsel et al., 1992). These data provide strong evidence that pulsatility is intrinsic to the GnRH neuron itself and does not require the coordination of synaptic signals from other cells. Moreover, the pulsatile nature of GnRH secretion is absolutely required for pituitary responsiveness to GnRH, as continuous infusion of GnRH results in desensitization due to a downregulation of GnRH receptors (Belchetz et al., 1978). Pharmacological interventions for patients with idiopathic hypogonadotropic hypogonadism (IHH) commonly use natural or synthetic GnRH analogs as a mechanism to restore fertility and induce ovulation. Patients with IHH fail to undergo sexual maturation, have impaired GnRH, LH, and FSH secretion, and are infertile. However, in order for the GnRH treatments to be successful in IHH patients, they must mimic the endogenous pattern of GnRH release.

The term GnRH 'pulse generator' was coined to describe the neuronal network involved in creating or modulating GnRH pulsatility. The fact that pulsatility is an intrinsic feature of each GnRH neuron raises the possibility that each GnRH neuron possesses an internal molecular clock regulating its activity. Thus, a fundamental question arises as to how these widely scattered, individually cycling GnRH neurons are then synchronized to produce a collective pulse of peptide release. One possibility is that there is direct cell : cell communication between GnRH neurons, e.g. via gap junctions, but there is very little anatomical evidence at the level of the perikarya to support this hypothesis. However, the dense clustering of GnRH axon terminals at the ME suggests that GnRH pulses could be synchronized through direct cellular contact at the synaptic terminals, rather than through the propagation of action potentials originating at the cell body. Another possibility is that there is some diffusible factor that selectively activates a population of cells independent of anatomical connectivity. Nitric oxide (NO) is one such factor that meets these criteria. Synthesized from L-arginine and oxygen by the enzyme nitric oxide synthase (NOS), NO is a diffusible gas capable of permeating cell membranes and activating second messenger signaling cascades, such as cGMP. Nitric oxide-producing cells are highly represented throughout the hypothalamus and are anatomically poised within close proximity to GnRH perikarya (Grossman et al., 1994; Herbison, Simonian, Norris, & Emson, 1996; Clasadonte, Poulain, Beauvillain, & Prevot, 2008). Although initial experiments using GT1-7 cell lines and hypothalamic explants indicated that NO was an important stimulatory factor for GnRH release (Lopez, Moretto, Merchenthaler, & Negro-Vilar, 1997; Lopez, Merchenthaler, Moretto, & Negro-Vilar, 1998), more recent evidence suggests exactly the opposite. Patch-clamp recordings measured directly from GnRH neurons in the transgenic GnRH-eGFP mouse showed that infusion of the NO precursor, L-arginine, significantly reduced spontaneous neuronal firing rates (Clasadonte et al., 2008), indicating that NO might be able to facilitate the coordinated firing of multiple GnRH neurons. Moreover, excess production of NO and its metabolites is cytotoxic and has been associated with decreased reproductive function during aging (Vernet, Bonavera, Swerdloff, Gonzalez-Cadavid, & Wang, 1998). Taken together, the data accumulated suggest that NO is an important regulator of GnRH secretion, but whether it is a synchronizing component of the GnRH pulse generator remains to be determined. Although the exact neural components and molecular mechanisms generating the pulsatile pattern are still unknown, the prevailing view is that the GnRH pulse generator requires a complex coordination of both intrinsic and extrinsic factors that collectively orchestrate the temporal signal that is required for reproductive success.

Over the years, several neurotransmitters (GABA, norepinephrine, glutamate); neuropeptides (opiods, galanin, melatonin, serotonin, neuropeptide Y, vasoactive intestinal peptide, leptin, prolactin); steroid hormones (P_4, estrogens, T, corticosterone); and neurotrophic factors (FGFs, insulin-like growth factor, transforming growth factor α and β) have been shown to have important, if not critical, roles in modulating GnRH secretion and pulsatility and have been reviewed extensively (Gore, 2002). The most recent player on the scene is the 54-amino-acid peptide kisspeptin (Kp), which is a product of the *KISS1* gene. Kisspeptin was first identified as a cognate ligand for the orphan G-protein-coupled receptor (GPCR), GPCR54 (Kotani et al., 2001). Shortly thereafter, a genetic screen of patients with IHH revealed that they all shared a common mutation in the GPCR54 gene (De Roux et al., 2003; Seminara et al., 2003). These two discoveries launched an exciting new area of research that has redefined basic concepts in the field of reproductive physiology. Intracerebroventricular or peripheral injections of Kp stimulate the release of LH and FSH in mice (Gottsch et al., 2004), rats (Irwig et al., 2004; Thompson et al., 2004; Navarro et al., 2005a; 2005b; Tovar et al., 2006), sheep (Messager et al., 2005), monkeys (Shahab et al., 2005; Plant, Ramaswamy, & Dipietro, 2006), and humans (Dhillo et al., 2005). These effects are blocked in animals pretreated with a GnRH antagonist and in the GPCR54-null mouse model, suggesting that the target of Kp is upstream from the anterior pituitary and that GPCR54 signaling is required (Gottsch et al., 2004; Irwig et al., 2004; Messager et al., 2005; Navarro et al., 2005b; Shahab et al., 2005). Both KISS1- and GPCR54-null mouse models have been generated, resulting in animals with the expected IHH phenotype; homozygote-null mice fail to undergo puberty, have low circulating levels of GTHs, and are infertile (Seminara et al., 2003; D'Anglemont de Tassigny et al., 2007; Lapatto et al., 2007). However, in all other respects, the animals appear phenotypically normal, suggesting that the KISS1 and GPCR54 genes are not critical for other physiological processes. In normal mice, Kp and GPCR54 are stably expressed in the brain at moderate to low levels prior to puberty, increase significantly during pubertal development, and then return to prepubertal levels during adulthood. This is consistent with the hypothesis that Kp is a primary regulator of pubertal onset by activating GnRH neurons. However, the role of Kp during other periods of dramatic reproductive changes, such as seasonal recrudescence, pregnancy, or menopause, remains to be fully elucidated. Moreover, it still unknown what factors regulate KISS1 and GPCR54 (now renamed KISS1R), whether there are additional targets for Kp in the brain or periphery, and whether there are multiple ligands for GPCR54. The landmark discovery of Kp and its role in reproduction is perhaps the most significant finding in the field of reproductive neuroendocrinology in this decade. Kisspeptin is the first identified neuroendocrine factor since the discovery of GnRH that is absolutely obligatory for reproductive function, although it is still upstream and ultimately acts through its immediate downstream target, GnRH, to achieve its effects on GTH release.

3. GONADOTROPIN-RELEASING HORMONE (GnRH) RECEPTORS

The superfamily of GPCRs share a common structure consisting of seven transmembrane α-helices connected by extra and intracellular loops, an extracellular N-terminal tail, and an intracellular C-terminal tail. Members of this superfamily are subclassified based on homology, ligand similarities, and the mechanisms of their intracellular signaling pathways. Gonadotropin-releasing hormone receptors belong to the GPCR subfamily A, which includes receptors for small molecule ligands such as peptides, opsins, and biogenic amines. Intracellular signaling patterns differ widely among GPCRs but, in general, ligand binding leads to an activation of one or more G-protein subunits. This in turn mobilizes effector proteins, such as kinases, to alter the intracellular concentration of second messenger signaling molecules (cGMP, cAMP, inositol phosphates, diacylglycerol, etc.).

Like the GnRH peptide, mammalian GnRHRs are highly conserved, with approximately 80% sequence homology between mouse, rat, sheep, cow, pig, and human (Kaiser, Zhao, Cardona, & Chin, 1992; Reinhart, Mertz, & Catt, 1992; Chi et al., 1993; Illing et al., 1993; Kakar, Rahe, & Neill, 1993; Perrin et al., 1993; Weesner & Matteri, 1994). Similarly, there are three known GnRHR types that are assumed to have evolved in parallel with the three main classes of GnRH ligands: types GnRHRI, GnRHRII, and GnRHRIII. The gene encoding GnRHRII has been identified in mammals, amphibians, and fishes. However, in many mammalian species, including humans, the GnRHRII gene is silenced or otherwise disrupted, preventing the translation of a functional protein (Stewart, Katz, Millar, & Morgan, 2009). Moreover, the GnRHRIII gene has only been identified in fish and amphibians (Millar et al., 2004). Therefore, the remainder of this section will focus on the type I GnRH receptor, which transduces the GnRH-I signal in gonadotropes.

The amino acid sequence of the mammalian GnRHRI receptor was first determined in mice and rats; however, the three-dimensional crystalline structure remains unresolved (Kaiser et al., 1992; Reinhart et al., 1992; Tsutsumi et al., 1992). Using classic biophysical and biochemical techniques, a basic archetype of the three-dimensional structure was generated from the amino acid sequence and largely validated based on comparisons with the recently achieved X-ray crystallography structure of the GPCR for bovine rhodopsin, also a member of subfamily A, coupled with mutational mapping of specific amino-acid residues (Davidson et al., 1994; Palczewski et al., 2000; Teller, Okada, Behnke, Palczewski, & Stenkamp, 2001; Soderhall, Polymeropoulos, Paulini, Gunther, & Kuhne, 2005).

Notably, mammalian GnRHRI lacks the intracellular C-terminal tail characteristic of most GPCRs, which is important for mediating rapid desensitization to agonists and agonist-induced receptor internalization through its recruitment of β-arrestin (Figure 2.3(a)). This unique structural feature challenges the assumption that the failure of chronically administered GnRH to elicit GTH secretion from the pituitary is due to receptor desensitization or downregulation. The N-terminal tail of GnRHRI contains only 35 amino acids, which, in addition to an absent C-terminal tail, makes them among the smallest GPCRs identified to date. The N-terminal tail contains one of four conserved cysteine residues, which are important for the formation of disulfide bonds on the extracellular face of GnRHRI and subsequent receptor stabilization. Disruption of disulfide bond formation reduces GnRH ligand binding, presumably due to disorganization of the ligand-binding domain. Ligand interactions with GPCRs often occur within a hydrophilic pocket that is created by a tight clustering of the seven transmembrane helices within the hydrophobic portion of the lipid bilayer. Gonadotropin-releasing hormone binding is accomplished through the interaction of the conserved C- and N-terminal ends of the mammalian GnRH peptide (Glu^1, His^2, Arg^8, and Gly^{10}) with specific amino acid residues on the hydrophilic face of the second (human: Asp^{98}, Trp^{101}, Asn^{102}) and third transmembrane helices (human: Lys^{121}) (Zhou et al., 1995; Davidson et al., 1996; Flanagan et al., 2000; Hoffmann, Ter Laak, Kuhne, Reilander, & Beckers, 2000). Site-directed mutagenesis studies have provided evidence that amino acid residues located on the third extracellular loop, connecting transmembrane helices VI and VII, are important for conferring ligand specificity for the mammalian form of GnRH, as mutation of Asp^{302} in the human GnRHRI decreases GnRH binding affinity (Fromme, Katz, Roeske, Millar, & Flanagan, 2001). It has been further demonstrated that the Arg^8 residue, which is unique to the mammalian and guinea pig forms of GnRH (see Table 2.1), interacts with Asp^{302}, thereby inducing a conformational change in the GnRH peptide, enabling a high affinity association with the ligand-binding pocket. Following ligand binding, GnRHRI undergoes a conformational change that allows for the activation of an intracellular heterotrimeric G-protein complex.

3.1. Gonadotropin-releasing Hormone (GnRH) Receptor Regulation in Gonadotropes

Gonadotropin synthesis and secretion induced by GnRH is dependent upon the expression of GnRHRI, which, in the pituitary, is restricted exclusively to gonadotropes. Therefore, the identification of the *cis*- and *trans*-acting factors that direct transcriptional activation of the GnRHRI promoter is critical for understanding the downstream events that lead to GTH regulation. The rat and mouse GnRHR promoters have been studied most thoroughly, primarily in the context of tumorigenic cell lines, such as αT3-1 and LβT2, which endogenously express GTHs. Although most of the regulatory regions have been mapped using the rat and mouse GnRHRI promoter, several of the *cis*-regulatory sites located within the proximal 500 base pairs of the promoter are highly homologous with the GnRHRI promoter in humans and sheep (Campion, Turzillo, & Clay, 1996; Pincas, Forrai, Chauvin, Laverriere, & Counis, 1998).

Specific expression of GnRHRI in gonadotropes is achieved through the coordinated actions of a three-part enhancer element located within the first 500 base pairs of the 5' flanking region. The first part of the enhancer element is a binding site for the orphan receptor steroidogenic factor-1 (SF-1) located at −244/−236 proximal to the transcription start site (Duval, Nelson, & Clay, 1997a). Steroidogenic factor-1 interacts with LIM homeodomain proteins to increase GnRHRI promoter activity, as well as to contribute to the differentiation of gonadotrope cell lineages in the developing pituitary (Savage, Yaden, Kiratipranon, & Rhodes, 2003; Granger et al., 2006). GnRHRI resides within membrane-associated microdomains called lipid rafts, consisting of regions enriched with sphingolipids and phospholipids (Navratil et al., 2003). Interestingly, SF-1 binds sphingosine and phospholipids, suggesting that another important function for SF-1 might be to help sequester GnRHRI to lipid rafts (Urs et al., 2007). The second part of the enhancer element is the binding sequence for activator protein-1 (AP-1), located at −336/−330. Activator protein-1 is a ubiquitous transcription factor formed by the heterodimerization of two immediate early gene products, c-fos and c-jun. The final component of the tripartite enhancer element that regulates the basal and cell-specific expression of the GnRHRI is referred to as GnRHR-activating sequence (GRAS), located at −391/−380 (Duval, Nelson, & Clay, 1997b). GnRHR-activating sequence interacts with multiple transcription factors including Smad4 and the forkhead protein FoxL2 (Ellsworth et al., 2003). Collectively, these three binding sites are required for proper expression of the GnRHRI in gonadotropes. These binding sites are highly conserved across many mammalian species, reflecting the critical functional role of these regulatory regions.

In addition to the *cis*-regulatory regions on the GnRHRI promoter needed for basal expression, there are several identified *trans*-regulatory proteins that serve to modulate the transient increase/decrease of GnRHRI that occurs during the estrous cycle, sexual maturation, and reproductive senescence. Notably, it is through its intrinsic pulsatility that the GnRH peptide mediates the temporal expression of GnRHRI. For example, frequent GnRH pulses favor increased expression of GnRHRI and LHβ subunit synthesis, whereas slow or continuous pulses downregulate GnRHRI (Katt, Duncan, Herbon, Barkan, & Marshall, 1985; Dalkin, Haisenleder, Ortolano, Ellis, & Marshall, 1989; Norwitz, Cardona, Jeong, & Chin, 1999). Gonadal steroid hormones, such as E_2 and P_4, have also been shown to modulate the expression of GnRHRI, especially during the estrous cycle (Gregg, Allen, & Nett, 1990; Gregg, Schwall, & Nett, 1991; McArdle, Schomerus, Groner, & Poch, 1992; Sakurai, Adams, & Adams, 1997; Duval et al., 2000; Cheng, Cheng, & Leung, 2001). In general, high levels of estrogens are correlated with increased GnRHRI expression, whereas the opposite is true for P_4. These findings are consistent with the necessity for increased GnRHRI expression just prior to the LH surge, when estrogen levels are highest. Two members of the TGFβ family of growth factors, activin and inhibin, directly regulate the expression of GnRHRI mRNA in rat pituitary cells (Braden, Farnworth, Burger, & Conn, 1990; Braden & Conn, 1992). Activin is endogenously produced in gonadotropes and regulates GnRHRI via an autocrine/paracrine mechanism. Moreover, activin directly increases the activity of the GnRHRI promoter by interacting with the GRAS element of the GnRHRI promoter in concert with a nearby region termed the downstream activin regulator element (DARE) (Cherrington et al., 2005; Cherrington, Farmerie, & Clay, 2006). Other transcription factors are implicated in regulating GnRHRI expression in gonadotropes, including pituitary adenylate cyclase-activating polypeptide (PACAP) and the basic helix-loop-helix proteins, NeuroD and Mash1 (Pincas, Laverriere, & Counis, 2001; Cherrington, Bailey, Diaz, & Mellon, 2008).

3.2. GnRH-I Receptor Signaling Pathways

G protein-coupled receptors initiate intracellular signaling cascades and activate downstream effectors when a ligand-bound receptor complex interacts with the G_α subunit of the inactive heterotrimeric G-protein complex. The inactive heterotrimeric complex consists of a guanosine diphosphate (GDP)-bound G_α subunit tightly associated with either a G_β or G_γ protein subunit. Ligand binding increases the affinity of the receptor for the α subunit, thereby promoting a nucleotide exchange of GDP with guanosine triphosphate (GTP) and a subsequent dissociation of the α subunit from the ternary complex. Following dissociation, the α subunit activates a distinct intracellular signaling pathway, depending on the specific subtype of the α subunit. The G_α subunit can be classified into four distinct subtype families based on sequence homology: G_s (G_s and G_{olf}); G_i (G_{tr}, G_{tc}, G_g, $G_{i1\text{-}3}$, G_o, G_z); G_q (G_q, G_{11}, G_{14}, $G_{15/16}$); and G_{12} (G_{12}, G_{13}). Gonadotropin-releasing hormone receptor agonists preferentially activate the $G_{q/11}$ subunits in gonadotropes, which initiates signaling through the protein kinase C (PKC) pathway by first stimulating the enzyme phospholipase C (PLCβ) to hydrolyze phosphatidylinositol-4,5 biphosphate (PIP_2) at the cell membrane into two second messenger products: inositol 1,4,5-trisphosphate (IP_3) and diacylglycerol (DAG). This sequence of events results in the mobilization of intracellular calcium from the endoplasmic reticulum by IP_3, and the activation of PKC by DAG (Figure 2.3). Studies in gonadotrope cell lines have shown that intracellular increases in calcium concentration is caused by the release of intracellular calcium stores and the opening of voltage-gated calcium channels (McArdle et al., 1995; 1996). These cellular events are responsible for the acute release of GTHs (Naor, Benard, & Seger, 2000). However, GnRH stimulates not only the synthesis and release of LH and FSH in the gonadotrope, but also promotes a host of cellular maintenance functions. This would suggest either that multiple divergent signaling pathways are activated by a single G_α subtype (i.e., $G_{q/11}$) or that multiple G_α subtypes can interact with GnRHRI. Evidence accumulated from the last decade indicates that at least some of the functions of GnRHRI agonists in gonadotropes can be attributed to receptor interactions with G_α subtypes other than $G_{q/11}$, such as G_s and G_i.

3.3. Gonadotropin-releasing Hormone (GnRH) Pulsatility Differentially Regulates Gonadotropins (GTHs)

Gonadotropin-releasing hormone binding of the GnRH receptor on gonadotropes signals the activation of a host of downstream factors that ultimately regulate the expression of the GTH subunits: common alpha-glycoprotein (Cgα = LHα, FSHα), LHβ, and FSHβ (Ferris & Shupnik, 2006).

Although GTH subunits are expressed at all times, GnRH pulse frequency specifies which type of GTH subunit is preferentially expressed and released (Kaiser, Jakubowiak, Steinberger, & Chin, 1997). A fast GnRH pulse frequency (once every 8–30 minutes) upregulates Cgα transcription, while an intermediate level of GnRH pulsatility (once every 30–60 minutes) favors LHβ subunit gene expression in mice. In contrast, a slow GnRH pulse frequency (once every two to three hours) results in the upregulation of the FSHβ subunit gene. Further, a high GnRH pulse frequency is responsible for the induction of the LH surge that occurs immediately prior to ovulation, while a slow-frequency GnRH pulse will preferentially cause the secretion of FSH (Ferris & Shupnik, 2006).

How does GnRH pulse frequency coordinate the preferential transcription of specific GTH subunits? Although this central question remains unanswered, a few possible mechanisms have been put forward. For example, changes in calcium concentration have been hypothesized to be responsible for the temporally regulated transcription of GTH genes. A likely calcium-responsive intermediary regulating intracellular calcium fluctuations is calmodulin kinase (CAMK) 2. Interestingly, studies in primary pituitary cultures and LβT2 immortalized cells have shown that CAMK2 phosphorylation is rapidly achieved following administration of GnRH (i.e., within two to five minutes); however, phosphorylated CAMK2 remains present for another 45 minutes (Haisenleder, Dalkin, Ortolano, Marshall, & Shupnik, 1991; Haisenleder, Ferris, & Shupnik, 2003). The slow decay of phosphorylated CAMK2 correlates well with the GnRH pulse intervals that favor transcription of Cgα and LHβ, and not FSHβ. In addition to CAMK2, GnRH pulses also activate the mitogen-activated protein kinase (MAPK) 1 pathway, which has been shown to activate Cgα and FSHβ transcription (Ferris & Shupnik, 2006). Thus, these two intracellular kinases, CAMK2 and MAPK1, initiate the transcription of different β-specific GTH subunits, LHβ and FSHβ respectively, yet they are both activated by a common stimulus: the GnRH pulse.

Interestingly, the immediate response to GnRH receptor binding is the activation of a negative feedback loop that prevents subsequent GnRH signaling. This process is thought to reset the GnRH/GnRH receptor signaling pathway in order to prepare for the next GnRH pulse. Normally, GPCRs are desensitized and internalized through the activation of their intracellular C-terminal sequence, which is absent in GnRH receptors (Willars et al., 1999). Consequently, GnRH receptor signaling desensitization is achieved through the upregulation of regulators of G-protein signaling-2 (*Rgs-2*), which shortens the activity of the GnRHRI-associated G-proteins (Cunningham, Waldo, Hollinger, Hepler, & Harden, 2001; Wurmbach, Yuen, Ebersole, & Sealfon, 2001; Kehrl & Sinnarajah, 2002).

Gonadotropin-releasing hormone signaling ultimately induces DNA-binding proteins to act on specific response elements on the promoters of GTH genes. For example, the promoter regions of Cgα, LHβ, and FSHβ contain response elements for the orphan nuclear receptor SF-1 (Savage et al., 2003), which has been proposed to regulate the steady-state expression of GTHs (Kaiser, Halvorson, & Chen, 2000; Salisbury, Binder, & Nilson, 2008). This hypothesis is consistent with the observation that SF-1-null mice are hypogonadal due to undetectable levels of Cgα, LHβ, and FSHβ (Zhao et al., 2001a; 2001b). Further, the LHβ promoter contains two early growth response protein-1 (EGR-1) binding sites that are required for its full activation. Coincidently, EGR-1 activation also increases the expression of the MAPK1 signaling pathway inhibitor DUSP4 (Roberson, Misra-Press, Laurance, Stork, & Maurer, 1995; Wurmbach et al., 2001), which could be a mechanism for GnRH signaling to selectively induce LHβ gene expression. Alternatively, GnRH signals through JUN N-terminal kinase to potentiate AP-1 activity, which in turn stimulates selective transcription of FSHβ (Salisbury et al., 2008). Together, these data suggest that GnRH may differentially regulate GTH expression through the selective activation of DNA-binding proteins.

4. CONCLUSIONS

The detection and isolation of GnRH ushered in a new era of rapid progress towards understanding how GTHs are regulated at the molecular and systems level. Despite these advances, the frequency of reproductive disorders and associated rates of infertility continue to climb (Ledger, 2009; Wilkes, Chinn, Murdoch, & Rubin, 2009). Infertility often stems from underlying medical conditions, such as obesity, diabetes, polycystic ovarian syndrome, or sexually transmitted diseases (Ledger, 2009; Livshits & Seidman, 2009; Wilkes & Murdoch, 2009). How these diseases contribute to the pathophysiology of reproductive dysfunction is not well understood, but might be resolved with a better understanding of the development and regulation of GnRH neurons. For instance, the precise developmental factors that dictate GnRH neuronal fate specification have yet to be determined, mainly due to a lack of unique markers for GnRH neurons that are present prior to the transcriptional activation of the GnRH gene. New techniques must be developed that will allow for the early identification and lineage-specific tracking of GnRH neurons in order to gain a better understanding of how the GnRH system is sculpted during early development. Further, hypothalamic releasing factors regulating GTH action other than GnRH have been proposed, such as gonadotropin-inhibiting hormone (GnIH) and FSH-releasing hormone; however, to date the precise sequence identification and biological actions of these additional factors have not been validated in mammals. The discovery of additional central regulators of GTHs could resolve some of the complexities surrounding the multifaceted regulatory signaling patterns of GnRH. Finally, the physiological cues dictating the onset of puberty, or, conversely, onset of menopause, remain unknown. Although recent evidence strongly supports the hypothesis that Kp is the primary regulator of pubertal onset in mammals, it remains to be determined what initiates Kp signaling. Overwhelming evidence would suggest that it is not one cue, but, rather, a host of precisely coordinated neuroendocrine signals that determine changes in reproductive competency across the lifespan. However, the fundamental questions of what factors coordinate the timing of these signals and how those factors are regulated remain unanswered. Similarly, it is unclear how this timing can be altered by environmental agents, pharmacological interventions, or early life experiences. Future research that finds answers to some of these essential questions will significantly increase our understanding of the mammalian reproductive system and lead to improved treatments for reproductive disorders.

ABBREVIATIONS

ACTH Corticotropin
AHA Anterior hypothalamic area
AP Anterior pituitary
AP-1 Activator protein-1
CAMK Calmodulin kinase
CG Chorionic gonadotropin
Cgα Common α-glycoprotein
CNS Central nervous system
CRF Corticotropin-releasing factor
DAG Diacylglycerol
DAPI 4',6-diamidino-2-phenylindole
DARE Downstream activin regulator element
DBB Diagonal band of Broca
DCC Deleted in colorectal cancer
dnFGFR Dominant-negative form of the fibroblast growth factor receptor
E Embryonic day
E_2 17β-estradiol
eGFP Enhanced green fluorescent protein
EGR-1 Early growth response protein-1
FGF Fibroblast growth factor
FGFR Fibroblast growth factor receptor
FSH Follicle-stimulating hormone
GABA Gamma-amino butyric acid
GAP Gonadotropin-releasing hormone-associated peptide
GDP Guanosine diphosphate
GnIH Gonadotropin-inhibiting hormone
GnRH Gonadotropin-releasing hormone
GnRHR Gonadotropin-releasing hormone receptor
GPCR G-protein-coupled receptor
GRAS GnRHR-activating sequence
GTH Gonadotropin
GTP Guanosine triphosphate
HPG Hypothalamus–pituitary–gonad
IHH Idiopathic hypogonadotropic hypogonadism
IP_3 Inositol triphosphate
Kp Kisspeptin
LH Luteinizing hormone
LHRH Luteinizing hormone-releasing hormone (i.e., GnRH)
LuRKO Luteinizing hormone receptor knockout
MAPK Mitogen-activated protein kinase
ME Median eminence
NELF Nasal embryonic luteinizing hormone-releasing hormone factor
NO Nitric oxide
NOS Nitric oxide synthase
OP Olfactory placodes
OVLT Organum vasculosum of the lamina terminalis
P_4 Progesterone
PACAP Pituitary adenylate cyclase-activating polypeptide
PD Pars distalis
PI Pars intermedia
PIP_2 Phosphatidylinositol-4,5 biphosphate
PKC Protein kinase C
PLCβ Phospholipase C
POA Preoptic area
pro-GnRH Progonadotropin-releasing hormone
PRV Ba2001 Psuedorabies virus
PT Pars tuberalis
Sey Small-eye
SF-1 Steroidogenic factor-1
T Testosterone
VNN Vomeronasal nerve
VNO Vomeronasal organ

REFERENCES

Amoss, M., Burgus, R., Blackwell, R., Vale, W., Fellows, R., & Guillemin, R. (1971). Purification, amino acid composition and N-terminus of the hypothalamic luteinizing hormone releasing factor (LRF) of ovine origin. *Biochem. Biophys. Res. Commun., 44*, 205–210.

Belchetz, P. E., Plant, T. M., Nakai, Y., Keogh, E. J., & Knobil, E. (1978). Hypophysial responses to continuous and intermittent delivery of hypothalamic gonadotropin-releasing hormone. *Science, 202*, 631–633.

Boehm, U., Zou, Z., & Buck, L. B. (2005). Feedback loops link odor and pheromone signaling with reproduction. *Cell, 123*, 683–695.

Bourguignon, J. P., & Franchimont, P. (1984). Puberty-related increase in episodic LHRH release from rat hypothalamus *in vitro*. *Endocrinology, 114*, 1941–1943.

Bousfield, G. R., Butnev, V. Y., Gotschall, R. R., Baker, V. L., & Moore, W. T. (1996). Structural features of mammalian gonadotropins. *Mol. Cell Endocrinol., 125*, 3–19.

Braden, T. D., & Conn, P. M. (1992). Activin-A stimulates the synthesis of gonadotropin-releasing hormone receptors. *Endocrinology, 130*, 2101–2105.

Braden, T. D., Farnworth, P. G., Burger, H. G., & Conn, P. M. (1990). Regulation of the synthetic rate of gonadotropin-releasing hormone receptors in rat pituitary cell cultures by inhibin. *Endocrinology, 127*, 2387–2392.

Cajal, S. (1911). *Histologie du systeme nerveux de l'homme et des vertebres.* Paris, France: Maloine.

Campbell, R. E., & Herbison, A. E. (2007). Definition of brainstem afferents to gonadotropin-releasing hormone neurons in the mouse using conditional viral tract tracing. *Endocrinology, 148*, 5884–5890.

Campbell, R. E., Han, S. K., & Herbison, A. E. (2005). Biocytin filling of adult gonadotropin-releasing hormone neurons *in situ* reveals extensive, spiny, dendritic processes. *Endocrinology, 146*, 1163–1169.

Campion, C. E., Turzillo, A. M., & Clay, C. M. (1996). The gene encoding the ovine gonadotropin-releasing hormone (GnRH) receptor: cloning and initial characterization. *Gene., 170*, 277–280.

Catt, K. J., Harwood, J. P., Clayton, R. N., Davies, T. F., Chan, V., Katikineni, M., et al. (1980). Regulation of peptide hormone receptors and gonadal steroidogenesis. *Recent Prog. Horm. Res., 36*, 557–662.

Cheng, K. W., Cheng, C. K., & Leung, P. C. (2001). Differential role of PR-A and -B isoforms in transcription regulation of human GnRH receptor gene. *Mol. Endocrinol., 15*, 2078–2092.

Cherrington, B. D., Bailey, J. S., Diaz, A. L., & Mellon, P. L. (2008). NeuroD1 and Mash1 temporally regulate GnRH receptor gene expression in immortalized mouse gonadotrope cells. *Mol. Cell. Endocrinol., 295*, 106–114.

Cherrington, B. D., Farmerie, T. A., & Clay, C. M. (2006). A specific helical orientation underlies the functional contribution of the activin responsive unit to transcriptional activity of the murine gonadotropin-releasing hormone receptor gene promoter. *Endocrine., 29*, 425–433.

Cherrington, B. D., Farmerie, T. A., Lents, C. A., Cantlon, J. D., Roberson, M. S., & Clay, C. M. (2005). Activin responsiveness of the murine gonadotropin-releasing hormone receptor gene is mediated by a composite enhancer containing spatially distinct regulatory elements. *Mol. Endocrinol., 19*, 898–912.

Chi, L., Zhou, W., Prikhozhan, A., Flanagan, C., Davidson, J. S., Golembo, M., et al. (1993). Cloning and characterization of the human GnRH receptor. *Mol. Cell Endocrinol., 91*, R1–R6.

Chuah, M. I., & Au, C. (1994). Olfactory cell cultures on ensheathing cell monolayers. *Chem. Senses, 19*, 25–34.

Chung, W. C. J., Moyle, S. S., & Tsai, P. S. (2008). Fibroblast growth factor 8 signaling through Fgf receptor 1 is required for the emergence of gonadotropin-releasing hormone neurons. *Endocrinology.*

Clasadonte, J., Poulain, P., Beauvillain, J. C., & Prevot, V. (2008). Activation of neuronal nitric oxide release inhibits spontaneous firing in adult gonadotropin-releasing hormone neurons: a possible local synchronizing signal. *Endocrinology, 149*, 587–596.

Crossley, P. H., & Martin, G. R. (1995). The mouse *Fgf8* gene encodes a family of polypeptides and is expressed in regions that direct outgrowth and patterning in the developing embryo. *Development, 121*, 439–451.

Cunningham, M. L., Waldo, G. L., Hollinger, S., Hepler, J. R., & Harden, T. K. (2001). Protein kinase C phosphorylates RGS2 and modulates its capacity for negative regulation of Galpha 11 signaling. *J. Biol. Chem., 276*, 5438–5444.

D'Anglemont de Tassigny, X., Fagg, L. A., Dixon, J. P., Day, K., Leitch, H. G., Hendrick, A. G., et al. (2007). Hypogonadotropic hypogonadism in mice lacking a functional *Kiss1* gene. *Proc. Natl. Acad. Sci. USA, 104*, 10714–10719.

Dalkin, A. C., Haisenleder, D. J., Ortolano, G. A., Ellis, T. R., & Marshall, J. C. (1989). The frequency of gonadotropin-releasing-hormone stimulation differentially regulates gonadotropin subunit messenger ribonucleic acid expression. *Endocrinology, 125*, 917–924.

Dalpathado, D. S., Irungu, J., Go, E. P., Butnev, V. Y., Norton, K., Bousfield, G. R., et al. (2006). Comparative glycomics of the glycoprotein follicle stimulating hormone: glycopeptide analysis of isolates from two mammalian species. *Biochemistry, 45*, 8665–8673.

Davidson, J. S., Flanagan, C. A., Becker, , II, Illing, N., Sealfon, S. C., & Millar, R. P. (1994). Molecular function of the gonadotropin-releasing hormone receptor: insights from site-directed mutagenesis. *Mol. Cell. Endocrinol., 100*, 9–14.

Davidson, J. S., McArdle, C. A., Davies, P., Elario, R., Flanagan, C. A., & Millar, R. P. (1996). Asn102 of the gonadotropin-releasing hormone receptor is a critical determinant of potency for agonists containing C-terminal glycinamide. *J. Biol. Chem., 271*, 15510–15514.

De Roux, N., Genin, E., Carel, J. C., Matsuda, F., Chaussain, J. L., & Milgrom, E. (2003). Hypogonadotropic hypogonadism due to loss of function of the KiSS1-derived peptide receptor GPR54. *Proc. Natl. Acad. Sci. USA., 100*, 10972–10976.

DeFalco, J., Tomishima, M., Liu, H., Zhao, C., Cai, X., Marth, J. D., et al. (2001). Virus-assisted mapping of neural inputs to a feeding center in the hypothalamus. *Science, 291*, 2608–2613.

Dellovade, T. L., Pfaff, D. W., & Schwanzel-Fukuda, M. (1998). The gonadotropin-releasing hormone system does not develop in Small-Eye (Sey) mouse phenotype. *Brain Res. Dev. Brain Res., 107*, 233–240.

Dhillo, W. S., Chaudhri, O. B., Patterson, M., Thompson, E. L., Murphy, K. G., Badman, M. K., et al. (2005). Kisspeptin-54 stimulates the hypothalamic–pituitary–gonadal axis in human males. *J. Clin. Endocrinol. Metab., 90*, 6609–6615.

Dudley, C. A., Vale, W., Rivier, J., & Moss, R. L. (1981). The effect of LHRH antagonist analogs and an antibody to LHRH on mating behavior in female rats. *Peptides, 2*, 393–396.

Duval, D. L., Farris, A. R., Quirk, C. C., Nett, T. M., Hamernik, D. L., & Clay, C. M. (2000). Responsiveness of the ovine gonadotropin-releasing hormone receptor gene to estradiol and gonadotropin-releasing hormone is not detectable *in vitro* but is revealed in transgenic mice. *Endocrinology, 141*, 1001–1010.

Duval, D. L., Nelson, S. E., & Clay, C. M. (1997a). A binding site for steroidogenic factor-1 is part of a complex enhancer that mediates expression of the murine gonadotropin-releasing hormone receptor gene. *Biol. Reprod., 56*, 160–168.

Duval, D. L., Nelson, S. E., & Clay, C. M. (1997b). The tripartite basal enhancer of the gonadotropin-releasing hormone (GnRH) receptor gene promoter regulates cell-specific expression through a novel GnRH receptor activating sequence. *Mol. Endocrinol., 11*, 1814–1821.

Ellsworth, B. S., Burns, A. T., Escudero, K. W., Duval, D. L., Nelson, S. E., & Clay, C. M. (2003). The gonadotropin releasing hormone (GnRH) receptor activating sequence (GRAS) is a composite regulatory element that interacts with multiple classes of transcription factors including Smads, AP-1 and a forkhead DNA binding protein. *Mol. Cell Endocrinol., 206*, 93–111.

Falardeau, J., Chung, W. C. J., Beenken, A., Raivio, T., Plummer, L., Sidis, Y., et al. (2008). Decreased FGF8 signaling causes deficiency of gonadotropin-releasing hormone in humans and mice. *J. Clin. Invest., 118*, 2822–2831.

Ferris, H. A., & Shupnik, M. A. (2006). Mechanisms for pulsatile regulation of the gonadotropin subunit genes by GNRH1. *Biol. Reprod., 74*, 993–998.

Flanagan, C. A., Rodic, V., Konvicka, K., Yuen, T., Chi, L., Rivier, J. E., et al. (2000). Multiple interactions of the Asp(2.61(98)) side chain of the gonadotropin-releasing hormone receptor contribute differentially to ligand interaction. *Biochemistry, 39*, 8133–8141.

Fromme, B. J., Katz, A. A., Roeske, R. W., Millar, R. P., & Flanagan, C. A. (2001). Role of aspartate7.32(302) of the human gonadotropin-releasing hormone receptor in stabilizing a high-affinity ligand conformation. *Mol. Pharmacol., 60*, 1280–1287.

Fujioka, H., Suzuki, M., Yamanouchi, K., Ohta, A., Nagashima, H., Kato, M., et al. (2003). Generation of transgenic rats expressing enhanced green fluorescent protein in gonadotropin-releasing hormone neurons. *J. Reprod. Dev., 49*, 523–529.

Garel, S., Marin, F., Grosschedl, R., & Charnay, P. (1999). Ebf1 controls early cell differentiation in the embryonic striatum. *Development, 126*, 5285–5294.

Gibson, M. J., Ingraham, L., & Dobrjansky, A. (2000). Soluble factors guide gonadotropin-releasing hormone axonal targeting to the median eminence. *Endocrinology, 141*, 3065–3071.

Gill, J. C., & Tsai, P. S. (2006). Expression of a dominant negative FGF receptor in developing GNRH1 neurons disrupts axon outgrowth and targeting to the median eminence. *Biol. Reprod., 74*, 463–472.

Gill, J. C., Moenter, S. M., & Tsai, P. S. (2004). Developmental regulation of gonadotropin-releasing hormone neurons by fibroblast growth factor signaling. *Endocrinology, 145*, 3830–3839.

Gorczyca, W., & Hardy, J. (1987). Arterial supply of the human anterior pituitary gland. *Neurosurgery, 20*, 369–378.

Gore, A. C. (2002). *GnRH: The Master Molecule of Reproduction.* Boston, MA: Kluwer Academic Publishers Group.

Gottsch, M. L., Cunningham, M. J., Smith, J. T., Popa, S. M., Acohido, B. V., Crowley, W. F., et al. (2004). A role for kisspeptins in the regulation of gonadotropin secretion in the mouse. *Endocrinology, 145*, 4073–4077.

Granger, A., Bleux, C., Kottler, M. L., Rhodes, S. J., Counis, R., & Laverriere, J. N. (2006). The LIM-homeodomain proteins Isl-1 and Lhx3 act with steroidogenic factor 1 to enhance gonadotrope-specific activity of the gonadotropin-releasing hormone receptor gene promoter. *Mol. Endocrinol., 20*, 2093–2108.

Green, J. D., & Harris, G. W. (1949). Observation of the hypophysio-portal vessels of the living rat. *J. Physiol., 108*, 359–361.

Gregg, D. W., Allen, M. C., & Nett, T. M. (1990). Estradiol-induced increase in number of gonadotropin-releasing hormone receptors in cultured ovine pituitary cells. *Biol. Reprod., 43*, 1032–1036.

Gregg, D. W., Schwall, R. H., & Nett, T. M. (1991). Regulation of gonadotropin secretion and number of gonadotropin-releasing hormone receptors by inhibin, activin-A, and estradiol. *Biol. Reprod., 44*, 725–732.

Grossman, A. B., Rossmanith, W. G., Kabigting, E. B., Cadd, G., Clifton, D., & Steiner, R. A. (1994). The distribution of hypothalamic nitric oxide synthase mRNA in relation to gonadotropin-releasing hormone neurons. *J. Endocrinol., 140*, R5–R8.

Guillemin, R., & Rosenberg, B. (1955). Humoral hypothalamic control of anterior pituitary: a study with combined tissue cultures. *Endocrinology, 57*, 599–607.

Guillemin, R., Hearn, W. R., Cheek, W. R., & Householder, D. E. (1957). Control of corticotrophin release: further studies with *in vitro* methods. *Endocrinology, 60*, 488–506.

Haisenleder, D. J., Dalkin, A. C., Ortolano, G. A., Marshall, J. C., & Shupnik, M. A. (1991). A pulsatile gonadotropin-releasing hormone stimulus is required to increase transcription of the gonadotropin subunit genes: evidence for differential regulation of transcription by pulse frequency *in vivo*. *Endocrinology, 128*, 509–517.

Haisenleder, D. J., Ferris, H. A., & Shupnik, M. A. (2003). The calcium component of gonadotropin-releasing hormone-stimulated luteinizing hormone subunit gene transcription is mediated by calcium/calmodulin-dependent protein kinase type II. *Endocrinology, 144*, 2409–2416.

Harris, G. W. (1948a). Electrical stimulation of the hypothalamus and the mechanism of neural control of the adenohypophysis. *J. Physiol., 107*, 418–429.

Harris, G. W. (1948b). Neural control of the pituitary gland. *Physiol. Rev., 28*, 139–179.

Herbison, A. E., Simonian, S. X., Norris, P. J., & Emson, P. C. (1996). Relationship of neuronal nitric oxide synthase immunoreactivity to GnRH neurons in the ovariectomized and intact female rat. *J. Neuroendocrinol., 8*, 73–82.

Hoffmann, S. H., Ter Laak, T., Kuhne, R., Reilander, H., & Beckers, T. (2000). Residues within transmembrane helices 2 and 5 of the human gonadotropin-releasing hormone receptor contribute to agonist and antagonist binding. *Mol. Endocrinol., 14*, 1099–1115.

Hsu, P., Yu, F., Feron, F., Pickles, J. O., Sneesby, K., & Mackay-Sim, A. (2001). Basic fibroblast growth factor and fibroblast growth factor receptors in adult olfactory epithelium. *Brain Res., 896*, 188–197.

Huhtaniemi, I., Zhang, F. P., Kero, J., Hamalainen, T., & Poutanen, M. (2002). Transgenic and knockout mouse models for the study of luteinizing hormone and luteinizing hormone receptor function. *Mol. Cell Endocrinol., 187*, 49–56.

Illing, N., Jacobs, G. F., Becker, , II, Flanagan, C. A., Davidson, J. S., Eales, A., et al. (1993). Comparative sequence analysis and functional characterization of the cloned sheep gonadotropin-releasing hormone receptor reveal differences in primary structure and ligand specificity among mammalian receptors. *Biochem. Biophys. Res. Commun., 196*, 745–751.

Irwig, M. S., Fraley, G. S., Smith, J. T., Acohido, B. V., Popa, S. M., Cunningham, M. J., et al. (2004). Kisspeptin activation of gonadotropin releasing hormone neurons and regulation of KiSS-1 mRNA in the male rat. *Neuroendocrinology, 80*, 264–272.

Itoh, N., & Ornitz, D. M. (2004). Evolution of the *Fgf* and *Fgfr* gene families. *Trends Genet., 20*, 563–569.

Kaiser, U. B., Halvorson, L. M., & Chen, M. T. (2000). Sp1, steroidogenic factor 1 (SF-1), and early growth response protein 1 (egr-1) binding sites form a tripartite gonadotropin-releasing hormone response element in the rat luteinizing hormone-beta gene promoter: an integral role for SF-1. *Mol. Endocrinol., 14*, 1235–1245.

Kaiser, U. B., Jakubowiak, A., Steinberger, A., & Chin, W. W. (1997). Differential effects of gonadotropin-releasing hormone (GnRH) pulse frequency on gonadotropin subunit and GnRH receptor messenger ribonucleic acid levels *in vitro*. *Endocrinology, 138*, 1224–1231.

Kaiser, U. B., Zhao, D., Cardona, G. R., & Chin, W. W. (1992). Isolation and characterization of cDNAs encoding the rat pituitary gonadotropin-releasing hormone receptor. *Biochem. Biophys. Res. Commun., 189*, 1645–1652.

Kakar, S. S., Rahe, C. H., & Neill, J. D. (1993). Molecular cloning, sequencing, and characterizing the bovine receptor for gonadotropin releasing hormone (GnRH). *Domest. Anim. Endocrinol., 10*, 335–342.

Katt, J. A., Duncan, J. A., Herbon, L., Barkan, A., & Marshall, J. C. (1985). The frequency of gonadotropin-releasing hormone stimulation determines the number of pituitary gonadotropin-releasing hormone receptors. *Endocrinology, 116*, 2113–2115.

Kawakami, M., Uemura, T., & Hayashi, R. (1982). Electrophysiological correlates of pulsatile gonadotropin release in rats. *Neuroendocrinology, 35*, 63–67.

Kawauchi, S., Shou, J., Santos, R., Hebert, J. M., McConnell, S. K., Mason, I., et al. (2005). Fgf8 expression defines a morphogenetic center required for olfactory neurogenesis and nasal cavity development in the mouse. *Development, 132*, 5211–5223.

Kehrl, J. H., & Sinnarajah, S. (2002). RGS2: a multifunctional regulator of G-protein signaling. *Int. J. Biochem. Cell. Biol., 34*, 432–438.

Kim, K., & Ramirez, V. D. (1986). *In-vitro* LHRH release from superfused hypothalamus as a function of the rat estrous cycle: effect of progesterone. *Neuroendocrinology, 42*, 392–398.

King, J. C., Tobet, S. A., Snavely, F. L., & Arimura, A. A. (1982). LHRH immunopositive cells and their projections to the median eminence and organum vasculosum of the lamina terminalis. *J. Comp. Neurol., 209*, 287–300.

Knobil, E., Plant, T. M., Wildt, L., Belchetz, P. E., & Marshall, G. (1980). Control of the rhesus monkey menstrual cycle: permissive role of hypothalamic gonadotropin-releasing hormone. *Science, 207*, 1371–1373.

Kotani, M., Detheux, M., Vandenbogaerde, A., Communi, D., Vanderwinden, J. M., Le Poul, E., et al. (2001). The metastasis suppressor gene KiSS-1 encodes kisspeptins, the natural ligands of the orphan G protein-coupled receptor GPR54. *J. Biol. Chem., 276*, 34631–34636.

Kramer, P. R., & Wray, S. (2000). Novel gene expressed in nasal region influences outgrowth of olfactory axons and migration of luteinizing hormone-releasing hormone (LHRH) neurons. *Genes Dev., 14*, 1824–1834.

Kramer, P. R., Krishnamurthy, R., Mitchell, P. J., & Wray, S. (2000). Transcription factor activator protein-2 is required for continued luteinizing hormone-releasing hormone expression in the forebrain of developing mice. *Endocrinology, 141*, 1823–1838.

Kumar, T. R. (2007a). Functional analysis of LHbeta knockout mice. *Mol Cell Endocrinol., 269*, 81–84.

Kumar, T. R. (2007b). Mouse models for gonadotropins: a 15-year saga. *Mol. Cell Endocrinol., 260-262*, 249–254.

Lapatto, R., Pallais, J. C., Zhang, D., Chan, Y. M., Mahan, A., Cerrato, F., et al. (2007). Kiss1-/- mice exhibit more variable hypogonadism than Gpr54-/- mice. *Endocrinology, 148*, 4927–4936.

Lawson, M. A., & Mellon, P. L. (1998). Expression of GATA-4 in migrating gonadotropin-releasing neurons of the developing mouse. *Mol. Cell Endocrinol., 140*, 157–161.

Ledger, W. L. (2009). Demographics of infertility. *Reprod. Biomed. Online* 18. *Suppl, 2*, 11–14.

Leung, P. C., & Armstrong, D. T. (1980). Interactions of steroids and gonadotropins in the control of steroidogenesis in the ovarian follicle. *Annu. Rev. Physiol., 42*, 71–82.

Livne, I., Gibson, M. J., & Silverman, A. J. (1993). Biochemical differentiation and intercellular interactions of migratory gonadotropin-releasing hormone (GnRH) cells in the mouse. *Dev. Biol., 159*, 643–656.

Livshits, A., & Seidman, D. S. (2009). Fertility issues in women with diabetes. *Womens Health (Lond Engl), 5*, 701–707.

Lopez, F. J., Merchenthaler, I. J., Moretto, M., & Negro-Vilar, A. (1998). Modulating mechanisms of neuroendocrine cell activity: the LHRH pulse generator. *Cell Mol. Neurobiol., 18*, 125–146.

Lopez, F. J., Moretto, M., Merchenthaler, I., & Negro-Vilar, A. (1997). Nitric oxide is involved in the genesis of pulsatile LHRH secretion from immortalized LHRH neurons. *J. Neuroendocrinol., 9*, 647–654.

Luscher, B., Mitchell, P. J., Williams, T., & Tjian, R. (1989). Regulation of transcription factor AP-2 by the morphogen retinoic acid and by second messengers. *Genes Dev., 3*, 1507–1517.

Ma, X., Dong, Y., Matzuk, M. M., & Kumar, T. R. (2004). Targeted disruption of luteinizing hormone beta-subunit leads to hypogonadism, defects in gonadal steroidogenesis, and infertility. *Proc. Natl. Acad. Sci. USA, 101*, 17294–17299.

McArdle, C. A., Forrest-Owen, W., Davidson, J. S., Fowkes, R., Bunting, R., Mason, W. T., et al. (1996). Ca2+ entry in gonadotrophs and alpha T3-1 cells: does store-dependent Ca2+ influx mediate gonadotropin-releasing hormone action? *J. Endocrinol., 149*, 155–169.

McArdle, C. A., Forrest-Owen, W., Willars, G., Davidson, J., Poch, A., & Kratzmeier, M. (1995). Desensitization of gonadotropin-releasing hormone action in the gonadotrope-derived alpha T3-1 cell line. *Endocrinology, 136*, 4864–4871.

McArdle, C. A., Schomerus, E., Groner, I., & Poch, A. (1992). Estradiol regulates gonadotropin-releasing hormone receptor number, growth and inositol phosphate production in alpha T3-1 cells. *Mol. Cell. Endocrinol., 87*, 95–103.

Means, A. R., Dedman, J. R., Tash, J. S., Tindall, D. J., Van Sickle, M., & Welsh, M. J. (1980). Regulation of the testis sertoli cell by follicle stimulating hormone. *Annu. Rev. Physiol., 42*, 59–70.

Merchenthaler, I., Gorcs, T., Setalo, G., Petrusz, P., & Flerko, B. (1984). Gonadotropin-releasing hormone (GnRH) neurons and pathways in the rat brain. *Cell Tissue Res., 237*, 15–29.

Messager, S., Chatzidaki, E. E., Ma, D., Hendrick, A. G., Zahn, D., Dixon, J., et al. (2005). Kisspeptin directly stimulates gonadotropin-releasing hormone release via G protein-coupled receptor 54. *Proc. Natl. Acad. Sci. USA, 102*, 1761–1766.

Meyers, E. N., Lewandoski, M., & Martin, G. R. (1998). An *Fgf8* mutant allelic series generated by Cre- and Flp-mediated recombination. *Nat. Genet., 18*, 136–141.

Millar, R. P., Lu, Z. L., Pawson, A. J., Flanagan, C. A., Morgan, K., & Maudsley, S. R. (2004). Gonadotropin-releasing hormone receptors. *Endocr. Rev., 25*, 235–275.

Moss, R. L., & McCann, S. M. (1975). Action of luteinizing hormone-releasing factor (lrf) in the initiation of lordosis behavior in the estrone-primed ovariectomized female rat. *Neuroendocrinology, 17*, 309–318.

Naor, Z., Benard, O., & Seger, R. (2000). Activation of MAPK cascades by G-protein-coupled receptors: the case of gonadotropin-releasing hormone receptor. *Trends Endocrinol. Metab., 11*, 91–99.

Navarro, V. M., Castellano, J. M., Fernandez-Fernandez, R., Tovar, S., Roa, J., Mayen, A., et al. (2005a). Effects of KiSS-1 peptide, the natural ligand of GPR54, on follicle-stimulating hormone secretion in the rat. *Endocrinology, 146*, 1689–1697.

Navarro, V. M., Castellano, J. M., Fernandez-Fernandez, R., Tovar, S., Roa, J., Mayen, A., et al. (2005b). Characterization of the potent luteinizing hormone-releasing activity of KiSS-1 peptide, the natural ligand of GPR54. *Endocrinology, 146*, 156–163.

Navratil, A. M., Bliss, S. P., Berghorn, K. A., Haughian, J. M., Farmerie, T. A., Graham, J. K., et al. (2003). Constitutive localization of the gonadotropin-releasing hormone (GnRH) receptor to low density membrane microdomains is necessary for GnRH signaling to ERK. *J. Biol. Chem., 278*, 31593–31602.

Norwitz, E. R., Cardona, G. R., Jeong, K. H., & Chin, W. W. (1999). Identification and characterization of the gonadotropin-releasing hormone response elements in the mouse gonadotropin-releasing hormone receptor gene. *J. Biol. Chem., 274*, 867–880.

Palczewski, K., Kumasaka, T., Hori, T., Behnke, C. A., Motoshima, H., Fox, B. A., et al. (2000). Crystal structure of rhodopsin: a G protein-coupled receptor. *Science, 289*, 739–745.

Perrin, M. H., Bilezikjian, L. M., Hoeger, C., Donaldson, C. J., Rivier, J., Haas, Y., et al. (1993). Molecular and functional characterization of GnRH receptors cloned from rat pituitary and a mouse pituitary tumor cell line. *Biochem. Biophys. Res. Commun., 191*, 1139–1144.

Pincas, H., Forrai, Z., Chauvin, S., Laverriere, J. N., & Counis, R. (1998). Multiple elements in the distal part of the 1.2 kb 5'-flanking region of the rat GnRH receptor gene regulate gonadotrope-specific expression conferred by proximal domain. *Mol. Cell Endocrinol., 144*, 95–108.

Pincas, H., Laverriere, J. N., & Counis, R. (2001). Pituitary adenylate cyclase-activating polypeptide and cyclic adenosine 3',5'-monophosphate stimulate the promoter activity of the rat gonadotropin-releasing hormone receptor gene via a bipartite response element in gonadotrope-derived cells. *J. Biol. Chem., 276*, 23562–23571.

Plant, T. M., Ramaswamy, S., & Dipietro, M. J. (2006). Repetitive activation of hypothalamic G protein-coupled receptor 54 with intravenous pulses of kisspeptin in the juvenile monkey (*Macaca mulatta*) elicits a sustained train of gonadotropin-releasing hormone discharges. *Endocrinology, 147*, 1007–1013.

Reinhart, J., Mertz, L. M., & Catt, K. J. (1992). Molecular cloning and expression of cDNA encoding the murine gonadotropin-releasing hormone receptor. *J. Biol. Chem., 267*, 21281–21284.

Roberson, M. S., Misra-Press, A., Laurance, M. E., Stork, P. J., & Maurer, R. A. (1995). A role for mitogen-activated protein kinase in mediating activation of the glycoprotein hormone alpha-subunit promoter by gonadotropin-releasing hormone. *Mol. Cell Biol., 15*, 3531–3539.

Roberts, J. L., Mani, S. K., Woller, M. J., Glucksman, M. J., & Wu, T. J. (2007). LHRH-(1-5): a bioactive peptide regulating reproduction. *Trends Endocrinol. Metab., 18*, 386–392.

Sakurai, H., Adams, B. M., & Adams, T. E. (1997). Concentration of GnRH receptor and GnRH receptor mRNA in pituitary tissue of orchidectomized sheep: effect of oestradiol, progesterone, and progesterone withdrawal. *J. Endocrinol., 152*, 91–98.

Salisbury, T. B., Binder, A. K., & Nilson, J. H. (2008). Welcoming beta-catenin to the gonadotropin-releasing hormone transcriptional network in gonadotropes. *Mol. Endocrinol., 22*, 1295–1303.

Savage, J. J., Yaden, B. C., Kiratipranon, P., & Rhodes, S. J. (2003). Transcriptional control during mammalian anterior pituitary development. *Gene., 319*, 1–19.

Schally, A. V., Arimura, A., Kastin, A. J., Matsuo, H., Baba, Y., Redding, T. W., et al. (1971). Gonadotropin-releasing hormone: one polypeptide regulates secretion of luteinizing and follicle-stimulating hormones. *Science, 173*, 1036–1038.

Schally, A. V., & Guillemin, R. (1959). Concentration of corticotropin releasing factor by chromatography on carboxymethylcellulose. *Proc. Soc. Exp. Biol. Med., 100*, 138–139.

Schwanzel-Fukuda, M., & Pfaff, D. W. (1989). Origin of luteinizing hormone-releasing hormone neurons. *Nature, 338*, 161–164.

Schwarting, G. A., Kostek, C., Bless, E. P., Ahmad, N., & Tobet, S. A. (2001). Deleted in colorectal cancer (DCC) regulates the migration of luteinizing hormone-releasing hormone neurons to the basal forebrain. *J. Neurosci., 21*, 911–919.

Seminara, S. B., Messager, S., Chatzidaki, E. E., Thresher, R. R., Acierno, J. S., Jr., Shagoury, J. K., et al. (2003). The *GPR54* gene as a regulator of puberty. *N. Engl. J. Med., 349*, 1614–1627.

Shahab, M., Mastronardi, C., Seminara, S. B., Crowley, W. F., Ojeda, S. R., & Plant, T. M. (2005). Increased hypothalamic GPR54 signaling: a potential mechanism for initiation of puberty in primates. *Proc. Natl. Acad. Sci. USA, 102*, 2129–2134.

Shipley, M. T., Murphy, A. Z., Rizvi, T. A., Ennis, M., & Behbehani, M. M. (1996). Olfaction and brainstem circuits of reproductive behavior in the rat. *Prog. Brain Res., 107*, 355–377.

Silverman, A. J., Wilson, R., Kesner, J. S., & Knobil, E. (1986). Hypothalamic localization of multiunit electrical activity associated with pulsatile LH release in the rhesus monkey. *Neuroendocrinology, 44*, 168–171.

Soderhall, J. A., Polymeropoulos, E. E., Paulini, K., Gunther, E., & Kuhne, R. (2005). Antagonist and agonist binding models of the human gonadotropin-releasing hormone receptor. *Biochem. Biophys. Res. Commun., 333*, 568–582.

Solov'ev, G. S., Bogdanov, A. V., Panteleev, S. M., & Yanin, V. L. (2008). Embryonic morphogenesis of the human pituitary. *Neurosci. Behav. Physiol., 38*, 829–833.

Spergel, D. J., Kruth, U., Shimshek, D. R., Sprengel, R., & Seeburg, P. H. (2001). Using reporter genes to label selected neuronal populations in transgenic mice for gene promoter, anatomical, and physiological studies. *Prog. Neurobiol., 63*, 673–686.

Spiess, J., Rivier, J., Rivier, C., & Vale, W. (1981). Primary structure of corticotropin-releasing factor from ovine hypothalamus. *Proc. Natl. Acad. Sci. USA, 78*, 6517–6521.

Stewart, A. J., Katz, A. A., Millar, R. P., & Morgan, K. (2009). Retention and silencing of prepro-GnRH-II and type II GnRH receptor genes in mammals. *Neuroendocrinology, 90*, 416–432.

Suter, K. J., Song, W. J., Sampson, T. L., Wuarin, J. P., Saunders, J. T., Dudek, F. E., et al. (2000). Genetic targeting of green fluorescent protein to gonadotropin-releasing hormone neurons: characterization of whole-cell electrophysiological properties and morphology. *Endocrinology, 141*, 412–419.

Teller, D. C., Okada, T., Behnke, C. A., Palczewski, K., & Stenkamp, R. E. (2001). Advances in determination of a high-resolution three-dimensional structure of rhodopsin, a model of G-protein-coupled receptors (GPCRs). *Biochemistry, 40*, 7761–7772.

Tennent, R., & Chuah, M. I. (1996). Ultrastructural study of ensheathing cells in early development of olfactory axons. *Brain Res. Dev. Brain Res., 95*, 135–139.

Thompson, E. L., Patterson, M., Murphy, K. G., Smith, K. L., Dhillo, W. S., Todd, J. F., et al. (2004). Central and peripheral administration of kisspeptin-10 stimulates the hypothalamic–pituitary–gonadal axis. *J. Neuroendocrinol., 16*, 850–858.

Tisay, K. T., & Key, B. (1999). The extracellular matrix modulates olfactory neurite outgrowth on ensheathing cells. *J. Neurosci., 19*, 9890–9899.

Tobet, S. A., & Schwarting, G. A. (2006). Minireview: recent progress in gonadotropin-releasing hormone neuronal migration. *Endocrinology, 147*, 1159–1165.

Tovar, S., Vazquez, M. J., Navarro, V. M., Fernandez-Fernandez, R., Castellano, J. M., Vigo, E., et al. (2006). Effects of single or repeated intravenous administration of kisspeptin upon dynamic LH secretion in conscious male rats. *Endocrinology, 147*, 2696–2704.

Trandafir, T., Sipot, C., & Froicu, P. (1990). On a possible neural ridge origin of the adenohypophysis. *Endocrinologie, 28*, 67–72.

Tsai, P. S., Werner, S., & Weiner, R. I. (1995). Basic fibroblast growth factor is a neurotropic factor in GT1 gonadotropin-releasing hormone neuronal cell lines. *Endocrinology, 136*, 3831–3838.

Tsutsumi, M., Zhou, W., Millar, R. P., Mellon, P. L., Roberts, J. L., Flanagan, C. A., et al. (1992). Cloning and functional expression of a mouse gonadotropin-releasing hormone receptor. *Mol. Endocrinol., 6*, 1163–1169.

Ulloa-Aguirre, A., & Timossi, C. (2000). Biochemical and functional aspects of gonadotropin-releasing hormone and gonadotropins. *Reprod. Biomed. Online, 1*, 48–62.

Ulloa-Aguirre, A., Maldonado, A., Damian-Matsumura, P., & Timossi, C. (2001). Endocrine regulation of gonadotropin glycosylation. *Arch. Med. Res., 32*, 520–532.

Ulloa-Aguirre, A., Timossi, C., Barrios-de-Tomasi, J., Maldonado, A., & Nayudu, P. (2003). Impact of carbohydrate heterogeneity in function of follicle-stimulating hormone: studies derived from *in vitro* and *in vivo* models. *Biol. Reprod., 69*, 379–389.

Urs, A. N., Dammer, E., Kelly, S., Wang, E., Merrill, A. H., Jr., & Sewer, M. B. (2007). Steroidogenic factor-1 is a sphingolipid binding protein. *Mol. Cell. Endocrinol., 265-266*, 174–178.

Vernet, D., Bonavera, J. J., Swerdloff, R. S., Gonzalez-Cadavid, N. F., & Wang, C. (1998). Spontaneous expression of inducible nitric oxide synthase in the hypothalamus and other brain regions of aging rats. *Endocrinology, 139*, 3254–3261.

Weesner, G. D., & Matteri, R. L. (1994). Rapid communication: nucleotide sequence of luteinizing hormone-releasing hormone (LHRH) receptor cDNA in the pig pituitary. *J. Anim. Sci., 72*, 1911.

Wetsel, W. C., Valenca, M. M., Merchenthaler, I., Liposits, Z., Lopez, F. J., Weiner, R. I., et al. (1992). Intrinsic pulsatile secretory activity of immortalized luteinizing hormone-releasing hormone-secreting neurons. *Proc. Natl. Acad. Sci. USA, 89*, 4149–4153.

Whitlock, K. E. (2005). Origin and development of GnRH neurons. *Trends Endocrinol. Metab., 16*, 145–151.

Wildt, L., Hausler, A., Marshall, G., Hutchison, J. S., Plant, T. M., Belchetz, P. E., et al. (1981). Frequency and amplitude of gonadotropin-releasing hormone stimulation and gonadotropin secretion in the rhesus monkey. *Endocrinology, 109*, 376–385.

Wilkes, S., & Murdoch, A. (2009). Obesity and female fertility: a primary care perspective. *J. Fam. Plann. Reprod. Health Care, 35*, 181–185.

Wilkes, S., Chinn, D. J., Murdoch, A., & Rubin, G. (2009). Epidemiology and management of infertility: a population-based study in UK primary care. *Fam. Pract., 26*, 269–274.

Willars, G. B., Heding, A., Vrecl, M., Sellar, R., Blomenrohr, M., Nahorski, S. R., et al. (1999). Lack of a C-terminal tail in the mammalian gonadotropin-releasing hormone receptor confers resistance to agonist-dependent phosphorylation and rapid desensitization. *J. Biol. Chem., 274*, 30146–30153.

Wilson, R. C., Kesner, J. S., Kaufman, J. M., Uemura, T., Akema, T., & Knobil, E. (1984). Central electrophysiologic correlates of pulsatile luteinizing hormone secretion in the rhesus monkey. *Neuroendocrinology, 39*, 256–260.

Wintermantel, T. M., Campbell, R. E., Porteous, R., Bock, D., Grone, H. J., Todman, M. G., et al. (2006). Definition of estrogen receptor pathway critical for estrogen positive feedback to gonadotropin-releasing hormone neurons and fertility. *Neuron., 52*, 271–280.

Wolfe, A., Divall, S., Singh, S. P., Nikrodhanond, A. A., Baria, A. T., Le, W. W., et al. (2008). Temporal and spatial regulation of CRE recombinase expression in gonadotropin-releasing hormone neurones in the mouse. *J. Neuroendocrinol., 20*, 909–916.

Wood, R. I., & Coolen, L. M. (1997). Integration of chemosensory and hormonal cues is essential for sexual behaviour in the male Syrian hamster: role of the medial amygdaloid nucleus. *Neuroscience, 78*, 1027–1035.

Wray, S. (2002). Development of gonadotropin-releasing hormone-1 neurons. *Front Neuroendocrinol., 23*, 292–316.

Wray, S., Grant, P., & Gainer, H. (1989). Evidence that cells expressing luteinizing hormone-releasing hormone mRNA in the mouse are derived from progenitor cells in the olfactory placode. *Proc. Natl. Acad. Sci. USA, 86*, 8132–8136.

Wray, S., Key, S., Qualls, R., & Fueshko, S. M. (1994). A subset of peripherin positive olfactory axons delineates the luteinizing hormone releasing hormone neuronal migratory pathway in developing mouse. *Dev. Biol., 166*, 349–354.

Wu, T. J., Glucksman, M. J., Roberts, J. L., & Mani, S. K. (2006a). Facilitation of lordosis in rats by a metabolite of luteinizing hormone releasing hormone. *Endocrinology, 147*, 2544–2549.

Wu, T. J., Glucksman, M. J., Roberts, J. L., & Mani, S. K. (2006b). Facilitation of lordosis in rats by a metabolite of luteinizing hormone releasing hormone (LHRH). *Endocrinology.*

Wu, T. J., Mani, S. K., Glucksman, M. J., & Roberts, J. L. (2005). Stimulation of luteinizing hormone-releasing hormone (LHRH) gene expression in GT1-7 cells by its metabolite, LHRH-(1-5). *Endocrinology, 146*, 280–286.

Wurmbach, E., Yuen, T., Ebersole, B. J., & Sealfon, S. C. (2001). Gonadotropin-releasing hormone receptor-coupled gene network organization. *J. Biol. Chem., 276*, 47195–47201.

Yoon, H., Enquist, L. W., & Dulac, C. (2005). Olfactory inputs to hypothalamic neurons controlling reproduction and fertility. *Cell, 123*, 669–682.

Zhao, L., Bakke, M., Krimkevich, Y., Cushman, L. J., Parlow, A. F., Camper, S. A., et al. (2001a). Hypomorphic phenotype in mice with pituitary-specific knockout of steroidogenic factor 1. *Genesis, 30*, 65–69.

Zhao, L., Bakke, M., Krimkevich, Y., Cushman, L. J., Parlow, A. F., Camper, S. A., et al. (2001b). Steroidogenic factor 1 (SF1) is essential for pituitary gonadotrope function. *Development, 128*, 147–154.

Zhou, W., Rodic, V., Kitanovic, S., Flanagan, C. A., Chi, L., Weinstein, H., et al. (1995). A locus of the gonadotropin-releasing hormone receptor that differentiates agonist and antagonist binding sites. *J. Biol. Chem., 270*, 18853–18857.

Chapter 3

Endocrine and Paracrine Regulation of Mammalian Spermatogenesis

Barry R. Zirkin*, Terry R. Brown*, Jonathan P. Jarow[†] and William W. Wright*

*Johns Hopkins Bloomberg School of Public Health, Baltimore, MD, USA, [†]Johns Hopkins School of Medicine, Baltimore, MD, USA

SUMMARY

Spermatogenesis is comprised of the mitotic divisions of spermatogonia to produce spermatocytes, the two meiotic divisions, and the differentiation of the resulting spermatids to spermatozoa. Additionally, the continuous production of male gametes requires the maintenance of a pool of stem spermatogonia as well as the regulated differentiation of a subset of these cells. During their development, germ cells associate with the somatic Sertoli cells. Sertoli cells produce proteins that are required for germ cell development, and, in turn, their functions are altered in response to the developing germ cells. Endocrine and paracrine factors regulate germ cell function and survival. In particular, testosterone (T), produced by Leydig cells in the testicular interstitial compartment, is required for spermatogenesis. Follicle-stimulating hormone (FSH) is required to establish normal numbers and proper functioning of Sertoli cells. The germ cells, the survival of which depends on adequate concentrations of T in the testis, do not respond directly to T. Rather, T function is mediated by the Sertoli cells, which express the androgen receptor (AR). Follicle-stimulating hormone receptors also are present on the Sertoli cells, and thus it is the Sertoli cells that transduce hormonal stimuli to coordinate germ cell differentiation.

1. OVERVIEW OF SPERMATOGENESIS

The mammalian testis is comprised of the seminiferous tubular and interstitial compartments. Within the seminiferous epithelium are the somatic Sertoli cells and the developing germ cells with which the Sertoli cells are associated. Spermatogenesis proceeds in an ordered fashion that includes diploid spermatogonial stem cells dividing mitotically to become spermatocytes, the two meiotic divisions producing haploid round spermatids, and the differentiation of the spermatids to elongated, mature sperm. In addition to their ability to differentiate, a subset of the spermatogonial cells are able to self-renew, thereby maintaining a pool of stem spermatogonia required for the continuous production of male gametes. The products of spermatogenesis, the spermatids, transform into mature sperm during spermiogenesis; sperm are released from Sertoli cells at spermiation and enter the epididymis, where they develop the ability for forward motility and fertility. Throughout the course of spermatogenesis, the proper balance of germ cell proliferation and germ cell death by apoptosis is critical. The seminiferous tubules apparently can support only a finite number of developing germ cells; it has been estimated that only 25–30% of potential germ cells succeed in differentiating to maturity, most of the remainder dying by apoptosis before entering meiosis (Russell, Chiarini-Garcia, Korsmeyer, & Knudson, 2002).

The Sertoli cells produce proteins that are required for spermatogenesis and, in turn, respond to the germ cells with which they are associated. The complex interaction between the Sertoli cells and the germ cells is crucial for spermatogenesis (Sharpe, 1994; Griswold & McLean, 2006; Kerr, Loveland, O'Bryan, & DeKretser, 2006).

A number of factors, both endocrine and paracrine, are known to regulate germ cell function and survival during spermatogenesis. Testosterone (T), produced by Leydig cells in the testicular interstitial compartment, is essential for spermatogenesis. Normal spermatogenesis also depends on follicle-stimulating hormone (FSH), which is required to establish normal numbers of and properly functioning Sertoli cells (Sairam & Krishnamurthy, 2001). Prolonged reduction in intratesticular T concentration results in significant germ cell loss, and can lead to transient infertility. This is the basis of male hormone-based contraception. In both rodents and humans, in which intratesticular levels of T (and therefore sperm count) are reduced, restoring T results in the restoration of normal spermatogenesis. The exact mechanism by which T regulates spermatogenesis is uncertain, in part because the germ cells, the survival of which depends on adequate concentrations of T in the testis, do not express the androgen receptor (AR).

Rather, germ cell apoptosis resulting from decreased intratesticular T is considered to be mediated by the somatic Sertoli cells, which express the AR. Follicle-stimulating hormone (FSH) receptors also are present on the Sertoli cells, and thus it is these cells that transduce hormonal stimuli to coordinate germ cell differentiation (O'Donnell, Meachem, Stanton, & McLachlan, 2006).

2. REGULATION OF STEM SPERMATOGONIA VIA GROWTH FACTORS AND PARACRINE INTERACTIONS

2.1. Characteristics of Stem Spermatogonia

Spermatogenesis consists of spermatogonial mitotic divisions; the entry of the ultimate products of these divisions, primary spermatocytes, into meiosis; two meiotic divisions resulting in the production of haploid round spermatids; and the differentiation of the round spermatids to sperm. The continuous production of male gametes, and thus male fertility, requires the maintenance of a pool of stem spermatogonia as well as the regulated differentiation of a subset of these cells into paired (A_{pr}) spermatogonia (Figure 3.1). The existence of stem spermatogonia was proven by the demonstration that a small fraction of spermatogonia has the capacity to restore spermatogenesis when transplanted into a germ cell-deficient testis (Ogawa, Dobrinski, Avarbock, & Brinster, 2000). Morphologically, stem spermatogonia have been defined in whole mounts of seminiferous tubules as single (A_s) spermatogonia (Huckins, 1978a). Hamra et al. (2004) identified 248 genes the expressions of which were at least two-fold higher in preparations of spermatogonia that were enriched in stem cells. Two of these genes encode the subunits of the receptor for glial-derived neurotrophic factor (GDNF)—RET (tyrosine protein binase subunit of the GDNF receptor) and GFRα1 (GDNF receptor-α1)—while another encodes the transcription factor promyelocytic leukemia zinc finger (PLZF) (Buaas et al., 2004; Naughton, Jain, Strickland, Gupta, & Milbrandt, 2006). However, unlike aligned A (A_{al}) and more mature spermatogonia (Figure 3.1), stem spermatogonia do not express cKit, the cell surface receptor for kit ligand/stem cell factor (De Rooij, 2001). As is the case with all stem cells, stem spermatogonia are scarce; there are approximately 30 000 per mouse testis (Ogawa, Ohmura, Yumura, Sawada, & Kubota, 2003). Guan et al. (2006) demonstrated that, *in vitro*, stem spermatogonia can form multipotent germline stem cells which, when placed in culture or when injected into a blastocyst, can give rise to cells in the endodermal, mesodermal, and ectodermal lineages. In contrast, in the testis, stem spermatogonia only form more stem or A_{pr} spermatogonia. The very infrequent exception to this rule is their formation of germ cell tumors of the testis (seminomas). Thus, the developmental potential of stem spermatogonial cells *in vivo* is tightly constrained, presumably by their niche. Division of the stem cells occurs at specific stages of the cycle of the seminiferous epithelium, suggesting that the niche also controls mitosis of stem spermatogonia. The proper functioning of their niche is fundamentally important to the replication and differentiation of stem spermatogonia, and therefore to male fertility. While it is apparent that the stem spermatogonial niche is created in part by Sertoli cells, it is not clear how the products of the Sertoli cells control stem spermatogonial differentiation and replication.

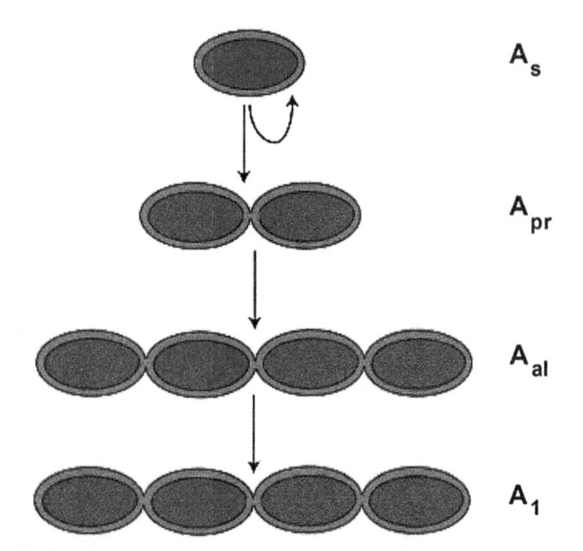

FIGURE 3.1 Self-renewing replication and differentiation of stem spermatogonia. Stem spermatogonia constitute a subset of the single (A_s) spermatogonia. When A_s spermatogonia divide, they form more of themselves or form paired (A_{pr}) spermatogonia that are joined to each other by an intracellular bridge. Replication of A_{pr} spermatogonia creates aligned (A_{al}) spermatogonia, which can form chains of up to 32 cells. A_s spermatogonia, A_{pr} spermatogonia, and A_{al} spermatogonia all are considered undifferentiated spermatogonia. A_1 spermatogonia are formed by the differentiation of A_{al} spermatogonia.

2.2. Replication and Differentiation of Stem Spermatogonia in their Niche

Mitosis of A_s spermatogonia can produce either two new A_s spermatogonia or A_{pr} spermatogonia that share an intracellular bridge and therefore remain in close proximity (Huckins, 1978b; Hamra et al., 2004). In the rat, most A_s cells cycle every 55–58 hours, but a small subset cycles approximately every 13 days (Huckins, 1971a; 1971b; 1971c). The numbers of replicating A_s spermatogonia differ depending on the stage of the cycle of the seminiferous epithelium (Figure 3.2). Thus, in both rats and mice, A_s spermatogonia replicate from stage 1 to the beginning of stage 5 (Huckins, 1971c; De Rooij, 2001). As the seminiferous epithelium progresses from stage 1 to stage 7, there is a gradual decrease in the percentages of replicating stem spermatogonia. Huckins (1971c) reported that, at stages 1, 3, and 4–8, 30%, 12%, and < 1% of the cells replicate, respectively. Differentiation of the stem spermatogonia occurs after they divide mitotically, with the progeny joined by cytoplasmic bridges as a consequence of incomplete cytokinesis (Huckins, 1978b). The differentiation of A_S spermatogonia into A_{pr} spermatogonia must be carefully regulated or the stem spermatogonial pool would be depleted (reviewed in De Rooij, 2001). The regulation of self-renewal and differentiation of stem spermatogonia may be linked. Thus, there is evidence that the number of more differentiated cells may influence or regulate stem spermatogonial replication and differentiation. For example, after irradiation-induced death of differentiated spermatogonia, the differentiation of A_S spermatogonia is repressed and the rate of mitotic division is increased (Dym & Clermont, 1970; Van Beek, Davids, Van de Kant, & De Rooij, 1984). This suggests a possible relationship between numbers of differentiated spermatogonia and the division and differentiation of stem spermatogonia. Observations such as these make it clear that the stem spermatogonial niche is a site of complex interrelationships among stem cells, differentiated spermatogenetic cells, and Sertoli cells.

2.3. Glial-derived Neurotrophic Factor (GDNF) and Other Growth Factors in the Stem Spermatogonia Niche

Sertoli cells secrete GDNF as well as other growth factors, and stem spermatogonia express RET and GFRα1 (Naughton et al., 2006). Analyses of the testes of mice that underexpress or overexpress GDNF demonstrate that GDNF is involved in regulating cell fate decisions of stem spermatogonia (Meng et al., 2000; De Rooij, 2001). For example, the reduced GDNF expression in adult $GDNF^{+/-}$ mice causes depletion of stem spermatogonia in some tubules. Additionally, when testes of newborn wild-type and $GDNF^{-/-}$ mice were transplanted into nude mice, spermatogenesis flourished in the wild-type testes for at least seven days, while only a few spermatogenetic cells remained in the $GDNF^{-/-}$ testes (Naughton et al., 2006). In contrast, when GDNF was overexpressed by a testis-specific promoter, A_S spermatogonia accumulated as masses within the seminiferous tubules (Meng et al., 2000). Taken together, these data from transgenic mice have led to the hypothesis that GDNF suppresses the differentiation of stem spermatogonia. It is not clear whether GDNF also stimulates the mitosis of stem spermatogonia, as it does for some neuronal progenitor cells (Chen, Slevin, Maley, & Gash, 2005). Meng et al. (2000) reported that, despite the fact that the overexpression of GDNF inhibited the differentiation of A_S spermatogonia, it had no effect on the rate of replication of these cells, and therefore that it is not a stem spermatogonia mitogen. However, Naughton et al. (2006) reached a different conclusion when they compared the replication of stem spermatogonia from testes of newborn wild-type and $GDNF^{-/-}$ mice that were transplanted into nude mice; replication of stem cells was clearly evident in the testes obtained from wild-type mice but was not evident in testes obtained from $GDNF^{-/-}$ mice. *In-vitro* studies also have been equivocal; whereas some studies identified

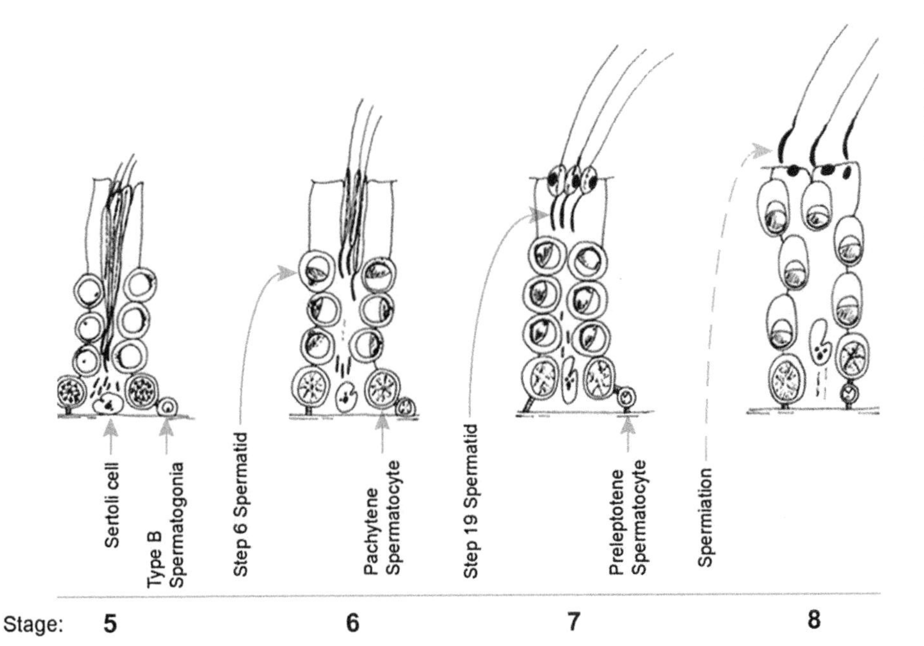

FIGURE 3.2 Four of the fourteen stages of the cycle of the rat seminiferous epithelium, illustrating the morphological changes in this epithelium as spermatogenic cells surrounding a Sertoli cell progress in synchrony from stages 5 to 8 of the cycle. Solid arrows in this figure point to specific cell types. The dashed arrow points to step 19 spermatids, the most mature step in the testis, that are being released from the Sertoli cell by the process of spermiation. These cells will enter the epididymis, where they further mature.

GDNF as essential for the replication of stem spermatogonia in culture and involved in stimulating the expression of cell cycle genes (Kubota, Avarbock, & Brinster, 2004; Hoffman, Braydich-Stolle, & Dym, 2005), in other studies (Guan et al., 2006) GDNF did not stimulate stem spermatogonial replication.

To date, only GDNF has been identified definitively as an *in-vivo* regulator of stem spermatogonia. However, it seems likely that other growth factors also are involved, since a number of growth factors and cytokines have been shown to be involved in the regulation of the replication and differentiation of other types of stem cell. For example, fibroblast growth factor-2 (FGF-2) has been shown to stimulate the replication of mouse neuronal stem cells isolated from the adult mouse forebrain, and insulin-like growth factor-1 (IGF-1) to promote their differentiation into neurons (Brooker et al., 2000). Fibroblast growth factor-2 also is a mitogen for stem and precursor cells of spinal cord neurons, and stimulates these cells to express autocrine regulators of their own replication and differentiation. These effects of FGF-2 are opposed by bone morphogenetic protein (BMP) and interferon gamma (Deleyrolle et al., 2006). The complete repertoire of growth factors regulating replication and differentiation of neuronal stem cells may be even larger. A recent model of the development of olfactory receptor neurons invokes the action of nine different cytokines and growth factors (Beites, Kawauchi, Crocker, & Calof, 2005). Fibroblast growth factor-2 also stimulates stem cells for blood, bone, and the intestinal epithelium (Allouche & Bikfalvi, 1995; Houchen, George, Sturmoski, & Cohn, 1999; Solchaga et al., 2005). The effect of FGF-2 on multiple types of stem cell raises the issue of whether, *in vivo*, it regulates stem spermatogonia. To date, this issue has not been addressed directly. However, FGF-2 is a known Sertoli cell product, and thus may be present in the stem spermatogonial niche (Mullaney & Skinner, 1992). Additionally, *in vitro,* it augments the ability of GDNF to stimulate an increase in the numbers of stem spermatogonia (Kubota et al., 2004; Hoffman et al., 2005).

Insulin-like growth factor-1 also stimulates the replication and/or differentiation of multiple stem cell lineages. Besides its effects on neuronal stem cells, IGF-1 stimulates stem cell-mediated muscle regeneration as well as the replication of cardiac stem cells (Brooker et al., 2000; Mourkiot & Rosenthal, 2005). As is the case with FGF-2, IGF-1 is produced by Sertoli cells and therefore is likely to be present in the stem spermatogonial niche (Dombrowicz et al., 1992). It is noteworthy, therefore, that Kubota et al. (2004) reported that IGF-1 augmented the effects of GDNF on the replication and differentiation of stem spermatogonia.

2.4. Regulation of Sertoli Cell Function by Differentiated Spermatogenetic Cells and the Implications of this Regulation for the Function of the Stem Spermatogonial Niche

A remarkable characteristic of mammalian spermatogenesis is that only specific types of spermatogonia, spermatocytes, and spermatids are found around a given Sertoli cell at a given time. Together, these germ cells progress in synchrony through the stages of the cycle of the seminiferous epithelium (Figure 3.2), whereas the Sertoli cells neither replicate nor die but their function changes, suggesting that, in species as diverse as humans and mice, stage-specific feedback from spermatogenetic cells affects Sertoli cell function (Suarez-Quian, Martinez-Garcia, Nistal, & Regadera, 1999; Wright, Smith, Kerr, & Charron, 2003). In the rat, the expressions of 137 genes by the Sertoli cells were found to vary at least four-fold with progression of the stages of the cycle; of these 137, the expressions of 26 varied at least 100-fold (Johnston et al., 2008). This indicates that stage-specific feedback from spermatogenetic cells has a widespread and profound effect on the function of the somatic Sertoli cells. The feedback is complex. For example, different spermatogenetic cells regulate different functions of Sertoli cells: pachytene spermatocytes and round spermatids regulate stage-specific expression of the cathepsin L gene, while preleptotene spermatocytes regulate urokinase activity (Penttila, Kaipia, Toppari, Parvinen, & Mali, 1994; Wright, Zabludoff, Penttilä, & Parvinen, 1995). Feedback from spermatogenetic cells also modulates the response of Sertoli cells to the one hormone that is essential for spermatogenesis: T (discussed in Section 3.4). The AR is expressed in a stage-specific manner, as are the expressions of the androgen-responsive genes *Rhox 5*, prostaglandin D_2 synthase, growth arrest-specific 6, and lipocalin 2 (Shan, Zhu, Bardin, & Hardy, 1995; Rao, Wayne, Meistrich, & Wilkinson, 2003; Eacker et al., 2007).

Given the diverse stage-specific effects of spermatogenetic cells on Sertoli cell function, it is possible that feedback from germ cells also affects the functions of Sertoli cells that contribute to the formation of the stem spermatogonial niche. This possibility is supported by the observation, noted above, that a significantly greater percentage of rat A_s spermatogonia replicate at stage 1 than at stage 7 of the cycle (Huckins, 1978a). Thus, as Sertoli cells promote the survival, development, and replication of differentiated spermatogenetic cells, these differentiated cells, in turn, feed back in a stage-specific manner to regulate specific Sertoli cell functions, including the synthesis of growth factors that act in the stem spermatogonial niche.

3. ENDOCRINE REGULATION OF SPERMATOGENESIS

3.1. Leydig Cells and Testosterone (T) Production

Spermatogenesis is regulated by the coordinated actions of the hypothalamus–pituitary–gonad (HPG) axis. Gonadotropin-releasing hormone (GnRH), a decapeptide neurohormone, is synthesized by the hypothalamus and secreted into the hypophysial portal bloodstream of the median eminence. Gonadotropin-releasing hormone binds to the GnRH receptor (GnRHR) in the anterior lobe of the pituitary and initiates the synthesis and secretion of luteinizing hormone (LH) and FSH. Luteinizing hormone is released in a pulsatile fashion in response to the pulsatile release of GnRH from the hypothalamus, and is delivered to the testes via the vascular system. Luteinizing hormone binds to its receptor on Leydig cells and stimulates the synthesis and secretion of T. Follicle-stimulating hormone also travels via the vascular system to the testis and acts on the Sertoli cells via the FSH receptors. Testosterone and its aromatized metabolite, 17β-estradiol (E_2), negatively feed back to the hypothalamus and pituitary to inhibit the release of GnRH and of LH/FSH. This negative feedback mechanism is crucial for maintaining physiological levels of hormones (O'Donnell et al., 2006).

Testosterone biosynthesis in Leydig cells is under the control of LH. After binding to the LH receptor, located on the Leydig cell surface, LH activates adenylyl cyclase and induces an increase of intracellular cyclic-3',5'-adenosine monophosphate (cAMP). This leads to the subsequent phosphorylation of key intermediates in the steroid biosynthetic pathway (Payne, 2007). The induction of cAMP leads to cholesterol rapidly moving into mitochondria via transporter protein (TSPO) (also known as peripheral benzodiazepine receptor (PBR)) and steroidogenic acute regulatory (StAR) protein (Hauet et al., 2005). This is followed by the conversion of cholesterol to pregnenolone in the mitochondria by P450 side-chain cleavage enzyme ($P450_{scc}$), and then the conversion of pregnenolone to T through a number of intermediates in the smooth endoplasmic reticulum (SER), catalyzed by steroidogenic enzymes including 3β-hydroxysteroid dehydrogenase (3β-HSD), 17-hydroxylase 17,20 lyase ($P450_{c17}$), and 17-ketosteroid reductase (Miller, 2008). There also are chronic effects of LH, in particular the LH-induced expression of several steroidogenic enzymes involved in T biosynthesis (Payne, 2007). Although T can be converted to various metabolites, it is the predominant androgen in the adult rat testis, constituting over 95% of androgen, with 5α-dihydrotestosterone (DHT) constituting only 5% (Turner et al., 1984). A similar relationship between T and DHT occurs in the human testis (Jarow & Zirkin, 2005).

3.2. Mechanism(s) by which Testosterone (T) Acts to Drive Spermatogenesis: The Role of Androgen Receptors (ARs)

As indicated above, spermatogenesis is a highly complex and precisely regulated process of cellular division and differentiation that is governed by two hormones: T and FSH (Roberts & Zirkin, 1991; Sharpe, 1994; McLachlan et al., 2002a; 2002b). Follicle-stimulating hormone acts directly on Sertoli cells via its membrane receptor; T is produced by Leydig cells within the testicular interstitial compartment in response to LH. It has long been known that T is essential for the initiation and maintenance of spermatogenesis. Recently published observations showing that mice with Sertoli cell-specific conditional knockout of the AR gene do not produce sperm provide definitive proof that AR expression in Sertoli cells is absolutely required for male fertility (Chang et al., 2004; De Gendt et al., 2004; Holdcraft & Braun, 2004). Studies in FSH-deficient genetic models and in animals following gonadotropin suppression confirm that T alone is sufficient for the initiation and maintenance of spermatogenesis in rodents (Awoniyi, Santulli, Chandrashekar, Schanbacher, & Zirkin, 1989a; Awoyi, Santulli, Sprando, Ewing, & Zirkin, 1989b; Awoniyi et al., 1990; Zirkin, 1998; Spaliviero, Jimenez, Allan, & Handelsman, 2004). As a result of their intimate association with the developing germ cells, Sertoli cells are regarded as the primary target cells within the testis that mediate the effects of T on spermatogenesis (Parvinen, 1982). Sertoli cells in rats, mice, and men exhibit highly concordant patterns of AR expression (Buzek & Sanborn, 1988; Bremner, Millar, Sharpe, & Saunders, 1994; Vornberger, Prins, Musto, & Suarez-Quian, 1994; Shan et al., 1995; Suarez-Quian et al., 1999; Zhu et al., 2000; Regadera et al., 2001; Zhou et al., 2002). During pubertal maturation, when spermatogenesis becomes an increasingly efficient process (Russell, Alger, & Nequin, 1987), there is an increase in AR expression in Sertoli cells (Buzek & Sanborn, 1988; Shan et al., 1995; Zhu et al., 2000). In adulthood, the pattern of AR expression in Sertoli cells is synchronized with the progression of the neighboring spermatogenetic cells through the stages of the cycle of the seminiferous epithelium (Bremner et al., 1994; Vornberger et al., 1994; Suarez-Quian et al., 1999; Regadera et al., 2001). Importantly, the stages with the highest levels of expression of the AR gene are the most sensitive to the loss of androgen stimulation (Sharpe et al., 1992; O'Donnell, McLachlan, Wreford, De Kretser, & Robertson, 1996; Saito, O'Donnell, MacLachlan, & Robertson, 2000). As discussed below, definitive proof that normal spermatogenesis and fertility require the expression of AR in Sertoli cells has recently emerged.

An implicit requirement for the AR in spermatogenesis is obvious from the defects in germ cell maturation that occur in the testes of androgen-insensitive humans, rats, and mice due to mutations in the X-linked AR gene (Lyon & Hawkes, 1970; Bardin, Bullock, Sherins, Mowszowicz, & Blackburn, 1973; Brown, 2002). Similar defects in spermatogenesis have been observed in AR gene knockout mice (Chang et al., 2004; De Gendt et al., 2004; Holdcraft & Braun, 2004), thus solidifying a central role for androgen action and AR in germ cell development. Over the years, the target cells affected by androgens in relation to the control of spermatogenesis have been debated, in part because previous biochemical and immunocytochemical techniques showed the presence of AR in each of Sertoli, peritubular myoid, germ, and Leydig cells in the adult testis (Isomaa, Parvinen, Janne, & Bardin, 1985; Buzek & Sanborn, 1988; Sar, Lubahn, French, & Wilson, 1990; Takeda, Chodak, Mutchnik, Nakamoto, & Chang, 1990; Bremner et al., 1994; Vornberger et al., 1994; Shan et al., 1995; Suarez-Quian et al., 1999; Zhu et al., 2000; Regadera et al., 2001; Zhou et al., 2002). With respect to germ cells as a direct site of androgen action, Lyon, Glenister, and Lamoreux (1975) showed that male mice that were chimeric for a normal X/Y genotype, and the androgen-insensitive testicular feminization (^{Tfm}X/Y) genotype, in which the AR gene was mutated, were able to produce offspring from the ^{Tfm}X/Y component. Similarly, when germ cells from ^{Tfm}X/Y mice were transplanted into the seminiferous tubules of azoospermic mice expressing a functional AR, complete and qualitatively normal donor-derived spermatogenesis was observed (Johnston, Russell, Friel, & Griswold, 2001). Therefore, it was concluded that germ cells do not require a functional AR for their maturation, and that the effects of androgens on spermatogenesis must be mediated via somatic cells in the testis.

To clarify how androgens affect spermatogenesis and to define the role of Sertoli cells as targets of androgen action, several groups used Cre/*loxP* technology to generate mice with a Sertoli cell-selective knockout of AR (Chang et al., 2004; De Gendt et al., 2004; Holdcraft & Braun, 2004). Mice in which exon 2 of the AR gene was floxed (i.e., sandwiched between two *lox P* sites) were crossed with mice engineered to express Cre recombinase under the control of the Sertoli cell-specific anti-Müllerian hormone (AMH) gene promoter. Compared to complete AR knockout (ARKO) mice, development of the urogenital tract and descent of the testes were normal in Sertoli cell-selective AR knockout (SCARKO) mice (Tan et al., 2005). Whereas ARKO mice exhibited a progressive reduction in Sertoli cell numbers/testis, SCARKO mice showed only minor changes, suggesting that androgen effects on Sertoli cell numbers are not mediated via direct action on Sertoli cells. The expressions of AMH, cyclin-dependent kinase inhibitor (p27^{kip1}), GATA-1 (a transcription factor involved in cell growth), and sulfated glycoprotein-2 (SGP-2) are normal in ARKO and SCARKO mice. The functional capacity of Sertoli cells to support spermatogonia is similar in SCARKO and wild-type mice, but reduced with age in ARKO mice. By contrast, severe deficits in the capacity of Sertoli cells to support all stages of germ cells were evident in adult ARKO and SCARKO mice. Spermatogenetic arrest occurred among primary spermatocytes at the diplotene stage, and SCARKO mice were infertile (Chang et al., 2004). The overall effect of an infertile phenotype in SCARKO mice was similar to that observed in ARKO mice except that the primary defect in spermatogenesis occurred at the pachytene stage in the latter mice. The stage-specific effects of AR gene knockout on spermatogenesis are consistent with stages 7–8 of the spermatogenetic cycle, the androgen-dependent stages (Figure 3.2). In SCARKO mice, gene expression analyses revealed reductions in the expression of genes such as cystatin-TE, claudin-11, Eppin, PCI tissue-type plasminogen activator, tubulin β3, procollagen type IV α6, brevican, Pem (Rhox5), fatty acid binding protein, and platelet derived growth factor-A, highlighting these as potentially androgen-regulated genes, though with functions that are uncertain (Tan et al., 2005; Eacker et al., 2007). Of note, an equal or greater number of genes were upregulated in SCARKO mice, among which were alcohol dehydrogenase-1, fatty acid desaturase-1, ferredoxin-1, cathepsin-H, angiotensinogen, and heat shock protein-1 (Eacker et. al., 2007). A significant number of the genes found to be misregulated in SCARKO mice were related to seminiferous tubular remodeling and to Sertoli cell junctional dynamics necessary to maintain the blood–testis barrier (Wang et al., 2006). In particular, increased levels of vimentin and decreased levels of laminin α5 in SCARKO mice contributed to the abnormal morphology of Sertoli cells and to loss of seminiferous epithelial integrity due to functional defects in the Sertoli cell cytoskeleton and their surrounding basement membrane in early postnatal life. Decreased expression of claudin-11, occludin, and gelsolin contributed to the defects in Sertoli cell junctional complex formation. Taken together, these studies unequivocally demonstrate that AR expression in Sertoli cells is absolutely required to support the meiotic and postmeiotic germ cell development necessary for normal spermatogenesis and fertility.

3.3. Androgen Receptor (AR) Expression is Stage-specific and Specific Stages of Spermatogenesis are Androgen-dependent

The morphologic association of germ cells and Sertoli cells within the seminiferous epithelium facilitates their direct cellular communication (Parvinen, 1982; Wright, 1993), with specific cohorts of germ cells associated with Sertoli cells to comprise the 14 stages of the cycle of the seminiferous epithelium in rats (Leblond & Clermont, 1952). Several genes, including AR, exhibit a cyclic pattern of expression in Sertoli cells, with maximal expression seen at various stages of the spermatogenetic cycle (Wright, 1993; Bremner et al., 1994; Vornberger et al., 1994; Eddy & O'Brien, 1998; Suarez-Quian et al., 1999; Regadera et al., 2001; Zhou et al., 2002). The androgen-dependent stages of spermatogenesis coincide with stages 7–8 (Figure 3.2), when AR levels in Sertoli cells are highest (Sharpe et al., 1992; O'Donnell et al., 1996; Saito et al., 2000; Spaliviero et al., 2004). These are the stages targeted by hormonal contraception in men (McLachlan et al., 2002a; 2002b; Kamischke & Nieschlag, 2003). Androgen withdrawal results in apoptosis of pachytene spermatocytes, sloughing of round spermatids, and failure of step 19 spermatids to be released at these stages (O'Donnell et al., 1996; Saito et al., 2000). Androgens also stimulate major changes in protein synthesis by seminiferous tubules at stages 7–8 (Sharpe et al., 1992). Taken together, these observations demonstrate that the androgen-dependent stages of germ cell maturation are coordinated with peak levels of AR expression in Sertoli cells. Thus, the molecular mechanisms that regulate stage-specific expression of AR in Sertoli cells are directly related to the biological actions of T in the process of germ cell maturation. The stage- and cell-specific regulation of AR gene transcription in Sertoli cells undoubtedly relies upon a subset of transcription factors and their coordinated interactions with 5'-upstream regulatory elements in the AR gene. As yet, there are very limited data, either *in vitro* or *in vivo*, regarding the transcription factors and *cis*-acting regulatory sites involved in the regulation of AR gene transcription.

3.4. Testosterone (T) and Spermatogenesis

In all mammalian species studied to date, T has been found to be the predominant intratesticular steroid. Turner et al. (1984) reported that, in the rat, the concentration of DHT within the testis is only 5% that of T, and that the T concentration is far in excess (25–30-fold greater) of its concentration in blood serum. Subsequent experimental studies showed that the normal total intratesticular T concentration (50–70 ng/ml (Turner et al., 1984)) is far higher than the concentration (20 ng/ml) shown experimentally to be *required* for the quantitative maintenance of spermatogenesis (Zirkin, Santulli, Awoniyi, & Ewing, 1989) or for the quantitative restoration of spermatogenesis in animals rendered azoospermic by T administration or by active immunization against GnRH (Awoniyi et al., 1989a; 1989b). The 20 ng/ml concentration, though less than 50% of the T concentration that is normally present within the testis, none-the-less is 10-fold higher than the total T concentration in normal rat blood serum (2 ng/ml). This also pertains in the human; the normal intratesticular T concentration is about 600 ng/ml and that in contracepted men about 50–100 ng/ml, the latter at least 10-fold higher than serum T concentration (Jarow, Chen, Rosner, Trentacoste, & Zirkin, 2001; Jarow & Zirkin, 2005).

As yet, the reason that intratesticular T concentration must be so high in order to maintain or restore spermatogenesis is unknown. One possibility is that much of the total T is bound and therefore unavailable to the Sertoli cell ARs. Androgen-binding protein (ABP), a secretory product of Sertoli cells that itself is regulated by T and FSH, is present in the mammalian testis (Joseph, 1994). Androgen-binding protein from the testis and human sex hormone-binding globulin (SHBG) from the liver share 68% identity of amino acid residues and have been shown to be closely related with respect to steroid-binding properties (Joseph, Hall, & French, 1987). Androgen-binding protein is widely considered to play an indirect role in spermatogenesis by sequestering androgens, thereby reducing their bioavailability. If it is true that the actions of androgens within the testes are affected by binding to ARs, androgens bound to ABP should have a reduced capability of initiating AR-mediated gene transcription. Roberts and Zirkin (1993) showed that, in rats made oligospermic as a result of receiving exogenous T, the concentration of intratesticular ABP was high enough to bind nearly 90% of the T. Additionally, T binding to ABP was shown to inhibit AR-mediated transcription (Roberts & Zirkin, 1993). These results suggested that ABP has the potential to prevent androgen-dependent transcription. Studies with transgenic mice have been consistent with this hypothesis. Thus, in studies of transgenic mice that overexpress ABP (Joseph, O'Brien, Sullivan, Becchis, & Tsuruta, 1997; Jeyaraj, Grossman, Weaver, & Petrusz, 2002), spermatogenesis was shown to begin normally despite the excess ABP, but progressive structural and functional abnormalities occurred in the testes in parallel with increasing amounts of ABP. As yet, however, it is not clear whether the effect of ABP on bioavailable androgen concentration can explain the relationship between intratesticular T concentration and the maintenance or restoration of spermatogenesis. Indeed, there is evidence that androgens may function through alternative mechanisms that complement the classical AR mechanism in Sertoli cells (e.g., rapid effects on cellular signaling (Walker, 2009)).

There have recently been efforts to conduct studies in the human that are comparable to those conducted in rats. Jarow et al. (2001) first described a minimally invasive percutaneous aspiration method for the collection of fluid from the human testis in sufficient quantity for reliable measurements to be made of intratesticular steroid content. As in the rat, there is a gradient between the concentration of T in the serum and the testis; in normal men, intratesticular T levels were found to be 100-fold higher than serum levels (Jarow & Zirkin, 2005). These results were consistent with previously conducted analyses of T concentration in human testicular fluids obtained from men undergoing open surgery (Swerdloff & Walsh, 1975; Adamopoulos et al., 1984; Sealey et al., 1988) or from testicular biopsies (Morse, Horike, Rowley, & Heller, 1973), invasive procedures that cannot be used for routine human studies. Analyses of intratesticular fluid using liquid chromatography tandem mass spectrometry (LC/MS/MS) revealed that intratesticular DHT levels are only 2% those of T (Zhao et al., 2004). Androgen bioavailability in the normal human testis, assessed by a recombinant protein mammalian cell-based bioassay (Raivio, Palvimo, Dunkel, Wickman, & Janne, 2001), was found to be about two-thirds that of the total intratesticular T, as measured by radioimmunoassay (RIA) (Jarow, Wright, Brown, Yan, & Zirkin, 2005). The bioactive androgen concentration in the testis, therefore, is still far greater than the androgen concentration in serum. Interestingly, the concentration of the known androgen-binding proteins within the human testis, sex hormone-binding globulin/androgen-binding protein, was found to be quite low—only about 4% that of intratesticular T (Jarow & Zirkin, 2005). Thus, the 30–40% reduction difference between total T and bioactive androgen concentrations within the testes cannot be explained solely by the binding of androgens to the known intratesticular androgen-binding proteins.

3.5. Male Hormonal Contraception

The administration to men of supraphysiological doses of T alone (World Health Organization (WHO), 1990; 1996), T in combination with a progestogen (Bebb et al., 1996; Handelsman, Conway, Howe, Turner, & Mackey, 1996; Wang et al., 2006), a GnRH antagonist (Pavlou et al., 1991; Behre, Nashan, Hubert, & Nieschlag, 1992; Swerdloff et al., 1998; Behre, Kliesch, Lemcke, Von Eckardstein, & Nieschlag, 2001), or a 5α-reductase inhibitor (Matthiesson et al., 2005) can suppress spermatogenesis reversibly (Liu, Swerdloff, Christenson, Handelsman, & Wang, 2006). Although azoospermia is desired and often is achieved with these formulations, significant numbers of men maintain rates of sperm production that are unacceptably high (WHO, 1996). Moreover, the success of hormonally induced azoospermia has been shown to vary with ethnic origin. For example, in trials using T-alone regimens, Asian men demonstrated far higher attainment of azoospermia than their non-Asian counterparts (Gu et al., 2003; Gui et al., 2004). Recently, Wang and colleagues reported that more than 90% of Asian men, but only 59% of Caucasian men, became severely oligospermic when administered T in contraceptive dosages (Wang et al., 2006). As yet, the biologic/mechanistic basis for individual or ethnic heterogeneity in response to hormonal contraception is not clear. One suggestion has been that there may be ethnic differences in spermatogenetic potential, with Asians having inherently lower daily sperm production than Caucasians, thereby predisposing Asians to greater responsiveness to contraceptive hormones (Johnson et al., 1998). It has been reported that there is greater suppression of gonadotropins by T in Asian than in Caucasian men (Wang et al., 2006), but there is competing evidence suggesting that the differential suppression of gonadotropins may not account for the variation in spermatogenetic suppression (Handelsman, Farley, Peregoudov, & Waites, 1995).

3.6. Male Infertility

Approximately 15% of couples fail to conceive after a year of unprotected intercourse, with a male factor the sole or contributory cause in approximately 50% of these couples (Thonneau et al., 1991). The possibility that some infertile men may derive benefit from increased intratesticular T was suggested by two studies reporting improvement in severely oligospermic men administered human chorionic gonadotropin (hCG) following varicocele repair (Dubin & Amelar, 1975; Mehan & Chehval, 1982). Approximately 25% of infertile men are diagnosed as idiopathic because no identifiable cause of their abnormal semen analyses is found. Clomiphene citrate, an antiestrogen, is often used to manage idiopathic infertility. Although most controlled studies have not shown efficacy of clomiphene citrate over placebo in men with idiopathic infertility, Wang, Chan, Wong, and Yeung (1983) observed a 36% pregnancy rate in couples when the male received 25 mg/day of clomiphene citrate compared to a 0% pregnancy rate in controls, suggesting that at least some men with idiopathic infertility may benefit from hormonal manipulation. Clomiphene citrate treatment raises serum T and FSH levels. It is possible that it also increases intratesticular T, though this has not been shown. In addition, a randomized study of men with idiopathic infertility by Adamopoulos, Nicopoulou, Kapolla, Karamertzanis, and Andreou (1997) demonstrated that the positive effect of antiestrogens could be enhanced by the additional administration of T. Thus, it is possible that at least a subset of men diagnosed as idiopathic would benefit from hormonal therapies designed to increase intratesticular bioavailable androgen concentration.

4. FUTURE DIRECTIONS

Although we have known for decades that T is absolutely required for spermatogenesis, we still do not understand how it functions, including whether its only actions are through ARs. We do not know why the intratesticular T concentration must be far in excess of its concentration in serum for normal spermatogenesis to occur. Just how stem spermatogonial cell self-renewal occurs, and how self-renewal and stem cell differentiation are regulated remain a mystery. Understanding these processes is critical because maintenance of a pool of functioning stem spermatogonia is required for the continuous production of male gametes. We still know little about how germ cells influence Sertoli cell function as they progress through their cyclical behavior, and how Sertoli cells influence their differentiation. The role of estrogens in the testis continues to be a mystery. The jury is still out on whether there is a future for hormone-based contraception in men. We also do not understand how environmental toxicants, particularly those to which the fetus is exposed, impact sperm formation, fertility, and progeny outcome. This is particularly critical because, increasingly, *in-vitro* methods are being used to achieve pregnancies. We need to understand the relationship between the quality of the spermatozoa being used for such procedures and outcomes beyond pregnancies. Finally, it seems possible that stem cells might be used to restore fertility to men, particularly young men, who have undergone chemotherapy for testicular or other cancers.

ACKNOWLEDGMENTS

We gratefully acknowledge the support by the Eunice Kennedy Shriver NICHD/NIH through cooperative agreement U54 HD055740 as part of the Specialized Cooperative Centers Program in Reproduction and Infertility Research.

ABBREVIATIONS

3β-HSD	3β-hydroxysteroid dehydrogenase
A_{al}	Aligned (spermatogonia)
ABP	Androgen-binding protein
AMH	Anti-Müllerian hormone
A_{pr}	Paired (spermatogonia)
AR	Androgen receptor
ARKO	Androgen receptor knockout
A_s	Single (spermatogonia)
BMP	Bone morphogenetic protein
cAMP	Cyclic-3',5'-adenosine monophosphate
ckit	Cell surface receptor for kit ligand/stem cell factor
DHT	5α-dihydrotestosterone
E_2	17β-estradiol
EGF	Epidermal growth factor
FGF-2	Fibroblast growth factor-2
FSH	Follicle-stimulating hormone
GATA	Transcriptional factor involved in cell growth
GDNF	Glial-derived neurotrophic factor
GFRα1	Glial-derived neurotrophic factor receptor-α1
GnRH	Gonadotropin-releasing hormone
GnRHR	Gonadotropin-releasing hormone receptor
hCG	Human chorionic gonadotropin
HPG	Hypothalamus–pituitary–gonad
IGF-1	Insulin-like growth factor-1
LC/MS/MS	Liquid chromatography tandem mass spectrometry
LH	Luteinizing hormone
$p27^{kip1}$	Cyclin-dependent kinase inhibitor
$P450_{C17}$	17-hydroxylase 17,20 lyase
$P450_{scc}$	Side chain-cleaving enzyme
PBR	Peripheral benzodiazepine receptor
PLZF	Promyelocytic leukemia zinc finger
RET	Tyrosine protein kinase subunit of the GDNF receptor
RIA	Radioimmunoassay
SCARKO	Sertoli cell-selective AR knockout
SER	Smooth endoplasmic reticulum
SGP-2	Sulfated glycoprotein-2
SHBG	Sex hormone-binding globulin
StAR	Steroidogenic acute regulatory
T	Testosterone
$^{Tfm}X/Y$	Androgen-insensitive testicular feminization (genotype)
TSPO	Transporter protein
X^{tmf}	Testicular feminization gene

REFERENCES

Adamopoulos, D., Lawrence, D. M., Vassilopoulos, P., Kapolla, N., Kontogeorgos, L., & McGarrigle, H. H. (1984). Hormone levels in the reproductive system of normospermic men and patients with oligospermia and varicocele. *J. Clin. Endocrinol. Metab., 59*, 447–452.

Adamopoulos, D. A., Nicopoulou, S., Kapolla, N., Karamertzanis, M., & Andreou, E. (1997). The combination of testosterone undecanoate with tamoxifen citrate enhances the effects of each agent given independently on seminal parameters in men with idiopathic oligozoospermia. *Fertil. Steril., 67*, 756–762.

Allouche, M., & Bikfalvi, A. (1995). The role of fibroblast growth factor-2 (FGF-2) in hematopoiesis. *Prog. Growth Factor Res., 6*, 35–48.

Awoniyi, C. A., Santulli, R., Chandrashekar, V., Schanbacher, B. D., & Zirkin, B. R. (1989a). Quantitative restoration of advanced spermatogenic cells in adult male rats made azoospermic by active immunization against luteinizing hormone or gonadotropin-releasing hormone. *Endocrinology, 125*, 1303–1309.

Awonyi, C. A., Santulli, R., Sprando, R. L., Ewing, L. L., & Zirkin, B. R. (1989b). Restoration of advanced spermatogenic cells in the experimentally regressed rat testis: quantitative relationship to testosterone concentration within the testis. *Endocrinology, 124*, 1217–1223.

Awonyi, C. A., Sprando, R. L., Santulli, R., Chandrashekar, V., Ewing, L. L., & Zirkin, B. R. (1990). Restoration of spermatogenesis by exogenously administered testosterone in rats made azoospermic by hypophysectomy or withdrawal of luteinizing hormone alone. *Endocrinology, 127*, 177–184.

Bardin, C. W., Bullock, L. P., Sherins, R. J., Mowszowicz, I., & Blackburn, W. R. (1973). Androgen metabolism and mechanism of action in male pseudohermaphroditism: a study of testicular feminization. *Rec. Prog. Horm. Res., 29*, 65–109.

Bebb, R. A., Anawalt, B. D., Christensen, R. B., Paulsen, C. A., Bremner, W. J., & Matsumoto, A. M. (1996). Combined administration of levonorgestrel and testosterone induces more rapid and effective suppression of spermatogenesis than testosterone alone: a promising male contraceptive approach. *J. Clin. Endocrinol. Metab., 81*, 757–762.

Behre, H. M., Kliesch, S., Lemcke, B., Von Eckardstein, S., & Nieschlag, E. (2001). Suppression of spermatogenesis to azoospermia by combined administration of GnRH antagonist and 19-nortestosterone cannot be maintained by this non-aromatizable androgen alone. *Hum. Reprod., 16*, 2570–2577.

Behre, H. M., Nashan, D., Hubert, W., & Nieschlag, E. (1992). Depot gonadotropin-releasing hormone agonist blunts the androgen-induced suppression of spermatogenesis in a clinical trial of male contraception. *J. Clin. Endocrinol. Metab., 74*, 84–90.

Beites, C. L., Kawauchi, S., Crocker, C. E., & Calof, A. L. (2005). Identification and molecular regulation of neural stem cells in the olfactory epithelium. *Exp. Cell. Res., 306*, 309–316.

Bremner, W. J., Millar, M. R., Sharpe, R. M., & Saunders, P. T. (1994). Immunohistochemical localization of androgen receptors in the rat testis: evidence for stage-dependent expression and regulation by androgens. *Endocrinology, 135*, 1227–1233.

Brooker, G. J., Kalloniatis, M., Russo, V. C., Murphy, M., Werther, G. A., & Bartlett, P. F. (2000). Endogenous IGF-1 regulates the neuronal differentiation of adult stem cells. *J. Neurosci. Res., 59*, 332–341.

Brown, T. R. (2002). Androgen resistance and disorders of androgenic metabolism. In J. D. Baxter, S. Melmed, & M. I. New (Eds.), *Genetics in Endocrinology* (pp. 709–738). Philadelphia, PA: Lippincott Williams and Wilkins.

Buaas, F. W., Kirsh, A. L., Sharma, M., McLean, D. J., Morris, J. L., Griswold, M. D., et al. (2004). Plzf is required in adult male germ cells for stem cell self-renewal. *Nat. Genet., 36*, 647–652.

Buzek, S. W., & Sanborn, B. W. (1988). Increase in testicular androgen receptor during sexual maturation in the rat. *Biol. Reprod., 39*, 39–49.

Chang, C., Chen, Y.-T., Yeh, S.-D., Xu, Q., Wang, R.-S., Guillou, F., et al. (2004). Infertility with defective spermatogenesis and hypotestosteronemia in male mice lacking the androgen receptor in Sertoli cells. *Proc. Natl. Acad. Sci. USA, 101*, 6876–6881.

Chen, Y., Ai, Y., Slevin, J. R., Maley, B. E., & Gash, D. M. (2005). Progenitor proliferation in the adult hippocampus and substantia nigra induced by glial cell line-derived neurotrophic factor. *Exp. Neurol., 196*, 87–95.

De Gendt, K., Swinnen, J. V., Saunders, P. T., Schoonjans, L., Dewerchin, M., Devos, A., et al. (2004). A Sertoli cell-selective knockout of the androgen receptor causes spermatogenic arrest in meiosis. *Proc. Natl. Acad. Sci. USA, 101*, 1327–1332.

Deleyrolle, L., Marchal-Victorion, S., Dromard, C., Fritz, V., Saunier, M., Sabourin, J. C., et al. (2006). Exogenous and fibroblast growth factor 2/ epidermal growth factor-regulated endogenous cytokines regulate neural precursor cell growth and differentiation. *Stem. Cells., 24*, 748–762.

De Rooij, D. G. (2001). Proliferation and differentiation of spermatogonial stem cells. *Reproduction, 121*, 347–354.

Dombrowicz, D., Hooghe-Peters, E. L., Gothot, A., Sente, B., Vanhaelst, L., Closset, J., et al. (1992). Cellular localization of IGF-I and IGF-II mRNAs in immature hypophysectomized rat testis and epididymis after *in vivo* hormonal treatment. *Arch. Int. Physiol. Biochim. Biophys., 100*, 303–308.

Dubin, L., & Amelar, R. D. (1975). Varicocelectomy as therapy in male infertility: a study of 504 cases. *J. Urology, 113*, 640–641.

Dym, M., & Clermont, Y. (1970). Role of spermatogonia in the repair of the seminiferous epithelium following x-irradiation of the rat testis. *Am. J. Anat., 128*, 265–282.

Eacker, S. M., Shima, J. E., Connolly, C. M., Sharma, M., Holdcraft, R. W., Griswold, M. D., et al. (2007). Transcriptional profiling of androgen receptor (AR) mutants suggests instructive and permissive roles of AR signaling in germ cell development. *Mol. Endocrinol., 21*, 895–907.

Eddy, E. M., & O'Brien, D. A. (1998). Gene expression during mammalian meiosis. *Curr. Top. Dev. Biol., 37*, 141–200.

Griswold, M. D., & McLean, D. (2006). In J. D. Neill (Ed.), *Knobil and Neill's Physiology of Reproduction* (3rd ed.). *The Sertoli. cell., Vol 1* (pp. 949–975) San Diego, CA: Elsevier.

Gu, Y. Q., Wang, X. H., Xu, D., Peng, L., Cheng, L. F., Huang, M. K., et al. (2003). A multicenter contraceptive efficacy study of injectable testosterone undecanoate in healthy Chinese men. *J. Clin. Endocrinol. Metab., 88*, 562–568.

Guan, K., Nayernia, K., Maier, L. S., Wagner, S., Dressel, R., Lee, J. H., et al. (2006). Pluripotency of spermatogonial stem cells from adult mouse testis. *Nature, 440*, 1199–1203.

Gui, Y. L., He, C. H., Amory, J. K., Bremner, W. J., Zheng, E. X., Yang, J., et al. (2004). Male hormonal contraception: suppression of spermatogenesis by injectable testosterone undecanoate alone or with levonorgestrel implants in Chinese men. *J. Androl., 25*, 720–727.

Hamra, F. K., Schultz, N., Chapman, K. M., Grellhesl, D. M., Cronkhite, J. T., Hammer, R. E., et al. (2004). Defining the spermatogonial stem cell. *Dev. Biol., 269*, 393–410.

Handelsman, D. J., Conway, A. J., Howe, C. J., Turner, L., & Mackey, M. A. (1996). Establishing the minimum effective dose and additive effects of depot progestin in suppression of human spermatogenesis by a testosterone depot. *J. Clin. Endocrinol. Metab., 81*, 4113–4121.

Handelsman, D. J., Farley, T. M., Peregoudov, A., & Waites, G. M. (1995). Factors in nonuniform induction of azoospermia by testosterone enanthate in normal men. World Health Organization Task Force on Methods for the Regulation of Male Fertility. *Fertil. Steril., 63*, 125–133.

Hauet, T., Yao, Z. X., Bose, H. S., Wall, C. T., Han, Z., Li, W., et al. (2005). Peripheral-type benzodiazepine receptor-mediated action of steroidogenic acute regulatory protein on cholesterol entry into Leydig cell mitochondria. *Mol. Endocrinol., 19*, 540–554.

Hofmann, M. C., Braydich-Stolle, L., & Dym, M. (2005). Isolation of male germ-line stem cells; influence of GDNF. *Dev. Biol., 279*, 114–124.

Holdcraft, R. W., & Braun, R. E. (2004). Androgen receptor function is required in Sertoli cells for the terminal differentitation of haploid spermatids. *Development, 131*, 459–467.

Houchen, C. W., George, R. J., Sturmoski, M. A., & Cohn, S. M. (1999). FGF-2 enhances intestinal stem cell survival and its expression is induced after radiation injury. *Am. J. Physiol., 276*, G249–258.

Huckins, C. (1971a). The spermatogonial stem cell population in adult rats. I. Their morphology, proliferation and maturation. *Anat. Rec., 169*, 533–557.

Huckins, C. (1971b). The spermatogonial stem cell population in adult rats. II. A radioautographic analysis of their cell cycle properties. *Cell Tissue Kinet., 4*, 313–334.

Huckins, C. (1971c). The spermatogonial stem cell population in adult rats. 3. Evidence for a long-cycling population. *Cell Tissue Kinet., 4*, 335–349.

Huckins, C. (1978a). The morphology and kinetics of spermatogonial degeneration in normal adult rats: an analysis using a simplified classification of the germinal epithelium. *Anat. Rec., 190*, 905–926.

Huckins, C. (1978b). Spermatogonial intercellular bridges in whole-mounted seminiferous tubules from normal and irradiated rodent testes. *Am. J. Anat., 153*, 97–121.

Isomaa, V., Parvinen, M., Janne, O. A., & Bardin, C. W. (1985). Nuclear androgen receptors in different stages of the seminiferous epithelial cycle and the interstitial tissue of rat testis. *Endocrinology, 116*, 132–137.

Jarow, J. P., Chen, H., Rosner, T. W., Trentacoste, S., & Zirkin, B. R. (2001). Assessment of the androgen environment within the human testis: minimally invasive method to obtain intratesticular fluid. *J. Androl., 22*, 640–645.

Jarow, J. P., Wright, W. W., Brown, T. R., Yan, X., & Zirkin, B. R. (2005). Bioactivity of androgens within the testes and serum of normal men. *J. Androl., 26*, 343–348.

Jarow, J. P., & Zirkin, B. R. (2005). The androgen microenvironment of the human testis and hormonal control of spermatogenesis. *Ann. N.Y. Acad. Sci., 1061*, 208–220.

Jeyaraj, D. A., Grossman, G., Weaver, C., & Petrusz, P. (2002). Dynamics of testicular germ cell proliferation in normal mice and transgenic mice overexpressing rat androgen-binding protein: a flow cytometric evaluation. *Biol. Reprod., 66*, 877–885.

Johnson, L., Barnard, J. J., Rodriguez, L., Smith, E. C., Swerdlof, R. S., Wang, X. H., et al. (1998). Ethnic differences in testicular structure and spermatogenic potential may predispose testes of Asian men to a heightened sensitivity to steroidal contraceptives. *J. Androl., 19*, 348–357.

Johnston, D. S., Russell, L. D., Friel, P. J., & Griswold, M. D. (2001). Murine germ cells do not require functional androgen receptors to complete spermatogenesis following spermatogonial stem cell transplantation. *Endocrinology, 142*, 2405–2408.

Johnston, D. S., Wright, W. W., Dicandeloro, P., Wilson, E., Kopf, G. S., & Jelinsky, S. A. (2008). Stage-specific gene expression is a fundamental characteristic of rat spermatogenic cells and Sertoli cells. *Proc. Natl. Acad. Sci. USA, 105*, 8315–8320.

Joseph, D. R. (1994). Structure, function, and regulation of androgen-binding protein/sex hormone-binding globulin. *Vitamins & Hormones, 49*, 197–280.

Joseph, D. R., Hall, S. H., & French, F. S. (1987). Rat and androgen-binding protein: evidence for identical subunits and amino acid sequence homology with human sex hormone-binding globulin. *Proc. Natl. Acad. Sci. USA, 84*, 339–343.

Joseph, D. R., O'Brien, D. A., Sullivan, P. M., Becchis, M., & Tsuruta, J. K. (1997). Overexpression of androgen-binding protein/sex hormone-binding globulin in male transgenic mice: tissue distribution and phenotypic disorders. *Biol. Reprod., 56*, 21–32.

Kamischke, A., & Nieschlag, E. (2003). Progress towards hormonal male contraception. *Trends Pharm. Sci., 25*, 49–57.

Kerr, J. B., Loveland, K. L., O'Bryan, M. K., & DeKretser, D. M. (2006). In J. D. Neill (Ed.), *Knobil's and Neill's Physiology of Reproduction* (3rd ed.). *Cytology of the testis and intrinsic control mechanisms, Vol 1* (pp. 827–948) San Diego, CA: Elsevier.

Kubota, H., Avarbock, M. R., & Brinster, R. L. (2004). Growth factors essential for self-renewal and expansion of mouse spermatogonial stem cells. *Proc. Natl. Acad. Sci. USA, 101*, 16489–16494.

Leblond, C. P., & Clermont, Y. (1952). Definition of the stages of the cycle of the seminiferous epithelium in the rat. *Ann. N.Y. Acad. Sci., 55*, 548–573.

Liu, P. Y., Swerdloff, R. S., Christenson, P. D., Handelsman, D. J., & Wang, C. (2006). Rate, extent, and modifiers of spermatogenic recovery after hormonal male contraception: an integrated analysis. *Lancet, 367*, 1412–1420.

Lyon, M. F., Glenister, P. H., & Lamoreux, M. L. (1975). Normal spermatozoa from androgen-resistant germ cells of chimaeric mice and the role of androgen in spermatogenesis. *Nature, 258*, 620–622.

Lyon, M. F., & Hawkes, S. G. (1970). X-linked gene for testicular feminization in the mouse. *Nature, 227*, 1217–1219.

Matthiesson, K. L., Stanton, P. G., O'Donnell, L., Meachem, S. J., Amory, J. K., Berger, R., et al. (2005). Effects of testosterone and levonorgestrel combined with a 5alpha-reductase inhibitor or gonadotropin-releasing hormone antagonist on spermatogenesis and intratesticular steroid levels in normal men. *J. Clin. Endocrinol. Metab., 90*, 5647–5655.

McLachlan, R. I., O'Donnell, L., Meachem, S. J., Stanton, P. G., De Kretser, D. M., Prati, K., et al. (2002a). Identification of specific sites of hormonal regulation in spermatogenesis in rats, monkeys and man. *Rec. Prog. Horm. Res., 57*, 149–179.

McLachlan, R. I., O'Donnell, L., Meachem, S. J., Stanton, P. G., De Kretser, D. M., Prati, K., et al. (2002b). Hormonal regulation of spermatogenesis in primates and man: insights for development of the male hormonal contraceptive. *J. Androl., 23*, 149–162.

Mehan, D. J., & Chehval, M. J. (1982). Human chorionic gonadotropin in the treatment of the infertile man. *J. Urol., 128*, 60–63.

Meng, X., Lindahl, M., Hyvonen, M. E., Parvinen, M., De Rooij, D. G., Hess, M. W., et al. (2000). Regulation of cell fate decision of undifferentiated spermatogonia by GDNF. *Science, 287*, 1489–1493.

Miller, W. L. (2008). Steroidogenic enzymes. *Endocr. Dev, 13*, 1–18.

Morse, H. C., Horike, N., Rowley, M. J., & Heller, C. G. (1973). Testosterone concentrations in testes of normal men: effects of testosterone propionate administration. *J. Clin. Endocrinol. Metab., 37*, 882–886.

Mourkiot, F., & Rosenthal, N. (2005). IGF-1, inflammation and stem cells: interactions during muscle regeneration. *Trends Immunol., 26*, 535–542.

Mullaney, B. P., & Skinner, M. K. (1992). Basic fibroblast growth factor (bFGF) gene expression and protein production during pubertal development of the seminiferous tubule: follicle-stimulating hormone-induced Sertoli cell bFGF expression. *Endocrinology, 131*, 2928–2934.

Naughton, C. K., Jain, S., Strickland, A. M., Gupta, A., & Milbrandt, J. (2006). Glial cell line-derived neurotrophic factor-mediated RET signaling regulates spermatogonial stem cell fate. *Biol. Reprod., 74*, 314–321.

O'Donnell, L., McLachlan, R. I., Wreford, N. G., De Kretser, D. M., & Robertson, D. M. (1996). Testosterone withdrawal promotes stage-specific detachment of round spermatids from rat seminiferous epithelium. *Biol. Reprod., 55*, 895–901.

O'Donnell, L., Meachem, S. J., Stanton, P. G., & McLachlan, R. I. (2006). In J. D. Neill (Ed.), *Knobil and Neill's Physiology of Reproduction* (3rd ed.). *Endocrine regulation of spermatogenesis, Vol 1* (pp. 1017–1069) San Diego, CA: Elsevier.

Ogawa, T., Dobrinski, I., Avarbock, M. R., & Brinster, R. L. (2000). Transplantation of male germ line stem cells restores fertility in infertile mice. *Nature Med., 6*, 29–34.

Ogawa, T., Ohmura, M., Yumura, Y., Sawada, H., & Kubota, Y. (2003). Expansion of murine spermatogonial stem cells through serial transplantation. *Biol. Reprod., 68*, 316–322.

Parvinen, M. (1982). Regulation of the seminiferous epithelium. *Endocr. Rev., 3*, 404–417.

Pavlou, S. N., Brewer, K., Farley, M. G., Lindner, J., Bastias, M. C., Rogers, B. J., et al. (1991). Combined administration of a gonadotropin-releasing hormone antagonist and testosterone in men induces reversible azoospermia without loss of libido. *J. Clin. Endocrinol. Metab., 73*, 1360–1369.

Payne, A. H. (2007). Steroidogenic enzymes in Leydig cells. In A. H. Payne, & M. P. Hardy (Eds.), *The Leydig Cell in Health and Disease* (pp. 157–172). Totowa, NJ: Human a Press.

Penttila, T. L., Kaipia, A., Toppari, J., Parvinen, M., & Mali, P. (1994). Localization of urokinase- and tissue-type plasminogen activator mRNAs in rat testes. *Mol. Cell. Endocrinol., 105*, 55–64.

Raivio, T., Palvimo, J. J., Dunkel, L., Wickman, S., & Janne, O. A. (2001). Novel assay for determination of androgen bioactivity in human serum. *J. Clin. Endocrinol. Metab., 86*, 1539–1544.

Rao, M. K., Wayne, C. M., Meistrich, M. L., & Wilkinson, M. F. (2003). Pem homeobox gene promoter sequences that direct transcription in a Sertoli cell-specific, stage-specific, and androgen-dependent manner in the testis *in vivo*. *Mol. Endocrinol., 17*, 223–233.

Regadera, J., Martinez-Garcia, F., Gonzalez-Peramato, P., Serrano, A., Nistal, M., & Suarez-Quian, C. (2001). Androgen receptor expression in Sertoli cells as a function of seminiferous tubule maturation in the human cryptorchid testis. *J. Clin. Endocrinol. Metab., 86*, 413–421.

Roberts, K. P., & Zirkin, B. R. (1991). Androgen regulation of spermatogenesis in the rat. *Ann. N.Y. Acad. Sci., 637*, 90–106.

Roberts, K. P., & Zirkin, B. R. (1993). Androgen binding protein inhibition of androgen dependent transcription explains the high minimal testosterone concentration required to maintain spermatogenesis in the rat. *Endocrine. J., 1*, 41–47.

Russell, L. D., Alger, L. E., & Nequin, L. G. (1987). Hormonal control of pubertal spermatogenesis. *Endocrinology, 120*, 1615–1632.

Russell, L. D., Chiarini-Garcia, H., Korsmeyer, S. J., & Knudson, C. M. (2002). Bax-dependent spermatogonia apoptosis is required for testicular development and spermatogenesis. *Biol. Reprod., 66*, 950–958.

Sairam, M. R., & Krishnamurthy, H. (2001). The role of follicle-stimulating hormone in spermatogenesis: lessons from knockout animal models. *Arch. Med. Res., 32*, 601–608.

Saito, K., O'Donnell, L., MacLachlan, R. I., & Robertson, D. M. (2000). Spermiation failure is a major contributor to early spermatogenic suppression caused by hormone withdrawal in adult rats. *Endocrinology, 141*, 2779–2785.

Sar, M., Lubahn, D. B., French, F. S., & Wilson, E. M. (1990). Immunohistochemical localization of the androgen receptor in rat and human tissues. *Endocrinology, 127*, 3180–3186.

Sealey, J. E., Goldstein, M., Pitarresi, T., Kudlak, T. T., Glorioso, N., Fiamengo, S. A., et al. (1988). Prorenin secretion from human testis: No evidence for secretion of active renin or angiotensinogen. *J. Clin. Endocrinol. Metab., 66*, 974–978.

Shan, L. X., Zhu, L. J., Bardin, C. W., & Hardy, M. P. (1995). Quantitative analysis of androgen receptor messenger ribonucleic acid in developing Leydig cells and Sertoli cells by *in situ* hybridization. *Endocrinology, 136*, 3856–3862.

Sharpe, R. M. (1994). Regulation of spermatogenesis. In E. Knobil, & J. D. Neill (Eds.), *The Physiology of Reproduction* (pp. 1363–1434). New York, NY: Raven Press.

Sharpe, R. M., Maddocks, S., Millar, M., Kerr, J. B., Saunders, P. T., & McKinnell, C. (1992). Testosterone and spermatogenesis. Identification of stage-specific, androgen-regulated proteins secreted by adult seminiferous tubules. *J. Androl., 13*, 172–184.

Solchaga, L. A., Penick, K., Porter, J. D., Goldberg, V. M., Caplan, A. I., & Welter, J. F. (2005). FGF-2 enhances the mitotic and chondrogenic potentials of human adult bone marrow-derived mesenchymal stem cells. *J. Cell. Physiol., 203*, 398–409.

Spaliviero, J. A., Jimenez, M., Allan, C. M., & Handelsman, D. J. (2004). Luteinizing hormone receptor-mediated effects on initiation of spermatogenesis in gonadotropin-deficient (hpg) mice are replicated by testosterone. *Biol. Reprod., 70*, 32–38.

Suarez-Quian, C. A., Martinez-Garcia, F., Nistal, M., & Regadera, J. (1999). Androgen receptor distribution in adult human testis. *J. Clin. Endocrinol. Metab., 84*, 350–358.

Swerdloff, R. S., & Walsh, P. C. (1975). Pituitary and gonadal hormones in patients with varicocele. *Fertil. Steril., 26*, 1006–1012.

Swerdloff, R. S., Bagatell, C. J., Wang, Anawalt, B. D., Berman, N., Steiner, B., et al. (1998). Suppression of spermatogenesis in man induced by Nal-Glu gonadotropin releasing hormone antagonist and testosterone enanthate (TE) is maintained by TE alone. *J. Clin. Endocrinol. Metab., 83*, 3527–3533.

Takeda, H., Chodak, G., Mutchnik, S., Nakamoto, T., & Chang, C. (1990). Immunohistochemical localization of androgen receptors with mono- and polyclonal antibodies to androgen receptor. *J. Endocrinol., 126*, 17–25.

Tan, K. A., De Gendt, K., Atanassova, N., Walker, M., Sharpe, R. M., Saunders, P. T. K., et al. (2005). The role of androgens in Sertoli cell proliferation and functional maturation: studies in mice with total or Sertoli cell-selective ablation of the androgen receptor. *Endocrinology, 146*, 2674–2683.

Thonneau, P., Marchand, S., Tallec, A., Ferial, M. L., Ducot, B., Lansac, J., et al. (1991). Incidence and main causes of infertility in a resident population (1,850,000) of three French regions (1988–1989). *Hum. Reprod., 6*, 811–816.

Turner, T. T., Jones, C. E., Howards, S. S., Ewing, L. L., Zegeye, B., & Gunsalus, G. L. (1984). On the androgen microenvironment of maturing spermatozoa. *Endocrinology, 115*, 1925–1932.

Van Beek, M. E., Davids, J. A., Van de Kant, H. J., & De Rooij, D. G. (1984). Response to fission neutron irradiation of spermatogonial stem cells in different stages of the cycle of the seminiferous epithelium. *Radiat. Res., 97*, 556–569.

Vornberger, W., Prins, G., Musto, N. A., & Suarez-Quian, C. A. (1994). Androgen receptor distribution in rat testis: new implications for androgen regulation of spermatogenesis. *Endocrinology, 134*, 2307–2316.

Walker, W. H. (2009). Molecular mechanisms of testosterone action in spermatogenesis. *Steroids, 74*, 602–607.

Wang, C., Chan, C. W., Wong, K. K., & Yeung, K. K. (1983). Comparison of the effectiveness of placebo, clomiphene citrate, mesterolone, pentoxifylline, and testosterone rebound therapy for the treatment of idiopathic oligospermia. *Fertil. Steril., 40*, 358–365.

Wang, C., Wang, X. H., Nelson, A. L., Lee, K. K., Cui, Y. G., Tong, J. S., et al. (2006). Levonorgestrel implants enhanced the suppression of spermatogenesis by testosterone implants: comparison between Chinese and non-Chinese men. *J. Clin. Endocrinol. Metab., 91*, 460–470.

Wang, R.-S., Yeh, S., Chen, L.-M., Lin, H.-Y., Zhang, C., Ni, J., et al. (2006). Androgen receptor in Sertoli cell is essential for germ nursery and junctional complex formation in mouse testes. *Endocrinology, 147*, 5624–5633.

World Health Organization (WHO) (1990). Contraceptive efficacy of testosterone-induced azoospermia in normal men. World Health Organization Task Force on methods for the regulation of male fertility. *Lancet, 336*, 955–959.

World Health Organization (WHO) (1996). Contraceptive efficacy of testosterone-induced azoospermia and oligozoospermia in normal men. *Fertil. Steril., 65*, 821–829.

Wright, W. W. (1993). Cellular interactions in the seminiferous epithelium. In C. Dejardins, & L. Ewing (Eds.), *Cell and Molecular Biology of the Testis* (pp. 377–399). New York, NY: Oxford University Press.

Wright, W. W., Smith, L., Kerr, C., & Charron, M. (2003). Mice that express enzymatically inactive cathepsin L exhibit abnormal spermatogenesis. *Biol. Reprod., 68*, 680–687.

Wright, W. W., Zabludoff, S. D., Penttilä, T.-L., & Parvinen, M. (1995). Germ cell–Sertoli cell interactions: Regulation by germ cells of the stage-specific expression of CP-2/Cathepsin l mRNA by Sertoli cells. *Dev. Genet., 6*, 104–113.

Zhao, M., Baker, S. D., Yan, X., Zhao, Y., Wright, W. W., Zirkin, B. R., et al. (2004). Simultaneous determination of steroid composition of human testicular fluid using liquid chromatography tandem mass spectrometry. *Steroids, 69*, 721–726.

Zhou, Q., Nie, R., Prins, G. S., Saunders, P. T., Katzenellenbogen, B. S., & Hess, R. A. (2002). Localization of androgen and estrogen receptors in adult male mouse reproductive tract. *J. Androl., 23*, 870–881.

Zhu, L. J., Hardy, M. P., Inigo, I. V., Huhtaniemi, I., Bardin, C. W., & Moo-Young, A. J. (2000). Effects of androgen on androgen receptor expression in rat testicular and epididymal cells: a quantitative immunohistochemical study. *Biol. Reprod., 63*, 368–376.

Zirkin, B. R. (1998). Spermatogenesis: its regulation by testosterone and FSH. *Semin. Cell Dev. Biol., 9*, 417–421.

Zirkin, B. R., Santulli, R., Awoniyi, C. A., & Ewing, L. L. (1989). Maintenance of advanced spermatogenic cells in the adult rat testis: quantitative relationship to testosterone concentration within the testis. *Endocrinology, 124*, 3043–3049.

Chapter 4

The Endocrinology of the Mammalian Ovary

David O. Norris and Kristin H. Lopez
University of Colorado at Boulder, Boulder, CO, USA

SUMMARY
This chapter emphasizes the roles of hormones and paracrine regulators that regulate, or are produced by, the mammalian ovary (including the oocyte, theca, granulosa, and interstitial cells). The regulation of developmental aspects of the ovary is discussed including embryogenesis, oogenesis, and folliculogenesis. Limited information on roles of specific genes is included. Factors affecting puberty (including melatonin and ghrelin as well as hormones from the hypothalamus, pituitary, and ovary) are discussed. The roles of hormones and paracrines during the ovarian cycle are separated into effects during the follicular phase, ovulation, and the luteal phase. Induced versus spontaneous ovulation is discussed as well as aspects of ovarian aging. A brief description of events during the ovarian cycles of rats, ewes, women, and elephants is provided. Finally, some suggestions are made for future research in this area.

1. INTRODUCTION

Understanding the structure, function, and regulatory control of the mammalian ovary is basic to any consideration of mammalian reproduction. In this chapter, we will briefly discuss the roles of hormones and other regulatory agents in the development of the ovary as well as in the processes of folliculogenesis, ovulation, and corpus luteum function. Several comprehensive treatments of the mammalian ovary have been published, and the interested reader is referred to these sources (e.g., Mossman & Duke, 1973; Jones, 1978; Leung & Adashi, 2004).

The cyclic reproductive events of female mammals are referred to as the estrous cycle and include a short period of heightened receptivity of the female for mating, which is called estrus. Only humans seem to lack a distinct estrous period. The role of the hypothalamic–pituitary–gonadal (HPG) axis in regulating the estrous cycle and ovarian function is described in Chapter 2, this volume. Briefly, in response to pulsatile hypothalamic gonadotropin-releasing hormone (GnRH) release, pituitary gonadotropins (GTHs) stimulate gamete maturation as well as steroidogenesis in females with release of ovarian estrogens, mainly 17β-estradiol (E_2) and lesser amounts of estrone (E_1) and estriol (E_3); androgens, primarily androstenedione (AND); and progestogens, mainly progesterone (P_4). These steroids may be released into the general circulation or may produce local ovarian effects. A variety of additional paracrine and autocrine factors are produced by the gonads in response to GTHs and influence steroidogenesis and oogenesis locally. Follicle growth in the ovary is controlled primarily by follicle-stimulating hormone (FSH), whereas induction of ovulation and formation of a corpus luteum from the postovulatory follicle occur as a result of luteinizing hormone (LH). Androgen synthesis is controlled primarily by LH, but conversion of androgens to estrogens by the enzyme aromatase ($P450_{aro}$) is under the control of FSH. In some species, prolactin (PRL) may play a role in regulating ovarian progestogen synthesis by the corpus luteum.

The hypothalamic centers regulating GnRH release are sensitive to circulating steroids that generally exert negative feedback on GnRH release. The exception to this pattern is the preovulatory positive feedback effect of estrogens on release of GnRH in females and subsequent stimulation of GTH release from the pituitary, as described in Chapter 2, this volume. Gonadal steroids, secreted as a result of the action of GTHs on somatic cells in the ovaries, also control differentiation and maintenance of accessory sexual organs such as the uterus and mammary glands.

2. EMBRYOGENESIS OF THE OVARY

The development of the ovary is best known for mice, as recently reviewed by Edson, Nagaraja, and Matzuk (2009), and the general pattern for mice appears to operate in most mammals. Very early in embryonic development, dedicated germline cells, the primordial germ cells (PGCs), arise within the yolk sac of the gastrula at embryonic day (E) 6.5 in

Hormones and Reproduction of Vertebrates, Volume 5—Mammals

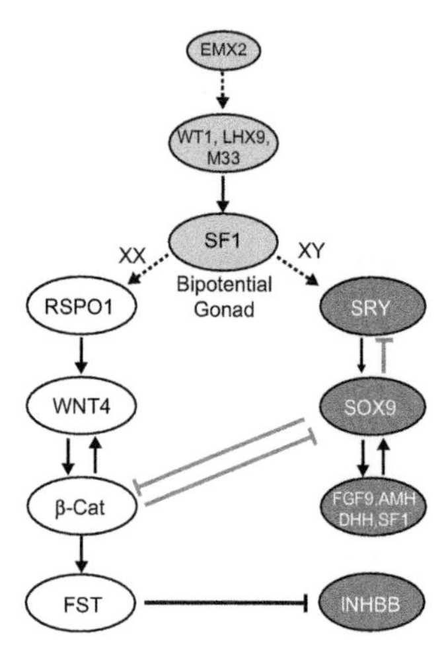

FIGURE 4.1 Sequence of gene activations in gonadal differentiation. β-catenin (β-Cat) activation can inhibit the male cascade. *Reprinted with permission from Edson, Nagaraja, and Matzuk (2009).*

mice. This process appears to be controlled by bone morphogenetic proteins (BMPs) secreted by the extra-embryonic ectoderm and the visceral endoderm (Lawson et al., 1999; see also Defalco & Capel, 2009). At about E10.5 in the mouse embryo, paired gonadal primordia consisting of two genital ridges arise from the intermediate mesoderm located on either side of the midline in close association with the transitory mesonephric kidneys. A bipotential gonad develops under the actions of a HOX gene (*Emx2*, empty spiracles homolog 2), *Wt1* (Wilms tumor 1 homolog), and eventually by activation of *Sf1* (steroidogenic factor 1). Prior to sexual differentiation of the gonads, PGCs begin a migration into the genital ridges from the base of the yolk sac. Chemoattractants released from the genital ridges, such as KIT ligand (also known as stem cell factor or steel factor; the product of the *Kit* gene in mice), direct the migration of PGCs from the yolk sac along the hindgut to the genital ridges. Once within the genital ridges, PGCs lose their ability to migrate and become remethylated, lagging behind the genome-wide remethylation in the somatic cells that occurred prior to gastrulation (Kafri et al., 1992). In the case of males, SF-1 activates the *Sry* gene on the Y-chromosome, causing the bipotential gonad to differentiate into a testis. In females, SF-1 activates a separate pathway, responsible for differentiation of an ovary (Figure 4.1). Therefore, formation of an ovary is not a simple 'default' that occurs in the absence of the *Sry* gene. Studies in both humans and mice indicate that production of β-catenin is both a major pro-ovarian and antitesticular factor that is activated within the ovarian gene cascade (Figure 4.1) and that in males *Sox9* is responsible for suppressing female development.

2.1. Oogenesis

The process of oogenesis begins with the conversion of PGCs to oogonia after they enter the genital ridge. Mitotic divisions of migrating PGCs as well as of oogonia greatly increase the size of the germ cell pool. Oogonial proliferation results in the formation of germ cell nests, clonal clusters of oogonia that are interconnected by intercellular bridges. Meiosis is initiated early in development and involves the entrance of oogonia into meiosis I to form primary oocytes. Recent evidence suggests that the signal initiating meiosis in the germ cells of the developing ovary is retinoic acid (RA) released from the adjacent mesonephros (Bowles et al., 2006; Bowles & Koopman, 2007). Although RA is also present in the developing testis at this stage, meiosis of male germ cells is prevented by the presence of an enzyme, $P450_{26B1}$ (CYP26B1), that catalyzes the degradation of RA. CYP26B1 is downregulated in the developing ovary, allowing the meiosis-inducing action of RA.

Oocytes that have entered meiosis do not complete the process and are arrested in diplotene of meiosis I prophase (the dictyate stage), in which they remain until just prior to ovulation. These primary oocytes, originally present in germ cell nests, become individually surrounded by a single layer of squamous pregranulosa cells to form primordial follicles, typically located near the surface of the ovary. Formation of primordial follicles requires the presence of a female germline-specific transcription factor, FIGLA (factor in the germline-α); female mice lacking the *Figla* gene fail to form primordial follicles and consequently are sterile (Soyal, Amleh, & Dean, 2000). Primordial follicles provide a reservoir of dormant follicles that may be activated at any time throughout the reproductive life of the female. Dormancy appears to be maintained by local inhibition, but the mechanisms maintaining dormancy and activating follicle growth are not understood (see reviews by Adhikari & Liu, 2009; Edson, Nagaraja, & Matzuk, 2009; Reddy, Zheng, & Liu, 2010; Richards & Pangas, 2010). Among the factors implicated in dormancy maintenance are anti-Müllerian hormone (AMH) and the tumor suppresser PTEN (phosphatase and tensin homolog deleted on chromosome 10) (Visser, De Jong, Laven, & Themmen, 2006; Broekmans et al., 2008).

In most mammals, it is assumed that there are no oogonia in the adult ovary because all of them enter meiosis and become primary oocytes prior to or shortly after birth. However, production of new oocytes and new follicle development have been reported after birth in mice as well as in certain primates (Johnson, Canning, Kaneko, Pru, & Tilly, 2004).

2.2. Folliculogenesis

Resting primordial follicles may enter the pool of growing follicles or simply fail to develop further. The neurotropin NT4 increases at the time that primordial follicles form in both mice and humans (see Anderson, Robinson, Brooks, & Spears, 2002). Follicular growth involves two steps: the

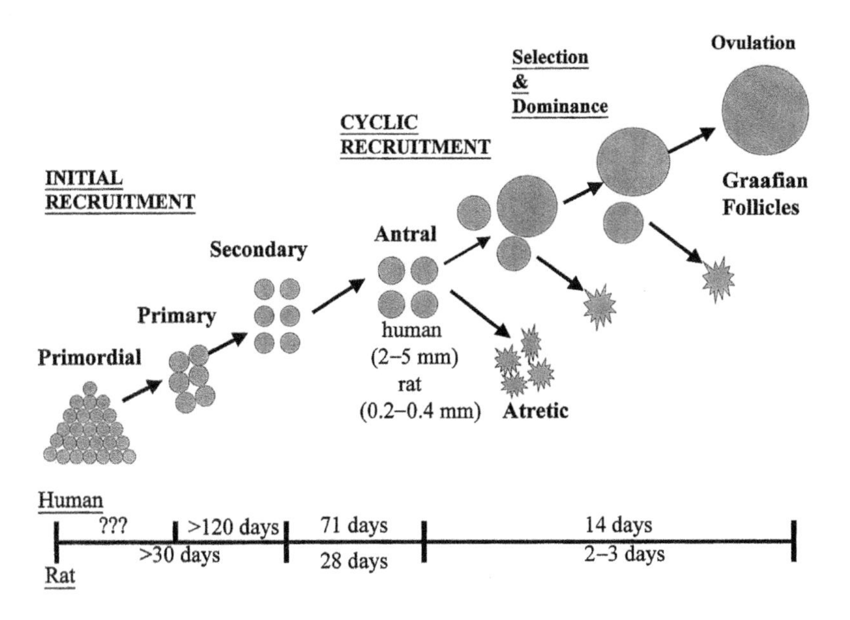

FIGURE 4.2 Developmental stages of the mammalian ovarian follicle. Primordial follicles constitute the pool of follicles available to advance to subsequent stages. Initial recruitment occurs throughout life or until primordial follicles are depleted. Cyclic recruitment begins at puberty in response to elevated follicle-stimulating hormone (FSH) and is repeated each ovarian cycle throughout the reproductive lifespan, providing a cohort of follicles available for selection for final growth and ovulation. The time required for each stage is shown for the human and rat ovarian follicle. *Reprinted with permission from McGee and Hsueh (2000).*

initial recruitment of primordial follicles into the growing follicular pool and the later cyclic recruitment of large growing follicles into the cohort that will compete for final maturation and ovulation (Figure 4.2) (McGee & Hsueh, 2000). Initial recruitment begins shortly after primordial follicles are formed and involves a steady trickle of these follicles into the preantral follicular pool. This initial recruitment begins in the juvenile animal or even before birth, depending on the species. Prior to puberty, follicles in the growing pool enlarge slowly, reaching the primary, secondary, or early antral phase before becoming atretic. Cyclic recruitment begins at puberty in response to the maturation of the HPG axis and continues cyclically throughout the reproductive lifespan of the female.

The process of folliculogenesis consists of a preantral phase independent of GTHs and an antral phase that is GTH-dependent (Figure 4.3). During the preantral phase, the pregranulosa cells of the primordial follicle differentiate into a single-layered sphere of cuboidal granulosa cells enclosed by a basement membrane and surrounding a primary oocyte to form a primary follicle. This process is stimulated by BMP-4 and BMP-7 (Knight & Glister, 2006). Later, proliferation of the granulosa forms a multicellular layer, and the follicle now is termed a secondary follicle. Granulosa cells function as nurse cells for the growing oocyte in each follicle. Communication among cells in the avascular granulosa is aided by gap junctions between adjacent granulosa cells and between the innermost follicular cells and the oocyte. The granulosa becomes surrounded by another layer of somatic cells, the theca, recruited from the ovarian stroma. Theca cells further differentiate into inner and outer layers, forming the endocrine theca interna and the connective tissue-like theca externa. The theca interna layer develops a rich supply of capillaries that do not cross the basement membrane and hence do not enter the granulosa layer.

In addition to their role as nurse cells, granulosa cells convert androgens secreted by the theca cells into estrogens. Organization of preantral follicles and follicle cell proliferation appear to involve paracrine cross-communication

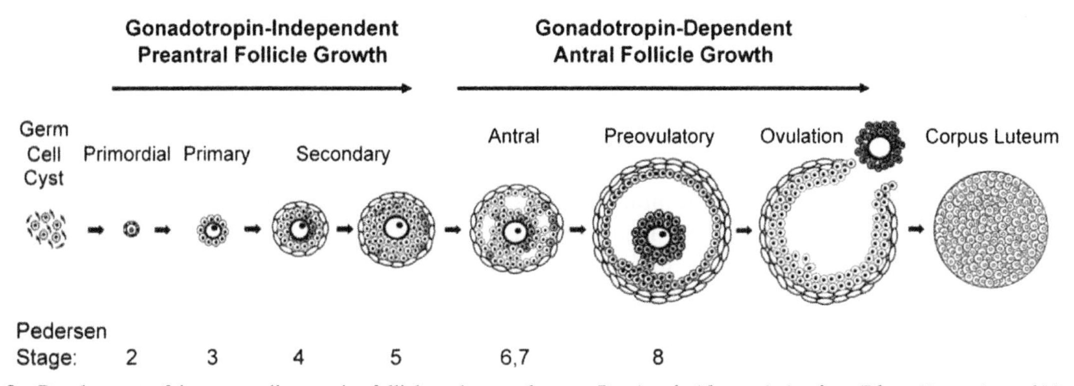

FIGURE 4.3 Development of the mammalian ovarian follicle and corpus luteum. *Reprinted with permission from Edson, Nagaraja, and Matzuk (2009).*

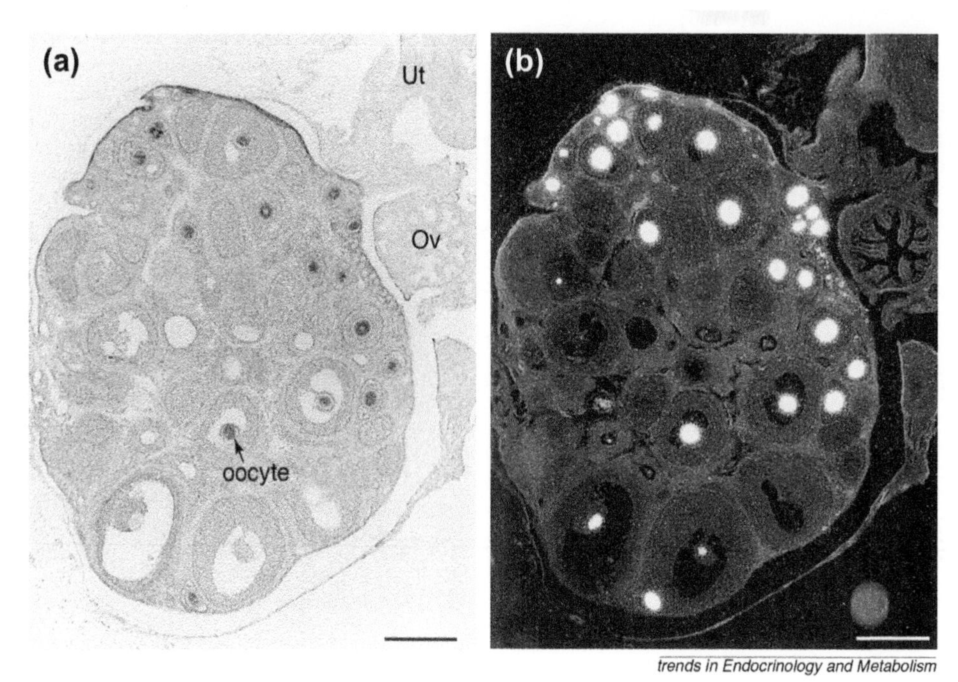

FIGURE 4.4 Oocyte expression of growth differentiation factor (GDF)-9. (a) Section of rat ovary showing follicles in various stages. (b) Same section showing distribution of GDF-9 in oocytes. *Reprinted with permission from Erickson and Shimasaki (2000).*

between the oocyte and follicular cells mediated by growth factors from both granulosa cells and the oocyte including fibroblast growth factors (FGFs), epidermal growth factors (EGFs), growth differentiation factor 9 (GDF-9), and neurotropins (e.g., nerve growth factor (NGF)). Informational molecules may also pass through the gap junctions connecting the oocyte and its surrounding granulosa cells.

The presence of FSH receptors on granulosa cells, and the ability to proliferate mitotically, are initiated by the oocyte through production of GDF-9 (see Figure 4.4). In *Gdf9*-knockout mice, the ovary develops normally but the follicles cannot respond to FSH (Erickson & Shimaski, 2000). Granulosa cell proliferation also is stimulated by insulin-like growth factor (IGF)-I and estrogens (Richards, Sharma, Falender, & Lo, 2002). Nuclear estrogen receptors (ERs), ERα and ERβ, are present in rats on granulosa cells and thecal cells of growing follicles, respectively (Richards et al., 1976), and ERα and/or ERβ knockout mice are infertile (see Edson et al., 2009), indicating the importance of local estrogen actions in the ovary.

Granulosa cells of the growing follicle secrete antral fluid, which accumulates in small pockets between the follicular cells. As the production of antral fluid increases, these pockets coalesce to form a single large fluid-filled cavity or antrum within the follicle of most mammals examined, and the follicle is then termed a 'tertiary' or 'antral' follicle. The antrum is surrounded by a relatively thin layer of mural granulosa cells. The oocyte is covered by a small mass of specialized granulosa cells, the cumulus oophorus, the cells of which are functionally distinct from those of the mural granulosa. The final meiotic maturation of the oocyte apparently is influenced by paracrine factors from cells of the cumulus oophorus prior to ovulation. The mature follicle is located just beneath the surface of the ovary, where it awaits the hormonal signal that will trigger ovulation. When the antral follicle has reached maximal size, it is termed a 'mature' follicle (also called a 'Graafian' follicle). Whereas the development of preantral follicles is largely regulated by intraovarian factors, antral follicles become dependent on GTHs for continued growth and survival.

2.2.1. Ovarian steroidogenesis

Under the influence of LH and FSH, growing ovarian follicles synthesize and release estrogens, predominantly E_2, into the general circulation. This synthesis is a cooperative effort between cells of the theca interna, which synthesize androgens (principally AND), and the granulosa, which convert AND to E_1 and then E_2. Synthesis of AND by theca cells of antral follicles is controlled by LH, whereas FSH increases $P450_{aro}$ levels in granulosa cells and conversion of AND to estrogens. Luteinizing hormone also stimulates production of steroidogenic acute regulatory (StAR) protein, which facilitates transport of cholesterol to the inner mitochondrial membrane, where it is converted to pregnenolone. Production of androgens by theca cells in preantral follicles is reduced by the absence of LH receptors as well as by production of activins and other regulators by granulosa cells. These paracrines inhibit AND synthesis (see Edson et al., 2009).

Ovarian estrogens stimulate proliferation and vascularization of the uterine endometrium to prepare it for accepting an implanting blastocyst, should pregnancy occur (see Chapter 5, this volume). Mammary gland development (thelarche) also is dependent on rising ovarian estrogens and often is the first sign of puberty in humans. Additionally, estrogens have important effects on bone metabolism and cardiovascular functions. Estrogens are also responsible for inducing the LH surge that causes ovulation (see Chapter 2, this volume).

In addition to its role in steroidogenesis and follicle growth, FSH stimulates the synthesis of inhibins, which feed back on the pituitary to selectively inhibit FSH release. Inhibins also act in an autocrine manner to inhibit $P450_{aro}$ activity in granulosa cells of the ovarian follicle. Conversely, the local production of activins increases $P450_{aro}$ activity. Inhibins and activins are members of the transforming growth factor-β (TGFβ) family. They also function as paracrines in the pituitary gland. The roles of activins and inhibins in reproduction have been reviewed by Bilezikijan et al., 2004; Muttukrishna, Tanetta, Groome, and Sargent, 2004; Bernard, Fortin, Wang, and Lamba, 2010.

During development, androgens and estrogens secreted by the gonads bring about permanent, irreversible organizational effects in the brain. These include transformation of the hypothalamic mechanisms that control GTH secretion from the female pattern of positive feedback resulting in the LH surge in adults to the steady male secretory pattern of GnRH secretion (for a detailed discussion see Chapter 1, this volume). Similarly, the development of male genitalia is an early organizational effect produced by androgens. Organizational effects cannot be reversed later by exposure to other gonadal steroids. In contrast, activational effects, such as specific reproductive behaviors, can be induced reversibly by exposure of adults to gonadal steroids. The type of behavior induced depends on the steroid applied and the organizational history of the individual but not on the genetic sex of the individual (see Figure 1, Chapter 1, this volume).

2.3. Puberty

The initial development of full hormonal and gametogenetic capability of ovaries and maturation of associated sex accessory structures is termed 'puberty,' whereas seasonal or annual development of similar reproductive events in postpubertal animals is termed 'recrudescence.' Both processes involve an increase in pulsatile secretion of GnRH by the hypothalamus, resulting in accelerated secretion of GTHs, which stimulate steroidogenesis in theca and granulosa cells as well as follicular growth. In wild animals, initiation of puberty or recrudescence is generally the result of photoperiodic cues operating through the pineal system, which allows for activation of GnRH secretion. Melatonin (MEL) secreted by the pineal gland has been linked to inhibition of reproduction at the hypothalamic level in numerous mammalian species as well as regulation of seasonal recrudescence in adults (see Revel, Masson-Pevet, Pevet, Mikkelsen, & Simmonneau, 2009). Further, nonsecreting pineal tumors that impair MEL secretion have been associated with central precocious puberty in humans (Macchi & Bruce, 2004; Srinivasan et al., 2009), further supporting the inhibitory role of the pineal. Additionally, MEL has been implicated as a protective antioxidant for mammalian gametes and during pregnancy (Reiter et al., 2009).

Evidence from numerous species supports a nutritional role in the activation of ovarian function in mammals (e.g., Urritia-Morales, et al., 2009). Estrogens have a positive effect on lipid storage (O'Sullivan, 2009). Puberty may be delayed in especially lean humans (see Donoso et al., 2010). Recent studies indicate that the peptide leptin, which is secreted by adipose cells, may influence GnRH release in the brain (see reviews by Casanueva & Dieguez, 1999; Mantzoros, 2000). Adipose cells secrete more leptin when lipid stores are greater, and leptin may provide a signal to the brain regarding the extent of fat storage and thus regulate hypothalamic secretion of GnRH. Puberty is delayed in ballet dancers who exhibit elevated adiponectin and reduced leptin (Donoso et al., 2010). Leptin also produces appetite suppression at the level of the hypothalamus. In contrast, ghrelin, which stimulates appetite, has been shown to inhibit GTH secretion in rodents (Fernandez-Fernandez et al., 2005).

3. THE OVARIAN CYCLE

The cyclical nature of ovarian function is the consequence of a genetically determined program that resides in the hypothalamus and in turn controls the secretion of GTHs from the pituitary. The length of the ovarian cycle is characteristic for each species, as is the number of tertiary follicles in a given cycle. The follicular phase is the portion of the cycle during which ovarian follicles undergo rapid growth, resulting in the development of one or more mature follicles, each containing an oocyte. The LH surge induces ovulation and rupturing of the follicle, resulting in release of the mature oocyte. The remains of each ruptured follicle, consisting mostly of granulosa cells, are transformed into a corpus luteum under the influence of LH. Ovulation marks the transition of the follicular phase to the luteal phase of the ovarian cycle, which lasts from a few days to weeks, depending upon the species. Some species, like humans, may begin another follicular phase during the later portion of the luteal phase, whereas other species enter a short inactive period (diestrus) before initiation of the next follicular phase. In monoestrous species, diestrus may last until the next breeding season, when a new follicular phase begins.

3.1. The Follicular Phase of the Ovarian Cycle

Following activation of the HPG axis at puberty, tertiary (antral) follicles respond to the increased levels of FSH and become dependent on the presence of FSH for continued growth and differentiation. At the beginning of the follicular phase of the ovarian cycle, several early antral follicles have developed to the stage at which they are capable of responding to rising levels of FSH. Gonadotropin stimulation initiates the cyclic recruitment of these antral follicles into final growth and maturation (McGee & Hsueh, 2000). However, only a limited, species-specific number of these follicles will reach preovulatory size. Thus, there is a process of follicular selection that results in a subset of available tertiary follicles achieving final maturation and ovulation in each cycle, while the remaining follicles die by atresia. The survival of these tertiary follicles appears to depend on sufficient access to FSH. Exposure of granulosa cells to FSH results in the induction of $P450_{aro}$ and the consequent increased synthesis of E_2. Feedback inhibition as a result of increased circulating E_2 diminishes pituitary secretion of GTHs, and the FSH-dependent cohort of antral follicles compete for the limited access to this hormone. The mechanisms involved in this follicular selection remain poorly understood, but it is likely that the more mature follicle(s) become dominant and escape atresia. In response to FSH, granulosa cells of maturing antral follicles acquire LH receptors. The acquisition of LH responsiveness may lessen the follicle's dependence on FSH (see Zeleznik, 2004, for a review of follicular selection). Local intrafollicular factors such as E_2 and growth factors such as IGF-I, EGF, bFGF, and members of the TGF-β superfamily, including inhibin, may have additional important roles in follicular selection (see Beg & Ginther, 2006).

Most of the follicles that initiate growth during a given ovarian cycle will exhibit apoptosis and undergo degeneration or atresia. There are different types of atresia observed in various mammals (see Rodgers & Irving-Rodgers, 2010). Some of the theca cells from these atretic follicles will remain steroidogenic and contribute to the interstitial gland of the ovary. The interstitial gland refers to the rather diffuse collection of cells of stromal and/or follicular origin that synthesize androgens (Duke, 1978).

Production of AND by the interstitial gland is stimulated by LH, and this production supplements the availability of AND from follicular theca cells for synthesis of estrogens by the granulosa cells. The interstitial gland also may be an important source of P_4 during pregnancy in some species (Duke, 1978).

3.2. Ovulation

Ovulation marks the end of the follicular phase and the beginning of the luteal phase in the ovary and is correlated with the appearance of estrous behavior. The progressive increase in circulating E_2 induces the LH surge (see Chapter 2, this volume for details) that causes ovulation of one or more follicles within a matter of hours. Surges in both LH and FSH occur in response to elevated GnRH prior to ovulation, but the magnitude of the LH surge greatly exceeds that of the FSH surge (see Chapter 2, this volume). Release of LH from the pituitary is enhanced selectively by the neuropeptide galanin, a peptide that is coreleased with GnRH just prior to the midcycle LH surge. Galanin has no effect on FSH release. Under the influence of FSH, the ovaries also secrete inhibin, which selectively reduces FSH release from the pituitary and contributes to the smaller FSH component in the midcycle GTH surge. In polyestrous species, the importance of the FSH surge may be related to initiation of follicle development for the next cycle. The number of follicles that reach maturity and ovulate is species-specific and relies on the levels of FSH receptors and the amount of FSH available.

The LH surge induces resumption of meiosis in mammalian primary oocytes. The oocytes are arrested in prophase of meiosis I for most of their existence—from days to decades, depending on the species. This protracted meiotic arrest is maintained by a high level of cyclic-3',5'-adenosine monophosphate (cAMP) within the oocyte. Growth of a follicle is accompanied by an accumulation within the oocyte of CDK1 and cyclin B, components of the heterodimer known as maturation-promoting factor (MPF), which induces the resumption of meiosis. High levels of cAMP prevent activation of the MPF complex, thus maintaining meiotic arrest. Nerve growth factor (NGF) also appears to be involved in ovulation (Dissen et al., 1966).

Oocytes within follicles, having reached the early antral stage (at which time sufficient levels of CDK1 and cyclin B have accumulated in the oocyte), are capable of resuming meiosis prematurely if removed from the follicle. Therefore, the follicular environment is important in maintaining meiotic arrest of meiotically competent antral oocytes. Luteinizing hormone acts on mural granulosa cells rather than directly on the oocyte, which, like the cumulus cells, lacks LH receptors. In response to LH, levels of cAMP within the oocyte fall, allowing the activation of MPF. The roles of oocyte and follicular cells in the resumption of meiosis are not completely understood (see Mehlmann, 2005; Edson et al., 2009 for reviews). The drop in cAMP is followed by germinal vesicle breakdown, the alignment of recombined homologous chromosomes at the metaphase plate, and the completion of the first meiotic division, resulting in the extrusion of the first polar body. Oocytes then enter meiosis II; in most mammalian species, meiosis

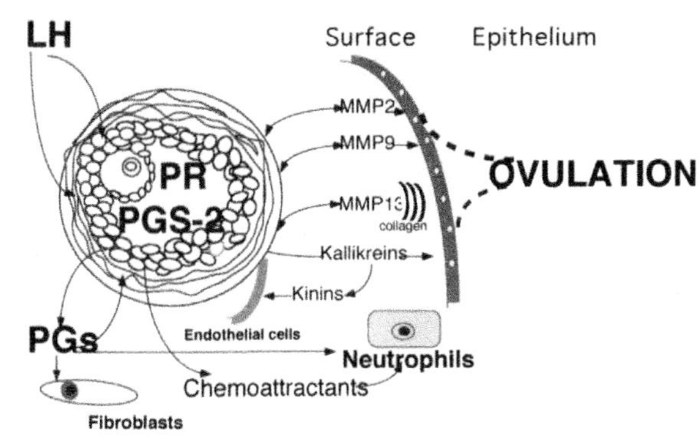

FIGURE 4.5 Mechanism of ovulation. Luteinizing hormone (LH) stimulates the production of progesterone receptors (PRs) and the enzyme prostaglandin synthase-2 (PGS-2), resulting in synthesis of prostaglandins (PGs) as well as collagenase enzymes (= matrix metalloproteinases (MMPs)) as well as the attraction of leukocytes (neutrophils). *Reprinted with permission from Richards, Russell, Robker, Dajee, and Alliston (1998).*

is arrested in metaphase II and will be completed only if fertilization occurs.

The physical mechanism by which LH causes its follicle to rupture and release the mature oocyte is not understood completely (see review by Edson et al., 2009). However, rupturing of the follicle occurs as a consequence of a series of events that begin at the avascular stigma located within the follicular wall (Figure 4.5). Luteinizing hormone activates the gene for production of locally available prostaglandin $F_{2\alpha}$ ($PGF_{2\alpha}$), which induces theca cells to produce collagenase, which in turn digests intracellular collagen, thus weakening the follicle wall. This hydrolysis of collagen is accompanied by an inflammatory response and release of another prostaglandin (PGE_2), which causes constriction of local blood vessels, resulting in local ischemia and cell death, weakening the follicle wall further. Hydrostatic pressure in the antral cavity is thought to cause the follicle wall to rupture at its weakest point, the stigma, and the ovum and surrounding cumulus oophorus (now called the corona radiata) are expelled. The actual cause of the increase in pressure is not understood. The synthesis of P_4 receptors (PRs) in the mural granulosa also results from LH exposure, and P_4 is essential for ovulation in mice, although its precise role is not defined (Lydon et al., 1995; Richards, Russell, Robker, Dajee, & Allison, 1998; Mulac-Jericevic & Conneely, 2004). The LH surge, in conjunction with factors supplied by the oocyte, results in expansion of the extracellular matrix within the cumulus oophorus. This LH-induced cumulus expansion is essential for ovulation, although cumulus cells, unlike mural granulosa cells, lack LH receptors. Luteinizing hormone appears to increase production of members of the EGF family, which in turn act upon the cumulus granulosa cells (Park et al., 2004). Numerous downstream genes have been implicated in the ovulatory process in mice (see Edson et al., 2009).

Some mammals are induced ovulators, including carnivores such as the domestic ferret (*Mustela putorius furo*), mink (*Mustela vison*), raccoon (*Procyon lotor*), and domestic cat (*Felis catus*); rodents (e.g, *Microtus californicus*); lagomorphs (e.g., *Sylvilagus* spp); artiodactyls (e.g., llama (*Lama glama*) and alpaca (*Vicugna pacos*)); at least one bat (lump-nosed bat (*Corynorhinus rafinesquei*)); and several insectivores (e.g., common shrew (*Sorex araneus*)) (see review by Bakker & Baum, 2000). Some other species are suspected to be induced ovulators, but most mammals are thought to be spontaneous ovulators. Copulation in induced ovulators causes a neural reflex that activates GnRH neurons and initiates an LH surge, leading to ovulation. In general, induced ovulators are relatively insensitive to positive feedback by estrogens, with the exception of the domestic cat (Bakker & Baum, 2000; Pelican et al., 2010). Preovulatory estrogens are responsible for inducing mating by induced ovulators and if mating occurs they will ovulate. Some spontaneously ovulating species, e.g., rats, mice, and cattle, also may ovulate following copulation. Although it has been suggested that humans may be induced to ovulate following copulation under special conditions (Clark & Zarrow, 1971), direct evidence is lacking.

3.3. The Luteal Phase of the Ovarian Cycle

In addition to causing ovulation, the LH surge prevents proliferation of mural granulosa cells of the ruptured follicle and causes them to differentiate into a corpus luteum (Fan et al., 2009). The follicular basement membrane loses its integrity, allowing invasion of blood vessels into the former granulosa layer followed by extensive angiogenesis. Theca interna cells, as well as ovarian stromal cells, may contribute to the corpus luteum (Stocco, Telleria, & Gibori, 2007). In some cases, anovulatory follicles may undergo luteinization and later function as accessory corpora lutea during pregnancy. In mice, premature luteinization of the late antral follicles is apparently prevented by the presence of inhibitory factors produced by the oocyte, possibly GFP9 (Pangas, Li, Robertson, & Matzuk, 2006).

Four types of corpora lutea have been described in mammals (Stocco, Telleria, & Gibori, 2007). The first is the corpus luteum, formed during an ovarian cycle that does not lead to pregnancy. The corpus luteum of pregnancy is found in all mammals and is essential for implantation and often the maintenance of pregnancy. The corpus luteum of pseudopregnancy has been described in rodents (see below), and the corpus luteum of lactation is found in species that ovulate right after giving birth.

Once formed, the corpus luteum begins secreting large quantities of P_4 along with lesser amounts of E_2 as well as other estrogens and progestogens. Circulating P_4 and E_2 are the major hormones involved in stimulating secretory activity of the uterine endometrium and maintaining its vascularity, respectively. If implantation occurs, P_4 is also important in reducing contractility of the uterine myometrium until the induction of contractions associated with parturition (see Chapter 7, this volume). Elevated P_4 and E_2 also inhibit hypothalamic GnRH secretion during the luteal phase, resulting in arrested follicular development. Ghrelin also is secreted by the corpus luteum in rats and humans, and ghrelin levels are positively correlated with peak activity of the corpus luteum (Garcia et al., 2007). In the case of the rat, this is true also when the corpus luteum is active during pregnancy. However, the precise role of ghrelin during the luteal phase is not understood.

In some species, corpora lutea are relatively short-lived and soon degenerate. In others, they may persist and require GTHs or other hormones to maintain their function. In sheep, e.g., PRL together with LH apparently stimulates steroid secretion by the corpus luteum, whereas in rodents only PRL is necessary to maintain the activity of the corpus luteum. In several species, preovulatory E_2 can produce a surge of PRL release that might be related to corpora luteal function. However, PRL has no role in corpus luteum function in primates and most other mammals that have been examined.

If pregnancy occurs, the corpus luteum plays an important role in gestation (see Chapter 6, this volume). If pregnancy is not initiated, the corpus luteum degenerates and the ovary enters a short or prolonged period of quiescence, depending on the species. Degeneration of the corpus luteum frees the hypothalamic GnRH centers from the inhibitory influence of estrogens and P_4, allowing resumption of follicular growth in polyestrous species. The length of the luteal phase and the life of the corpus luteum is programmed for each species. In nonprimates, uterine endometrial cells produce $PGF_{2\alpha}$, which is primarily responsible for corpus luteum degeneration, and this synthesis is dependent upon the actions of P_4 and estrogens as well as oxytocin on the uterine cells (for details see Figure 5.1, Chapter 5, this volume). A pattern of luteolysis in primates is beginning to emerge. As the corpus luteum secretes P_4, it also begins producing a variant of the LH receptor that does not bind LH but readily binds human chorionic gonadotropin (hCG), produced by the conceptus (Dickinson, Stewart, Myers, Millar, & Duncan, 2009). There is a reduction in cholesterol uptake (Bogan & Hennebold, 2010). Further, locally produced $PGF_{2\alpha}$ may play a role in primate luteolysis (Priyanka, Jayaram, Sridaran, & Medhamurthy, 2009) by reducing StAR protein, which is responsible for transporting cholesterol to the mitochondrial inner membrane, where it is converted to pregnenolone and ultimately to P_4 (see Devoto et al., 2009). These effects would result in a reduction of steroidogenesis. An accompanying increase in collagenase enzymatic activity could be responsible in part for the structural degeneration of the corpus luteum (see Devoto et al., 2009).

4. OVARIAN LIFESPAN AND REPRODUCTIVE AGING

Few animals live beyond their peak of reproductive activity in nature due to predation, disease, or other environmentally related phenomena. In contrast, life after reproductive age is a common occurrence in human females, in whom the follicle supply is largely depleted and the ovary becomes refractory to GTHs, a condition called menopause. Cycles of these women become irregular and eventually they cease to ovulate and menstruate; this is accompanied by a marked depression in circulating levels of estrogens as well as of adrenal androgens such as dehydroepiandrosterone (DHEA). Consequently, there is a marked rise in GTHs. Menopause is accompanied by vaginal atrophy, hot flashes or flushes, often changes in libido, and accelerated bone resorption, which may lead to calcium deficiency syndromes such as osteopenia and osteoporosis. Transition from the premenopausal (active reproduction) to postmenopausal (nonreproductive) state usually is gradual over several years. Many studies have shown that heart disease and other cardiovascular disorders increase exponentially in postmenopausal women, although the separate effects of chronological aging and the menopausal transition remain to be clarified, and deaths due to cardiac disease are several-fold greater than for uterine and breast cancer combined.

Estrogen replacement therapy (ERT), usually in combination with a progestogen, alleviates many of the symptoms in postmenopausal women including prevention of bone resorption when combined with calcium and vitamin D supplements. Estrogen therapies have been associated both with improvement (e.g., Peganini-Hill, 2001; Turgeon, McDonnell, Martin, & Wise, 2004) and reduction (Sherwin, 2002) in cognitive skills. However, ERT has become controversial because of links to increased incidence of breast cancer and lack of improvement in other areas, based largely on the results of the Women's Health Initiative (see Lemay, 2002; Rossouw et al., 2002; Turgeon, Carr, Maki, Mendelsohn, & Wise, 2006). However, one must consider the populations of women involved, when treatment is begun, the specific regimens used, and the choices of replacement steroids not evaluated fully in the Women's Health Initiative (see Turgeon et al., 2006; Studd, 2010).

5. OVARIAN EVENTS AND REPRODUCTIVE CYCLES OF SELECTED MAMMALS

The examples of ovarian cycles described here were selected to emphasize some of the diversity that exists among mammals. Additional information on ovarian events

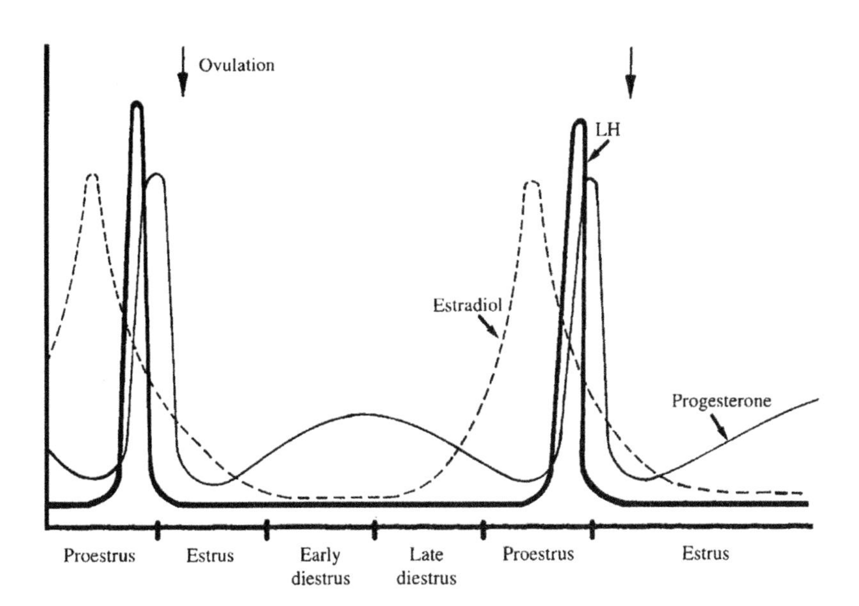

FIGURE 4.6 Hormone levels during the ovarian cycle of the four-day-cycling rat. LH, luteinizing hormone. *Reprinted with permission from Norris (2007).*

can be found in the accompanying chapters in this volume describing reproductive cycles in marsupials (Chapter 10), wild rodents (Chapter 11), bats (Chapter 12), and primates (Chapter 13).

5.1. The Four-day-cycling Rat

The female laboratory rat is polyestrous and an induced ovulator. The ovarian cycle lasts only four or five days if females do not become pregnant (Figure 4.6). The rat cycle is separable into proestrus (one day), estrus (one day), and diestrus (two or three days) and is cued closely to environmental events. Typically, several follicles develop and ovulate during each cycle, resulting in litters of 8–12 offspring and formation of multiple corpora lutea in the postovulatory ovary.

Plasma E_2 peaks on the morning of proestrus and stimulates an LH surge accompanied by a small surge in FSH. The GTH surge occurs on the afternoon of proestrus and ovulation occurs a few hours after midnight on the day of estrus. Ovulation is followed rapidly by secretion of P_4 from several short-lived corpora lutea. Several follicles usually mature simultaneously, and multiple ovulations commonly occur. Estrus lasts about 9 to 15 hours, during which time the female is highly receptive to the male. During proestrus, E_2 causes development of cornified cells in the superficial layers of the vagina, and their presence in vaginal smears characterizes estrus. As a result of the short-lived nature of the corpora lutea, the ovarian interstitial cells provide most of the P_4 and 20α-dihydroprogesterone (20α-DHP) secreted from the rat ovary.

In the four-day-cycling rat, the third and fourth days are termed diestrus I and diestrus II. A short transitional period, metestrus, is often described between estrus and diestrus, during which the female is no longer receptive to the male but some cornified cells still appear in smears prepared from the vaginal mucosa. Vaginal smears during diestrus lack cornified cells and exhibit a predominance of leukocytes in the smear, reflecting a reduction in E_2 levels.

Mating stimulates the GnRH surge followed by ovulation; consequently, the corpora lutea form and, together with the interstitial ovarian tissue, secrete significant amounts of P_4 and 20α-DHP. If fertilization and subsequent implantation do not occur, the mated female will not return immediately to proestrus but will exhibit pseudopregnancy, which may last for several days (see Chapter 5, this volume). Pseudopregnancy is often accompanied by PRL-induced effects on lactation and behavior. If fertilization was successful and implantation does occur, the corpora lutea and interstitial tissue continue to secrete progestogens.

5.2. The Ewe

Sheep estrous cycles occur seasonally, and the duration of one complete ovarian cycle is 16–17 days (Figure 4.7). A ewe may return to proestrus at least once if fertilization does not occur. During the follicular phase (= proestrus), there is a marked increase in estrogen and androgen levels. A surge of plasma LH occurs about 12 hours after the peak E_2 level, although AND may be responsible for inducing estrous behavior, acting as a substrate for E_2 synthesis in the brain. Ovulation of usually a single follicle occurs about 24 hours after the LH surge, and a corpus luteum soon forms under the influence of LH. Low levels of LH following ovulation and an E_2-induced surge of PRL stimulate the corpus luteum to secrete P_4. The uterine endometrium synthesizes a luteolytic prostaglandin ($PGF_{2\alpha}$) under the influence of P_4 that causes degeneration

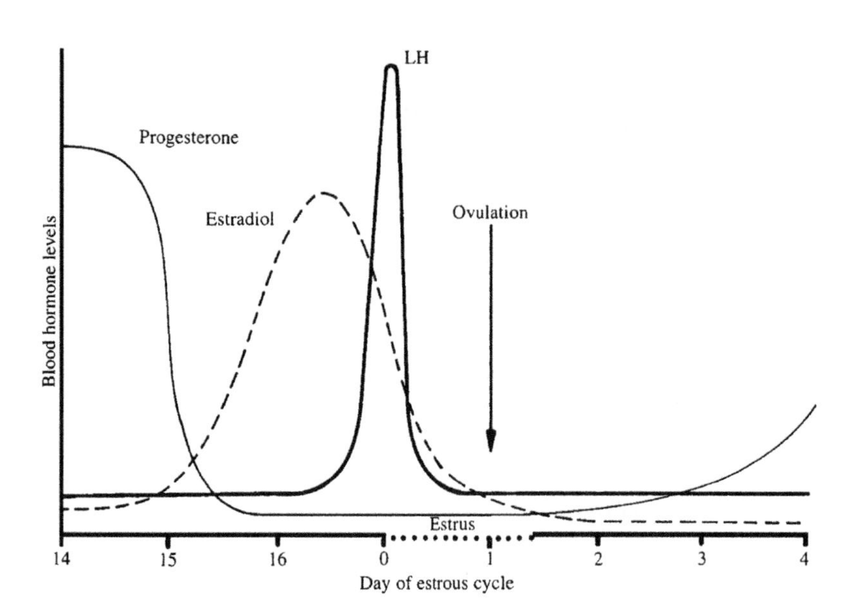

FIGURE 4.7 Hormonal levels during the ovarian cycle of the ewe. LH, luteinizing hormone. *Reprinted with permission from Norris (2007).*

of the corpus luteum and resumption of proestrus if fertilization does not occur.

5.3. Women

Humans are unusual among mammals in that they lack both a seasonal reproductive pattern and an estrous cycle. The ovarian cycle is synchronized with maturation of the uterine endometrium in preparation for successful implantation. In the absence of pregnancy, endometrial tissue degrades and is lost during an overt menstrual flow; thus, the human female reproductive cycle, including changes in the hypothalamus, pituitary, ovary, uterus, and oviducts, is known as the 'menstrual cycle.'

The length of the ovarian cycle, averaging about 28 days, is determined by the period of GTH-dependent antral follicular growth (follicular phase) as well as the lifespan of the corpus luteum (luteal phase) (Figure 4.8). Variations in cycle length exist among women and within women, depending on stage of the reproductive lifespan as well as body weight, physical activity, stress, and other physiological factors. These variations predominantly affect the length of the follicular phase, which is more variable than that of the luteal phase, as the lifespan of the corpus luteum in the absence of pregnancy is relatively consistent, at 14 days. In a 'typical' 28-day cycle, ovulation occurs at midcycle.

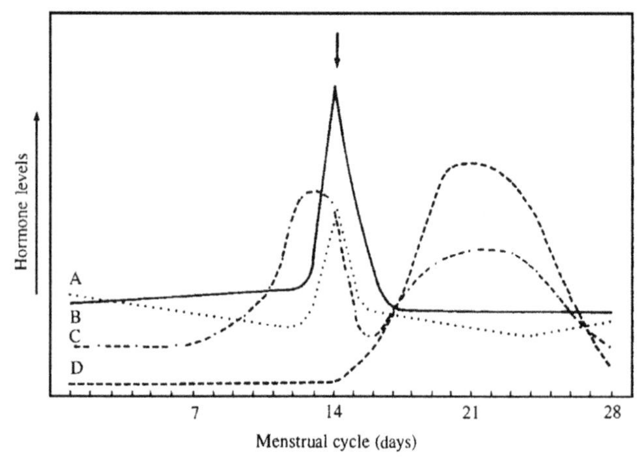

FIGURE 4.8 Hormonal levels during the ovulatory cycle of women. A, follicle-stimulating hormone (FSH); B, luteinizing hormone (LH); C, estradiol (E_2); D, progesterone (P_4). *Reprinted with permission from Bolander (1989).*

Follicular dynamics in the human ovary have been described and reviewed by Gougeon (1996). In the fetal ovary, primordial follicles begin to appear at midgestation, with each ovary containing approximately 300 000−500 000 follicles. Initial recruitment from this resting pool of follicles begins soon after their formation, but atresia of all growing follicles and many primordial follicles reduces the follicular pool by about half by the time of birth. Atresia of follicles that emerge from the primordial pool continues until puberty, when rising GTH levels rescue one follicle per cycle. After a follicle is recruited into preantral growth, its development to the early antral stage takes up to a year, spanning seven or more ovarian cycles (Figure 4.2). Preovulatory growth is much more rapid (two weeks). Like most other primates, humans are monovular. Although several follicles in each ovary begin FSH-dependent growth, only one normally attains dominance in preparation for ovulation and is 'rescued'; the others become atretic.

Follicular luteinization and increased progestogen production begin within an hour of the LH surge, followed by a final spurt of follicular growth to about 25 mm diameter caused by antral expansion, and ovulation occurs 24−36 hours postsurge. The human corpus luteum is the major source of P_4, which maintains endometrial receptivity to implantation during the luteal phase (for a review of endocrine and molecular mechanisms involved in luteal function in the human see Devoto et al., 2009). If pregnancy occurs, the death of the corpus luteum is prevented by hCG, produced by the syncytiotrophoblast of the early blastocyst (Baird, Weinberg, McConnaughey, & Wilcox, 2003). Levels of hCG diminish by the end of the first trimester, and the placenta subsequently becomes the primary source of progestogens essential for the maintenance of pregnancy. Ovarian cyclicity normally resumes within four months following delivery but can be further delayed in women who breastfeed (McNeilly, 2001), especially if feeding bouts are frequent (Konner & Worthman, 1980).

5.4. The Elephant

Asian (*Elephas maximus*) and African (*Loxodonta africana*) elephants are seasonal breeders with similar four-to-five-year birth intervals in females that are a consequence of a relatively long gestation period and extended lactation with repression of folliculogenesis (Hodges, 1998). Females are more likely to mate with older males that exhibit "musth" a period of increased activity and mating behaviors. In the Asian elephant, a forthcoming ovulation is advertised by the excretion of (Z)-7-dodecen-1-yl acetate in urine, which arouses interest by males. A male that is in musth secretes fluid from his temporal glands and dribbles strong urine about that attracts females whose ovaries are in the follicular phase.

The ovarian cycle is about 16 weeks in duration and ovulation occurs during a one-week period of estrus near the middle of the cycle. Unlike the characteristic mammalian pattern of GTH secretion, there are two successive LH surges approximately two to three weeks apart (Figure 4.9). The first LH surge possibly stimulates accessory corpora luteal development and the second LH surge stimulates ovulation. Typically, only one follicle ovulates as twins are rare and a single offspring is produced following a 22-month gestation. Elephants produce large amounts of 5α-DHP and 5α-pregnan-3-ol-20-one. 5α-dihydroprogesterone has the strongest affinity for the PR and may be the only physiologically important progestogen in elephants. Following birth of the offspring, the female elephant enters a prolonged period of lactation during which GTH release is suppressed and the ovary remains quiescent.

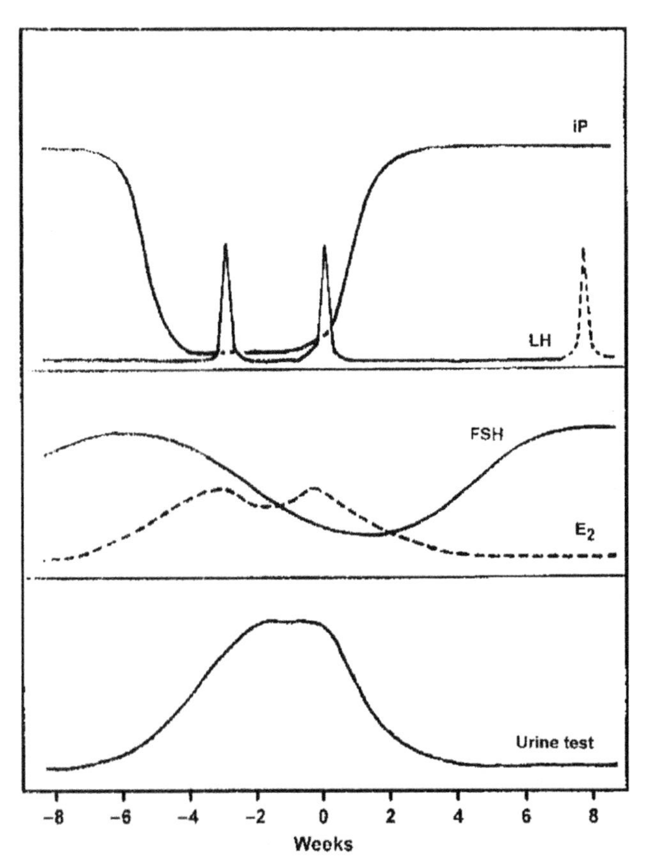

FIGURE 4.9 Hormone levels during the ovarian cycle of elephants. Only the second luteinizing hormone (LH) surge resulted in ovulation. E_2, estradiol; iP, inter-luteal period; LH, luteinizing hormone. *Reprinted with permission from Hodges (1998).*

6. FUTURE RESEARCH NEEDS

Despite many decades of research, several fundamental aspects of mammalian ovarian physiology remain incompletely understood. These include the molecular regulation of the mitotic/meiotic transition, meiotic arrest, and its resumption in female gametes; an understanding of the factors involved in the selection of preovulatory follicles; the regulation of litter size; and mechanisms controlling follicular and corpus luteal death. Although we have learned a great deal about the genes responsible for directing early ovarian formation, we need to learn much more about the specific roles of genes that regulate follicle development and atresia, follicular selection, arrest and reactivation of meiosis, and formation and lifespan of the corpus luteum. Studies of genetically engineered mice are providing valuable information identifying the genes involved in these processes, but the initiation of these gene cascades is less known. Much of our understanding of the mammalian ovary has emerged from work on rodents, domestic species, and humans, but we need more information from additional species to identify similarities and differences with the mouse models.

ABBREVIATIONS

20α-DHP	20α-dihydroprogesterone
5α-DHP	5α-dihydroprogesterone
AMH	Anti-Müllerian hormone
AND	Androstenedione

BMP	Bone morphogenetic protein
cAMP	Cyclic-3',5'-adenosine monophosphate
DHEA	Dehydroepiandrosterone
E	Embryonic day
E_1	Estrone
E_2	17β-estradiol
E_3	Estriol
EGF	Epidermal growth factor
Emx2	Empty spiracles homolog 2
ER	Estrogen receptor
ERT	Estrogen replacement therapy
FGF	Fibroblast growth factor
FIGLA	Female germline-specific transcription factor
FSH	Follicle-stimulating hormone
GDF-9	Growth differentiation factor 9
GnRH	Gonadotropin-releasing hormone
GTH	Gonadotropin
hCG	Human chorionic gonadotropin
HPG	Hypothalamic–pituitary–gonad
IGF	Insulin-like growth factor
LH	Luteinizing hormone
MEL	Melatonin
MPF	Maturation-promoting factor
NGF	Nerve growth factor
P_4	Progesterone
$P450_{aro}$	Aromatase
PGC	Primordial germ cell
PGE_2	Prostaglandin E_2
$PGF_{2\alpha}$	Prostaglandin $F_{2\alpha}$
PR	Progesterone receptor
PRL	Prolactin
PTEN	Phosphatase and tensin homolog deleted on chromosome 10
RA	Retinoic acid
SF-1	Steroidogenic factor 1 gene
StAR	Steroidogenic acute regulatory protein
TGFβ	Transforming growth factor-β
Wt1	Wilms' tumor 1 homolog

REFERENCES

Adhikari, D., & Liu, K. (2009). Molecular mechanisms underlying the activation of mammalian primordial follicles. *Endo. Revs., 30*, 438–464.

Anderson, R. A., Robinson, L. L. L., Brooks, J., & Spears, N. (2002). Neurotropins and their receptions are expressed in the human fetal ovary. *J. Clin. Endocrinol. Metab., 87*, 890–897.

Baird, D. D., Weinberg, C. R., McConnaughey, D. R., & Wilcox, A. J. (2003). Rescue of the corpus luteum in human pregnancy. *Biol. Reprod., 68*, 448–456.

Bakker, J., & Baum, M. J. (2000). Neuroendocrine regulation of GnRH release in induced ovulators. *Fron. Neuroendocrinol., 21*, 220–262.

Beg, M. A., & Ginther, J. (2006). Follicle selection in cattle and horses: role of intrafollicular factors. *Reproduction, 132*, 365–377.

Bernard, D. J., Fortin, J., Wang, Y., & Lamba, P. (2010). Mechanisms of FSH synthesis: what we know, what we don't know, and why do we care? *Fert. Steril., 93*, 2465–2485.

Bilezikjian, L. M., Blount, A. L., Leal, A. M., Donaldson, C. J., Fischer, W. H., & Vale, W. W. (2004). Autocrine/paracrine regulation of pituitary function by activin, inhibin and follistatin. *Mol. Cell. Endocrinol., 225*, 29–36.

Bogan, R. L., & Hennebold, J. D. (2010). The reverse cholesterol transport system as a potential mediator of luteolysis in the primate corpus luteum. *Reproduction, 139*, 163–176.

Bowles, J., & Koopman, P. (2007). Retinoic acid, meiosis and germ cell fate in mammals. *Development, 134*, 3401–3411.

Bowles, J., Knight, D., Smith, C., Wilhelm, D., Richman, J., Mamiya, S., et al. (2006). Retinoid signaling determines germ cell fate in mice. *Science, 312*, 596–600.

Broekmans, F. J., Visser, J. A., Laven, J. S. E., Broer, S. L., Thermmen, A. P. N., & Fauser, B. C. (2008). Anti-Müllerian hormone and ovarian dysfunction. *Trends Endocrinol. Metab., 19*, 340–347.

Cansanueva, F. F., & Dieguez, C. (1999). Neuroendocrine regulation and actions of leptin. *Fron. Neuroendocrinol., 20*, 317–363.

Clark, J. H., & Zarrow, M. X. (1971). Influence of copulation on time of ovulation in women. *Amer. J. Obstet. Gynecol., 109*, 1083–1085.

Defalco, T., & Capel, B. (2009). Gonadal morphogenesis in vertebrates: divergent means to a convergent end. *Annu. Rev. Cell Devel. Biol., 25*, 457–482.

Devoto, L., Fuentes, A., Koken, P., Cespedes, P., Palomino, A., Pommer, R., et al. (2009). The human corpus luteum: life cycle and function in natural cycles. *Fertil. Steril., 92*, 1067–1079.

Dickinson, R. E., Stewart, A. J., Myers, M., Millar, R. P., & Duncan, W. C. (2009). Differential expression of luteinizing hormone receptor splice variants in human luteal cells: implications for luteolysis. *Endocrinology, 150*, 2873–2881.

Dissen, G. A., Hill, D. F., Costa, M. E., Les Dees, C. W., Lara, H. E., & Ojeda, S. R. (1996). A role for trkA nerve growth factor receptors in mammalian ovulation. *Endocrinology, 137*, 198–209.

Donoso, M. A., Muñoz-Calvo, M. T., Barrios, V., Garrido, G., Hawkins, F., & Argente, J. (2010). Increased circulating adiponectin and decreased leptin/soluble leptin receptor ratio throughout puberty in female ballet dancers: association with body composition and the delay in puberty. *Eur. J. Endocrinol., 162*, 905–911.

Duke, K. L. (1978). Nonfollicular ovarian components. In R. E. Jones (Ed.), *The Vertebrate Ovary* (pp. 563–582). New York, NY: Plenum Press.

Edson, M. A., Nagaraja, A. K., & Matzuk, M. M. (2009). The mammalian ovary from genesis to revelation. *Endo. Revs., 30*, 624–712.

Erickson, G. F., & Shimasaki, S. (2000). The role of the oocyte in folliculogenesis. *Trends Endocrinol. Metab., 11*, 193–198.

Fan, H. Y., Liu, Z., Shimada, M., Sterneck, E., Johnson, P. F., Hedrick, S. M., et al. (2009). MPAK3/1 (ERK1/2) in ovarian granulosa cells are essential for female fertility. *Science, 324*, 938–941.

Fernandez-Fernandez, R., Tena-Sempere, M., Navarro, V. M., Barreiro, M. L., Casrtellano, J. M., Aguilar, E., & Pinella, L. (2005). Effects of ghrelin upon gonadotropin-releasing hormone and gonadotropin secretion in adult female rats: *in-vivo* and *in-vitro* studies. *Neuroendocrinology, 82*, 245–255.

Garcia, M. C., Lopez, M., Alvarez, C. V., Casanueva, F., Tena-Sempere, M., & Dieguez, C. (2007). Role of ghrelin in reproduction. *Reproduction, 133*, 531–540.

Gougeon, A. (1996). Regulation of ovarian follicular development in primates: facts and hypotheses. *Endocr. Rev., 17*, 121–155.

Hodges, J. K. (1998). Endocrinology of the ovarian cycle and pregnancy in the Asian (*Elephas maximus*) and African (*Loxodonta africana*) elephant. *Anim. Reprod. Sci., 53*, 3–18.

Johnson, J., Canning, J., Kaneko, T., Pru, J. K., & Tilly, J. L. (2004). Germline stem cells and follicular renewal in the postnatal mammalian ovary. *Nature, 428*, 145–150.

Jones, R. E. (1978). *The Vertebrate Ovary.* New York, NY: Plenum Press.

Kafri, T., Ariel, M., Brandeis, M., Shemer, R., Urven, L., McCarrey, J., et al. (1992). Developmental pattern of gene-specific DNA methylation in the mouse embryo and germ line. *Genes Dev., 6*, 705–714.

Knight, P. G., & Glister, C. (2006). TGF-β superfamily members and ovarian follicle development. *Reproduction, 132*, 191–206.

Konner, M., & Worthman, C. (1980). Nursing frequency, gonadal function, and birth spacing among !Kung hunter-gatherers. *Science, 277*, 788–791.

Lawson, K. A., Dunn, N. R., Roelen, B. A., Zeinstra, L. M., Davis, A. M., et al. (1999). Bmp4 is required for the generation of primordial germ cells in the mouse embryo. *Genes Dev., 13*, 424–436.

Lemay, A. (2002). The relevance of the Women's Health Initiative results on combined hormone replacement therapy in clinical practice,. *J. Obstet. Gynaecol. Can., 24*, 711–715.

Leung, C. K., & Adashi, E. Y. (2004). *The Ovary* (2nd ed.). San Diego, CA: Academic Press.

Lydon, J. P., DeMayo, F. J., Funk, C. R., Mani, S. K., Hughes, A. R., Montgomery, C. A., Jr., et al. (1995). Mice lacking progesterone receptor exhibit pleiotropic reproductive abnormalities. *Genes Dev., 9*, 2266–2278.

Macchi, M. M., & Bruce, J. N. (2004). Human pineal physiology and functional significance of melatonin. *Fron. Neuroendocrinol., 25*, 177–195.

Mantzoros, C. S. (2000). Role of leptin in reproduction. *Ann. NY Acad. Sci., 900*, 174–183.

McGee and, E. A., & Hsueh, A. J. W. (2000). Initial and cyclic recruitment of ovarian follicles. *Endocr, Rev., 21*, 200–214.

McNeilly, A. S. (2001). Lactational control of reproduction. *Reprod. Fertil. Dev., 13*, 508–590.

Mehlmann, L. M. (2005). Stops and starts in mammalian oocytes: recent advances in understanding the regulation of meiotic arrest and oocyte maturation. *Reproduction, 130*, 791–798.

Mossman, H. W., & Duke, K. L. (1973). *Comparative Morphology of the Mammalian Ovary.* Madison WI: University of Wisconsin Press.

Mulac-Jericevic, B., & Conneely, O. M. (2004). Reproductive tissue selective actions of progesterone receptors. *Reproduction, 128*, 139–146.

Muttukrishna, S., Tanetta, D., Groome, N., & Sargent, I. (2004). Activin and follistatin in female reproduction. *Mol. Cell. Endocrinol., 225*, 45–56.

O'Sullivan, A. J. (2009). Does estrogen allow women to store more fat more efficiently? A biological advantage for fertility and gestation. *Obes. Rev., 10*, 168–177.

Paganini-Hill. (2001). Hormone replacement therapy and stroke: risk protection or no effect? *Maturitas, 38*, 243–261.

Pangas, S. A., Li, X., Robertson, E. J., & Matzuk, M. M. (2006). Premature luteinization and cumulus cell defects in ovarian-specific smad4 knockout mice. *Mol. Endocrinol, 20*, 1406–1422.

Park, J.-Y., Su, Y.-Q., Ariga, M., Law, E., Jin, S.-L. C., & Conti, M. (2004). EGF-like growth factors as mediators of LH action in the ovulatory follicle. *Science, 303*, 682–684.

Pelican, K. M., Spindler, R. E., Pukazhenthi, B. S., Wildt, D. E., Ottinger, M. A., & Howard, J.-G. (2010). Progestin exposure before gonadotropin stimulation improves embryo development after *in-vitro* fertilization in the domestic cat. *Biol. Reprod.* DOI: 10.1095/biolrepod.109.083360.

Priyankas, S., Jayaram, P., Sridaran, R., & Medhamurthy, R. (2009). Genome-wide gene analysis reveals a dynamic interplay between luteotropic and luteolytic factors in the regulation of corpus luteum function in the bonnet monkey (*Macaca radiata*). *Endocrinology, 150*, 1473–1484.

Reddy, P., Zheng, W., & Liu, K. (2010). Mechanisms maintaining the dormancy and survival of mammalian primordial follicles. *Trends Endocrinol. Metab., 21*, 96–103.

Reiter, R. J., Tan, D. X., Manchester, L. C., Paredes, S. D., Mayo, J. C., & Sainz, R. M. (2009). Melatonin and reproduction revisted. *Biol. Reprod., 81*, 445–456.

Revel, F. G., Masson-Pevet, M., Pevet, P., Mikkelsen, J. D., & Simonneaux, V. (2009). Melatonin controls seasonal breeding by a network of hypothalamic targets. *Neuroendocrinology, 90*, 1–14.

Richards, J. S., & Pangas, S. A. (2010). The ovary: basic biology and clinical implications. *J. Clin. Invest., 120*, 963–972.

Richards, J. S., Ireland, J. J., Rao, M. C., Bernath, G. A., Midgley, A. R., Jr., & Reichart, L. E., Jr. (1976). Ovarian follicular development in the rat: hormone receptor regulation by estradiol, follicle-stimulating hormone and luteinizing hormone. *Endocrinology, 99*, 1562–1570.

Richards, J. S., Russell, D. L., Robker, R. L., Dajee, M., & Alliston, T. N. (1998). Molecular mechanisms of ovulation and luteinization. *Mol. Cell. Endocrinol., 145*, 47–54.

Richards, J. S., Sharma, S. C., Falender, A. E., & Lo, Y. K. (2002). Expression of FKHR, FKHRL1, and AFX genes in the rodent ovary: evidence for regulation by IGF-I, estrogen, and the gonadotropins. *Mol. Endocrinol., 16*, 580–599.

Rodgers, R. J., & Irving-Rodgers, H. F. (2010). Morphological classification of bovine ovarian follicles. *Reproduction, 139*, 309–318.

Rossouw, J. E., Anderson, G. L., Prentice, R. L., La Croix, A. Z., Kooperberg, C., Stefanick, M. L., et al. (2002). Risks and benefits of estrogen plus progestin in health postmenopausal women: principal results from the Women's Health Initiative randomized controlled study. *JAMA, 288*, 321–333.

Sherwin, B. B. (2002). Estrogen and cognitive aging in women. *Trends Pharmacol. Sci., 23*, 527–534.

Soyal, S., Amleh, A., & Dean, J. (2000). FIGalpha, a germ cell-specific transcription factor required for ovarian follicle formation. *Development, 2000*, 4645–4654.

Srinivasan, V., Spence, W. D., Pandi-Perumal, S. R., Zakharia, R., Bhatnagar, K. P., & Brzezinski, A. (2009). Melatonin and human reproduction: shedding light on the darkness. *Gynecol. Endocrinol., 25*, 779–785.

Stocco, C., Telleria, C., & Gibori, G. (2007). The molecular control of corpus luteum formation, function, and regression. *Endo. Revs., 28*, 117–149.

Studd, J. (2010). Ten reasons to be happy about hormone replacement therapy: a guide for patients. *Menopause Int., 16*, 44–46.

Turgeon, J. L., Carr, M. C., Maki, P. M., Mendelsohn, M. E., & Wise, P. M. (2006). Complex actions of sex steroids in adipose tissue, the cardiovascular system, and brain: insights from basic science and clinical studies. *Endo. Rev., 27*, 575–605.

Turgeon, J. L., McDonnell, D. P., Martin, K. A., & Wise, P. M. (2004). Hormone therapy: physiological complexity belies therapeutic simplicity. *Science, 304*, 1269–1273.

Urrutia-Morales, J., Meza-Herrrero, C. A., Escobar-Medinda, F. J., Gamez-Vazquez, H. G., Ramirez-Andrade, B. M., Diaz-Gomez, M. O., et al. (2009). Relative roles of photoperiodic and nutritional cues in modulating ovarian activity in goats. *Reprod. Biol., 9*, 283–294.

Visser, J. A., De Jong, F. H., Laven, J. S. E., & Themmen, A. P. N. (2006). Anti-Müllerian hormone: a new marker for ovarian function. *Reproduction, 131*, 1–9.

Zeleznik, A. J. (2004). The physiology of follicle selection. *Reprod. Biol. Endocrinol., 2*, 31.

Chapter 5

Hormones and Pregnancy in Eutherian Mammals

Fuller W. Bazer and Thomas E. Spencer
Texas A&M University, Texas, TX, USA

SUMMARY

The establishment and maintenance of pregnancy requires that eutherian mammals maintain a functional corpus luteum (CL) for production of progesterone (P_4), the hormone of pregnancy. This is accomplished by species-specific mechanisms as pregnancy recognitions signals, including chorionic gonadotropin in primates, estrogens in pigs, lactogenic hormones in rodents, and interferon tau in ruminants. Uterine receptivity to implantation is P_4-dependent; however, a prerequisite for implantation in all species studied is loss of expression of receptors for P_4 and estrogens by uterine epithelia. Thus, P_4 acts via P_4 receptor (PR)-positive uterine stromal cells that express one or more progestamedins—e.g., fibroblast growth factors-7 and -10 and/or hepatocyte growth factor—that act via their respective receptors on uterine epithelia and trophectoderm to regulate expression of genes in concert with other hormones such as estrogens, chorionic gonadotropin, interferons, and lactogenic hormones. Progesterone is permissive to actions of other hormones that affect implantation and placentation, which are required for the successful growth and development of the conceptus (the embryo/fetus and its associated placental membranes) and birth of viable offspring.

1. INTRODUCTION

The endocrinology of recurring estrous/menstrual cycles and pregnancy in primates, ruminants, swine, horses, cats, dogs, and rodents is discussed in this chapter with emphasis on hormonal signaling for maternal recognition of pregnancy and the endocrinology of pregnancy. The estrous cycle of subprimate species is uterine-dependent, because the uterus is the source of prostaglandin $F_{2\alpha}$ ($PGF_{2\alpha}$), the luteolytic hormone responsible for functional and structural regression of the ovarian corpus luteum (CL) in the absence of an appropriate maternal recognition of pregnancy signal. In primates, however, the menstrual cycle is uterine-independent, as luteolytic $PGF_{2\alpha}$ originates from an intra-ovarian source. During the peri-implantation period of pregnancy, maternal recognition of pregnancy signals from the conceptus (embryo and associated developing placental membranes) to the maternal system are either antiluteolytic—i.e., they prevent uterine release of luteolytic $PGF_{2\alpha}$—or they are luteotropic—i.e., they act directly on the CL to prevent luteolysis. The functional lifespan of the CL is controlled by timing of release of $PGF_{2\alpha}$ from the uterus and/or ovary, while maternal recognition of pregnancy signals from the conceptus may act in either a paracrine or endocrine manner to prevent uterine and/or ovarian release of luteolytic $PGF_{2\alpha}$. Following successful signaling for maternal recognition and establishment of pregnancy, progesterone (P_4) from the CL, oxytocin (OXT) from the pituitary, and a number of hormones from the placenta (e.g., P_4, chorionic gonadotropin, placental lactogens, etc.) are required for the maintenance of pregnancy and birth of viable offspring.

1.1. Corpus Luteum (CL) Formation and Regression

Mammals are classified as spontaneous ovulators (e.g., ruminants, pigs, horses, and primates), induced ovulators (e.g., rabbits and cats), or spontaneous ovulators with induced CL (mice and rats). With the exception of primates, the estrous cycle is dependent on the uterine production of luteolytic $PGF_{2\alpha}$, which is responsible for functional (cessation of P_4 production) and structural regression of the CL (Stouffer, 1988). In ruminants, e.g., luteolytic pulses of $PGF_{2\alpha}$ are released by the uterine luminal epithelium (LE) and superficial glandular epithelia (sGE), which express OXT receptors (OXTR) and prostaglandin synthase-2 (PTGS2), the rate-limiting enzyme used in the synthesis of prostaglandins. As illustrated in Figure 5.1, the uterine luteolytic mechanism requires sequential effects of P_4, estradiol (E_2), and OXT acting through their respective receptors on the uterine LE/sGE to stimulate the accumulation of phospholipids, the expression of OXTR, and the release of OXT-induced luteolytic pulses of $PGF_{2\alpha}$ (Spencer

Hormones and Reproduction of Vertebrates, Volume 5—Mammals

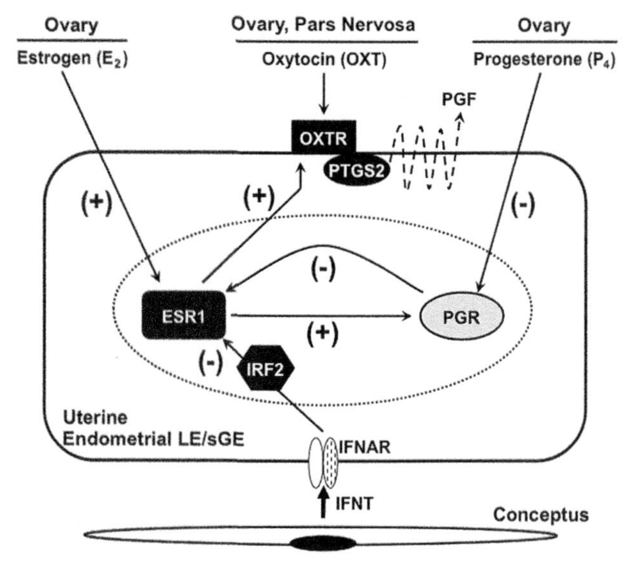

FIGURE 5.1 Schematic diagram illustrating the hormonal regulation of the endometrial luteolytic mechanism and the antiluteolytic effects of the conceptus on the ovine uterine endometrium, which appears to be applicable to other ruminant species such as the cow and the nanny goat (Spencer, Johnson, Bazer, Burghardt, & Palmarini, 2007; Bazer, Burghardt, Johnson, Spencer, & Wu, 2008). During estrus and metestrus, the expression of oxytocin receptors (OXTR) by the uterine luminal and superficial ductal glandular epithelia (LE/sGE) increase in response to estrogens from the ovarian follicles, which first stimulate the expression of estrogen receptor alpha (ERα) and estrogens act via ERα to increase OXTR. Progesterone receptors (PRs) are expressed by the LE/sGE during metestrus and diestrus, but low systemic levels of progesterone are insufficient to act via PGR to suppress ERα and OXTR gene expression. During early diestrus, endometrial ERα and estrogens are low, but progesterone levels begin to increase with the formation of the corpus luteum (CL). Progesterone (P_4) acts through the PR to suppress ERα and OXTR synthesis for 8 to 10 days. Continuous exposure of the endometrium to progesterone eventually downregulates PR gene expression in the endometrial LE/sGE by days 11 to 12 of the estrous cycle. The loss of PR terminates the progesterone block to ERα and OXYR formation. Thus, ERα appears between days 11 and 12 postestrus and is closely followed by increases in OXTR on days 13 and 14. The increase in OXTR expression is facilitated by increasing the secretion of estrogens by ovarian follicles. In both cyclic and pregnant ewes, OXY is released from the posterior pituitary and ovarian CL, beginning on day 9. In cyclic ewes, OXT binds to OXTR on the LE/sGE and increases the release of luteolytic pulses of prostaglandin $F_{2\alpha}$ (PGF) to regress the CL through a PTGS2-dependent pathway. In pregnant ewes, interferon tau (IFNT) is synthesized and secreted by the elongating conceptus, beginning on day 10 of pregnancy. Interferon tau binds to type I IFN receptors (IFNAR) on the endometrial LE/sGE and inhibits transcription of the *ESR1* gene through a signaling pathway involving interferon regulatory factor-2 (IRF2). These antiluteolytic actions of IFNT on the *ERα* gene prevent OXTR formation, thereby maintaining the CL and the P_4 production required for the establishment and maintenance of pregnancy.

& Bazer, 2002). Sheep are seasonally polyestrous and exhibit recurring estrous cycles of 16 to 17 days in length during the period from August to February each year. The estrous cycle is defined as the interval from the day of onset of sexual receptivity to mating (estrus) to the onset of the next period of sexual receptivity. At the onset of estrus (day 0), estrogens from mature Graafian follicles increase their expression of uterine estrogen receptor alpha (ERα (also abbreviated to ESR1)), P_4 receptor (PR), and OXTR, respectively. During diestrus, P_4 from the CL (1) stimulates the accumulation of phospholipids in the LE/sGE, which liberates arachidonic acid for synthesis and secretion of $PGF_{2\alpha}$; (2) acts via PR to 'block' expression of ERα and OXTR in uterine LE/sGE for 10 to 12 days; and (3) downregulates PR expression between days 11 and 13 to release the 'P_4 block' to expression of ERα and then OXTR by the LE/sGE necessary for OXT-induced release of luteolytic pulses of $PGF_{2\alpha}$ from the uterine endometrium and consequent CL regression. The promoter region of the ovine *OXTR* gene contains transcription factor Sp1 (Sp1) response elements that mediate responsiveness to occupied ERα, which increases transcription of *OXYR* in the LE/sGE. Importantly, continuous exposure of the uterus to P_4 for 10 to 12 days downregulates PR in the endometrial LE/sGE to allow a rapid increase in ERα in those cells beginning on day 13 and OXTR to increase from days 14 to 17 in cyclic ewes. The mechanism responsible for P_4-induced downregulation of PR is not known. In ewes, pulsatile release of OXY from the posterior pituitary gland and CL induces release of luteolytic pulses of $PGF_{2\alpha}$ from the uterine LE/sGE on days 15 and 16, to induce luteolysis.

1.2. Pregnancy Recognition Signals for Maintenance of the Corpus Luteum (CL)

Pregnancy is established and maintained in subprimate mammals in response to signaling from the conceptus and/or anterior pituitary to the uterus and/or CL to exert antiluteolytic effects that prevent uterine release of luteolytic $PGF_{2\alpha}$ or act as a luteotropic signal to directly stimulate development and function of the CL. Although the luteolytic mechanism in primates remains unclear, chorionic gonadotropin (CG) is the luteotropin required to extend the lifespan of the CL and is produced by peri-implantation of primate blastocysts to signal maternal recognition of pregnancy.

1.3. Luteinizing Hormone (LH), Chorionic Gonadotropin (CG), and their Receptor (LHCGR)

The information in this section is taken from recent reviews (see Rao, 2001; Ziecik, Kaczmarek, Blitek, Kowalczyk, Li, & Rahmanc, 2007). Luteinizing hormone (LH), a dimeric glycoprotein, shares a common alpha subunit with follicle-stimulating hormone (FSH), thyroid-stimulating hormone (TSH), and CG. However, the beta subunits of LH (LHβ) and CG (CGβ) are unique and confer biological activities and immunological identity, but both act via LH or a common receptor (LHCGR). Binding of LH or CG to LHCGR activates a G-protein bound to the cytoplasmic domain of the LHCGR, which detaches from LHCGR to activate the cyclic-3',5'-adenosine monophosphate (cAMP) system and cAMP-dependent protein kinases (PKs). The PKs are tetramers with two regulatory and two catalytic subunits. Upon binding cAMP by the regulatory units, the catalytic units are released to initiate phosphorylation of proteins that then activate genes that effect biological responses. Luteinizing hormone acts on mature Graafian follicles to induce ovulation and luteinization of theca and granulosa cells, which form the CL and produce P_4. Basal output of LH from the anterior pituitary is permissive for maintenance of CL function in most species. However, CG from the trophectoderm/chorion is the maternal recognition of pregnancy signal in primates. In addition, LH and CG bind to LHCGR expressed in tissues of the female (myometrium, endometrium, oviduct, broad ligament, uterine blood vessels, cervix) and male (epididymis, prostate) reproductive tracts, as well as the mammary gland, placenta, umbilical cord, brain, adrenal glands, skin, and lymphocytes. Inactivation of *Lhcgr* in by gene targeting revealed effects during both fetal development and in adult mice, as follows: (1) fetal mice were viable and experienced normal sexual differentiation of gonads and the reproductive tract; (2) there was a lack of redundancy for LH signaling for ovarian follicular growth beyond the antral stage and ovum maturation; (3) it was found that initiation of pregnancy requires LH signaling; and (4) it was found that LH signaling is required to maintain expression of ERα and steroidogenic acute regulatory (StAR) protein in the ovaries and testes of fetal and adult mice.

In pregnant pigs, P_4 is known to inhibit contractions of the uterine myometrium, but other functions of LH are to act directly to maintain a quiescent myometrium and to act directly or indirectly to affect the production of prostaglandins (PGs), particularly PGE_2, which stimulates uterine blood flow (Ziecik et al., 2007).

1.4. Progesterone (P_4) and Progesterone Receptors (PRs)

A maternal recognition of pregnancy signal is required in most mammals to protect the CL from luteolytic hormones and ensure that it continues to produce P_4, the hormone of pregnancy, required for a uterine environment supportive of implantation, placentation, and development of viable offspring, as well as mammogenesis and lactogenesis for continued nutritional support of the newborn during the neonatal and postnatal periods (see Bazer, Burghardt, Johnson, Spencer, & Wu, 2008; Bazer, Spencer, & Johnson, 2009). Uterine receptivity to implantation of blastocysts is dependent on P_4, which is permissive to the actions of type I interferons (IFNA, IFNB, IFNT, IFND, IFNW) and the single known type II interferon (IFNG), CG, lactogenic hormones such as prolactin (PRL), and chorionic somatomammotropin hormone (CSH1; also known as placental lactogen). An interesting paradox is that downregulation of PR and ERα by uterine epithelia appears to be a prerequisite for uterine receptivity to implantation, the expression of genes for secretory proteins, and the selective transport of molecules that support conceptus development into the uterine lumen. Downregulation of PR is associated with a loss of expression of proteins on the uterine LE/sGE, such as cell surface-associated mucin-1 (MUC1), which would otherwise interfere with implantation. Further, P_4 acts via PR-positive uterine stromal cells to induce expression of progestamedins; i.e., fibroblast growth factor-7 (FGF7) and -10 (FGF10) and perhaps hepatocyte growth factor (HGF; also known as met proto-oncogene), which exert paracrine effects on uterine epithelia and conceptus trophectoderm that express receptors for FGF7 and FGF10 (fibroblast growth factor receptor-2 splice variant (FGFR2IIIb)) and HGF (met proto-oncogene (MET)). In sheep and perhaps other ruminants, a select set of genes expressed by the uterine LE/sGE are P_4-induced and interferon tau (IFNT)-stimulated; however, a fundamental unanswered question concerns whether the actions of progestamedins and IFNs on uterine epithelia involve nonclassical cell signaling pathways that are independent of PR and signal transducers and activators of transcription (STAT1), such as mitogen-activated protein kinases (MAPKs) and phosphoinositide-3 kinase (PI3K) (Plantanias, 2005). Indeed, IFNT activates cell signaling pathways to differentially affect gene expression in the uterine LE/sGE compared to glandular epithelium (GE) and stromal cells in sheep. Expression of interferon regulatory factor-2 (IRF2), a potent transcriptional repressor, is limited to the uterine LE/sGE and inhibits expression of many IFN-stimulated genes (ISGs) and ERα, but IRF2 is not expressed in GE or stromal cells, which allows for induction of ISGs by IFNT.

The caveat, relative to the actions of P_4, is that P_4 can mediate effects independently of nuclear PRs, which are absent in the uterine epithelia throughout pregnancy. Based on work in primates and sheep, the effects of P_4 on the epithelia are mediated by progestamedins such as FGF10, FGF7, and HGF. In addition, membrane receptors for P_4 (mPRs) may also mediate rapid nongenomic responses to P_4. Known mPRs have been studied most thoroughly in fishes, primarily sea trout, and in a few species of mammal. They include P_4 receptor membrane components-1 and -2 (PGRMC1 and PGRMC2); membrane progesterone receptors a, b, and c (MPRA, MPRB, and MPRC); and hypothetical RNA-binding protein (RDA288) (see Nillson et al., 2006; Thomas, 2008; Gellerson, Fernandes, & Brosens, 2009). MPRA, MPRB, and MPRC are members of the seven-transmembrane P_4 adiponectin Q receptor (PAQR) family, which are coupled directly to G-proteins that activate pertussis-sensitive inhibitory G-proteins (G_i) and downregulate adenylyl cyclase activity. The known biological effects of P_4 that are mediated via mPR in fish include induction of oocyte maturation and hypermotility of sperm, but they also affect some uterine functions and GnRH secretion in rodents (Gellerson et al., 2009). The PGRMC1 and PGRMC2 proteins have predicted motifs for interactions with tyrosine kinase via SH2 domain-containing phosphatase anchor protein-1 and SH3 domain protein-5, based on their amino acid sequences. PGRMC1 mediates the antiapoptotic effects of P_4 in rat granulosa cells and perhaps P_4 induction of the acrosome reaction in mammalian sperm. The mPR subtypes detected in fishes and mammals have a cytoplasmic localization for progestin binding and G-protein coupling for cell signaling. These subtypes are expressed in different tissues and have different patterns of expression, which suggests different physiological functions in the female reproductive tract. For example, *MPRA* mRNA is expressed in the ovary, amnion, chorion and uterine endometrium and myometrium, kidney, adrenal gland, mammary gland, hypothalamus, pituitary, brain, and lymphocytes of mammals. PGRMC1 and PGRMC2 are also expressed in tissues of the female reproductive tract, particularly the ovary. However, the physiological relevance of mPGRs in mediating the actions of P_4 in reproductive organs is largely unknown.

1.5. Relaxin (RLX)

This section summarizes known biological effects of relaxin (RLX) based on a recent review by Sherwood (2004). Relaxin was first studied in guinea pigs as a hormone from the CL that induced relaxation of connective tissue in the cervix and prevented spontaneous myometrial contractions in rodents and pigs, but these same biological responses to RLX have not been reported for primates or ruminants. Relaxin is a member of a family of peptide hormones that includes insulin and the insulin-like growth factors (IGFs). Relaxin upregulates cAMP by binding to its G-protein-coupled receptors, relaxin/insulin-like family peptide receptor-1 (RXFP1 (or LGR7)) or -2 (RXFP2 (or LGR8)) to induce a G-protein signaling pathway, as well as a tyrosine kinase pathway that inhibits expression of phosphodiesterase (PDE4A), an enzyme that degrades cAMP. Relaxin-induced widening of the birth canal in guinea pigs is due to increased collagen extensibility, likely mediated by the actions of matrix metalloproteinases that are modulated by tissue inhibitors of metalloproteinase. Relaxin-induced connective tissue remodeling also occurs in the uterus, amnion, and ovarian follicles. The response of skin fibroblasts is associated with vascular changes that allow metabolic heat generated by both mother and conceptus to be dissipated efficiently in rats.

There is evidence that secretion of RLX early in pregnancy supports implantation of blastocysts in rats, pigs, and humans. In women and marmoset monkeys, production of RLX by the CL is positively associated with successful pregnancy outcomes due to RLX-induced decidualization of uterine stromal cells and the expression of several genes associated with the implantation of blastocysts, including genes for metalloproteinases, insulin-like growth factor binding protein-1 (IGFBP1), vascular endothelial growth factor (VEGF), and PRL. In marmoset monkeys, RLX from the CL causes thickening and extensive vascularization of the uterus, as well as effects on the stromal and epithelial cells. Because RLX deficits are compensated for in mice by steroid hormones and perhaps other agents, *Rlx*-knockout mice do not exhibit an altered phenotype. In rats, RLX is also positively associated with increased blood volume in pregnancy as it (1) resets the vasopressin response in the brain to hyperosmolarity; (2) increases rates of glomerular filtration in the kidney; (3) stimulates vasodilation and neoangiogenesis in conjunction with increasing nitric oxide synthase (NOS) activity to generate nitric oxide (NO); and (4) induces expression of VEGF and FGF2 . Relaxin may affect plasma osmolality during pregnancy in rats and mice by stimulating water intake and decreasing secretion of vasopressin.

Relaxin, particularly in concert with E_2, promotes growth of the uterine myometrium and endometrium in nonpregnant rodents and pigs, which is associated with the dilation of small arteries and/or veins and increased content of water, protein, collagen, glycogen, DNA, IGFs, IGF-binding proteins, connexins, E-cadherin, VEGF, and tissue inhibitor of matrix metalloproteinases (TIMPs). However, P_4 inhibits acute RLX-induced increases in uterine wet weight and collagen content in ovariectomized prepubertal rats, but enhances uterotropic effects of RLX in ovariectomized gilts. In pigs, the effects of RLX on the uterus may be important in the accommodation of conceptuses. With respect to the uterine myometrium, RLX reduces the frequency and amplitude of myometrial contractions in rats, mice, and pigs.

The cardiovascular system of all mammals must adapt to an increase in plasma volume, cardiac output, and heart rate, as well as a decrease in blood pressure and vascular resistance during pregnancy. Relaxin contributes to these adaptations through effects on the kidney, vasculature, and heart by (1) increasing both the glomerular filtration rate and renal blood flow; (2) reducing myogenic reactivity of small arteries and vasoconstrictive responses to angiotensin II, and increasing endothelin type B receptors on vascular endothelial by increasing NO production; and (3) increasing cardiac output to accommodate the demands of the pregnant uterus for high rates of perfusion.

Studies of *Rlx* knockout and relaxin receptor (*Rxfp1* or *Lgr7*) knockout mice revealed a requirement for RLX for the normal delivery of pups. Although there is no evidence that RLX is required for parturition in humans or pigs, there is evidence to indicate that exogenous RLX facilitates birth in rats, mice, and pigs by stimulating growth and remodeling of the cervix, vagina, and interpubic ligaments, as well as increasing the extensibility of the cervix. Relaxin is thought to decrease crosslinking among collagen fibers, hydration, dry weight, and content of glycosaminoglycans (dermatan sulfate, heparan sulfate, hyaluronic acid) in the connective tissue of the cervix. In particular, increases in the accumulation of dermatan sulfate among collagen fibrils either reduce crosslinking or prevent their aggregation to increase the extensibility of the cervix. Similar events likely occur in the pubic ligaments and the vagina during parturition.

Relaxin also affects the connective tissue components of the mammary glands of pregnant pigs, rats, and mice; however, the major effect is to stimulate development of the nipples in rodents. The nipples of mammary glands of RLX-deficient mice are so small that pups are unable to suckle. In pigs, nipples of mammary glands are smaller in RLX-deficient females, but piglets can still suckle. The effect of RLX on the development of the nipples of mammary glands is independent of the effects of estrogens.

1.6. Chorionic Somatomammotropin Hormone (CSH1)

Chorionic somatomammotropin hormone (CSH1) is similar in structure and function to growth hormone (GH1) and affects the metabolic state of the mother during pregnancy to facilitate the transfer of nutrients to the conceptus (Soares, Konno, & Khorshed Alam, 2007; Ben-Jonathan, LaPensee, & LaPensee, 2008). Temporal patterns of secretion of maternal CSH1 are similar during pregnancy for humans, sheep, goats, rats, mice, and cattle, as concentrations increase to peak values in the last trimester and then decrease abruptly near the time of parturition (Figure 5.2). With the exception of cows, concentrations of CSH1 are higher in fetal than maternal blood and, in all species, are positively related to the amount of placental tissue. Biological activities of CSH1 are mediated via homodimers of PRL receptors (PRLRs) and/or heterodimers of PRLR and GH1 receptors (GHR), including luteotropic effects on CL in some species, stimulation of fetal–placental growth, mammogenesis, and lactogenesis. The luteotropic effect of CSH1 and PRL is established in rodents (mice and rats). Chorionic somatomammotropin hormone is also involved in the partitioning of nutrients to support fetal development through stimulation of amino acid uptake and glyconeogenesis in fetal–placental tissues. Chorionic somatomammotropin hormone also stimulates lobulo-alveolar growth in mammary glands through its mitogenic activity and stimulation of the synthesis of milk proteins. It also decreases concentrations of urea nitrogen and increases concentrations IGF1, nonesterified fatty acids, and glucose in serum, with little effect on lipolysis, and it decreases maternal insulin sensitivity to allow increased concentrations of glucose in maternal serum and decreased utilization of maternal glucose by maternal tissues that favor transport of glucose and other nutrients to the fetal circulation. In humans, CSH1 influences both fat and carbohydrate metabolism and may contribute to the diabetogenic state of women during pregnancy.

2. THE MENSTRUAL CYCLE, LUTEOLYSIS, AND PREGNANCY IN PRIMATES (*HOMO SAPIENS* AND *MACACA MULATTA*)

2.1. The Menstrual Cycle

The mechanism(s) responsible for luteolysis in primates is not established, but several possibilities have been identified (see Stouffer, 1988). Estradiol may act directly on the luteal cells of primates to initiate luteolysis or suppress the LH required for luteal maintenance; however, evidence for this E_2-mediated luteolytic pathway is controversial. Similarly, intraovarian sources of $PGF_{2\alpha}$ may act directly on the primate CL to initiate luteolysis as inhibitors of $PGF_{2\alpha}$ synthesis do not extend CL lifespan, whereas intraluteal infusion of $PGF_{2\alpha}$, but not other prostaglandins, causes premature luteolysis in monkeys. Therefore, $PGF_{2\alpha}$ may be an intraovarian paracrine or autocrine hormone that acts in concert with other intraovarian peptides/proteins to cause luteolysis of CL in primates. Corpus luteum regression is clearly uterine-independent as hysterectomy does not alter cyclic ovarian function in either women or monkeys.

2.2. Pregnancy

Maternal recognition of pregnancy signaling extends CL function throughout pregnancy or until the time of the luteal–placental shift, when production of P_4 by the placenta is adequate to support pregnancy (Fazleabas, Kim, & Strakova, 2004). Primate embryos enter the uterus on day three or four postovulation at the morula stage, develop into hatched blastocysts, and begin implantation to the uterine LE on days seven to nine postovulation in macaques and humans or days 11 to 12 in marmoset monkeys. Chorionic gonadotropin produced by primate blastocysts signals maternal recognition of pregnancy through its luteotropic actions via LHCGR on the luteal cells. Circulating concentrations of CG, first detected around the time of implantation in all primates, increase to peak values in the

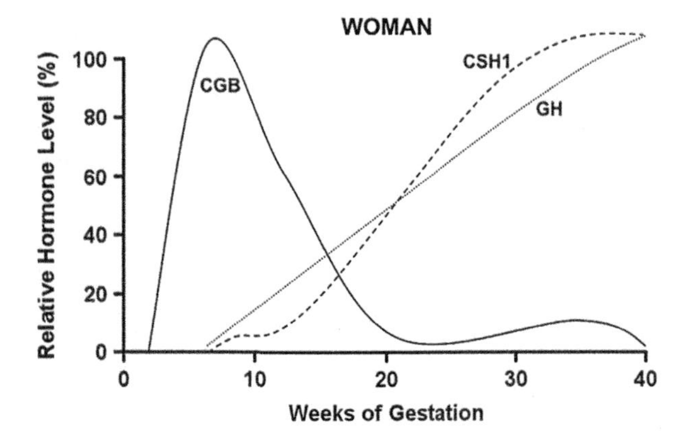

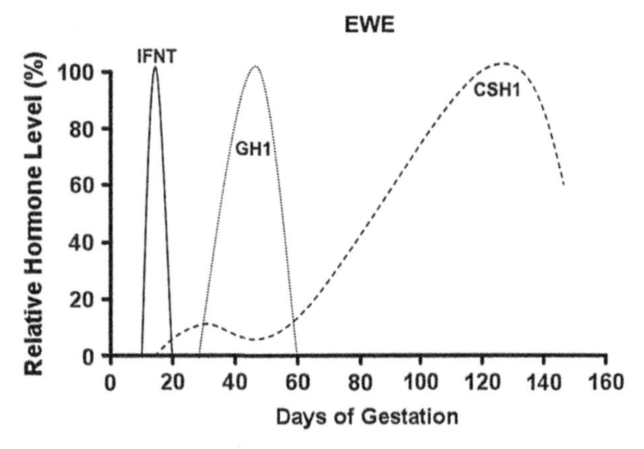

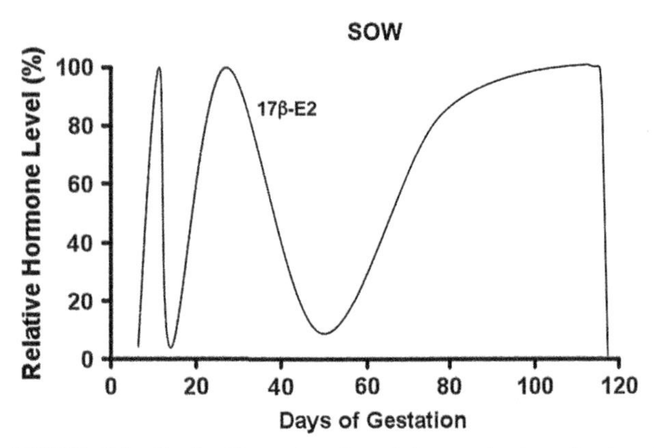

FIGURE 5.2 Graphs illustrating the relative production of selected hormones by the placentae of women, ewes, and sows during gestation. CG, chorionic gonadotropin; CSH1, chorionic somatomammotropin (placental lactogen); E_2, 17β-estradiol; GH1, growth hormone; IFNT, interferon tau.

first trimester and then decrease during late gestation in humans. However, circulating levels are reduced to near the limits of detection by midpregnancy in some primates whereas CG does not peak until midpregnancy in other species and may not decline until just before parturition in others (see Figure 5.3). In rhesus monkeys, 'rescue' of the CL occurs on day nine postovulation, when CG is detectable in utero–ovarian venous blood, but not peripheral blood, probably due to dilution effects. There is evidence for *CGβ* mRNA in eight-cell human embryos and hatched blastocysts of the rhesus monkey and LHCGR in primate endometria.

The production of CG by human trophoblasts may be regulated by gonadotropin-releasing hormone (GnRH) from the uterus. Receptors for GnRH (GnRHR) are detectable in placental tissue, as well as inhibin, activin, and P_4 from the ovary and/or placenta. Importantly, GnRH agonists enhance and GnRH1 antagonists suppress CG secretion. Primate blastocysts also produce inhibin and steroids that may affect CG secretion. In many primates, CG production decreases at the time of the luteal–placental shift in P_4 production. Both passive and active immunization of primates with modified forms of CG result in infertility, but do not prevent primates from continuing normal menstrual cycles. Further, exogenous CG increases P_4 production and extends CL lifespan in women and monkeys. Thus, CG alone appears to be the maternal recognition of pregnancy signal in nonhuman primates and in humans.

There are other aspects of uterine receptivity to implantation and pregnancy in humans that must be appreciated (see Fazleabas et al., 2004). First, PRs are limited to endometrial decidual tissue, which is responsive to P_4. Second, CG acts via LHCGR on the uterine epithelia and stromal cells to induce decidualization of stromal cells, which (1) secrete PRL and IGFBP1; (2) increase edema; (3) express alpha smooth muscle actin and PTGS2; (4) increase angiogenesis and blood flow; and (5) express leukemia inhibitory factor (LIF). Interleukin-1 beta (IL1B) from trophoblasts also increases the expression of IGFBP1 and the decidualization of endometrial stromal cells. During the window of implantation in humans, P_4 stimulates development of uterine glands and secretory activity by GE and downregulates ERα in uterine epithelia. Thereafter, ERα and PR are restricted to the basalis zone of the endometrium prior to implantation while PRs remain abundant in the uterine stromal cells. The window of implantation in humans is characterized by the expression of integrin heterodimers $\alpha_1\beta_1$, $\alpha_4\beta_1$, and $\alpha_v\beta_3$, which can bind fibronectin, vitronectin, thrombospondin, von Willebrand factor, bone sialoprotein-1, and secreted phosphoprotein-1 (SPP1, or osteopontin). Binding of L-selectin to $\alpha_4\beta_1$ is associated with the establishment of connections between invading conceptus trophectoderm and maternal vasculature that extends to placentation, while $\alpha_v\beta_3$ and SPP1 complexes localize to pinopodes of endometrial LE to serve as a marker of implantation. Glycodelin from uterine glands may facilitate implantation, but LIF and calcitonin are required for implantation, at least in mice (see Fazleabas et al., 2004).

Secretions from uterine GE increase in response to P_4, likely acting via PR in stromal cells that express

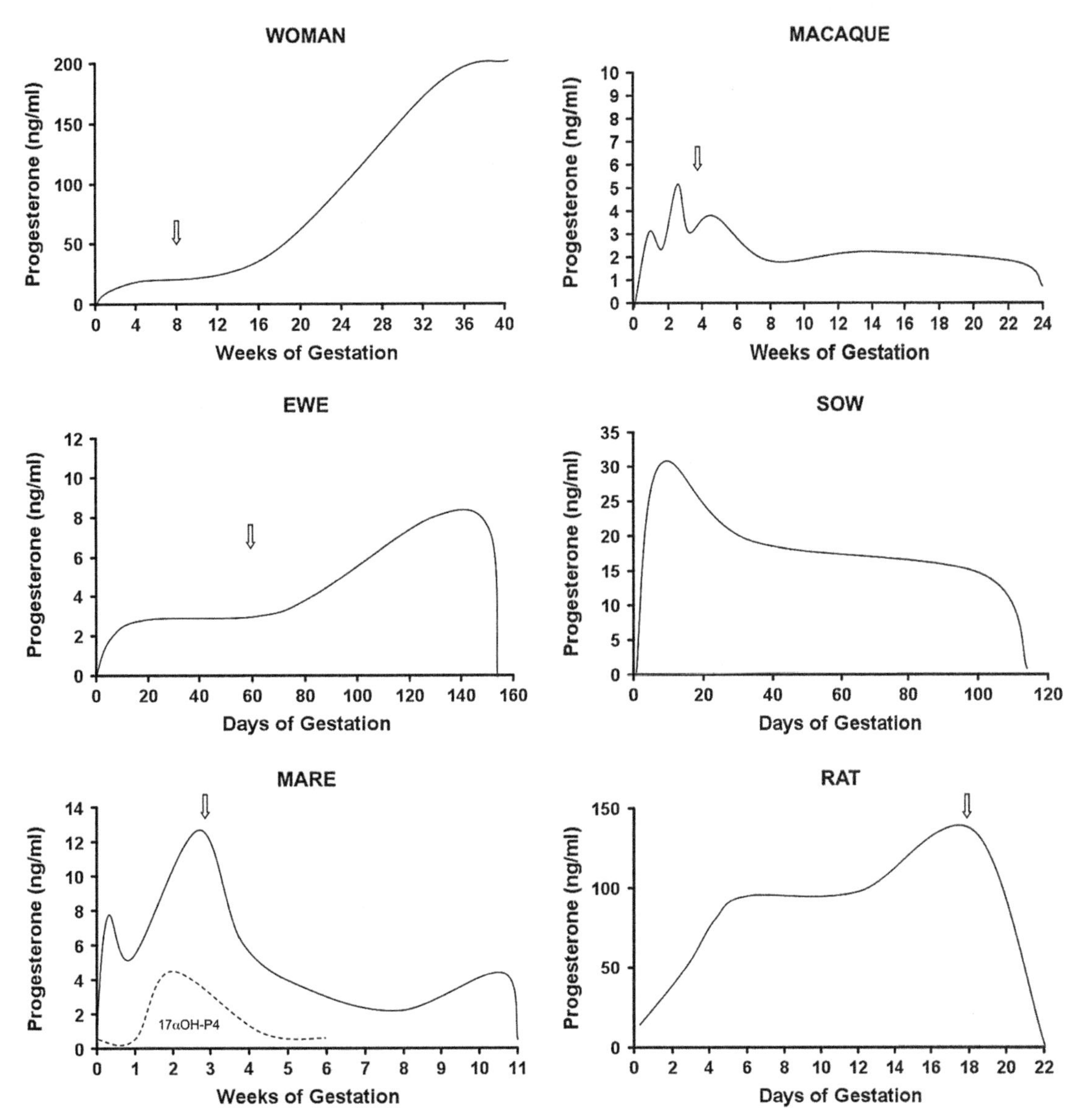

FIGURE 5.3 Patterns of change in concentrations of circulating progesterone throughout pregnancy in different mammals. The white arrow indicates the time in pregnancy when ovariectomy does not cause abortion because production of P_4 by the placenta is sufficient for the maintenance of pregnancy. *Adapted from Leroy-Heinrichs and Ribbon (1989) and Leymarie and Martal (1991).*

progestamedins such as FGF7, FGF10, and HGF (Slayden & Keater, 2007). Proteins in these secretions include uteroglobin, histone A2, spermidine/spermine acetyltransferase-2, secretory leukocyte protease inhibitor, and metallothionein. Stromal cells of humans and macaque also secrete proprotein convertase-6 (PC6) at the implantation site. However, P_4 suppresses the expression of transforming growth factor-β (TGFβ), matrix metalloproteinase-11, proenkephalins, cysteine/glycine-rich protein-2, collagen type $VII_{\alpha 1}$ and frizzle-related protein-4, while FGF7 has antiapoptotic effects on uterine GE (Figure 5.4).

Human extravillous and villous trophoblasts from first and third trimester human pregnancies produce IFNA and IFNB, which may (1) regulate the proliferation of trophoblast or uterine cells; (2) exert immunosuppressive effects by suppressing mitogen-induced proliferation of T- and B-cells; (3) protect the conceptus from viral infections; (4) regulate cellular differentiation and expression of cell surface antigens; (5) stimulate expression of ϵ-globin, a component of embryonic hemoglobin; and (6) suppress the expression of proto-oncogenes such as epidermal growth factor receptors (EGFRs) and colony-stimulating factor-1 receptor (CSFR1) to affect trophoblast growth and differentiation (Aboagye-Mathiesen, Toth, Zdravkovic, & Ebbesen, 1995). Interferon-stimulated genes upregulated in human endometrial stromal cells treated with human

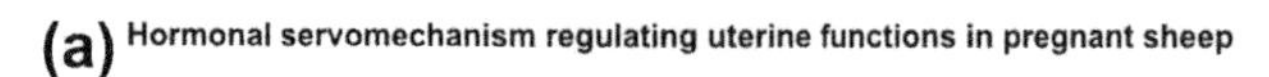

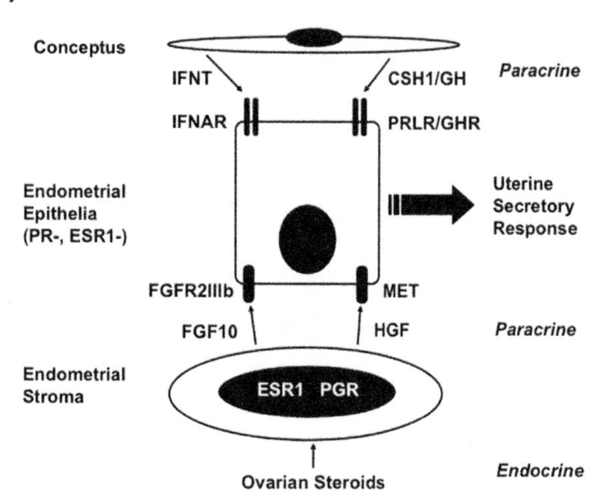

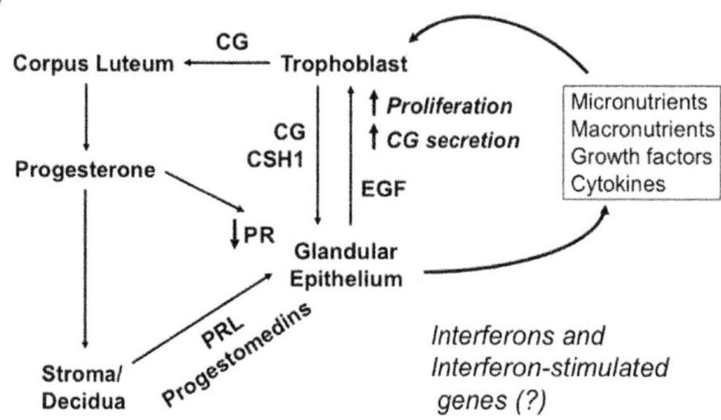

FIGURE 5.4 Summary of the concept of a servomechanism involving sequential interactions between ovarian progesterone (P_4) and placental hormones for the regulation of endometrial functions and conceptus growth and development. (a) Ovarian steroid hormone regulation of endometrial function and uterine response to placental hormones during pregnancy in sheep (Spencer & Bazer, 2004). During early pregnancy, continuous exposure of the endometrium to P_4 downregulates P_4 receptors (PRs) in all endometrial epithelia. Indeed, the endometrial epithelia in the ovine uterus are negative for ERα and PR during most of gestation. Therefore, the endocrine effects of ovarian steroids, particularly progesterone, on endometrial function are promulgated by paracrine-acting growth factors produced by stromal cells, which are PR-positive. These progestomedins and estramedins, growth factors from uterine stromal cells that mediate the actions of estrogens, include stromal fibroblast growth factor (FGF)-10 and hepatocyte growth factor (HGF), which activate their receptors, fibroblast growth factor receptor-2 splice variant (FGFR2IIIb) and met proto-oncogene (MET), expressed only on uterine epithelia and the conceptus trophectoderm. Further, stromal-derived growth factors appear to be permissive for the effects of placental hormones, i.e., IFNT, CSH1, and GH1, to activate their receptors and cellular functions, including secretion of proteins and other molecules required for the growth and development of the conceptus and placentation. (b) A summary of a potential servomechanism by which the human conceptus may stimulate the activity of the uterine glandular epithelial cells to meet the requirements for the establishment and maintenance of pregnancy through at least the first trimester of gestation. *Adapted from Burton, Jauniaux, & Charnock-Jones (2007).* Similarly to the servomechanism described in [a] for ewes, P_4, along with hormones from the uterine decidua and placenta, regulates uterine gland secretions to provide nutrients and growth factors that, in turn, regulate placental hormone production and development. CG, chorionic gonadotropin; CSH1, chorionic somatomammotropin; EGF, epidermal growth factor; ERα, estrogen receptor alpha; FGF10, fibroblast growth factor-10; FGFR2IIIb, fibroblast growth factor receptor-2; GH, growth hormone; GHR, growth hormone receptor; HGF, hepatocyte growth factor; MET, met proto-oncogene (hepatocyte growth factor receptor); IFNAR, type I interferon receptor; IFNT, interferon tau; PGR, progesterone receptor; PRL, prolactin; PRLR, prolactin receptor.

trophoblast conditioned medium include *DDX58*, *GBP1*, *GBP2*, *HSXIAPF1*, *IFIH1*, *IFIT1*, *IFIT2*, *IFIT3*, *IFIT5*, *IFI35*, *IRF1*, *ISG15*, *MIC*, *MX1*, *MX2*, *NMI*, *OAS1*, *OAS2*, *OAS3*, *PLSCR1*, *RSAD2*, *RTP4*, *SERPING1*, *STAT1*, and *STAT2* (Hess et al., 2007). During the peri-implantation period of pregnancy, increases in *B2M*, *GBP1*, *IFI27*, *IRF1*, and *ISG15* occur in the endometria of humans, baboons, domestic animals, and laboratory animals. Guanylate binding protein-1 (GBP1), a guanosine triphosphatase, is a marker of uterine receptivity to implantation, but its function is unknown. Myxovirus resistance-1 protein (MX1), and also guanosine triphosphate (GTP)ases, may protect against viral infection. The shift in endometrial production from $PGF_{2\alpha}$ to PGE_2 is associated with implantation in humans and may be due to actions of interferons that suppress P_4-regulated production of $PGF_{2\alpha}$ but not PGE_2, as occurs in response to IFNT in ruminants.

The luteal–placental shift in the production of P_4 from the CL to the placenta is accompanied by increases in CG secretion stimulated by GnRH1, EGF, activin, IL1 and IL6, macrophage colony-stimulating factor, γ-aminobutyric acid, retinoic acid, glucocorticoids, and dehydroepiandrosterone, but inhibited by TGFβ, dopamine, inhibin, P_4, IGF1, testosterone, and estrogens. The placentae of primates can produce all hormones of the hypothalamic–pituitary–gonadal axis that regulate LH and FSH secretion, as well as those that regulate the production of P_4, estrogens, CSH1, and RLX during gestation. An excellent review of these hormones with emphasis on CG, PRL, and CSH1 has been published (Ben-Jonathan et al., 2008). The major points from that review are summarized in the following sections.

During the first trimester of pregnancy, PRs are expressed by uterine decidual cells and uterine GEs express abundant receptors for LHCGR. Chorionic gonadotropin is a luteotropin, but it also stimulates the expression of PTGS2 by uterine epithelia. Receptors for PRL are also expressed by the uterine GE and decidualized stromal cells during early pregnancy. Long-term exposure of the uterine GE to P_4 stimulates decidual cells via the PR to secrete PRL and the syncytiotrophoblast to secrete CSH1. Both PRL and CSH1 stimulate uterine GE, in concert with CG, to regulate the secretory activity of the uterine GE, which supplies nutrients and other factors to the conceptus (Burton, Jauniaux, & Charnock-Jones, 2007). In addition, local paracrine signals within the decidua influence the accumulation of natural killer (NK) immune cells in the vicinity of the uterine GE during early pregnancy. The NK cells secrete a variety of cytokines when activated, as well as EGF. Thus, interactions between extravillous trophoblast and NK cells may signal the presence of a conceptus to the endometrial GE in humans and other primates. Deficiencies in uterine GE secretions may result in early pregnancy failure. The uterine glands secrete glycodelin A and MUC1, both of which are lower in women who fail to maintain pregnancy. Failure to secrete adequate amounts of these key proteins may result from failure of the uterine GE to downregulate PR in women with a luteal phase defect. At present, it is not known whether miscarriages result from development of an incomplete cytotrophoblast shell due to inadequate EGF-induced proliferation or abnormal immunological interactions within decidual cells through a lack of modulation by glycodelin A.

There is no evidence that either PRL or CSH1 is luteotropic for primate CL. Rather, pituitary LH and CG are responsible for the development and maintenance of CL for the production of the P_4 required for pregnancy. However, from midpregnancy to term, the human placenta is the primary source of P_4 and estrogens. The human uterine decidua produces PRL during pregnancy, but its functions with respect to decidual tissue or chorioamnion are not established. However, PRL and CSH1 affect transport mechanisms for water, electrolytes, and nutrients across the chorioamnion (see Bazer, Goldstein, & Barron, 1981), and PRL stimulates the production of prostaglandins and fetal lung development. Prolactin receptors are expressed by decidual cells, the chorionic cytotrophoblast, the amniotic epithelium, and syncytiotrophoblasts. Secretion of CSH1 and GH1 by human placentae is highest between weeks 6 and 30 of gestation. Chorionic somatomammotropin hormone increases maternal appetite and food intake, glucose uptake, insulin secretion, insulin sensitivity, and lipolysis for the mobilization and utilization of maternal free fatty acids.

During the course of pregnancy, the mammary gland initially grows, due primarily to lobulo-alveolar differentiation (mammogenesis) and then stage I (milk protein synthesis) and stage II (milk secretion) lactogenesis. During pregnancy, mammogenesis and lactogenesis occur in response to the actions of E_2, P_4, PRL, CSH1, insulin, GH1, corticosteroids, and thyroid hormones. Stage 1 lactogenesis begins midgestation with progressive expression of genes for milk constituents including casein, whey acidic protein, and lactalbumin. Stage 2 lactogenesis occurs during the prepartum period, when there are abundant levels of colostrum proteins being secreted, and the secretion of milk is initiated. Finally, RLX, at least in rodents, stimulates development of the nipples of the mammary gland to facilitate suckling by the newborn (Sherwood, 2004).

3. ESTROUS CYCLES, LUTEOLYSIS, PSEUDOPREGNANCY, DELAYED IMPLANTATION, AND PREGNANCY IN SUBPRIMATE MAMMALS

3.1. Rodents: Rat (*Rattus norvegicus*), Mouse (*Mus musculus*), Guinea Pig (*Cavia porcellus*)

3.1.1. The estrous cycle and luteolysis

Laboratory rodents are nonseasonal, spontaneously ovulating, polyestrous mammals having an estrous cycle of four to five days in length that includes proestrus (12 to 14 hours), estrus (25 to 27 hours), metestrus (6 to 8 hours) and diestrus (55 to 57 hours) (see Freeman, Smith, Nazian, & Neill, 1974; Soares et al., 2007). For rats, mice, and hamsters, the CL secretes P_4 for two days, but becomes fully functional only after vaginal stimulation. The absence of a true luteal phase in nonmated rodents allows for recurrent estrous cycles of short duration and frequent opportunities for mating and the establishment of pregnancy. During the estrous cycle of the rat, newly formed CL are maintained through metestrus of the following cycle and then regress due to apoptosis of luteal cells, degeneration of blood vessels, infiltration of leukocytes, and increased 20-alpha-hydroxysteroid dehydrogenase (20α-HSD). Although $PGF_{2\alpha}$ is luteolytic in rodents, its mechanism of action is not known. The CL of cyclic rodents secretes P_4, which is rapidly metabolized by 20α-HSD to 20α-hydroxyprogesterone (20α-OHP_4), an inactive metabolite that can not support a uterine decidual reaction for implantation or pregnancy. As secretion of luteal 20α-OHP_4 declines during diestrus, there is onset of proestrus and then estrus and ovulation.

3.1.2. Pseudopregnancy and pregnancy

The gestation period for rats, mice, and hamsters is 20 to 22 days and the functional CL must produce P_4 through day 17 (see Soares et al., 2007). Thus, maternal recognition of pregnancy in rodents involves signaling for the formation of a functional CL that can produce P_4 to support the uterine decidual reaction, implantation, and pregnancy. Sterile mating or cervical stimulation of rodents during estrus results in the establishment of functional CL that secrete P_4 for 12 to 14 days—a period known as pseudopregnancy. Extension of CL lifespan beyond day 12 requires the presence of viable conceptuses within the uterus. The establishment and maintenance of pregnancy requires two endocrine events. First, mating elicits diurnal and nocturnal surges of PRL from the pituitary that increase LHCGR on luteal cells for the formation of a CL and the suppression of 20α-HSD activity in the CL to prevent conversion of P_4 to 20α-OHP_4. The second endocrine event required for maintenance of pregnancy beyond day 12 in rodents is dependent on implantation, conceptus development, and production of lactogenic hormones by the uterine decidua and placentae. These members of the lactogenic family of hormones take over from pituitary PRL as luteotropic hormones that act on CL to maintain production of P_4 throughout gestation.

Interferons (IFNA, IFNB, IFNG) are produced by the placentae of mice as early as day seven of pregnancy and several ISGs are expressed by cells in implantation sites (Austin et al., 2003; Bany & Cross, 2006). The expression of both IFNA and interferon-regulated gene-1 (IRG1) increase in the uteri of pregnant rats between days one and four and then decrease following implantation. Further, trophoblast giant cells express IFNA, which, in turn, induces expression of ISG15 during the peri-implantation period in mice. Microarray analyses revealed upregulation of ISGs, as well as downregulation of genes for histocompatibility class 2, T region locus 23, IRF6, and major histocompatibility complex (MHC) class I and class II at implantation sites in mice. The expression of IFNB also increased at implantation sites when E_2 was used to induce implantation, and IFNG is expressed by the uterine LE and GE, trophoblast cells, and degenerating metrial gland cells of mice. However, most of the IFNG in the decidua of mice at midgestation is from uterine NK cells, which effect changes in the uterine spiral arteries. Thus, IFNA and IFNB, as well as IFNG, bind to type I and type II interferon receptors, respectively, to initiate unique cell-signaling pathways that likely modify gene expression at implantation sites in rodents to directly affect conceptuses or to exert indirect effects on blood vessels and decidual cells.

3.1.3. Rodent decidual prolactin (PRL)-like proteins and placental lactogens

Implantation in rodents involves transformation of stromal cells into decidual cells: a P_4-dependent process termed 'decidualization.' In turn, decidual cells, characterized as small, loosely packed fibroblast-like and binucleate cells, produce PRL, alpha-2-macroglobulin, and growth factors, while antimesometrial decidua formed by giant polyploid cells secrete hormones that include follistatin, activin, TGFβ, and a number of PRL family proteins including PRL-like protein B (PRLPB) and decidual PRL-like protein (dPRP) (Soares et al., 2007). Hormones produced by the antimesometrial decidua sustain luteal production of P_4 and inhibit aromatization of androgens to estradiol by ovarian follicles (Gu et al., 1995). The main luteotropic hormone of the decidua is PRL, which maintains secretion of P_4. The dPRP also binds receptors on mesometrial decidual cells and upregulates expression of alpha-2-macroglobulin. Various forms of lactogenic hormone are produced by the placentae of rodents, but the primary roles of CSH1 are (1) maintenance of P_4 secretion by the CL during pregnancy; (2) development of the mammary gland during pregnancy; and (3) control of maternal metabolic functions. Various isoforms of placental lactogens and PRL, along with P_4, E_2, corticosteroids, insulin, and GH1, are also responsible for adipogenesis and glucose and lipid metabolism, as well as mammogenesis and lactogenesis (Soares et al., 2007; Ben-Jonathan et al., 2008). The lactogenic hormones in rodents exert effects on the maternal immune system, hematopoiesis, angiogenesis, and metabolism during pregnancy (Soares et al., 2007).

3.2. Ruminants

3.2.1. The estrous cycle and luteolysis

Ruminants, e.g., cattle, sheep, and goats, are spontaneously ovulating, polyestrous mammals with recurring estrous cycles of 21, 17, and 20 days, respectively (see Spencer, Burghardt, Johnson, & Bazer, 2004; Spencer, Johnson, Bazer, Burghardt, & Palmarini, 2007). Domestic cattle (*Bos taurus* and *Bos indicus*) are not seasonal breeders, but sheep (*Ovis aries*) and goats (*Capra hircus*) are short-day breeders with regular estrous cycles from late summer through midwinter. Sheep have been studied in greatest detail with respect to endocrine regulation of the estrous cycle and pregnancy; however, the basic mechanisms for luteolysis and maternal recognition of pregnancy are very similar for ewes, cows, and goats.

In ewes, day zero of estrus is designated as the day of the onset of sexual receptivity for mating and the beginning of the estrous cycle. Estrus lasts about 30 hours and spontaneous ovulation occurs about 30 hours after the onset of estrus, in response to an estrogen-induced ovulatory surge of LH and FSH from the anterior pituitary at the onset of estrus. Metestrus (days one to four of the estrous cycle) is characterized by luteinization of the theca and granulosa cells of the ovarian follicle under the influence of LH to form the CL, which secretes P_4. Diestrus (days 4 to 14 of the estrous cycle) is when the CL reaches its maximum size and secretion of P_4. Near the end of diestrus, PRs in the uterine epithelia are downregulated by P_4 to allow increases in ERα and then OXTR, necessary for OXY-induced secretion of luteolytic $PGF_{2\alpha}$ and luteolysis. The onset of proestrus begins when the CL is fully regressed and the ovarian follicles begin producing significant amounts of E_2. A dominant follicle(s) is selected for ovulation and the onset of estrus marks the beginning of the next estrous cycle. The estrous cycle of ruminants is dependent on the uterine endometrium for production of luteolytic $PGF_{2\alpha}$. During diestrus, P_4 increases phospholipid stores and prostaglandin synthase in the uterine epithelia, which are necessary for the mobilization of arachidonic acid by phospholipase A2 and its conversion by PTGS2 to $PGF_{2\alpha}$

during late diestrus. Importantly, exposure of the uterus to P_4 for 10 to 12 days downregulates PR, which, in turn, allows ERα and OXTR expression, initially by the LE/sGE and then by the GE and stromal cells. These are key events in the activation of the luteolytic mechanism for endometrial production of luteolytic $PGF_{2\alpha}$. Following upregulation of ERα and OXTR, E_2 induces phospholipase A2 to mobilize arachidonic acid for conversion to $PGF_{2\alpha}$ and OXT is released in a pulsatile manner from the CL and posterior pituitary to act via OXTR to induce pulsatile release of luteolytic $PGF_{2\alpha}$, which culminates in regression of the CL on day 16. If ewes are hysterectomized during the active life of the CL, luteolysis does not occur and CL lifespan is prolonged to about five months; i.e., similar to the duration of normal pregnancy.

Intimate intertwining of the uterine branch of the ovarian vein and the ovarian artery allows for the countercurrent exchange of luteolytic $PGF_{2\alpha}$ from the uterine venous drainage directly to the ovarian artery for delivery to the CL for induction of luteolysis. Luteolysis does not require a decrease in LH receptors or withdrawal of basal LH support from the CL, but the CL must be exposed to five pulses of PGF over a 24-hour period to undergo complete luteolysis. Luteolytic pulses of $PGF_{2\alpha}$ on days 15 and 16 in cyclic ewes are preceded by increases in circulating levels of E_2. Experimentally, administration of E_2 to cyclic ewes on day 9 or 10 of the estrous cycle sequentially upregulates *ESR1* mRNA, ESR1 protein, and then *OXYR* mRNA and protein in the endometrial LE/sGE, and then luteolysis. Luteolytic effects of $PGF_{2\alpha}$ on the CL have been attributed to (1) a decrease in luteal blood flow; (2) a reduction in LH receptors; (3) uncoupling of the LH receptor from adenylyl cyclase; (4) activation of protein kinase C; (5) an influx of high levels of calcium and/or; (6) activation of a cytotoxic cascade. Prostaglandin $F_{2\alpha}$ may also act in concert with endothelin to cause (1) vasoconstriction of the blood vessels to the CL; (2) inhibition of the expression of LHCGR and StAR inhibits P_4 production; and (3) recruitment of the immune cells that produce cytokines (IL1B, TNFA, IFNG), which stimulate production of NO and apoptosis of luteal cells.

3.2.2. Pregnancy

Hormones from the pituitary (PRL), ovary (P_4), and placenta (CSH1 and GH1), as well as fetal adrenal (glucocorticoids) and pancreas (insulin), stimulate mammogenesis, lactogenesis, and uterine functions supportive of conceptus development in ruminants (see Guilbault, Thatcher, Collier, & Wilcox, 1985; Kann, Delobelle-Deroide, Belair, Gertler, & Djiane, 1999). Relaxin is not known to have a role in parturition or mammogenesis in cattle, but this has not been established in sheep or goats.

The antiluteolytic signal for maternal recognition of pregnancy in ruminants is IFNT, produced by mononuclear trophectoderm cells of the conceptus (Spencer et al., 2007; Bazer et al., 2008). The effect of IFNT in abrogating the uterine luteolytic mechanism for maternal recognition of pregnancy signaling is similar, if not identical, for ewes, cows, and nanny goats. Interferon tau, a type I IFN, is secreted by the trophectoderm of peri-implantation ruminant conceptuses and has potent antiviral, antiproliferative, and immunomodulatory activities, as well as a unique role as the maternal recognition of pregnancy signal in ruminants. Interferon tau is produced in massive amounts by the conceptus trophectoderm as it undergoes morphological transition from spherical to tubular and filamentous forms during the peri-implantation period. Secreted forms of ovine IFNT are not glycosylated, whereas bovine IFNT is glycosylated and caprine IFNT may be either glycosylated or nonglycosylated. The structural relatedness of IFNT in ruminants is based on evidence that trophoblastic vesicles from sheep extend the interestrous interval when placed into the uterine lumen of cattle, and ovine IFNT suppresses OXT-induced $PGF_{2\alpha}$ production by bovine and caprine uteri.

All type I IFNs bind a common receptor composed of two subunits, IFNAR1 and IFNAR2, to induce cell signaling via the classical Janus-activated kinases (JAKs) and STAT1 pathway. However, several nonclassical ISGs are P_4-induced and IFNT-stimulated in the ovine uterine LE/sGE, which lack both PR and STAT1. Therefore, the actions of P_4 and IFNT may operate through the combined effects of a progestamedin(s) (e.g., FGF10 in sheep) and a nonclassical cell signaling pathway(s) for IFNT, such as MAPK and PI3K pathways, to affect gene expression and uterine receptivity to implantation (Plantanias, 2005; Bazer et al., 2009). Restriction of the expression of classical ISGs by the ovine uterine LE/sGE is due, at least in part, to the expression of interferon regulatory factor-2 (IRF2), a potent transcriptional repressor, in those cells, but IRF2 is not expressed in the uterine GE and stromal cells.

Secretion of ovine IFNT begins on about day 10 and increases as conceptuses undergo morphological changes from spherical (312 ng/ml uterine flush) to tubular (1380 ng) and filamentous (4455 ng) forms between days 12 and 13 of pregnancy. Successful transfer of conceptuses to cyclic ewes can occur as late as day 12; i.e., 48 to 72 hours prior to the luteolytic period. Thus, IFNT acts to abrogate the uterine luteolytic mechanism and prevent pulsatile release of $PGF_{2\alpha}$. Intrauterine infusions of IFNT alone from days 11 to 15 of the estrous cycle prevent luteolysis and extend the CL lifespan. The antiluteolytic effects of IFNT primarily occur on the uterine LE/sGE in ewes. Interferon tau silences transcription of the *ESR1* gene and, therefore, ESR1-dependent expression of the *OXYR* gene in the uterine LE/sGE to prevent development of the endometrial luteolytic mechanism, which requires pulsatile release of $PGF_{2\alpha}$. However, basal production of $PGF_{2\alpha}$ is actually higher in pregnant than cyclic ewes due to the continued expression of PTGS2 in the uterine LE/sGE. Further, silencing ESR1 expression by IFNT prevents E_2 action via ESR1 from inducing PR in endometrial epithelia. The absence of PR in uterine epithelia is required for the expression of a unique set of P_4-induced and IFNT-stimulated genes in the ovine uterine LE/sGE during early pregnancy. For most, if not all, actions of IFNT on the uterus, P_4 is the permissive hormone. That is, uterine receptivity to implantation is P_4-dependent, but is preceded by loss of expression of PR and ERα by uterine epithelia; the loss of PR is a prerequisite for the expression of several ISGs in ewes. Thus, P_4 likely acts on PR-positive stromal cells to increase the expression of a progestamedin(s) in ewes, likely FGF10, that exerts paracrine effects on uterine epithelia and the conceptus trophectoderm, which express its receptor FGFR2IIIb. In sheep, classical ISGs (e.g., IFN-stimulated gene-15, mouse myxovirus resistance-1 and 2', 5' oligoadenylate synthase) are induced by IFNT only in the uterine GE, the stroma, and the immune cells that do not express IRF2. As the ovine uterine LE/sGE lacks PR and STAT1, IFNT is unable to affect gene transcription through the classical JAK-STAT1 cell signaling pathway and must activate gene transcription through alternate cell-signaling pathways such as MAPK and PI3K in the uterine LE/sGE.

3.2.3. Cows

Bovine IFNT is secreted between days 12 and 38 of pregnancy and activates mechanisms that prevent uterine secretion of luteolytic pulses of $PGF_{2\alpha}$ (see Thatcher et al., 2001). As in ewes, the available evidence indicates that neither exogenous E_2 nor OXT stimulate uterine release of $PGF_{2\alpha}$. Therefore, it appears that *ESR1* and *OXTR* mRNAs are either less abundant or not responsive to E_2 and OXT in the endometria of pregnant as compared to cyclic cows, and in cows that received intrauterine injections of ovine IFNT. In any event, bovine IFNT abrogates uterine production of luteolytic pulses of $PGF_{2\alpha}$.

3.2.4. Goats

Caprine IFNT is secreted between days 16 and 21 of gestation to prevent pulsatile release of luteolytic $PGF_{2\alpha}$ (Newton, Ott, Woldesenbet, Shelton, & Bazer, 1996). Intrauterine injections of ovine IFNT in nanny goats extend CL lifespan.

3.2.5. Sheep

In sheep, the establishment and maintenance of pregnancy requires integration of endocrine and paracrine signals from the ovary, conceptus, and uterus (Spencer & Bazer, 2004). Superficial implantation and placentation occur between days 15 and 60 of pregnancy as the uterus grows and remodels to accommodate the development and growth of the conceptus in the last trimester of pregnancy. In addition, caruncles and cotyledons develop and interdigitate to form placentomes, which increase in vascularity, and the intercaruncular endometrial glands grow substantially during pregnancy to secrete increasing amounts of histotroph, which is transported across the areolae of the chorioallantois into the fetal–placental circulation. During gestation, endometrial gland hyperplasia occurs between days 15 and 50 and is followed by hypertrophy for maximal production of histotroph after day 60. During this period, the pregnant ovine uterus is exposed sequentially to E_2, P_4, IFNT, CSH1, and placental GH1, which regulate endometrial gland morphogenesis and differentiated functions. The binucleate cells of the chorion secrete CSH1 from day 16 of pregnancy, which occurs simultaneous with initiation of the expression of uterine milk proteins (UTMP) (serpin family of serine protease inhibitors) and SPP1 (an extracellular matrix protein), which are excellent markers of endometrial GE differentiation and secretory capacity. In maternal serum, CSH1 is detectable by day 50 and peaks between days 120 and 130 of gestation. It can bind either homodimers of PRLR or heterodimers of PRLR and GHR to transducer cell signaling. In the ovine uterus, PRLRs are specifically expressed in the GE. Increasing levels of CSH1 are associated with hyperplasia and hypertrophy of the uterine GE, as well as increased production of UTMP, SPP1, and other components of histotroph that support growth and development of the conceptus. Sequential exposure of the pregnant ovine endometrium to E_2, P_4, IFNT, CSH1, and placental GH1 constitutes a 'servomechanism' that activates and maintains endometrial remodeling, secretory function, and uterine growth during gestation. The chronic effect of P_4 is to downregulate epithelial PRs to allow the expression of proteins by the uterine GE, while CSH1 and GH1 increase the number of endometrial glands and levels of expression of genes encoding for proteins such as UTMP and SPP1. The net effect in ewes is a developmentally programmed sequence of events, mediated by specific paracrine-acting factors at the conceptus–endometrial interface, that stimulate both intercaruncular endometrial remodeling and differentiated function of uterine GE for increased production of histotroph to support fetal–placental growth.

3.3. Swine (*Sus domestica*)

3.3.1. The estrous cycle and luteolysis

Pigs are a spontaneously ovulating, polyestrous, litter-bearing species that reach puberty between four and nine months of age depending on breed, nutritional status, and housing conditions (see Anderson, 1993). For gilts (non-parous) and sows (parous), estrous cycle length averages 21 days (range 18–24 days). Estrus lasts 40 to 60 hours with gilts exhibiting shorter periods of estrus than sows. Estrus, a behavioral response to estrogens from ovarian follicles, also stimulates a protracted ovulatory surge of LH, resulting in ovulation about 44 hours after the onset of estrus. The number of follicles ovulated ranges from 8 to 10 at the pubertal estrus to more than 20 for sows. Levels of circulating P_4 increase from 2 to 4 ng/ml at metestrus (days three to five) to 20 to 40 ng/ml at mid-diestrus (days 6 to 14). In the absence of pregnancy, uterine release of luteolytic $PGF_{2\alpha}$ occurs between days 16 and 18 to regress the CL and allow a new wave of follicles to develop and secrete estrogens during proestrus (days 18 to the onset of estrus) if fertilization is not successful. Circulating concentrations of estrogens reach peak values about two days prior to the onset of estrus. Concentrations of PRL in serum are greatest during proestrus and estrus while levels of RLX from the CL are low during estrus and highest during diestrus. Just prior to and during estrus, the vulva swells and becomes pink to red. Females exhibit restlessness at this time and mount other females or stand to be mounted, and females in estrus seek out the boar for mating. Females in estrus exhibit lordosis, characterized by ears laid back along the neck and legs locked in response to mounting or hand pressure on their back. The presence of the boar facilitates estrus detection in females due to the presence of pheromones in the saliva and preputial diverticulum of the sheath of the penis of boars.

Endocrine requirements for luteolysis in pigs are not well defined; however, luteolysis occurs during late diestrus and proestrus following stimulation of the uterine endometrium by P_4 for 10 to 12 days to allow the accumulation of phospholipids and necessary enzymes for the production of luteolytic pulses of PGF (Bazer, 1989; Ludwig et al., 1998; Spencer et al., 2004). Hysterectomy extends CL function to about 120 days due to the removal of the uterine source of $PGF_{2\alpha}$. Further, the CLs of pigs are refractory to luteolytic effects of $PGF_{2\alpha}$ until days 12–13 of the estrous cycle due to low numbers of luteal $PGF_{2\alpha}$ receptors. In ruminants, OXT from the CL and posterior pituitary act via uterine OXTR to elicit pulsatile release of $PGF_{2\alpha}$; however, the CLs of pigs contain very low amounts of OXT and vasopressin. Thus, the role(s), if any, of these neuropeptides in luteolysis in pigs is not known. The uterine endometrium is a source of OXY in pigs, but its role is not known. The administration of exogenous OXT decreases the interestrous interval in gilts when administered between days 10 and 16 postestrus, but not when administered to ovary-intact hysterectomized gilts, suggesting that the effect of OXT is uterine-dependent. The endometrium of pigs contains receptors for OXT and lysine vasopressin, but only responds to OXT with increased secretion of $PGF_{2\alpha}$, whereas both peptides stimulate inositol phosphate turnover indicative of stimulation of the protein kinase C and calcium-calmodulin kinase cell-signaling pathways. Oxytocin stimulates phospholipase C activity and phosphatidylinositol hydrolysis increases intracellular concentrations of calcium and diacylglycerol, which activate protein kinase C and calcium-calmodulin kinase to activate phospholipase A_2 and stimulate release of arachidonic acid and increased production of $PGF_{2\alpha}$ in pigs. Although concentrations of OXT increase in the peripheral circulation during luteolysis, OXT-induced increases in circulating concentrations of 13,14-dihydro,15-keto- $PGF_{2\alpha}$ (PGFM), the inactive metabolite of $PGF_{2\alpha}$, are lower in pregnant than cyclic gilts or gilts induced into pseudopregnancy by injection of exogenous E_2 from days 11–15 postestrus. Prostaglandins, however, are critical for the establishment of pregnancy in pigs, as inhibition of PTGS2 results in pregnancy failure and concentrations of PGFM in the circulation are elevated in pregnant gilts, beginning on day 12.

3.3.2. Pregnancy

After hatching from the zona pellucida, pig blastocysts expand and undergo morphological transition to spherical (10 to 15 mm diameter), tubular (15 by 50 mm), and filamentous (1 by 100–200 mm) forms between days 10 and 12 of pregnancy and achieve a length of 800 to 1000 mm between days 12 and 15 of pregnancy (see Bazer, 1992). During rapid elongation of the trophectoderm, it produces estrogens, as well as IFNG and IFND. The pregnancy recognition signal is E_2, produced by the conceptus from days 11 and 12 to day 15 of pregnancy. It directs secretion of $PGF_{2\alpha}$ away from the uterine vasculature and into the uterine lumen (exocrine secretion), where it is sequestered and metabolized to prevent luteolysis. In nonpregnant gilts, $PGF_{2\alpha}$ is released from the uterine endometrium into the venous drainage (endocrine secretion) to be transported to the CL to induce luteolysis. Estrogens released from the conceptus also modulate the uterine gene expression responsible for endometrial remodeling for implantation between days 13 and 25 of gestation. Both SPP1 and FGF7 are induced by estrogens in the uterine LE to affect trophectoderm growth and development, as well as adhesion to the LE for signal transduction and cell migration during the peri-implantation period. The trophectoderm also secretes IL1B and E_2 modulates uterine responses to IL1B (Geisert & Yelich, 1997). On day 15 of pregnancy, IFNG and IFND colocalize to perinuclear membranes typically occupied by endoplasmic reticulum and Golgi apparatus, as well as cytoplasmic vesicles within clusters of trophectoderm cells along the endometrial LE. Their expression is accompanied by the appearance of zona occludens-1 (ZO1), a marker of epithelial tight junctions, on their basal aspect, suggesting changes in endometrial polarity. These IFNs apparently have no known antiluteolytic effects on the CL, but they do stimulate secretion of PGE_2 by the uterus.

Interactions of E_2 and IFNs regulate cell-type specific expression of multiple genes in the uterine endometrium, which suggests a complex interplay between the endometrium and conceptus for pregnancy recognition and implantation in pigs (see Ziecik et al., 2007; Bazer et al., 2008). In pigs, implantation is noninvasive and the placenta is epitheliochorial. Genes induced in the LE by E_2 include SPP1, FGF7, aldo-keto reductase family 1 member B1 (aldose reductase or AKR1B1), cluster of differentiation 24 (CD24), neuromedin-β (NMβ), STAT1, and IRF2. Interferon regulatory factor-2, a potent inhibitor of the transcription of ISGs, is induced in the uterine LE of pigs by E_2. Genes expressed in the presence of IRF2 are likely to have roles in the establishment of pregnancy, including release of histotroph from uterine epithelia into the uterine lumen to affect cell proliferation, and the attachment and development of conceptus trophectoderm. In addition, IFND and IFNG may affect blastocyst attachment to the LE by remodeling uterine epithelia to affect polarity and stimulate production of PGE_2. Expression of the IRF2 repressor by the porcine uterine LE/sGE restricts expression of most ISGs to the endometrial stroma and GE, including STAT1, STAT2, IRF1, MX1, swine leukocyte antigens (SLAs) 1–3 and 6–8, and beta 2 microglobulin (B2M). The pregnancy-specific roles of these ISGs are not known, but they may (1) affect decidual/stromal remodeling to protect the fetal semiallograft from immune rejection; (2) limit conceptus invasion into the endometrium; and/or (3) stimulate development of the uterine vasculature. Because IFNG can initiate endometrial vascular development, conceptus-derived IFNG in pigs may facilitate vascular changes for hematotropic support of developing conceptuses.

The roles of PGs in the pig uterus during pregnancy remain to be clarified; however, inhibitors of PG synthesis are detrimental to the establishment and maintenance of pregnancy (see Ziecik et al., 2007). Further, $PGF_{2\alpha}$ and PGE_2 are present in greater amounts in the uterine lumen of pregnant than cyclic gilts. There is evidence that $PGF_{2\alpha}$ from the uterus is taken up by the mesometrium and transferred back to the uterus in arterial blood by a countercurrent system for conversion to its inactive metabolite, PGFM. Further, PGE_2 synthase, $PGF_{2\alpha}$ synthase, carbonyl reductase/prostaglandin 9-ketoreductase, and PGE_2 synthase : $PGF_{2\alpha}$ synthase ratios are higher in the CL of pregnant than cyclic gilts, but not between the CL on ovaries ipsilateral and contralateral to the pregnant uterine horn. Therefore, compounds from the conceptus appear to be transported within the to both ovaries to enhance CL maintenance and function. An integral part of the maternal recognition of pregnancy signaling events may involve the lipid signaling system consisting of $PGF_{2\alpha}$, PGE_2, and lysophosphatidic acid (LPA3), with PGE_2 synthase in the trophoblast and endometrium decreasing $PGF_{2\alpha}$ synthase and carbonyl reductase/prostaglandin 9-ketoreductase in conceptuses favoring PGE_2 production to support uterine functions, CL maintenance, and the establishment and maintenance of pregnancy. Expression of LPA3 during pregnancy may affect the establishment and maintenance of pregnancy in pigs as it is critical for embryo migration and spacing in mice and appears to be similarly important in pigs to enhance successful implantation and placentation.

Estrogens are secreted from pig conceptuses between days 10 and 15 for pregnancy recognition, but also increase the expression of growth factors such as IGF1 and FGF7, which in turn act on the trophectoderm to stimulate cell proliferation and development (see Spencer et al., 2004; Bazer et al., 2008). Insulin-like growth factor-1 is expressed by the uterine glands of cyclic and pregnant pigs and IGF1 receptors (IGF1Rs) are expressed in the endometrium and conceptuses, suggesting paracrine and autocrine actions of IGF1. Fibroblast growth factor 7, an established paracrine mediator of hormone-regulated epithelial growth and differentiation, is expressed by the uterine LE and is most abundant between days 12 and 15 of the estrous cycle and pregnancy. Fibroblast growth factor 7 binds to and activates FGF2IIIb expressed by the uterine epithelia and conceptus trophectoderm. Estradiol increases FGF7 expression and FGF7, in turn, increases cell proliferation, phosphorylated FGFR2IIIb, the mitogen-activated protein kinase cascade, and expression of urokinase-type plasminogen activator, a marker for trophectoderm cell differentiation. From about day 20 of pregnancy, FGF7 is expressed by the uterine GE in pigs and may continue to affect uterine epithelia and conceptus development (Johnson, Burghardt, Ka, & Bazer, unpublished results).

In addition to the increase in the secretion of estrogens between days 11 and 15 of pregnancy for maternal recognition of pregnancy, concentrations of estrogens increase in fetal fluids and maternal blood between days 15 and 30, decrease to day 60, and then increase again between day 60 of gestation and term (114 days). The increase in estrogens between days 20 and 30 is associated with increases in the expression of endometrial receptors for PRL, rapid accumulation of allantoic fluid, elongation of the chorioallantoic membranes during placentation, and increased uterine blood flow. Between days 30 and 75 of gestation, placentation is completed and uterine gland activity is greatest in terms of the transport of histotroph across the placental areolae into the fetal–placental circulation and the accumulation in allantoic fluid. Uterine blood flow increases dramatically between days 35 and 100 of gestation along with increased uptake of amino acids and glucose across the uterus and interareolear areas of the placenta. Uterine and placental development are affected by E_2, P_4, PRL, and

RLX, but details of their individual effects in pigs are limited. However, PRL and P_4 affect the secretory activity of the uterine GE and placental transport of nutrients in pigs (see Goldstein, Bazer, Spellacy, & Buhi, 1976; Knight, Bazer, Thatcher, Franke, & Wallace, 1977; Bazer et al., 1981; Young, Kraeling, & Bazer, 1990). In terms of mammogenesis and lactogenesis in pigs, E_2, P_4, PRL, and possibly RLX affect mammogenesis between days 20 and 30, stage I lactogenesis (around day 90), and stage II lactogenesis (day 110) of the 114-day period of gestation (Kensinger, Collier, Bazer, Ducsay, & Becker, 1980).

3.4. Horses (*Equus ferus caballus*)

3.4.1. The estrous cycle and luteolysis

Mares are seasonally polyestrous, with the onset of cyclicity occurring in late winter in the northern hemisphere (see Irvine, 1995). Ovarian activity is inhibited by decreasing day length (photoperiod), but often lags several months after the solstice. The effects of photoperiod appear to be regulated by melatonin from the pineal gland. Mares typically exhibit estrous cycles of 21 to 22 days, although variability in length of estrus and diestrus is common, particularly during transition into and out of the breeding season. Mares have the highest levels of fertility between May and July; however, pregnancy lasts 11 months and the goal among breeders is for mares to foal near January 1 (the arbitrary birth date for all foals born in a given calendar year). Artificial lengthening of the photoperiod, specifically delaying sunset, is often used in an attempt to hasten onset of the breeding season. Advancing sunrise has not been effective in altering the breeding season. Estrus lasts from three to seven days and is accompanied by swelling and reddening of the vulva in response to estrogens from ovarian follicles. During estrus the vulva becomes covered with a thin mucus layer and the cervix dilates several centimeters. The mare frequently assumes a posture associated with urination; i.e., the tail head is raised and deviated and the clitoris exposed repeatedly by rhythmic contractions of the vulva (termed 'winking'). Ovulation occurs after the oocyte undergoes the first meiotic division, typically 24 to 48 hours before the end of behavioral estrus. Ovulation is preceded by an increase and then decrease in concentrations of LH in serum over a period of up to 10 days. Multiple ovulations are not uncommon, but secondary ovulations typically occur within 48 hours of the initial ovulation. Additional ovulations during the luteal phase have been reported, but the physiological basis for them has not been defined. Secretions of FSH and LH increase in parallel, with the initial increase during late estrus/early diestrus and a second increase during mid-diestrus.

Progesterone production begins about 24 hours after ovulation and is maximal between days 6 and 18–20 postestrus. Concentrations of P_4 range from 4 to 8 ng/ml during diestrus, but decline rapidly in the nonpregnant mare at the end of diestrus in response to uterine release of $PGF_{2\alpha}$ and luteolysis. The uterine endometrium releases luteolytic $PGF_{2\alpha}$, but neither the pattern of release required for luteolysis nor endocrine regulation of the uterine production of luteolytic $PGF_{2\alpha}$ are clearly established. However, cervical stimulation results in the release of OXT via the Ferguson reflex and stimulates uterine secretion of $PGF_{2\alpha}$, as does administration of exogenous OXT. It is assumed that the combined effects of P_4, E_2, and OXT are responsible for activation of the luteolytic mechanism in mares, and the CLs of mares are responsive to the luteolytic effects of $PGF_{2\alpha}$ after day 5 postovulation.

3.4.2. Pregnancy

The equine conceptus produces an unknown factor(s) that inhibits uterine release of luteolytic $PGF_{2\alpha}$ (see Sharp, McDowell, Weithenauer, & Thatcher, 1989). In cycling mares, concentrations of $PGF_{2\alpha}$ in uterine venous plasma and uterine flushings increase between days 14 and 16, when luteolysis occurs and concentrations of P_4 in plasma decline. Receptors for $PGF_{2\alpha}$ on luteal cells are abundant between day 14 of the estrous cycle and estrus, as well as day 18 of pregnancy. The equine conceptus migrates between the two uterine horns until fixation on day 18 of pregnancy to activate an antiluteolytic mechanism, as amounts of $PGF_{2\alpha}$ in uterine fluids and uterine venous plasma are reduced and PGFM in peripheral plasma is not released in a pulsatile pattern in pregnant mares. Further, the presence of the conceptus abrogates endometrial production of $PGF_{2\alpha}$ in response to both cervical stimulation and exogenous OXT, indicating the absence of or a reduction in the expression of endometrial OXTR in pregnant mares. Equine conceptuses produce increasing amounts of E_2 between days 8 and 20 of gestation; however, attempts to prolong CL lifespan in mares by injections of E_2 have yielded variable results. The equine conceptus also secretes proteins of 400, 65, and 50 kDa between days 12 and 14 of pregnancy, as well as IFND (Cochet, Vaiman, & Lefevre, 2009), but their role(s) in pregnancy recognition are not known. Equine conceptuses secrete IFND, but there are no reports of its function in mares.

3.5. Rabbits (*Oryctolagus cuniculus*)

3.5.1. The ovarian cycle and estrus

The rabbit doe is polyestrous from puberty at three to five months to the end of her reproductive life, at 12 to 36 months (see Miller & Pawlak, 1994; Ramirez & Beyer, 1994). Domestic rabbits reproduce all year when exposed to greater than 12 hours of light per day. After puberty, mature follicles persist in the ovary for 7 to 10 days before undergoing atresia and being replaced by another wave of follicles that secrete high levels of both E_2 and inhibin. Estradiol stimulates sexual receptivity and inhibin suppresses additional follicular growth by blocking FSH secretion. During this period of growth and atresia of follicles, does are receptive to mating and exhibit lordosis when mounted by does or bucks. Does ovulate in response to mating with sterile bucks and then experience a 16- to 18-day period of pseudopregnancy that is analogous to diestrus. The vulva of receptive does is red and swollen, and the does exhibit 'chinning' behavior.

The rabbit, an induced ovulator, ovulates in response to mating or stimulation of the perineal or vaginal area. This elicits a neural input into the hypothalamus for secretion of GnRH and a subsequent ovulatory surge of LH and ovulation 9 to 12 hours postcoitum (pc). The ovulatory surge of LH increases cholesterol mobilization and production of 20α-OHP_4 by ovarian interstitial cells within 10 to 60 minutes pc; it peaks at four to six hours and returns to below basal levels by nine to twelve hours pc. Concentrations of LH and FSH are highest at one to two and two to three hours pc, respectively, and both return to premating levels within six to twelve hours pc. Circulating concentrations of both LH and FSH are low throughout pregnancy except for a second peak of FSH on days one to two of pregnancy. Concentrations of testosterone and E_2 in plasma increase about three-fold between 90 and 120 minutes pc and return to below basal levels by 12 hours pc. At mating, concentrations of PRL in plasma decrease transiently, return to basal levels in about 30 minutes, and increase again three to four days pc and remain elevated for up to two-thirds of gestation. In addition to its role in lactation, PRL stimulates steroidogenesis in the rabbit ovary. Concentrations of OXY in the blood increase in response to mating, parturition, and lactation. There is a second release of OXY about five hours pc and it may stimulate uterine and oviductal contractions responsible for sperm transport.

Regression of the CL occurs at the end of pregnancy or pseudopregnancy in response to $PGF_{2\alpha}$ produced by the ovary and/or endometrium, or in response to injections. The uterus synthesizes prostaglandins; however, hysterectomy delays, but does not completely block, luteolysis in pseudopregnant does. The CL also synthesizes $PGF_{2\alpha}$, PGE_2, and 6-keto-prostaglandin $F_{1\alpha}$, suggesting intraovarian mechanisms for luteolysis. The increase in uterine PGFs may initiate luteal regression, because levels of P_4 in serum decline prior to increases in PGFs. Purified PGE-reductase from the CL of pseudopregnant rabbits possesses both PGE-9-keto-reductase and 20α-HSD activity, suggesting that both P_4 and PGE_2 are substrates and that PG production and steroid metabolism are tightly linked in the luteolytic cascade in rabbits. The CL from pseudopregnant does becomes responsive to exogenous $PGF_{2\alpha}$ around day 12, whereas the CL of pregnant does is responsive to the luteolytic effects of $PGF_{2\alpha}$ as early as day 7 but is unresponsive by day 15. Thus, the conceptus affects the CL prior to maternal recognition of pregnancy on day 12 and renders the CL more resistant to $PGF_{2\alpha}$ after maternal recognition of pregnancy (Marcinkiewicz, Moy, & Bahr, 1992). The basis for the altered sensitivity of the CL to $PGF_{2\alpha}$ is not known. There is also participation of immune cells in luteolysis in rabbits (Nariai, Kanayama, Endo, & Tsukise, 1995).

3.5.2. Pregnancy

The rabbit has a duplex uterus, each with a cervix, so the male deposits semen into the anterior vagina so that sperm can be transported into each uterine horn to fertilize ova, which are ovulated from each ovary approximately 10 hours pc (Browning, Keyes, & Wolf, 1980). Fertilization occurs at the ampullary–isthmic junction of the oviduct one to two hours postovulation and embryos enter the uterus on day 3 and blastocysts undergo implantation on day 7. Rabbits have a hemochorial placenta with discoid villous distribution. The placenta is not a source of P_4 so a CL is required for production of sufficient P_4 to maintain pregnancy to term. Following sterile mating, a CL forms and persists for 14 to 16 days, a period known as pseudopregnancy. For both pseudopregnant and pregnant does, P_4 increases from 1 to 2 ng/ml on day 2 pc to 12–20 ng/ml between days six and eight pc. Between days eight and ten pc, P_4 profiles of pregnant and pseudopregnant does diverge as levels decline rapidly to basal levels between days 16 and 18 of pseudopregnancy. Pregnant does exhibit elevated levels of P_4 until three to four days prior to parturition, which occurs between days 28 and 36 postcoitum. Circulating levels of P_4 and E_2 are not different between pregnant and pseudopregnant does until after implantation (Browning et al., 1980).

Maternal recognition of pregnancy in rabbits occurs between days 10 and 12 pc in response to E_2 and an unidentified placental luteotropin (see Keyes, Kostyo, & Towns, 1994). Luteal cells contain LH receptors; however, LH does not stimulate P_4 production *in vivo*. Rather, estrogen exerts its luteotropic effect by uncoupling P_4 production from cAMP. If E_2 is withdrawn from does, exogenous CG stimulates luteal cAMP and both CG and cAMP stimulate P_4 production. The luteotropic effect of the placenta does not result from increased concentrations or affinity of luteal ERα for E_2. Rabbit placentae secrete immunoreactive GnRH-like activity, which appears to act locally on the uterus but not directly on luteal cells (Nowak & Bahr, 1987). A putative 6 to 8 kDa placental luteotropic factor also has been reported to enhance P_4 production by cultured luteal cells alone or in conjunction with E_2 (Gadsby, 1989). Further, a 12 to 14 kDa rabbit placental luteotropin that is acidic, trypsin-, and heat-sensitive also has been reported (Marcinkiewicz & Bahr, 1993). This factor stimulated production of P_4 by luteal explants in the presence of E_2; however, 200 μg/ml of conceptus protein was necessary to achieve a modest increase in P_4 production. Rabbit placental giant cells contain immunoreactive CG and cytotrophoblast cells contain immunoreactive CSH1/PRL. However, the effects of CSH1 and PRL on luteal cells are not known (Grunder, Hrabe de Angelis, & Kirchner, 1994). It is known that circulating levels of E_2 increase, P_4 levels decrease, and PRL levels increase about two days prepartum. Prolactin influences nest-building behavior but mainly affects lactogenesis and milk production, whereas E_2 and P_4 are only temporally associated with nest-building (Negatu & McNitt, 2002). As for other species, E_2 and P_4 also affect mammogenesis and lactogenesis.

3.6. Domestic Cats (*Felis catus*)

3.6.1. The estrous cycle

The domestic cat is a seasonally polyestrous, induced ovulator with onset of cyclicity occurring as early as four months and as late as 21 months of age depending on breed, photoperiod, and level of nutrition (Tsutsui & Stabenfeldt, 1993). Cats are long-day breeders, so litters are generally born in the spring and summer months in the northern hemisphere, but cats exhibit sexual activity throughout the year in the tropics. In the absence of mating, estrus lasts two to ten days (average seven days), with females vocalizing and exhibiting 'treading' with hind legs, rubbing against objects, and accepting males for mating. Interestrous periods of 3 to 14 days (average 10 days) separate periods of estrus and are characterized by low circulating concentrations of E_2 and nonreceptivity to mating. With growth of ovarian follicles there is increased secretion of E_2 over a two- to three-day period prior to estrus, with circulating levels of 40−100 pg/ml at estrus. Queens ovulate mature follicles in response to mating-induced activation of a neuroendocrine reflex, which releases GnRH and an ovulatory surge of LH as early as five minutes pc that peaks at 20 minutes and returns to basal levels by 60 minutes. However, cats sometimes ovulate in the absence of cervical stimulation and physical contact with other cats. The ovulatory surge of LH varies in amplitude (10 to 100 ng/ml) and duration (1 to 24 hours), based on the number of copulations. Multiple matings on successive days of estrus ensure ovulation in all queens, but less than 50% of queens in estrus ovulate after a single copulation (see Wildt, Chan, Seager, & Chakraborty, 1981).

The CL of domestic cats is resistant to exogenous $PGF_{2\alpha}$ even late in gestation. However, production of $PGF_{2\alpha}$ by the fetal−placental unit and endometrium increases during the last half of pregnancy, reaches a plateau around day 45, and increases sharply just before parturition. Administration of exogenous $PGF_{2\alpha}$ to cats after day 40 of gestation induces abortion, but the mechanism is not known (Tsutsui & Stabenfeldt, 1993).

3.6.2. Pregnancy

Cats have a bipartite uterus and the male deposits semen in the anterior vagina at ejaculation (see Tsutsui & Stabenfeldt, 1993). Ovulation occurs 25 to 50 hours pc and frequent matings reduce the time to ovulation. Fertilization takes place in the oviduct up to 48 hours after ovulation, and embryos enter the uterus at the blastocyst stage four to six days postovulation. Blastocysts hatch from the zona pellucida on day 11 and implantation occurs on days 12 to 13 of pregnancy. The cat has an endotheliochorial-type placenta with zonary villous distribution. Following mating, concentrations of P_4 in plasma increase to 15 to 90 ng/ml between days 10 and 40 of pregnancy and days 13 to 30 of pseudopregnancy. Pseudopregnancy typically lasts 40 days whereas the length of gestation averages 63 to 65 days but ranges from 56 to 71 days. By day 30, circulating levels of P_4 are higher in pregnant than pseudopregnant queens. Physical activity, litter size, and nutritional status all affect the duration of gestation in queens.

The placenta does not produce sufficient P_4 to maintain pregnancy, as ovariectomy on day 45 results in a rapid decline in circulating P_4 and abortion within six to nine days (Verstegen et al., 1993). Prolactin levels increase during the last trimester of gestation to peak values at parturition (5−10 ng/ml) and remain elevated during lactation in response to suckling stimulus. Prolactin is considered an important luteotropin in late gestation. Relaxin, produced by the fetal−placental unit, increases to 5−10 ng/ml plasma during the second half of gestation and, acting in concert with P_4, maintains a quiescent uterus but later facilitates parturition by softening the connective tissues of the pelvis. Following parturition, queens experience anestrus during lactation and resume cycling two to three weeks after weaning kittens. A comprehensive review of the endocrinology of domestic and nondomestic felids indicates differences among species, including changes in circulating levels of the various hormones of pregnancy (see Brown, 2006). Brown's review highlights the variability in endocrine profiles among domestic and nondomestic felids, including the type of ovulation (spontaneous vs. induced), steroid metabolism, seasonal effects on reproduction, adrenal responses to husbandry practices, and ovarian responses to exogenous gonadotropins and steroids.

3.7. Domestic Dogs (*Canis lupus familiaris*)

3.7.1. The ovarian cycle and estrus

The female dog, or bitch, is monoestrous and spontaneously ovulates once or twice per year from 7 to 12 months of age to reproductive senescence (Concannon, 1993). Follicles grow and secrete E_2 and inhibin at the end of anestrus in association with a transient increase in LH and slight reduction in FSH just prior to proestrus. Ovarian inhibin suppresses FSH secretion during proestrus as E_2 increases to between 50 and 100 pg/ml. Proestrus lasts about seven days, but ranges from three days to three weeks. A bloody discharge from the vagina is common during proestrus, and the vaginal and perineal areas increase in size and turgidity. An LH surge (5–20 ng/ml) at the onset of estrus is accompanied by a decrease in circulating levels of E_2, increasing concentrations of P_4, and increased sexual receptivity. The GnRH-induced LH surge lasts two to three days and concentrations of FSH peak shortly after the LH surge and return to basal levels in one to two days. Bitches ovulate about 48 hours after the LH surge. During anestrus, concentrations of LH are low and pulsatile while levels of FSH are five to ten times higher than during proestrus. Oocytes are at the germinal vesicle stage when ovulated, reach metaphase II in the oviduct, and are fertilizable for two to three days as spermatozoa are viable in the female reproductive tract for six to seven days. Fertilization occurs four days after the LH surge, morulae/blastocysts enter the uterus on days 9 to 10, and implantation occurs on days 16–18 after the LH surge.

The length of the luteal phase is similar in pseudopregnant and pregnant bitches and concentrations of P_4 reach 15 to 90 ng/ml between days 10 and 40 post-LH surge before declining to term at 64 to 66 days of gestation (Concannon, 1993). Circulating concentrations of E_2 increase during the last half of gestation and likely stimulate development of the mammary glands. Concentrations of PRL increase in the serum of pregnant bitches between days 30 and 40 of gestation and peak at one to two days prior to parturition (50–60 ng/ml). Prolactin is luteotropic as suppression of PRL secretion causes luteolysis after day 30 of pregnancy. Prolactin may regulate the length of anestrus by affecting the secretion of FSH and/or LH or the responsiveness of the ovary to these gonadotropins (Jeffcoate, 1993). Relaxin, a pregnancy-specific hormone in bitches, is produced by both the ovary and uterus during the last trimester of gestation and it remains detectable in the maternal blood for one to two months postpartum (Steinetz, Goldsmith, Hasan, & Lust, 1990). The specific functions of RLX in the bitch are not known, but RLX may affect uterine functions, cervical dilation at parturition, and mammary development, as reported for rodents (see Sherwood, 2004).

Luteolysis is protracted, as concentrations of P_4 decrease gradually at the end of the luteal phase and remains at less than 1 ng/ml until the next follicular phase. However, luteolysis occurs rapidly immediately postpartum due to the absence of the effects of LH and PRL. Frequent administration of exogenous $PGF_{2\alpha}$ is luteolytic in the bitch, but hysterectomy does not prolong CL lifespan, which suggests that luteolysis is uterine-independent (Concannon & McCann, 1989).

3.7.2. Pregnancy

Oocytes are fertilized two to five days after ovulation in the bitch and, on day 10, blastocysts enter the uterus, where they are free-floating until hatching and implantation occur around day 16 (Concannon & McCann, 1989). The dog has an endotheliochorial placenta with zonary villous distribution. The CL is the primary source of P_4 as both ovariectomy and hypophysectomy at any stage of pregnancy results in abortion. Since the CLs of pregnancy and pseudopregnancy have similar lifespans, a pregnancy recognition signal does not seem to be required for CL maintenance and pregnancy. A recent comprehensive review of the endocrinology of pregnancy in the bitch indicates that much is yet to be learned about reproduction in dogs (Verstegen-Onclin & Verstegen, 2008). The concentrations of P_4 in plasma are similar in pregnant and pseudopregnant bitches prior to implantation, but P_4 levels increase postimplantation in parallel with increases in circulating levels of RLX. Secretions of PRL and RLX increase in parallel independently of the effects of E_2, produced by the CL. Prolactin is essential for CL maintenance and function, whereas basal concentrations of LH from the anterior pituitary may be permissive to CL function during pregnancy. Luteolysis precedes parturition due to the release of $PGF_{2\alpha}$ from the uterus, beginning 36 hours prepartum, but the role of estrogens in parturition is not known. As with the other species, it is assumed that E_2, P_4, PRL, RLX, insulin, IGF1, and glucocorticoids are required for mammogenesis and lactogenesis in dogs.

4. SUMMARY AND CONCLUSIONS

Comparative aspects of reproduction in various mammals indicate that the mechanisms responsible for recurring estrous cycles and the transition from recurring estrous cycles to pregnancy are highly variable and may involve signaling from the anterior pituitary, conceptus, uterus, and/or ovaries. The common theme, however, is that each mechanism for pregnancy recognition serves to ensure maintenance of the CL and the continued production of P_4 to support pregnancy, mammogenesis, and lactogenesis. The strategies for CL maintenance range from the direct effects of a product of the conceptus (e.g., CG) or anterior pituitary (e.g., PRL) in primates and rodents; to anti-luteolytic effects of conceptus signals, which abrogate the luteolytic mechanism in the uterus (e.g., IFNT and E_2) in domestic ruminants, horses, and pigs; to a situation in which the lengths of diestrus and CL maintenance (pseudopregnancy and pregnancy) are of approximately equivalent lengths without a requirement for a luteotropin or antiluteolytic hormone (e.g., dog, cat). Progesterone, the permissive hormone of pregnancy, is required for uterine functions supportive of pregnancy; this is due, in part, to its suppression of PR in uterine epithelia, but also to the stimulation of the secretion of progestamedins by the uterine stromal cells to affect the function of uterine epithelia and the trophectoderm, chorioamnion, or chorioallantois, depending on the species. It is also clear that one or more IFNs are produced by the trophectoderm and/or resident immune cells to act independently or in concert with P_4/progestamedins to affect the expression of genes critical to uterine functions that, in turn, affect conceptus development and the birth of viable offspring. It is also important to note that, in most species that have been examined in detail, P_4 not only prepares the uterus for pregnancy but is also central to the development of the uterine luteolytic mechanism, which ensures luteolysis in the event that pregnancy is not established. This area of research will continue to be exciting as advances provide strategies for enhancing the reproductive efficiency and reproductive health of humans and animals.

5. FUTURE RESEARCH

There is much known about the endocrinology of the estrous cycle and novel mechanisms responsible for conceptus–endometrial interactions for the establishment and maintenance of pregnancy. However, there are many mechanisms yet to be discovered. A fundamental gap in our knowledge concerns the requirement for the loss of expression of PR by endometrial epithelia as a prerequisite for implantation, the expression of genes for secretory proteins, and the selective transport of molecules into the uterine lumen to support conceptus growth and development for the successful establishment of pregnancy. Clarification of what regulates the downregulation of PR and the identification of progestamedins or estramedins unique to each species is necessary in terms of understanding mechanisms whereby these sex steroids regulate uterine epithelia individually and in concert with IFNs, lactogenic hormones, and prostaglandins, for successful pregnancy outcomes. Comparative reproductive biology is necessary to advance our understanding of these mechanisms. For example, the ewe is a proven model for research in terms of understanding the roles of IFNs during the peri-implantation period because the trophectoderm or immune cells, as the sites of implantation of most, if not all, mammals, are now known to express type I and/or type II IFNs. Thus, IFN-stimulated genes are among the most highly upregulated genes in human decidualized stromal cells treated with a trophoblast-conditioned medium and in the uteri of domestic and laboratory animals. Understanding the effects of ovine IFNT on gene expression in the uterus will advance our understanding of novel mechanisms whereby P_4 and IFNs directly or indirectly act on the cells of the reproductive system to induce ISGs critical to the establishment and maintenance of pregnancy in mammals. Similarly, understanding the roles of novel endogenous retroviruses in reproductive tissues will advance our understanding of their roles in implantation, placentation, and the endocrinology of pregnancy. This knowledge is essential for translational research into strategies to enhance reproductive efficiencies and reproductive health in humans and animals.

ABBREVIATIONS

20α-HSD 20-alpha-hydroxysteroid dehydrogenase
20α-OHP$_4$ 20α-hydroxyprogesterone
2-AG 2-arachidonyl glycerol
AEA Arachidonoyl ethanloamine
AKR1B1 Aldo-keto reductase family 1 member B1
AVP Arginine vasopressin
B2M Beta 2 microglobulin
cAMP Cyclic-3',5'-adenosine monophosphate
CD24 Cluster of differentiation 24
CG Chorionic gonadotropin
CL Corpus luteum
CNS Central nervous system
CSFR1 Colony-stimulating factor-1 receptor
CSH1 Chorionic somatomammotropin hormone
dPRP Decidual prolactin-like protein
E_2 Estradiol
EGF Epidermal growth factor
EGFR Epidermal growth factor receptor
ERα Estrogen receptor alpha
FGF Fibroblast growth factor
FGFR2IIIb Fibroblast growth factor receptor-2 splice variant
FSH Follicle-stimulating hormone
GBP1 Guanylate binding protein-1
GE Glandular epithelium
GH1 Growth hormone
GHR Growth hormone receptor
GnRH Gonadotropin-releasing hormone
GnRHR Gonadotropin-releasing hormone receptor
GRP54 Kisspeptin receptor
GTP Guanosine triphosphate
hCG Human chorionic gonadotropin
HGF Hepatocyte growth factor
HPA Hypothalamic–pituitary–adrenal
HPG Hypothalamic–pituitary–gonadal
IFN Interferon
IGF Insulin-like growth factor
IGF1R Insulin-like growth factor-1 receptor
IGFBP Insulin-like growth factor binding protein

IL	Interleukin
IRF	Interferon regulatory factor
IRG	Interferon-regulated gene
ISG	Interferon-stimulated gene
JAK	Janus-activated kinase
LE	Luminal epithelium
LGR	See RXFP
LH	Luteinizing hormone
LHCGR	Luteinizing hormone/chorionic gonadotropin receptor
LIF	Leukemia inhibitory factor
LPA3	Lysophosphatidic acid
LPS	Lipopolysaccharide
LSA	Lysophosphatidic acid
MAPK	Mitogen-activated protein kinase
MET	Met proto-oncogene
MHC	Major histocompatibility complex
mPR	Membrane progesterone receptor
mPOA	Medial preoptic area
MPR	Membrane progesterone receptor
MUC1	Mucin, surface-associated
MX1	Myxovirus resistance-1 protein
NK	Natural killer
NMβ	Neuromedin-β
NO	Nitric oxide
NOS	Nitric oxide synthase
NPY	Neuropeptide Y
NTS	Nucleus tratus solitarius
OXT	Oxytocin
OXTR	Oxytocin receptor
P_4	Progesterone
PAQR	P_4 adiponectin Q receptor
pc	Postcoitum
PC6	Proprotein convertase-6
PDE4A	Phosphodiesterase
PG	Prostaglandin
PGE_2	Prostaglandin E_2
$PGF_{2\alpha}$	Prostaglandin $F_{2\alpha}$
PGFM	13,14-dihydro,15-keto- $PGF_{2\alpha}$
PGFR	PGF2α receptor
PI3K	Phosphoinositide-3 kinase
PK	Protein kinase
PLPB	Prolactin-like protein B
PR	Progesterone receptor
PRL	Prolactin
PRLPB	Prolactin-like protein B
PRLR	Prolactin receptor
PTGS2	Prostaglandin synthase-2
PVN	Paraventricular nucleus
RLX	Relaxin
RXFP	Relaxin/insulin-like family peptide receptor
sGE	Superficial glandular epithelia
SLA	Swine leukocyte antigen
Sp1	Transcription factor Sp1
SPP1	Secreted phosphoprotein-1
StAR	Steroidogenic acute regulatory
STAT1	Signal transducers and activators of transcription
TGFβ	Transforming growth factor-β
TIMP	Tissue inhibitor of matrix metalloproteinase
TNFα	Tissue necrosis factor
TSH	Thyroid-stimulating hormone
UTMP	Uterine milk proteins
VEGF	Vascular endothelial growth factor
VMH	Ventromedial hypothalamus
ZO1	Zona occludens-1

REFERENCES

Aboagye-Mathiesen, G., Toth, F. D., Zdravkovic, M., & Ebbesen, P. (1995). Human trophoblast interferons: production and possible roles in early pregnancy. *Early Pregnancy, 1*, 41–53.

Anderson, L. L. (1993). Pigs. In E. S. E. Hafez (Ed.), *Reproduction in Farm Animals* (pp. 343–360). Philadelphia, PA: Lea & Febiger.

Austin, K. J., Bany, B. M., Belden, E. L., Rempel, L. A., Cross, J. C., & Hansen, T. R. (2003). Interferon-stimulated gene-15 (Isg15) expression is up-regulated in the mouse uterus in response to the implanting conceptus. *Endocrinology, 144*, 3107–3113.

Bany, B. M., & Cross, J. C. (2006). Post-implantation mouse conceptuses produce paracrine signals that regulate the uterine endometrium undergoing decidualization. *Develop. Biol., 294*, 445–456.

Bazer, F. W. (1989). Establishment of pregnancy in sheep and pigs. *Reprod. Fertil. Develop., 1*, 237–242.

Bazer, F. W. (1992). Mediators of maternal recognition of pregnancy in mammals. *Proc. Soc. Exp. Biol. Med., 199*, 373–384.

Bazer, F. W., Burghardt, R. C., Johnson, G. A., Spencer, T. E., & Wu, G. (2008). Interferons and progesterone for establishment and maintenance of pregnancy: Interactions among novel cell signaling pathways. *Reprod. Biol., 8*, 179–211.

Bazer, F. W., Goldstein, M. H., & Barron, D. H. (1981). Water and electrolyte transport by pig chorioallantois. In L. Mastroianni, J. D. Biggers, & W. A. Sadler (Eds.), *Fertilization and Embryonic Development* In Vitro (pp. 289–321). New York, NY: Plenum Publishing Corp.

Bazer, F. W., Spencer, T. E., & Johnson, G. A. (2009). Interferons and uterine receptivity. *Semin. Reprod. Med., 27*, 90–102.

Ben-Jonathan, N., LaPensee, C. R., & LaPensee, E. W. (2008). What can we learn from rodents about prolactin in humans? *Endocrine Rev., 29*, 1–41.

Brown, J. L. (2006). Comparative endocrinology of domestic and nondomestic felids. *Theriogenology, 66*, 25–36.

Browning, J. Y., Keyes, P. L., & Wolf, R. C. (1980). Comparison of serum progesterone, 20 alpha-dihydroprogesterone, and estradiol-17 beta in pregnant and pseudopregnant rabbits: evidence for postimplantation recognition of pregnancy. *Biol. Reprod., 23*, 1014–1019.

Burton, G. J., Jauniaux, E., & Charnock-Jones, D. S. (2007). Human early placental development: potential roles of the endometrial glands. *Placenta, 28*(Suppl. A), S64–69.

Cochet, M., Vaiman, D., & Lefèvre, F. (2009). Novel interferon delta genes in mammals: cloning of one gene from the sheep, two genes expressed by the horse conceptus and discovery of related sequences in several taxa by genomic database screening. *Gene, 433*, 88–99.

Concannon, P. W. (1993). Biology of gonadotrophin secretion in adult and prepubertal female dogs. *J. Reprod. Fertil., 47*(Suppl.), 3–27.

Concannon, P. W., & McCann, J. P. (1989). Biology and endocrinology of ovulation, pregnancy and parturition in the dog. *J. Reprod. Fertil., (Suppl.), 39*, 3–25.

Fazleabas, A. T., Kim, J. J., & Strakova, Z. (2004). Implantation: embryonic signals and the modulation of the uterine environment—a review. *Placenta, 25*, S26–S31.

Freeman, M. E., Smith, M. S., Nazian, S. J., & Neill, J. D. (1974). Ovarian and hypothalamic control of the daily surges of prolactin secretion during pseudopregnancy. *Endocrinology, 94*, 875–882.

Gadsby, J. E. (1989). Control of corpus luteum function in the pregnant rabbit. *J. Reprod. Fertil., 37*(Suppl.), 45–54.

Geisert, R. D., & Yelich, J. V. (1997). Regulation of conceptus development and attachment in pigs. *J. Reprod. Fertil. Suppl., 52*, 133–149.

Gellersen, B., Fernandes, M. S., & Brosens, J. J. (2009). Non-genomic progesterone actions in female reproduction. *Human Reprod. Update, 15*, 119–138.

Goldstein, M. H., Bazer, F. W., Spellacy, W. N., & Buhi, W. C. (1976). Stimulation of active transport across porcine and human placentae by human placental lactogen. *Gynecol. Invest., 7*, 58.

Grunder, C., Hrabe de Angelis, M., & Kirchner, C. (1994). Chorionic gonadotropin-like proteins in the placental giant cells of the rabbit. *Cell Tissue Res., 278*, 573–578.

Gu, Y., Srivastava, R. K., Ou, J., Krett, N. L., Mayo, K. E., & Gibori, G. (1995). Cell-specific expression of activin and its two binding proteins in the rat decidua: role of α_2-macroglobulin and follistatin. *Endocrinology, 136*, 3815–3822.

Guilbault, L. A., Thatcher, W. W., Collier, R. J., & Wilcox, C. J. (1985). Periparturient endocrine changes of conceptus and maternal units in holstein heifers bearing genetically different conceptuses. *J. Anim. Sci., 61*, 1505–1515.

Hess, A. P., Hamilton, A. E., Talbi, S., Dosiou, C., Nyegaard, M., Nayak., N., et al. (2007). Decidual stromal cell response to paracrine signals from the trophoblast: amplification of immune and angiogenic modulators. *Biol. Reprod., 76*, 102–117.

Irvine, C. H. G. (1995). The nonpregnant mare: a review of some current research and of the last 25 years of endocrinology. In: "Equine Reproduction IV" (F.W. Bazer and D.C. Sharp, Eds.), Biol Reprod Monograph Series Vol. 1, pp 343–360.

Jeffcoate, I. A. (1993). Endocrinology of anoestrous bitches. *J. Reprod. Fertil., 47*(Suppl.), 69–76.

Kann, G., Delobelle-Deroide, A., Belair, L., Gertler, A., & Djiane, J. (1999). Demonstration of *in vivo* mammogenic and lactogenic effects of recombinant ovine placental lactogen and mammogenic effect of recombinant ovine GH in ewes during artificial induction of lactation. *J. Endocrinol., 160*, 365–377.

Kensinger, R. S., Collier, R. J., Bazer, F. W., Ducsay, C. A., & Becker, H. N. (1980). Nucleic acid, metabolic and histological changes in gilt mammary tissue during pregnancy and lactogenesis. *Biol. Reprod., 22*, 1168–1180.

Keyes, P. L., Kostyo, J. L., & Towns, R. (1994). The autonomy of the rabbit corpus luteum. *J. Endocrinol., 143*, 423–431.

Knight, J. W., Bazer, F. W., Thatcher, W. W., Franke, D. E., & Wallace, H. D. (1977). Conceptus development in intact and unilaterally hysterectomized–ovariectomized gilts: Interrelations among hormonal status, placental development, fetal fluids and fetal growth. *J. Anim. Sci., 44*, 620–637.

Leroy-Heinrichs, W., & Ribbon, W. E. (1989). Endocrinology of pregnancy. In S. A. Brodie, & K. Veland (Eds.), *Endocrine Disorders of Pregnancy* (pp. 65–80). East Norwak, CT: Appleton and Lange.

Leymarie, P., & Martal, J. (1991). The corpus luteum from cycle to gestation. In C. Thibault, M. C. Levasseur, & R. H. F. Hunter (Eds.), *Reproduction in Mammals and Man* (pp. 413–433). Paris, France: Ellipses.

Ludwig, T. E., Sun, B. C., Carnahan, K. G., Uzumcu, M., Yelich, J. V., Geisert, R. D., et al. (1998). Endometrial responsiveness to oxytocin during diestrus and early pregnancy in pigs is not controlled solely by changes in oxytocin receptor population density. *Biol. Reprod., 58*, 769–777.

Marcinkiewicz, J. L., & Bahr, J. M. (1993). Identification and preliminary characterization of luteotropic activity in the rabbit placenta. *Biol. Reprod., 48*, 403–408.

Marcinkiewiz, J. L., Moy, E. S., & Bahr, J. M. (1992). Change in responsiveness of rabbit corpus luteum to prostaglandin F-2α during pregnancy and pseudopregnancy. *J. Reprod. Fertil., 94*, 305–310.

Miller, J. B., & Pawlak, C. M. (1994). Characterization and physiological variation in prostaglandin, prostacyclin, and thromboxane synthesis by corpora lutea, non-luteal and uterine tissues during pseudopregnancy in the rabbit. *Life Sci., 54*, 341–353.

Nariai, K., Kanayama, K., Endo, T., & Tsukise, A. (1995). Effects of splenectomy on luteolysis in pseudopregnant rabbits. *J. Vet. Med. Sci., 57*, 503–505.

Negatu, Z., & McNitt, J. I. (2002). Hormone profiles and nest-building behavior during the periparturient period in rabbit does. *Anim. Reprod. Sci., 72*, 125–135.

Newton, G. R., Ott, T. L., Woldesenbet, S., Shelton, A. H., & Bazer, F. W. (1996). Biochemical and immunological properties of related small ruminant trophoblast interferons. *Theriogenology, 46*, 703–716.

Nowak, R. A., & Bahr, J. M. (1987). Secretion of a gonadotrophin-releasing hormone-(GnRH)like factor by the rabbit fetal placenta *in vitro*. *Placenta, 8*, 299–304.

Platanias, L. C. (2005). Mechanisms of type-I- and type-II-interferon-mediated signalling. *Nature Rev. Immunol., 5*, 375–386.

Ramirez, V. D., & Beyer, C. (1994). In E. Knobil, & J. D. Neill (Eds.), *The Physiology of Reproduction. The neuroendocrine control of the rabbit ovarian cycle, Vol 2* (pp. 585–612). New York: Raven Press.

Rao, C. V. (2001). Multiple novel roles of luteinizing hormone. *Fertil. Steril., 76*, 1097–1100.

Sharp, D. C., McDowell, K. J., Weithenauer, J., & Thatcher, W. W. (1989). The continuum of events leading to maternal recognition of pregnancy in mares. *J. Reprod. Fertil., (Suppl.), 37*, 101–107.

Sherwood, D. O. (2004). Relaxin's physiological roles and other diverse actions. *Endocrine Rev., 25*, 205–234.

Slayden, O. D., & Keater, C. S. (2007). Role of progesterone in nonhuman primate implantation. *Semin. Reprod. Med., 25*, 418–430.

Soares, M. J., Konno, T., & Khorshed Alam, S. K. M. (2007). The prolactin family: effectors of pregnancy-dependent adaptations. *Trends Endocrinol. Metabol., 18*, 114–121.

Spencer, T. E., & Bazer, F. W. (2002). Biology of progesterone action during pregnancy recognition and maintenance of pregnancy. *Frontiers in Biosciences, 7*, 1879–1898.

Spencer, T. E., & Bazer, F. W. (2004). Uterine and placental factors regulating conceptus growth in domestic animals. *J. Anim. Sci., 82 E-Suppl.* E4–13.

Spencer, T. E., Burghardt, R. C., Johnson, G. A., & Bazer, F. W. (2004). Conceptus signals for establishment and maintenance of pregnancy. *Anim. Reprod. Sci., 82-83*, 537–550.

Spencer, T. E., Johnson, G. A., Bazer, F. W., Burghardt, R. C., & Palmarini, M. (2007). Pregnancy recognition and conceptus implantation in domestic ruminants: roles of progesterone, interferons and endogenous retroviruses. *Reprod. Fertil. Develop., 19*, 65–78.

Steinetz, B. G., Goldsmith, L. T., Hasan, S. H., & Lust, G. (1990). Diurnal variation of serum progesterone, but not relaxin, prolactin, or estradiol-17 beta in the pregnant bitch. *Endocrinology, 127*, 1057–1063.

Stouffer, R. L. (1988). Perspectives on the corpus luteum of the menstrual cycle and early pregnancy. *Semin. Reprod. Endocrinol., 6*, 103–113.

Thatcher, W. W., Guzeloglu, A., Mattos, R., Binelli, M., Hansen, T. R., & Pru, J. K. (2001). Uterine–conceptus interactions and reproductive failure in cattle. *Theriogenology, 56*, 1435–1450.

Thomas, P. (2008). Characteristics of membrane progestin receptor alpha (mPRa) and progesterone membrane receptor component 1 (PGMRC1) and their roles in mediating rapid progestin actions. *Frontiers in Neuroendocrinol., 29*, 292–312.

Tsutsui, T., & Stabenfeldt, G. H. (1993). Biology of ovarian cycles, pregnancy and pseudopregnancy in the domestic cat. *J. Reprod. Fertil., (Suppl.), 47*, 29–35.

Verstegen, J. P., Onclin, K., Silva, L. D. M., Wouters-Ballman, P., Delahaut, P., & Ectors, F. (1993). Regulation of progesterone during pregnancy in the cat: studies on the roles of corpora lutea, placenta and prolactin secretion. *J. Reprod. Fertil., 47*(Suppl.), 165–173.

Verstegen-Onclin, K., & Verstegen, J. (2008). Endocrinology of pregnancy in the dog: a review. *Theriogenology, 70*, 291–299.

Wildt, D. E., Chan, S. Y. W., Seager, S. W. J., & Chakraborty, P. K. (1981). Ovarian activity, circulating hormones, and sexual behavior in the cat. I. Relationships during the coitus-induced luteal phase and the estrous period without mating. *Biol. Reprod., 25*, 15–28.

Young, K. H., Kraeling, R. R., & Bazer, F. W. (1990). Effect of pregnancy and exogenous ovarian steroids on endometrial prolactin receptor ontogeny and uterine secretory response in pigs. *Biol. Reprod., 43*, 592–599.

Ziecik, A. J., Kaczmarek, M. M., Blitek, A., Kowalczyk, A. E., Li, X., & Rahmanc, N. A. (2007). Novel biological and possible applicable roles of LH/hCG receptor. *Mol. Cell. Endocrinol., 269*, 51–60.

Chapter 6

The Comparative Physiology of Parturition in Mammals: Hormones and Parturition in Mammals

I. Ross Young,* Marilyn B. Renfree,† Sam Mesiano,** Geoff Shaw,† Graham Jenkin†† and Roger Smith***

*Monash University, Clayton, Victoria, Australia, †The University of Melbourne, Victoria, Australia, **Case Western Reserve University, Cleveland, OH, USA, ††Monash Institute of Medical Research (MIMR), Clayton, Victoria, Australia, ***The University of Newcastle, NSW, Australia

SUMMARY

The physiology of parturition in mammals is highly diverse and almost species specific. A range of examples is presented from the marsupial Tammar wallaby (Macropus eugenii) to the primate Homo sapiens. In some species, such as the sheep, parturition is initiated by the fetus while in others, for example the goat, parturition is initiated by the mother. In the human and other great apes parturition appears to be regulated by placental production of steroids from both fetal (estriol) and maternal (estradiol) origin. The role of progesterone related steroids and estrogens varies dramatically amongst species. In a number of mammals progesterone withdrawal initiates labor while in others a rise in estrogens is the precipitant. However, in other species, such as the Tammar, neither estrogens or progestagens appear to have any role. In all species examined to date prostaglandins play a role in the process of parturition although the regulation of increases in prostaglandins varies with differing prostaglandin synthases involved in different species and differences in the tissues of origin and sites of action. The enormous variation in the physiology of parturition is related to the high rates of perinatal death of both mother and offspring and the consequent evolutionary pressures that are exerted on this process.

1. INTRODUCTION

Much of our knowledge of the physiology of parturition in mammals has been gained from studies of only a few species, notably the sheep, pig, rat, rabbit, some nonhuman primates, and the human. From the perspective of comparative physiology, this is a poor coverage as these species fall into only four of the 28 orders and only half a dozen or so of the ~5000 species of mammals. Moreover, the fundamental mechanism underlying the initiation of parturition is not established in any species, and considerable variation exists among the species that have been investigated. Rather than surveying the physiology of parturition species by species, this contribution will attempt to identify mechanisms that vary widely among species and those that are well-conserved. This classical comparative approach explicitly assumes that highly conserved mechanisms are fundamentally important because they have remained unchanged in the face of changing selective pressures during the course of evolution. While the comparative approach on its own can tell us what is fundamentally important about the process we are investigating, our understanding can be expanded by simultaneous consideration of the phylogenetic relationships among the taxa we are studying. In doing so, we make an assumption that fundamental, conserved processes have evolved earlier and are more widespread among distantly related taxa than are processes that exhibit wider variations (Somero, 2000). In this chapter we shall apply this style of reasoning to the following questions: What determines gestation length? How are fetal maturation and birth synchronized? How are the uterotonic mechanisms activated?

Parturition is fundamental to viviparity, but even a cursory overview of the mammals makes it apparent that there is huge variation. This review will draw heavily from studies in three species: the tammar wallaby (tammar), the sheep, and the human. Tammars are marsupials with a relatively brief gestation (the organogenesis phase lasts barely eight days). The highly altricial neonate is barely 1/10 000 of maternal weight, with many organ systems poorly developed, but it climbs to the pouch and finds a teat where it then spends around nine months of lactation completing its development. Sheep have more extensive development *in utero* and the relatively large precocial young are able to stand and run minutes after birth; the young gain independence after a relatively short lactation. Humans too give birth to large young. However, their large

Hormones and Reproduction of Vertebrates, Volume 5—Mammals

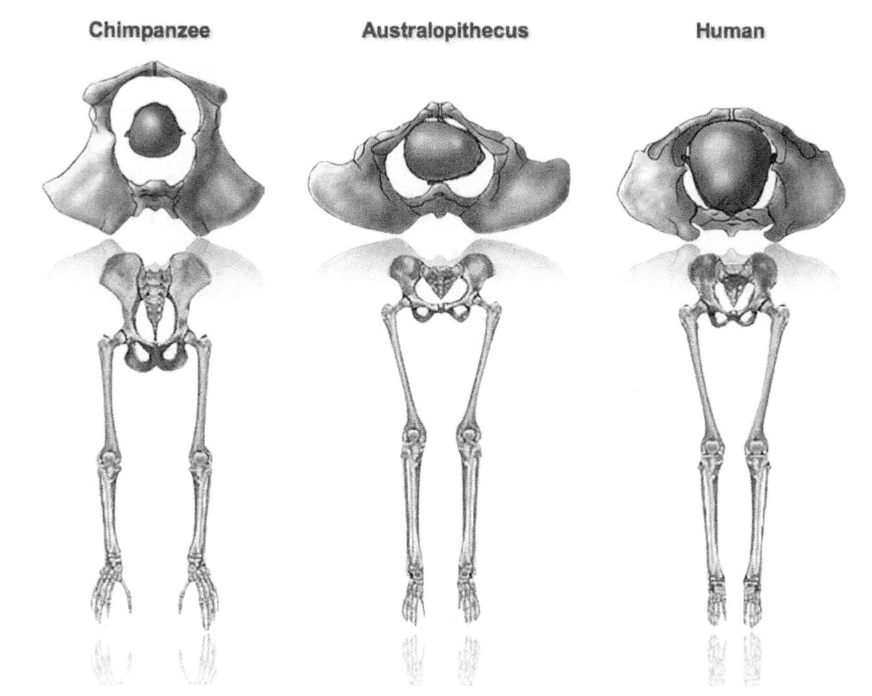

FIGURE 6.1 In the chimpanzee, as in all extant primates except *Homo sapiens*, the fetal head passes through the maternal pelvis with the face in an anterior position facing the pubic symphysis. Upright posture as present in Australopithecus was associated with a narrowing of the anterior to posterior diameter of the pelvis, forcing the fetus at delivery to rotate to traverse the birth canal. Further changes to the pelvis in *Homo sapiens* require a series of rotations to enable the large head and wide shoulders of the human fetus to negotiate the birth passage, placing a severe constraint on the size of the fetus and especially the fetal brain at the time of birth. See color plate section.

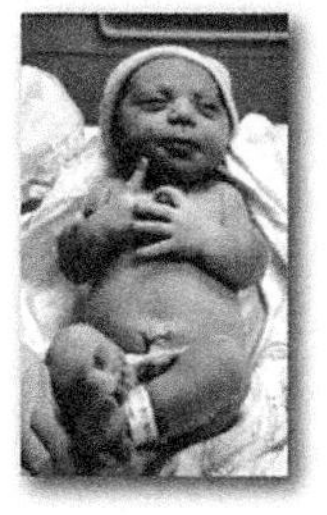

FIGURE 6.2 Various mammals adopt different strategies regarding pre- and postnatal development. Precocial species have relatively long gestational periods and deliver relatively mature, independent young; altricial species have shorter gestational periods and deliver immature, dependent young. Primates in general follow the precocial pattern, giving birth to relatively independent young after a long gestational period; however, humans have a complex pattern consisting of a long gestational period with delivery of an immature offspring that is highly dependent for an extended period of time. The human pattern seems to be a consequence of changes required for upright posture and possession of a large brain. The human brain is relatively large compared to total body size at birth and continues to grow rapidly after birth. The relatively narrow pelvis of primates has become narrowed with the adoption of upright posture (see Figure 6.1) in Australopithecus and modern humans, providing a physical limit on the size of the fetal brain that can traverse the birth canal. See color plate section.

brains provide a challenge at birth when the relatively large head must fit through a pelvis designed for bipedalism (Figure 6.1). Consequently, humans have acquired the trait of secondary altriciality, giving birth to a very immature neonate in which the brain continues to grow postnatally to a much greater extent than observed in other primates (Figure 6.2). The late maturation of the brain means that human offspring require prolonged maternal care during an extended lactation and beyond, before they become independent.

Failure of birth means failure to pass on genes, with likely death of both mother and offspring, so there has been heavy Darwinian selection of the processes of parturition. Because of this pressure, failsafe systems may be present with multiple pathways to fetal expulsion. Whichever of these pathways is first triggered becomes the normal 'mechanism'. The redundancy in the pathway provides a fertile ground for evolution as secondary paths can be independently developed and eventually subsume the role of the previously dominant path, which may then become superfluous and be lost from the genome. Thus, as species have evolved, the trigger used may have changed—so long as one pathway is effective the genes will be passed on and different 'mechanisms' become normal. However, the various trigger mechanisms tend to converge on a smaller set of final processes that stimulate uterine contractions (notably oxytocin (OXY) and prostaglandin $F_{2\alpha}$ ($PGF_{2\alpha}$)), remove inhibition (notably via regulation of progesterone (P_4) action and nitric oxide synthase (NOS)) and facilitate delivery (e.g., cervical ripening, facilitated by hormones such as relaxin and prostaglandin E_2 (PGE_2)). Clearly these processes need to be synchronized in order to effect birth, and so a key question becomes how this synchronization is achieved and how this is coordinated with fetal development, so that birth occurs when the fetus is developmentally ready for delivery.

2. WHAT DETERMINES GESTATION LENGTH?

The duration of gestation for each species is remarkably constant. This suggests that some time-measuring or counting process may be involved. If so, the process may involve counting the number of cell divisions that have occurred since the time of syngamy, although experimental confirmation of this is lacking. Alternatively, the process may involve measurement of the time elapsed since syngamy, which would require counting the repetitions of an isochronous process, as occurs in clocks. This may occur in the mother, the placenta, or the fetus. If the putative time-keeping mechanism resides in the mother, the isochronous process may be the external photoperiod or some physiological rhythm linked to it, such as a circadian rhythm in the concentration of a metabolic (e.g., plasma glucose), endocrine (e.g., plasma melatonin (MEL) or cortisol (F)), or behavioral (e.g., activity or feeding) parameter. The recent discovery of genes and proteins that may transduce photic information into physiological signals in the mother may lead to an improved understanding of the mechanisms underlying the establishment of 24-hour rhythms (Gekakis et al., 1995; Shearman et al., 1997; Steeves et al., 1999; Zheng et al., 1999; Hida et al., 2000); however, no link has been made to a circadian cycle-counting mechanism, which would be expected if these phenomena form the basis of a clock that keeps count of days. In any case, empirical evidence suggests that photoperiodic or circadian information is not a necessary determinant of gestation length in primates and rats, although it modulates the time of day when births occur (Bosc, 1990; Honnebier et al., 1991; Rowland et al., 1991; Honnebier & Nathanielsz, 1994). The ovine fetus can signal its own birth after destruction of the fetal optic nerves and suprachiasmatic nuclei (SCN) (Poore et al., 1999), and maternal pinealectomy abolishes the

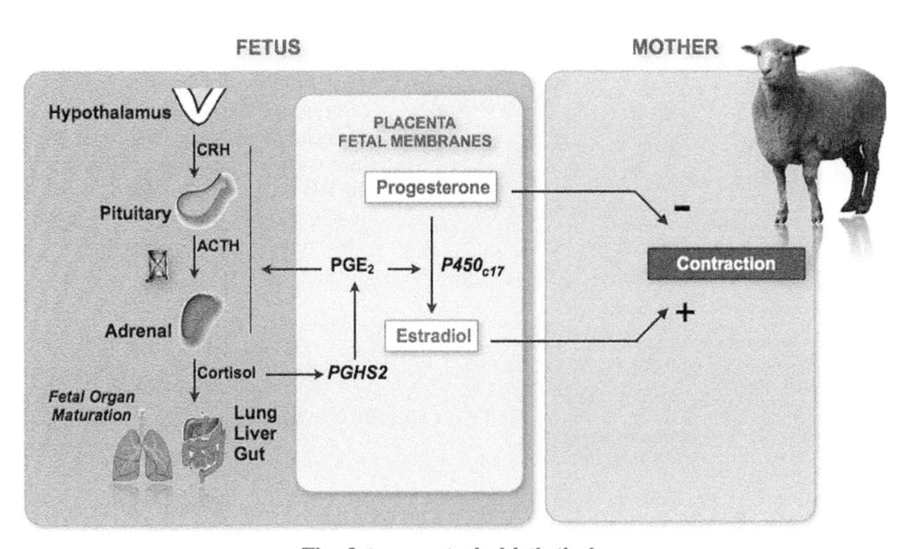

FIGURE 6.3 In the sheep, maturation of the fetal hypothalamus appears to regulate the timing of birth. ACTH, corticotropin; CRH, corticotropin-releasing hormone; $P450_{C17}$, 17-hydroxylase 17,20 lyase; PGE_2, prostaglandin E_2; PGHS2, prostaglandin synthase-2. See color plate section.

diurnal MEL rhythm but does not affect the duration of gestation (McMillen & Nowak, 1989). These observations suggest that the recognized components of the fetal or maternal biological clocks do not affect gestation length in the sheep.

In species such as the sheep, where the signal to end gestation arises within the fetus (Figure 6.3), it is difficult to conceive of a time-counting mechanism operating from the time of conception, except at the cellular level. The structures we associate with generating or marking the diurnal rhythms (the SCN and pineal gland) do not differentiate until the conceptus is several weeks old. The problem becomes deeper when we consider species exhibiting embryonic diapause, a state of suspended development in which the blastocyst remains quiescent in the uterus awaiting a maternal signal to recommence growth. In the tammar, embryos may develop continuously with a gestation length of ~27 days or, more usually, with an 11-month period of embryonic diapause when the gestation is a year. In the latter case, the sum of the periods before diapause and after the signal to end diapause is the same as the uninterrupted gestation length plus a few days required to transmit and receive the maternal signal to reactivate the embryo (Renfree, 1994). However, the timing appears to be related to the developmental stage of the embryo, because there is a remarkably uniform period of 19 days from the time of a P_4 pulse (which occurs four to six days after removal of pouch young) to birth, whereas the time from fertilization, reactivation, or artificially induced reactivation of the blastocyst is variable (Renfree, 1994). It is therefore unlikely that a fetal clock-like function is involved in determining gestation length in this marsupial.

In tammars one might argue for a clock based on the lifespan of the corpus luteum, which undergoes luteolysis with a corresponding fall in plasma P_4 at birth (reviewed in Shaw & Renfree, 2006). At this time there is normally also a rise in plasma 17β-estradiol (E_2) in concert with the start of postpartum estrus (Tyndale-Biscoe et al., 1988). Falling P_4 and rising E_2 are known mechanisms that can activate uterotonic mechanisms in some eutherian mammals. However, these changes in P_4 and E_2 in tammars are not important for birth, since not all females have a rise in E_2 and postpartum estrus (Shaw & Renfree, 1984), and, when folliculogenesis is blocked by passive immunization, preventing a rise in E_2 (Short, Flint, & Renfree, 1985), the timing of birth is not affected. Similarly, birth timing is unaffected if P_4 levels near term are manipulated (Shaw, 2006). Moreover, in many nonmacropodid marsupial species, luteolysis or rising E_2 are not normally closely associated with parturition (Tyndale-Biscoe & Renfree, 1987).

As a time-measuring mechanism does not appear to determine gestation length, we may consider the possibility that the conceptus of each species develops at a genetically determined rate and that birth occurs in response to a signal given when the fetus (or litter) attains sufficient size or maturity. Such a signal may be transduced by the mother (Figure 6.4) (e.g., uterine volume), the fetus (e.g., nutrient restriction; developmental transition), or the placenta (e.g., increased fetal demand for nutrients). In all of these cases, the duration of gestation would ultimately depend on the genetically determined developmental or growth rate of the conceptus, and not on any objective measurement of time. Uterine volume, once considered a possible determinant of gestation length, now seems unlikely to be so, although larger uterine volume (e.g., in multiple vs. singleton ovine or human pregnancy) is associated with slightly shorter gestation length. In the case of multiple ovine pregnancy, birth occurs much closer to the time expected for singleton pregnancies (145−150 days) than to the time (about 120 days) when the combined fetal mass reaches that of a term singleton pregnancy. Further, dysfunction of the fetal pituitary or adrenals in sheep is associated with prolonged gestation and continued fetal growth. In this case, the ewe does not deliver the fetus but eventually dies of starvation because the uterine volume

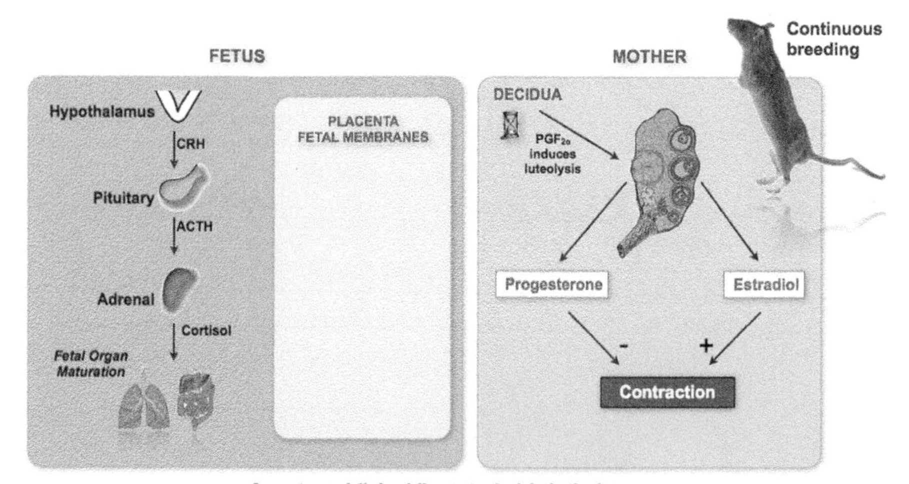

FIGURE 6.4 In rodents, the timing of birth appears to be controlled by events in the maternal tissues leading to luteolysis and a fall in progesterone (P_4) levels. ACTH, corticotropin; CRH, corticotropin-releasing hormone; PGF_2, prostaglandin F_2. See color plate section.

compromises her ability to take in sufficient food (Van Kampen & Ellis, 1972). Thus, increased uterine volume appears insufficient to trigger birth in the sheep. In the human, while twin gestations deliver prematurely at a much higher rate than singletons, this appears to be related more to placental hormone secretion than to increased uterine volume or myometrial tension (Smith et al., 2009).

Persuasive arguments have been made that the rapid accretion of fetal tissue in late gestation outstrips the ability of the placenta to supply substrates at the necessary rate, with the result that fetal substrate concentrations actually fall, activating the fetal hypothalamo–pituitary–adrenal (HPA) axis, which, in at least some species, provides the signal for birth. An alternative hypothesis suggests that the supply of fetal nutrients is maintained by adaptation of the placenta to the increased fetal demand, but that one of the placental responses to that increased demand is the secretion of a factor, hypothetically PGE_2, that activates the fetal HPA axis. Evidence that brain PGE_2 modulates the preparturient rise in corticotropin (ACTH) in the fetal sheep has recently been published (Gersting et al., 2008). Both these hypotheses are subject to the objection that substrate limitation would evolve gradually and it is difficult to reconcile this feature with the precision that normally characterizes the timing of birth.

In marsupials, the term fetus is tiny and places almost no metabolic demands on the mother—e.g., a 5000 g female tammar has a single 0.4 g fetus at term (Tyndale-Biscoe & Renfree, 1987). However, most marsupials have a relatively noninvasive yolk sac placenta, so it is possible that parturition occurs when the limit of placental transfer is reached. When parturition in tammars was prevented by administration of indomethacin, dead postmature fetuses 20% heavier than normal were recovered from the uterus two days after the normal birth time (Renfree, 1994). Whether the fetal death was attributable to toxic effects of indomethacin or to fetal growth beyond what the placenta could supply is not clear. However, it is hard to argue that such a limitation occurs in another marsupial group, the bandicoots, which develop an invasive chorioallantoic placenta in the last day or so of the 12.5 day gestation (Padykula & Taylor, 1976), yet the term fetus weighs under 250 mg and is at a relatively earlier stage of development than the tammar neonate.

Since it is difficult to find conclusive support for time-keeping mechanisms or fetal size as fundamental determinants of the onset of parturition, other mechanisms must be considered. One possibility is that the mechanism for the timing of birth is encoded in the fetal genome and is activated when certain prerequisite developmental events have occurred. The details of such a putative mechanism are unknown, but may be elucidated with further research, just as the molecular bases for other, previously mysterious, processes in the early development of the embryo are being discovered.

Human pregnancy lasts approximately 38 weeks after conception, with minor variations among different racial groups (Patel et al., 2004). In humans, the timing of birth is associated with development of the placenta and, in particular, placental synthesis of corticotropin-releasing hormone (CRH) (McLean et al., 1995). An association between levels of maternal plasma CRH, which is of placental origin, and the timing of birth has been found in several large cohort studies (Hobel et al., 1999; Leung et al., 1999; Holzman et al., 2001; Inder et al., 2001; Leung et al., 2001; Ellis et al., 2002; Sandman et al., 2006; Torricelli et al., 2007). Maternal plasma CRH increases exponentially as pregnancy advances, peaking at the time of delivery. In women destined to deliver preterm, the exponential increase is more rapid, whereas in women destined to deliver after the estimated date of delivery the rise is slower (Figure 6.5) (McLean et al., 1995; Torricelli et al., 2006). These findings suggest that a placental clock determines the timing of delivery in humans (McLean et al., 1995).

Production of CRH by the placenta is restricted to primates (Robinson et al., 1989; Smith et al., 1999; Bowman et al., 2001) and even within the order of primates the pattern of production varies considerably. In New and Old World

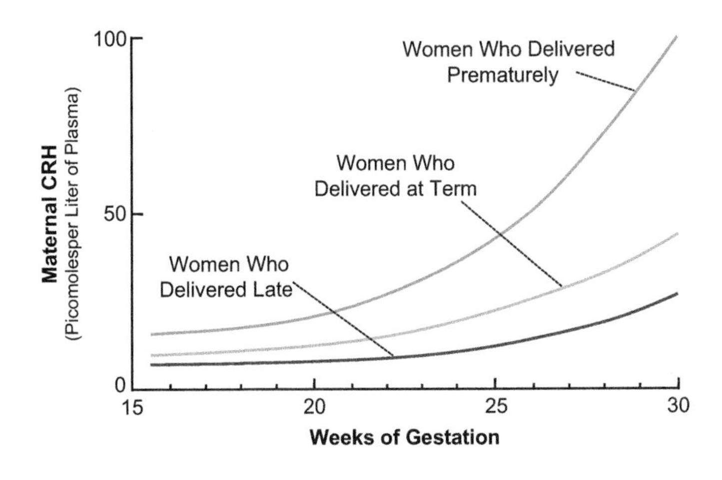

FIGURE 6.5 Placentally produced corticotropin-releasing hormone (CRH) can be measured in maternal plasma in humans. The concentration of CRH in maternal plasma follows an exponential curve that peaks at the time of delivery. In women who deliver prematurely (less than 37 completed weeks of gestation), the rate of rise of the exponential curve is greater than that observed in those that deliver at term (37–40 weeks), which is greater than that of those who deliver late (more than 40 weeks). See color plate section.

monkeys there is a pronounced peak in CRH production in midgestation, but only in the great apes is there an exponential production similar to the rise in maternal CRH in humans (Smith et al., 1993; Smith et al., 1999). Humans and other great apes also produce a circulating binding protein for CRH (CRHBP). At the end of pregnancy, CRHBP levels fall, thereby increasing the bioavailability of CRH (Linton et al., 1993; McLean et al., 1995).

Glucocorticoids stimulate expression of the *CRH* gene and production of CRH by the placenta (Robinson et al., 1988; Cheng et al., 2000). In turn, CRH stimulates ACTH production by the pituitary and this ACTH causes release of F by the adrenal cortex. This arrangement permits a positive feed-forward system that has been shown by mathematical modeling to mimic the changes observed in human pregnancy (Emanuel et al., 1994). Placental CRH production is also modified by estrogens, P_4, and nitric oxide (NO), which are inhibitory, and by a range of neuropeptides, including adrenalin, noradrenalin, and acetylcholine, which are stimulatory (Petraglia et al., 1989; Ni et al., 1997; 2002; 2004). In each individual woman, levels of placental CRH in maternal blood follow an exponential function to produce a particular trajectory for that pregnancy so that each woman maintains her position relative to median levels throughout the pregnancy. That is, if a woman has relatively low levels early in pregnancy compared to her peers, she will also have relatively low levels late in pregnancy. Small changes in the exponential function describing CRH production at the beginning of pregnancy lead to large differences in concentrations between different women later in pregnancy. Given the large variations between individuals, it is likely that the rate of rise of maternal concentrations of CRH, rather than absolute levels, is the signal for the onset of labor (Leung et al., 2001; McGrath et al., 2002). African Americans have lower maternal plasma CRH concentrations than other racial groups studied to date, although among African American women CRH concentrations correlate with the timing of birth (Holzman et al., 2001) and African American women have relatively high rates of preterm birth.

Thus, in humans and other great apes, it seems likely that the length of gestation is determined by the maturation of the placenta and the developmental regulation of expression of the CRH gene in the placental tissue modified by factors that impinge on this process through stress-related production of glucocorticoids. The observed bias of human deliveries to the night time (Lindow et al., 2000) suggests that diurnal rhythm in other factors, such as adrenal steroid production, provides a fine tuning to the actual time of delivery.

The underlying factor determining gestation length in most species remains elusive, and we should remain open to the possibility that these factors may differ between species and that in all species the control may be multifactorial.

3. HOW ARE FETAL MATURATION AND PARTURITION SYNCHRONIZED?

The timely birth of suitably mature fetuses requires that some mechanism synchronizes fetal maturation with the maternal mechanisms that effect the birth, including preparatory (uterotropic) and expulsive (uterotonic) phases. In some species, including ruminants and the tammar, the synchronizing factor is glucocorticoid secreted by the fetal adrenal cortex.

In tammars, the fetal adrenals almost double in size in the last two days of gestation, and their F content increases 10-fold, then falls postpartum (Renfree 1994; Ingram, Shaw, & Renfree, 1999). Cortisol concentrations in fetal blood and allantoic and yolk-sac fluid follow a similar pattern. Adrenals from term fetuses secrete F *in vitro*, and F synthesis is stimulated by ACTH and by PGE_2 (Ingram et al., 1999). Administration of exogenous glucocorticoids on day 24 precipitates premature parturition 22 hours later on day 25, one day earlier than normal (Shaw, Renfree, & Fletcher, 1996). Glucocorticoids probably also synchronize organ development with parturition. For example, lung surfactant is controlled by glucocorticoids and is activated by the time of birth, even though the lungs are only at the early stages of the terminal air-sac phase in tammars (Renfree, 1994). This rise in F is therefore clearly a fetal signal for parturition in the tammar.

The rising glucocorticoid concentrations that characterize the late gestation fetal sheep not only accelerate the maturation of organ systems required for extrauterine life, but induce placental enzymes (especially 17α-hydroxylase-C17,20 lyase ($P450_{c17}$)) that increase the synthesis of estrogens (potent uterotropic factors) at the expense of P_4. These changes, in turn, lead to synthesis of uterotonins such as prostaglandins (PGs), with consequent activation of the myometrium and delivery.

The F concentrations in ovine fetal plasma increase exponentially over the last month of the 147-day gestation, a phenomenon that has come to be known as the F surge (Poore et al., 1998). The role of increased glucocorticoid concentrations in accelerating the maturation of multiple organ systems needed for postnatal survival is well established. An example is the respiratory system, where glucocorticoid is needed for structural and functional maturation of the lungs, a feature that has led to the prophylactic use of glucocorticoid treatment in cases of threatened premature delivery in humans. This treatment helps to mature the lung and reduce the incidence and severity of respiratory distress syndrome in the neonate. The obligate requirement for prenatal glucocorticoid exposure was underlined when it was shown that neonatal mice lacking the CRH gene cannot survive birth unless they receive glucocorticoid via their mothers *in utero* (Cole et al., 1995; Muglia et al., 1995).

It is important to establish the precise role of glucocorticoids in fetal maturation, including whether they act to initiate or simply accelerate the differentiation of cells in the organs of interest. As a result of the mechanism of action of glucocorticoid receptors on the genome of the target cell, there is a clear potential for these hormones to affect gene expression and thus to serve as initiators of cellular differentiation. Evidence acquired to date in the hematopoietic system suggests that, rather than acting in this way, glucocorticoids accelerate maturational change, which is initiated by other means (Wintour et al., 1985; Zitnik et al., 1995; Wessely et al., 1997). Another issue requiring resolution is the degree of prenatal glucocorticoid exposure that is optimal. As a result of the approximately 30-fold increase in F concentration observed in fetal sheep in the month before birth, it might be assumed that this degree of exposure is required. None-the-less, the F concentrations of human fetuses appear to rise more slowly and to a more modest extent, bringing into question the degree of F exposure needed in this species. Infants born with defective F synthesis, e.g., those affected by congenital adrenal hyperplasia, have low plasma glucocorticoid concentrations but do not usually come to clinical attention because of lung immaturity (New & Speiser, 1986; White & Speiser, 2000). This observation suggests that the degree of glucocorticoid exposure required for organ maturation in humans may be quite small. However, studies of fetal HPA activity in humans (measured by the levels of estriol (E_3) in the maternal circulation) suggest that late in the third trimester maternal F crosses the placenta and enters the fetal circulation (Patrick et al., 1979). This may represent an adaptive trait to ensure that fetal organ systems are matured in situations where fetal glucocorticoid production is compromised.

In addition to its maturational role, the F surge also constitutes the signal for parturition in the sheep and other ruminants (Liggins et al., 1967; Liggins, 1968; Bassett & Thorburn, 1969) (and even in the tammar, which has a ruminant-type physiology (Ingram et al., 1999)). Glucocorticoids do not serve this role in all species, but an increase in fetal glucocorticoid concentration occurs prior to parturition in all species examined to date, indicating that it plays an essential role in late gestation, presumably that of promoting organ maturation. In species such as the mouse, horse, and human, which do not rely on a fetal F signal to initiate birth, the possibility still exists for a causal relationship between fetal glucocorticoid-mediated maturation and the birth process. The link involves one of the ubiquitous uterotonins, PGE_2. This eicosanoid is secreted in increasing amounts by the placenta or fetal membranes in many species during the prelude to labor and is a potent activator of the fetal HPA axis (Louis et al., 1976; Hollingworth et al., 1995). Thus, even in species that do not rely on glucocorticoid as the signal for birth, glucocorticoid concentrations in fetal plasma may increase as a consequence of the parturient mechanisms. The often poor outcome for prematurely born humans suggests it may be of little practical importance, although, in this case, structural immaturity of the lung and other vital organs may limit their ability to respond to glucocorticoids and it is important to extrapolate cautiously from pathological phenomena to normal physiology.

On balance, the universality of the fetal F surge preceding normal labor at term suggests that it constitutes a fundamental signal. It is clearly linked to maturational processes in the fetus, although not universally to the initiation of parturition.

4. HOW DOES THE FETUS SIGNAL THE INITIATION OF LABOR?

The role of fetally derived glucocorticoids in initiating birth in ruminants is well established. In species such as the sheep, in which the placenta is the principal source of P_4 for pregnancy maintenance, glucocorticoids induce the expression of enzymes that redirect the steroidogenic pathway to favor the production of estrogens at the expense of P_4 (Anderson et al., 1975; Steele et al., 1976). The resultant increase in the estrogen/P_4 ratio causes several uterotropic changes and activates prostaglandin G/H synthase in the placenta, leading to myometrial activation and labor (Flint et al., 1974; Liggins, 1974; McLaren et al., 1996; 2000). Thus, the prepartum F surge is accepted as the fetal signal for labor in the sheep and related species as well as in the tammar, and probably in other macropodid marsupials (see Figures 6.3 and 6.4).

4.1. A More Complicated Case: The Corpus Luteum-Dependent Species

The goat, a close relative of the sheep, presents an intriguing and significant difference from the sheep. In this species, as in the sheep, the fetal F surge is indispensible for the initiation of labor (Currie & Thorburn, 1977a; 1977b; Flint et al., 1978; Ford et al., 1998), yet the corpus luteum and not the placenta is the principal source of P_4. This condition is termed corpus luteum dependence because removal of the corpus luteum results in loss of the pregnancy. To effect a decrease in P_4 secretion, the fetal glucocorticoid signal must be transmitted to this remote site and induce luteolysis. It seems extremely unlikely that fetal F could cross the placenta, be mixed with maternal blood in the maternal circulatory system, and reach the corpus luteum in sufficient concentration to induce luteolysis directly. A more likely scenario is that F is the signal for some change in placental endocrine function leading to luteolysis. Induction of the placental $P450_{c17}$ enzyme may act, as it does in the sheep, to redirect steroidogenesis from P_4 to E_2 synthesis (Flint et al., 1978). The caprine placenta synthesizes as much P_4 as the ovine placenta, but it does not secrete sufficient P_4 to maintain pregnancy because the P_4 is metabolized before it can be secreted (Sheldrick et al., 1980; 1981). As fetal glucocorticoid secretion induces placental $P450_{c17}$ activity in this species (Flint et al., 1978), the induction of this enzyme may lead to the synthesis of E_2, using either pregnenolone or P_4 as its substrate. Exogenous E_2 can induce luteolysis in nonpregnant does and labor in pregnant does (Currie et al., 1976), although the mechanism by which it does so is not fully established.

When maternally administered E_2 was used to induce parturition in does, the latency to parturition was about three days and peripheral plasma P_4 concentrations started to decrease about 40 hours after the commencement of the E_2 infusion. Utero–ovarian venous $PGF_{2\alpha}$ concentrations subsequently increased markedly in association with labor; however, episodes of $PGF_{2\alpha}$ secretion were seen earlier, about 30 hours into the infusion period, and these may have induced luteolysis (Currie et al., 1976). Consistent with this interpretation, direct infusion of $PGF_{2\alpha}$ into the utero–ovarian vein induced labor in 30–36 hours, after a precipitous decrease in plasma P_4 concentrations during the first six hours of $PGF_{2\alpha}$ infusion (Currie & Thorburn, 1973). These data suggest that, in the normal doe at term, an increase in endogenous E_2 secretion could induce PG synthesis, leading to luteolysis and labor, although experimental corroboration of this hypothesis has been difficult to obtain. When larger numbers of does were studied after a longer period of postoperative equilibration, it became apparent that P_4 decreases over the last ten days of gestation with a more rapid decline in the last three days. This latter change coincides with an increase in E_2 concentration as expected, but no evidence of increased pulsatile or mean $PGF_{2\alpha}$ secretion was found until the final, labor-associated increase on the day of birth (Ford et al., 1998; 1999).

Further studies are needed to clarify the nature of the luteolytic signal(s) in the goat. This information may have considerable economic importance as the physiology of parturition in the goat parallels that in cattle. Increasingly, large dairy herds in developed countries use synchronized, artificially assisted breeding programs with consequent synchronization of calving. None-the-less, there is still considerable variability in calving dates, resulting in significant costs to producers in terms of the human workload needed to minimize and manage obstetrical mishaps. Improved understanding of the normal physiology of parturition in corpus luteum-dependent species may permit more effective strategies for manipulation of calving dates with economic benefits to producers and improved welfare for their stock.

4.2. The Role of the Fetal Hypothalamo–pituitary–adrenal (HPA) Axis in the Initiation of Parturition

Although the sheep and the goat differ in the way in which fetally derived F signals the induction of the uterotonic mechanisms, there is general agreement that, in these and most other ruminant species, the fetal F surge is indeed the signal for birth. In keeping with this view, exogenous glucocorticoid induces labor in both species, whether it is administered to the fetus or the mother. In species other than the even-toed ungulates (Artiodactyla), the role of the fetal HPA axis in the initiation of parturition is less clear. For example, in the horse, an odd-toed ungulate (Perissodactyla), the fetal F profile increases only in the last 48 hours before delivery and maternally administered glucocorticoid does not induce labor as it does in sheep and goats (Silver, 1990; Silver & Fowden, 1991). Similarly, there is little evidence for glucocorticoids as inducers of parturition in rodents. Mice lacking a functional CRH gene have atrophic adrenal cortices, yet are born at the normal time (Muglia et al., 1995) and the timing of birth in guinea pigs is unaffected by ablation of the fetal pituitary (Donovan & Peddie, 1973) or maternal administration of dexamethasone (Donovan & Peddie, 1973; Illingworth et al., 1974). Data from other species are equivocal and it is probably unsafe to assume that fetal glucocorticoids play a causative role in parturition in eutherians generally.

Since fetal glucocorticoids signal birth in only a restricted range of eutherian species, it is noteworthy that there is strong evidence for this phenomenon in a marsupial species, the tammar. The marsupial lineage diverged from that of the Eutheria during the Cretaceous period between 148 and 125 million years ago (Luo et al., 2003; Bininda-Emonds et al., 2007), before the major radiation that gave rise to the Artiodactyla and most other extant eutherian orders. Notwithstanding the distance of their phylogenetic relationship, the tammar conforms to the artiodactyl pattern in that the concentration of F in fetal fluids increases in the last two days of the ~27 day gestation period (Ingram et al., 1999). Exogenous glucocorticoid induces premature birth with maternal endocrine profiles closely matching those associated with spontaneous birth (Renfree & Shaw, 1996; Shaw & Renfree, 2001), and inhibition of glucocorticoid synthesis may delay delivery (Renfree, 1994; Shaw et al., 1996). Whether this physiological similarity between two widely separated groups is an example of convergent evolution or represents the persistence of a primitive condition that has been replaced in many other eutherian taxa is open to debate. In either case, the induction of labor as a result of the ubiquitous prepartum increase in fetal F is not a highly conserved mechanism across taxa and therefore appears not to be a fundamental one.

In primates, the fetal adrenal cortex contains a specific functional compartment known as the fetal zone, which produces the C_{19} androgen dehydroepiandrosterone sulfate (DHEA-S) in response to ACTH. The fetal zone grows disproportionately from around the 10th week of gestation. Dehydroepiandrosterone sulfate is 16-hydroxylated in the fetal liver to form 16-OH-DHEA-S, which is then converted by sulfatase and aromatase into E_3 in the placenta. As the 16-hydroxylase enzyme is only expressed by the fetal liver, the production of E_3 by the placenta is reflective of the steroidogenic activity of the fetal HPA axis. Some DHEA-S, after removal of the sulfate, is also converted to estrone (E_1) and E_2 by the placenta. Thus, the estrogens of human pregnancy (E_1, E_2, and E_3) are produced by the placenta from C_{19} substrate provided by the fetal zone of the fetal adrenal cortex, although production of E_2 and E_1 also occurs from maternal adrenal DHEAs.

Placental CRH is also released into the fetus. Although the concentrations of CRH in the fetal circulation are lower than in the maternal circulation, they still rise with advancing gestation (Nodwell et al., 1999). In the fetus, CRH receptors are present in the pituitary (Asa et al., 1991) and on fetal zone cells in the adrenal cortex (Smith et al., 1998). Based on the known effects of CRH on the HPA axis, it is thought that stimulation of the fetal pituitary by CRH increases ACTH production and consequently the synthesis of F by the fetal adrenal gland and maturation of the fetal lung. In turn, the rising F concentrations in the fetus further stimulate placental CRH production. This leads to an exponential increase in placental CRH production and is reflected by increasing levels in the maternal and fetal circulations. Corticotropin-releasing hormone may have direct effects on fetal organ maturation as well as promoting the steroidogenic activity of the fetal HPA axis. In baboons, CRH directly stimulates fetal lung development and strongly induces surfactant phospholipid synthesis (Emanuel et al., 2000) but it is not clear if this occurs in humans. Interestingly, CRH also directly affects fetal zone function by interacting with specific receptors on fetal zone cells that lead to increased DHEA-S production (Smith et al., 1998). The fetal zone of the adrenal involutes rapidly after delivery of the placenta in association with the decline in circulating CRH levels, which is consistent with the idea that placental factors, such as CRH, maintain the fetal zone.

In New and Old World monkeys the situation is more complex as peak fetal zone size occurs at midgestation. In the New and Old World monkeys, E_1 and E_2 are the dominant estrogens of pregnancy. The two estrogens, however, follow very different patterns: E_2 peaks in midgestation while estrone increases progressively to peak at parturition. Maternal plasma CRH concentrations in these species peak at midgestation (Smith et al., 1993), suggesting that in monkeys CRH drives E_2 production

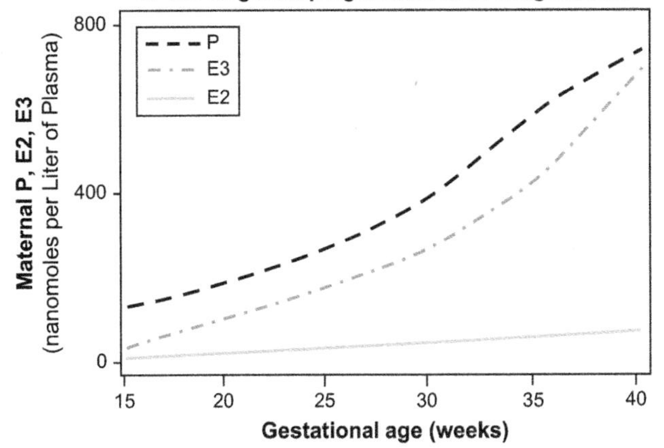

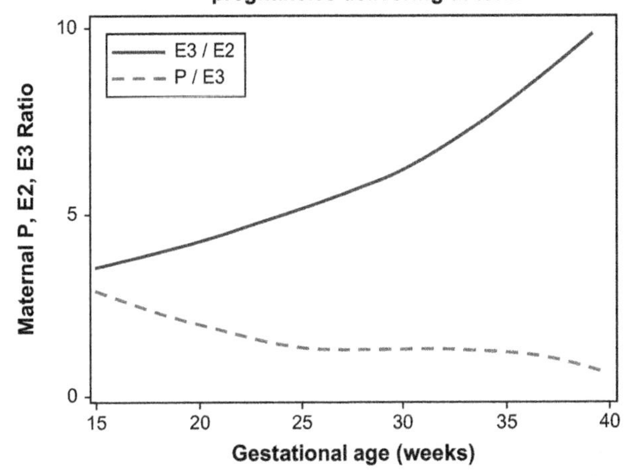

FIGURE 6.6 In human pregnancy, estriol (E_3) derived from fetal adrenal dehydroepiandrosterone sulfate (DHEA-S) increases as pregnancy advances under the drive from the exponential increase in placental corticotropin-releasing hormone (CRH) production. This leads to an increase in the ratio of E_3 to 17β-estradiol (E_2) (which is derived from maternal sources). The increase in E_3 also increases the ratio of E_3 to progesterone (P_4) in maternal plasma. Together, the changes in these ratios increase the biologically effective estrogen action at estrogen receptors and change the balance of estrogenic to progestational actions to favor estrogen and promote the onset of labor. See color plate section.

rather than the E_3 produced from adrenal precursors stimulated by CRH in the great apes (Smith et al., 1999). These contrasting patterns in monkeys and great apes are likely related to the complex actions of the different estrogens at the estrogen receptor (ER). Individually, E_1, E_2, and E_3 are all effective agonists at the ER but act as mutual antagonists at equimolar concentrations (Melamed et al., 1997). In the great apes, in which the major estrogens are E_3 and E_2, the exponential increase in placental CRH drives a late pregnancy increase in E_3 derived from fetal adrenal androgens, which leads to a molar excess of E_3 over E_2 and the estrogenic environment required for labor (Figure 6.6).

In the setting of fetal death the drop in E_3 likely leads to an excess of E_2 and the onset of labor. In the monkeys, a midgestation rise in E_2 opposes the actions of E_1 at the level of the estrogen nuclear receptor. The subsequent fall in E_2 as E_1 is maintained allows a single estrogen (E_1) to become dominant, creating an estrogenic environment that coincides with the onset of labor.

Thus, the complex functional interaction between placental CRH and the fetal HPA axis represents a potential link between the putative placental clock, reflected by placental CRH and the production of glucocorticoids needed for fetal organ maturation and estrogens needed for the process of parturition.

4.3. Signals from the Fetal Lungs

Birth should not occur until the fetal lungs (and other organ systems that are needed for extrauterine life) are sufficiently mature that the newborn is capable of air breathing. As described above, in most species F produced by the fetal adrenals stimulates lung maturation and in some animals it ensures that the timing of birth is synchronized with the functional development of organ systems (especially the lungs) that the newborn will need to survive outside the uterus. Recent studies in rodents (Condon et al., 2004) have indicated that the fetal lungs contribute to the parturition process. In this species, surfactant protein-A (SP-A), a critical lung surfactant protein, produced by the fetal lung epithelium is secreted into the amniotic fluid where it activates fetal macrophages and causes them to migrate to the myometrium. In the myometrium the increased production of cytokines by the fetal macrophages induces activation of the transcription factor NFκB in the myometrial cells, which opposes the transcriptional activity of nuclear P_4 receptors (PRs), leading to functional P_4 withdrawal (see Section X) and parturition (Condon et al., 2004). This interaction ensures that fetal lung maturation, indicated by SP-A production, and birth timing are coordinated. However, evidence for a similar system is not apparent in the human. The trafficking of fetal macrophages in humans and mice differs substantially. However, this does not discount the possibility that SP-A, or some other signaling factor produced by the developing fetal lung, affects myometrial contractility. *In-vitro* studies have shown that SP-A may activate the inflammatory process by interacting with cell surface receptors CD14 and toll-like receptor (TLR)-2 (TLR2) (Murakami et al., 2002; Sato et al., 2003) and TLR4 and MD-2 on macrophages (Yamada et al., 2006). In addition, some studies suggest that SP-A interacts directly with human myometrial cells via a specific receptor to increase prostaglandin production and contractile capacity (Garcia-Verdugo et al., 2008). Thus, SP-A produced by the human fetal lungs may directly affect myometrial contractility and the capacity for the development of an inflammatory response in the gestational tissues, especially the fetal membranes and perhaps the myometrium. This may impact on the sensitivity of the gestational tissues to inflammation-induced parturition. An interesting twist to this paradigm is the observation that the fetal membranes express SP-A in response to glucocorticoid and that SP-A induces PG synthesis by chorionic trophoblasts (Sun et al., 2006).

In summary, the synchronization of fetal maturation with birth is probably achieved through increased fetal glucocorticoid secretion in the prelude to labor. This is also the signal for labor in a minority of species. In the sheep, the glucocorticoid signal is transduced by the placenta and the chain of events leading from this step to myometrial activation is well defined. In corpus luteum-dependent ruminants, the F signal is also transduced by the placenta, but the mechanism linking this step to P_4 withdrawal requires further definition. In most other species glucocorticoids do not provide the signal for labor, but their concentrations in fetal plasma may be increased in response to spillover of uterotonins such as PGE_2 from placental tissues into the umbilical vasculature.

5. HOW ARE THE UTEROTONIC MECHANISMS ACTIVATED?

During gestation, the uterus grows to accommodate the increasing mass of its contents and, as term approaches, it prepares for labor by increasing its ability to contract and expel those contents. The preparatory changes are described as uterotropic, while the mechanisms that effect myometrial activation and contraction are called uterotonic. Clearly, both types of mechanism are important for the successful delivery of healthy young and frequently they are coordinately regulated, so it is somewhat artificial to separate them. The following sections concentrate on the uterotonic factors but also refer to uterotropic mechanisms.

5.1. The Sex Steroids: Progesterone (P_4) and Estrogens

In common laboratory and domestic species such as the rat, mouse, rabbit, sheep, goat, cow, and dog, P_4 is required for the maintenance of pregnancy. It is secreted into the maternal circulation from the corpus luteum of pregnancy and, in some species, the placenta takes over this role at some stage in gestation. The high P_4 concentrations achieved during pregnancy generally favor myometrial quiescence. Progesterone is thought to achieve this effect via the classic genomic mode of steroid hormone action whereby its interaction with nuclear PRs leads to their activation as transcription factors that modulate the expression of specific genes, the products of which promote a relaxed myometrial phenotype. In addition, some studies have indicated that P_4 also binds to membrane-associated PRs (mPRs) and that this directly affects intracellular signaling cascades that hyperpolarize myometrial cells, thus opposing the generation of electrical and mechanical activity within the tissue (Kuriyama & Suzuki, 1976; Parkington, 1983), although there are long-standing technical difficulties in establishing this with certainty (reviewed in Finn & Porter, 1975). In contrast to the relaxatory actions of P_4, estrogens (mainly E_3 in the human) promote labor and delivery by favoring the synthesis of contractile proteins in myometrial cells and enhancing electrical coupling through connexin 43 and the expression of OXY receptors, leading to well-propagated, coordinated contractions. Thus, the general consensus is that through most of pregnancy the influence of P_4 dominates to maintain the uterus in a relaxed and quiescent state with the cervix closed and rigid. At parturition, the relaxatory actions of P_4 are withdrawn and the stimulatory actions of estrogens prevail to transform the myometrium to a highly contractile and excitable state and promote cervical softening. In this context the key event in the hormonal control of parturition is the mechanism by which P_4 withdrawal is mediated and controlled and how this is coordinated with increased estrogenic drive to the myometrium and cervix. This event appears to be fundamental across multiple viviparous species but with variability existing in the physiology and biochemistry of P_4 withdrawal and estrogen activation.

In many species, the maternal plasma P_4 concentration decreases shortly before the onset of labor, and P_4 withdrawal before term results in premature labor (see Figure 6.7).

In others, such as the guinea pig and human, no such marked decrease in plasma P_4 occurs, yet preterm treatment with antiprogestogens induces premature labor, at least

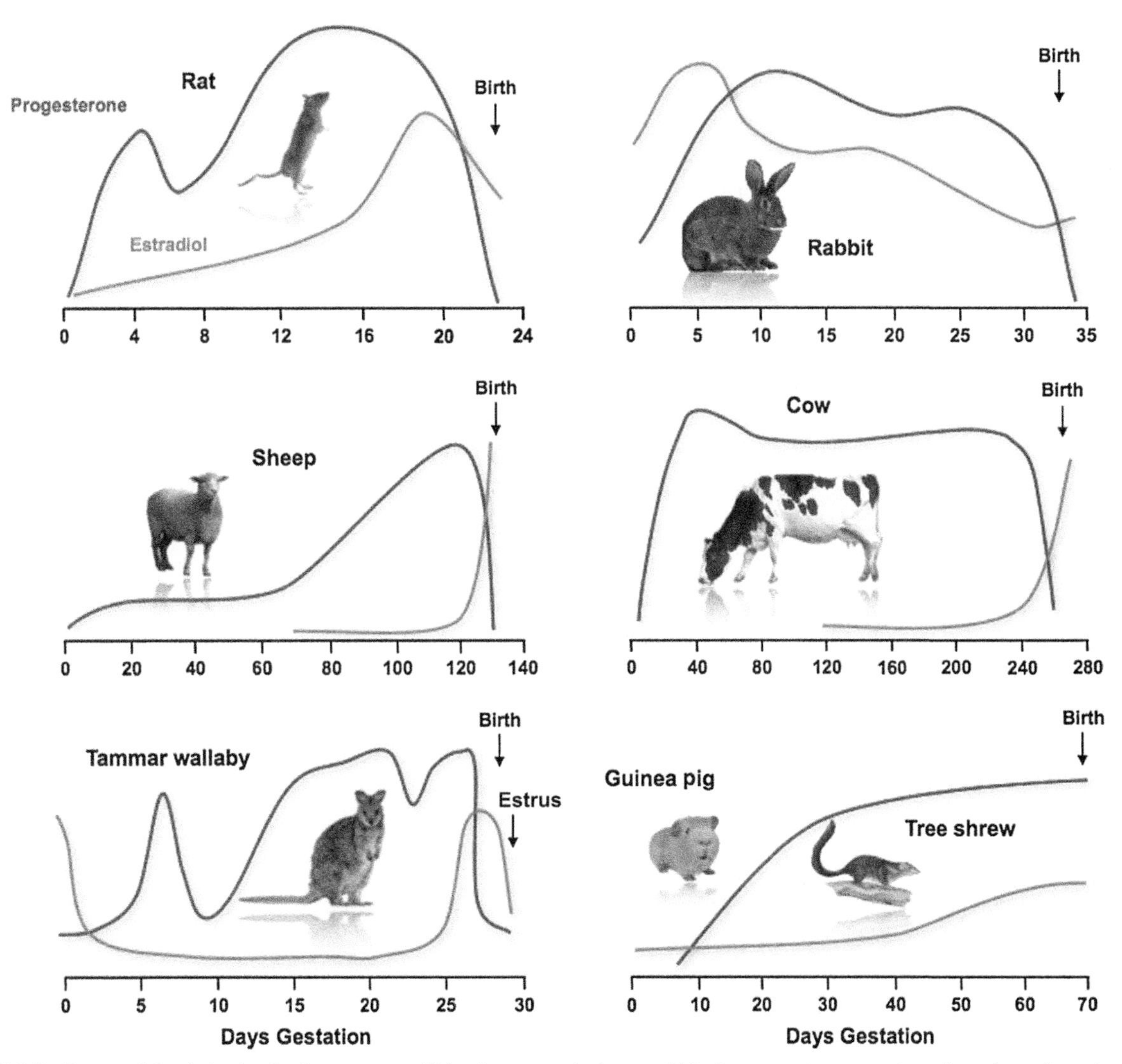

FIGURE 6.7 Pattern of circulating levels of progesterone (P_4) and estrogens in the rat, rabbit, sheep, cow, tammar, guinea pig, and tree shrew throughout gestation. See color plate section.

under some conditions. This apparent paradox may be resolved in the case of the guinea pig by the observation that, although the circulating steroid concentrations do not change, the abundance of PRs decreases while ERs increase near term (Glasier & Hobkirk, 1993). This suggests that functional P_4 withdrawal in species with persistent exposure of the uterus to high P_4 levels is mediated by decreased myometrial P_4 responsiveness. Myometrial P_4 responsiveness is determined primarily by the extent of expression of specific PRs and in particular the nuclear PRs. The fact that labor is initiated by administration of specific PR antagonists (e.g., mifepristone (RU486)) reflects the importance of nuclear PR-mediated P_4 actions for the maintenance of pregnancy. Moreover, it shows that inhibition of this signaling pathway is sufficient to trigger the full parturition cascade. Based on this reasoning, it is hypothesized that human parturition involves a functional nuclear PR-mediated P_4 withdrawal whereby genomic P_4 responsiveness in myometrial cells is abrogated (Mesiano et al., 2002).

The human PR exists as two major forms: the full-length PR-B and the truncated (by 164 N-terminal amino acids) PR-A (Kastner et al., 1990). Many *in-vitro* studies have led to the concept that PR-B is the principal mediator of P_4 actions, whereas PR-A decreases P_4 responsiveness by repressing the transcriptional activity of PR-B (Haluska et al., 2002; Mesiano, 2004; Merlino et al., 2007). Thus, genomic P_4 responsiveness may be related to the PR-A/PR-B ratio in target cells. This has led to the PR-A/PR-B hypothesis, which posits that P_4 withdrawal in human parturition is mediated by an increase in the myometrial PR-A/PR-B ratio. In support of this hypothesis, several studies have found that PR-A levels (and the PR-A/PR-B ratio) increase in the human pregnancy myometrium in association with advancing gestation and the onset of labor (Mesiano et al., 2002; Merlino et al., 2007). Interestingly, a similar labor-associated increase in the myometrial PR-A/PR-B protein ratio has been observed in the rhesus monkey (Haluska et al., 2002), a species that also lacks a systemic P_4 withdrawal at parturition.

Other mechanisms also may facilitate functional P_4 withdrawal at parturition. The transcriptional activity of the PRs requires specific coregulators that form the nPR-induced transcriptional complex. These include cAMP (cyclic-3',5'-adenosine monophosphate)-response-element-binding protein (CREB)-binding protein and steroid receptor coactivators 2 and 3 (Condon et al., 2003). Studies in human term myometrium have shown that the levels of these critical coactivators decrease with the onset of labor (Condon et al., 2003) and as such the capacity for ligand-activated PR-B to control gene expression could be compromised. Progesterone may also be metabolized to products with diminished progestin activity. For example, at the time of labor, the potent relaxatory steroid 5β-dihydroprogesterone decreases along with a reduction in steroid 5β-reductase expression and activity (Sheehan et al., 2005). The transcription/activity of factor NFκB in myometrial cells may also be important in blocking the action of P_4 at the receptor level (Kalkhoven

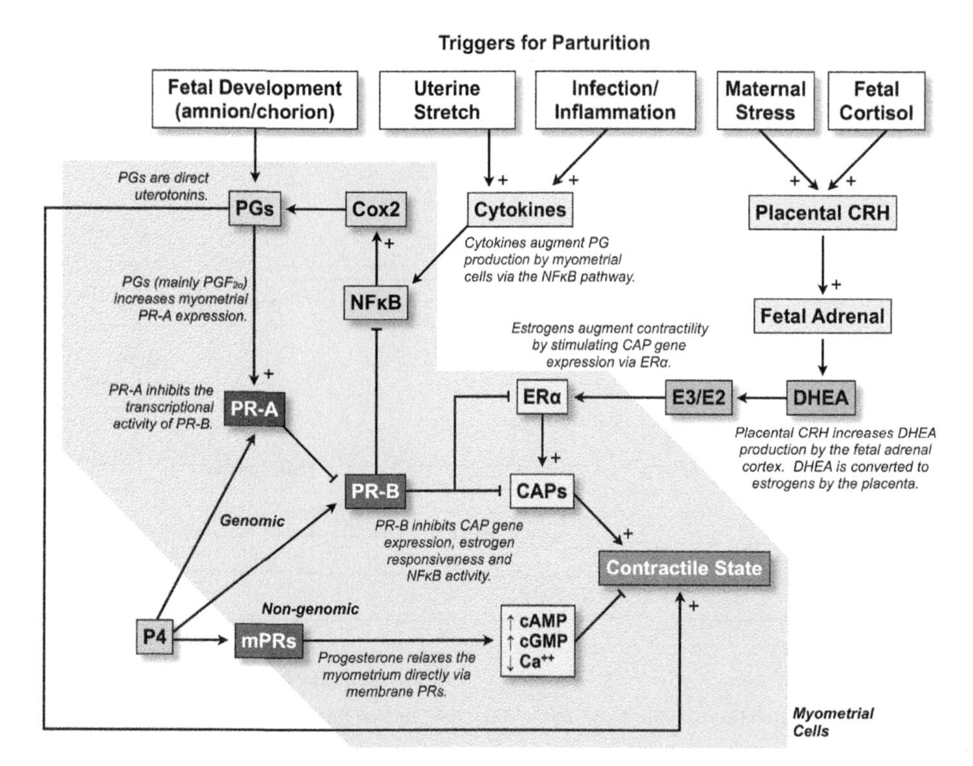

FIGURE 6.8 The onset of labor in humans. Placental production of corticotropin-releasing hormone (CRH) leads to a change in the balance of estriol (E_3) to 17β-estradiol (E_2) and of E_3 to progesterone (P_4). Contraction-associated proteins (CAPs) include connexin 43. Ca^{++}; cAMP, cyclic-3',5'-adenosine monophosphate; cGMP; CAP, contraction-associated protein; ER, estrogen receptor; IκBα; NFκB; PG, prostaglandin; PGRMC; PR, progesterone receptor. See color plate section.

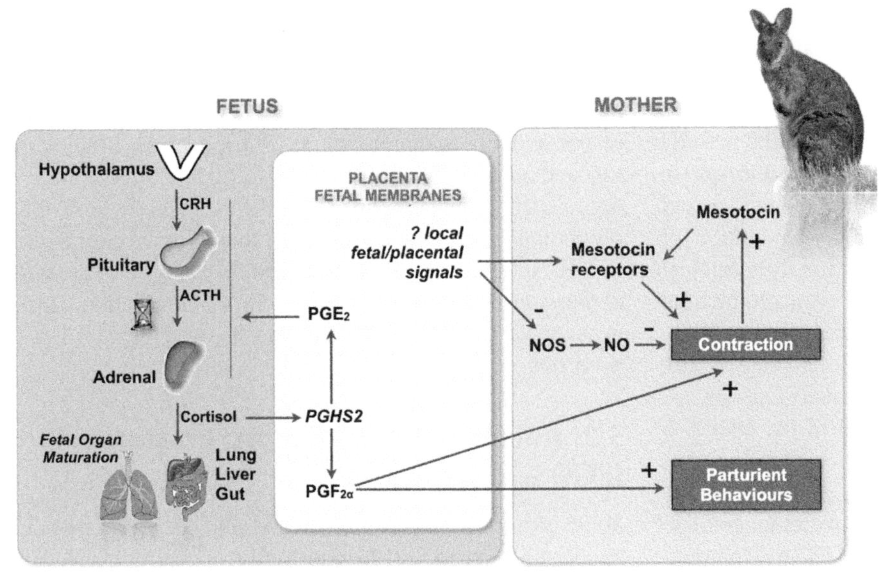

FIGURE 6.9 In the Tammar fetal production of cortisol is a dominant signal to the onset of parturition. Cortisol promotes placental transcription of the PGHS2 gene leading to increased prostaglandin production. Both PGE_2 and $PGF_{2\alpha}$ are produced within the placenta. PGE_2 feeds back to promote ACTH release from the fetal pituitary promoting the rapid rise in fetal cortisol that precedes parturition. $PGF_{2\alpha}$ is released from the placenta into the maternal circulation where it promotes parturient behaviours and myometrical contraction. Other placental factors may stimulate nitric oxide synthase (NOS) which promote myometrial relaxation during pregnancy. Mesotocin from the mother also plays a role in promoting myometrial contraction. The increase in mesotocin and $PGF_{2\alpha}$ prior to delivery occurs as a brief peak (approximately one hour). See color plate section.

et al., 1996; Lindstrom & Bennett, 2005). This may be a critical mechanism for infection-induced preterm labor, since NFκB is a major mediator of the inflammatory response; one of its actions in myometrial cells may be to block the transcriptional activity of PR-B. Taken together, the current data indicate that in human parturition the P_4 block to the onset of labor is removed by specific biochemical events in myometrial cells that abrogate the capacity of ligand-activated PR-B to promote relaxation (see Figure 6.8).

In the horse, circulating P_4 concentrations are low in late gestation, while high concentrations of 5α-reduced P_4 metabolites are found (Holtan et al., 1975; 1991). Inhibition of P_4 synthesis with the 3β-hydroxysteroid dehydrogenase (3β-HSD) inhibitor, epostane, does not induce premature labor in mares even though plasma P_4 concentrations fall (Fowden & Silver, 1987). Interestingly, in those studies epostane also inhibited estrogen synthesis, complicating the interpretation of the outcomes, as the estrogen/P_4 ratio may be more important than the absolute concentrations of the individual steroids in the equine. Estrogens are synthesized by the equine placenta using as the principal substrate DHEA-S derived from the fetal gonads (Raeside et al., 1979). If the fetal gonads are removed, the estrogen/P_4 ratio falls dramatically but the timing of parturition is unaffected, suggesting that even the estrogen/P_4 ratio is relatively unimportant, although there is some contradictory evidence (Haluska & Currie, 1988). Considered together, these findings suggest that the classical roles of the sex steroids in the physiology of parturition may not apply in the horse. Similarly, there is no prepartum change in P_4 in the tammar (Ward & Renfree, 1984; Tyndale-Biscoe, Hinds, & Horn, 1988; Renfree, 1994). Indeed, in this and related species the duration of gestation and the estrous cycle are the same and the P_4 and estrogen profiles during pregnancy do not differ. Neither blocking P_4 nor increasing it via exogenous injection, or removal of estrogens, has any influence on the timing of parturition (Ward & Renfree, 1984; Short et al., 1985). Further comparative studies are needed to determine whether P_4 or related progestogens play an obligatory role in the maintenance of pregnancy across the mammalian subclasses and orders.

Estrogens promote protein synthesis, tissue growth, and blood flow in the uterus, and, in the common laboratory and domestic species, they favor the synthesis of uterotonins, including prostaglandins. They also promote the expression of receptors for uterotonins (PGs and OXY) in the myometrium. As such, they oppose P_4 action, and the simultaneous prepartum decrease in circulating P_4 and increase in estrogen concentrations in the sheep are regulated coordinately through the activation of placental $P450_{c17}$.

While it is generally true that estrogens predispose the myometrium to contract, they are not essential for parturition in all species. Antagonism of estrogen actions delays but does not abolish the birth process in rats (Fang et al., 1996) and pregnancy and parturition proceed in women when placental aromatase is lacking (Shozu et al., 1991; 1992). Similarly, exogenous estrogens do not induce or accelerate labor in humans, and other data indicating that estrogens are not essentially linked to the onset of human labor have been reviewed previously (Liggins & Thorburn, 1994). Despite these findings, studies in the rhesus monkey suggest that increased estrogen synthesis due to increased availability of C_{19} precursors may precipitate parturition (Giussani et al., 1996; Mecenas et al., 1996; Nathanielsz et al., 1998). Recent work in the human suggests a critical role for the ratio of E_3 to E_2, with a marked rise in E_3 occurring just prior to delivery that is correlated with increasing concentrations of CRH (Smith et al., 2009). This finding indicates a physiological link between the activity of the fetal HPA axis and the timing of parturition in primates. A unique feature of the primate fetal adrenal cortex is the fetal zone and its abundant production of DHEA-S. Dehydroepiandrosterone sulfate is the principal source of C_{19} steroids for placental estrogen production. The principal placental estrogen of human pregnancy is E_3. As the placenta lacks the 16-hydroxylase enzyme, it can only produce E_3 from a 16-hydroxylated C_{19} steroid precursor (Diczfalusy et al., 1964). Most of the DHEA-S produced by the fetal zone is converted to 16α-hydroxy-DHEA-S by the fetal liver and to a lesser extent within the adrenal itself (Seron-Ferre, 1981). The maternal adrenal also makes a substantial contribution to DHEA-S production but the maternal tissues lack the 16-hydroxylase enzyme. In the placenta, the sulfatase enzyme removes the sulfate moiety from DHEA-S and 16α-hydroxy-DHEA-S, producing DHEA, which is aromatized to E_1 and E_2, and 16α-hydroxy-DHEA, which is aromatized to E_3. Thus, E_3 production is exclusively from fetal sources while E_2 and E_1 may originate from either maternal or fetal precursors. In primates, the placenta lacks $P450_{c17}$ throughout gestation; therefore, P_4 production does not decline at the end of pregnancy as it does in sheep.

Androgens may be important for the initiation of primate parturition. Infusion of androstenedione (AND) into pregnant rhesus monkeys late in gestation increased maternal estrogens and nocturnal OXY concentrations and induced cervical dilation and normal parturition (Mecenas et al., 1996). Thus, in primates, androgen, produced by the fetal adrenals as a source of aromatizable substrate for estrogen synthesis by the placenta, may be the link between fetus and mother in the initiation of parturition. However, removal of the fetus (fectectomy) in baboons does not prolong retention of the placenta, and administration of E_2 prevents placental delivery and prolongs gestation in fetectomized baboons, indicating that in this species estrogens produced by the feto–placental unit may actually inhibit parturition. However, the interpretation of these data needs to be conducted in the light of the likely role of changing ratios of different estrogens close to the onset of labor in primates. The questionable importance of estrogen for parturition in horses and wallabies (Shaw & Renfree, 1984) already has been discussed.

In human pregnancy, maternal estrogen levels begin to increase from around the 10th week of gestation and continue to be elevated up to delivery. Thus, for most of pregnancy, the myometrium is exposed to large amounts of estrogens, yet it remains in a relaxed and quiescent state. This apparent refractoriness may be attributable to the mutual antagonism of E_3 and E_2, and to the inhibition of myometrial ERα expression by P_4. Thus, P_4 may not only inhibit the expression of contraction-associated genes, such as connexin 43, but also decrease myometrial estrogen responsiveness by inhibiting ERα expression. Studies of PR and ERα expression in the term human pregnancy myometrium support this hypothesis and suggest that functional P_4 withdrawal mediated by increased expression of PR-A (and the PR-A/PR-B ratio) leads to increased expression of ERα. In this scenario the circulating estrogens could promote contractility once the P_4 block has been removed.

5.2. Inflammatory Mediators: Prostaglandins (PGs) and Cytokines

Some of the changes seen in uterine and cervical tissues in the period surrounding the onset of labor resemble changes that typically occur in response to tissue damage. These include tissue edema, neutrophil infiltration, and expression of chemical mediators of the inflammatory reaction, of which cytokines and PGs are key players (Kelly, 1996). The case for inflammatory mediators as effectors in parturition is strengthened by the observation that stimuli such as mechanical or chemical trauma, or infection, which commonly induce the expression of these agents, are associated with premature delivery and are often effective as abortifacients.

In considering the roles of proinflammatory agents in parturition, it is useful to distinguish between their effects on the body of the uterus and the cervix. The body of the uterus consists principally of the endometrium and the myometrium, while the cervix is unlike either of these tissues, being composed mainly of collagenous connective tissue. Cytokines and PGs are implicated in the activation of the myometrium, but they play an equally important role in cervical ripening. This process is characterized by hydration of the ground substance, together with enzymatic unlinking and loss of orientation of the collagen fibers that compose the bulk of the cervical tissue, with consequent softening and increased compliance of the cervix. This allows transformation of the entire structure from an unyielding, tubular one into a compliant, effaced opening allowing the safe exit of the fetus (Calder & Greer, 1990; Denison et al., 1999).

Among the cytokines, interleukins (ILs)-1, -6, and -8 and tumor necrosis factor (TNF)-α are implicated in cervical ripening and in the prelude to myometrial activation, although they do not exert uterotonic effects directly (Romero et al., 1991; Romero & Tartakovsky 1992; Elliott et al., 1998; Denison et al., 1999). Interestingly, evidence suggests that IL-6 is secreted in response to tissue damage but acts to mitigate the inflammatory response (Wu et al., 2004). The neutrophils that infiltrate the cervical tissue are probably responsible for the enzymatic unlinking of collagen fibers although fibroblasts that are also abundant express collagenases and may participate in this process (Calder & Greer, 1990; Kelly, 1996).

In the myometrium, PGs induce muscle contraction. Prostaglandin E_2 does not elicit contraction in all species or at all stages of gestation, but $PGF_{2\alpha}$ is secreted into the uterine veins during labor in all species studied to date including marsupials, and appears to act as a universal myometrial contractant. Accordingly, disruption of the genes encoding the inducible PG synthase (PGHS)-2 or the PGF receptor is associated with disordered labor in mice (Dinchuk et al., 1995; Morham et al., 1995; Sugimoto et al., 1997). Similarly, direct infusion of $PGF_{2\alpha}$ (and, in some species, PGE_2) induces labor whereas pharmacological inhibition of PGHS-2 inhibits it (Poore et al., 1999). Proinflammatory cytokines in the parturient process appear to promote induction of PGHS-2, the expression and activity of which increase at the onset of labor with consequent increases in PG production by intrauterine tissues. Positive feedback of PGs on cytokine expression also has been demonstrated (Denison et al., 1999). Although it may not cause uterine contraction in all species, PGE_2 is secreted in large amounts from the placenta in late gestation in many species and is effective in inducing cervical ripening. Thus, uterine activity and cervical compliance are affected simultaneously by the complementary actions of $PGF_{2\alpha}$ and PGE_2. These effects seem to apply in all species studied to date.

The sites of PG synthesis within the uterus and the means by which these cytokines reach the myometrium have been the subject of investigations. Contrary to earlier expectations, the myometrium does not generally exhibit increased expression of PGHS-1 or -2 at labor in the sheep (McLaren et al., 1996; Gibb et al., 2000; McLaren et al., 2000) or the rat (Myatt et al., 1994), raising questions about where these PGs are synthesized and how they reach the target tissue. In the sheep, the only tissue showing increased expression of PGHS-2 at spontaneous or glucocorticoid-induced labor appears to be the placenta (McLaren et al., 1996; 2000). Prostaglandins may reach the myometrium from the 80–100 placentomes by diffusion, either directly through the basal plates of the placentomes or from the many small veins draining the placentomes. Thus, in the sheep, the signaling pathway from the fetus to the placenta via glucocorticoids and from the placenta to the myometrium via PGs appears reasonably straightforward. In humans, the situation is much less clear. The human fetal HPA axis probably plays an important role in normal parturition but is not essential in signaling the onset of labor (reviewed in Liggins & Thorburn, 1994) and the principal site of PG production is believed to be the amnion (Sadovsky et al., 2000), although PGHS-2 increases in the human myometrium at labor (Chan et al., 2002; Bisits et al., 2005). The predominant location of PGHS-2 in a fetal structure (the amnion) may reflect the need for a fetal signal to initiate labor, but the amnion is relatively distant from the site of PG action in the myometrium, and the PGs would need to diffuse through the chorion containing the dehydrogenase that inactivates them. In the tammar, PGs accumulate in the yolk sac fluid, and the placenta appears to be the site of synthesis (Shaw, 1983; Shaw & Bell, 1999) These PGs are critical for the induction of normal birth behavior as well as for uterine contractility (Hinds et al., 1990; Shaw, 1990) (see Figure 6.9).

Data from studies of human myometrial cells suggest that PGs play a critical role in the induction of parturition by inducing P_4 withdrawal via their induction of PR-A expression (Madsen et al., 2004). Several observations provide clues as to the hormonal regulators of myometrial PR expression. Prime candidates include locally produced immune/inflammatory cytokines, particularly PGE_2 and $PGF_{2\alpha}$, that may be involved in normal term birth and infection-associated preterm birth. In women, exogenous PGs induce the full parturition cascade at all stages of pregnancy (Robins & Mann, 1975; Jain & Mishell, 1994). However, in contrast to the potent and almost instantaneous uterotonic actions of PGs when administered to women already in active labor, the induction of labor in a quiescent uterus by exogenous PGs occurs after a latency of 15–20 hours (Robins & Mann, 1975; Jain & Mishell, 1994). This likely represents the time needed for the myometrium to transform to a contractile state. Thus, PGs may first act as uterotropins to transform the myometrium to a contractile phenotype, and then exert potent uterotonic actions once the contractile state is established. Consistent with this hypothesis is that, in a human myometrial cell line, $PGF_{2\alpha}$ increases the PR-A/PR-B mRNA ratio by specifically increasing expression of PR-A but not PR-B expression (Madsen et al., 2004). This finding suggests that increased local $PGF_{2\alpha}$ production and/or increased $PGF_{2\alpha}$ accessibility to the myometrium from the amnion through decreased PG dehydrogenase in the chorion could initiate labor by modulating myometrial PR expression to induce functional P_4 withdrawal. Indeed normal labor at term is associated with increased PG production by the fetal membranes and decidua (Mitchell et al., 1995) and abnormal parturition, particularly infection/inflammation-associated preterm labor, could involve the same process. In this regard, the link between the local immune/inflammatory response within the gestational tissues and myometrial PR expression becomes an important issue. Activation of the NFκB transcription factor complex (a key mediator of the inflammatory response) stimulates expression of PRs in myometrial cells (Condon et al., 2006). Thus, current data suggest that local immune/inflammatory cytokines, and in particular $PGF_{2\alpha}$, converge on myometrial PR expression to modulate P_4 responsiveness and potentially initiate labor by inducing functional P_4 withdrawal. This may be a central pathway in the hormonal control of human parturition.

5.3. Oxytocin (OXY)

The data currently available are consistent with a role for OXY in the second stage of labor in all or most mammalian species, while its putative role in the onset of labor in any species awaits confirmation. During established labor, OXY is secreted episodically from the posterior pituitary of most if not all species in response to cervico–vaginal stretch and acts on the myometrium to increase the frequency and force of contractions. Physiological to supraphysiological concentrations of exogenously administered OXY are used clinically to induce labor in women at term. However, the role of OXY appears more subtle than that of PGs since it does not normally induce preterm labor and, even at term, it is unreliable unless some procedure inducing tissue damage (such as amniotomy) is also performed. Similarly, pharmacological inhibition of OXY action may delay but does not abolish delivery in rats and guinea pigs (Chan & Chen, 1992; Schellenberg, 1995), and OXY-null mice deliver spontaneously at the expected time (Nishimori et al., 1996; Young et al., 1996).

The expression of OXY receptors in the pregnant uterus is increased at term in humans, rats, and all other species studied to date (Alexandrova & Soloff, 1980a; 1980b; 1980c; Fuchs et al., 1982). In the tammar, a similar increase has been reported in the concentration of receptors for mesotocin (8-Ile-oxytocin (MST)), the tammar equivalent of OXY (Parry et al., 1996; Renfree et al., 1996; Parry et al., 1997; Sebastian et al., 1998), and blocking MST inhibits parturition (Renfree et al., 1996). The consistency of this phenomenon across taxa suggests a fundamental role for OXY, although it does not argue specifically for a role in the initiation of labor. The induction of OXY receptors (OXY-R) at term in the rat can be inhibited by blocking PGHS (Chan, 1980), and this treatment is associated with prolonged gestation and inefficient uterine contractions (Chan & Chen, 1992). In the same study, specific OXY-R blockade did not prolong gestation but interfered with the process of delivery. Together, these observations suggest that PGs are positive regulators of OXY-R expression, that the uterotonic action of PGs depends partly on OXY action, and that PGs, but not OXY, are involved in the initiation of labor.

Oxytocin also stimulates PG release from the endometrium of the pregnant rat and the OXY-R type mediating this response may be pharmacologically differentiated from the myometrial receptors responsible for the utertonic actions of oxytocin (Chan, 1980; Chan & Chen, 1992; Chan et al., 1993). These observations suggest the existence of a positive feedback control mechanism with PGHS as the controlling variable and endometrial OXY-R expression as the controlled variable. Evidence from other species is less complete than that obtained in rats, but is broadly consistent with this hypothesis.

The OXY gene is expressed in the uterus itself as well as the hypothalamus, raising the possibility that OXY synthesized within the uterus could participate in the initiation or the accomplishment of parturition (Mitchell et al., 1998). The expression of the OXY gene as measured by mRNA concentration peaks some three to four days before labor in the rat uterus, although this timing raises questions about the role and mechanism of action of OXY in the initiation of birth (Fang et al., 1996). The prepartum increase in OXY gene expression in the rat uterus depends on estrogen action, being abolished by the ER antagonist tamoxifen. Significantly, even when OXY gene expression was suppressed by tamoxifen, labor was eventually initiated, although it was delayed. This delay was associated with suppression of the tissue PGE_2 concentration immediately before the expected time of delivery, and a restoration of this parameter by the time of labor. These observations are consistent with a complex positive regulatory interaction between estrogens, PGHS, and the uterine OXY system, and are consistent with the proposed role of PGs as the final common effector in myometrial activation. They also suggest regulation of PGHS expression or activity by factors other than estrogens and OXY since normal PGE_2 concentrations were reached despite significant downregulation of uterine ERs and OXY gene expression. This action could be exerted by cytokines, as discussed previously.

In the tammar, uterine contractility is mediated by a decrease in NOS and NO in conjunction with an increase in MST, both probably stimulated by local fetal–placental signals in concert with the rapid PG pulse at parturition (Ingram, Renfree, & Shaw, 2001). In the pig, the OXY gene is expressed in the myometrium and the endometrium, although the tissue mRNA concentration does not increase as labor approaches. These data lend mixed support for the notion that intrauterine OXY gene expression is an important step for the initiation of labor.

5.4. Electrical Conductivity of the Myometrium

It has long been known that propagation of action potentials through the myometrium of rabbits is inhibited by high P_4 concentrations and that the 'P_4-blocked' uterus is characterized by a reduction in both the amplitude of contraction and intrauterine pressure (Csapo, 1977). These phenomena were confirmed in the tammar (Ingram et al., 2001; Shaw, 2001) and sheep (Lye & Porter, 1978) for uterine contractions induced by both oxytocin and $PGF_{2\alpha}$, although the explanation for this phenomenon at the cellular or tissue level was not established at that time. It was subsequently shown that the numbers of gap junctions on myometrial cells of guinea pigs increase at parturition (Garfield et al.,

1982) and that these are permeable to ions, intracellular signaling molecules, and dyes. They also constitute high conductance pathways to facilitate electrical coupling between adjacent cells and provide an anatomical basis for the increased conductivity of myometrial tissue obtained from pregnant animals at term, relative to other times.

The gap junctions are formed by oligomers of a protein called connexin 43. The expression of connexin 43 increases at term or labor in the myometrium of the human (Garfield & Hayashi, 1981; Chow & Lye, 1994; Sparey et al., 1999), sheep (McNutt et al., 1994), rabbit (Nnamani et al., 1994), and rat (Orsino et al., 1996); is positively regulated by estrogens in the rat (Lefebvre et al., 1995) and cow (Doualla-Bell et al., 1995); and is inhibited by P_4 in the rat (Orsino et al., 1996; Risek & Gilula, 1996; Ou et al., 1997). In the human, connexin 43 expression correlates with ER-A expression in the myometrium (Bisits et al., 2005) and explanted cells increase their expression of connexin 43 in response to exogenous estrogens (Kilarski et al., 2000). Progesterone may not regulate connexin 43 in the human myometrium (Geimonen et al., 1998; Kilarski et al., 2000). The human myometrium shows greater expression of connexin 43 in the fundal than the cervical region, suggesting that contractions may be propagated from the former zone to the latter, which would be expected to aid expulsion of the fetus (Sparey et al., 1999). It is highly likely that connexin 43 and other gap junction subunit proteins are important for parturition in humans but the factors regulating their expression in humans require elucidation. The distribution and regulation of gap junction proteins appears not to have been studied systematically beyond the standard laboratory and farmed species, but the consistency of the scenario across those that have been studied suggests that these proteins play a fundamental and generalized role in the physiology of parturition.

6. CONCLUSIONS

1. There appears to be a general requirement for antenatal exposure of the fetus to glucocorticoid, which promotes the maturational changes needed for extrauterine life.
2. The prepartum increase in glucocorticoid concentrations in fetal plasma constitutes the signal for myometrial activation in only a few species, although not all these species are closely related.
3. Progesterone is required for the maintenance of uterine quiescence and its action is withdrawn at term by decreased secretion or receptor downregulation in many, but not all, species.
4. Estrogens generally favor myometrial activation.
5. Although there is considerable interspecific diversity in the ultimate determinants of parturition, there is much more consistency when it comes to the effector mechanisms at the level of the myometrium. Prostaglandin $F_{2\alpha}$ and PGE_2 are universally involved in direct activation of the myometrium and also in preparation of the cervix and other components of the birth canal. The importance of gap junction proteins has not been assessed in many species, but it may be a widespread feature of uterine activation. Oxytocin plays a role in inducing uterine contractions in most, if not all, species, although this role is not responsible for the initiation of parturition.
6. The physiology of parturition in humans is not well represented by any animal model and application of the methods of comparative physiology suggests that we should exercise caution in extrapolating knowledge gained in laboratory, domesticated, or captive animal species to humans. This caution applies increasingly as one considers mechanisms more distant from the level of the myometrium.

ABBREVIATIONS

16-OH DHEA-S	16-Hydroxy dehydroepiandrosterone
3β-HSD	3β-hydroxysteroid dehydrogenase
ACTH	Corticotropin
AND	Androstenedione
cAMP	Cyclic-3',5'-adenosine monophosphate
CAP	Contraction-associated protein
CREB	cAMP-response-element-binding protein
CRH	Corticotropin-releasing hormone
CRHBP	Circulating binding protein for CRH
DHEA-S	Dehydroepiandrosterone sulfate
E_1	Estrone
E_2	17β-estradiol
E_3	Estriol
ER	Estrogen receptor
F	Cortisol
HPA	Hypothalamo–pituitary–adrenal
IL	Interleukin
MEL	Melatonin
mPR	Membrane-associated progesterone receptor
MST	Mesotocin
NO	Nitric oxide
NOS	Nitric oxide synthase
OXY	Oxytocin
OXY-R	Oxytocin receptor
P_4	Progesterone
$P450_{C17}$	17-hydroxylase 17,20 lyase
PG	Prostaglandin
PGE_2	Prostaglandin E_2
$PGF_{2\alpha}$	Prostaglandin $F_{2\alpha}$
PGHS	Prostaglandin synthase
PR	Progesterone receptor
RU486	Mifepristone
SCN	Suprachiasmatic nucleus
SP-A	Surfactant protein-A
TLR	Toll-like receptor
TNF	Tumor necrosis factor

REFERENCES

Alexandrova, M., & Soloff, M. S. (1980). Oxytocin receptors and parturition. I. Control of oxytocin receptor concentration in the rat myometrium at term. *Endocrinology, 106*, 730–735.

Alexandrova, M., & Soloff, M. S. (1980). Oxytocin receptors and parturition. II. Concentrations of receptors for oxytocin and estrogen in the gravid and nongravid uterus at term. *Endocrinology, 106*(3), 736–738.

Alexandrova, M., & Soloff, M. S. (1980). Oxytocin receptors and parturition. III. Increases in estrogen receptor and oxytocin receptor concentrations in the rat myometrium during prostaglandin F2 alpha-induced abortion. *Endocrinology, 106*(3), 739–743.

Anderson, A. B., Flint, A. P., et al. (1975). Mechanism of action of glucocorticoids in induction of ovine parturition: effect on placental steroid metabolism. *J. Endocrinol., 66*(1), 61–70.

Asa, S. L., Kovacs, K., et al. (1991). Human fetal adenohypophysis: morphologic and functional analysis *in vitro*. *Neuroendocrinology, 53* (6), 562–572.

Bassett, J. M., & Thorburn, G. D. (1969). Foetal plasma corticosteroids and the initiation of parturition in sheep. *J. Endocrinol., 44*(2), 285–286.

Bininda-Emonds, O. R., Cardillo, M., et al. (2007). The delayed rise of present-day mammals. *Nature, 446*(7135), 507–512.

Bisits, A. M., Smith, R., et al. (2005). Inflammatory aetiology of human myometrial activation tested using directed graphs. *PLoS Comput. Biol., 1*(2), 132–136.

Bosc, M. J. (1990). Photoperiodic regulation of the time of birth in rats: involvement of circadian endogenous mechanisms. *Physiol. Behav., 48*(3), 441–446.

Bowman, M. E., Lopata, A., et al. (2001). Corticotropin-releasing hormone-binding protein in primates. *Am. J. Primatol., 53*(3), 123–130.

Calder, A. A., & Greer, I. A. (1990). Prostaglandins and the biological control of cervical function. *Reprod. Fertil. Dev., 2*(5), 459–465.

Chan, E. C., Fraser, S., et al. (2002). Human myometrial genes are differentially expressed in labor: a suppression subtractive hybridization study. *J. Clin. Endocrinol. Metab., 87*(6), 2435–2441.

Chan, W. Y. (1980). The separate uterotonic and prostaglandin-releasing actions of oxytocin. Evidence and comparison with angiotensin and methacholine in the isolated rat uterus. *J. Pharmacol. Exp. Ther., 213* (3), 575–579.

Chan, W. Y., & Chen, D. L. (1992). Myometrial oxytocin receptors and prostaglandin in the parturition process in the rat. *Biol. Reprod., 46* (1), 58–64.

Chan, W. Y., Chen, D. L., et al. (1993). Oxytocin receptor subtypes in the pregnant rat myometrium and decidua: pharmacological differentiations. *Endocrinology, 132*(3), 1381–1386.

Cheng, Y. H., Nicholson, R. C., et al. (2000). Glucocorticoid stimulation of corticotropin-releasing hormone gene expression requires a cyclic adenosine 3',5'-monophosphate regulatory element in human primary placental cytotrophoblast cells. *J. Clin. Endocrinol. Metab., 85*(5), 1937–1945.

Chow, L., & Lye, S. J. (1994). Expression of the gap junction protein connexin-43 is increased in the human myometrium toward term and with the onset of labor. *Am. J. Obstet. Gynecol., 170*(3), 788–795.

Cole, T. J., Blendy, J. A., et al. (1995). Targeted disruption of the glucocorticoid receptor gene blocks adrenergic chromaffin cell development and severely retards lung maturation. *Genes. Dev., 9*(13), 1608–1621.

Condon, J. C., Hardy, D. B., et al. (2006). Up-regulation of the progesterone receptor (PR)-C isoform in laboring myometrium by activation of nuclear factor-kappaB may contribute to the onset of labor through inhibition of PR function. *Mol. Endocrinol., 20*(4), 764–775.

Condon, J. C., Jeyasuria, P., et al. (2004). Surfactant protein secreted by the maturing mouse fetal lung acts as a hormone that signals the initiation of parturition. *Proc. Natl. Acad. Sci. USA, 101*(14), 4978–4983.

Condon, J. C., Jeyasuria, P., et al. (2003). A decline in the levels of progesterone receptor coactivators in the pregnant uterus at term may antagonize progesterone receptor function and contribute to the initiation of parturition. *Proc. Natl. Acad. Sci. USA, 100*(16), 9518–9523.

Csapo, A. I. (1977). The 'see-saw' theory of parturition. *Ciba. Found. Symp.,* (47), 159–210.

Currie, W. B., & Thorburn, G. D. (1973). Induction of premature parturition in goats by prostaglandin F2 administered into the uterine vein. *Prostaglandins, 4*(2), 201–214.

Currie, W. B., & Thorburn, G. D. (1977a). The fetal role in timing the initiation of parturition in the goat. *Ciba. Found. Symp.,* (47), 49–72.

Currie, W. B., & Thorburn, G. D. (1977b). Parturition in goats: studies on the interactions between the foetus, placenta, prostaglandin F and progesterone before parturition, at term or at parturition induced prematurely by corticotrophin infusion of the foetus. *J. Endocrinol., 73*(2), 263–278.

Currie, W. B., Cox, R. I., et al. (1976). Release of prostaglandin F, regression of corpora lutea and induction of premature parturition in goats treated with estradiol-17beta. *Prostaglandins, 12*(6), 1093–1103.

Denison, F. C., Calder, A. A., et al. (1999). The action of prostaglandin E2 on the human cervix: stimulation of interleukin 8 and inhibition of secretory leukocyte protease inhibitor. *Am. J. Obstet. Gynecol., 180*(3 Pt 1), 614–620.

Diczfalusy, E., Barr, M., et al. (1964). Oestriol metabolism in an anencephalic monster. *Acta. Endocrinol. (Copenh), 46*, 511–524.

Dinchuk, J. E., Car, B. D., et al. (1995). Renal abnormalities and an altered inflammatory response in mice lacking cyclooxygenase II. *Nature, 378*(6555), 406–409.

Donovan, B. T., & Peddie, M. J. (1973). Foetal hypothalamic and pituitary lesions, the adrenal glands and abortion in the guinea pig. In R. S. Comline, K. W. Cross, G. S. Dawes, & P. W. N. Nathanielsz (Eds.), *Foetal and Neonatal Physiology* (pp. 603–605). Cambridge, UK: Cambridge University Press.

Doualla-Bell, F., Lye, S. J., et al. (1995). Differential expression and regulation of connexin-43 and cell–cell coupling in myocytes from the circular and longitudinal layers of bovine myometrium. *Endocrinology, 136*(12), 5322–5328.

Elliott, C. L., Kelly, R. W., et al. (1998). Regulation of interleukin 8 production in the term human placenta during labor and by antigestagens. *Am. J. Obstet. Gynecol., 179*(1), 215–220.

Ellis, M. J., Livesey, J. H., et al. (2002). Plasma corticotropin-releasing hormone and unconjugated estriol in human pregnancy: gestational patterns and ability to predict preterm delivery. *Am. J. Obstet. Gynecol., 186*(1), 94–99.

Emanuel, R. L., Robinson, B. G., et al. (1994). Corticotrophin releasing hormone levels in human plasma and amniotic fluid during gestation. *Clin. Endocrinol., (Oxf), 40*(2), 257–262.

Emanuel, R. L., Torday, J. S., et al. (2000). Direct effects of corticotropin-releasing hormone and thyrotropin-releasing hormone on fetal lung explants. *Peptides, 21*(12), 1819–1829.

Fang, X., Wong, S., et al. (1996). Relationships among sex steroids, oxytocin, and their receptors in the rat uterus during late gestation and at parturition. *Endocrinology, 137*(8), 3213–3219.

Finn, C. A., & Porter, D. G. (1975). *The Uterus*. London, UK: Elek.

Flint, A. P., Anderson, A. B., et al. (1974). Control of utero–ovarian venous prostaglandin F during labour in the sheep: acute effects of vaginal and cervical stimulation. *J. Endocrinol., 63*(1), 67–87.

Flint, A. P., Kingston, E. J., et al. (1978). Initiation of parturition in the goat: evidence for control by foetal glucocorticoid through activation of placental C21-steroid 17alpha-hydroxylase. *J. Endocrinol., 78*(3), 367–378.

Ford, M. M., Thorburn, G. D., et al. (1999). Pulsatile output of prostaglandin F(2alpha) does not increase around the time of luteolysis in the pregnant goat. *Biol. Reprod., 61*(2), 411–415.

Ford, M. M., Young, I. R., et al. (1998). Fetal and maternal endocrine changes approaching parturition in the goat: lack of evidence for prostaglandins E2 and F2alpha as signals for luteolysis. *Biol. Reprod., 58*(4), 1065–1070.

Fowden, A. L., & Silver, M. (1987). Effects of inhibiting 3 beta-hydroxysteroid dehydrogenase on plasma progesterone and other steroids in the pregnant mare near term. *J. Reprod. Fertil. Suppl., 35*, 539–545.

Fuchs, A. R., Fuchs, F., et al. (1982). Oxytocin receptors and human parturition: a dual role for oxytocin in the initiation of labor. *Science, 215*(4538), 1396–1398.

Garcia-Verdugo, I., Tanfin, Z., et al. (2008). Surfactant protein A signaling pathways in human uterine smooth muscle cells. *Biol. Reprod., 79*(2), 348–355.

Garfield, R. E., & Hayashi, R. H. (1981). Appearance of gap junctions in the myometrium of women during labor. *Am. J. Obstet. Gynecol., 140* (3), 254–260.

Garfield, R. E., Daniel, E. E., et al. (1982). Changes of gap junctions in myometrium of guinea pig at parturition and abortion. *Can. J. Physiol. Pharmacol., 60*(3), 335–341.

Geimonen, E., Boylston, E., et al. (1998). Elevated connexin-43 expression in term human myometrium correlates with elevated c-Jun expression and is independent of myometrial estrogen receptors. *J. Clin. Endocrinol. Metab., 83*(4), 1177–1185.

Gekakis, N., Saez, L., et al. (1995). Isolation of timeless by PER protein interaction: defective interaction between timeless protein and long-period mutant PERL. *Science, 270*(5237), 811–815.

Gersting, J., Schaub, C. E., et al. (2008). Inhibition of brain prostaglandin endoperoxide synthase-2 prevents the preparturient increase in fetal adrenocorticotropin secretion in the sheep fetus. *Endocrinology, 149* (8), 4128–4136.

Gibb, W., Sun, M., et al. (2000). Localization of prostaglandin synthase type-1 (PGHS-1) mRNA and prostaglandin synthase type-2 (PGHS-2) mRNA in ovine myometrium and endometrium throughout gestation. *J. Endocrinol., 165*(1), 51–58.

Giussani, D. A., Jenkins, S. L., et al. (1996). Daily and hourly temporal association between delta4-androstenedione-induced preterm myometrial contractions and maternal plasma estradiol and oxytocin concentrations in the 0.8 gestation rhesus monkey. *Am. J. Obstet. Gynecol., 174*(3), 1050–1055.

Glasier, M. A., & Hobkirk, R. (1993). Nuclear receptors for progesterone and estradiol in the guinea pig uterine compartment during gestation. *Steroids, 58*(10), 478–483.

Haluska, G. J., & Currie, W. B. (1988). Variation in plasma concentrations of oestradiol-17 beta and their relationship to those of progesterone, 13,14-dihydro-15-keto-prostaglandin F-2 alpha and oxytocin across pregnancy and at parturition in pony mares. *J. Reprod. Fertil., 84*(2), 635–646.

Haluska, G. J., Wells, T. R., et al. (2002). Progesterone receptor localization and isoforms in myometrium, decidua, and fetal membranes from rhesus macaques: evidence for functional progesterone withdrawal at parturition. *Journal of the Society For Gynecologic Investigation, 9*(3), 125–136.

Hida, A., Koike, N., et al. (2000). The human and mouse Period1 genes: five well-conserved E-boxes additively contribute to the enhancement of mPer1 transcription. *Genomics, 65*(3), 224–233.

Hinds, L. A., Tyndale-Biscoe, C. H., et al. (1990). Effects of prostaglandin and prolactin on luteolysis and parturient behaviour in the non-pregnant tammar, *Macropus eugenii. J. Reprod. Fertil., 88*(1), 323–333.

Hobel, C. J., Dunkel-Schetter, C., et al. (1999). Maternal plasma corticotropin-releasing hormone associated with stress at 20 weeks' gestation in pregnancies ending in preterm delivery. *Am. J. Obstet. Gynecol., 180*(1 Pt 3), S257–S263.

Hollingworth, S. A., Deayton, J. M., et al. (1995). Prostaglandin E2 administered to fetal sheep increases the plasma concentration of adrenocorticotropin (ACTH) and the proportion of ACTH in low molecular weight forms. *Endocrinology, 136*(3), 1233–1240.

Holtan, D. W., Houghton, E., et al. (1991). Plasma progestagens in the mare, fetus and newborn foal. *J. Reprod. Fertil. Suppl., 44*, 517–528.

Holtan, D. W., Nett, T. M., et al. (1975). Plasma progestagens in pregnant mares. *J. Reprod. Fertil. Suppl.*, (23), 419–424.

Holzman, C., Jetton, J., et al. (2001). Second trimester corticotropin-releasing hormone levels in relation to preterm delivery and ethnicity. *Obstet. Gynecol., 97*(5 Pt 1), 657–663.

Honnebier, M. B., & Nathanielsz, P. W. (1994). Primate parturition and the role of the maternal circadian system. *Eur. J. Obstet. Gynecol. Reprod. Biol., 55*(3), 193–203.

Honnebier, M. B., Jenkins, S. L., et al. (1991). Temporal structuring of delivery in the absence of a photoperiod: preparturient myometrial activity of the rhesus monkey is related to maternal body temperature and depends on the maternal circadian system. *Biol. Reprod., 45*(4), 617–625.

Illingworth, D. V., Challis, J. R., et al. (1974). Parturition in the guinea-pig; plasma levels of steroid hormones, steroid-binding proteins, and oxytocin, and the effect of corticosteroids, prostaglandins and adrenocorticotrophin. *J. Endocrinol., 63*(3), 557–570.

Inder, W. J., Prickett, T. C., et al. (2001). The utility of plasma CRH as a predictor of preterm delivery. *J. Clin. Endocrinol. Metab., 86*, 5706–5710.

Ingram, J. N., Renfree, M. B., & Shaw, G. (2001). Spontaneous contractility and nitric oxide sensitivity in gravid and non-gravid myometrium during pregnancy in a marsupial (*Macropus eugenii*). *Endocrinology, 142*, 2244–2251.

Ingram, J. N., Shaw, G., & Renfree, M. B. (1999). Cortisol in fetal fluids and the fetal adrenal at parturition in the tammar wallaby (*Macropus eugenii*). *Biol. Reprod., 60*(3), 651–655.

Jain, J. K., & Mishell, D. R., Jr. (1994). A comparison of intravaginal misoprostol with prostaglandin E2 for termination of second-trimester pregnancy. *N. Engl. J. Med., 331*(5), 290–293.

Kalkhoven, E., Wissink, S., et al. (1996). Negative interaction between the RelA(p65) subunit of NF-kappaB and the progesterone receptor. *J. Biol. Chem., 271*(11), 6217–6224.

Kastner, P., Krust, A., et al. (1990). Two distinct estrogen-regulated promoters generate transcripts encoding the two functionally different human progesterone receptor forms A and B. *EMBO J., 9*(5), 1603–1614.

Kelly, R. W. (1996). Inflammatory mediators and parturition. *Rev. Reprod., 1*(2), 89–96.

Kilarski, W. M., Hongpaisan, J., et al. (2000). Effect of progesterone and oestradiol on expression of connexin43 in cultured human myometrium cells. *Folia. Histochem. Cytobiol., 38*(1), 3–9.

Kuriyama, H., & Suzuki, H. (1976). Changes in electrical properties of rat myometrium during gestation and following hormonal treatments. *J. Physiol., 260*(2), 315–333.

Lefebvre, D. L., Piersanti, M., et al. (1995). Myometrial transcriptional regulation of the gap junction gene, connexin-43. *Reprod. Fertil. Dev., 7*(3), 603–611.

Leung, T. N., Chung, T. K., et al. (1999). Elevated mid-trimester maternal corticotrophin-releasing hormone levels in pregnancies that delivered before 34 weeks. *Br. J. Obstet. Gynaecol., 106*(10), 1041–1046.

Leung, T. N., Chung, T. K., et al. (2001). Rate of rise in maternal plasma corticotrophin-releasing hormone and its relation to gestational length. *Bjog., 108*(5), 527–532.

Liggins, G. C. (1968). Premature parturition after infusion of corticotrophin or cortisol into foetal lambs. *J. Endocrinol., 42*(2), 323–329.

Liggins, G. C. (1974). Parturition in the sheep and the human. *Basic Life Sci., 4*(PT. B), 423–443.

Liggins, G. C., & Thorburn, G. D. (1994). *Initiation of parturition. Marshall's Physiology of Reproduction.* G. E. Lamming (ed.). London, UK: Chapman & Hall. 863–1002.

Liggins, G. C., Kennedy, P. C., et al. (1967). Failure of initiation of parturition after electrocoagulation of the pituitary of the fetal lamb. *Am. J. Obstet. Gynecol., 98*(8), 1080–1086.

Lindow, S. W., Jha, R. R., et al. (2000). 24 hour rhythm to the onset of preterm labour. *Bjog, 107*(9), 1145–1148.

Lindstrom, T. M., & Bennett, P. R. (2005). The role of nuclear factor kappa B in human labour. *Reproduction, 130*(5), 569–581.

Linton, E. A., Perkins, A. V., et al. (1993). Corticotropin releasing hormone-binding protein (CRH-BP): plasma levels decrease during the third trimester of normal human pregnancy. *J. Clin. Endocrinol. Metab., 76*(1), 260–262.

Louis, T. M., Challis, J. R., et al. (1976). Rapid increase of foetal corticosteroids after prostaglandin E2. *Nature, 264*(5588), 797–799.

Luo, Z. X., Ji, Q., et al. (2003). An Early Cretaceous tribosphenic mammal and metatherian evolution. *Science, 302*(5652), 1934–1940.

Lye, S. J., & Porter, D. G. (1978). Demonstration that progesterone 'blocks' uterine activity in the ewe *in vivo* by a direct action on the myometrium. *J. Reprod. Fertil., 52*(1), 87–94.

Madsen, G., Zakar, T., et al. (2004). Prostaglandins differentially modulate progesterone receptor-A and -B expression in human myometrial cells: evidence for prostaglandin-induced functional progesterone withdrawal. *J. Clin. Endocrinol. Metab., 89*(2), 1010–1013.

McGrath, S., McLean, M., et al. (2002). Maternal plasma corticotropin-releasing hormone trajectories vary depending on the cause of preterm delivery. *Am. J. Obstet. Gynecol., 186*(2), 257–260.

McLaren, W. J., Young, I. R., et al. (1996). Expression of prostaglandin G/H synthase-1 and -2 in ovine amnion and placenta following glucocorticoid-induced labour onset. *J. Endocrinol., 151*(1), 125–135.

McLaren, W. J., Young, I. R., et al. (2000). Localisation and temporal changes in prostaglandin G/H synthase-1 and -2 content in ovine intrauterine tissues in relation to glucocorticoid-induced and spontaneous labour. *J. Endocrinol., 165*(2), 399–410.

McLean, M., Bisits, A., et al. (1995). A placental clock controlling the length of human pregnancy [see comments]. *Nat. Med., 1*(5), 460–463.

McMillen, I. C., & Nowak, R. (1989). Maternal pinealectomy abolishes the diurnal rhythm in plasma melatonin concentrations in the fetal sheep and pregnant ewe during late gestation. *J. Endocrinol., 120*(3), 459–464.

McNutt, C. M., Nicholson, B. J., et al. (1994). ACTH-induced preterm labour in the ewe is associated with increased mRNA and protein levels of myometrial gap junction protein, connexin-43. *J. Endocrinol., 141*(2), 195–202.

Mecenas, C. A., Giussani, D. A., et al. (1996). Production of premature delivery in pregnant rhesus monkeys by androstenedione infusion. *Nat. Med., 2*(4), 443–448.

Melamed, M., Castano, E., et al. (1997). Molecular and kinetic basis for the mixed agonist/antagonist activity of estriol. *Mol. Endocrinol., 11* (12), 1868–1878.

Merlino, A. A., Welsh, T. N., et al. (2007). Nuclear progesterone receptors in the human pregnancy myometrium: evidence that parturition involves functional progesterone withdrawal mediated by increased expression of progesterone receptor-A. *J. Clin. Endocrinol. Metab., 92*(5), 1927–1933.

Mesiano, S. (2004). Myometrial progesterone responsiveness and the control of human parturition. *J. Soc. Gynecol. Investig., 11*(4), 193–202.

Mesiano, S., Chan, E. C., et al. (2002). Progesterone withdrawal and estrogen activation in human parturition are coordinated by progesterone receptor A expression in the myometrium. *J. Clin. Endocrinol. Metab., 87*(6), 2924–2930.

Mitchell, B. F., Fang, X., et al. (1998). Oxytocin: a paracrine hormone in the regulation of parturition? *Rev. Reprod., 3*(2), 113–122.

Mitchell, M. D., Romero, R. J., et al. (1995). Prostaglandins and parturition. *Reprod. Fertil. Dev., 7*(3), 623–632.

Morham, S. G., Langenbach, R., et al. (1995). Prostaglandin synthase 2 gene disruption causes severe renal pathology in the mouse. *Cell, 83* (3), 473–482.

Muglia, L., Jacobson, L., et al. (1995). Corticotropin-releasing hormone deficiency reveals major fetal but not adult glucocorticoid need. *Nature, 373*(6513), 427–432.

Murakami, S., Iwaki, D., et al. (2002). Surfactant protein A inhibits peptidoglycan-induced tumor necrosis factor-alpha secretion in U937 cells and alveolar macrophages by direct interaction with toll-like receptor 2. *J. Biol. Chem., 277*(9), 6830–6837.

Myatt, L., Langdon, G., et al. (1994). Identification and changes in concentrations of prostaglandin H synthase (PGHS) isoforms in rat myometrium at parturition. *Prostaglandins, 48*(5), 285–296.

Nathanielsz, P. W., Jenkins, S. L., et al. (1998). Local paracrine effects of estradiol are central to parturition in the rhesus monkey. *Nat. Med., 4* (4), 456–459.

New, M. I., & Speiser, P. W. (1986). Genetics of adrenal steroid 21-hydroxylase deficiency. *Endocr. Rev., 7*(3), 331–349.

Ni, X., Chan, E. C., et al. (1997). Nitric oxide inhibits corticotropin-releasing hormone exocytosis but not synthesis by cultured human trophoblasts. *J. Clin. Endocrinol. Metab., 82*(12), 4171–4175.

Ni, X., Hou, Y., et al. (2004). Progesterone receptors A and B differentially modulate corticotropin-releasing hormone gene expression through a cAMP regulatory element. *Cell Mol. Life Sci., 61*(9), 1114–1122.

Ni, X., Nicholson, R. C., et al. (2002). Estrogen represses whereas the estrogen-antagonist ICI 182780 stimulates placental CRH gene expression. *J. Clin. Endocrinol. Metab., 87*(8), 3774–3778.

Nishimori, K., Young, L. J., et al. (1996). Oxytocin is required for nursing but is not essential for parturition or reproductive behavior. *Proc. Natl. Acad. Sci. USA, 93*(21), 11699–11704.

Nnamani, C., Godwin, A., et al. (1994). Regulation of cell–cell communication mediated by connexin 43 in rabbit myometrial cells. *Biol. Reprod., 50*(2), 377–389.

Nodwell, A., Carmichael, L., et al. (1999). Placental release of corticotrophin-releasing hormone across the umbilical circulation of the human newborn. *Placenta., 20*(2-3), 197–202.

Orsino, A., Taylor, C. V., et al. (1996). Connexin-26 and connexin-43 are differentially expressed and regulated in the rat myometrium throughout late pregnancy and with the onset of labor. *Endocrinology, 137*(5), 1545–1553.

Ou, C. W., Orsino, A., et al. (1997). Expression of connexin-43 and connexin-26 in the rat myometrium during pregnancy and labor is differentially regulated by mechanical and hormonal signals. *Endocrinology, 138*(12), 5398–5407.

Padykula, H. A., & Taylor, J. M. (1976). Ultrastructural evidence for loss of the trophoblastic layer in the chorioallantoic placenta of Australian bandicoots (Marsupialia: Peramelidae). *Anat. Rec., 186*(3), 357–385.

Parkington, H. C. (1983). Electrical properties of the costo-uterine muscle of the guinea-pig. *J. Physiol., 335*, 15–27.

Parry, L. J., Bathgate, R. A., et al. (1997). Evidence for a local fetal influence on myometrial oxytocin receptors during pregnancy in the tammar wallaby (*Macropus eugenii*). *Biol. Reprod., 56*(1), 200–207.

Parry, L. J., Guymer, F. J., et al. (1996). Release of an oxytocic peptide at parturition in the marsupial, *Macropus eugenii. J. Reprod. Fertil., 107* (2), 191–198.

Patel, R. R., Steer, P., et al. (2004). Does gestation vary by ethnic group? A London-based study of over 122,000 pregnancies with spontaneous onset of labour. *Int. J. Epidemiol., 33*(1), 107–113.

Patrick, J., Challis, J., et al. (1979). Circadian rhythms in maternal plasma cortisol, estrone, estradiol, and estriol at 34 to 35 weeks' gestation. *Am. J. Obstet. Gynecol., 135*(6), 791–798.

Petraglia, F., Sutton, S., et al. (1989). Neurotransmitters and peptides modulate the release of immunoreactive corticotropin-releasing factor from cultured human placental cells. *Am. J. Obstet. Gynecol., 160*(1), 247–251.

Poore, K. R., Canny, B. J., et al. (1999). Adrenal responsiveness and the timing of parturition in hypothalamo–pituitary disconnected ovine foetuses with and without constant adrenocorticotrophin infusion. *J. Neuroendocrinol., 11*(5), 343–349.

Poore, K. R., Young, I. R., et al. (1998). Studies on the role of ACTH in the regulation of adrenal responsiveness and the timing of parturition in the ovine fetus. *J. Endocrinol., 158*(2), 161–171.

Poore, K. R., Young, I. R., et al. (1999). Efficacy of the selective prostaglandin synthase type 2 inhibitor nimesulide in blocking basal prostaglandin production and delaying glucocorticoid-induced premature labor in sheep. *Am. J. Obstet. Gynecol., 180*(5), 1244–1253.

Raeside, J. I., Liptrap, R. M., et al. (1979). A precursor role for DHA in a feto–placental unit for oestrogen formation in the mare. *J. Reprod. Fertil. Suppl.*, (27), 493–497.

Renfree, M. B. (1994). *Endocrinology of pregnancy, parturition and lactation in marsupials Marshall's Physiology of Reproduction.* G. E. Lamming (ed.). London, UK: Chapman & Hall. 677–751.

Renfree, M. B., & Shaw, G. S. (1996). Reproduction of a marsupial: from uterus to pouch. *Anim. Reprod. Sci., 42*, 393–404.

Renfree, M. B., Parry, L. J., et al. (1996). Infusion with an oxytocin receptor antagonist delays parturition in a marsupial. *J. Reprod. Fertil., 108*(1), 131–137.

Risek, B., & Gilula, N. B. (1996). Gap junction regulation during preterm labor in the rat: multiple effects of the antiprogesterone RU486. *Biol. Reprod., 55*(3), 525–535.

Robins, J., & Mann, L. I. (1975). Midtrimester pregnancy termination by intramuscular injection of a 15-methyl analogue of prostaglandin F2 alpha. *Am. J. Obstet. Gynecol., 123*(6), 625–631.

Robinson, B. G., Arbiser, J. L., et al. (1989). Species-specific placental corticotropin releasing hormone messenger RNA and peptide expression. *Mol. Cell. Endocrinol., 62*(2), 337–341.

Robinson, B. G., Emanuel, R. L., et al. (1988). Glucocorticoid stimulates expression of corticotropin-releasing hormone gene in human placenta. *Proc. Natl. Acad. Sci. USA, 85*(14), 5244–5248.

Romero, R., & Tartakovsky, B. (1992). The natural interleukin-1 receptor antagonist prevents interleukin-1-induced preterm delivery in mice. *Am. J. Obstet. Gynecol., 167*(4 Pt 1), 1041–1045.

Romero, R., Mazor, M., et al. (1991). Systemic administration of interleukin-1 induces preterm parturition in mice. *Am. J. Obstet. Gynecol., 165*(4 Pt 1), 969–971.

Rowland, D. L., Wagonblast, A. L., et al. (1991). Timing of parturition in the rat: an analysis of successive births. *Chronobiologia., 18*(1), 31–38.

Sadovsky, Y., Nelson, D. M., et al. (2000). Effective diminution of amniotic prostaglandin production by selective inhibitors of cyclooxygenase type 2. *Am. J. Obstet. Gynecol., 182*(2), 370–376.

Sandman, C. A., Glynn, L., et al. (2006). Elevated maternal cortisol early in pregnancy predicts third trimester levels of placental corticotropin releasing hormone (CRH): priming the placental clock. *Peptides, 27* (6), 1457–1463.

Sato, S., Sugiyama, M., et al. (2003). Toll/IL-1 receptor domain-containing adaptor inducing IFN-beta (TRIF) associates with TNF receptor-associated factor 6 and TANK-binding kinase 1, and activates two distinct transcription factors, NF-kappa B and IFN-regulatory factor-3, in the Toll-like receptor signaling. *J. Immunol., 171*(8), 4304–4310.

Schellenberg, J. C. (1995). The effect of oxytocin receptor blockade on parturition in guinea pigs. *J. Clin. Invest., 95*(1), 13–19.

Sebastian, L. T., De Matteo, L., et al. (1998). Mesotocin receptors during pregnancy, parturition and lactation in the tammar wallaby. *Anim. Reprod. Sci., 51*(1), 57–74.

Seron-Ferre, M., & Jaffe, R. B. (1981). The fetal adrenal gland. *Annu. Rev. Physiol., 43*, 141–162.

Shaw, G. (1983). Effect of PGF-2 alpha on uterine activity, and concentrations of 13, 14-dihydro-15-keto-PGF-2 alpha in peripheral plasma during parturition in the tammar wallaby (*Macropus eugenii*). *J. Reprod. Fertil., 69*(2), 429–436.

Shaw, G. (1990). Control of parturient behaviour by prostaglandin F-2 alpha in the tammar wallaby (*Macropus eugenii*). *J. Reprod. Fertil., 88*(1), 335–342.

Shaw, G., & Renfree, M. B. (2006). Parturition and perfect prematurity: birth in marsupials. *Aust. J. Zool., 54*, 139–149.

Shaw, G., G.H., & Bell, E. C. (1999). Production of prostaglandin f2alpha and its metabolite by endometrium and yolk sac placenta in late gestation in the tammar wallaby, *Macropus eugenii. Biol. Reprod., 60* (3), 611–614.

Shaw, G., & Renfree, M. B. (1984). Concentrations of oestradiol-17 beta in plasma and corpora lutea throughout pregnancy in the tammar, *Macropus eugenii. J. Reprod. Fertil., 72*(1), 29–37.

Shaw, G., & Renfree, M. B. (2001). Fetal control of parturition in marsupials. *Reprod. Fert. Develop., 13*, 653–659.

Shaw, G. R., Renfree, M. B., & Fletcher, T. P. (1996). A role for glucocorticoids in parturition in a marsupial, *Macropus eugenii. Biol. Reprod., 54*, 728–733.

Shearman, L. P., Zylka, M. J., et al. (1997). Two period homologs: circadian expression and photic regulation in the suprachiasmatic nuclei. *Neuron., 19*(6), 1261–1269.

Sheehan, P. M., Rice, G. E., et al. (2005). 5 Beta-dihydroprogesterone and steroid 5 beta-reductase decrease in association with human parturition at term. *Mol. Hum. Reprod., 11*(7), 495–501.

Sheldrick, E. L., Ricketts, A. P., et al. (1980). Placental production of progesterone in ovariectomized goats treated with a synthetic progestagen to maintain pregnancy. *J. Reprod. Fertil., 60*(2), 339–348.

Sheldrick, E. L., Ricketts, A. P., et al. (1981). Placental production of 5 beta-pregnane-3 alpha,20 alpha-diol in goats. *J. Endocrinol., 90*(2), 151–158.

Short, R. V., Flint, A. P. F., & Renfree, M. B. (1985). Influence of passive immunisation against gonadotrophin releasing hormone on pregnancy and parturition in the tammar wallaby. *J. Reprod. Fert., 75*, 567–575.

Shozu, M., Akasofu, K., et al. (1992). Mutations in the aromatase gene and their effect [letter; comment]. *Fertility and Aterility, 58*(5), 1084–1085.

Shozu, M., Akasofu, K., et al. (1991). A new cause of female pseudohermaphroditism: placental aromatase deficiency. *Journal of Clinical Endocrinology and Metabolism, 72*(3), 560–566.

Silver, M. (1990). Prenatal maturation, the timing of birth and how it may be regulated in domestic animals. *Exp. Physiol., 75*(3), 285–307.

Silver, M., & Fowden, A. L. (1991). Response of the fetal adrenal to ACTH in the foal near term. *Journal of Endocrinology, 131*(Suppl). Abstract 69.

Smith, R., Chan, E. C., et al. (1993). Corticotropin-releasing hormone in baboon pregnancy. *J. Clin. Endocrinol. Metab., 76*(4), 1063–1068.

Smith, R., Mesiano, S., et al. (1998). Corticotropin-releasing hormone directly and preferentially stimulates dehydroepiandrosterone sulfate secretion by human fetal adrenal cortical cells. *J. Clin. Endocrinol. Metab., 83*(8), 2916–2920.

Smith, R., Smith, J. I., et al. (2009). Patterns of plasma corticotropin-releasing hormone, progesterone, estradiol, and estriol change and the onset of human labor. *J. Clin. Endocrinol. Metab., 94*(6), 2066–2074.

Smith, R., Wickings, E. J., et al. (1999). Corticotropin-releasing hormone in chimpanzee and gorilla pregnancies. *J. Clin. Endocrinol. Metab., 84*(8), 2820–2825.

Somero, G. N. (2000). Unity in diversity: a perspective on the methods, contributions, and future of comparative physiology. *Annu. Rev. Physiol., 62*, 927–937.

Sparey, C., Robson, S. C., et al. (1999). The differential expression of myometrial connexin-43, cyclooxygenase-1 and -2, and Gs alpha proteins in the upper and lower segments of the human uterus during pregnancy and labor. *J. Clin. Endocrinol. Metab., 84*(5), 1705–1710.

Steele, P. A., Flint, A. P., et al. (1976). Activity of steroid C-17,20 lyase in the ovine placenta: effect of exposure to foetal glucocorticoid. *J. Endocrinol., 69*(2), 239–246.

Steeves, T. D., King, D. P., et al. (1999). Molecular cloning and characterization of the human CLOCK gene: expression in the suprachiasmatic nuclei. *Genomics, 57*(2), 189–200.

Sugimoto, Y., Yamasaki, A., et al. (1997). Failure of parturition in mice lacking the prostaglandin F receptor. *Science, 277*(5326), 681–683.

Sun, K., Brockman, D., et al. (2006). Induction of surfactant protein A expression by cortisol facilitates prostaglandin synthesis in human chorionic trophoblasts. *J. Clin. Endocrinol. Metab., 91*(12), 4988–4994.

Torricelli, M., Giovannelli, A., et al. (2007). Labor (term and preterm) is associated with changes in the placental mRNA expression of corticotrophin-releasing factor. *Reprod. Sci., 14*(3), 241–245.

Torricelli, M., Ignacchiti, E., et al. (2006). Maternal plasma corticotrophin-releasing factor and urocortin levels in post-term pregnancies. *Eur. J. Endocrinol., 154*(2), 281–285.

Tyndale-Biscoe, C. H., & Renfree, M. B. (1987). *Reproductive Physiology of Marsupials*. Cambridge University Press.

Tyndale-Biscoe, C. H., Hinds, L. A., & Horn, C. A. (1988). Fetal role in the control of parturition in the tammar, Macropus eugenii. *J. Reprod. Fertil., 82*(2), 419–428.

Van Kampen, K. R., & Ellis, L. C. (1972). Prolonged gestation in ewes ingesting *Veratrum californicum*: morphological changes and steroid biosynthesis in the endocrine organs of cyclopic lambs. *J. Endocrinol., 52*(3), 549–560.

Ward, K. L., & Renfree, M. B. (1984). Effects of progesterone on parturition in the tammar, *Macropus eugenii. J. Reprod. Fertil., 72*(1), 21–28.

Wessely, O., Deiner, E. M., et al. (1997). The glucocorticoid receptor is a key regulator of the decision between self-renewal and differentiation in erythroid progenitors. *Embo. J., 16*(2), 267–280.

White, P. C., & Speiser, P. W. (2000). Congenital adrenal hyperplasia due to 21-hydroxylase deficiency. *Endocr. Rev., 21*(3), 245–291.

Wintour, E. M., Smith, M. B., et al. (1985). The role of fetal adrenal hormones in the switch from fetal to adult globin synthesis in the sheep. *J. Endocrinol., 104*(1), 165–170.

Wu, W. X., Ma, X. H., et al. (2004). Prostaglandin mediates premature delivery in pregnant sheep induced by estradiol at 121 days of gestational age. *Endocrinology, 145*(3), 1444–1452.

Yamada, C., Sano, H., et al. (2006). Surfactant protein A directly interacts with TLR4 and MD-2 and regulates inflammatory cellular response. Importance of supratrimeric oligomerization. *J. Biol. Chem., 281*(31), 21771–21780.

Young, W. S., 3rd, Shepard, E., et al. (1996). Deficiency in mouse oxytocin prevents milk ejection, but not fertility or parturition. *J. Neuroendocrinol., 8*(11), 847–853.

Zheng, B., Larkin, D. W., et al. (1999). The mPer2 gene encodes a functional component of the mammalian circadian clock. *Nature, 400*(6740), 169–173.

Zitnik, G., Peterson, K., et al. (1995). Effects of butyrate and glucocorticoids on gamma- to beta-globin gene switching in somatic cell hybrids. *Mol. Cell Biol., 15*(2), 790–795.

Chapter 7

Stress and Reproduction in Mammals

Lynda Uphouse
Texas Woman's University, Denton, TX, USA

SUMMARY

Stress is associated with reproductive modification due to interaction between the stress-responsive hypothalamic–pituitary–adrenal (HPA) axis and the hypothalamic–pituitary–gonadal (HPG) axis. The mechanisms responsible for this interaction include neuronal and peripheral events that are translated into alterations of gonadotropin secretion from the hypothalamus. Free-ranging mammals are routinely exposed to the environmental challenges of food scarcity, disease, and predation, all of which reduce both behavioral and physiological indices of reproduction. Although classically viewed as very different in their means of suppression of reproduction, stressors share multiple commonalities. While the 'fight or flight' response to a potential predator seems distinct from the 'sick behavior' evoked by disease, both provoke similar compromises in reproductive function. In this chapter, food restriction, infection, and other stressors will be discussed with an emphasis on their effect on the release of pituitary luteinizing hormone, their behavioral impact, and the mechanisms responsible for their reproductive action.

1. INTRODUCTION

Successful reproduction in mammals, as in other vertebrates, requires coordination among behavioral and physiological processes so that the act of mating occurs coincident with physiological readiness to produce offspring. In the male, successful reproduction requires the maintenance of spermatogenesis and appropriate mating activity. In the female, ovulation and mating must occur close enough in time to ensure viability of both ova and sperm prior to fertilization. The events required for implantation and successful gestation must properly proceed, and postpartum mammalian offspring must receive nurturing so that they survive to reproduce themselves. Stress can interfere with any or all of these reproductive events. In many excellent reviews, the effects of stress on neuroendocrine events critical for ovulation and spermatogenesis have been discussed (Rivest & Rivier, 1995; Sapolsky, Romero, & Munck, 2000; Tilbrook, Turner, & Clarke, 2000; Haddad, Saade, & Safieh-Garabedian, 2002; Wade & Jones, 2003; Wingfield & Sapolsky, 2003; Breen & Karsch, 2006; Hill, Elmquist, & Elias, 2008). However, behavioral readiness to mate is also vulnerable to disruption by stress (Avitsur & Yirmiya, 1999; Wade & Jones, 2003). Acute stressors may interfere with behavioral patterns without disrupting ovulation or spermatogenesis but, without the act of mating, no offspring will be produced. Therefore, consideration of the impact of stress on reproduction must also include its effects on mating rituals.

Reproductive activity is coordinated through the hypothalamic–pituitary–gonadal (HPG) axis, with gonadotropin-releasing hormone (GnRH) neurons functioning to integrate information from hormonal and neural feedback loops (Evans, 1999). Release of GnRH into the median eminence and its transport through the hypothalamic–hypophysial portal system controls the secretion of follicle-stimulating hormone (FSH) and luteinizing hormone (LH) from the anterior pituitary. Follicle-stimulating hormone and LH act on the gonads to influence the maturation of gametes and, in the female, the rupture of mature follicles. Examination of the effects of stress on the reproductive axis has relied heavily upon the measurement of LH and GnRH, leading to the current conception that effects of stress on reproduction ultimately involve effects on GnRH neurons (Rivest & Rivier, 1995; Wade & Jones, 2004; Breen, Billings, Wagenmaker, Wessinger, & Karsch, 2005). The necessary and sufficient stress-produced signals that mediate these effects on GnRH neurons are less clear and probably include multiple neural and hormonal pathways that differ depending on the nature of the stimulus, the species, the strain, and hormonal or social status (Windle, Wood, Lightman, & Ingram, 1998; Jeong, Jacobson, Widmaier, & Majzoub, 1999; Bekris, Antoniou, Daskas, & Papadopoulou-Daifoti, 2005; Breen et al., 2005; Machatschke, Wallner, & Dittami, 2006; Avitsur et al., 2007).

The precise location of GnRH neurons varies with species (King & Anthony, 1984; Merchenthaler, Gorcs,

Setalo, Petrusz, & Flerko, 1984; Palkovits, 1986). For example, in rodents, GnRH neurons are most prominent in the medial preoptic area (mPOA), while, in primates, many GnRH neurons are located near the ventrolateral hypothalamus in the tuberal region. This differential location of GnRH neurons, relative to the blood–brain barrier, may account for reported differences between species in their response to systemic treatment with stress-inducing compounds.

Although sexual behavior and physiological readiness to mate are coordinated by the release of GnRH and HPG activation, the neural sites controlling GnRH release and sexual behavior are not identical (Pfaff & Modianos, 1985; Flanagan-Cato & McEwen, 1995; Hull & Dominguez, 2007). Sexual behavior includes three components: arousal, motivation, and consummation (Sodersten, 1981; Erskine, 1989; Hull & Dominguez, 2007). Appetitive behaviors (often associated with sexual motivation) include those behavioral patterns that are designed to attract the attention of a potential mate and engage him/her in pursuit or interaction. Consummatory behavior includes the actual act of mating (mounting, intromission, ejaculation by the male or assumption of an appropriate mating posture—e.g., lordosis, lateral displacement—in the female).

In most studies of the effects of stress on sexual behavior, consummatory behavior has been examined. In female rodents, the ventromedial nucleus of the hypothalamus (VMH) is a critical site for the consummatory act of lordosis (Pfaff & Modianos, 1985), but many brain areas contribute to the complete behavioral female repertoire (Cottingham & Pfaff, 1986; Flanagan-Cato, Lee, & Calizo, 2006). In male rodents, the mPOA is a critical site for integration of sensory and motor events required for copulation, while additional areas contribute to the entire mating sequence (Hull & Dominguez, 2007). In primates, in which sexual behavior is less dependent on physiological readiness to mate, the factors controlling sexual activity are more complex.

Similarly to the effects of stressors on GnRH, sexual behavior declines after prolonged stress and the effects of acute stressors are more variable. Female mammals, which usually bear the greater reproductive burden, are more likely to be affected by acute stress than are males. However, since some degree of arousal is critical for sexual behavior to occur, it is often difficult to determine when arousing stimuli have become stressful.

A major caveat intrinsic to the study of stress and reproduction is a difficulty in defining exactly what is and is not a stressor. It is generally agreed that stressors activate the hypothalamic–pituitary–adrenal (HPA) axis and the sympathetic nervous system (Kubovcakova et al., 2004; Tilbrook et al., 2000; Von Borell, Dobson, & Prunier, 2007). Hypothalamic–pituitary–adrenal responses include an increase in corticotropin-releasing hormone (CRH) and arginine vasopressin (AVP) from the parvocellular division of the paraventricular nucleus (PVN) of the hypothalamus (Herman & Cullinan, 1997; Engler, Redei, & Kola, 1999; Carrasco & Van de Kar, 2003). Corticotropin-releasing hormone, secreted into the hypophysial portal system, stimulates anterior pituitary corticotropes to secrete corticotropin (ACTH) (Gibbs & Vale, 1982; Antoni, 1986) and other peptides (e.g., β-endorphin, α-melanocortin-stimulating hormone), which are derived by differential processing of the pro-opiomelanocortin precursor (Engler, Pham, Fullerton, Clarke, & Funder, 1989). Arginine vasopressin synergizes with CRH to increase ACTH release (C. Rivier, J. Rivier, Mormede, & Vale, 1984), and, at least in some species, can modulate ACTH in the absence of CRH (Jeong et al., 1999; Muglia, Bethin, Jacobson, Vogt, & Majzoub, 2000). Corticotropin stimulates the adrenal cortex to secrete glucocorticoids (cortisol in primates and corticosterone (CORT) in rodents), which increase glucose availability (Breen & Karsch, 2006), and glucocorticoids provide negative feedback to the HPA axis, thereby limiting HPA activation (Dallman et al., 1987). The sympathetic nervous system is activated by brainstem regions including the locus coeruleus (LC) (Van Bockstaele, Colago, & Valentino, 1996; 1998).

Collectively, these responses mobilize energy stores in preparation for an adaptive (e.g., defense/escape) response to the stressor. Reduced reproductive function during/after stress prevents a waste of valuable energetic resources under conditions in which individual and/or offspring survival is questionable. Therefore, a decline in reproduction is a logical outcome of the stress response.

Classically, it has been tempting to operationally define the occurrence of stress by HPA activation and an increase in glucocorticoids. This has led to a plethora of research in which individual members of the stress response have been examined for their impact on reproductive function. However, while CRH synaptic contacts are present at GnRH neurons (MacLusky, Naftolin, & Leranth, 1988), where they could directly modulate GnRH release, CRH neurons and their projections also exist outside the PVN in areas that are not directly involved in the HPA axis (Merchenthaler, 1984; Van Bockstaele et al., 1996; 1998; Reyes, Valentino, Xu, & Van Bockstaele, 2005). Corticotropin-releasing hormone neurons project to midbrain and hindbrain areas that integrate peripheral and central nervous system (CNS) responses to stress and provide critical links between stressful stimuli and the reproductive axis (Valentino & Van Bockstaele, 2008).

Moreover, while it is true that events generally agreed to be stressful do, in fact, activate the HPA axis, HPA activation occurs under conditions that are not generally thought to be stressful. Most arousing stimuli (e.g., presence of a mate, availability of food, novel environment) will activate the HPA axis without disrupting reproduction

(Auger, Moffatt, & Blaustein, 1997; Hennessy, Zate, & Maken, 2008) and may produce differential effects on behavior (Tonissaar et al., 2008). Therefore, while many mechanisms leading to stress-induced disruption of reproductive function have been identified, the relationships among these mechanisms continue to emerge.

In freely roaming species, the most obvious threats to the individual are predation, undernutrition or malnutrition, and disease. Severe and long-term undernutrition have long been recognized to interfere with reproduction. Similarly, when disease is prevalent, reproductive fitness declines. In the laboratory, food restriction and immune challenges have been used to examine the consequences of stress on reproductive function. However, stress-induced reproductive decline may also occur in healthy individuals in the presence of a plentiful food supply if the individual is subjected to repeated episodes of environmental challenge. In the laboratory, a variety of stimuli including footshock, restraint, crowding, or unpredictable disturbance have been used to simulate these conditions. Collectively, these studies have led to a general consensus that prolonged experience with such stimuli can reduce reproductive fitness. However, there are also many examples where resistance to stress-induced reproductive suppression occurs. This resistance is especially evident in seasonal breeders such as the grey seal (*Halichoerus gyrpus*) or Arctic ground squirrel (*Spermophilus parryi plesius*), where there is a limited reproductive timespan and where failure to reproduce offspring may create an especially high burden in terms of species fitness (Wingfield & Sapolsky, 2003). Escape from stress-induced reproductive suppression may also occur in species such as olive baboons (*Papio anubis*) with a clear dominance hierarchy and where the dominant individual begets most of the offspring (Tilbrook et al., 2000; Wingfield & Sapolsky, 2003; Lidgard, Boness, Bowen, & McMillan, 2008). In such cases, subordinate individuals exhibit stress-induced reproductive suppression while the dominant individual remains reproductively competent under a variety of adverse conditions (Lidgard et al., 2008). Sapolsky reported that resistance of the dominant male baboon resulted from a smaller effect of stress so that testosterone (T) levels were less affected in dominant individuals (Sapolsky, 1985; 1986). However, dominance seems to be associated with lower stress responses only if dominance is accompanied by additional factors such as social support and good coping mechanisms (Wingfield & Sapolsky, 2003).

In the following discussion, the effects of food restriction, immune challenge, and other stressors (e.g., immobilization, footshock, etc.) on GnRH, LH, and sexual behavior will be examined. Classically, these stressors have been categorized as metabolic or nonmetabolic, metabolic or neurogenic, physical or psychosocial, interoceptive or exteroceptive, with the stimulus source and/or transduction pathway as the defining characteristic. Clearly, the behavioral 'fight or flight' response to a potential predator seems distinct from the 'sick behavior' evoked by disease, so that such categorizations are useful in defining specific pathways whereby information is transmitted to the reproductive system. In recent years, considerable commonalities have emerged. Nevertheless, the following discussion is grouped according to the probable sources of stress (e.g., absence of food supply, presence of infection, or external challenge) that are most likely to be encountered by freely roaming mammals.

2. EFFECTS OF FOOD RESTRICTION

Reduced availability of the food supply is one of the most common stressors likely to impact reproduction in freely roaming mammals (Wade & Schneider, 1992; Wade & Jones, 2003). The effects of food restriction vary with the degree and duration of food restriction and with the type of food restriction (e.g., general caloric restriction or reduction of select dietary components) (Wade & Jones, 2003; 2004) and there is a considerable range between types of food restriction that do and do not affect reproduction (Hoyenga & Hoyenga, 1982). What is reproductively tolerable may vary across species or even within species with the season (in seasonal breeders) and/or degree of hormonal priming (Tropp & Markus, 2001). Nevertheless, if the food supply is severely limited for a prolonged time period, reproduction is unlikely to occur (Jones & Lubbers, 2001). Reproduction is an energy-expensive endeavor, so that reproductive activities are reduced when individuals are forced to choose between individual survival and reproduction (K. B. Hoyenga & K. T. Hoyenga, 1982; Schneider, Zhou, & Blum, 2000). Hence, most investigators would agree that long-term food restriction and the resulting metabolic consequences have a negative influence on fertility.

Although males and females show changes in reproductive function during periods of food restriction (Tena-Sempere & Barreiro, 2002; Fernandez-Fernandez et al., 2006; Martin et al., 2007), female mammals (who generally bear the greater energy burden for offspring production) are more likely to be affected (K. B. Hoyenga & K. T. Hoyenga, 1982; Wade & Jones, 2004; Martin et al., 2007; Martin, Qasim, & Reilly, 2008). Females respond to food deprivation with a greater increase in CORT than do males (Martin et al., 2007) and, in females, estradiol accentuates the decline in LH that follows fasting (Cagampang, Maeda, Tsukamura, Ohkura, & Ota, 1991; Nagatani et al., 1996a).

In female mammals, long-term food restriction disrupts estrous cyclicity (Wade & Schneider, 1992; Tropp & Markus, 2001; Williams, Helmreich, Parfitt, Caston-Balderrama, & Cameron, 2001); females become anestrus (Schneider et al., 1998; Schneider, Blum, & Wade, 2000), fail to ovulate (Wade, Schneider, & Li, 1996), and do not show sexual receptivity (Dickerman, Li, & Wade, 1993; Gill & Rissman, 1997; Jones, Pick, & Wade, 2002). In males and females, after long-term food restriction, most components (e.g., brain, pituitary, gonads) of the HPG axis are disrupted (Kinoshita, Moriyama, Tsukamura, & Maeda, 2003; Wade & Jones, 2004; Martin et al., 2007). Gonadotropic hormones (e.g., LH and FSH) are reduced (Cagampang, Maeda, Yokoyama, & Ota, 1990; Henry, Goding, Tilbrook, Dunshea, & Clarke, 2001; Kiyma et al., 2004; Wade & Jones, 2004), gonadal hormones decline (Martin et al., 2007), and hypogonadism results. The precise hormonal changes vary but generally reflect a disruption of the neuroendocrine profiles characteristic of a healthy, reproductively competent mammal. The effects of food deprivation are not permanent because reproductive function resumes when food availability increases (Jones & Lubbers, 2001; Wade & Jones, 2004).

While it makes intuitive sense that severe and long-term energy depletion would produce a decline in reproductive function, food restriction can interfere with reproductive processes long before the threat to individual survival is imminent (Jones & Lubbers, 2001; Wade & Jones, 2004). In females, short-term food restriction, even if relatively transient, can disrupt estrous cyclicity, reduce lordosis behavior, and decrease ovulation (Schneider, Goldman, Leo, & Rosen, 1997; Schneider et al., 1998; Schneider, Hall, & Wade, 1997). While the reproductive system of males is also affected by short-term food restriction (Dallman et al., 1999), a transient decrease in spermatogenesis is unlikely to substantially reduce male reproductive fitness.

The rapidity of the reproductive suppression following food restriction has led investigators to conclude that organisms are capable of sensing minute-by-minute availability of internal energy resources (Schneider et al., 2000a; Wade & Jones, 2003; Wade and Jones, 2004). The mechanisms whereby this information about energy status is communicated to reproductively relevant brain areas has been the subject of intensive investigation and several excellent reviews are available (e.g., see Dallman et al., 1999; Wade and Jones, 2003; 2004). Ultimately, however, the hypothalamus is the final point of integration of the various metabolic signals, with food restriction reducing LH through an influence on GnRH neurons.

Although the critical communication link(s) between food restriction and reproduction remain undetermined, the number of possible signal molecules is vast and classically has included such candidates as glucose and insulin (Dallman et al., 1999; Wade & Jones, 2004). With the emergence of leptin, ghrelin, cholecystokinin, and a variety of other gut peptides (all of which have been implicated in reproductive suppression after food restriction), the number of candidates has substantially increased (Giacobini et al., 2004; Sahu, 2004; Stanley, Wynne, McGowan, & Bloom, 2005; Wynne, Stanley, McGowan, & Bloom, 2005; Ahima & Lazar, 2008; Sajapitak et al., 2008). However, regardless of the signal molecule, metabolic signals are conveyed to the hypothalamus, where they directly or indirectly influence GnRH neurons (Wade & Jones, 2003; 2004; Martin et al., 2007; Hill et al., 2008; Martin et al., 2008).

Leptin initially received the most attention as a contributor to the effects of food restriction on reproductive functioning. A threshold amount of leptin is required for the maintenance of reproductive functioning (Hill et al., 2008), and leptin levels are responsive to food restriction. Even acute fasting can reduce serum leptin, and considerable evidence has accumulated to link leptin with effects of food restriction on reproduction (Schneider et al., 1998; Schneider et al., 2000b; Wade & Jones, 2004; Hill et al., 2008). Leptin attenuates the food-deprivation-induced decline in LH and fertility in females and attenuates the food-deprivation-induced decline in plasma T in males (Nagatani et al., 1998; Wade & Jones, 2004; Hill et al., 2008). However, long-term elevation of leptin may itself reduce T by a direct inhibitory effect at the testes (Sirotkin et al., 2008). Nevertheless, even though leptin can reverse fasting-induced anestrus in female hamsters (Schneider et al., 1997b; 1998; 2000b), leptin does not restore estrous cyclicity when fuel oxidation is also reduced (Schneider et al., 1998; Schneider & Zhou, 1999). Moreover, leptin does not restore sexual behavior in ovariectomized, hormone-treated, food-deprived hamsters unless somewhat 'heroic' concentrations of estrogens are administered (Wade & Jones, 2004; Wade, Lempicki, Panicker, Frisbee, & Blaustein, 1997).

In addition, reproductive function (both LH pulsatility and estrous behavior) is reinstated rapidly when refeeding occurs, whereas leptin levels do not increase until later, and LH pulsatility is reduced by short-term food restriction, which does not alter levels of leptin (Jones & Lubbers, 2001; Wade & Jones, 2004). Therefore, in spite of considerable emphasis on leptin as a potential link between food restriction and reproductive function (Schneider et al., 2000b), it cannot be the only key in this interesting puzzle. Nevertheless, leptin may still be a prominent player so that leptin's role in fasting-induced declines in LH cannot be ignored.

Leptin's effects on food intake are integrated in part through leptin receptors located on hypothalamic neurons, which contribute to the regulation of food intake (Cunningham et al., 1999; Zigman & Elmquist, 2003; Sahu, 2004). However, GnRH neurons do not appear to express leptin receptors (Hakansson, Brown, Ghilardi, Skoda, & Meister, 1998; Cunningham, Clifton, & Steiner, 1999) so that information carried by leptin must be communicated indirectly to GnRH neurons. A possible candidate for such communication is the Kiss-1 neuron

(Castellano et al., 2005; 2008; Ramaswamy, Guerriero, Gibbs, & Plant, 2008).

Kisspeptin is the gene product of the *Kiss-1* gene and binds to the GRP54 receptor, a G-protein-coupled receptor located on GnRH neurons (Irwig et al., 2004; Messager et al., 2005), and is a potent regulator of GnRH neurons (Gottsch, Clifton, & Steiner, 2004; Irwig et al., 2004; Roa, Aguilar, Dieguez, Pinilla, & Tena-Sempere, 2008; Smith et al., 2008). Kiss-1 neurons may translate many metabolic cues, including leptin, to GnRH neurons (Castellano et al., 2008; Hill et al., 2008; Martin et al., 2008; Ramaswamy et al., 2008; Roa et al., 2008; Smith et al., 2008). Food deprivation reduces Kiss-1 mRNA (Castellano et al., 2005; Luque, Kineman, & Tena-Sempere, 2007; Brown, Imran, Ur, & Wilkinson, 2008) and kisspeptin can restore the fasting-induced decline in LH (Castellano et al., 2005; 2008). Kisspeptin is effective even in the presence of leptin antibodies, leading to the suggestion that the actions of kisspeptin occur downstream from the effects of leptin (Castellano et al., 2005). A role for Kiss-1 neurons in the response to multiple metabolic challenges has led investigators to conclude that kisspeptin may be a critical integrator of metabolic cues to GnRH neurons (Castellano et al., 2008; Hill et al., 2008; Martin et al., 2008; Roa et al., 2008).

Brainstem areas, including the vagal-innervated area postrema/nucleus tractus solitarius (NTS), are important in detecting and translating a variety of items of metabolic fuel information to the hypothalamus (Estacio, Yamada, Tsukamura, Hirunagi, & Maeda, 1996; Murahashi et al., 1996; Kinoshita et al., 2003; Wade & Jones, 2004; Thammacharoen, Lutz, Geary, & Asarian, 2008). Brainstem areas communicate this information through a catecholaminergic and/or neuropeptide Y (NPY)-containing pathway that directly or indirectly modulates CRH and consequently reduces LH (Maeda, Cagampang, Coen, & Tsukamura, 1994; Tsukamura, Nagatani, Cagampang, Kawakami, & Maeda, 1994; Nagatani et al., 1996b; I'Anson, Sundling, Roland, & Ritter, 2003; Kinoshita et al., 2003; Wade & Jones, 2004). An important role for CRH in the suppression of LH during food restriction is evidenced by the ability of a variety of CRH receptor antagonists to attenuate reproductive effects of food restriction or negative energy balance (Maeda et al., 1994; Tsukahara, Tsukamura, Foster, & Maeda, 1999; Seymour, Dettloff, Jones, & Wade, 2005; Vulliemoz, Xiao, Xia-Zhang, Rivier, & Ferin, 2008). However, lesions of the area postrema/NTS block the effect of food restriction on hamster lordosis behavior but do not prevent estrous cycle disruption following all types of energy restriction (Wade & Jones, 2004). Moreover, antagonists for different CRH receptors (CRH-R1 and CRH-R2) provide different profiles of protection against food-restriction-induced declines in cyclicity and behavior (Seymour et al., 2005). Therefore, overlapping but separate pathways may be involved in the effect of food restriction on GnRH release and on sexual behavior (Wade & Jones, 2004; Seymour et al., 2005).

Recent evidence has accumulated to implicate endocannabinoids in the stress response (Shimizu & Yokotani, 2008) and as players in the control of food intake and reproductive decline following food restriction (Pagotto, Marsicano, Cota, Lutz, & Pasquali, 2006; Carr, Elliot, & Brown, 2008). 2-Arachidonyl glycerol (2-AG) and arachidonoyl ethanolamine (AEA) are the two endogenous endocannabinoids that have received the most attention (Wang, Dey, & Maccarrone, 2006). Leptin modulates endocannabinoid levels in the hypothalamus (Hill et al., 2008) and 2-AG increases during fasting, declines upon refeeding, and returns to normal in satiated animals (Hanus et al., 2003). In females, endocannabinoids reduce LH, probably via modulation of various neurotransmitters that in turn modulate LH release; however, both pituitary and hypothalamic sites may be involved (Olah, Milloh, & Wenger, 2008), and in males endocannabinoids reduce LH, decrease T production, and suppress spermatogenesis (Pagotto et al., 2006; Wang et al., 2006). Endocannabinoids have begun to emerge as possible intermediaries of a variety of stressors in vertebrates (Coddington, Lewis, Rose, & Moore, 2007; Denver, 2007).

3. EFFECTS OF STRESS ON IMMUNITY AND REPRODUCTION

In response to injury or infection, a widespread immune/inflammatory response occurs that shifts the organism's energy away from nonessential activities and toward responses that combat the infection or repair the injury. Immune activation also elicits a variety of symptoms such as fever, anorexia, and reduced activity that are associated with sickness and that are mediated by peripheral and central cytokines (Rivest & Rivier, 1995; Avitsur & Yirmiya, 1999; Haddad et al., 2002; Goshen & Yirmiya, 2008). Such defensive activities are generally accompanied by a reduction in reproductive activity. Not unlike the response to undernutrition, female sexual behavior is more likely to be reduced during illness than is male sexual behavior (Avitsur, Donchin, Barak, Cohen, & Yirmiya, 1995; Avitsur & Yirmiya, 1999). For females, where mating has a high probability of pregnancy, sexual behavior can potentially increase the female's energy burden and reduce her own chances of survival. The probability of bringing the pregnancy to term is also reduced if the mother is ill, so it would be more adaptive for the female to simply avoid the possibility of pregnancy. Immune/inflammatory responses play a role in pregnancy avoidance by (1) reducing the female's motivation for mating, (2) reducing the male's attraction for the female, and (3) disrupting the hormonal events required for successful reproduction.

In laboratory animals, the reproductive consequences of immune-mediated responses have been examined after (1) administration of a variety of individual pathogens; (2) treatment with the endotoxin lipopolysaccharide (LPS), a glycolipid present in the outer membrane of Gram-negative bacteria (Raetz, 1990); or (c) treatment with the cytokines themselves (Rivest & Rivier, 1995; Haddad et al., 2002; Goshen & Yirmiya, 2008). The endotoxin LPS increases peripheral and central cytokines (Nathan, 1987; Ebisui et al., 1992; Koldzic-Zivanovic et al., 2006), such as interleukins (ILs), tumor necrosis factor (TNF-α), and interferon. Although not identical in effect, administration of these compounds elicits many of the 'sick behaviors' of infection (for review, see Busbridge & Grossman, 1991; Rivest & Rivier, 1995; Avitsur & Yirmiya, 1999).

Sexual behavior after immune challenge has been studied primarily in rodents. In rats, peripheral or central administration of LPS or IL-1β reduced female proceptivity, lordosis behavior, and the female's preference for spending time with the male (Yirmiya, Avitsur, Donchin, & Cohen, 1995; Avitsur, Pollak, & Yirmiya, 1997). Since female rat sexual behavior is tightly linked with ovulatory events, any compound that disrupts normal estrous cyclicity will ultimately prevent sexual behavior. However, LPS and IL-1β also reduced female sexual behavior in ovariectomized rats when gonadal hormones were controlled by exogenous administration (Avitsur et al., 1995; Avitsur & Yirmiya, 1999). Consistent with ideas that animals choose to mate with individuals that offer greater potential for reproductive success (K. B. Hoyenga & K. T. Hoyenga, 1982; Husak & Moore, 2008), endotoxin/cytokine-treated females were less attractive to males than their healthy counterparts (reviewed in Avitsur & Yirmiya, 1999).

Comparable treatments to male rats had no effect on male sexual behavior (Avitsur & Yirmiya, 1999) and, interestingly, the exhibition of 'sickness behavior' after a low dose of IL-1β was less evident when the male was in the presence of an estrous female (Avitsur & Yirmiya, 1999). Therefore, males could overcome some of the behavioral effects of the immune/inflammatory challenge to engage in sexual behavior. However, laboratory conditions do not simulate those in the wild, in which males must compete for the attention of receptive females. In addition, T production may be affected by cytokine treatment (Rivest & Rivier, 1995), so that reproductive fitness may decline in spite of normal male sexual behavior. Moreover, although female rats in the laboratory did not appear to differentiate between control and cytokine-treated males (Avitsur & Yirmiya, 1999), in a natural setting, such equality may not be present.

In female rats, immune/inflammatory compounds disrupt the reproductive cycle, leading to persistent vaginal estrus and the induction of a pseudopregnancy-like state (Rivier & Erickson, 1993). Depending on the stage of the reproductive cycle when the initial treatment occurs, LPS/IL-1β interrupts ovulation and/or disrupts luteolysis in a variety of mammalian species (Rivier & Erickson, 1993; Nappi & Rivest, 1997; Battaglia, Krasa, Padmanabhan, Viguie, & Karsch, 2000) and a potentially long-lasting effect of endotoxin after discontinuation of treatment was reported in rhesus monkeys (Xiao et al., 2007). An LPS challenge, administered during the luteal phase, decreased corpus luteum formation during the endotoxin challenge and for at least two additional cycles after termination of treatment (Xiao, Xia-Zhang, Vulliemoz, Rivier, & Ferin, 2007).

Sites of action of immune/inflammatory compounds can include peripheral effects directly on the gonads, effects on sensory input to the CNS (probably mediated through vagus afferents), or direct central effects (Rivest & Rivier, 1995; Goshen & Yirmiya, 2008). Most evidence is consistent with a central mechanism that results in a decrease in GnRH release and consequent decrease in LH (Rivest & Rivier, 1995; Haddad et al., 2002). Lipopolysaccharide and cytokines reduce LH in ovariectomized females (with or without hormonal priming) and block the preovulatory LH surge in intact females (P. Kalra, Sahu, & S. Kalra, 1990; Rivest & Rivier, 1995; Haddad et al., 2002). Immune/inflammatory compounds disrupt estradiol's positive feedback on the preovulatory LH surge (Battaglia et al., 1999); reduce GnRH release from GnRH neurons (Rivest & Rivier, 1993a; 1993b; Battaglia et al., 1997; Debus et al., 2002); reduce GnRH receptor mRNA in the pituitary (Nappi & Rivest, 1997); and, in sheep, reduce pituitary sensitivity to GnRH pulsatility (Battaglia et al., 1997; Debus et al., 2002). In cultured primary hypothalamic neurons or in immortalized GnRH neurons, IL-1β reduces GnRH release (Igaz et al., 2006) and *in-vivo* treatment reduces GnRH mRNA in GnRH neurons (Rivest & Rivier, 1993a). Additional evidence for an effect on GnRH neurons is derived from findings that LPS or IL-1β inhibit the expression of *c-fos* that normally occurs in GnRH neurons (Gore & Roberts, 1995) on the day of proestrus (Rivest, Lee, Attardi, & Rivier, 1993; Nappi & Rivest, 1997).

In rats, most evidence implicates a primary effect of endotoxin/cytokine within the central nervous system (Rivest & Rivier, 1995). Lateral ventricle infusion of IL-1β, IL-1α, or LPS reduced LH in castrated rats, blocked the proestrus LH surge in intact females, and led to persistent diestrus (Rivier & Vale, 1990; Ebisui et al., 1992; Rivest et al., 1993; Rivier, 1993). In contrast, systemic treatment with IL-1α or IL-1β had no effect on LH (Rivier & Vale, 1990; Rivest & Rivier, 1993a). However, LPS reduced LH following either systemic or central administration (Ebisui et al., 1992; Rivest & Rivier, 1995), but the effects could result from an LPS-induced increase in central IL-1α and IL-1β. Lipopolysaccharide increases levels of IL-1α and IL-1β in the CNS and effects of LPS can be partially blocked by anti-IL-1α or anti-IL-1β antibodies (Ebisui et al., 1992).

In contrast to rats, the immune/inflammatory reduction in LH in sheep has been attributed to both central and pituitary sites. Intravenous administration of endotoxin reduced the amplitude of LH pulses and also reduced the pulsatility of GnRH release into the portal system (Battaglia et al., 1997; Debus et al., 2002). When endogenous GnRH release was blocked by estrogens and the pituitary response to exogenous GnRH was examined during endotoxin treatment, the pituitary LH response to GnRH was reduced (Williams, Harris, Battaglia, Viguie, & Karsch, 2001). Therefore, in sheep, both hypothalamic and pituitary sites appear to be involved. However, the relative importance of the two sites may vary depending on hormonal priming (Williams et al., 2001a). The HPA response to endotoxin and IL-1α is increased by gonadectomy and reduced by estradiol or T replacement (Xia-Zhang, Xiao, & Ferin, 1995; Puder, Freda, Goland, & Wardlaw, 2001).

Mechanisms involved in the endotoxin/cytokine-induced decrease in LH include activation of CRH, increased prostaglandins, or elevations in a variety of neurally active compounds (Haddad et al., 2002; Goshen & Yirmiya, 2008). Endotoxin/cytokines activate the HPA axis, leading to increased CRH release into the hypothalamic–hypophysial portal system, increased ACTH release from the pituitary, and increased secretion of glucocorticoids and progesterone (P_4) from the adrenal (Rivest & Rivier, 1995; Battaglia et al., 1998; Haddad et al., 2002; Tonelli et al., 2002; Beishuizen & Thijs, 2003). Any or all of these secretions could influence the reproductive system. However, CRH has received the most attention because of its recognized suppression of LH (Rivest & Rivier, 1995). In ovariectomized ewes, endotoxin-induced declines in GnRH pulses and reductions in LH temporally coincided with an increase in CRH and AVP in hypothalamic–hypophysial portal blood (Battaglia et al., 1998). In ovariectomized rhesus monkeys, IL-1α increased cortisol and reduced LH and both effects were blocked by either a CRH (Feng et al., 1991) or an AVP antagonist (Shalts, Feng, & Ferin, 1992). Moreover, in intact rhesus monkeys, a CRH receptor antagonist, astressin B, blocked the increase in CRH, prevented the decline in LH in response to LPS, and accelerated the normal luteal cyclicity after discontinuation of the endotoxin challenge (Xiao et al., 2007). While the effect on luteal function may have included an effect at the ovary, the astressin B reversal of endotoxin effect on cortisol and LH was interpreted to reflect an important role for CRH in the CNS.

In rats, however, increased CRH does not appear to be required for the decline in LH after endotoxin/cytokine treatment. Corticotropin-releasing hormone antagonists do not block the effects of IL-1β on LH, even though effects on ACTH are reduced (Bonavera, S. Kalra, & P. Kalra, 1993; Rivier & Rivest, 1993; Haddad et al., 2002). In addition, lesioning of the PVN does not prevent IL-1β from reducing LH (Rivest & Rivier, 1991). Therefore, at least in rats, immune/inflammatory compounds can reduce LH independent of activation of the HPA axis (Rivest & Rivier, 1995). However, under nonexperimental conditions, there is little doubt that increased CRH could be a contributing factor to endotoxin/cytokine-induced reproductive decline.

As prostaglandins mediate many of the 'sick behaviors' of the immune/inflammatory response and are increased in the brain in response to inflammatory compounds (Cao, Matsumura, Yamagata, & Watanabe, 1997; Komaki, Arimura, & Koves, 1992), prostaglandins have been implicated as contributors to the reproductive decline evident during illness (Rivest & Rivier, 1995; Harris et al., 2000; Matsuwaki, Kayasuga, Yamanouchi, & Nishihara, 2006). Prostaglandin inhibitors reduce the ability of LPS/cytokines to decrease either LH or female sexual behavior (Avitsur, Weidenfeld, & Yirmiya, 1999; Harris et al., 2000). However, there is evidence for both positive and negative regulation of prostaglandins on reproduction (Ojeda, Negro-Vilar, & McCann, 1981; Rage, Lee, Ma, & Ojeda, 1997), possibly related to the variable effects of individual cytokines in different brain regions, and effects may vary with the type of hormonal priming (Rage et al., 1997; Williams et al., 2001a; Mouihate & Pittman, 2003).

Cytokines can signal to the CNS via neural mechanisms mediated through vagal input to brainstem areas and ultimately hypothalamic areas (Goshen & Yirmiya, 2008), and gonad to CNS communication is also present (Gerendai, Toth, Boldogkoi, Medveczky, & Halasz, 2000a; 2000b). There is evidence that a reduction in gonadal steroids could occur independently of the HPG axis. After IL-1β treatment in rats, a reduction in the testicular response to human chorionic gonadotropin (hCG) occurs (Turnbull & Rivier, 1995; Turnbull & Rivier, 1997b; Ogilvie & Rivier, 1998; Selvage & Rivier, 2003); this reduction is mediated by a pathway including the PVN, central amygdala, NTS, LC, and cortex with brainstem catecholamine neurons implicated as key players in the response to IL-1β (Gerendai et al., 2000a; Lee, Miselis, & Rivier, 2002; Selvage & Rivier, 2003). Although destruction of the PVN did not abolish the inhibitory effects of IL-1β on LH, it did abolish the effect of IL-1β on the responsivity of the testes to hCG (Selvage, Lee, Parsons, Seo, & Rivier, 2004), and microinfusion of IL-1β into the pons near the LC mimicked the effect of ventricular infusion of the cytokine on the testicular response to hCG (Selvage, Parsons, & Rivier, 2006). However, lesion of the LC did not prevent the ability of IL-1β to reduce the response to hCG (Selvage et al., 2006). Therefore, the exact mechanism is not known, but catecholamine involvement is likely. Catecholamines may be especially important players in the ability of peripheral cytokines to communicate with the CNS (Goshen & Yirmiya, 2008).

In summary, immune activation leads to reductions in reproductive functioning that are especially evident in female mammals. Multiple mechanisms exist whereby information can be relayed from the immune system to the CNS. Depending on the nature of the immune challenge and its location, the precise mechanisms may vary. The overall result, however, is a decline in the activity of GnRH neurons in the hypothalamus and, at least in some cases, a reduction in pituitary sensitivity to the GnRH released into the hypothalamic–hypophysial portal system. This diversion from reproductive energy provides an adaptive mechanism that would, in the longer term, enhance the fitness value of the individual.

4. EFFECTS OF OTHER STRESSORS

Other sources of stress are often categorized as physical (e.g., footshock, cold, heat), psychological (resulting from prior 'learned' association), or social (resulting from altered social interaction) (Pacak & Palkovits, 2001). Although noted for the induction of stressor-specific neurochemical and neuroendocrine responses (Dubrovsky, 2000; Thrivikraman, Nemeroff, & Plotsky, 2000; Pacak & Palkovits, 2001), if sufficiently long or intense, most such stressors will affect male or female reproduction by reducing LH and steroidogenesis. Therefore, in the following discussion, effects of such physical, psychological, and social stressors are collectively discussed.

4.1. Females

Disruption of female estrous cyclicity following chronic stress was reported in the 1970s (Collu, Tache, & Ducharme, 1979) and has been confirmed by many investigators (Axelson, 1987; Briski & Sylvester, 1988; Marchlewska-Koj, Pochron, Galewicz-Sojecka, & Galas, 1994; Machatschke et al., 2006). Chronic immobilization for six hours per day for 3–42 days reduced serum LH and produced persistent vaginal diestrus in intact female rats (Collu et al., 1979). Similarly, chronic unpredictable emotional and physical stress over a 14-day period led to constant diestrus in intact, previously cycling female rats (Rodriguez Echandia, Gonzalez, Cabrera, & Fracchia, 1988), and eight weeks of forced swimming and running disrupted estrous cyclicity (Axelson, 1987). In contrast, Anderson et al. (1996) found no disruption of estrous cyclicity after 14 days of intermittent footshock. However, since the stressor also produced only small changes in ACTH and CORT, these findings illustrate the variability in response dependent on the nature and/or intensity of the stressor.

Since sexual behavior of subprimate mammals is dependent on gonadal hormones, a suppression of the HPG axis after chronic stress would be expected to lead, also, to a decline in sexual behavior. Hence, in most cases, sexual behavior has not been examined after chronic stress, but, in at least one incidence, sexual behavior was examined after eight weeks of forced exercise (Axelson, 1987). Sexual behavior was unaffected and females bred normally. However, by the time sexual receptivity was examined, estrous cyclicity had begun to recover. Such recovery from the estrous cycle disruption may reflect neural adaptation to the repeated stress, as has been reported for a pharmacological stressor (Uphouse, Hensler, Sarkar, & Grossie, 2006).

To avoid the confounding effects of the estrous cycle disruption, investigators have examined effects of chronic stressors in ovariectomized rats where sexual receptivity is induced by exogenous hormonal priming. When hormonal priming is suboptimal (e.g., leads to low levels of sexual receptivity), various types of subchronic or chronic stress have been reported to increase, rather than decrease, female sexual behavior (Williams, McGinnis, & Lumia, 1992; Gorzalka, Hanson, & Brotto, 1998; Brotto, Hanson, & Gorzalka, 1999; Hanson & Gorzalka, 1999). A potential role for adrenal secretions was initially implicated by findings that chronic treatment with CORT could enhance sexual behavior of ovariectomized, adrenalectomized rats (Hanson & Gorzalka, 1999). However, effects of chronic CORT vary with the nature of the hormonal priming and the presence or absence of an intact adrenal (DeCatanzaro & Gorzalka, 1980; Gorzalka & Moe, 1994; Gorzalka et al., 1998). In female rats, an increase in adrenal CORT and P_4 precedes the ovulatory LH surge and is important in the facilitation of female rat sexual behavior (Plas-Roser & Aron, 1977). Chronic adrenalectomy disrupts the normal facilitation of lordosis behavior that occurs on the day of proestrus (Plas-Roser & Aron, 1977) and removal of the adrenal influences the effect of estradiol priming in ovariectomized rats (DeCatanzaro & Gorzalka, 1980). However, adrenal effects may be more important for the facilitation of proceptive behavior rather than the onset of lordosis (DeCatanzaro & Gorzalka, 1980).

Although early emphasis was placed on the possibility that adrenal glucocorticoids were responsible for chronic effects of stress on female lordosis behavior, adrenal P_4 may be equally or more important (Plas-Roser & Aron, 1981; Auger et al., 1997). In rats, increased P_4 from the adrenal is involved in facilitation of lordosis in normally cycling female rats (Plas-Roser & Aron, 1981; Gorzalka & Moe, 1994) and in regulating estrous cyclicity (Barfield & Lisk, 1974). In ovariectomized female rats primed only with estradiol, adrenalectomy increased sexual receptivity but reduced proceptivity (Hanson & Gorzalka, 1999), a behavior predominantly influenced by P_4 (Sodersten, 1981).

Both CORT and P_4 undergo metabolism to a variety of neurosteroids (e.g., allopregnanolone and deoxycorticosterone) that can modulate γ-aminobutyric acid A (GABAA) receptor function, reduce anxiety, increase social interaction, and influence female rat lordosis behavior (Kubli-Garfias, 1990; Frye, 2001; Mellon & Griffin, 2002). Many stressors increase neurosteroids (Paul & Purdy, 1992; Biggio, Concas, Follesa, Sanna, & Serra, 2007; Frye, 2007; Girdler & Klatzkin, 2007); mating increases neurosteroids (Frye & Bayon, 1999); and P_4 metabolites enhance female rodent sexual behavior (Frye, 2001). Therefore, facilitatory effects of repeated stress probably include a variety of mechanisms that are related to normal estrous cycle events that are conducive to female rat sexual behavior. However, facilitative effects of chronic stress reported in rats (Hanson & Gorzalka, 1999) may be unique because of the importance of adrenal secretions to the normal control of sexual receptivity in this species. It would be important to determine whether facilitative effects of chronic stress are also present in other species. Moreover, facilitative effects of chronic stress only have been reported under artificial conditions (e.g., in ovariectomized rats with hormonal manipulations) or in intact females when the female's estrous cycle was not disrupted by the stressor. In a freely roaming species, in which chronic stress may be more debilitating than when present in the laboratory environment, it is unlikely that any enhancement of reproductive function would occur.

In contrast to general agreement that long-term experience with stress leads to reproductive dysfunction, there is considerable variability in the effects of acute stress on sexual behavior (Wingfield & Sapolsky, 2003). One source of variability is the measure of sexual behavior examined. The full complement of sexual behavior includes motivation (arousal) to mate, approach to a potential mate, and the consummatory act. Depending on the testing condition, stress may disrupt one of these elements without suppressing another. The behavioral variability also reflects an interaction between the nature of the testing paradigm and which additional behavioral options are available (Kavaliers & Choleris, 2001; Uphouse et al., 2008). Nevertheless, the behavioral response to an acute stressor may be a critically important determinant of reproductive success. For female mammals, with their greater sensitivity to stress (Kavaliers & Choleris, 2001), even a transient disruption of the behavioral profile could alter the temporal relationship between ovulation and the presence of sperm (Pedersen, 2007).

Acute stress has been reported to decrease, increase, or have no effect on reproduction in female rats. In hormonally primed, ovariectomized rats, the effect of the stressor varied with the type of stressor used, the time after ovariectomy, and with the presence or absence of estrogens (Briski & Sylvester, 1988). In intact females, acute stressors have minimal effects on female sexual behavior unless the stressor is applied near the ovulatory surge of LH (Truitt et al., 2003; Donadio et al., 2007). However, when applied near the ovulatory surge, a variety of acute stressors (restraint or ether) increased CORT; decreased, but did not prevent, the LH surge that afternoon; slightly reduced lordosis behavior; and reduced the number of oocytes in the ovary (Donadio et al., 2007). Therefore, acute stress near the time of sexual receptivity has the potential to influence female reproductive behavior.

Effects of acute mild stress vary with the type of hormonal priming. Five minutes of restraint produced a robust, but transient, decline in lordosis behavior of ovariectomized rats primed only with estradiol (Truitt et al., 2003; White & Uphouse, 2004), whereas no effects were present in rats primed with estradiol and P_4. Hormonal priming also altered the effect of restraint on the female's sexual motivation, as measured by preference for spending time with a sexually active male. However, effects of restraint also varied with the testing paradigm. The female's preference for interaction with a male showed a prolonged reduction when the female had the opportunity to escape to a dark 'burrow' (Uphouse, Selvamani, Lincoln, Morales, & Comeaux, 2005). In contrast, there was no effect of restraint when escape from a central area allowed the female to choose between a sexually active male or an ovariectomized female (Uphouse et al., 2008). These findings suggest that the effects of acute stress on the female's probability of mating will depend not only on the specifics of the stressor but also on the behavioral options available. Similarly to effects of acute stress on behavior, effects of stressors on LH vary with hormonal priming. For example, in ovariectomized rats without hormone priming, a 15-minute restraint reduced LH release but restraint increased LH release in estradiol-primed rats (Briski & Sylvester, 1988).

4.2. Males

In rats, male sexual behavior is measured by approach and interaction with a receptive female and is usually quantified in terms of latencies to mounting, intromission, and ejaculation and/or frequency of ejaculation (Hull & Dominguez, 2007). A variety of 'stressful' stimuli delay or reduce intromission and ejaculation frequency while often leaving mounting behavior intact (D'Aquila, Peana, Carboni, & Serra, 2000; Gronli et al., 2005). After two weeks of chronic stress, male rats showed a reduced frequency of mounts and intromissions, but other behaviors were also reduced (Gronli et al., 2005). Locomotor activity and consumption of sucrose (used as a hedonic measure) both declined (Gronli et al., 2005). These findings and others suggest that male sexual motivation may be less affected by stress than are the male's motor or hedonic responses (Bekris et al., 2005). In the study by D'Aguila (2000), rats were successful at mounting but never completed mating beyond the mounting stage.

In general, chronic stress appears to have smaller effects in males than in females, but footshock has been reported to reduce LH in castrated male rats (Rivier, Rivier, & Vale, 1986). A decline in LH would be expected to produce a consequent reduction in T. However, castration (removing the gonadal source of T) results in a gradual rather than a precipitous decline in sexual activity (Hull & Dominguez, 2007). Decrements of sexual behavior following chronic stress could, therefore, result from long-term gonad-suppressive effects of stress, perhaps due to a chronic elevation in glucocorticoids (which can reduce steroidogenesis in Leydig cells (Agular, Vinggaard, & Vind, 1992; Payne & Youngblood, 1995)), but declines during acute stress might have little impact. However, male sexual activity increases both plasma CORT and T, and effects of stress are not always correlated with levels of T (Retana-Marquez, Bonilla-Jaime, & Velazquez-Moctezuma, 1998; Retana-Marquez, Bonilla-Jaime, Vazquez-Palacios, Martinez-Garcia, & Velazquez-Moctezuma, 2003) and administration of T did not reduce effects of chronic cold-water stress on the sexual behavior of castrated male rats (Retana-Marquez et al., 2003). Nevertheless, variability in T levels and/or response to stress (Hull & Dominguez, 2007) may be important to individual differences in the reproductive effects of chronic or acute stress on male sexual behavior.

As for female sexual behavior, the degree of sexual disruption is dependent on the intensity and duration of the stressor (Pednekar, Mulgaonker, & Mascarenhas, 1993; Retana-Marquez et al., 2003) or behavioral testing conditions. For example, sexual stimulation associated with mating reduced effects of stress on male reproduction (Lemaire, Taylor, & Mormede, 1997). Effects of a familiar chronic stressor on male sexual behavior may not be evident, but effects may emerge in the presence of an additional challenge (Gorzalka & Hanson, 1998). This is consistent with evidence that one stressor can sensitize the system to the effects of another stressor (Berga, 2008). Prior stress also may reduce the male's response to a variety of rewards other than sexual reward (Bekris et al., 2005). The degree of control the individual has over the stressor can also influence the behavioral response. When rats were subjected to 12 days of either inescapable or escapable footshock, sexual behavior was reduced in rats given inescapable, but not escapable, footshock (Holmer, Rodman, Helmreich, & Parfitt, 2003). Moreover, the environmental milieu shapes the male's behavioral response to the stressor so that the male's dominance position (Monder, Sakai, Miroff, Blanchard, & Blanchard, 1994) or social status (Lidgard et al., 2008) can alter the response to stress. For example, in Mongolian gerbils, social stress reduced marking frequency of subordinate, but not dominant, males (Yamaguchi, Kikusui, Takeuchi, Yoshimura, & Mori, 2005).

The relative resistance of dominant males to stress-induced reproductive decline is well known (Sapolsky, 1982; Sapolsky, 1986; Tilbrook et al., 2000; Lidgard et al., 2008). In capture and release projects, dominant males often show capture-induced increases in both glucocorticoids and T, whereas subordinate males increase glucocorticoids but do not increase T in response to stress (Lidgard et al., 2008). Moreover, the glucocorticoid response is of high magnitude and short duration in dominant males, while subordinate males show a more prolonged elevation of glucocorticoids. Similarly, in olive baboon colonies, males with higher reproductive success show a higher T response to stress while males with lower reproductive success show a reduced T response (Lidgard et al., 2008). Dominant males, therefore, may be able to remain reproductively competent in spite of chronic stress.

Depending on the nature of the stressor and its intensity, acute stress may have no or only marginal effects on male sexual activity (Retana-Marquez, Salazar, & Velazquez-Moctezuma, 1996; Berton, Durand, Aguerre, Mormede, & Chaouloff, 1999; Retana-Marquez et al., 2003). Mating-induced elevations of T may be reduced by acute stress (Retana-Marquez et al., 2003) but still have only minor effects on male sexual activity. In fact, sexually sluggish or naïve rats may even increase copulatory behavior after a mild or intermittent stress (Barfield & Sachs, 1968; Caggiula & Eibergen, 1969; Crowley, Popolow, & Ward, 1973). Nevertheless, in spite of the relative resistance of male sexual behavior to stress, acute severe stressful stimuli can still reduce male reproductive function (Sapolsky et al., 2000; Tilbrook, Turner, & Clarke, 2002; Wingfield & Sapolsky, 2003; Lidgard et al., 2008). However, the degree to which the decline in sexual behavior reflects a specific sexual response is questionable. Since sexual behavior requires some degree of arousal, it has been argued that mild stressors function to enhance attention to stimuli while more severe stressful stimuli may produce avoidance behavior (Smith, Stewart, & Pfaus, 1997). Therefore, the behavioral effect of stress on male sexual behavior may reflect a tradeoff between attention and avoidance.

5. ROLE OF HPA ACTIVATION (GLUCOCORTICOIDS AND CRH)

Exogenous administration of glucocorticoids or CRH has been used to determine whether either or both are causally related to the chronic stress-induced decline in reproductive function. Prolonged administration of glucocorticoids reduces LH in males and females of a variety of species (Baldwin, 1979; Adams, Sakurai, & Adams, 1999; Sapolsky et al., 2000; Breen et al., 2005; Breen & Karsch, 2006). In females, prolonged elevation of glucocorticoids reduces the preovulatory LH surge and blocks ovulation (Turner, Hemsworth, Canny, & Tilbrook, 1999; Tilbrook et al., 2000; Breen & Karsch, 2004; Breen et al., 2005). In males, prolonged elevation decreases LH, disrupts steroidogenesis, and reduces T (Neumann, 1984; Orr & Mann, 1992; Weber et al., 2000). Therefore, a sustained increase in glucocorticoids will inevitably reduce reproductive function.

In female ewes, there is convincing evidence that exogenous CORT (in the absence of stress) can reduce plasma LH levels by reducing pituitary responsiveness to GnRH (Debus et al., 2002; Breen & Karsch, 2004; Breen et al., 2005). Similarly, in an ovine pituitary cell culture, cortisol reduced the response to GnRH without decreasing either GnRH receptor protein or pituitary LH content (Breen et al., 2008) and the type II glucocorticoid receptor (GR) has been implicated in the direct pituitary effect of cortisol (Breen et al., 2004; 2008). However, during the follicular stage of female sheep, both hypothalamic and pituitary sites are important in the cortisol-induced suppression of LH (Breen et al., 2004), and gender differences and/or hormonal condition (Adams et al., 1999; Daley, Sakurai, Adams, & Adams, 1999; Wagenmaker et al., 2008a) may influence the relative importance of hypothalamic or pituitary sites to the effects of cortisol. For example, cortisol transiently reduced the increase in GnRH receptor and its mRNA that is induced by estradiol (Adams et al., 1999). This pituitary response to exogenous glucocorticoids contrasts with the effects of psychosocial stress, where the hypothalamus appears to be essential for the suppression of LH (Wagenmaker, Breen, Oakley, Tilbrook, & Karsch, 2008b).

In contrast to ideas that glucocorticoid elevation might account for stress-induced suppression of LH, other investigators have suggested that acute elevations of glucocorticoids are beneficial to reproductive function (Kilen et al., 1996; McGillivray, Thackray, Coss, & Mellon, 2007). Recent reports provide a plausible mechanism whereby glucocorticoids could maintain LH secretion during a stressful experience (Matsuwaki, Watanabe, Suzuki, Yamanouchi, & Nishihara, 2003; Matsuwaki et al., 2006). Ovariectomized and adrenalectomized (ovx/adx) rats were treated with TNF-α, LPS, or 2-deoxyglucose, or were subjected to one hour of restraint. Adrenalectomy enhanced the effect of TNF-α on LH, and exogenous CORT restored LH and reversed the TNF-α-induced induction of *c-fos* in the PVN (Matsuwaki et al., 2003). Similarly, LPS, 2-deoxyglucose, or a one-hour restraint all reduced LH in ovx/adx rats, and exogenous CORT restored LH pulsatility. In the PVN, stress increased COX-2, an enzyme required for prostaglandin synthesis (Feng et al., 1993) in ovx/adx rats and the increase was abolished by CORT. Indomethacin, used to reduce prostaglandin synthesis, mimicked the effects of CORT by reducing the effects of the stressors on LH (Matsuwaki et al., 2006). Glucocorticoids are known to reduce induction of prostaglandins by decreasing induction of COX-2 (Tanabe & Tohnai, 2002). From these findings, it was suggested that CORT, by blocking the induction of COX-2, reduced prostaglandin synthesis in the PVN and thereby maintained LH secretion in spite of the stressor (Matsuwaki et al., 2006).

This latter suggestion is especially interesting given increasing evidence that a variety of stressors increase cytokines in the brain (Haddad et al., 2002). Since prostaglandins have been classically associated with the effects of cytokines (Harris et al., 2000) and since glucocorticoids are appreciated for their anti-inflammatory actions (Goshen & Yirmiya, 2008), such an intersection with stress effects on the reproductive system is not unreasonable.

Reports of the effects of CORT on sexual behavior are not consistent. Direct injection of CORT into the brain inhibited female rodent sexual behavior (DeCatanzaro, 1987) and CORT inhibited sexual behavior in female pigs (Matsuwaki et al., 2006). In addition, CORT rapidly, but transiently, inhibited the preference of female mice for male odors (Kavaliers & Ossenkopp, 2001). However, CORT or its metabolites, administered intravenously, increased lordosis behavior in suboptimally hormone-primed ovariectomized rats (Kubli-Garfias, 1990) and 13 days of treatment with CORT increased the sexual behavior of ovariectomized female rats (Gorzalka, Brotto, & Hong, 1999). In male rats, ejaculatory behavior may be reduced by chronic treatment with CORT (Gorzalka & Hanson, 1998; Gorzalka, Hanson, & Hong, 2001) but other investigators found no effect of daily CORT on either T or copulatory performance (Retana-Marquez et al., 1998).

Therefore, collective evidence does not identify glucocorticoids as major regulators of the stress-induced suppression of either LH or sexual behavior. More evidence implicates CRH as a participant. As CRH neurons make synaptic contact with GnRH neurons (MacLusky et al., 1988), this peptide has received considerable emphasis as a mediator of the stress-induced reproductive decline. A potential role of CRH in stress-induced suppression of LH revolves around three major observations: (1) most stressors increase CRH; (2) CRH mimics many of the effects of stressors on reproductive function; and (3) CRH antagonists prevent some of the effects of stress on reproduction. As has been discussed in the sections on specific stressors, undernutrition, immune challenge, and physical/psychological stress all increase CRH. In addition, there is considerable evidence to suggest that increases in CRH reduce both LH (Rivier & Vale, 1984; Rivier et al., 1986; Maeda et al., 1994; Roozendaal, Swarts, Wiegant, &

Mattheij, 1995; Argiolas, 1999; Li et al., 2004) and sexual behavior (Sirinathsinghji, Rees, Rivier, & Vale, 1983; Sirinathsinghji, 1985; Dornan & Malsbury, 1989; Almeida, Hassan, & Holsboer, 1993; Keen-Rhinehart et al., 2008). In addition, overexpression of CRH in transgenic mice completely eliminated sexual activity in female mice, but only reduced (and did not eliminate) sexual activity in male mice (Heinrichs et al., 1997).

However, CRH does not reduce LH or sexual behavior under all conditions and may even, in some cases, stimulate LH. In the female musk shrew, increased HPA activation occurs in conjunction with ovulation and prior to sexual behavior and is critical for the induction of female sexual activity (Schiml & Rissman, 2000). In this species, infusion of CRH into the lateral ventricle led to an increase in mating behavior (Schiml & Rissman, 2000). Therefore, CRH elevation and/or elevations of glucocorticoids will not always inhibit sexual behavior or the HPG axis (Wang, Huether, Doble, & Liptrap, 1986; Retana-Marquez et al., 1998).

Similarly, the effects of antagonists of CRH action are not consistent across stressors or species. In primates, effects of IL-1α (Feng et al., 1991), LPS (Xiao et al., 2007), or insulin hypoglycemia (Chen et al., 1996) on LH were blocked (or reduced) by a CRH receptor antagonist. In rats, fasting (Maeda et al., 1994) or footshock-induced (Rivier et al., 1986) decrements in LH were blocked by alpha-helical CRH-(9-41). However, effects of IL-1β on LH were not blocked by a CRH receptor antagonist, and lesioning of the PVN did not block effects of IL-1β (Rivier et al., 1986; Rivier & Rivest, 1993) or of footshock on LH (Rivier & Rivest, 1993). In CRH-deficient mice, both restraint and food withdrawal still reduced LH so CRH clearly was not required (Jeong et al., 1999). Independently of whether or not CRH receptor antagonists blocked the effects of stress on LH, HPA activation was still blocked with CRH receptor antagonists.

Therefore, while there may be species- and stressor-specific differences in the importance of CRH, at least from the PVN, CRH does not appear to be required for a stress-induced decline in LH. The PVN is essential for stress-induced HPA activation but not for stress-induced decrements in LH. Consequently stress-induced HPA activation and HPG suppression appear to be dissociable (Van Vugt, Piercy, Farley, Reid, & Rivest, 1997).

In addition to CRH pathways from the PVN to the median eminence, CRH neurons and terminals exist in other brain areas in which the effects of stress may be integrated. Moreover, there are four different CRH-related peptides (CRH, urocortins (Unc I, Unc II, and Unc III)), and two different receptors (CRH-R1 and CRH-R2) upon which these peptides act (Bale & Vale, 2004). Corticotropin-releasing hormone receptor-1 seems to be primarily responsible for the activation of pituitary corticotropes and the release of ACTH (Turnbull & Rivier, 1997a; Turnbull, Vaughan, J. Rivier, Vale, & C. Rivier, 1999; Preil et al., 2001) but the role of CRH-R1 may vary with the type of stressor since inflammatory stress was able to produce HPA activation even in mice that were deficient in CRH-R1 receptors (Turnbull et al., 1999a). Corticotropin-releasing hormone receptor-2 may play the more important role in the decline in LH after restraint stress (C. Rivier, Grigoriadis, & J. Rivier, 2003; Li, Bowe, Lightman, & O'Byrne, 2005). Corticotropin-releasing hormone receptor-1 and CRH-R2 mRNAs are differentially distributed in the brain (Lovenberg, Chalmers, Liu, & De Souza, 1995; Van Pett et al., 2000) so that excitation of different neuronal pathways and CRH-like peptides may be involved in the effects of stress on the HPA and HPG axes (Li et al., 2005). Corticotropin-releasing hormone receptor-1 has a more global distribution with high density in the pituitary while CRH-R2 is more restricted, located in limbic-associated areas and midbrain/brainstem regions such as the raphe and NTS (Van Pett et al., 2000). Corticotropin-releasing hormone has a higher affinity for CRH-R1 than for CRH-R2, while Unc I has equal affinity for both receptors (Chen, Lewis, Perrin, & Vale, 1993; Lovenberg et al., 1995; Hauger et al., 2003). Unc II and Unc III are considered as endogenous ligands for CRH-R2 and have little affinity for CRH-R1. Therefore, reproductive effects originally attributed to CRH may actually involve one of these other CRH-like peptides (Jeong et al., 1999). Consistent with this idea is the finding that lateral ventricle administration of Unc II can suppress LH (Li et al., 2005) and that CRH-R2 and CRH-1 antagonists are differentially effective in reducing effects of CRH on female sexual behavior (Seymour et al., 2005).

6. IMPLICATIONS FOR WILDLIFE

Most research designed to evaluate the effects of stress on reproduction has been conducted in laboratory animals or in domestic animals living in a confined area. While intrusive mechanistic approaches cannot be applied to natural populations, behavioral observations, analyses of glucocorticoid levels in hair or fecal boli, and comparisons of populations under different environmental conditions lead to the same basic conclusions as those derived from laboratory studies (Tilbrook et al., 2000; Kavaliers & Choleris, 2001; Wingfield & Sapolsky, 2003; Rasmussen, Ganswindt, Douglas-Hamilton, & Vollrath, 2008; Van Meter et al., 2008). In the shorter term, an inverse association between stress and reproduction favors survival of the species in spite of a reduced fitness at the moment. The continuous interactions among environmental challenges and reproductive performance maintain the population at a level that can be efficiently supported by the environment (Wingfield & Sapolsky, 2003; Lidgard et al., 2008). However, the nature and intensity of stressors differ in laboratory and natural settings (Wiedenmayer, 2004). In the laboratory, individual or a small combination of stressful events are used, whereas, in the wild, it is the impacts of multiple stressors that ultimately determine the reproductive success of a particular generation (Wiedenmayer, 2004). Species that are well-adapted to their environment will maintain population density at a level consistent with the availability of food and space required for species viability. Under conditions of drought, extremes in temperature, or increased predator activity, reduced reproductive activity will conserve energy for later reproduction when offspring have a greater probability of surviving to reproductive age. Under extreme environmental conditions (e.g., extended drought) or in the introduction of a pathogen, the survival of the species can be at risk, but, in the absence of these extremes, the interplay between stress and reproduction allows for an adaptive pattern of reproductive activity that is consistent with the species' history with environmental challenge.

Novel and subtle stressors that are not part of the species' history may lead to a change in population dynamics. Familiar examples come from the release of pesticides into the environment (Saradha & Mathur, 2006). However, even a subtle association with humans may impact population dynamics (Martin & Reale, 2008). For example, chipmunks trapped in areas with more human activity were more 'docile' and had higher exploration levels than did chipmunks trapped distant from the location frequented by humans. The question raised by Martin and Reale was whether experience with humans had shaped chipmunk temperament or whether individual chipmunk temperament had influenced their habitat choice. In either case, the local population behavior changed as a consequence of proximity to humans. It is, therefore, possible that subtle changes in temperament of wild populations may ultimately influence their reproductive response to stressors. There are multiple examples in domestic animals where selection for one trait (e.g., increased production or tameness) leads to correlated selection of other traits so that association with humans may indirectly (and unintentionally) alter the characteristics of the freely roaming species and their reproductive response to stress (Stark, Banks, & Vargus, 2004) (see also Chapter 14, this volume for evidence of subtle stressors that may be altering reproductive functions in wildlife).

7. CONCLUSIONS

Although different types of stressor have historically been viewed as distinct in their effects on the CNS and consequences for reproduction, the current picture is that all such stressors may have more commonalities than differences. With an examination of stressors such as food restriction, infection, and social/psychological stressors, multiple commonalities emerge. Reproductive consequences depend on (1) intensity and duration, (2) hormonal status, (3) prior history, (4) social status, and (5) coping mechanisms such as social support or the opportunity to avoid the stressor.

Although central mechanisms vary depending on the modality whereby the stimulus is received and the context in which the stressor occurs, except in 'emergency' situations, it is less the particular attribute of the stressor than the individual's perception of the stimulus that determines the ultimate reproductive outcome (Sapolsky et al., 2000; Breuner, Patterson, & Hahn, 2008; Crockford, Wittig, Whitten, Seyfarth, & Cheney, 2008). The 'perceived attribute' of the stimulus involves CNS sites beyond the HPA and HPG axis and incorporates learning and adaptation into the equation of stress. The noradrenergic system emerges as a participant in the receipt of metabolic and immune information from the periphery, in the receipt of 'cognitive' information from the forebrain, and as a regulator of the autonomic nervous system (Valentino & Van Bockstaele, 2008). Thus, a phylogenetically older stress system has been incorporated as a recipient of newer, more cognitively related stressors, allowing for context-dependency of response without reinvention of effector mechanisms.

Multiple neurotransmitters contribute to the effect of stress and their 'perception.' Glutamate, GABA, serotonin, and norepinephrine have classically been implicated as key players in the response to stress, in the regulation of reproduction, and in their interaction (Dobson, Ghuman, Prabhakar, & Smith, 2003; Leonard, 2005; Bethea, Centeno, & Cameron, 2008; Durand, Pampillo, Caruso, & Lasaga, 2008; Ferguson, Latchford, & Samson, 2008). However, more recently, the endocannabinoids (Pagotto et al., 2006; Wang et al., 2006; Coddington et al., 2007; Denver, 2007; Gorzalka, Hill, & Hillard, 2008; Gorzalka, Morrish, & Hill, 2008) and prostaglandins (Madrigal et al., 2003; Garcia-Bueno et al., 2005; Garcia-Bueno & Leza, 2008), as well as gaseous molecules such as nitric oxide and carbon monoxide (Turnbull, Kim, Lee, & Rivier, 1998; Lee, Kim, & Rivier, 1999; Kim & Rivier, 2000) have emerged as important participants. Because of their localized, paracrine action, these molecules have the potential to translate the context of the stressor to the reproductive system.

Therefore, while different types of stressor have the potential to influence reproductive functioning through common final pathways, the issue of what is and what is not a stressor remains unresolved. In his now classical 1985 paper, Bronson (1985) wrote, 'If they promote reproductive success, we classify them under the rubric of 'arousal'; if they act to the detriment of reproduction we evoke the classic concept of 'stress'' (p. 18). Thus, we have come full-circle to recognizing that the reproductive distinction between stressful and nonstressful stimuli may be subtle, localized, and dependent on the individual's perspective.

Finally, most mechanistic hypotheses regarding stress and reproduction have relied upon laboratory investigations. Although generalization to wild mammalian populations is often assumed, it is important to note that such generalizations may not translate directly to the non-laboratory environment.

ACKNOWLEDGEMENTS

Work supported by NIH HD28419 and GM 55380. Appreciation is expressed to Ms. Jutatip Guptarak for reading an earlier version of the chapter.

ABBREVIATIONS

2-AG	2-Arachidonyl glycerol
ACTH	Corticotropin
AEA	Arachidonoyl ethanolamine
AVP	Arginine vasopressin
CNS	Central nervous system
CORT	Corticosterone
CRH	Corticotropin-releasing hormone
CRH-R	Corticotropin-releasing hormone receptor
FSH	Follicle-stimulating hormone
GABAA	γ-aminobutyric acid A
GnRH	Gonadotropin-releasing hormone
GR	Glucocorticoid receptor
hCG	Human chorionic gonadotropin
HPA	Hypothalamic–pituitary–adrenal
HPG	Hypothalamic–pituitary–gonadal
IL	Interleukin
LC	Locus coeruleus
LH	Luteinizing hormone
LPS	Lipopolysaccharide
mPOA	Medial preoptic area
NPY	Neuropeptide Y
NTS	Nucleus tractus solitarius
ovx/adx	Ovariectomized and adrenalectomized
P_4	Progesterone
PVN	Paraventricular nucleus
T	Testosterone
TNF-α	Tumor necrosis factor
Unc	Urocortin
VMH	Ventromedial nucleus of the hypothalamus

REFERENCES

Adams, T. E., Sakurai, H., & Adams, B. M. (1999). Effect of stress-like concentrations of cortisol on estradiol-dependent expression of gonadotropin-releasing hormone receptor in orchidectomized sheep. *Biol. Reprod., 60*, 164–168.

Agular, B. M., Vinggaard, A. M., & Vind, C. (1992). Regulation by dexamethasone of the 3 beta-hydroxysteroid dehydrogenase activity in adult rat Leydig cells. *J. Steroid. Biochem. Mol. Biol., 43*, 565–571.

Ahima, R. S., & Lazar, M. A. (2008). Adipokines and the peripheral and neural control of energy balance. *Mol. Endocrinol., 22*, 1023–1031.

Almeida, O. F., Hassan, A. H., & Holsboer, F. (1993). Intrahypothalamic neuroendocrine actions of corticotropin-releasing factor. *Ciba. Found. Symp., 172*, 151–169, discussion 169–72.

Anderson, S. M., Saviolakis, G. A., Bauman, R. A., Chu, K. Y., Ghosh, S., & Kant, G. J. (1996). Effects of chronic stress on food acquisition, plasma hormones, and the estrous cycle of female rats. *Physiol. Behav., 60*, 325–329.

Antoni, F. A. (1986). Hypothalamic control of adrenocorticotropin secretion: advances since the discovery of 41-residue corticotropin-releasing factor. *Endocr. Rev., 7*, 351–378.

Argiolas, A. (1999). Neuropeptides and sexual behaviour. *Neurosci. Biobehav. Rev., 23*, 1127–1142.

Auger, A. P., Moffatt, C. A., & Blaustein, J. D. (1997). Progesterone-independent activation of rat brain progestin receptors by reproductive stimuli. *Endocrinology, 138*, 511–514.

Avitsur, R., & Yirmiya, R. (1999). The immunobiology of sexual behavior: gender differences in the suppression of sexual activity during illness. *Pharmacol. Biochem. Behav., 64*, 787–796.

Avitsur, R., Donchin, O., Barak, O., Cohen, E., & Yirmiya, R. (1995). Behavioral effects of interleukin-1 beta: modulation by gender, estrus cycle, and progesterone. *Brain Behav. Immun., 9*, 234–241.

Avitsur, R., Kinsey, S. G., Bidor, K., Bailey, M. T., Padgett, D. A., & Sheridan, J. F. (2007). Subordinate social status modulates the vulnerability to the immunological effects of social stress. *Psychoneuroendocrinology, 32*, 1097–1105.

Avitsur, R., Pollak, Y., & Yirmiya, R. (1997). Different receptor mechanisms mediate the effects of endotoxin and interleukin-1 on female sexual behavior. *Brain Res., 773*, 149–161.

Avitsur, R., Weidenfeld, J., & Yirmiya, R. (1999). Cytokines inhibit sexual behavior in female rats: II. Prostaglandins mediate the suppressive effects of interleukin-1beta. *Brain Behav. Immun., 13*, 33–45.

Axelson, J. F. (1987). Forced swimming alters vaginal estrous cycles, body composition, and steroid levels without disrupting lordosis behavior or fertility in rats. *Physiol. Behav., 41*, 471–479.

Baldwin, D. M. (1979). The effect of glucocorticoids on estrogen-dependent luteinizing hormone release in the ovariectomized rat and on gonadotropin secretin in the intact female rat. *Endocrinology, 105*, 120–128.

Bale, T. L., & Vale, W. W. (2004). CRF and CRF receptors: role in stress responsivity and other behaviors. *Annu. Rev. Pharmacol. Toxicol., 44*, 525–557.

Barfield, M. A., & Lisk, R. D. (1974). Relative contributions of ovarian and adrenal progesterone to the timing of heat in the 4-day cyclic rat. *Endocrinology, 94*. 571-571.

Barfield, R. J., & Sachs, B. D. (1968). Sexual behavior: stimulation by painful electrical shock to skin in male rats. *Science, 161*, 392–393.

Battaglia, D. F., Beaver, A. B., Harris, T. G., Tanhehco, E., Viguie, C., & Karsch, F. J. (1999). Endotoxin disrupts the estradiol-induced luteinizing hormone surge: interference with estradiol signal reading, not surge release. *Endocrinology, 140*, 2471–2479.

Battaglia, D. F., Bowen, J. M., Krasa, H. B., Thrun, L. A., Viguie, C., & Karsch, F. J. (1997). Endotoxin inhibits the reproductive neuroendocrine axis while stimulating adrenal steroids: a simultaneous view from hypophyseal portal and peripheral blood. *Endocrinology, 138*, 4273–4281.

Battaglia, D. F., Brown, M. E., Krasa, H. B., Thrun, L. A., Viguie, C., & Karsch, F. J. (1998). Systemic challenge with endotoxin stimulates corticotropin-releasing hormone and arginine vasopressin secretion into hypophyseal portal blood: coincidence with gonadotropin-releasing hormone suppression. *Endocrinology, 139*, 4175–4181.

Battaglia, D. F., Krasa, H. B., Padmanabhan, V., Viguie, C., & Karsch, F. J. (2000). Endocrine alterations that underlie endotoxin-induced disruption of the follicular phase in ewes. *Biol. Reprod., 62*, 45–53.

Beishuizen, A., & Thijs, L. G. (2003). Endotoxin and the hypothalamo–pituitary–adrenal (HPA) axis. *J. Endotoxin. Res., 9*, 3–24.

Bekris, S., Antoniou, K., Daskas, S., & Papadopoulou-Daifoti, Z. (2005). Behavioural and neurochemical effects induced by chronic mild stress applied to two different rat strains. *Behav. Brain Res., 161*, 45–59.

Berga, S. L. (2008). Stress and reprodution: a tale of false dichotomy? *Endocrinology, 149*, 867–888.

Berton, O., Durand, M., Aguerre, S., Mormede, P., & Chaouloff, F. (1999). Behavioral, neuroendocrine and serotonergic consequences of single social defeat and repeated fluoxetine pretreatment in the Lewis rat strain. *Neuroscience, 92*, 327–341.

Bethea, C. L., Centeno, M. L., & Cameron, J. L. (2008). Neurobiology of stress-induced reproductive dysfunction in female macaques. *Mol. Neurobiol., 38*, 199–230.

Biggio, G., Concas, A., Follesa, P., Sanna, E., & Serra, M. (2007). Stress, ethanol, and neuroactive steroids. *Pharmacol. Ther., 116*, 140–171.

Bonavera, J. J., Kalra, S. P., & Kalra, P. S. (1993). Mode of action of interleukin-1 in suppression of pituitary LH release in castrated male rats. *Brain Res., 612*, 1–8.

Breen, K. M., & Karsch, F. J. (2004). Does cortisol inhibit pulsatile luteinizing hormone secretion at the hypothalamic or pituitary level? *Endocrinology, 145*, 692–698.

Breen, K. M., & Karsch, F. J. (2006). New insights regarding glucocorticoids, stress and gonadotropin suppression. *Front Neuroendocrinol., 27*, 233–245.

Breen, K. M., Billings, H. J., Wagenmaker, E. R., Wessinger, E. W., & Karsch, F. J. (2005). Endocrine basis for disruptive effects of cortisol on preovulatory events. *Endocrinology, 146*, 2107–2115.

Breen, K. M., Davis, T. L., Doro, L. C., Nett, T. M., Oakley, A. E., Padmanabhan, V., et al. (2008). Insight into the neuroendocrine site and cellular mechanism by which cortisol suppresses pituitary responsiveness to gonadotropin-releasing hormone. *Endocrinology, 149*, 767–773.

Breen, K. M., Stackpole, C. A., Clarke, I. J., Pytiak, A. V., Tilbrook, A. J., Wagenmaker, E. R., et al. (2004). Does the type II glucocorticoid receptor mediate cortisol-induced suppression in pituitary responsiveness to gonadotropin-releasing hormone? *Endocrinology, 145*, 2739–2746.

Breuner, C. W., Patterson, S. H., & Hahn, T. P. (2008). In search of relationships between the acute adrenocortical response and fitness. *Gen. Comp. Endocrinol., 157*, 288–295.

Briski, K. P., & Sylvester, P. W. (1988). Effect of specific acute stressors on luteinizing hormone release in ovariectomized and ovariectomized estrogen-treated female rats. *Neuroendocrinology, 47*, 194–202.

Bronson, F. H. (1985). Mammalian reproduction: an ecological perspective. *Biol. Reprod., 32*, 1–26.

Brotto, L. A., Hanson, L. A., & Gorzalka, B. B. (1999). Nefazodone attenuates the stress-induced facilitation of wet dog shaking behaviour but not the facilitation of sexual behaviour in female rats. *Eur. J. Pharmacol., 381*, 101–104.

Brown, R. E., Imran, S. A., Ur, E., & Wilkinson, M. (2008). KiSS-1 mRNA in adipose tissue is regulated by sex hormones and food intake. *Mol. Cell. Endocrinol., 281*, 64–72.

Busbridge, N. J., & Grossman, A. B. (1991). Stress and the single cytokine: interleukin modulation of the pituitary-adrenal axis. *Mol. Cell. Endocrinol., 82*, C209–C214.

Cagampang, F. R., Maeda, K. I., Tsukamura, H., Ohkura, S., & Ota, K. (1991). Involvement of ovarian steroids and endogenous opioids in the fasting-induced suppression of pulsatile LH release in ovariectomized rats. *J. Endocrinol., 129*, 321–328.

Cagampang, F. R., Maeda, K., Yokoyama, A., & Ota, K. (1990). Effect of food deprivation on the pulsatile LH release in the cycling and ovariectomized female rat. *Horm. Metab. Res., 22*, 269–272.

Caggiula, A. R., & Eibergen, R. (1969). Copulation of virgin male rats evoked by painful peripheral stimulation. *J. Comp. Physiol. Psychol., 69*, 414–419.

Cao, C., Matsumura, K., Yamagata, K., & Watanabe, Y. (1997). Involvement of cyclooxygenase-2 in LPS-induced fever and regulation of its mRNA by LPS in the rat brain. *Am. J. Physiol., 272*, R1712–R1725.

Carr, T., Elliot, D., & Brown, A. (2008). Endocannabinoids, metabolic regulation, and the role of diet. *Nutrition Research, 28*, 641–650.

Carrasco, G. A., & Van de Kar, L. D. (2003). Neuroendocrine pharmacology of stress. *Eur. J. Pharmacol., 463*, 235–272.

Castellano, J. M., Navarro, V. M., Fernandez-Fernandez, R., Nogueiras, R., Tovar, S., Roa, J., et al. (2005). Changes in hypothalamic KiSS-1 system and restoration of pubertal activation of the reproductive axis by kisspeptin in undernutrition. *Endocrinology, 146*, 3917–3925.

Castellano, J. M., Roa, J., Luque, R. M., Dieguez, C., Aguilar, E., Pinilla, L., et al. (2008). KiSS-1/kisspeptins and the metabolic control of reproduction: Physiologic roles and putative physiopathological implications. *Peptides.*

Chen, M. D., Ordog, T., O'Byrne, K. T., Goldsmith, J. R., Connaughton, M. A., & Knobil, E. (1996). The insulin hypoglycemia-induced inhibition of gonadotropin-releasing hormone pulse generator activity in the rhesus monkey: roles of vasopressin and corticotropin-releasing factor. *Endocrinology, 137*, 2012–2021.

Chen, R., Lewis, K. A., Perrin, M. H., & Vale, W. W. (1993). Expression cloning of a human corticotropin-releasing-factor receptor. *Proc. Natl. Acad. Sci. USA, 90*, 8967–8971.

Coddington, E., Lewis, C., Rose, J. D., & Moore, F. L. (2007). Endocannabinoids mediate the effects of acute stress and corticosterone on sex behavior. *Endocrinology, 148*, 493–500.

Collu, R., Tache, Y., & Ducharme, J. R. (1979). Hormonal modifications induced by chronic stress in rats. *J. Steroid. Biochem., 11*, 989–1000.

Cottingham, S. L., & Pfaff, D. (1986). Interconnectedness of steroid hormone-binding neurons: existence and implications. In D. Ganten, & D. Pfaff (Eds.), *Current Topics in Neuroendocrinology, Vol. 7* (pp. 223–249). Berlin, Germany: Springer.

Crockford, C., Wittig, R. M., Whitten, P. L., Seyfarth, R. M., & Cheney, D. L. (2008). Social stressors and coping mechanisms in wild female baboons (*Papio hamadryas ursinus*). *Horm. Behav., 53*, 254–265.

Crowley, W. R., Popolow, H. B., & Ward, O. B., Jr. (1973). From dud to stud: copulatory behavior elicited through conditioned arousal in sexually inactive male rats. *Physiol. Behav., 10*, 391–394.

Cunningham, M. J., Clifton, D. K., & Steiner, R. A. (1999). Leptin's actions on the reproductive axis: perspectives and mechanisms. *Biol. Reprod., 60*, 216–222.

D'Aquila, P. S., Peana, A. T., Carboni, V., & Serra, G. (2000). Exploratory behaviour and grooming after repeated restraint and chronic mild stress: effect of desipramine. *Eur. J. Pharmacol., 399*, 43–47.

Daley, C. A., Sakurai, H., Adams, B. M., & Adams, T. E. (1999). Effect of stress-like concentrations of cortisol on gonadotroph function in orchidectomized sheep. *Biol. Reprod., 60*, 158–163.

Dallman, M. F., Akana, S. F., Bhatnagar, S., Bell, M. E., Choi, S., Chu, A., et al. (1999). Starvation: early signals, sensors, and sequelae. *Endocrinology, 140*, 4015–4023.

Dallman, M. F., Akana, S. F., Jacobson, L., Levin, N., Cascio, C. S., & Shinsako, J. (1987). Characterization of corticosterone feedback regulation of ACTH secretion. *Ann. NY. Acad. Sci., 512*, 402–414.

Debus, N., Breen, K. M., Barrell, G. K., Billings, H. J., Brown, M., Young, E. A., et al. (2002). Does cortisol mediate endotoxin-induced inhibition of pulsatile luteinizing hormone and gonadotropin-releasing hormone secretion? *Endocrinology, 143*, 3748–3758.

DeCatanzaro, D. (1987). Alteration of estrogen-induced lordosis through central administration of corticosterone in adrenalectomized-ovariectomized rats. *Neuroendocrinology, 46*, 468–474.

DeCatanzaro, D., & Gorzalka, B. B. (1980). Effects of dexamethasone, corticosterone, and ACTH on lordosis in ovariectomized and adrenalectomized-ovariectomized rats. *Pharmacol. Biochem. Behav., 12*, 201–206.

Denver, R. J. (2007). Endocannabinoids link rapid, membrane-mediated corticosteroid actions to behavior. *Endocrinology, 148*, 490–492.

Dickerman, R. W., Li, H. Y., & Wade, G. N. (1993). Decreased availability of metabolic fuels suppresses estrous behavior in Syrian hamsters. *Am. J. Physiol., 264*, R568–R572.

Dobson, H., Ghuman, S., Prabhakar, S., & Smith, R. (2003). A conceptual model of the influence of stress on female reproduction. *Reproduction, 125*, 151–163.

Donadio, M. V., Kunrath, A., Corezola, K. L., Franci, C. R., Anselmo-Franci, J. A., Lucion, A. B., et al. (2007). Effects of acute stress on the day of proestrus on sexual behavior and ovulation in female rats: participation of the angiotensinergic system. *Physiol. Behav., 92*, 591–600.

Dornan, W. A., & Malsbury, C. W. (1989). Neuropeptides and male sexual behavior. *Neurosci. Biobehav. Rev., 13*, 1–15.

Dubrovsky, B. (2000). The specificity of stress responses to different nocuous stimuli: neurosteroids and depression. *Brain Res. Bull., 51*, 443–455.

Durand, D., Pampillo, M., Caruso, C., & Lasaga, M. (2008). Role of metabotropic glutamate receptors in the control of neuroendocrine function. *Neuropharmacology, 55*, 577–583.

Ebisui, O., Fukata, J., Tominaga, T., Murakami, N., Kobayashi, H., Segawa, H., et al. (1992). Roles of interleukin-1 alpha and -1 beta in endotoxin-induced suppression of plasma gonadotropin levels in rats. *Endocrinology, 130*, 3307–3313.

Engler, D., Pham, T., Fullerton, M. J., Clarke, I. J., & Funder, J. W. (1989). Evidence for an ultradian secretion of adrenocorticotropin, beta-endorphin and alpha-melanocyte-stimulating hormone by the ovine anterior and intermediate pituitary. *Neuroendocrinology, 49*, 349–360.

Engler, D., Redei, E., & Kola, I. (1999). The corticotropin-release inhibitory factor hypothesis: a review of the evidence for the existence of inhibitory as well as stimulatory hypophysiotropic regulation of adrenocorticotropin secretion and biosynthesis. *Endocr. Rev., 20*, 460–500.

Erskine, M. S. (1989). Solicitation behavior in the estrous female rat: a review. *Horm. Behav., 23*, 473–502.

Estacio, M. A., Yamada, S., Tsukamura, H., Hirunagi, K., & Maeda, K. (1996). Effect of fasting and immobilization stress on estrogen receptor immunoreactivity in the brain in ovariectomized female rats. *Brain Res., 717*, 55–61.

Evans, J. J. (1999). Modulation of gonadotropin levels by peptides acting at the anterior pituitary gland. *Endocr. Rev., 20*, 46–67.

Feng, L., Sun, W., Xia, Y., Tang, W. W., Chanmugam, P., Soyoola, E., et al. (1993). Cloning two isoforms of rat cyclooxygenase: differential regulation of their expression. *Arch. Biochem. Biophys., 307*, 361–368.

Feng, Y. J., Shalts, E., Xia, L. N., Rivier, J., Rivier, C., Vale, W., et al. (1991). An inhibitory effects of interleukin-1a on basal gonadotropin release in the ovariectomized rhesus monkey: reversal by a corticotropin-releasing factor antagonist. *Endocrinology, 128*, 2077–2082.

Ferguson, A. V., Latchford, K. J., & Samson, W. K. (2008). The paraventricular nucleus of the hypothalamus—a potential target for integrative treatment of autonomic dysfunction. *Expert. Opin. Ther. Targets, 12*, 717–727.

Fernandez-Fernandez, R., Martini, A. C., Navarro, V. M., Castellano, J. M., Dieguez, C., Aguilar, E., et al. (2006). Novel signals for the integration of energy balance and reproduction. *Mol. Cell Endocrinol., 254-255*, 127–132.

Flanagan-Cato, L. M., & McEwen, B. S. (1995). Pattern of Fos and Jun expression in the female rat forebrain after sexual behavior. *Brain Res., 673*, 53–60.

Flanagan-Cato, L. M., Lee, B. J., & Calizo, L. H. (2006). Co-localization of midbrain projections, progestin receptors, and mating-induced fos in the hypothalamic ventromedial nucleus of the female rat. *Horm. Behav., 50*, 52–60.

Frye, C. A. (2001). The role of neurosteroids and non-genomic effects of progestins and androgens in mediating sexual receptivity of rodents. *Brain Res. Brain Res. Rev., 37*, 201–222.

Frye, C. A. (2007). Progestins influence motivation, reward, conditioning, stress, and/or response to drugs of abuse. *Pharmacol. Biochem. Behav., 86*, 209–219.

Frye, C. A., & Bayon, L. E. (1999). Mating stimuli influence endogenous variations in the neurosteroids 3alpha,5alpha-THP and 3alpha-Diol. *J. Neuroendocrinol., 11*, 839–847.

Garcia-Bueno, B., & Leza, J. C. (2008). Inflammatory/anti-inflammatory mechanisms in the brain following exposure to stress. *Rev. Neurol., 46*, 675–683.

Garcia-Bueno, B., Madrigal, J. L., Lizasoain, I., Moro, M. A., Lorenzo, P., & Leza, J. C. (2005). Peroxisome proliferator-activated receptor gamma activation decreases neuroinflammation in brain after stress in rats. *Biol. Psychiatry., 57*, 885–894.

Gerendai, I., Toth, I. E., Boldogkoi, Z., Medveczky, I., & Halasz, B. (2000a). Central nervous system structures labelled from the testis using the transsynaptic viral tracing technique. *J. Neuroendocrinol., 12*, 1087–1095.

Gerendai, I., Toth, I. E., Boldogkoi, Z., Medveczky, I., & Halasz, B. (2000b). CNS structures presumably involved in vagal control of ovarian function. *J. Auton. Nerv. Syst., 80*, 40–45.

Giacobini, P., Kopin, A. S., Beart, P. M., Mercer, L. D., Fasolo, A., & Wray, S. (2004). Cholecystokinin modulates migration of gonadotropin-releasing hormone-1 neurons. *J. Neurosci., 24*, 4737–4748.

Gibbs, D. M., & Vale, W. (1982). Presence of corticotropin releasing factor-like immunoreactivity in hypophysial portal blood. *Endocrinology, 111*, 1418–1420.

Gill, C. J., & Rissman, E. F. (1997). Female sexual behavior is inhibited by short- and long-term food restriction. *Physiol. Behav., 61*, 387–394.

Girdler, S. S., & Klatzkin, R. (2007). Neurosteroids in the context of stress: implications for depressive disorders. *Pharmacol. Ther., 116*, 125–139.

Gore, A. C., & Roberts, J. L. (1995). Regulation of gonadotropin-releasing hormone gene expression in the rat during the luteinizing hormone surge. *Endocrinology, 136*, 889–896.

Gorzalka, B. B., & Hanson, L. A. (1998). Sexual behavior and wet dog shakes in the male rat: regulation by corticosterone. *Behav. Brain Res., 97*, 143–151.

Gorzalka, B. B., & Moe, I. V. (1994). Adrenal role in proceptivity and receptivity induced by two modes of estradiol treatment. *Physiol. Behav., 55*, 29–34.

Gorzalka, B. B., Brotto, L. A., & Hong, J. J. (1999). Corticosterone regulation of 5-HT2A receptor-mediated behaviors: attenuation by melatonin. *Physiol. Behav., 67*, 439–442.

Gorzalka, B. B., Hanson, L. A., & Brotto, L. A. (1998). Chronic stress effects on sexual behavior in male and female rats: mediation by 5-HT2A receptors. *Pharmacol. Biochem. Behav., 61*, 405–412.

Gorzalka, B. B., Hanson, L. A., & Hong, J. J. (2001). Ketanserin attenuates the behavioural effects of corticosterone: implications for 5-HT (2A) receptor regulation. *Eur. J. Pharmacol., 428*, 235–240.

Gorzalka, B. B., Hill, M. N., & Hillard, C. J. (2008a). Regulation of endocannabinoid signaling by stress: implications for stress-related affective disorders. *Neurosci. Biobehav. Rev., 32*, 1152–1160.

Gorzalka, B. B., Morrish, A. C., & Hill, M. N. (2008b). Endocannabinoid modulation of male rat sexual behavior. *Psychopharmacology (Berl), 198*, 479–486.

Goshen, I., & Yirmiya, R. (2008). Interleukin-1 (IL-1): A central regulator of stress responses. *Front Neuroendocrinol.* doi: 10.1016/j.yfrne.2008.10.001.

Gottsch, M. L., Clifton, D. K., & Steiner, R. A. (2004). Galanin-like peptide as a link in the integration of metabolism and reproduction. *Trends Endocrinol. Metab., 15*, 215–221.

Gronli, J., Murison, R., Fiske, E., Bjorvatn, B., Sorensen, E., Portas, C. M., et al. (2005). Effects of chronic mild stress on sexual behavior, locomotor activity and consumption of sucrose and saccharine solutions. *Physiol. Behav., 84*, 571–577.

Haddad, J. J., Saade, N. E., & Safieh-Garabedian, B. (2002). Cytokines and neuro-immune-endocrine interactions: a role for the hypothalamic–pituitary–adrenal revolving axis. *J. Neuroimmunol., 133*, 1–19.

Hakansson, M. L., Brown, H., Ghilardi, N., Skoda, R. C., & Meister, B. (1998). Leptin receptor immunoreactivity in chemically defined target neurons of the hypothalamus. *J. Neurosci., 18*, 559–572.

Hanson, L. A., & Gorzalka, B. B. (1999). The influence of corticosterone on serotonergic stereotypy and sexual behavior in the female rat. *Behav. Brain Res., 104*, 27–35.

Hanus, L., Avraham, Y., Ben-Shushan, D., Zolotarev, O., Berry, E. M., & Mechoulam, R. (2003). Short-term fasting and prolonged semi-starvation have opposite effects on 2-AG levels in mouse brain. *Brain Res., 983*, 144–151.

Harris, T. G., Battaglia, D. F., Brown, M. E., Brown, M. B., Carlson, N. E., Viguie, C., et al. (2000). Prostaglandins mediate the endotoxin-induced suppression of pulsatile gonadotropin-releasing hormone and luteinizing hormone secretion in the ewe. *Endocrinology, 141*, 1050–1058.

Hauger, R. L., Grigoriadis, D. E., Dallman, M. F., Plotsky, P. M., Vale, W. W., & Dautzenberg, F. M. (2003). International Union of Pharmacology. XXXVI. Current status of the nomenclature for receptors for corticotropin-releasing factor and their ligands. *Pharmacol. Rev., 55*, 21–26.

Heinrichs, S. C., Min, H., Tamraz, S., Carmouche, M., Boehme, S. A., & Vale, W. W. (1997). Anti-sexual and anxiogenic behavioral consequences of corticotropin-releasing factor overexpression are centrally mediated. *Psychoneuroendocrinology, 22*, 215–224.

Hennessy, M. B., Zate, R., & Maken, D. S. (2008). Social buffering of the cortisol response of adult female guinea pigs. *Physiol. Behav., 93*, 883–888.

Henry, B. A., Goding, J. W., Tilbrook, A. J., Dunshea, F. R., & Clarke, I. J. (2001). Intracerebroventricular infusion of leptin elevates the secretion of luteinising hormone without affecting food intake in long-term food-restricted sheep, but increases growth hormone irrespective of bodyweight. *J. Endocrinol., 168*, 67–77.

Herman, J. P., & Cullinan, W. E. (1997). Neurocircuitry of stress: central control of the hypothalamo–pituitary–adrenocortical axis. *Trends Neurosci., 20*, 78–84.

Hill, J. W., Elmquist, J. K., & Elias, C. F. (2008). Hypothalamic pathways linking energy balance and reproduction. *Am. J. Physiol. Endocrinol. Metab., 294*, E827–E832.

Holmer, H. K., Rodman, J. E., Helmreich, D. L., & Parfitt, D. B. (2003). Differential effects of chronic escapable versus inescapable stress on male syrian hamster (*Mesocricetus auratus*) reproductive behavior. *Horm. Behav., 43*, 381–387.

Hoyenga, K. B., & Hoyenga, K. T. (1982). Gender and energy balance: sex differences in adaptations for feast and famine. *Physiol. Behav., 28*, 545–563.

Hull, E. M., & Dominguez, J. M. (2007). Sexual behavior in male rodents. *Horm. Behav., 52*, 45–55.

Husak, J. F., & Moore, I. T. (2008). Stress hormones and mate choice. *Trends Ecol. Evol., 23*, 532–534.

I'Anson, H., Sundling, L. A., Roland, S. M., & Ritter, S. (2003). Immunotoxic destruction of distinct catecholaminergic neuron populations disrupts the reproductive response to glucoprivation in female rats. *Endocrinology, 144*, 4325–4331.

Igaz, P., Salvi, R., Rey, J. P., Glauser, M., Pralong, F. P., & Gaillard, R. C. (2006). Effects of cytokines on gonadotropin-releasing hormone (GnRH) gene expression in primary hypothalamic neurons and in GnRH neurons immortalized conditionally. *Endocrinology, 147*, 1037–1043.

Irwig, M. S., Fraley, G. S., Smith, J. T., Acohido, B. V., Popa, S. M., Cunningham, M. J., et al. (2004). Kisspeptin activation of gonadotropin releasing hormone neurons and regulation of KiSS-1 mRNA in the male rat. *Neuroendocrinology, 80*, 264–272.

Jeong, K. H., Jacobson, L., Widmaier, E. P., & Majzoub, J. A. (1999). Normal suppression of the reproductive axis following stress in corticotropin-releasing hormone-deficient mice. *Endocrinology, 140*, 1702–1708.

Jones, J. E., & Lubbers, L. S. (2001). Suppression and recovery of estrous behavior in Syrian hamsters after changes in metabolic fuel availability. *Am. J. Physiol. Regul. Integr. Comp. Physiol., 280*, R1393–R1398.

Jones, J. E., Pick, R. R., & Wade, G. N. (2002). Food deprivation inhibits estrous behavior in hormone-treated Syrian hamsters despite elevated estradiol levels. *Horm. Behav., 41*, 316–320.

Kalra, P. S., Sahu, A., & Kalra, S. P. (1990). Interleukin-1 inhibits the ovarian steroid-induced luteinizing hormone surge and release of hypothalamic luteinizing hormone-releasing hormone in rats. *Endocrinology, 126*, 2145–2152.

Kavaliers, M., & Choleris, E. (2001). Antipredator responses and defensive behavior: ecological and ethological approaches for the neurosciences. *Neurosci. Biobehav. Rev., 25*, 577–586.

Kavaliers, M., & Ossenkopp, K. P. (2001). Corticosterone rapidly reduces male odor preferences in female mice. *Neuroreport, 12*, 2999–3002.

Keen-Rhinehart, E., Michopoulos, V., Toufexis, D. J., Martin, E. I., Nair, H., Ressler, K. J., et al. (2008). Continuous expression of corticotropin-releasing factor in the central nucleus of the amygdala emulates the dysregulation of the stress and reproductive axes. *Mol. Psychiatry.* doi: 10.1038/mp.2008.91.

Kilen, S. M., Szabo, M., Strasser, G. A., McAndrews, J. M., Ringstrom, S. J., & Schwartz, N. B. (1996). Corticosterone selectively increases follicle-stimulating hormone beta-subunit messenger ribonucleic acid in primary anterior pituitary cell culture without affecting its half-life. *Endocrinology, 137*, 3802–3807.

Kim, C. K., & Rivier, C. L. (2000). Nitric oxide and carbon monoxide have a stimulatory role in the hypothalamic–pituitary–adrenal response to physico–emotional stressors in rats. *Endocrinology, 141*, 2244–2253.

King, J. C., & Anthony, E. L. (1984). LHRH neurons and their projections in humans and other mammals: species comparisons. *Peptides, 1* (5 Suppl.), 195–207.

Kinoshita, M., Moriyama, R., Tsukamura, H., & Maeda, K. I. (2003). A rat model for the energetic regulation of gonadotropin secretion: role of the glucose-sensing mechanism in the brain. *Domest. Anim. Endocrinol., 25*, 109–120.

Kiyma, Z., Alexander, B. M., Van Kirk, E. A., Murdoch, W. J., Hallford, D. M., & Moss, G. E. (2004). Effects of feed restriction on reproductive and metabolic hormones in ewes. *J. Anim. Sci., 82*, 2548–2557.

Koldzic-Zivanovic, N., Tu, H., Juelich, T. L., Rady, P. L., Tyring, S. K., Hudnall, S. D., et al. (2006). Regulation of adrenal glucocorticoid synthesis by interleukin-10: a preponderance of IL-10 receptor in the adrenal zona fasciculata. *Brain Behav. Immun., 20*, 460–468.

Komaki, G., Arimura, A., & Koves, K. (1992). Effect of intravenous injection of IL-1 beta on PGE2 levels in several brain areas as determined by microdialysis. *Am. J. Physiol., 262*, E246–E251.

Kubli-Garfias, C. (1990). Chemical structure of corticosteroids and its relationship with their acute induction of lordosis in the female rat. *Horm. Behav., 24*, 443–449.

Kubovcakova, L., Tybitanclova, K., Sabban, E. L., Majzoub, J., Zorad, S., Vietor, I., et al. (2004). Catecholamine synthesizing enzymes and their modulation by immobilization stress in knockout mice. *Ann. NY. Acad. Sci., 1018*, 458–465.

Lee, S., Kim, C. K., & Rivier, C. (1999). Nitric oxide stimulates ACTH secretion and the transcription of the genes encoding for NGFI-B, corticotropin-releasing factor, corticotropin-releasing factor receptor type 1, and vasopressin in the hypothalamus of the intact rat. *J. Neurosci., 19*, 7640–7647.

Lee, S., Miselis, R., & Rivier, C. (2002). Anatomical and functional evidence for a neural hypothalamic–testicular pathway that is independent of the pituitary. *Endocrinology, 143*, 4447–4454.

Lemaire, V., Taylor, G. T., & Mormede, P. (1997). Adrenal axis activation by chronic social stress fails to inhibit gonadal function in male rats. *Psychoneuroendocrinology, 22*, 563–573.

Leonard, B. E. (2005). The HPA and immune axes in stress: the involvement of the serotonergic system. *Eur. Psychiatry, 20*(Suppl 3), S302–S306.

Li, X. F., Bowe, J. E., Lightman, S. L., & O'Byrne, K. T. (2005). Role of corticotropin-releasing factor receptor-2 in stress-induced suppression of pulsatile luteinizing hormone secretion in the rat. *Endocrinology, 146*, 318–322.

Li, X. F., Bowe, J. E., Mitchell, J. C., Brain, S. D., Lightman, S. L., & O'Byrne, K. T. (2004). Stress-induced suppression of the gonadotropin-releasing hormone pulse generator in the female rat: a novel neural action for calcitonin gene-related peptide. *Endocrinology, 145*, 1556–1563.

Lidgard, D. C., Boness, D. J., Bowen, W. D., & McMillan, J. I. (2008). The implications of stress on male mating behavior and success in a sexually dimorphic polygynous mammal, the grey seal. *Horm. Behav., 53*, 241–248.

Lovenberg, T. W., Chalmers, D. T., Liu, C., & De Souza, E. B. (1995). CRF2 alpha and CRF2 beta receptor mRNAs are differentially distributed between the rat central nervous system and peripheral tissues. *Endocrinology, 136*, 4139–4142.

Luque, R. M., Kineman, R. D., & Tena-Sempere, M. (2007). Regulation of hypothalamic expression of KiSS-1 and GPR54 genes by metabolic factors: analyses using mouse models and a cell line. *Endocrinology, 148*, 4601–4611.

Machatschke, I. H., Wallner, B., & Dittami, J. (2006). Impact of social environment on female chimpanzee reproductive cycles. *Horm. Behav., 50*, 126–131.

MacLusky, N. J., Naftolin, F., & Leranth, C. (1988). Immunocytochemical evidence for direct synaptic connections between corticotrophin-releasing factor (CRF) and gonadotrophin-releasing hormone (GnRH)-containing neurons in the preoptic area of the rat. *Brain Res., 439*, 391–395.

Madrigal, J. L., Garcia-Bueno, B., Moro, M. A., Lizasoain, I., Lorenzo, P., & Leza, J. C. (2003). Relationship between cyclooxygenase-2 and nitric oxide synthase-2 in rat cortex after stress. *Eur. J. Neurosci., 18*, 1701–1705.

Maeda, K., Cagampang, F. R., Coen, C. W., & Tsukamura, H. (1994). Involvement of the catecholaminergic input to the paraventricular nucleus and of corticotropin-releasing hormone in the fasting-induced suppression of luteinizing hormone release in female rats. *Endocrinology, 134*, 1718–1722.

Marchlewska-Koj, A., Pochron, E., Galewicz-Sojecka, A., & Galas, J. (1994). Suppression of estrus in female mice by the presence of conspecifics or by foot shock. *Physiol. Behav., 55*, 317–321.

Martin, B., Pearson, M., Kebejian, L., Golden, E., Keselman, A., Bender, M., et al. (2007). Sex-dependent metabolic, neuroendocrine, and cognitive responses to dietary energy restriction and excess. *Endocrinology, 148*, 4318–4333.

Martin, J. G., & Reale, D. (2008). Animal temperament and human disturbance: implications for the response of wildlife to tourism. *Behav. Processes, 77*, 66–72.

Martin, S. S., Qasim, A., & Reilly, M. P. (2008). Leptin resistance: a possible interface of inflammation and metabolism in obesity-related cardiovascular disease. *J. Am. Coll. Cardiol., 52*, 1201–1210.

Matsuwaki, T., Kayasuga, Y., Yamanouchi, K., & Nishihara, M. (2006). Maintenance of gonadotropin secretion by glucocorticoids under stress conditions through the inhibition of prostaglandin synthesis in the brain. *Endocrinology, 147*, 1087–1093.

Matsuwaki, T., Watanabe, E., Suzuki, M., Yamanouchi, K., & Nishihara, M. (2003). Glucocorticoid maintains pulsatile secretion of luteinizing hormone under infectious stress condition. *Endocrinology, 144*, 3477–3482.

McGillivray, S. M., Thackray, V. G., Coss, D., & Mellon, P. L. (2007). Activin and glucocorticoids synergistically activate follicle-stimulating hormone beta-subunit gene expression in the immortalized LbetaT2 gonadotrope cell line. *Endocrinology, 148*, 762–773.

Mellon, S. H., & Griffin, L. D. (2002). Synthesis, regulation, and function of neurosteroids. *Endocr. Res., 28*, 463.

Merchenthaler, I. (1984). Corticotropin releasing factor (CRF)-like immunoreactivity in the rat central nervous system. Extra-hypothalamic distribution. *Peptides 5 Suppl, 1*, 53–69.

Merchenthaler, I., Gorcs, T., Setalo, G., Petrusz, P., & Flerko, B. (1984). Gonadotropin-releasing hormone (GnRH) neurons and pathways in the rat brain. *Cell Tissue Res., 237*, 15–29.

Messager, S., Chatzidaki, E. E., Ma, D., Hendrick, A. G., Zahn, D., Dixon, J., et al. (2005). Kisspeptin directly stimulates gonadotropin-releasing hormone release via G protein-coupled receptor 54. *Proc. Natl. Acad. Sci. USA, 102*, 1761–1766.

Monder, C., Sakai, R. R., Miroff, Y., Blanchard, D. C., & Blanchard, R. J. (1994). Reciprocal changes in plasma corticosterone and testosterone in stressed male rats maintained in a visible burrow system: evidence for a mediating role of testicular 11 beta-hydroxysteroid dehydrogenase. *Endocrinology, 134*, 1193–1198.

Mouihate, A., & Pittman, Q. J. (2003). Neuroimmune response to endogenous and exogenous pyrogens is differently modulated by sex steroids. *Endocrinology, 144*, 2454–2460.

Muglia, L. J., Bethin, K. E., Jacobson, L., Vogt, S. K., & Majzoub, J. A. (2000). Pituitary–adrenal axis regulation in CRH-deficient mice. *Endocr. Res., 26*, 1057–1066.

Murahashi, K., Bucholtz, D. C., Nagatani, S., Tsukahara, S., Tsukamura, H., Foster, D. L., et al. (1996). Suppression of luteinizing hormone pulses by restriction of glucose availability is mediated by sensors in the brain stem. *Endocrinology, 137*, 1171–1176.

Nagatani, S., Bucholtz, D. C., Murahashi, K., Estacio, M. A., Tsukamura, H., Foster, D. L., et al. (1996a). Reduction of glucose availability suppresses pulsatile luteinizing hormone release in female and male rats. *Endocrinology, 137*, 1166–1170.

Nagatani, S., Guthikonda, P., Thompson, R. C., Tsukamura, H., Maeda, K. I., & Foster, D. L. (1998). Evidence for GnRH regulation by leptin: leptin administration prevents reduced pulsatile LH secretion during fasting. *Neuroendocrinology, 67*, 370–376.

Nagatani, S., Tsukamura, H., Murahashi, K., Bucholtz, D. C., Foster, D. L., & Maeda, K. (1996b). Paraventricular norepinephrine release mediates glucoprivic suppression of pulsatile luteinizing hormone secretion. *Endocrinology, 137*, 3183–3186.

Nappi, R. E., & Rivest, S. (1997). Effect of immune and metabolic challenges on the luteinizing hormone-releasing hormone neuronal system in cycling female rats: an evaluation at the transcriptional level. *Endocrinology, 138*, 1374–1384.

Nathan, C. F. (1987). Secretory products of macrophages. *J. Clin. Invest., 79*, 319–326.

Neumann, F. (1984). Effects of drugs and chemicals on spermatogenesis. *Arch. Toxicol. Suppl., 7*, 109–117.

Ogilvie, K., & Rivier, C. (1998). The intracerebroventricular injection of interleukin-1beta blunts the testosterone response to human chorionic gonadotropin: role of prostaglandin- and adrenergic-dependent pathways. *Endocrinology, 139*. 3088-1995.

Ojeda, S. R., Negro-Vilar, A., & McCann, S. M. (1981). Role of prostaglandins in the control of pituitary hormone secretion. *Prog. Clin. Biol. Res., 74*, 229–247.

Olah, M., Milloh, H., & Wenger, T. (2008). The role of endocannabinoids in the regulation of luteinizing hormone and prolactin release. Differences between the effects of AEA and 2AG. *Mol. Cell. Endocrinol., 286*, S36–S40.

Orr, T. E., & Mann, D. R. (1992). Role of glucocorticoids in the stress-induced suppression of testicular steroidogenesis in adult male rats. *Horm. Behav., 26*, 350–363.

Pacak, K., & Palkovits, M. (2001). Stressor specificity of central neuro-endocrine responses: implications for stress-related disorders. *Endocr. Rev., 22*, 502–548.

Pagotto, U., Marsicano, G., Cota, D., Lutz, B., & Pasquali, R. (2006). The emerging role of the endocannabinoid system in endocrine regulation and energy balance. *Endocr. Rev., 27*, 73–100.

Palkovits, M. (1986). Afferents onto neuroendocrine cells. In D. Ganten, & D. Pfaff (Eds.), *Current Topics in Neuroenocrinology, Vol. 7* (pp. 197–222). Berlin, Germany: Springer-Verlag.

Paul, S. M., & Purdy, R. H. (1992). Neuroactive steroids. *Faseb J, 6*, 2311–2322.

Payne, A. H., & Youngblood, G. L. (1995). Regulation of expression of steroidogenic enzymes in Leydig cells. *Biol. Reprod., 52*, 217–225.

Pedersen, L. J. (2007). Sexual behaviour in female pigs. *Horm. Behav., 52*, 64–69.

Pednekar, J. R., Mulgaonker, V. K., & Mascarenhas, J. F. (1993). Effect of intensity and duration of stress on male sexual behaviour. *Indian J. Exp. Biol., 31*, 638–640.

Pfaff, D. W., & Modianos, D. (1985). Neural mechanisms of female reproductive behavior. In R. W. Goy (Ed.), *Handbook of Behavioral Neurobiology* (pp. 423–493). New York, NY: Plenum Press.

Plas-Roser, S., & Aron, C. (1977). New data concerning the control by the adrenals of sexual receptivity in the rat. *Physiol. Behav., 19*, 57–60.

Plas-Roser, S., & Aron, C. (1981). Stress related effects in the control of sexual receptivity and in the secretion of progesterone by the adrenals in cyclic female rats. *Physiol. Behav., 27*, 261–264.

Preil, J., Muller, M. B., Gesing, A., Reul, J. M., Sillaber, I., Van Gaalen, M. M., et al. (2001). Regulation of the hypothalamic–pituitary–adrenocortical system in mice deficient for CRH receptors 1 and 2. *Endocrinology, 142*, 4946–4955.

Puder, J. J., Freda, P. U., Goland, R. S., & Wardlaw, S. L. (2001). Estrogen modulates the hypothalamic–pituitary–adrenal and inflammatory cytokine responses to endotoxin in women. *J. Clin. Endocrinol. Metab., 86*, 2403–2408.

Raetz, C. R. (1990). Biochemistry of endotoxins. *Annu. Rev. Biochem., 59*, 129–170.

Rage, F., Lee, B. J., Ma, Y. J., & Ojeda, S. R. (1997). Estradiol enhances prostaglandin E2 receptor gene expression in luteinizing hormone-releasing hormone (LHRH) neurons and facilitates the LHRH response to PGE2 by activating a glia-to-neuron signaling pathway. *J. Neurosci., 17*, 9145–9156.

Ramaswamy, S., Guerriero, K. A., Gibbs, R. B., & Plant, T. M. (2008). Structural interactions between kisspeptin and GnRH neurons in the mediobasal hypothalamus of the male rhesus monkey (*Macaca mulatta*) as revealed by double immunofluorescence and confocal microscopy. *Endocrinology, 149*, 4387–4395.

Rasmussen, H. B., Ganswindt, A., Douglas-Hamilton, I., & Vollrath, F. (2008). Endocrine and behavioral changes in male African elephants: linking hormone changes to sexual state and reproductive tactics. *Horm. Behav., 54*, 539–548.

Retana-Marquez, S., Bonilla-Jaime, H., Vazquez-Palacios, G., Martinez-Garcia, R., & Velazquez-Moctezuma, J. (2003). Changes in masculine sexual behavior, corticosterone and testosterone in response to acute and chronic stress in male rats. *Horm. Behav., 44*, 327–337.

Retana-Marquez, S., Bonilla-Jaime, H., & Velazquez-Moctezuma, J. (1998). Lack of effect of corticosterone administration on male sexual behavior of rats. *Physiol. Behav., 63*, 367–370.

Retana-Marquez, S., Salazar, E. D., & Velazquez-Moctezuma, J. (1996). Effect of acute and chronic stress on masculine sexual behavior in the rat. *Psychoneuroendocrinology, 21*, 39–50.

Reyes, B. A., Valentino, R. J., Xu, G., & Van Bockstaele, E. J. (2005). Hypothalamic projections to locus coeruleus neurons in rat brain. *Eur. J. Neurosci., 22*, 93–106.

Rivest, S., & Rivier, C. (1991). Influence of the paraventricular nucleus of the hypothalamus in the alteration of neuroendocrine functions induced by intermittent footshock or interleukin. *Endocrinology, 129*, 2049–2057.

Rivest, S., & Rivier, C. (1993a). Centrally injected interleukin-1 beta inhibits the hypothalamic LHRH secretion and circulating LH levels via prostaglandins in rats. *J. Neuroendocrinol., 5*, 445–450.

Rivest, S., & Rivier, C. (1993b). Interleukin-1 beta inhibits the endogenous expression of the early gene c-fos located within the nucleus of LH-RH neurons and interferes with hypothalamic LH-RH release during proestrus in the rat. *Brain Res., 613*, 132–142.

Rivest, S., & Rivier, C. (1995). The role of corticotropin-releasing factor and interleukin-1 in the regulation of neurons controlling reproductive functions. *Endocr. Rev., 16*, 177–199.

Rivest, S., Lee, S., Attardi, B., & Rivier, C. (1993). The chronic intracerebroventricular infusion of interleukin-1 beta alters the activity of the hypothalamic–pituitary–gonadal axis of cycling rats. I. Effect on LHRH and gonadotropin biosynthesis and secretion. *Endocrinology, 133*, 2424–2430.

Rivier, C. (1993). Neuroendocrine effects of cytokines in the rat. *Rev. Neurosci., 4*, 223–237.

Rivier, C., & Erickson, G. (1993). The chronic intracerebroventricular infusion of interleukin-1 beta alters the activity of the hypothalamic–pituitary–gonadal axis of cycling rats. II. Induction of pseudopregnant-like corpora lutea. *Endocrinology, 133*, 2431–2436.

Rivier, C., & Rivest, S. (1993). Mechanisms mediating the effects of cytokines on neuroendocrine functions in the rat. *Ciba. Found. Symp., 172*, 204–220, discussion 220–5.

Rivier, C., & Vale, W. (1984). Influence of corticotropin-releasing factor on reproductive functions in the rat. *Endocrinology, 114*, 914–921.

Rivier, C., & Vale, W. (1990). Cytokines act within the brain to inhibit luteinizing hormone secretion and ovulation in the rat. *Endocrinology, 127*, 849–856.

Rivier, C. L., Grigoriadis, D. E., & Rivier, J. E. (2003). Role of corticotropin-releasing factor receptors type 1 and 2 in modulating the rat adrenocorticotropin response to stressors. *Endocrinology, 144*, 2396–2403.

Rivier, C., Rivier, J., Mormede, P., & Vale, W. (1984). Studies of the nature of the interaction between vasopressin and corticotropin-releasing factor on adrenocorticotropin release in the rat. *Endocrinology, 115*, 882–886.

Rivier, C., Rivier, J., & Vale, W. (1986). Stress-induced inhibition of reproductive functions: role of endogenous corticotropin-releasing factor. *Science, 231*, 607–609.

Roa, J., Aguilar, E., Dieguez, C., Pinilla, L., & Tena-Sempere, M. (2008). New frontiers in kisspeptin/GPR54 physiology as fundamental gatekeepers of reproductive function. *Front Neuroendocrinol., 29*, 48–69.

Rodriguez Echandia, E. L., Gonzalez, A. S., Cabrera, R., & Fracchia, L. N. (1988). A further analysis of behavioral and endocrine effects of unpredictable chronic stress. *Physiol. Behav., 43*, 789–795.

Roozendaal, M. M., Swarts, H. J., Wiegant, V. M., & Mattheij, J. A. (1995). Effect of restraint stress on the preovulatory luteinizing hormone profile and ovulation in the rat. *Eur. J. Endocrinol., 133*, 347–353.

Sahu, A. (2004). Minireview: a hypothalamic role in energy balance with special emphasis on leptin. *Endocrinology, 145*, 2613–2620.

Sajapitak, S., Iwata, K., Shahab, M., Uenoyama, Y., Yamada, S., Kinoshita, M., et al. (2008). Central lipoprivation-induced suppression of luteinizing hormone pulses is mediated by paraventricular catecholaminergic inputs in female rats. *Endocrinology, 149*, 3016–3024.

Sapolsky, R. M. (1982). The endocrine stress-response and social status in the wild baboon. *Horm. Behav., 16*, 279–292.

Sapolsky, R. M. (1985). Stress-induced suppression of testicular function in the wild baboon: role of glucocorticoids. *Endocrinology, 116*, 2273–2278.

Sapolsky, R. M. (1986). Stress-induced elevation of testosterone concentration in high ranking baboons: role of catecholamines. *Endocrinology, 118*, 1630–1635.

Sapolsky, R. M., Romero, L. M., & Munck, A. U. (2000). How do glucocorticoids influence stress responses? Integrating permissive, suppressive, stimulatory, and preparative actions. *Endocr. Rev., 21*, 55–89.

Saradha, B., & Mathur, P. (2006). Effect of environmental contaminants on male reproduction. *Environmental Toxicology and Pharmacology, 21*, 34–42.

Schiml, P. A., & Rissman, E. F. (2000). Effects of gonadotropin-releasing hormones, corticotropin-releasing hormone, and vasopressin on female sexual behavior. *Horm. Behav., 37*, 212–220.

Schneider, J. E., & Zhou, D. (1999). Interactive effects of central leptin and peripheral fuel oxidation on estrous cyclicity. *Am. J. Physiol., 277*, R1020–R1024.

Schneider, J. E., Blum, R. M., & Wade, G. N. (2000a). Metabolic control of food intake and estrous cycles in syrian hamsters. I. Plasma insulin and leptin. *Am. J. Physiol. Regul. Integr. Comp. Physiol., 278*, R476–R485.

Schneider, J. E., Goldman, M. D., Leo, N. A., & Rosen, M. E. (1997a). Central vs. peripheral metabolic control of estrous cycles in Syrian hamsters. II. Glucoprivation. *Am. J. Physiol., 272*, R406–R412.

Schneider, J. E., Goldman, M. D., Tang, S., Bean, B., Ji, H., & Friedman, M. I. (1998). Leptin indirectly affects estrous cycles by increasing metabolic fuel oxidation. *Horm. Behav., 33*, 217–228.

Schneider, J. E., Hall, A. J., & Wade, G. N. (1997b). Central vs. peripheral metabolic control of estrous cycles in Syrian hamsters. I. Lipoprivation. *Am. J. Physiol., 272*, R400–R405.

Schneider, J. E., Zhou, D., & Blum, R. M. (2000b). Leptin and metabolic control of reproduction. *Horm. Behav., 37*, 306–326.

Selvage, D. J., & Rivier, C. (2003). Importance of the paraventricular nucleus of the hypothalamus as a component of a neural pathway between the brain and the testes that modulates testosterone secretion independently of the pituitary. *Endocrinology, 144*, 594–598.

Selvage, D. J., Lee, S. Y., Parsons, L. H., Seo, D. O., & Rivier, C. L. (2004). A hypothalamic–testicular neural pathway is influenced by brain catecholamines, but not testicular blood flow. *Endocrinology, 145*, 1750–1759.

Selvage, D. J., Parsons, L., & Rivier, C. (2006). Role played by brainstem neurons in regulating testosterone secretion via a direct neural pathway between the hypothalamus and the testes. *Endocrinology, 147*, 3070–3075.

Seymour, P. L., Dettloff, S. L., Jones, J. E., & Wade, G. N. (2005). Corticotropin-releasing factor receptor subtypes mediating nutritional suppression of estrous behavior in Syrian hamsters. *Am. J. Physiol. Regul. Integr. Comp. Physiol., 289*, R418–R423.

Shalts, E., Feng, Y. J., & Ferin, M. (1992). Vasopressin mediates the interleukin-1 alpha-induced decrease in luteinizing hormone secretion in the ovariectomized rhesus monkey. *Endocrinology, 131*, 153–158.

Shimizu, T., & Yokotani, K. (2008). Bidirectional roles of the brain 2-arachidonoyl-sn-glycerol in the centrally administered vasopressin-induced adrenomedullary outflow in rats. *Eur. J. Pharmacol., 582*, 62–69.

Sirinathsinghji, D. J. (1985). Modulation of lordosis behaviour in the female rat by corticotropin releasing factor, beta-endorphin and gonadotropin releasing hormone in the mesencephalic central gray. *Brain Res., 336*, 45–55.

Sirinathsinghji, D. J., Rees, L. H., Rivier, J., & Vale, W. (1983). Corticotropin-releasing factor is a potent inhibitor of sexual receptivity in the female rat. *Nature, 305*, 232–235.

Sirotkin, A. V., Chrenkova, M., Nitrayova, S., Patras, P., Darlak, K., Valenzuela, F., et al. (2008). Effects of chronic food restriction and treatments with leptin or ghrelin on different reproductive parameters of male rats. *Peptides, 29*, 1362–1368.

Smith, P. M., Chambers, A. P., Ho, W., Hopf, C., Sharkey, K. A., & Ferguson, A. V. (2008). The subfornical organ: a CNS site for actions of circulating leptin. *Am. J. Physiol. Regul. Integr. Comp. Physiol.*

Smith, W. J., Stewart, J., & Pfaus, J. G. (1997). Tail pinch induces fos immunoreactivity within several regions of the male rat brain: effects of age. *Physiol. Behav., 61*, 717–723.

Sodersten, P. (1981). Estradiol-progesterone interactions in the reproductive behavior of female rats. In D. Ganten, & D. Pfaff (Eds.), *Current Topics in Neuroendocrinology: Actions of Progesterone on the Brain* (pp. 141–174). New York, NY: Springer-Verlag.

Stanley, S., Wynne, K., McGowan, B., & Bloom, S. (2005). Hormonal regulation of food intake. *Physiol. Rev., 85*, 1131–1158.

Stark, J. D., Banks, J. E., & Vargas, R. (2004). How risky is risk assessment: the role that life history strategies play in susceptibility of species to stress. *Proc. Natl. Acad. Sci. USA, 101*, 732–736.

Tanabe, T., & Tohnai, N. (2002). Cyclooxygenase isozymes and their gene structures and expression. *Prostaglandins Other Lipid Mediat, 68-69*, 95–114.

Tena-Sempere, M., & Barreiro, M. L. (2002). Leptin in male reproduction: the testis paradigm. *Mol. Cell. Endocrinol., 188*, 9–13.

Thammacharoen, S., Lutz, T. A., Geary, N., & Asarian, L. (2008). Hindbrain administration of estradiol inhibits feeding and activates estrogen receptor-alpha-expressing cells in the nucleus tractus solitarius of ovariectomized rats. *Endocrinology, 149*, 1609–1617.

Thrivikraman, K. V., Nemeroff, C. B., & Plotsky, P. M. (2000). Sensitivity to glucocorticoid-mediated fast-feedback regulation of the hypothalamic–pituitary–adrenal axis is dependent upon stressor specific neurocircuitry. *Brain Res., 870*, 87–101.

Tilbrook, A. J., Turner, A. I., & Clarke, I. J. (2000). Effects of stress on reproduction in non-rodent mammals: the role of glucocorticoids and sex differences. *Rev. Reprod., 5*, 105–113.

Tilbrook, A. J., Turner, A. I., & Clarke, I. J. (2002). Stress and reproduction: central mechanisms and sex differences in non-rodent species. *Stress, 5*, 83–100.

Tonelli, L., Kramer, P., Webster, J. I., Wray, S., Listwak, S., & Sternberg, E. (2002). Lipopolysaccharide-induced oestrogen receptor regulation in the paraventricular hypothalamic nucleus of Lewis and Fischer rats. *J. Neuroendocrinol., 14*, 847–852.

Tonissaar, M., Herm, L., Eller, M., Koiv, K., Rinken, A., & Harro, J. (2008). Rats with high or low sociability are differently affected by chronic variable stress. *Neuroscience, 152*, 867–876.

Tropp, J., & Markus, E. J. (2001). Effects of mild food deprivation on the estrous cycle of rats. *Physiol. Behav., 73*, 553–559.

Truitt, W., Harrison, L., Guptarak, J., White, S., Hiegel, C., & Uphouse, L. (2003). Progesterone attenuates the effect of the 5-HT1A receptor agonist, 8-OH-DPAT, and of mild restraint on lordosis behavior. *Brain Res., 974*, 202–211.

Tsukahara, S., Tsukamura, H., Foster, D. L., & Maeda, K. I. (1999). Effect of corticotropin-releasing hormone antagonist on oestrogen-dependent glucoprivic suppression of luteinizing hormone secretion in female rats. *J. Neuroendocrinol., 11*, 101–105.

Tsukamura, H., Nagatani, S., Cagampang, F. R., Kawakami, S., & Maeda, K. (1994). Corticotropin-releasing hormone mediates suppression of pulsatile luteinizing hormone secretion induced by activation of alpha-adrenergic receptors in the paraventricular nucleus in female rats. *Endocrinology, 134*, 1460–1466.

Turnbull, A., & Rivier, C. (1995). Brain-periphery connections: do they play a role in mediating the effect of centrally injected interleukin-1 beta on gonadal function? *Neuroimmunomodulation, 2*, 224–235.

Turnbull, A. V., & Rivier, C. (1997a). Corticotropin-releasing factor (CRF) and endocrine responses to stress: CRF receptors, binding protein, and related peptides. *Proc. Soc. Exp. Biol. Med., 215*, 1–10.

Turnbull, A. V., & Rivier, C. (1997b). Inhibition of gonadotropin-induced testosterone secretion by the intracerebroventricular injection of interleukin-1 beta in the male rat. *Endocrinology, 138*, 1008–1013.

Turnbull, A. V., Kim, C. K., Lee, S., & Rivier, C. L. (1998). Influence of carbon monoxide, and its interaction with nitric oxide, on the adrenocorticotropin hormone response of the normal rat to a physico–emotional stress. *J. Neuroendocrinol., 10*, 793–802.

Turnbull, A. V., Smith, G. W., Lee, S., Vale, W. W., Lee, K. F., & Rivier, C. (1999a). CRF type I receptor-deficient mice exhibit a pronounced pituitary–adrenal response to local inflammation. *Endocrinology, 140*, 1013–1017.

Turnbull, A. V., Vaughan, J., Rivier, J. E., Vale, W. W., & Rivier, C. (1999b). Urocortin is not a significant regulator of intermittent electrofootshock-induced adrenocorticotropin secretion in the intact male rat. *Endocrinology, 140*, 71–78.

Turner, A. I., Hemsworth, P. H., Canny, B. J., & Tilbrook, A. J. (1999). Inhibition of the secretion of LH in ovariectomised pigs by sustained but not repeated acute elevation of cortisol in the absence but not the presence of oestradiol. *J. Endocrinol., 163*, 477–486.

Uphouse, L., Hensler, J. G., Sarkar, J., & Grossie, B. (2006). Fluoxetine disrupts food intake and estrous cyclicity in Fischer female rats. *Brain Res., 1072*, 79–90.

Uphouse, L., Hiegel, C., Sarkar, J., Hurlburt, J., Templeton, C., Guptarak, J., et al. (2008). Female gonadal hormones, mild restraint, and male preference. *Pharmacol. Biochem. Behav., 90*, 758–762.

Uphouse, L., Selvamani, A., Lincoln, C., Morales, L., & Comeaux, D. (2005). Mild restraint reduces the time hormonally primed rats spend with sexually active males. *Behav. Brain Res., 157*, 343–350.

Valentino, R. J., & Van Bockstaele, E. (2008). Convergent regulation of locus coeruleus activity as an adaptive response to stress. *Eur. J. Pharmacol., 583*, 194–203.

Van Bockstaele, E. J., Colago, E. E., & Valentino, R. J. (1996). Corticotropin-releasing factor-containing axon terminals synapse onto catecholamine dendrites and may presynaptically modulate other afferents in the rostral pole of the nucleus locus coeruleus in the rat brain. *J. Comp. Neurol., 364*, 523–534.

Van Bockstaele, E. J., Colago, E. E., & Valentino, R. J. (1998). Amygdaloid corticotropin-releasing factor targets locus coeruleus dendrites: substrate for the co-ordination of emotional and cognitive limbs of the stress response. *J. Neuroendocrinol., 10*, 743–757.

Van Meter, P. E., French, J. A., Dloniak, S. M., Watts, H. E., Kolowski, J. M., & Holekamp, K. E. (2008). Fecal glucocorticoids reflect socio–ecological and anthropogenic stressors in the lives of wild spotted hyenas. *Horm. Behav.* doi: 10.1016/j.yhbeh.2008.11.001.

Van Pett, K., Viau, V., Bittencourt, J. C., Chan, R. K., Li, H. Y., Arias, C., et al. (2000). Distribution of mRNAs encoding CRF receptors in brain and pituitary of rat and mouse. *J. Comp. Neurol., 428*, 191–212.

Van Vugt, D. A., Piercy, J., Farley, A. E., Reid, R. L., & Rivest, S. (1997). Luteinizing hormone secretion and corticotropin-releasing factor gene expression in the paraventricular nucleus of rhesus monkeys following cortisol synthesis inhibition. *Endocrinology, 138*, 2249–2258.

Von Borell, E., Dobson, H., & Prunier, A. (2007). Stress, behaviour and reproductive performance in female cattle and pigs. *Horm. Behav., 52*, 130–138.

Vulliemoz, N. R., Xiao, E., Xia-Zhang, L., Rivier, J., & Ferin, M. (2008). Astressin B, a nonselective corticotropin-releasing hormone receptor antagonist, prevents the inhibitory effect of ghrelin on luteinizing hormone pulse frequency in the ovariectomized rhesus monkey. *Endocrinology, 149*, 869–874.

Wade, G. N., & Jones, J. E. (2003). Lessons from experimental disruption of estrous cycles and behaviors. *Med Sci Sports Exerc, 35*, 1573–1580.

Wade, G. N., & Jones, J. E. (2004). Neuroendocrinology of nutritional infertility. *Am. J. Physiol. Regul. Integr. Comp. Physiol., 287*, R1277–R1296.

Wade, G. N., & Schneider, J. E. (1992). Metabolic fuels and reproduction in female mammals. *Neurosci. Biobehav. Rev., 16*, 235–272.

Wade, G. N., Lempicki, R. L., Panicker, A. K., Frisbee, R. M., & Blaustein, J. D. (1997). Leptin facilitates and inhibits sexual behavior in female hamsters. *Am. J. Physiol., 272*, R1354–R1358.

Wade, G. N., Schneider, J. E., & Li, H. Y. (1996). Control of fertility by metabolic cues. *Am. J. Physiol., 270*, E1–19.

Wagenmaker, E. R., Breen, K. M., Oakley, A. E., Pierce, B. N., Tilbrook, A. J., Turner, A. I., et al. (2008a). Cortisol interferes with the estradiol-induced surge of luteinizing hormone in the ewe. *Biol. Reprod.* doi: 10.1095/biolreprod.108.074252.

Wagenmaker, E. R., Breen, K. M., Oakley, A. E., Tilbrook, A. J., & Karsch, F. J. (2008b). Psychosocial stress inhibits amplitude of gonadotropin-releasing hormone pulses independent of cortisol action on the type II glucocorticoid receptor. *Endocrinology.* doi: 10.1210/en.2008.0757.

Wang, H., Dey, S. K., & Maccarrone, M. (2006). Jekyll and hyde: two faces of cannabinoid signaling in male and female fertility. *Endocr. Rev., 27*, 427–448.

Wang, L. X., Huether, P., Doble, E., & Liptrap, R. M. (1986). Effect of cortisol on luteinizing hormone release during sexual behavior in the boar. *Can. J. Vet. Res., 50*, 540–542.

Weber, M. A., Groos, S., Hopfl, U., Spielmann, M., Aumuller, G., & Konrad, L. (2000). Glucocorticoid receptor distribution in rat testis during postnatal development and effects of dexamethasone on immature peritubular cells *in vitro*. *Andrologia, 32*, 23–30.

White, S., & Uphouse, L. (2004). Estrogen and progesterone dose-dependently reduce disruptive effects of restraint on lordosis behavior. *Horm. Behav., 45*, 201–208.

Wiedenmayer, C. P. (2004). Adaptations or pathologies? Long-term changes in brain and behavior after a single exposure to severe threat. *Neurosci. Biobehav. Rev., 28*, 1–12.

Williams, C. Y., Harris, T. G., Battaglia, D. F., Viguie, C., & Karsch, F. J. (2001a). Endotoxin inhibits pituitary responsiveness to gonadotropin-releasing hormone. *Endocrinology, 142*, 1915–1922.

Williams, G. W., McGinnis, M. Y., & Lumia, A. R. (1992). The effects of olfactory bulbectomy and chronic psychosocial stress on serum glucocorticoids and sexual behavior in female rats. *Physiol. Behav., 52*, 755–760.

Williams, N. I., Helmreich, D. L., Parfitt, D. B., Caston-Balderrama, A., & Cameron, J. L. (2001b). Evidence for a causal role of low energy availability in the induction of menstrual cycle disturbances during strenuous exercise training. *J. Clin. Endocrinol. Metab., 86*, 5184–5193.

Windle, R. J., Wood, S. A., Lightman, S. L., & Ingram, C. D. (1998). The pulsatile characteristics of hypothalamo–pituitary–adrenal activity in female Lewis and Fischer 344 rats and its relationship to differential stress responses. *Endocrinology, 139*, 4044–4052.

Wingfield, J. C., & Sapolsky, R. M. (2003). Reproduction and resistance to stress: when and how. *J. Neuroendocrinol., 15*, 711–724.

Wynne, K., Stanley, S., McGowan, B., & Bloom, S. (2005). Appetite control. *J. Endocrinol., 184*, 291–318.

Xia-Zhang, L., Xiao, E., & Ferin, M. (1995). A 5-day estradiol therapy, in amounts reproducing concentrations of the early-mid follicular phase, prevents the activation of the hypothalamo–pituitary–adrenal axis by interleukin-1 alpha in the ovariectomized rhesus monkey. *J. Neuroendocrinol., 7*, 387–392.

Xiao, E., Xia-Zhang, L., Vulliemoz, N., Rivier, J., & Ferin, M. (2007). Astressin B, a corticotropin-releasing hormone receptor antagonist, accelerates the return to normal luteal function after an inflammatory-like stress challenge in the rhesus monkey. *Endocrinology, 148*, 841–848.

Yamaguchi, H., Kikusui, T., Takeuchi, Y., Yoshimura, H., & Mori, Y. (2005). Social stress decreases marking behavior independently of testosterone in Mongolian gerbils. *Horm. Behav., 47*, 549–555.

Yirmiya, R., Avitsur, R., Donchin, O., & Cohen, E. (1995). Interleukin-1 inhibits sexual behavior in female but not in male rats. *Brain Behav. Immun., 9*, 220–233.

Zigman, J. M., & Elmquist, J. K. (2003). Minireview: From anorexia to obesity—the yin and yang of body weight control. *Endocrinology, 144*, 3749–3756.

Chapter 8

Behavioral Neuroendocrinology of Reproduction in Mammals

Jin Ho Park* and Emilie F. Rissman†

** University of Massachusetts, Boston, MA, USA, † University of Virginia, Charlottesville, VA, USA*

SUMMARY

This chapter provides a broad overview of the neuroendocrinological mechanisms underlying mammalian reproduction, with a focus on discoveries made in the past two decades, integrating both male and female reproductive behavior. Although a wide variety of mammalian species are covered when reviewing the neuroendocrinology of reproductive behavior, most of our understanding of this field has been based on rodent animal models, and the majority of studies reviewed in this chapter reflect this bias. After providing a brief description of male and female sexual behavior (MSB/FSB) and the neural circuitry underlying those behaviors, the roles of gonadal steroids, neurotransmitters, and neuropeptides in both MSB/FSB are discussed. In addition, potential gonadal steroid-independent mechanisms governing sexual behavior in both males and females are covered. Lastly, we discuss recent findings concerning the genetic underpinnings of MSB/FSB, concluding with a discussion of the neuroendocrinology of human sexual behavior.

1. INTRODUCTION

The mechanisms that underlie sexual behavior range from the social composition of the group to coactivators that bind with steroid receptors to stimulate transcription and translation of new proteins in the brain. In this chapter, 'sexual behavior' refers to behaviors associated with courtship, mating, and copulation, and not those associated with maternal behavior, territoriality, lactation, etc. We attempt to integrate male and female sexual behavior (MSB/FSB) in mammals and compare the mechanisms that underlie each. This is not an easy task because few laboratories study both sexes, and the approaches used to study behavior in the two sexes are historically different. Males have been examined from two perspectives: one that deals with the mechanics of copulation and another that asks about motivational aspects of copulation and, more recently, how this is associated with reward. Female sexual behavior traditionally has been used as a bioassay to map steroid-dependent spinal cord to brain circuits. The most stereotyped aspect of the behavior, lordosis, is the rigid posture that female rodents assume in response to stimulation of the hind quarters. The motivational aspects of FSB have been examined only recently in detail. One common aspect of these two fields is the reliance on a few strains of inbred laboratory rats as model subjects. There are good reasons for this choice. Rats breed well in captivity, they are large enough to collect blood for hormone sampling, and their brains are large enough to allow for lesion and implantation work. The behaviors themselves are easy to identify and the amount of individual variation is small enough to permit statistical power with group sizes of about 10. However, species variation constitutes a rich source of information about the types of mechanism that are general vs. those that are specifically adapted to selected environmental niches. Moreover, in the postgenomic era, many behavioral researchers have turned to mice, since the genome is complete, and the power of genetic engineering can be applied to this question. We will largely present the data collected in rats, but also will discuss species differences as well as review data collected using mouse strains. The material we have selected to review deals with the activational impact of hormones on adult sexual behaviors and not the developmental factors that organize the brain (see Chapter 1, this volume).

2. DESCRIPTION OF MALE AND FEMALE SEXUAL BEHAVIOR (MSB/FSB)

Sexual behavior includes courtship behaviors that are displayed by both male and female rats. Some forms of aggression in males are also courtship behaviors since they allow a male to gain access to a female; e.g., fighting with other males to establish a breeding territory. Males and females often explore to find potential mates. These are considered motivational (also referred to as appetitive) behaviors. In most of the species studied in the laboratory,

Hormones and Reproduction of Vertebrates, Volume 5—Mammals

there are three distinguishable behavioral components to the act of male copulation: mounting, intromission, and ejaculation (also referred to as performance or consummatory behaviors). Mounting occurs when the male assumes a copulatory position, with his front paws on the back of the female, but does not necessarily include insertion of the penis into the female's vagina. Intromissions occur when the penis enters the vagina during a mount. An intromission is defined as an ejaculation when it includes the expulsion of semen from the distal urethra. The temporal arrangements of the components of a mating bout ending in ejaculation have species-typical features. For example, in rats, males mount and perform many intromissions that end in a single ejaculation, defined by a deeper, longer thrust and a slow, relaxed dismount, often accompanied by a lifting of the forepaws off the female prior to withdrawal. Male mice tend to perform more bouts of mounting and intromissions before a final bout that includes an ejaculation. In Syrian hamsters (*Mesocricetus auratus*), ejaculation can be difficult to detect and is generally characterized by a subtle change in the rate of intromissions. In mice and musk shrews, the male grasps the female and remains completely still for ~10–30 seconds after the ejaculation. Another aspect of MSB is the postejaculatory interval, in which the male enters a period of sexual quiescence. This interval can be as short as 30 seconds in the Syrian hamster, about 5–15 minutes in rats, or as long as hours to days in other species including some commonly used mouse strains (Dewsbury, 1972).

Female sexual behavior traditionally includes three components: attractivity, proceptivity, and receptivity. Attractivity is the stimulus value of a female for a given male. In general, males of most mammalian species will spend more time in close proximity to females that are ready to display sexual behavior over those that are not so inclined (reviewed in Beach, 1976; Wallen, 1990). Behavior may also increase the attractiveness of females. Females that solicit copulations have a higher stimulus value for males than females that do not. Female proceptive behaviors in rats include darts, hops, and crouches. Females of many nonhuman primates present to males by backing their hindquarters toward the males and males spend more time with females that engage in this behavior (Dixon, Everitt, Herbert, Rugman, & Scruton, 1973). Another proceptive behavior observed in a wide range of species is alternating approaches and withdrawals. In addition, females of many species, including farm animals, also demonstrate mounting behavior accompanied by thrusting pelvic movements in which females may mount either males or females (Beach, 1942; Afonso, Bablekis, & Pfaus, 2006; Afonso, Woehrling, & Pfaus, 2006). In goats and musk shrews, tail wagging is a proceptive behavior and it occurs directly prior to the first attempted mount by the male (Rissman, Clendenon, & Krohmer, 1990; Katz, 2007). Receptivity can be defined as those female reactions that are necessary and sufficient for complete copulation. Receptivity is indicated by a species-specific mating posture in many mammals examined (one exception is humans). Rabbits (*Oryctolagus cuniculus*) and mares (*Equus ferus caballus*) move their tails to one side to allow intromissions. Musk shrews (*Suncus murinus*) hunch their backs up and slow down to allow males a single intromission per mount. Female rodents display the lordosis posture in response to tactile stimulation by the male.

In the laboratory, sexual behavior is typically observed in pairs containing one male and one female; however, in nature, mating occurs in a group context in some species. One of the best examples of the difference that the social context can make comes from work in Rhesus monkeys (*Macaca mulatta*). When Rhesus monkeys are tested for FSB in pairs, the females allow males to mate during most of the menstrual cycle. On the other hand, when mating is examined in the context of groups, typically containing a stable population of females and a few males, females restrict their sexual behaviors to the ovulatory portion of the cycle (Wallen, 1990; 2005). Rodents typically do not allow mating at times other than the evening of proestrus. However, when rats are tested in seminatural group settings, receptive female rats are able to pace their interactions with males (McClintock & Adler, 1978; McClintock, 1987). Paced mating is a repertoire of behaviors displayed by a sexually receptive female rat in which, when given the opportunity, the female will approach and withdraw from the male rat, controlling the timing of the receipt of mounts, intromissions, and ejaculations. Even in pair tests, if females are able to escape from males, they pace copulatory behavior so that the average interintromission interval is around three minutes as compared to around one minute in a standard mating arena where escape is not possible (McClintock & Adler, 1978; Erskine, 1985). Recent work in mice suggests that females also can pace their mating bouts (Johansen, Clemens, & Nunez, 2008). Functionally, paced mating provides more efficient activation of a progestational reflex, which in turn facilitates pregnancy. In standard nonpaced mating tests, ~10 intromissions are required to induce luteal function and ensure successful pregnancy, whereas, in paced mating tests, only around five intromissions are necessary (Erskine, 1985). Males need fewer intromissions to ejaculate in paced as compared with standard laboratory mating tests (Erskine, 1985). Paced mating also induces more expression of c-fos in neurons that are part of the neural circuitry underlying FSB (Erskine & Hanrahan, 1997). Pacing behavior is dependent upon both estradiol (E_2) and progesterone (P_4) to induce sexual receptivity, and both hormones play an important role in its neural control (Brandling-Bennett, Blasberg, & Clark, 1999).

3. THE ROLE OF GONADAL STEROIDS IN SEXUAL BEHAVIOR

In most animals, reproductive behavior and gamete production are synchronized; thus, it is no surprise that gonadal steroids promote the display of a myriad of social behaviors in both sexes. In females that have an estrous cycle, expression of sexual receptivity occurs after E_2 and P_4 levels have reached their zenith. Early studies by Beach (1948) and Young (1961) described the dependence of female reproductive behavior on E_2 and P_4. These two steroids work sequentially and synergistically and the most pronounced FSB is seen when both E_2 and P_4 are given; E_2 alone can trigger lower levels of receptivity in ovariectomized (OVX) females (Davidson, Rodgers, Smith, & Bloch, 1968; Pfaff, 1970), but P_4 alone is without effect. The effects of estrogens can be facilitated by administration of P_4 concentrations well below physiological concentrations (Sodersten, 1985), and more complete displays of female reproductive behavior are observed when an OVX E_2-primed female is administered P_4. Thus, since E_2 treatment alone is sufficient for FSB under certain circumstances, P_4 is not necessary for the expression of FSB; in fact, not all females require P_4 for sexual behavior, particularly induced ovulators such as shrews, voles, and rabbits. Musk shrews are an interesting case. At the time that mating begins, plasma levels of testosterone (T) are 10-fold higher than E_2 in females (Rissman & Crews, 1988), yet the effective steroid is still E_2, which is produced via aromatase enzyme ($P450_{aro}$) in the brain (Veney & Rissman, 1998; 2000).

Gonadal steroids are essential in modulating a myriad of male social behaviors that contribute to successful reproduction and survival across a wide range of mammalian species, including humans (reviewed in Young, 1961; Davidson, Camargo, & Smith, 1979; Salmimies, Kockott, Pirke, Vogt, & Schill, 1982; Hull, Wood & McKenna, 2006). In males, the major steroid secreted by the testes is T, which can be converted by $P450_{aro}$ to E_2 or to the nonaromatizable androgen, 5α-dihydrotestosterone (DHT) by 5α-reductase. Both enzymes exist in the brain; thus, there is the potential for regionally specific conversion of T to E_2 or DHT. While some species-specificity exists, in general, the activation of the receptors of both androgens and estrogens is required for complete MSB (reviewed in Hull et al., 2006).

In agreement with these observations, orchidectomy (ORCH) or OVX diminishes expression of both performance and motivational aspects of sexual behaviors in the vast majority of mammalian species. For example, in almost all males of several well-studied species of rodent and more primitive mammals, such as musk shrews (Rissman, 1987), the ejaculatory response disappears within a few weeks of ORCH (Young, 1961). This decline in sexual behavior is reversible with T treatment, which typically restores sexual behavior within 5–10 days (Beach & Holz-Tucker, 1949; Davidson, 1966a; McGinnis & Dreifuss, 1989; Putnam, Du, Sato, & Hull, 2001). Likewise, OVX females are not receptive to male mounting attempts; however, periodic injections or silastic capsules that contain E_2 restore many aspects of sexual receptivity (Whalen & Nakayama, 1965; Gerall & Dunlap, 1973; Beach & Orndoff, 1974; Parsons, MacLusky, Krieger, McEwen, & Pfaff, 1979).

In female rats, P_4 is necessary for proceptive behavior (Fadem, Barfield, & Whalen, 1979; Tennent, Smith, & Davidson, 1980), and P_4 regulates the termination of receptivity. Progesterone facilitates species-specific effects on FSB in rodents (Ball, 1941; Zucker, 1968). After the initial increase in P_4 secreted from the Graafian follicles, P_4 levels remain elevated after ovulation in several species due to increased secretion by the transient corpora lutea while estrogens decline. A metabolite of DHT, 3α-androstanediol, also plays an important role in the termination of sexual receptivity in rodents (Erskine, Hippensteil, & Kornberg, 1992). Estradiol-induced sexual receptivity in several rodent species can be inhibited by DHT, and, although the precise mechanisms by which DHT inhibits reproductive processes are unclear, there is strong evidence that DHT's effects may be mediated by metabolism to, or enhanced production of, 3α-androstanediol (Dohanich & Clemens, 1983; Erskine, MacLusky, & Baum, 1985). The mechanism by which 3α-androstanediol has effects on lordosis is unclear, but, interestingly, its effects on termination may be acting at $GABA_A$/benzodiazepine receptor complexes independent of intracellular androgen receptors (ARs) (Frye, Duncan, Basham, & Erskine, 1996; Frye, 2001). Another metabolite of DHT, 3β-androstanediol, is a high-affinity ligand for estrogen receptor (ER) β. Although this has not been tested directly, mutant female mice that do not express ERβ have stronger lordosis responses than wild-type (WT) littermates (Kudwa & Rissman, 2003) and they remain in estrus for a longer time during their cycle (Ogawa et al., 1999a), suggesting a potential role of ERβ in the termination of lordosis.

Although the cycle of T secretion by the testes is not as predictable as ovarian steroid secretion, T is released episodically, sometimes with a circadian rhythm superimposed on higher frequency ultradian pulses (e.g., a three-hour T pulse approximately twice a day in rats) (Sodersten, Eneroth, & Pettersson, 1983). Testosterone remains at low basal values except for several fleeting surges each day in mice (Coquelin & Desjardins, 1982) and rats (Coquelin & Desjardins, 1982; Ellis & Desjardins, 1982). In Syrian hamsters, T concentrations are elevated for about four hours beginning shortly prior to dark onset (Pieper & Lobocki, 2000). Assessment of androgenic influences on MSB of rodents has, with very few exceptions, involved either daily injections of pure or esterified hormones or implantation of constant-release capsules that generate supraphysiological and/or constantly elevated T concentrations. The episodic ultradian rhythms of T secretion of intact rodents are not mimicked by replacement procedures that chronically elevate circulating hormone concentrations (Damassa, Smith, Tennent, & Davidson, 1977; Arteaga-Silva et al., 2005).

To determine the minimum amount of T necessary to maintain or restore MSB, Park et al. (2007) utilized an infusion system that allowed for the control of the amount and duration of the delivery of T systemically. A daily four-hour infusion of T, which was dissolved in ethanol, allowing for rapid clearance of T in the circulation soon after delivery, maintained MSB in castrated Syrian hamsters (Park et al., 2007). In another study, Piekarski et al. (2009) found that once-daily subcutaneous T (dissolved in ethanol) injections both maintained and restored MSB. Testosterone injected once every four or seven days maintained MSB in most hamsters, despite long intervals between injections during which circulating T was undetectable or well below physiological concentrations (Piekarski et al., 2009). The demonstration that relatively brief, infrequent elevations of T are sufficient to support MSB provides a useful model to assess the neuroendocrine basis of MSB and raises the possibility that infrequent low-dose androgen-replacement protocols may restore sexual behavior to hypogonadal men without inducing some of the negative side effects associated with chronic high-dose treatments.

4. NEURAL CIRCUITRY REGULATING SEXUAL BEHAVIOR

4.1. Chemosensory Inputs to Male Sexual Behavior (MSB)

The reliance of these behaviors on steroid hormones provides an access point both for the neural circuitry of the behaviors and, as we describe later, to discover the molecular mechanisms that activate the behaviors. In male hamsters, chemoinvestigatory behaviors are all but eliminated when the accessory portion of the olfactory bulb or the vomeronasal organ are lesioned (reviewed in Ballard & Wood, 2007). This discovery allowed several laboratories to map motivational circuitry in males (reviewed in Chapter 9, this volume; see Figure 9.1 within that chapter). A comprehensive overview of the neuroanatomical basis of rat MSB is reviewed in Hull, Wood, & McKenna (2006), and the following account highlights a few features relevant to this chapter. The medial preoptic area (mPOA), the posterior dorsal medial nucleus of the amygdala (dmAMG), and the bed nucleus of the stria terminalis (BNST) are all essential areas in this network that regulate MSB. The dmAMG receives projections from the main and accessory olfactory bulbs (Davis, Macrides, Youngs, Schneider, & Rosene, 1978; Lehman & Winans, 1982), and sends its projections to the mPOA, BNST, and hypothalamus (Kevetter & Winans, 1981; Maragos, Newman, Lehman, & Powers, 1989; Gomez & Newman, 1992). Within the neural network, the primary integration site for MSB in most mammalian species is the mPOA (Everitt, 1990; Robertson et al., 1991; Baum & Everitt, 1992; Newman, 1999; Riters & Ball, 1999; Wood & Williams, 2001b; Ferris et al., 2004). The mPOA mediates copulatory behavior by sending and receiving indirect input from every sensory modality, and connects reciprocally with other brain areas known to be important for the expression of MSB, including the dmAMG and BNST. When the mPOA is lesioned, MSB is disrupted (Ginton & Merari, 1977; Leedy & Hart, 1986; Bailhache, Surlemont, & Balthazart, 1993; Hoshina, Takeo, Nakano, Sato, & Sakuma, 1994; Paredes, Tzschentke, & Nakach, 1998; Balthazart, Stamatakis, Bacola, Absil, & Dermon, 2001). Moreover, the mPOA is rich with ARs and ERs in rodents (Simerly, Chang, Muramatsu, & Swanson, 1990; Greco, Edwards, Zumpe, & Clancy, 1998) and in humans (Fernandez-Guasti, Kruijver, Fodor, & Swaab, 2000; Osterlund, Gustafsson, Keller, & Hurd, 2000; Osterlund, Keller, & Hurd, 2000; Hestiantoro & Swaab, 2004; Hrabovszky et al., 2004). Testosterone and E_2 implants into the mPOA facilitate sexual activity in castrated male rodents (Davis & Barfield, 1979; Tang & Sisk, 1991; Nyby, Matochik, & Barfield, 1992; Wood, 1996; Wood & Coolen, 1997; Wood & Williams, 2001a). In addition, the mPOA contains several neurotransmitters and their receptors, which are critical for expression of copulation (Simerly, Gorski, & Swanson, 1986; Canteras, Simerly, & Swanson, 1992) and will be reviewed later in this chapter.

In nonhuman primates, no differences in c-fos protein in the mPOA were noted in males that ejaculated compared with unmated males. In other regions, such as the ventromedial hypothalamus (VMH), arcuate nucleus (ARC), lateral mammillary area, and BNST fewer c-fos-ir neurons were noted in mated males compared to unmated males, (Michael, Clancy, & Zumpe, 1999). These results are contradictory to the rodent literature in which more c-fos-ir neurons are found in sexually experienced males when compared to naïve males. This may reflect taxonomic differences between primates and other species; greater activation in feral species that have constant sensory stimulation of conspecifics vs. domesticated, inbred species; or possibly deactivation of neural activity in these specific areas by mating in male primates.

4.2. Neural Circuitry Regulating Lordosis

The study of lordosis, the reflexive female mating posture permitting penile insertion by the male, has been essential for our understanding of the neuroendocrinology of FSB, and the neural circuitry underlying lordosis has been mapped out (reviewed in detail in Micevych, Eckersell, Brecha, & Holland, 1997; Pfaff, Sakuma, Kow, Lee, & Easton, 2006) (Figure 8.1). The following section is a summary of the neural circuitry underlying lordosis, highlighting the major neural areas crucial for expression of FSB.

Lordosis requires hormonal priming, as described above, and is displayed in response to stimulation of the flanks and perineal region, normally provided by a copulating male. This sensory information ascends in the anterolateral columns of the spinal cord and distributes in certain regions of the medullary reticular formation, lateral vestibular nucleus, and midbrain central gray (MCG). The sensory cues are integrated at the hypothalamic module where estrogen-facilitated neuronal activity regulates the expression of lordosis. On the motor side of the circuit, neurons descending from the midbrain periaqueductal gray activate certain medullary reticular spinal neurons which, in turn, connect with lateral vestibulospinal neurons that control the deep back muscles (lateral longissimus and transversospinalis) that execute lordosis behavior. If the spinal cord is cut at any point where key motor neurons exit, females cannot exhibit lordosis (Pfaff et al., 2006).

At the top of the neural circuit of lordosis, sensory cues are integrated with steroid signals in several brain sites to mediate lordosis: the VMH, MCG, and the mPOA. Electrolytic lesions of the VMH typically reduce the frequency of lordosis (Kennedy, 1964; Yamanouchi, 1980; Clark, Pfeifle, & Edwards, 1981). Implantation of both E_2 and P_4 directly into the VMH is both necessary and sufficient to induce lordosis after ovariectomy (Rubin & Barfield, 1983). Tract tracing in this area has led to the discovery that efferent fibers from the VMH project to the MCG region, and destruction of these fibers leads to a reduction in lordosis (Sakuma & Pfaff, 1979). Destruction of the midbrain ascending ventral noradrenergic bundle (VNAB) completely abolishes lordosis (Hansen, Stanfield, & Everitt, 1980; 1981). Neurons in the MCG receive direct input from the VMH, and the MCG is an important integrative area. The MCG is also a major brain area that

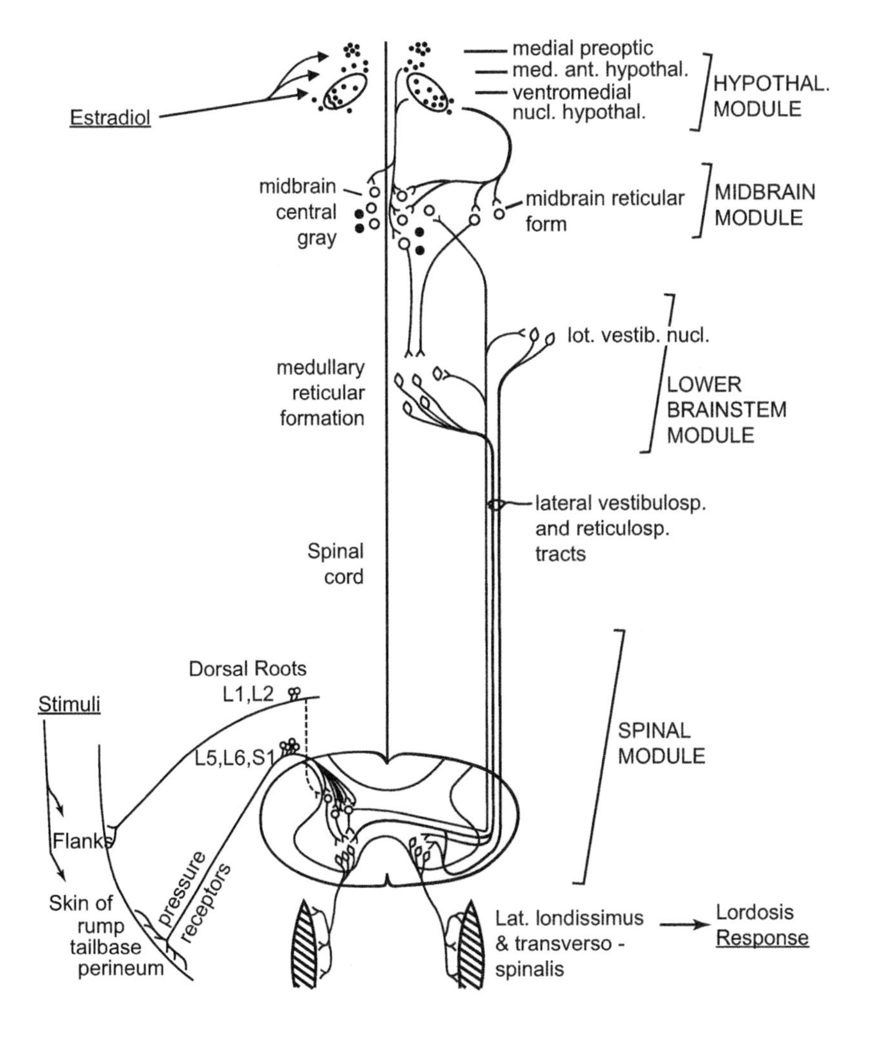

FIGURE 8.1 The neural circuitry of lordosis behavior, separated into structural and functional neural modules. *Reproduced from Lee, Devidze, Pfaff, & Zhou (2006)*

regulates pain sensitivity, anxiety, blood pressure, heart rate, and urination; the MCG will block activity related to competing bodily functions in order for the neural activity relevant to lordosis to occur or will block the propagation of impulses specific to lordosis when appropriate (Bandler & Shipley, 1994).

As discussed in the previous section, the mPOA is an integral site for MSB. In the female, the mPOA is generally considered inhibitory to FSB, but a subset of mPOA neurons may also facilitate some components of FSB. An excitotoxic lesion of the mPOA releases some of the inhibition on lordosis and also diminishes proceptivity (Hoshina et al., 1994), while electrical stimulation of the mPOA disrupts the lordosis reflex without diminishing proceptivity (Takeo, Chiba, & Sakuma, 1993; Hoshina et al., 1994). Neuronal activity in the mPOA implicates different subsets of mPOA neurons in the regulation of individual behavioral components (Hasegawa & Sakuma, 1993). Neurons within the mPOA contain both ERα and ERβ, and local implants of E_2 induce receptivity in OVX female Syrian hamsters (Floody, Blinn, Lisk, & Vomachka, 1987). These results collectively seem to indicate that different components of FSB are regulated separately by different subsets of mPOA neurons, probably under the influence of estrogens.

5. STEROID RECEPTORS AND SEXUAL BEHAVIOR

5.1. Males

Activation of MSB usually involves steroids binding to intracellular receptors. Both AR and ER α and ERβ play various roles in MSB. All three steroid receptors are widely but selectively distributed throughout the central nervous system and are present in areas important for MSB (reviewed in Hull, 2006). Distribution of steroid receptors is widespread within neurons, including the nucleus, cytoplasm, and possibly the cell membrane, allowing steroids to mediate a variety of physiological actions (Blaustein, Lehman, Turcotte, & Greene, 1992; Milner et al., 2005). In most cases, once bound to the appropriate ligand, steroid receptors are translocated to DNA and act as transcription factors to promote synthesis of new proteins, which in turn is assumed to promote MSB (Yahr & Ulibarri, 1987; McGinnis & Kahn, 1997). Further, once in the cell nucleus, receptors bind additional coregulatory proteins that also confer specificity of steroid actions (McKenna, Lanz, & O'Malley, 1999). Steroids also have nongenomic effects that occur more rapidly, and these effects are mediated by membrane receptors that alter ion channel activity and second-messenger systems (reviewed in Kelly, Qiu, & Ronnekleiv, 2003).

The fact that T can be converted to E_2 *in situ* led to the development of the influential 'aromatization hypothesis,' which speculated that, at the receptor level, actions of T in the brain may be mediated largely by ERs (Naftolin, Ryan, Davies, Petro, & Kuhn, 1975). In laboratory rats and mice, exogenous E_2 is sufficient to reinstate most copulatory behavior in ORCH males. Systemic administration of aromatase inhibitors (Beyer, Morali, Naftolin, Larsson, & Perez, 1976; Bonsall, Clancy, & Michael, 1992; Vagell & McGinnis, 1997; Roselli, Cross, Poonyagariyagorn, & Stadelman, 2003) or ER antagonists (Beyer et al., 1976) to castrates inhibited or prevented the subsequent restoration of copulation by T administration. However, E_2 alone is not sufficient to maintain all aspects of MSB, as E_2-treated ORCH rats have greater difficulty achieving the ejaculatory reflex when compared with castrates treated with T or with a combination of E_2 and DHT (reviewed in Hull, 2006). In a number of species, DHT alone is sufficient to maintain or restore male copulatory behavior in rabbits (Agmo & Sodersten, 1975), guinea pigs (*Cavia porcellus*) (Alsum & Goy, 1974; Butera & Czaja, 1989), deer mice (*Peromyscus maniculatus*) (Clemens & Pomerantz, 1982), rhesus monkeys (Phoenix, 1974; Michael, Zumpe, & Bonsall, 1986), and laboratory mice (Luttge & Hall, 1973); although, in ORCH male rats, DHT alone was not sufficient to restore or maintain copulation (McDonald et al., 1970; Whalen & Luttge, 1971; Beyer, Larsson, Perez-Palacios, & Morali, 1973). In Syrian hamsters, comparable doses of DHT alone were sufficient to maintain MSB (Powers, Bergondy, & Matochik, 1985; Romeo, Cook-Wiens, Richardson, & Sisk, 2001), but not sufficient to restore MSB after long-term ORCH. These species- and strain-specific differences underscore the fact that model organisms may not yield all the answers.

Orchidectomy leads to a 10-fold decrease in the expression of AR and a 50% decrease in both ERα and ERβ mRNA levels; treatment with T propionate or DHT propionate restores AR to precastration levels and treatment with estradiol benzoate increases ERα and ERβ levels back to those of control mice (Arteaga-Silva, Rodriguez-Dorantes, Baig, & Morales-Montor, 2007). In addition, the density of steroid receptor coactivators varies between discrete brain regions and is influenced by circulating concentrations of steroids. Thus, steroids can contribute not only to functional specialization of brain areas, but also to cell-specificity of steroid receptor-mediated signaling in the brain (Meijer, Steenbergen, & De Kloet, 2000; Mitev, Wolf, Almeida, & Patchev, 2003). Recent studies have shown that nuclear steroid receptors can become transcriptionally active in the absence of steroid hormones (Denner, Weigel, Maxwell, Schrader, & O'Malley, 1990; Power, Mani, Codina, Conneely, & O'Malley, 1991; Auger, 2001; 2004), and these studies will be reviewed in more depth in a later section.

Work with various genetically engineered mice has helped establish the functions of AR, ERα, and ERβ in MSB. Androgen receptor knockout (ARKO) mice show less masculine sexual behavior than their WT control counterparts, suggesting a role for ARs in MSB (Sato et al., 2004). However, these mice have female-like external genitalia that certainly influence their behavior. Male mice with a testicular feminization mutation (Tfm), caused by a spontaneous mutation in the AR gene, also display impaired MSB similar to ARKO mice (Ono, Geller, & Lai, 1974), but the impairment in mounts and thrusts is reversed after E_2 replacement in adulthood (Bodo & Rissman, 2007). This finding indicates that AR activation during development is of little importance for the expression of MSB in mice, or that sufficient ER stimulation can overcome a lack of AR activation. Interestingly, masculinization of other sex-related behaviors, such as partner preference in mice, does appear to rely on the presence of a functional AR during development, as Tfm males act like WT females and show no partner preferences for females or their odors (Bodo & Rissman, 2007). Unlike in Tfm mice, partner preference is masculinized in Tfm male rats (Sherins & Bardin, 1971; Bardin, Bullock, Sherins, Mowszowicz, & Blackburn, 1973; Naess et al., 1976; Charest et al., 1991; Hamson, Csupity, Ali, & Watson, 2005).

Estrogen receptor-α plays a more substantial role in MSB in mice than ERβ. Male mice lacking the ERα (ERα-knockout (ERαKO) mice) mount occasionally, but perform few mounts with intromissions and rarely display ejaculations (Ogawa, Lubahn, Korach, & Pfaff, 1997; Wersinger et al., 1997; Ogawa et al., 1998b; Dominguez-Salazar, Bateman, & Rissman, 2004). Male ERαKO mice have higher levels of T than WT mice, yet castrated ERαKO males treated with either T (Wersinger & Rissman, 2000) or DHT (Ogawa et al., 1998b) do not display ejaculations. Systemic administration with both T and apomorphine, a DA agonist, reinstates MSB in ERαKO mice (Wersinger & Rissman, 2000); however, when ERαKO mice are treated with apomorphine via intracerebroventricular infusion, other components of MSB (but not ejaculations) are restored (Burns-Cusato, Scordalakes, & Rissman, 2004). Male mice with the aromatase enzyme knockout (ArKO) do not display much sexual behavior in adulthood (Sato et al., 2004). However, either E_2 treatment during the early neonatal period or prior to testing in adulthood can restore much of the MSB (Bakker, Honda, Harada, & Balthazart, 2004; Toda et al., 2001). The role of ERβ in MSB is less substantial than ERα, as there were no deficits in adult mating behavior of ERβ-knockout mice (Ogawa et al., 1999a). However, these males have delayed behavioral puberty compared with WT littermates, as indicated by the attainment of their first ejaculation at an older age (Temple, Scordalakes, Bodo, Gustafsson, & Rissman, 2003).

5.2. Females

The effects of E_2 and P_4 on FSB are largely mediated by the genomic action of their receptors, and estrogens regulate the expression of these receptors. Estrogenic facilitation of the lordosis response requires new transcription and protein synthesis. Protein synthesis inhibitors infused directly into the hypothalamus eliminate lordosis in female rats (Meisel & Pfaff, 1984). The distribution of ERα is fairly limited in the brain; it is present in the POA, hypothalamus, some components of the limbic system, and at lower levels in the midbrain, hindbrain, and spinal cord. The distribution is highly conserved phylogenically, with a few caveats. When the two receptors are colocalized within the same cell, they can dimerize *in vitro*, suggesting yet another way in which they can interact (Pettersson, Grandien, Kuiper, & Gustafsson, 1997).

In the mouse, ERα, not ERβ, is required for the expression of FSB (Kudwa & Rissman, 2003). Studies on female ERαKO mice have clearly demonstrated that receptivity is inhibited when the ERα gene is not functional (Ogawa et al., 1998a; Rissman, Early, Taylor, Korach, & Lubahn, 1997; reviewed in Rissman, Wersinger, Taylor, & Lubahn, 1997). Recent work with ligand-specific agonists for ERα and ERβ shows that, in rats, the activation of ERα, not ERβ, also is essential for lordosis (Mazzucco, Walker, Pawluski, Lieblich, & Galea, 2008). In addition, use of siRNA, delivered to the VMH, to selectively knock down ERα in the VMH also depresses lordosis (Musatov, Chen, Pfaff, Kaplitt, & Ogawa, 2006). In contrast to female ERαKO mice, female mice lacking functional ERβ receptors exhibit normal lordosis behaviors, demonstrating that ERβ receptors are not necessary for the expression of FSB (Krege et al., 1998; Ogawa et al., 1999b). Ovariectomized female mice primed with E_2 and P_4 that had both the ERα and ERβ genes disrupted failed to display lordosis (Kudwa & Rissman, 2003). Ovariectomized ERβKO mice hormonally primed may be qualitatively even more receptive than their WT counterparts, and the length of time in estrus was extended (Kudwa & Rissman, 2003).

The ERα primes for the induction of P_4 receptor (PR) in the VMH (Kudwa & Rissman, 2003). Female mice that lack the functional P_4 receptor knockout (PRKO), like ERαKO mice, are infertile (Lydon, DeMayo, Conneely, & O'Malley, 1996). Ovariectomized PRKO mice do not display lordosis after treatment with E_2 alone or E_2 with P_4 (Lydon et al., 1996). This work correlates with a previous study that demonstrated that giving an infusion of PR antisense into the VMH of E_2-primed female rats led to a significant decrease in the expression of both lordosis behavior and proceptive behavior (Ogawa, Olazabal, Parhar, & Pfaff, 1994).

Of course, ERs induce transcription of genes other than PR and certainly many of these must be involved in sexual behavior. Estradiol needs to be given many hours in advance of sexual behavior and its genomic actions, via the ERs, have been emphasized. The cascade hypothesis (reviewed in Pfaff et al., 2006) suggests that specific and discrete transcriptional and translational events occur within the neurons in the VMH as a result of initial ER binding, and that these events are required for later estrogen-dependent and -independent events to occur. Continuous ER occupation is not required, but, rather, estrogens must occupy receptors at specific critical times in order for lordosis to be expressed.

In order to investigate the effects of estrogens on gene expression in female mice, microarray is beginning to be used (Mong et al., 2003a). One gene differentially regulated by E_2 treatment is lipocalin-type prostaglandin D synthase (L-PGDS); mRNA levels were upregulated in the basomedial hypothalamus whereas they were downregulated in the mPOA. Lipocalin-type prostaglandin D synthase catalyzes the conversion of prostaglandin H (PGH_2) to prostoglandin D_2 (PGD_2) a neuromodulator involved in a variety of functions, including sleep, pain, and odor responses. The decreased levels of L-PGDS mRNA in the mPOA suggested a role in the regulation of arousal, since the mPOA is a suspected site of action for the sleep-promoting effects of PGD_2 (Ueno, Ishikawa, Nakayama, & Hayaishi, 1982). When L-PGDS mRNA function was knocked down in the mPOA, activity levels were increased; moreover, lordosis behavior was observed in OVX mice that were not E_2-primed (Mong, Devidze, Goodwillie, & Pfaff, 2003). These results demonstrated that reduction of L-PGDS in the mPOA can mimic the action of estrogens, contributing to the general arousal that may lead to successful expression of FSB. More recently, Jasnow, Mong, Romeo, and Pfaff (2007) used microarray to assess rapid (two-hour) vs. longer-term (24-hour) effects of a single E_2 treatment on mRNA in the amygdalae of female mice. Two hours after E_2 treatment, 44 mRNAs were significantly up- or downregulated and at 24 hours only 13 transcripts were affected by E_2. Another interesting gene, CaMKIIα, was examined with rtPCR and western blots and was confirmed to decrease in message and protein two hours but not 24 hours after E_2 administration (Jasnow et al., 2007). This approach is just beginning to be used. The more candidate genes identified, the more likely we are to decipher the genetic network that underlies these behaviors.

6. LIGAND-INDEPENDENT ACTIVATION OF STEROID RECEPTORS

Interestingly, both ERs and PRs can be activated in the absence of their respective ligands, and there is clear evidence that there is 'crosstalk' between neurotransmitters and steroid receptors. The first report of ligand-independent activation of steroid hormone receptors found that both the PRs and the ERs could be activated *in vitro* by activation of dopamine (DA) release (Power, Lydon, Conneely, & O'Malley, 1991). A DA D1 receptor agonist, but not a DA D2 receptor agonist, mimicked the effects of P_4 in facilitating FSB in rats (Mani, Allen, Clark, Blaustein, & O'Malley, 1995). The facilitatory effects of DA were blocked by a PR antagonist (Mani et al., 1995; Auger, Moffatt, & Blaustein, 1996). The D5 receptor is present in the female rat VMH, and has been implicated in FSB (Mani et al., 2000). In the D5 receptor KO mouse, lordosis persists when E_2 and P_4 are given but the general DA agonist apomorphine was not able to induce lordosis in the D5KO as it did in the wild type mouse (Kudwa, Dominguez-Salazar, Cabrera, Sibley, & Rissman, 2005).

Growth factors, such as epidermal growth factor (EGF) and insulin-like growth factor-1 (IGF-1) (Ignar-Trowbridge et al., 1992; Aronica & Katzenellenbogen, 1993; Klotz et al., 2002); protein and peptide hormones (Waring & Turgeon, 1992; Ma et al., 1994b; Demay, De Monti, Tiffoche, Vaillant, & Thieulant, 2001); neurotransmitter agonists (Power et al., 1991b); and numerous activators of particular intracellular signaling pathways (Aronica & Katzenellenbogen, 1993; Cho & Katzenellenbogen, 1993; Joel, Traish, & Lannigan, 1995; Kato et al., 1995; Bunone, Briand, Miksicek, & Picard, 1996; Patrone et al., 1996; Trowbridge, Rogatsky, & Garabedian, 1997; Martin et al., 2000; Schreihofer, Resnick, Lin, & Shupnik, 2001) activate either PRs and/or ERs *in vitro*. There are a few studies that have investigated the relationship between estrogens and growth factors in FSB. Growth factors such as EGF (reviewed in Plata-Salaman, 1991; Ma, Dissen, Merlino, Coquelin, & Ojeda, 1994), IGF-1 (Garcia-Sergura, Perez, Pons, Rejas, & Torres-Aleman, 1991; Etgen, Gonzalez-Flores, & Todd, 2006), and other growth factor ligands (Kaser, Lakshmanan, & Fisher, 1992; reviewed in Pfaff, Schwartz-Giblin, McCarthy, & Kow, 1994) are widely distributed in the mammalian central nervous system, including the cerebral cortex, hippocampus, substantia nigra, and mPOA.

Crosstalk between ERs and IGF-1 receptors (IGF-1Rs) mediates changes in the brain and in peripheral reproductive tissues. ERα, ERβ, and IGF-1R are colocalized in neurons and glia in various brain regions, including the hypothalamus and mPOA (Cardona-Gomez, DonCarlos, & Garcia-Segura, 2000), and E_2 and IGF-1 can act independently or synergistically to stimulate signaling pathways in the brain, including mitogen-activated protein kinase (MAPK) and phosphatidylinositol-3-kinase (PI3K) (reviewed in Daftary & Gore, 2005). Although there are no reports of IGF-1 independently activating ERs in the absence of estrogens *in vivo*, there are data suggesting that the combined activation of these downstream signaling pathways underlies the neuroendocrine regulation of FSB, as evidenced by abolished E_2 facilitation of lordosis after intracerebroventricular (ICV) infusions of both MAPK and PI3K inhibitors (Etgen & Acosta-Martinez, 2003).

There is also support for ligand-independent activation of steroid hormone receptors that occurs *in vivo* (Mani, Blaustein, & O'Malley, 1997; Auger, 2001; Klotz et al., 2002; Blaustein, 2003; Ciana et al., 2003). In particular, ER-dependent transcription may be activated *in vivo* in a ligand-independent manner by EGF to produce short latency (one to four hours) facilitation of female reproductive behavior in OVX rats that have not been primed with E_2 (Apostolakis, Garai, Lohmann, Clark, & O'Malley, 2000). Clearly, steroid hormone receptors can be activated by a variety of signaling pathways, including neurotransmitters and intracellular signaling systems, through a process that does not require binding to cognate ligands.

7. RELEVANT NEUROTRANSMITTERS AND NEUROPEPTIDES

7.1. Females

Both classical neurotransmitter systems and some novel neuropeptides work along with steroid hormones to regulate sexual behaviors (Tables 8.1 and 8.2). However, the

TABLE 8.1 Effects of neurotransmitters and neuropeptides on male sexual behavior (MSB). + indicates facilitation of MSB; − indicates inhibition of MSB; +/− indicates agent may have either facilitory or inhibitory effects on MSB

Agent (receptor type)	Effects on male sexual behavior (MSB)	Remarks	References
α-melanocyte-stimulating hormone	+	Increases erections	Argiolas et al., 2000; Giuliano, 2004
Acetylcholine	+/−	Increases number of ejaculations	Maeda et al., 1990; Zarrindast et al., 1994
Corticotropin-releasing factor	−	Inhibitory	Sirinathsinghji, 1987
Dopamine	+	Facilitation in both gonad-intact and castrated rats	Malmnas, 1976; Scaletta & Hull, 1990; Wersinger & Rissman, 2000; also reviewed in Hull & Dominguez, 2006
Gamma-aminobutryic acid	−	Impaired copulation	Agmo et al., 1997
Gonadotropin- releasing hormone	+/−	Inconsistent results	Moss et al., 1975; Myers & Baum, 1980
Nitric oxide	+	Generally facilitory	Benelli et al., 1995; Giuliano et al., 2003; Hull et al., 1994
Norepinephrine	+/−	Inconsistent results	Clark et al., 1984; Smith et al., 1987a; 1987b; Tallentire et al., 1996
Opiates	+/−	Mixed, dose-dependent effects	Agmo & Paredes, 1988; Gomez-Marrero et al., 1988; Pfaus & Wilkins, 1995; Leyton & Stewart, 1996
Oxytocin	+/−	Mixed, dose-dependent effects on MSB	Witt, 1995
Prolactin	+/−	Inhibitory; inconsistent results	Doherty et al., 1986; Shrenker & Bartke, 1987
Serotonin (5-HT1A)	+/−	Facilitation of ejaculation in rats, inhibititory in mice	Ahlenius & Larsson, 1984b; Fernandez-Guasti et al., 1992; Matuszewich et al., 1999; Popova & Amstislavskaya, 2002; Rodriguez-Manzo et al., 2002
Serotonin (5-HT1B)	−	Inhibitory in both rats and mice	Hillegaart & Ahlenius, 1998; Rodriguez-Manzo et al., 2002; also reviewed in Seagraves, 1990; Rosen et al., 1999;
Serotonin (5-HT2C)	+/−	Facilitation of erection and inhibition of ejaculation	Foreman et al., 1989; Pomerantz et al., 1993
Vasopressin	+	Increases erections	Gozes et al., 1989

TABLE 8.2 Effects of neurotransmitters and neuropeptides on female sexual behavior (FSB). + indicates facilitation of FSB; − indicates inhibition of FSB; +/− indicates agent may have either facilitory or inhibitory effects on FSB

Agent (receptor type)	Effects on female sexual behavior (FSB)	References
β-endorphin	−	Wiesner & Moss, 1984
α-melanocyte-stimulating hormone	+	Shadiack et al., 2007; Pfaus et al., 2004
Acetylcholine	+	Clemens et al., 1981; Dohanich & Clemens, 1981
Cholecystokinin	+/−	Bloch et al., 1987
Corticotropin-releasing factor	−	Sirinathsinghji et al., 1983
Dopamine	+/−	Everitt et al., 1974
Gamma-aminobutryic acid	+	McGinnis et al., 1980; McCarthy et al., 1990; 1991
Glutamate	−	Kow et al., 1985
Gonadotropin-releasing hormone-I	+	Moss & McCann, 1973; Pfaff, 1973; Moss & Foreman, 1976; Kendrick & Dixson, 1985; Meredith & Howard, 1992; Fernandez-Fewell & Meredith, 1995
Gonadotropin-releasing hormone-II	+	Temple et al., 2003a
Neuropeptide Y	−	Clark et al., 1985
Norepinephrine (α_1)	+	Foreman & Moss, 1978; Crowley, 1986; Etgen et al., 1992; Kalra, 1993; Herbison, 1997; Etgen et al., 2001; reviewed in Meyerson et al., 1985
Oxytocin	+	Caldwell, 1992; Pedersen & Boccia, 2006; Baskerville & Douglas, 2008; Neumann, 2008
Prolactin	+	Harlan et al., 1983
Serotonin (5-HT_{1A})	−	Reviewed in Mendelson, 1992
Serotonin (5-HT_2)	+	Reviewed in Gorzalka et al., 1990; Mendelson, 1992
Substance P	+	Dornan et al., 1987

Modified from Kow, Paden, and Pfaff (1981).

way in which sex steroids act on these systems is only partly understood (Pfaff et al., 2006; reviewed in Fabre-Nys, 1998). As mentioned previously, DA may effect some of its actions via the PR, but most of the neurotransmitters carry out their actions on their own receptor subtypes. Generally, neurotransmitters that enhance neuronal excitation are increased in the brain by steroids and increase sexual behavior. For lordosis, examples include norepinephrine (NE), DA, acetylcholine (ACh), and oxytocin (OXY). Conversely, neurotransmitters that inhibit neuronal excitation, such as opiates and serotonin (5-HT), are inhibitory to lordosis. However, many neurotransmitter systems have dual or mixed effects on sexual behavior.

7.1.1. Norepinephrine (NE)

The monoamine neurotransmitter NE has been implicated as a neurochemical mediator of both lordosis behavior and the preovulatory luteinizing hormone (LH) surge (Crowley, 1986; Etgen, Ungar, & Petitti, 1992; Kalra, 1993; Herbison, 1997; Etgen, Ansonoff, & Quesada, 2001). Direct infusion of NE to the mPOA or the VMH can activate lordosis in E_2-treated female rats (Foreman & Moss, 1978; Fernandez-Guasti, Larsson, & Beyer, 1985). Administration of aDA β-hydroxylase inhibitor, which leads to decreased levels of NE, results in decreased expression of lordosis in female guinea pigs (Nock & Feder, 1979). Depletion of hypothalamic NE by lesions of the ventral NE bundle abolishes the expression of lordosis behavior in E_2 and P_4-primed female rats (Hansen et al., 1980). In addition, E_2 and P_4 can influence the strength of NE synaptic communication by regulating NE release in the VMH (reviewed in Etgen, Chu, Fiber, Karkanias, & Morales, 1999).

The NE-mediated facilitation of FSB is mediated specifically through the α_1-adrenergic receptors and not the α_2-adrenergic. Pretreatment with phenoxybenzamine, a specific α_1-adrenergic antagonist, results in decreased expression of lordosis in OVX E_2-treated female guinea pigs (Nock & Feder, 1984). The NE-mediated facilitation of FSB can be blocked by systemic administration of prazosin, a α_1-adrenergic antagonist, but not by antagonists for the α_2-adrenergic receptor (Fernandez-Guasti et al., 1985; Etgen, Vathy, Petitti, Ungar, & Karkanias, 1990). In addition, direct infusion of selective α_1-adrenergic agonists, but not an α_2-adrenergic agonist, facilitated lordosis (Kow, Weesner, & Pfaff, 1992). In agreement with the results that facilitation of FSB is mediated through NE are electrophysiological studies of hypothalamic tissue slices that reveal activation of α_1-adrenergic receptors in the VMH leading to excitation of neuronal activity, whereas activation of α_2-adrenergic receptors leads to inhibition (Kow et al., 1992). Converging lines of evidence strongly suggest that NE released from the ventral noradrenergic bundle excites neurons in the VMH by binding specifically to

α_1-adrenergic receptors, which in turn leads to the facilitation of steroid-primed FSB.

7.1.2. Acetylcholine (ACh)

ACh plays an important role in estrogen-facilitated lordosis behavior. Direct bilateral infusions of a cholinergic agonist, carbachol, into the mPOA or VMH facilitates lordosis in OVX rats treated with E_2 benzoate alone as soon as 15 minutes after infusion (Clemens, Dohanich, & Witcher, 1981; Dohanich & Clemens, 1981). Infusion of carbachol into the mesencephalic reticular formation or frontal cortex failed to activate lordosis, suggesting that the lordosis effect of cholinergic stimulation may be related to the ability of brain areas to concentrate estrogens. Infusions of carbachol into the midbrain central gray in OVX E_2-treated female rats also facilitated FSB, indicating that some aspects of the cholinergically facilitated hypothalamic outputs are routed through the midbrain central gray (Richmond & Clemens, 1986). Lastly, administration of either the cholinergic antagonist atropine sulfate or the ACh synthesis inhibitor hemicholinium leads to a decline in lordosis behavior (Clemens & Dohanich, 1980).

7.1.3. Gonadotropin-releasing hormone (GnRH)

Gonadotropin-releasing hormone (GnRH) is a hypothalamic decapeptide that promotes the secretion of LH and FSH, which in turn stimulate steroid production by the gonads (see Chapter 2, this volume). Clearly, via this indirect role, GnRH is involved in facilitating mammalian reproductive behavior. In addition to terminals in the median eminence, GnRH is released in many brain regions, where it has neurotransmitter actions. Via these direct neural actions, GnRH stimulates reproductive behavior in several species (Moss & McCann, 1973; Pfaff, 1973; Moss & Foreman, 1976; Kendrick & Dixson, 1985; Fernandez- Meredith & Howard, 1992; Fewell & Meredith, 1995). In rats, GnRH facilitates lordosis behavior (Moss & McCann, 1973; Pfaff, 1973). Of interest, a metabolite of GnRH-I, GnRH1-5, can enhance lordosis despite the fact that it cannot bind to the GnRH receptor (Roberts, Mani, Woller, Glucksman, & Wu, 2007). In ewes, GnRH can extend the duration of receptivity and a GnRH antagonist can reduce sexual behavior when given along with E_2 and P_4 (Fabre-Nys & Gelez, 2007). In OVX mice treated with E_2 and P_4, GnRH causes a small enhancement of lordosis (Kauffman, Buenzle, Fraley, & Rissman, 2005).

Gonadotropin-releasing hormone-II, has seven of ten amino acids in common with GnRH-I, and is present in the brains of all representative vertebrate classes that have been studied (Millar, Pawson, Morgan, Rissman, & Lu, 2008). In mammals, the majority of GnRH-II cell bodies reside in the midbrain, with few cells present in hypothalamic and extrahypothalamic regions (Dellovade, King, Millar, & Rissman, 1993; Kasten et al., 1996; White, Eisen, Kasten, & Fernald, 1998). In the musk shrew, the majority of GnRH-II terminals are present in the medial habenula (Rissman, Alones, Craig-Veit, & Millam, 1995). Temple, Millar, and Rissman (2003) proposed that GnRH-II, acting as a neurotransmitter, regulates reproduction according to a female's energetic status. Reproduction is an energetically costly process in female mammals. Female musk shrews that are food-restricted for 48 hours exhibit a significant decline in mating behaviors; interestingly, central administration of GnRH-II, but not GnRH-I, rapidly reverses the inhibitory effects of food restriction on female musk shrew sexual behavior. Gonadotropin-releasing hormone-II also promotes sexual behavior in underfed female mice. Pretreatment with the type 1 GnRH antagonist, Antide, did not prevent GnRH-II from promoting mating behavior (Kauffman & Rissman, 2004; Kauffman et al., 2005). These results suggest that GnRH-II's behavioral actions are mediated through the type 2 GnRH receptor.

7.1.4. Opiates

Lordosis behavior is suppressed in OVX E_2-primed female rats after intraventricular infusion of β-endorphin (Wiesner & Moss, 1984). Because opioid peptides regulate a variety of behavioral functions, studies were conducted to ensure that the suppression of FSB was due to direct action of β-endorphin rather than potentially confounding generalized behavioral suppressions (Wiesner & Moss, 1986a; 1986b). Direct infusion of naloxone, a μ-opioid antagonist, to the MCG resulted in the facilitation of FSB (Sirinathsinghji, 1984). Opiates also act at the spinal level of the female mating circuitry, as intrathecal injections of morphine in OVX E_2-primed female rats inhibited the lordosis response whereas treatment with naloxone facilitated the lordosis response (Wiesenfeld-Hallin & Sodersten, 1984).

7.2. Neurotransmitters that Act in Both Sexes

There are a few agents, other than steroids, that can affect both male and female sexual behavior (Tables 8.1 and 8.2), including DA, 5-HT, OXY, and α-melanocyte-stimulating hormone (MSH). Dopamine is particularly interesting since it works both in the VMH and POA and also is a key regulator of the general reward pathway. Sex is a rewarding behavior, and considering it in this light helps us to outline commonalities between regulation of sex and other addictive behaviors such as drug taking and overeating in humans (Sato, Schulz, Sisk, & Wood, 2008).

7.2.1. Dopamine (DA)

The neurotransmitter DA has been shown to facilitate MSB. The first indication that DA may play a significant role in human sexual behavior resulted from case studies in which patients with Parkinson's disease who were administered L-3, 4-dihydroxyphenylalanine (L-DOPA) often reported increased libido and greater sexual potency (Barbeau, 1969; Bowers, Van Woert, & Davis, 1971). The role of DA in facilitating male sexual function is well documented in rodents (reviewed in Hull, 2006). Three regionally specific groups of DA neurons have been implicated in the regulation of MSB. These include the mesolimbic DA neurons, projecting from the midbrain ventral tegmental area to the nucleus accumbens, the incertohypothalamic DA neurons projecting from the dorsal-posterior hypothalamus to the POA, and the tuberoinfundibular DA neurons projecting from the arcuate nucleus to the median eminence (reviewed in Moore & Lookingland, 1995).

In rats, systemic DA agonists facilitate MSB (Bitran & Hull, 1987; Melis & Argiolas, 1995; Giuliano & Allard, 2001), while DA antagonists block behavior (Ahlenius & Larsson, 1984a; Agmo & Fernandez, 1989; Pfaus & Phillips, 1989; Agmo & Picker, 1990; Ahlenius & Larsson, 1990). Specifically, the nonselective DA agonist apomorphine partially restored copulation in short- and long-term sexually experienced ORCH rats (Malmnas, 1976; Scaletta & Hull, 1990). In addition, the D2 receptor family agonists, 7-hydroxy-2(di-*n* propylamino) tetralin (7-OH-DPAT), 2-amino-4,5,6,7-tetrahydro-6-propylamino-benzothiazol dihydrochloride) (SND 919), and 6-allyl-2-amino-5,6,7,8-tetrahydro-4H-thiazolo[4,5d]azepine dihydrochloride (B-HT 920) facilitate ejaculation in male rats (Ferrari & Giuliani, 1995; 1996; Giuliani & Ferrari, 1996). A DA D1 agonist, SKF 81292, facilitated copulation in DA-deficient mice (Szczypka, Zhou, & Palmiter, 1998); a D2 agonist (quinpirole) inhibited copulation, but this may have been due to locomotor confounds caused by high doses of the agonist. Male ERαKO mice, which usually show little copulatory behavior, copulated normally after receiving systemic injections of apomorphine (Wersinger & Rissman, 2000).

Dopamine release in the mPOA is essential for activation of adult MSB (Hull, Du, Lorrain, & Matuszewich, 1997). Sex steroids modulate DA activity at various levels of DA transmission. Testosterone upregulates nitric oxide synthase (NOS) in the mPOA (Hadeishi & Wood, 1996; Putnam, Sato, Riolo, & Hull, 2005), which in turn increases basal extracellular DA levels (Hanbauer, Wink, Osawa, Edelman, & Gally, 1992) mediating the female-stimulated mPOA DA release essential for male copulation (Sato, Horita, Kurohata, Adachi, & Tsukamoto, 1998). Dopamine neurons are direct targets for estrogens, and estrogens stimulate neurite extension/branching and the expression of tyrosine hydroxylase, a rate-limiting enzyme in DA synthesis (Pasqualini, Olivier, Guibert, Frain, & Leviel, 1995; Kuppers, Ivanova, Karolczak, & Beyer, 2000). In the septum, nonclassical estrogenic actions mediated through membrane receptors coupled to diverse intracellular signaling cascades regulate DA release (Mermelstein, Becker, & Surmeier, 1996).

Dopamine is associated with rewarding aspects of FSB. One way to assess reward is by conditioned place preferences (CPP) associated with mating. In female rats, paced mating-induced CPP is correlated with DA released in the nucleus accumbens (Jenkins & Becker, 2003). In female Syrian hamsters, D2 blockade can inhibit CPP associated with mating (Meisel, Joppa, & Rowe, 1996). Females treated with 6-hydroxy-dopamine to selectively lesion the DA inputs to the nucleus accumbens do not show the experience-facilitated increase in receptivity seen in controls (Meisel & Mullins, 2006).

As mentioned previously, DA can activate steroid receptors in a ligand-independent manner and influence FSB. In the PRKO, female mice without a functional PR do not display receptivity when E_2 treatment is combined with either P_4 or DA agonists (Mani, Reyna, Chen, Mulac-Jericevic, & Conneely, 2006; Mani, Portillo, & Reyna, 2009). In a set of studies using rats, mice, and PR knockout mice, Mani and colleagues have shown that DA affects both phosphatases and kinases. Thus 'crosstalk' may be caused by convergence of signaling pathways within cells containing steroid receptors, and this in turn will likely turn out to be a common theme for steroid receptor nongenomic function.

7.2.2. Serotonin (5-HT)

Serotonin is generally inhibitory to sexual behavior. Nine groups of 5-HT neurons (B1–9) are located in the medulla, pons, and midbrain, and axons from these nuclei ascend and descend to nearly every part of the brain and spinal cord. Men who take selective serotonin reuptake inhibitors (SSRI) commonly report decreased libido and difficulty achieving orgasm or ejaculation as a side effect (reviewed in Seagraves, 1990; Rosen, Lane, & Menza, 1999). The physiological mechanisms that mediate these effects are unclear, but they may include increased prolactin (PRL) secretion, anticholinergic effects, decreased nitric oxide production (Rosen et al., 1999), or decreased penile sensitivity (Yilmaz, Tatlisen, Turan, Arman, & Ekmekcioglu, 1999).

Administration of the SSRI fluoxetine to male rats (Taylor et al., 1996; Cantor, Binik, & Pfaus, 1999; Frank, Hendricks, & Olson, 2000) and clomipramine, a 5-HT reuptake inhibitor, to hamsters (Boscarino & Parfitt, 2002) resulted in decreased sexual motivation and ejaculation. Injections of the 5-HT precursor, 5-hydroxytryptophan, increased intromission and ejaculation latencies and the number of intromissions in rats (Ahlenius & Larsson, 1991). Complementary to these studies, p-chlorophenylalanine, a drug that depletes 5-HT, or lesions of the medial raphe nucleus, a major source of 5-HT in the brain, facilitate MSB (Salis & Dewsbury, 1971; Albinsson, Andersson, Andersson, Vega-Matuszczyk, & Larsson, 1996).

There are several subtypes of 5-HT receptors, and, interestingly, activation of different 5-HT receptor subtypes leads to either the facilitation or inhibition of MSB and also seems to be species-dependent (see Table 8.1). As expected, activation of most of the 5-HT receptor subtypes results in deficits of MSB. In rats, the 5-HT_{1B} agonist anpirtoline inhibited ejaculation, and its effects were blocked by selective 5-HT_{1B} antagonists (Hillegaart & Ahlenius, 1998). 5-HT_{1B} knockout mice were less debilitated by several serotonergic drugs than were WT mice, again suggesting an inhibitory effect of the 5-HT_{1B} receptor (Rodriguez-Manzo, Lopez-Rubalcava, Hen, & Fernandez-Guasti, 2002).

Contrary to 5-HT's mainly inhibitory effects on sexual behavior, stimulation of one receptor subtype, the 5-HT_{1A} receptor, facilitates ejaculation. Administration of 8-hydroxy-2-(di-n propylamino) tetralin (8-OH-DPAT), a 5-HT_{1A} agonist, decreased both the time and number of intromissions preceding ejaculation in rats (Ahlenius & Larsson, 1984b). The facilitative effects of 8-OH-DPAT may be mediated in part through increasing extracellular DA in the mPOA (Lorrain, Matuszewich, & Hull, 1998). Postsynaptic receptors, rather than inhibition of 5-HT release, seem to mediate the effects of 8-OH-DPAT. Lesions of serotonergic nuclei did not alter the effects of 8-OH-DPAT (Fernandez-Guasti & Escalante, 1991). Injections of 8-OH-DPAT into the mPOA and nucleus accumbens, areas where all 5-HT_{1A} receptors are postsynaptic, facilitated MSB (Fernandez-Guasti, Escalante, Ahlenius, Hillegaart, & Larsson, 1992; Matuszewich et al., 1999). The facilitative effects of 8-OH-DPAT are dependent upon normal levels of T and on sexual experience. In castrates with subnormal T replacement, 8-OH-DPAT did not facilitate MSB, and, in those with threshold T concentrations, only sexually experienced males copulated after 8-OH-DPAT administration (Rowland & Houtsmuller, 1998). In summary, while activation of 5-HT_{1A} facilitates ejaculation, activation of 5-HT_{1B} receptors leads to the inhibition of ejaculation.

Interestingly, two studies in mice reported not only inhibitory effects of 5-HT_{1B} agonists, but also of the 5-HT_{1A} agonist (8-OH-DPAT), which is contrary to studies in rats and is another example of species-specific effects of 8-OH-DPAT (Popova & Amstislavskaya, 2002; Rodriguez-Manzo et al., 2002). Similarly to 5-HT_{2B} receptors, activation of 5-HT_{2C} receptors leads to inhibition of ejaculation; however, activation of 5-HT_{2C} also may lead to facilitation of erection, as shown in rhesus monkeys (Pomerantz, Hepner, & Wertz, 1993). Lastly, 5-HT_2 receptors also inhibit copulation in rats (Foreman, Hall, & Love, 1989).

In females, 5-HT can either facilitate or inhibit FSB depending on which receptor subtype is activated (reviewed in Gorzalka, Mendelson, & Watson, 1990; Mendelson, 1992). Facilitation of FSB through 5-HT is mediated mainly by the 5-HT_2 receptors (Hunter, Hole, & Wilson, 1985; Mendelson & Gorzalka, 1985; Wilson & Hunter, 1985; Mendelson & Gorzalka, 1986). The inhibitory effect of 5-HT on lordosis is mediated by the postsynaptic 5-HT_{1A} (Mendelson & Pfaus, 1989). Treatment with a 5-HT_{1A} agonist systemically or directly infused into the VMH led to the inhibition of lordosis (Uphouse, Montanez, Richards-Hill, Caldarola-Pastuszka, & Droge, 1991). When treated with 5-HT precursors, lordosis is inhibited in female rats (Meyerson, 1964; 1975), while SSRIs decrease FSB in female rats (Everitt & Fuxe, 1977). Conversely, FSB is facilitated after inhibition of 5-HT synthesis (Zemlan, Ward, Crowley, & Margules, 1973) and treatment with 5-HT antagonists (Ward, Crowley, & Zemlan, 1975). Lesions of 5-HT fibers in the hypothalamus led to the facilitation of lordosis behavior in E_2-treated female rats (Luine, Frankfurt, Rainbow, Biegon, & Azmitia, 1983). In contrast, transplantation of fetal cells containing 5-HT into the hypothalamus led to the inhibition of FSB (Luine, Renner, Frankfurt, & Azmitia, 1984).

7.2.3. Oxytocin (OXY)

Oxytocin, the neurohypophysial peptide well known for its hormonal role in lactation and parturition (see Chapter 6, this volume), is present in both females and males and is thought to be involved in several functions, such as memory, learning, and affiliative and sociosexual behaviors including penile erection and copulation (reviewed in Argiolas & Melis, 2004; Baskerville & Douglas, 2008). Oxytocin facilitates erectile function and male sexual behavior in mice, rats, rabbits, and monkeys, and possibly in humans (reviewed in Argiolas & Melis, 2004). Central administration of OXY-induced penile erection (Argiolas, Melis, & Gessa, 1985) and improved copulatory behavior, as evidenced by shorter ejaculation and postejaculatory interval latencies (Arletti, Bazzani, Castelli, & Bertolini, 1985) in male rats. The proerectile effect of OXY is T-dependent; it is abolished by castration and restored with exogenous T treatment (reviewed in Argiolas & Melis, 1995). Oxytocin also improves sexual behavior in dominant, but not in subordinate, male squirrel monkeys (*Saimiri sciureus*), and, when given an OXY antagonist, the facilitative effects were prevented (Winslow & Insel, 1991). In contrast, OXY is reported to reduce sexual activity in the male prairie vole (*Microtus ochrogaster*) (Mahalati, Okanoya, Witt, & Carter, 1991).

In many female rodent species, OXY has actions on anxiety, social memory, social bonding, and maternal behavior and thus its role in sexual behavior may in part be mediated by other processes (Baskerville & Douglas, 2008; Neumann, 2008). Social interaction alone induces OXY release within various brain regions (Landgraf & Neumann, 2004) and mice lacking OXY or its receptor display impaired social and affiliative behavior (Takayanagi et al., 2005), further supporting the idea of a general permissive role for OXY in FSB. When given with E_2 and P_4, OXY enhances proceptive behavior and lordosis in rats (Caldwell, 1992; Pedersen & Boccia, 2006). Estrogens can induce an increase in the synthesis of OXY (Robinson, 1974; Robinson, Haluszczak, Wilkins, Huellmantel, & Watson, 1977), and the promoter of the OXY gene has estrogen response elements (Richard & Zingg, 1990). Oxytocin mRNA levels increase in response to elevated estrogen concentrations, as evidenced by increased levels of OXY mRNA levels during the estrous phase of the female rat's cycle (Van Tol, Van den Buuse, De Jong, & Burbach, 1988). In both the female rat and mouse, ERβ is colocalized with OXY, producing neurons in the paraventricular nucleus (PVN) of the hypothalamus (Patisaul, Scordalakes, Young, & Rissman, 2003). Treatment with E_2 and P_4 increased OXY mRNA expression in WT but not ERβKO females in the PVN, suggesting that ERβ is necessary for the regulation of the expression of OXY.

Oxytocin infusions can facilitate lordosis when delivered to several brain sites, including the VMH (Schulze & Gorzalka, 1991), whereas treatment with OXY antagonists leads to inhibition of FSB (Caldwell, Barakat, Smith, Hruby, & Pedersen, 1990; Witt & Insel, 1991). Administration of antisense oligonucleotides directed against the OXY receptor mRNA into the VMH of E_2-primed female rats also decreases the lordosis response (McCarthy, Kleopoulos, Mobbs, & Pfaff, 1994). In prairie voles, OXY facilitates social bonding but also enhances female receptivity when given during natural estrus (Cushing & Carter, 1999).

7.2.4. α-melanocyte-stimulating hormone (MSH, or melanocortin)

One more recent addition to the list of neurotransmitters that influence sexual behavior is melanocortin, as a variety of neuropeptide hormones, including MSH and corticotropin (ACTH), have been shown to cause penile erection following ICV administration (Argiolas, Melis, Murgia, & Schioth, 2000; Giuliano, 2004). Normally thought of for its role in pigmentation, subcutaneous administration of several synthetic analogs of MSH (melanotan-II (MT-II) and bremelanotide (PT-141)) causes spontaneous penile erection in humans (Wessells et al., 1998; 2000). Melanotan-II also enhanced copulatory behavior and augmented erections in WT but not MC-4 receptor knockout mice (Van der Ploeg et al., 2002). Intranasal administration of PT-141 elicited erections in rats and in men with erectile dysfunction (Molinoff, Shadiack, Earle, Diamond, & Quon, 2003). In female rats allowed to pace their interactions with males, PT-141 increased proceptive hopping and darting behavior with no effect on lordosis (Pfaus, Shadiack, Van Soest, Tse, & Molinoff, 2004). Preclinical data support the assertion that PT-141 acts on melanocortin receptors (MCRs) in the hypothalamus to increase sexual arousal or desire (Shadiack, Sharma, Earle, Spana, & Hallam, 2007). In 2008, development of PT-141 was discontinued because the US Food and Drug Administration expressed concerns regarding the side effects of increased blood pressure. Recently, development of PT-141 has been reconsidered, as an alternative administration (subcutaneous administration vs. intranasal) seems to circumvent the serious side effects. More clinical studies are needed to examine the efficacy and safety of these melanocortin agonists in humans to more fully address its therapeutic relevance and benefit.

8. GONAD-INDEPENDENT SEXUAL BEHAVIOR

As humans of both sexes engage in sexual behavior after gonadectomy, and in the case of women well after menopause, there has been some interest in identification of other species with this interesting characteristic. In general, mammalian FSB is dependent upon the gonadal steroids secreted by the ovaries (reviewed in Pfaff, 1980; Barfield, Rubin, Glaser, & Davis, 1982; Feder, 1984; Pfaff et al., 1994). Humans, rhesus monkeys, bonobos (*Pan paniscus*), rabbits, and horses are notable exceptions (reviewed in Beach, 1948; De Waal, 1995; Crowell-Davis, 2007). It is difficult to generalize based on so few species but it is striking that these animals all reside in social groups. One hypothesis about the evolution of 'silent' ovulation in humans and horses is that sexual activity is beneficial for group maintenance (Crowell-Davis, 2007). Little experimental work has been done on these animals; however, the leading hypothesis for gonadal steroid-independent mating in mares and rabbits is that androgens produced by the adrenals may be converted to estrogens in the brain (Beyer, Hoffman, & Gonzalez-Flores, 2007; Crowell-Davis, 2007). This would create a situation similar to that described in the musk shrew (Rissman, 1990).

The extent of the reduction in MSBs after castration varies both between individuals and within species (reviewed in Larsson, 1979). Striking interindividual variation in the response to castration has been documented in rats (Davidson, 1966b), dogs (Beach, 1970), cats (Rosenblatt & Aronson, 1958), rhesus monkeys (Phoenix, Slob, & Goy, 1973), and humans (Zverina, Hampl, Sulocava, & Starka, 1990). Because most of our knowledge concerning the physiology of MSB has been gleaned from animal models that require gonads, we remain largely ignorant of the underlying mechanisms governing gonadal steroid-independent MSB. In most of the well-studied rodents, testicular steroids are essential and castration eliminates copulatory behavior; however, there are two notable exceptions. In addition to the Siberian hamster (Park et al., 2004), retention of the ejaculatory reflex long after castration has been reported for the hybrid B6D2F1 mouse (Manning & Thompson, 1976b; McGill & Manning, 1976).

B6D2F1 hybrid male mice are produced by crossing C57BL/6J females with DBA/2J males, and D2B6F1 males are produced by the reciprocal cross. Some individuals in both crosses continue to copulate after castration, but the numbers are larger in the B6D2F1 than the D2B6F1 hybrids, suggesting a potential interaction with a sex-linked gene (McGill & Manning, 1976). Approximately 30% of castrated B6D2F1 hybrid males retain the ejaculatory reflex in mating tests conducted as long as 25 weeks after castration (herein after referred to as 'persistent copulators;' those that did not display this behavioral phenotype will be referred to as 'nonpersistent copulators') (Clemens et al., 1988a).

Steroids secreted by the adrenal glands do not play a role in persistent copulation, as adrenalectomy did not prevent copulation after long-term castration (Thompson, McGill, McIntosh, & Manning, 1976). Interestingly, variability exists in the expression of MSB among those identified as persistent copulators (Wee & Clemens, 1989). The underlying mechanisms mediating this variability of the persistence of MSB within the persistent copulators has yet to be investigated. Hormonal characterization of the hybrid B6D2F1 mouse failed to reveal anything exceptional (Quadagno, McGill, Yellon, & Goldman, 1979; Coquelin, 1991; Sinchak, Roselli, & Clemens, 1996). Plasma T, E_2, and DHT concentrations and hypothalamic nuclear ERs were reduced to the same extent in the hybrid mice that continued or ceased to copulate after castration (Clemens et al., 1988a; Sinchak et al., 1996). Further, total ER binding was similar between persistent copulators and nonpersistent copulators, and the activity of the aromatase enzyme that converts T to E_2 was uniformly low in the mPOA of castrated males, regardless of their ability to mate (Clemens et al., 1988a; Sinchak et al., 1996).

Sexual experience prior to castration in adulthood facilitates retention of MSB in ORCH mice (Manning & Thompson, 1976a), cats (Rosenblatt & Aronson, 1958), and Syrian hamsters (Bunnell & Kimmel, 1965; Lisk & Heimann, 1980). Prior sexual experience renders neural substrates less dependent on androgens (Hart, 1974). However, prior sexual experience does not promote retention of sexual behavior in gonadectomized rats (Rabedeau & Whalen, 1959; Bloch, Babcock, Gorski, & Micevych, 1987) and dogs (Hart, 1968; Beach, 1970). In contrast to these studies, one study reports that significantly more sexually experienced male rats maintained copulatory behavior one month longer after ORCH as compared to nonexperienced males (Pfaus, Kippin, & Centeno, 2001). Notably, similar patterns of activation in brain areas essential for copulatory behavior were found in male rats that had sexual experience and those that were just presented with a nonvolatile odor that had been previously associated via classical conditioning with a sexually receptive female, demonstrating that the mechanisms underlying MSB also may be activated conditionally with experience (Pfaus & Heeb, 1997).

The potential role of prior copulatory experience has been investigated in the postcastration retention of sexual behavior in the Siberian hamster (Costantini et al., 2007). Postoperative sexual experience was not necessary for the persistence of copulatory behavior in long-term castrated Siberian hamsters (Costantini et al., 2007). Preoperative sexual experience did, however, increase the incidence of postcastration sexual behavior, but was not necessary for postcastration retention of sexual behavior, as 25% of the males demonstrated the complete repertoire of copulatory behavior after castration (Costantini et al., 2007). Similar to Siberian hamsters, precastration sexual experience was not necessary for expression of MSB after castration in B6D2F1 hybrid mice (Manning & Thompson, 1976a), nor was weekly behavioral testing after castration necessary for the retention of ejaculatory reflexes (Wee & Clemens, 1989). Interestingly, when tested in their home cage, 75% of the castrated males retained the ability to ejaculate, but there was a significant decrease in the percentage of castrated males displaying male copulatory behavior when tested in a test arena, indicating that environmental factors impact this behavioral phenotype in some of the castrated hybrid mice.

9. GENETICS OF SEXUAL BEHAVIOR

One useful avenue for pursuit of novel genes involved in MSB is the use of various KO mice. Despite the negatives associated with the lifelong silencing of selected genes, these animals have been successfully used to tentatively identify new players in MSB. One other avenue that could be utilized is a genetic examination of strain differences in mouse MSB, which are well documented (McGill & Blight, 1963; McGill & Tucker, 1964; Clemens et al., 1988b; Burns-Cusato et al., 2004; Dominguez-Salazar et al., 2004). These differences are often attributed to genetics but few studies have examined this aspect. Here we summarize a few select studies that employ transgenic mice to examine the roles of various genes in MSB (see Tables 8.3 and 8.4 for an overview).

TABLE 8.3 Effects of specific targeted gene disruption on male sexual behavior (MSB) in mice

Genotype	Sexual behavior	References
Activin receptor II KO	Deficits in MSB	Ma et al., 2005
Amyloid precursor protein (APP) overexpressing mice	Normal MSB	Vloeberghs et al., 2007
Androgen receptor KO	Deficits in MSB	Bakker et al., 2004; Sato et al., 2004; Kudwa et al., 2007
BCL2-associated X protein (BAX) KO	Deficits in MSB	Jyotika et al., 2007
β1-3 N-acetylglucosaminyltransferase-1 (B3GNT1) KO	Deficits in MSB	Biellmann et al., 2008
Carboxypeptidase E	Deficits in MSB	Srinivasan et al., 2004
Cyclic nucleotide-gated channel alpha2 (CNGA2) KO	Deficits in MSB	Mandiyan et al., 2005
D-aspartate KO	Deficits in MSB	Huang et al., 2006; Weil et al., 2006
Dopamine D5 receptor KO (D5KO)	Normal MSB	Kudwa et al., 2005
Estrogen receptor α knockout (ERαKO)	Deficits in MSB	Ogawa et al., 1997; Wersinger et al., 1997; Ogawa et al., 1998b; Dominguez-Salazar et al., 2004
Estrogen Receptor β KO	Normal MSB	Ogawa et al., 1999a
ETS-domain transcription factor pea3 (PEA3) KO	Deficits in MSB	Laing et al., 2000
G protein-coupled receptor 54 (GPR54) KO	Normal MSB	Kauffman et al., 2007
Glu-R containing AMPA receptors KO	Deficits in MSB	Shimshek et al., 2006
Gonadotropin-releasing hormone-deficient hypogonadal (hpg) mice	Deficits in MSB	Livne et al., 1992
Growth hormone KO	Deficits in MSB	Bartke et al., 1999
Growth hormone overexpressing mice	Deficits in MSB	Bartke et al., 1999
Heme oxygenase-2 KO	Deficits in MSB	Burnett et al., 1998
Histamine (HCD) KO	Deficits in MSB	Par et al., 2003
M2 muscarinic receptor KO	Deficits in MSB	Wang et al., 2008
Monoamine oxidase type A KO	Normal MSB	Popova et al., 2001
N-methyl-D-aspartate (NMDA) receptor KO	Normal MSB	Shimshek et al., 2006
Oxytocin KO	Normal MSB	Nishimori et al., 1996
Paternally expressed 3 (PEG3) KO	Normal MSB, but MSB does not change with sexual experience	Swaney et al., 2007
Progesterone receptor KO	Normal MSB	Schneider et al., 2005
Progranulin gene KO	Deficits in MSB	Kayasuga et al., 2007
Serotonin 1B receptor KO	Deficits in MSB	Rodriguez-Manzo et al., 2002
Steroidogenic factor 1 (Sf1, also called AD4BP) KO	Deficits in MSB; lack gonads and adrenals	Parker et al., 2002
Testicular feminization mutation (Tfm)	Deficits in MSB	Ono et al., 1974; Bodo & Rissman, 2007
Transforming growth factor β 1 KO (TGFB1 KO)	Deficits in MSB; able to mount, deficits in intromission behavior, and unable to ejaculate	Ingman & Robertson, 2007; Ingman et al., 2009

KO, knockout.

TABLE 8.4 Effects of specific targeted gene disruption on female sexual behavior (FSB) in mice

Genotype	Sexual behavior	References
Connexin 43 (CX43) KO	Deficits in FSB; reduced lordosis behavior relative to wild-type controls	Gulinello & Etgen, 2005
DARPP-32 (a dopamine-and adenosine 3′,5′-monophosphate (cAMP)-regulated phosphoprotein) KO	Facilitative effect of P blocked	Mani et al., 2000
Estrogen receptor α KO	Deficits in FSB	Rissman et al., 1997a; Ogawa et al., 1998a; reviewed in Rissman et al., 1997b
Estrogen receptor β KO	Exhibit normal lordosis behaviors	Krege et al., 1998; Ogawa et al., 1999b
Estrogen receptor α KO and estrogen receptor β KO	Deficits in FSB	Kudwa & Rissman, 2003
γ-aminobutyric acid (GABA(B1)) KO	Deficits in FSB	Catalano et al., 2005
Gonadotropin-releasing hormone-deficient hypogonadal (hpg) mice	Deficits in FSB	Ward & Charlton, 1981
GPR54 KO	Normal FSB	Kauffman et al., 2007
Growth hormone KO	Deficits in FSB	Danilovich et al., 1999
Growth hormone overexpressing mice	Deficits in FSB	Bartke et al., 1999
μ-opiod receptor KO	Hormonal treatment facilitates lordosis, but reduced sexual receptivity vs. WT controls	Sinchak et al., 2005
Nescient helix loop helix 2 (NHLH2) KO	Deficits in FSB	Johnson et al., 2004
Oxytocin KO	Impaired social and affiliative behaviors	Takayanagi et al., 2005
Pituitary adenylate cyclase-activating polypeptide (PACAP) KO	Deficits in FSB	Shintani et al., 2002
Progesterone receptor isoform A KO	Deficits in FSB	Lydon et al., 1995; White et al., 2007
Progesterone receptor KO	Sterile and do not display lordosis after hormone treatment	Lydon et al., 1996
Thyroid transcription factor 1 (TTF1) KO	Delayed puberty, reduced reproductive capacity, and a short reproductive span	Mastronardi et al., 2006
Transforming growth factor beta 1 KO (TGFB1 KO)	Deficits in FSB; able to mate and produce offspring, but significantly reduced compared to controls	Ingman & Robertson, 2008

In addition to steroid receptors, ionotropic glutamate receptors, specifically the GluR-B subunit, containing α-amino-3-hydroxyl-5-methyl-4-isoxazole-propionate (AMPA) receptors in GnRH neurons and other limbic system neurons are also essential for MSB (Shimshek et al., 2006). Male mice lacking GluR-B-containing AMPA receptors are poor breeders and demonstrated decreased mounting compared to WT controls. Activin receptors also seem to play an important role in MSB. Male mice lacking the activin receptor type II display multiple reproductive behavioral deficits, including delayed initiation of copulation, reduced mounting and intromission frequencies, and increased mounting, intromission, and ejaculation latencies (Ma, Reyna, Mani, Matzuk, & Kumar, 2005). Interestingly, male mice lacking this particular receptor also have decreased NOS activity in the mPOA but not in the rest of the hypothalamus or cortex, suggesting that activins are upstream regulators of NOS activity within the mPOA and play an important role in MSB.

Several studies have investigated the role of sex chromosomes in MSB. One study investigated gonadally intact congenic mice possessing Y chromosomes from two different common mouse strains (DBA/2J and C57BL/10J). A small, but significant, difference in the percentage that displayed mounting behavior with a receptive female was reported (Shrenker & Maxson, 1983). In another study using a mouse cross in which gonadal and chromosome sex were uncoupled, XY^{-}*Sry* males (which have a Y^{-} chromosome with a spontaneous deletion of the *Sry* gene, plus an autosomal *Sry* transgene that induces development of testes) and XX*Sry* males did not differ in MSB but differed from normal XY males in latency to thrust (De Vries et al., 2002).

Sex chromosome aneuploidy has a significant impact on MSB. In men, its occurrence is fairly common with incidence rates as high as 1 : 500 in both 47,XXY and 47,XYY individuals (Rives, Simeon, Milazzo, Barthelemy, & Mace, 2003; Simpson et al., 2003). Behavioral and psychological assessments of men with an additional Y chromosome (e.g., 47,XYY) suggest a tendency towards increased emotional instability and increased sexual drive (Theilgaard, 1984). Men with Klinefelter Syndrome (47,XXY) typically report low sex drive, possibly due to low endogenous T levels (Sorensen, Nielsen, Froland, & Johnsen, 1979; Yoshida et al., 1997), although some men with Klinefelter Syndrome report that low libido persists even after androgen replacement (Stewart, Bailey, Netley, & Park, 1990; Winter, 1991). To assess the role of sex chromosome aneuploidy on sexual behavior, one study used mice with six male genotypes: XY, XYY^{-}, XX*Sry*, XXY^{-}*Sry*, XY*Sry*, and XYY^{-}*Sry* (Park et al., 2008). These mice were generated by aneuploid dams (XXY^{-}) and sires that had a Sry transgene (XY*Sry*). In order to eliminate possible differences in levels of T, all of the subjects were castrated and received T implants prior to tests for MSB. Mice with an additional copy of the Y^{-} chromosome (XYY^{-}) had shorter latencies to intromit and achieve ejaculations than XY males, indicating that the additional Y^{-} chromosome enhanced several components of MSB. In a comparison of the four genotypes bearing the *Sry* transgene, males with two copies of the X chromosome (XX*Sry* and XXY^{-}*Sry*) had longer latencies to mount and thrust than males with only one copy of the X chromosome (XY*Sry* and XYY^{-}*Sry*), suggesting that increased X chromosome dosage leads to deficits in several components of MSB. These results suggest novel roles for sex chromosome genes in MSB, and future studies utilizing these types of mice need to be conducted in order to determine whether Y genes are expressed in brain regions controlling MSB, whether expression of Y genes is tightly correlated with the expression of the behavior, and how the Y genes interact with androgens to influence social behaviors.

10. NEUROENDOCRINOLOGY OF HUMAN SEXUAL BEHAVIOR

10.1. Men

In general, a pronounced decrease in T concentration correlates with a decrease in MSB in human males. Sexual function as well as circulating T levels decline with age in men (see reviews in Stanworth & Jones, 2008; Tostain & Blanc, 2008). Hypogonadal men reported a failure to obtain fantasy-induced erections, suggesting a significant impact of T on libido and sexual fantasies (Bancroft, 1989). In a few studies on the sexual consequences of castration in men, results indicate that T plays a more modulatory role for male sexual activity in humans than in other mammals. Castration in men treated for testicular tuberculosis led to failed ejaculation in some whereas libido and erectile potency were maintained in others (Hammond, 1943). Of men castrated for 'treatment' of sexual offenses, 50 to 67% reported rapid loss of sexual desire and interest, whereas for the remaining men sexual activity waned gradually (Heim & Hursch, 1979).

As in rodent models, variability in T levels and its relation to MSB exists between individual human males. Some men reportedly engaged in sexual intercourse as long as a decade after bilateral orchidectomy without androgen replacement therapy (Heim & Hursch, 1979; Weinberger, Sreenivasan, Garrick, & Osran, 2005). In a population of 101 healthy men, no correlation between T levels and libido, sexual thoughts, or fantasies was noted (Bagatell, Heiman, Rivier, & Bremner, 1994). In another study of healthy males ranging in age from 21 to 82, T levels and self-reported sexual function did not vary across age groups (Rowland, Greenleaf, Dorfman, & Davidson, 1993). In hypogonadal men, graded doses of T did not increase their libido (Cunningham, Hirshkowitz, Korenman, & Karacan, 1990; Bhasin, 1992).

It should be noted that one significant difference between men castrated in adulthood and hypogonadal men is that the latter group has had minimal or no exposure to androgens since birth. It seems that a disruption in exposure to T during the critical periods may lead to a decrease in responsiveness of neural and peripheral tissues to T and more T is necessary to reactivate dormant copulatory systems. Thus, increased sexual dysfunction observed in hypogonadal men may be influenced by the absence of early exposure to androgens. Two meta-analyses of placebo-controlled studies of T therapy on sexual function in men concluded that T treatment moderately improved sexual dysfunction in men with low circulating T (Isidori et al., 2005; Bolona et al., 2007). In addition, the effects of T on erectile function, but not libido, were inversely related to mean baseline T concentrations (Isidori et al., 2005). However, large-scale, long-term, randomized controlled studies to formally investigate the efficacy and potential side effects of both short- and long-term T replacement therapy in symptomatic men with reduced T levels and sexual dysfunction still need to be conducted. It remains to be elucidated whether the large individual differences of sexual function observed in normal men are due to gonadal steroid-independent mechanisms, differential degrees of sensitivity to T, organizational actions of steroids, or aspects of sexual reward.

Studies utilizing techniques such as immunocytochemistry to help map the neural regulation of MSB in humans are not practical. However, with the advent of brain imaging techniques, it is now possible to begin to determine which neural regions are active during human sexual behavior. Using Positron emission tomography (PET), blood flow changes in the ventral tegmental area, cerebellum, and the lateral corpus striatum in men were scanned during erection and ejaculation (Holstege et al., 2003). Interestingly and counter to what would have been predicted, the mPOA, BNST, and AMG did not have increased blood flow in sexually active men compared to controls. Before firm conclusions can be made from the results of this study, several variables must be taken into account. First, PET is a low-sensitivity technique, and thus may not reveal subtle differences in blood flow. Second, the stimuli used to stimulate sexual arousal may not have been optimal, and variability in the allotted time that subjects were allowed to view the sexually arousing stimuli may have been too long (or too short) to capture differences in brain vasculature.

Several magnetic imaging studies have been performed on men while they were viewing sexually arousing images, and, in contrast to the PET study, some studies have demonstrated visual regions showing signs of activation in the AMG and hypothalamus, including the mPOA (Mouras et al., 2003; Hamann, Herman, Nolan, & Wallen, 2004a). Other studies have revealed a sex difference in human brain activity during exposure to putative male vs. female pheromones: male subjects displayed cellular activity in the mPOA after exposure to putative female pheromones whereas females did not show a response in this region to either substance (Savic, 2002; 2005).

10.2. Women

Women are capable of engaging in sexual behavior at any time during the menstrual cycle, during pregnancy and lactation, and after menopause. Although it seems FSB is thus emancipated from hormonal control and from the fertile phase of the reproductive cycle, the question of whether women are affected by hormones has not been answered adequately. Although many facets of the neuroendocrine basis of FSB in rodents have been elucidated, female sexual arousal disorder (FSAD) is not fully understood. The National Health and Social Life Survey, which was conducted in 1999, showed the prevalence of sexual dysfunction to be relatively high (43% of 1749 women surveyed in the US) compared to that of American men (31%) (Laumann, Paik, & Rosen, 1999). Some factors that adversely affect sexual function in women include age, status of relationship, menopausal status, and hysterectomy (L. Zussman, S. Zussman, Sunley, & Bjornson, 1981). Menopausal transition doubles the prevalence of sexual dysfunction from 42 to 88% (Dennerstein, Randolph, Taffe, Dudley, & Burger, 2002).

Functional magnetic resonance imaging (fMRI) technology has been utilized to map out the brain areas associated with FSB. Females were presented with arousing visual materials while fMRI was used to determine whether and which cerebral regions were affected (Park et al., 2001). The inferior frontal lobe, cingulate gyrus, insula gyrus, corpus callosum, thalamus, caudate nucleus, globus pallidus, and inferior temporal lobe were all significantly activated when women were watching erotic films compared to nonerotic films; several of these areas overlap with those previously described in men (Stoleru et al., 1999). Greater arousal as indicated by significantly greater activation in several of these areas was found in male subjects when compared with females, specifically in the AMG and hypothalamus (Karama et al., 2002; Hamann, Herman, Nolan, & Wallen, 2004b). One interesting study utilizing fMRI demonstrated that vaginal–cervical self-stimulation in women with complete spinal cord injury increased activity in the region of the nucleus of the solitary tract, which is the brainstem nucleus to which the vagus nerves project, suggesting that the vagus nerve can convey genital sensory input directly to the brain in women with spinal cord injury, thus completely bypassing the injured spinal pathways (Komisaruk et al., 2004).

Current pharmacological strategies for treating FSAD include treatment with androgens such as a T patch, topically applied prostaglandin E1 (PGE1; also know as alprostadil), or phosphodiesterase 5 (PDE-5) inhibitors; however, none of these strategies have been FDA-approved for treating FSAD. Estrogen therapy also improves several sexual components, including clitoral sensitivity, orgasm rates, sexual desire level, and sexual activity (Sarrel, 1990; Nathorst-Boos, Wiklund, Mattsson, Sandin, & Von Schoultz, 1993). In postmenopausal women, exogenous estrogens will increase vaginal blood flow and lubrication and prevent vaginal atrophy associated with lowered estrogen levels. However, the effects of estrogens on sexual desire have not been elucidated (Myers, Dixen, Morrissette, Carmichael, & Davidson, 1990). Women receiving methyltestosterone had higher bioavailable testosterone and lower sex hormone-binding globulin (SHBG) than those receiving esterified estrogens. They also reported an improvement in their sexual desire, suggesting a possible relation between bioavailable T, SHBG, and libido (Lobo, Rosen, Yang, Block, & Van Der Hoop, 2003). There are data showing that exogenous T therapy increases sexual desire, sexual responsiveness, and the frequency of sexual activity in postmenopausal women (reviewed in Hubayter & Simon, 2008). In reproductive-age women, T is produced by the adrenal glands, the ovaries, and from peripheral conversion of other circulating androgens to T. As women age, androgen production from the adrenal glands declines. In postmenopausal women, the ovary continues to secrete premenopausal levels of T, but it produces less androstenedione (AND), which leads to a decline in serum T levels. However, there is no clear association between T levels and sexual function, and a myriad of interrelated factors, both biologic and psychosocial, affect sexual desire in postmenopausal women, complicating the determination of whether decreased sexual desire is due to a hypoandrogenic state (Dennerstein et al., 2002; Davis, Davison, Donath, & Bell, 2005). Before treatment recommendations are made concerning T, more studies must be conducted. It is still unclear whether T affects libido via the AR or the ERs, and most data in other animals, including rhesus monkeys, suggests the latter is most likely (Zehr, Maestripieri, & Wallen, 1998).

Reviews of published studies investigating the potential use of PDE-5 inhibitors for women with sexual dysfunction reveal that there is no conclusive or consistent benefit of sildenafil (although some women may show significant improvements on some measures) (Basson, McInnes, Smith, Hodgson, & Koppiker, 2002; J. Berman, L. Berman, Toler, Gill, & Haughie, 2003; Mayer, Stief, Truss, & Uckert, 2005; Shields & Hrometz, 2006). In addition to treating erectile dysfunction in men, clinical trials have been conducted to evaluate the safety and efficacy of bremelanotide to treat FSAD (Diamond et al., 2006; Safarinejad, 2008). These clinical trials are based on data from animal models that suggest melanocortin agonists selectively and strongly activate central mechanisms involved in sexual arousal in female rodents that may be analogous to central mechanisms of arousal, motivation, and desire in human females (Pfaus et al., 2004). Specifically, subcutaneous injection of bremelanotide to hormonally primed female rats selectively increased solicitation behaviors towards male rats, without affecting expression of lordosis behavior (Pfaus et al., 2004). Another melanocortin agonist, MT-II, also enhanced proceptive sexual behaviors in the female rat (Rossler et al., 2006). In two separate randomized double-blind placebo-controlled studies aiming to determine the effect of bremelanotide on sexual response, women reported increased arousal and desire, suggesting that the melanocortin system is a viable drug target for the treatment of FSAD (Diamond et al., 2006; Safarinejad, 2008). However, further studies will be required before final conclusions are drawn.

The role of OXY in human FSB has yet to be determined. Plasma OXY concentrations were elevated in both men and women during the time of orgasm, and remained elevated for at least five minutes after orgasm (Carmichael et al., 1987). In another study, OXY concentrations were elevated one minute after orgasm but declined to baseline by five minutes postorgasm in women (Blaicher et al., 1999). There is one case study that reports increased sexual response after OXY administration (Anderson-Hunt & Dennerstein, 1994); however, no adequate double-blind trials have been conducted to confirm this observation (Anderson-Hunt & Dennerstein, 1995). Based on the human literature, it is difficult to draw conclusions on OXY and its role in sexual arousal, and the functional significance of the elevated OXY levels during orgasm remains to be elucidated.

11. SUMMARY

A tremendous amount of work has been done on the neuroendocrine control of reproductive behavior; however, data are still lacking to address and treat important sexual and reproductive problems in humans. For example, younger men who are hypogonadal or older men with reduced androgen secretion contend with reduced libido, and, unfortunately, the optimal T treatments are controversial and most have undesirable side effects. Although there have been recent advances in the field of neuropharmacology of human MSB, such as the advent of sildenafil (Viagra), vardenafil hydrochloride (Levitra), and tadalafil (Cialis) for the treatment of erectile dysfunction (Montorsi, Corbin, & Phillips, 2004; reviewed in Doggrell, 2007), many still report loss of libido, sexual desire, and intimacy (Rousseau, Dupont, Labrie, & Couture, 1988; Arai et al., 1999; Al-Abany, Steineck, Agren Cronqvist, & Helgason, 2000; Sanchez-Ortiz et al., 2000; Vale, 2000; Incrocci, Slob, & Levendag, 2002; Anastasiadis et al., 2003; Brucker & Cella, 2003; Lee, Huang, & Lu, 2003; Miller, Bissonette, Bahnson, Wilson, & Theodorescu, 2003; Hisasue et al., 2004). In addition, many postmenopausal women report decreased sexual desire and/or difficulty achieving orgasm, and we lack effective corrective hormone treatments. Fortunately, outstanding recent advances in genetics and molecular biology make possible exciting new approaches to understanding the neuroendocrine underpinnings of sexuality and hold the potential for developing new therapies for sexual behavior dysfunction.

ABBREVIATIONS

5-HT Serotonin
7-OH-DPAT 7-hydroxy-2(di-n propylamino) tetralin
8-OH-DPAT 8-hydroxy-2-(di-n propylamino) tetralin
ACh Acetylcholine
ACTH Corticotropin
AMG Amygdala
AMPA α-amino-3-hydroxyl-5-methyl-4-isoxazole-propionate
AND Androstenedione
AR Androgen receptor

ARC	Arcuate nucleus
ARKO	Androgen receptor knockout
ArKO	Aromatase enzyme knockout
B-HT 920	6-allyl-2-amino-5,6,7,8-tetrahydro-4H-thiazolo[4,5d] azepine dihydrochloride
BNST	Bed nucleus of the stria terminalis
CPP	Conditioned place preferences
DA	Dopamine
DHT	5α-dihydrotestosterone
dmAMG	Dorsal medial nucleus of the amygdala
E_2	Estradiol
EGF	Epidermal growth factor
ER	Estrogen receptor
ERαKO	ERα knockout
ERβKO	ERβ knockout
fMRI	Functional magnetic resonance imaging
FSAD	Female sexual arousal disorder
FSB	Female sexual behavior
FSH	Follicle-stimulating hormone
GnRH	Gonadotropin-releasing hormone
ICV	Intracerebroventricular
IGF	Insulin-like growth factor
IGF-1R	Insulin-like growth factor-1 receptor
L-DOPA	L-3, 4- dihydroxyphenylalanine
LH	Luteinizing hormone
L-PGDS	Lipocalin-type prostaglandin D synthase
MAPK	Mitogen-activated protein kinase
MCG	Midbrain central gray
MCR	Melanocortin receptor
mPOA	Medial preoptic area
MSB	Male sexual behavior
MSH	α-melanocyte-stimulating hormone
NE	Norepinephrine
NOS	Nitric oxide synthase
ORCH	Orchidectomy
OVX	Ovariectomized
OXY	Oxytocin
P_4	Progesterone
$P450_{aro}$	Aromatase enzyme
PDE-5	Phosphodiesterase 5
PET	Positron emission tomography
PG	Prostaglandin
PGD_2	Prostaglandin D_2
PGE1	Prostaglandin E1 (also known as alprostadil)
PGH_2	Prostaglandin H
PI3K	Phosphatidylinositol-3-kinase
PR	Progesterone receptor
PRKO	Progesterone receptor knockout
PRL	Prolactin
PT-141	Bremelanotide
PVN	Paraventricular nucleus
SHBG	Sex hormone-binding globulin
SND 919	2-amino-4,5,6,7-tetrahydro-6-propylamino-benzthiazol dihydrochloride
SSRI	Selective serotonin reuptake inhibitors
T	Testosterone
Tfm	Testicular feminization mutation
VMH	Ventromedial hypothalamus
VNAB	Ventral noradrenergic bundle
WT	Wild-type

REFERENCES

Afonso, V. M., Bablekis, V., & Pfaus, J. G. (2006a). Sensory mediation of female-male mounting in the rat: II. Role of tactile and conspecific cues. *Physiol. Behav., 87*, 863–869.

Afonso, V. M., Woehrling, A., & Pfaus, J. G. (2006b). Sensory mediation of female–male mounting in the rat: I. Role of olfactory cues. *Physiol. Behav., 87*, 857–862.

Agmo, A., & Fernandez, H. (1989). Dopamine and sexual behavior in the male rat: a reevaluation. *J. Neural. Transm., 77*, 21–37.

Agmo, A., & Paredes, R. (1988). Opioids and sexual behavior in the male rat. *Pharmacol. Biochem. Behav., 30*, 1021–1034.

Agmo, A., & Picker, Z. (1990). Catecholamines and the initiation of sexual behavior in male rats without sexual experience. *Pharmacol. Biochem. Behav., 35*, 327–334.

Agmo, A., & Sodersten, P. (1975). Sexual behaviour in castrated rabbits treated with testosterone, oestradiol, dihydrotestosterone or oestradiol in combination with dihydrotestosterone. *Journal of Endocrinology, 67*, 327–332.

Agmo, A., Paredes, R. G., Sierra, L., & Garces, I. (1997). The inhibitory effects on sexual behavior and ambulatory activity of the mixed GABAA/GABAB agonist progabide are differentially blocked by GABA receptor antagonists. *Psychopharmacology (Berl), 129*, 27–34.

Ahlenius, S., & Larsson, K. (1984a). Apomorphine and haloperidol-induced effects on male rat sexual behavior: no evidence for actions due to stimulation of central dopamine autoreceptors. *Pharmacol. Biochem. Behav., 21*, 463–466.

Ahlenius, S., & Larsson, K. (1984b). Lisuride, LY-141865, and 8-OH-DPAT facilitate male rat sexual behavior via a non-dopaminergic mechanism. *Psychopharmacology (Berl), 83*, 330–334.

Ahlenius, S., & Larsson, K. (1990). Effects of selective dopamine D1 and D2 antagonists on male rat sexual behavior. *Experientia, 46*, 1026–1028.

Ahlenius, S., & Larsson, K. (1991). Opposite effects of 5-methoxy-N, N-di-methyl-tryptamine and 5-hydroxytryptophan on male rat sexual behavior. *Pharmacol. Biochem. Behav., 38*, 201–205.

al-Abany, M., Steineck, G., Agren Cronqvist, A. K., & Helgason, A. R. (2000). Improving the preservation of erectile function after external beam radiation therapy for prostate cancer. *Radiotherapy and Oncology, 57*, 201–206.

Albinsson, A., Andersson, G., Andersson, K., Vega-Matuszczyk, J., & Larsson, K. (1996). The effects of lesions in the mesencephalic Raphe systems on male rat sexual behavior and locomotor activity. *Behav. Brain Res., 80*, 57–63.

Alsum, P., & Goy, R. W. (1974). Actions of esters of testosterone, dihydrotestosterone, or estradiol on sexual behavior in castrated male guinea pigs. *Hormones and Behavior, 5*, 207–217.

Anastasiadis, A. G., Sachdev, R., Salomon, L., Ghafar, M. A., Stisser, B. C., Shabsigh, R., et al. (2003). Comparison of health-related quality of life and prostate-associated symptoms after primary and salvage cryotherapy for prostate cancer. *Journal of Cancer Research and Clinical Oncology, 129*, 676–682.

Anderson-Hunt, M., & Dennerstein, L. (1994). Increased female sexual response after oxytocin. *B MJ, 309*, 929.

Anderson-Hunt, M., & Dennerstein, L. (1995). Oxytocin and female sexuality. *Gynecol. Obstet. Invest., 40*, 217–221.

Apostolakis, E. M., Garai, J., Lohmann, J. E., Clark, J. H., & O'Malley, B. W. (2000). Epidermal growth factor activates reproductive behavior independent of ovarian steroids in female rodents. *Mol. Endocrinol., 14*, 1086–1098.

Arai, Y., Okubo, K., Aoki, Y., Maekawa, S., Okada, T., Maeda, H., et al. (1999). Patient-reported quality of life after radical prostatectomy for prostate cancer. *International journal of urology, 6*, 78–86.

Argiolas, A., & Melis, M. R. (1995). Oxytocin-induced penile erection. Role of nitric oxide. *Adv. Exp. Med. Biol., 395*, 247–254.

Argiolas, A., & Melis, M. R. (2004). The role of oxytocin and the paraventricular nucleus in the sexual behaviour of male mammals. *Physiol. Behav., 83*, 309–317.

Argiolas, A., Melis, M. R., & Gessa, G. L. (1985). Intraventricular oxytocin induces yawning and penile erection in rats. *Eur. J. Pharmacol., 117*, 395–396.

Argiolas, A., Melis, M. R., Murgia, S., & Schioth, H. B. (2000). ACTH- and alpha-MSH-induced grooming, stretching, yawning and penile erection in male rats: site of action in the brain and role of melanocortin receptors. *Brain Res. Bull., 51*, 425–431.

Arletti, R., Bazzani, C., Castelli, M., & Bertolini, A. (1985). Oxytocin improves male copulatory performance in rats. *Horm. Behav., 19*, 14–20.

Aronica, S. M., & Katzenellenbogen, B. S. (1993). Stimulation of estrogen receptor-mediated transcription and alteration in the phosphorylation state of the rat uterine estrogen receptor by estrogen, cyclic adenosine monophosphate, and insulin-like growth factor-I. *Mol. Endocrinol., 7*, 743–752.

Arteaga-Silva, M., Marquez-Villanueva, Y., Martinez-Garcia, R., Hernandez-Gonzalez, M., Bonilla-Jaime, H., & Retana-Marquez, S. (2005). Effects of hormonal replacement with androgens and estrogens on male sexual behavior and plasma levels of these steroids in gonadectomized golden hamsters (*Mesocricetus auratus*). *Physiology and Behavior, 85*, 571–580.

Arteaga-Silva, M., Rodriguez-Dorantes, M., Baig, S., & Morales-Montor, J. (2007). Effects of castration and hormone replacement on male sexual behavior and pattern of expression in the brain of sex-steroid receptors in BALB/c AnN mice. *Comp. Biochem. Physiol. A Mol. Integr. Physiol., 147*, 607–615.

Auger, A. P. (2001). Ligand-independent activation of progestin receptors: relevance for female sexual behaviour. *Reproduction, 122*, 847–855.

Auger, A. P. (2004). Steroid receptor control of reproductive behavior. *Hormones and Behavior, 45*, 168–172.

Auger, A. P., Moffatt, C. A., & Blaustein, J. D. (1996). Reproductively-relevant stimuli induce Fos-immunoreactivity within progestin receptor-containing neurons in localized regions of female rat forebrain. *J. Neuroendocrinol., 8*, 831–838.

Bagatell, C. J., Heiman, J. R., Rivier, J. E., & Bremner, W. J. (1994). Effects of endogenous testosterone and estradiol on sexual behavior in normal young men. *The Journal of clinical endocrinology and metabolism, 78*, 711–716.

Bailhache, T., Surlemont, C., & Balthazart, J. (1993). Effects of neurochemical lesions of the preoptic area on male sexual behavior in the Japanese quail. *Brain research bulletin, 32*, 273–283.

Bakker, J., Honda, S., Harada, N., & Balthazart, J. (2004). Restoration of male sexual behavior by adult exogenous estrogens in male aromatase knockout mice. *Horm. Behav., 46*, 1–10.

Ball, J. (1941). Effect of progesterone upon sexual excitability in the female monkey. *Psychol. Bull., 38*, 533–534.

Ballard, C. L., & Wood, R. I. (2007). Partner preference in male hamsters: steroids, sexual experience and chemosensory cues. *Physiol. Behav., 91*, 1–8.

Balthazart, J., Stamatakis, A., Bacola, S., Absil, P., & Dermon, C. R. (2001). Effects of lesions of the medial preoptic nucleus on the testosterone-induced metabolic changes in specific brain areas in male quail. *Neuroscience, 108*, 447–466.

Bancroft, J. (1989). *Human Sexuality and its Problems*. Edinburgh, UK: Churchill Livingston.

Bandler, R., & Shipley, M. T. (1994). Columnar organization in the midbrain periaqueductal gray: modules for emotional expression? *Trends Neurosci., 17*, 379–389.

Barbeau, A. (1969). L-dopa therapy in Parkinson's disease: a critical review of nine years' experience. *Can. Med. Assoc. J., 101*, 59–68.

Bardin, C. W., Bullock, L. P., Sherins, R. J., Mowszowicz, I., & Blackburn, W. R. (1973). Androgen metabolism and mechanism of action in male pseudohermaphroditism: a study of testicular feminization. *Recent Prog. Horm. Res., 29*, 65–109.

Barfield, R. J., Rubin, B. S., Glaser, J. H., & Davis, P. G. (1982). Sites of action of ovarian hormones in the regulation of oestrous responsiveness in rats. In J. Balthazart, E. Prove, & R. Gilles (Eds.), *Hormones and Behaviour in Higher Vertebrates* (pp. 2–18). Berlin, Germany: Springer-Verlag.

Bartke, A., Chandrashekar, V., Turyn, D., Steger, R. W., Debeljuk, L., Winters, T. A., et al. (1999). Effects of growth hormone overexpression and growth hormone resistance on neuroendocrine and reproductive functions in transgenic and knock-out mice. *Proc. Soc. Exp. Biol. Med., 222*, 113–123.

Baskerville, T. A., & Douglas, A. J. (2008). Interactions between dopamine and oxytocin in the control of sexual behaviour. *Prog. Brain Res., 170*, 277–290.

Basson, R., McInnes, R., Smith, M. D., Hodgson, G., & Koppiker, N. (2002). Efficacy and safety of sildenafil citrate in women with sexual dysfunction associated with female sexual arousal disorder. *J. Womens Health Gend. Based Med., 11*, 367–377.

Baum, M. J., & Everitt, B. J. (1992). Increased expression of c-fos in the medial preoptic area after mating in male rats: role of afferent inputs from the medial amygdala and midbrain central tegmental field. *Neuroscience, 50*, 627–646.

Beach, F. A. (1942). Importance of progesterone to induction of sexual receptivity in spayed femal rats. *Proc. Soc. Exp. Biol. Med, 51*, 369–371.

Beach, F. A. (1948). *Hormones and Behavior*. New York, NY: Paul B. Hoeber.

Beach, F. A. (1970). Coital behavior in dogs. VI. Long-term effects of castration upon mating in the male. *Journal of comparative and physiological psychology, 70*, 1–32.

Beach, F. A. (1976). Sexual attractivity, proceptivity, and receptivity in female mammals. *Horm. Behav., 7*, 105–138.

Beach, F. A., & Holz-Tucker, A. M. (1949). Effects of different concentrations of androgen upon sexual behavior in castrated male rats. *Journal of comparative and physiological psychology, 42*, 433–453.

Beach, F. A., & Orndoff, R. K. (1974). Variation in the responsiveness of female rats to ovarian hormones as a function of preceding hormonal deprivation. *Horm. Behav., 5*, 201–205.

Benelli, A., Bertolini, A., Poggioli, R., Cavazzuti, E., Calza, L., Giardino, L., et al. (1995). Nitric oxide is involved in male sexual behavior of rats. *Eur. J. Pharmacol., 294*, 505–510.

Berman, J. R., Berman, L. A., Toler, S. M., Gill, J., & Haughie, S. (2003). Safety and efficacy of sildenafil citrate for the treatment of female sexual arousal disorder: a double-blind, placebo controlled study. *J. Urol., 170*, 2333–2338.

Beyer, C., Hoffman, K. L., & Gonzalez-Flores, O. (2007). Neuroendocrine regulation of estrous behavior in the rabbit: similarities and differences with the rat. *Horm. Behav., 52*, 2–11.

Beyer, C., Larsson, K., Perez-Palacios, G., & Morali, G. (1973). Androgen structure and male sexual behavior in the castrated rat. *Hormones and Behavior, 7*, 99–108.

Beyer, C., Morali, G., Naftolin, F., Larsson, K., & Perez, p. (1976). Effect of some antiestrogens and aromatase inhibitors on androgen induced sexual behavior in castrated male rats. *Hormones and Behavior, 7*, 353–363.

Bhasin, S. (1992). Clinical review 34: Androgen treatment of hypogonadal men. *The Journal of clinical endocrinology and metabolism, 74*, 1221–1225.

Biellmann, F., Henion, T. R., Burki, K., & Hennet, T. (2008). Impaired sexual behavior in male mice deficient for the beta1-3 N-acetylglucosaminyltransferase-I gene. *Mol. Reprod. Dev., 75*, 699–706.

Bitran, D., & Hull, E. M. (1987). Pharmacological analysis of male rat sexual behavior. *Neurosci. Biobehav. Rev., 11*, 365–389.

Blaicher, W., Gruber, D., Bieglmayer, C., Blaicher, A. M., Knogler, W., & Huber, J. C. (1999). The role of oxytocin in relation to female sexual arousal. *Gynecol. Obstet. Invest., 47*, 125–126.

Blaustein, J. D. (2003). Progestin receptors: neuronal integrators of hormonal and environmental stimulation. *Ann. NY. Acad. Sci., 1007*, 238–250.

Blaustein, J. D., Lehman, M. N., Turcotte, J. C., & Greene, G. (1992). Estrogen receptors in dendrites and axon terminals in the guinea pig hypothalamus. *Endocrinology, 131*, 281–290.

Bloch, G. J., Babcock, A. M., Gorski, R. A., & Micevych, P. E. (1987). Cholecystokinin stimulates and inhibits lordosis behavior in female rats. *Physiology and Behavior, 39*, 217–224.

Bodo, C., & Rissman, E. F. (2007). Androgen receptor is essential for sexual differentiation of responses to olfactory cues in mice. *The European journal of neuroscience, 25*, 2182–2190.

Bolona, E. R., Uraga, M. V., Haddad, R. M., Tracz, M. J., Sideras, K., Kennedy, C. C., et al. (2007). Testosterone use in men with sexual dysfunction: a systematic review and meta-analysis of randomized placebo-controlled trials. *Mayo Clinic Proceedings, 82*, 20–28.

Bonsall, R. W., Clancy, A. N., & Michael, R. P. (1992). Effects of the nonsteroidal aromatase inhibitor, Fadrozole, on sexual behavior in male rats. *Hormones and Behavior, 26*, 240–254.

Boscarino, B. T., & Parfitt, D. B. (2002). Chronic oral administration of clomipramine decreases sexual behavior in the male Syrian hamster (Mesocricetus auratus). *Physiol. Behav., 75*, 361–366.

Bowers, M. B., Jr., Van Woert, M., & Davis, L. (1971). Sexual behavior during L-dopa treatment for Parkinsonism. *Am. J. Psychiatry, 127*, 1691–1693.

Brandling-Bennett, E. M., Blasberg, M. E., & Clark, A. S. (1999). Paced mating behavior in female rats in response to different hormone priming regimens. *Horm. Behav., 35*, 144–154.

Brucker, P. S., & Cella, D. (2003). Measuring self-reported sexual function in men with prostate cancer. *Urology, 62*, 596–606.

Bunnell, B. N., & Kimmel, M. E. (1965). Some effects of copulatory experience on postcastration mating behavior in the male hamster. *Psychonomic science, 3*, 179–180.

Bunone, G., Briand, P. A., Miksicek, R. J., & Picard, D. (1996). Activation of the unliganded estrogen receptor by EGF involves the MAP kinase pathway and direct phosphorylation. *Embo. J., 15*, 2174–2183.

Burnett, A. L., Johns, D. G., Kriegsfeld, L. J., Klein, S. L., Calvin, D. C., Demas, G. E., et al. (1998). Ejaculatory abnormalities in mice with targeted disruption of the gene for heme oxygenase-2. *Nat. Med., 4*, 84–87.

Burns-Cusato, M., Scordalakes, E. M., & Rissman, E. F. (2004). Of mice and missing data: what we know (and need to learn) about male sexual behavior. *Physiol. Behav., 83*, 217–232.

Butera, P. C., & Czaja, J. A. (1989). Effects of intracranial implants of dihydrotestosterone on the reproductive physiology and behavior of male guinea pigs. *Hormones and Behavior, 23*, 424–431.

Caldwell, J. D. (1992). Central oxytocin and female sexual behavior. *Ann. NY. Acad. Sci., 652*, 166–179.

Caldwell, J. D., Barakat, A. S., Smith, D. D., Hruby, V. J., & Pedersen, C. A. (1990). A uterotonic antagonist blocks the oxytocin-induced facilitation of female sexual receptivity. *Brain Res., 512*, 291–296.

Canteras, N. S., Simerly, R. B., & Swanson, L. W. (1992). Connections of the posterior nucleus of the amygdala. *The Journal of comparative neurology, 324*, 143–179.

Cantor, J. M., Binik, Y. M., & Pfaus, J. G. (1999). Chronic fluoxetine inhibits sexual behavior in the male rat: reversal with oxytocin. *Psychopharmacology (Berl), 144*, 355–362.

Cardona-Gomez, G. P., DonCarlos, L., & Garcia-Segura, L. M. (2000). Insulin-like growth factor I receptors and estrogen receptors colocalize in female rat brain. *Neuroscience, 99*, 751–760.

Carmichael, M. S., Humbert, R., Dixen, J., Palmisano, G., Greenleaf, W., & Davidson, J. M. (1987). Plasma oxytocin increases in the human sexual response. *J. Clin. Endocrinol. Metab., 64*, 27–31.

Catalano, P. N., Bonaventura, M. M., Silveyra, P., Bettler, B., Libertun, C., & Lux-Lantos, V. A. (2005). GABA(B1) knockout mice reveal alterations in prolactin levels, gonadotropic axis, and reproductive function. *Neuroendocrinology, 82*, 294–305.

Charest, N. J., Zhou, Z. X., Lubahn, D. B., Olsen, K. L., Wilson, E. M., & French, F. S. (1991). A frameshift mutation destabilizes androgen receptor messenger RNA in the Tfm mouse. *Mol. Endocrinol., 5*, 573–581.

Cho, H., & Katzenellenbogen, B. S. (1993). Synergistic activation of estrogen receptor-mediated transcription by estradiol and protein kinase activators. *Mol. Endocrinol., 7*, 441–452.

Ciana, P., Raviscioni, M., Mussi, P., Vegeto, E., Que, I., Parker, M. G., et al. (2003). *In-vivo* imaging of transcriptionally active estrogen receptors. *Nat. Med., 9*, 82–86.

Clark, A. S., Pfeifle, J. K., & Edwards, D. A. (1981). Ventromedial hypothalamic damage and sexual proceptivity in female rats. *Physiol. Behav., 27*, 597–602.

Clark, J. T., Kalra, P. S., & Kalra, S. P. (1985). Neuropeptide Y stimulates feeding but inhibits sexual behavior in rats. *Endocrinology, 117*, 2435–2442.

Clark, J. T., Smith, E. R., & Davidson, J. M. (1984). Enhancement of sexual motivation in male rats by yohimbine. *Science, 225*, 847–849.

Clemens, L. G., & Dohanich, G. P. (1980). Inhibition of lordotic behavior in female rats following intracerebral infusion of anticholinergic agents. *Pharmacol. Biochem. Behav., 13*, 89–95.

Clemens, L. G., & Pomerantz, S. M. (1982). Testosterone acts as a prohormone to stimulate male copulatory behavior in male deer mice (*Peromyscus maniculatus bairdi*). *Journal of comparative and physiological psychology, 96*, 114–122.

Clemens, L. G., Dohanich, G. P., & Witcher, J. A. (1981). Cholinergic influences on estrogen-dependent sexual behavior in female rats. *J. Comp. Physiol. Psychol., 95*, 763–770.

Clemens, L. G., Wee, B. E., Weaver, D. R., Roy, E. J., Goldman, B. D., & Rakerd, B. (1988a). Retention of masculine sexual behavior following castration in male B6D2F1 mice. *Physiology & behavior, 42*, 69–76.

Clemens, L. G., Wee, B. E., Weaver, D. R., Roy, E. J., Goldman, B. D., & Rakerd, B. (1988b). Retention of masculine sexual behavior following castration in male B6D2F1 mice. *Physiol. Behav., 42*, 69–76.

Coquelin, A. (1991). Persistent sexual behavior in castrated, recombinant inbred mice. *Biology of reproduction, 45*, 680–684.

Coquelin, A., & Desjardins, C. (1982). Luteinizing hormone and testosterone secretion in young and old male mice. *Am. J. Physiol., 243*, E257–E263.

Costantini, R. M., Park, J. H., Beery, A. K., Paul, M. J., Ko, J. J., & Zucker, I. (2007). Post-castration retention of reproductive behavior and olfactory preferences in male Siberian hamsters: role of prior experience. *Hormones and Behavior, 51*, 149–155.

Crowell-Davis, S. L. (2007). Sexual behavior of mares. *Horm. Behav., 52*, 12–17.

Crowley, W. R. (1986). Reproductive neuroendocrine regulation in the female rat by central catecholamine-neuropeptide interactions: a local control hypothesis. *Ann. NY. Acad. Sci., 474*, 423–436.

Cunningham, G. R., Hirshkowitz, M., Korenman, S. G., & Karacan, I. (1990). Testosterone replacement therapy and sleep-related erections in hypogonadal men. *The Journal of clinical endocrinology and metabolism, 70*, 792–797.

Cushing, B. S., & Carter, C. S. (1999). Prior exposure to oxytocin mimics the effects of social contact and facilitates sexual behaviour in females. *J. Neuroendocrinol., 11*, 765–769.

Daftary, S. S., & Gore, A. C. (2005). IGF-1 in the brain as a regulator of reproductive neuroendocrine function. *Exp. Biol. Med. (Maywood), 230*, 292–306.

Damassa, D. A., Smith, E. R., Tennent, B., & Davidson, J. M. (1977). The relationship between circulating testosterone levels and male sexual behavior in rats. *Hormones and Behavior, 8*, 275–286.

Danilovich, N., Wernsing, D., Coschigano, K. T., Kopchick, J. J., & Bartke, A. (1999). Deficits in female reproductive function in GH-R-KO mice; role of IGF-I. *Endocrinology, 140*, 2637–2640.

Davidson, J. M. (1966a). Characteristics of sex behaviour in male rats following castration. *Anim. Behav., 14*, 266–272.

Davidson, J. M. (1966b). Characteristics of sex behaviour in male rats following castration. *Animal behaviour, 14*, 266–272.

Davidson, J. M., Camargo, C. A., & Smith, E. R. (1979). Effects of androgen on sexual behavior in hypogonadal men. *The Journal of clinical endocrinology and metabolism, 48*, 955–958.

Davidson, J. M., Rodgers, C. H., Smith, E. R., & Bloch, G. J. (1968). Stimulation of female sex behavior in adrenalectomized rats with estrogen alone. *Endocrinology, 82*, 193–195.

Davis, B., Macrides, F., Youngs, W., Schneider, S., & Rosene, D. (1978). Efferents and centrifugal afferents of the main and accessory olfactory bulbs in the hamster. *Brain Res. Bull., 3*, 59–72.

Davis, P. G., & Barfield, R. J. (1979). Activation of masculine sexual behavior by intracranial estradiol benzoate implants in male rats. *Neuroendocrinology, 28*, 217–227.

Davis, S. R., Davison, S. L., Donath, S., & Bell, R. J. (2005). Circulating androgen levels and self-reported sexual function in women. *Jama, 294*, 91–96.

De Vries, G. J., Rissman, E. F., Simerly, R. B., Yang, L. Y., Scordalakes, E. M., Auger, C. J., et al. (2002). A model system for study of sex chromosome effects on sexually dimorphic neural and behavioral traits. *J. Neurosci., 22*, 9005–9014.

De Waal, F. B. (1995). Bonobo sex and society. *Sci Am, 272*, 82–88.

Dellovade, T. L., King, J. A., Millar, R. P., & Rissman, E. F. (1993). Presence and differential distribution of distinct forms of immunoreactive gonadotropin-releasing hormone in the musk shrew brain. *Neuroendocrinology, 58*, 166–177.

Demay, F., De Monti, M., Tiffoche, C., Vaillant, C., & Thieulant, M. L. (2001). Steroid-independent activation of ER by GnRH in gonadotrope pituitary cells. *Endocrinology, 142*, 3340–3347.

Denner, L. A., Weigel, N. L., Maxwell, B. L., Schrader, W. T., & O'Malley, B. W. (1990). Regulation of progesterone receptor-mediated transcription by phosphorylation. *Science, 250*, 1740–1743.

Dennerstein, L., Randolph, J., Taffe, J., Dudley, E., & Burger, H. (2002). Hormones, mood, sexuality, and the menopausal transition. *Fertil. Steril. 77. Suppl, 4*, S42–S48.

Dewsbury, D. A. (1972). Patterns of copulatory behavior in male mammals. *Q. Rev. Biol., 47*, 1–33.

Diamond, L. E., Earle, D. C., Heiman, J. R., Rosen, R. C., Perelman, M. A., & Harning, R. (2006). An effect on the subjective sexual response in premenopausal women with sexual arousal disorder by bremelanotide (PT-141), a melanocortin receptor agonist. *J. Sex. Med., 3*, 628–638.

Dixon, A. F., Everitt, G. J., Herbert, J., Rugman, S. J., & Scruton, D. M. (1973). Hormonal and other determinants of sexual attractiveness and receptivity in rhesus and talapoin monkeys. In *IVth International Congress of Primatology, Vol 2* (pp. 36–63). Basel, Switzerland: Karger.

Doggrell, S. (2007). Do vardenafil and tadalafil have advantages over sildenafil in the treatment of erectile dysfunction? *International journal of impotence research, 19*, 281–295.

Dohanich, G. P., & Clemens, L. G. (1981). Brain areas implicated in cholinergic regulation of sexual behavior. *Horm. Behav., 15*, 157–167.

Dohanich, G. P., & Clemens, L. G. (1983). Inhibition of estrogen-activated sexual behavior by androgens. *Horm. Behav., 17*, 366–373.

Doherty, P. C., Baum, M. J., & Todd, R. B. (1986). Effects of chronic hyperprolactinemia on sexual arousal and erectile function in male rats. *Neuroendocrinology, 42*, 368–375.

Dominguez-Salazar, E., Bateman, H. L., & Rissman, E. F. (2004). Background matters: the effects of estrogen receptor alpha gene disruption on male sexual behavior are modified by background strain. *Horm. Behav., 46*, 482–490.

Dornan, W. A., Malsbury, C. W., & Penney, R. B. (1987). Facilitation of lordosis by injection of substance P into the midbrain central gray. *Neuroendocrinology, 45*, 498–506.

Ellis, G. B., & Desjardins, C. (1982). Male rats secrete luteinizing hormone and testosterone episodically. *Endocrinology, 110*, 1618–1627.

Erskine, M. S. (1985). Effects of paced coital stimulation on estrus duration in intact cycling rats and ovariectomized and ovariectomized-adrenalectomized hormone-primed rats. *Behav. Neurosci., 99*, 151–161.

Erskine, M. S., & Hanrahan, S. B. (1997). Effects of paced mating on c-fos gene expression in the female rat brain. *J. Neuroendocrinol., 9*, 903–912.

Erskine, M. S., Hippensteil, M., & Kornberg, E. (1992). Metabolism of dihydrotestosterone to 3 alpha-androstanediol in brain and plasma: effect on behavioural activity in female rats. *J. Endocrinol., 134*, 183–195.

Erskine, M. S., MacLusky, N. J., & Baum, M. J. (1985). Effect of 5 alpha-dihydrotestosterone on sexual receptivity and neural progestin receptors in ovariectomized rats given pulsed estradiol. *Biol. Reprod., 33*, 551–559.

Etgen, A. M., & Acosta-Martinez, M. (2003). Participation of growth factor signal transduction pathways in estradiol facilitation of female reproductive behavior. *Endocrinology, 144*, 3828–3835.

Etgen, A. M., Ansonoff, M. A., & Quesada, A. (2001). Mechanisms of ovarian steroid regulation of norepinephrine receptor-mediated signal transduction in the hypothalamus: implications for female reproductive physiology. *Horm. Behav., 40*, 169–177.

Etgen, A. M., Chu, H. P., Fiber, J. M., Karkanias, G. B., & Morales, J. M. (1999). Hormonal integration of neurochemical and sensory signals governing female reproductive behavior. *Behav. Brain Res., 105*, 93–103.

Etgen, A. M., Gonzalez-Flores, O., & Todd, B. J. (2006). The role of insulin-like growth factor-I and growth factor-associated signal transduction pathways in estradiol and progesterone facilitation of female reproductive behaviors. *Front Neuroendocrinol., 27*, 363–375.

Etgen, A. M., Ungar, S., & Petitti, N. (1992). Estradiol and progesterone modulation of norepinephrine neurotransmission: Implications for the regulation of female reproductive behavior. *J. Neuroendocrinol., 4*, 255–271.

Etgen, A. M., Vathy, I., Petitti, N., Ungar, S., & Karkanias, G. B. (1990). Ovarian steroids, female reproductive behavior, and norepinephrine neurotransmission in the hypothalamus. *Comparative Physiology, 8*, 116–128.

Everitt, B. J. (1990). Sexual motivation: a neural and behavioural analysis of the mechanisms underlying appetitive and copulatory responses of male rats. *Neurosci. Biobehav. Rev., 14*, 217–232.

Everitt, B. J., & Fuxe, K. (1977). Serotonin and the sexual behavior of female rats: effects of hallucinogenic indolealkylamines and phenylethylamines. *Neurosci. Lett., 4*, 213–220.

Everitt, B. J., Fuxe, K., & Hokfelt, T. (1974). Inhibitory role of dopamine and 5-hydroxytryptamine in the sexual behaviour of female rats. *Eur. J. Pharmacol., 29*, 187–191.

Fabre-Nys, C. (1998). Steroid control of monoamines in relation to sexual behaviour. *Rev. Reprod., 3*, 31–41.

Fabre-Nys, C., & Gelez, H. (2007). Sexual behavior in ewes and other domestic ruminants. *Horm. Behav., 52*, 18–25.

Fadem, B. H., Barfield, R. J., & Whalen, R. E. (1979). Dose-response and time-response relationships between progesterone and the display of patterns of receptive and proceptive behavior in the female rat. *Horm. Behav., 13*, 40–48.

Feder, H. H. (1984). Hormones and sexual behavior. *Annu. Rev. Psychol, 35*, 165–200.

Fernandez-Fewell, G. D., & Meredith, M. (1995). Facilitation of mating behavior in male hamsters by LHRH and AcLHRH5-10: interaction with the vomeronasal system. *Physiol. Behav., 57*, 213–221.

Fernandez-Guasti, A., & Escalante, A. (1991). Role of presynaptic serotonergic receptors on the mechanism of action of 5-HT1A and 5-HT1B agonists on masculine sexual behaviour: physiological and pharmacological implications. *J. Neural. Transm. Gen. Sect., 85*, 95–107.

Fernandez-Guasti, A., Escalante, A. L., Ahlenius, S., Hillegaart, V., & Larsson, K. (1992). Stimulation of 5-HT1A and 5-HT1B receptors in brain regions and its effects on male rat sexual behaviour. *Eur. J. Pharmacol., 210*, 121–129.

Fernandez-Guasti, A., Kruijver, F. P., Fodor, M., & Swaab, D. F. (2000). Sex differences in the distribution of androgen receptors in the human hypothalamus. *The Journal of comparative neurology, 425*, 422–435.

Fernandez-Guasti, A., Larsson, K., & Beyer, C. (1985). Potentiative action of alpha- and beta-adrenergic receptor stimulation in inducing lordosis behavior. *Pharmacol. Biochem. Behav., 22*, 613–617.

Ferrari, F., & Giuliani, D. (1995). Sexual attraction and copulation in male rats: effects of the dopamine agonist SND 919. *Pharmacol. Biochem. Behav., 50*, 29–34.

Ferrari, F., & Giuliani, D. (1996). Behavioral effects induced by the dopamine D3 agonist 7-OH-DPAT in sexually-active and -inactive male rats. *Neuropharmacology, 35*, 279–284.

Ferris, C. F., Snowdon, C. T., King, J. A., Sullivan, J. M., Jr., Ziegler, T. E., Olson, D. P., et al. (2004). Activation of neural pathways associated with sexual arousal in non-human primates. *Journal of magnetic resonance imaging, 19*, 168–175.

Floody, O. R., Blinn, N. E., Lisk, R. D., & Vomachka, A. J. (1987). Localization of hypothalamic sites for the estrogen-priming of sexual receptivity in female hamsters. *Behav. Neurosci., 101*, 309–314.

Foreman, M. M., & Moss, R. L. (1978). Role of hypothalamic alpha and beta adrenergic receptors in the control of lordotic behavior in the ovariectomized-estrogen primed rat. *Pharmacol. Biochem. Behav., 9*, 235–241.

Foreman, M. M., Hall, J. L., & Love, R. L. (1989). The role of the 5-HT2 receptor in the regulation of sexual performance of male rats. *Life Sci., 45*, 1263–1270.

Frank, J. L., Hendricks, S. E., & Olson, C. H. (2000). Multiple ejaculations and chronic fluoxetine: effects on male rat copulatory behavior. *Pharmacol. Biochem. Behav., 66*, 337–342.

Frye, C. A. (2001). The role of neurosteroids and non-genomic effects of progestins and androgens in mediating sexual receptivity of rodents. *Brain Res. Brain. Res. Rev., 37*, 201–222.

Frye, C. A., Duncan, J. E., Basham, M., & Erskine, M. S. (1996). Behavioral effects of 3 alpha-androstanediol. II: Hypothalamic and preoptic area actions via a GABAergic mechanism. *Behav. Brain Res., 79*, 119–130.

Garcia-Sergura, L., Perez, J., Pons, S., Rejas, M., & Torres-Aleman, I. (1991). Localization of insulin-like growth factor I (IGF-1)-like immunoreactivity in the developing and adult rat brain. *Brain Res., 560*, 167–174.

Gerall, A. A., & Dunlap, J. L. (1973). The effect of experience and hormones on the initial receptivity in female and male rats. *Physiol. Behav., 10*, 851–854.

Ginton, A., & Merari, A. (1977). Long range effects of MPOA lesion on mating behavior in the male rat. *Brain research, 120*, 158–163.

Giuliani, D., & Ferrari, F. (1996). Differential behavioral response to dopamine D2 agonists by sexually naive, sexually active, and sexually inactive male rats. *Behav. Neurosci., 110*, 802–808.

Giuliano, F. (2004). Control of penile erection by the melanocortinergic system: experimental evidences and therapeutic perspectives. *J. Androl., 25*, 683–691.

Giuliano, F., & Allard, J. (2001). Dopamine and male sexual function. *Eur. Urol., 40*, 601–608.

Giuliano, F., Bernabe, J., Alexandre, L., Niewoehner, U., Haning, H., & Bischoff, E. (2003). Pro-erectile effect of vardenafil: *in-vitro* experiments in rabbits and *in-vivo* comparison with sildenafil in rats. *Eur. Urol., 44*, 731–736.

Gomez-Marrero, J., Feria, M., & Mas, M. (1988). Stimulation of opioid receptors suppresses penile erectile reflexes and seminal emission in rats. *Pharmacol. Biochem. Behav., 31*, 393–396.

Gomez, D., & Newman, S. (1992). Differential projections of the anterior and posterior regions of the medial amygdaloid nucleus in the Syrian hamster. *J. Comp. Neurol., 317*, 195–218.

Gorzalka, B. B., Mendelson, S. D., & Watson, N. V. (1990). Serotonin receptor subtypes and sexual behavior. *Ann. NY. Acad. Sci., 600*, 435–444, discussion 445–456.

Gozes, I., Meltzer, E., Rubinrout, S., Brenneman, D. E., & Fridkin, M. (1989). Vasoactive intestinal peptide potentiates sexual behavior: inhibition by novel antagonist. *Endocrinology, 125*, 2945–2949.

Greco, B., Edwards, D. A., Zumpe, D., & Clancy, A. N. (1998). Androgen receptor and mating-induced fos immunoreactivity are co-localized in limbic and midbrain neurons that project to the male rat medial preoptic area. *Brain Research, 781*, 15–24.

Gulinello, M., & Etgen, A. M. (2005). Sexually dimorphic hormonal regulation of the gap junction protein, CX43, in rats and altered female reproductive function in CX43+/- mice. *Brain Res., 1045*, 107–115.

Hadeishi, Y., & Wood, R. I. (1996). Nitric oxide synthase in mating behavior circuitry of male Syrian hamster brain. *Journal of neurobiology, 30*, 480–492.

Hamann, S., Herman, R. A., Nolan, C. L., & Wallen, K. (2004a). Men and women differ in amygdala response to visual sexual stimuli. *Nature neuroscience, 7*, 411–416.

Hamann, S., Herman, R. A., Nolan, C. L., & Wallen, K. (2004b). Men and women differ in amygdala response to visual sexual stimuli. *Nat. Neurosci., 7*, 411–416.

Hammond, T. (1943). The function of the testes after puberty. *British journal of urology, 6*, 128–141.

Hamson, D. K., Csupity, A. S., Ali, F. T., & Watson, N. V. (2005). Mating behavior and CNS morphology in rats carrying the testicular feminization mutation. *Soc. Neurosci., 320*, 19.

Hanbauer, I., Wink, D., Osawa, Y., Edelman, G. M., & Gally, J. A. (1992). Role of nitric oxide in NMDA-evoked release of [3H]-dopamine from striatal slices. *Neuroreport, 3*, 409–412.

Hansen, S., Stanfield, E. J., & Everitt, B. J. (1980). The role of ventral bundle noradrenergic neurones in sensory components of sexual behaviour and coitus-induced pseudopregnancy. *Nature, 286*, 152–154.

Hansen, S., Stanfield, E. J., & Everitt, B. J. (1981). The effects of lesions of lateral tegmental noradrenergic neurons on components of sexual behavior and pseudopregnancy in female rats. *Neuroscience, 6*, 1105–1117.

Harlan, R. E., Shivers, B. D., & Pfaff, D. W. (1983). Midbrain microinfusions of prolactin increase the estrogen-dependent behavior, lordosis. *Science, 219*, 1451–1453.

Hart, B. L. (1968). Role of prior experience in the effects of castration on sexual behavior of male dogs. *Journal of comparative and physiological psychology, 66*, 719–725.

Hart, B. L. (1974). Gonadal androgen and sociosexual behavior of male mammals: a comparative analysis. *Psychological bulletin, 81*, 383–400.

Hasegawa, T., & Sakuma, Y. (1993). Developmental effect of testosterone on estrogen sensitivity of the rat preoptic neurons with axons to the ventral tegmental area. *Brain Res., 611*, 1–6.

Heim, N., & Hursch, C. J. (1979). Castration for sex offenders: Treatment or punishment? A review and critique of recent European literature. *Archives of sexual behavior, 8*, 281–304.

Herbison, A. E. (1997). Noradrenergic regulation of cyclic GnRH secretion. *Rev. Reprod., 2*, 1–6.

Hestiantoro, A., & Swaab, D. F. (2004). Changes in estrogen receptor-alpha and -beta in the infundibular nucleus of the human hypothalamus are related to the occurrence of Alzheimer's disease neuropathology. *The Journal of clinical endocrinology and metabolism, 89*, 1912–1925.

Hillegaart, V., & Ahlenius, S. (1998). Facilitation and inhibition of male rat ejaculatory behaviour by the respective 5-HT1A and 5-HT1B receptor agonists 8-OH-DPAT and anpirtoline, as evidenced by use of the corresponding new and selective receptor antagonists NAD-299 and NAS-181. *Br. J. Pharmacol., 125*, 1733–1743.

Hisasue, S., Kato, R., Takahashi, A., Masumori, N., Itoh, N., Miyao, N., et al. (2004). Erectile function following external beam radiotherapy for clinically organ-confined or locally advanced prostate cancer. *Japanese journal of clinical oncology, 34*, 269–273.

Holstege, G., Georgiadis, J. R., Paans, A. M., Meiners, L. C., Van der Graaf, F. H., & Reinders, A. A. (2003). Brain activation during human male ejaculation. *Journal of neuroscience, 23*, 9185–9193.

Hoshina, Y., Takeo, T., Nakano, K., Sato, T., & Sakuma, Y. (1994). Axon-sparing lesion of the preoptic area enhances receptivity and diminishes proceptivity among components of female rat sexual behavior. *Behavioural brain research, 61*, 197–204.

Hrabovszky, E., Kallo, I., Steinhauser, A., Merchenthaler, I., Coen, C. W., Petersen, S. L., et al. (2004). Estrogen receptor-beta in oxytocin and vasopressin neurons of the rat and human hypothalamus: Immunocytochemical and *in-situ* hybridization studies. *The Journal of comparative neurology, 473*, 315–333.

Huang, A. S., Beigneux, A., Weil, Z. M., Kim, P. M., Molliver, M. E., Blackshaw, S., et al. (2006). D-aspartate regulates melanocortin formation and function: behavioral alterations in D-aspartate oxidase-deficient mice. *J. Neurosci., 26*, 2814–2819.

Hubayter, Z., & Simon, J. A. (2008). Testosterone therapy for sexual dysfunction in postmenopausal women. *Climacteric, 11*, 181–191.

Hull, E., Wood, RI, McKenna, KE. (2006). Neurobiology of male sexual behavior. In J.D. Neill, et al., (Eds.), *Knobil and Neill's Physiology of Reproduction, Third Edition* (pp. 1729–1824), New York.

Hull, E. M., & Dominguez, J. M. (2006). Getting his act together: roles of glutamate, nitric oxide, and dopamine in the medial preoptic area. *Brain Res., 1126*, 66–75.

Hull, E. M., Du, J., Lorrain, D. S., & Matuszewich, L. (1997). Testosterone, preoptic dopamine, and copulation in male rats. *Brain Res. Bull., 44*, 327–333.

Hull, E. M., Lumley, L. A., Matuszewich, L., Dominguez, J., Moses, J., & Lorrain, D. S. (1994). The roles of nitric oxide in sexual function of male rats. *Neuropharmacology, 33*, 1499–1504.

Hunter, A. J., Hole, D. R., & Wilson, C. A. (1985). Studies into the dual effects of serotonergic pharmacological agents on female sexual behaviour in the rat: preliminary evidence that endogenous 5HT is stimulatory. *Pharmacol. Biochem. Behav., 22*, 5–13.

Ignar-Trowbridge, D. M., Nelson, K. G., Bidwell, M. C., Curtis, S. W., Washburn, T. F., McLachlan, J. A., et al. (1992). Coupling of dual signaling pathways: epidermal growth factor action involves the estrogen receptor. *Proc. Natl. Acad. Sci. USA, 89*, 4658–4662.

Incrocci, L., Slob, A. K., & Levendag, P. C. (2002). Sexual (dys)function after radiotherapy for prostate cancer: a review. *International journal of radiation oncology, biology, physics, 52*, 681–693.

Ingman, W. V., & Robertson, S. A. (2007). Transforming growth factor-beta1 null mutation causes infertility in male mice associated with testosterone deficiency and sexual dysfunction. *Endocrinology, 148*, 4032–4043.

Ingman, W. V., & Robertson, S. A. (2008). Mammary gland development in transforming growth factor beta1 null mutant mice: systemic and epithelial effects. *Biol. Reprod., 79*, 711–717.

Ingman, W. V., McGrath, L. M., Breed, W. G., Musgrave, I. F., Robker, R. L., & Robertson, S. A. (2009). The mechanistic basis for sexual dysfunction in male transforming growth factor beta1 null mutant mice. *J. Androl.*

Isidori, A. M., Giannetta, E., Gianfrilli, D., Greco, E. A., Bonifacio, V., Aversa, A., et al. (2005). Effects of testosterone on sexual function in men: results of a meta-analysis. *Clinical endocrinology, 63*, 381–394.

Jasnow, A. M., Mong, J. A., Romeo, R. D., & Pfaff, D. W. (2007). Estrogenic regulation of gene and protein expression within the amygdala of female mice. *Endocrine, 32*, 271–279.

Jenkins, W. J., & Becker, J. B. (2003). Female rats develop conditioned place preferences for sex at their preferred interval. *Horm. Behav., 43*, 503–507.

Joel, P. B., Traish, A. M., & Lannigan, D. A. (1995). Estradiol and phorbol ester cause phosphorylation of serine 118 in the human estrogen receptor. *Mol. Endocrinol., 9*, 1041–1052.

Johansen, J. A., Clemens, L. G., & Nunez, A. A. (2008). Characterization of copulatory behavior in female mice: evidence for paced mating. *Physiol. Behav., 95*, 425–429.

Johnson, S. A., Marin-Bivens, C. L., Miele, M., Coyle, C. A., Fissore, R., & Good, D. J. (2004). The Nhlh2 transcription factor is required for female sexual behavior and reproductive longevity. *Horm. Behav., 46*, 420–427.

Jyotika, J., McCutcheon, J., Laroche, J., Blaustein, J. D., & Forger, N. G. (2007). Deletion of the Bax gene disrupts sexual behavior and modestly impairs motor function in mice. *Dev. Neurobiol., 67*, 1511–1519.

Kalra, S. P. (1993). Mandatory neuropeptide-steroid signaling for the preovulatory luteinizing hormone-releasing hormone discharge. *Endocr. Rev., 14*, 507–538.

Karama, S., Lecours, A. R., Leroux, J. M., Bourgouin, P., Beaudoin, G., Joubert, S., et al. (2002). Areas of brain activation in males and females during viewing of erotic film excerpts. *Hum. Brain Mapp., 16*, 1–13.

Kaser, M., Lakshmanan, J., & Fisher, D. (1992). Comparison between epidermal growth factor, transforming growth factor-a and EGF receptor levels in regions of adult rat brain. *Br. Res. Mol. Brain Res., 16*, 316–322.

Kasten, T. L., White, S. A., Norton, T. T., Bond, C. T., Adelman, J. P., & Fernald, R. D. (1996). Characterization of two new preproGnRH mRNAs in the tree shrew: first direct evidence for mesencephalic GnRH gene expression in a placental mammal. *Gen. Comp. Endocrinol., 104*, 7–19.

Kato, S., Endoh, H., Masuhiro, Y., Kitamoto, T., Uchiyama, S., Sasaki, H., et al. (1995). Activation of the estrogen receptor through phosphorylation by mitogen-activated protein kinase. *Science, 270*, 1491–1494.

Katz, L. S. (2007). Sexual behavior of domesticated ruminants. *Horm. Behav., 52*, 56–63.

Kauffman, A. S., & Rissman, E. F. (2004). The evolutionarily conserved gonadotropin-releasing hormone II modifies food intake. *Endocrinology, 145*, 686–691.

Kauffman, A. S., Buenzle, J., Fraley, G. S., & Rissman, E. F. (2005). Effects of galanin-like peptide (GALP) on locomotion, reproduction, and body weight in female and male mice. *Horm. Behav., 48*, 141–151.

Kauffman, A. S., Park, J. H., McPhie-Lalmansingh, A. A., Gottsch, M. L., Bodo, C., Hohmann, J. G., et al. (2007). The kisspeptin receptor GPR54 is required for sexual differentiation of the brain and behavior. *J. Neurosci., 27*, 8826–8835.

Kayasuga, Y., Chiba, S., Suzuki, M., Kikusui, T., Matsuwaki, T., Yamanouchi, K., et al. (2007). Alteration of behavioural phenotype in mice by targeted disruption of the progranulin gene. *Behav. Brain. Res., 185*, 110–118.

Kelly, M. J., Qiu, J., & Ronnekleiv, O. K. (2003). Estrogen modulation of G-protein-coupled receptor activation of potassium channels in the central nervous system. *Annals of the New York Academy of Sciences, 1007*, 6–16.

Kendrick, K. M., & Dixson, A. F. (1985). Luteinizing hormone releasing hormone enhances proceptivity in a primate. *Neuroendocrinology, 41*, 449–453.

Kennedy, G. C. (1964). Hypothalamic control of the endocrine and behavioural changes associated with oestrus in the rat. *J. Physiol., 172*, 383–392.

Kevetter, G., & Winans, S. (1981). Connections of the corticomedial amygdala in the golden hamster. I. Efferents of the vomeronasal amygdala. *J. Comp. Neurol., 197*, 81–98.

Klotz, D. M., Hewitt, S. C., Ciana, P., Raviscioni, M., Lindzey, J. K., Foley, J., et al. (2002). Requirement of estrogen receptor-alpha in insulin-like growth factor-1 (IGF-1)-induced uterine responses and *in-vivo* evidence for IGF-1/estrogen receptor cross-talk. *J. Biol. Chem., 277*, 8531–8537.

Komisaruk, B. R., Whipple, B., Crawford, A., Liu, W. C., Kalnin, A., & Mosier, K. (2004). Brain activation during vaginocervical self-stimulation and orgasm in women with complete spinal cord injury: fMRI evidence of mediation by the vagus nerves. *Brain Res., 1024*, 77–88.

Kow, L. M., Harlan, R. E., Shivers, B. D., & Pfaff, D. W. (1985). Inhibition of the lordosis reflex in rats by intrahypothalamic infusion of neural excitatory agents: evidence that the hypothalamus contains separate inhibitory and facilitatory elements. *Brain Res., 341*, 26–34.

Kow, L. M., Paden, C. M., & Pfaff, D. W. (1981). Dispensability of spinal monoaminergic systems in mediating the lordosis reflex of the female rat. *Pharmacol. Biochem. Behav., 14*, 707–711.

Kow, L. M., Weesner, G. D., & Pfaff, D. W. (1992). Alpha 1-adrenergic agonists act on the ventromedial hypothalamus to cause neuronal excitation and lordosis facilitation: electrophysiological and behavioral evidence. *Brain. Res., 588*, 237–245.

Krege, J. H., Hodgin, J. B., Couse, J. F., Enmark, E., Warner, M., Mahler, J. F., et al. (1998). Generation and reproductive phenotypes of mice lacking estrogen receptor beta. *Proc. Natl. Acad. Sci. USA, 95*, 15677–15682.

Kudwa, A. E., & Rissman, E. F. (2003). Double oestrogen receptor alpha and beta knockout mice reveal differences in neural oestrogen-mediated progestin receptor induction and female sexual behaviour. *J. Neuroendocrinol., 15*, 978–983.

Kudwa, A. E., Boon, W. C., Simpson, E. R., Handa, R. J., & Rissman, E. F. (2007). Dietary phytoestrogens dampen female sexual behavior in mice with a disrupted aromatase enzyme gene. *Behav. Neurosci., 121*, 356–361.

Kudwa, A. E., Dominguez-Salazar, E., Cabrera, D. M., Sibley, D. R., & Rissman, E. F. (2005). Dopamine D5 receptor modulates male and female sexual behavior in mice. *Psychopharmacology (Berl), 180*, 206–214.

Kuppers, E., Ivanova, T., Karolczak, M., & Beyer, C. (2000). Estrogen: a multifunctional messenger to nigrostriatal dopaminergic neurons. *Journal of neurocytology, 29*, 375–385.

Laing, M. A., Coonrod, S., Hinton, B. T., Downie, J. W., Tozer, R., Rudnicki, M. A., et al. (2000). Male sexual dysfunction in mice bearing targeted mutant alleles of the PEA3 ets gene. *Mol. Cell Biol., 20*, 9337–9345.

Landgraf, R., & Neumann, I. D. (2004). Vasopressin and oxytocin release within the brain: a dynamic concept of multiple and variable modes of neuropeptide communication. *Front Neuroendocrinol., 25*, 150–176.

Larsson, K. (1979). Features of the neuroendocrine regulation of masculine sexual behavior. In C. Beyer (Ed.), *Endocrine Control of Sexual Behavior* (pp. 77–163). New York, NY: Raven Press.

Laumann, E. O., Paik, A., & Rosen, R. C. (1999). Sexual dysfunction in the United States: prevalence and predictors. *Jama, 281*, 537–544.

Lee, A. W., Devidze, N., Pfaff, D. W., & Zhou, J. (2006). Functional genomics of sex hormone-dependent neuroendocrine systems: specific and generalized actions in the CNS. *Prog. Brain Res., 158*, 243–272.

Lee, Y. H., Huang, J. K., & Lu, C. M. (2003). The impact on sexual function after nerve sparing and non-nerve sparing radical retropubic prostatectomy. *Journal of the Chinese Medical Association, 66*, 13–18.

Leedy, M. G., & Hart, B. L. (1986). Medial preoptic-anterior hypothalamic lesions in prepubertal male cats: effects on juvenile and adult sociosexual behaviors. *Physiology & behavior, 36*, 501–506.

Lehman, M., & Winans, S. (1982). Vomeronasal and olfactory pathways to the amygdala controlling male hamster sexual behavior: autoradiographic and behavioral analyses. *Brain Res., 240*, 27–41.

Leyton, M., & Stewart, J. (1996). Acute and repeated activation of male sexual behavior by tail pinch: opioid and dopaminergic mechanisms. *Physiol. Behav., 60*, 77–85.

Lisk, R. D., & Heimann, J. (1980). The effects of sexual experience and frequency of testing on retention of copulatory behavior following castration in the male hamster. *Behavioral and neural biology, 28*, 156–171.

Livne, I., Silverman, A. J., & Gibson, M. J. (1992). Reversal of reproductive deficiency in the hpg male mouse by neonatal androgenization. *Biol. Reprod., 47*, 561–567.

Lobo, R. A., Rosen, R. C., Yang, H. M., Block, B., & Van Der Hoop, R. G. (2003). Comparative effects of oral esterified estrogens with and without methyltestosterone on endocrine profiles and dimensions of sexual function in postmenopausal women with hypoactive sexual desire. *Fertil. Steril., 79*, 1341–1352.

Lorrain, D. S., Matuszewich, L., & Hull, E. M. (1998). 8-OH-DPAT influences extracellular levels of serotonin and dopamine in the medial preoptic area of male rats. *Brain Res., 790*, 217–223.

Luine, V. N., Frankfurt, M., Rainbow, T. C., Biegon, A., & Azmitia, E. (1983). Intrahypothalamic 5,7-dihydroxytryptamine facilitates feminine sexual behavior and decreases [3H]imipramine binding and 5-HT uptake. *Brain Res., 264*, 344–348.

Luine, V. N., Renner, K. J., Frankfurt, M., & Azmitia, E. C. (1984). Facilitated sexual behavior reversed and serotonin restored by Raphe nuclei transplanted into denervated hypothalamus. *Science, 226*, 1436–1439.

Luttge, W. G., & Hall, N. R. (1973). Androgen-induced agonistic behavior in castrated male Swiss-Webster mice: comparison of four naturally occurring androgens. *Behavioral biology, 8*, 725–732.

Lydon, J. P., DeMayo, F. J., Conneely, O. M., & O'Malley, B. W. (1996). Reproductive phenotpes of the progesterone receptor null mutant mouse. *J. Steroid Biochem. Mol. Biol., 56*, 67–77.

Lydon, J. P., DeMayo, F. J., Funk, C. R., Mani, S. K., Hughes, A. R., Montgomery, C. A., Jr., et al. (1995). Mice lacking progesterone receptor exhibit pleiotropic reproductive abnormalities. *Genes Dev., 9*, 2266–2278.

Ma, X., Reyna, A., Mani, S. K., Matzuk, M. M., & Kumar, T. R. (2005). Impaired male sexual behavior in activin receptor type II knockout mice. *Biol. Reprod., 73*, 1182–1190.

Ma, Y., Dissen, G., Merlino, G., Coquelin, A., & Ojeda, S. (1994a). Overexpression of a human transforming growth factor- (TGF) transgene reveals a dual antagonistic role of TGF in female sexual development. *Endocrinology, 135*, 1392–1400.

Ma, Z. Q., Santagati, S., Patrone, C., Pollio, G., Vegeto, E., & Maggi, A. (1994b). Insulin-like growth factors activate estrogen receptor to control the growth and differentiation of the human neuroblastoma cell line SK-ER3. *Mol. Endocrinol., 8*, 910–918.

Maeda, N., Matsuoka, N., & Yamaguchi, I. (1990). Septohippocampal cholinergic pathway and penile erections induced by dopaminergic and cholinergic stimulants. *Brain Res., 537*, 163–168.

Mahalati, K., Okanoya, K., Witt, D. M., & Carter, C. S. (1991). Oxytocin inhibits male sexual behavior in prairie voles. *Pharmacol. Biochem. Behav., 39*, 219–222.

Malmnas, C. O. (1976). The significance of dopamine, versus other catecholamines, for L-dopa induced facilitation of sexual behavior in the castrated male rat. *Pharmacology, biochemistry, and behavior, 4*, 521–526.

Mandiyan, V. S., Coats, J. K., & Shah, N. M. (2005). Deficits in sexual and aggressive behaviors in Cnga2 mutant mice. *Nat. Neurosci., 8*, 1660–1662.

Mani, S. K., Allen, J. M., Clark, J. H., Blaustein, J. D., & O'Malley, B. W. (1995). Steroid hormone and neurotransmitter-induced rat sexual behavior: addendum. *Science, 268*, 1833.

Mani, S. K., Blaustein, J. D., & O'Malley, B. W. (1997). Progesterone receptor function from a behavioral perspective. *Horm. Behav., 31*, 244–255.

Mani, S. K., Fienberg, A. A., O'Callaghan, J. P., Snyder, G. L., Allen, P. B., Dash, P. K., et al. (2000). Requirement for DARPP-32 in progesterone-facilitated sexual receptivity in female rats and mice. *Science, 287*, 1053–1056.

Mani, S. K., Portillo, W., & Reyna, A. (2009). Steroid hormone action in the brain: cross talk between signalling pathways. *J. Neuroendocrinol.*

Mani, S. K., Reyna, A. M., Chen, J. Z., Mulac-Jericevic, B., & Conneely, O. M. (2006). Differential response of progesterone receptor isoforms in hormone-dependent and -independent facilitation of female sexual receptivity. *Mol. Endocrinol., 20*, 1322–1332.

Manning, A., & Thompson, M. L. (1976a). Postcastration retention of sexual behaviour in the male BDF1 mouse: the role of experience. *Anim. Behav., 24*, 523–533.

Manning, A., & Thompson, M. L. (1976b). Postcastration retention of sexual behaviour in the male BDF1 mouse: the role of experience. *Animal Behavior, 24*, 523–533.

Maragos, W., Newman, S., Lehman, M., & Powers, J. (1989). Neurons of origin and fiber trajectory of amygdalofugal projections to the medial preoptic area in Syrian hamsters. *J. Comp. Neurol., 280*, 59–71.

Martin, M. B., Franke, T. F., Stoica, G. E., Chambon, P., Katzenellenbogen, B. S., Stoica, B. A., et al. (2000). A role for Akt in mediating the estrogenic functions of epidermal growth factor and insulin-like growth factor I. *Endocrinology, 141*, 4503–4511.

Mastronardi, C., Smiley, G. G., Raber, J., Kusakabe, T., Kawaguchi, A., Matagne, V., et al. (2006). Deletion of the Ttf1 gene in differentiated neurons disrupts female reproduction without impairing basal ganglia function. *J. Neurosci., 26*, 13167–13179.

Matuszewich, L., Lorrain, D. S., Trujillo, R., Dominguez, J., Putnam, S. K., & Hull, E. M. (1999). Partial antagonism of 8-OH-DPAT'S effects on male rat sexual behavior with a D2, but not a 5-HT1A, antagonist. *Brain Res., 820*, 55–62.

Mayer, M., Stief, C. G., Truss, M. C., & Uckert, S. (2005). Phosphodiesterase inhibitors in female sexual dysfunction. *World J. Urol., 23*, 393–397.

Mazzucco, C. A., Walker, H. A., Pawluski, J. L., Lieblich, S. E., & Galea, L. A. (2008). ERalpha, but not ERbeta, mediates the expression of sexual behavior in the female rat. *Behav. Brain Res., 191*, 111–117.

McCarthy, M. M., Kleopoulos, S. P., Mobbs, C. V., & Pfaff, D. W. (1994). Infusion of antisense oligodeoxynucleotides to the oxytocin receptor in the ventromedial hypothalamus reduces estrogen-induced sexual receptivity and oxytocin receptor binding in the female rat. *Neuroendocrinology, 59*, 432–440.

McCarthy, M. M., Malik, K. F., & Feder, H. H. (1990). Increased GABAergic transmission in medial hypothalamus facilitates lordosis but has the opposite effect in preoptic area. *Brain Res., 507*, 40–44.

McCarthy, M. M., Masters, D. B., Fiber, J. M., Lopez-Colome, A. M., Beyer, C., Komisaruk, B. R., et al. (1991). GABAergic control of receptivity in the female rat. *Neuroendocrinology, 53*, 473–479.

McClintock, M. K. (1987). A functional approach to the behavioral endocrinology of rodents. In D. Crews (Ed.), *Psychobiology of Reproductive Behavior: An Evolutionary Perspective* (pp. 176–203). Englewood Cliffs, NJ: Prentice Hall.

McClintock, M. K., & Adler, N. T. (1978). The role of the female during copulation in wild and domestic Norway rats (*Rattus norvegicus*). *Behaviour, 67*, 67–96.

McDonald, P., Beyer, C., Newton, F., Brien, B., Baker, R., Tan, H. S., et al. (1970). Failure of 5alpha-dihydrotestosterone to initiate sexual behaviour in the castrated male rat. *Nature, 227*, 964–965.

McGill, T. E., & Blight, W. C. (1963). Effects of genotype on the recovery of sex drive in the male mouse. *J. Comp. Physiol. Psychol., 56*, 887–888.

McGill, T. E., & Manning, A. (1976). Genotype and retention of the ejaculatory reflex in castrated male mice. *Animal Behavior, 24*, 507–518.

McGill, T. E., & Tucker, G. R. (1964). Genotype and sex drive in intact and in castrated male mice. *Science, 145*, 514–515.

McGinnis, M. Y., & Dreifuss, R. M. (1989). Evidence for a role of testosterone-androgen receptor interactions in mediating masculine sexual behavior in male rats. *Endocrinology, 124*, 618–626.

McGinnis, M. Y., & Kahn, D. F. (1997). Inhibition of male sexual behavior by intracranial implants of the protein synthesis inhibitor anisomycin into the medial preoptic area of the rat. *Hormones and Behavior, 31*, 15–23.

McGinnis, M. Y., Gordon, J. H., & Gorski, R. A. (1980). Influence of gamma-aminobutyric acid on lordosis behavior and dopamine activity in estrogen primed spayed female rats. *Brain Res., 184*, 179–197.

McKenna, N. J., Lanz, R. B., & O'Malley, B. W. (1999). Nuclear receptor coregulators: cellular and molecular biology. *Endocrine reviews, 20*, 321–344.

Meijer, O. C., Steenbergen, P. J., & De Kloet, E. R. (2000). Differential expression and regional distribution of steroid receptor coactivators SRC-1 and SRC-2 in brain and pituitary. *Endocrinology, 141*, 2192–2199.

Meisel, R. L., & Mullins, A. J. (2006). Sexual experience in female rodents: cellular mechanisms and functional consequences. *Brain Res., 1126*, 56–65.

Meisel, R. L., & Pfaff, D. W. (1984). RNA and protein synthesis inhibitors: effects on sexual behavior in female rats. *Brain Res. Bull., 12*, 187–193.

Meisel, R. L., Joppa, M. A., & Rowe, R. K. (1996). Dopamine receptor antagonists attenuate conditioned place preference following sexual behavior in female Syrian hamsters. *Eur. J. Pharmacol., 309*, 21–24.

Melis, M. R., & Argiolas, A. (1995). Dopamine and sexual behavior. *Neurosci. Biobehav. Rev., 19*, 19–38.

Mendelson, S. D. (1992). A review and reevaluation of the role of serotonin in the modulation of lordosis behavior in the female rat. *Neurosci Biobehav Rev, 16*, 309–350.

Mendelson, S. D., & Gorzalka, B. B. (1985). A facilitatory role for serotonin in the sexual behavior of the female rat. *Pharmacol. Biochem. Behav., 22*, 1025–1033.

Mendelson, S. D., & Gorzalka, B. B. (1986). Serotonin type 2 antagonists inhibit lordosis behavior in the female rat: reversal with quipazine. *Life Sci., 38*, 33–39.

Mendelson, S. D., & Pfaus, J. G. (1989). Level searching: a new assay of sexual motivation in the male rat. *Physiol Behav, 45*, 337–341.

Meredith, M., & Howard, G. (1992). Intracerebroventricular LHRH relieves behavioral deficits due to vomeronasal organ removal. *Brain Res. Bull., 29*, 75–79.

Mermelstein, P. G., Becker, J. B., & Surmeier, D. J. (1996). Estradiol reduces calcium currents in rat neostriatal neurons via a membrane receptor. *The Journal of Neuroscience, 16*, 595–604.

Meyerson, B. J. (1964). Central nervous monoamines and hormone induced estrus behaviour in the spayed rat. *Acta. Physiol. Scand. Suppl, SUPPL, 241*, 1–32.

Meyerson, B. J. (1975). Drugs and sexual motivation in the female rat. In M. Sandier, & G. L. Gessa (Eds.), *Sexual behavior, pharmacology and biochemistry*. New York, NY: Raven Press.

Meyerson, B. J., Malmnas, C. O., & Everitt, B. J. (1985). In D. a. Adler, Pfaff, & R. W. Goy (Eds.), *Handbook of Behavioral Neurobiology. Neuropharmacology, neurotransmitters and sexual behavior in mammals, Vol. 7* (pp. 495–536). New York, NY: Plenum Press.

Micevych, P. E., Eckersell, C. B., Brecha, N., & Holland, K. L. (1997). Estrogen modulation of opioid and cholecystokinin systems in the limbic—hypothalamic circuit. *Brain Res. Bull., 44*, 335—343.

Michael, R. P., Clancy, A. N., & Zumpe, D. (1999). Effects of mating on c-fos expression in the brains of male macaques. *Physiology & behavior, 66*, 591—597.

Michael, R. P., Zumpe, D., & Bonsall, R. W. (1986). Comparison of the effects of testosterone and dihydrotestosterone on the behavior of male cynomolgus monkeys (*Macaca fascicularis*). *Physiology & behavior, 36*, 349—355.

Millar, R. P., Pawson, A. J., Morgan, K., Rissman, E. F., & Lu, Z. L. (2008). Diversity of actions of GnRHs mediated by ligand-induced selective signaling. *Front Neuroendocrinol., 29*, 17—35.

Miller, N. L., Bissonette, E. A., Bahnson, R., Wilson, J., & Theodorescu, D. (2003). Impact of a novel neoadjuvant and adjuvant hormone-deprivation approach on quality of life, voiding function, and sexual function after prostate brachytherapy. *Cancer, 97*, 1203—1210.

Milner, T. A., Ayoola, K., Drake, C. T., Herrick, S. P., Tabori, N. E., McEwen, B. S., et al. (2005). Ultrastructural localization of estrogen receptor beta immunoreactivity in the rat hippocampal formation. *The Journal of comparative neurology, 491*, 81—95.

Mitev, Y. A., Wolf, S. S., Almeida, O. F., & Patchev, V. K. (2003). Developmental expression profiles and distinct regional estrogen responsiveness suggest a novel role for the steroid receptor coactivator SRC-1 as discriminative amplifier of estrogen signaling in the rat brain. *The FASEB journal: official publication of the Federation of American Societies for Experimental Biology, 17*, 518—519.

Molinoff, P. B., Shadiack, A. M., Earle, D., Diamond, L. E., & Quon, C. Y. (2003). PT-141: a melanocortin agonist for the treatment of sexual dysfunction. *Ann. NY. Acad. Sci., 994*, 96—102.

Mong, J. A., Devidze, N., Frail, D. E., O'Connor, L. T., Samuel, M., Choleris, E., et al. (2003a). Estradiol differentially regulates lipocalin-type prostaglandin D synthase transcript levels in the rodent brain: evidence from high-density oligonucleotide arrays and *in-situ* hybridization. *Proc. Natl. Acad. Sci. USA, 100*, 318—323.

Mong, J. A., Devidze, N., Goodwillie, A., & Pfaff, D. W. (2003b). Reduction of lipocalin-type prostaglandin D synthase in the preoptic area of female mice mimics estradiol effects on arousal and sex behavior. *Proc. Natl. Acad. Sci. USA, 100*, 15206—15211.

Montorsi, F., Corbin, J., & Phillips, S. (2004). Review of phosphodiesterases in the urogenital system: new directions for therapeutic intervention. *The journal of sexual medicine, 1*, 322—336.

Moore, K. E., & Lookingland, K. J. (1995). Dopaminergic neuronal systems in the hypothalamus. In F. E. Bllom, & D. J. Kupfer (Eds.), *Psychopharmacology: the fourth generation of progress* (pp. 245—246). New York, NY: Raven Press.

Moss, R. L., & Foreman, M. M. (1976). Potentiation of lordosis behavior by intrahypothalamic infusion of synthetic luteinizing hormone-releasing hormone. *Neuroendocrinology, 20*, 176—181.

Moss, R. L., & McCann, S. M. (1973). Induction of mating behavior in rats by luteinizing hormone-releasing factor. *Science, 181*, 177—179.

Moss, R. L., McCann, S. M., & Dudley, C. A. (1975). Releasing hormones and sexual behavior. *Prog. Brain Res., 42*, 37—46.

Mouras, H., Stoleru, S., Bittoun, J., Glutron, D., Pelegrini-Issac, M., Paradis, A. L., et al. (2003). Brain processing of visual sexual stimuli in healthy men: a functional magnetic resonance imaging study. *Neuroimage, 20*, 855—869.

Musatov, S., Chen, W., Pfaff, D. W., Kaplitt, M. G., & Ogawa, S. (2006). RNAi-mediated silencing of estrogen receptor {alpha} in the ventromedial nucleus of hypothalamus abolishes female sexual behaviors. *Proc. Natl. Acad. Sci. USA, 103*, 10456—10460.

Myers, B. M., & Baum, M. J. (1980). Facilitation of copulatory performance in male rats by naloxone: effects of hypophysectomy, 17 alpha-estradiol, and luteinizing hormone releasing hormone. *Pharmacol. Biochem. Behav., 12*, 365—370.

Myers, L. S., Dixen, J., Morrissette, D., Carmichael, M., & Davidson, J. M. (1990). Effects of estrogen, androgen, and progestin on sexual psychophysiology and behavior in postmenopausal women. *J. Clin. Endocrinol. Metab., 70*, 1124—1131.

Naess, O., Haug, E., Attramadal, A., Aakvaag, A., Hansson, V., & French, F. (1976). Androgen receptors in the anterior pituitary and central nervous system of the androgen "insensitive" (Tfm) rat: correlation between receptor binding and effects of androgens on gonadotropin secretion. *Endocrinology, 99*, 1295—1303.

Naftolin, F., Ryan, K. J., Davies, I. J., Petro, Z., & Kuhn, M. (1975). The formation and metabolism of estrogens in brain tissues. *Adv. Biosci., 15*, 105—121.

Nathorst-Boos, J., Wiklund, I., Mattsson, L. A., Sandin, K., & Von Schoultz, B. (1993). Is sexual life influenced by transdermal estrogen therapy? A double blind placebo controlled study in postmenopausal women. *Acta. Obstet. Gynecol. Scand., 72*, 656—660.

Neumann, I. D. (2008). Brain oxytocin: a key regulator of emotional and social behaviours in both females and males. *J. Neuroendocrinol., 20*, 858—865.

Newman, S. W. (1999). The medial extended amygdala in male reproductive behavior. A node in the mammalian social behavior network. *Annals of the New York Academy of Sciences, 877*, 242—257.

Nishimori, K., Young, L. J., Guo, Q., Wang, Z., Insel, T. R., & Matzuk, M. M. (1996). Oxytocin is required for nursing but is not essential for parturition or reproductive behavior. *Proc. Natl. Acad. Sci. USA, 93*, 11699—11704.

Nock, B., & Feder, H. H. (1979). Noradrenergic transmission and female sexual behavior of guinea pigs. *Brain Res., 166*, 369—380.

Nock, B., & Feder, H. H. (1984). Alpha 1-noradrenergic regulation of hypothalamic progestin receptors and guinea pig lordosis behavior. *Brain Res., 310*, 77—85.

Nyby, J., Matochik, J. A., & Barfield, R. J. (1992). Intracranial androgenic and estrogenic stimulation of male-typical behaviors in house mice (*Mus domesticus*). *Hormones and Behavior, 26*, 24—45.

Ogawa, S., Chan, J., Chester, A. E., Gustafsson, J. A., Korach, K. S., & Pfaff, D. W. (1999a). Survival of reproductive behaviors in estrogen receptor beta gene-deficient (betaERKO) male and female mice. *Proceedings of the National Academy of Sciences of the United States of America, 96*, 12887—12892.

Ogawa, S., Chan, J., Chester, A. E., Gustafsson, J. A., Korach, K. S., & Pfaff, D. W. (1999b). Survival of reproductive behaviors in estrogen receptor beta gene-deficient (betaERKO) male and female mice. *Proc. Natl. Acad. Sci. USA, 96*, 12887—12892.

Ogawa, S., Eng, V., Taylor, J., Lubahn, D. B., Korach, K. S., & Pfaff, D. W. (1998a). Roles of estrogen receptor-alpha gene expression in reproduction-related behaviors in female mice. *Endocrinology, 139*, 5070—5081.

Ogawa, S., Lubahn, D. B., Korach, K. S., & Pfaff, D. W. (1997). Behavioral effects of estrogen receptor gene disruption in male mice.

Proceedings of the National Academy of Sciences of the United States of America, 94, 1476–1481.

Ogawa, S., Olazabal, U. E., Parhar, I. S., & Pfaff, D. W. (1994). Effects of intrahypothalamic administration of antisense DNA for progesterone receptor mRNA on reproductive behavior and progesterone receptor immunoreactivity in female rat. *J. Neurosci., 14*, 1766–1774.

Ogawa, S., Washburn, T. F., Taylor, J., Lubahn, D. B., Korach, K. S., & Pfaff, D. W. (1998b). Modifications of testosterone-dependent behaviors by estrogen receptor-alpha gene disruption in male mice. *Endocrinology, 139*, 5058–5069.

Ono, S., Geller, L. N., & Lai, E. V. (1974). TfM mutation and masculinization versus feminization of the mouse central nervous system. *Cell, 3*, 235–242.

Osterlund, M. K., Gustafsson, J. A., Keller, E., & Hurd, Y. L. (2000a). Estrogen receptor beta (ERbeta) messenger ribonucleic acid (mRNA) expression within the human forebrain: distinct distribution pattern to ERalpha mRNA. *The Journal of clinical endocrinology and metabolism, 85*, 3840–3846.

Osterlund, M. K., Keller, E., & Hurd, Y. L. (2000b). The human forebrain has discrete estrogen receptor alpha messenger RNA expression: high levels in the amygdaloid complex. *Neuroscience, 95*, 333–342.

Par, G., Szekeres-Bartho, J., Buzas, E., Pap, E., & Falus, A. (2003). Impaired reproduction of histamine deficient (histidine-decarboxylase knockout) mice is caused predominantly by a decreased male mating behavior. *Am. J. Reprod. Immunol., 50*, 152–158.

Paredes, R. G., Tzschentke, T., & Nakach, N. (1998). Lesions of the medial preoptic area/anterior hypothalamus (MPOA/AH) modify partner preference in male rats. *Brain Research, 813*, 1–8.

Park, J. H., Burns-Cusato, M., Dominguez-Salazar, E., Riggan, A., Shetty, S., Arnold, A. P., et al. (2008). Effects of sex chromosome aneuploidy on male sexual behavior. *Genes Brain Behav., 7*, 609–617.

Park, J. H., Paul, M. J., Butler, M. P., Villa, P., Burke, M., Kim, D. P., et al. (2007). Short duration testosterone infusions maintain male sex behavior in Syrian hamsters. *Horm. Behav., 52*, 169–176.

Park, J. H., Takasu, N., Alvarez, M. I., Clark, K., Aimaq, R., & Zucker, I. (2004). Long-term persistence of male copulatory behavior in castrated and photo-inhibited Siberian hamsters. *Hormones and Behavior, 45*, 214–221.

Park, K., Kang, H. K., Seo, J. J., Kim, H. J., Ryu, S. B., & Jeong, G. W. (2001). Blood-oxygenation-level-dependent functional magnetic resonance imaging for evaluating cerebral regions of female sexual arousal response. *Urology, 57*, 1189–1194.

Parker, K. L., Rice, D. A., Lala, D. S., Ikeda, Y., Luo, X., Wong, M., et al. (2002). Steroidogenic factor 1: an essential mediator of endocrine development. *Recent Prog. Horm. Res., 57*, 19–36.

Parsons, B., MacLusky, N. J., Krieger, M. S., McEwen, B. S., & Pfaff, D. W. (1979). The effects of long-term estrogen exposure on the induction of sexual behavior and measurements of brain estrogen and progestin receptors in the female rat. *Horm. Behav., 13*, 301–313.

Pasqualini, C., Olivier, V., Guibert, B., Frain, O., & Leviel, V. (1995). Acute stimulatory effect of estradiol on striatal dopamine synthesis. *Journal of neurochemistry, 65*, 1651–1657.

Patisaul, H. B., Scordalakes, E. M., Young, L. J., & Rissman, E. F. (2003). Oxytocin, but not oxytocin receptor, is rRegulated by oestrogen receptor beta in the female mouse hypothalamus. *J. Neuroendocrinol., 15*, 787–793.

Patrone, C., Ma, Z. Q., Pollio, G., Agrati, P., Parker, M. G., & Maggi, A. (1996). Cross-coupling between insulin and estrogen receptor in human neuroblastoma cells. *Mol. Endocrinol., 10*, 499–507.

Pedersen, C. A., & Boccia, M. L. (2006). Vasopressin interactions with oxytocin in the control of female sexual behavior. *Neuroscience, 139*, 843–851.

Pettersson, K., Grandien, K., Kuiper, G. G., & Gustafsson, J. A. (1997). Mouse estrogen receptor beta forms estrogen response element-binding heterodimers with estrogen receptor alpha. *Mol. Endocrinol., 11*, 1486–1496.

Pfaff, D. (1970). Nature of sex hormone effects on rat sex behavior: specificity of effects and individual patterns of response. *J. Comp. Physiol. Psychol., 73*, 349–358.

Pfaff, D. W. (1973). Luteinizing hormone-releasing factor potentiates lordosis behavior in hypophysectomized ovariectomized female rats. *Science, 182*, 1148–1149.

Pfaff, D. W. (1980). *Estrogens and Brain Function: Neural Analysis of a Hormone-Controlled Mammalian Reproductive Behavior.* Berlin: Springer-Verlag.

Pfaff, D.W., Sakuma, Y., Kow, L.-M., Lee, A.W.L., and Easton, A. (2006). Hormonal, neural, and genomic mechanisms for female reproductive behaviors, motivation, and arousal. In J.D. Neill, et al., (Eds.), Knobil and Neill's Physiology of Reproduction, Third Edition (pp. 1825–1920), New York, NY: Raven Press

Pfaff, D. W., Schwartz-Giblin, S., McCarthy, M. M., & Kow, L. M. (1994). Cellular and molecular mechanisms of female reproductive behaviors. In E. Knobel, & J. D. Neill (Eds.), *The physiology of reproduction, Vol. 2* (pp. 107–220). New York, NY: Raven Press.

Pfaus, J. G., & Heeb, M. M. (1997). Implications of immediate-early gene induction in the brain following sexual stimulation of female and male rodents. *Brain research bulletin, 44*, 397–407.

Pfaus, J. G., Kippin, T. E., & Centeno, S. (2001). Conditioning and sexual behavior: a review. *Hormones and Behavior, 40*, 291–321.

Pfaus, J. G., & Phillips, A. G. (1989). Differential effects of dopamine receptor antagonists on the sexual behavior of male rats. *Psychopharmacology (Berl), 98*, 363–368.

Pfaus, J. G., & Wilkins, M. F. (1995). A novel environment disrupts copulation in sexually naive but not experienced male rats: reversal with naloxone. *Physiol. Behav., 57*, 1045–1049.

Pfaus, J. G., Shadiack, A., Van Soest, T., Tse, M., & Molinoff, P. (2004). Selective facilitation of sexual solicitation in the female rat by a melanocortin receptor agonist. *Proc. Natl. Acad. Sci. USA, 101*, 10201–10204.

Phoenix, C. H. (1974). Effects of dihydrotestosterone on sexual behavior of castrated male rhesus monkeys. *Physiology & behavior, 12*, 1045–1055.

Phoenix, C. H., Slob, A. K., & Goy, R. W. (1973). Effects of castration and replacement therapy on sexual behavior of adult male rhesuses. *Journal of comparative and physiological psychology, 84*, 472–481.

Piekarski, D. J., Routman, D. M., Schoomer, E. E., Driscoll, J. R., Park, J. H., Butler, M. P., et al. (2009). Infrequent low dose testosterone treatment maintains male sexual behavior in Syrian hamsters. *Horm. Behav., 55*, 182–189.

Pieper, D. R., & Lobocki, C. A. (2000). Characterization of serum dehydroepiandrosterone secretion in golden hamsters. *Proc. Soc. Exp. Biol. Med., 224*, 278–284.

Plata-Salaman, C. (1991). Review: epidermal growth factor and the nervous system. *Peptides, 12*, 653–663.

Pomerantz, S. M., Hepner, B. C., & Wertz, J. M. (1993). 5-HT1A and 5-HT1C/1D receptor agonists produce reciprocal effects on male sexual behavior of rhesus monkeys. *Eur. J. Pharmacol., 243*, 227–234.

Popova, N. K., & Amstislavskaya, T. G. (2002). Involvement of the 5-HT(1A) and 5-HT(1B) serotonergic receptor subtypes in sexual arousal in male mice. *Psychoneuroendocrinology, 27*, 609–618.

Popova, N. K., Skrinskaya, Y. A., Amstislavskaya, T. G., Vishnivetskaya, G. B., Seif, I., & De Meier, E. (2001). Behavioral characteristics of mice with genetic knockout of monoamine oxidase type A. *Neurosci. Behav. Physiol., 31*, 597–602.

Power, R. F., Lydon, J. P., Conneely, O. M., & O'Malley, B. W. (1991a). Dopamine activation of an orphan of the steroid receptor superfamily. *Science, 252*, 1546–1548.

Power, R. F., Mani, S. K., Codina, J., Conneely, O. M., & O'Malley, B. W. (1991b). Dopaminergic and ligand-independent activation of steroid hormone receptors. *Science, 254*, 1636–1639.

Powers, J. B., Bergondy, M. L., & Matochik, J. A. (1985). Male hamster sociosexual behaviors: effects of testosterone and its metabolites. *Physiol. Behav., 35*, 607–616.

Putnam, S. K., Du, J., Sato, S., & Hull, E. M. (2001). Testosterone restoration of copulatory behavior correlates with medial preoptic dopamine release in castrated male rats. *Hormones and Behavior, 39*, 216–224.

Putnam, S. K., Sato, S., Riolo, J. V., & Hull, E. M. (2005). Effects of testosterone metabolites on copulation, medial preoptic dopamine, and NOS-immunoreactivity in castrated male rats. *Hormones and Behavior, 47*, 513–522.

Quadagno, D. M., McGill, T. E., Yellon, S. M., & Goldman, B. D. (1979). Neither non-contact exposure nor mating affect serum LH and FSH in male B6D2F1 house mice. *Physiology & behavior, 22*, 191–192.

Rabedeau, R. G., & Whalen, R. E. (1959). Effects of copulatory experience on mating behavior in the male rat. *Journal of comparative and physiological psychology, 52*, 482–484.

Richard, S., & Zingg, H. H. (1990). The human oxytocin gene promoter is regulated by estrogens. *J. Biol. Chem., 265*, 6098–6103.

Richmond, G., & Clemens, L. G. (1986). Evidence for involvement of midbrain central gray in cholinergic mediation of female sexual receptivity in rats. *Behav. Neurosci., 100*, 376–380.

Rissman, E. F. (1987). Gonadal influences on sexual behavior in the male musk shrew (*Suncus murinus*). *Hormones and Behavior, 21*, 132–136.

Rissman, E. F. (1990). The musk shrew, *Suncus murinus*, a unique animal model for the study of female behavioral endocrinology. *J. Exp. Zool. Suppl, 4*, 207–209.

Rissman, E. F., & Crews, D. (1988). Hormonal correlates of sexual behavior in the female musk shrew: the role of estradiol. *Physiol. Behav., 44*, 1–7.

Rissman, E. F., Alones, V. E., Craig-Veit, C. B., & Millam, J. R. (1995). Distribution of chicken-II gonadotropin-releasing hormone in mammalian brain. *J. Comp. Neurol., 357*, 524–531.

Rissman, E. F., Clendenon, A. L., & Krohmer, R. W. (1990). Role of androgens in the regulation of sexual behavior in the female musk shrew. *Neuroendocrinology, 51*, 468–473.

Rissman, E. F., Early, A. H., Taylor, J. A., Korach, K. S., & Lubahn, D. B. (1997a). Estrogen receptors are essential for female sexual receptivity. *Endocrinology, 138*, 507–510.

Rissman, E. F., Wersinger, S. R., Taylor, J. A., & Lubahn, D. B. (1997b). Estrogen receptor function as revealed by knockout studies: neuroendocrine and behavioral aspects. *Horm. Behav., 31*, 232–243.

Riters, L. V., & Ball, G. F. (1999). Lesions to the medial preoptic area affect singing in the male European starling (*Sturnus vulgaris*). *Hormones and Behavior, 36*, 276–286.

Rives, N., Simeon, N., Milazzo, J. P., Barthelemy, C., & Mace, B. (2003). Meiotic segregation of sex chromosomes in mosaic and non-mosaic XYY males: case reports and review of the literature. *Int. J. Androl., 26*, 242–249.

Roberts, J. L., Mani, S. K., Woller, M. J., Glucksman, M. J., & Wu, T. J. (2007). LHRH-(1-5): a bioactive peptide regulating reproduction. *Trends Endocrinol. Metab., 18*, 386–392.

Robertson, G. S., Pfaus, J. G., Atkinson, L. J., Matsumura, H., Phillips, A. G., & Fibiger, H. C. (1991). Sexual behavior increases c-fos expression in the forebrain of the male rat. *Brain Research, 564*, 352–357.

Robinson, A. G. (1974). Elevation of plasma neurophysin in women on oral contraceptives. *J. Clin. Invest., 54*, 209–212.

Robinson, A. G., Haluszczak, C., Wilkins, J. A., Huellmantel, A. B., & Watson, C. G. (1977). Physiologic control of two neurophysins in humans. *J. Clin. Endocrinol. Metab., 44*, 330–339.

Rodriguez-Manzo, G., Lopez-Rubalcava, C., Hen, R., & Fernandez-Guasti, A. (2002). Participation of 5-HT(1B) receptors in the inhibitory actions of serotonin on masculine sexual behaviour of mice: pharmacological analysis in 5-HT(1B) receptor knockout mice. *Br. J. Pharmacol., 136*, 1127–1134.

Romeo, R. D., Cook-Wiens, E., Richardson, H. N., & Sisk, C. L. (2001). Dihydrotestosterone activates sexual behavior in adult male hamsters but not in juveniles. *Physiol. Behav., 73*, 579–584.

Roselli, C. E., Cross, E., Poonyagariyagorn, H. K., & Stadelman, H. L. (2003). Role of aromatization in anticipatory and consummatory aspects of sexual behavior in male rats. *Hormones and Behavior, 44*, 146–151.

Rosen, R. C., Lane, R. M., & Menza, M. (1999). Effects of SSRIs on sexual function: a critical review. *J. Clin. Psychopharmacol., 19*, 67–85.

Rosenblatt, J., & Aronson, L. R. (1958). The decline of sexual behavior in male cats after castration with special reference to the role of prior sexual experience. *Behaviour, 12*, 285–338.

Rossler, A. S., Pfaus, J. G., Kia, H. K., Bernabe, J., Alexandre, L., & Giuliano, F. (2006). The melanocortin agonist, melanotan II, enhances proceptive sexual behaviors in the female rat. *Pharmacol. Biochem. Behav., 85*, 514–521.

Rousseau, L., Dupont, A., Labrie, F., & Couture, M. (1988). Sexuality changes in prostate cancer patients receiving antihormonal therapy combining the antiandrogen flutamide with medical (LHRH agonist) or surgical castration. *Archives of sexual behavior, 17*, 87–98.

Rowland, D. L., & Houtsmuller, E. J. (1998). 8-OH-DPAT interacts with sexual experience and testosterone to affect ejaculatory response in rats. *Pharmacol. Biochem. Behav., 60*, 143–149.

Rowland, D. L., Greenleaf, W. J., Dorfman, L. J., & Davidson, J. M. (1993). Aging and sexual function in men. *Archives of sexual behavior, 22*, 545–557.

Rubin, B. S., & Barfield, R. J. (1983). Induction of estrous behavior in ovariectomized rats by sequential replacement of estrogen and progesterone to the ventromedial hypothalamus. *Neuroendocrinology, 37*, 218–224.

Safarinejad, M. R. (2008). Evaluation of the safety and efficacy of bremelanotide, a melanocortin receptor agonist, in female subjects with arousal disorder: a double-blind placebo-controlled, fixed dose, randomized study. *J. Sex Med., 5*, 887–897.

Sakuma, Y., & Pfaff, D. W. (1979). Facilitation of female reproductive behavior from mesensephalic central gray in the rat. *Am. J. Physiol., 237*, R278–R284.

Salis, P. J., & Dewsbury, D. A. (1971). p-chlorophenylalanine facilitates copulatory behaviour in male rats. *Nature, 232*, 400–401.

Salmimies, P., Kockott, G., Pirke, K. M., Vogt, H. J., & Schill, W. B. (1982). Effects of testosterone replacement on sexual behavior in hypogonadal men. *Arch. Sex. Behav., 11*, 345–353.

Sanchez-Ortiz, R. F., Broderick, G. A., Rovner, E. S., Wein, A. J., Whittington, R., & Malkowicz, S. B. (2000). Erectile function and quality of life after interstitial radiation therapy for prostate cancer. *International journal of impotence research 12. Suppl, 3*, S18–S24.

Sarrel, P. M. (1990). Sexuality and menopause. *Obstet. Gynecol., 75*, 26S–30S, discussion 31S–35S.

Sato, S. M., Schulz, K. M., Sisk, C. L., & Wood, R. I. (2008). Adolescents and androgens, receptors and rewards. *Horm. Behav., 53*, 647–658.

Sato, T., Matsumoto, T., Kawano, H., Watanabe, T., Uematsu, Y., Sekine, K., et al. (2004). Brain masculinization requires androgen receptor function. *Proc. Natl. Acad. Sci. USA, 101*, 1673–1678.

Sato, Y., Horita, H., Kurohata, T., Adachi, H., & Tsukamoto, T. (1998). Effect of the nitric oxide level in the medial preoptic area on male copulatory behavior in rats. *The American journal of physiology, 274*, R243–R247.

Savic, I. (2002). Sex differentiated hypothalamic activation by putative pheromones. *Molecular psychiatry, 7*, 335–336.

Savic, I. (2005). Brain imaging studies of the functional organization of human olfaction. *Chemical senses* 30. *Suppl, 1*, i222–i223.

Scaletta, L. L., & Hull, E. M. (1990). Systemic or intracranial apomorphine increases copulation in long-term castrated male rats. *Pharmacology, biochemistry, and behavior, 37*, 471–475.

Schneider, J. S., Burgess, C., Sleiter, N. C., DonCarlos, L. L., Lydon, J. P., O'Malley, B., et al. (2005). Enhanced sexual behaviors and androgen receptor immunoreactivity in the male progesterone receptor knockout mouse. *Endocrinology, 146*, 4340–4348.

Schreihofer, D. A., Resnick, E. M., Lin, V. Y., & Shupnik, M. A. (2001). Ligand-independent activation of pituitary ER: dependence on PKA-stimulated pathways. *Endocrinology, 142*, 3361–3368.

Schulze, H. G., & Gorzalka, B. B. (1991). Oxytocin effects on lordosis frequency and lordosis duration following infusion into the medial pre-optic area and ventromedial hypothalamus of female rats. *Neuropeptides, 18*, 99–106.

Seagraves, R. T. (1990). Effects of psychotropic drugs on human erection and ejaculation. *Arch. Gen. Psychiatry, 46*, 275–284.

Shadiack, A. M., Sharma, S. D., Earle, D. C., Spana, C., & Hallam, T. J. (2007). Melanocortins in the treatment of male and female sexual dysfunction. *Curr. Top. Med. Chem., 7*, 1137–1144.

Sherins, R. J., & Bardin, C. W. (1971). Preputial gland growth and protein synthesis in the androgen-insensitive male pseudohermaphroditic rat. *Endocrinology, 89*, 835–841.

Shields, K. M., & Hrometz, S. L. (2006). Use of sildenafil for female sexual dysfunction. *Ann. Pharmacother., 40*, 931–934.

Shimshek, D. R., Bus, T., Grinevich, V., Single, F. N., Mack, V., Sprengel, R., et al. (2006). Impaired reproductive behavior by lack of GluR-B containing AMPA receptors but not of NMDA receptors in hypothalamic and septal neurons. *Mol. Endocrinol., 20*, 219–231.

Shintani, N., Mori, W., Hashimoto, H., Imai, M., Tanaka, K., Tomimoto, S., et al. (2002). Defects in reproductive functions in PACAP-deficient female mice. *Regul. Pept., 109*, 45–48.

Shrenker, P., & Bartke, A. (1987). Effects of hyperprolactinaemia on male sexual behaviour in the golden hamster and mouse. *J. Endocrinol., 112*, 221–228.

Shrenker, P., & Maxson, S. C. (1983). The genetics of hormonal influences on male sexual behavior of mice and rats. *Neurosci. Biobehav. Rev., 7*, 349–359.

Simerly, R. B., Chang, C., Muramatsu, M., & Swanson, L. W. (1990). Distribution of androgen and estrogen receptor mRNA-containing cells in the rat brain: an *in-situ* hybridization study. *The Journal of comparative neurology, 294*, 76–95.

Simerly, R. B., Gorski, R. A., & Swanson, L. W. (1986). Neurotransmitter specificity of cells and fibers in the medial preoptic nucleus: an immunohistochemical study in the rat. *The Journal of comparative neurology, 246*, 343–363.

Simpson, J. L., De la Cruz, F., Swerdloff, R. S., Samango-Sprouse, C., Skakkebaek, N. E., Graham, J. M., Jr., et al. (2003). Klinefelter syndrome: expanding the phenotype and identifying new research directions. *Genet. Med., 5*, 460–468.

Sinchak, K., Roselli, C. E., & Clemens, L. G. (1996). Levels of serum steroids, aromatase activity, and estrogen receptors in preoptic area, hypothalamus, and amygdala of B6D2F1 male house mice that differ in the display of copulatory behavior after castration. *Behavioral neuroscience, 110*, 593–602.

Sinchak, K., Shahedi, K., Dewing, P., & Micevych, P. (2005). Sexual receptivity is reduced in the female mu-opioid receptor knockout mouse. *Neuroreport, 16*, 1697–1700.

Sirinathsinghji, D. J. (1984). Modulation of lordosis behavior of female rats by naloxone, beta-endorphin and its antiserum in the mesencephalic central gray: possible mediation via GnRH. *Neuroendocrinology, 39*, 222–230.

Sirinathsinghji, D. J. (1987). Inhibitory influence of corticotropin releasing factor on components of sexual behaviour in the male rat. *Brain Res., 407*, 185–190.

Sirinathsinghji, D. J., Whittington, P. E., Audsley, A., & Fraser, H. M. (1983). beta-Endorphin regulates lordosis in female rats by modulating LH-RH release. *Nature, 301*, 62–64.

Smith, E. R., Lee, R. L., Schnur, S. L., & Davidson, J. M. (1987a). Alpha 2-adrenoceptor antagonists and male sexual behavior: I. Mating behavior. *Physiol. Behav., 41*, 7–14.

Smith, E. R., Lee, R. L., Schnur, S. L., & Davidson, J. M. (1987b). Alpha 2-adrenoceptor antagonists and male sexual behavior: II. Erectile and ejaculatory reflexes. *Physiol Behav, 41*, 15–19.

Sodersten, P. (1985). In D. Ganten, & P.D.W.. (Eds.), *Current topics in neuroendocrinology. Estradio-progesterone interactions in the reproductive behavior of female rats, Vol. 5*. Berlin, Germany: Springer-Verlag.

Sodersten, P., Eneroth, P., & Pettersson, A. (1983). Episodic secretion of luteinizing hormone and androgen in male rats. *J. Endocrinol., 97*, 145–153.

Sorensen, K., Nielsen, J., Froland, A., & Johnsen, S. G. (1979). Psychiatric examination of all eight adult males with the karyotype 46, XX diagnosed in Denmark till 1976. *Acta. Psychiatr. Scand., 59,* 153–163.

Srinivasan, S., Bunch, D. O., Feng, Y., Rodriguiz, R. M., Li, M., Ravenell, R. L., et al. (2004). Deficits in reproduction and pro-gonadotropin-releasing hormone processing in male Cpefat mice. *Endocrinology, 145,* 2023–2034.

Stanworth, R. D., & Jones, T. H. (2008). Testosterone for the aging male; current evidence and recommended practice. *Clin. Interv. Aging, 3,* 25–44.

Stewart, D. A., Bailey, J. D., Netley, C. T., & Park, E. (1990). Growth, development, and behavioral outcome from mid-adolescence to adulthood in subjects with chromosome aneuploidy: the Toronto Study. *Birth Defects Orig. Artic. Ser., 26,* 131–188.

Stoleru, S., Gregoire, M. C., Gerard, D., Decety, J., Lafarge, E., Cinotti, L., et al. (1999). Neuroanatomical correlates of visually evoked sexual arousal in human males. *Arch. Sex Behav., 28,* 1–21.

Swaney, W. T., Curley, J. P., Champagne, F. A., & Keverne, E. B. (2007). Genomic imprinting mediates sexual experience-dependent olfactory learning in male mice. *Proc. Natl. Acad. Sci. USA, 104,* 6084–6089.

Szczypka, M. S., Zhou, Q. Y., & Palmiter, R. D. (1998). Dopamine-stimulated sexual behavior is testosterone dependent in mice. *Behav. Neurosci., 112,* 1229–1235.

Takayanagi, Y., Yoshida, M., Bielsky, I. F., Ross, H. E., Kawamata, M., Onaka, T., et al. (2005). Pervasive social deficits, but normal parturition, in oxytocin receptor-deficient mice. *Proc. Natl. Acad. Sci. USA, 102,* 16096–16101.

Takeo, T., Chiba, Y., & Sakuma, Y. (1993). Suppression of the lordosis reflex of female rats by efferents of the medial preoptic area. *Physiol. Behav., 53,* 831–838.

Tallentire, D., McRae, G., Spedding, M., Clark, R., & Vickery, B. (1996). Modulation of sexual behaviour in the rat by a potent and selective alpha 2-adrenoceptor antagonist, delequamine (RS-15385-197). *Br. J. Pharmacol., 118,* 63–72.

Tang, Y. P., & Sisk, C. L. (1991). Testosterone in MPOA elicits behavioral but not neuroendocrine responses in ferrets. *Brain research bulletin, 26,* 373–378.

Taylor, G., Bardgett, M., Csernansky, J., Early, T., Haller, J., Scherrer, J., et al. (1996). Male reproductive systems under chronic fluoxetine or trimipramine treatment. *Physiol. Behav., 59,* 479–485.

Temple, J. L., Millar, R. P., & Rissman, E. F. (2003a). An evolutionarily conserved form of gonadotropin-releasing hormone coordinates energy and reproductive behavior. *Endocrinology, 144,* 13–19.

Temple, J. L., Scordalakes, E. M., Bodo, C., Gustafsson, J. A., & Rissman, E. F. (2003b). Lack of functional estrogen receptor beta gene disrupts pubertal male sexual behavior. *Hormones and Behavior, 44,* 427–434.

Tennent, B. J., Smith, E. R., & Davidson, J. M. (1980). The effects of estrogen and progesterone on female rat proceptive behavior. *Horm. Behav., 14,* 65–75.

Theilgaard, A. (1984). A psychological study of the personalities of XYY- and XXY-men. *Acta. Psychiatr. Scand. Suppl, 315,* 1–133.

Thompson, M. L., McGill, T. E., McIntosh, S. M., & Manning, A. (1976). Effects of adrenalectomy on the sexual behaviour of castrated and intact BDF1 mice. *Animal Behavior, 24,* 519–522.

Toda, K., Okada, T., Takeda, K., Akira, S., Saibara, T., Shiraishi, M., et al. (2001). Oestrogen at the neonatal stage is critical for the reproductive ability of male mice as revealed by supplementation with 17beta-oestradiol to aromatase gene (Cyp19) knockout mice. *J. Endocrinol., 168,* 455–463.

Tostain, J. L., & Blanc, F. (2008). Testosterone deficiency: a common, unrecognized syndrome. *Nat. Clin. Pract. Urol., 5,* 388–396.

Trowbridge, J. M., Rogatsky, I., & Garabedian, M. J. (1997). Regulation of estrogen receptor transcriptional enhancement by the cyclin A/Cdk2 complex. *Proc. Natl. Acad. Sci. USA, 94,* 10132–10137.

Ueno, R., Ishikawa, Y., Nakayama, T., & Hayaishi, O. (1982). Prostaglandin D2 induces sleep when microinjected into the preoptic area of conscious rats. *Biochem. Biophys. Res. Commun., 109,* 576–582.

Uphouse, L., Montanez, S., Richards-Hill, R., Caldarola-Pastuszka, M., & Droge, M. (1991). Effects of the 5-HT1A agonist, 8-OH-DPAT, on sexual behaviors of the proestrous rat. *Pharmacol. Biochem. Behav., 39,* 635–640.

Vagell, M. E., & McGinnis, M. Y. (1997). The role of aromatization in the restoration of male rat reproductive behavior. *Journal of neuroendocrinology, 9,* 415–421.

Vale, J. (2000). Erectile dysfunction following radical therapy for prostate cancer. *Radiotherapy and oncology, 57,* 301–305.

Van der Ploeg, L. H., Martin, W. J., Howard, A. D., Nargund, R. P., Austin, C. P., Guan, X., et al. (2002). A role for the melanocortin 4 receptor in sexual function. *Proc. Natl. Acad. Sci. USA, 99,* 11381–11386.

Van Tol, H. H., Van den Buuse, M., De Jong, W., & Burbach, J. P. (1988). Vasopressin and oxytocin gene expression in the supraoptic and paraventricular nucleus of the spontaneously hypertensive rat (SHR) during development of hypertension. *Brain Res., 464,* 303–311.

Veney, S. L., & Rissman, E. F. (1998). Co-localization of estrogen receptor and aromatase enzyme immunoreactivities in adult musk shrew brain. *Horm. Behav., 33,* 151–162.

Veney, S. L., & Rissman, E. F. (2000). Steroid implants in the medial preoptic area or ventromedial nucleus of the hypothalamus activate female sexual behaviour in the musk shrew. *J. Neuroendocrinol., 12,* 1124–1132.

Vloeberghs, E., Van Dam, D., Franck, F., Staufenbiel, M., & De Deyn, P. P. (2007). Mood and male sexual behaviour in the APP23 model of Alzheimer's disease. *Behav. Brain Res., 180,* 146–151.

Wallen, K. (1990). Desire and ability: hormones and the regulation of female sexual behavior. *Neurosci. Biobehav. Rev., 14,* 233–241.

Wallen, K. (2005). Hormonal influences on sexually differentiated behavior in nonhuman primates. *Front Neuroendocrinol., 26,* 7–26.

Wang, H., Liang, S., Burgdorf, J., Wess, J., & Yeomans, J. (2008). Ultrasonic vocalizations induced by sex and amphetamine in M2, M4, M5 muscarinic and D2 dopamine receptor knockout mice. *PLoS ONE, 3,* e1893.

Ward, B. J., & Charlton, H. M. (1981). Female sexual behaviour in the GnRH deficient, hypogonadal (hpg) mouse. *Physiol. Behav., 27,* 1107–1109.

Ward, I. L., Crowley, W. R., & Zemlan, F. P. (1975). Monoaminergic mediation of female sexual behavior. *J. Comp. Physiol. Psychol., 88,* 53–61.

Waring, D. W., & Turgeon, J. L. (1992). A pathway for luteinizing hormone releasing-hormone self-potentiation: cross-talk with the progesterone receptor. *Endocrinology, 130,* 3275–3282.

Wee, B. E., & Clemens, L. G. (1989). Environmental influences on masculine sexual behavior in mice. *Physiology & behavior, 46,* 867–872.

Weil, Z. M., Huang, A. S., Beigneux, A., Kim, P. M., Molliver, M. E., Blackshaw, S., et al. (2006). Behavioural alterations in male mice lacking the gene for D-aspartate oxidase. *Behav. Brain Res., 171*, 295–302.

Weinberger, L. E., Sreenivasan, S., Garrick, T., & Osran, H. (2005). The impact of surgical castration on sexual recidivism risk among sexually violent predatory offenders. *The journal of the American Academy of Psychiatry and the Law, 33*, 16–36.

Wersinger, S. R., & Rissman, E. F. (2000). Dopamine activates masculine sexual behavior independent of the estrogen receptor alpha. *The Journal of Neuroscience, 20*, 4248–4254.

Wersinger, S. R., Sannen, K., Villalba, C., Lubahn, D. B., Rissman, E. F., & De Vries, G. J. (1997). Masculine sexual behavior is disrupted in male and female mice lacking a functional estrogen receptor alpha gene. *Horm. Behav., 32*, 176–183.

Wessells, H., Fuciarelli, K., Hansen, J., Hadley, M. E., Hruby, V. J., Dorr, R., et al. (1998). Synthetic melanotropic peptide initiates erections in men with psychogenic erectile dysfunction: double-blind, placebo controlled crossover study. *J. Urol., 160*, 389–393.

Wessells, H., Gralnek, D., Dorr, R., Hruby, V. J., Hadley, M. E., & Levine, N. (2000). Effect of an alpha-melanocyte stimulating hormone analog on penile erection and sexual desire in men with organic erectile dysfunction. *Urology, 56*, 641–646.

Whalen, R. E., & Luttge, W. G. (1971). Testosterone, androstenedione and dihydrotestosterone: effects on mating behavior of male rats. *Hormones and Behavior, 2*, 117–125.

Whalen, R. E., & Nakayama, K. (1965). Induction of oestrous behaviour: facilitation by repeated hormone treatments. *J. Endocrinol., 33*, 525–526.

White, M. M., Sheffer, I., Teeter, J., & Apostolakis, E. M. (2007). Hypothalamic progesterone receptor-A mediates gonadotropin surges, self priming and receptivity in estrogen-primed female mice. *J. Mol. Endocrinol., 38*, 35–50.

White, R. B., Eisen, J. A., Kasten, T. L., & Fernald, R. D. (1998). Second gene for gonadotropin-releasing hormone in humans. *Proc. Natl. Acad. Sci. USA, 95*, 305–309.

Wiesenfeld-Hallin, Z., & Sodersten, P. (1984). Spinal opiates affect sexual behaviour in rats. *Nature, 309*, 257–258.

Wiesner, J. B., & Moss, R. L. (1984). Beta-endorphin suppression of lordosis behavior in female rats; lack of effect of peripherally-administered naloxone. *Life Sci., 34*, 1455–1462.

Wiesner, J. B., & Moss, R. L. (1986a). Behavioral specificity of beta-endorphin suppression of sexual behavior: differential receptor antagonism. *Pharmacol. Biochem. Behav., 24*, 1235–1239.

Wiesner, J. B., & Moss, R. L. (1986b). Suppression of receptive and proceptive behavior in ovariectomized, estrogen-progesterone-primed rats by intraventricular beta-endorphin: studies of behavioral specificity. *Neuroendocrinology, 43*, 57–62.

Wilson, C. A., & Hunter, A. J. (1985). Progesterone stimulates sexual behaviour in female rats by increasing 5-HT activity on 5-HT2 receptors. *Brain Res., 333*, 223–229.

Winslow, J. T., & Insel, T. R. (1991). Social status in pairs of male squirrel monkeys determines the behavioral response to central oxytocin administration. *J. Neurosci., 11*, 2032–2038.

Winter, J. S. (1991). *Androgen therapy in Klinefelter syndrome during adolescence.* Wiley-Liss, for the March of Dimes Birth Defects Foundation. Birth Defects. New York, NY: Original Article Series.

Witt, D. M. (1995). Oxytocin and rodent sociosexual responses: from behavior to gene expression. *Neurosci. Biobehav. Rev., 19*, 315–324.

Witt, D. M., & Insel, T. R. (1991). A selective oxytocin antagonist attenuates progesterone facilitation of female sexual behavior. *Endocrinology, 128*, 3269–3276.

Wood, R. I. (1996). Estradiol, but not dihydrotestosterone, in the medial amygdala facilitates male hamster sex behavior. *Physiology & behavior, 59*, 833–841.

Wood, R. I., & Coolen, L. M. (1997). Integration of chemosensory and hormonal cues is essential for sexual behaviour in the male Syrian hamster: role of the medial amygdaloid nucleus. *Neuroscience, 78*, 1027–1035.

Wood, R. I., & Williams, S. J. (2001a). Steroidal control of male hamster sexual behavior in Me and MPOA: effects of androgen dose and tamoxifen. *Physiol. Behav., 72*, 727–733.

Wood, R. I., & Williams, S. J. (2001b). Steroidal control of male hamster sexual behavior in Me and MPOA: effects of androgen dose and tamoxifen. *Physiology & behavior, 72*, 727–733.

Yahr, P., & Ulibarri, C. (1987). Polyadenylated and nonadenylated messenger RNA and androgen control of sexual behavior and scent marking in male gerbils. *Hormones and Behavior, 21*, 53–64.

Yamanouchi, K. (1980). Inhibitory and facilitatory neural mechanisms involved in the regulation of lordosis behavior in female rats: effects of dual cuts in the preoptic area and hypothalamus. *Physiol. Behav., 25*, 721–725.

Yilmaz, U., Tatlisen, A., Turan, H., Arman, F., & Ekmekcioglu, O. (1999). The effects of fluoxetine on several neurophysiological variables in patients with premature ejaculation. *J. Urol., 161*, 107–111.

Yoshida, A., Miura, K., Nagao, K., Hara, H., Ishii, N., & Shirai, M. (1997). Sexual function and clinical features of patients with Klinefelter's syndrome with the chief complaint of male infertility. *Int. J. Androl., 20*, 80–85.

Young, W. (1961). The hormones and mating behavior. In W. Young (Ed.), *Sex and Internal Secretions* (pp. 1173–1239). Baltimore, MD: Williams and Wilkins.

Zarrindast, M. R., Mamanpush, S. M., & Rashidy-Pour, A. (1994). Morphine inhibits dopaminergic and cholinergic induced ejaculation in rats. *Gen. Pharmacol., 25*, 803–808.

Zehr, J. L., Maestripieri, D., & Wallen, K. (1998). Estradiol increases female sexual initiation independent of male responsiveness in rhesus monkeys. *Horm. Behav., 33*, 95–103.

Zemlan, F. P., Ward, I. L., Crowley, W. R., & Margules, D. L. (1973). Activation of lordotic responding in female rats by suppression of serotonergic activity. *Science, 179*, 1010–1011.

Zucker, I. (1968). Biphasic effects of progesterone on sexual receptivity in the female guinea pig. *J. Comp. Physiol. Psychol., 65*, 472–478.

Zussman, L., Zussman, S., Sunley, R., & Bjornson, E. (1981). Sexual response after hysterectomy-oophorectomy: recent studies and reconsideration of psychogenesis. *Am. J. Obstet. Gynecol., 140*, 725–729.

Zverina, J., Hampl, R., Sulocava, J., & Starka, L. (1990). Hormonal status and sexual behaviour of 16 men after surgical castration. *Urological, nephrological, and andrological sciences, 62*, 55–58.

Chapter 9

Pheromones and Reproduction in Mammals

Aras Petrulis
Georgia State University, Atlanta, GA, USA

SUMMARY
Many mammalian species, especially nocturnal animals, use pheromonal signals to coordinate reproduction by altering the physiology and behavior of both sexes. Pheromones can prime reproductive physiology so that individuals become sexually mature and active at times when mating is most probable and suppress it when it is not. Once in reproductive condition, odor cues produced and deposited by both males and females are used to find and select individuals for mating. In most cases, the production, dissemination, and appropriate responses to these social odors are modulated heavily by gonadal sex steroids, which thereby link pheromonal communication with the broader reproductive context. Mechanistically, pheromones are perceived by both main and accessory (vomeronasal) olfactory systems, with the importance of each system tied more strongly to the nature of the stimulus rather than to the response. More centrally, the vast majority of pheromonal responses are mediated by corticomedial amygdala connections with forebrain structures irrespective of chemosensory modality. Despite the importance of social odors in mammals, many details of pheromonal communication differ even among closely related nocturnal species and defy easy categorization.

1. INTRODUCTION

1.1. Scope

Chemical signals, often termed 'pheromones,' are a prominent and often critical means of reproductive communication for most mammalian species. Most mammals produce, disseminate, and respond to social odors in many contexts, including those surrounding reproduction, parent—offspring interactions, and territorial/dominance relationships (Brown & Macdonald, 1985). This chapter will focus on the most direct links between social odors and reproduction; i.e., the effects of pheromones on reproductive physiology and behavior. The depth and breadth of the existing literature on pheromones and reproduction preclude the review of pheromone-influenced behaviors or physiological processes that are further removed from copulation. So, while territorial acquisition/defense (Gosling, Roberts, Peter, Slater, & Timothy, 2001; Hurst & Beynon, 2004) and maternal behavior (Kendrick et al., 1997) are critical for reproductive success and involve pheromonal communication, work in these areas will not be covered. Further, a truly comprehensive review that covers each mammalian order is not possible here and so the focus will be on species for which there is the most information. Unfortunately, this will generally limit discussion to companion, farmed, and laboratory species; the interested reader is directed to several texts that provide a more comprehensive treatment of odor communication in mammals: Brown and Macdonald (1985) and relevant sections in Wyatt (2003), Mason, LeMaster, and Müller-Schwarze (2005), and Hurst, Beynon, Roberts, and Wyatt (2008).

1.2. Pheromones

Before asking what role pheromones play in mammalian reproduction, one must define what is meant by the term 'pheromone.' It was used initially to describe the conspecific odor cues that elicit behavioral and physiological responses of insects and was defined as 'substances which are secreted to the outside by an individual and received by a second individual of the same species, in which they release a specific reaction, e.g., a definite behavior or a developmental process' Karlson & Luscher, 1959 (p. 55); see also Wyatt, 2009). This led to the designations 'releaser pheromones' for odors that affected behavior and 'primer pheromones' for those that affected developmental/physiological processes. While useful as a general description, the problematic extension of the term to mammalian biology was recognized early by researchers, who noted that mammalian reproductive behavior and physiology are not rigid but are, instead, flexible, context-dependent, and modifiable by experience (McClintock, 2002). Moreover,

Hormones and Reproduction of Vertebrates, Volume 5—Mammals

the term 'pheromone' has been taken to refer to one or a small number of species-specific compounds that have significant activity on a specific behavior or physiological process. Despite some successes in isolating specific behaviorally active compounds (primarily in rodents), these individual substances have rarely been as effective as the full odor and, in several cases, are not species-specific (Ingersoll & Launay, 1986; Rasmussen, Lee, Roelofs, Zhang, & Daves, 1996; Gelez & Fabre-Nys, 2004; Zhang, Sun, Zhang, & Feng, 2008). Indeed, some of the identified compounds in social odors have multiple behavioral and physiological functions (Novotny, 2003), suggesting that the releaser vs. primer distinction is not meaningful at the stimulus level. Moreover, many fine distinctions made by animals on the basis of social odors (such as individuality, familiarity, kin) are unlikely to be mediated by the presence or absence of one or a very small number of unique compounds and may, instead, require processing of an odor mosaic or blend (Johnston, 2008). Consequently, in this chapter, the use of the term 'pheromone' will be limited to specific compounds with significant physiological or behavioral activity and other, more neutral terms such as 'chemosignal' or 'social odor' will be used in all other circumstances.

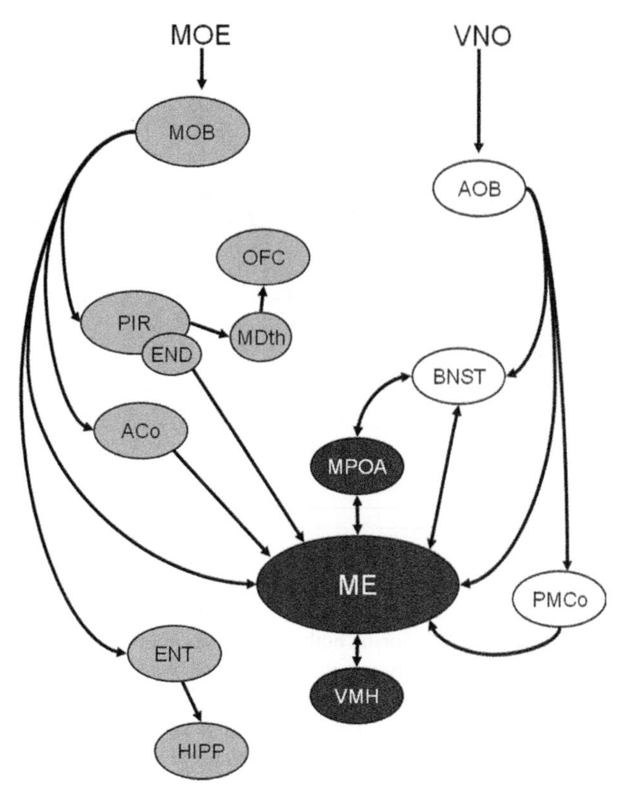

FIGURE 9.1 Abbreviated schematic of main olfactory (MOS; grey) and accessory olfactory or vomeronasal (AOS; white) systems along with integrative areas (black). For clarity, only unidirectional olfactory bulb connections are presented and several areas without known relevance to social odor processing are omitted. AOB, accessory olfactory bulb; acAMG, anterior cortical amygdala; BNST, bed nucleus of the stria terminalis; END, endopiriform nucleus; ENT, entorhinal cortex; HIPP, hippocampus; mAMG, medial amygdala; MDth, mediodorsal thalamus; MOB, main olfactory bulb; MOE, main olfactory epithelium; mPOA, medial preoptic area; OFC, orbitofrontal cortex; PIR, piriform cortex; pcAMG, posteromedial cortical amygdala; VMH, ventromedial hypothalamus; VNO, vomeronasal organ.

1.3. Chemosensory Systems

The basic anatomy of odor-processing circuits is well-defined and consists primarily of the main olfactory system (MOS) and the accessory olfactory (or vomeronasal) system (AOS); it is reviewed in detail elsewhere (Halpern & Martinez-Marcos, 2003; Zufall & Leinders-Zufall, 2007). These two systems are largely separate, having both separate sensory neuron populations in the nasal cavity and segregated representations in the olfactory bulbs as well as significant separation in postbulbar structures (Figure 9.1). Interconnections between the two systems occur primarily in the medial amygdala (mAMG) (Martinez-Marcos, 2009). The mAMG and connected areas also contain many neurons that are responsive to gonadal steroids (Wood, 1997). In general, perception of both male and female odors elicits increased immediate−early gene expression, an indirect measure of neural activity, in neurons within most parts of the AOS and MOS of both sexes; the greatest sex differences are observed in the preoptic area (POA) and hypothalamic region (Baum & Kelliher, 2008).

1.4. Organization

This review will explore the physiological/developmental and behavioral effects of exposure to opposite-sex odors separately, even though the same odor source may cause both physiological and behavioral changes. Within these sections, responses by males and females also will be treated separately and an effort will be made in each section to provide a basic understanding of (1) the phenomenon in question, (2) its species diversity, (3) how relevant signals and responses are modulated by hormones and other factors, (4) the sensory and neural mechanisms involved, and (5) possible linkages to ecological relevance.

2. SOCIAL ODORS AND PHYSIOLOGY

2.1. Responses by Females

2.1.1. Puberty acceleration

The presence of male odors prior to reproductive maturity accelerates the onset of puberty (first day of estrus) in juvenile female house mice (*Mus musculus*) (Vandenbergh, 1969), meadow voles (*Microtus pennsylvanicus*) (Baddaloo & Clulow, 1981), Siberian hamsters (*Phodopus sungorus*) (Reasner & Johnston, 1988), prairie voles (*Microtus ochrogaster*) (Carter, Getz, Gavish, McDermott, & Arnold, 1980), pine voles (*Microtus pinetorum*) (Lepri & Vandenbergh, 1986), and sheep (*Ovis aries*) (Knight, Lindsay, & Oldham, 1975). It is, however, the tactile interactions between male and female plus male urinary cues that lead to the greatest acceleration of puberty (Bronson & Maruniak, 1975). In mice, this pubertal acceleration occurs in response to soiled bedding or urine produced by reproductively active, dominant male mice (Vandenbergh, 1969); urine from juvenile, subordinate, or castrated males does not alter female reproductive physiology (Lombardi, Vandenbergh, & Whitsett, 1976; Drickamer & Murphy, 1978). The ability to accelerate puberty is androgen-dependent, as testosterone (T) injections can restore activity to castrated males' or even to diestrous females' urine. This suggests that androgens may stimulate the production of a common set of odor compounds irrespective of sex or even species. Indeed, sexual maturation of female mice can be hastened by exposure to male rat urine (Colby & Vandenberg, 1974). On the other hand, odors from any reproductively active individual may be effective in accelerating puberty. For example, urine from singly housed pregnant, lactating, or estrous adult females also accelerates puberty in juvenile female mice (Drickamer & Hoover, 1979); this stimulatory effect in mice is eliminated by group-housing stimulus females and is absent from androgenized females (Cowley & Pewtress, 1986). Indeed, urine from close relatives is equally effective at accelerating puberty as urine from unfamiliar animals (Drickamer, 1984).

The puberty-accelerating chemosignals in male mouse urine are present in bladder urine and survive removal of male preputial glands (Colby & Vandenberg, 1974). Initial characterizations of these stimuli suggested that the protein fraction was critical for biological action (Vandenbergh, Whitsett, & Lombardi, 1975), but it now appears that these major urinary proteins (MUPs) bind to several hydrophobic, low-molecular-weight (LMW) compounds that accelerate puberty (Novotny, Ma, Wiesler, & Zidek, 1999). These MUPs may normally protect these molecules from oxidation as well as acting as a reservoir for the slow release of volatile pheromones and possibly aiding in the transport to and interaction of volatiles with chemosensory structures, rather than acting directly as pheromones (Novotny et al., 1999b; Utsumi et al., 1999; Hurst & Beynon, 2004 but see Mucignat-Caretta, Caretta, & Cavaggioni, 1995). These volatile compounds have been identified as 2-sec-butyl dihydrothiazole (SBT), 3, 4-dehydro-exo-brevicomin (DHB), 6-hydroxy-6-methyl-3-heptanone (HMH), and two farnesenes (Novotny et al., 1999a; 1999b). Other urinary compounds such as isobutylamine and isoamylamine may also accelerate puberty (Nishimura, Utsumi, Yuhara, Fujitani, & Iritani, 1989). The occurrence of these volatile pheromones matches the conditions under which urine will accelerate puberty; they are only found in the urine of dominant adult male mice and are T-dependent (Novotny et al., 1999a). Similarly, MUPs are more abundant in male urine and show some dependence on circulating androgen levels (Armstrong, Robertson, Cheetham, Hurst, & Beynon, 2005). The puberty-acceleration substances found in the urine of estrous, pregnant, or lactating females partially overlap with those found in male urine (Jemiolo, Andreolini, Xie, Wiesler, & Novotny, 1989).

Puberty-accelerating odors increase the release of luteinizing hormone (LH) from the pituitary in juvenile female mice within 30 minutes of exposure (Bronson & Desjardins, 1974), which, through increases in gonadal estrogen secretion, induces puberty (Sisk & Foster, 2004). This effect is eliminated by removing the olfactory bulbs (OBX) (Zarrow, Estes, Denenberg, & Clark, 1970) or damaging the vomeronasal organ (VNO) system (Kaneko, Debski, Wilson, & Whitten, 1980; Lomas & Keverne, 1982). Indeed, puberty-accelerating pheromones alter activity in VNO receptor neurons with high specificity and sensitivity (Leinders-Zufall et al., 2000), with different pheromones activating different VNO receptor populations (Brennan, Schellinck, & Keverne, 1999; Boschat et al., 2002). Although the main olfactory epithelium (MOE) can detect HMH (Trinh & Storm, 2003), females with MOE damage still accelerate their puberty in response to male urine, but interestingly not to urine from pregnant or lactating females (Drickamer, 1986). Both the MOS and AOS provide input via the mAMG to the hypothalamic gonadotropin-releasing hormone (GnRH) cells that control release of LH. Medial amygdala neurons presynaptic to GnRH cells show increased immediate—early gene activity in response to male urine and α-farnesene, suggesting that these amygdala neurons mediate the effects of male pheromones on puberty acceleration (Boehm, Zou, & Buck, 2005; Yoon, Enquist, & Dulac, 2005).

Based on the short lifespan of wild mice, any advancement of reproduction in the presence of breeding opportunity, e.g. indicated by cues from opposite-sex individuals, would seem to be an adaptive response. This does not appear to be the case, however, as early pubertal onset induced by males or male urine reduces survivorship and leads to fewer litters with fewer offspring per litter in wild female mice (Drickamer, 1988). Although the production of puberty-accelerating pheromones is present in wild male mice (Massey & Vandenbergh, 1981) and does increase population density (Drickamer & Mikesic, 1990), it appears that female reproductive success may increase by avoiding pubertal acceleration under natural conditions.

2.1.2. Puberty delay

Pubertal onset can be delayed by several days in single-sex, group-housed female mice (Colby & Vandenberg, 1974), pine voles (Lepri & Vandenbergh, 1986), and California voles (*Microtus californicus*) (Rissman & Johnston, 1985) exposed to urine from group-housed females. Unlike puberty acceleration, delay of sexual maturation in response to grouped-female mouse urine is equivalent to that produced by group housing itself (Drickamer, 1977), indicating that social odors are the major social factor retarding puberty in mice. Puberty delay is insensitive to the genetic relationship between recipient and donor in mice (Drickamer, 1984) but may be important in vole species, where mothers may suppress their daughters' reproduction (Rissman & Johnston, 1985; Lepri & Vandenbergh, 1986).

Bladder urine from both singly and group-housed females delays sexual maturation in juvenile female mice (McIntosh & Drickamer, 1977). However, treating bladder urine with homogenates of urethral tissue from singly housed female mice reversed its pubertal delay properties, indicating that singly housed females normally produce perigenital substances that suppress puberty-delay odors. These delay odors are produced by reproductively active, juvenile and ovariectomized (OVX) group-housed females (Drickamer, 1977; Drickamer, McIntosh, & Rose, 1978), suggesting minimal involvement of the female reproductive system in production of the odor. In contrast, adrenalectomy eliminates production of the puberty-delay odor (Drickamer & McIntosh, 1980). Jemiolo and Novotny (1994) have identified 2,5-dimethylpyrazine (DMP) as the active pheromone in group-housed female urine that delays female sexual maturation. 2,5-dimethylpyrazine appears to be a particularly strong pheromone in that it is active by itself and independent of additional substances present in urine (Novotny, Jemiolo, Harvey, Wiesler, & Marchlewska-Koj, 1986).

Main olfactory epithelium damage does not impair puberty-delay in response to group-housed female urine (Drickamer, 1986). Although not directly tested, the VNO is likely required for response to the puberty-delay pheromone, as DMP increases cellular activity in VNO receptor neurons (Sam et al., 2001).

The relationship between puberty inhibition and group housing in laboratory settings suggests that delay of reproduction might exist under natural conditions of high population density and related increases in resource competition. To test this hypothesis, isolated areas were stocked with wild-caught mice and urine collected from animals trapped across a one-year period. Urine from females in populations experiencing rapid growth and dense populations delayed puberty in juvenile female laboratory mice whereas urine from stable, less dense populations did not (Massey & Vandenbergh, 1980). Conversely, application of urine from group-housed females to mice living in large outdoor enclosures resulted in reduced population density compared to control conditions (Drickamer & Mikesic, 1990). Taken together, pheromone-induced delay of reproductive maturity likely occurs under natural conditions.

2.1.3. Inhibition of ovarian cyclicity

Exposure to group-housed female odors can suppress estrus cyclicity by lengthening the diestrous phase in adult female mice (Champlin, 1971). Although ovarian hormones modulate the production of the cycle-suppressive odor (Clee, Humphreys, & Russell, 1975), the adrenal glands appear to be critical for its production (Ma, Miao, & Novotny, 1998). Indeed, DMP, the same adrenal-sensitive compound that delays puberty, also suppresses estrus in female mice (Ma et al., 1998). Estrus suppression by odors from grouped-housed female mice is dependent on a VNO-mediated increase in prolactin (PRL) secretion (Reynolds & Keverne, 1979). Estrus suppression can be alleviated by reducing PRL levels in group-housed female mice, and induced in singly housed females by elevating PRL levels.

2.1.4. Facilitation of ovarian cyclicity

The presence of odors from adult males can promote ovulation in group-housed mice (Marsden & Bronson, 1964) or anovulatory brown rats (*Rattus norvegicus*) (Johns, Feder, Komisaruk, & Mayer, 1978) through increases in LH and decreases in PRL secretion (Keverne & De la Riva, 1982). This pro-ovulatory 'male effect' also has been reported in induced-ovulator species such as prairie voles (Carter et al., 1980), gray short-tailed opossums (*Monodelphis domestica*) (Fadem, 1987), and domestic sheep and goats (*Capra aegagrus hircus*) (Gelez & Fabre-Nys, 2004). However, in other species, such as pine voles, exposure to males themselves, rather than their odors, is required for reproductive activation (Solomon, Vandenbergh, Wekesa, & Barghusen, 1996). Normally only reproductively active males generate odors that induce ovulation, and this ability depends on T in mice (Bronson & Whitten, 1968), Siberian hamsters (Dodge, Kristal, & Badura, 2002), prairie voles (Carter et al., 1980), and goats (Iwata et al., 2000); T treatment can even cause female mice to produce these odors (Bronson & Whitten, 1968).

In mice, estrus-accelerating pheromones appear to be the volatile compounds SBT and DHB within bladder urine (Bronson & Whitten, 1968; Jemiolo, Harvey, & Novotny, 1986), but other male-specific components have not been tested. In goats and sheep, the male pheromones that elicit LH release and ovulation also appear to be volatile (Gelez & Fabre-Nys, 2004). In sheep, the ram pheromone appears to be a blend of 1, 2-hexadecanediol, 1, 2-octadecanediol, and fatty acids (Cohen-Tannoudji, Einhorn, & Signoret, 1994), whereas in goats substances derived from 4-ethyl octanoic acid may be critical (Murata et al., 2009). However, not all estrus-accelerating pheromones are volatile; direct contact with non-volatile odors is required for reproductive activation in female prairie voles (Carter et al., 1980).

Species differences exist in the sensory systems that mediate estrus acceleration or induction. In female rats (Johns et al., 1978), opossums (Jackson & Harder, 1996), and prairie voles (Lepri & Wysocki, 1987), AOS damage prevents or reduces induction of estrus. In contrast, VNO removal does not block onset of mating in meadow voles (Meek, Lee, Rogers, & Hernandez, 1994) or odor induction of LH release in sheep (Cohen-Tannoudji, Lavenet, Locatelli, Tillet, & Signoret, 1989). In ewes, damaging the MOE impairs LH secretion in response to male odors, implicating the MOS in this response (Gelez & Fabre-Nys, 2004). Indeed, inactivation of the anterior cortical amygdala (acAMG), but not the mAMG, eliminates the endocrine response to male odors in ewes (Gelez, Archer, Chesneau, Magallon, & Fabre-Nys, 2004). Interestingly, OBX (Cohen-Tannoudji, Locatelli, & Signoret, 1986) has no effect on LH released by exposure to the male itself, indicating the importance of somatosensory cues in LH release (Bakker & Baum, 2000).

2.1.5. Ovarian synchrony

In addition to the well-studied phenomenon of male-induced ovarian synchrony described above, several reports indicate that rats and humans may show chemosignal-induced synchronization of their ovulatory cycles. In an early study, McClintock (1971) presented evidence that cohabitating female humans synchronize their menstrual cycles over time and postulated that this was mediated by chemosignals. Similarly, housing female rats together or allowing the sharing of airborne cues was reported to increase the number of animals showing estrous cycle synchrony (McClintock & Adler, 1978; McClintock, 1984). Further research suggested that synchrony may be brought about by two pheromones: a phase-advancing signal produced by follicular-stage females and a phase-delay signal produced by ovulatory females (Schank & McClintock, 1992). Such a pheromonal effect was reported for humans (Stern & McClintock, 1998) and has been inferred in earlier studies reporting synchrony in humans (Weller & Weller, 1993). Many studies, however, have not found evidence of cycle synchrony in humans (Schank, 2001b) or even in rats (Schank, 2001a) or hamsters (Schank, 2000). At best, cycle synchrony may only occur in very restricted and contextually specific circumstances (McClintock, 2002); at worst, the phenomenon is a methodological and statistical artifact (Schank, 2001b).

2.1.6. Pregnancy blockage

Exposure of recently inseminated female mice to novel males induces a high rate of pregnancy failure and early resumption of estrous cycles, but not if the male is the one that mated with the female (Bruce, 1969). This selective pregnancy block is mediated by chemosensory cues, as urine from novel males alone can disrupt pregnancy (Dominic, 1966). The ability of male urine to induce pregnancy disruption depends on T, as urine from gonadectomized or juvenile male mice is ineffective in blocking pregnancy, whereas T replacement, even in females, restores the ability of urine to disrupt pregnancy (Dominic, 1965). Although male urine is as potent as the presence of a male in disrupting pregnancy in mice (Dominic, 1966), pine voles (Schadler, 1981), and prairie voles (Smale, 1988), this is not the case for field (*Microtus agrestis*) and meadow voles (Milligan, 1976). Even in mice, the genetic strain of subjects and odor donors as well as the specifics of experimental testing conditions determine the strength of odor-specific effects (De Catanzaro et al., 1995; De Catanzaro, Muir, Sullivan, & Boissy, 1999)

The constituents of male mouse urine that induce pregnancy disruption are currently undefined but are known to require contact (De Catanzaro et al., 1995). More recently, however, the LMW, but not the high-molecular-weight (HMW), fraction of unfamiliar male urine has been shown to block pregnancy in some animals, suggesting that the individual information is carried by volatile components of urine (Peele, Salazar, Mimmack, Keverne, & Brennan, 2003). Nevertheless, the LMW fraction of unfamiliar male urine was not as effective as unfractionated urine in blocking pregnancy unless recombined with the HMW component. This may indicate that volatile signals may benefit from binding to MUPs for transport to, and interaction with, chemosensory receptors. Similarly, volatiles might also bind to major histocompatibility complex (MHC) gene products that are known to contribute to odor individuality (Yamazaki et al., 1983; Spehr et al., 2006). The volatile components of an individual's odor profile that allow a female to recognize familiar or unfamiliar males are unknown but do not involve the volatile male mouse pheromones SBT, DHB, or farnesenes (Brennan et al., 1999; Zacharias, De Catanzaro, & Muir, 2000).

Pregnancy disruption by urine from unfamiliar males is the result of a VNO-mediated neuroendocrine reflex (Brennan & Keverne, 1997). Removal of the VNO prevents the urine of unfamiliar males from disrupting early pregnancy, whereas destruction of the MOE does not (Lloyd-Thomas & Keverne, 1982). However, recent evidence suggests that inhibition of MOS processing via dopamine action in the main olfactory bulb (MOB) may prevent odor-induced disruption of late pregnancy (Serguera, Triaca, Kelly-Barrett, Banchaabouchi, & Minichiello, 2008). Normally, perception of male urinary chemosignals by the VNO increases LH secretion as well as decreasing PRL secretion by the anterior pituitary via the mAMG (Marchlewska-Koj & Jemiolo, 1978; Li, Kaba, Saito, &

Seto, 1990). In an unmated female, this drop in PRL, along with increased LH secretion, can accelerate puberty and induce estrus. In a mated female, however, this pheromone-induced drop in PRL eliminates support for the corpora lutea and thereby pregnancy (Bellringer, Pratt, & Keverne, 1980). Indeed, urine from unfamiliar males only disrupts pregnancy if given during the daily, postmating, elevations of PRL (Rosser, Remfry, & Keverne, 1989). To prevent the mating male's urine from disrupting his own mate's pregnancy, the female forms a memory of the stud male that effectively disrupts accessory olfactory bulb (AOB) output to the mAMG in response to his urinary cues (Brennan, Kaba, & Keverne, 1990). The AOB itself is the site of this suppressive memory of the stud odor and is mediated by mating-induced release of norepinephrine, which ultimately restructures cellular responses in the AOB (Brennan & Keverne, 1997). Increased dopaminergic activity in the MOB also impairs olfactory sensitivity during late pregnancy and thereby limits behavioral investigation and processing of strange males' odors (Serguera et al., 2008).

Although the prevention of pregnancy blockade by the sire would seem to benefit both the stud male and the dam, attempts at demonstrating male-induced pregnancy termination in the field have produced equivocal results. An initial study using outside enclosures reported that prairie voles do show evidence of pregnancy block in more natural conditions (Heske & Nelson, 1984). However, in larger enclosures that better resemble the actual size of vole home ranges, rapid turnover of males by trapping and introducing new males does not lead to population-level changes in pregnancy in grey-tailed voles (*Microtus canicaudus*) (De la Maza, Wolff, & Lindsey, 1999) and leads to only minor effects in prairie voles (Mahady & Wolff, 2002).

2.2. Responses by Males

2.2.1. Reproductive development

In contrast to the well-developed literature on odor effects on female reproductive physiology, there is considerably less information about similar effects in males (Koyama, 2004). What is known suggests that the development and function of male reproductive physiology is also sensitive to odor exposure. For example, the development of puberty in male mice can be delayed by exposure to male urine, grouped-female urine, and DMP, a puberty-delay pheromone derived from the urine of grouped females (Jemiolo & Novotny, 1994). In California voles, odors from mothers suppress their sons' pubertal onset more strongly than odors from their fathers (Rissman & Johnston, 1985). Interestingly, exposure to unrelated, nongrouped females does not advance puberty in male mice (Maruniak, Coquelin, & Bronson, 1978).

2.2.2. Sperm allocation

Exposure to social odors can alter male gonadal physiology in adulthood. For example, dominant, but not subordinate, male mice display increased sperm density when housed with female bedding, suggesting that exposure to female odors increases spermatogenesis (Koyama & Kamimura, 2000). Sperm allocation can also be adaptively regulated in odor contexts that signal increased sperm competition: more sperm is ejaculated by meadow voles during copulation in the presence of another male's odor due to increased movement of sperm to the vas deferens prior to copulation (Delbarco-Trillo & Ferkin, 2004).

2.2.3. Hormone release

Upon exposure to conspecific female odors, adult male mice (Macrides, Bartke, & Dalterio, 1975), Syrian hamsters (*Mesocricetus auratus*) (Macrides, Bartke, Fernandez, & D'Angelo, 1974), Siberian hamsters (Anand, Turek, & Horton, 2004), rats (Bonilla-Jaime, Vazquez-Palacios, Arteaga-Silva, & Retana-Marquez, 2006), and common marmosets (*Callithrix jacchus*) (Ziegler, Schultz-Darken, Scott, Snowdon, & Ferris, 2005) release LH within 15–30 minutes, followed by a dramatic peak in circulating androgens. This response rapidly habituates to repeated presentations of the same female odors and is increased by presentation of cues from novel females (Coquelin & Bronson, 1979), indicating that female novelty strongly modulates this response.

The reflexive hormone surge to female odors can be elicited by both sexually naïve and sexually experienced male Syrian hamsters (Pfeiffer & Johnston, 1994) and mice (Maruniak & Bronson, 1976) but not by sexually naïve rats (Bonilla-Jaime et al., 2006), indicating significant species differences in unconditioned physiological responses. Nevertheless, LH/androgen responses to the presentation of the entire female survive combined AOS and MOS lesions in sexually experienced male Syrian hamsters (Pfeiffer & Johnston, 1994), suggesting that, even in species with unconditioned responses to odors, hormonal responses can become conditioned to nonodor copulatory cues. Indeed, pairing of neutral odor stimuli with mating causes male rats to later produce LH/androgen surges in response to the previously neutral odor (Graham & Desjardins, 1980).

Surprisingly, the reproductive state of the female has minimal impact on her ability to produce odors that induce LH/androgen surges (Johnston & Bronson, 1982). The chemical identity of the female pheromones eliciting LH/androgen surges are unknown but, in mice, appear to be LMW molecules that may be bound to MUPs (Singer, Clancy, Macrides, Agosta, & Bronson, 1988).

Removing the VNO eliminates or reduces LH/androgen surges in both sexually experienced and naïve males in response to female odors but not to the female herself; damage to the MOE is without effect (Wysocki, Katz, & Bernhard, 1983; Coquelin, Clancy, Macrides, Noble, & Gorski, 1984; Pfeiffer & Johnston, 1994). Combined lesions of the VNO and MOE do eliminate surges in response to females in sexually naïve Syrian hamsters, but not sexually experienced males, indicating that chemosensory function is required for LH/androgen response prior to the animal's initial copulatory experience (Pfeiffer & Johnston, 1994).

The function of the reflexive LH/androgen surge has been elusive but evidence suggests that it may facilitate male copulatory behavior through reduction of anxiety or changes in penile reflexes (Nyby, 2008).

3. SOCIAL ODORS AND BEHAVIOR

3.1. Responses by Females

3.1.1. Attraction and investigation

Postpubertal female Syrian hamsters (Johnston, 1979), rats (Carr, Loeb, & Dissinger, 1965), mice (Hurst, 1990), meadow voles (Ferkin & Zucker, 1991), and domestic ferrets (*Mustela putorius furo*) (Kelliher & Baum, 2002) are preferentially attracted to the odors of reproductively active male conspecifics independent of sexual experience. Female attraction to male odors in mice, but not Syrian hamsters (Maras & Petrulis, 2008a), develops through a process that requires contact with male odors (Moncho-Bogani, Lanuza, Hernandez, Novejarque, & Martinez-Garcia, 2002; Ramm, Cheetham, & Hurst, 2008).

Despite expectations of hormone dependency, female Syrian hamsters are attracted to male odors across their entire estrous cycle and during lactation but not during pregnancy (Johnston, 1979; Eidson, Maras, Epperson, & Petrulis, 2007). Female meadow voles are attracted to male odors across pregnancy and also during lactation (Ferkin & Johnston, 1995). Indeed, female mice and Syrian hamsters continue to display preferences for male odors following gonadectomy (Moncho-Bogani, Lanuza, Lorente, & Martinez-Garcia, 2004; Eidson et al., 2007). In contrast, opposite-sex odor preference of female rats (Xiao, Kondo, & Sakuma, 2004), ferrets (Woodley & Baum, 2003), and meadow voles (Ferkin, Gorman, & Zucker, 1991) decreases following ovariectomy and is reinstated by estradiol (E_2) or T injections. This suggests significant species differences in the hormone sensitivity of female attraction to male odors.

In many species, female preference for males is strongly modulated by chemosensory indicators of male quality such as high androgen levels (Kempenaers, Peters, & Foerster, 2008) that can be indexed by androgen-sensitive odors and marking behavior (Thiessen & Rice, 1976). Indeed, gonadectomy decreases attractiveness of male odors in meadow voles and is reversed by gonadal steroid treatment (Ferkin & Johnston, 1993). Attractiveness of male odors is also modulated by other cues of reproductive fitness such as social status, diet, and pathogen load. For example, female house mice prefer the odors of dominant males over those of subordinates (Mossman & Drickamer, 1996), and meadow voles prefer odors of males fed on high protein diets, a normally scarce and valuable resource (Ferkin, Sorokin, Johnston, & Lee, 1997). Female mice also decrease their preference for odors from males subclinically infected with parasites (Kavaliers & Colwell, 1995).

In several species, breeding is biased by having mates that differ from oneself at the MHC locus, a cluster of genes critical for self/nonself immune recognition, as a means of avoiding inbreeding and/or increasing genetic diversity (Penn, Damjanovich, & Potts, 2002). Although female mice can distinguish between the volatile odors of males differing only at the MHC locus (Singer, Beauchamp, & Yamazaki, 1997), they may not prefer odors from dissimilar MHC males (Ehman & Scott, 2001). Moreover, MUP variation, rather than that of MHC, in wild-type male mice appears to regulate female preference (Cheetham et al., 2007; Thom et al., 2008).

In addition to factors altering the intrinsic attractiveness of male odor, female interest is also modulated by the spatial pattern of odor deposition indicative of a territory owner. Female mice and Syrian hamsters prefer odors from males they have previously encountered and those that have counter-marked a competitor's odors (Hurst & Beynon, 2004; Johnston, 2008).

Treatment with the mouse pheromones SBT, DHB, hexadecanol and hexadecyl acetate, and (methylthio) methanethiol increases the attractiveness of castrate male urine to females (Jemiolo, Alberts, Sochinski-Wiggins, Harvey, & Novotny, 1985; Lin, Zhang, Block, & Katz, 2005; Zhang, Liu, Zhang, & Sun, 2008). However, these volatile molecules are ineffective or less effective in water, implying that other constituents of urine such as MUPs are also important for maximal attraction. In addition, two other attractant compounds, α- and β-farnesene, are found in the preputial glands of intact males (Jemiolo, Xie, & Novotny, 1991). However, attraction to these compounds at normal concentrations requires sexual experience, which suggests that they are not the major sexually attractive components of mouse urine.

Female preference for male odors requires both the AOS and the MOS, although the relative contribution of these two systems varies with the nature of the odor stimulus. Vomeronasal organ removal in female Syrian hamsters (Petrulis, Peng, & Johnston, 1999), mice (Keller, Pierman, Douhard, Baum, & Bakker, 2006), domestic pigs (*Sus scrofa*) (Dorries, Adkins-Regan, & Halpern, 1997), and ferrets (Woodley, Cloe, Waters, & Baum, 2004) does not impair their preference for volatile male odors. Attraction to distant male odors is, instead, dependent on the MOS; impairing the MOE eliminates preference for volatile male odors in female mice (Keller et al., 2006b) and ferrets (Kelliher & Baum, 2001). In contrast, female attraction to nonvolatile components of male odor is reduced by AOS damage in mice (Keller et al., 2006b), Syrian hamsters (Petrulis et al., 1999), ferrets (Woodley et al., 2004), and opossums (Zuri & Halpern, 2005). Taken together, this suggests that the AOS is not critical for attraction of females to distal male odors but is needed for close and persistent investigation of male odors and for the development of conditioned preference for male volatiles (Lanuza et al., 2008).

Female preference for male volatile odors is eliminated by mAMG damage in female Syrian hamsters (Petrulis & Johnston, 1999) and rats (Kondo & Sakuma, 2005), with further hypothalamic processing dependent on the species investigated. Specifically, lesions of the ventromedial hypothalamus (VMH), but not the medial preoptic area (mPOA), reduce the preference of female ferrets for investigating anesthetized males (Robarts & Baum, 2007), whereas mPOA lesions eliminate female rats' preference for male volatile odors (Xiao, Kondo, & Sakuma, 2005). Other potential forebrain structures, such as the entorhinal cortex (ENT) and orbitofrontal cortex (OFC), are not required for female attraction to male odors (Petrulis, DeSouza, Schiller, & Johnston, 1998; Petrulis, Peng, & Johnston, 2000). Unlike other motivated behaviors, preference for male odors is either independent of, or suppressed by, dopaminergic and opiate neural systems (Lanuza et al., 2008).

3.1.2. Scent marking

Female mice (Rich & Hurst, 1999), Syrian hamsters (Johnston, 1977), domestic rabbits (*Oryctolagus cuniculus*) (Gonzalez-Mariscal, Melo, Zavala, & Beyer, 1990), and rats (Birke, 1984) scent-mark in response to male odors and vary their marking across reproductive states to advertise impending receptivity. The best-studied case of female reproductive marking is vaginal marking by Syrian hamsters, a stereotyped behavior that deposits a sex-attractant vaginal secretion on the substrate (Been & Petrulis, 2008). Vaginal marking peaks on the night before behavioral receptivity but vanishes during the onset of sexual receptivity, consistent with its role in attracting widely dispersed males for mating (Johnston, 1977; Gattermann et al., 2001). This dramatic rise in vaginal marking during behavioral proestrus appears to be mediated by rising E_2 levels, and the rapid switch between vaginal marking and receptive behavior is likely due to the subsequent rise in progesterone (P_4) (Lisk & Nachtigall, 1988). The striking hormonally mediated changes in vaginal marking across the estrous cycle appear to be regulated by neurons in the mPOA and VMH; E_2 implants in these regions reinstate vaginal marking in OVX females (Takahashi, Lisk, & Burnett, 1985). However, the VMH may not be critical for vaginal marking as lesions of this structure do not eliminate vaginal marking (Floody, 2002), whereas lesions of the mPOA do (Malsbury, Kow, & Pfaff, 1977).

Vaginal marking by female Syrian hamsters is stimulated by male conspecific odors, specifically those from sexually dimorphic flank glands (Petrulis & Johnston, 1997), and inhibited by female odors (Johnston, 1977; Johnston & Brenner, 1982), with most marking directed toward odors of unrelated males (Heth, Todrank, & Johnston, 1998). The preferential marking toward male odors appears to require early experience with odors from male siblings (Maras & Petrulis, 2008a) similar to the development of odor preference in female mice (Moncho-Bogani et al., 2002). This chemosensory regulation of vaginal marking is mediated primarily by the MOS, as VNO removal has minimal effect on vaginal marking whereas MOE destruction reduces the behavior (Johnston, 1992; Petrulis et al., 1999). Main olfactory system connections with the mAMG appear to mediate vaginal marking responses to odors as mAMG lesions dramatically reduce vaginal marking to both male and female odors (Petrulis & Johnston, 1999). However, mAMG-lesioned females still mark more in response to male odors than to female odors and continue to show cyclic variation in their vaginal marking, indicating that the mAMG is not critical for hormonal control or differential marking. Other MOS structures such as the ENT and OFC do not mediate vaginal marking (Petrulis et al., 1998; 2000).

3.2. Responses by Males

3.2.1. Attraction and investigation

Adult males of many species approach female scent from a distance and engage in prolonged investigation of these odors (Beauchamp & Beruter, 1973; Johnston, 1974; Brown, 1978). Males use the arrangement, spatial patterning, and freshness of female scent marks to bias their search toward nearby mating partners (Johnston, 2008) and toward areas associated with female odors (Pankevich, Cherry, & Baum, 2006) or where the presence of estrous females is anticipated (Ferkin, Combs, DelBarco-Trillo, Pierce, & Franklin, 2008), even in the face of predator cues (Kavaliers, Choleris, & Colwell, 2001). There is species variation in the optimal stimulus that elicits male investigation. In species such as the Syrian hamster, in which females are dispersed and scent-mark to signal impending receptivity, males are equally attracted to vaginal secretion from females across reproductive conditions (Johnston, 1974; Kwan & Johnston, 1980; Macrides, Singer, Clancy,

Goldman, & Agosta, 1984). In other species, such as rats, where interactions with females are more common, males are more attracted to urine odors from estrous females than from diestrous females (Lydell & Doty, 1972). This differential attraction to estrous odors is dependent on sexual experience in male rats, mice, and dogs (*Canis lupus familiaris*) (Stern, 1970; Doty & Dunbar, 1974; Hayashi & Kimura, 1974), which suggests, more broadly, that aspects of odor preference are learned during adulthood. In support of this idea, male mice and Syrian hamsters can be conditioned to reduce investigation of female odors by pairing the odors with gastrointestinal distress; however, this conditioning is rapidly extinguished by social experience (Johnston & Zahorik, 1975; Kay & Nyby, 1992).

Although putative sexual-attractant compounds have been identified in females of several species (Zhang et al., 2005; Zhang, Rao, Sun, Zhao, & Qin, 2007; Zhang et al., 2008b) there are very few cases in which actual attractant pheromones have been isolated. One prominent case is the identification of (Z)-7-dodecenyl acetate as the major attractant compound found in the urine of estrous female Asian elephants (Rasmussen et al., 1996). Another well-known attempt was the putative characterization of dimethyl disulfide (DMDS) as the attractant pheromone within hamster vaginal secretion. Dimethyl disulfide is more attractive than several other volatile components within the secretion and is half as attractive as the entire secretion when tested in an animal's home cage (Singer et al., 1976; O'Connell, Singer, Macrides, Pfaffmann, & Agosta, 1978). However, when presented in a neutral arena, DMDS is much less effective than volatiles from vaginal secretion in attracting male hamsters (Petrulis & Johnston, 1995). Moreover, unlike vaginal secretion, no sex difference in investigation was evident toward DMDS and attraction to it was not dependent on gonadal hormones. Thus, in a context more similar to how male hamsters might find vaginal secretion, DMDS by itself does not meet the definition of a sex-attractant pheromone.

Adult male attraction to female odors appears to be largely due to increased circulating levels of T in males, as castration eliminates this attraction and can be reversed by T treatment (Stern, 1970; Gregory, Engel, & Pfaff, 1975) or by the combination of its estrogenic and androgenic metabolites (Powers, Bergondy, & Matochik, 1985; Steel & Hutchison, 1986). Treatments with estrogens or non-aromatizable androgens alone can be sufficient to increase attraction to female odors in ferrets (Woodley & Baum, 2003) but are either insufficient or suboptimal for restoring normal odor preference in male rats (Xiao et al., 2004), hamsters (Powers et al., 1985; Steel & Hutchison, 1986), meadow voles (Ferkin & Gorman, 1992), and mice (Bean, Nyby, Kerchner, & Dahinden, 1986).

Preference for receptive female odors is eliminated by OBX in male Syrian hamsters (Murphy & Schneider, 1970) and rats (Edwards, Griffis, & Tardivel, 1990) but the relative contribution of the AOS and MOS to odor preference depends on access to the odor cue. If only volatile chemosignals are available, damage to the MOE, but not to the VNO, impairs opposite-sex odor attraction in male Syrian hamsters (Powers & Winans, 1973; O'Connell & Meredith, 1984), mice (Pankevich, Baum, & Cherry, 2004), and ferrets (Kelliher & Baum, 2001). If contact is allowed, damage to either system can decrease attraction to vaginal secretion in sexually inexperienced male Syrian hamsters (Powers, Fields, & Winans, 1979; Pfeiffer & Johnston, 1994) and mice (Pankevich et al., 2004), but not in sexually experienced Syrian hamsters (O'Connell & Meredith, 1984). Similarly, VNO removal in sexually inexperienced guinea pigs (*Cavia porcellus*) leads to rapid extinction of urine investigative behavior (Beauchamp, Martin, Wysocki, & Wellington, 1982).

Centrally, lesions of either the anterior mAMG (amAMG) or posterior mAMG (pmAMG), but not the posteromedial cortical amygdala (pcAMG) (Maras & Petrulis, 2008b), eliminate preference for female odors in sexually naïve male Syrian hamsters, albeit in different ways (Maras & Petrulis, 2006). Anterior mAMG damage dramatically increases investigation of both male and female odors whereas males with pmAMG lesions show reduced investigation specifically toward female odors. This suggests that amAMG functions to evaluate the significance of social stimuli whereas the pmAMG generates undifferentiated attraction to social odors. Outputs from the mAMG to the bed nucleus of the stria terminalis (BNST) (Powers, Newman, & Bergondy, 1987; Edwards, Walter, & Liang, 1996) and mPOA (Hurtazo & Paredes, 2005; Alekseyenko, Waters, Zhou, & Baum, 2007) may be critical for driving odor attraction. Other structures may also be involved in male attraction toward female odors. For example, in Syrian hamsters, OFC lesions eliminate preference for female odor (Sapolsky & Eichenbaum, 1980) and lesions to the ENT eliminate a male hamster's preference for investigating odors from novel females (Petrulis & Eichenbaum, 2003).

3.2.2. Vocalizations

Male rats (Geyer & Barfield, 1978), mice (Nyby, Wysocki, Whitney, & Dizinno, 1977), and guinea pigs (Eisthen, Wysocki, & Beauchamp, 1987) produce ultrasonic vocalizations (USVs) in response to female, but not male, odors. Production of USVs in mice is eliminated by gonadectomy and restored, and even induced in females, by T or its metabolites (Nyby, Dizinno, & Whitney, 1977; Bean et al., 1986). Exposure to a rapidly oxidized component of female urine is necessary to induce USV in sexually naïve male mice without rapid extinction of the response and will support conditioned USV response to previously neutral stimuli (Sipos, Kerchner, & Nyby, 1992). Transient inhibition of USVs to odors also can be conditioned by pairing odors with gastric distress, although this effect is easily extinguished (Kay & Nyby, 1992).

Olfactory bulbectomy in males eliminates USVs to female odors and similar effects are observed after VNO removal, especially in sexually naïve animals (Bean, 1982; Wysocki, Nyby, Whitney, Beauchamp, & Katz, 1982; Eisthen et al., 1987) as well as after MOE damage (Sipos, C. Wysocki, Nyby, L. Wysocki, & Nemura, 1995). Both VNO and MOE signals ultimately work through steroid-sensitive regions in the mPOA to trigger USVs, as T or E_2 implants in the mPOA restore USVs in castrated male mice (Nyby, Matochik, & Barfield, 1992).

3.3. Copulatory behavior

Masculine copulatory behavior in rodents is largely regulated by chemosensory cues, with the absolute requirement for odor input varying between species and with sexual experience. For example, OBX eliminates copulatory behavior independently of sexual experience in Syrian hamsters (Murphy & Schneider, 1970) and mice (Rowe & Edwards, 1972), whereas in rats and guinea pigs OBX has more variable or contextually specific effects on copulation and can be ameliorated by sexual experience (Beauchamp, Magnus, Shmunes, & Durham, 1977; Edwards et al., 1990; 1996). In sexually experienced Syrian hamsters, female vaginal secretions are sufficient for eliciting mounting when applied to the anogenital area of anesthetized males (Murphy, 1973; Johnston, 1975). This copulation-promoting function of vaginal secretions may be a unique feature of Syrian hamster biology as attempts at conditioning copulation to artificial odors have not been successful in this species (Macrides, Clancy, Singer, & Agosta, 1984). Although volatiles from vaginal secretions can induce low levels of mounting, the mounting stimulus is primarily in the nonvolatile, HMW fraction (O'Connell & Meredith, 1984; Singer, Clancy, Macrides, & Agosta, 1984). The major active component of this fraction was identified as a protein ('aphrodisin') that is produced within the vaginal tract (Briand, Trotier, & Pernollet, 2004). Like MUPs, aphrodisin normally binds a number of water-insoluble volatile ligands and may act as a carrier molecule; the pure protein does not promote mounting. Although sufficient for mounting of surrogate females, vaginal secretion is not necessary for copulatory behavior in sexually experienced male hamsters (Johnston, 1986) and aversive conditioning of it does not greatly impair copulation (Johnston, Zahorik, Immler, & Zakon, 1978).

Both the AOS and MOS are involved in regulating copulatory behavior of male rodents, but the relative importance of each system varies across species. Although damage to the MOE does not impair copulatory behavior in Syrian hamsters (Powers & Winans, 1973), it eliminates copulation in mice, independent of experience (Mandiyan, Coats, & Shah, 2005; Keller, Douhard, Baum, & Bakker, 2006) and greatly reduces erections produced by rats in response to volatile odors from conspecific females (Kondo, Tomihara, & Sakuma, 1999). On the other hand, VNO damage eliminates copulation in 20–30% of male hamsters (Powers & Winans, 1975) but has a less pronounced effect in male rats and mice (Clancy, Coquelin, Macrides, Gorski, & Noble, 1984; Saito & Moltz, 1986) and no effect in guinea pigs (Beauchamp et al., 1982). The absence of copulatory behavior in some VNO-lesioned male hamsters is not observed following sexual experience (Meredith, 1986) and can be reversed by exposure to vaginal secretion prior to testing (Westberry & Meredith, 2003). In mice, functional disorganization of the VNO induces inappropriate mounting behavior toward male conspecifics (Leypold et al., 2002; Stowers, Holy, Meister, Dulac, & Koentges, 2002), but this effect is not observed in mice following VNO removal, suggesting that impairment of VNO function may be more disruptive than VNO removal (Pankevich et al., 2004). These species differences may be due, in part, to the nature of the chemosensory stimuli that facilitate copulation in different species. For example, in Syrian hamsters, the procopulatory aspects of vaginal secretion are largely nonvolatile and may therefore depend more on the VNO (Halpern & Martinez-Marcos, 2003; Zufall & Leinders-Zufall, 2007), whereas in mice the vaginal odors that promote mating (Hayashi & Kimura, 1974) may be primarily volatile.

More centrally, lesions of the AMG, especially the amAMG, eliminate copulatory behavior and greatly reduce anogenital investigation by male hamsters (Lehman, Winans, & Powers, 1980). In rats, damage to the pmAMG, and to a lesser extent the amAMG, impairs noncontact erections to volatile odors from females (Kondo, Sachs, & Sakuma, 1998; Kondo & Sachs, 2002). Connections from the mAMG to the BNST appear to be critical for anogenital investigation of females (Lehman, Powers, & Winans, 1983) and noncontact erections in response to female odors (Liu, Salamone, & Sachs, 1997). Steroidal effects on chemoinvestigatory behavior during mating are mediated by the mAMG and mPOA, as T or E_2 implants in either of these regions reinstate anogenital investigation and copulation in sexually experienced, castrated Syrian hamsters (Wood & Newman, 1995; Wood, 1996).

4. CONCLUSIONS

It is clear that social odors have profound and oftentimes obligatory roles in mammalian reproduction, especially in nocturnal species. The effects of these signals are to coordinate reproductive physiology and behavior and to adapt these responses to prevailing environmental conditions. For example, chemosensory signals can accelerate or decelerate maturation and adult function of the hypothalamic–pituitary–gonadal (HPG) axis of recipients such that increased HPG functioning responds to the appearance of mating partners, and decreases in HPG function occur in reproductively suboptimal conditions. Similarly, perception of social odors from opposite-sex conspecifics leads reproductively active adults to initiate behaviors that increase the likelihood of contacting mates, such as chemosensory investigation, scent marking, and vocalizations. Even after contact occurs, copulatory behavior in both sexes may still be dependent on chemosensory social cues, perhaps indicating that these cues provide a more detailed evaluation of the mating partner. The coordination of reproductive physiology and behavior by social odors is most evident by the observation that gonadal steroids not only regulate odor-guided reproductive behavior but also the production, release, and deposition of social odors themselves.

In any discussion of mammalian chemosensory communication, the inevitable question is raised: does pheromonal communication occur in humans? Based on the dominance of visual and auditory cues in human communication and the complexity of human social behavior, we would not expect social odors to play a major role in human reproduction. Indeed, they do not. Although adult humans do produce odors and can discriminate between different categories of individuals, such as males and females or kin and nonkin, there is no evidence that adult social behavior is altered by these odors (Wysocki & Preti, 2004). The most direct evidence supporting a behavioral role for human social odors comes from two studies claiming increased sexual attractiveness by wearing proprietary derivatives of human odors (Cutler, Friedmann, & McCoy, 1998; McCoy & Pitino, 2002). Unfortunately, a statistical reanalysis of these data indicated no significant attractant effect in either study (Winman, 2004). Although a component of human male sweat (androstadienone (AND)) may positively bias a female's perception of male attractiveness, this does not translate into increased selection of these men as dating partners (Saxton, Lyndon, Little, & Roberts, 2008). Other, more indirect, evidence has attempted to link preferences for odors of opposite-sex individuals that are more (Jacob, McClintock, Zelano, & Ober, 2002) or less (Wedekind, Seebeck, Bettens, & Paepke, 1995) similar at MHC loci to population-level marriage patterns and have found equivocal results (Chaix, Cao, & Donnelly, 2008; Havlicek & Roberts, 2009). Although human behavior may not be altered by exposure to human odors, increased positive mood has been reported in response to AND by women (McClintock, 2002). These alterations of female mood are heavily context-specific; they are, e.g., only observed if the experimenter is male (Lundstrom & Olsson, 2005). Additionally, exposure to AND increases sympathetic system arousal in women in a dose-dependent fashion (Jacob, Hayreh, & McClintock, 2001; Wyart et al., 2007). Even though the reported phenomenon of menstrual synchrony may be a methodological artifact, prolonged exposure to extracts of male axillary odor does advance female LH surges by a modest amount (Preti, Wysocki, Barnhart, Sondheimer, & Leyden, 2003). Combined with the preponderance of evidence that adult humans do not have a functioning VNO (Meredith, 2001), the existing evidence suggests that, unlike many other mammals, social odors play only a minor role in human reproductive physiology and behavior.

ABBREVIATIONS

acAMG Anterior cortical amygdala
amAMG Anterior medial amygdala
AND Androstenedione
AOB Accessory olfactory bulb
AOS Accessory olfactory (or vomeronasal) system
BNST Bed nucleus of the stria terminalis
DHB 3, 4-dehydro-exo-brevicomin
DMDS Dimethyl disulfide
DMP 2,5-dimethylpyrazine
E_2 Estradiol
ENT Entorhinal cortex
GnRH Gonadotropin-releasing hormone
HMH 6-hydroxy-6-methyl-3-heptanone
HMW High-molecular-weight
HPG Hypothalamic–pituitary–gonadal
LH Luteinizing hormone
LMW Low-molecular-weight
mAMG Medial amygdala
MHC Major histocompatibility complex
MOB Main olfactory bulbectomy
MOE Main olfactory epithelium
MOS Main olfactory system
mPOA Medial preoptic area
MUP Major urinary protein
OBX Olfactory bulbectomy
OFC Orbitofrontal cortex
OVX Ovariectomized
P_4 Progesterone
pcAMG Posteromedial cortical amygdala
pmAMG Posterior medial amygdala
POA Preoptic area
PRL Prolactin
SBT 2-sec-butyl dihydrothiazole
T Testosterone
USV Ultrasonic vocalization
VMH Ventromedial hypothalamus
VNO Vomeronasal organ

REFERENCES

Alekseyenko, O. V., Waters, P., Zhou, H., & Baum, M. J. (2007). Bilateral damage to the sexually dimorphic medial preoptic area/anterior hypothalamus of male ferrets causes a female-typical preference for and a hypothalamic Fos response to male body odors. *Physiol. Behav., 90*, 438–449.

Anand, S., Turek, F. W., & Horton, T. H. (2004). Chemosensory stimulation of luteinizing hormone secretion in male Siberian hamsters (*Phodopus sungorus*). *Biol. Reprod., 70*, 1033–1040.

Armstrong, S. D., Robertson, D. H., Cheetham, S. A., Hurst, J. L., & Beynon, R. J. (2005). Structural and functional differences in isoforms of mouse major urinary proteins: a male-specific protein that preferentially binds a male pheromone. *Biochem. J., 391*, 343–350.

Baddaloo, E. G. Y., & Clulow, F. V. (1981). Effects of the male on growth, sexual maturation, and ovulation of young female meadow voles, *Microtus pennsylvanicus*. *Can. J. Zool., 59*, 415–421.

Bakker, J., & Baum, M. J. (2000). Neuroendocrine regulation of GnRH release in induced ovulators. *Front. Neuroendocrinol., 21*, 220–262.

Baum, M. J., & Kelliher, K. R. (2008). Complementary roles of the main and accessory olfactory systems in mammalian mate recognition. *Annu. Rev. Physiol.*

Bean, N. J. (1982). Olfactory and vomeronasal mediation of ultrasonic vocalizations in male mice. *Physiol. Behav., 28*, 31–37.

Bean, N. J., Nyby, J., Kerchner, M., & Dahinden, Z. (1986). Hormonal regulation of chemosignal-stimulated precopulatory behaviors in male housemice (*Mus musculus*). *Horm. Behav., 20*, 390–404.

Beauchamp, G. K., & Beruter, J. (1973). Source and stability of attractive components in guinea pig (*Cavia porcellus*) urine. *Behav. Biol., 9*, 43–47.

Beauchamp, G. K., Magnus, J. G., Shmunes, N. T., & Durham, T. (1977). Effects of olfactory bulbectomy on social behavior of male guinea pigs (*Cavia porcellus*). *J. Comp. Physiol. Psychol., 91*, 336–346.

Beauchamp, G. K., Martin, I. G., Wysocki, C. J., & Wellington, J. L. (1982). Chemoinvestigatory and sexual behavior of male guinea pigs following vomeronasal organ removal. *Physiol. Behav., 29*, 329–336.

Been, L. and Petrulis, A. (2008). The neurobiology of sexual solicitation: vaginal marking in female syrian hamsters (*Mesocricetus auratus*). In *Chemical Signals in Vertebrates 11*, pp. 231–239.

Bellringer, J. F., Pratt, H. P., & Keverne, E. B. (1980). Involvement of the vomeronasal organ and prolactin in pheromonal induction of delayed implantation in mice. *J. Reprod. Fertil., 59*, 223–228.

Birke, L. I. (1984). Effects of estradiol and progesterone on scent-marking behavior of female rats. *Horm. Behav., 18*, 95–98.

Boehm, U., Zou, Z., & Buck, L. B. (2005). Feedback loops link odor and pheromone signaling with reproduction. *Cell, 123*, 683–695.

Bonilla-Jaime, H., Vazquez-Palacios, G., Arteaga-Silva, M., & Retana-Marquez, S. (2006). Hormonal responses to different sexually related conditions in male rats. *Horm. Behav., 49*, 376–382.

Boschat, C., Pelofi, C., Randin, O., Roppolo, D., Luscher, C., Broillet, M. C., et al. (2002). Pheromone detection mediated by a V1r vomeronasal receptor. *Nat. Neurosci., 5*, 1261–1262.

Brennan, P. A., & Keverne, E. B. (1997). Neural mechanisms of mammalian olfactory learning. *Prog. Neurobiol., 51*, 457–481.

Brennan, P., Kaba, H., & Keverne, E. B. (1990). Olfactory recognition: a simple memory system. *Science, 250*, 1223–1226.

Brennan, P. A., Schellinck, H. M., & Keverne, E. B. (1999). Patterns of expression of the immediate–early gene egr-1 in the accessory olfactory bulb of female mice exposed to pheromonal constituents of male urine. *Neuroscience, 90*, 1463–1470.

Briand, L., Trotier, D., & Pernollet, J. C. (2004). Aphrodisin, an aphrodisiac lipocalin secreted in hamster vaginal secretions. *Peptides, 25*, 1545–1552.

Bronson, F. H., & Desjardins, C. (1974). Relationships between scent marking by male mice and the pheromone-induced secretion of the gonadotropic and ovarian hormones that accompany puberty in female mice. *Adv. Behav. Biol., 11*, 157–178.

Bronson, F. H., & Maruniak, J. A. (1975). Male-induced puberty in female mice: evidence for a synergistic action of social cues. *Biol. Reprod., 13*, 94–98.

Bronson, F. H., & Whitten, W. K. (1968). Oestrus-accelerating pheromone of mice: assay, androgen-dependency and presence in bladder urine. *J. Reprod. Fertil., 15*, 131–134.

Brown, R. E. (1978). Hormonal control of odor preferences and urine-marking in male and female rats. *Physiol. Behav., 20*, 21–24.

Brown, R. E., & Macdonald, D. W. (1985). *Social odours in mammals.* New York, NY: Clarendon Press.

Bruce, H. M. (1969). Pheromones and behavior in mice. *Acta. Neurol. Psychiatr. Belg., 69*, 529–538.

Carr, W. J., Loeb, L. S., & Dissinger, M. L. (1965). Responses of Rats to Sex Odors. *J. Comp. Physiol. Psychol., 59*, 370–377.

Carter, C. S., Getz, L. L., Gavish, L., McDermott, J. L., & Arnold, P. (1980). Male-related pheromones and the activation of female reproduction in the prairie vole (*Microtus ochrogaster*). *Biol. Reprod., 23*, 1038–1045.

Chaix, R., Cao, C., & Donnelly, P. (2008). Is mate choice in humans MHC-dependent? *PLoS Genet., 4*. e1000184.

Champlin, A. K. (1971). Suppression of oestrus in grouped mice: the effects of various densities and the possible nature of the stimulus. *J. Reprod. Fertil., 27*, 233–241.

Cheetham, S. A., Thom, M. D., Jury, F., Ollier, W. E., Beynon, R. J., & Hurst, J. L. (2007). The genetic basis of individual-recognition signals in the mouse. *Curr. Biol., 17*, 1771–1777.

Clancy, A. N., Coquelin, A., Macrides, F., Gorski, R. A., & Noble, E. P. (1984). Sexual behavior and aggression in male mice: involvement of the vomeronasal system. *J. Neurosci., 4*, 2222–2229.

Clee, M. D., Humphreys, E. M., & Russell, J. A. (1975). The suppression of ovarian cyclical activity in groups of mice, and its dependence on ovarian hormones. *J. Reprod. Fertil., 45*, 395–398.

Cohen-Tannoudji, J., Einhorn, J., & Signoret, J. P. (1994). Ram sexual pheromone: first approach of chemical identification. *Physiol. Behav., 56*, 955–961.

Cohen-Tannoudji, J., Lavenet, C., Locatelli, A., Tillet, Y., & Signoret, J. P. (1989). Non-involvement of the accessory olfactory system in the LH response of anoestrous ewes to male odour. *J. Reprod. Fertil., 86*, 135–144.

Cohen-Tannoudji, J., Locatelli, A., & Signoret, J. P. (1986). Non-pheromonal stimulation by the male of LH release in the anoestrous ewe. *Physiol. Behav., 36*, 921–924.

Colby, D. R., & Vandenberg, J. G. (1974). Regulatory effects of urinary pheromones on puberty in the mouse. *Biol. Reprod., 11*, 268–279.

Coquelin, A., & Bronson, F. H. (1979). Release of luteinizing hormone in male mice during exposure to females: habituation of the response. *Science, 206*, 1099–1101.

Coquelin, A., Clancy, A. N., Macrides, F., Noble, E. P., & Gorski, R. A. (1984). Pheromonally induced release of luteinizing hormone in male mice: involvement of the vomeronasal system. *J. Neurosci., 4*, 2230–2236.

Cowley, J. J., & Pewtress, R. K. (1986). Post-parturition and late lactation urine: long-term effects of exposure on mouse activity and sniffing behaviour. *Reprod. Nutr. Dev., 26*, 113–126.

Cutler, W. B., Friedmann, E., & McCoy, N. L. (1998). Pheromonal influences on sociosexual behavior in men. *Arch. Sex Behav., 27*, 1–13.

De Catanzaro, D., Muir, C., Sullivan, C., & Boissy, A. (1999). Pheromones and novel male-induced pregnancy disruptions in mice: exposure to conspecifics is necessary for urine alone to induce an effect. *Physiol. Behav., 66*, 153–157.

De Catanzaro, D., Wyngaarden, P., Griffiths, J., Ham, M., Hancox, J., & Brain, D. (1995). Interactions of contact, odor cues, and androgens in strange-male-induced early pregnancy disruptions in mice (*Mus musculus*). *J. Comp. Psychol., 109*, 115–122.

De la Maza, H. M., Wolff, J. O., & Lindsey, A. (1999). Exposure to strange adults does not cause pregnancy disruption or infanticide in the gray-tailed vole. *Behav. Ecol. Sociobiol., 45*, 107–113.

Delbarco-Trillo, J., & Ferkin, M. H. (2004). Male mammals respond to a risk of sperm competition conveyed by odours of conspecific males. *Nature, 431*, 446–449.

Dodge, J. C., Kristal, M. B., & Badura, L. L. (2002). Male-induced estrus synchronization in the female Siberian hamster (*Phodopus sungorus sungorus*). *Physiol. Behav., 77*, 227–231.

Dominic, C. J. (1965). The origin of the pheromones causing pregnancy block in mice. *J. Reprod. Fertil., 10*, 469–472.

Dominic, C. J. (1966). Observations on the reproductive pheromones of mice. I. Source. *J. Reprod. Fertil., 11*, 407–414.

Dorries, K. M., Adkins-Regan, E., & Halpern, B. P. (1997). Sensitivity and behavioral responses to the pheromone androstenone are not mediated by the vomeronasal organ in domestic pigs. *Brain Behav. Evol., 49*, 53–62.

Doty, R. L., & Dunbar, I. (1974). Attraction of beagles to conspecific urine, vaginal and anal sac secretion odors. *Physiol. Behav., 12*, 825–833.

Drickamer, L. C. (1977). Delay of sexual maturation in female house mice by exposure to grouped females or urine from grouped females. *J. Reprod. Fertil., 51*, 77–81.

Drickamer, L. C. (1984). Urinary chemosignals from mice (*Mus musculus*): acceleration and delay of puberty in related and unrelated young females. *J. Comp. Psychol., 98*, 414–420.

Drickamer, L. C. (1986). Peripheral anosmia affects puberty-influencing chemosignals in mice: donors and recipients. *Physiol. Behav., 37*, 741–746.

Drickamer, L. C. (1988). Long-term effects of accelerated or delayed sexual maturation on reproductive output in wild female house mice (*Mus musculus*). *J. Reprod. Fertil., 83*, 439–445.

Drickamer, L. C., & Hoover, J. E. (1979). Effects of urine from pregnant and lactating female house mice on sexual maturation of juvenile females. *Dev. Psychobiol., 12*, 545–551.

Drickamer, L. C., & McIntosh, T. K. (1980). Effects of adrenalectomy on the presence of a maturation-delaying pheromone in the urine of female mice. *Horm. Behav., 14*, 146–152.

Drickamer, L. C., & Mikesic, D. G. (1990). Urinary chemosignals, reproduction, and population size for house mice (*Mus domesticus*) living in field enclosures. *J. Chem. Ecol., 16*, 2955–2968.

Drickamer, L. C., McIntosh, T. K., & Rose, E. A. (1978). Effects of ovariectomy on the presence of a maturation-delaying pheromone in the urine of female mice. *Horm. Behav., 11*, 131–137.

Drickamer, L. C., & Murphy, R. X., Jr. (1978). Female mouse maturation: effects of excreted and bladder urine from juvenile and adult males. *Dev. Psychobiol., 11*, 63–72.

Edwards, D. A., Griffis, K. T., & Tardivel, C. (1990). Olfactory bulb removal: effects on sexual behavior and partner-preference in male rats. *Physiol. Behav., 48*, 447–450.

Edwards, D. A., Walter, B., & Liang, P. (1996). Hypothalamic and olfactory control of sexual behavior and partner preference in male rats. *Physiol. Behav., 60*, 1347–1354.

Ehman, K. D., & Scott, M. E. (2001). Urinary odour preferences of MHC congenic female mice, *Mus domesticus*: implications for kin recognition and detection of parasitized males. *Anim. Behav., 62*, 781–789.

Eidson, L. N., Maras, P. M., Epperson, E., & Petrulis, A. (2007). Female hamster preference for odors is not regulated by circulating gonadal hormones. *Physiol. Behav., 91*, 134–141.

Eisthen, H. L., Wysocki, C. J., & Beauchamp, G. K. (1987). Behavioral responses of male guinea pigs to conspecific chemical signals following neonatal vomeronasal organ removal. *Physiol. Behav., 41*, 445–449.

Fadem, B. H. (1987). Activation of estrus by pheromones in a marsupial: stimulus control and endocrine factors. *Biol. Reprod., 36*, 328–332.

Ferkin, M. H., & Gorman, M. R. (1992). Photoperiod and gonadal hormones influence odor preferences of the male meadow vole, *Microtus pennsylvanicus*. *Physiol. Behav., 51*, 1087–1091.

Ferkin, M. H., & Johnston, R. E. (1995). Effects of pregnancy lactation and postpartum estrus on odor signals and the attraction to odors in female meadow voles, *Microtus pennsylvanicus*. *Anim. Behav., 49*, 1211–1217.

Ferkin, M. H., & Johnston, R. E. (1993). Roles of gonadal hormones in control of five sexually attractive odors of meadow voles (*Microtus pennsylvanicus*). *Horm. Behav., 27*, 523–538.

Ferkin, M. H., & Zucker, I. (1991). Seasonal control of odour preferences of meadow voles (*Microtus pennsylvanicus*) by photoperiod and ovarian hormones. *J. Reprod. Fertil., 92*, 433–441.

Ferkin, M. H., Gorman, M. R., & Zucker, I. (1991). Ovarian hormones influence odor cues emitted by female meadow voles, *Microtus pennsylvanicus*. *Horm. Behav., 25*, 572–581.

Ferkin, M. H., Combs, A., delBarco-Trillo, J., Pierce, A. A., & Franklin, S. (2008). Meadow voles, *Microtus pennsylvanicus*, have the capacity to recall the "what", "where", and "when" of a single past event. *Anim. Cogn., 11*, 147–159.

Ferkin, M. H., Sorokin, E. S., Johnston, R. E., & Lee, C. J. (1997). Attractiveness of scents varies with protein content of the diet in meadow voles. *Animal. Behaviour, 53*, 133–141.

Floody, O. R. (2002). Time course of VMN lesion effects on lordosis and proceptive behavior in female hamsters. *Horm. Behav., 41*, 366–376.

Gattermann, R., Fritzsche, P., Neumann, K., Al-Hussein, I., Kayser, A., Abiad, M., et al. (2001). Notes on the current distribution and the ecology of wild golden hamsters (*Mesocricetus auratus*). *J. Zool., 254*, 359–365.

Gelez, H., & Fabre-Nys, C. (2004). The "male effect" in sheep and goats: a review of the respective roles of the two olfactory systems. *Horm. Behav., 46*, 257–271.

Gelez, H., Archer, E., Chesneau, D., Magallon, T., & Fabre-Nys, C. (2004). Inactivation of the olfactory amygdala prevents the endocrine response to male odour in anoestrus ewes. *Eur. J. Neurosci., 19*, 1581–1590.

Geyer, L. A., & Barfield, R. J. (1978). Influence of gonadal hormones and sexual behavior on ultrasonic vocalization in rats: I. Treatment of females. *J. Comp. Physiol. Psychol., 92*, 438–446.

Gonzalez-Mariscal, G., Melo, A. I., Zavala, A., & Beyer, C. (1990). Variations in chin-marking behavior of New Zealand female rabbits throughout the whole reproductive cycle. *Physiol. Behav., 48*, 361–365.

Gosling, L. M., Roberts, S. C., Peter, J. B., Slater, J. S. R. C. T. S., & Timothy, J. R. (2001). Scent-marking by male mammals: Cheat-proof signals to competitors and mates. In *Advances in the Study of Behavior, vol. 30* (pp. 169–217). Academic Press.

Graham, J. M., & Desjardins, C. (1980). Classical conditioning: induction of luteinizing hormone and testosterone secretion in anticipation of sexual activity. *Science, 210*, 1039–1041.

Gregory, E., Engel, K., & Pfaff, D. (1975). Male hamster preference for odors of female hamster vaginal discharges: studies of experiential and hormonal determinants. *J. Comp. Physiol. Psychol., 89*, 442–446.

Halpern, M., & Martinez-Marcos, A. (2003). Structure and function of the vomeronasal system: an update. *Prog. Neurobiol., 70*, 245–318.

Havlicek, J., & Roberts, S. C. (2009). MHC-correlated mate choice in humans: A review. *Psychoneuroendocrinology, 34*, 497–512.

Hayashi, S., & Kimura, T. (1974). Sex-attractant emitted by female mice. *Physiol. Behav., 13*, 563–567.

Heske, E. J., & Nelson, R. J. (1984). Pregnancy interruption in *Microtus ochrogaster*: laboratory artifact or field phenomenon? *Biol. Reprod., 31*, 97–103.

Heth, G., Todrank, J., & Johnston, R. E. (1998). Kin recognition in golden hamsters: evidence for phenotype matching. *Anim. Behav., 56*, 409–417.

Hurst, J. L. (1990). Urine marking in populations of wild house mice *Mus domesticus* Rutty. III. Communication between the sexes. *Anim. Behav., 40*, 233–243.

Hurst, J. L., & Beynon, R. J. (2004). Scent wars: the chemobiology of competitive signalling in mice. *Bioessays, 26*, 1288–1298.

Hurst, J. L., Beynon, R. J., Roberts, S. C., & Wyatt, T. D. (2008). *Chemical Signals in Vertebrates 11*. New York, NY: Springer.

Hurtazo, H. A., & Paredes, R. G. (2005). Olfactory preference and Fos expression in the accessory olfactory system of male rats with bilateral lesions of the medial preoptic area/anterior hypothalamus. *Neuroscience, 135*, 1035–1044.

Ingersoll, D. W., & Launay, J. (1986). Murine aggression induced by a boar chemosignal: a stimulus presentation dependency. *Physiol. Behav., 36*, 263–269.

Iwata, E., Wakabayashi, Y., Kakuma, Y., Kikusui, T., Takeuchi, Y., & Mori, Y. (2000). Testosterone-dependent primer pheromone production in the sebaceous gland of male goat. *Biol. Reprod., 62*, 806–810.

Jackson, L. M., & Harder, J. D. (1996). Vomeronasal organ removal blocks pheromonal induction of estrus in gray short-tailed opossums (*Monodelphis domestica*). *Biol. Reprod., 54*, 506–512.

Jacob, S., Hayreh, D. J., & McClintock, M. K. (2001). Context-dependent effects of steroid chemosignals on human physiology and mood. *Physiol. Behav., 74*, 15–27.

Jacob, S., McClintock, M. K., Zelano, B., & Ober, C. (2002). Paternally inherited HLA alleles are associated with women's choice of male odor. *Nat. Genet., 30*, 175–179.

Jemiolo, B., Alberts, J., Sochinski-Wiggins, S., Harvey, S., & Novotny, M. (1985). Behavioural and endocrine responses of female mice to synthetic analogues of volatile compounds in male urine. *Anim. Behav., 33*, 1114–1118.

Jemiolo, B., Andreolini, F., Xie, T. M., Wiesler, D., & Novotny, M. (1989). Puberty-affecting synthetic analogs of urinary chemosignals in the house mouse, *Mus domesticus*. *Physiol. Behav., 46*, 293–298.

Jemiolo, B., Harvey, S., & Novotny, M. (1986). Promotion of the Whitten effect in female mice by synthetic analogs of male urinary constituents. *Proc. Natl. Acad. Sci. USA, 83*, 4576–4579.

Jemiolo, B., & Novotny, M. (1994). Inhibition of sexual maturation in juvenile female and male mice by a chemosignal of female origin. *Physiol. Behav., 55*, 519–522.

Jemiolo, B., Xie, T. M., & Novotny, M. (1991). Socio–sexual olfactory preference in female mice: attractiveness of synthetic chemosignals. *Physiol. Behav., 50*, 1119–1122.

Johns, M. A., Feder, H. H., Komisaruk, B. R., & Mayer, A. D. (1978). Urine-induced reflex ovulation in anovulatory rats may be a vomeronasal effect. *Nature, 272*, 446–448.

Johnston, R. E. (1974). Sexual attraction function of golden hamster vaginal secretion. *Behav. Biol., 12*, 111–117.

Johnston, R. E. (1975). Sexual excitation function of hamster vaginal secretion. *Anim. Learn. Behav., 3*, 161–166.

Johnston, R. E. (1977). The causation of two scent-marking behaviour patterns in female hamsters (*Mesocricetus auratus*). *Anim. Behav., 25*, 317–327.

Johnston, R. E. (1979). Olfactory preferences, scent marking, and "proceptivity" in female hamsters. *Horm. Behav., 13*, 21–39.

Johnston, R. E. (1986). Effects of female odors on the sexual behavior of male hamsters. *Behav. Neural. Biol., 46*, 168–188.

Johnston, R. E. (1992). Vomeronasal and/or olfactory mediation of ultrasonic calling and scent marking by female golden hamsters. *Physiol. Behav., 51*, 437–448.

Johnston, R.E. (2008). Individual odors and social communication: individual recognition, kin recognition, and scent over-marking. In *Advances in the Study of Behavior*, vol. Volume 38 (ed. Brockmann, H. J., Roper, T. J., Naguib, M., Wagne-Edwards, C. E., Barnard, C., & Mitani, J. C.), pp. 439–505: Academic Press.

Johnston, R. E., & Brenner, D. (1982). Species-specificity of scent marking in hamsters. *Behavioral and Neural Biology, 35*, 46–55.

Johnston, R. E., & Bronson, F. (1982). Endocrine control of female mouse odors that elicit luteinizing hormone surges and attraction in males. *Biol. Reprod., 27*, 1174–1180.

Johnston, R. E., & Zahorik, D. M. (1975). Taste aversions to sexual attractants. *Science, 189*, 893–894.

Johnston, R. E., Zahorik, D. M., Immler, K., & Zakon, H. (1978). Alterations of male sexual behavior by learned aversions to hamster vaginal secretion. *J. Comp. Physiol. Psychol., 92*, 85–93.

Kaneko, N., Debski, E. A., Wilson, M. C., & Whitten, W. K. (1980). Puberty acceleration in mice. II. Evidence that the vomeronasal organ is a receptor for the primer pheromone in male mouse urine. *Biol. Reprod., 22*, 873–878.

Karlson, P., & Luscher, M. (1959). 'Pheromones': a new term for a class of biologically active substances. *Nature, 183*, 55–56.

Kavaliers, M., & Colwell, D. D. (1995). Discrimination by female mice between the odours of parasitized and non-parasitized males. *Proc. Biol. Sci., 261*, 31–35.

Kavaliers, M., Choleris, E., & Colwell, D. D. (2001). Brief exposure to female odors "emboldens" male mice by reducing predator-induced behavioral and hormonal responses. *Horm. Behav., 40*, 497–509.

Kay, E., & Nyby, J. (1992). LiCl aversive conditioning has transitory effects on pheromonal responsiveness in male house mice (*Mus domesticus*). *Physiol. Behav., 52*, 105–113.

Keller, M., Douhard, Q., Baum, M. J., & Bakker, J. (2006a). Sexual experience does not compensate for the disruptive effects of zinc sulfate-lesioning of the main olfactory epithelium on sexual behavior in male mice. *Chem. Senses, 31*, 753–762.

Keller, M., Pierman, S., Douhard, Q., Baum, M. J., & Bakker, J. (2006b). The vomeronasal organ is required for the expression of lordosis behaviour, but not sex discrimination in female mice. *Eu.r J. Neurosci., 23*, 521–530.

Kelliher, K., & Baum, M. (2002). Effect of sex steroids and coital experience on ferrets' preference for the smell, sight and sound of conspecifics. *Physiol. Behav., 76*, 1–7.

Kelliher, K. R., & Baum, M. J. (2001). Nares occlusion eliminates heterosexual partner selection without disrupting coitus in ferrets of both sexes. *J. Neurosci., 21*, 5832–5840.

Kempenaers, B., Peters, A., & Foerster, K. (2008). Sources of individual variation in plasma testosterone levels. *Philo. Trans. Royal Soci. B: Biol. Sci., 363*, 1711–1723.

Kendrick, K. M., Da Costa, A. P., Broad, K. D., Ohkura, S., Guevara, R., Levy, F., et al. (1997). Neural control of maternal behaviour and olfactory recognition of offspring. *Brain Res. Bull., 44*, 383–395.

Keverne, E. B., & De la Riva, C. (1982). Pheromones in mice: reciprocal interaction between the nose and brain. *Nature, 296*, 148–150.

Knight, T. W., Lindsay, D. R., & Oldham, C. M. (1975). Proceedings: the influence of rams on the fertility of the ewe. *J. Reprod Fertil., 43*, 377–378.

Kondo, Y., & Sachs, B. D. (2002). Disparate effects of small medial amygdala lesions on noncontact erection, copulation, and partner preference. *Physiol. Behav., 76*, 443–447.

Kondo, Y., & Sakuma, Y. (2005). The medial amygdala controls the coital access of female rats: a possible involvement of emotional responsiveness. *Jpn. J. Physiol., 55*, 345–353.

Kondo, Y., Sachs, B. D., & Sakuma, Y. (1998). Importance of the medial amygdala in rat penile erection evoked by remote stimuli from estrous females. *Behav. Brain Res., 91*, 215–222.

Kondo, Y., Tomihara, K., & Sakuma, Y. (1999). Sensory requirements for noncontact penile erection in the rat. *Behav. Neurosci., 113*, 1062–1070.

Koyama, S. (2004). Primer effects by conspecific odors in house mice: a new perspective in the study of primer effects on reproductive activities. *Horm. Behav., 46*, 303–310.

Koyama, S., & Kamimura, S. (2000). Influence of social dominance and female odor on the sperm activity of male mice. *Physiol. Behav., 71*, 415–422.

Kwan, M., & Johnston, R. E. (1980). The role of vaginal secretion in hamster sexual behavior: males' responses to normal and vaginectomized females and their odors. *J. Comp. Physiol. Psychol., 94*, 905–913.

Lanuza, E., Novejarque, A., Martinez-Ricos, J., Martinez-Hernandez, J., Agustin-Pavon, C., & Martinez-Garcia, F. (2008). Sexual pheromones and the evolution of the reward system of the brain: the chemosensory function of the amygdala. *Brain Res. Bull., 75*, 460–466.

Lehman, M. N., Powers, J. B., & Winans, S. S. (1983). Stria terminalis lesions alter the temporal pattern of copulatory behavior in the male golden hamster. *Behav. Brain Res., 8*, 109–128.

Lehman, M. N., Winans, S. S., & Powers, J. B. (1980). Medial nucleus of the amygdala mediates chemosensory control of male hamster sexual behavior. *Science, 210*, 557–560.

Leinders-Zufall, T., Lane, A. P., Puche, A. C., Ma, W., Novotny, M. V., Shipley, M. T., et al. (2000). Ultrasensitive pheromone detection by mammalian vomeronasal neurons. *Nature, 405*, 792–796.

Lepri, J. J., & Vandenbergh, J. G. (1986). Puberty in pine voles, *Microtus pinetorum*, and the influence of chemosignals on female reproduction. *Biol. Reprod., 34*, 370–377.

Lepri, J. J., & Wysocki, C. J. (1987). Removal of the vomeronasal organ disrupts the activation of reproduction in female voles. *Physiol. Behav., 40*, 349–355.

Leypold, B. G., Yu, C. R., Leinders-Zufall, T., Kim, M. M., Zufall, F., & Axel, R. (2002). Altered sexual and social behaviors in trp2 mutant mice. *Proc. Natl. Acad. Sci. USA, 99*, 6376–6381.

Li, C. S., Kaba, H., Saito, H., & Seto, K. (1990). Neural mechanisms underlying the action of primer pheromones in mice. *Neuroscience, 36*, 773–778.

Lin, D. Y., Zhang, S. Z., Block, E., & Katz, L. C. (2005). Encoding social signals in the mouse main olfactory bulb. *Nature, 434*, 470–477.

Lisk, R. D., & Nachtigall, M. J. (1988). Estrogen regulation of agonistic and proceptive responses in the golden hamster. *Horm. Behav., 22*, 35–48.

Liu, Y. C., Salamone, J. D., & Sachs, B. D. (1997). Lesions in medial preoptic area and bed nucleus of stria terminalis: differential effects on copulatory behavior and noncontact erection in male rats. *J. Neurosci., 17*, 5245–5253.

Lloyd-Thomas, A., & Keverne, E. B. (1982). Role of the brain and accessory olfactory system in the block to pregnancy in mice. *Neuroscience, 7*, 907–913.

Lomas, D. E., & Keverne, E. B. (1982). Role of the vomeronasal organ and prolactin in the acceleration of puberty in female mice. *J. Reprod. Fertil., 66*, 101–107.

Lombardi, J. R., Vandenbergh, J. G., & Whitsett, J. M. (1976). Androgen control of the sexual maturation pheromone in house mouse urine. *Biol. Reprod., 15*, 179–186.

Lundstrom, J. N., & Olsson, M. J. (2005). Subthreshold amounts of social odorant affect mood, but not behavior, in heterosexual women when tested by a male, but not a female, experimenter. *Biol. Psychol., 70*, 197–204.

Lydell, K., & Doty, R. L. (1972). Male rat of odor preferences for female urine as a function of sexual experience, urine age, and urine source. *Horm. Behav., 3*, 205–212.

Ma, W., Miao, Z., & Novotny, M. V. (1998). Role of the adrenal gland and adrenal-mediated chemosignals in suppression of estrus in the house mouse: the lee-boot effect revisited. *Biol. Reprod., 59*, 1317–1320.

Macrides, F., Bartke, A., & Dalterio, S. (1975). Strange females increase plasma testosterone levels in male mice. *Science, 189*, 1104–1106.

Macrides, F., Bartke, A., Fernandez, F., & D'Angelo, W. (1974). Effects of exposure to vaginal odor and receptive females on plasma testosterone in the male hamster. *Neuroendocrinology, 15*, 355–364.

Macrides, F., Clancy, A. N., Singer, A. G., & Agosta, W. C. (1984a). Male hamster investigatory and copulatory responses to vaginal discharge: an attempt to impart sexual significance to an arbitrary chemosensory stimulus. *Physiol. Behav., 33*, 627–632.

Macrides, F., Singer, A. G., Clancy, A. N., Goldman, B. D., & Agosta, W. C. (1984b). Male hamster investigatory and copulatory responses to vaginal discharge: relationship to the endocrine status of females. *Physiol. Behav., 33*, 633–637.

Mahady, S., & Wolff, J. (2002). A field test of the Bruce effect in the monogamous prairie vole (*Microtus ochrogaster*). *Behav. Ecol. Sociobiol., 52*, 31–37.

Malsbury, C. W., Kow, L. M., & Pfaff, D. W. (1977). Effects of medial hypothalamic lesions on the lordosis response and other behaviors in female golden hamsters. *Physiol. Behav., 19*, 223–237.

Mandiyan, V. S., Coats, J. K., & Shah, N. M. (2005). Deficits in sexual and aggressive behaviors in Cnga2 mutant mice. *Nat Neurosci, 8*, 1660–1662.

Maras, P. M., & Petrulis, A. (2006). Chemosensory and steroid-responsive regions of the medial amygdala regulate distinct aspects of opposite-sex odor preference in male Syrian hamsters. *Eur. J. Neurosci., 24*, 3541–3552.

Maras, P. M., & Petrulis, A. (2008a). Olfactory experience and the development of odor preference and vaginal marking in female Syrian hamsters. *Physiol. Behav., 94*, 545–551.

Maras, P. M., & Petrulis, A. (2008b). The posteromedial cortical amygdala regulates copulatory behavior, but not sexual odor preference, in the male Syrian hamster (*Mesocricetus auratus*). *Neuroscience, 156*, 425–435.

Marchlewska-Koj, A., & Jemiolo. (1978). Evidence for the involvement of dopaminergic neurons in the pregnancy block effect. *Neuroendocrinology, 26*, 186–192.

Marsden, H. M., & Bronson, F. H. (1964). Estrous synchrony in mice: alteration by exposure to male urine. *Science, 144*, 1469.

Martinez-Marcos, A. (2009). On the organization of olfactory and vomeronasal cortices. *Prog. Neurobiol., 87*, 21–30.

Maruniak, J. A., & Bronson, F. H. (1976). Gonadotropic responses of male mice to female urine. *Endocrinology, 99*, 963–969.

Maruniak, J. A., Coquelin, A., & Bronson, F. H. (1978). The release of LH in male mice in response to female urinary odors: characteristics of the response in young males. *Biol. Reprod., 18*, 251–255.

Mason, R. T., LeMaster, M. P., & Müller-Schwarze, D. (2005). *Chemical Signals in Vertebrates 10*. Boston, MA: Springer.

Massey, A., & Vandenbergh, J. G. (1980). Puberty delay by a urinary cue from female house mice in feral populations. *Science, 209*, 821–822.

Massey, A., & Vandenbergh, J. G. (1981). Puberty acceleration by a urinary cue from male mice in feral populations. *Biol. Reprod., 24*, 523–527.

McClintock, M. K. (1971). Menstrual synchrony and suppression. *Nature, 229*, 244–245.

McClintock, M. K. (1984). Estrous synchrony: modulation of ovarian cycle length by female pheromones. *Physiol. Behav., 32*, 701–705.

McClintock, M. K. (2002). Pheromones, odors and vasanas: the neuroendocrinology of social chemosignals in humans and animals. In D. W. Pfaff, A. Arnold, A. Cardoso, D. Blake, S. Newcomer, & R. Quimby (Eds.), *Hormones, Brain and Behavior* (pp. 797–870). New York, NY: Elsevier.

McClintock, M. K., & Adler, N. T. (1978). Induction of persistent estrus by airborne chemical communication among female rats. *Horm. Behav., 11*, 414–418.

McCoy, N. L., & Pitino, L. (2002). Pheromonal influences on sociosexual behavior in young women. *Physiol. Behav., 75*, 367–375.

McIntosh, T. K., & Drickamer, L. C. (1977). Excreted urine, bladder urine, and the delay of sexual maturation in female house mice. *Anim. Behav., 25*, 999–1004.

Meek, L. R., Lee, T. M., Rogers, E. A., & Hernandez, R. G. (1994). Effect of vomeronasal organ removal on behavioral estrus and mating latency in female meadow voles (*Microtus pennsylvanicus*). *Biol. Reprod., 51*, 400–404.

Meredith, M. (1986). Vomeronasal organ removal before sexual experience impairs male hamster mating behavior. *Physiol. Behav., 36*, 737–743.

Meredith, M. (2001). Human vomeronasal organ function: a critical review of best and worst cases. *Chem. Senses, 26*, 433–445.

Milligan, S. R. (1976). Pregnancy blocking in the vole, *Microtus agrestis*. I. Effect of the social environment. *J. Reprod. Fertil., 46*, 91–95.

Moncho-Bogani, J., Lanuza, E., Hernandez, A., Novejarque, A., & Martinez-Garcia, F. (2002). Attractive properties of sexual pheromones in mice: innate or learned? *Physiol. Behav., 77*, 167–176.

Moncho-Bogani, J., Lanuza, E., Lorente, M. J., & Martinez-Garcia, F. (2004). Attraction to male pheromones and sexual behaviour show different regulatory mechanisms in female mice. *Physiol. Behav., 81*, 427–434.

Mossman, C. A., & Drickamer, L. C. (1996). Odor preferences of female house mice (*Mus domesticus*) in seminatural enclosures. *J. Comp. Psychol., 110*, 131–138.

Mucignat-Caretta, C., Caretta, A., & Cavaggioni, A. (1995). Acceleration of puberty onset in female mice by male urinary proteins. *J. Physiol., 486*(Pt 2), 517–522.

Murata, K., Wakabayashi, Y., Kitago, M., Ohara, H., Watanabe, H., Tamogami, S., et al. (2009). Modulation of gonadotrophin-releasing hormone pulse generator activity by the pheromone in small ruminants. *J. Neuroendocrinol., 21*, 346–350.

Murphy, M. R. (1973). Effects of female hamster vaginal discharge on the behavior of male hamsters. *Behav. Biol., 9*, 367–375.

Murphy, M. R., & Schneider, G. E. (1970). Olfactory bulb removal eliminates mating behavior in the male golden hamster. *Science, 167*, 302–304.

Nishimura, K., Utsumi, K., Yuhara, M., Fujitani, Y., & Iritani, A. (1989). Identification of puberty-accelerating pheromones in male mouse urine. *J. Exp. Zool., 251*, 300–305.

Novotny, M., Jemiolo, B., Harvey, S., Wiesler, D., & Marchlewska-Koj, A. (1986). Adrenal-mediated endogenous metabolites inhibit puberty in female mice. *Science, 231*, 722–725.

Novotny, M. V. (2003). Pheromones, binding proteins and receptor responses in rodents. *Biochem. Soc. Trans., 31*, 117–122.

Novotny, M. V., Jemiolo, B., Wiesler, D., Ma, W., Harvey, S., Xu, F., et al. (1999a). A unique urinary constituent, 6-hydroxy-6-methyl-3-heptanone, is a pheromone that accelerates puberty in female mice. *Chem. Biol., 6*, 377–383.

Novotny, M. V., Ma, W., Wiesler, D., & Zidek, L. (1999b). Positive identification of the puberty-accelerating pheromone of the house mouse: the volatile ligands associating with the major urinary protein. *Proc. Biol. Sci., 266*, 2017–2022.

Nyby, J., Dizinno, G., & Whitney, G. (1977a). Sexual dimorphism in ultrasonic vocalizations of mice (*Mus musculus*): gonadal hormone regulation. *J. Comp. Physiol. Psychol., 91*, 1424–1431.

Nyby, J., Matochik, J. A., & Barfield, R. J. (1992). Intracranial androgenic and estrogenic stimulation of male-typical behaviors in house mice (*Mus domesticus*). *Horm. Behav., 26*, 24–45.

Nyby, J., Wysocki, C. J., Whitney, G., & Dizinno, G. (1977b). Pheromonal regulation of male mouse ultrasonic courtship (*Mus musculus*). *Anim. Behav., 25*, 333–341.

Nyby, J. G. (2008). Reflexive testosterone release: a model system for studying the nongenomic effects of testosterone upon male behavior. *Front. Neuroendocrinol., 29*, 199–210.

O'Connell, R. J., & Meredith, M. (1984). Effects of volatile and nonvolatile chemical signals on male sex behaviors mediated by the main and accessory olfactory systems. *Behav. Neurosci., 98*, 1083–1093.

O'Connell, R. J., Singer, A. G., Macrides, F., Pfaffmann, C., & Agosta, W. C. (1978). Responses of the male golden hamster to mixtures of odorants identified from vaginal discharge. *Behav. Biol., 24*, 244–255.

Pankevich, D. E., Baum, M. J., & Cherry, J. A. (2004). Olfactory sex discrimination persists, whereas the preference for urinary odorants from estrous females disappears in male mice after vomeronasal organ removal. *J. Neurosci., 24*, 9451–9457.

Pankevich, D. E., Cherry, J. A., & Baum, M. J. (2006). Accessory olfactory neural Fos responses to a conditioned environment are blocked in male mice by vomeronasal organ removal. *Physiol. Behav., 87*, 781–788.

Peele, P., Salazar, I., Mimmack, M., Keverne, E. B., & Brennan, P. A. (2003). Low molecular weight constituents of male mouse urine mediate the pregnancy block effect and convey information about the identity of the mating male. *Eur. J. Neurosci., 18*, 622–628.

Penn, D. J., Damjanovich, K., & Potts, W. K. (2002). MHC heterozygosity confers a selective advantage against multiple-strain infections. *Proc. Natl. Acad. Sci. USA, 99*, 11260–11264.

Petrulis, A., DeSouza, I., Schiller, M., & Johnston, R. E. (1998). Role of frontal cortex in social odor discrimination and scent-marking in female golden hamsters (*Mesocricetus auratus*). *Behav. Neurosci., 112*, 199–212.

Petrulis, A., & Eichenbaum, H. (2003). The perirhinal–entorhinal cortex, but not the hippocampus, is critical for expression of individual recognition in the context of the Coolidge effect. *Neuroscience, 122*, 599–607.

Petrulis, A., & Johnston, R. E. (1995). A reevaluation of dimethyl disulfide as a sex attractant in golden hamsters. *Physiol. Behav., 57*, 779–784.

Petrulis, A., & Johnston, R. E. (1997). Causes of scent marking in female golden hamsters (*Mesocricetus auratus*): specific signals or classes of information? *J. Comp. Psychol., 111*, 25–36.

Petrulis, A., & Johnston, R. E. (1999). Lesions centered on the medial amygdala impair scent-marking and sex-odor recognition but spare discrimination of individual odors in female golden hamsters. *Behav. Neurosci., 113*, 345–357.

Petrulis, A., Peng, M., & Johnston, R. E. (1999). Effects of vomeronasal organ removal on individual odor discrimination, sex-odor preference, and scent marking by female hamsters. *Physiol. Behav., 66*, 73–83.

Petrulis, A., Peng, M., & Johnston, R. E. (2000). The role of the hippocampal system in social odor discrimination and scent-marking in female golden hamsters (*Mesocricetus auratus*). *Behav. Neurosci., 114*, 184–195.

Pfeiffer, C. A., & Johnston, R. E. (1994). Hormonal and behavioral responses of male hamsters to females and female odors: roles of olfaction, the vomeronasal system, and sexual experience. *Physiol. Behav., 55*, 129–138.

Powers, J. B., Bergondy, M. L., & Matochik, J. A. (1985). Male hamster sociosexual behaviors: effects of testosterone and its metabolites. *Physiol. Behav., 35*, 607–616.

Powers, J. B., Fields, R. B., & Winans, S. S. (1979). Olfactory and vomeronasal system participation in male hamsters attraction to female vaginal secretions. *Physiol. Behav., 22*, 77–84.

Powers, J. B., Newman, S. W., & Bergondy, M. L. (1987). MPOA and BNST lesions in male Syrian hamsters: differential effects on copulatory and chemoinvestigatory behaviors. *Behav. Brain Res., 23*, 181–195.

Powers, J. B., & Winans, S. S. (1973). Sexual behavior in peripherally anosmic male hamsters. *Physiol. Behav., 10*, 361–368.

Powers, J. B., & Winans, S. S. (1975). Vomeronasal organ: critical role in mediating sexual behavior of the male hamster. *Science, 187*, 961–963.

Preti, G., Wysocki, C. J., Barnhart, K. T., Sondheimer, S. J., & Leyden, J. J. (2003). Male axillary extracts contain pheromones that affect pulsatile secretion of luteinizing hormone and mood in women recipients. *Biol. Reprod., 68*, 2107–2113.

Ramm, S. A., Cheetham, S. A., & Hurst, J. L. (2008). Encoding choosiness: female attraction requires prior physical contact with individual male scents in mice. *Proc. Biol. Sci., 275*, 1727–1735.

Rasmussen, L. E., Lee, T. D., Roelofs, W. L., Zhang, A., & Daves, G. D., Jr. (1996). Insect pheromone in elephants. *Nature, 379*, 684.

Reasner, D. S., & Johnston, R. E. (1988). Acceleration of reproductive development in female Djungarian hamsters by adult males. *Physiol. Behav., 43*, 57–64.

Reynolds, J., & Keverne, E. B. (1979). The accessory olfactory system and its role in the pheromonally mediated suppression of oestrus in grouped mice. *J. Reprod. Fertil., 57*, 31–35.

Rich, T. J., & Hurst, J. L. (1999). The competing countermarks hypothesis: reliable assessment of competitive ability by potential mates. *Anim. Behav., 58*, 1027–1037.

Rissman, E. F., & Johnston, R. E. (1985). Female reproductive development is not activated by male California voles exposed to family cues. *Biol. Reprod., 32*, 352–360.

Robarts, D. W., & Baum, M. J. (2007). Ventromedial hypothalamic nucleus lesions disrupt olfactory mate recognition and receptivity in female ferrets. *Horm. Behav., 51*, 104–113.

Rosser, A. E., Remfry, C. J., & Keverne, E. B. (1989). Restricted exposure of mice to primer pheromones coincident with prolactin surges blocks pregnancy by changing hypothalamic dopamine release. *J. Reprod. Fertil., 87*, 553–559.

Rowe, F. A., & Edwards, D. A. (1972). Olfactory bulb removal: influences on the mating behavior of male mice. *Physiol. Behav., 8*, 37–41.

Saito, T. R., & Moltz, H. (1986). Copulatory behavior of sexually naïve and sexually experienced male rats following removal of the vomeronasal organ. *Physiol. Behav., 37*, 507–510.

Sam, M., Vora, S., Malnic, B., Ma, W., Novotny, M. V., & Buck, L. B. (2001). Odorants may arouse instinctive behaviours. *Nature, 412*, 142.

Sapolsky, R. M., & Eichenbaum, H. (1980). Thalamocortical mechanisms in odor-guided behavior. II. Effects of lesions of the mediodorsal thalamic nucleus and frontal cortex on odor preferences and sexual behavior in the hamster. *Brain Behav. Evol., 17*, 276–290.

Saxton, T. K., Lyndon, A., Little, A. C., & Roberts, S. C. (2008). Evidence that androstadienone, a putative human chemosignal, modulates women's attributions of men's attractiveness. *Horm. Behav., 54*, 597–601.

Schadler, M. H. (1981). Postimplantation abortion in pine voles (*Microtus pinetorum*) induced by strange males and pheromones of strange males. *Biol. Reprod., 25*, 295–297.

Schank, J. C. (2000). Can pseudo entrainment explain the synchrony of estrous cycles among golden hamsters (*Mesocricetus auratus*)? *Horm. Behav., 38*, 94–101.

Schank, J. C. (2001a). Do Norway rats (*Rattus norvegicus*) synchronize their estrous cycles? *Physiol. Behav., 72*, 129–139.

Schank, J. C. (2001b). Menstrual-cycle synchrony: problems and new directions for research. *J. Comp Psychol, 115*, 3–15.

Schank, J. C., & McClintock, M. K. (1992). A coupled-oscillator model of ovarian-cycle synchrony among female rats. *J. Theor. Biol., 157*, 317–362.

Serguera, C., Triaca, V., Kelly-Barrett, J., Banchaabouchi, M. A., & Minichiello, L. (2008). Increased dopamine after mating impairs olfaction and prevents odor interference with pregnancy. *Nat. Neurosci., 11*, 949–956.

Singer, A. G., Agosta, W. C., O'Connell, R. J., Pfaffmann, C., Bowen, D. V., & Field, F. H. (1976). Dimethyl disulfide: an attractant pheromone in hamster vaginal secretion. *Science, 191*, 948–950.

Singer, A. G., Beauchamp, G. K., & Yamazaki, K. (1997). Volatile signals of the major histocompatibility complex in male mouse urine. *Proc. Natl. Acad. Sci. USA, 94*, 2210–2214.

Singer, A. G., Clancy, A. N., Macrides, F., & Agosta, W. C. (1984). Chemical studies of hamster vaginal discharge: male behavioral responses to a high molecular weight fraction require physical contact. *Physiol. Behav., 33*, 645–651.

Singer, A. G., Clancy, A. N., Macrides, F., Agosta, W. C., & Bronson, F. H. (1988). Chemical properties of a female mouse pheromone that stimulates gonadotropin secretion in males. *Biol. Reprod., 38*, 193–199.

Sipos, M. L., Kerchner, M., & Nyby, J. G. (1992). An ephemeral sex pheromone in the urine of female house mice (*Mus domesticus*). *Behav. Neural. Biol., 58*, 138–143.

Sipos, M. L., Wysocki, C. J., Nyby, J. G., Wysocki, L., & Nemura, T. A. (1995). An ephemeral pheromone of female house mice: perception via the main and accessory olfactory systems. *Physiol. Behav., 58*, 529–534.

Sisk, C. L., & Foster, D. L. (2004). The neural basis of puberty and adolescence. *Nat. Neurosci., 7*, 1040–1047.

Smale, L. (1988). Influence of male gonadal hormones and familiarity on pregnancy interruption in prairie voles. *Biol. Reprod., 39*, 28–31.

Solomon, N. G., Vandenbergh, J. G., Wekesa, K. S., & Barghusen, L. (1996). Chemical cues are necessary but insufficient for reproductive activation of female pine vole (*Microtus pinetorum*). *Biol. Reprod., 54*, 1038–1045.

Spehr, M., Kelliher, K. R., Li, X. H., Boehm, T., Leinders-Zufall, T., & Zufall, F. (2006). Essential role of the main olfactory system in social recognition of major histocompatibility complex peptide ligands. *J. Neurosci., 26*, 1961–1970.

Steel, E., & Hutchison, J. B. (1986). Olfactory recognition in the male hamster: effect of non-aromatizable androgens, 17 beta-hydroxy-17 alpha-methyl-estra-4,9,11-triene-3-one (R 1881) and 5 alpha-dihydrotestosterone, in combination with oestrogen. *J. Endocrinol., 110*, 525–531.

Stern, J. J. (1970). Responses of male rats to sex odors. *Physiol. Behav., 5*, 519–524.

Stern, K., & McClintock, M. K. (1998). Regulation of ovulation by human pheromones. *Nature, 392*, 177–179.

Stowers, L., Holy, T. E., Meister, M., Dulac, C., & Koentges, G. (2002). Loss of sex discrimination and male–male aggression in mice deficient for TRP2. *Science, 295*, 1493–1500.

Takahashi, L. K., Lisk, R. D., & Burnett, A. L., 2nd (1985). Dual estradiol action in diencephalon and the regulation of sociosexual behavior in female golden hamsters. *Brain Res., 359*, 194–207.

Thiessen, D., & Rice, M. (1976). Mammalian scent gland marking and social behavior. *Psychol. Bull., 83*, 505–539.

Thom, M. D., Stockley, P., Jury, F., Ollier, W. E., Beynon, R. J., & Hurst, J. L. (2008). The direct assessment of genetic heterozygosity through scent in the mouse. *Curr. Biol., 18*, 619–623.

Trinh, K., & Storm, D. R. (2003). Vomeronasal organ detects odorants in absence of signaling through main olfactory epithelium. *Nat. Neurosci., 6*, 519–525.

Utsumi, M., Ohno, K., Kawasaki, Y., Tamura, M., Kubo, T., & Tohyama, M. (1999). Expression of major urinary protein genes in the nasal glands associated with general olfaction. *J. Neurobiol., 39*, 227–236.

Vandenbergh, J. G. (1969). Male odor accelerates female sexual maturation in mice. *Endocrinology, 84*, 658–660.

Vandenbergh, J. G., Whitsett, J. M., & Lombardi, J. R. (1975). Partial isolation of a pheromone accelerating puberty in female mice. *J. Reprod. Fertil., 43*, 515–523.

Wedekind, C., Seebeck, T., Bettens, F., & Paepke, A. J. (1995). MHC-dependent mate preferences in humans. *Proc. Biol. Sci., 260*, 245–249.

Weller, L., & Weller, A. (1993). Human menstrual synchrony: a critical assessment. *Neurosci. Biobehav. Rev., 17*, 427–439.

Westberry, J. M., & Meredith, M. (2003). Pre-exposure to female chemosignals or intracerebral GnRH restores mating behavior in naive male hamsters with vomeronasal organ lesions. *Chem. Senses, 28*, 191–196.

Winman, A. (2004). Do perfume additives termed human pheromones warrant being termed pheromones? *Physiol. Behav., 82*, 697–701.

Wood, R. I. (1996). Estradiol, but not dihydrotestosterone, in the medial amygdala facilitates male hamster sex behavior. *Physiol. Behav., 59*, 833–841.

Wood, R. I. (1997). Thinking about networks in the control of male hamster sexual behavior. *Horm. Behav., 32*, 40–45.

Wood, R. I., & Newman, S. W. (1995). The medial amygdaloid nucleus and medial preoptic area mediate steroidal control of sexual behavior in the male Syrian hamster. *Horm. Behav., 29*, 338–353.

Woodley, S. K., & Baum, M. J. (2003). Effects of sex hormones and gender on attraction thresholds for volatile anal scent gland odors in ferrets. *Horm. Behav., 44*, 110–118.

Woodley, S. K., Cloe, A. L., Waters, P., & Baum, M. J. (2004). Effects of vomeronasal organ removal on olfactory sex discrimination and odor preferences of female ferrets. *Chem. Senses, 29*, 659–669.

Wyart, C., Webster, W. W., Chen, J. H., Wilson, S. R., McClary, A., Khan, R. M., et al. (2007). Smelling a single component of male sweat alters levels of cortisol in women. *J. Neurosci., 27*, 1261–1265.

Wyatt, T. D. (2003). *Pheromones and Animal Behaviour: Communication by Smell and Taste*. Cambridge, UK; New York, NY: Cambridge University Press.

Wyatt, T. D. (2009). Fifty years of pheromones. *Nature, 457*, 262–263.

Wysocki, C. J., & Preti, G. (2004). Facts, fallacies, fears, and frustrations with human pheromones. *Anat. Rec. A. Discov. Mol. Cell Evol. Biol., 281*, 1201–1211.

Wysocki, C. J., Katz, Y., & Bernhard, R. (1983). Male vomeronasal organ mediates female-induced testosterone surges in mice. *Biol. Reprod., 28*, 917–922.

Wysocki, C. J., Nyby, J., Whitney, G., Beauchamp, G. K., & Katz, Y. (1982). The vomeronasal organ: primary role in mouse chemosensory gender recognition. *Physiol. Behav., 29*, 315–327.

Xiao, K., Kondo, Y., & Sakuma, Y. (2004). Sex-specific effects of gonadal steroids on conspecific odor preference in the rat. *Horm. Behav., 46*, 356–361.

Xiao, K., Kondo, Y., & Sakuma, Y. (2005). Differential regulation of female rat olfactory preference and copulatory pacing by the lateral septum and medial preoptic area. *Neuroendocrinology, 81*, 56–62.

Yamazaki, K., Beauchamp, G. K., Wysocki, C. J., Bard, J., Thomas, L., & Boyse, E. A. (1983). Recognition of H-2 types in relation to the blocking of pregnancy in mice. *Science, 221*, 186–188.

Yoon, H., Enquist, L. W., & Dulac, C. (2005). Olfactory inputs to hypothalamic neurons controlling reproduction and fertility. *Cell, 123*, 669–682.

Zacharias, R., De Catanzaro, D., & Muir, C. (2000). Novel male mice disrupt pregnancy despite removal of vesicular-coagulating and preputial glands. *Physiol. Behav., 68*, 285–290.

Zarrow, M. X., Estes, S. A., Denenberg, V. H., & Clark, J. H. (1970). Pheromonal facilitation of ovulation in the immature mouse. *J. Reprod. Fertil., 23*, 357–360.

Zhang, J. X., Liu, Y. J., Zhang, J. H., & Sun, L. (2008a). Dual role of preputial gland secretion and its major components in sex recognition of mice. *Physiol. Behav., 95*, 388–394.

Zhang, J. X., Rao, X. P., Sun, L., Zhao, C. H., & Qin, X. W. (2007). Putative chemical signals about sex, individuality, and genetic background in the preputial gland and urine of the house mouse (*Mus musculus*). *Chem. Senses, 32*, 293–303.

Zhang, J. X., Soini, H. A., Bruce, K. E., Wiesler, D., Woodley, S. K., Baum, M. J., et al. (2005). Putative chemosignals of the ferret (*Mustela furo*) associated with individual and gender recognition. *Chem. Senses, 30*, 727–737.

Zhang, J. X., Sun, L., Zhang, J. H., & Feng, Z. Y. (2008b). Sex- and gonad-affecting scent compounds and 3 male pheromones in the rat. *Chem. Senses, 33*, 611–621.

Ziegler, T. E., Schultz-Darken, N. J., Scott, J. J., Snowdon, C. T., & Ferris, C. F. (2005). Neuroendocrine response to female ovulatory odors depends upon social condition in male common marmosets, *Callithrix jacchus*. *Horm. Behav., 47*, 56–64.

Zufall, F., & Leinders-Zufall, T. (2007). Mammalian pheromone sensing. *Curr. Opin. Neurobiol., 17*, 483–489.

Zuri, I., & Halpern, M. (2005). Modification of odor investigation and discrimination in female opossums (*Monodelphis domestica*) following the ablation of the accessory olfactory bulbs. *Behav. Neurosci., 119*, 612–621.

Chapter 10

Reproductive Endocrinology of Prototherians and Metatherians

Bronwyn M. McAllan
The University of Sydney, Sydney, NSW, Australia

SUMMARY
Reproduction in prototherians (monotremes) and metatherians (marsupials) is typified by the production of altricial young, and associated with some differences in the endocrinology of reproduction when compared to eutherians ('true' mammals, often called placental mammals). Thc corpus luteum secretes progesterone (P_4) in monotremes and marsupials and, as gestation in the marsupial does not extend beyond the length of an estrous cycle, the preluteal phase of the estrous cycle has been extended. Parturition is poorly understood in monotremes, but more is known for marsupials. In macropods (kangaroos), plasma P_4 falls, and estradiol, associated with the postpartum estrus, rises. Relaxin may prepare the cervix and vaginae for birth, and prostaglandins and mesotocin are essential for the maternal contractions involved in delivery of the neonate. During the last two days of gestation, fetal adrenal glucocorticoid production is increased. In male marsupials, development of testicular tissue is controlled by androgens, but not all sexual differentiation in the neonate depends on testicular hormones. The steroid androstenediol may have a previously unrecognized role in virilization. Little is known about regulation of testicular function in monotremes. Seasonal reproduction is controlled by pineal function in those marsupials in which regulation has been investigated.

1. THE PROTOTHERIANS AND METATHERIANS: THE 'OTHER' MAMMALS

Prototherians (monotremes) and metatherians (marsupials) differ from eutherians ('true' mammals, often called placental mammals) in their mode of reproduction. Prototherians and metatherians share with eutherians the mammalian features of hair, a lower jaw consisting of two dentary bones, three middle ear bones, and suckling their young using mammary tissue modified from apical glands of the skin (Blackburn, 1993; Augee, Carrick, Grant, & Temple-Smith, 2008). Studies on the lactational tissues indicate that the biochemical, ultrastructural, developmental, and histological similarities of the mammary glands and mammary secretions of extant monotremes, marsupials, and eutherians provide convincing evidence that lactation had a common origin that predated the divergence of these groups (Oftedal, 2002). Ontogenetic studies, in conjunction with information on the placentation of metatherians and eutherians, have demonstrated that the common ancestor of the Theria was still oviparous, similar to the extant monotremes, and that viviparity occurred independently in the phylogeny of metatherians and eutherians (Zeller, 1999).

1.1. Prototherians: Monotremes

The subclass Prototheria is represented today by only one group of mammals, the order Monotremata (monotremes). Monotremes diverged from other mammal lineages approximately 200 million years ago, and there are only five extant species of monotreme: the platypus (*Ornithorhynchus anatinus*), the short-beaked echidna (*Tachyglossus aculeatus*), and three species of long-beaked echidna (*Zaglossus* spp.). The platypus is found only in Australia, long-beaked echidnas are found only in Papua New Guinea, with short-beaked echidnas found throughout Australia and Papua New Guinea (Augee et al., 2008). Monotremes can be identified anatomically from other mammalian subclasses by their retention of interclavicle, coracoid, and precoracoid bones in the shoulder girdle, and ectopterygoid bones at the base of the skull (Augee et al., 2008). Monotremes have a true cloaca, whereas marsupials do not: marsupials have a simple external opening for excretory and reproductive functions (urogenital sinus) (Temple-Smith & Grant, 2001). All male monotremes have a bilateral venom system associated with a spur on the ankle connected by a long duct to a gland in the upper leg, although only the system of the platypus is particularly toxic (Augee et al., 2008).

The monotremes differ in their mode of reproduction from other mammals by laying eggs and also producing

Hormones and Reproduction of Vertebrates, Volume 5—Mammals

milk for their altricial young (Temple-Smith & Grant, 2001). The soft-shelled eggs of monotremes are laid after early embryological development has occurred *in utero*, with a gestation period of 20–23 days for short-beaked echidnas (Beard & Grigg, 2000). The eggs are then incubated for about 10 days by the female in her shallow pouch area, and are held there by the female curling around the egg (Temple-Smith & Grant, 2001; Nicol & Andersen, 2006). Monotremes feed their young through lactational pores, which duct directly to the surface of the shallow pouch area (Blackburn, 1993). Lactation occurs for three to four months for the platypus and five to six months for the short-beaked echidna (Beard & Grigg, 2000; Temple-Smith & Grant, 2001).

1.2. Metatherians: Marsupials

Marsupial mammals diverged from placental mammals about 125–130 million years ago, although some debate exists about the exact timing, with dates of between 125–180 million years ago suggested (Oftedal, 2002). There are about 270 extant marsupial mammals, which are found in South America, Papua New Guinea, and Australia. A major difference in the biology between the marsupials and other mammalian groups is that marsupial neonates are born in an extremely altricial state after a relatively short gestation period, and development occurs postnatally in the pouch ('lactation over gestation' option) (Renfree, 1983; 2006).

Marsupials do have a functional placenta, with most having a choriovitelline placenta and bandicoots having an invasive chorioallantoic placenta (Freyer, Zeller, & Renfree, 2003; 2007). Thus, the term 'placentals' for the eutherian placental mammals is perhaps a misnomer. Marsupials were named because they have a pouch (marsupium); however, not all marsupial females have a pouch, with many of the didelphids having only a pouch area, and no males have a pouch, or indeed any mammary remnants (unlike eutherian males) (Renfree, Pask, & Shaw, 2001). Gestational length for marsupials can be as short as 10.5 days (*Sminthopsis macroura*, the stripe-faced dunnart (Selwood & Woolley, 1991)), with some didelphids and dasyurids having gestational lengths of 12–16 days (see McAllan, 2003), but most species have a gestational length of about 27–35 days (e.g., *Macropus eugenii*, the tammar wallaby (Tyndale-Biscoe & Renfree, 1987)). Marsupial neonates weigh as little as 5 mg (*Tarsipes rostratus*, the honey possum (Renfree, 1979)), and all are undeveloped, with most organ development occurring postbirth, during the long, frequently complex lactational period in the pouch. The altricial developmental state at birth includes the brain, with olfactory, somatosensory, and brainstem regulatory regions well developed, but with other cerebral and limbic regions very undeveloped (Reynolds et al., 1985; Gemmell & Nelson, 1988; Nelson, Knight, & Kinghorn, 2003). The marsupial neonate has relatively well-developed digestive, respiratory, and circulatory systems, although lower limbs and musculoskeletal systems are poorly developed with the exception of those needed to climb to the pouch after birth (Shaw & Renfree, 2006). Marsupials are born retaining a functional mesonephric kidney, and the scrotum and primordial mammary glands are present at birth, even though the testes and ovaries do not develop until well after birth (Shaw, Renfree, & Short, 1990; Shaw & Renfree, 2006).

2. ENDOCRINE CONTROL OF SEXUAL DIFFERENTIATION IN MONOTREMES AND MARSUPIALS

2.1. Monotremes

Few details are known about sexual differentiation in monotremes, although recent studies have shown that the differentiation of the gonads is determined after birth (Wallis, Waters, & Graves, 2008). However, the mechanisms for sexual differentiation in monotremes have not been investigated, although the genetic control of sex determination has been well studied (Graves, 1990; Wallis et al., 2008). Most studies on early development in monotremes have concentrated on embryonic morphogenesis and organogenesis (Hughes & Hall, 1998; Ferner, Zeller, & Renfree, 2009).

2.2. Marsupials

The ovaries and testes of marsupials are undifferentiated at birth, which fact has provoked a significant amount of work on the origins of sex determination in therians. The pouch and scrotal sac can be observed at birth, and this is believed to be genetically determined, with one X chromosome determining the development of a scrotal sac, and two X chromosomes promoting the development of a pouch (O, Short, Renfree, & Shaw, 1988). In contrast, eutherians require a functional fetal testis for the development of the scrotum from the undifferentiated urogenital area (O et al., 1988). In the tammar, or tammar wallaby, differentiation of the mammary primordia, gubernaculum, and processus vaginalis occurs early in fetal development under genetic control, whereas development of the vas deferentia, epididymides, prostate, male urethra, and phallus is under the control of androgens (Renfree, Wilson, Short, Shaw, & George, 1992; Butler, Harry, Deakin, Cooper, & Renfree, 1998; Shaw et al., 2000; Wilson, Leihy, Shaw, & Renfree, 2002). Hormonal manipulation of the developing neonates and young marsupials in the pouch, notably the tammar, indicates that the development of the prostate and phallus is promoted in the male by the androgen 5α-androstane-3 α, 17β-diol (Shaw et al., 2000).

In a series of experiments, the precise role of androgens in determining the male phenotype was determined. Androgen secretion from the testes of the tammar begins on day two postpartum, when the testis cords have differentiated (Renfree et al., 1992; Renfree, O, Short, & Shaw, 1996). The testes of the grey short-tailed opossum (*Monodelphis domestica*) are differentiated in 50% of males on the day of birth, and are differentiated in all males on day two postpartum (Xie, Mackay, Ullmann, Gilmore, & Payne, 1996). Leydig cells are evident by day three postpartum; however, testosterone (T) secretion is low until four weeks after birth, although 3β-hydroxysteroid dehydrogenase (3β-HSD) activity is found in adrenals and gonads of both male and female grey short-tailed opossums from birth (Russell et al., 2003; Xie et al., 1998). In tammars, the testicular concentration of T rises from day two postpartum to plateau from day 10 until about day 40 postpartum, after which point it drops until reaching basal levels at day 70 postpartum (still in the pouch) (Renfree et al., 1992; 1996). In females, absence of androgens results in development of the female urogenital system (Renfree et al., 1992; 1996).

The ontogeny of gonadal development after birth is best-studied in the family Dasyuridae. In both males and females, development of the gonads takes several months, with five to six months recorded for several species of marsupial including the carnivorous *Antechinus* spp. and *Sminthopsis* spp. (Woolley, 1966; Taylor & Horner, 1970; Wilson, 1986). In many other seasonal species, including rodents, gonadal development takes place in only six to eight weeks (Woolley, 1966; Taylor & Horner, 1970; Wilson, 1986). Oocytes in juvenile female *Antechinus* spp. are small and have a single layer of granulosa cells around the oocyte. As the year progresses towards the breeding season, secondary oocytes are seen, with many layers of granulosa cells around the oocyte, and small antra develop within the oocytes, eventually joining to become single large antra of the Graafian follicles. During development of the follicles, the female external morphological features are relatively unchanged until after the winter solstice, when secondary oocytes mature into Graafian follicles. At this point, the urogenital sinus enlarges and the pouch becomes easily discernable, with thinning hair and guard hairs developing. The pouch continues to develop during the mating season in late winter or early spring, ovulation, and during pregnancy, such that by parturition the pouch ridges form, with pronounced guard hairs, and the pouch itself is hairless and shiny; lactational tissue has developed. The external morphological changes parallel reproductive hormone cycles in the agile antechinus, *Antechinus agilis* (Kerr & Hedger, 1983; Hinds & Selwood, 1990).

3. THE OVARIAN CYCLE IN MONOTREMES AND MARSUPIALS

3.1. Monotremes

Little information is known about the reproductive cycles of monotremes, partly because of their enigmatic behavior but also because, as protected species, they are not easily available for determining reproductive status. Monotremes are not sexually mature until their second year of life, and are seasonal breeders (Temple-Smith & Grant, 2001; Nicol, Andersen, & Jones, 2005). In short-beaked echidnas during the nonbreeding season (September to May in Tasmania, Australia), mean plasma progesterone (P_4) concentrations of 0.18 ± 0.12 ng/mL (0.77 ± 0.69 mmol/L) have been recorded, and during the breeding season (June to August in Tasmania) the mean plasma P_4 concentrations are above 0.5 ng/mL (1.59 mmol/L) (Nicol et al., 2005). Peak plasma concentrations coincided with the last part of pregnancy in these echidnas (Nicol et al., 2005). The corpus luteum of the platypus is believed to secrete P_4 from the follicular epithelial cells, and P_4 acts on the uterine epithelium to promote uterine secretions (Hughes & Carrick, 1978; Gemmell, 1995).

3.2. Marsupials

The estrous cycle overlaps with pregnancy in many of the marsupials that have been studied; thus, ovarian cycles and pregnancy will be considered together. In most marsupials, ovulation is spontaneous, estradiol (E_2) is the main circulating estrogen, and P_4 falls prior to parturition (Tyndale-Biscoe & Renfree, 1987; Hinds & Selwood, 1990). Ovulation induced by the presence of males has been reported in the grey short-tailed opossum (Fadem, 1985; Hinds, Reader, Wernberg-Möller, & Saunders, 1992), the greater bilby (*Macrotis lagotis*) (Ballantyne, Matson, Noakes, Nicolson, & Johnston, 2009), and the brush-tailed bettong (*Bettongia penicillata*; also known as the woylie or brush-tailed rat-kangaroo) (Hinds & Smith, 1992). However, there is also evidence for a postlactational estrus independent of male stimuli in the grey short-tailed opossum (Jackson & Harder, 2000). For this opossum and woylie, the sensory contact of auditory and olfactory cues can be enough to stimulate estrous cycles and estrus activity (Fadem, 1985; 1987; Hinds & Smith, 1992; Hinds et al., 1992a), but the greater bilby requires the presence of males; auditory and olfactory cues alone are not sufficient to induce ovulation (Ballantyne et al., 2009). In the woolly opossum (*Caluromys philander*), the female does ovulate spontaneously; however, cycles are increased and regularized by the presence of males (Perret & Ben M'Barek, 1991). The koala (*Phascolarctos cinereus*) is an obligatory reflex ovulator, requiring the act of copulation to promote the luteal phase of the cycle (Johnston, McGowan, O'Callaghan, Cox, & Nicolson, 2000; Johnston, O'Callaghan, Nilsson, Tzipori, & Curlewis, 2004), although the western quoll (or chuditch; *Dasyurus geoffroii*) may also be a reflex ovulator (Stead-Richardson, Bradshaw, Bradshaw, & Gaikhorst, 2001).

In marsupials, estrous cycles exhibit three broad patterns (Tyndale-Biscoe & Renfree, 1987). The first pattern occurs in many didelphids, dasyurids, and phalangerids, and has a luteal phase lasting up to 60% of the cycle length and a follicular phase following, which leads into the next estrus and ovulation (Figure 10.1). Pregnancy is completed within the luteal phase, and lactation suppresses the later follicular phase. Regression of the corpus luteum coincides with parturition, and, if lactation occurs, the corpus luteum of pregnancy slowly disappears or can become a persisting corpus albicans (Smith & How, 1973; Shorey & Hughes, 1973; Tyndale-Biscoe & Renfree, 1987). The second pattern is seen in the peramelids and thylacomyids, in which parturition occurs during the luteal phase and the corpora lutea remain large throughout early lactation when the follicular phase is suppressed (Gemmell, 1981; 1984; Ballantyne et al., 2009) (Figure 10.1). The third pattern is exhibited by most of the macropods, in which the luteal phase lasts for about 90% of the cycle and the follicular phase is not suppressed by birth and subsequent lactation. Indeed, gestation occupies most of the luteal phase, and there is a postpartum estrus and ovulation (Figure 10.1). If this is followed by lactation, the corpus luteum from the postpartum estrus does not regress but stays quiescent, persisting for as long as there is a young in the pouch (Tyndale-Biscoe & Renfree, 1987; Hinds, Fletcher, & Rodger, 1996). The estrous cycle of macropods is uninterrupted by pregnancy and parturition, and the cessation of lactation will initiate a resumption of the estrous cycle (Sharman, 1970; Lee, Woolley, & Braithwaite, 1982; Tyndale-Biscoe & Renfree, 1987).

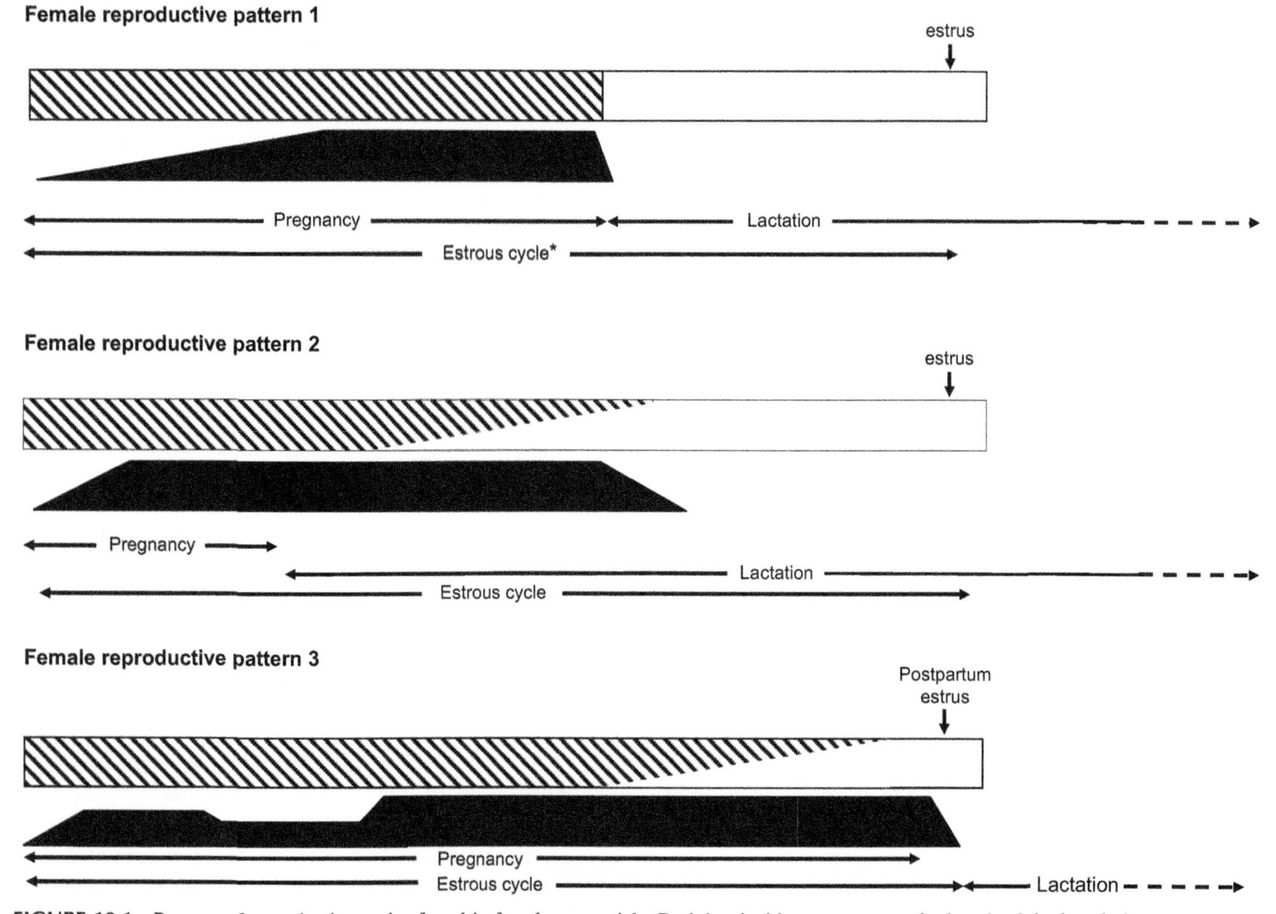

FIGURE 10.1 Patterns of reproductive cycles found in female marsupials. Dark hatched boxes represent the length of the luteal phase; open boxes represent the follicular phase. Black boxes represent the patterns of progesterone secretion; insufficient data are available for consistent representation of estradiol data. Relative length of pregnancy and the lactational phases are indicated by arrows. Length of lactation is variable, so the extension of lactation is indicated by a hatched arrow. Asterisk indicates a representative cycle for this group, where, if pregnancy is not followed by lactation, the estrous cycle is repeated. In dasyurid females, where the life history is typified by complete male mortality at the end of the mating period, females are monovular and thus no further cycles can occur in that breeding season. The length of pregnancy ranges from 10.5–32 days in type 1 females, 12.5–17 days in type 2 females, and 26–37 days in type 3 females.

3.2.1. Type 1: Short gestation, long luteal phase

The brown antechinus (marsupial 'shrew;' *Antechinus stuartii*), a dasyurid, exhibits the first pattern. If mating occurs in female *A. stuartii*, spontaneous ovulation usually is followed by pregnancy. Spontaneous ovulation also can be followed by a pseudopregnancy if females are unmated or matings are unsuccessful (Woolley, 1966; McAllan, Joss, & Firth, 1991). The pseudopregnant female undergoes enlargement of the shallow pouch, formation of the pouch ridge, and lengthening of the guard hairs in exactly the same way as in the pregnant female. If birth does not occur, the pouch slowly regresses to its previous state (Woolley, 1966; McAllan et al., 1991). Plasma P_4 concentrations in pregnant and pseudopregnant females are indistinguishable and rise from about 4 $ng{\cdot}mL^{-1}$ (12.7 $mmol{\cdot}L^{-1}$) at ovulation to 11 $ng{\cdot}mL^{-1}$ (35 $mmol{\cdot}L^{-1}$) at days 15–22 of gestation (Hinds & Selwood, 1990). At this time, corpora lutea are large and it is likely that pouch development depends on corpora luteal activity rather than the influences of pregnancy. In *Antechinus* spp. the histological changes that occur in the gravid and nongravid uteri correlate with the developmental rate of the embryo, circulating P_4 concentrations, and formation of the corpus luteum (Cruz & Selwood, 1993). The corpus luteum controls the rate of embryonic development, and P_4 regulation of embryonic development is mediated by the uterus, and the interactions are believed to be coordinated by embryonic signaling (Woolley, 1966; Hinds & Selwood, 1990; Cruz & Selwood, 1993).

3.2.2. Type 2: Short gestation, prolonged luteal phase

The second pattern is exhibited by the peramelids and thylacomyids, having a short gestation period, typically about 12–13 days, which is dominated by the luteal phase (Gemmell, 1981; 1984; 1989; Ballantyne et al., 2009). The corpus luteum persists well into the lactational period, and continues to secrete P_4 until at least day 40 of lactation (Gemmell, 1989; Ballantyne et al., 2009). The common bandicoot (*Isoodon macrourus*) demonstrates unusual P_4 cycles, with plasma P_4 concentrations increasing following ovulation and then dropping slightly after parturition, but remaining raised for at least four days postpartum (at 10–12 $ng{\cdot}mL^{-1}$) and declining slowly until they reach basal levels at about day 19 (3–5 $ng{\cdot}mL^{-1}$) (Gemmell, 1985; 1995). It was initially thought that the raised P_4 concentrations depend on the number of young, with plasma P_4 remaining elevated until after day 19 in females suckling larger litters and returning to basal by day 19 if there was only one young, suggesting that the suckling stimulus maintains the corpus luteum in bandicoots (Gemmell, 1984). However, more recent studies show that corpora luteal function is not influenced by lactation, nor do corpora lutea influence lactation (Gemmell, 1989). Similar patterns of P_4 secretion have been observed in the greater bilby, in which cycles of changing P_4 concentrations are present during the lactational period (Ballantyne et al., 2009). Bandicoots are also unusual in that the chorioallantois is fused with the uterine epithelium. The chorioallantois is absent from the placentae of other marsupials. In peramelids, the chorioallantois may provide some nutrient transfer as well as perhaps providing prostaglandins (PGs), necessary for induction of birth (Gemmell, 2002; Freyer et al., 2003).

3.2.3. Type 3: Long gestation, delayed luteal phase

The third pattern is exhibited by the macropods, and has been studied extensively in the tammar. Here, the luteal phase occupies up to 90% of the cycle, and pregnancy occurs during the luteal phase. In the last third of the luteal phase there is new follicular development so that a postpartum estrus and ovulation can occur immediately after the birth of the young (Tyndale-Biscoe & Renfree, 1987; Hinds et al., 1996). In contrast to dasyurids, the endometrium of the gravid uterus in many macropods is morphologically different from the contralateral nongravid uterus (Renfree, 1972; Shaw & Rose, 1979; Renfree, 1983; Rose, 1992). This is not the case for the Virginia opossum (*Didelphis virginiana*) (Fleming & Harder, 1981), in which the uterine differences are believed to reflect a maternal recognition of pregnancy (Merchant, 1976; Renfree, 1983; Rose, 1992).

Lactation in type 3 marsupials prevents the conceptus from developing beyond the blastocyst stage and the diapausing embryo does not reactivate until the previous young is weaned (Renfree & Shaw, 2000). This embryonic diapause is exhibited by 22 macropods and three possums from three families (the feathertail glider (*Acrobates pygmaeus*); the mountain pygmy possum (*Burramys parvus*); and the honey possum) (Renfree & Shaw, 2000). Embryonic diapause is facultative in most species, but some show seasonal diapause as part of the yearly cycle, where the changing photoperiod promotes the cessation of embryonic diapause (Renfree & Shaw, 2000). In all species with a type 3 pattern, the corpus luteum is quiescent during diapause (Renfree & Shaw, 2000).

In both the tammar and the quokka (*Setonix brachyurus*), after ovulation a transient rise in E_2 occurs, coinciding with growth of the corpus luteum and the blastocyst (Flint & Renfree, 1982; Bradshaw & Bradshaw, 1992). A second rise in E_2 occurs then just before parturition and the postpartum estrus (Flint & Renfree 1982). Plasma P_4 concentrations rise during pregnancy and drop rapidly at parturition (Hinds & Tyndale-Biscoe, 1982;

Tyndale-Biscoe, Hinds, Horn, & Jenkin, 1983; Harder, Hinds, Horn, & Tyndale-Biscoe, 1985). Similar plasma P_4 changes have been found in other macropodid marsupials (Shorey & Hughes, 1973; Cake, Owen, & Bradshaw, 1980; Curlewis, White, & Loudon, 1987; Fletcher, 1989a; Muths & Hinds, 1996). These changes in circulating P_4 are related to the presence of the corpus luteum and are independent of pregnancy. This phenomenon is also seen in the dasyurid *A. stuartii* (type 1 pattern), whose ovulation is spontaneous, where pregnant and pseudopregnant females both exhibit rises in circulating P_4 following ovulation (Hinds & Selwood, 1990).

In other marsupials besides the macropods, the corpus luteum is essential for development of the embryo beyond the blastocyst stage, or is essential for parturition (Bryant & Rose, 1986; Tyndale Biscoe & Renfree, 1987). Progesterone is secreted by the corpus luteum and rises in the second half of the estrous cycle in the didelphids, dasyurids, phalangerids, vombatids, and macropods (Shorey & Hughes, 1973; Hinds & Tyndale-Biscoe, 1982; Tyndale Biscoe & Renfree, 1987; Hinds, 1989; Hinds & Selwood, 1990; West, Galloway, Shaw, Trounson, & Paris, 2004). Not all these marsupials have a postpartum estrus, nor do they exhibit embryonic diapause, and the P_4 concentrations resolve to basal concentrations soon after parturition.

As well as the corpus luteum, the Graafian follicle has a role in the cycling of P_4 and the postpartum estrous in the macropods (Shorey & Hughes, 1973; Sutherland, Evans, & Tyndale-Biscoe, 1980; Harder et al., 1985; Fletcher & Renfree, 1988; Fletcher, 1989a; Hinds et al., 1992b). Follicular development in eutherians is biphasic, with the development of the oocyte and its associated follicular tissues developing concurrently. However, in the second part, follicular development is more pronounced, with the granulosa and thecal cells undergoing significant mitotic activity and the antral cavity accumulating fluid (Hinds et al., 1996). Marsupials also have biphasic cycles, although, unlike eutherians, the oocytes continue to grow after formation of the follicular antrum in the four phylogenetically diverse species that have been studied closely: the brush-tailed possum (*Trichosurus vulpecula*), the woylie, the tammar, and the grey short-tailed opossum (Rodger, Giles, & Mate, 1992). Oocytes reach diameters two to three times those seen in eutherians (Rodger et al., 1992a). The final development of follicles and the Graafian follicle in particular is dependent on pituitary influences, and this is best described for the tammar, where hypophysectomy or immunization against gonadotropin-releasing hormone (GnRH) results in failure of the follicles to mature (Short, Flint, & Renfree, 1985; Hinds et al., 1996).

In marsupials with the type 3 reproductive pattern, P_4 continues to rise until birth, when there is a rapid decrease in circulating P_4 concentrations (Hinds & Tyndale-Biscoe, 1982; Tyndale-Biscoe et al., 1983; Harder, Hinds, Horn, & Tyndale-Biscoe, 1984; Shaw & Renfree, 1984; Harder et al., 1985). This decrease is followed by an increase in circulating E_2 within six hours of the P_4 decline, and an E_2 peak 12 hours postpartum (Renfree et al., 1984; Shaw & Renfree, 1984; Harder et al., 1985). The developing Graafian follicle, not the corpus luteum, is the source of the E_2, and peripheral and antral E_2 increases proportionally with the increasing size of the follicle (Harder et al., 1984; Shaw & Renfree, 1984). Estradiol concentrations are highest in the utero–ovarian circulation draining the Graafian follicle and not that draining the corpus luteum, which in macropods is in the opposite ovary to that of the Graafian follicle (Harder et al., 1984). Estradiol feeds back to the pituitary to induce the preovulatory surge of luteinizing hormone (LH), and removal of the Graafian follicle prevents the preovulatory surge of E_2 and subsequent LH surge 16 hours postpartum, but does not prevent parturition (Harder et al., 1985). In the tammar wallaby, plasma E_2 is low, with the exception of estrus and at around day five of the estrous cycle, when E_2 peaks at about $15\,pg{\cdot}mL^{-1}$ before declining to about $5-10\,pg{\cdot}mL^{-1}$ for the remainder of the estrous cycle, dropping further before parturition (Shaw & Renfree 1984; Harder et al., 1985). Estradiol concentrations over the duration of the cycle have been measured in a few other marsupials, with values known for the Virginia opossum and the brush-tailed possum. Estradiol values increase in the proestrous part of the cycle and decline at parturition in the Virginia opossum, peaking at $23 \pm 2\,pg{\cdot}mL^{-1}$ and decreasing to $9 \pm 1\,pg{\cdot}mL^{-1}$ at their lowest (Harder & Fleming, 1981). In the brush-tailed possum, E_2 values are low, except for around estrus, presumably originating from the preovulatory follicles (Eckery et al., 2002a).

Analysis of fecal steroids has allowed examination of the estrous cycle from rare and/or wild animals. Fecal E_2 is raised around the time of ovulation in the red-tailed phascogale (*Phascogale calura*) (Foster, Caton, Thomas, Cox, & Taggart, 2008), the honey possum (Oates, F. Bradshaw, S. Bradshaw, Stead-Richardson, & Philippe, 2007), and in the follicular phase of the estrous cycle for the Tasmanian devil (*Sarcophilus harrisii*) (Hesterman, Jones, Schwarzenberger, 2008a) and the spotted-tailed quoll (*Dasyurus maculatus*) (Hesterman, Jones, Schwarzenberger, 2008b). Plasma E_2 concentrations are raised in the follicular part of the estrous cycle in the southern hairy-nosed wombat (*Lasiorhinus latifrons*) and probably peak at ovulation (Finlayson et al., 2006).

The data for cyclic changes in circulating E_2 concentrations are sketchy for marsupials, with much more information available for the role that P_4 plays in the development of the uterus, maintenance of pregnancy, and facilitation of embryonic development. In the brown antechinus, P_4 increases parallel to the development of the corpus luteum. Moreover, the stage of embryonic development appears to parallel that of corpus luteum formation and positively correlates with plasma P_4 levels (Hinds & Selwood, 1990). This phenomenon also has been observed in other species with the type 1 reproductive pattern, including the stripe-faced dunnart, in which the plasma P_4 profile is biphasic, peaking during unilaminar blastocyst expansion and on the day of implantation. Periods of rapid embryonic development are associated with increasing plasma P_4 concentrations and maternal body weight (Menkhorst, Hinds, & Selwood, 2009). A transient P_4 peak during unilaminar blastocyst formation occurs in some species and is higher in the pregnant than the nonpregnant quokka (Cake et al., 1980), the tammar (Hinds & Tyndale-Biscoe, 1982), and the eastern quoll (*Dasyurus viverrinus*) (Hinds, 1989).

4. CONTROL OF OVARIAN CYCLES IN MONOTREMES AND MARSUPIALS

4.1. Monotremes

Little is known about the control of ovarian cycles in female monotremes. Indeed, even the length of gestation is uncertain in the platypus and that of the short-beaked echidna is still debated (Temple-Smith & Grant, 2001; Holland & Jackson, 2002).

4.2. Marsupials

Control of ovarian cycles is known for a few marsupials, including the kowari (*Dasyuroides byrneii*), the brush-tailed possum, the stripe-faced dunnart, the grey-short-tailed opossum (all type 1 reproductive pattern), and the tammar (type 3 reproductive pattern). Plasma LH remains low for most species until behavioral estrus, when there is a sharp rise. In tammars, there is a short-lived LH surge followed one to two days later by ovulation (Sutherland et al., 1980; Harder et al., 1985) and a similar pattern is seen in the brush-tailed possum with a short-lived LH surge coincident with behavioral estrus one day before ovulation (Shorey & Hughes, 1973; Crawford et al., 1999). In contrast, the pattern of LH secretion in the kowari rises in the first part of the ovarian cycle and shows no surge prior to ovulation (Fletcher, 1989b).

In female tammars, basal concentrations of LH are low (0.94 ng·mL^{-1}) with pulses of low magnitude (1.3 ng·mL^{-1}) and low frequency (Hinds, Diggle, & Tyndale-Biscoe, 1992). Nonluteal ovarian feedback is necessary for LH secretion from the pituitary, and E_2 promotes the LH surge (Hinds et al., 1992b; Rudd, Short, McFarlane, & Renfree, 1999). Removal of the sucking stimulus in tammars affects LH pulse frequency but there are no seasonal differences in pulses of LH (Hinds et al., 1992b). In the brush-tailed possum, mRNA coding for LH receptors are found in the granulosa cells of the ovary, first appearing at the time of antrum formation in the theca interna, and are also present in the interstitial tissue, although the tissue does not appear to respond to LH stimulation (Eckery et al., 2002b; Haydon, Juengel, Thomson, & Eckery, 2008).

Data for the role of follicle-stimulating hormone (FSH) in the reproductive cycle of female marsupials are patchy. In tammars, the pattern of FSH release is not pulsatile (Hinds et al., 1992b). Circulating FSH does not change when the sucking stimulus of young is removed, but concentrations are higher during seasonal quiescence (Hinds et al., 1992b). Concentrations of FSH in the plasma of female tammars are low but increase around the time of estrus, coinciding with an increase in follicular development (Evans, Tyndale-Biscoe, & Sutherland, 1980). Treatment of both cycling and noncycling female tammars and brush-tailed possums with exogenous eutherian GTHs (porcine LH and FSH in females primed with pregnant mare serum gonadotropin (PMSG)) increases the number of developing follicles and ovulation sites (Glazier & Molinia, 1998; Molinia, Gibson, Smedley, & Rodger, 1998). The response is seasonal in brush-tailed possums (Glazier, 1998), but not in tammars (Molinia et al., 1998). Nonluteal ovarian tissue is essential for the negative-feedback effects on both LH and FSH secretion in the tammar (Hinds et al., 1992b). The FSH receptor (FSH-R) gene has been cloned and characterized in this species and found to share 94% amino acid similarity with human FSH-R, and is expressed in both the adult testis and ovary (Mattiske, Pask, Shaw, & Shaw, 2002). This would suggest a similar function for this gene in both marsupials and eutherians.

Specific sites for the FSH receptor include the granulosa cells of healthy follicles containing at least two complete layers of cells in brush-tailed possums (Eckery et al., 2002b). In the brush-tailed possum, expression of the FSH-R was observed in granulosa cells of follicles shortly after they had begun to grow, was limited to these granulosa cells, and was active in granulosa cells of follicles shortly after antrum formation (Eckery et al., 2002b). Similarly to eutherian mammals, multiple ovulatory follicles can be induced to develop in the brush-tailed possum by administration of FSH, suggesting a key role for this hormone in the regulation of follicular growth (McLeod, Hunter, Crawford, & Thompson, 1999). Plasma FSH concentrations fall progressively over the period of preovulatory follicle development and rise again after ovulation (Crawford et al., 1999).

For all other species, data are sketchy, with only some information on the use of LH and FSH available from studies promoting the breeding of rare and endangered marsupials. The dunnarts (*Sminthopsis crassicaudata* and *S. macroura*) have been used to establish the best methods for artificial reproductive technology for some of the rare dasyurids such as the northern quoll (*Dasyurus hallucatus*) and the Tasmanian devil, the latter of which currently is under significant conservation threat from a combination of facial tumor disease and land clearing (Hesterman et al., 2008a; 2008b). Luteinizing hormone and/or FSH stimulation have been used in both of these dunnart species in efforts to harvest healthy follicles for reproductive technologies (Menkhorst, Ezard, & Selwood, 2007; Czarny, Garnham, Harris, & Rodger, 2009). Similar studies have been performed on the southern hairy-nosed wombat and common wombat (*Vombatus ursinus*) in an effort to save the northern hairy-nosed wombat (*Lasiorhinus krefftii*), which is critically endangered (West et al., 2004; McDonald et al., 2006; Druery et al., 2007; West et al., 2007).

Other data are provided by the stimulation of ovarian cycles by GnRH or its agonists to promote artificial insemination of rare and endangered species, with GnRH agonists promoting normal luteal phases in the koala (Allen et al., 2008). Treatment with equine GnRH also promotes normal estrous cycling in the stripe-faced dunnart (Menkhorst et al., 2007) and the fat-tailed dunnart (*S. crassicaudata*) (Rodger, Breed, & Bennett, 1992), although this is dose-dependent (Rodger et al., 1992b). In the eastern grey kangaroo (*Macropus giganteus*), implants of potent GnRH agonists suppress reproduction, presumably because of the initial hyperstimulation of LH and FSH secretion, followed by downregulation of pituitary activities (Herbert, Trigg, & Cooper, 2006). Immunization against GnRH prevents follicular growth and development of the luteal phase in the tammar (Short et al., 1985), indicating a similar role for GnRH as that found for reproduction in eutherians.

5. HORMONAL REGULATION OF BIRTH IN MONOTREMES AND MARSUPIALS

5.1. Monotremes

Little is known about the regulation of parturition in monotremes; indeed, birth has rarely been observed (Temple-Smith & Grant, 2001). A pouch area develops in short-beaked echidnas, and females will stay with the young in a burrow during both incubation and the early part of lactation (Morrow & Nicol, 2009). Recently it was confirmed that these echidnas use torpor (a controlled reduction in body temperature) during the reproductive season, which would explain the variety of gestation periods reported in this species (Temple-Smith & Grant, 2001; Morrow & Nicol, 2009).

5.2. Marsupials

In marsupials, successful delivery of live young depends on the presence of the corpus luteum as well as both the anterior and posterior pituitaries; however, similarly to many eutherian mammals, the fetus may determine the time of birth (see Tyndale-Biscoe & Renfree 1987; see also Chapter 6, this volume). It is known that birth in the tammar wallaby is determined by the minute (400 mg) fetus, and that the timing of this signal is determined by the fetal genotype (Shaw & Renfree, 2001). A fetal signal activates a nitric oxide (NO)−guanylate cyclase system in the myometrium that may maintain myometrial inactivity, and this is downregulated at term (Ingram, Shaw, & Renfree, 2001). Nitric oxide is a paracrine smooth muscle relaxant that can be demonstrated to promote the cascade that enables gravid uterine contractile quiescence, and the effects are localized, with only the tissue surrounding the fetus affected (Ingram et al., 2001).

Parturition has been studied in a number of marsupial species, and the hormones involved are essentially the same as those involved in eutherian mammals, although some marsupials have mesotocin (8-ile-oxytocin) (MST) instead of oxytocin (OXY), as seen in prototherians, many marsupials, and eutherians (Sebastian, De Matteo, Shaw, & Renfree, 1998; Gemmell, 2002). The roles of P_4, E_2, relaxin, prolactin (PRL), PGs, and OXY/MST in late pregnancy and parturition are well understood in the tammar. Around the time of parturition, plasma P_4 falls and E_2 associated with the postpartum estrus typically rises, although neither hormone is essential for the timing of birth. Experimental evidence indicates the independence of birth from the estrous cycle and hormone circulation (Hinds & Tyndale-Biscoe, 1982; see Shaw & Renfree 2001). Relaxin, however, may loosen the connective tissue of the cervix and vaginae for birth, and relaxin concentrations in the corpora lutea of tammars increases significantly in the four days prior to birth (Parry, Bathgate, Shaw, & Renfree, 1997). However, some tissue softening still occurs around the time of birth, and this may be due to high P_4 concentrations prior to birth (Hinds & Tyndale-Biscoe, 1982). It would appear that preparation of the uterine connective tissues for the birth process relies on the presence of both relaxin and the presence of prebirth elevated P_4.

Promotion of parturition occurs via the production of PGs and receptors for MST. In the tammar, PG production in the gravid endometrium rises during the last two days of gestation and this parallels increased placental PG synthesis. There is also a pregnancy-specific upregulation of OXY receptors in the gravid myometrium that increases the responsiveness of the gravid uterus to MST (Shaw & Renfree, 2001; 2006). In the Brazilian mouse opossum (or white-bellied opossum) (*Didelphis albiventris*), the sensitivity to OXY is upregulated by the presence of P_4 (Ladosky & Calixto, 1984). Both PGs and MST (or OXY) are essential for the maternal contractions involved in delivery of the neonate (Shaw & Renfree, 2001; 2006).

The fetal signal is probably related to glucocorticoid production. During the last two days of gestation in the tammar wallaby, fetal adrenal glucocorticoid production increases, promoting lung maturation and surfactant production and ultimately triggering labor (Ingram, Shaw, & Renfree, 1999). In the last few days of pregnancy in the tammar, the fetal adrenal gland doubles in size in preparation for birth (Ingram et al., 1999). Cortisol production by the fetal adrenal can be stimulated *in vitro* by corticotropin (ACTH) and PGE_2, which suggests that the *in-vivo* increase may be stimulated by release of ACTH from the fetal hypothalamus–pituitary axis and also from PGE_2 originating from the placenta (Ingram et al., 1999). Fetal adrenal activity is believed to be important for later maturation purposes (Shaw & Renfree, 2006), especially for the neonatal lung, which, while capable of producing surfactant and of functioning, is so undeveloped that at birth 95% of respiratory exchange occurs through the skin (Frappell & McFarlane, 2006).

6. EMBRYONIC DIAPAUSE AND GESTATIONAL DEVELOPMENTAL ARREST

Another feature of marsupial reproduction is delay of production of young after fertilization in many species. The delay can be a part of embryonic diapause, which is under endocrine control, or as a developmental arrest, the control of which is less clear, but occurs due to slowing of post-blastocyst development and is believed to be due in part to temperature changes experienced by the mother (Renfree & Shaw, 2000). Torpor use by both reproductive monotremes (Nicol et al., 2005; Morrow & Nicol, 2009) and marsupials (Geiser & Masters, 1994; Geiser, McAllan, & Brigham, 2005) can contribute to gestational delays. Similar observations have been reported in temperate-zone bats (see Chapter 13, this volume).

6.1. Embryonic Diapause

Embryonic diapause occurs in eight orders of eutherians and one order of marsupial, and is controlled by two methods: either seasonal cycles of melatonin (MEL) secretion from the pineal gland (seasonal diapause) or by the presence of a suckling stimulus (lactational diapause). In the marsupial species in which seasonal diapause has been investigated, the regulation occurs via the pineal gland, where the secretion of MEL changes with the length of the day, and this allows timekeeping information to be processed by the animal. In the two best-studied species, the tammar and Bennett's wallaby (*Macropus rufogriseus*), the breeding season begins at the summer solstice, when ovulation occurs, followed by pregnancy and birth and then a postpartum estrus and fertilization, during which time, if pouch young are removed, the luteal and embryonic quiescences are terminated and birth can occur during lactational diapause. During lactational diapause the suckling stimulus promotes PRL release, which inhibits the growth of the corpus luteum and thus the P_4-induced activation of the embryo and also development of the follicle in the opposite ovary (Loudon, Brinklow, Gulland, Boyle, & Flint, 1990; Brinklow & Loudon, 1993). After the winter solstice, removal of pouch young does not reactivate the diapausing embryo, and this failure to reactivate is mediated by the presence of increasing levels of MEL, which promotes the inhibition of the corpus luteum by PRL (McConnell, Tyndale-Biscoe, & Hinds, 1986; Brinklow & Loudon, 1989; 1993; Renfree & Shaw, 2000). There are some differences in embryonic diapause between macropods and possums, with diapause occurring at the unilaminar blastocyst stage in possums but with continued slow growth of the blastocyst (Ward, 1990; Renfree & Shaw, 2000). It is unclear what mechanisms are involved in the control of embryonic diapause in possums (Renfree & Shaw, 2000).

6.2. Gestational Developmental Arrest

In many dasyurid species, gestational length has been described as "variable" and part of this variability in published records is due to the developmental arrest of the embryo. Gestational length in the agile antechinus is confirmed at 27 days but there are several periods when developmental arrest can occur (Selwood, 1980; 1981). During development, the first and second cleavage divisions are rapid, but can be followed by a developmental arrest of up to four days (Selwood, 1980). Another period of arrest has been seen from days 11 to 13 or 14, when the embryo is at the unilaminar blastocyst stage.

Until day 14, formation of the corpus luteum is incomplete and plasma P_4 is low (Hinds & Selwood, 1990). However, a modest increase in plasma P_4 is seen at day four, while maturation of the corpus luteum is marked by a sustained increase in P_4 secretion. Thus, an increase in plasma P_4 level is required at days 4 and 14 to allow the embryo to continue developing. The uterus is required to mediate these changes in developmental rate, although it is likely that an embryonic signal is also required (Cruz & Selwood, 1993). Cultured embryos in a stage of developmental arrest cannot resume development when P_4 is administered without the influences of the uterus (Hinds & Selwood, 1990). Periods of slow or arrested development coincident with low P_4 have been observed in other marsupials with long gestation periods for their size (Hinds & Selwood, 1990). Another interrupter of gestational development is torpor, which has been seen in other pregnant dasyurids including the stripe-faced dunnart (*S. macroura*) (Geiser et al., 2005) and the mulgara (*Dasycercus cristicauda*) (Geiser & Masters, 1994).

7. REGULATION OF THE TESTICULAR CYCLE IN MONOTREMES AND MARSUPIALS

7.1. Monotremes

Little is known about the regulation of testicular function in monotremes, including the testicular cycle and hormone profiles. Plasma androgens rise during the breeding season in the platypus (New, Jabukowski, Stone, & Jones, 1998; Nicol et al., 2005), and seasonal cycles of plasma androgens have been determined in the short-beaked echidna (Nicol et al., 2005; Johnston et al., 2007). There is no apparent diurnal cycle in T secretion, and T is secreted in response to GnRH agonists and human chorionic gonadotropin (hCG) (Johnston et al., 2007). One study found that, while echidna plasma taken before hCG stimulation showed no evidence of a seasonal change in T secretion, blood samples collected from the same animals four hours after hCG injection revealed significant changes in mean T concentration, depending on the season (Johnston et al., 2007). During the breeding season of the short-beaked echidna (June–August), sexually mature males had T concentrations in excess of $0.2\,ng{\cdot}mL^{-1}$ and were found in mating groups at some time during this period (Nicol et al., 2005). Short-beaked echidnas hibernate from October to May, when mean T levels are $0.09 \pm 0.05\,ng{\cdot}mL^{-1}$ (peak $= 4.62\,ng{\cdot}mL^{-1}$) (Nicol et al., 2005).

Monotremes lack seminal vesicles, and the presence of a functional prostate is controversial (Temple-Smith & Grant, 2001). There are no seasonal changes observed in the purported prostate of the platypus (Temple-Smith & Grant, 2001); some nominative prostatic tissue is present in the short-beaked echidna, and the activity of this tissue is androgen-dependent (Jones, Stone, & Zupp, 1992). The secretions of the nominative prostate epithelium are unlikely to contribute significantly to the fluid component of monotreme seminal plasma (Temple-Smith & Grant, 2001). There is also one pair of androgen-dependent bulbourethral (Cowper's) glands, which are seasonal in their activity, and they secrete a clear, viscous fluid drained by single ducts into the membranous portion of the urethra, as also seen in both marsupials and eutherians (Temple-Smith & Grant, 2001).

7.2. Marsupials

Development of the testicular tissue in marsupials is controlled by androgens, although not all sexual differentiation in the neonate depends on testicular hormones. While maintenance of the adult male accessory reproductive tract is androgen-dependent, sperm production appears to be less reliant on raised circulating androgen concentrations. The main circulating androgen secreted by male marsupials is T, although 5α-dihydrotestosterone (DHT) is also present (Allen & Bradshaw, 1980; McDonald, Lee, Bradley, & Than, 1981; Inns, 1982; Curlewis & Stone, 1985; Gemmell, Johnston, & Barnes, 1985; Wilson, Renfree, Auchus, Pask, & Shaw, 2009).

In seasonally breeding marsupials, the annual cycle of the male reproductive organs shows three distinct patterns. The first is represented by many of the dasyurid marsupials including those from the genus *Antechinus*, *Phascogale*, *Dasyurus*, and putatively *Didelphis*, in which puberty occurs during the first year of life, followed by spontaneous irreversible failure of spermatogenesis, a brief mating period, and then a physiologically cataclysmic death (known as male 'die-off') (Kerr & Hedger, 1983; Wilson & Bourne, 1984; Bradley, 2003; Woods & Hellgren, 2003; McAllan, 2009) (Figure 10.2). A second pattern is seen in species such as the ring-tailed possum (*Pseudocheirus peregrinus*) (Hughes, Thomson, & Owen, 1965; Phillips, McKinnon, Keeley, & Johnston, 2008), eastern quoll (Fletcher, 1985; Bryant, 1986), and greater glider (*Petauroides volans*) (Smith, 1969), in which there is an annual cycle of growth and regression of the testis with a complete cessation of spermatogenesis outside the breeding season (Figure 10.2). There are also seasonal changes in the weights of the prostate and epididymis (Smith 1969; Fletcher, 1985). The third group includes many of the macropods (Inns, 1982; Tyndale-Biscoe & Renfree, 1987), the brush-tailed possum (Gemmell, Cepon, & Barnes, 1986), the bandicoot (Todhunter & Gemmell, 1987), and the dunnarts (*S. crassicaudata* and *S. macroura*) (McAllan, 2003). Spermatogenesis continues throughout the year, and testis and epididymis weights do not show seasonal variation; however, there is a large seasonal change in the size of the prostate and Cowper's (bulbourethral) glands with maximal weight coincident with the mating season (Inns, 1982; Tyndale-Biscoe & Renfree, 1987; McAllan, 2003) (Figure 10.2).

7.2.1. Type 1: Seasonal reproductive cycle followed by complete spermatogenetic collapse

The uncoupling of T and spermatogenetic development can be seen best in male dasyurid marsupials, especially the semelparous (once in a lifetime reproduction) dasyurids from the genera *Antechinus* and *Phascogale*. Testosterone significantly affects the morphology of the male *Antechinus* reproductive tract. In eutherians, androgens are believed to be essential for promoting spermatogenesis and for early germ cell development and later spermiogenesis (McLachlan, Wredford, O'Donnell, De Kretser, & Robertson, 1996). Many marsupials display a marked effect of T on accessory reproductive tract growth (sternal glands, bulbourethral glands, and prostate), without any change in size

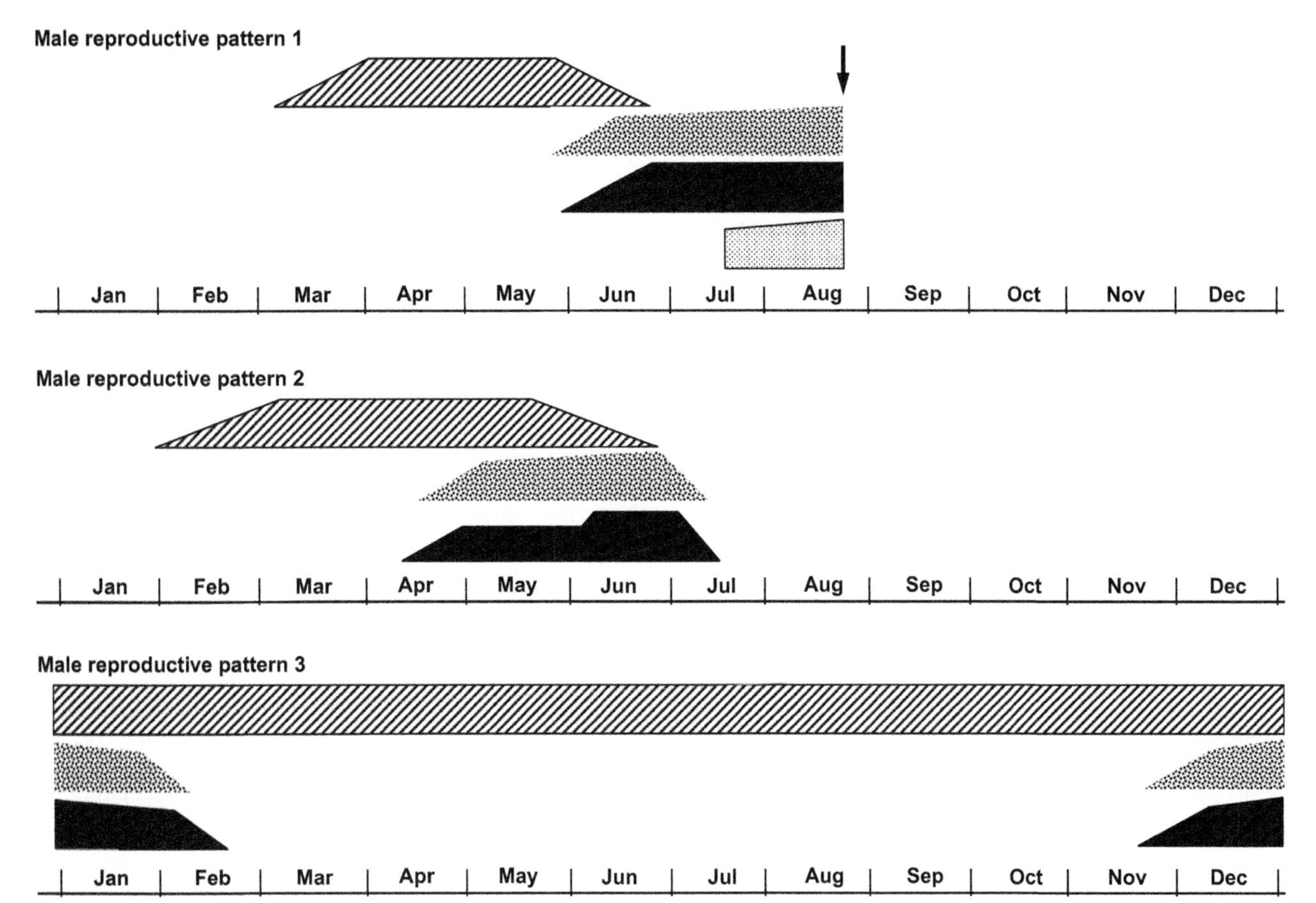

FIGURE 10.2 Patterns of reproductive cycles found in male marsupials. Diagonally striped boxes represent the period of the year during which spermatogenesis occurs; heavily speckled boxes represent the time of year that the accessory glands are enlarged and functional; black boxes represent the time of the year that plasma testosterone concentrations are raised. The lightly speckled box, which is present only for reproductive pattern 1, represents the time of year that plasma cortisol concentrations are raised in males. The arrow indicates the time of year of the abrupt male mortality seen in males exhibiting reproductive pattern 1. Males from patterns 2 and 3 usually live for several breeding seasons, and the reproductive pattern is repeated each year for these males. The schematic representation for pattern 3 is for the macropods such as *Macropus eugenii* and *Macropus rufogriseus*, which have shorter reproductive seasons than other species that exhibit this pattern.

or mass of the testes or effect on the spermatogenetic cycle (the brush-tailed possum (Cook, McDonald, & Gibson, 1978; Gemmell et al., 1986); the tammar (Inns, 1982; Jones, Stone, Hinds, & Setchell, 1988); the agile antechinus (Kerr & Hedger, 1983); the swamp antechinus (*Antechinus minimus*) (Wilson & Bourne, 1984); Bennett's wallaby (Curlewis, 1991); and the brown antechinus (McAllan, 1998)). Only in the type 1 male pattern is spermatogenesis completely uncoupled from the reproductive season. Spermatogenesis in *A. stuartii* has ceased by the time plasma T concentrations start to rise (McAllan, 1998). Testicular T concentrations may be 10 times higher than peripheral levels, which are sufficient to maintain testicular activity in rats (Rommerts, 1988; Walker & Cheung, 2005). Thus, in marsupials that display no relationship between peak plasma T levels and spermatogenesis, spermatogenetic activity may be regulated locally by androgens released from mature Leydig cells (McAllan, 1998).

In the agile antechinus, the gonad appears undifferentiated at three days postpartum and well-differentiated by four to five months of age (Taggart, Johnson, & Temple-Smith, 1993). The progression of spermatogenetic stages in the seminiferous tubules is synchronized within each population from about five months of age onwards (Kerr & Hedger, 1983). In *A. agilis*, plasma androgen concentrations become detectable at about five months of age (McDonald et al., 1981; Kerr & Hedger, 1983). Similar observations have been made in other studies on *Antechinus* (Woolley, 1966; Inns, 1975; McAllan, Roberts, & O'Shea, 1997).

Despite its surprising lack of effect on spermatogenetic activity in *A. stuartii*, an increase in plasma T concentration is coincident with the observation of male secondary sexual characteristics, including sternal gland secretions (sebaceous glands found lying over the sternum that secrete a musky sticky product in some marsupials), increase in the size of the bulbourethral glands and prostate, and pigmentation of the testes (Woolley, 1966; McAllan et al., 1997). Testosterone administration to immature males promotes the development of secondary sexual characteristics and also causes hypertrophy in cells in the caudal end of the epididymis but not the caput epithelial cells, significant alterations in mass and morphology of the prostate and bulbourethral glands, and development of a mature penis within 10 days of T injections (McAllan, 1998).

During the lifecycle of male *A. agilis A. stuartii,* circulating T begins to increase in late June (approximately the winter solstice) until it reaches an eight-fold peak in August (Bradley, McDonald, & Lee, 1980; Kerr & Hedger, 1983). Prior to the winter solstice, circulating T is basal (Bradley et al., 1980; Kerr & Hedger, 1983). The increase in circulating T coincides with increasingly aggressive, territorial behavior (Wood, 1970; Braithwaite, 1974; Bradley, McDonald, & Lee, 1976). Testosterone has significant anabolic effects in *Antechinus*, promoting an increase in body mass and a concomitant increase in food consumption as circulating T concentrations increase (McAllan et al., 1997; McAllan, Roberts, & O'Shea, 1998a). Besides the musculoskeletal system, T also affects other body systems, notably the renal system (McAllan et al., 1998a) and the accessory reproductive tract (McAllan, 1998). The accessory reproductive tract increases in mass as a result of T administration (McAllan, 1998).

In male, but not female, *A. stuartii*, renal structure and function exhibit seasonal changes, and many of the changes are correlated with circulating concentrations of T (McAllan, Roberts, & O'Shea, 1996; 1998b). In males, glomerular filtration rate (GFR) declines in July and August, coincident with the increase in circulating T (McAllan et al., 1998b). Testosterone administration also significantly decreases GFR. Outside the breeding season, the GFR of males is similar to that seen in female *A. stuartii* and to that found for other small mammals (McAllan et al., 1998b; Goldstein & Newland, 2004). Circulating T also may increase hematocrit in *A. stuartii* (McAllan et al., 1998a). Other studies have found a decrease in hematocrit in male *A. agilis* during the breeding season (Cheal, Lee, & Barnett, 1976; Barker, Beveridge, Bradley, & Lee, 1978), mostly due to blood losses as a result of gastric ulceration. Gastric ulceration is an effect of relatively high circulating free cortisol in males, neither of which occurs in females (McAllan et al., 1998a). Sex differences in hematocrit are known in humans and in brush-tailed possums but are considered uncommon in all mammals (Barnett, How, & Humphreys, 1979).

Seasonal rises in T are found in macropodid and phalangerid marsupials (see Section 7.2.3); however, the seasonal changes in T are associated with significant changes in free cortisol in several dasyurid marsupials. The increase of plasma T concentration in *Antechinus* and Phascogale males, which continues from June until death in August/September (Bradley et al., 1980; Bradley, 1987; 1990), is coincident with changes in the binding capacity of corticosteroid-binding globulins (CBGs) in the blood. When circulating CBGs decline, free or albumin-bound corticosteroids are more effective (Bradley et al., 1976; Lee, Bradley, & Braithwaite, 1977; Bradley et al., 1980). These effects also have been observed in the yellow-footed antechinus (*Antechinus flavipes*), the dusky antechinus (*Antechinus swainsonii*), and the red-tailed phascogale (McDonald, Lee, Than, & Martin, 1986; Bradley, 1987; 1990; 2003) but not in iteroparous (repeated reproductive bouts per lifetime) dasyurid marsupials such as *S. crassicaudata* (McDonald et al., 1981).

In the *Antechinus* genus and in other semelparous dasyurids studied, circulating plasma cortisol levels increase dramatically just prior to and during the breeding season (Bradley et al., 1980; McDonald et al., 1981; Bradley 1987; 2003). These relatively high plasma cortisol concentrations are causally related to male postbreeding mortality (Braithwaite 1974; Bradley, McDonald, & Lee, 1975; Barker et al., 1978; Bradley et al., 1980; Bradley 1987; 2003). The continuing rise in circulating cortisol levels is believed to be the result of a combination of a T-induced fall in CBG levels in the plasma and the failure of the glucocorticoid negative feedback mechanism mediated through the hypothalamus–pituitary–adrenal (HPA) axis (Bradley et al., 1976; 1980; McDonald et al., 1986; Bradley 1987; 2003). Cortisol administration to prebreeding males also causes a dose-dependent increase in mortality within five to six weeks of injection (Bradley et al., 1975; 1976; McAllan et al., 1997; McAllan, 1998).

A distinguishing feature of male senescence in *Antechinus* spp. is the complete collapse of the germinal epithelium of the testes before the males go through 'die-off' (Woolley, 1966; Inns, 1975; Kerr & Hedger, 1983; Wilson & Bourne, 1984; McAllan et al., 1997). Unlike many other marsupial and eutherian mammals, the spermatogenetic cycle in all *Antechinus* studied has only four stages dominating the seminiferous tubules at any point in time, not six (Woolley, 1966; Taylor & Horner, 1970; Inns, 1975; Calaby & Taylor, 1981; Kerr & Hedger, 1983; Wilson & Bourne, 1984; McAllan et al., 1997). Only one sequence is present in the seminiferous tubules at any point in time, differing from all other mammals studied, in which at least four to six stages are present either in adjacent tubules or in wave-like progressions along the same tubule when males are mature and ready for breeding (Woolley, 1966; Taylor & Horner, 1970; Inns, 1975; Calaby & Taylor, 1981; Kerr & Hedger, 1983; Wilson & Bourne, 1984; McAllan et al., 1997). Spermatogenesis is complete at the winter solstice, long before the breeding season begins, when plasma T concentrations are barely detectable. Male semelparity is independent of circulating T concentrations (Kerr & Hedger, 1983; McAllan et al., 1997; McAllan, Westman, & Joss, 2002). Germ cell maturation follows a similar progression to that in other mammals until May, after which depletion of the spermatogonia and early primary spermatocytes indicate spermatogenetic failure (Kerr & Hedger, 1983). In August, at the peak of circulating plasma T levels, the seminiferous tubules have collapsed because of the depletion of the germ cells, and sperm have disappeared from the lumen of each seminiferous tubule in *A. agilis* (Kerr & Hedger, 1983). Recruitment of sperm into the epididymis from the testes has ceased by mid-August (Taggart & Temple-Smith, 1990).

In other type 1 male marsupials, the spermatogenetic cycle and associated hormonal changes are not as dramatic as in *Antechinus* spp. The seasonal cycle of testicular activity in the Virginia opossum is well known (Winegarner, 1982); however, the possibility that they exhibit the semelparous life history has only recently been investigated (Woods & Hellgren, 2003). Unlike in the dasyurids, male mortality in this species is not associated with raised plasma cortisol concentrations (Woods & Hellgren, 2003). Other didelphids also may demonstrate a type 1 reproductive pattern, but data are sparse; species examined include the Brazilian gracile mouse opossum (*Gracilinanus microtarsus*) (Martins, Bonato, Da Silva, & Dos Reis, 2006), the short-tailed opossum (*Monodelphis dimidiata*) (Pine, Dalby, & Matson, 1985), and the grey slender mouse opossum (*Marmosa incana*) (Lorini, Oliveira, & Persson, 1994). The evidence for the latter two species is not derived from survival rate estimates but rather from the seasonal distribution of weights of specimens taken in the field (Pine et al., 1985; Lorini et al., 1994). The Brazilian gracile mouse opossum appears to undergo partial semelparity, where not all males succumb to physiological collapse and death at the end of the breeding season (Martins et al., 2006). Partial semelparity also has been observed in some of the dasyurids including the dibbler (*Parantechinus apicalis*) and the northern quoll (*D. hallucatus*) (see McAllan, 2009 for discussion).

7.2.2. Type 2: Seasonal reproductive cycle with testicular involution

The type 2 marsupial male reproductive cycle is most similar to that seen for many of the northern hemisphere rodents, in which a distinct seasonal reproductive period and the complete involution of the reproductive tract during the reproductive quiescent period is observed (Bronson, 1985). Testicular involution is followed by reduction in size and secretory activity of the accessory reproductive tract (Hughes et al., 1965; Smith, 1969; Fletcher, 1985; Bryant, 1986). For most of these species, histological examination definitively excludes year-round spermatogenetic activity, and associated circulating hormone levels are unknown. Plasma T has been reported only for the eastern quoll, with values ranging from 0.5–0.7 $ng \cdot mL^{-1}$ in the nonbreeding season to 5.0 $\pm$ 0.3 $ng \cdot mL^{-1}$ in the breeding season (Bryant, 1986).

7.2.3. Type 3: Seasonal reproductive cycle with continuous spermatogenetic activity

The third type of reproductive cycle is perhaps the most common for marsupials, with spermatogenetic activity continuing throughout the year and the accessory reproductive tract and plasma T concentrations showing seasonal patterns of change. Usually, however, the testis and epididymal masses remain unchanged throughout the year (Gilmore, 1969; Inns, 1982; Perret & Atramentowicz, 1989; Curlewis, 1991). For the tammar, seasonal changes have been described in some detail, with the histological cycles of both the testes and accessory reproductive tract determined in conjunction with seasonal cycles of circulating T (Inns, 1982). Testosterone concentrations range from 1–3 $ng \cdot mL^{-1}$ in the nonbreeding season to 6.0 $ng \cdot mL^{-1}$ in the breeding season and increases in the masses of the prostate gland and the bulbourethral glands (Cowper's glands) occur in parallel with increases in circulating T (Inns, 1982). Similar patterns are seen in the white-bellied opossum, including seasonal changes in Leydig cell volumes (De Quieroz & Nogueira, 1992; De Quieroz, Da Silva, & Nogueira, 1995).

In the brush-tailed possum, there are seasonal changes of plasma T and DHT, with both rising significantly during the breeding season (Curlewis & Stone 1985a; 1985b), although a recent study found no significant seasonal changes in these same hormones (Fink, McLeod, Assinder, Parry, & Nicholson, 2005). In conjunction with raised T and DHT are some of the largest changes in accessory reproductive tracts seen in any mammal, with the prostate enlarging to as much as 8 g/kg of body mass (Gilmore, 1969; Fink et al., 2005). The prostate of the brush-tailed possum also contains many MST receptors, and prostatic MST concentrations have been found to be highest immediately before the increases in prostate mass associated with the autumn and spring breeding periods. At this time, MST receptors were also present in the prostatic capsule in addition to those present in the glandular tissue (Fink et al., 2005). Mesotocin is present in the Leydig cells of brush-tailed possum testes (Bathgate et al., 1995) and in the prostate of the common bandicoot (Gemmell & Sernia, 1989). In the tammar prostate, a MST receptor with high sequence homology to the eutherian OXY receptor has been identified and this receptor has a similar distribution to that seen in the primate prostate, being present in the outer stromal capsule and the glandular epithelial cells (Parry & Bathgate, 1998).

Similar patterns have been found in Bennett's wallaby, in which there are seasonal changes in plasma T concentrations, from basal concentrations of $<1\ ng{\cdot}mL^{-1}$ to peaks of $4 \pm 1\ ng{\cdot}mL^{-1}$ in the breeding season (Curlewis, 1991). However, testes masses also showed a small but significant seasonal cycle in this species, although spermatogenesis was continuous (Curlewis, 1991). In the bandicoot, there is an increase in testicular interstitial cell volume but not testis mass during the time when T secretion is high (Todhunter & Gemmell, 1987). Associated with the changes in T levels in Bennett's wallaby are changes in pituitary PRL and LH, with both peaking before the T surge at the beginning of the breeding season and remaining high throughout the breeding season, although plasma circulating values remain low, presumably because of the pulsate nature of PRL and LH (Curlewis, 1991).

8. CONTROL OF TESTICULAR FUNCTION IN MONOTREMES AND MARSUPIALS

8.1. Monotremes

As for ovarian cycles in female monotremes, little is known about the regulation of testicular cycles by pituitary hormones in male monotremes, with only seasonal cycles of gonadal hormones occasionally described (Nicol et al., 2005).

8.2. Marsupials

Similarly to eutherians, testis function is regulated by LH in male marsupials, and although little is known about the role of FSH in testis function. Several studies have determined the role of LH in the stimulation of T production from the Leydig cells, mostly using the tammar (Hearn, 1975; Lincoln, 1978), but also the kowari (Fletcher, 1989b), the eastern quoll (Bryant, 1992), the Virginia opossum (Tyndale-Biscoe & Renfree, 1987), and Bennett's wallaby (Curlewis, 1991). Similarly to studies on females, examination of the role of GnRH is limited to stimulation of gonadal function for conservation biology. Gonadotropin-releasing hormone (or GnRH agonists) promotes T production in male koalas (Allen, McKinnon, Lisle, D'Occhio, & Johnston, 2006; Allen et al., 2008), the tammar (Herbert et al., 2004), the ring-tailed possum (Phillips et al., 2008), and the Virginia opossum (Johnston et al., 2008).

9. SUMMARY AND CONCLUSIONS

Many important aspects of monotreme reproduction remain unknown, with breeding in captivity remaining a rare and cryptic event. Data are so sparse that we cannot describe the reproductive cycles in females, nor fully describe the seasonal changes in the endocrinology of male marsupials. It is hoped that with the publication of the platypus genome more can be discovered about this unusual group of mammals. While more is known about the endocrinology of reproduction in marsupials, it is clear that there are many gaps in our knowledge. While more is known about the endocrine control of sexual differentiation and development, less is known about the regulation of reproduction. For example, we know little about the role that FSH plays in the control of reproduction in male marsupials, and, although data are better for females, information is patchy. Of particular concern is the lack of knowledge of the reproductive endocrinology of South American marsupials, where even basic information, e.g. regarding reproductive cycles, are poorly known for most species. The information is urgently needed for conservation purposes, especially for captive breeding programs.

ABBREVIATIONS

3β-HSD	3β-hydroxysteroid dehydrogenase
ACTH	Corticotropin
CBG	Corticosteroid-binding globulin
DHT	5α-dihydrotestosterone
E_2	Estradiol
FSH	Follicle-stimulating hormone
FSH-R	Follicle-stimulating hormone receptor
GFR	Glomerular filtration rate
GnRH	Gonadotropin-releasing hormone
hCG	Human chorionic gonadotropin
HPA	Hypothalamus–pituitary–adrenal
LH	Luteinizing hormone
MEL	Melatonin
MST	Mesotocin
NO	Nitric oxide
OXY	Oxytocin
P_4	Progesterone
PG	Prostaglandin
PMSG	Pregnant mare serum gonadotropin
PRL	Prolactin
T	Testosterone

REFERENCES

Allen, N. T., & Bradshaw, S. D. (1980). Diurnal variation in plasma concentrations of testosterone, 5α-dihydrotestosterone, and corticosteroids in the Australian brush-tailed possum, *Trichosurus vulpecula* (Kerr). *General and Comparative Endocrinology, 40*, 455–458.

Allen, C. D., Burridge, M., Chafer, M. L., Nicholson, V. N., Iago, S. C., Booth, R. J., et al. (2008). Control of the koala (*Phascolarctos cinereus*) anterior pituitary–gonadal axis with analogues of GnRH. *Reproduction, Fertility and Development, 20*, 598–605.

Allen, C. D., McKinnon, A. J., Lisle, A. T., D'Occhio, M. J., & Johnston, S. J. (2006). Use of a GnRH agonist and hCG to obtain an index of testosterone secretory capacity in the koala (*Phascolarctos cinereus*). *Journal of Andrology, 27*, 720–724.

Augee, M. L., Carrick, F. N., Grant, T. R., & Temple-Smith, P. D. (2008). Order Monotremata: Platypus and echidnas. In S. Van Dyck, & R. Strahan (Eds.), *The Mammals of Australia* (3rd ed.). (pp. 30–31) Sydney, Australia: Reed New Holland (New Holland Publishers Pty Ltd).

Ballantyne, K., Matson, P., Noakes, N., Nicolson, V., & Johnston, S. D. (2009). Reproductive physiology of the female greater bilby (*Macrotis lagotis* Thylacomyidae): evidence for a male-induced luteal phase. *Reproduction, Fertility and Development, 21*, 274–282.

Barker, I. K., Beveridge, I., Bradley, A., & Lee, A. K. (1978). Observations on spontaneous stress-related mortality among males of the dasyurid marsupial *Antechinus stuartii* Macleay. *Australian Journal of Zoology, 26*, 435–447.

Barnett, J. L., How, R. A., & Humphreys., W. F. (1979). Blood parameters in natural populations of *Trichosurus* species (Marsupialia: Phalangeridae) I. Age, sex and seasonal variation in *T. caninus* and *T. vulpecula*. *Australian Journal of Zoology, 27*, 913–926.

Bathgate, R. A. D., Parry, L. J., Fletcher, T. P., Shaw, G., Renfree, M. B., Gemmell, R. T., et al. (1995). Comparative aspects of oxytocin-like hormones in marsupials. *Advances in Experimental and Medical Biology, 395*, 639–655.

Beard, L. A., & Grigg, G. C. (2000). Reproduction in the short-beaked echidna, *Tachyglossus aculeatus*: Field observations at an elevated site in south-east Queensland. *Proceedings of the Linnean Society of New South Wales, 122*, 89–99.

Blackburn, D. G. (1993). Lactation—historical patterns and potential for manipulation. *Journal of Dairy Science, 76*, 3195–3212.

Bradley, A. J. (1987). Stress and mortality in the red-tailed phascogale, *Phascogale calura* (Marsupialia: Dasyuridae). *General and Comparative Endocrinology, 67*, 85–100.

Bradley, A. J. (1990). Failure of glucocorticoid feedback during breeding in the male red-tailed phascogale, *Phascogale calura* (Marsupialia: Dasyuridae). *Journal of Steroid Biochemistry and Molecular Biology, 37*, 155–163.

Bradley, A. J. (2003). Stress, hormones and mortality in small carnivorous marsupials. In M. Jones, C. Dickman, & M. Archer (Eds.), *Predators with Pouches: The Biology of Carnivorous Marsupials*. CSIRO publishers.

Bradley, A. J., McDonald, I. R., & Lee, A. K. (1975). Effects of Exogenous cortisol on mortality of a dasyurid marsupial. *Journal of Endocrinology, 66*, 281–282.

Bradley, A. J., McDonald, I. R., & Lee, A. K. (1976). Corticosteroid-binding globulin and mortality in a dasyurid marsupial. *Journal of Endocrinology, 70*, 323–324.

Bradley, A. J., McDonald, I. R., & Lee, A. K. (1980). Stress and mortality in a small marsupial (*Antechinus stuartii*, Macleay). *General and Comparative Endocrinology, 40*, 188–200.

Bradshaw, F. J., & Bradshaw, S. D. (1992). Estradiol-17ß secretion by the newly formed corpus luteum of a macropodid marsupial, the quokka (*Setonix brachyurus*). *General and Comparative Endocrinology, 87*, 425–435.

Braithwaite, R. W. (1974). Behavioural changes associated with the population cycle of *Antechinus stuartii* (Marsupialia). *Australian Journal of Zoology, 22*, 45–62.

Brinklow, B. R., & Loudon, A. S. I. (1989). Effects of exogenous prolactin and bromocriptine on the seasonal reproductive quiescence in the Bennett's wallaby, *Macropus rufogriseus rufogriseus*. *Journal of Endocrinology, 120*, 189–193.

Brinklow, B. R., & Loudon, A. S. I. (1993). Evidence for a circannual cycle of reproduction and prolactin in a seasonally reproducing macropodid marsupial, the Bennett's wallaby, *Macropus rufogriseus rufogriseus*. *Journal of Reproduction and Fertility, 98*, 625–630.

Bronson, F. H. (1985). Mammalian reproduction—an ecological Perspective. *Biology of Reproduction, 32*, 1–26.

Bryant, S. L. (1986). Seasonal variation of plasma testosterone in a wild population of male eastern quoll, *Dasyurus viverrinus* (Marsupialia: Dasyuridae) from Tasmania. *General and Comparative Endocrinology, 64*, 75–79.

Bryant, S. L. (1992). Testosterone-LH Response and Episodic Secretion in the male marsupial, *Dasyurus viverrinus*. *General and Comparative Endocrinology, 87*, 410–415.

Bryant, S. L., & Rose, R. W. (1986). Growth and role of the corpus luteum throughout delayed gestation in the potoroo, *Potorous tridactylus*. *Journal of Reproduction and Fertility, 7*, 409–414.

Butler, C. M., Harry, J. L., Deakin, J. E., Cooper, D. W., & Renfree, M. B. (1998). Development of the expression of the androgen receptor during virilization of the urogenital system of a marsupial. *Biology of Reproduction, 59*, 725–732.

Calaby, J. H., & Taylor, J. M. (1981). Reproduction in two marsupial-mice, *Antechinus bellus* and *A. bilarni* (Dasyuridae), of tropical Australia. *Journal of Mammalogy, 62*, 329–334.

Cake, M. H., Owen, F. J., & Bradshaw, S. D. (1980). Difference in concentration of progesterone in plasma between pregnant and non-pregnant quokkas (*Setonix brachyurus*). *Journal of Endocrinology, 84*, 153–158.

Cheal, P. D., Lee, A. K., & Barnett, J. L. (1976). Changes in the haematology of *Antechinus stuartii* (Marsupialia) and their association with male mortality. *Australian Journal of Zoology, 24*, 299–311.

Cook, B., McDonald, I. R., & Gibson, W. R. (1978). Prostatic function in the brush-tailed possum, *Trichosurus vulpecula*. *Journal of Reproduction and Fertility, 53*, 369–375.

Crawford, J. L., McLeod, B. J., Thompson, E. G., Hurst, P. R., Colbourne, L. E., Lun, S., et al. (1999). Plasma gonadotropin concentrations in the cyclic female brushtail possum (*Trichosurus vulpecula*). *General and Comparative Endocrinology, 116*, 73–80.

Cruz, Y. P., & Selwood, L. (1993). Uterine histology of the dasyurid marsupial, *Antechinus stuartii*: relationship with differentiation of the embryo. *Journal of Reproduction and Fertility, 99*, 237–242.

Curlewis, J. D. (1991). Seasonal changes in the reproductive organs and plasma and pituitary hormone content of the male Bennett's wallaby (*Macropus rufogriseus rufogriseus*). *Journal of Zoology London, 223*, 223–231.

Curlewis, J. D., & Stone, G. M. (1985a). Some effects of breeding season and castration on the prostate and epididymis of the brushtail possum, *Trichosurus vulpecula*. *Australian Journal of Biological Sciences, 38*, 313–326.

Curlewis, J. D., & Stone, G. M. (1985b). Peripheral androgen levels in the male brush-tail possum (*Trichosurus vulpecula*). *Journal of Endocrinology, 105*, 63–70.

Curlewis, J. D., White, A. S., & Loudon, A. S. I. (1987). The onset of seasonal quiescence in the Bennett's wallaby (*Macropus rufogriseus rufogriseus*). *Journal of Reproduction and Fertility, 80*, 119–124.

Czarny, N. A., Garnham, J. I., Harris, M. S., & Rodger, J. C. (2009). Comparison of the production, quality, and *in-vitro* maturation capacity of oocytes from untreated cycling and intermediate phase equine serum gonadotropin-treated fat-tailed dunnarts (*Sminthopsis crassicaudata*). *Reproduction, 138*, 23–31.

De Queiroz, G. F., & Nogueira, J. C. (1992). Duration of the cycle of the seminiferous epithelium and quantitative histology of the testis of the South American white-belly opossum (*Didelphis albiventris*), Marsupialia. *Reproduction Fertility and Development, 4*, 213–222.

De Queiroz, G. F., da Silva, M. N. F., & Nogueira, J. C. (1995). Testicular sperm reserve and plasma testosterone levels of the South American white-belly opossum (*Didelphis albiventris*), Marsupialia. *Mammalia, 59*, 255–261.

Druery, G. V., Shimmin, G. A., Taggart, D. A., Temple-Smith, P. T., Breed, W. G., McDonald, C. H., et al. (2007). Ovarian follicular superstimulation and oocyte maturation in the anoestrous southern hairy-nosed wombat (*Lasiorhinus latifrons*). *Animal Reproduction Science, 99*, 363–376.

Eckery, D. C., Juengel, J. L., Whale, L. J., Thomson, B. P., Lun, S., & McNatty, K. P. (2002a). The corpus luteum and interstitial tissue in a marsupial, the brushtail possum (*Trichosurus vulpecula*). *Molecular and Cellular Endocrinology, 191*, 81–87.

Eckery, D. C., Lun, S., Thomason, B. P., Chie, W. N., Moore, L. G., & Juengel, J. L. (2002b). Ovarian expression of messenger RNA encoding the receptors for luteinizing hormone and follicle-stimulating hormone in a marsupial, the brushtail possum (*Trichosurus vulpecula*). *Biology of Reproduction, 66*, 1310–1317.

Evans, S. M., Tyndale-Biscoe, C. H., & Sutherland, R. L. (1980). Control of gonadotropin secretion in the female tammar wallaby (*Macropus eugenii*). *Journal of Endocrinology, 86*, 13–23.

Fadem, B. H. (1985). Evidence for the activation of female reproduction by males in a marsupial, the gray short-tailed opossum *(Monodelphis domesticus)*. *Biology of Reproduction, 33*, 112–116.

Fadem, B. H. (1987). Activation of estrus by males in a marsupial: stimulus control and endocrine factors. *Biology of Reproduction, 36*, 328–332.

Ferner, K., Zeller, U., & Renfree, M. B. (2009). Lung development of monotremes: evidence for the mammalian morphotype. *The Anatomical Record, 292*, 190–201.

Fink, J. W., McLeod, B. J., Assinder, S. J., Parry, L. J., & Nicholson, H. D. (2005). Seasonal changes in mesotocin and localization of its receptor in the prostate of the brushtail possum (*Trichosurus vulpecula*). *Biology of Reproduction, 72*, 470–478.

Finlayson, G. R., Shimmin, G. A., Taggart, D. A., Skinner, J. F., Gilmore, A., & Paris, M. C. J. (2006). Oestrous cycle of captive southern hairy-nosed wombats (*Lasiorhinus latifrons*) in South Australia, Australia. *Animal Reproduction Science, 95*, 295–306.

Fleming, M. W., & Harder, J. D. (1981). Uterine histology and reproductive cycles in pregnant and non-pregnant opossums, *Didelphis virginiana*. *Journal of Reproduction and Fertility, 63*, 21–24.

Fletcher, T. P. (1985). Aspects of reproduction in the male eastern quoll, *Dasyurus viverrinus* (Shaw) (Marsupialia: Dasyuridae), with notes on polyoestry in the female. *Australian Journal of Zoology, 33*, 101–110.

Fletcher, T. P. (1989a). Plasma progesterone and body weight in the pregnant and non-pregnant kowari, *Dasyuroides byrnei* (Marsupialia: Dasyuridae). *Reproduction, Fertility and Development, 1*, 65–74.

Fletcher, T. P. (1989b). Luteinizing hormone in the kowari, *Dasyuroides byrnei* (Marsupialia: Dasyuridae), during the oestrous cycle and pregnancy, and the effects of gonadectomy in male and female. *Reproduction, Fertility and Development, 1*, 55–64.

Fletcher, T. P., & Renfree, M. B. (1988). Effects of corpus luteum removal on progesterone, oestradiol-17ß and LH in early pregnancy of the tammar, *Macropus eugenii*. *Journal of Reproduction and Fertility, 83*, 185–191.

Flint, A. P., & Renfree, M. B. (1982). Oestradiol-17β in the blood during seasonal reactivation of the diapausing blastocyst in a wild population of tammar wallabies. *Journal of Endocrinology, 95*, 293–300.

Foster, W. K., Caton, W., Thomas, J., Cox, S., & Taggart, D. A. (2008). Timing of births and reproductive success in captive red-tailed phascogales, *Phascogale calura*. *Journal of Mammalogy, 89*, 1136–1144.

Freyer, C., Zeller, U., & Renfree, M. B. (2003). The marsupial placenta: a phylogenetic analysis. *Journal of Experimental Zoology A, 299*, 59–77.

Freyer, C., Zeller, U., & Renfree, M. B. (2007). Placental function in two distantly related marsupials. *Placenta, 28*, 249–257.

Frappell, P. B., & McFarlane, P. M. (2006). Development of the respiratory system in marsupials. *Respiratory Physiology and Neurobiology, 154*, 252–267.

Geiser, F., & Masters, P. (1994). Torpor in relation to reproduction in the mulgara, *Dasycercus cristicauda*, (Dasyuridae: Marsupialia). *Journal of Thermal Biology, 19*, 33–40.

Geiser, F., McAllan, B. M., & Brigham, R. M. (2005). Daily torpor in a pregnant dunnart (*Sminthopsis macroura* Dasyuridae: Marsupialia). *Mammalian Biology, 70*, 117–121.

Gemmell, R. T. (1981). The role of the corpus luteum of lactation in the marsupial *Isoodon macrourus* (Marsupialia: Peramelidae). *General and Comparative Endocrinology, 44*, 13–19.

Gemmell, R. T. (1984). Plasma concentrations of progesterone and 13,14-dihydro-15-keto-prostaglandin $F_{2\alpha}$ during regression of the corpora lutea of lactation in the bandicoot, *Isoodon macrourus*. *Journal of Reproduction and Fertility, 72*, 295–299.

Gemmell, R. T. (1985). The effect of prostaglandin $F_{2\alpha}$ analog on the plasma concentration of progesterone in the bandicoot, *Isoodon macrourus* (Marsupialia: Peramelidae) during lactation. *General and Comparative Endocrinology, 57*, 405–410.

Gemmell, R. T. (1989). The persistence of the corpus luteum of pregnancy into lactation in the marsupial bandicoot *Isoodon macrourus*. *General and Comparative Endocrinology, 75*, 355–362.

Gemmell, R. T. (1995). A comparative study of the corpus luteum. *Reproduction, Fertility and Development, 7*, 303–312.

Gemmell, R. T. (2002). Induction of birth in the bandicoot (*Isoodon macrourus*) with prostaglandin and oxytocins. *Reproduction, 123*, 301–306.

Gemmell, R. T., & Nelson, J. (1988). Ultrastructure of the olfactory system of three newborn marsupial species. *Anatomical Record, 221*, 655–662.

Gemmell, R. T., & Sernia, C. (1989). The localization of oxytocin and mesotocin in the reproductive tract of the male marsupial bandicoot *Isoodon macrourus*. *General and Comparative Endocrinology, 75*, 103–109.

Gemmell, R. T., Johnston, G., & Barnes, A. (1985). Seasonal variation in plasma testosterone concentration in male marsupial bandicoot *Isoodon macrourus* in captivity. *General and Comparative Endocrinology, 59*, 184–191.

Gemmell, R. T., Cepon, G., & Barnes, A. (1986). Weekly variations in body weight and plasma testosterone concentrations in the captive male possum, *Trichosurus vulpecula*. *General and Comparative Endocrinology, 62*, 1–7.

Gilmore, D. P. (1969). Seasonal reproductive periodicity in the male Australian Brush-tailed possum, *(Trichosurus vulpecula)*. *Journal of Zoology (London), 157*, 75–98.

Glazier, A. M. (1998). Seasonal variation in ovarian response to pregnant mares serum gonadotrophin in the brushtail possum (*Trichosurus vulpecula*). *Reproduction, Fertility and Development, 10*, 499–503.

Glazier, A. M., & Molinia, F. C. (1998). Improved method of superovulation in monovulatory brushtail possums (*Trichosurus vulpecula*) using pregnant mares serum gonadotrophin-luteinizing hormone. *Journal of Reproduction and Fertility, 113*, 191–195.

Goldstein, D. L., & Newland, S. (2004). Water balance and kidney function in the least shrew (*Cryptotis parva*). *Comparative Biochemistry and Physiology A-Molecular & Integrative Physiology, 139*, 71–76.

Graves, J. A. M. (1990). Sex Chromosome Function in Marsupials and Monotremes. *Australian Journal of Zoology, 37*, 409–410.

Harder, J. D., & Fleming, M. W. (1981). Estradiol and progesterone profiles indicate a lack of endocrine recognition of pregnancy in the opossum. *Science, 212*, 1400–1402.

Harder, J. D., Hinds, L. A., Horn, C. A., & Tyndale-Biscoe, C. H. (1984). Oestradiol in follicular fluid and in utero–ovarian venous and peripheral plasma during parturition and postpartum oestrus in the tammar, *Macropus eugenii. Journal of Reproduction and Fertility, 72*, 551–558.

Harder, J. D., Hinds, L. A., Horn, C. A., & Tyndale-Biscoe, C. H. (1985). Effects of removal in late pregnancy of the corpus luteum, Graafian follicle or ovaries on plasma progesterone, oestradiol, LH, parturition and post-partum oestrus in tammar wallaby, *Macropus eugenii. Journal of Reproduction and Fertility, 75*, 449–459.

Haydon, L. J., Juengel, J. L., Thomson, B. P., & Eckery, D. C. (2008). Expression of mRNAs encoding oestrogen receptor (ER) α and ERβ, androgen receptor and progesterone receptor during gonadal and follicular development in the marsupial brushtail possum (*Trichosurus vulpecula*). *Reproduction, Fertility and Development, 20*, 335–349.

Hearn, J. P. (1975). The role of the pituitary in the reproduction of the male tammar wallaby, *Macropus eugenii. Journal of Reproduction and Fertility, 42*, 399–402.

Herbert, C. A., Trigg, T. E., Renfree, M. B., Shaw, G., Eckery, D. C., & Cooper, D. W. (2004). Effects of a gonadotropin-releasing hormone agonist implant on reproduction in a male marsupial, *Macropus eugenii. Biology of Reproduction, 70*, 1836–1842.

Herbert, C. A., Trigg, T. E., & Cooper, D. W. (2006). Fertility control in eastern grey kangaroos using the GnRH agonist deslorelin. 1. Effects on reproduction. *Wildlife Research, 33*, 41–46.

Hesterman, H., Jones, S. M., & Schwarzenberger, F. (2008a). Reproductive endocrinology of the largest Dasyurids: Characterization of ovarian cycles by plasma and fecal steroid monitoring. Part I. The Tasmanian devil (*Sarcophilus harrisii*). *General and Comparative Endocrinology, 155*, 245–254.

Hesterman, H., Jones, S. M., & Schwarzenberger, F. (2008b). Reproductive endocrinology of the largest Dasyurids: Characterization of ovarian cycles by plasma and fecal steroid monitoring. Part II. The spotted-tailed quoll (*Dasyurus maculatus*). *General and Comparative Endocrinology, 155*, 245–254.

Hinds, L. A. (1989). Plasma progesterone through pregnancy and the estrus cycle in the eastern quoll, *Dasyurus viverrinus. General and Comparative Endocrinology, 75*, 110–117.

Hinds, L. A., & Selwood, L. (1990). Plasma progesterone concentrations during pregnancy in the dasyurid marsupial, *Antechinus stuartii*: Relationship with differentiation in the embryo. *Reproduction, Fertility and Development, 2*, 61–70.

Hinds, L. A., & Smith, M. J. (1992). Evidence from plasma progesterone concentrations for male-induced ovulation in the brush-tailed bettong, *Bettongia penicillata. Journal of Reproduction and Fertility, 95*, 291–302.

Hinds, L. A., & Tyndale-Biscoe, C. H. (1982). Plasma progesterone levels in the pregnant and non-pregnant tammar, *Macropus eugenii. Journal of Endocrinology, 93*, 99–107.

Hinds, L. A., Reader, M., Wernberg-Mőller, S., & Saunders, N. (1992a). Hormonal evidence for induced ovulation *Monodelphis domesticus. Journal of Reproduction and Fertility, 95*, 302–312.

Hinds, L. A., Diggle, P. J., & Tyndale-Biscoe, C. H. (1992b). Effects of the ovary, sucking stimulus and season on the pattern of LH and FSH release in the female tammar *Macropus eugenii. Reproduction, Fertility and Development, 4*, 25–34.

Hinds, L. A., Fletcher, T. P., & Rodger, J. C. (1996). Hormones of oestrus and ovulation and their manipulation in marsupials. *Reproduction, Fertility and Development, 8*, 661–672.

Holland, N., & Jackson, S. M. (2002). Reproductive behaviour and food consumption associated with captive breeding in the platypus *Ornithorhyncus anatinus. Journal of Zoology (London), 256*, 279–288.

Hughes, R. L., & Carrick, F. N. (1978). Reproduction in female monotremes. *Australian Zoologist, 20*, 233–253.

Hughes, R. L., & Hall, L. S. (1998). Early development and embryology of the platypus. *Philosophical Transactions: Biological Sciences, 353* (1372), 1101–1114.

Hughes, R. L., Thomson, J. A., & Owen, W. H. (1965). Reproduction in natural populations of the Australian ringtail possum *Pseudocheirus peregrinus* (Marsupialia: Phalangeridae), in Victoria. *Australian Journal of Zoology, 13*, 383–406.

Ingram, J. L., Shaw, G., & Renfree, M. B. (1999). Cortisol in fetal fluids and the fetal adrenal at parturition in the tammar wallaby (*Macropus eugenii*). *Biology of Reproduction, 60*, 651–655.

Ingram, J. L., Shaw, G., & Renfree, M. B. (2001). Differential contractility and nitric oxide sensitivity in the gravid and non-gravid myometrium during late pregnancy in a marsupial. *Endocrinology, 142*, 2244–2251.

Inns, R. W. (1975). Some seasonal changes in *Antechinus flavipes* (Marsupialia: Dasyuridae). *Australian Journal of Zoology, 24*, 523–531.

Inns, R. W. (1982). Seasonal changes in the accessory reproductive system and plasma testosterone levels of the male tammar wallaby, *Macropus eugenii*, in the wild. *Journal of Reproduction and Fertility, 66*, 675–680.

Jackson, L. M., & Harder, J. D. (2000). Evidence for the spontaneous postlactational estrus in gray short-tailed opossums, *Monodelphis domestica. Biology of Reproduction, 62*, 1823–1827.

Johnston, S. D., Camacho, F. C., Carrillo, L., Guy, N., Govea, J., Martinez, O., et al. (2008). The development of a testosterone stimulation test in the Virginia opossum (*Didelphis virginiana*) and its use in evaluating deslorelin contraception. *Reproduction Fertility and Development, 20*, 563–569.

Johnston, S. D., McGowan, M. R., O'Callaghan, P., Cox, R., & Nicolson, V. (2000). Studies of the oestrous cycle, oestrus and pregnancy in the koala (*Phascolarctos cinereus*). *Journal of Reproduction and Fertility, 120*, 49–57.

Johnston, S. D., Nicolson, V., Madden, C., Logie, S., Pyne, M., Roser, A., et al. (2007). Assessment of reproductive status in male echidnas. *Animal Reproduction Science, 97*, 114–127.

Johnston, S. D., O'Callaghan, P., Nilsson., K., Tzipori, G., & Curlewis, J. D. (2004). Semen-induced luteal phase and identification of a LH surge in the koala (*Phascolarctos cinereus*). *Reproduction, 128*, 629–634.

Jones, R. C., Stone, G. M., Hinds, L. A., & Setchell, B. P. (1988). Distribution of 5α-reductase in the epididymis of the tammar wallaby (*Macropus eugenii*) and dependence of the epididymis on systemic testosterone and luminal fluids from the testis. *Journal of Reproduction and Fertility, 83*, 779–783.

Jones, R. C., Stone, G. M., & Zupp, J. (1992). Reproduction in the male echidna. In M. L. Augee. (Ed.), *Platypus and Echidnas* (pp. 115–126). Sydney, Australia: Royal Zoological Society of New South Wales.

Kerr, J. B., & Hedger, M. P. (1983). Spontaneous spermatogenic failure in the marsupial mouse *Antechinus stuartii* Macleay (Dasyuridae: Marsupialia). *Australian Journal of Zoology, 31*, 445–466.

Ladosky, W., & Calixto, S. L. (1984). Influence of estrogen and progesterone on the uterine Sensitivity *in vitro* to neuropituitary hormones in the Brazilian marsupial *Didelphis albiventris*: comparison with lactating animals. *General and Comparative Endocrinology, 53*, 69–77.

Lee, A. K., Bradley, A. J., & Braithwaite, R. W. (1977). Corticosteroid levels and male mortality in *Antechinus stuartii*. In B. Stonehouse, & D. Glimore (Eds.), *The Biology of Marsupials* (pp. 209–220). Baltimore, MD: University Park Press.

Lee, A. K., Woolley, P. A., & Braithwaite, R. W. (1982). Life history strategies of dasyurid marsupials. In M. Archer (Ed.), *Carnivorous Marsupials* (pp. 1–11). Sydney, Australia: Royal Zoological Society of New South Wales.

Lincoln, G. (1978). Plasma testosterone profiles in male macropodid marsupials. *Journal of Endocrinology, 77*, 347–351.

Lorini, M. L., Oliveira, J. A., & Persson, V. G. (1994). Annual age structure and reproductive patterns in *Marmosa incana* (Lund, 1841) (Didelphidae, Marsupialia). *Mammalian Biology, 59*, 65–73.

Loudon, A. S. I., Brinklow, B. R., Gulland, F. D., Boyle, J., & Flint, A. P. F. (1990). Roles of prolactin and the uterus in the control of luteal regression in the Bennett's wallaby (*Macropus rufogriseus rufogriseus*). *Reproduction Fertility and Development, 2*, 71–78.

Martins, E. G., Bonato, V., da Silva, C. Q., & Dos Reis, S. F. (2006). Partial semelparity in the neotropical didelphid marsupial *Gracilinanus microtarsus*. *Journal of Mammalogy, 87*, 915–920.

Mattiske, D., Pask, A. J., Shaw, J. M., & Shaw, G. (2002). Structure and expression of the follicle-stimulating hormone receptor gene in a marsupial, *Macropus eugenii*. *Molecular Reproduction and Development, 63*, 24–31.

McAllan, B. M. (1998). Effect of testosterone and cortisol administration on the reproductive tract of male *Antechinus stuartii* (Marsupialia). *Journal of Reproduction and Fertility, 112*, 199–209.

McAllan, B. M. (2003). Timing of reproduction in carnivorous marsupials. In Menna Jones, Chris Dickman, & Mike Archer (Eds.), *Predators with Pouches: The Biology of Carnivorous Marsupials* (pp. 147–164). CSIRO publishers.

McAllan, B. M. (2009). Reproductive parameters of post "die-off" male *Antechinus flavipes* and *Antechinus stuartii* (Dasyuridae: Marsupialia). *Australian Mammalogy, 31*, 17–23.

McAllan, B. M., Joss, J. M. P., & Firth, B. T. (1991). Phase delay of the natural photoperiod alters reproductive timing in the marsupial *Antechinus stuartii*. *Journal of Zoology (London), 225*, 633–646.

McAllan, B. M., Roberts, J. R., & O'Shea, T. (1996). Seasonal changes in the renal morphometry of *Antechinus stuartii* (Marsupialia: Dasyuridea). *Australian Journal of Zoology, 44*, 337–354.

McAllan, B. M., Roberts, J. R., & O'Shea, T. (1997). The seasonal changes in the reproductive anatomy of male *Antechinus stuartii* (Marsupialia: Dasyuridae). *Journal of Morphology, 231*, 266–275.

McAllan, B. M., Roberts, J. R., & O'Shea, T. (1998a). The effects of cortisol and testosterone on renal function in male *Antechinus stuartii* (Marsupialia). *Journal of Comparative Physiology [B], 168*, 248–256.

McAllan, B. M., Roberts, J. R., & O'Shea, T. (1998b). Seasonal changes in glomerular filtration rate in *Antechinus stuartii* (Marsupialia: Dasyuridae). *Journal of Comparative Physiology [B], 168*, 41–49.

McAllan, B. M., Westman, W., & Joss, J. M. P. (2002). The seasonal reproductive cycle of a marsupial, *Antechinus stuartii*: effects of oral administration of melatonin. *General and Comparative Endocrinology, 128*, 82–90.

McLeod, B. J., Hunter, M. G., Crawford, J. L., & Thompson, E. G. (1999). Follicle development in cyclic, anoestrous and FSH-treated brushtail possums *Trichosurus vulpecula*. *Animal Reproduction Science, 57*, 217–227.

McConnell, S. J., Tyndale-Biscoe, C. H., & Hinds, L. A. (1986). Change in duration of elevated concentrations of melatonin is the major factor in photoperiod response of the tammar, *Macropus eugenii*. *Journal of Reproduction and Fertility, 77*, 623–632.

McDonald, C. H., Taggart, D. A., Breed, W. G., Druery, G. V., Shimmin, G. A., Finlayson, G. R., et al. (2006). Effect of exogenous gonadotrophins on ovarian morphology and oocyte maturation in the southern hairy nosed wombat *Lasiohinus latifrons* during the breeding season. *Reproduction Fertility and Development, 18*, 477–494.

McDonald, I. R., Lee, A. K., Bradley, A. J., & Than, K. A. (1981). Endocrine changes in dasyurid marsupials with differing mortality Patterns. *General and Comparative Endocrinology, 44*, 292–301.

McDonald, I. R., Lee, A. K., Than, K. A., & Martin, R. W. (1986). Failure of glucocorticoid feedback in males of a population of small marsupials (*Antechinus swainsonii*) during the period of mating. *Journal of Endocrinology, 108*, 63–68.

McLachlan, R. I., Wredford, N. G., O'Donnell, L., De Kretser, D. M., & Robertson., D. M. (1996). The endocrine regulation of spermatogenesis: independent roles for testosterone and FSH. *Journal of Endocrinology, 148*, 1–9.

Menkhorst, E., Ezard, N., & Selwood, L. (2007). Induction of ovulation and natural oestrous cycling in the stripe-faced dunnart, *Sminthopsis macroura*. *Reproduction, 133*, 495–502.

Menkhorst, E., Hinds, L. A., & Selwood, L. (2009). Progesterone concentration in the marsupial *Sminthopsis macroura*: relationship with the conceptus, uterine glandular regeneration, and body weight. *Reproduction, 137*, 107–117.

Merchant, J. C. (1976). Breeding biology of the agile wallaby, *Macropus agilis* (Gould) (Marsupialia: Macropodidae), in captivity. *Australian Wildlife Research, 3*, 93–103.

Molinia, F. C., Gibson, R. J., Smedley, M. A., & Rodger, J. C. (1998). Further observations of the ovarian response of the tammar wallaby *Macropus eugenii* to exogenous gonadotrophins: an improved method for superovulation using FSH/LH. *Animal Reproduction Science, 53*, 253–263.

Morrow, G., & Nicol, S. C. (2009). Cool sex? Hibernation and reproduction overlap in the echidna. *PLoS One, 4*(6), 1–5.

Muths, E., & Hinds, L. A. (1996). Circulating levels of prolactin and progesterone in a wild population of red kangaroos (*Macropus rufus*) Marsupialia: Macropodidae. *General and Comparative Endocrinology, 101*, 317–322.

Nelson, J., Knight, R. M., & Kinghorn, C. (2003). Perinatal sensory and motor development in marsupials with special reference to the northern quoll, *Dasyurus hallucatus*. In M. Jones, C. Dickman, & M. Archer (Eds.), *Predators with Pouches: The Biology of Carnivorous Marsupials* (pp. 205–217). CSIRO publishers.

New, N. P., Jabukowski, J. M., Stone, G. M., & Jones, R. C. (1998). Seasonal pattern of androgen secretion in the male platypus, *Ornithorhynchus anatinus*, in the Upper Barnard River, New South Wales. *Australian Mammalogy, 20*, 8–9.

Nicol, S., & Andersen, N. A. (2006). Body temperature as an indicator of egg-laying in the echidna, *Tachyglossus aculeatus*. *Journal of Thermal Biology, 31*, 483–490.

Nicol, S., Andersen, N. A., & Jones, S. M. (2005). Seasonal variations in reproductive hormones in free-ranging echidnas (*Tachyglossus aculeatus*): Interaction between reproduction and hibernation. *General and Comparative Endocrinology, 144*, 204–210.

O, W. S., Short, R. V., Renfree, M. B., & Shaw, G. (1988). Primary genetic control of somatic sexual differentiation in a mammal. *Nature, 331*, 716–717.

Oates, J. E., Bradshaw, F. J., Bradshaw, S. D., Stead-Richardson, E. J., & Philippe, D. L. (2007). Reproduction and embryonic diapause in a marsupial: Insights from captive female honey possums, *Tarsipes rostratus* (Tarsipedidae). *General and Comparative Endocrinology, 150*, 445–461.

Oftedal, O. T. (2002). The mammary gland and its origin during synapsid evolution. *Journal of Mammary Gland Biology and Neoplasia, 7*, 225–252.

Parry, L. J., & Bathgate, R. A. D. (1998). Mesotocin receptor gene and protein expression in the prostate gland, but not testis, of the tammar wallaby, *Macropus eugenii. Biology Reproduction, 59*, 1101–1107.

Parry, L. J., Bathgate, R. A. D., Shaw, G., & Renfree, M. B. (1997). Evidence for a local fetal influence on myometrial oxytocin receptors during pregnancy in the tammar wallaby (*Macropus eugenii*). *Biology of Reproduction, 56*, 200–207.

Perret, M., & Atramentowicz, M. (1989). Plasma concentrations of progesterone and testosterone in captive woolly opossums (*Caluromys philander*). *Journal of Reproduction and Fertility, 85*, 31–41.

Perret, M., & Ben M'Barek, S. (1991). Male influence on oestrous cycles in the female woolly opossum, *Caluromys philander*. *Journal of Reproduction and Fertility, 91*, 557–566.

Pine, R. H., Dalby, P. L., & Matson, J. O. (1985). Ecology, postnatal development, morphometrics, and taxonomic status of the short-tailed opossum, *Monodelphis dimidiata*, an apparently semelparous annual marsupial. *Annals of Carnegie Museum, 54*, 195–231.

Phillips, D. J., McKinnon, A., Keeley, T., & Johnston, S. D. (2008). Testosterone secretion, testicular histology and the cryopreservation of cauda epididymidal spermatozoa in the common ringtail possum (*Pseudocheirus peregrinus*). *Reproduction Fertility and Development, 20*, 391–401.

Renfree, M. B. (1972). Influence of the embryo on the marsupial uterus. *Nature, 240*, 475–477.

Renfree, M. B. (1979). Initiation of development of diapausing embryo by mammary denervation during lactation in a marsupial. *Nature, 278*, 549–551.

Renfree, M. B. (1983). Marsupial reproduction: the choice between placentation and lactation. In C. A. Finn (Ed.), *Oxford Reviews of Reproductive Biology 5* (pp. 1–29). Oxford, UK: Clarendon Press.

Renfree, M. B. (2006). Life in the pouch: womb with a view. *Reproduction Fertility and Development, 18*, 721–734.

Renfree, M. B., & Shaw, G. (2000). Diapause. *Annual Reviews in Physiology, 62*, 353–375.

Reufree, M. B., Flint, A. P. F., Green, S. W., & Heap, R. B. (1984). Ovarian steroid metabolism and estrogens in the corpus luteum of the tammar wallaby. *J. Endocrin., 101*, 23–24.

Renfree, M. B., Wilson, J., Short, R. V., Shaw, G., & George, F. (1992). Steroid hormone content of the gonads of the tammar wallaby during sexual differentiation. *Biology of Reproduction, 47*, 644–647.

Renfree, M. B., O, W.S., Short, R. V., & Shaw, G. (1996). Sexual differentiation of the urogenital system of the fetal and neonatal tammar wallaby. *Anatomy and Embryology, 194*, 111–134.

Renfree, M. B., Pask, A. J., & Shaw, G. (2001). Sex down under: the differentiation of sexual dimorphisms during marsupial development. *Reproduction Fertility and Development, 13*, 679–690.

Reynolds, M. L., Cavanaugh, M. E., Dziegielewska, K. M., Hinds, L. A., Saunders, N. R., & Tyndale-Biscoe, C. H. (1985). Postnatal development of the telencephalon of the tammar wallaby (*Macropus eugenii*): An accessible model of neocortical differentiation. *Anatomy and Embryology, 173*, 81–94.

Rodger, J. C., Giles, I., & Mate, K. E. (1992a). Unexpected oocyte growth after follicular antrum formation in four marsupial species. *Journal of Reproduction and Fertility, 96*, 755–763.

Rodger, J. C., Breed, W. G., & Bennett, J. H. (1992b). Gonadotrophin-induced oestrus and ovulation in the polyovulatory marsupial *Sminthopsis crassicaudata*. *Reproduction Fertility and Development, 4*, 145–152.

Rommerts, F. F. G. (1988). How much androgen is required for maintenance of spermatogenesis? *Journal of Endocrinology, 116*, 7–9.

Rose, R. W. (1992). Maternal recognition of pregnancy in the Tasmanian bettong, Bettongia gaimardi (Marsupialia: Macropodoidea). *Reproduction Fertility and Development, 4*, 35–41.

Rudd, C. D., Short, R. V., McFarlane, J. R., & Renfree, M. B. (1999). Sexual differentiation of oestradiol-LH positive feedback in a marsupial. *Journal of Reproduction and Fertility, 115*, 269–274.

Russell, A. J., Gilmore, D. P., Mackay, S., Ullmann, S. L., Baker, P. J., & Payne, A. P. (2003). The role of androgens in development of the scrotum in the grey short-tailed Brazilian opossum (*Monodelphis domestica*). *Anatomy and Embryology, 206*, 381–389.

Sebastian, L. T., De Matteo, L., Shaw, G., & Renfree, M. B. (1998). Mesotocin receptors during pregnancy, parturition and lactation in the tammar wallaby. *Animal Reproduction Science, 51*, 57–74.

Selwood, L. (1980). A timetable of embryonic development of the dasyurid marsupial, *Antechinus stuartii* (Macleay). *Australian Journal of Zoology, 28*, 649–668.

Selwood, L. (1981). Delayed embryonic development in the dasyurid marsupial, *Antechinus stuartii*. *Journal of Reproduction and Fertility Supplement, 29*, 79–82.

Selwood, L., & Woolley, P. A. (1991). A timetable of embryonic development and ovarian and uterine development in the strip-faced dunnart *Sminthopsis macroura* (Marsupialia: Dasyuridae). *Journal of Reproduction and Fertility, 91*, 213–227.

Sharman, G. B. (1970). Reproductive physiology of marsupials. *Science, 167*, 1221–1228.

Shaw, G., & Renfree, M. B. (1984). Concentrations of oestradiol-17ß in plasma and corpora lutea throughout pregnancy in the tammar, *Macropus eugenii*. *Journal of Reproduction and Fertility, 72*, 29–37.

Shaw, G., & Renfree, M. B. (2001). Fetal control of parturition in marsupials. *Reproduction Fertility and Development, 13*, 653–659.

Shaw, G., & Renfree, M. B. (2006). Parturition and perfect prematurity: birth in marsupials. *Australian Journal of Zoology, 54*, 139–149.

Shaw, G., Renfree, M. B., & Short, R. V. (1990). Primary genetic control of sexual differentiation in marsupials. *Australian Journal of Zoology, 37*, 443–450.

Shaw, G., Renfree, M. B., Leihy, M. W., Shackleton, C. H. L., Roitman, E., & Wilson, J. D. (2000). Prostate formation in a marsupial is mediated by the androgen 5α-androstane-3 α, 17β-diol. *Proceedings of the National Academy of Sciences USA, 97*, 12256–12259.

Shaw, G., & Rose, R. W. (1979). Delayed gestation in the potoroo *Potorous tridactylus* (Kerr). *Australian Journal of Zoology, 27*, 901–912.

Shorey, C. D., & Hughes, R. L. (1973). Cyclical changes in the uterine endometrium and peripheral plasma concentrations of progesterone in the marsupial *Trichosurus vulpecula*. *Australian Journal of Zoology, 21*, 1–19.

Short, R. V., Flint, A. P. F., & Renfree, M. B. (1985). Influence of passive immunization against GnRH on pregnancy and parturition in the tammar wallaby, *Macropus eugenii*. *Journal of Reproduction and Fertility, 75*, 567–575.

Smith, R. F. C. (1969). Studies on the marsupial glider, *Schoinobates volans* (Kerr). 1. Reproduction. *Australian Journal of Zoology, 17*, 625–636.

Smith, M. J., & How, C. A. (1973). Reproduction in the mountain possum, *Trichosurus caninus* (Ogilby) in captivity. *Australian Journal of Zoology, 21*, 321–329.

Stead-Richardson, E. J., Bradshaw, S. D., Bradshaw, F. J., & Gaikhorst, G. (2001). Monitoring the oestrous cycle of the chuditch (*Dasyurus geoffroii*) (Marsupialia: Dasyuridae): non-invasive analysis of faecal oestradiol-17β. *Australian Journal of Zoology, 49*, 183–193.

Sutherland, R. L., Evans, S. M., & Tyndale-Biscoe, C. H. (1980). Macropodid marsupial luteinizing hormone: validation of assay procedures and changes in plasma levels during the oestrous cycle in the female tammar wallaby (*Macropus eugenii*). *Journal of Endocrinology, 86*, 1–12.

Taggart, D. A., & Temple-Smith, P. D. (1990). Effects of breeding season and mating on total number and distribution of spermatozoa in the epididymis of the brown marsupial mouse, *Antechinus stuartii*. *Journal of Reproduction and Fertility, 88*, 81–91.

Taggart, D. A., Johnson, J., & Temple-Smith, P. D. (1993). Testicular and epididymal development in the brown marsupial mouse, *Antechinus stuartii* (Dasyuridae, Marsupialia). *Anatomy and Embryology, 188*, 87–100.

Taylor, J. M., & Horner, B. E. (1970). Gonadal activity in the marsupial mouse, *Antechinus bellus*, with notes on other species of the genus (Marsupialia: Dasyuridae). *Journal of Mammalogy, 51*, 659–668.

Temple-Smith, P. D., & Grant, T. R. (2001). Uncertain Breeding: A short history of reproduction in Monotremes. *Reproduction Fertility and Development, 13*, 487–497.

Todhunter, R., & Gemmell, R. T. (1987). Seasonal changes in the reproductive tract of the male marsupial bandicoot *Isoodon macrourus*. *Journal of Anatomy, 154*, 173–186.

Tyndale-Biscoe, H., & Renfree, M. B. (1987). *Reproductive Physiology of Marsupials*. Cambridge, UK: Cambridge University Press.

Tyndale-Biscoe, H., Hinds, L. A., Horn, C. A., & Jenkin, G. (1983). Hormonal changes at oestrus, parturition and post-partum oestrus in the tammar wallaby (*Macropus eugenii*). *Journal of Endocrinology, 96*, 155–161.

Walker, W. H., & Cheng, J. (2005). FSH and testosterone signaling in Sertoli cells. *Reproduction, 130*, 15–28.

Wallis, M. C., Waters, P. D., & Graves, J. A. M. (2008). Sex determination in mammals—before and after the evolution SRY. *Cellular and Molecular Life Sciences, 65*, 3185–3192.

Ward, S. J. (1990). Reproduction in the western pygmy-possum, *Cercartetus concinnus* (Marsupialia: Burramyidae), with notes on reproduction of some other small possum species. *Australian Journal of Zoology, 38*, 423–438.

West, M., Galloway, D., Shaw, J., Trounson, A., & Paris, M. C. J. (2004). The oestrous cycle of the common wombat, *Vombatus ursinus* in Victoria, Australia. *Reproduction, Fertility and Development, 16*, 336–349.

West, M., Lacham-Kaplan, O., Cleary, M., Galloway, D., Shaw, J., Trounson, A. O., et al. (2007). *In vitro* maturation and intracytoplasmic sperm injection of oocytes collected from hormonally stimulated common wombats, *Vombatus ursinus*. *Animal Reproduction Science, 98*, 311–321.

Wilson, B. A. (1986). Reproduction in the female dasyurid *Antechinus minimus maritimus* (Marsupialia: Dasyuridae). *Australian Journal of Zoology, 34*, 189–197.

Wilson, B. A., & Bourne, A. R. (1984). Reproduction in the male dasyurid *Antechinus minimus minimus* (Marsupialia: Dasyuridae). *Australian Journal of Zoology, 32*, 311–318.

Wilson, J. D., Leihy, M. W., Shaw, G., & Renfree, M. B. (2002). Androgen physiology: unsolved problems at the millennium. *Molecular and Cellular Endocrinology, 198*, 1–5.

Wilson, J. D., Renfree, M. B., Auchus, R. J., Pask, A. J., & Shaw, G. (2009). Formation of 5α-reduced androgens in the testes and urogenital tract of the grey short-tailed opossum, *Monodelphis domestica*. *Reproduction, Fertility and Development, 21*, 649–654.

Winegarner, M. S. (1982). Seasonal changes in the reproductive tract of the male opossum *Didelphis virginiana* Kerr in Florida. *American Midland Naturalist, 107*, 258–261.

Wood, D. H. (1970). An ecological study of *Antechinus stuartii* (Marsupialia) in a south-east Queensland rain forest. *Australian Journal of Zoology, 18*, 185–207.

Woods, H. A., & Hellgren, E. C. (2003). Seasonal changes in the physiology of male Virginia opossums (*Didelphis virginiana*). *Physiological and Biochemical Zoology, 76*, 406–417.

Woolley, P. (1966). Reproduction in *Antechinus* spp. and other dasyurid marsupials. *Symposia of the Zoological Society of London, 15*, 281–294.

Xie, Q., Mackay, S., Ullmann, S. L., Gilmore, D. P., & Payne, A. P. (1996). Testis development in the opossum *Monodelphis domestica*. *Journal of Anatomy, 189*, 393–406.

Xie, Q., Mackay, S., Ullmann, S. L., Gilmore, D. P., Payne, A. P., & Gray, C. (1998). Post natal development of Leydig cells in the opossum (*Monodelphis domestica*): an immunohistological and endocrinological study. *Biology of Reproduction, 58*, 664–669.

Zeller, U. (1999). Mammalian reproduction: Origin and evolutionary transformations. *Zoologischer Anzeiger, 238*, 117–131.

Chapter 11

Hormones and Reproductive Cycles in Rodents

Karen L. Bales and Caroline M. Hostetler
University of California, Davis, CA, USA

SUMMARY
In this chapter we examine the ecological and social factors affecting reproductive cycles in rodents, with concentration on field data and unusual species whenever possible. We present the literature regarding the effects of food availability, photoperiod, water availability, temperature, and social cues, and include a brief review of multiannual population cycles. Special attention is paid to the comparative literature and to sex differences. We also discuss new directions in the field (e.g., the effects of personality on reproductive success in rodents), and make suggestions for future research.

1. INTRODUCTION

'Reproductive cycles in rodents' could be interpreted in a number of different ways; it could relate to the female estrous cycle, male reproductive cycles, seasonal breeding, hibernation and emergence, multiannual cycles of animal density, etc. The understanding of how and why these cycles exist requires integration of both ultimate and proximate factors, and many fields including ecology, evolution, natural history, physiology, and neuroscience. The goal of this chapter is to review the current literature on rodent cycles, with a focus on those questions not normally covered in a review of the female rat estrous cycle. Estrous cycles in the most common laboratory rodents, *Rattus* and *Mus*, have been described excellently elsewhere in the literature (Long & Evans, 1922). Therefore, here we concentrate on more unusual rodent species, and on field data where possible.

In his classic paper on the ecology of mammalian reproduction, Bronson (1985) identified both a lack of information on small-bodied mammals (such as rodents), especially those that live in the tropics, and a need for a unified understanding of how the environment affects reproduction. The most important ecological cues for mammalian reproduction in the wild are food availability, temperature, rainfall, photoperiodicity, and social factors (Bronson, 1985; 1989). These are still the topics of most interest to behavioral ecologists that study mammal reproduction. Tropical and temperate rodent reproduction might be expected to vary on a number of parameters including cycles, seasonality, hibernation, litter size, etc. However, the variability of reproductive strategies *within* the tropics is nearly as wide as that between tropic and temperate zones, rendering generalizations almost impossible to make (Bronson, 1989). In particular, although ambient temperature may vary less throughout the year in the tropics, rainfall may vary and result in changing food availability and seasonal reproduction (Bronson, 1988). In this chapter, we will examine the relationship of these proximate cues to reproductive cycles in rodents.

2. FOOD AVAILABILITY

Food availability is clearly a primary determinant of the ability to reproduce. Bronson (1985; 1989) references the results of many studies in which food supplementation of wild populations improved reproduction. Food availability often interacts with photoperiod to influence reproductive readiness. In the marsh rice rat (*Oryzomys palustris*), adult males kept on short days showed no effect of food restriction on testes, seminal vesicles, or testosterone (T) levels; however, on long days food restriction reduced the testes and seminal vesicle masses (Edmonds, Riggs, & Stetson, 2003). Food restriction in California mice (*Peromyscus californicus*) interacts with photoperiod and temperature (Demas & Nelson, 1998) to affect male reproductive function. Under long days, food restriction reduced testes mass at both mild and low temperatures whereas, under short days, food restriction reduced testes mass only at low temperatures.

Not every study has shown rodent reproduction to be constrained by food availability. In a field experiment on wild house mice (*Mus musculus*) in Australia, Ylonen and

Hormones and Reproduction of Vertebrates, Volume 5—Mammals

colleagues supplemented certain farms with high-protein food pellets. However, they found no differences in any reproductive parameters between supplemented and non-supplemented farms (Ylonen, Jacob, Runcie, & Singleton, 2003). The authors suggest that, in the arid environment of Australia, water rather than food may be the factor limiting reproduction. In field voles (*Microtus agrestis*), food supplementation increased reproduction in the winter, but not in the summer (Fey, Banks, & Korpimaki, 2008). Reproduction of female Cape ground squirrels (*Xeris inauris*) was compared at a high resource site and a low resource site. Length of gestation, interbirth intervals, juvenile body mass, juvenile emergence, and litter size did not differ between the two sites (Pettitt, Waterman, & Wheaton, 2008). The relationship between reproduction and overall food availability may be obscured if reproduction is triggered only by a specific food, such as beech mast in fat dormice (*Myoxis glis*) (Pilastro, Tavecchia, & Marin, 2003).

2.1. Neuroendocrine Mechanisms Regulating the Reproductive Response to Food Availability

In the last decade, and even in the last few years, our knowledge of the neuroendocrine mechanisms by which food availability regulates reproduction has grown exponentially. Previously, there was a consensus that nutritional infertility in mammals was due to low levels of fat stores (the 'critical body fat hypothesis') (Frisch & MacArthur, 1974), and that cycles would be restored when sufficient levels of fat were achieved. However, it has become apparent that the mechanisms for energy-related regulation of feeding and secretion of the pituitary hormone luteinizing hormone (LH) are more rapid than the time required for a response to increase stored body fat in females (Wade & Jones, 2003; Furman & Wade, 2007), and also different from those affecting sexual behavior. The current 'metabolic fuels' hypothesis suggests that both reproductive function and feeding behavior are regulated by short-term changes in glucose availability (Wade & Jones, 2003; Schneider, 2004). Recent discoveries also have included several new peripheral hormones that appear to modulate hunger and satiety (e.g., leptin, ghrelin, obestatin) (Bouret, Draper, & Simerly, 2004; Alexe, Syridou, & Petridou, 2006; Cummings, 2006) via actions on central neurons expressing CART (cocaine- and amphetamine-related transcript), POMC (pro-opiomelanocortin), AgRP (agouti-related peptide), and NPY (neuropeptide Y). These are also candidate hormones to regulate motivational interactions between hunger and sexual behavior.

Gonadotropin-releasing hormone (GnRH) from the hypothalamus stimulates the release of LH from the pituitary, which in turn stimulates androgen synthesis in the ovary and eventually causes ovulation. Gonadotropin-releasing hormone also stimulates release of follicle-stimulating hormone (FSH), which in turn regulates oocyte growth and the synthesis of estrogens from androgens. The hypothalamus responds quickly to a reduction in available glucose, which is the primary metabolic cue that inhibits GnRH release (Wade & Jones, 2003; Schneider, 2004). Refeeding of fasted animals can also quickly reinstate gonadotropin (GTH) release. The mechanism of detection for glucose levels is likely in the area postrema of the hindbrain (Schneider & Zhu, 1994; Ritter, Dinh, & Zhang, 2000), which has direct neural connections to the frontal cortex (Sawchenko & Swanson, 1982a; 1982b; Loewy, 1990). In contrast to these quick effects on LH, it takes extended treatment with glucose inhibitors, or extended fasting, to affect estrous behaviors, and these effects likely involve hormone modulators. Corticotropin-releasing hormone (CRH), and the related hormone urocortin, may play a critical role in the suppression of estrous behavior during food deprivation (Wade & Jones, 2003), although they are not the only hormonal candidates (Schneider, 2004).

Leptin, a hormone produced by fat cells (Friedman & Halaas, 1998), has been postulated to act as a long-term signal of energy balance and to decrease feeding, and is undeniably an important developmental signal for neurons of the arcuate nucleus of the hypothalamus, a key nucleus involved in feeding behavior (Bouret et al., 2004; Simerly, 2008). However, there are no leptin receptors on GnRH neurons, so effects on that system must be mediated indirectly, perhaps by kisspeptin (Chan & Mantzoros, 2001; Smith, Acohido, Clifton, & Steiner, 2006a; Tena-Sempere, 2007) (see Chapter 2, this volume). Ghrelin, on the other hand, is produced in the brain and in the stomach and is orexigenic (Van der Lely, Tschop, Heiman, & Ghigo, 2004). Centrally administered ghrelin can suppress GnRH (Furuta, Funabashi, & Kimura, 2001). It is also possible that high levels of ghrelin associated with low food availability could inhibit growth of embryos and reduce litter size (Kawamura et al., 2003). Both leptin and ghrelin also may have direct effects on the gonads (Fernandez-Fernandez et al., 2006; Tena-Sempere, 2007).

Many field studies rely on postmortem measurements of testicular or ovarian organ mass to detect changes in reproduction due to food availability. However, even at a level of food restriction that does not result in reduction of testicular mass, male prairie voles (*Microtus ochrogaster*) showed increases in immuno-labeled GnRH neuronal cell bodies from the medial septum to the anterior hypothalamus, as well as increased labeling of fibers in the median eminence, indicating a reduction in the release of GnRH (Kriegsfeld, Ranalli, Trasy, & Nelson, 2001). Testicular regression in response to limited food restriction and photoperiod may be mediated by apoptosis (Young, Zirkin, & Nelson, 2000).

2.1.1. Developmental effects of food availability

Food restriction of the dam, or of the developing young, can have lifelong effects. In juvenile male marsh rice rats that were food-restricted to 60% of ad lib from age three to eight weeks, seminal vesicle and testes masses were reduced relative to ad lib and 80% of ad lib feeding (Edmonds et al., 2003). Daughters of wild adult house mice (*M. musculus*) that were food-restricted did not produce fewer litters and showed no difference in birth or weaning weights, but did wean slightly fewer pups in their second litter (Meikle & Westberg, 2001a). Sons of food-deprived mothers, in contrast, had lighter seminal vesicles and preputial glands, but only if they had undergone previous aggressive encounters (Meikle & Westberg, 2001b). They were less likely to be dominant in aggressive encounters, although it was not necessary for the male to be the loser in the encounter in order to experience the reduction in organ weight.

2.1.2. Hibernation

Rodents that hibernate, such as ground squirrels and marmots, have developed a number of physiological adaptations, primarily the ability to drop body temperature and metabolic rate (Lyman, Willis, Malan, & Wang, 1982). These animals require specialized lipid storage and utilization mechanisms for each phase of their annual cycle, including deposition of fat prior to hibernation, a period (torpor) in which temperature and metabolism drop, and re-emergence. Torpor is broken up with periodic short bursts of arousal. Successful hibernation is dependent on reserves of both white and brown adipose tissue. Lipid metabolism changes drastically, in an apparently programmed way, during each phase (Dark, 2005).

There are fascinating sex differences in the way that hibernation interacts with reproduction. In most species, females and nonreproductive (juvenile) males emerge from hibernation at their lowest body mass, when food becomes environmentally available. In contrast, reproductive males (which usually have larger white adipose tissue reserves) emerge several weeks earlier, before food is available, fully reproductively prepared and with spermatogenesis complete (Michener, 1992; Dark, 2005; Strauss, Mascher, Palme, & Millesi, 2007). The termination of hibernation appears to be mediated by rises in T, which occur across periodic arousals (Barnes, Kretzmann, Zucker, & Licht, 1988). Termination of hibernation is also dependent upon reserves of white adipose tissue (Dark, 2005). The choice of whether or not to reproduce as a juvenile can be an important life-history determinant of reproductive success; juvenile males that attempt to breed in their first year can be involved in more aggressive interactions and end the breeding season at lower body masses than nonreproductive juveniles (Strauss et al., 2007).

In hibernating rodents, food availability interacts with timing mechanisms. One of the most fascinating questions concerns how the animal's internal circadian clock adapts to deep hibernation. A recent study showed that the clock 'stops ticking,' or, more specifically, that an enzyme crucial to the regulation of melatonin (MEL) stops exhibiting circadian changes, during hibernation (Revel et al., 2007b). Clearly, there is regional brain activity specific to different phases of hibernation that regulates torpor, periodic arousals, and emergence (Bratincsak et al., 2007), including continued activity of the suprachiasmatic nucleus (SCN) or 'biological clock.'

This is a quickly moving area of research, with recent advances both in molecular techniques (Andrews, 2007) and in technology used to study hibernation in the field (Long, Hut, & Barnes, 2007). Long and colleagues used light-sensitive radiotransmitters and implanted, temperature-sensitive dataloggers to record body temperature and activity data in wild arctic ground squirrels (*Spermophilus parryii*).

3. PHOTOPERIOD

3.1. Seasonal Breeding

Seasonal breeding is a characteristic of many nonequatorial small mammals (Bronson, 1989). The energetic costs of reproduction, particularly for females, contribute to the restriction of gestation and births to spring months, when food availability is high. As such, natural selection may favor the ability to predict seasonal changes, in particular the onset of winter (Prendergast, Kriegsfeld, & Nelson, 2001). Photoperiod is the most reliable cue for time of year, and many animals use daylight to regulate reproduction. Among the rodent taxa identified as seasonally breeding are Syrian hamsters (*Mesocricetus auratus*) (Petterborg, Vaughan, Johnson, Champney, & Reiter, 1984), Mongolian gerbils (*Meriones unguiculatus*) (Petterborg et al., 1984), white-footed mice (*Peromyscus leucopus*) (Lynch, Heath, & Johnston, 1981; Petterborg & Reiter, 1982), deer mice (*Peromyscus maniculatus*) (Dark, Johnston, Healy, & Zucker, 1983) meadow voles (*Microtus pennsylvanicus*) (Dark & Zucker, 1983), bank voles (*Clethrionomys glareolus*) (Gockel & Ruf, 2001; Kotula-Balak et al., 2003), prairie voles (Grocock, 1981), California voles (*Microtus californicus*) (Nelson, Dark, & Zucker, 1983), marsh rice rats (Edmonds & Stetson, 1993; Edmonds, Riggs, & Masden, 2005), woodchucks (*Marmota monax*) (Concannon, Roberts, Graham, & Tennant, 1998; Concannon, Castracane, Rawson, & Tennant, 1999), collared lemmings (*Dicrostonyx groenlandicus*) (Millar, 2001), Indian palm squirrels (*Funambulus pennanti*) (Kanwar & Chaudhry, 1976), and Cape dune mole-rat (*Bathyergus suillus*) (Hart, O'Riain, Jarvix, & Bennett, 2006b). It is important to note that any number of environmental clues may influence reproductive responses to seasonal changes. Although many of the listed species respond to photoperiod as a seasonal cue, not all do.

Most laboratory studies reviewed here, unless otherwise noted, involve animals housed under one or both of two day length conditions: short day length (SD), or long day length (LD), roughly recreating winter and summer light conditions, respectively. However, it has been argued that this paradigm lacks ecological validity, as animals under natural conditions are exposed to gradually shifting day lengths (Prendergast, 2005). In the wild, gonadal regression is anticipated by decreasing day lengths that are still longer than the typical SD light period. Gradual day-length transitions are more quick and effective at inducing changes in reproductive function than abrupt transfer between SD and LD (Gorman, 1995; Gorman & Zucker, 1997). In order to anticipate seasonal changes, individuals must maintain a 'photoperiodic history' based on the relative changes of day length over time. For adults, the acquisition period is approximately 8–14 days (Stetson, Ray, Creyaufmiller, & Horton, 1989). Studying species under more naturalistic, gradual changes in day lengths is increasing in popularity and is a promising area of study for this field (Prendergast, 2005).

One other issue to note is the recurrent confusion regarding the common and scientific names for the Siberian (*Phodopus sungorus)* and Djungarian (*Phodopus campbelli)* hamsters (Steinlechner, 1998). Due to the rise to species status of a former subspecies (*campbelli*), many publications have referred to *P. sungorus* as the Djungarian hamster. In this chapter, we use the current accepted common name for the species we are discussing, rather than the common name that was used in the original article. In general, *P. sungorus* or Siberian hamsters are a widely used seasonally breeding species, while *P. campbelli* or Djungarian hamsters are not as seasonally responsive.

3.2. Neural Mechanisms for Photoperiodic Information

Information regarding photoperiod is transmitted via the retinal–hypothalamic–pineal pathway. During daylight, photons interact with the retina, sending an inhibitory signal via retinal ganglion cells to the SCN (Reiter, 1981). The SCN signals the pineal gland via a multisynaptic pathway (Revel et al., 2007a). First, the SCN projects to the paraventricular nucleus (PVN) of the hypothalamus, which then projects to the intermediolateral cells of the spinal cord. The next projection is to the superior cervical ganglion, which finally projects to the pineal gland. At night, when inhibition from the SCN is lifted, norepinephrine is released at the pineal gland (Axelrod, 1974). This norepinephrine signal induces the synthesis and release of MEL (Reiter, 1991). This nocturnal synthesis of MEL is consistent across all vertebrates (Gern, Duvall, & Nervina, 1986), acting as an 'endocrine code of day length' (Malpaux, Thiery, & Chemineau, 1999). Melatonin circulates throughout the organism's body, exposing effectively all cells to this chemical signal of circadian rhythm (Reiter, 1991).

The duration of nocturnal secretion of MEL from the pineal gland is critical for eliciting seasonally appropriate responses in physiology and behavior across many mammalian species (Bartness, Powers, Hastings, Bittman, & Goldman, 1993). Duration is the salient feature of the MEL signal for reproductive regulation, rather than amplitude or circadian phase (Goldman, Darrow, & Yogev, 1984; Gorman, 2003), although the timing of MEL release may also be a significant factor for regulating reproductive physiology (Gunduz & Stetson, 2001). In rodents, high concentrations of melatonin receptors (MEL-Rs) are found in the hypothalamus (specifically, the SCN and dorsomedial nucleus), median eminence, thalamus (reuniens nucleus and PVN), and the pars tuberalis of the pituitary (Carlson, Weaver, & Reppert, 1991; Weaver, Provencio, Carlson, & Reppert, 1991). Melatonin modulates pulsatile GnRH secretion from the hypothalamus, leading to corresponding changes in LH on female and male reproductive physiology and behavior. A key enzyme for MEL synthesis, arylalkylamine N-acetyltransferase (AANAT), is under much more complex intracellular regulation in the photoresponsive Syrian hamster than in laboratory rats or mice (Simonneaux, Sinitskaya, Salingre, Garidou, & Pevet, 2006). However, MEL does not act directly on GnRH neurons, and the specific mechanisms underlying photoperiodic control of GnRH neuronal activity are not understood, although several promising candidates are being investigated (as discussed below).

Early laboratory research on photoperiodic control of reproductive physiology was conducted on white-footed mice, and is discussed below. Subsequent studies in other species have revealed similar gonadal, endocrine, and neurobiological responses to SD (Prendergast et al., 2001). Melatonin was identified early on as the key signaler for photoperiod responsiveness of reproductive cycles in the white-footed mouse (Petterborg, Richardson, & Reiter, 1981; Lynch, Heath, & Margolis, 1982; Hall & Lynch, 1985; Carlson, Zimmermann, & Lynch, 1989). White-footed mice housed under SD have more immunoreactive GnRH cell bodies in the olfactory peduncle, anterior hypothalamus, and preoptic area when compared to animals under LD conditions (Glass, 1986), as well as increased GnRH levels in the medial basal and anterior hypothalamus (Glass, Ferreira, & Deaver, 1988), indicating inhibition of GnRH release and subsequent suppression of pituitary/gonadal functions. Short day length housing is also associated with decreased circulating LH and T, as well as LH in the pituitary (Glass et al., 1988).

Testicular regression under SD has been associated with testicular apoptosis (Young, Zirkin, & Nelson, 1999; reviewed in Young & Nelson, 2001). Vascular endothelial growth factor (VEGF) also has been implicated in SD-induced testicular regression (Young & Nelson, 2000); VEGF functions as a paracrine regulator of blood vessel growth and maintenance. Under SD conditions, VEGF levels are lowered and this decrease precedes testicular regression in a number of measures (Young & Nelson, 2000). Short day length conditions also trigger changes in the expression of a number of genes involved in angiogenesis and apoptosis (Pyter, Hotchkiss, & Nelson, 2005).

3.3. Kisspeptin

The neuropeptide kisspeptin (the product of the *KiSS1* gene) and its receptor GPR54 have emerged as important regulators of negative feedback on GTH secretion. Kisspeptin is a potent stimulator of GnRH release from the hypothalamus, and sex steroids negatively feed back on *KiSS1* gene expression (as reviewed in Smith, Clifton, & Steiner, 2006b; Popa, Clifton, & Steiner, 2008). Recent work on Siberian (*Phodopus sungorus*) and Syrian hamsters suggests that kisspeptin differentially regulates the reproductive axis of seasonally breeding animals. The two hypothalamic nuclei that are important for GTH regulation by kisspeptin and GPR54 are the anteroventral periventricular nucleus and the arcuate nucleus (ARC) (Smith & Clarke, 2007). Siberian and Syrian hamsters show different kisspeptin expression patterns in these hypothalamic areas in response to photoperiod (Greives et al., 2007; Mason et al., 2007; Revel et al., 2007a), but both species show changes in cell labeling in those areas associated with photoresponsive reproductive quiescence. In the Siberian hamster, kisspeptin expression is influenced by both gonadal dependent (i.e., feedback) and independent (i.e., photoperiodic) mechanisms (Greives et al., 2008a). The downregulation of *KiSS1* expression in the ARC of Syrian hamsters under SD is mediated through MEL but not through sex steroid feedback (Revel et al., 2007a; Simonneaux et al., 2009).

Further species differences are found with responses to exogenous kisspeptin. Chronic administration of exogenous kisspeptin induces testicular regrowth in SD-regressed Syrian hamsters (Revel et al., 2006a), but not Siberian hamsters (Greives, Kriegsfeld, & Demas, 2008c). There is also evidence for sex differences in kisspeptin responsivity in the Siberian hamster. Kisspeptin administration to adult male Siberian hamsters induces elevated circulating levels of LH, regardless of reproductive status (Greives et al., 2007). This suggests that, in males, GnRH neurons are equally sensitive to kisspeptin, regardless of reproductive status, and that seasonal regression may be associated with a decrease in kisspeptin. In contrast, reproductively quiescent female Siberian hamsters show a reduced response to exogenous kisspeptin during SD compared to LD females, suggesting a decrease in kisspeptin sensitivity under SD conditions (Mason et al., 2007). Additionally, *KiSS1* and *GPR54* mRNA are both upregulated in the ovaries of female Siberian hamsters during induced recrudescence, suggesting a function for these proteins in ovulation and ovarian recrudescence (Shahed & Young, 2009).

3.4. RFamide-related Peptide (RFRP)

Another peptide of interest in seasonally breeding rodents is RFamide-related peptide (RFRP), which is considered to be the mammalian homolog of the avian gonadotropin-inhibiting hormone (GnIH) (RFRP is often referred to as GnIH in the literature). Whereas kisspeptin upregulates the hypothalamus–pituitary–gonadal (HPG) axis, RFRP functions as a downregulator (as reviewed in Kriegsfeld, 2006; Greives, Kriegsfeld, Bentley, Tsutsui, & Demas, 2008). RFamide-related peptide has inhibitory effects on GTH synthesis and release, and is differentially expressed based on season (Revel, Saboureau, Pevet, Simonneaux, & Mikkelsen, 2008). Both Siberian and Syrian hamsters show a decrease in RFRP mRNA and RFRP-immunoreactive cell bodies of the medial basal hypothalamus (MBH) under SD conditions, compared to animals maintained under LD photoperiods, showing a seasonal effect but in the opposite direction of that predicted if RFMP acts as a reproductive downregulator (Revel et al., 2008). Photoperiod had no effect on RFRP expression in the Wistar rat. Pinealectomized Syrian hamsters do not show photoperiodic responses of RFRP, and chronic MEL administration to LD animals reduced RFRP expression to that of SD animals. Therefore, photoperiod differences in RFRP appear to be under MEL-dependent regulation, and provide another potential mechanism for seasonal regulation of the reproductive axis (Revel et al. (2008), but see Ebling and Luckman (2008) for a criticism of the methods and interpretation of the Revel et al. study).

3.5. Thyroid Hormones

Regulation of thyroid hormones also appears to play an important species-specific role in photoperiodic regulation of physiology in the Indian palm squirrel (Haldar, Shavali, & Singh, 1992), collared lemmings (Gower, Nagy, & Stetson, 1996), woodchucks (Concannon et al., 1999), Syrian hamster (Revel, Saboureau, Pevet, Mikkelsen, & Simonneaux, 2006; Yasuo, Yoshimura, Ebihara, & Korf, 2007), and Siberian hamster (Barrett et al., 2007; Freeman,

Teubner, Smith, & Prendergast, 2007). The Siberian hamster utilizes type III deiodinase (D3), which catabolizes the thyroid hormones thyroxine (T_4) and triiodothyronine (T_3), of which T_3 is the more potent (Barrett et al., 2007). An increase in D3 levels in SD leads to a decrease in T_3, and a subsequent suppression of the reproductive axis. Watanabe et al. (2007) found that this increase in D3 spikes within the first few weeks after transfer to SD housing, but prolonged SD housing is associated with a decrease in D3 and spontaneous recrudescence. Syrian hamsters differentially regulate the type II deiodonase (D2), which converts T_4 into T_3 (Revel et al., 2006b; Yasuo et al., 2007). Type III deiodinase levels are decreased under SD, leading to less available T_3 in Syrian hamsters. Thus, both species have less available bioactive T_3 under SD, but the molecular mechanisms differ. In Syrian hamsters, T_3 levels are suppressed by chronic administration of MEL, suggesting a MEL-mediated mechanism for regulation of D2 activity (Watanabe et al., 2007).

3.6. Pubertal Timing

Delayed sexual maturation under naturally decreasing day length or laboratory SD conditions has been noted in a number of species, including the cotton rat (*Sigmodon hispidus*) (Johnston & Zucker, 1979), prairie vole (Nelson, 1985), montane vole (*Microtus montanus*) (Horton, 1984), field vole (Grocock, 1981), white-footed mouse (Johnston & Zucker, 1980; Forger & Zucker, 1985), deer mouse (Whitsett & Lawton, 1982; Nelson, Marinovic, Moffatt, Kriegsfeld, & Kim, 1997), grasshopper mouse (*Onychomys leucogaster*) (Frost & Zucker, 1983), Siberian hamster (Hoffman, 1978; Yellon & Goldman, 1984), Syrian hamster (Cherry, 1987; Ebling, 1994), and marsh rice rat (Edmonds & Stetson, 1993). Among these species, the effects of photoperiod on pubertal timing may greatly vary. For example, comparative studies have consistently shown that reproductive development in juvenile Siberian hamsters is more sensitive to SD than in the Djungarian hamster (*Phodopus campbelli*) (Ebling, 1994; Timonin, Place, Wanderi, & Wynne-Edwards, 2006).

Species may vary in the relative sensitivity of juveniles and adults to photoperiod. For example, although juvenile cotton rats delay sexual maturation in response to SD, adults of this species are generally insensitive to SD (Johnston & Zucker, 1979). Similarly, maturing male prairie voles are also more sensitive to SD conditions than adult males (Nelson, 1985). Both juvenile and adult males under SD show a decrease in testes and seminal vesicle mass, but only juvenile males suffer a loss in fertility relative to LD animals (Nelson, 1985). Other species, such as the white-footed mouse, show a similarity of effects of MEL or photoperiod on each sexual maturation and gonadal regression within a population (Petterborg & Reiter, 1982; Carlson et al., 1989).

In the field, pubertal timing is associated with the seasonal timing of birth: individuals born in the spring tend to undergo puberty and breed within their birth season, whereas those individuals born later in the summer will delay puberty until the spring of the following season. Under simulated natural photoperiod, cohorts of Siberian hamster males and females born at various intervals before and following the summer solstice showed pubertal timing consistent with the field literature (Butler, Trumbull, Turner, & Zucker, 2007; Butler et al., 2007b). Specifically, those born before the simulated summer solstice reached puberty within the summer season, and the animals born six weeks after the solstice delayed puberty until the next spring. The autumnal gonadal regression and spring recrudescence was synchronized by all animals, regardless of time of birth.

Recent studies suggest that, in addition to postnatal photoperiod, prenatal photoperiod may provide seasonal information that can influence the timing of pubertal development in Siberian hamsters (Weaver, Keohan, & Reppert, 1987; Tuthill et al., 2005; Shaw & Goldman, 2007), and to a lesser extent in Syrian hamsters (Beery, Paul, Routman, & Zucker, 2008). The acquisition period for photoperiodic history for fetuses is also highly sensitive, requiring only two to three days compared to eight to fourteen days in adults (Stetson et al., 1989; Davies & Norman, 2002). Even in the domestic guinea pig (*Cavia aperea f. porcellus*), a nonseasonally reproducing species, the prenatal and early postnatal photoperiod and temperature can influence gonadal steroids and onset of puberty (Bauer, Womastek, Dittami, & Huber, 2008).

3.7. Intraspecies Variation in Photoperiod Responsiveness

Initial reports on intraspecific variability in sensitivity to photoperiodic regulation of reproduction focused on population differences, typically emphasizing latitudinal factors. In the case of the white-footed mouse, reproductive regression is generally more responsive to photoperiod in individuals from northern populations than in those from southern latitudes (Lynch et al., 1981; Carlson et al., 1989). Specifically, individuals from the Georgia population are thought to be insensitive to the effects of MEL (Lynch, Sullivan, Heath, & Tamarkin, 1982; Carlson et al., 1989). Deer mice derived from a population in Manitoba, Canada, showed more reproductive sensitivity to SD than those from South Dakota or Chihuahua, Mexico (Dark et al., 1983).

In many species, intraspecific variation in photoresponsiveness is maintained either within populations or laboratory colonies (see Prendergast et al., 2001 for a comprehensive review and list of species). Variation in responsiveness has been observed within wild populations, such as meadow voles (Kerbeshian, Bronson, & Bellis, 1994). In the early 1980s, laboratory studies on intraspecies differences began to focus more on individual 'responders' and 'nonresponders,' rather than overall population differences. Initial reports that geographically specific sensitivities are maintained in laboratory stocks suggested a genetic regulation of photoresponsiveness in the white-footed mouse (Gram, Heath, Wichman, & Lynch, 1982). Subsequent crossbreeding and laboratory selection experiments support genetic control of reproductive photoresponsiveness in this species (Heideman & Bronson, 1991; Wichman & Lynch, 1991). A genetic basis for phenotypic variation in photoresponsiveness also has been noted in a number of other rodent populations, including deer mice (Desjardins, Bronson, & Blank, 1986), field voles (Spears & Clark, 1988), and Siberian hamsters (Lynch, Lynch, & Kliman, 1989). It has been argued that individual differences in photoperiod responsiveness can affect seasonal breeding in such a way as to influence population density fluctuations that occur across multiple years (Nelson, 1987).

3.8. Spontaneous Recrudescence

Anticipating seasonal changes is crucial to an individual's ability to synchronize reproduction with resource availability (Prendergast, 2005). External cues such as photoperiod, food availability, and temperature may provide seasonal information regarding the oncoming winter, initiating gonadal regression. In contrast, internal signals drive gonadal recrudescence in anticipation of spring, independently of photoperiod (Prendergast, 2005). Initiation of spontaneous recrudescence—regrowth of reproductive physiology in the absence of external cues—normally occurs in midwinter, allowing litters to be born in spring.

Under laboratory conditions, animals exposed to prolonged periods of SD will spontaneously initiate reproductive function (Young, Zirkin, & Nelson, 2001). For example, in white-footed mice, testicular regression is maximal at 14 weeks of SD, but by 26 weeks there is no difference in testicular mass between SD and LD subjects (Young et al., 2001). Siberian hamsters will initiate gonadal growth after four to five months of continuous SD housing, with restoration of reproductive function within one to two months (Prendergast, 2005). Bittman (1978) showed that recrudescence follows a loss of target tissue sensitivity to inhibitory MEL signaling. This testicular recrudescence is driven by a decrease in testicular apoptosis (Young et al., 2001), and changes in genes involved in angiogenesis (Pyter et al., 2005).

There are substantial heritable variability and sex differences in the timing of recrudescence (Prendergast, Renstrom, & Nelson, 2004). Males initiate gonadal regrowth approximately five weeks earlier than females, consistent with the greater maturation time of spermatogenesis relative to follicular maturation and ovulation. Males also show a stronger heritability for the timing of recrudescence, and the high variability in this trait may confer differential benefits depending on annual variations in the onset of spring.

3.9. Female Reproduction and Sex Differences in Response to Photoperiod

Ovarian and uterine weights are larger in female white-footed mice under LD vs. SD (Spanel-Borowski, Petterborg, & Reiter, 1983). The ovarian morphology of Siberian hamsters differs in a number of characteristics depending on photoperiod conditions, including primordial follicles (Place, Tuthill, Schoomer, Tramontin, & Zucker, 2004; Timonin et al., 2006; Kabithe & Place, 2008), matrix metalloproteinases (Salverson McMichael, Sury, Shahed, & Young, 2008), and follicular apoptosis (Moffatt-Blue, Sury, & Young, 2006).

Prairie vole females may be relatively nonresponsive to day length changes, as Nelson (1985) found that neither the incidence of pregnancy nor the occurrence or timing of postpartum estrus was affected in SD compared to LD individuals. Female collared lemmings are also resistant to photoperiod-induced gonadal regression relative to males (Weil, Bowers, & Nelson, 2007). In Syrian hamsters, females initiate gonadal regression earlier than males (Beery, Trumbull, Tsao, Costantini, & Zucker, 2007).

3.10. Unmasking Techniques and Nonphotoresponsive Species

A number of interventions may 'unmask' photoresponsivity (i.e., reveal a latent or vestigial photoresponsive tendency) in species that normally show no reproductive response to shifts in photoperiod. These techniques include prepubertal olfactory bulbectomy, prolonged food restriction, and neonatal exposure to androgens. Olfactory bulbectomy prior to puberty unmasks photoperiod-induced reproductive suppression in laboratory rats and mice (Reiter, 1969; Nelson & Zucker, 1981; Nelson, Bamat, & Zucker, 1982; Petterborg et al., 1984; Nelson, Moffatt, & Goldman, 1994). Under normal conditions, the rat olfactory bulb provides a tonic stimulatory signal for gonadotropin secretion, which is eliminated by bulbectomy (Pieper & Newman, 1999).

It is interesting to note that olfactory bulbectomy in a photosensitive species can have the opposite effect. In the Syrian hamster, olfactory bulbectomy prevents SD-induced gonadal regression, and the olfactory bulb has a tonic inhibitory influence on the HPG axis (Pieper, Tang, Lipski, Subramanian, & Newman, 1984; Clancy, Goldman, Bartke, & Macrides, 1986). Olfactory bulbectomy does not affect the HPG axis of Siberian hamsters, however, suggesting that the olfactory bulb does not play a significant role in the GTH response to photoperiod in this species (Prendergast, Pyter, Galang, & Kay, 2008).

The cane mouse (*Zygodontomys brevicauda*) is a tropical rodent that breeds throughout the year (Heideman & Bronson, 1990). Unlike other nonphotoresponsive species previously discussed, attempts to unmask vestigial photoresponsiveness (via olfactory bulbectomy, prolonged food restriction, or neonatal T exposure) have no effect on reproductive measures (Bronson & Heideman, 1992).

4. WATER AVAILABILITY

The availability of water, particularly in arid environments, may sometimes be a more salient reproductive cue than food availability or photoperiod (Ylonen et al., 2003). In the desert environment of the golden spiny mouse (*Acomys russatus*), water content in plants is lower at the end of the summer. In the laboratory, golden spiny mice were given a diet higher either in salt or arginine vasopressin (AVP) (also known as antidiuretic hormone), which is released in times of thirst in order to aid water retention. Both of these treatments led to decreases in uteri and body masses in females (Shanas & Haim, 2004). The Mongolian gerbil, another desert rodent, displayed a complete halt in reproduction when water-deprived. Water deprivation also resulted in reduced weights of the testes, seminal vesicles, and ventral scent glands. Females had reduced adrenals and ventral scent glands, but no changes in ovarian or uterine weight (Yahr & Kessler, 1975).

Water restriction in another desert rodent, the hopping mouse (*Notomys alexis*), resulted in lower body weight, reduced ovary and uterine weights, and longer estrus periods due to a disestrus prolonged phase (Breed, 1975). Vaginal perforation was delayed in water-deprived females (Breed, 1976). In this study, there were several variables that showed individual variation in response to water restriction; for instance, 7 of the 16 water-deprived females showed regular estrous cycles while the rest did not.

A field study in which supplementary water was provided to three desert rodents (*Demodillus auricularis*, *Gerbillarus paeba*, and *Rhabdomys pumilio*) found mixed effects (Christian, 1979). The two seasonally breeding rodents (*G. paeba* and *R. pumilio*), when given extra water, extended their breeding seasons longer into the hotter portion of the year, while *D. auricularis* did not respond to the experimental treatment.

A similar experiment carried out in California with wild house mice had very interesting consequences. A population that received supplementary water was larger and had higher survival rates than a control population during the dry season. However, the population then declined rapidly, despite supplemental food, in response to increased predation (Newsome, Stendell, & Myers, 1976).

A population of *Peromyscus truei* in California was given supplemental water for 5.3 weeks during the midsummer period. Males from the watered group had longer seminal vesicles and more sperm in the epididymides. Females from the watered group were more likely to have perforate vaginae and to have corpora lutea, as well as larger vaginal openings (Bradford, 1975). However, none of the females was pregnant, indicating that water supply was not the only factor limiting summer reproduction in this species.

Shaw's jird (*Meriones shawi*), a desert rodent native to Egypt, showed a response to free-water restriction in the laboratory but limited changes in the reproductive system. Although water restriction reduced body weight and white fat stores, it did not affect testes or uterine weights (El-Bakry, Zahran, & Bartness, 1999).

Water availability is also crucial to rodents living in less extreme environments. In deer mice, which are temperate-zone rodents living in a mesic environment, water restriction reduced testicular mass and spermatogenesis. Interestingly, the response within the water-restricted group varied from becoming aspermic (22 of 62 animals), to displaying normal epididymal sperm numbers (12 of 62 animals), suggesting considerable individual variation (Nelson & Desjardins, 1987). The water restriction in this study did not affect food intake.

In seasonally breeding California mice, water restriction reduced seminal vesicle and epididymal masses, as well as plasma prolactin (Nelson, Gubernick, & Blom, 1995). Neither photoperiod nor availability of green food affected any reproductive variables, suggesting that water is the salient cue for this species. In contrast, California voles show reproductive regression during short days as well as a response to reduced water. Both can be overridden by extra green food (Nelson et al., 1983). In prairie voles, water restriction also reduced body mass and epididymal and seminal vesicle masses (Nelson, Frank, Bennett, & Carter, 1989).

These results from outbred, relatively 'wild' laboratory rodents stand in contrast to results from inbred strains of mice. In a comparison of a highly inbred (CF1) strain to F4 descendants of wild house mice, Nelson found that the inbred strain showed little responsiveness to water restriction; there was a reduction in seminal vesicle mass but no loss in spermatogenesis. The close descendants of wild mice showed a decline in body mass, a reduction in spermatogenesis, and significantly lighter testes, epididymides, and seminal vesicles (Nelson, 1988). This difference could be the result of selection for ability to breed in the laboratory (reproductive responsiveness) or to higher adiposity in the inbred strain (Nelson, 1988).

5. TEMPERATURE

Temperature may provide a salient signal for both seasonal and opportunistic breeders. Although the role of ambient temperature in reproduction in rodents is not well understood, a number of field and laboratory studies suggest that it is a significant environmental cue for certain species.

Within species, optimal temperatures may vary based on geographic origin (Bronson & Pryor, 1983). Deer mice from wild-caught stocks were paired and housed in temperatures ranging from extreme cold (−6°C) to extreme heat (34°C), and successful births recorded. Reproductive success was sensitive to temperature depending on geographical origin. Somewhat paradoxically, deer mice from a southern population (Texas) were most successful at colder temperatures, and northern deer mice (Alberta) had maximal breeding success at room temperature. Ruf, Korytko, Stieglitz, Lavenburg, and Blank (1997) found no additional effects of temperature on reproductive measures of male deer mice placed under SD conditions. The role of temperature in this species remains unclear.

Interactions of temperature and photoperiod on reproduction have been observed in a number of species including deer mice (Desjardins & Lopez, 1983), Siberian hamsters (Steinlechner, Stieglitz, Ruf, Heldmaier, & Reiter, 1991), guinea pigs (Bauer et al., 2008), and prairie voles (Nelson, Frank, Smale, & Willoughby, 1989; Kriegsfeld, Ranalli, Bober, & Nelson, 2000; Kriegsfeld, Trasy, & Nelson, 2000). In male prairie voles, photoperiod interacts with ambient temperature such that the combination of low temperatures and SD leads to regressed reproductive systems, as assessed by a number of physiological and anatomical measures (Nelson et al., 1989b; Kriegsfeld et al., 2000a). Reproductive regression under these environmental conditions appears to be mediated by alterations to the GnRH neuronal system (Kriegsfeld et al., 2000a; 2000b). Low temperatures also may interact with day length to influence pubertal timing in the Siberian hamster (Steinlechner et al., 1991). Temperature can also interact with food availability. In a laboratory study of wild house mice, only females that were both food-restricted and held under cold conditions showed reduced numbers of surviving litters (Marsteller & Lynch, 1987).

The African grass rat (*Arvicanthis niloticus*) is a Saharan desert rodent that maintains breeding capability during the dry season (Sicard, Fuminier, Maurel, & Boissin, 1993). Wild-caught males were tested for reproductive activity under varying temperature and humidity conditions. There was an interaction of temperature and humidity, such that testicular activity was inhibited with high temperatures (30−35°C) combined with low humidity and stimulated by high humidity combined with low temperatures (20−22°C) (Sicard et al., 1993). Other combinations of temperature and humidity led to intermediate testicular growth. This environmental regulation of reproduction in this species suggests that they have an opportunistic breeding strategy. Results from females suggest a very similar reproductive strategy, with spontaneous ovulation and no seasonal anoestrus (Sicard, Maurel, Fuminier, & Boissin, 1994).

A path analysis examining the effects of environmental factors on reproduction in wild female tuco-tucos (*Ctenomys talarum*) found that high temperatures had a negative, direct relationship with proportion of both pregnant and lactating females, perhaps indicating a negative effect of heat stress on female reproduction. However, the negative effects of high temperature were lessened by an indirect positive effect of temperature on plant-water content, which is favorable for reproduction (Fanjul, Zenuto, & Busch, 2006). This study demonstrates the importance of field work on environmental regulation of reproductive variables, as indirect effects that are controlled under laboratory conditions may still significantly contribute to breeding patterns.

6. SOCIAL CUES

Many of the studies of the effects of social environment on reproduction in laboratory rats and mice (e.g., the Lee-Boot, Vandenbergh, Whitten, and Bruce effects) are discussed in Chapter 9 of this volume. Here we focus on social regulation of reproductive cycles in wild populations and less traditional laboratory models. Given the species-specific nature of reproductive responses to the social environment, we will discuss each species individually.

6.1. Reproductive Suppression in Social Mole-rats

The African mole-rats (family Bathyergidae) are subterranean rodents displaying the entire range of social systems, from solitary to eusocial. Colonies of all social mole-rats include a single breeding female and one to three breeding males (depending on species), with all other sexually mature individuals in the colony failing to reproduce. Although this reproductive suppression of 'subordinates' is a common characteristic of all social mole-rats, the underlying physiological and behavioral mechanisms driving this unequal distribution of reproduction are highly species-specific (Faulkes & Bennett, 2001). The naked mole-rat (*Heterocephalus glaber*) and Damaraland mole-rat (*Cryptomys damarensis*) are the only mammals known to display eusociality, a highly ordered social system characterized by reproductive division of labor (i.e., reproductive and nonreproductive 'castes'), cooperative care of young, and overlapping generations (Wilson, 1971). Many studies on the reproductive biology of mole-rat species are conducted with wild-caught animals, contributing ecological validity to our understanding of behavioral endocrinology in these species.

6.1.1. *Naked mole-rats (*Heterocephalus glaber*)*

Reproduction in naked mole-rats is restricted to a single breeding female and two to three breeding males, with nonbreeding individuals that are organized into castes (Sherman, Jarvis, & Alexander, 1991). Colonies average 70–80 individuals, although as many as 295 individuals have been found in a single colony (Sherman et al. 1991). Physiological suppression of reproduction in both sexes is a characteristic of this species (Sherman et al., 1991; Faulkes & Abbott, 1997). It is generally thought that subordinates are reproductively suppressed by the breeding queen via dominance and aggression, rather than pheromonal or other olfactory means (as discussed below).

Ovulation is suppressed and circulating LH is lower in both wild and captive nonbreeding female naked mole-rats (Faulkes, Abbott, & Jarvis, 1990). Nonbreeding females also show a reduced LH response to low doses of exogenous GnRH, an effect reversed by GnRH priming. This suggests that these females may have inhibited GnRH release from the hypothalamus, leading to a reduction in pituitary GnRH receptors (Faulkes, Abbott, Jarvis, & Sherriff, 1990). Removing nonbreeding females from the natal group reverses their reproductive suppression when these females are either singly housed or exposed to a nonbreeding male.

Both captive and wild-caught nonbreeding males produce viable, mature gametes (Faulkes, Abbott, & Jarvis, 1991). However, nonbreeding males have reduced levels of urinary T and plasma LH, as well as a blunted LH response to exogenous GnRH when compared to breeding males (Faulkes et al., 1991). An experimental study with captive colonies supports male reproductive suppression, as nonbreeding males removed from their colony and singly housed show a marked increased in T and LH (Faulkes & Abbott, 1991), as well as an increase in body weight, testes size (corrected for body size), number of spermatozoa, and sperm motility (Faulkes, Trowell, Jarvis, & Bennett, 1994). When nonbreeding males that have been singly housed are introduced to a nonbreeding female, there is a partial restoration of endocrine suppression, except for an increase in T during the early follicular phase of the female's cycle (Faulkes & Abbott, 1991). These studies suggest that reproductive suppression in male mole-rats is regulated by the breeding female of the colony.

Similarly to males, nonbreeding females that are removed from the natal colony will commence ovarian cycling, suggesting a removal of reproductive suppression. This induction of ovarian cycling has been shown under a number of laboratory conditions, including single housing, housing with a nonbreeding male, and single housing with regular social contact with nonbreeding members of the natal colony (Faulkes & Abbott, 1993; Smith, Faulkes, & Abbott, 1997). Additionally, daily rotation of the bedding of nest, food, and toilet chambers of the natal colony in any of these housing conditions does not appreciably affect reproductive measures, suggesting that olfactory cues from the natal nest do not play a significant role in reproductive suppression in females of this species (Faulkes & Abbott, 1993; Smith et al., 1997). Reproductive suppression may be reversed following the death or removal of the breeding female, further supporting a role for the breeding female in controlling reproduction of other colony members (Clarke & Faulkes, 1997).

Holmes et al. (2007) investigated neurobiological correlates of breeding status in this species and found differences in regions of the brain that are implicated in social behavior in a number of other rodents species. Specifically, the bed nucleus of the stria terminalis, PVN, and medial amygdala are all larger in breeders compared to nonbreeders, and there are more cell bodies in the ventromedial hypothalamus of breeders. These differences were seen in both males and females. Breeders also have more AVP-immunoreactive cells in the dorsomedial hypothalamus (DMH) when compared to subordinates, and this is particularly pronounced in females (Rosen, De Vries, Goldman, Goldman, & Forger, 2007). As discussed in Section 4, AVP is involved in water balance, but also plays a key role in social and reproductive behaviors in mammalian species. Interestingly, AVP-ir (immunoreactivity) in the DMH is increased in female golden hamsters during lactation, suggesting a specific role of AVP in the DMH for reproduction that would be consistent with reproductive demands on breeding mole-rats (Delville, Conklin, & Ferris, 1995).

Although reproductive suppression in the natal nest is a defining characteristic of this eusocial species, a subset of reproductively viable and morphologically distinct males have been identified in captive populations (Oriain, Jarvis, & Faulkes, 1996). These individuals, unlike their nonbreeding siblings, have elevated circulating LH levels and will attempt to copulate with foreign conspecifics (but, importantly, not the queen of their natal colony). These reproductively active individuals also have a higher dispersal rate.

6.1.2. *Damaraland mole-rats (*Cryptomys damarensis*)*

The Damaraland mole-rats live in social colonies of approximately 40 individuals, with breeding restricted to a single female and one or two reproductive males, whom she actively selects as her partners. Nonreproductive individuals are divided into two worker castes (Bennett &

Jarvis, 1988; Jacobs, Bennett, Jarvis, & Crowe, 1991). Subordinate females are physiologically suppressed, whereas subordinate males are not. The predominant mechanism of suppression is inbreeding avoidance, rather than behavioral or chemical signaling from the queen (Bennett, Faulkes, & Molteno, 1996; Clarke, Miethe, & Bennett, 2001).

Nonbreeding males in this species do not undergo complete physiological suppression (Bennett, 1994). Faulkes et al. (1994) found no differences in the gamete viability (number of sperm or sperm motility) between wild-caught breeding and nonbreeding males, but Maswanganye, Bennett, Brinders, & Cooney (1999) found that breeding males have a great proportion of motile sperm than subordinates, and subordinates also have a higher percentage of immature sperm in the epididymis. Breeding status does not affect urinary or circulating T (Bennett, 1994; Maswanganye et al., 1999). Luteinizing hormone levels in males also are not affected, as both basal levels and response to exogenous GnRH do not differ between wild-caught males of different breeding statuses (Bennett, Jarvis, Faulkes, & Millar, 1993). Taken together, these studies support the idea that nonbreeding males in this species are not physiologically suppressed (Bennett, 1994).

In contrast to males, nonbreeding female Damaraland mole-rats do display physiological reproductive function. In an 18-month study of a captive colony, nonreproductive females were anovulatory due to halted follicular development (Bennett, 1994). The authors proposed that unruptured luteinized follicles may produce low levels of circulating progesterone (P_4) that are still sufficient for negative feedback, preventing an LH surge and subsequent ovulation. Another possible mechanism for reproductive suppression is reduced GnRH sensitivity, as evidenced by the significantly lower basal circulating LH levels as well as a blunted LH response to GnRH in wild-caught nonbreeding females (Bennett et al., 1993). Although there are no differences in the distribution, morphology, or size of GnRH neurons in nonbreeding females compared to breeding females, GnRH release appears to be inhibited in these animals (Molteno, Kallo, Bennett, King, & Coen, 2004). Although levels of GnRH in the brain were the same between breeding and nonbreeding females, nonbreeding subjects had reduced GnRH immunoreactivity in the median eminence and pituitary stalk. Differences in GTH secretion between breeding and nonbreeding females are independent of endogenous opioids or gonadal steroids (Molteno & Bennett, 2000; 2002).

Physiological suppression is maintained while the breeding female remains in the colony. Reproductive suppression is not mediated by the breeding female's behavior or pheromones (Clarke et al., 2001). The consistently replicated ability of the introduction of a novel, nonrelated male to induce ovulation in these previously nonbreeding females suggests a key role for an incest avoidance strategy (Cooney & Bennett, 2000; Clarke et al., 2001). Sibling pairs will not engage in sexual behavior or mating, although nonsibling pairings will successfully mate and conceive (Bennett et al., 1996).

6.1.3. *Mashona mole-rats (*Cryptomys darlingi*)*

Mashona mole-rats live in colonies of five to nine individuals with a single breeding pair that reproduces throughout the year (Bennett, Jarvis, & Cotterill, 1994). Behavioral inhibition is the dominant mechanism for Mashona mole-rats (*Cryptomys darlingi*) (Bennett, Faulkes, & Spinks, 1997). Nonbreeding females in this species have mature ovarian morphology, and nonbreeding males show evidence of spermatogenesis (Bennett et al., 1994). There are no differences in either basal or GnRH-challenged levels of LH between breeders and nonbreeders of either sex (Bennet et al., 1997). Rather than physiological suppression, socially induced infertility may be caused by obligate incest avoidance in this species.

6.1.4. *Ansell's mole-rats (*Fukomys anselli*)*

Ansell's mole-rats (formerly *Cryptomys anselli*) maintain reproductive suppression through incest avoidance mechanisms and kin recognition. If the breeding female dies or is removed from a colony, no other nonbreeding female will replace her, and reproduction in the colony will permanently cease. Little is known about the long-term consequences of this sterility on the colony, as well as dispersal behavior in this species. The loss of a breeding male, however, does not lead to colony sterility, as he may be replaced by older sons (Burda, 1995). Females in this species do not appear to be physiologically suppressed, as subordinates express normal early follicular development, although ovulation does not occur (Willingstorfer, Burda, & Winckler, 1998). Burda (1995) showed that individual recognition plays a key role in reproductive suppression in common mole-rat colonies. Ansell's mole-rats can discriminate conspecific odors based on relatedness (Heth, Todrank, Begall, Wegner, & Burda, 2004). Such strong incest-avoidance mechanisms in a colony of highly related individuals may contribute to colony sterility following the loss of the dominant breeding female.

6.1.5. *Common mole-rats (*Cryptomys hottentotus hottentotus*)*

Common mole-rats live in colony groups up to 14 individuals with an average of five members (Bennett, 1989). These groups include one breeding pair and their nonreproductive offspring, who stay in the natal colony until conditions favor dispersal (Spinks, Bennett, Faulkes, &

Jarvis, 2000). The common mole-rat is also among the two seasonally breeding species of social mole-rat (DuToit, Bennett, Katz, Kallo, & Coen, 2006).

Male common mole-rats are not physiologically suppressed (Spinks, VanderHorst, & Bennett, 1997). The LH response to administration of exogenous GnRH in wild-caught common mole-rats did not differ based on breeding status (Spinks et al., 2000). Neuroanatomical measures of the GnRH system in females of this species do not differ based on reproductive status (DuToit et al., 2006b). Taken together, these data strongly point to a lack of active physiological suppression of either sex.

Seasonal breeding in common mole-rats is behaviorally induced, as no aspect of reproductive physiology is affected by season in either wild-caught males (as measured by spermatogenesis, sperm quality, and testicular anatomical and histological measures (Spinks et al., 1997)) or females (as measured by ovarian histology (Spinks, Bennett, & Jarvis, 1999)). The inactive breeding season is also a period of dispersal opportunity, and maintaining reproductive activity may reduce intersexual aggression, thus aiding in dispersal and outbreeding success (Spinks et al., 1997).

6.1.6. *Highveld mole-rats (*Cryptomys hottentotus pretoriae*)*

Colonies of highveld mole-rats consist of an average of 10 individuals, including a single breeding female and one or two breeding males (Moolman, Bennett, & Schoeman, 1998). Highveld mole-rats are the second of the social mole-rat species that are seasonal breeders, although the dominant male and female do not display physiological suppression in the nonbreeding season (Van der Walt, Bennett, & Schoeman, 2001).

Subordinate female highveld mole-rats demonstrate some physiological suppression in that follicular development is normal but ovulation is inhibited (Van der Walt et al., 2001). Neuroanatomical evidence suggests that subordinate female highveld mole-rats have inhibition of GnRH release. This inhibition does not appear to be related to endogenous opioids, as administration of the opioid antagonist naloxone did not affect LH levels acutely, or under acute or long-term GnRH challenge (DuToit, Bennett, Gutjahr, & Coen, 2006). In contrast, subordinate males of this species do not display physiological suppression in the natal colony, suggesting behavioral incest avoidance as the driving force for reproductive suppression (Van Rensburg, Bennett, Van der Merwe, Schoeman, & Brinders, 2003). Additionally, males in this seasonally breeding species maintain reproductive activity throughout the year, consistent with the dispersal strategy seen in common mole-rats (Van Rensburg, Bennett, Van der Merwe, & Schoeman, 2002).

6.2. Reproductive Suppression in Solitary Mole-rats

6.2.1. *Cape dune mole-rats (*Bathyergus suillus*)*

The Cape dune mole-rat is among the solitary mole-rat species, and shows seasonal reproductive suppression (Hart et al., 2006b). Basal levels of LH are lower during the nonbreeding season in both sexes (Hart, O'Riain, Jarvis, & Bennett, 2006a). However, LH response to a GnRH challenge does not vary in either sex by season, suggesting that the GnRH system is not differentially regulated based on season (Hart et al., 2006a). A recent neuroanatomical study found no difference in the number or distribution of hypothalamic GnRH cell bodies between seasons (Hart et al., 2008). The cell body size of GnRH neurons is smaller in females during the nonbreeding season, but males show no size difference across seasons. It is possible that conserving the integrity of the GnRH system allows opportunistic breeding in this species, when opportunities arise outside the normal breeding season.

6.2.2. *Cape mole-rats (*Georychus capensis*)*

The Cape mole-rat (*Georychus capensis*) is a solitary and seasonally breeding species (Bennett & Jarvis, 1988; Oosthuizen & Bennett, 2007). Similarly to the Cape dune mole-rat, the Cape mole-rat does not show seasonal differences in GnRH regulation (Oosthuizen & Bennett, 2007). Females of this species are induced ovulators (Van Sandwyk & Bennett, 2005). Taken together, these studies are consistent with opportunistic breeding in this species.

6.3. Reproductive Suppression and Kin Recognition in Other Rodent Species

6.3.1. *Cactus mice (*Peromyscus eremicus*)*

In one of the earliest experimental studies of reproductive suppression in the natal group, Skryja (1978) studied reproductive success of weanling female cactus mice that were housed with an unrelated male, both parents, or their father. Females housed with both parents showed a marked reproductive inhibition. However, females paired with their father only showed no inhibition in reproduction, suggesting that mothers in this species suppress reproduction in daughters.

6.3.2. *Prairie voles (*Microtus ochrogaster*)*

Females of this species are induced ovulators, and first estrus is induced by detecting a chemosignal in male urine (Carter, Getz, Gavish, McDermott, & Arnold, 1980). Exposure to a gonadally intact male or application of intact male urine to the upper lip of a female will lead to an increase in uterine weight (Carter et al. 1980). Olfactory access to male urine induces an LH surge in female prairie voles, and thus initiates ovulation (Dluzen, Ramirez, Carter, & Getz, 1981). Females must actively obtain this chemosignal by sniffing the male genital region, as it is not transmitted via the air. Young female prairie voles do not engage in sniffing the anogenital region of familiar males; this functions as an internal behavioral control for inbreeding avoidance (Carter et al., 1980; McGuire & Getz, 1991). The urine of another female suppresses reproduction in female prairie voles (Getz, Dluzen, & McDermott, 1983), although this can be overcome with repeated exposure to male urine (Hofmann & Getz, 1988). There are fewer available studies on male prairie voles, although laboratory studies have consistently shown reproductive suppression in the natal nest (Batzli, Getz, & Hurley, 1977).

Estrous females show a preference to mate with unfamiliar males; however, in the absence of an unfamiliar male they may mate with a familiar, related male, although the mating may not be successful (McGuire & Getz, 1991). Mating with the father is also inhibited by the presence of the mother, although Wolff, Dunlap, and Ritchhart (2001) found no mother-driven reproductive suppression of daughters toward unfamiliar males.

6.3.3. *Pine voles (*Microtus pinetorum*)*

Female pine voles are anovulatory when housed with their mother in the natal group (Solomon, Brant, Callahan, & Steinly, 2001). Behavioral mechanisms appear to play a larger role than chemical cues in the reproductive suppression of female pine voles in their natal nest (Brant, Schwab, Vanderbergh, Schaefer, & Solomon, 1998). In a laboratory study, breeding males were removed from groups and replaced with either their soiled bedding or an unfamiliar male. Exposure to the father's soiled bedding did not affect female reproduction, suggesting that familiar male chemical cues are not important for suppression in this species. Agonistic behavior from the mother directed toward the daughter did decrease reproduction in daughters, suggesting a behavioral regulation of reproductive suppression (Brant et al., 1998).

6.3.4. *Mongolian gerbils (*Meriones unguiculatus*)*

Reproductive suppression in the natal nest has been observed in both the field and laboratory (Payman & Swanson, 1980; Agren, 1981; Clark & Galef, 2001; Saltzman, Ahmed, Fahimi, Wittwer, & Wegner, 2006). Inhibition of female reproduction occurs in the presence of an older, reproductive female but can also occur between sisters (Clark & Galef, 2001). Both male and female Mongolian gerbils are reproductively inhibited by cohabitation with their parents; however, females (but not males) are further suppressed by the presence of younger siblings (Saltzman et al., 2009). This may be a protective mechanism for pup care, as females in this species become infanticidal in conjunction with reproductive maturity (Saltzman et al., 2006; 2009).

6.3.5. *Alpine marmots (*Marmota marmota*) and field studies of other species*

Social groups in the cooperatively breeding alpine marmot include a dominant, territorial pair, their offspring, and additional unrelated subordinates. Reproductive suppression in wild populations has been observed in both sexes and is behaviorally mediated. Ovulation is not suppressed in this species, and subordinate females do become impregnated (Hacklander, Mostl, & Arnold, 2003). The dominant female's direct agonistic behavior toward gestating subordinates and the severity of these encounters are negatively associated with relatedness to the subordinate. Repeated aggression from the dominant female leads to an increase in glucocorticoids and decreased P_4 in the subordinate, and eventual loss of pregnancy. Thus, females in this species behaviorally mediate reproductive suppression through a negative effect of stress on the HPG axis (Hacklander et al., 2003). Although a direct behavioral effect of dominant males on subordinates has not been shown, one study found that subordinates that are not related to the dominant male are reproductively suppressed, whereas subordinate 'sons' are not (Arnold & Dittami, 1997). These 'non-sons' also sustain more injuries from fights, suggesting that they may be reproductively inhibited by the physiological effects of social stress.

In a 40-year study of wild yellow-bellied marmots (*Marmota flaviventris*), onset of reproduction was delayed due to reproductive suppression by older, reproductive females (Oli & Armitage, 2003). Population density drives reproductive suppression in wild populations of capybara (*Hydrochoerus hydrochaeris*) (Maldonado-Chaparro & Blumstein, 2008).

6.4. Spontaneous vs. Induced (Reflex) Ovulators

As evidenced from the details above, one significant source of variation in female rodent cycling is the type of ovulatory cycle displayed. Ovulatory cycles in virgin females are typically either *spontaneous* or *male-induced* (Zarrow & Clark, 1968; Jochle, 1973; Sawrey & Dewsbury, 1985; Komisaruk & Steinman 1986; Dewsbury, 1989). Spontaneous ovulators display cycles for the first time without need for exposure to a male. Induced ovulators require exposure to a male, including either odors, contact, or copulation (Dewsbury, 1989).

This division does not always hold strictly. *Rattus* and *Mus* are the classic laboratory examples of spontaneous estrous cycles; however, vaginal stimulation clearly leads to ovulation of additional eggs (Zarrow & Clark, 1968) and cohabitation with a male can hasten ovulation (Jochle, 1973). Djungarian hamsters (*Phodopus campbelli*) show induced components of a spontaneous estrous cycle (Erb, Edwards, Jenkins, Mucklow, & Wynne-Edwards, 1993). It is most likely that many species have some characteristics of both spontaneous and induced ovulation, and that this is really a continuum (Weir, 1974).

Although spontaneous and induced ovulation have been studied for many decades, the evolutionary and ecological significance of this phenomenon in the life history of different species is still a very current topic in many different taxa, including both primates and carnivores (Bercovitch & Ziegler, 2002; Lariviere & Ferguson, 2003). Hypotheses suggest that induced ovulation should be advantageous when (1) females are solitary and need to be able to ovulate quickly upon finding a fertile male or (2) reproduction is seasonal and thus ovulation is constricted to occur at an optimal time.

Table 11.1 shows details on spontaneous vs. induced ovulators including their social systems. While a full phylogenetic analysis is beyond the scope of this paper, there are several clear patterns. First of all, some taxa include all spontaneous or all induced ovulators (e.g., *Microtus*). In the case of *Microtus*, all species are induced ovulators but some have extremely divergent social systems, as well as divergent neural patterns of peptides subserving social systems (Getz, Carter, & Gavish, 1981; Witt, Carter, & Insel, 1991; Insel & Shapiro, 1992; Salo, Shapiro, & Dewsbury, 1993; Insel, Wang, & Ferris, 1994). An example of a taxon in which the species are all spontaneous ovulators, but that also have divergent social systems, is *Peromyscus* (Bradley & Terman, 1979; Dewsbury, 1988; Gubernick, 1988). Although induced ovulation is theoretically most advantageous for females that do not live continuously with available males, it appears that induced ovulation is not sufficiently disadvantageous for social species (or vice versa) to have been lost due to selection in *Microtus*. Alternatively, these reproductive strategies may not have yet had time to diverge in this taxon. In contrast, species of the family Bathyergidae appear to have evolved different modes of ovulation, with the two most eusocial species displaying spontaneous ovulation while the other social and solitary species do not. It is apparent that, while sociality may affect ovulation patterns, there is also a large role for phylogenetic constraints, and probably for seasonality or aseasonality of breeding, in many taxa.

Males also have adopted reproductive strategies to succeed in inducing ovulation in females. In particular, males of many species have developed penile spines in order to provide more extensive stimulation of the female (Zarrow & Clark, 1968; Parag, Bennett, Faulkes, & Bateman, 2006). In the Bathyergidae, there is reduced penile ornamentation in spontaneously ovulating species (Parag et al., 2006). Other families such as the Sciuridae, which contain mostly reflex ovulators, have penile spines (Zarrow & Clark, 1968).

6.5. Postpartum and Lactational Estrus

Another source of variation in rodent female reproductive cycling, which can directly affect maternal and other social behavior, is the presence or absence of postpartum and lactational estrus (Gilbert, 1984; Dewsbury, 1989). Postpartum estrus is an estrus occurring less than two days after giving birth, whereas lactational estrus occurs more than two days after giving birth but while the female is still lactating (Gilbert, 1984). Either allows the female to be pregnant and lactating at the same time, thus shortening the time until the birth of her next litter.

There are a number of differences in copulatory behavior between a cycling or male-induced estrus, and postpartum estrus. For instance, males mating with females in postpartum estrus required more intromissions and achieved fewer ejaculations, and the period of female receptivity was shorter (Dewsbury, 1989; Witt, Carter, Chayer, & Adams, 1990). Postpartum female meadow voles received fewer intromissions, with shorter intervals in between, and a shorter time for total copulation than did females in male-induced estrus (DelBarco-Trillo & Ferkin, 2007). Bank vole females in postpartum estrus approached males more frequently than did virgin females or females in induced estrus (Marchlewska-Koj, Kruczek, & Olejnicjak, 2003). Following copulation, these females in male-induced estrus have higher P_4 levels than females in postpartum estrus (Carter et al., 1989). Females of some species in postpartum estrus also do not display the pregnancy blockage ('Bruce effect') of a strange male (Dewsbury, 1989).

Another crucial difference is that, during the postpartum estrus, there are young pups that require maternal (and sometimes paternal) care. The presence of pups significantly shortens postpartum estrus duration in prairie voles (Witt et al., 1990). In Mongolian gerbils, another biparental species, females were the primary caregivers for the pups during the postpartum estrus, while the males were mainly seeking copulations. Whether pups are born to a female in postpartum estrus or not can have long-term effects on their own behavior. In laboratory-housed Norway rats (*Rattus norvegicus*), older pups did not stop nursing upon the birth of pups produced by a postpartum mating, and continued to spend time in the nest, despite aggression from the mother before and after the birth of the second litter (Gilbert, Burgoon, Sullivan, & Adler, 1983). Adult male and diestrus female Wistar rats reared with younger siblings were less

TABLE 11.1 Type of ovulation (induced or spontaneous) and social system for representative rodent species other than Muridae

Family	Species	Ovulation	Social System
Ctenomyidae	***Ctenomys talarum* (tuco-tuco)**[a]	**Induced**	**Solitary**
Bathyergidae	*Heterocephalus glaber* (naked mole-rat)[b]	Spontaneous	Eusocial/CB
	Cryptomys hottentotus hottentotus (common mole-rat)[c]	Induced	Eusocial/CB
	Cryptomys hottentotus pretoriae (highveld mole-rat)[d]	Induced	Social
	Cryptomys hottentotus natalensis[e]	Induced	Social
	Cryptomys damarensis (Damaraland mole-rat)[f]	Spontaneous	Eusocial/CB
	Fukomys anselli (Ansell's mole-rat)[g]	Induced	Eusocial/CB
	Georychus capensis (Cape mole-rat)[h]	Induced	Solitary
Cricetidae	*Microtus ochrogaster* (prairie vole)[i]	Induced	Monogamous/CB
	Microtus pennsylvanicus (meadow vole)[j]	Induced	Polygynous
	Microtus montanus (montane vole)[k]	Induced	Polygynous
	Microtus pinetorum (pine vole)[j]	Induced	Monogamous/CB
	Microtus californicus (California vole)[l]	Induced	Monogamous/CB
	Peromyscus maniculatus (deer mouse)[m]	Spontaneous	Polygynous/CB
	Peromyscus californicus (California mouse)[n]	Spontaneous	Monogamous
	Peromyscus leucopus (white-footed mouse)[o]	Spontaneous	Polygynous/CB
	Mesocricetus auratus (Syrian golden hamster)[p]	Spontaneous	Polygynous
	Meriones unguiculatus (Mongolian gerbil)[q]	Spontaneous	Monogamous/CB
	Phodopus campbelli (Djungarian hamster)[r]	Spontaneous (but see text)	Monogamous
	Phodopus sungorus (Siberian hamster)[r]	Spontaneous	Solitary
Caviidae	*Cavia porcellus* (domesticated guinea pig)[a]	Spontaneous	Polygynous
	Cavia aperea (wild guinea pigs)[s]	Spontaneous	Polygynous
	Galea musteloides (yellow-toothed cavy)[s]	Induced	Promiscuous
	Hydrochoerus hydrochoerus (capybara)[t]	Spontaneous	Social
Sciuridae	*Spermophilus tridecemlineatus* (ground squirrel)[u]	Induced	Solitary
	Marmota monax (woodchuck)[v]	Induced	Solitary
	Xerus inauris (Cape ground squirrel)[w]	Spontaneous	Promiscuous

CB, cooperative breeders.
[a]*Weir (1974); Fanjul and Zanuto (2008).*
[b]*Jarvis (1991).*
[c]*Spinks, Bennett, and Jarvis (1999).*
[d]*Malherbe, Schoeman, and Bennett (2004).*
[e]*Jackson and Bennett (2005).*
[f]*Snyman, Jackson, and Bennett (2006).*
[g]*Willingstorfer, Burda, and Winckler (1998).*
[h]*Van Sandwyk and Bennett (2005).*
[i]*Carter, Getz, Gavish, McDermott, and Arnold (1980); Carter, Witt, Schneider, Harris, and Volkening (1987).*
[j]*Sawrey and Dewsbury (1985); Taylor, Salo, and Dewsbury (1992).*
[k]*Gray, Davis, Zerylnick, and Dewsbury (1974).*
[l]*Kenney, Hartung, and Dewsbury (1979).*
[m]*Bradley and Terman (1979).*
[n]*Gubernick (1988).*
[o]*Clark (1938).*
[p]*Carter (1973).*
[q]*Barfield and Beeman (1968).*
[r]*Wynne-Edwards and Lisk (1987); Erb, Edwards, Jenkins, Mucklow, and Wynne-Edwards (1993).*
[s]*Touma, Palme, and Sachser (2001); Asher, De Oliveira, and Sachser (2004).*
[t]*Alvarez and Kravetz (2006).*
[u]*Foster (1934).*
[v]*Hikim, Woolf, Bartke, and Amador (1992).*
[w]*Bouchie, Bennett, Jackson, and Waterman (2006).*

anxious than those not reared with younger siblings (Uriarte, Ferreira, Rosa, Sebben, & Lucion, 2008).

In a review of these strategies in rodents, Gilbert (1984) found relatively few species that did not display a postpartum estrus. In contrast, lactational estrus was relatively uncommon. Among species with lactational estrus were Syrian hamsters, California mice and deer mice (but not white-footed mice, a closely related species), pocket gophers (*Geomyidae* spp.), and most Sciuridae (squirrels). Only three species were identified by Gilbert as definitively lacking both postpartum and lactational estrus: the Syrian hamster, mountain beaver (*Aplodontia rufa*), and eastern gray squirrel (*Sciurus carolinensis*). One explanation for the lack of a postpartum estrus in hibernating species is that the cycle allows the female to raise only one litter before needing to deposit fat for her winter hibernation. A further variation in female reproductive tactics, mating at postpartum estrus but delaying implantation, was found in a number of species (Gilbert, 1984).

7. SMALL RODENT POPULATION CYCLES

Population fluctuations in which voles and lemmings, once every three to five years, produce a large excess of animals followed by a population crash, have fascinated ecologists for nearly a century (Elton, 1924), and are still producing much debate (Stenseth, 1999). These multiannual cycles have been hypothesized to be dependent on interactions with food sources (Turchin & Batzli, 2001) and predators (Hanski, Henttonen, Korpimaki, Oksanen, & Turchin, 2001; Sundell, 2006), and clearly rely on both direct and delayed density-dependent interactions (Stenseth, 1999).

Although this question has engaged both theoreticians and empiricists, climate change may be having significant effects on ongoing natural observations (Kausrad et al., 2008). Kausrad and colleagues showed that, over a 38-year period, changes in climate in a Norwegian habitat reduced the amount of subnivean space (a space under the snow in which lemmings and other small rodents can move freely and be protected from predators). Since 1994, there have been no observed rodent population peaks.

8. OTHER ECOLOGICAL FACTORS AFFECTING REPRODUCTION

8.1. Personality and Individual Differences

Personality (temperament, or behavioral syndromes), defined as consistent individual differences in behavior across multiple measures, is being studied increasingly in animals, although studies are still rare in natural populations (Capitanio, 1999; Sih, Bell, Johnson, & Ziemba, 2004). A study of North American red squirrels (*Tamiasciurus hudsonicus*) found that the effects of female personality on reproductive success varied across years; in some years, more aggressive females had more surviving offspring, but in other years the opposite relationship was true (Boon, Reale, & Boutin, 2007). Changes in food supply across years, especially due to masting of the white spruce, might account for these differences (Boon et al., 2007). Temperament may provide one basis for individual differences and differing responses to heterogeneous environments.

8.2. Parasite Load

Parasite load may affect reproduction in rodent populations. Treatment of white-footed mice and deer mice with the drug ivermectin (designed to remove intestinal nematodes) resulted in lesser seasonal population declines than in control populations; when combined with food supplementation, population declines were eliminated completely (Pedersen & Greives, 2008). A different study of white-footed mice found similar effects of antihelminthic treatment in that mice continued to reproduce during normal times of reproductive hiatus, and demonstrated increased body condition, growth rate, and survival. Food supplementation did not have the same positive effects on reproduction (Vandegrift, Raffel, & Hudson, 2008).

8.3. Compounds Available in Green Plants

The presence of certain dietary plant compounds may serve as a signal for reproductive readiness. 6-methoxy-2-benzoxazolinone (6-MBOA) is a compound found in young plants (a byproduct of spontaneous breakdown due to injury of plant tissue (Berger, Negus, Sanders, & Gardner, 1981)). Male prairie voles whose mothers were fed a diet with added 6-MBOA displayed increased testicular mass and higher epididymal and testicular spermatogenesis, while female prairie voles were not affected by maternal exposure (Nelson, 1991). For a review of endocrine-active phytochemicals affecting reproduction of rodents and other vertebrates, see Vajda and Norris (2005).

9. CONCLUSIONS

The study of reproductive cycles in rodents is entering its second century. As it does so, there are clearly certain topics that present themselves as important, some of which have never been fully addressed and some of which are new areas of research.

9.1. Species Differences

Several previous works have emphasized the need for additional comparative data (Bronson, 1985; Schneider, 2004). Bronson in particular pointed out the lack of studies on tropical rodents, which make up a large proportion of the world's mammals. This situation has improved since 1985, with data on many desert rodents as well as tropical populations of rodents such as the Bathyergidae. However, our understanding of these species is still limited and requires further study.

9.2. Sex Differences

It is often said that females are more responsive to energetic constraints on reproduction than males. What are the proximate mechanisms for this responsiveness? How do the observed sex differences in responses to hormones such as estrogens and leptin factor into this?

9.3. Individual Differences

In several of the studies cited above, we have noted when there was heterogeneity of response to a treatment. What is this variation based on? One possibility is personality (Boon et al., 2007). Other possibilities include genetic variability and differential early environment. One tactic in examining individual differences is to use selective breeding experiments to produce individuals that are responsive, or unresponsive, to a given stimulus (Reilly, Oum, & Heideman, 2006).

9.4. Field vs. Laboratory Studies

The majority of experimental studies on rodent reproductive cycles are still done in the laboratory; more controlled field experiments will be valuable additions to our knowledge. Studies of rodent reproductive responses to changing habitat (edge vs. center, fragmentation effects, etc.) may become increasingly important as more rodent habitat is lost (Wilder & Meikle, 2005).

ABBREVIATIONS

6-MBOA 6-methoxy-2-benzoxazolinone
AANAT Arylalkylamine N-acetyltransferase
AgRP Agouti-related peptide
ARC Arcuate nucleus
AVP Arginine vasopressin
CART Cocaine- and amphetamine-related transcript
CRH Corticotropin-releasing hormone
D2 Type II deiodonase
D3 Type III deiodinase
DMH Dorsomedial hypothalamus
FSH Follicle-stimulating hormone
GnIH Gonadotropin-inhibiting hormone
GnRH Gonadotropin-releasing hormone
GPR54 Kisspeptin receptor
GTH Gonadotropin
HPG Hypothalamus–pituitary–gonadal
ir Immunoreactivity
LD Long day length
LH Luteinizing hormone
MBH Medial basal hypothalamus
MEL Melatonin
MEL-R Melatonin receptor
NPY Neuropeptide Y
P_4 Progesterone
POMC Pro-opiomelanocortin
PVN Paraventricular nucleus
RFRP RFamide-related peptide
SCN Suprachiasmatic nucleus
SD Short day length
T Testosterone
T_3 Triiodothyronine
T_4 Thyroxine
VEGF Vascular endothelial growth factor

REFERENCES

Agren, G. (1981). 2 Laboratory experiments on inbreeding avoidance in the Mongolian gerbil. *Behav. Proc., 6,* 291–297.

Alexe, D.-M., Syridou, G., & Petridou, E. T. (2006). Determinants of early life leptin levels and later life degenerative outcomes. *Clin. Med. Res., 4,* 326–335.

Alvarez, M. R., & Kravetz, F. O. (2006). Reproductive performance of capybaras (*Hydrochoerus hydrochaeris*) in captivity under different management systems in Argentina. *Anim. Res., 55,* 153–164.

Andrews, M. T. (2007). Advances in molecular biology of hibernation in mammals. *BioEssays., 29,* 431–440.

Arnold, W., & Dittami, J. (1997). Reproductive suppression in male alpine marmots. *Anim. Behav., 53,* 53–66.

Asher, M. R., De Oliveira, E. S., & Sachser, N. (2004). Social system and spatial organization of wild guinea pigs (*Cavia aperea*) in a natural population. *J. Mammal., 85,* 788–796.

Axelrod, J. (1974). The pineal gland: a neurochemical transducer. *Science, 184,* 1341–1348.

Barfield, M. A., & Beeman, E. A. (1968). The oestrus cycle in the Mongolian gerbil, *Meriones unguiculatus*. *J. Reprod. Fert., 17,* 247–251.

Barnes, B. M., Kretzmann, M., Zucker, I., & Licht, P. (1988). Plasma androgen and gonadotropin levels during hibernation and testis maturation in golden-mantled ground squirrels. *Biol. Reprod., 38,* 616–622.

Barrett, P., Ebling, F. J., Schuhler, S., Wilson, D., Ross, A. W., Warner, A., et al. (2007). Hypothalamic thyroid hormone catabolism acts as a gatekeeper for the seasonal control of body weight and reproduction. *Endocrinology, 148,* 3608–3617.

Bartness, T. J., Powers, J. B., Hastings, M. H., Bittman, E. L., & Goldman, B. D. (1993). The timed infusion paradigm for melatonin delivery: what has it taught us about the melatonin signal, its reception, and the photoperiodic control of seasonal responses? *J. Pineal Res., 15,* 161–190.

Batzli, G. O., Getz, L. L., & Hurley, S. S. (1977). Suppression of growth and reproduction of microtine rodents by social factors. *J. Mammal., 58,* 583–591.

Bauer, B., Womastek, I., Dittami, J., & Huber, S. (2008). The effects of early environmental conditions on the reproductive and somatic development of juvenile guinea pigs (*Cavia aperea f. porcellus*). *Gen. Comp. Endocrinol., 155,* 680–685.

Beery, A. K., Paul, M. J., Routman, D. M., & Zucker, I. (2008). Maternal photoperiodic history affects offspring development in Syrian hamsters. *J. Biol. Rhythms, 23,* 445–455.

Beery, A. K., Trumbull, J. J., Tsao, J. M., Costantini, R. M., & Zucker, I. (2007). Sex differences in the onset of seasonal reproductive quiescence in hamsters. *Proc. Biol. Sci., 274,* 281–286.

Bennett, N. C. (1989). The social structure and reproductive biology of the common mole rat, *Cryptomys hottentotus hottentotus*, and remarks on the trends in reproduction and sociality in the family Bathyergidae. *J. Zool., 219*, 45–59.

Bennett, N. C. (1994). Reproductive suppression in social *Cryptomys damarensis* colonies—a lifetime of socially-induced sterility in males and females (Rodentia, Bathyergidae). *J. Zool., 234*, 25–39.

Bennett, N. C., & Jarvis, J. U. M. (1988). The reproductive-biology of the cape mole-rat, *Georychus-capensis* (Rodentia, Bathyergidae). *J. Zool., 214*, 95–106.

Bennett, N. C., Faulkes, C. G., & Molteno, A. J. (1996). Reproductive suppression in subordinate, non-breeding female Damaraland mole-rats: two components to a lifetime of socially induced infertility. *Proc. Biol. Sci., 263*, 1599–1603.

Bennett, N. C., Faulkes, C. G., & Spinks, A. C. (1997). LH responses to single doses of exogenous GnRH by social Mashona mole-rats: a continuum of socially induced infertility in the family Bathyergidae. *Proc. Biol. Sci., 264*, 1001–1006.

Bennett, N. C., Jarvis, J. U., Faulkes, C. G., & Millar, R. P. (1993). LH responses to single doses of exogenous GnRH by freshly captured Damaraland mole-rats, *Cryptomys damarensis*. *J. Reprod. Fert., 99*, 81–86.

Bennett, N. C., Jarvis, J. U. M., & Cotterill, F. P. D. (1994). The colony structure and reproductive-biology of the afrotropical Mashona mole-rat, *Crymptomys darlingi*. *J. Zool., 234, 477–487.*

Bercovitch, F. B., & Ziegler, T. E. (2002). Current topics in primate socioendocrinology. *Ann. Rev. Anthro., 31*, 45–67.

Berger, P. J., Negus, N. C., Sanders, E. H., & Gardner, P. D. (1981). Chemical triggering of reproduction in *Microtus montanus*. *Science, 214*, 69–70.

Bittman, E. L. (1978). Hamster refractoriness: the role of insensitivity of pineal target tissues. *Science, 202*, 648–650.

Boon, A. K., Reale, D., & Boutin, S. (2007). The interaction between personality, offspring fitness and food abundance in North American red squirrels. *Ecol. Lett., 10*, 1094–1104.

Bouchie, L., Bennett, N. C., Jackson, T., & Waterman, J. M. (2006). Are Cape ground squirrels (*Xerus inauris*) induced or spontaneous ovulators? *J. Mammal., 87*, 60–66.

Bouret, S. G., Draper, S. J., & Simerly, R. B. (2004). Trophic action of leptin on hypothalamic neurons that regulate feeding. *Science, 304*, 108–110.

Bradford, D. F. (1975). The effects of an artificial water supply on free-living *Peromyscus truei*. *J. Mammal., 56*, 705–707.

Bradley, E. L., & Terman, C. R. (1979). Ovulations in *Peromyscus maniculatus bairdi* under laboratory conditions. *J. Mammal., 60*, 543–549.

Bradley, E. L., & Terman, C. R. (1979). Ovulation in *Peromyscus maniculatus bairdi* under laboratory conditions. *J. Mammal., 60*, 543–549.

Brant, C. L., Schwab, T. M., Vanderbergh, J. G., Schaefer, R. L., & Solomon, N. G. (1998). Behavioural suppression of female pine voles after replacement of the breeding male. *Anim. Behav., 55*, 615–627.

Bratincsak, A., McMullen, D., Miyake, S., Toth, Z., Hallenbeck, J. M., & Palkovits, M. (2007). Spatial and temporal activation of brain regions in hibernation: *c-fos* expression during the hibernation bout in thirteen-lined ground squirrel. *J. Comp. Neurol., 505*, 443–458.

Breed, W. G. (1975). Environmental factors and reproduction in the female hopping mouse, *Notomys alexis*. *J. Reprod. Fert., 45*, 273–281.

Breed, W. G. (1976). Effect of environment on ovarian activity of wild hopping mice (*Notomys alexis*). *J. Reprod. Fert., 47*, 395–397.

Bronson, F. H. (1985). Mammalian reproduction: an ecological perspective. *Biol. Reprod., 32*, 1–26.

Bronson, F. H. (1988). Mammalian reproductive strategies: genes, photoperiod, and latitude. *Reprod. Nutr. Develop., 28*, 335–347.

Bronson, F. H. (1989). *Mammalian Reproductive Biology.* Chicago, IL: University of Chicago Press.

Bronson, F. H., & Heideman, P. D. (1992). Lack of reproductive photoresponsiveness and correlative failure to respond to melatonin in a tropical rodent, the cane mouse. *Biol. Reprod., 46*, 246–250.

Bronson, F. H., & Pryor, S. (1983). Ambient temperature and reproductive success in rodents living at different latitudes. *Biol. Reprod., 29*, 72–80.

Burda, H. (1995). Individual recognition and incest avoidance in eusocial common mole-rats rather than reproductive suppression by parents. *Cell. Mol. Life Sci., 51*, 411–413.

Butler, M. P., Trumbull, J. J., Turner, K. W., & Zucker, I. (2007a). Timing of puberty and synchronization of seasonal rhythms by simulated natural photoperiods in female Siberian hamsters. *Amer. J. Phys. Reg. Int. Comp. Phys., 293*, R413–R420.

Butler, M. P., Turner, K. W., Park, J. H., Butler, J. P., Trumbull, J. J., Dunn, S. P., et al. (2007b). Simulated natural day lengths synchronize seasonal rhythms of asynchronously born male Siberian hamsters. *Amer. J. Phys. -Reg. Int. Comp. Phys., 293*, R402–R412.

Capitanio, J. P. (1999). Personality dimensions in adult male rhesus macaques: Prediction of behaviors across time and situation. *Amer. J. Primatol., 47*, 299–320.

Carlson, L. L., Weaver, D. R., & Reppert, S. M. (1991). Melatonin receptors and signal transduction during development in Siberian hamsters (*Phodopus sungorus*). *Dev. Brain Res., 59*, 83–88.

Carlson, L. L., Zimmermann, A., & Lynch, G. R. (1989). Geographic differences for delay of sexual maturation in *Peromyscus leucopus*: effects of photoperiod, pinealectomy, and melatonin. *Biol. Reprod., 41*, 1004–1013.

Carter, C. S. (1973). Stimuli contributing to decrement in sexual receptivity of female golden hamsters *(Mesocricetus auratus). Anim. Behav., 21*, 827–834.

Carter, C. S., Getz, L. L., Gavish, L., McDermott, J. L., & Arnold, P. (1980). Male-related pheromones and the activation of female reproduction in the prairie vole (*Microtus ochrogaster*). *Biol. Reprod., 23*, 1038–1045.

Carter, C. S., Getz, L. L., Gavish, L., McDermott, J. L., & Arnold, P. (1980). Male-related pheromones and the activation of female reproduction in the prairie vole (*Microtus ochrogaster*). *Biol. Reprod., 23*, 1038–1045.

Carter, C. S., Witt, D. M., Manock, S. R., Adams, K. A., Bahr, J. M., & Carlstead, K. (1989). Hormonal correlates of sexual behavior and ovulation in male-induced and postpartum estrus in female prairie voles. *Physiol. Behav., 46*, 941–948.

Carter, C. S., Witt, D. M., Schneider, J., Harris, Z. L., & Volkening, D. (1987). Male stimuli are necessary for female sexual behavior and uterine growth in prairie voles (*Microtus ochrogaster*). *Horm. Behav., 21*, 74–82.

Chan, J. L., & Mantzoros, C. S. (2001). Leptin and the hypothalamic–pituitary regulation of the gonadotropin–gonadal axis. *Pituitary., 4*, 87–92.

Cherry, J. A. (1987). The effect of photoperiod on development of sexual behavior and fertility in golden hamsters. *Physiol. Behav., 39*, 521–526.

Christian, D. P. (1979). Comparative demography of three Namib desert rodents: responses to the provision of supplementary water. *J. Mammal., 60*, 679–690.

Clancy, A. N., Goldman, B. D., Bartke, A., & Macrides, F. (1986). Reproductive effects of olfactory bulbectomy in the Syrian hamster. *Biol. Reprod., 35*, 1202–1209.

Clark, F. H. (1938). Age of sexual maturity in mice of the genus *Peromyscus. J. Mammal., 19*, 230–234.

Clark, M. M., & Galef, B. G. (2001). Socially induced infertility: familial effects on reproductive development of female Mongolian gerbils. *Anim. Behav., 62*, 897–903.

Clarke, F. M., & Faulkes, C. G. (1997). Dominance and queen succession in captive colonies of the eusocial naked mole-rat, *Heterocephalus glaber. Proc. Royal Soc. London B, 264*, 993–1000.

Clarke, F. M., Miethe, G. H., & Bennett, N. C. (2001). Reproductive suppression in female Damaraland mole-rats *Cryptomys damarensis*: dominant control or self-restraint? *Proc. Royal Soc. London B, 268*, 899–909.

Concannon, P. W., Castracane, V. D., Rawson, R. E., & Tennant, B. C. (1999). Circannual changes in free thyroxine, prolactin, testes, and relative food intake in woodchucks, *Marmota monax. Amer. J. Phys., 277*, R1401–R1409.

Concannon, P. W., Roberts, P., Graham, L., & Tennant, B. C. (1998). Annual cycle in LH and testosterone release in response to GnRH challenge in male woodchucks *(Marmota monax). J. Reprod. Fert., 114*, 299–305.

Cooney, R., & Bennett, N. C. (2000). Inbreeding avoidance and reproductive skew in a cooperative mammal. *Proc. Royal Soc. London B, 267*, 801–806.

Cummings, D. E. (2006). Ghrelin and the short- and long-term regulation of appetite and body weight. *Physiol. Behav., 89*, 71–84.

Dark, J. (2005). Annual lipid cycles in hibernators: integration of physiology and behavior. *Ann. Rev. Nutr., 25*, 469–497.

Dark, J., & Zucker, I. (1983). Short photoperiods reduce winter energy requirements of the meadow vole, *Microtus pennsylvanicus. Physiol. Behav., 31*, 699–702.

Dark, J., Johnston, P. G., Healy, M., & Zucker, I. (1983). Latitude of origin influences photoperiodic control of reproduction of deer mice (*Peromyscus maniculatus*). *Biol. Reprod., 28*, 213–220.

Davies, M. J., & Norman, R. J. (2002). Programming and reproductive functioning. *Trends Endocrinol. Metab., 13*, 386–392.

DelBarco-Trillo, J., & Ferkin, M. H. (2007). Female meadow voles, *Microtus pennsylvanicus*, experience a reduction in copulatory behavior during postpartum estrus. *Ethology, 113*, 466–473.

Delville, Y., Conklin, L. S., & Ferris, C. F. (1995). Differential expression of vasopressin receptor binding in the hypothalamus during lactation in golden hamsters. *Brain Res., 689*, 147–150.

Demas, G. E., & Nelson, R. J. (1998). Photoperiod, ambient temperature, and food availability interact to affect reproductive and immune function in adult male deer mice. *J. Biol. Rhythm., 13*, 253–262.

Desjardins, C., & Lopez, M. J. (1983). Environmental cues evoke differential responses in pituitary–testicular function in deer mice. *Endocrinology, 112*, 1398–1406.

Desjardins, C., Bronson, F. H., & Blank, J. L. (1986). Genetic selection for reproductive photoresponsiveness in deer mice. *Nature, 322*, 172–173.

Dewsbury, D. A. (1988). Sperm competition in deer mice (*Peromyscus maniculatus bairdi*): effects of cycling versus postpartum estrus and delays between matings. *Behav. Ecol. Sociobiol., 22*, 251–256.

Dewsbury, D. A. (1989). Modes of estrus induction as a factor in studies of the reproductive behavior of rodents. *Neurosci. Biobehav. Rev., 14*, 147–155.

Dluzen, D. E., Ramirez, V. D., Carter, C. S., & Getz, L. L. (1981). Male vole urine changes luteinizing hormone-releasing hormone and norepinephrine in female olfactory bulb. *Science, 212*, 573–575.

DuToit, L., Bennett, N. C., Gutjahr, G. H., & Coen, C. W. (2006a). Reproductive suppression in subordinate female highveld mole-rats (*Cryptomys hottentotus pretoriae*): no role for endogenous opioid peptides. *Physiol. Behav., 87*, 897–902.

DuToit, L., Bennett, N. C., Katz, A. A., Kallo, I., & Coen, C. W. (2006b). Relations between social status and the gonadotrophin-releasing hormone system in females of two cooperatively breeding species of African mole-rats, *Cryptomys hottentotus hottentotus* and *Cryptomys hottentotus pretoriae*: neuroanatomical and neuroendocrinological studies. *J. Comp. Neurol., 494*, 303–313.

Ebling, F. J. P. (1994). Photoperiodic differences during development in the dwarf hamsters *Phodopus sungorus* and *Phodopus campbelli. Gen. Comp. Endocrinol., 95*, 475–482.

Ebling, F. J. P., & Luckman, S. M. (2008). RFamide-related peptide: Another sexy peptide? *Endocrinology, 149*, 899–901.

Edmonds, K., Riggs, L., & Masden, T. (2005). Effects of photoperiod, melatonin, and the pineal gland on compensatory gonadal hypertrophy during postnatal development in the marsh rice rat (*Oryzomys palustris*). *Zoolog. Sci., 22*, 763–774.

Edmonds, K. E., & Stetson, M. H. (1993). Effect of photoperiod on gonadal maintenance and development in the marsh rice rat (*Oryzomys palustris*). *Gen. Comp. Endocrinol., 92*, 281–291.

Edmonds, K. E., Riggs, L., & Stetson, M. H. (2003). Food availability and photoperiod affect reproductive development and maintenance in the marsh rice rat (*Oryzomys palustris*). *Physiol. Behav., 78*, 41–49.

El-Bakry, H. A., Zahran, W. M., & Bartness, T. J. (1999). Control of reproductive and energetic status by environmental cues in a desert rodent, Shaw's jird. *Physiol. Behav., 66*, 657–666.

Elton, C. S. (1924). Periodic fluctuations in numbers of animals: their causes and effects. *Brit. J. Exp. Biol., 2*, 119–163.

Erb, G. E., Edwards, H. E., Jenkins, K. L., Mucklow, L. C., & Wynne-Edwards, K. E. (1993). Induced components in the spontaneous ovulatory cycle of the Djungarian hamster (*Phodopus campbelli*). *Physiol. Behav., 54*, 955–959.

Erb, G. E., Edwards, H. E., Jenkins, K. L., Mucklow, L. C., & Wynne-Edwards, K. E. (1993). Induced components in the spontaneous ovulatory cycle of the Djungarian hamster (*Phodopus campbelli*). *Physiol. Behav., 54*, 955–959.

Fanjul, M. S., & Zenuto, R. R. (2008). Copulatory pattern of the subterranean rodent *Ctenomys talarum. Mammalia, 72*, 102–108.

Fanjul, M. S., Zenuto, R. R., & Busch, C. (2006). Seasonality of breeding in wild tuco-tucos *Ctenomys talarum* in relation to climate and food availability. *Acta Theriol., 51*, 283–293.

Faulkes, C. G., & Abbott, D. H. (1991). Social control of reproduction in breeding and non-breeding male naked mole-rats (*Heterocephalus glaber*). *J. Reprod. Fertil., 93*, 427–435.

Faulkes, C. G., & Abbott, D. H. (1993). Evidence that primer pheromones do not cause social suppression of reproduction in male and female naked mole-rats (*Heterocephalus glaber*). *J. Reprod. Fertil., 99*, 225–230.

Faulkes, C. G., & Abbott, D. H. (1997). Proximate mechanisms regulating a reproductive dictatorship: a single dominant female controls male and female reproduction in colonies of naked mole-rats. In N. G. Solomon, & J. A. French (Eds.), *Cooperative Breeding in Mammals* (pp. 302–334). Cambridge, UK: Cambridge University Press.

Faulkes, C. G., & Bennett, N. C. (2001). Family values: group dynamics and social control of reproduction in African mole-rats. *Trends Ecol. Evol., 16*, 184–190.

Faulkes, C. G., Abbott, D. H., & Jarvis, J. U. (1990a). Social suppression of ovarian cyclicity in captive and wild colonies of naked mole-rats, *Heterocephalus glaber. J. Reprod. Fertil., 88*, 559–568.

Faulkes, C. G., Abbott, D. H., & Jarvis, J. U. (1991). Social suppression of reproduction in male naked mole-rats, *Heterocephalus glaber. J. Reprod. Fertil., 91*, 593–604.

Faulkes, C. G., Abbott, D. H., Jarvis, J. U., & Sherriff, F. E. (1990b). LH responses of female naked mole-rats, *Heterocephalus glaber*, to single and multiple doses of exogenous GnRH. *J. Reprod. Fertil., 89*, 317–323.

Faulkes, C. G., Trowell, S. N., Jarvis, J. U., & Bennett, N. C. (1994). Investigation of numbers and motility of spermatozoa in reproductively active and socially suppressed males of two eusocial African mole-rats, the naked mole-rat (*Heterocephalus glaber)* and the Damaraland mole-rat (*Cryptomys damarensis*). *J. Reprod. Fert., 100*, 411–416.

Fernandez-Fernandez, R., Martini, A. C., Navarro, V. M., Castellano, J. M., Dieguez, C., Aguilar, E., et al. (2006). Novel signals for the integration of energy balance and reproduction. *Mol. Cell. Endocrinol., 254-255*, 127–132.

Fey, K., Banks, P. B., & Korpimaki, E. (2008). Voles on small islands: effects of food limitation and alien predation. *Oecologia, 157*, 419–428.

Forger, N. G., & Zucker, I. (1985). Photoperiodic regulation of reproductive development in male white-footed mice (*Peromyscus leucopus*) born at different phases of the breeding season. *J. Reprod. Fert., 73*, 271–278.

Foster, M. A. (1934). The reproductive cycle in the female ground squirrel, *Citellus tridecemlineatus* (Mitchell). *Amer. J. Anat., 54*, 487–510.

Freeman, D. A., Teubner, B. J., Smith, C. D., & Prendergast, B. J. (2007). Exogenous T3 mimics long day lengths in Siberian hamsters. *Am. J. Physiol. Regul. Integr. Comp. Physiol., 292*, R2368–R2372.

Friedman, J. M., & Halaas, H. (1998). Leptin and regulation of body weight in mammals. *Nature, 395*, 763–770.

Frisch, R. E., & MacArthur, J. W. (1974). Menstrual cycles: fatness as a determinant of minimum weight for height necessary for their maintenance or onset. *Science, 185*, 949–951.

Frost, D., & Zucker, I. (1983). Photoperiod and melatonin influence seasonal gonadal cycles in the grasshopper mouse (*Onychomys leucogaster*). *J. Reprod. Fert., 69*, 237–244.

Furman, M., & Wade, G. N. (2007). Animal models in the study of nutritional infertility. *Curr. Opin. Endocrinol. Diab. Obes., 14*, 475–481.

Furuta, M., Funabashi, T., & Kimura, F. (2001). Intracerebroventricular administration of ghrelin rapidly suppresses pulsatile luteinizing hormone secretion in ovariectomized rats. *Biochem. Biophys. Res. Comm., 288*, 780–785.

Gern, W. A., Duvall, D., & Nervina, J. M. (1986). Melatonin—a discussion of its evolution and actions in vertebrates. *Amer. Zoologist., 26*, 985–996.

Getz, L. L., Carter, C. S., & Gavish, L. (1981). The mating system of the prarie vole *Microtus ochrogaster*: field and laboratory evidence for pair-bonding. *Behav. Ecol. Sociobiol., 8*, 189–194.

Getz, L. L., Dluzen, D., & McDermott, J. L. (1983). Suppression of reproductive maturation in male-stimulated virgin female *Microtus* by a female urinary chemosignal. *Behav. Proc., 8*, 59–64.

Gilbert, A. N. (1984). Postpartum and lactational estrus: a comparative analysis in Rodentia. *J. Comp. Psychol., 98*, 232–245.

Gilbert, A. N., Burgoon, D. A., Sullivan, K. A., & Adler, N. T. (1983). Mother-weanling interactions in Norway rats in the presence of a successive litter produced by postpartum mating. *Physiol. Behav., 30*, 267–271.

Glass, J. D. (1986). Gonadotropin-releasing hormone neuronal system of the white-footed mouse, *Peromyscus leucopus. Neuroendocrinology, 43*, 220–229.

Glass, J. D., Ferreira, S., & Deaver, D. R. (1988). Photoperiodic adjustments in hypothalamic amines, gonadotropin-releasing hormone, and beta-endorphin in the white-footed mouse. *Endocrinology, 123*, 1119–1127.

Gockel, J., & Ruf, T. (2001). Alternative seasonal reproductive strategies in wild rodent populations. *J. Mammal., 82*, 1034–1046.

Goldman, B. D., Darrow, J. M., & Yogev, L. (1984). Effects of timed melatonin infusions on reproductive development in the Djungarian hamster (*Phodopus sungorus*). *Endocrinology, 114*, 2074–2083.

Gorman, M. R. (1995). Seasonal adaptations of Siberian hamsters.1. Accelerated gonadal and somatic development in increasing versus static long-day lengths. *Biol. Reprod., 53*, 110–115.

Gorman, M. R. (2003). Melatonin implants disrupt developmental synchrony regulated by flexible interval timers. *J. Neuroendocrinol., 15*, 1084–1094.

Gorman, M. R., & Zucker, I. (1997). Environmental induction of photononresponsiveness in the Siberian hamster, *Phodopus sungorus. Amer. J. Physiol. Reg. Int. Comp. Phys., 41*, R887–R895.

Gower, B. A., Nagy, T. R., & Stetson, M. H. (1996). Influence of photoperiod, time, and sex on hormone concentrations in collared lemmings (*Dicrostonyx groenlandicus*). *Gen. Comp. Endocrinol., 101*, 53–62.

Gram, W. D., Heath, H. W., Wichman, H. A., & Lynch, G. R. (1982). Geographic variation in *Peromyscus leucopus*: short-day induced reproductive regression and spontaneous recrudescence. *Biol. Reprod., 27*, 369–373.

Gray, G. D., Davis, H. N., Zerylnick, M., & Dewsbury, D. A. (1974). Oestrus and induced ovulation in montane voles. *J. Reprod. Fert., 38*, 193–196.

Greives, T. J., Humber, S. A., Goldstein, A. N., Scotti, M. A., Demas, G. E., & Kriegsfeld, L. J. (2008a). Photoperiod and testosterone interact to drive seasonal changes in kisspeptin expression in Siberian hamsters (*Phodopus sungorus*). *J. Neuroendocrinol., 20*, 1339–1347.

Greives, T. J., Kriegsfeld, L. J., & Demas, G. E. (2008c). Exogenous kisspeptin does not alter photoperiod-induced gonadal regression in Siberian hamsters (*Phodopus sungorus*). *Gen. Comp. Endocrinol., 156*, 552–558.

Greives, T. J., Kriegsfeld, L. J., Bentley, G. E., Tsutsui, K., & Demas, G. E. (2008b). Recent advances in reproductive neuroendocrinology: a role for RFamide peptides in seasonal reproduction? *Proc. Biol. Sci., 275*, 1943–1951.

Greives, T. J., Mason, A. O., Scotti, M. A., Levine, J., Ketterson, E. D., Kriegsfeld, L. J., & Demas, G. E. (2007). Environmental control of kisspeptin: implications for seasonal reproduction. *Endocrinology, 148*, 1158–1166.

Grocock, C. A. (1981). Effect of different photoperiods on testicular weight changes in the vole, *Microtus agrestis*. *J. Reprod. Fert., 62*, 25–32.

Gubernick, D. J. (1988). Reproduction in the California mouse, *Peromyscus californicus*. *J. Mammal., 69*, 860.

Gubernick, D. J. (1988). Reproduction in the California mouse, *Peromyscus californicus*. *J. Mammal., 69*, 857–860.

Gunduz, B., & Stetson, M. H. (2001). A test of the coincidence and duration models of melatonin action in Siberian hamsters: the effects of 1-hr melatonin infusions on testicular development in intact and pinealectomized prepubertal *Phodopus sungorus*. *J. Pineal Res., 30*, 97–107.

Hacklander, K., Mostl, E., & Arnold, W. (2003). Reproductive suppression in female Alpine marmots, *Marmota marmota*. *Anim. Behav., 65*, 1133–1140.

Haldar, C., Shavali, S. S., & Singh, S. (1992). Photoperiodic response of pineal–thyroid axis of the female Indian palm squirrel, *Funambulus pennanti*. *J. Neural. Transm. Gen. Sect., 90*, 45–52.

Hall, E. S., & Lynch, G. R. (1985). Two daily melatonin injections differentially induce nonshivering thermogenesis and gonadal regression in the mouse (*Peromyscus leucopus*). *Life Sci., 37*, 783–788.

Hanski, I., Henttonen, H., Korpimaki, E., Oksanen, L., & Turchin, P. (2001). Small rodent dynamics and predation. *Ecology, 82*, 1505–1520.

Hart, L., Bennett, N. C., Kalamatianos, T., Oosthuizen, M. K., Jarvis, J. U., O'Riain, M. J., et al. (2008). Neuroanatomical investigation of the gonadotrophin-releasing hormone 1 system in the seasonally breeding Cape dune mole-rat, *Bathyergus suillus*. *Brain Res. Bull.*

Hart, L., O'Riain, M. J., Jarvis, J. U., & Bennett, N. C. (2006a). The pituitary potential for opportunistic breeding in the Cape dune mole-rat, *Bathyergus suillus*. *Physiol. Behav., 88*, 615–619.

Hart, L., O'Riain, M. J., Jarvix, J. U. M., & Bennett, N. C. (2006b). Is the Cape dune mole-rat, *Bathyergus suillus* (Rodentia: Bathyergidae), a seasonal or aseasonal breeder? *J. Mammal., 87*, 1078–1085.

Heideman, P. D., & Bronson, F. H. (1990). Photoperiod, melatonin secretion, and sexual maturation in a tropical rodent. *Biol. Reprod., 43*, 745–750.

Heideman, P. D., & Bronson, F. H. (1991). Characteristics of a genetic polymorphism for reproductive photoresponsiveness in the white-footed mouse (*Peromyscus leucopus*). *Biol. Reprod., 44*, 1189–1196.

Heth, G., Todrank, J., Begall, S., Wegner, R. E., & Burda, H. (2004). Genetic relatedness discrimination in eusocial *Cryptomys anselli* mole-rats, Bathyergidae, Rodentia. *Folia Zool., 53*, 269–278.

Hikim, A. P. S., Woolf, A., Bartke, A., & Amador, A. G. (1992). Further observations on estrus and ovulation in woodchucks (*Marmota momax*) in captivity. *Biol. Reprod., 46*, 10–16.

Hoffmann, K. (1978). Effects of short photoperiods on puberty, growth and molt in Djungarian Hamster (*Phodopus sungorus*). *J. Reprod. Fert., 54*, 29–35.

Hofmann, J. E., & Getz, L. L. (1988). Multiple exposures to adult males and reproductive activation of virgin female *Microtus ochrogaster*. *Behav. Proc., 17*, 57–61.

Holmes, M. M., Rosen, G. J., Jordan, C. L., De Vries, G. J., Goldman, B. D., & Forger, N. G. (2007). Social control of brain morphology in a eusocial mammal. *Proc. Natl. Acad. Sci. USA, 104* (25), 10548–10552.

Horton, T. H. (1984). Growth and reproductive development of male *Microtus montanus* is affected by the prenatal photoperiod. *Biol. Reprod., 31*, 499–504.

Insel, T. R., & Shapiro, L. E. (1992). Oxytocin receptor distribution reflects social organization in monogamous and polygamous voles. *PNAS, 89*, 5981–5985.

Insel, T. R., Wang, Z. X., & Ferris, C. F. (1994). Patterns of brain vasopressin receptor distribution associated with social organization in microtine rodents. *J. Neurosci., 14*, 5381–5392.

Jackson, C. R., & Bennett, N. C. (2005). Is the Natal mole-rat (*Cryptomys hottentotus natalensis*) a spontaneous or induced ovulator? *J. Mammal., 86*, 1–6.

Jacobs, D. S., Bennett, N. C., Jarvis, J. U. M., & Crowe, T. M. (1991). The colony structure and dominance hierarchy of the Damaraland mole-rat, *Cryptomys damarensis* (Rodentia: Bathyergidae), from Namibia. *J. Zool., 224*, 553–576.

Jarvis, J. U. M. (1991). Reproduction of naked mole-rats. In P. W. Sherman, J. U. M. Jarvix, & R. D. Alexander (Eds.), *The Biology of the Naked Mole-Rat* (pp. 384–425). Princeton, NJ: Princeton University Press.

Jochle, W. (1973). Coitus-induced ovulation. *Contraception, 7*, 523–564.

Johnston, P. G., & Zucker, I. (1979). Photoperiodic influences on gonadal development and maintenance in the cotton rat, *Sigmodon hispidus*. *Biol. Reprod., 21*, 1–8.

Johnston, P. G., & Zucker, I. (1980). Photoperiodic regulation of reproductive development in white-footed mice (*Peromyscus leucopus*). *Biol. Reprod., 22*, 983–989.

Kabithe, E. W., & Place, N. J. (2008). Photoperiod-dependent modulation of anti-Mullerian hormone in female Siberian hamsters, *Phodopus sungorus*. *Reproduction, 135*, 335–342.

Kanwar, K. C., & Chaudhry, V. (1976). Seasonal reproductive periodicity and thyroid fluctuations in female Indian palm squirrel, *Funambulus pennanti*. *Acta. Biol. Med. Ger., 35*, 1643–1649.

Kausrad, K. L., Mysterud, A., Steen, H., Vik, J. O., Ostbye, E., Cazelles, B., et al. (2008). Linking climate change to lemming cycles. *Nature, 456*, 93–98.

Kawamura, K., Sato, N., Fukuda, J., Kodama, H., Kumegai, J., Tanikawa, H., et al. (2003). Ghrelin inhibits the development of mouse preimplantation embryos *in vitro*. *Endocrinology, 144*, 2623–2633.

Kenney, A. M., Hartung, T. G., & Dewsbury, D. A. (1979). Copulatory behavior and the initiation of pregnancy in California voles (*Microtus californicus*). *Brain Behav. Evol., 16*, 176–191.

Kerbeshian, M. C., Bronson, F. H., & Bellis, E. D. (1994). Variation in reproductive photoresponsiveness in a wild population of meadow voles. *Biol. Reprod., 50*, 745–750.

Komisaruk, B. R., & Steinman, J. L. (1986). Genital stimulation as a trigger for neuroendocrine and behavioral control of reproduction. *Ann. NY Acad. Sci., 474*, 64–75.

Kotula-Balak, M., Slomczynska, M., Fraczek, B., Bourguiba, S., Tabarowski, Z., Carreau, S., et al. (2003). Complementary approaches demonstrate that cellular aromatization in the bank vole testis is related to photoperiod. *Eur. J. Histochem., 47*, 55–62.

Kriegsfeld, L. J. (2006). Driving reproduction: RFamide peptides behind the wheel. *Horm. Behav., 50*, 655–666.

Kriegsfeld, L. J., Ranalli, N. J., Bober, M. A., & Nelson, R. J. (2000a). Photoperiod and temperature interact to affect the GnRH neuronal system of male prairie voles (*Microtus ochrogaster*). *J. Biol. Rhythms, 15*, 306–316.

Kriegsfeld, L. J., Ranalli, N. J., Trasy, A. G., & Nelson, R. J. (2001). Food restriction affects the gonadotropin releasing hormone neuronal system of male prairie voles (*Microtus ochrogaster*). *J. Neuroendocrinol., 13*, 791–798.

Kriegsfeld, L. J., Trasy, A. G., & Nelson, R. J. (2000b). Temperature and photoperiod interact to affect reproduction and GnRH synthesis in male prairie voles. *J. Neuroendocrinol., 12*, 553–558.

Lariviere, S., & Ferguson, S. H. (2003). Evolution of induced ovulation in North American carnivores. *J. Mammal., 84*, 937–947.

Loewy, A. D. (1990). Central autonomic pathways. In A. D. Loewy, & K. M. Spyer (Eds.), *Central Regulation of Autonomic Functions* (pp. 88–103). New York, NY: Oxford University Press.

Long, J. A., & Evans, H. M. (1922). The oestrus cycle in the rat and its associated phenomena. *Memoirs of the University of California, 6*, 1–148.

Long, R. A., Hut, R. A., & Barnes, B. M. (2007). Simultaneous collection of body temperature and activity data in burrowing mammals: a new technique. *J. Wildlife Manag., 71*, 1375–1379.

Lyman, C. P., Willis, J. S., Malan, A., & Wang, L. C. H. (1982). *Hibernation and Torpor in Mammals and Birds*. New York, NY: Academic Press.

Lynch, G. R., Heath, H. W., & Johnston, C. M. (1981). Effect of geographical origin on the photoperiodic control of reproduction in the white-footed mouse, *Peromyscus leucopus*. *Biol. Reprod., 25*, 475–480.

Lynch, G. R., Heath, H. W., & Margolis, D. J. (1982a). Daily melatonin injections: their usefulness in understanding photoperiodism in *Peromyscus leucopus*. *Int. J. Biometeorol., 26*, 305–309.

Lynch, G. R., Lynch, C. B., & Kliman, R. M. (1989). Genetic analyses of photoresponsiveness in the Djungarian hamster, *Phodopus sungorus*. *J. Comp. Physiol. [A], 164*, 475–481.

Lynch, G. R., Sullivan, J. K., Heath, H. W., & Tamarkin, L. (1982b). Daily melatonin rhythms in photoperiod sensitive and insensitive white-footed mice (*Peromyscus leucopus*). *Prog. Clin. Biol. Res., 92*, 67–73.

Maldonado-Chaparro, A., & Blumstein, D. T. (2008). Management implications of capybara (*Hydrochoerus hydrochaeris*) social behavior. *Biol. Cons., 141*, 1945–1952.

Malherbe, G. P., Schoeman, A. S., & Bennett, N. C. (2004). Is the highveld mole-rat *Cryptomys hottentotus pretoriae* (Rodentia: Bathyergidae) an induced or spontaneous ovulator? *J. Zool., 263*, 159–165.

Malpaux, B., Thiery, J. C., & Chemineau, P. (1999). Melatonin and the seasonal control of reproduction. *Reprod. Nutr. Dev., 39*, 355–366.

Marchlewska-Koj, A., Kruczek, M., & Olejnicjak, P. (2003). Mating of bank voles (*Clethrionomys glareolus*) modified by hormonal and social factors. *Mammal. Biol., 68*, 144–152.

Marsteller, F. A., & Lynch, C. B. (1987). Reproductive responses to variation in temperature and food supply by mouse mice. I. Mating and pregnancy. *Biol. Reprod., 37*, 838–843.

Mason, A. O., Greives, T. J., Scotti, M. A., Levine, J., Frommeyer, S., Ketterson, E. D., et al. (2007). Suppression of kisspeptin expression and gonadotropic axis sensitivity following exposure to inhibitory day lengths in female Siberian hamsters. *Horm. Behav., 52*, 492–498.

Maswanganye, K. A., Bennett, N. C., Brinders, J., & Cooney, R. (1999). Oligospermia and azoospermia in non-reproductive male Damaraland mole-rats *Cryptomys damarensis* (Rodentia: Bathyergidae). *J. Zool., 248*, 411–418.

McGuire, B., & Getz, L. L. (1991). Responses of young female prairie voles (*Microtus ochrogaster*) to nonresident males: implications for population regulation. *Can. J. Zool., 69*, 1348–1355.

Meikle, D., & Westberg, M. (2001a). Maternal nutrition and reproduction of daughters in wild house mice (*Mus musculus*). *Reproduction, 122*, 437–442.

Meikle, D., & Westberg, M. (2001b). Social dominance rank and accessory sex glands in wild adult male house mice born to food-deprived mothers. *Physiol. Behav., 72*, 359–364.

Michener, G. R. (1992). Sexual differences in over-winter torpor patterns of Richardson's ground squirrels in natural hibernacula. *Oecologia, 89*, 396–407.

Millar, J. S. (2001). On reproduction in lemmings. *Ecoscience, 8*, 145–150.

Moffatt-Blue, C. S., Sury, J. J., & Young, K. A. (2006). Short photoperiod-induced ovarian regression is mediated by apoptosis in Siberian hamsters (*Phodopus sungorus*). *Reproduction, 131*, 771–782.

Molteno, A. J., & Bennett, N. C. (2000). Anovulation in non-reproductive female Damaraland mole-rats (*Cryptomys damarensis*). *J. Reprod. Fertil., 119*, 35–41.

Molteno, A. J., & Bennett, N. C. (2002). Social suppression in nonreproductive female Damaraland mole-rats, *Cryptomys damarensis*: no apparent role for endogenous opioid peptides. *Horm. Behav., 41*, 115–125.

Molteno, A. J., Kallo, I., Bennett, N. C., King, J. A., & Coen, C. W. (2004). A neuroanatomical and neuroendocrinological study into the relationship between social status and the GnRH system in cooperatively breeding female Damaraland mole-rats, *Cryptomys damarensis*. *Reproduction, 127*, 13–21.

Moolman, M., Bennett, N. C., & Schoeman, A. S. (1998). The social structure and dominance hierarchy of the highveld mole- *rat Cryptomys hottentotus pretoriae* (Rodentia: Bathyergidae). *J. Zool., 246*, 193–201.

Nelson, R. J. (1985). Photoperiod influences reproduction in the prairie vole (*Microtus ochrogaster*). *Biol. Reprod., 33*, 596–602.

Nelson, R. J. (1987). Photoperiod-nonresponsive morphs—a possible variable in microtine population density fluctuations. *Am. Nat., 130*, 350–369.

Nelson, R. J. (1988). Restricted water intake influences male reproduction in two strains of house mice (*Mus musculus*). *Physiol. Behav., 43*, 217–221.

Nelson, R. J. (1991). Maternal diet influences reproductive development in male prairie vole offspring. *Physiol. Behav., 50*, 1063–1066.

Nelson, R. J., & Desjardins, C. (1987). Water availability affects reproduction in deer mice. *Biol. Reprod., 37*, 257–260.

Nelson, R. J., & Zucker, I. (1981). Photoperiodic control of reproduction in olfactory-bulbectomized rats. *Neuroendocrinology, 32*, 266–271.

Nelson, R. J., Bamat, M. K., & Zucker, I. (1982). Photoperiodic regulation of testis function in rats: mediation by a circadian mechanism. *Biol. Reprod., 26*, 329–335.

Nelson, R. J., Dark, J., & Zucker, I. (1983). Influence of photoperiod, nutrition and water availability on reproduction of male California voles (*Microtus californicus*). *J. Reprod. Fert., 69*, 673–677.

Nelson, R. J., Frank, D., Bennett, S. A., & Carter, C. S. (1989a). Simulated drought influences reproduction in male prairie voles. *Physiol. Behav., 46*, 849–852.

Nelson, R. J., Frank, D., Smale, L., & Willoughby, S. B. (1989b). Photoperiod and temperature affect reproductive and nonreproductive functions in male prairie voles (*Microtus ochrogaster*). *Biol. Reprod., 40*, 481–485.

Nelson, R. J., Gubernick, D. J., & Blom, J. M. C. (1995). Influence of photoperiod, green food, and water availability on reproduction in male California mice (*Peromyscus californicus*). *Physiol. Behav., 57*, 1175–1180.

Nelson, R. J., Marinovic, A. C., Moffatt, C. A., Kriegsfeld, L. J., & Kim, S. (1997). The effects of photoperiod and food intake on reproductive development in male deer mice (*Peromyscus maniculatus*). *Physiol. Behav., 62*, 945–950.

Nelson, R. J., Moffatt, C. A., & Goldman, B. D. (1994). Reproductive and nonreproductive responsiveness to photoperiod in laboratory rats. *J. Pineal Res., 17*, 123–131.

Newsome, A. E., Stendell, R. C., & Myers, J. H. (1976). Free-watering a wild population of house mice: a test of an Australian hypothesis in California. *J. Mammal., 57*, 677–686.

Oli, M. K., & Armitage, K. B. (2003). Sociality and individual fitness in yellow-bellied marmots: insights from a long-term study (1962-2001). *Oecologia, 136*, 543–550.

Oosthuizen, M. K., & Bennett, N. C. (2007). LH responses to single doses of exogenous GnRH in the Cape mole rat (*Georychus capensis*): the pituitary potential for opportunistic breeding. *J. Zool., 271*, 198–202.

Oriain, M. J., Jarvis, J. U. M., & Faulkes, C. G. (1996). A dispersive morph in the naked mole-rat. *Nature, 380*, 619–621.

Parag, A., Bennett, N. C., Faulkes, C. G., & Bateman, P. W. (2006). Penile morphology of African mole rats (Bathyergidae): structural modification in relation to mode of ovulation and degree of sociality. *J. Zool., 270*, 323–329.

Payman, B. C., & Swanson, H. H. (1980). Social influence on sexual maturation and breeding in the female Mongolian gerbil (*Meriones unguiculatus*). *Anim. Behav., 28*. 528–&.

Pedersen, A. B., & Greives, T. J. (2008). The interaction of parasites and resources cause crashes in a wild mouse population. *J. Anim. Ecol., 77*, 370–377.

Petterborg, L. J., & Reiter, R. J. (1982). Effect of photoperiod and pineal indoles on the reproductive system of young female white-footed mice. *J. Neural Transm., 55*, 149–155.

Petterborg, L. J., Richardson, B. A., & Reiter, R. J. (1981). Effect of long or short photoperiod on pineal melatonin content in the white-footed mouse, *Peromyscus leucopus*. *Life Sci., 29*, 1623–1627.

Petterborg, L. J., Vaughan, M. K., Johnson, L. Y., Champney, T. H., & Reiter, R. J. (1984). Modification of testicular and thyroid function by chronic exposure to short photoperiod: a comparison in four rodent species. *Comp. Biochem. Physiol. A Comp. Physiol., 78*, 31–34.

Pettitt, B. A., Waterman, J. M., & Wheaton, C. J. (2008). Assessing the effects of resource availability and parity on reproduction in female Cape ground squirrels: resources do not matter. *J. Zool., 276*, 291–298.

Pieper, D. R., & Newman, S. W. (1999). Neural pathway from the olfactory bulbs regulating tonic gonadotropin secretion. *Neurosci. Biobehav. Rev., 23*, 555–562.

Pieper, D. R., Tang, Y. K., Lipski, T. P., Subramanian, M. G., & Newman, S. W. (1984). Olfactory bulbectomy prevents the gonadal regression associated with short photoperiod in male golden hamsters. *Brain Res., 321*, 183–186.

Pilastro, A., Tavecchia, G., & Marin, G. (2003). Long living and reproduction skipping in the fat dormouse. *Ecology, 84*, 1784–1792.

Place, N. J., Tuthill, C. R., Schoomer, E. E., Tramontin, A. D., & Zucker, I. (2004). Short day lengths delay reproductive aging. *Biol. Reprod., 71*, 987–992.

Popa, S. M., Clifton, D. K., & Steiner, R. A. (2008). The role of kisspeptins and GPR54 in the neuroendocrine regulation of reproduction. *Ann. Rev. Physiol., 70*, 213–238.

Prendergast, B. J. (2005). Internalization of seasonal time. *Horm. Behav., 48*, 503–511.

Prendergast, B. J., Kriegsfeld, L. J., & Nelson, R. J. (2001). Photoperiodic polyphenisms in rodents: neuroendocrine mechanisms, costs, and functions. *Quart. Rev. Biol., 76*, 293–325.

Prendergast, B. J., Pyter, L. M., Galang, J., & Kay, L. M. (2008). Reproductive responses to photoperiod persist in olfactory bulbectomized Siberian hamsters (*Phodopus sungorus*). *Behav. Brain Res.*

Prendergast, B. J., Renstrom, R. A., & Nelson, R. J. (2004). Genetic analyses of a seasonal interval timer. *J. Biol. Rhythms, 19*, 298–311.

Pyter, L. M., Hotchkiss, A. K., & Nelson, R. J. (2005). Photoperiod-induced differential expression of angiogenesis genes in testes of adult *Peromyscus leucopus*. *Reproduction, 129*, 201–209.

Reilly, S. J., Oum, R., & Heideman, P. D. (2006). Phenotypic plasticity of reproductive traits in response to food availability and photoperiod in white-footed mice (*Peromyscus leucopus*). *Oecologia, 150*, 373–382.

Reiter, R. J. (1969). Relationships among the pineal gland, the photoperiod, olfaction and reproduction in female rats. *J. Reprod. Fertil., 18*, 152.

Reiter, R. J. (1981). The mammalian pineal gland: structure and function. *Am. J. Anat., 162*, 287–313.

Reiter, R. J. (1991). Melatonin: the chemical expression of darkness. *Mol. Cell Endocrinol., 79*, C153–C158.

Revel, F. G., Ansel, L., Klosen, P., Saboureau, M., Pevet, P., Mikkelsen, J. D., et al. (2007a). Kisspeptin: a key link to seasonal breeding. *Rev. Endocr. Metab. Disord., 8*, 57–65.

Revel, F. G., Herwig, A., Garidou, M.-L., Dardente, H., Menet, J. S., Masson-Pevet, M., et al. (2007b). The circadian clock stops ticking during deep hibernation in the European hamster. *PNAS, 104*, 13816–13820.

Revel, F. G., Saboureau, M., Masson-Pevet, M., Pevet, P., Mikkelsen, J. D., & Simonneaux, V. (2006a). Kisspeptin mediates the photoperiodic control of reproduction in hamsters. *Curr. Biol., 16*, 1730–1735.

Revel, F. G., Saboureau, M., Pevet, P., Mikkelsen, J. D., & Simonneaux, V. (2006b). Melatonin regulates type 2 deiodinase gene expression in the Syrian hamster. *Endocrinology, 147*, 4680–4687.

Revel, F. G., Saboureau, M., Pevet, P., Simonneaux, V., & Mikkelsen, J. D. (2008). RFamide-related peptide gene is a melatonin-driven photoperiodic gene. *Endocrinology, 149*, 902–912.

Ritter, S., Dinh, T. T., & Zhang, Y. B. (2000). Localization of hindbrain glucoreceptive sites controlling food intake and blood glucose. *Brain Res., 856*, 37–47.

Rosen, G. J., De Vries, G. J., Goldman, B. D., Goldman, B. D., & Forger, N. G. (2007). Distribution of vasopressin in the brain of the eusocial naked mole-rat. *J. Comp. Neurol., 500*, 1093–1105.

Ruf, T., Korytko, A. I., Stieglitz, A., Lavenburg, K. R., & Blank, J. L. (1997). Phenotypic variation in seasonal adjustments of testis size, body weight, and food intake in deer mice: role of pineal function and ambient temperature. *J. Comp. Physiol. [B], 167*, 185–192.

Salo, A. L., Shapiro, L., & Dewsbury, D. A. (1993). Affiliative behavior in different species of voles (*Microtus*). *Psych. Report., 72*, 316–318.

Saltzman, W., Ahmed, S., Fahimi, A., Wittwer, D. J., & Wegner, F. H. (2006). Social suppression of female reproductive maturation and infanticidal behavior in cooperatively breeding Mongolian gerbils. *Horm. Behav., 49*, 527–537.

Saltzman, W., Thinda, S., Higgins, A. L., Matsumoto, W. R., Ahmed, S., McGeehan, L., et al. (2009). Effects of siblings on reproductive maturation and infanticidal behavior in cooperatively breeding Mongolian gerbils. *Devel. Psychobiol., 51*, 60–72.

Salverson, T. J., McMichael, G. E., Sury, J. J., Shahed, A., & Young, K. A. (2008). Differential expression of matrix metalloproteinases during stimulated ovarian recrudescence in Siberian hamsters (*Phodopus sungorus*). *Gen. Comp. Endocrinol., 155*, 749–761.

Sawchenko, P. E., & Swanson, L. W. (1982a). Immunohistochemical identification of neurons in the paraventricular nucleus of the hypothalamus that project to the medulla or to the spinal cord in the rat. *J. Comp. Neurol., 205*, 260–272.

Sawchenko, P. E., & Swanson, L. W. (1982b). The organization of noradrenergic pathways from the brainstem to the paraventricular and supraoptic nuclei in the rat. *Brain Res. Rev., 4*, 275–325.

Sawrey, D. K., & Dewsbury, D. A. (1985). Control of ovulation, vaginal estrus, and behavioral receptivity in voles (*Microtus*). *Neurosci. Biobehav. Rev., 9*, 563–571.

Sawrey, D. K., & Dewsbury, D. A. (1985). Control of ovulation, vaginal estrus, and behavioral receptivity in voles (*Microtus*). *Neurosci. Biobehav. Rev., 9*, 563–571.

Schneider, J. E. (2004). Energy balance and reproduction. *Physiol. Behav., 81*, 289–317.

Schneider, J. E., & Zhu, Y. (1994). Caudal brain stem plays a role in metabolic control of estrous cycles in Syrian hamsters. *Brain Res., 661*, 70–74.

Shahed, A., & Young, K. A. (2009). Differential ovarian expression of KiSS-1 and GPR-54 during the estrous cycle and photoperiod induced recrudescence in Siberian hamsters (*Phodopus sungorus*). *Mol. Reprod. Dev., 76*(5), 444–452.

Shanas, U., & Haim, A. (2004). Diet salinity and vasopressin as reproduction modulators in the desert-dwelling golden spiny mouse (*Acomys russatus*). *Physiol. Behav., 81*, 645–650.

Shaw, D., & Goldman, B. D. (2007). Developmental changes in male Siberian hamsters (*Phodopus sungorus*) exposed to different gestational and postnatal photoperiods. *J. Pineal Res., 43*, 25–34.

Sherman, P. W., Jarvis., J. U. M., & Alexander, R. D. (1991). *The Biology of the Naked Mole-Rat*. Princeton, NJ: Princeton University Press.

Sicard, B., Fuminier, F., Maurel, D., & Boissin, J. (1993). Temperature and water conditions mediate the effects of day length on the breeding cycle of a Sahelian rodent, *Arvicanthis niloticus*. *Biol. Reprod., 49*, 716–722.

Sicard, B., Maurel, D., Fuminier, F., & Boissin, J. (1994). Climate, trophic factors, and breeding patterns of the Nile grass rat (*Arvicolus niloticus solatus*): a 5-year study in the Sahelian region of Burkina Faso (formerly Upper Volta). *Can. J. Zool., 72*, 201–214.

Sih, A., Bell, A. M., Johnson, J. C., & Ziemba, R. E. (2004). Behavioral syndromes: An integrative overview. *Quart. Rev. Biol., 79*, 241–277.

Simerly, R. B. (2008). Hypothalamic substrates of metabolic imprinting. *Physiol. Behav., 94*, 79–89.

Simonneaux, V., Ansel, L., Revel, F. G., Klosen, P., Pevet, P., & Mikkelsen, J. D. (2009). Kisspeptin and the seasonal control of reproduction in hamsters. *Peptides, 30*, 146–153.

Simonneaux, V., Sinitskaya, N., Salingre, A., Garidou, M. L., & Pevet, P. (2006). Rat and Syrian hamster: two models for the regulation of AANAT gene expression. *Chronobiol. Int., 23*, 351–359.

Skryja, D. D. (1978). Reproductive inhibition in female cactus mice (*Peromyscus eremicus*). *J. Mammal., 59*, 543–550.

Smith, J. T., & Clarke, I. J. (2007). Kisspeptin expression in the brain: catalyst for the initiation of puberty. *Rev. Endocr. Metab. Disord., 8*, 1–9.

Smith, J. T., Acohido, B. V., Clifton, D. K., & Steiner, R. A. (2006a). KISS-1 neurons are direct targets for leptin in the ob/ob mouse. *J. Neuroendocrinol., 18*, 298–303.

Smith, J. T., Clifton, D. K., & Steiner, R. A. (2006b). Regulation of the neuroendocrine reproductive axis by kisspeptin-GPR54 signaling. *Reproduction, 131*, 623–630.

Smith, T. E., Faulkes, C. G., & Abbott, D. H. (1997). Combined olfactory contact with the parent colony and direct contact with nonbreeding animals does not maintain suppression of ovulation in female naked mole-rats (*Heterocephalus glaber*). *Horm. Behav., 31*, 277–288.

Snyman, P. C., Jackson, C. R., & Bennett, N. C. (2006). Do dispersing non-reproductive Damaraland mole-rats *Cryptomys damarensis* (Rodentia: Bathyergidae) exhibit spontaneous or induced ovulation? *Physiol. Behav., 87*, 88–94.

Solomon, N. G., Brant, C. L., Callahan, P. A., & Steinly, B. A. (2001). Mechanisms of reproductive suppression in female pine voles (*Microtus pinetorum*). *Reproduction, 122*, 297–304.

Spanel-Borowski, K., Petterborg, L. J., & Reiter, R. J. (1983). Morphological and morphometric changes in the ovaries of white-footed mice (*Peromyscus leucopus*) following exposure to long or short photoperiod. *Anat. Rec., 205*, 13–19.

Spears, N., & Clark, J. R. (1988). Selection in field voles (*Microtus agrestis*) for gonadal growth under short photoperiods. *J. Anim. Ecol., 57*, 61–70.

Spinks, A. C., Bennett, N. C., & Jarvis, J. U. M. (1999). Regulation of reproduction in female common mole-rats (*Cryptomys hottentotus hottentotus*): the effects of breeding season and reproductive status. *J. Zool., 248*, 161–168.

Spinks, A. C., Bennett, N. C., & Jarvis, J. U. M. (1999). Regulation of reproduction in female common mole-rats (*Cryptomys hottentotus hottentotus*): the effects of breeding season and reproductive status. *J. Zool., 248*, 161–168.

Spinks, A. C., Bennett, N. C., Faulkes, C. G., & Jarvis, J. U. M. (2000). Circulating LH levels and the response to exogenous GnRH in the common mole-rat: implications for reproductive regulation in this social, seasonal breeding species. *Horm. Behav., 37*, 221–228.

Spinks, A. C., VanderHorst, G., & Bennett, N. C. (1997). Influence of breeding season and reproductive status on male reproductive characteristics in the common mole-rat, *Cryptomys hottentotus hottentotus*. *J. Reprod. Fert., 109*, 79–86.

Steinlechner, S. (1998). Djungarian hamster and/or Siberian hamster: who is who? *Eur. Pineal Soc. NEWS, 38*, 7–11.

Steinlechner, S., Stieglitz, A., Ruf, T., Heldmaier, G., & Reiter, R. J. (1991). Integration of environmental signals by the pineal gland and its significance for seasonality in small mammals. In F. Fraschini, & R. J. Reiter (Eds.), *Role of Melatonin and Pineal Peptides in Neuroimmunomodulations* (pp. 159–163). New York, NY: Plenum Press.

Stenseth, N. C. (1999). Population cycles in voles and lemmings: density dependence and phase dependence in a stochastic world. *Oikos, 87*, 427–461.

Stetson, M. H., Ray, S. L., Creyaufmiller, N., & Horton, T. H. (1989). Maternal transfer of photoperiodic information in Siberian hamsters. II. The nature of the maternal signal, time of signal transfer, and the effect of the maternal signal on peripubertal reproductive development in the absence of photoperiodic input. *Biol. Reprod., 40*, 458–465.

Strauss, A., Mascher, E., Palme, R., & Millesi, E. (2007). Sexually mature and immature yearling male European ground squirrels: a comparison of behavioral and physiological parameters. *Horm. Behav., 52*, 646–652.

Sundell, J. (2006). Experimental tests of the role of predation in the population dynamics of voles and lemmings. *Mamm. Rev., 36*, 107–141.

Taylor, S. A., Salo, A. L., & Dewsbury, D. A. (1992). Estrus induction in four species of voles (*Microtus*). *J. Comp. Psychol., 106*, 366–373.

Tena-Sempere, M. (2007). Roles of ghrelin and leptin in the control of reproductive function. *Neuroendocrinology, 86*, 241.

Timonin, M. E., Place, N. J., Wanderi, E., & Wynne-Edwards, K. E. (2006). *Phodopus campbelli* detect reduced photoperiod during development but, unlike *Phodopus sungorus*, retain functional reproductive physiology. *Reproduction, 132*, 661–670.

Touma, C., Palme, R., & Sachser, N. (2001). Different types of oestrous cycle in two closely related South American rodents (*Cavia aperea* and *Galea musteloides*) with different social and mating systems. *Reproduction, 121*, 791–801.

Turchin, P., & Batzli, G. O. (2001). Availability of food and the population dynamics of arvicoline rodents. *Ecology, 82*, 1521–1534.

Tuthill, C. R., Freeman, D. A., Butler, M. P., Chinn, T., Park, J. H., & Zucker, I. (2005). Perinatal influences of melatonin on testicular development and photoperiodic memory in Siberian hamsters. *J. Neuroendocrinol., 17*, 483–488.

Uriarte, N., Ferreira, A., Rosa, X. F., Sebben, V., & Lucion, A. B. (2008). Overlapping litters in rats: effects on maternal behavior and offspring emotionality. *Physiol. Behav., 93*, 1061–1070.

Vajda, A. M., & Norris, D. O. (2005). Endocrine-active phytochemicals: Environmental signaling contact and mechanisms. In D. O. Norris, & J. A. Carr (Eds.), *Endocrine Disruption Biological Bases for Health Effects in Wildlife and Humans* (pp. 390–423). New York, NY: Oxford University Press.

Van der Lely, A. J., Tschop, M., Heiman, M. L., & Ghigo, E. (2004). Biological, physiological, pathophysiological, and pharmacological aspects of ghrelin. *Endocrine Rev., 25*, 426–457.

Van der Walt, L., Bennett, N. C., & Schoeman, S. (2001). Reproductive suppression and pituitary sensitivity to exogenous GnRH in the highveld mole-rat (*Cryptomys hottentotus pretoriae*). *J. Zool., 254*, 177–184.

Van Rensburg, L. J., Bennett, N. C., Van der Merwe, M., & Schoeman, A. S. (2002). Seasonal reproduction in the highveld mole-rat, *Cryptomys hottentotus pretoriae* (Rodentia: Bathyergidae). *Can. J. Zool., 80*, 810–820.

Van Rensburg, L. J., Bennett, N. C., Van der Merwe, M., Schoeman, A. S., & Brinders, J. (2003). Are non-reproductive male highveld mole-rats, *Cryptomys hottentotus pretoriae* physiologically suppressed while in the confines of the natal colony? *J. Zool. Soc. London, 260*, 73–78.

Van Sandwyk, J. H. D. T., & Bennett, N. C. (2005). Do solitary, seismic signalling Cape mole-rats (*Georychus capensis*) demonstrate spontaneous or induced ovulation? *J. Zool., 267*, 75–80.

Van Sandwyk, J. H.d. T., & Bennett, N. C. (2005). Do solitary, seismic signaling Cape mole-rats (*Georychus capensis*) demonstrate spontaneous or induced ovulation? *J. Zool., 267*, 75–80.

Vandegrift, K. J., Raffel, T. R., & Hudson, P. J. (2008). Parasites prevent summer breeding in white-footed mice, *Peromyscus leucopus*. *Ecology, 89*, 2251–2258.

Wade, G. N., & Jones, J. F. (2003). Lessons from experimental disruption of estrous cycles and behaviors. *Med. Sci. Sports Exer., 35*, 1573–1580.

Watanabe, T., Yamamura, T., Watanabe, M., Yasuo, S., Nakao, N., Dawson, A., et al. (2007). Hypothalamic expression of thyroid hormone-activating and -inactivating enzyme genes in relation to photorefractoriness in birds and mammals. *Amer. J. Physiol. Regul. Integr. Comp. Physiol., 292*, R568–R572.

Weaver, D. R., Keohan, J. T., & Reppert, S. M. (1987). Definition of a prenatal sensitive period for maternal–fetal communication of day length. *Am. J. Physiol., 253*, E701–E704.

Weaver, D. R., Provencio, I., Carlson, L. L., & Reppert, S. M. (1991). Melatonin receptors and signal transduction in photorefractory Siberian hamsters (*Phodopus sungorus*). *Endocrinology, 128*, 1086–1092.

Weil, Z. M., Bowers, S. L., & Nelson, R. J. (2007). Photoperiod alters affective responses in collared lemmings. *Behav. Brain Res., 179*, 305–309.

Weir, B. J. (1974). Reproductive biology of hystricomorph rodents. In I. W. Rowland, & B. J. Weir (Eds.), *The Biology of Hystricomorph Rodents* (pp. 265–301). New York, NY: Academic Press.

Weir, B. J. (1974). Reproductive characteristics of hystricomorph rodents. In I. W. Rowlands, & B. J. Weir (Eds.), *The Biology of Hystricomorph Rodents* (pp. 265–301). New York, NY: Academic Press.

Whitsett, J. M., & Lawton, A. D. (1982). Social stimulation of reproductive development in male deer mice housed on a short-day photoperiod. *J. Comp. Physiol. Psych., 96*, 416–422.

Wichman, H. A., & Lynch, C. B. (1991). Genetic variation for seasonal adaptation in *Peromyscus leucopus*: nonreciprocal breakdown in a population cross. *J. Hered., 82*, 197–204.

Wilder, S. M., & Meikle, D. B. (2005). Reproduction, foraging, and the negative density-area relationship of a generalist rodent. *Oecologia, 144*, 391–398.

Willingstorfer, W. J., Burda, H., & Winckler, J. (1998). Ovarian growth and folliculogenesis in breeding and nonbreeding females of a social rodent, the Zambian common mole-rat, *Cryptomys* sp. *J. Morphol., 237*, 33–41.

Willingstorfer, W.-J., Burda, H., & Winckler, J. (1998). Ovarian growth and folliculogenesis in breeding and non-breeding females of a social rodent, the Zambian mole-rat, *Cryptomys* sp. *J. Morphol., 237*, 33–41.

Wilson, E. O. (1971). *The Insect Societies*. Cambridge, MA: Harvard University Press.

Witt, D. M., Carter, C. S., & Insel, T. R. (1991). Oxytocin receptor binding in female prairie voles—Endogenous and exogenous estradiol stimulation. *J. Neuroendocrinol., 3*, 155–161.

Witt, D. M., Carter, C. S., Chayer, R., & Adams, K. (1990). Patterns of behaviour during postpartum oestrus in prairie voles, *Microtus ochrogaster*. *Anim. Behav., 39*, 528–534.

Wolff, J. O., Dunlap, A. S., & Ritchhart, E. (2001). Adult female prairie voles and meadow voles do not suppress reproduction in their daughters. *Behav. Proc., 55*, 157–162.

Wynne-Edwards, K. E., & Lisk, R. D. (1987). Male—female interactions across the female estrous cycle: a comparison of two species of dwarf hamster (*Phodopus campbelli* and *Phodopus sungorus*). *J. Comp. Psychol., 101*, 335–344.

Yahr, P., & Kessler, S. (1975). Suppression of reproduction in water-deprived Mongolian gerbils (*Meriones unguiculatus*). *Biol. Reprod., 12*, 249–254.

Yasuo, S., Yoshimura, T., Ebihara, S., & Korf, H. W. (2007). Temporal dynamics of type 2 deiodinase expression after melatonin injections in Syrian hamsters. *Endocrinology, 148*, 4385–4392.

Yellon, S. M., & Goldman, B. D. (1984). Photoperiod control of reproductive development in the male Djungarian hamster (*Phodopus sungorus*). *Endocrinology, 114*, 664–670.

Ylonen, H., Jacob, J., Runcie, M. J., & Singleton, G. R. (2003). Is reproduction of the Australian house mouse (*Mus domesticus*) constrained by food? A large-scale field experiment. *Oecologia, 135*, 372–377.

Young, K. A., & Nelson, R. J. (2000). Short photoperiods reduce vascular endothelial growth factor in the testes of *Peromyscus leucopus*. *Amer. J. Physiol. Reg. Int. Comp. Physiol., 279*, R1132–R1137.

Young, K. A., & Nelson, R. J. (2001). Mediation of seasonal testicular regression by apoptosis. *Reproduction, 122*, 677–685.

Young, K. A., Zirkin, B. R., & Nelson, R. J. (1999). Short photoperiods evoke testicular apoptosis in white-footed mice (*Peromyscus leucopus*). *Endocrinology, 140*, 3133–3139.

Young, K. A., Zirkin, B. R., & Nelson, R. J. (2000). Testicular regression in response to food restriction and short photoperiod in white-footed mice (*Peromyscus leucopus*) is mediated by apoptosis. *Biol. Reprod., 62*, 347–354.

Young, K. A., Zirkin, B. R., & Nelson, R. J. (2001). Testicular apoptosis is down-regulated during spontaneous recrudescence in white-footed mice (*Peromyscus leucopus*). *J. Biol. Rhythms, 16*, 479–488.

Zarrow, M. X., & Clark, J. H. (1968). Ovulation following vaginal stimulation in a spontaneous ovulator and its implications. *J. Endocrinol., 40*, 343–352.

Chapter 12

Hormones and Reproductive Cycles in Bats

Amitabh Krishna* and Kunwar P. Bhatnagar†
*Banaras Hindu University, Varanasi, India, †University of Louisville, Louisville, KY, USA

SUMMARY

Bats belonging to the order Chiroptera exhibit numerous reproductive specializations. Prolonged sperm retention in the epididymis; asynchrony between gametogenesis, accessory sex gland function, and copulation; reproductive tract asymmetries; and menstruation are topics of great interest. Gametogenesis in bats is prolonged and exhibits unique patterns of hormonal changes. Reproductive activities in bats are diverse and unusually sensitive to environmental changes, resulting in a variety of reproductive delays, such as delayed ovulation, implantation, and development, all coinciding with winter dormancy. The hypothalamic–pituitary regulation of reproductive processes is poorly known. Hormones such as insulin, leptin, and melatonin may interact to regulate the amount of body fat and the reproductive cycle. As there is a huge diversity of bats, it has been hard to find one specific factor responsible for controlling the various reproductive strategies. Studies on vespertilionid bats suggest that unique reproductive features of bats are strongly linked to winter dormancy (heterothermy), fat deposition, and low basal metabolic rate.

1. INTRODUCTION

Bats belong to the order Chiroptera, which is the second largest order of mammals. They are grouped into two suborders: Megachiroptera (megabats) and Microchiroptera (microbats). There are approximately 175 species of megabat and 790 species of microbat. Megabats belong to one family, the Pteropodidae, which are known as 'Old World fruit bats' or 'flying foxes;' they are confined to Old World tropics and feed mainly on fruit, flower, nectar, and pollen. These bats are generally larger than microbats in size and do not echolocate, with the exception of the genus *Rousettus*. The 17 families of microbats are a diverse group and do echolocate. Thus, the Microchiroptera trap insects skillfully using echolocation in flight. The Microchiroptera are distributed in tropical climates as well as in temperate and subtropical regions, where insects are not available in sufficient numbers throughout the year. Thus, bats have evolved mechanisms to ensure that the birth of their young is timed to give the greatest chance of survival, both for the females and their young. Restriction of food during the winter months has been overcome by the occurrence of torpor over this period and reliance on deposition of subcutaneous fat laid down during the summer or autumn months. For bats living at higher latitudes (temperate zone), the summer is short and the young must be born early to ensure full development before winter. Following gestation and lactation, both mothers and young must accumulate sufficient fat to be used for hibernation during the winter. Bats solve this problem through reproductive asynchrony, delayed ovulation, sperm storage in the male or female genital tract, delayed fertilization, delayed implantation or delayed development, and reproductive synchrony with the environment.

Bat populations are in decline worldwide. Effective conservation strategies require detailed knowledge of their reproduction. This chapter summarizes various aspects of bat reproduction to provide a generalized picture of reproductive strategies and underlying endocrine mechanisms. Knowledge of various reproductive events and their controlling mechanisms may ultimately be central to the conservation of vulnerable bat populations.

1.1. The Hypothalamus–pituitary–gonadal (HPG) Axis in Bats

The hypothalamus–pituitary–gonadal (HPG) axis of bats is similar to that of other mammals described in this volume (see Chapters 2–5, this volume). Briefly, the main stimulant from the hypothalamus to the gonadotropic cells causing synthesis and release of gonadotropins (GTHs) from the pituitary is gonadotropin-releasing hormone (GnRH). The two GTHs, luteinizing hormone (LH) and follicle-stimulating hormone (FSH), cooperatively stimulate gamete production and steroidogenesis by the gonads. The gonadal steroids (androgens, estrogens, and progestogens)

Hormones and Reproduction of Vertebrates, Volume 5—Mammals

in turn stimulate reproductive ducts and glands as well as various sex accessory structures, and provide feedback to the hypothalamus and pituitary. In males, the Leydig or interstitial cells are primarily responsible for androgen production, including testosterone (T), 5α-dihydrotestosterone (DHT), and androstenedione (AND). The seminiferous tubules contain spermatogenetic cells that result in sperm production and Sertoli cells that produce estrogens and numerous other factors involved in sperm production and maturation. In females, the ovaries consist of growing follicles, each containing an oocyte. The follicle wall consists of androgen-producing thecal cells surrounding estrogen-producing granulosa cells. The estrogens produced include estrone (E_1), 17β-estradiol (E_2), and estriol (E_3). Following ovulation and release of the mature oocyte, the collapsed follicle wall is converted into a corpus luteum (CL) by LH and proceeds to secrete estrogens and progesterone (P_4). The number of follicles that mature and the number of resulting corpora lutea is specific for each mammalian species. The CL contributes to the establishment and maintenance of pregnancy to varying extents, depending on the species. The placenta and the fetus also produce numerous hormones that are necessary for the maintenance of pregnancy (see Chapter 6, this volume) and the pituitary, ovaries, placenta, and fetus all may play roles in parturition, depending on the species (see Chapter 7, this volume). Additionally, prolactin (PRL) produced by the pituitary plays a variety of species-specific roles. The particular roles of these various hormones in bats are discussed in the following sections.

2. BAT REPRODUCTIVE CYCLES

2.1. Patterns of Reproductive Cycles

Early knowledge of bat reproduction was based mainly on temperate-zone species. Tropical bats have received little attention in spite of the fact that bats are primarily tropical in distribution (Koopman & Jones, 1970). In general, breeding patterns in tropical bats are diversified, in contrast to the rather stereotyped annual cycle of temperate-zone bats.

2.1.1. Reproduction in temperate-zone bats

In temperate-zone bats, reproductive activities are closely regulated by ambient temperature and food availability (insects) (Wilson, 1979), with parturition and lactation occurring during the periods of optimal food availability. Hibernation and arousal exert a profound influence on the physiology of reproduction in these temperate-zone species. These bats are seasonally monoestrous (Racey, 1982). Significant features of special interest in temperate-zone hibernating bats are reproductive delays (delayed ovulation, prolonged storage of sperm, and/or delayed embryonic development). According to the timing of reproductive events, the unusual reproductive activity of temperate-zone hibernating bats can be classified into two types.

2.1.1.1. Reproductive pattern type I

In this reproductive pattern, a brief proestrus occurs in mid to late summer followed by estrus and subsequent copulation in late summer or early autumn. Bats typically enter hibernation soon thereafter. Sperm remain stored in a viable state in the female reproductive tract and ovulation, fertilization, and gestation take place after arousal in the following spring (Figure 12.1). This pattern was described by Benecke (1879) in *Pipistrellus pipistrellus* and *Plecotus auritus* and by Fries (1879) in several species including *Rhinolophus hipposideros*, *P. auritus*, and *Vespertilio murinus*. The classical studies of Rollinat and Trouessart (1895; 1897) on reproduction in *V. murinus* and *Rhinolophus ferrumequinum* clearly demonstrated that sperm remain stored in the uterus during winter for over six months from October to April. Ovulation and fertilization take place at the beginning of the spring, five to ten days after the bats emerge from hibernation. This pattern of reproduction has since been reported in numerous species of vespertilionid bats from temperate latitudes throughout the world (Wimsatt, 1945; Sluiter & Bels, 1951; Pearson, Koford, & Pearson, 1952; Uchida & Mori, 1977; Oh, Mori, & Uchida, 1985; Son, Yoon, Mori, & Uchida, 1987).

A variation of the type I reproductive pattern is exhibited by certain temperate-zone species and involves mating, ovulation, and fertilization in the spring (Hartman & Cuyler, 1927). This is an exceptional pattern exhibited by bats that failed to copulate in the previous autumn. Guthrie (1933) provided the only clear-cut evidence of copulation in the spring in several American microchiropterans. The occurrence of active spermatogenesis was seen not only in autumn but also in spring in several species—e.g., *P. pipistrellus*, *Eptesicus fuscus*, *Myotis lucifugus*, *Rhinolophus* sp., and *Hipposideros* sp. (Caffier & Kolbow, 1934)—providing strong circumstantial evidence for the occurrence of copulation in the spring.

2.1.1.2. Reproductive pattern type II

This type of reproductive pattern is exhibited by members of the genus *Miniopterus* (family Vespertilionidae) and the genus *Rhinolophus* (family Rhinolophidae). Gonadal recrudescence is initiated during the summer; estrus and copulation occur in the autumn, followed immediately by ovulation, fertilization, and initial embryogenesis (Figure 12.1). The bats then enter hibernation in a pregnant condition, and there may be retardation of embryonic

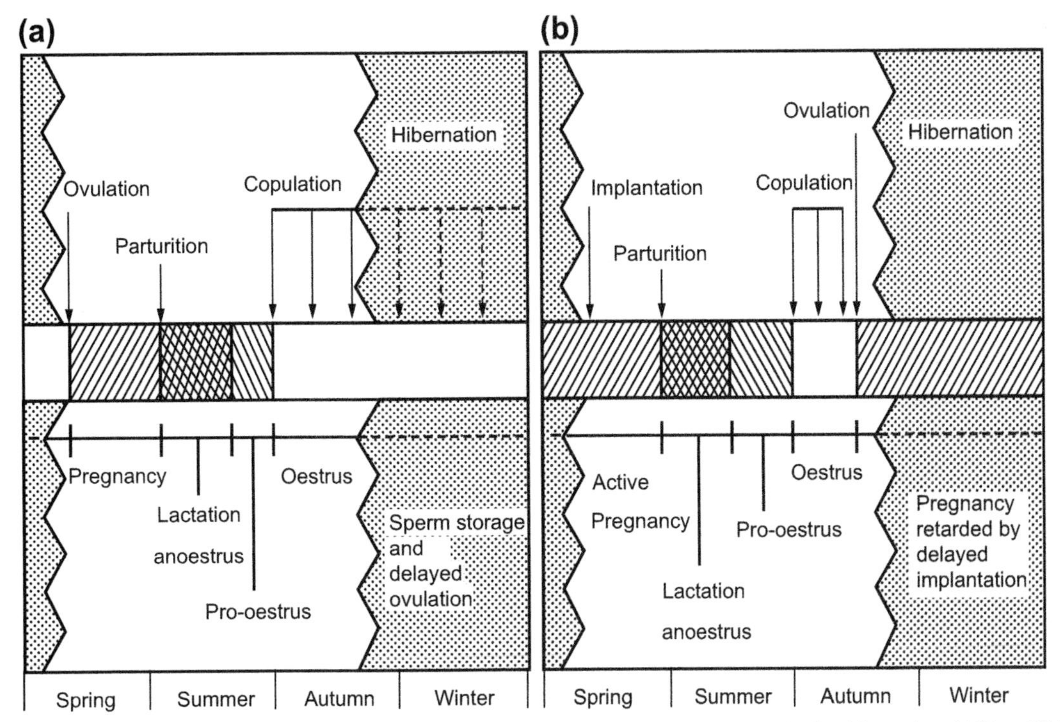

FIGURE 12.1 Reproductive patterns in temperate-zone (TZ) bats. (a) Type TZ-I: delayed ovulation until after hibernation. (b) Type TZ-II: ovulation and fertilization in autumn; embryogenesis delayed until hibernation. *Adapted from Oxberry (1979).*

development during hibernation. Normal growth resumes after arousal in the spring. This pattern was noted by Courrier (1927) in the European species *Miniopterus schreibersii* and later confirmed by Planel, Guilhem, and Soleilhavoup (1961) and Peyre and Herlant (1963a; 1963b). A similar pattern has been reported in Australian *M. schreibersii* and *Miniopterus australis* (Dwyer, 1963a; 1963b; Richardson, 1977) and in the phyllostomid *Macrotus californicus* (Bradshaw, 1961).

2.1.2. Reproduction in tropical bats

In contrast to bats living in cold and temperate climates, information on the breeding habits of tropical/subtropical bats is very limited. Ovulation and fertilization ensue immediately after mating. The reproductive cycles of tropical bats are varied and numerous. There is growing evidence that, for the majority of bats in the tropical zone, reproductive activities are associated with rainfall (e.g., Ramakrishna & Rao, 1977; Wilson, 1979; Happold & Happold, 1990a; Cumming & Bernard, 1997) and the associated food availability (e.g., Krishna & Dominic, 1983; Willig, 1985; Heideman, Deoraj, & Bronson 1992; Kofron, 1997), or photoperiod (e.g., Racey, 1978; Heideman & Bronson, 1994). There are many vespertilionid bats in the subtropical zone that show patterns of reproduction similar to temperate-zone vespertilionid bats and exhibit delayed ovulation, prolonged storage of sperm in the female genital tract, and asynchronized gamete production by males and females (see Sections 3.1, 3.2, 5.1.2.3 & 6.1.1). The reproductive patterns manifested by tropical bats fall into four broad categories: seasonal monoestry, bimodal polyestry, seasonal polyestry, and aseasonal polyestry (Fleming, Hopper, & Wilson, 1972; Wilson, 1973).

2.1.2.1. Seasonal monoestry

Seasonal monoestry involves a single breeding period in a year. This pattern exhibits three variations.

- Copulation takes place late in autumn and is followed by fertilization and pregnancy, with delivery of young in the following spring. This has been reported in *Megaderma lyra* (Gopalakrishna, 1950), *Megaderma spasma* (Ramakrishna, 1951), *Hipposideros fulvus fulvus* (Patil, 1968), and *Pteropus giganteus* (Sahastrabudhe, 1971).
- Copulation in spring is immediately followed by fertilization and pregnancy. This has been reported in *M. australis* in the New Hebrides (Baker & Bird, 1936), *Taphozous georgianus* in central Queensland, Australia (Jolly, 1990), *Pipestrellus rusticus* in South Africa (Van der Merwe & Rautenbach, 1990), *Scotophilus wroughtoni* (Gopalakrishna, 1947), and several species of bat inhabiting central and western India (Brosset, 1962a; 1962b; 1962c; 1963).

- In this extended seasonal monoestrous pattern, proestrus occurs in mid to late autumn, followed by delayed ovulation and subsequent copulation in late autumn or early winter. Bats typically undergo winter dormancy. Sperm remain stored in a viable state in the female reproductive tract, and ovulation, fertilization, and gestation take place after arousal the following spring; e.g., *Scotophilus heathi* (Krishna & Dominic, 1981).

2.1.2.2. Bimodal polyestry

Bimodal polyestry involves a restricted breeding season with two pregnancies in quick succession, as seen in *Artibeus*, *Carollia*, *Phyllostomus*, and *Chiroderma*. An interesting variation in this pattern is shown by *Artibeus jamaicensis*, which undergo delayed development during this period (Fleming, 1971). Several neotropical phyllostomids (Fleming, 1971; Fleming et al., 1972; Wilson, 1973) and small pteropodids such as *Rousettus aegyptiacus* (Mutere, 1968a), *Rousettus leschenaulti* (Gopalakrishna, 1964), and *Cynopterus sphinx* (Krishna & Dominic, 1983; 1984) also exhibit this pattern. The emballonurid Indian species *Taphozous longimanus* (Krishna & Dominic, 1982a; 1982b) and *Coleura afra* in Kenya (McWilliam, 1987a), and the molossid species *Tadarida condylura* in Africa (Happold & Happold, 1989) and *Molossus fortis* in Puerto Rico (Krutzsch & Crichton, 1985), also breed twice in quick succession.

2.1.2.3. Seasonal polyestry

There is continuous breeding in some bats during most of the year, with a brief period of sexual inactivity, e.g. as seen in *Pipistrellus mimus* (Krishna, 1985), *Myotis nigricans* (Wilson & Findley, 1970; Wilson, 1973), and *Tadarida pumila* (Van der Merwe, 1987; Happold & Happold, 1989). Polyestry may be an adaptation of bats for producing more than one young per breeding cycle and has been documented in tropical insectivorous and frugivorous bats.

2.1.2.4. Aseasonal polyestry

Aseasonal polyestry is the continuous asynchronous breeding pattern as seen in some tropical members of the families Desmodontidae (Wimsatt & Trapido, 1952), Phyllostomatidae (Tamsitt & Valdivieso, 1965a; 1965b), Pteropodidae, Emballonuridae, and Molossidae (Mutere, 1968b; Anciaux de Faveaux, 1973). According to Gopalakrishna (1955), *T. longimanus* breeds throughout the year in central India. *Desmodus rotundus*, a sanguivorous bat, *Molosus ater* (Marques, 1986), and *T. pumila* (Mutere, 1973) may show this pattern due to aseasonal food supply.

3. REPRODUCTIVE DELAYS IN BATS

Female bats are unique mammalian models for studies on reproductive delay. They exhibit a variety of reproductive delays and generally have a much longer pregnancy than one would predict on the basis of their body size. Four types of reproductive delay occur in female bats: delayed ovulation, delayed fertilization, delayed implantation, and delayed development. Delayed ovulation and delayed fertilization (by sperm stored in the uterus) are uncommon in mammals but are the norm in the majority of temperate-zone vespertilionid and rhinolophid bats (Oxberry, 1979; Racey, 1982; Crichton, 2000) as well as in tropical/subtropical vespertilionid species (Myers, 1977; Krutzsch, 1979; Dominic & Krishna, 1989; Singh & Krishna, 1997). While much is known about delayed implantation among mammals (Flint, Renfree, & Weir, 1981), the postimplantation delay in development is the least common (Burns, Baker, & Bleier, 1972; Burns & Wallace, 1975; Burns & Easley, 1977; Burns, 1981; Heideman, Cummings, & Heaney, 1993).

3.1. Delayed Ovulation

During hibernation/winter dormancy in heterothermic bats, ovarian activity apparently arrests and thus final maturation and ovulation of the Graafian follicles are postponed for several months. Delayed ovulation means that a species-specific number of specialized Graafian follicles persist in the ovary throughout hibernation, and ovulation is postponed or delayed until spring (Oxberry, 1979). This unique feature of bats has evolved as a result of the superimposition of the period of hibernation on the season of reproduction. Delayed ovulation was first described in hibernating members of rhinolophid (Matthews, 1937; Gaisler, 1966) and vespertilionid bats (Courrier, 1924) from Europe. This has since been reported in numerous species of rhinolophid and vespertilionid bats from temperate latitudes throughout the world (Wimsatt, 1944; Sluiter & Bels, 1951; Pearson, Koford, & Pearson, 1952; Uchida & Mori, 1977; Oh, Mori, & Uchida, 1985; Son et al., 1987). The maximal period of delayed ovulation reported is five to seven months in *Myotis ricketti* (Wang, Liang, Racey, Wang, & Zhang, 2008).

The Graafian follicles that are present in the ovaries of these bats during the period of delayed ovulation have a small antrum, with each ovum being surrounded by a large multilayered cumulus, the cells of which are hypertrophied and vacuolated (Oxberry, 1979). Histochemical and/or ultrastructural studies have established that this vacuolation stores either glycogen (*Myotis lucifugus* (Wimsatt & Kallen, 1957; Wimsatt & Parks, 1966); *Pipistrellus abramus* (Uchida & Mori, 1977)) or lipids (*R. f. nippon* (Oh et al., 1985)) as nutrient sources during the prolonged survival of the Graafian follicles.

Glycogen may be an important nutrient for the Graafian follicle during hibernation (Oxberry, 1979). Ultrastructural studies of Wimsatt and Parks (1966) on the Graafian follicles of temperate-zone *M. lucifugus* show that glycogen stored in the granulosa cells is transferred to oocytes by follicle cell processes that pass through the zona pellucida and penetrate the oocyte surface. A number of nonhibernating tropical vespertilionid bats also show the presence of Graafian follicles that exhibit a similar specialization (Medway, 1972; Gopalakrishna, Madhvan, Thakur, & Rajgopal, 1974; Kitchner & Halse, 1978; Krishna, 1985). The mechanism of ovulatory delay may involve a depressing effect of hibernation on the synthesis and release of LH as well as on LH-responsiveness, but not on FSH (Herlant, 1956a; 1968; Wimsatt, 1960). Racey (1982) speculated that the obstacle to ovulation in hibernating bats did not appear to originate in the ovary but was probably due to low pituitary secretion of LH. Intense LH and FSH immunoreactivity was seen in *S. heathi* during both delayed ovulation and breeding periods (Singh & Krishna, 1996b). Anthony et al. (1989) reported enhanced hypothalamic tissue content but low secretion of GnRH in *M. lucifugus* during hibernation (preovulatory period) but increased GnRH secretion during the postovulatory period. Parallel reductions in hypothalamic and pituitary neural lobe GnRH content during the periovulatory period support the hypothesis that the neural lobe component of the system contributes to the control of GTH secretion in this species.

The ovary of *S. heathi*, a seasonally monoestrous subtropical vespertilionid bat, shows many large antral follicles during the month of November but ovulation is delayed until early March (Krishna & Singh, 1992). All attempts to induce ovulation with human chorionic gonadotropin (hCG) during this period were unsuccessful (Singh & Krishna, 1992). Further studies suggested that increased AND produced by the ovary from November to February might be responsible for delayed ovulation by desensitizing the surviving follicles to gonadotropin stimulation, making the follicles unsuitable for ovulation (Abhilasha & Krishna, 1996; Chanda, Abhilasha, & Krishna, 2006). A similar association between hyperandrogenism and anovulation is known to exist in patients with polycystic ovary syndrome (PCOS). Recently, more data have accumulated suggesting a close similarity between PCOS and the ovarian morphology and endocrinological features of the bat during the period of delayed ovulation (Table 12.1) (Chanda, Abhilasha, & Krishna, 2003; Srivastava & Krishna, 2006).

TABLE 12.1 Comparison of polycystic ovarian syndrome (PCOS) in women and delayed ovulation in *Scotophilus heathi*

Signs/Symptoms	Polycystic Ovarian Syndrome (PCOS)	*Scotophilus heathi* During Delayed Ovulation
Anovulation	+	+
Enlarged ovaries with multiple small follicles	+	+
Hyperthecosis	+	+
Hyperandrogenism	+	+
Hyperinsulinemia	+	+
Insulin resistance	+	+
Luteinizing hormone (LH) hypersecretion	+	*
Follicle-stimulating hormone (FSH) insufficiency	+	*
Increased abdominal adiposity	+	+

*Data not available.

3.2. Delayed Fertilization

Delayed fertilization/prolonged sperm storage is an essential component of the reproductive cycle in almost all rhinolophid and vespertilionid bats undergoing hibernation/winter dormancy (Racey, 1982; Crichton, 2000). These species display a clear dissociation between male and female gamete production. Spermatogenesis occurs during summer and early autumn and epididymides are full of sperm from winter until spring. Ovulation occurs on arousal from hibernation (Wimsatt, 1969). In these bats, mating activity commences in the autumn and sperm are then stored in the female genital tract throughout the winter hibernating period and subsequently may be used in the fertilization of ova released at ovulation after arousal in the spring (Wimsatt, 1969; Racey, 1975). The nature of the specializations that have evolved in these bats to permit prolonged sperm survival is still poorly understood. Sperm storage was thought to be an adaptation for overwintering and a specialization related to hibernation, since females of a nonhibernating population of *Myotis austroriparius* did not appear to store sperm (Rice, 1957). However, tropical vespertilionids store viable sperm (Gopalakrishna & Madhavan, 1971; Medway, 1972; Gopalakrishna & Madhavan, 1978; Krishna & Dominic, 1978; Krishna, 1984), suggesting that sperm storage was not a preadaptation for hibernation in temperate regions.

The maximal period of sperm storage is 198 days in female *Nyctalus noctula* (Racey, 1973). In general, tropical bats store sperm for much shorter periods as compared with the hibernating temperate-zone bats. The interval between copulation and ovulation is up to five weeks in *S. heathi* (Krishna & Dominic, 1978) and 30–35 days in *P. mimus* (Krishna, 1984). The site of female sperm storage varies, from the periovarian spaces (Uchida & Mori, 1987) to the vagina, depending on the species (Racey, 1975). In some vespertilionid bats, *viz. Chalinolobus gouldii* (Kitchener, 1975) and *M. s. fuliginosus*, the storage site is limited to the uterotubal junction, where intact sperm are found in the diverticulae formed by the mucosal fold (Mori & Uchida, 1980). The sperm are stored in the uterotubal junction in *Myotis daubentoni* (Racey, 1975), *M. lucifugus*, and *Myotis velifer* (Krutzsch, Crichton, & Nagle, 1982). In *Rhinolophus hipposideros*, *R. ferrumequinum*, *R. f. Nippon*, and *Myotis nattereri*, the principal site of sperm storage is the oviduct (Racey, 1975; Mori, Oh, & Uchida, 1982).

Perpendicular orientation of sperm, with their heads towards the reproductive tract epithelium, is characteristic of all sperm-storing bat species so far investigated, suggesting that such relationships are an integral part of the mechanism of prolonged storage of sperm (Figure 12.2) (Racey, 1979; Uchida & Mori, 1987). This technique has also been used in other vertebrates that store sperm, such as snakes (Hoffman & Wimsatt, 1972), birds (Van Krey, Ogasawara, & Pangborn, 1967), fishes (Jalabert & Billard, 1969), and mammals (Racey, 1975). It was suggested that the epithelial cells of the storage site actively provide sperm with nutrients for their prolonged survival; e.g., glycogen in the uterine epithelium of *P. abramus*, *P. pipistrellus*, *M. lucifugus*, *M. velifer*, and *S. heathi* (Nakano, 1928; Racey, 1975; Crichton, Krutzsch, & Wimsatt, 1981; Krishna, 1997). Further, association of sperm with the epithelium of the storage organ probably protects them from various degradative factors (e.g., muscular movements of the uterus). Certain enzymes of glycolysis have been located in the sperm as well as in the uterus (Racey, 1975). Beside nutritional support, another factor required for long-term maintenance of sperm might be a mechanism that acts to suppress sperm metabolism, thereby reducing demand for available substrates. In this regard, Hunter, Barker, Johnson, Fahning, and Schultz (1971) and Hunter, Johnson, Barker, Fahning, and Schultz (1971) observed that the seminal fluid of *M. lucifugus* contains a protein that is inhibitory to uterine motility and the phagocytic system.

It appears that the ultimate control of the mechanism underlying sperm storage may be hormonal. Racey (1972) ovariectomized *P. pipistrellus* during hibernation, causing detrimental effects on the survival of stored sperm after two weeks as compared with control bats. Circulating levels of androgens that are required for the maintenance of sperm integrity and viability in the male are known for only a few bat species. Circulating AND may be present in elevated levels in females storing sperm (Crichton, 2000). Androgen

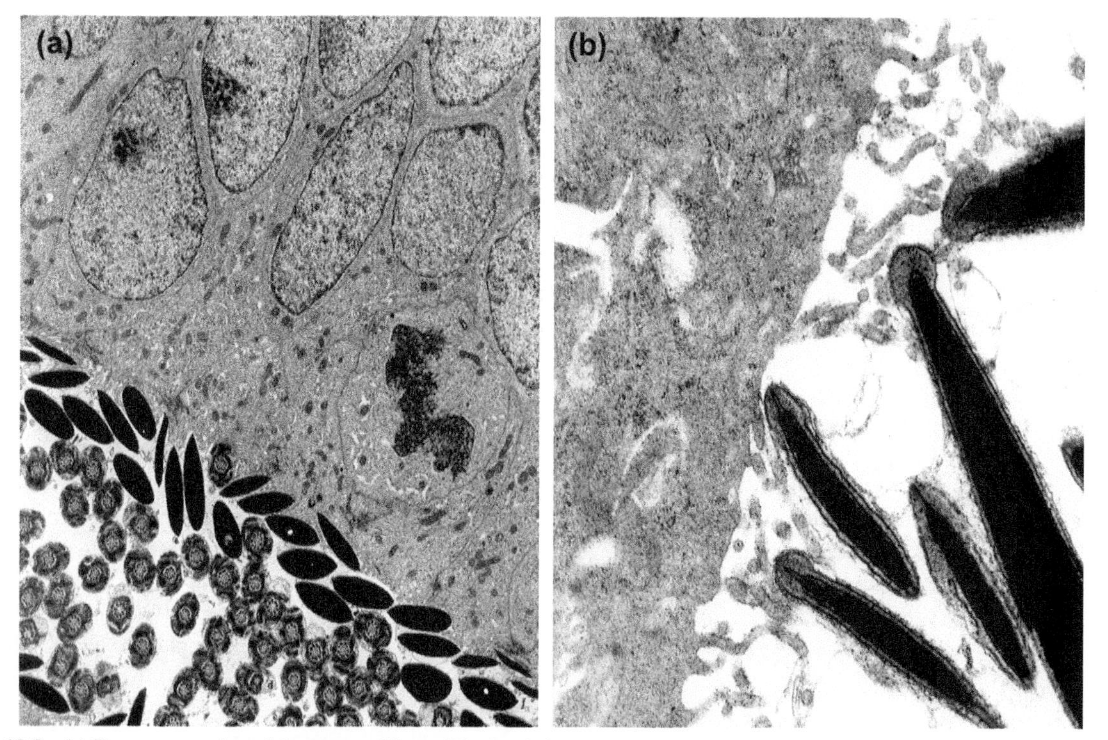

FIGURE 12.2 (a) Transverse section of the uterus of *Scotophilus heathi* in January showing the arrangement of the sperm with their heads oriented towards the uterine epithelium. (b) Electron micrograph of the uterus of *S. heathi* in February showing close contact of the sperm with the microvilli of the ciliated epithelial cells.

receptors (ARs) and androgen-binding proteins (ABPs) are present in the epithelial cell lining of the uterotubal junction (storage organ) during the period of sperm storage in *S. heathi* (Figure 12.3) (Roy & Krishna, 2010). It is suggested that androgens create a unique microenvironment (e.g., secretion of ABP within the lumen of the uterotubal junction) that facilitates prolonged sperm storage.

Sperm storage may increase fecundity in bats. With immediate availability of sperm, females can fertilize eggs whenever the opportunity arises or in rapid succession (Neubaum & Wolfner, 1999). Female sperm storage enables mating and ovulation to occur at widely separated intervals (Briskie & Montogomeric, 1993). When sperm are stored, the frequency of matings can be minimized. Mating requires time for courtship and copulation; it may disrupt feeding and increase exposure to predation, disease, and parasite transmission (Thornhill & Alcock, 1983; Hunter, Petrie, Otronen, Birkhead, & Moller, 1993). According to Racey (1982), sperm storage may function to synchronize parturition and lactation in species in which copulation is relatively asynchronous. Krutzsch et al. (1982) and Uchida and Mori (1987) have suggested that prolonged sperm storage among bats is closely related to the delayed sperm capacitation, which may be considered as an adaptive characteristic for sperm in bats.

3.3. Delayed Implantation

This phenomenon is not a common ordinal reproductive strategy in bats (Fleming, 1971), and mainly occurs in hibernating members of the genus *Miniopterus* (Mead, 1993; Krishna, 1999). These species differ from other hibernating vespertilionids in that copulation, ovulation, and fertilization occur in quick succession and the female enters hibernation in the pregnant state but implantation is delayed until the spring arousal. *Eidolon helvum* (Mutere, 1965) and *Rhinolophus rouxi* (Ramakrishna & Rao, 1977) demonstrate delayed implantation of several months duration. The period of delayed implantation almost coincides with that of winter hibernation (Planel et al., 1961; Peyre

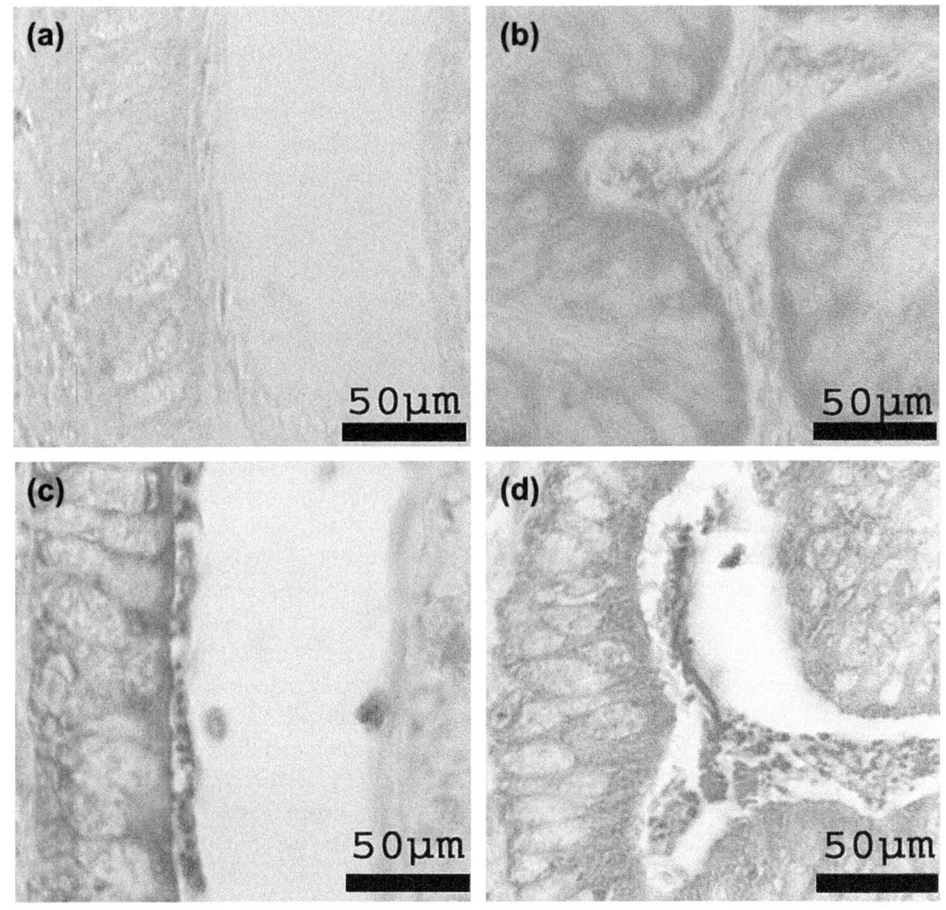

FIGURE 12.3 (a) Immunolocalization of androgen receptors (ARs) during December showing mild staining. (b) Immunolocalization of ARs during January showing strong immunostaining. (c) Immunolocalization of androgen-binding protein (ABP) during December showing mild staining in the cytoplasm of the luminal epithelial cells. (d) ABP in the epithelial lining of the uterotubal junction of *Scotophilus heathi* during the period of sperm storage. See color plate section

& Herlant, 1963a; 1963b; 1967; Richardson, 1977; Kimura & Uchida, 1983). Wimsatt (1969; 1975) proposed that the delay could be due to lowered cellular metabolism during hibernation. However, Peyre and Herlant (1963a; 1963b; 1967) speculated that an arrest in LH secretion following ovulation coupled with insufficient PRL is responsible for delayed implantation. Delayed implantation is not always a temperate zone phenomenon, but is found in various *Miniopterus* spp. distributed in the subtropical and tropical regions (Richardson, 1977; McWilliam, 1988; Bernard, Cotteril, & Forgusson, 1996). *M. schreibersii* showed an inactive CL and low plasma P_4 and LH concentrations during the time of delayed implantation (Bernard & Bojarski, 1994).

Torpor associated with hibernation may be responsible for suppression of luteal activity and consequently delay in implantation. Treatment of pregnant bats with exogenous PRL for 10 days during the early stages of delayed implantation resulted in the activation of luteal activity, high plasma P_4 concentrations, and the initiation of implantation. Treatment of bats with hCG during this period also induced similar changes in luteal activity and plasma P_4 concentrations but failed to initiate implantation. Treatment with exogenous P_4 had no influence on luteal activity or the initiation of implantation. Implantation occurs as day length increases in all bats that exhibit delayed implantation, suggesting that increasing day length may be the environmental cue that triggers implantation (Racey, 1982). Prolactin may be the pituitary hormone that mediates photoperiodic control of delayed implantation in *M. schreibersii*, where the period of delayed implantation lengthens with increasing latitude (Crichton, Seamark, & Krutzsch, 1989).

3.4. Delayed Development

Postimplantation delay in development is the least common form of reproductive delay in mammals and is described only in the Chiroptera (Bradshaw, 1962; Fleming, 1971; Bernard & Meester, 1982; Heideman, 1988; 1989; Singh & Krishna, 1997). The first report was in the phyllostomid bat *M. californicus* (Bradshaw, 1962). Later, Bleier (1975) showed an eight-month gestation period for *M. californicus* with 4.5 months of very slow embryonic growth. Delayed development also occurs in *A. jamaicensis* (Fleming, 1971) and in several other species (Table 12.2) (Bernard & Meester, 1982; Krishna & Dominic, 1982b; Heideman, 1989; Heideman & Powell, 1998).

A decline in food availability and low ambient temperature may affect embryonic development in bats (Racey, 1982; Mead, 1993; Bernard, 1994). Delayed development can be broadly categorized into two types: temperature-dependent and temperature-independent. In some species, such as *P. pipistrellus*, *Myotis myotis*, and *Corynorhinus* (or *Plecotus*) *rafinesquei*, reduced environmental temperature significantly lengthens gestation by slowing embryonic development (Eisentraut, 1937; Pearson et al., 1952; Racey, 1969). During such periods the female hibernates and gestation is prolonged approximately to the duration of hibernation (Racey, 1969; Racey & Swift, 1981). Since this form of embryonic diapause may or may not occur, depending upon environmental conditions, it is referred to as facultative delayed development. Delayed development lasting about four to five months always occurs in the California leaf-nosed bat (*M. californicus*) and is independent of temperature and food conditions. This is referred to as an obligate form of delayed development. The fruit bat (*Haplonycteris fischeri*) from the Philippines exhibits a nine-month delay in embryonic development, which is presumably of the obligate type (Heideman, 1989).

Bradshaw (1962) found that neither food provided in excess nor elevated temperature affected the embryonic growth rate in *M. californicus*. They did find lowered plasma thyroxine (T_4) concentration during part of the delay and suggested that hypothyroidism may be the cause of delay. Later, Burns and Wallace (1975) failed to increase the rate of embryonic development in this animal by treating with T_4 for up to 40 days. Burns and Wallace (1975) described biphasic peaks in plasma E_1 and E_2 during development, the first peak corresponding to the early part of the delay and the second to the period after delay. Burns and Easley (1977) found a similar biphasic pattern for plasma P_4. They speculated that the elevated steroid hormones might be required for implantation and suggested that the lowered hormone levels might be responsible for the delay. Crichton, Hoyer, and Krutzsch (1990) suggested that delayed development in *M. californicus* might be due to delayed luteal cell development and resultant nutritional incompetence of the uterus through inadequate steroid stimulation. Low circulating levels of P_4 suggested that the CL of embryonic diapause in *M. californicus* secretes fewer steroids than after the onset of active embryogenesis (Crichton et al., 1990). In this species, luteal cells increase in size as transition from delayed to normal development occurs (Crichton and Krutzsch, 1985). The role of the CL in developmental delay has also been examined in *M. schreibersii* (Crichton et al., 1989) and *C. sphinx* (Meenakumari, Banerjee, & Krishna, 2009), in which luteal cells are less active and the number of steroidogenic organelles in luteal cytoplasm are greatly reduced compared with normal embryogenesis. At the end of developmental delay, mitochondria, the smooth endoplasmic reticulum, and lipid droplets increase markedly in the luteal cells (Crichton et al., 1989). The low concentration of plasma P_4 recorded by Van der Merwe and Van Aarde (1989) in *M. s. natalensis* during the period of relatively slower embryonic development is in agreement with the depressed values reported for

TABLE 12.2 A partial list of the bats showing delayed embryonic development

Sl. No	Name of species	Family	Duration of delay	Embryonic Stage	Place	Nature of study	Reference
1	*Macrotus californicus*	Phyllostomidae	4.5 months	Newly implanted blastocyst	Arizona USA	Histology Embryo & CL, P_4, E_2, T_3, T_4	Bradshaw, 1962
2	*Artibeus jamaicensis*	Phyllostomidae	3 months	Newly implanted blastocyst	Panama Neotropical Mexico	Histology embryo	Flemming, 1971
3	*Miniopterus schreibersii*	Vespertilionidae	2.5-3.5 in Australia 5 months in France	Gastrulation	Japan, South Australia	Histilogy Embryo & CL, P_4	Crichton *et al.*, 1989
4	*Haplonecteris fischeri*	Pteropodidae	8 months	Gastrulation	Philippines	Histology Embryo & CL	Heideman, 1989
5	*Otopteopus cartilagonodus*	Pteropodidae	About 30 days	Primitive streak	Luzon Island, Philippines	Histology, Embryo	Heideman *et al.*, 1993
6	*Carollia perspicillata*	Phyllostomidae	41-110 days	Primitive streak	Trinidad, West Indies	Histology, Embryo	Rasweiler and Bradwaik, 1997
7	*Ptenochirus jagori*	Pteropodidae	5 months young female	Gastrulation	Phillipinee Island	Histology, Embryo	Heideman and Powell, 1998
8	*Cynopterus sphinx*	Pteropodidae	About 25 days	Gastrulation	Varanasi, India	Uterine diameter, CL, P4, E2, Embryo Histology	Meenakumari and Krishna, 2005

Delayed embryonic development is also implicated in *Rhinolophus hipposideros* (Kolb, 1950); *Natalus stramineus* (Mitchell, 1965); *Pipistrellus pipistrellus* (Racey and Swift, 1981); *Taphozous longimanus* (Krishna and Dominic, 1982b); *Miniopterus schreibersii* (Kimura and Uchida, 1983); *Hipposideros caffer* (Bernard and Meester, 1982) and *Rhinolophus ferrumequinum* (McOwat and Andrews, 1995).

the Japanese long-fingered bat (*M. s. fuliginosus*) (Kimura, Takeda, & Uchida, 1987). Studies on *C. sphinx* showed a marked decline in the steroidogenic enzyme activity (side chain cleavage enzyme, $P450_{scc}$) and steroidogenic acute regulatory (StAR) activity in the CL during the delayed period (Meenakumari et al., 2009). Insufficient PRL and GTHs may be responsible for suboptimal luteal cell development and consequently failure of the luteal cells to secrete sufficient hormone during delay (Kimura et al., 1987; Crichton et al., 1990; Meenakumari et al., 2009). Similar observations have been made for *M. californicus* (Richardson, 1979) during delayed development as well as for *M. schreibersii* during delayed implantation in Japan (Mikami et al., 1988a) and France (Peyre & Herlant, 1963b).

The period of retarded embryonic development in *C. sphinx* coincides with changes in various metabolic factors, especially those favoring fat accumulation (Banerjee, Meenakumari, & Krishna, 2007). Reduced ambient temperature during early winter leads to a decline in body temperature. Succinate dehydrogenase (SDH) activity may be responsible for decreased oxygen consumption and, consequently, a decline in basal metabolic rate (BMR) during this time. This period of low BMR, body temperature, and SDH activity may be responsible for retarded embryonic development during early winter (November–December). The energy saved by suppressing embryonic development and depressing body temperature may be utilized for the accumulation of fat.

4. HORMONES AND REPRODUCTION IN BATS

Very little is known of the endocrine mechanisms underlying bat reproduction, especially at the level of the hypothalamus and pituitary. Earlier studies on the bat pituitary focused primarily on temperate-zone hibernating bats and relied initially on tinctorial and histochemical techniques for identification of the various secretory cell types of the pars distalis. These studies have been reviewed in great detail by Richardson (1979). The most extensive investigations on the chiropteran pituitary are those of Herlant (1953; 1956a; 1956b; 1963; 1964; 1968), who described variations in the pituitary cytology in *M. myotis* throughout the annual reproductive cycle.

4.1. Hypothalamic Regulation of Pituitary Function

The presence of GnRH in the hypothalamus was first demonstrated in *M. lucifugus* (King, Anthony, Gustafson, & Damassa, 1984). Subsequently, sites of GnRH synthesis and secretion have been studied immunocytochemically in the vespertilionid bats *M. schreibersii* (Mikami, Chiba, Taniguchi, Kubokawa, & Ishii, 1988; Fernandez, Muniz, Gragera, & Martinez-Rodriguez, 1992) and *Eptesicus fuscus* (Oelschläger & Northcutt, 1992). In all the bats the neuronal cell bodies responsible for GnRH synthesis are widely distributed within the hypothalamus and in surrounding forebrain areas. The vespertilionid bat *M. lucifugus* has only mammalian GnRH (GnRH-I); chicken gonadotropin-releasing hormone (GnRH-II) is not present (King, Steneveld, Curlewis, Rissman, & Millar, 1994).

In *M. lucifugus*, the authenticity of GnRH-immunoreactive material in the neural lobe has been established by high performance liquid chromatography (HPLC) (Anthony, Wu, Bruhn, & Jackson, 1987). The principal sites of GnRH secretion in vespertilionid bats were shown immunocytochemically within the lower infundibular stem and the neural lobe. The majority of GnRH neuronal perikarya in the vespertilionid bats *M. lucifugus* (King et al., 1984), *M. schreibersii* (Mikami et al., 1988b; Fernandez et al., 1992), and *E. fuscus* (Oeschläger & Northcutt, 1992) are located in the medial basal hypothalamus (MBH), particularly in the region of the arcuate nucleus (ARC), and their distribution resembles that in higher primates (Parker, Neaves, & Porter, 1980; Silverman et al., 1982). Gonadotropin-releasing hormone neuronal cell bodies are preferentially distributed in the medial preoptic area (mPOA) in the hypothalamus (Witkin, Paden, & Silverman, 1982) and in some fibers terminating in the upper infundibular stem.

With respect to GnRH transport within the hypothalamic–pituitary complex, bats and rats seem to exemplify two extreme and opposite conditions. In rats, the majority of the fibers terminate at the level of the median eminence (ME), making this species heavily dependent on long portal routes of delivery. Bats, on the other hand, rely on short portal and possibly nonvascular routes, as GnRH fibers extend to more distal sites of secretion. Locations of GnRH terminal fields are representative of various mammalian groups, and primates and carnivores appear to fall between these two extremes, with long and short portal vessels contributing more equally to GnRH transport (Anthony, King, & Stopa, 1984).

4.2. Seasonal Dynamics of the Gonadotropin-releasing Hormone (GnRH) Neuronal System

Seasonal changes of the GnRH neuronal system in relation to reproductive function have been investigated in only a few species of bat. Kawamoto, Kurahashi, and Hayashi (1998) demonstrated seasonal changes in the hypothalamic GnRH neurons of the rhinolophid bat *R. ferrumequinum*. Gonadotropin-releasing hormone-immunoreactive (ir) neurons in this species were concentrated in the mPOA and MBH. This distribution seems to be a characteristic feature of rhinolophid bats. In vespertilionids, GnRH-immunoreactive neurons were mainly found in the MBH. Numbers of GnRH-ir neurons in the MBH vary depending on reproductive state (Anthony et al., 1989; Fernandez et al., 1992; Anthony, 2000). These findings suggest that GnRH neurons located in the MBH are the main perikarya of GnRH terminals in the ME and play a central role in the seasonal changes of GTH secretion in both rhinolophid and vespertilionid bats. Studies on rhinolophid bats revealed sex differences in the number of GnRH neurons in the mPOA in early June to late July. The number of ir neurons located in the mPOA was significantly smaller in females than in males. Enhanced GnRH neuronal activity in the mPOA from May to August may be related to female-specific reproductive events; i.e., pregnancy, parturition, and lactation periods. Mikami et al. (1988b) reported that GnRH immunoreactive perikarya were larger in hibernating male and female Japanese bats (*M. schreibersii*) than in active (pregnant and lactating) females. Fernandez et al. (1992) found that the perikarya of pregnant and lactating bats were smaller and fewer in number than those in hibernating and preovulatory individuals of *M. schreibersii* in Spain.

Anthony et al. (1989) examined the frequency and distribution of GnRH perikarya during the pre- and postovulatory periods of *M. lucifugus* using light and electron microscopic immunocytochemistry (ICC), and examined GnRH tissue contents by radioimmunoassay. The results showed a significant decline in the total number of ir perikarya as well as the amount of hypothalamic GnRH content in postovulatory compared to preovulatory females. In female *M. lucifugus*, GnRH neurons innervate both the ME and pars nervosa (PN) (King et al., 1984). Gonadotropin-releasing hormone immunoreactivity in the PN was chemically identical with that in the hypothalamus (Anthony et al., 1987b) and GnRH-ir was reduced significantly at both sites during the periovulatory period (Anthony et al., 1989). This reduction in GnRH associated with ovulation thus suggests that both the ME and PN contribute to the control of GTH secretion in this species.

In seasonally breeding mammals, gonadal activity depends especially on photoperiod. Animals exposed to short-day photoperiods in captivity or to naturally declining day lengths show a decline in gonadal function concomitant with a decrease in the release of hypothalamic GnRH (Kawamoto, Tanaka, & Hayashi, 2000) and of GTHs

(Pickard & Silverman, 1979; Niklowitz, Khan, Bergman, Hoffmann, & Nieschlag, 1989). In mammals, the effect of photoperiod is mediated by the nocturnal secretion of melatonin (MEL) from the pineal gland. The exact mechanism of neuroendocrine regulation by photoperiod or pineal hormone that causes physiological adjustment remains unknown (Malpaux, Migaud, Tricoire, & Chemineau, 2001).

4.3. Gonadotropins

The presence of two distinct types of gonadotropic cell (FSH and LH cells) in the pars distalis of several bat species has been reported by earlier workers using cytochemical, immunocytochemical, and ultrastructural techniques (Anthony & Gustafson, 1984; Mikami et al., 1988a; Bhiwgade, Akolkar, Menon, Manekar, & Senad, 1989). In a study of *S. heathi* using antisera against human βFSH and human βLH on consecutive sections of the pituitary, it was found that some of the cells react with both antisera. Other cells reacted either with βFSH or βLH antisera. Thus, this study of *S. heathi* suggests the occurrence of at least three populations of gonadotropic cells in the hypophysis, *viz.* those that contain both LH and FSH and those that contain either LH or FSH only. This is in agreement with results in *M. californicus* (Richardson, 1979) and *M. lucifugus* (Anthony & Gustafson, 1984).

Gonadotropin cells show marked differences in morphology and distribution from other cell types. In most mammals studied so far, including bats, GTH cells in both sexes are scattered individually or in small groups throughout the pars distalis (Richardson, 1979, Singh & Krishna, 1994a). Aggregation of GTH cells within the pars distalis of *M. lucifugus* is located adjacent to portions of the infundibular stalk and neural lobe, which contain GnRH fibers (Anthony et al. 1984). The portion of the pars distalis that is enriched in GTH cells in *M. lucifugus* may be analogous to the pars tuberalis of bats and other mammals (Mikami et al., 1988a); accounts of the pars tuberalis are conflicting among different species of bats studied (Richardson, 1981). This distribution may reflect a common tendency for GTH cells to be clustered near points of entry of portal blood. Portal vessels have been reported to enter the pars distalis in the region of the pars tuberalis in bats (Herlant, 1953; Anthony & Gustafson, 1984).

Cytological characteristics of GTH cells are described in *M. lucifugus* (Anthony, 1987), and similar features also have been noticed in the tropical vespertilionid bat, *S. heathi* (Singh & Krishna, 1994a). Gonadotropin cells are generally large, oval or round, and sometimes irregularly shaped, and contain irregularly shaped electron-dense secretary granules ranging from 350–500μm in diameter. These granules are scattered throughout the cytoplasm. The nucleus is typically irregular in shape. The well-developed Golgi complex contains immature secretory granules in bats from spring and summer. The rough endoplasmic reticulum (RER) exhibits striking seasonal variation in morphology, and is inconspicuous in bats in autumn and winter. The rough endoplasmic reticulum in GTH cells was well developed in bats collected in April, existing as extensive arrays of parallel and flattened cisternae. Mitochondria were proportionately larger and elongated at that time of year (April). During summer (June–August), the GTH cells of males contained extremely dilated RER with homogenous electron-dense material in the cisternae.

This seasonal variation in ultrastructural features suggests particularly high levels of synthetic and secretory activity in the GTH cells of males during spring and summer. Active protein synthesis in spring is indicated by the extensive development of the RER, accompanied by the presence of immature secretory granules in the Golgi complex. Even more striking is the external dilation in summer of the cisternae of the RER, which is similar to gonadotropes of castrated rats (Farquhar & Rinehart, 1954). This suggests increased synthesis and secretion of GTHs. Seasonal changes in intensity of LHβ- and FSHβ-ir and the size of GTH cells have been reported for numerous species including *M. californicus*, *M. lucifugus*, *M. schreibesii*, *S. heathi*, *Taphozous melanopogon*, and *R. ferrumequinum* (Richardson 1979; 1981; Anthony & Gustafson, 1984; Anthony, 1987; Badwaik, 1988; Mikami et al., 1988a; Bernard, Bojarski, & Millar, 1991a; Bernard, Bojarski, & Millar 1991b; Singh & Krishna 1994a; 1996b; Kawamoto et al., 2000), and these variations are greater in female than in male *M. lucifugus* (Anthony, 1987). The percentage of the pars distalis occupied by LH and FSH cells exhibits significant variation during the reproductive cycle in female *S. heathi* (Figure 12.4 and Table 12.3) (Singh & Krishna, 1996b). Many of the bat species studied so far showed conspicuous hypertrophy and granulation of gonadotrophs but weak LHβ- and FSHβ-ir during the reproductively active period, as demonstrated in *M. lucifugus* (Anthony & Gustafson, 1984), *S. heathi* (Singh & Krishna, 1996b), *R. ferrumequinum* (Kawamoto et al., 2000), and other seasonally breeding mammals such as the ferret (*Mustela putorius*) (Holmes, 1963), vole (*Microtus agrestis*) (Clarke & Forsyth, 1964), etc. A significant decline in the volume percentage of LHβ-ir occurs following ovulation and also during pregnancy, as demonstrated in *S. heathi* (Singh & Krishna, 1996b). Degranulation of LHβ-ir cells occurs during the periovulatory period in *M. myotis* (Herlant, 1956a), *M. californicus* (Richardson, 1981), and *M. schreibersii* (Mikami et al. 1988a), as well as in the rat (*Rattus norvegicus*) (Merchant, 1974; Blake, 1980) and sow (*Sus domestica*) (Herlant & Ectors, 1969). Follicle-stimulating hormone-β-ir became undetectable during

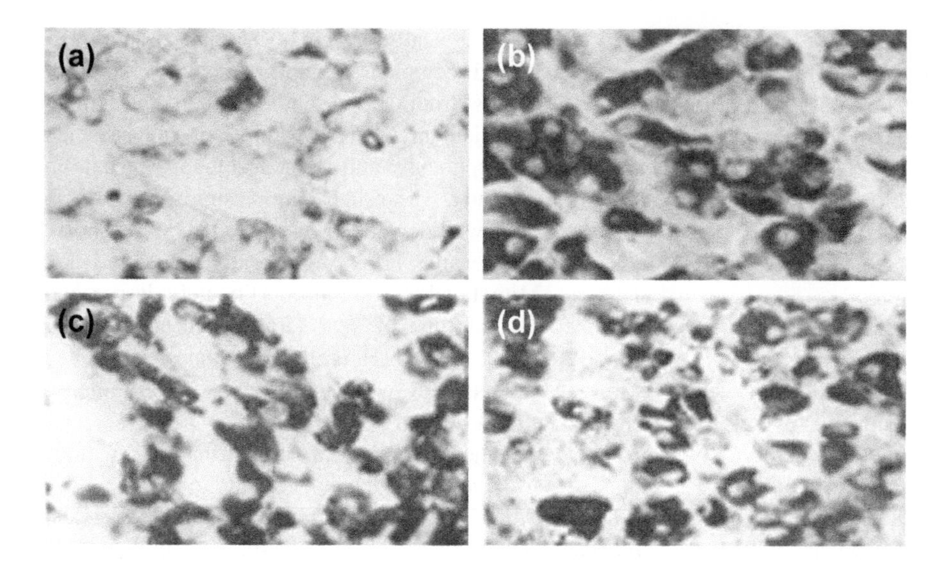

FIGURE 12.4 Transverse sections of the anterior pituitary of female *Scotophilus heathi* during different stages of the reproductive cycle. Sections are stained immunocytochemically with antiserum against human βLH. Magnification 275×. (a) During quiescence (August), showing weak staining. (b) During recrudescence, showing highly granulated LH cells. (c) During winter dormancy (December), showing granulated cells. (d) During ovulation in March, showing degranulated LH cells. *Adapted from Singh and Krishna (1996b).*

pregnancy of all of the above bat species. During the period of prolonged survival of Graafian follicles or delayed ovulation during hibernation, both LH- and FSH-ir cells were numerous in the pars distalis. This suggests that LH and FSH are apparently stored within these cells throughout the period of hibernation; unfortunately, ICC cannot provide information on hormone turnover. A reciprocal relationship between the immunoreactivity and cell size of gonadotropes in bats of both sexes may suggest seasonal differences in the dynamics of secretion, storage, and synthesis of LH and FSH hormones during the annual reproductive cycle (Kawamoto, 2003).

The maximal level of LH secretion and responsiveness of gonadotropes to GnRH in male *R. ferrumequinum* (Kawamoto et al., 2000) coincided with spermatogenetic activity and mating, but relatively low levels were found in other phases (the postarousal and hibernating periods). Changes in the rate of FSH secretion were similar to those for LH secretion. Treatment with GnRH significantly increased the total amount of GTH secreted from the GTH cells throughout the annual reproductive cycle, except for LH in the postarousal period and FSH in the hibernation period. The effect of GnRH on the secretion of both hormones was markedly enhanced during the spermatogenetic period. Thus, the responsiveness of GTH cells to GnRH also shows seasonal variation, which may be closely associated with the number of GnRH receptors on GTH cells and/or with the increased response of GTH cells to GnRH, caused by the action of steroids (Kawakami & Winters, 1999). During hibernation, the majority of GTH cells were quiescent, but a slight secretory response of these cells to GnRH was detected at low incubation temperatures (10°C) (Kawamoto et al., 2000). These results suggest that GnRH receptors and subsequent GnRH-mediated GTH secretion decline under heterothermic conditions.

Bernard et al. (1991b) calculated an index of pituitary LHβ-ir cells with circulating LH levels in female *M. schreibersii*. Changes in the LHβ-ir cell index throughout the year were remarkably consistent with changes in plasma LH level, both being high during follicular development and peaking just before ovulation. They were very low during early delayed implantation, rising again

TABLE 12.3 Seasonal variation in mean percentage area of the pars distalis occupied by luteinizing hormone (LH), follicle-stimulating hormone (FSH), and prolactin (PRL) cells of adult female *Scotophilus heathi*

Cell Types/ Reproductive Stages	Quiescence	Recrudescence	Winter Dormancy	Breeding	Postovulatory period	Pregnancy
Luteinizing hormone (LH)	4.24 ± 2.28	9.88 ± 1.56	8.95 ± 1.79	11.53 ± 1.39	2.94 ± 0.37	7.58 ± 2.28
Follicle-stimulating hormone (FSH)	3.71 ± 0.45	7.91 ± 1.22	9.12 ± 1.64	10.90 ± 2.21	6.04 ± 1.29	6.33 ± 1.25
Prolactin (PRL)	17.52 ± 1.74	20.84 ± 2.08	16.86 ± 3.25	21.94 ± 4.32	*	30.34 ± 6.38

Data not available.

coincident with normal pregnancy. The only discrepancy occurred in the months of January and February, when circulating LH levels were relatively low but the immunoreactivity index was high. This discrepancy suggests that during this period LH is being stored in the pituitary with only a minimal amount released. This study by Bernard et al. (1991a) provides a close agreement between ICC parameters and circulating hormone levels.

Luteinizing hormone levels in *M. lucifugus* are low in late summer, prior to the autumn mating period, until midhibernation during the period of delayed ovulation, but rise significantly in spring, prior to ovulation (Anthony & Gustafson, 1984; Canney & Butler, 1987). Pituitary storage of LH occurs throughout hibernation in *S. heathi* (Singh & Krishna, 1996b), with high circulating androgen levels during the period of delayed ovulation (Abhilasha & Krishna, 1996), suggesting that negative feedback is responsible for preventing LH secretion at this time in *M. lucifugus* and *S. heathi* as well as in other hibernating species.

4.4. Prolactin (PRL)

The presence of mammosomatotropes (bihormonal cells that secrete both PRL and growth hormone) has been well documented in the pars distalis of various mammals, including bats (Li, Stefaneanu, Kovacs, Horvath, & Smyth, 1993; Ishibashi & Shiino, 1989). In *P. abramus*, mammosomatotropes are found throughout the annual cycle, while mammotropes (lactotropes) containing pure PRL granules appear only during pregnancy and lactation. The role of mammosomatotropes is not completely understood. Mammotropes show significant seasonal changes in bats from both temperate and tropical zones (Kawamoto, 2003). In male bats, seasonal variation is not apparent (Kawamoto, 2003).

Prolactin has a widely recognized role in lactation but studies also link this pituitary hormone to the regulation of seasonal reproductive delay phenomena in some mammals (Anthony, 2000). In mustelid carnivores, the period of delayed implantation coincides with short days, and increased pineal MEL activity is believed to be responsible for suppressing PRL release. Prolactin secretion is necessary to support the critical function of the CL; thus, inhibition of PRL delays implantation in mustelid carnivores (Curlewis, 1992).

Prolactin-ir and the number of lactotropes in the pituitary of the vespertilionid bat *M. schreibersii* decreased from March through June, during follicular development, and the early part of the delayed implantation period, which coincided with low circulating PRL levels (Bojarski, 1993; Bernard & Bojarski, 1994). Increased serum PRL levels in July and August coincided with the end of delay and the period of implantation, and lactotropic cell activity and circulating PRL continued to increase as embryonic development proceeded. Increased size and/or abundance of PRL-ir cells are found during pregnancy in *S. heathi* (Singh & Krishna, 1996b) and similar cytological evidence of increased activity is seen in pregnant *M. myotis* (Muniz, Jimenez, Gragera, Fernandez, & Rua, 1991). Further, circulating PRL levels are higher in the female megachiropteran bat *Pteropus poliocephalus* during pregnancy and lactation periods but are low in the autumn during the breeding season (O'Brien, Curlewis, & Martin, 1996). Increasing synthesis and release of PRL after the winter solstice may activate the CL and create an endocrine environment that is conducive for implantation in this species. Increasing day length may be the environmental cue releasing PRL that mediates implantation. Although treatment with exogenous PRL induces high plasma P_4 concentrations and implantation, the initiation of implantation is not due to P_4 alone (Bernard & Bojarski, 1994), and the route by which PRL induces implantation remains unknown.

M. schreibersii in early delayed implantation collected in May and treated with daily subcutaneous PRL (0.01 mg) for 10 days showed CL activation and the onset of implantation (Bernard & Bojarski, 1994). However, bats in the same study treated with hCG (1 iu) showed only stimulation of steroidogenesis but no evidence of implantation. In *S. heathi*, PRL treatment delayed ovulation and caused a decline in the size and abundance as well as the PRL-ir content of lactotropes during winter dormancy (Figure 12.5) (Singh & Krishna, 1996b).

5. HORMONES AND REPRODUCTION IN MALE BATS

Only fragmentary information has accumulated concerning the reproductive biology of male Chiroptera and rarely has a repertoire of analytical tools common to current basic investigation been employed. Most of the hibernating male bats exhibit prolonged sperm retention in the epididymis, asynchrony between gametogenesis and the accessory reproductive organs, and asynchrony of the periods of spermatogenesis and copulation (Wimsatt, 1960; Gustafson, 1979; 1987). Collectively, these findings suggest that the reproductive patterns operative in hibernating bats are markedly different from those of other mammals. However, in most instances nonhibernating bats demonstrate reasonable synchrony between male and female reproductive processes. Sperm are most often produced and accessory sex glands active at a time consistent with the onset of the estrous cycle in the female, and copulation, ovulation, and fertilization are usually contemporary events (Krutzsch, 1979). There are exceptions, wherein sperm are stored in the epididymis (Gopalakrishna & Bhatia, 1980) of nonhibernating species;

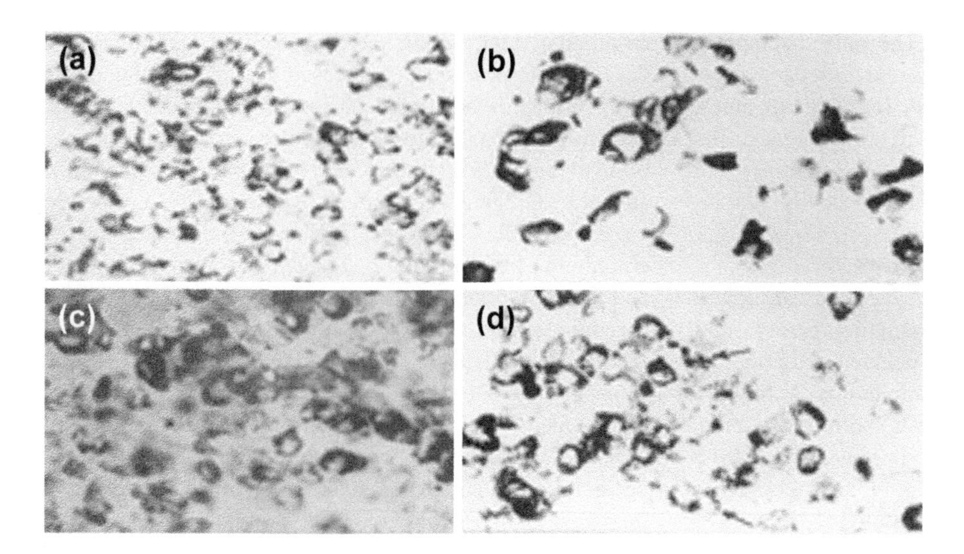

FIGURE 12.5 Transverse sections of the anterior pituitary of female *Scotophilus heathi* during different stages of the reproductive cycle. Sections are stained immunocytochemically with antiserum against human prolactin. Magnification 375×. (a) During quiescence (August), showing weak staining. (b) During the breeding phase (February), showing highly granulated cells. (c) During pregnancy (April), showing granulated cells. (d) During winter dormancy (December), showing degranulation.

e.g., *T. georgianus* of tropical Australia exhibits an extreme asynchrony of primary and secondary reproductive function (Jolly & Blackshaw, 1987).

Only limited studies have been undertaken on the testicular activities of bats, particularly in relation to unique asynchronous features and their regulation by endocrine and paracrine factors. Previous studies on the reproductive endocrinology of male bats have been limited to a few measurements of plasma T, AND, and localization of gonadotropes (Martin & Bernard, 2000).

5.1. Reproductive Patterns in Male Bats

Male bats show a wide range of annual patterns of reproduction, but they may be broadly classified into the following two categories: patterns in temperate-zone bats and patterns in tropical/subtropical-zone bats.

5.1.1. Temperate-zone male bats

Reproductive cycles in temperate-zone species are linked with hibernation and arousal. The timing of reproductive events in temperate zone hibernating bats can be classified into the following two subtypes.

5.1.1.1. Reproductive pattern subtype I

In this pattern, male reproductive cycles are interrupted annually by hibernation so that secondary sexual activities extend into and throughout the period of dormancy. This subtype corresponds to the pattern of delayed ovulation seen in females. In these species, although the increase in testicular weight, spermatogenesis, and the activity of Leydig cells occur in summer, the development of accessory organs and mating behavior is delayed until autumn, by which time spermatogenetic tissue regresses, epididymides are full of sperm, and the accessory sex organs show maximal development. Such asynchrony in activity between the testes and accessory sex organs is unique and is noted in many vespertilionid bats—e.g., *M. lucifugus*, *N. noctula*, and *Eptesicus (Vespadelus) vulturnus*—and rhinolophid bats—e.g., *R. ferrumequinum* and *Rhinolophus capensis* (Gustafson, 1979; Bernard, 1986; Martin & Bernard, 2000).

5.1.1.2. Reproductive pattern subtype II

In this pattern, male reproductive cyclicity is characterized by synchronized activity of the testes and accessory sex organs. The male reproductive cycle does not overlap with the period of winter dormancy/hibernation, because the increases in spermatogenesis, testicular mass, circulating T levels, and activity of the accessory sex organs are synchronized and completed mostly before hibernation. Both seminiferous tubules and accessory sex glands are inactive during winter dormancy. The most extensively studied representative of this pattern is *M. schreibersii* (Gustafson, 1979; Bernard et al., 1991a) and corresponds to the female pattern of delayed implantation or delayed development. For several American microchiropterans that failed to copulate in the previous autumn, mating occurred in the spring (Guthrie, 1933). The occurrence of active spermatogenesis both during the autumn (November) and spring (March) is demonstrated in some species; e.g., *P. pipistrellus*, *E. fuscus*, and *M. lucifugus* (Caffier & Kolbow, 1934).

5.1.2. Tropical/subtropical-zone bats

Studies of male reproductive cycles have revealed that a wide-ranging variation in reproductive patterns exists within closely related tropical/subtropical bats, even though the different species reside in nearly ecologically and geographically identical sites. Reproductive patterns in tropical/subtropical male bats can be organized into the following eight patterns based on frequency and/or duration of the spermatogenetic cycle, duration of sperm storage in epididymides, and the timing of mating and secretory cycle of male accessory sex glands together with their relationship with the female gametic cycle.

The following patterns vary with latitude, and some bats are seasonally bimodal at one locality whereas at another locality they are continuously polyestrous; e.g., *T. longimanus* at Varanasi vs. Nagpur (Gopalakrishna, 1955; Krishna & Dominic, 1982a). Some emballonurid bats are monoestrous in parts of their range and polyestrous in others (Krutzsch, 2000); e.g., *Hipposideros caffer* is monoestrous in some parts of its range but bimodally or continuously polyestrous in other parts (Bernard & Cumming, 1997).

5.1.2.1. Single restricted seasonal breeding (seasonal monoestry in females)

Male and female gametic cycles are synchronized. Testicular hypertrophy, gametogenesis, and epididymal enlargement are closely synchronized with the hypertrophy and secretory cycle of the accessory sex glands. Changes in male reproductive tissues usually occur in synchrony with estrus. Examples: *Lasiurus ego* (Myers, 1977); *S. wroughtonii* (Gopalakrishna, 1948); *Tylonycteris pachypus* and *Tylonycteris robustula* (Medway, 1972); *M. australis* (Baker & Bird, 1936).

5.1.2.2. Single extended seasonal breeding

Males and females exhibit nearly synchronized gamete cycles. Spermatogenesis commences in late summer and epididymides contain sperm from autumn to summer. Examples: *Nycteris thebaica* (Bernard, 1976; 1982); *M. lyra* (Ramakrishna, 1951; Ramaswamy, 1961; Pal, 1977; 1984).

5.1.2.3. Single extended asynchronized breeding

Both males and females exhibit single asynchronized breeding. Spermatogenesis commences in autumn with mating during winter. Sperm persists in epididymides for a prolonged period. Females may also store sperm. Ovulation occurs in early spring followed by fertilization and pregnancy. Examples: *S. heathi* (Krishna & Singh, 1992); *Rhinopoma kinneari* (Anand Kumar, 1965); *Rhinopoma hardwickei hardwickei* (Karim & Banerjee, 1985).

5.1.2.4. Single seasonal breeding with sperm stored throughout the year

Sperm are present in the testes and epididymides for an extended period although females are monoestrous. Epididymides store sperm throughout the year. Mating may occur for extended an period. Examples: *Balantiopteryx plicata* (Lopez-Forment, 1981); *H. fischeri* (Heideman, 1988); *T. melanopogon* (Brosset, 1962b; Khaparde, 1976); *Taphozous nudiventris*, *T. georgianus*, *T. hilli* (Kitchener, 1976; 1980; 1983; Jolly & Blackshaw, 1987).

5.1.2.5. Seasonal bimodal breeding

The male reproductive cycle is in synchrony with the female breeding cycle (twice in a year) and they undergo biseasonal testicular hypertrophy. Examples: *Epomophorus anurus* (Okia, 1974); *R. aegyptiacus* (Mutere, 1968a).

5.1.2.6. Seasonal bimodal breeding with sperm stored throughout the year

Males remain sexually active throughout the year with no variation in the testes, and the epididymides contain sperm all year round. Females in these species breed twice a year. Examples: *Cynopterus minutus* (Kofron, 1997); *C. sphinx* (Krishna & Dominic, 1983; 1984).

5.1.2.7. Multiple seasonal breeding (seasonal female polyestry)

Some bats show multiple (two or more) breeding bouts during most of the year with a brief period of reproductive quiescence, as seen in *P. mimus* (Krishna, 1985), *M. nigricans* (Wilson & Findley, 1970; Wilson, 1973), *Eonycteris spelaea* (Beck & Boo-Liat, 1973), and *T. pumila* (Van der Merwe, 1987; Happold & Happold, 1989). In *P. mimus*, the male bat also exhibits multiple (two or more) peaks of spermatogenic activity (Krishna, 1985).

5.1.2.8. Continuous breeding throughout the year (aseasonal polyestry)

Continuous but asynchronous breeding occurs in some tropical members of the families Desmodontidae (*D. rotundus* (Wimsatt & Trapido, 1952)), Phyllostomatidae (Tamsitt & Valdivieso, 1965a; 1965b), Pteropodidae, Emballonuridae, and Molossidae (Mutere, 1968a; 1968b; Anciaux de Faveaux, 1973). The best known of the virtually continuous-breeding molossids are *Chaerephon* (or *Tadarida*) *hindei* from Uganda (Marshall & Corbet, 1959) and *T. pumila* from Ghana (McWilliam, 1987a; 1987b). Males of these bats show spermatogenesis throughout the year.

5.2. Morphological Changes in Male Reproductive Tissue

Histological and ultrastructural methods have generally been used for the assessment of annual variations in Leydig cells for the majority of bats studied (Gustafson, 1976; Ohata, 1979; Loh & Gemell, 1980; Bernard, 1986; Bernard et al., 1991b). Two patterns of Leydig cell activity have been described among vespertilionid bats. In the first type, Leydig cells are active during spermatogenesis and remain active during winter hibernation ('Pipistrellus pattern'), while, in the second, the Leydig cells are active during spermatogenesis but regress before winter hibernation ('Myotis pattern') (Gustafson, 1979). Data on plasma T values suggest elevated T concentration during winter in the pipistrelle bat (*P. pipistrellus*) and depressed concentration of plasma T during winter in the little brown bat (*M. lucifugus*) and Cape horseshoe bat (*R. capensis*) (Racey & Tam, 1974; Gustafson, 1979; Bernard, 1986; Gustafson; 1987). Based upon the morphology of Leydig cells and ICC of P450 side chain cleavage enzyme and plasma levels of AND and T, *S. heathi* exhibits a unique pattern of Leydig cell activity. In *S. heathi*, plasma T and AND levels are elevated during active spermatogenesis, but only T is suppressed whereas AND remains high during the winter (Krishna & Singh, 1997).

The testes deliver sperm via the epididymal duct system and ductus deferens to the ampullary glands and then to the lumen of the prostatic urethra and onwards. The cauda epididymides may store sperm for various durations. Spermatogenesis in hibernating bats takes place during summer, with initial copulation at the onset of winter. During winter there is no spermatogenetic activity but sperm are stored in the cauda epididymides, the accessory glands remain active, and further copulation may occur (Gustafson, 1979; Racey, 1979; 1982). The interfemoral caudae epididymides are often enclosed in a heavily pigmented sheath (tunica vaginalis), the role of which is not known, though it is speculated to be thermoregulatory and related to sperm viability (Krutzsch, 2000).

The morphology of the accessory sex glands of Chiroptera shows great diversity, and surprisingly wide differences are found in closely related species. These glands (prostate, seminal vesicle, ampullary, urethral, and Cowper's) contribute various substances (fructose, citric acid, sialic acid, and zinc) to the ejaculate (Rajalakshmi & Prasad, 1970; Krutzsch, Watson, & Lox, 1976; Mokkapati & Dominic, 1976) and their central duct seems to be a temporary storage site for sperm in some species (e.g., *M. schreibersii* (Krutzsch & Crichton, 1990b)). The accessory sex glands undergo seasonal changes in size and activity, presumably in response to circulating androgens (Krishna & Singh, 1997; Krutzsch, 2000). These glands hypertrophy during the reproductively active phase. Studies have reported the presence of neutral mucins and protein-bound sialomucins (Gadegone & Sapkal, 1983) and fructose (Krutzsch et al., 1976; Krishna & Dominic, 1982a) in the seminal vesicles. Interesting reports of Hunter et al. (1971a; 1971b) have described a protein in the seminal vesicles of *M. lucifugus* that, when injected into other mammals, diminished smooth muscle contractile frequency as well as decreasing circulating lymphocytes. The authors speculated that this protein may block sperm transport and alter the phagocytic system so as to allow prolonged sperm storage in the female tract. It is presumed that accessory sex gland secretions in certain bat species produce, in conjunction with uterine secretion, a vaginal plug (Matthews, 1941; Martin et al., 1995). The seminal vesicle of *R. leschenaulti* (Gopalakrishna & Karim, 1974) and *C. sphinx* (Mokkapati & Dominic, 1977; Krishna & Dominic, 1984) showed the presence of spherical bodies (secretory blebs). These bodies are also found in the vagina of females for several days after copulation and, unlike other secretions from the accessory sex glands, do not form a vaginal plug. Such rounded bodies are also described in the prostate of certain marsupials and insectivores (Mokkapati & Dominic, 1977).

5.3. Spermatogenesis and its Regulation

The process of spermatogenesis in bats has been studied in only a few bat species: *R. kinneari* (Singwi & Lall, 1983), *P. giganteus* (David & Lall, 1985), *P. poliocephalus* (McGuckin & Blackshaw, 1987a; 1987b), *R. leschenaulti* (Saidapur & Patil, 1992), and *Pteropus vampyrus* and *Rhinolophus cornutus* (Morigaki et al., 2001). Further, only a few studies have investigated the hormonal control of spermatogenesis (Kawamoto, 2003). Generally, spermatogenesis in bats is prolonged and varies from three to four months between the first signs of growth of the seminiferous tubules to the release of sperm from the tubules. In contrast, the duration of spermatogenesis varies from 35 to 74 days in other mammals (Monesi, 1972).

In rhinolophid bats (e.g., *R. ferrumequinum*), the concentrations of FSH receptors (specific binding per unit tissue weight) in the testes vary throughout the annual reproductive cycle. Follicle-stimulating hormone receptors are primarily localized in Sertoli cells (Simoni, Gromoll, & Nieschlag, 1997). The decrease in the concentration of FSH binding sites per unit weight basis in summer (spermatogenetic period) could be due to an increased number of germ cells as compared to Sertoli cell number. The elevated concentration of binding sites during the mating period is due to a rapid decrease of testicular germ cells, spermatocytes, and spermatids, because there is some evidence that the absolute number of Sertoli cells per testis is stable after puberty and does not vary throughout the seasons in mammals (Steinberger & Steinberger, 1977; Sinha Hikim, Bartke, & Russell, 1988). Nevertheless, the concentration of FSH binding sites and total binding sites in the testis, and FSH-stimulated production of cAMP, were very low during hibernation. These results suggest that, in the testis in the hibernating period, not only was the number of FSH binding sites per Sertoli cell extremely decreased, but that consequently the activity of the adenylate cyclase system led by the FSH-receptor complex was much reduced.

Luteinizing hormone and FSH secretion and the responsiveness of gonadotropes to GnRH as assessed by cell immunoblot assay were markedly increased in the spermatogenetic period compared to other periods. Seasonal changes in FSH secretion correlate with concentration of FSH binding sites. Therefore, testicular FSH receptors might be upregulated in response to elevated circulating FSH levels in bats. After arousal from hibernation, elevated FSH secretion may be required for the initiation of testicular recrudescence and the concomitant increase in testicular weight (Schatt, De Geyter, Kliesch, Nieschlag, & Bergman, 1995).

5.4. Hormonal Changes in Males

Seasonally reproductive species in the tropical zone generally exhibit increased plasma T concentrations during the breeding season, and there is a corresponding synchronous stimulation of spermatogenesis and sex accessory gland functions (Table 12.4). A close correlation of increased circulating T concentration with peak spermatogenesis has been widely reported in most tropical bats (Singh & Krishna, 2000), suggesting a direct role of T in spermatogenesis. Curiously, in the majority of hibernating vespertilionid and rhinolophid bats, activation of the sex accessory glands and the onset of breeding are not synchronized with the period of peak spermatogenesis and plasma T levels, indicating temporal dissociation of the exocrine and endocrine activities of the testis (see Table 12.4).

The endocrine changes during reproductive asynchrony in male bats have been investigated most comprehensively In this species, there is a significant increase in testicular size together with increases in spermatogenesis, Leydig cell function, plasma T levels, and steroid-binding globulin (SBG) levels during arousal from hibernation in spring. Maximal T levels coincide with peak spermatogenesis, but little increase in accessory sex gland size occurs after peak spermatogenesis, when plasma T and SBG levels are declining (Gustafson & Shemesh, 1976). Increased accessory sex gland activity and libido are maintained throughout hibernation in *M. lucifugus*, when T levels are minimal. Gustafson and Damassa (1984; 1985) suggest that SBG reduces free T, thereby inhibiting hypertrophy of accessory sex organs.

In several microchiropteran and megachiropteran species, male reproductive cyclicity is characterized by synchronized activity of the testes and accessory organs; e.g., in *M. schreibersii* (Gustafson, 1979; Bernard et al., 1991a), the pteropodid bats *P. poliocephalus* and *Pteropus scapulatus* (McGuckin & Blackshaw, 1991), and *T. longimanus* (Singh & Krishna, 2000). In these bat species, all the male reproductive activities, such as the increase in spermatogenesis, testicular mass, circulating T levels, and activity of the accessory glands, are synchronized. After copulation, these activities decrease, resulting in a regression of testicular mass and accessory organs and a decline in circulating androgen levels. The range of the circulating T and AND concentration shown in Table 12.4 is higher than that reported for most other mammalian species (Gustafson & Shemesh, 1976; Jolly & Blackshaw, 1989; Singh & Krishna, 2000), especially in most of the vespertilionid bats that exhibit prolonged sperm storage.

Sperm storage has been reported in various families of Chiroptera: Emballonuridae (Jolly & Blackshaw, 1987); Molossidae (Racey, 1974; Krutzsch & Crichton, 1987); *Hipposideros* (Gopalakrishna & Bhatia, 1980); and Rhinolophidae (Karim & Banerjee, 1985). Thus, in view of the ubiquity with which sperm storage appears in the Chiroptera, sperm longevity (persistence) in the reproductive tracts of male and female bats may be an inherent characteristic of this mammalian order.

Although it is well known that LH is the major stimulus for initiation and maintenance of steroidogenesis by Leydig cells, earlier studies in the bat *S. heathi* along with a number of other studies mainly on the rat have suggested a vital role for FSH in the control of testicular endocrine functions (Chen, Payne, & Kelch, 1976; Gaytan, Bellido, Morales, Van Rooijen, & Aguilar 1995). Follicle-stimulating hormone is capable of stimulating androgen production *in vitro* from the testes of *S. heathi* (Krishna & Singh, 1999) (Table 12.5). Interestingly, a differential effect of FSH on T and AND production was observed during recrudescence and winter dormancy. During recrudescence, FSH was not very effective in stimulating AND production as compared to the same dose (1 μg/ml) of LH, whereas during winter dormancy FSH was as effective as LH in stimulating AND production. Follicle-stimulating hormone was more effective in stimulating T production *in vitro* during recrudescence than during winter dormancy. The higher responsiveness to FSH during winter dormancy may be responsible for maintaining high AND concentration in *S. heathi* and for maintaining copulatory behavior.

6. HORMONES AND REPRODUCTION OF FEMALE BATS

The Chiroptera are noteworthy in that anatomical and/or functional asymmetries of the female reproductive organs are far more frequent and profound than those recorded for any other mammalian order. Female reproductive cycles can be categorized according to unique specializations such as delayed ovulation/sperm storage, delayed implantation/development, and ovarian asymmetry/postpartum estrus. In the vespertilionid and rhinolophid bats (Oxberry, 1979; Racey, 1982), initial mating is a prehibernation event and

TABLE 12.4 Circulating testosterone (T) and androstenedione (AND) in male bats during different phases of the reproductive cycle

Species	Mean Plasma Testosterone (T) or Androstenedione (AND) levels in ng/ml					References
	Reproductive Cycle	Active Accessory Sex Gland	Spermatogenesis	Nonbreeding Phase	Sperm Storage	
Myotis lucifugus	T: < 0.1–60	T: 59	T: 60	T: < 0.1	T: < 0.1	Gustafson and Shemesh (1976)
Nyctalus noctula	T: 30–102	T: 45	T:102	T: 30	T: 30	Racey (1974)
Pipistrellus pipistrellus	T: 90–81	T: 120	T: 81	T:90	T: 120	Racey and Tam (1974)
Vespadelus vulturnus	AND: 20–100	n/a	AND: 100	AND: 20	AND: 10–20	Tidemann (1993)
Nyctophilus geoffroyi	T: 80–254 pg/testis	n/a	T: 254 pg/testis	T: 80 pg/testis	T: ~100 pg/testis	Hosken et al. (1998)
Mormopterus planiceps	T & AND: 0.2–5		T: 4	T: < 1	T: 8	Krutzsch and Crichton (1987)
Rhinolophus capensis	T: 0.2–6.2	T: 1.5	T: 6.2	T: 2–3	T: 0.2–1.2	Bernard (1986)
Taphozous georgianus	T: 0.36–2.10 AND: 0.61–3.66	T:2.1	*	*	*	Jolly and Blackshaw (1989)
Miniopterus schreibersii	T: 27.0 & 0.1–0.2	T: 10.3 (1–15)	T: 15	T: 2.17	n/a	Krutzsch and Crichton (1990); Bernard et al. (1991b);
Macrotus californicus	T: 0.5–2.7	T: 3.0–2.0	T: 2.7	T: 0.5	n/a	Krutzsch et al. (1976)
Pteropus poliocephalus	T: 2.18–46.3	T: 15.0–20.0	T: 46.3 AND: ~16.0	T: 2.18 AND: ~4.0	n/a	McGuckin and Blackshaw (1987b)
Pteropus scapulatus	T: 0.5–2.6	T: 1.0–2.0	T: 2.6	T: 0.5	n/a	McGuckin (1988)
Epomops franqueti	T: 0.1–0.3	n/a	T: 0.3	T: 0.1	n/a	Ifuta et al. (1988)
Myotis ricketti	T: 0.03–16.0	T: 0.37	T: 16.0	T: 0.37	T: 0.23–16	Wang et al. (2008)
Scotophilus heathi	T: 0.66–19.95 AND: 7.3–165.20	T: 9.54 AND: 19.93	T: 19.95 AND: 165.31	T: 0.66–5.49 AND: 7.3–16.90	T: 19.95–4.49 AND: 165.31–7.3	Krishna and Singh (1997)
Taphozous longimanus	T: 2–36 AND: 1.5–22	T: 36 AND: 22	T: 36 AND: 22	T: 2–7 AND: 7	T: 2–36 AND: 1.5–10	Singh and Krishna (2000)

Partially adopted from Martin and Bernard (2000).
Data not available

TABLE 12.5 The effect of luteinizing hormone (LH) and follicle-stimulating hormone (FSH) on androstenedione (AND) and testosterone (T) synthesis *in vitro* in the testes[!] of *Scotophilus heathi*

	Mean Androstenedione (AND) (ng/ml)		Mean Testosterone (T) (ng/ml)	
Treatment	Recrudescence	Winter Dormancy	Recrudescence	Winter Dormancy
Control	6.15 ± 0.41	2.30 ± 0.15	6.98 ± 2.44	0.91 ± 0.06
b. LH (1 μg/ml)	31.08 ± 5.96[a]	6.40 ± 0.57	22.44 ± 6.95[a]	8.34 ± 2.84[a]
c. LH (10 μg/ml)	6.40 ± 0.11[b]	6.49 ± 1.29	9.54 ± 0.93b	*
d. FSH (1 μg/ml)	10.52 ± 2.12[b]	15.47 ± 2.19[a]	16.21 ± 5.85	2.84 ± 0.39
e. FSH (10 μg/ml)	21.40 ± 8.99[a]	14.17 ± 5.58[a]	12.33 ± 0.22	3.0 ± 0.11
f. LH + FSH (1 μg each)	16.41 ± 0.98[b]	5.03 ± 0.12	9.78 ± 3.59[b]	6.58 ± 3.14[a]

Values are mean ± standard error of the mean (SEM).
[a]*Statistically significant ($p < 0.05$) vs. control values.*
[b]*Statistically significant ($p < 0.05$) vs. LH (1 μg/ml) values.*
**Data not available.*
[!]*One testis in two halfs per tube were cultered for two hours.*

sperm remain alive in the female reproductive tract over winter until ovulation the following spring. The majority of these bats undergo hibernation induced by low ambient temperature and food deprivation, which in turn acts to retard the progress of reproductive events, resulting in delayed ovulation or delayed embryonic development (Oxberry, 1979; Racey, 1982).

6.1. Anatomy and Physiology of the Female Reproductive Tract

The female reproductive system in bats includes paired ovaries, which are the source of germ cells or ova and hormones that regulate sexual development. The paired fallopian tubes transport sperm to the egg and transport fertilized ova to the uterus. The uterus receives sperm arriving from the vagina and is the site of the development of the fetus. Within this basic framework, the female reproductive tract of bats exhibits many anatomical and functional specializations. The uteri of bats vary considerably from duplex to bicornuate to simplex (Figure 12.6). As in most mammals, the bat ovary is divided into an outer cortex and an inner medulla. The cortex contains the follicles, CL, interstitial gland cells, atretic follicles, and stromal cells. The medulla contains the intraovarian blood vessels, lymphatic vessels, nerves, medullary cords, interstitial gland cells, and the ovarian rete. A surface epithelium envelops the exterior of the cortex.

6.1.1. Reproductive asymmetry

The Chiroptera present an extremely high incidence of female reproductive asymmetry in which one or the other ovary and the ipsilateral or contralateral uterine horn are nonfunctional. Asymmetry in bats is always associated with the monotocous habit (producing a single offspring at birth). In the relatively few polytocous species (producing two or more offspring at birth), the female reproductive organs are bilaterally symmetrical. The partial or complete dominance of one ovary, the alteration of successive ovulation, differential stimulation of the oviducts and uterine horns, and/or the dominance of one uterine horn are far more frequent and profound in bats than has been noted for any other mammalian order. Wimsatt (1979) classified the reproductive asymmetries in female bats into five types:

6.1.1.1. Alternating dominance

Both ovaries and uterine cornua are morphologically and physiologically normal with ovulation alternating from the right or left ovary; postpartum estrus. For example, *P. giganteus*, *C. sphinx*, and *T. longimanus* (Krishna & Dominic, 1982a; 1982b; 1983).

6.1.1.2. Dextral dominance

The right uterine cornu dominates as implantation occurs in the right cornu; ovulation usually occurs from the right ovary. For example, the molossid bat *M. ater* ovulates only from the right ovary; the smaller left ovary has a relatively less developed blood supply (Rasweiler, 1988). Crichton and Krutzsch (1985) suggested that the dominant right ovary or dextral dominance in *M. californicus* might be due to differences in the levels of GTH receptors between the two ovaries. Ovulation was demonstrated from the left ovary in *M. californicus* following experimental removal of the right ovary (Bleier & Ehteshami, 1981). A similar response was not observed in *Molossus* sp. after removal of the dominant right ovary (Wimsatt, 1975). Jerrett (1979)

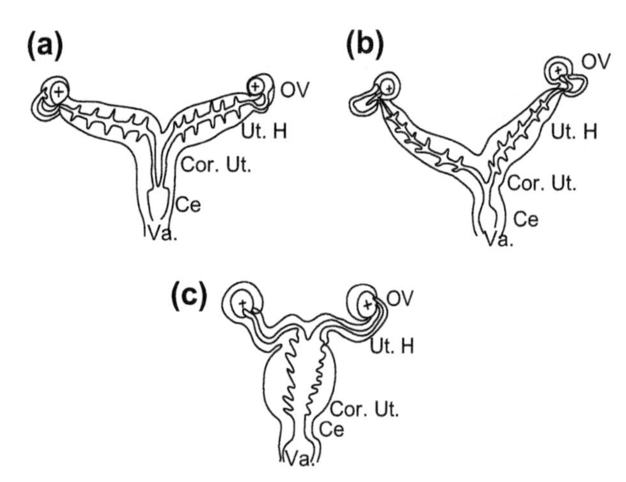

FIGURE 12.6 Types of female genital tract found in bats. (a) Simplex; (b) bicornuate; (c) duplex. OV, ovary; Ut. H., uterine horn; Cor. Ut., corpus uteri; C, cervix; Va, vagina.

noted an increase in the weight of the left ovary, but could not induce ovulation following removal of the dominant right ovary in *T. brasiliensis*.

6.1.1.3. Sinistral dominance

The left uterine cornu dominate, implantation occurs in the left uterine cornu, and ovulation usually occurs from the left ovary. For example, in *M. lyra* (Badwaik & Gopalakrishna, 1990), 70% of the females show sinistral dominance (Karim, 1973).

6.1.1.4. Contralateral dominance

Two species of the genus *Miniopterus* employ a contralateral dominance pattern in which ovulation occurs from the left ovary and the early embryo then undergoes transuterine migration and implantation occurs in the right cornu (Medway, 1971; Richardson, 1977; Bernard, 1980).

6.1.1.5. Symmetric dominance

Both ovaries and both uterine cornua are functional and generally polytocous (twin embryo in opposite cornu).

6.1.2. *Unilateral endometrial involvement*

Generally, reproductive asymmetry in bats is expressed as a unilateral dominance of the right or left ovary, the right or left uterus, or both in various combinations. Unlike birds, dextral asymmetry predominates in bats but a few also show ipsilateral and contralateral dominance (Wimsatt, 1979). The bats of the families Pteropodidae, Phyllostomidae, Emballonuridae, etc. generally show a nonrandom alteration of ovulation between the right and left ovaries in successive cycles. An intriguing characteristic of this pattern is the development of a localized progestational reaction of the endometrium at the distal end of the cornu on the side of the ovulating ovary. Several hypotheses have been proposed to explain how preferential stimulation of the ipsilateral oviduct and/or uterine horn might be accomplished in bats (Bonilla & Rasweiler, 1974; Rasweiler, 1978; Wimsatt, 1979; Rasweiler & Badwaik, 2000). This unilateral effect in bats may be due to the existence of special vascular arrangements that would enable an active ovary to exert a local, unilateral influence on the ipsilateral uterine horn or oviduct. Such an arrangement could be one that would permit a countercurrent exchange of steroid hormones or other diffusible substances between ovarian or uterine venous and/or lymphatic vessels and arterial branches supplying the ipsilateral cornu and/or oviduct. It is also possible that the hormones in the follicular fluid released at the time of ovulation may have a local effect upon the ipsilateral ducts, oviduct, and uterus. The unilateral nature of this action implies the existence of a local pathway whereby the endocrine products of the functional ovary directly influence the ipsilateral uterine cornu. Marshall (1953) first observed the close juxtaposition of ovaries and uterine cornual extremities in the giant fruit bat (*P. giganteus*). Space restriction is a possible explanation for the asymmetry patterns observed in Chiroptera. Alternation of ovulation may facilitate the rapid re-establishment of a pregnancy after either a reproductive failure or parturition. The ability to alternate ovulation appears to be essential in order for the females to conceive successfully at postpartum estrus (Rasweiler, 1982). It has been suggested that, although sperm transport is inhibited on the recently parturient side of the tract, it can take place normally through the previously nongravid uterus (Rasweiler, 1982). Similarly, by alternating ovulations the bat can successfully establish a pregnancy on the previously nongravid side, while involution of the postparturient horn is still in progress.

The progestational reaction is localized to a small segment near the cranial end of the uterine cornua in *C. sphinx* during the oviductal journey of the embryo. The progestational changes are augmented in the cornu on the side of ovulation, whereas the contralateral cornu reverts to an anestrus condition after implantation of the blastocyst. Unilateral progestational changes at the cranial part of the uterus have been reported in *P. giganteus* (Marshall, 1953), *T. longimanus* (Gopalakrishna & Murthy, 1960; Krishna & Dominic, 1982a), *R. leschenaulti* (Gopalakrishna & Karim, 1971), *Noctilio albiventris* (Rasweiler, 1978), *Pteropteryx kappleri* (Rasweiler, 1982), and *H. fischeri* (Heideman, 1989). The progestational swelling occurs due to local proliferation of the uterine endometrium and uterine gland and becomes macroscopically visible prior to the arrival of the blastocyst. The progestational reaction in *C. sphinx* may preclude movement of the blastocyst past the zone of expansion (Heideman, 1989).

6.1.3. Menstruation

The process of menstruation has been demonstrated in five species of bat belonging to the families Phyllostomidae (*Glossophaga soricina* and *C. perspicillata*), Molossidae (*M. ater*), Desmodontidae (*D. rotundus*), and Vespertilionidae (*M. ricketti*) (Rasweiler, 1970; 1991; Rasweiler & De Bonilla, 1992; Rasweiler & Badwaik, 2000; Wang et al., 2008). Menstruation in these bats involves necrosis and sloughing of the superficial lamina functionalis of the endometrium with associated bleeding. However, no external signs of menstruation have been noted, because most of the menstrual debris undergoes dissolution within the uterine cavity. Generally menstruation has been observed in bats during the periovulatory (immediately before or after ovulation) period. Thus, the menstruation process in bats appears to be more closely associated with the time of estrus and ovulation, as occurs in canids, than with events in the menstrual cycle, as in catarrhine primates.

Although the endocrine control of the uterine cycle has not been fully worked out for these menstruating bats, in *G. soricina*, *C. perspicillata*, *D. rotundus*, and *M. ater*, regressing CL are usually present in the ovaries at the time of menstruation.

6.2. Ovarian Morphology

In mammals, including bats, the ovary is covered with a connective tissue bursa and divided into an outer cortex and inner medulla. In adults the cortex contains follicles and CL along with interstitial cells, degenerating follicles and CL, and stroma cells. The medulla contains interstitial cells, intraovarian blood and lymphatic vessels, nerves, medullary cords, and the ovarian rete. A surface or 'germinal' epithelium envelops the exterior of the cortex. In the phyllostomid *G. soricina*, primordial follicles are congregated in a cup-shaped zone on the medial side of the ovary (Rasweiler, 1972). Primordial follicles have a similar restricted distribution in several other phyllostomid bats—e.g., *Leptonycteris curasoae* (Hood & Smith, 1983) – and pteropodid bats – e.g., *Otopteropus cartilagonodus*, *P. jagori*, and *Pteropus* spp. (Heideman et al., 1993; Pow & Martin, 1994; 1995; Heideman & Powell, 1998).

The type and relative abundance of thecal/interstitial gland cells (TICs) varies among bats (Rasweiler & Badwaik, 2000). Many of the vespertilionid, molossid, and emballonurid bats have ovaries that typically possess an abundance of stromal TICs. This differs markedly from the situation in the noctilionids and other bat species. In *C. perspicillata* and *N. albiventris*, TICs are not distinctive in the ovary during the periovulatory phase or early pregnancy. In *G. soricina*, a small number of interstitial cells (ICs) are sometimes evident adjacent to atretic follicles, but not in the ovarian stroma. In the New Zealand short-tailed bat (*Mystacina tuberculata*), TICs are abundant in the functional left ovary but not in the right ovary. The ovary of *T. longimanus* shows a predominance of TICs in the stroma, especially during the period of recrudescence and winter dormancy (Singh & Krishna, 2002). The increase in TICs in the ovarian stroma of *T. longimanus* during these reproductive phases could not be due to either increased differentiation of stroma cells into TICs or accumulation of TICs derived from thecal cells of atretic follicles (Singh & Krishna, 2002). Thecal/interstitial gland cells are a major steroidogenic component of the ovary in many insectivorous bats (Abhilasha & Krishna, 1998).

One of the most striking cytological features of the ovary of *S. heathi* is the dominance of TICs. The majority of the interfollicular tissue of the ovary of *S. heathi* consists of a mass of TICs composed of the histologically similar thecal cells and ICs. The ultrastructural features of TICs are the presence of organelles typical of a steroidogenic cell; i.e., a granular endoplasmic reticulum, mitochondria with tubular cristae, and abundant and large lipid droplets (Singh & Krishna, 1994b). Immunoreactive, $P450_{scc}$, 3β-hydroxysteroid dehydrogenase (3β-HSD), and 17α-hydroxylase ($P450_{17\alpha}$) enzymes are present in TICs of the ovarian stroma. Thus, the high androgen production by the ovary of *S. heathi* may be the result of increased differentiation of the stromal cells into thecal ICs. Similar extensive development of TICs has been reported in *Otomops martiensseni*, *M. ater*, and *Rhinolophus curale* (Kayanja & Mutere, 1975; Rasweiler, 1988; Tsvetkov & Takeva, 1988). The ovary of female spotted hyenas with masculinization of external genitalia also shows hypertrophied ICs as the major source of androgen secretion (Lindeque, Skinner, & Miller, 1986). Similarly, an increased mass of theca-like ICs in the ovarian stroma has been demonstrated to be the main site of excess androgen production in women with PCOS (Erickson, Magoffin, Dyer, & Hofeditz, 1985).

6.2.1. Follicular development

Ovarian follicular development in bats can be of two major patterns. In pattern I, follicular development is prolonged; Graafian follicles survive for much longer than is usual during reproductive cycles of other mammals and ovulation may be delayed for several months. This occurs in the majority of vespertilionid and rhinolophid bats. In pattern II, development of follicles and ovulation occur in quick succession within a brief period.

6.2.1.1. Pattern I: Prolonged survival of Graafian follicles

In most of the vespertilionid bats, several waves of follicular growth and development occur (Wimsatt, 1944).

A number of antral follicles form in the autumn as a result of waves of follicular growth and development. These large antral follicles or Graafian follicles form during early winter and survive for a prolonged period (Oxberry, 1979; Krishna & Dominic 1982c). During hibernation, these Graafian follicles remained unchanged (Wimsatt & Kallen, 1957; Krishna & Abhilasha, 2000a). The static appearance of the Graafian follicles (Oxberry, 1979) and studies showing the ineffectiveness of pituitary hormone in inducing ovulation in torpid bats (Wimsatt, 1960; Singh & Krishna, 1992) led to the suggestion that ovarian activity in vespertilionid bats is arrested during hibernation/winter dormancy. Morphologically, these follicles show hypertrophy and hyperplasia of granulosa cells of the discus proligerus (cumulus oophorus), resulting in reduction of the antrum and accumulation of glycogen, specializations not shown in other mammals (Oxberry, 1979; Son et al., 1987; Singh & Krishna, 1996a). *S. heathi* exhibits two waves of follicular development (Krishna & Abhilasha, 2000a). The first wave is initiated prior to winter dormancy and may be responsible for the development of nonovulatory Graafian follicles and thus the delay in ovulation. Hypertrophy of granulosa cells and accumulation of glycogen may be responsible for the decline in proliferation as well as apoptosis of granulosa cells, and in the development of specialized Graafian follicles, as shown in some vespertilionid bats. The second wave of follicular development following the winter dormancy period is characterized by active proliferation of granulosa cells and recruitment of Graafian follicles for ovulation. Rhinolophid bats show some variation from the pattern followed by vespertilionid bats. Wimsatt (1969) noted that, while Graafian follicles are present in the ovaries of the European rhinolophids during hibernation (Matthews, 1937; Gaisler, 1966), no-one had demonstrated that the follicle destined for ovulation persists throughout that period. The second wave of Graafian follicles occurs shortly after the end of hibernation (Gaisler, 1966). In the Japanese *R. ferrumequinum*, the situation is different in that a single large Graafian follicle is maintained in the ovary throughout hibernation (Oh et al., 1985). Although many lipid droplets were observed in the oocyte and cumulus cells of such follicles, abundant glycogen was not found in this bat. The maximal period of delayed ovulation was reported to be five months for *R. ferrumequinum nippon* (Oh et al., 1985) and about five to seven months for *M. ricketti* (Wang et al., 2008).

6.2.1.2. Pattern II: Unilateral follicular development and ovulation

With the exception of vespertilionid and rhinolophid bats, the majority of the bats show a brief period of estrus with mating, ovulation, and fertilization occurring in quick succession. During each cycle, a single large Graafian follicle (about 350 μm in diameter) develops that contains a small cumulus oophorus, large antrum, and two to three layers of membrane granulose. The ICs are generally few and not very prominent. Endocrine activity of the ovary is restricted to thecal cells and granulosa layers. These bats are generally monotocous or monovular and mono- or polyestrous. The majority of these species breed twice each year and have both ovaries functional (symmetric dominance). As bats find it difficult to successfully carry more than a single conceptus, these bats exhibit alternation of successive ovulation between two ovaries. The postpartum estrus always occurs in the ovary contralateral to that which supported the first or earlier pregnancy of the breeding season.

Many monovular bats have only one functional ovary. In dextral dominant molossid bats, the ovary contains numerous small follicles at different stages of development and atresia. A single large Graafian follicle develops before ovulation (350 μm in diameter). The left ovary contains numerous follicles up to 300 μm in diameter, but no Graafian follicles or CL. Hypertrophied ICs are prominent in the left ovary, show steroidogenic enzymes, and appear to perform only endocrine functions (Jerrett, 1979).

6.3. Ovarian Steroidogenesis and Peripheral Hormone Concentrations

Reproductive events in female bats follow one of the two basic sequences. In the type I sequence, estrus and mating are initiated in late summer or autumn and these bats typically undergo hibernation. Sperm are stored in the female's genital tract until spring, when ovulation, fertilization, and gestation take place. Thus, these bats exhibit the unique feature of delayed ovulation. In the type II sequence, females mate in the autumn, followed immediately by ovulation, fertilization, and early embryonic development prior to hibernation. These bats exhibit delayed embryonic development during hibernation and exhibit dextral or sinistral reproductive asymmetry.

Ovarian steroidogenesis and/or circulating steroid concentrations have been reported for only a few bat species. Hormonal changes during the reproductive cycle are best known in bats exhibiting the type I sequence, with delayed ovulation (e.g., *S. heathi*, *Antrozous pallidus*, *M. lucifugus*), as compared to bats exhibiting the type II sequence, with delayed embryonic development (e.g., *M. schreibersii*, *M. californicus*, *C. sphinx*).

The ovary of *S. heathi* secretes unusually large amounts of androgens, both AND and T, during the period of recrudescence and the early phase of ovulatory delay (Abhilasha & Krishna, 1996). Thecal/interstitial gland cells surrounding the follicles show intense $P450_{scc}$ and $P450_{17\alpha}$ activities and thus may be secreting large amounts of androgens. Overproduction of androgens may inhibit follicular growth and consequently selection of a dominant follicle. In female *S. heathi*, high circulating AND may be responsible for development of nonovulatory antral follicles. Circulating AND and T decline to very low values during the period when mating and ovulation normally take place. *S. heathi* females show two peaks of circulating E_2 during the reproductive cycle, coinciding closely with two waves of recruitment and development of antral follicles (Krishna & Abhilasha, 2000a). The concentration of E_2 declines during delayed ovulation, suggesting that antral follicles in the ovary of *S. heathi* during the period of delayed ovulation may not be steroidogenically active.

Similarly, in the emballonurid bat *T. longimanus*, a high level of circulating AND may be responsible for suppressing follicular maturation and ovulation, leading to prolongation of the follicular phase and delayed ovulation (Singh & Krishna, 2002). Increased plasma AND in *T. longimanus* coincides with increased development of ICs in the ovary. This suggests that increased development or differentiation of ICs may be related to high circulating AND and this increased androgen level may be responsible for slowing follicular development and delaying ovulation in these bats. Insulin and/or insulin-like growth factor-1 (IGF-1) may be involved in inducing high AND synthesis by the ovary (Doval & Krishna, 1998).

In *A. pallidus*, the ovaries contain a developed Graafian follicle and relatively low circulating P_4 levels during the period of prolonged estrus during winter, whereas E_2 levels increased temporarily towards the end of estrus (Oxberry, 1979). Low P_4 seems to be essential for the maintenance of delayed ovulation. Although the concentration of circulating androgens was not estimated in this species, the presence of extensive ICs together with strong expression of 3β-HSD and 17β -hydroxysteroid dehydrogenase (17β-HSD) in the stroma suggest that the ovaries of *A. pallidus* have the potential to produce high androgen levels during delayed ovulation. After arousal from hibernation, circulating E_2 levels appear to rise temporarily at ovulation and then decline immediately after ovulation. This might be due to fresh recruitment of the antral follicle for ovulation. At the early stage of pregnancy, P_4 levels decline briefly and then increase considerably to reach a peak prior to parturition. Changes in these steroid levels are similar in other species showing delayed ovulation, such as *M. lucifugus* (Buchanan & Young Lai, 1986; 1988) and *C. gouldii* (Hosken, O'Shea, & Blackberry, 1996).

Variations in these patterns of reproduction are found in certain species including the vespertilionid bat *M. schreibersii*, the phyllostomid bat *M. californicus*, and the pteropodid bat *C. sphinx*. In these bats, estrus and mating occur in autumn, followed immediately by ovulation, fertilization, and early embryonic development. These bats then enter hibernation in the pregnant condition and are characterized by a prolonged period of slow embryonic development. In general, circulating P_4 levels are low during the period of delayed or slow embryonic development and not significantly different from those of nonpregnant females. Significantly elevated P_4 levels are observed during subsequent active fetal development, accompanied by development of the placenta, reaching peak levels just prior to parturition (Martin & Bernard, 2000).

Delayed development has been described in several bat species but little evidence has been provided for or against the role of low P_4. *C. sphinx* have relatively low circulating plasma P_4 and E_2 for a prolonged period coinciding with the period of slow embryonic development during the first pregnancy. This suggests that slow embryonic development during the first pregnancy might be due to reduced synthesis of P_4 and E_2. The majority of bat species with delayed development also show low P_4 levels during delayed development, as compared with the period of normal embryogenesis (Krishna, 1999). *M. californicus* shows low plasma P_4 during delayed development and its level increases significantly at the resumption of normal embryonic development (Crichton et al., 1990). *M. schreibersii* also has low circulating P_4 during delayed implantation (Van der Merwe & Van Aarde, 1989; Bernard et al., 1991a; 1991b). This may be a general phenomenon related to delayed development, as depressed P_4 levels are also seen during developmental delay in the striped skunk (*Mephitis mephitis*) (Wade-Smith, Richmond, Mead, & Taylor, 1980).

In *A. jamaicensis*, CL cells are smaller during delayed development, and Fleming (1971) suggested that reduced secretion of P_4 from the CL might cause the delay. During the delay period, CL cells were found to be less active, and the number of steroidogenic organelles was greatly reduced as compared with normal embryogenesis in *M. schreibersii* (Crichton et al., 1989). Bernard et al. (1991a) observed a decrease in CL volume during embryonic diapause in the natal clinging bat (*M. s. natalensis*). Kimura et al. (1987) reported a decrease in CL volume and low P_4 levels in the Japanese long-fingered bat (*M. s. fuliginosus*) during hibernation. In the roe deer (*Capreolus capreolus*), which has an active CL during implantation delay (Aitkin, 1981), circulating P_4 levels remain low during the delay compared to levels when implantation occurs (Sempere, 1977).

The two major sources for P_4 in bats are the CL and the placenta, with the relative importance of these two organs differing both temporally within species and between species. Additional sources of P_4 are the ovarian TICs and the adrenal gland (Martin & Bernard, 2000). The most detailed study of circulating P_4 levels is available for *M. lucifugus* and *M. schreibersii*, in which the placenta takes over P_4 synthesis from the CL in the final third of pregnancy (Buchanan & Young Lai, 1988; Van Aarde, Van der Merwe, & Skinner, 1994).

A high baseline level for serum P_4 (~9.00 ng/ml) in nonpregnant *C. sphinx* is similar to other bat species (Table 12.6) such as *M. schreibersii* (~7 ng/ml), *A. pallidus*, and *T. brasiliensis mexicana* (~5.0 ng/ml) (Jerrett, 1979; Oxberry, 1979; Kimura et al. 1987), suggesting that P_4 is secreted by ovarian nonluteal tissue (Jerrett, 1979) and/or by the adrenal cortex, as in the ewe (McCracken & Baird, 1969) and women (Baird, 1976). Strong 3β-HSD-ir has been reported in the thecal layer of growing follicles in nonpregnant bats (Meenakumari et al., 2009). Serum P_4 levels rise significantly at ovulation and remain high during early pregnancy. This increase in P_4 is probably necessary for early embryonic development and implantation. Wang, Rider, Heap, and Feinstein (1984) showed that P_4 antibody administered to mice after coitus not only prevented implantation but blocked embryonic development at the morula stage. Several bats that do not undergo delayed implantation, such as *T. b. mexicana* (Jerrett, 1979), *A. pallidus* (Oxberry, 1979), and *P. pipistrellus* (Racey & Swift, 1981), show an increase in serum P_4 level following ovulation. Serum P_4 levels in *C. sphinx* rise again during midpregnancy and attain peak levels in the late stage of pregnancy. Progesterone is higher during mid- and late pregnancy than during early pregnancy. A substantial rise in serum P_4 level coincides with the most rapid rate of fetal growth. The prepartum decline in serum P_4 shown in some bat species (Buchanan & Young Lai, 1986) does not occur in *C. sphinx*, although the possibility of a prepartum fall in P_4 near the time of parturition cannot be ruled out. During the period of delayed development in *C. sphinx*, circulating P_4 is much higher (~20 ng/ml) than that reported in other mammals during the early stage of embryonic development. This contradicts the hypothesis that a decline in P_4 secretion by the CL is one of the major causes for delayed embryonic development. It is also possible that *C. sphinx* might require a much higher level of P_4 for the post-implantational development. High maternal serum P_4 during the postpartum period suggests a role for P_4 in lactation in *C. sphinx*.

TABLE 12.6 Circulating progesterone (P_4) levels in female bats

Species	Range of Progesterone Concentration	References
Tadarida brasiliensis mexicana	106.0 ng/ml	Jerret (1979)
Antrozous pallidus	~4.00 to 53.00 ng/ml	Oxberry (1979)
Pipistrellus pipistrellus	10.00 to 40.00 mol/L	Racey and Swift (1981)
Myotis lucifugus	6.00 to 45.00 ng/ml	Buchanan and Young Lai (1986)
Miniopterus schreibersii fulginosus	4.70 to 11.80 ng/ml	Kimura et al. (1987)
Mormopherus planiceps	1.50 to 37.00 ng/ml	Crichton and Krutzsch (1987)
Myotis lucifugus	5.00 to 65.00 ng/ml	Currie et al. (1988)
Miniopterus schreibersii natalensis	2.01 to 181.30 ng/ml	Van der Merwe and Van Aarde (1989)
Macrotus californicus	11.24 to 108.0 ng/ml	Crichton et al. (1990)
Miniopterus schreibersii	1.54 to 64.82 ng/ml	Bernard et al. (1991a)
Pteropus poliocephalus	0.13 to 800.0 ng/ml	Towers and Martin (1995)
Pteropus (Chalinolobous) gouldii	0.50 to 13.00 ng/ml	Hosken et al. (1996)
Taphozous longimanus	5.71 to 88.57 ng/ml	Singh and Krishna (2002)
Cynopterus sphinx	8.67 to 116.7 ng/ml	Meenakumari et al. (2009)

Adapted from Meenakumari (2006).

6.4. The Corpus Luteum (CL)

The pattern of growth and survival of the CL of pregnancy varies widely among chiropteran species. In two African bats, *Nycteris lutea* and *Triaenops afer* (Matthews, 1941), as well as in *C. rafinesquei* (Pearson et al., 1952) and *P. mimus* (Krishna, 1985), the CL is short-lived and disappears following implantation early in pregnancy. In some Australian bats, *viz. C. gouldii* (Kitchener, 1975), *Chalinolobus morio* (Kitchener & Coster, 1981), and *Eptesicus regulus* (Kitchener & Halse, 1978), even though the CL persists up to late pregnancy, degeneration starts well before parturition in late midpregnancy. The CL of pregnancy in *S. heathi* (Krishna & Dominic, 1988), *T. longimanus*, and *C. sphinx* (Meenakumari et al., 2009) resembles that of *T. georgianus* (Kitchener, 1973) and *D. r. murinus* (Wimsatt & Trapido, 1952) in its large size, rapid development, and persistence until parturition. Gopalakrishna (1969) noted that in *R. leschenaulti* the CL of pregnancy persists for several months after parturition. Interestingly, in *R. ferrumequinum* (Matthews, 1937) and *H. fulvus* (Patil, 1974), the CL becomes everted early in its development. In these bats the CL attains maximal size by implantation but soon undergoes regression. In *G. soricina*, ovulation occurs alternatively between the left and right

TABLE 12.7 Seasonal changes in body mass, accumulated fat mass, insect availability, body temperature, and circulating leptin and insulin concentrations for *Scotophilus heathi*

Calendar Month	Body Mass Female (gm ± SEM)	Accumulated Fat Mass (gm ± SEM)	Insects (Number Captured/ Night)**	Body Temperature (°C)	Leptin (ng/ml ± SEM)	Insulin (Pmol/L ± SEM)
July	30.08 ± 0.5	*	94	29.6 ± 0.3	6.8 ± 0.2	84.8 ± 5.3
August	30.08 ± 0.2	*	259	29.2 ± 0.2	6.3 ± 0.2	102.5 ± 6.3
September	30.08 ± 0.4	0.9 ± 0.1	650	29.4 ± 0.3	6.8 ± 0.3	104.3 ± 5.6
October	33.0 ± 0.7	3.9 ± 0.4	742	27.6 ± 0.4	6.7 ± 0.2	104.3 ± 7.5
November	45.5 ± 0.5	11.4 ± 0.8	102	22.7 ± 0.4	14.7 ± 1.4	137.8 ± 7.4
December	42.3 ± 1.1	4.6 ± 0.2	26	18.4 ± 0.5	15.6 ± 0.6	168.3 ± 7.4
January	37.7 ± 0.9	2.6 ± 0.3	20	16.5 ± 0.4	8.6 ± 0.4	*
February	31.8 ± 0.9	*	10	22.1 ± 0.5	7.3 ± 0.3	146.0 ± 5.5
March	33.5 ± 0.3	*	12	24.2 ± 0.5	7.1 ± 0.1	122.5 ± 14.8
April	32.1 ± 0.6	*	40	29.9 ± 0.4	7.0 ± 1.1	121.3 ± 8.6
May	36.4 ± 0.6	*	56	31.7 ± 0.3	7.4 ± 0.3	165.8 ± 2.8
June	43.3 ± 0.7	*	72	32.4 ± 0.6	11.7 ± 0.2	180.0 ± 31.9

Adapted partially from Srivastava and Krishna (2008).
SEM, standard error of the mean.
**Data not available*
***For details see Srivastava and Krishna (2008).*

ovaries and the CL persists for two cycles, the formation of the second CL and the regression of the previous corpus being controlled independently of each other (Rasweiller, 1972). Bonilla and Rasweiler (1974) describe the formation of secondary CL resulting from additional ovulations in *C. perspicillata* during early pregnancy.

The morphological and steroidogenic changes in the CL have been studied extensively in bat species showing developmental delay (Crichton et al., 1989; Meenakumari et al., 2009). In *A. jamaicensis*, luteal cells were smaller during the period of delayed development. Based on this observation, Fleming (1971) suggested that reduced P_4 might cause the delay. During the delay period, CL cells were less active and the number of steroidogenic organelles was greatly reduced as compared with CL cells during embryogenesis in *M. schreibersii* (Crichton et al., 1989).

6.5. Induced Ovulation

Several attempts have been made to induce ovulation during delayed ovulation in female vespertilionid bats. The reproductive cycle of vespertilionid bats is characterized by a prolonged survival of one to many large antral follicles, and final maturation and ovulation of the mature follicle is delayed for several months (Wimsatt, 1969; Oxberry, 1979). Attempts to induce ovulation during the period of delayed ovulation in female vespertilionid bats have yielded contradictory results (Racey, 1982). Earlier attempts to induce ovulation in hibernating vespertilionid bats exhibiting delayed ovulation were centered around arousal from hibernation alone or arousal accompanied by administration of GTHs (Racey, 1976) or some steroids (Herlant, 1954; Ramaswami & Anand Kumar, 1966). Hibernating animals are generally unresponsive to hormonal stimulation and only those aroused will ovulate (Wimsatt, 1960). However, increased metabolism and elevated body temperature are not sufficient in themselves to induce ovulation in hibernating bats (Wimsatt, 1960). Wimsatt (1960) succeeded in inducing ovulation by treatment with GTHs or with homogenized bat pituitary in bats maintained at room temperature. Although some studies succeeded in inducing ovulation during the delay period with GTHs, there is marked variation in the dose of hormones administered to animals of approximately the same size (e.g., dose of hCG ranging from 1.5 iu to 100 iu) reported (Racey, 1982). Guthrie and Smith (1940) and Smith (1951) were unable to induce ovulation in *Myotis grisescens* by hCG during August to early September, even though the ovary contains only antral follicles at this time.

Failure to induce ovulation with GTHs in vespertilionid bats during the period when the ovary contains a number of large antral follicles is unusual and intriguing. In order to understand the mechanism of ovulation suppression in vespertilionid bats during the period of delayed ovulation, effects of hCG, pregnant mare serum gonadotropin (PMSG), FSH, and a GnRH agonist on ovarian morphology and steroid concentration were investigated in *S. heathi*. Treatment with any of these agents failed to induce ovulation, although these hormones produced a high degree of ovarian stimulation. Both hCG and PMSG induced ovarian enlargement, intense hyperemia, marked changes in the TICs, development of several antral follicles, and a varying degree of abnormalities in the oocytes of most of the antral follicles. Treatment with hCG, PMSG, or GnRH agonist increased plasma AND whereas E_2 levels decreased. Administration of FSH caused regression of TICs and pyknosis of granulosa cells in the majority of antral follicles. Follicle-stimulating hormone did not enhance AND levels, however. The failure of hormonal treatments to induce ovulation during the period of delayed ovulation may be due to a seasonal desensitization of ovarian follicles in *S. heathi*. As the numerous TICs are capable of producing a very high circulating concentration of AND, a somewhat more complicated approach is required for induction of ovulation in these bats.

7. ENVIRONMENTAL FACTORS AND ADDITIONAL HORMONES AFFECTING REPRODUCTION IN BATS

7.1. The Roles of Hibernation, Fat Accumulation, and Leptin

Charles Darwin (1896) was the first to suggest that fertility in animals might be affected by food availability, such that fertility is decreased during times when nutritional supplies are scarce and is restored when adequate nutrition becomes available. Many studies have since demonstrated that nutritional status has a profound effect on reproductive functions. The fact that animals breed seasonally in environments where the variation in food availability is severe further supports the role of nutrition in reproductive processes.

Body mass or body fat is suggested to play a role in transmitting information about the nutritional status of the body to the reproductive axis (Bronson & Manning, 1991). Reproduction is an energy- and nutrient-consuming process. The timing of reproduction is critical in many bat species living in temperate/subtropical regions. Most of these species are reproductively quiescent during winter dormancy/hibernation, when ambient temperatures decline and food availability is typically low, and they resume reproductive activities in the early spring/summer, when the ambient temperature and food supply are more favorable. This suggests that the reproductive neuroendocrine system is highly sensitive to changes in nutrition (Woods, Figelewiez, Lattermann, Schwartz, & Porte, 1990). The sensitivity of the reproductive axis to nutrition would be advantageous, ensuring that fertility is impaired at times when energy resources are not sufficient for high-energy-requiring reproductive processes such as pregnancy and lactation.

Preparation for hibernation in some mammals involves the deposition of fat reserves prior to the onset of food shortage, which in turn provides the sole source of energy during the prolonged winter fast (Young, 1976; Mrosovsky, 1985). Several investigators have documented increases in body mass during the prehibernation period in several vespertilionid bats as well as in some other mammals (Kenagy, 1987; Boswell, Woods, & Kenagy, 1994; Abhilasha & Krishna, 1997; Kunz, Wrazen, & Burnett, 1998). Increases in body mass during the prehibernation period have been attributed largely to deposition of white adipose tissue (the major energy storage depot for fat). White adipose tissue serves a variety of functions in mammals, but for most species its primary value resides in its capacity to store energy as insurance against future energy needs, such as those encountered during winter scarcity of food (Hudson & White, 1985; Bronson, Heideman, & Kerbeshian, 1991). Several species of bat from temperate as well as tropical regions display annual cycles of food intake and deposition of white adipose tissue associated with the use of fat as fuel during hibernation/winter dormancy (Ewing, Studier, & O'Farrell, 1970; Abhilasha & Krishna, 1997; Kunz et al., 1998; Speakman & Rowland, 1999; Kronfeld-Schor, Richardson, Silvia, Kunz, & Widmaier, 2000). During the active season, bats increase their food intake, gain body mass, and accumulate large fat stores. Fat deposition is a characteristic feature of many microchiropteran bats, which undergo hibernation/winter dormancy. Studies of the vespertilionid bats *M. lucifugus* and *S. heathi* showed that the rapid deposition of fat reserves depends upon seasonal hyperphagia, lowering of body temperature, and seasonal production of leptin and insulin (Widmaier, Long, Cadigan, Gurgel, & Kuntz, 1997; Srivastava & Krishna, 2008) (Table 12.7). This is followed by a drastic reduction of food intake and loss of body mass, a state that is maintained throughout hibernation. Although the timing of autumn fattening and subsequent hibernation appears to be similar in temperate-zone bats and rodents, there are striking differences in the reproductive patterns of these two mammalian groups (Kunz et al., 1998). The majority of hibernating bats typically mate and initiate reproductive processes before hibernation, whereas rodents mate following arousal from hibernation. These differences reflect the fact that bats, in contrast to rodents, have evolved mechanisms for sperm storage, prolonged survival of Graafian follicles, prolonged retention of blastocysts, etc., during hibernation (Wimsatt, 1969; Fenton, 1984; Hosken, 1997). The reproductive success of bats is strongly linked to the quality and quantity of fat deposited (Kunz et al., 1998). Several recent studies have shown that successful hibernation in rodents not only depends on the amount of fat but also on the relative ratio of dietary fatty acids (Frank, 1991). The rapid deposition of fat reserves in autumn/early winter observed in bats reflects the selection of insects high in unsaturated fatty acids, seasonal hyperphagia, relaxation of thermoregulation, selection of roost sites that promote torpor, and/or the seasonal production of insulin, leptin, or neuropeptide Y (Widmaier et al., 1997; Speakman & Rowland, 1999; Bartness, Demas, & Song, 2002).

A close association between adiposity and reproductive processes is demonstrated in *M. lucifugus* and *S. heathi* (Kunz et al., 1998; Chanda et al., 2003). Changes in body mass because of fat accumulation and mobilization and changes in reproductive activity are intimately interdependent in *S. heathi* (Srivastava & Krishna, 2007; 2008). Mating success of male bats appears to be greater in autumn following a prolonged period of spermatogenesis and fat deposition. Recent evidence suggests that male bats that mate during hibernation may have the greatest reproductive success (Hosken, 1997). Successful ovulation following hibernation appears to depend on stored fat reserves.

Since its discovery in 1994, leptin, a hormone synthesized by adipose cells, has been recognized as a major physiological regulator of adiposity in mammals (Zhang et al., 1994; Sahu, 2004; McMillen, Edwards, Duffield, & Muhlhausler, 2006; Morris & Rui, 2009). Leptin is the product of the Ob (obese) gene in mice (Zhang et al., 1994; Masuzaki et al., 1997). The production of leptin is modulated by insulin, glucocorticoids, β-adrenergic receptor agonists, and fasting (Trayhum et al., 1998). Available evidence suggests that the role of leptin is to provide key afferent information to the CNS or peripheral tissues concerning the amount and distribution of body fat (Benoit, Clegg, Seeley, & Woods, 2004). In humans, serum leptin levels are highly correlated with the percentage of body fat, and decrease in response to weight loss (Maffei et al., 1995; Considine et al., 1996; Weigle et al., 1997). Increasing adiposity is accompanied by insulin resistance and hyperinsulinemia (Kopelman, 1994), and suggests the possibility of an interaction between insulin and leptin. The positive correlation of both leptin and insulin levels with body mass (Figure 12.7) observed in *S. heathi* suggests that both leptin and insulin might be linked to the development and maintenance of adiposity in *S. heathi*, as shown in rodents (Maffei et al., 1995). The relationship of leptin to reproduction in bats has been examined in *M. lucifugus* (Widmaier et al., 1997), *E. fuscus* (Kunz et al. 1999), *S. heathi* (Srivastava & Krishna, 2007), and *T. longimanus* (Singh, Krishna, & Bhatnagar, 2007). In *M. lucifugus* and *E. fuscus*, leptin levels rose during pregnancy and declined following parturition. During lactation, reduced leptin was significantly correlated with body weight in *E. fuscus*, but not in *M. lucifugus*. Leptin was also found to be significantly reduced in acutely fasted (16 hours) animals. These studies support the suggestion that leptin may play a role in reproduction and the regulation of energy balance in bats.

Increased leptin levels suppress ovulation in rats (Duggal et al., 2000). Leptin levels increase prior to winter dormancy in *S. heathi*, simultaneously with an increase in body fat. Increased leptin levels in *S. heathi* may be responsible for suppressed ovarian activity leading to delayed ovulation (Srivastava & Krishna, 2007). This finding may explain why bats exhibit reproductive delays

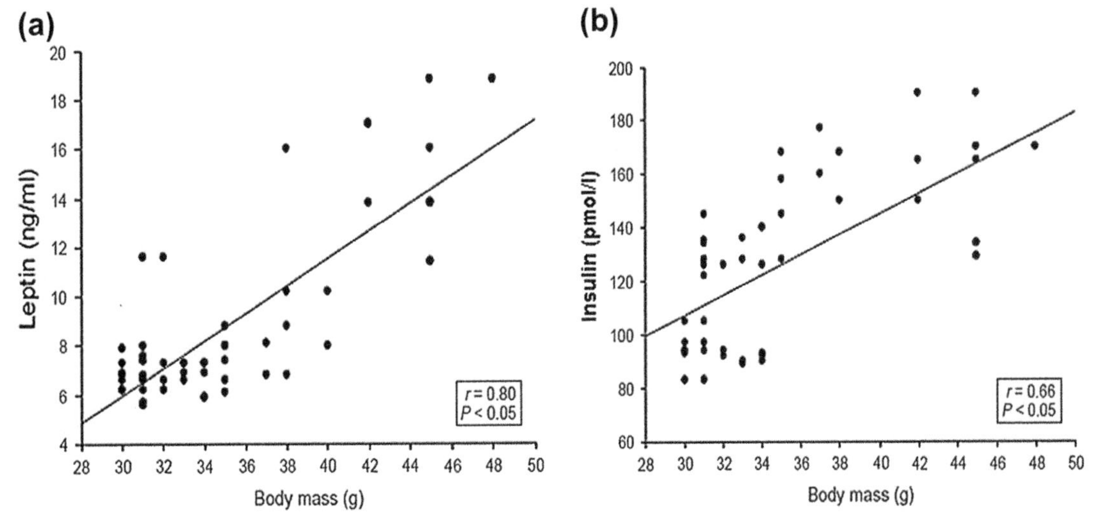

FIGURE 12.7 (a) Correlation between circulating leptin levels and body mass in *Scotophilus heathi*. (b) Correlation between circulating insulin levels and body mass of *S. heathi*.

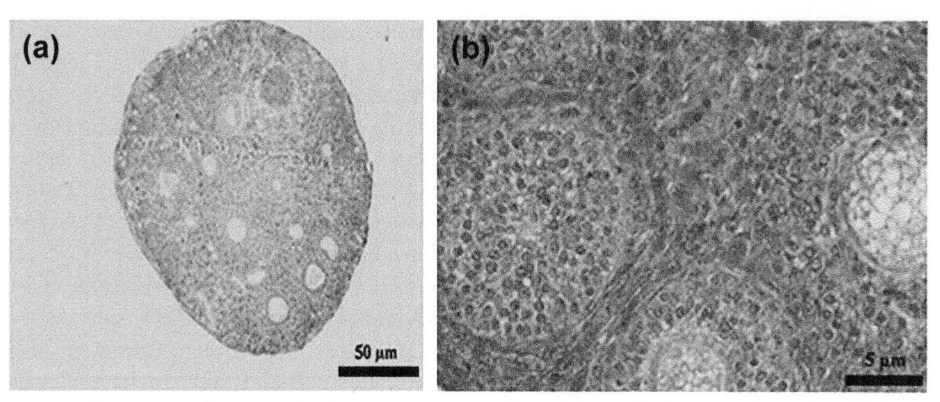

FIGURE 12.8 Immunocytochemical localization of leptin receptors in the ovaries of *Scotophilus heathi*. (a) Note the presence of intense immunostaining in the interfollicular area of the ovary. (b) Higher view demonstrating the presence of intense immunostaining of the leptin receptor, mainly in the thecal interstitial cells. Granulosa cells showed no immunostaining. See color plate section.

corresponding with the period of fat deposition. Ovulation in *S. heathi* occurs in late February/early March, when fat reserves are completely mobilized, and a decline in leptin during this period may be important for the fresh recruitment and maturation of preovulatory follicles (Krishna & Abhilasha, 2000a). That leptin directly affects the ovary in bats is suggested by the presence of receptors in ovaries of *S. heathi* during the period of delayed ovulation (Figure 12.8) (Srivastava, 2008) as well as leptin receptors and leptin receptor mRNA in human, mouse, rat, and pig ovaries (Karlsson et al., 1997; Ruiz-Cortes et al., 2000; Kikuchi et al. 2001). The way in which leptin affects ovarian activity is not clearly known.

7.2. Environmental Factors Regulating Reproduction

It has been suggested that bats routinely use predictive cues to time reproduction (Beasley & Zucker, 1984; McGuckin & Blackshaw, 1992; Heideman & Bronson, 1994), rather than following an opportunistic strategy of reproductive timing (Heideman, 1995). In most temperate-zone vespertilionid bats, prolonged hibernation during winter restricts the reproductive processes to the summer (Bernard, 1989). Spermatogenesis in these species is completed over a short period of time between late summer and early autumn, prior to the beginning of hibernation. Copulation occurs just prior to and during hibernation; sperm are stored in the female genital tract until spring, when ovulation and fertilization occur (Racey, 1982). In certain temperate-zone vespertilionid bats, active spermatogenesis occurs in summer and early fall but also in spring, soon after arousal from winter hibernation (Caffier & Kolbow, 1934).

Little is known about the environmental cues used by tropical bats to time seasonal reproduction. Annual changes in photoperiod are used by many temperate-zone mammals to time their reproduction, and there is some evidence that temperate and subtropical bat populations also may rely upon photoperiod as a seasonal cue (Beasley, Smale, & Smith, 1984; McGuckin & Blackshaw, 1985; 1988; 1992). However, many of the tropical bats studied so far are not reproductively photoresponsive (Racey, 1978; Heideman, 1995). A tropical population of *Anoura goeffroyi* from 10°N latitude (Heideman et al., 1992; Heideman & Bronson, 1994) and a subtropical/tropical population of *Pteropus scapultatus* from about 25°S latitude (O'Brien, Curlewis, & Martin, 1993) apparently do not use photoperiod to regulate their reproduction seasonally.

It is often impossible to characterize a specific pattern of reproduction in species with a wide distribution (Bernard & Cumming, 1997). In a tropical vespertilionid, *S. wroughtoni*, spermatogenesis was shown to occur during late winter and early spring, prior to the time of ovulation (Gopalakrishna, 1947). In *S. heathi* from Trichur, India (10°N, 76°E), spermatogenesis occurs from October to December (Madhavan, 1981). Copulation in this species in Trichur occurs mid-November, and ovulation occurs by the end of December. The reproductive cycle of *S. heathi* at a subtropical zone (Varanasi, India: 25°N, 83°E) seems to be peculiar as it differs from some but resembles other aspects of the temperate zone hibernating members of the family Vespertilionidae. Spermatogenesis in *S. heathi* at Varanasi is prolonged and extended over the period from autumn (November) to spring (March), with testes becoming quiescent during the winter period. The major differences between the reproductive patterns reported for temperate zone vespertilionids and *S. heathi* at Varanasi lie in the length of winter dormancy. Thus, in vespertilionid bats, the spermatogenetic cycle appears to depend on climate and temperature, and therefore varies with climatic zone. There appears to be a gradual shift in spermatogenetic cycle from summer and early autumn in temperate zones to autumn and spring in subtropical zones and to spring only in tropical zone bats (Figure 12.9). A gradual change in reproductive pattern from multiple to single

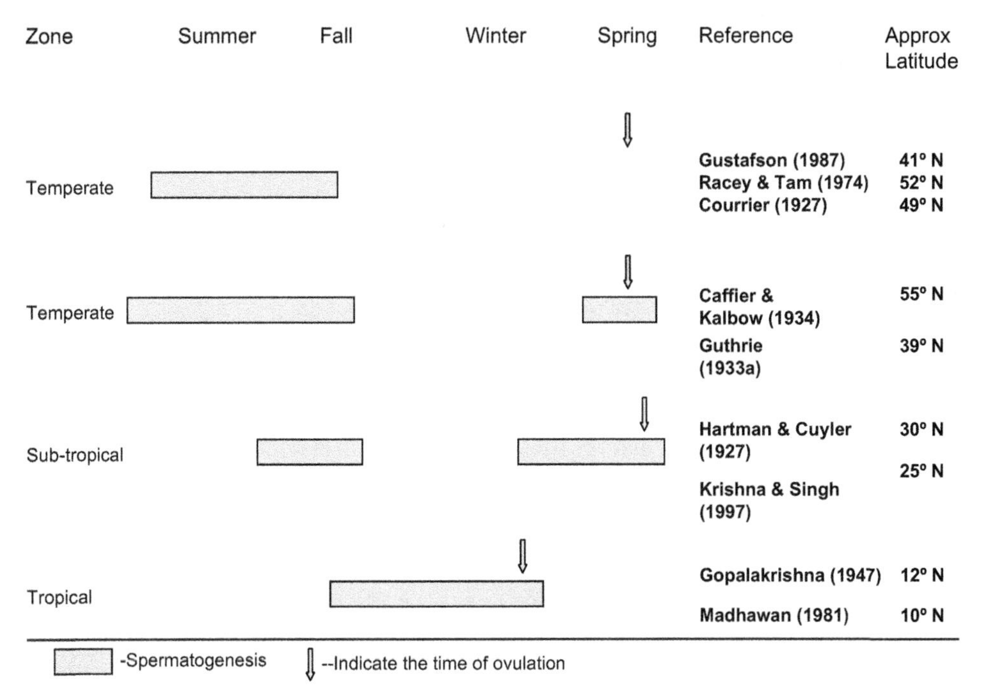

FIGURE 12.9 Summary of the spermatogenetic cycle of vespertilionid bats from different temperature zones.

breeding along a gradient of increasing latitude has also been demonstrated among molossid bat species on the African continent (Happold & Happold, 1990a).

There is abundant correlational evidence suggesting that rainfall is one of the primary causes of seasonal birth peaks in tropical bats, probably through its effects on food production (reviewed by Racey, 1982; Heideman, 1995). At temperate latitudes, spermatogenesis corresponds with the peak in insect availability and foraging opportunity during summer (Speakman & Racey, 1986; Entwistle, 1994). Similar patterns have been noted in tropical species, in which spermatogenesis corresponds with periods of rainfall and food availability (Jones, 1972; McWilliam, 1987b; 1988). The pattern of rainfall and the severity of intervening dry seasons in different geographic regions may be responsible for different reproductive patterns found in bats; e.g., bimodal birth corresponds with bimodal rainfall, such as in *Micropteropus pusillus* (Thomas & Marshall, 1984); monoestry with short unimodal rains, e.g. in *Pipistrellus nanus* (Happold & Happold, 1990a; 1990b); and bimodal or multimodal polyestry with an extended period of unimodal rainfall, e.g. *Epomops bultikoferi* (Kofron & Chapman, 1994). These observations suggest that food supply may be a key factor in determining the timing of reproductive cycles in males. Endogenous reproductive rhythms are common in long-lived seasonally breeding mammals (reviewed by Bronson & Heideman, 1994) and have also been reported in bats from the temperate zone (Beasley & Zucker, 1984), a tropical phyllostomid (*A. geoffroyi* (Heideman & Bronson, 1994)), and other tropical bat species (Haussler, Moller, & Schmidt, 1981; Taft & Handley, 1991).

Mammals in seasonal environments have evolved two general strategies of reproductive timing: opportunism and reliance on environmental cue(s) (Bronson, 1985). Opportunism results in seasonal reproduction simply because favorable conditions often occur seasonally (Heideman & Bronson, 1992). Opportunistic breeders mate whenever environmental conditions are good, whereas seasonal breeders restrict mating to a time that results in optimal conditions for rearing of the young—the most physiologically demanding event of the reproductive cycle. However, opportunism may be rare as a seasonal strategy in bats because the relatively long gestation period of bats (2–11.5 months) makes it unlikely that a female could produce a follicle, ovulate, mate, and then reach the most demanding stages of reproduction (lactation and weaning) before the good conditions have passed. Instead, most seasonally reproducing bats probably react to environmental cues that predict oncoming periods when climatic, dietary, and other conditions permit successful reproduction.

Marshall (1942) was the first to emphasize the dependence of the breeding season in many species of bat on external environmental factors. The environment can act on bat reproduction in two fundamentally different ways (Heideman, 2000). First, extreme physical conditions or starvation can directly alter reproduction (Tuttle & Stevenson, 1982). Second, the environment can affect reproduction via specific neuroendocrine pathways that have evolved to play a role in the optimal timing of different stages of reproduction (Beasley & Zucker, 1984). No seasonal cycle can persist in the absence of environmental regulation and this implies that most mammals including bats are under some form of environmental regulation. Environmental factors are known to affect the timing of reproductive maturation, sexual receptivity and mating, ovulation and fertilization, implantation, rate of post-implantation growth, embryo loss, birth, gonadal regression, and gonadal recrudescence in a variety of bat species (Racey, 1982; Tuttle & Stevenson, 1982; Bronson & Heideman, 1994). Specific environmental factors affect the timing of ovulation and fertilization (temperature), gonadal recrudescence (photoperiod), rates of post-implantation growth (temperature, food, and possibly stress), and embryo loss (stress) in bats. Seasonal environmental changes appear to be both cause and cue for the characteristic reproductive cycle in many bats. Although relatively few tropical mammals have been tested for reproductive responses to variation in photoperiod, the available data suggest that reproductive photoresponsiveness may be present in some populations of mammals above 13° latitude, but absent in populations from latitudes of 10° or less (Heideman & Bronson, 1994).

In general, photoperiod exerts effects on the mammalian reproductive system via a complex pathway (Tamarkin, Baird, & Almeida, 1985; Turek & Van Cauter, 1994). The suprachiasmatic nucleus (SCN) is the central circadian oscillator controlling the daily rhythms of mammals, and it can maintain a free-running circadian rhythm even in the absence of a daily light cycle. Light is detected by retinal photoreceptors, the output of which is directed through fibers in the retinohypothalamic tract to the SCN. The SCN provides input to the hypothalamus and from there neuronal connections extend to the superior cervical ganglia, which in turn act on the pineal gland via sympathetic adrenergic neurons. This pathway has not been studied in detail in bats, but it appears to be fairly similar across the different orders of mammals and hence applicable to bats (Heideman, 2000).

The pineal gland secretes large amounts of MEL at night and mediates the effect of photoperiod on reproduction in pallid bats (*A. pallidus*) (Beasley & Zucker, 1984; Beasley et al., 1984), as it does in rodents (Bronson & Heideman, 1994). In *Pteropus geddiei*, Baker and Bird (1936) noted that the testes increase in size when the length of the day and the number of hours of sunshine were increasing. In *M. australis*, mating occurs at the time of year when the days are beginning to get longer (Baker & Bird, 1936). O'Farrell and Studier (1975) reported that the time of emergence of *Myotis thysanodes* and *M. lucifugus* from their roost is directly correlated with the time of sunset and appears to be a photoperiodic response. A weak reproductive response to photoperiod was reported in *P. poliocephalus* by O'Brien, Curlewis, and Martin (1991) and McGuckin and Blackshaw (1992), and in *C. sphinx* by Haldar and Alipreeta (2001). There are a few species of seasonally breeding bats for which there is experimental evidence that photoperiod is not an important environmental cue, such as *Pteropus rodricensis* (Cheke & Dahl, 1981; Carroll, 1988) and *A. geoffroyi* (Heideman & Bronson, 1994).

Although heterothermy has evolved independently in several mammalian groups (Kayser, 1961; Cade, 1964) the ability to become torpid facultatively in response to adverse environmental factors during pregnancy seems to be limited to bats. Eisentraut (1937) was first to note the effect of changes in environmental temperature on the rate of fetal development in heterothermic bats. Kolb (1950) reported a delay of three weeks, compared with the previous year, in the birth of the first young in a colony of lesser horseshoe bats (*R. hipposideros*) with three weeks of cold weather in May. Later, Racey (1973) confirmed experimentally that the length of gestation in heterothermic bats might be profoundly affected by environmental conditions. Pearson et al. (1952) studied the effect of variation in environmental temperature on the marked variability in their estimates of the gestation period (56 to 100 days) of *C. rafinesquei* examined in different localities and in different years, and concluded that maternal torpor primarily extends the preimplantation stage of development. In many bat species, the period of torpor induced by low ambient temperature and food deprivation acts to arrest or retard the progress of embryonic development (Racey, 1982).

Racey and Swift (1981) found that the duration of gestation of *P. pipistrellus* in England in 1978 was 10 days shorter than the gestation length in 1979, in which there was a two-week period of cold temperature and few trappable insects during the early part of gestation. Studies of *R. ferrumequinum* in England showed a strong relationship between cold temperature and gestation length, with cold temperatures acting by slowing postimplantation development (Ransome & McOwat, 1994; McOwat & Andrews, 1995). Audet and Fenton (1988) presented evidence that *E. fuscus* was forced to enter torpor during colder springs in southern Canada, thus extending gestation and delaying parturition. Krishna and Dominic (1982b) reported a 19-day difference in duration of gestation in the two annual pregnancies in *T. longimanus*. The longer gestation period occurs during the colder months of the year, and lowered body temperature might explain the reproductive delay.

The observation of delayed embryonic development in *M. californicus*, a nonhibernating species, suggests that torpidity due to low ambient temperature may not always cause delay. Photoperiod has been implicated in the timing of implantation in several species of mustelid such as the spotted skunk (*Spilogale putorius latiforns*), in which exposure to 14-hour light : 10-hour dark in winter causes implantation to take place an average of 49 days earlier than in controls (Mead, 1971). Experimentally increasing the daily photoperiod resulted in premature implantation in mink (*Mustela vision*) (Pearson & Enders, 1944). Delayed implantation also occurs in *M. schreibersii*. In this species the gestation length increases with latitude (Racey, 1982), the range of which extends from 37°S to 45° N (Richardson, 1977). Implantation of blastocysts occurs during February at 45° N and during August at 37°S, when the day length is 11–12 hours and increasing, suggesting that this may be the stimulus causing implantation (Racey, 1981). Some species of bat in the tropics also exhibit delayed embryonic development, e.g. *E. helvum* (Mutere, 1965) and *A. jamaicensis* (Fleming, 1971) during the height of the rainy season. Whatever the species-specific mechanism, the ecological result of embryonic diapause is one of synchronization. It effectively lengthens the active gestation period, which allows mating to occur and young to be born at times of the year optimal for that species.

7.3. Non-Hypothalamus–pituitary–gonad (HPG) Hormones Affecting Reproduction in Bats

7.3.1. The pineal gland

The pineal gland generally serves as the intermediary between photoperiod and the reproductive system through production of the hormone MEL. The effect of MEL upon the HPG axis was shown for the first time by Wurtman, Axlrod, and Chu (1963). Melatonin may have either anti- or progonadotropic effects, depending upon the circumstances. Bats roost in darkness during the day and are active at night except during winter. The quantity of light directly determines the functional status of the pineal, and darkness enhances pineal activity (Kappers, 1976). Bats, particularly vespertilionids that undergo winter dormancy, may present a different cyclic pattern of MEL secretion than non-hibernating species.

A survey of 88 bat species has firmly established that the pineal organ is not only well-developed in Chiroptera, but it is as complex an organ as in rodents. A few species of bat have been reported to lack a pineal (Bhatnagar, Frahm, & Stephan, 1986; 1990) but it may be very small and concealed, as is the case for *A. geoffroyi* (Heideman et al., 1996). Interestingly, the largest pineal gland (in relation to body size) in any mammal occurs in the megachiropteran bat *Dobsonia proedatrix* (Bhatnagar et al., 1990). Night-time serum MEL levels in *A. geoffroyi* (50–100 pg/ml) are within the range reported for other mammals (Heideman & Bronson, 1990; Yellon & Hilliker, 1994). McGuckin and Blackshaw (1992) reported serum MEL in *P. poliocephalus* of less than 90 pg/ml during the day and 100–300 pg/ml during the night. In the pallid bat (*A. pallidus*), MEL treatment stimulates spermatogenesis but apparently does not influence androgen secretion (Beasley et al., 1984). In *S. heathi*, MEL during winter dormancy suppresses LH-induced T production by the testis *in vitro* via the suppression of 17β-HSD activity (Figure 12.10) (Singh & Krishna, 1995). Haldar and Alipreeta (2001) experimentally showed an inverse relationship between pineal and gonadal activity in *C. sphinx*. *P. poliocephalus* produces a short-duration nocturnal rise in MEL under short nights and a long-duration nocturnal rise in MEL under long nights (McGuckin & Blackshaw, 1992), suggesting that the reproductive effects of photoperiod may be mediated by MEL in bats. Finally, Haldar and Ghosh (1993) reported that MEL injections inhibited gonadal recrudescence in both males and females of *C. sphinx*.

The MEL signal may be interpreted differently by the reproductive axis at different times of the year and under different physiological conditions (Singh & Krishna, 1995). Melatonin inhibited ovarian LH-induced T synthesis during winter dormancy and breeding due to a direct effect of MEL on 17β-HSD (Singh & Krishna, 1995), but showed no significant effect during quiescence and recrudescence phases in *S. heathi* (Abhilasha & Krishna, 1999). In contrast, MEL stimulated LH-induced ovarian T production during October (Srivastava, 2008). The effects of MEL on steroidogenic enzymes have been shown in the rat (Ng & Lo, 1988; Mandal, Ghosh, & Biswas, 1990) and it appears that MEL may have different effects on different enzymes. The increased circulating MEL level in *S. heathi* during winter inhibits ovulation through increased AND synthesis, but suppression of E_2 (Srivastava & Krishna, 2010a). Recent studies of *C. sphinx* showed increased circulating MEL and MEL receptors in the ovary during

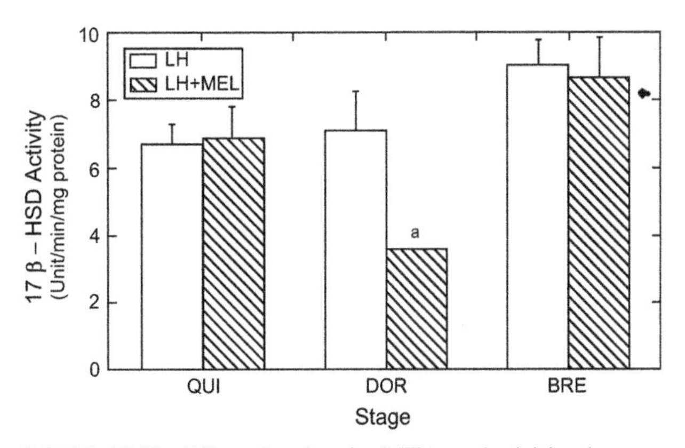

FIGURE 12.10 Effect of melatonin (MEL) on luteinizing hormone (LH)-induced 17β-hydroxysteroid dehydrogenase (HSD) activity in the testis of *Scotophilus heathi, in vitro*. Note that MEL treatment suppressed 17β HSD activity during winter dormancy, but not during other phases of the reproductive cycle of *S. heathi*. *Adapted from Singh and Krishna (1995).*

December as compared with their levels during April (Haldar, Yadav, & Alipreeta, 2006; Banerjee, Meenakumari, Udin, & Krishna, 2009). It is suggested that increased MEL during December in *C. sphinx* may be responsible for suppressed P_4 production by the ovary, resulting in the delayed embryonic development (Banerjee et al., 2009; Meenakumari et al., 2009).

Melatonin also is implicated in the regulation of seasonal adiposity and reproductive activity. However, the mechanism by which MEL regulates seasonal changes in physiology, metabolism, and reproduction is not very clear, and contradictory data exist. Daily administration of MEL to middle-aged rats restores nocturnal plasma MEL to youthful levels, decreases body weight, and suppresses visceral fat and plasma leptin levels (Wolden-Hanson et al., 2000). A recent study in a bat, *S. heathi*, showed that MEL interferes with glucose homeostasis by modulating intracellular glucose transport and may protect bats from hypoglycemia during winter dormancy (Srivastava & Krishna, 2010b). In addition, pinealectomy causes severe hyperinsulinemia, hyperleptinemia, and accumulation of triglycerides in the liver in normal and type 2 diabetic rats (Canpolat et al., 2001; Nishida, Sato, Murai, & Nakagawa, 2003). In contrast to these observations, Mustonen et al. (2004) reported that exogenous MEL implants in female mink elevate plasma leptin levels. Thus, hormones such as insulin, leptin, and MEL may interact to regulate the amount of body fat in bats.

The specific site of action of MEL remains unclear, and the possibility exists that MEL may act at a variety of peripheral and/or central sites, including the brain (Glass, 1988), pituitary gland (Martin & Sattler, 1979), and gonads (Ellis, 1972; Cohen, Roselle, & Chabner, 1978; Abhilasha & Krishna, 1999). Photoperiodic modulation of the secretion of LH and FSH is critical to the reproductive rhythms of both long- and short-day breeders. The most widely accepted mechanism is MEL's ability to modify the frequency of discharge in the hypothalamic GnRH pulse generator (Robinson, 1987). Melatonin binding sites are reported to be present in the SCN of the hypothalamus (Bubenik, Brown, & Grota, 1976). High-affinity (MEL-R1) and low-affinity (MEL-R2) receptors have been described. Three MEL-R-1s have been cloned (MEL-R1a, MEL-R1b, and MEL-R1c) (Reppert, Weaver, Ebisawa, Mahle, & Kolakowski, 1996) but their localization in the reproductive system of bats requires further investigation, particularly in relation to various delay phenomena.

7.3.2. The thyroid

Thyroid hormones are known to have diverse effects on the growth, development, differentiation, and metabolism of nearly all tissues. The thyroid is essential for the normal maintenance of reproductive function (Maruo et al., 1992; Goldman, Dirnfeld, Abramovici, & Kraiem, 1993). Both hyper- and hypothyroidism result in reproductive dysfunction and impaired fertility in many animals (Werner, 1969), and subclinical hypothyroidism often causes infertility in humans (Maruo, Katayama, Barnea, & Mochizuki, 1992). Hagino (1971) reported that, in thyroidectomized rats, ovulation was erratic and occurred in a small percentage of animals, and only a small number of ova were produced. The role of the thyroid gland in seasonal reproduction was first suggested more than 50 years ago when it was found that thyroidectomized starlings failed to undergo the annual decline in gonadal function (Woitkewitsch, 1940). Subsequently, the requirement of the thyroid gland to permit a normal transition from the reproductively active to inactive condition has been described in several seasonally breeding mammalian species (Shi & Barrell, 1992; Jannini, Ulisse, & D'Armiento, 1995).

Studies of thyroid activity (Table 12.8) are mainly restricted to temperate-zone species that exhibit prolonged hibernation (Sadler & Tyler, 1960a; 1960b; Burns et al., 1972; Velicky & Titlbach, 1972; 1974; Ifuta, Gevaerts, & Kuhn, 1988; Damassa, Gustafson, Kwiecinski, & Gagin, 1995). Thyroid activity in these species increases in the spring and declines in the autumn, remaining quiescent throughout the hibernation period. A slight difference was noticed in *M. lucifugus* (Damassa et al., 1995), in which plasma thyroxine levels were lowest at the beginning of hibernation and attained maximal values at the time of arousal. There is a positive correlation between thyroid and testes activity from quiescence to winter dormancy but not during the phases from breeding to postbreeding in *S. heathi* (Krishna & Singh, 1998). Studies in *T. longimanus* support the hypothesis that thyroid activity is necessary for transition from the breeding to the nonbreeding phase, as in seasonally breeding mammals (Webster et al., 1991). In *T. longimanus*, both triiodothyronine (T_3) and T_4 concentrations increased dramatically in late winter. This is similar to the reports on hibernating ground squirrels such as *Spermophilus tridecemlineatus* (Hudson & Wang, 1979) and *Spermophilus richardsoni* (Magnus & Henderson, 1988), and the bat *M. lucifugus* (Damassa et al., 1995). An increase in thyrotropin-releasing hormone (TRH) in response to cold might be the reason for the changes in T_3 and T_4 concentration during hibernation (Ranta, 1975).

Circulating T_4 in female nonhibernating *M. californicus* varies between 26 and 64 nM during different stages of the reproductive cycle (Burns et al., 1972). These values are lower than those reported in normal women (117 to 168 nM), but within the range of serum T_4 levels for other eutherian mammals (Hulbert & Augee, 1982). Similar circulating T_4 concentrations were reported in the hibernating vespertilionid bat *M. lucifugus* (Kwiecinski, Damassa, & Gustafson, 1986), the pteropodid bat *E. franqueti* (Ifuta et al., 1988), the subtropical vespertilionid bat

TABLE 12.8 A partial listing of bat species in which thyroid function has been implicated with reproductive cycles

Species	Observation	References
Antrozous pallidus	Thyroid active in hibernation	Sadler & Tyler (1960a)
Tadarida mexicana; Macrotus californicus	Not responsive to ambient temperatures of 22–37°C	Sadler & Tyler (1960b)
Antrozous pallidus	Thyroid active in hibernation through the sixth day at 4°C	Sadler & Tyler (1960b)
Rhinolophus hipposideros; Plecotus auritus	Ultrastructural differences in follicular and parafollicular cells seen during hibernation	Velicky & Titlbach (1972)
Myotis myotis (male, female)	Thyroid in active period showed pronounced changes in the number and size of organelles	Velicky & Titlbach (1974)
Macrotus waterhousii (male, female)	T_4 low during early pregnancy	Burns et al. (1972)
Epomops franqueti	T_3 low in nonpregnant bats and T_4 does not change	Ifuta et al. (1988)
Myotis lucifugus (male)	T_4 increased during hibernation	Damassa et al. (1995)
Scotophilus heathi (male)	T_4 maximal during postbreeding season	Krishna & Singh (1998)
Taphozous longimanus (male)	T_3 and T_4 low during early pregnancy	Singh et al. (2002)

T_3, triiodothyronine; T_4, thyroxine. *Adapted from Singh, Krishna, and Bhatnagar (2002).*

S. heathi (Krishna & Singh, 1998), the emballonurid bat *T. longimanus* (Singh, Krishna, & Bhatnagar, 2002), and the pteropodid bat *C. sphinx* (Banerjee et al., 2007).

Maternal thyroid activity declines during the early stage of pregnancy in *M. californicus* (Burns et al., 1972). Triiodothyronine and T_4 levels in *T. longimanus* decline during the early stage of a first pregnancy, causing a more prolonged gestation, but not during the second pregnancy (Singh et al., 2002), and might be the cause of delayed embryonic development. This was not the case, however, in *C. sphinx* (Banerjee et al., 2007).

7.3.3. The adrenal cortex

The adrenal cortex plays an important role in metabolism and in many other vital functions of the body such as reproduction, stress, and behavior. Seasonal variations in the hypothalamus–pituitary–adrenal (HPA) axis of bats have mostly been inferred from morphological evidence such as changes in adrenal weight, cytology, and histochemistry. Limited reports are available on the adrenal glands of bats. The adrenal cortex in males of the temperate-zone vespertilionid bat (*M. lucifugus*) remains active during hibernation, possibly more active than it is during the nonhibernating interval (Gustafson & Belt, 1981). Krutzsch and Hess (1961) showed that the adrenal ascorbic acid concentration (a measure of adrenal activity) in *M. lucifugus* was higher in active bats than in hibernating ones. In a histological and histochemical study of another hibernating bat, *M. schreibersii*, Planel et al. (1961) presented data indicating that the zona fasciculata and the zona reticularis are more developed during the hibernation period.

Measurements of adrenal function in bats have been made in a limited number of species (Table 12.9), but results indicate that there are both seasonal and stress-related changes in adrenocortical activity. Microchiropteran bats generally show more variable levels of circulating glucocorticoids. Circulating glucocorticoid levels in *M. lucifugus* and *S. heathi* show a distinct seasonal cycle (Gustafson & Belt, 1981; Krishna, Singh, Doval, & Chanda, 1998). Plasma corticosterone (CORT) levels in *S. heathi* were observed during the period of fat accumulation, and an increase in CORT concentration was seen during mobilization of depot fat, suggesting that CORT may have a catabolic effect on fat accumulation (Krishna et al., 1998). *M. lucifugus*, on the other hand, exhibited lower levels of cortisol during the months of activity and higher levels during hibernation. Corticosterone in *S. heathi* also may be required to stimulate feeding activity (Green, Wilkinson, & Woods, 1992). Circulating cortisol levels in *M. lucifugus* were lowest during the months of activity (1.3 μM) and highest during hibernation (2.6 μM). The adrenal gland of *M. lucifugus* appears to regulate metabolism during hibernation (Gustafson & Belt, 1981). The cortisol levels of *M. lucifugus* are the highest reported for any mammalian hibernators. Cortisol, the principal glucocorticoid in captive *Pteropus hypomelanus*, *P. vampyrus*, and *P. pumulus*, was extremely high in these species, ranging from 102 to 590 ng/ml (Widmaier & Kunz, 1993; Widmaier, Harmer, Sulak, & Kunz, 1994).

The zona fasciculata and zona reticularis of the adrenal cortex in *M. schreibersii* are cytologically more developed during hibernation than during the summer active period (Planel et al., 1961). Morphological studies on the adrenal gland of *Vesperugo savi* and *Vesperugo piccolo* also correlated with the high level of glucocorticoid during

TABLE 12.9 Glucocorticoid concentrations in selected bat species assayed during reproductive cycles

Species	Cortisol (ng/ml ± SEM)	Corticosterone (ng/ml ± SEM)	References
Suborder: Megachiroptera; Family: Pteropodidae			
Pteropus vampyrus	569	102	Widmaier and Kunz (1993)
Pteropus hypomelanus	483	121	Widamaier et al. (1994)
Rousettus aegyptiacus	*	36	Widmaier and Kunz (1993)
Pteropus pumilus	394	55	Widamaier et al. (1994)
Suborder: Microchiroptera; Family: Phyllostomidae			
Carollia perspicillata	*	34	Widamaier et al. (1994)
Artbeus jamaicansis	58	26	Widamaier et al. (1994)
Family: Vespertilionidae			
Myotis lucifugus Morning	14	85	Widamaier et al. (1994)
Midday	49	52	Widamaier et al. (1994)
Evening	170	NP	Widamaier et al. (1994)
Myotis lucifugus			
Early hibernation (January)	1114	*	Gustafson and Belt (1981)
Late hibernation (April)	500	*	Gustafson and Belt (1981)
Active (August)	350	*	Gustafson and Belt (1981)
Scotophilus heathi (male) Quiescent (August)	*	186.7	Krishna et al. (1998)
Active prehibernation (November)	*	58.6	Krishna et al. (1998)
Winter dormancy (December)	*	100.34	Krishna et al. (1998)
Breeding (February)	*	117.2	Krishna et al. (1998)
Postbreeding (April)	*	293.3	Krishna et al. (1998)
Scotophilus heathi (female) Active feeding (August–September)	*	240–260	Krishna and Bahuguna (2002)
Fat deposition (October–December)	*	130–140	Krishna and Bahuguna (2002)
Fat depletion (January–February)	*	130–145	Krishna and Bahuguna (2002)
First pregnancy	*	130–145	Krishna and Bahuguna (2002)
Second pregnancy (March–May)	*		Krishna and Bahuguna (2002)

SEM, standard error of the mean.
NP, Not present in detectable amount.
Data not available.

hibernation. Interestingly, in *M. lucifugus* and *M. septentrionalis*, very high levels of AND were found during spermatogenesis (~1000 ng/ml) and throughout the period of hibernation (200–400 ng/ml) coincident with the period of maximal adrenocortical activation. Whether AND is of testicular or adrenal origin and the way in which it influences reproduction in bats showing various seasonal reproductive adaptations remain to be determined. Recent studies of *M. lucifugus* demonstrated that glucocorticoids may be involved in increased feeding and/or fat deposition during pregnancy and the prehibernation period (Reeder, Kosteczko, Kunz, & Widmaier, 2004).

7.3.4. The pancreas

The endocrine hormone-producing cells of the pancreas are organized into 'islets' (1–2% of pancreatic mass) dispersed among the pancreatic exocrine acinar cells. Although pancreatic islets in mammals produce a number of hormones, studies in bats have focused on insulin and glucagon only. Insulin secretion is primarily stimulated by hyperglycemia, and insulin controls cellular storage and metabolism of ingested metabolic fuels. The highest serum insulin concentration in *S. heathi* was observed during November, the period of maximal body weight and fat content (Doval & Krishna, 1998). Thus, overfeeding induces an increase in serum insulin in *S. heathi* that may facilitate the accumulation of fat (Forbes, Brown, Welle, & Underwood, 1989). A decline in serum insulin was observed in *S. heathi* during winter dormancy, when a decline in body weight occurred due to gradual depletion of body fat. Exposure of *S. heathi* ovaries *in vitro* to a supraphysiological dose of insulin during November significantly augmented the production of AND (Krishna & Abhilasha, 2000b) (Table 12.10). This finding suggests that insulin amplifies the LH-dependent increase in androgen synthesis by the ovary in *S. heathi*. In mammals in general, the major physiological stimulus for glucagon secretion is hypoglycemia. Similarly, in bats, glycogen may be responsible for a rise in blood glucose by stimulating hepatic glycogenolysis and gluconeogenesis.

Plasma levels of glucose in three species of non-hibernating bat (*P. vampyrus*, *P. hypomelanus*, and *R. aegyptiacus*) showed the normal mammalian range (80–100 mg/dl), with no difference between males and females (Widmaier & Kunz, 1993). In another study on *M. lucifugus*, pancreatic protein content showed a decrease of 21% from nonhibernation to the prehibernation state, which further decreased by 50% during hibernation. Insulin and glucagon content in the pancreas increased from the nonhibernating state to the prehibernating state and during the first half of hibernation. Persistent elevation of glucagon occurred during hibernation; this is perhaps related to energy homeostasis during fasting. Seasonal changes in serum insulin levels in *S. heathi* varied markedly, as did body fat and serum leptin. Serum insulin and leptin exhibited two peaks in females, one at the beginning of hibernation when body fat was maximal and another during pregnancy. A significant decline in serum insulin was observed after hibernation. There was a close association in *S. heathi* between changes in body mass related to fat deposition and mobilization with the changes in reproductive activity, suggesting that they are interdependent. Hyperinsulinemia (HI) and insulin resistance (IR) are associated with obesity in women (Conway & Jacobs, 1993), and perhaps insulin has a role in the hormonal changes associated with adiposity in bats. Increased circulating levels of both insulin and glucose in *S. heathi* (Figure 12.11) suggest HI and the possibility of IR conditions before the fat accumulation during the prehibernation period. Thus, the development of HI and IR may be required for fat accumulation in *S. heathi*. Therefore, the metabolic and morphological changes in the pancreatic cells of bats during hibernation may be associated with the adaptive changes for energy storage needed to counter the forthcoming adverse environmental conditions of hibernation.

8. CONCLUSIONS AND FUTURE DIRECTIONS

Although bats are the most numerous mammals on earth after rodents, knowledge on the hormonal regulation of their reproductive cycles is limited to a few species, mostly microchiropteran. Bats demonstrate numerous unique features in reproductive processes. Only limited attempts have been made towards the understanding of mechanism (s) responsible for these unique reproductive phenomena. The neuroendocrine regulation of reproductive processes is poorly known and limited to only a few microchiropterans. Lack of pure bat pituitary hormones and their antisera are in part responsible for the failure to elucidate many central regulatory neuroendocrine mechanisms. It has been suggested that delayed ovulation may result from long-loop negative feedback by steroids on the HPG axis. Although

TABLE 12.10 **The effect of insulin on basal and hCG-induced androstenedione production by *Scotophilus heathi* ovaries *in vitro***

Treatment	Control	hCG	Insulin (IU)			Insulin (IU) + hcG (0.1μg)		
Dose		(0.1 μg)	0.4	1.0	2.0	0.4	1.0	2.0
Values	12.7±1.5	79.7^{a}±13.2	58.0^{a}±3.8,	86.5^{a}±14.3,	116.3^{a}±11.6	68.9^{a}±5.9,	95.0^{a}±16.2,	122.2^{ab}±10.2

Adapted from Doval and Krishna (1998)
Values are means ± SEM of four samples.
^{a}*Values are significantly different (P<0.05) compared with control.*
^{ab}*Values are significantly different (P<0.05) compared with the hCG-treated groups.*

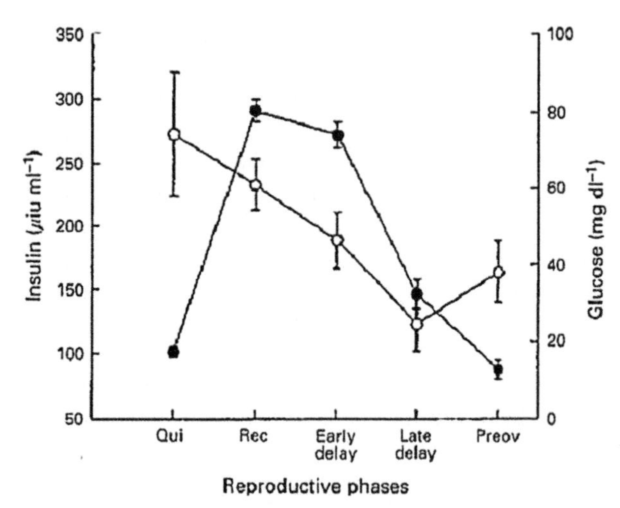

FIGURE 12.11 Circulating glucose (●) and insulin (○) concentrations in *Scotophilus heathi* during different reproductive phases. Qui, quiescence; Rec, recrudescence; Early delay, early phase of ovulatory delay; Late delay, late phase of ovulatory delay; Preov, preovulatory period. Values are mean ± standard error of the mean (SEM) of five samples. The presence of a high level of glucose as well as insulin during recrudescence suggests a condition of insulin resistance. *Adapted from Doval and Krishna (1998).*

the environmental changes affecting PRL synthesis remain to be investigated, the possible involvement of MEL in regulating PRL release as well as seasonal variations in GnRH neurons need to be examined.

Our fragmentary information on the anatomy and physiology of male bat reproduction, based upon studies of a few microchiropterans, mostly vespertilionids, reveals two most interesting features: prolonged retention of sperm in epididymides and asynchrony between gametogenesis and accessory reproductive gland function. In the majority of bats, the spermatogenetic cycle is generally restricted during summer and varies with the climatic zone. Despite this restricted period of spermatogenesis, mating in most bat species is spread over an extended period. Whether prolonged sperm storage in bats is due to inherent unique characteristics of the gametes and/or the relatively high circulating level of androgens in bats requires detailed investigation.

The morphological and physiological changes that occur in female reproductive organs during delayed ovulation are now well demonstrated in several vespertilionid and rhinolophid bats, although detailed study of the ovarian cycle requires investigation beyond vespertilionid and rhinolophid bats. Reproductive tract asymmetries, unilateral endometrial reaction, menstruation, follicular and luteal development, and induced ovulation are all topics of great interest that need further study.

Low ambient temperature and food deprivation are two major factors that appear to be responsible for suppression of reproductive events, leading to reproductive delay phenomena (of ovulation, implantation, and/or development) in bats. The roles suggested for lipid and glycogen storage as well as the involvement of metabolic hormones such as leptin and insulin need additional study. Roles for thyroid hormones and MEL have been implied in the regulation of seasonal adiposity and reproductive activity, and more work is needed in these areas.

Finally, detailed studies of more bat species are needed to better understand bat reproductive cycles. A particular emphasis of considerable importance should be placed on species that breed in temperate zones vs. tropical and semitropical environments, in order to better understand the mechanisms responsible for reproductive adaptations related to hibernation.

Bats are useful and necessary for the human environment as they are an integral part of the food chain. As natural habitats of bats are increasingly threatened by human activities, conserving natural bat populations should be a primary goal. Knowledge of the reproductive characteristics of this unique and ecologically important group of mammals is essential for their conservation. The great diversity of bat species means that there are many different adaptations and variations in life history strategies, with possibly unique hormonal components to be investigated. Much research remains to be done before the endocrine control of the reproductive cycles of bats comes out of the dark.

ABBREVIATIONS

17β-HSD	17β-hydroxysteroid dehydrogenase
3β-HSD	3β-hydroxysteroid dehydrogenase
5-HT	Serotonin
ABP	Androgen-binding protein
AND	Androstenedione
AR	Androgen receptor
ARC	Arcuate nucleus
BMR	Basal metabolic rate
CBG	Corticosteroid-binding globulin
CL	Corpus luteum
CORT	Corticosterone
DHT	5α-dihydrotestosterone
E_1	Estrone
E_2	17β-estradiol
E_3	Estriol
FSH	Follicle-stimulating hormone
GH	Growth hormone
GnIH	Gonadotropin-inhibiting hormone
GnRH	Gonadotropin-releasing hormone
GnRH-I	Mammalian gonadotropin-releasing hormone
GnRH-II	Chicken gonadotropin-releasing hormone
GTH	Gonadotropin
hCG	Human chorionic gonadotropin
HI	Hyperinsulinemia

HPA Hypothalamus–pituitary–adrenal
HPG Hypothalamus–pituitary–gonadal
HPLC High performance liquid chromatography
IC Interstitial cell
ICC Immunocytochemistry
IGF-1 Insulin-like growth factor-1
ir Immunoreactive
IR Insulin resistance
LH Luteinizing hormone
MBH Medial basal hypothalamus
ME Median eminence
MEL Melatonin
MEL-R1 High-affinity melatonin receptor
MEL-R2 Low-affinity melatonin receptor
mPOA Medial preoptic area
P_4 Progesterone
$P450_{scc}$ Sidechain-cleaving enzyme
$P450_{17\alpha}$ 17α-hydroxylase
PCOS Polycystic ovary syndrome
PMSG Pregnant mare serum gonadotropin
PN Pars nervosa
PRL Prolactin
RER Rough endoplasmic reticulum
SBG Steroid-binding globulin
SCN Suprachiasmatic nucleus
SDH Succinate dehydrogenase
StAR Steroidogenic acute regulatory
T Testosterone
T_3 Triiodothyronine
T_4 Thyroxine
TICs Thecal/interstitial gland cells
TRH Thyrotropin-releasing hormone

REFERENCES

Abhilasha, & Krishna, A. (1996). High androgen production by ovarian thecal-interstitial cells: a mechanism for delayed ovulation in a tropical vespertilionid bat, *Scotophilus heathi*. *J. Reprod. Fertil., 106*, 207–211.

Abhilasha, & Krishna, A. (1997). Adiposity and androstenedione production in relation to delayed ovulation in the Indian bat, *Scotophilus heathi*. *Comp. Biochem. Physiol., 116C*, 97–101.

Abhilasha, & Krishna, A. (1998). Immunolocalization of cytochrome P450 side chain cleavage, 17-alpha-hydroxylase and aromatase in the ovary of vespertilionid bat (*Scotophilus heathi*) during different phases of ovulatory delay. *Biol. Res., 31*, 49–57.

Abhilasha, & Krishna, A. (1999). Ovarian steroidogenesis in the vespertilionid bat, *Scotophilus heathi*: role of melatonin. *Acta Chiropterologica, 1*, 223–230.

Aitken, R. J. (1981). Aspects of delayed implantation in the roe deer (C *apreolus capreolus*). *J. Reprod. Fertil. (suppl.), 29*, 83–95.

Anand Kumar, T. C. (1965). Reproduction in the rat-tailed bat *Rhinopoma kinneari*. *J. Zool. Lond., 147*, 147–155.

Anciaux de Faveaux, M. (1973). Essai de synthese sur la reproduction de chiropeteres d'Afrique (region faunistique Ethiopienne). *Period Biol., 75*, 195–199.

Anthony, E. L. P. (1987). The role of the anterior pituitary and the hyothalamus in controlling reproductive cycles in bats. In M. B. Fenton, P. Racey, & J. M. V. Rayner (Eds.), *Recent Advances in the Study of Bats* (pp. 421–439). London, UK: Cambridge University Press.

Anthony, E. L. P. (2000). Endocrinology of reproduction in bats: central control. In E. G. Crichton, & P. H. Krutzsch (Eds.), *Reproductive biology of bats* (pp. 1–26). London, UK: Academic Press.

Anthony, E. L. P., & Gustafson, A. W. (1984). Seasonal variations in pituitary LH-gonadotropes of the hibernating bat *Myotis lucifugus lucifugus*: An immunohistochemical study. *Am. J. Anat., 170*, 101–115.

Anthony, E. L. P., Wu, P., Bruhn, T. O., & Jackson, I. M. D. (1987). Characterization of LH-RH immunoreactivity in mammalian pituitary neural lobe by HPLC. *Brain Res., 424*, 258–263.

Anthony, E. L. P., King, J. C., & Stopa, E. G. (1984). Immunocytochemical localization of LHRH in the median eminence, infundibular stalk, and neurohypophysis. Evidence for multiple sites or releasing hormone secretion in humans and other mammals. *Cell Tissue Res., 236*, 5–14.

Anthony, E. L. P., Weston, P. J., Montivilo, J. A., Bruhn, T. O., Neel, K., & King, J. C. (1989). Dynamic aspects of the LHRH system associated with ovulation in the little brown bat (*Myotis lucifugus*). *J. Reprod. Fertil., 87*, 671–686.

Audet, D., & Fenton, M. B. (1988). Heterothermy and the use of torpor by the bat *Eptesicus fuscus* (Chiroptera: Vespertilionidae): a field study. *Phys. Zool., 61*, 197–204.

Badwaik, N. K. (1988). Cytology and seasonal changes of pituitary of the emballonurid bat, *Taphozous melanopogon* (temnick). *Proc. Indian Acad. Sci. (Anim. Sci.), 97*, 479–489.

Badwaik, N. K., & Gopalakrishna, A. (1990). Asymmetry of the female genitilia in some Indian bats. *Trends Life Sci., 5*, 11–17.

Baird, D. T. (1976). Ovarian steroid secretion and metabolism in women. In V. H. T. James, M. Serio, & G. Giusti (Eds.), *The Endocrine function of the Human Ovary* (pp. 125–133). London, UK: Academic Press.

Baker, J. R., & Bird, T. F. (1936). The seasons in a tropical rain-forest (New Hebrides): Part 4. Insectivorous bats. (Vespertilionidae and Rhinolophidae). *J. Linn. Soc. Zool., 40*, 143–161.

Banerjee, A., Meenakumari, K. J., & Krishna, A. (2007). Relationship between delayed embryonic development and metabolic factors and fat deposition in fruit bat *Cynopterus sphinx*. *Reprod. Fertil. Develop., 19*, 626–633.

Banerjee, A., Meenakumari, K. J., Udin, S., & Krishna, A. (2009). Melatonin regulates delayed embryonic development in the short-nosed fruit bat, *C. sphinx*. *Reproduction, 138*(6), 935–944.

Bartness, T. J., Demas, G. E., & Song, C. K. (2002). Seasonal changes in adiposity: the roles of the photoperiod, melatonin and other hormones, and sympathetic nervous system. *Exp. Biol. Med., 227*, 363–376.

Beasley, L. J., & Zucker, I. (1984). Photoperiod influences the annual reproductive cycle of the male pallid bat (*Antrozous pallidus*). *J. Reprod. Fertil., 70*, 567–573.

Beasley, L. J., Smale, L., & Smith, E. R. (1984). Melatonin influences the reproductive physiology of male pallid bats. *Biol. Reprod., 30*, 300–305.

Beck, A. J., & Boo-Liat, L. (1973). Reproductive biology of *Eonycteris spelaea*, Dobson (Megachiroptera) in West Malaysia. *Acta Trop., 30*, 251–260.

Benecke, B. (1879). Ueber Reifung und Befruchtung des Eies bei den Fledermausen. *Zool. Anz., 2*, 304–305.

Benoit, S. C., Clegg, D. J., Seeley, R. J., & Woods, S. C. (2004). Insulin and leptin as adiposity signals. *Recent Prog. Horm. Res., 59*, 267–285.

Bernard, R. T. F. (1976). Some aspects of the dynamics of a bat population in Natal, South Africa. *Honours thesis.* South Africa: University of Natal.

Bernard, R. T. F. (1980). Reproductive cycles of *Miniopterus schreibersii natalensis* (Kuhl, 1819) and *Miniopterus fraterculus* (Thomas and Schwann, 1906). *Ann. Trans. Mus., 32*, 55–64.

Bernard, R. T. F. (1982). Female reproductive cycle of *Nycteris thebaica* (Microchiroptera) from Natal, South Africa. *Zeitschrift fur Saugertierkunde, 47*, 12–18.

Bernard, R. T. F. (1986). Seasonal changes in plasma testosterone concentrations and Leydig cell and accessory gland activity in the Cape horseshoe bat (*Rhinolophus capensis*). *J. Reprod. Fertil., 78*, 413–422.

Bernard, R. T. F. (1989). The adaptive significance of reproductive delay phenomenon in some South African Microchiroptera. *Mammal. Rev., 19*(1), 27–34.

Bernard, R. T. F. (1994). Reproductive synchrony and annual variation in foetal growth rate in the long fingered bat *(Miniopterus schreibersii). J. Zool. Lond., 232*, 485–490.

Bernard, R. T. F., & Bojarski, C. (1994). Effects of prolactin and hCG treatment on luteal activity and the conceptus during delayed implantation in Schreibers' long-fingered bat *(Miniopterus schreibersii). J. Reprod. Fertil., 100*, 359–365.

Bernard, R. T. F., & Cummings, G. S. (1997). African bats: evolution of reproductive patterns and delays. *Q. Rev. Biol., 72*, 253–274.

Bernard, R. T. F., & Meester, J. A. J. (1982). Female reproduction and the female reproductive cycle of *Hipposideros caffer* (Sundevall, 1846) in Natal, South Africa. *Ann. Transvaal. Mus., 33*, 131–144.

Bernard, R. T. F., Bojarski, C., & Millar, R. P. (1991a). Plasma progesterone and luteinizing hormone concentrations and the role of the corpus luteum and LH gonadotrophs in the control of delayed implantation in Schreiber's long-fingered bat (*Miniopterus schreibersii*). *J. Reprod. Fertil., 93*, 31–42.

Bernard, R. T. F., Bojarski, C., & Millar, R. P. (1991b). Pattern of Leydig cells and LH gonadotroph activity and plasma testosterone concentration in the seasonally reproducing schreibers long fingered bat *(Miniopterus schreibersii). J. Reprod. Fertil., 91*, 479–492.

Bernard, R. T. F., Cotteril, F. P. D., & Forgusson, R. A. (1996). On the occurrence of a short period of delayed implantation in Schreibers' long-fingered bat (*Miniopterus schreibersii*) from tropical latitude in Zimbabwe. *J. Zool., Lond., 238*, 13–22.

Bhatnagar, K. P., Frahm, H. D., & Stephan, H. (1986). The pineal organ of bats: a comparative morphological and volumetric investigation. *J. Anat., 147*, 143–161.

Bhatnagar, K. P., Frahm, H. D., & Stephan, H. (1990). The megachiropteran pineal organ: A comparative morphological and volumetric investigation with special emphasis on the remarkably large pineal of *Dobsonia praedatrix. J. Anat., 168*, 143–166.

Bhiwgade, D. A., Akolkar, V. V., Menon, S. N., Manekar, A. P., & Senad, D. G. (1989). Ultrastructural and functional characteristics of anterior pituitary cells in the Indian fruit bat, *Rousettus leschenaulti* (Desmarest). *Acta Anat., 135*, 129–141.

Blake, C. A. (1980). Correlative study of changes in the morphology of the LH-gonadotroph and anterior pituitary gland LH secretion during the 4-days rat estrous cycle. *Biol. Reprod., 23*, 1097–1108.

Bleier, W. J. (1975). Early embryology and implantation in the California leaf- nosed bat, *Macrotus californicus. Anat. Rec., 182*, 237–254.

Bleier, W. J., & Ehteshami, M. (1981). Ovulation following unilateral ovariectomy in the California leaf-nosed bat (*Macrotus californicus*). *J. Reprod. Fertil., 63*, 181–183.

Bojarski, C. (1993). Seasonal changes in pituitary and plasma prolactin concentrations, and the role of prolactin in the control of delayed implantation in female *Miniopterus schreibersii. PhD Thesis.* South Africa: Rhodes University.

Bonilla, H., & de., and Rasweiler, J. J. (1974). Breeding activity, preimplantation development, and oviduct histology of the short-tailed fruit bat, *Carollia*, in captivity. *Anat. Rec., 179*, 385–404.

Boswell, T., Woods, S. C., & Kenagy, G. J. (1994). Seasonal changes in body mass, insulin and glucocortioids of free-living golden-mantled ground squirrels. *Gen. Comp. Endocrinol., 96*, 339–346.

Bradshaw, G. V. R. (1961). Le cycle de reproduction de *Macrotus californicus* (Chiroptera, Phyllostomidae). *Mammalia, 25*, 117–119.

Bradshaw, G. V. R. (1962). Reproductive cycle of the California leaf-nosed bat, *Macrotus californicus. Science, 136*, 645–646.

Briskie, J. V., & Montgomeric, R. (1993). Patterns of sperm storage in relation to sperm competition in passerine birds. *Condor, 95*, 442–454.

Bronson, F. H. (1985). Mammalian reproduction: an ecological perspective. *Biol. Reprod., 32*, 1–26.

Bronson, F. H., & Heideman, P. D. (1994). Seasonal regulation of reproduction in mammals. In E. Knobil, & J. D. Neill (Eds.), *The Physiology of Reproduction* (pp. 541–583). New York, NY: Raven.

Bronson, F. H., & Manning, J. M. (1991). The energetic regulation of ovulation: a realistic role for body fat. *Biol. Reprod., 44*, 945–950.

Bronson, F. H., Heideman, P. D., & Kerbeshian, M. C. (1991). Lability of fat stores in peripubertal wild house mice. *J. Comp. Physiol., 161B*, 15–18.

Brosset, A. (1962a). La reproduction des chiropteres de l'ouest et du centre de, l'Inde. *Mammalia, 26*, 176–213.

Brosset, A. (1962b). The bats of Central and Western India. Pt. I. *J. Bombay Nat. Hist. Soc., 59*, 1–57.

Brosset, A. (1962c). The bats of Central and Western India Pt.2. *J. Bombay Nat Hist Soc., 59*, 583–624.

Brosset, A. (1963). The bats of Central and Western India. Pt. 4. *J. Bomaby Nat. Hist. Soc., 60*, 337–355.

Bubenik, G. A., Brown, G. M., & Grota, L. J. (1976). Immunohistochemical localization of melatonin in the rat Harderian gland. *J. Histochem. Cytochem., 24*, 1173–1177.

Buchanan, G. D., & Young Lai, E. V. (1986). Plasma progesterone levels during pregnancy in the little brown bat *Myotis lucifugus* (Vespertilionidae). *Biol. Reprod., 34*, 878–884.

Buchanan, G. D., & Young Lai, E. V. (1988). Plasma progesterone concentrations in female little brown bats (*Myotis lucifugus*) during hibernation. *J. Reprod. Fertil., 83*, 59–65.

Burns, J. M. (1981). Aspects of endocrine control of delay phenomenon in bats with special emphasis on delayed development. *J. Reprod. Fertil. Suppl., 29*, 61–66.

Burns, J. M., & Easley, R. G. (1977). Hormonal control of delayed development in the California leaf-nosed bat, *Macrotus californicus*. III Changes in plasma progesterone during pregnancy. *Gen. Comp. Endocrinol., 32*, 163–166.

Burns, J. M., & Wallace, W. E. (1975). Hormonal control of delayed development in *Macrotus waterhousii*. II. Radioimmunoassay of plasma estrone and estradiol 17 β during pregnancy. *Gen. Comp. Endocrinol., 25*, 529–533.

Burns, J. M., Baker, R. J., & Bleier, W. J. (1972). Hormonal control of delayed development in *Macrotus waterhousii*. 1. Changes in plasma thyroxine during pregnancy and lactation. *Gen. Comp. Endocrinol., 18*, 54–58.

Case, T. J. (1964). The evolution of torpidity in rodents. Suomal. *Tiedeakat. Toim., Ser. A, 71*, 77–112.

Caffier, P., & Kolbow, H. (1934). Anatomisch-physiologische genitalstudien an Fleudermausen zur Klarung der therapeutischen Sexualhormonwirkung. *Z. Geburtsch. Gynäk., 108*, 185–235.

Canney, D. H., & Butler, V. V. R. (1987). Plasma and pituitary luteinizing hormone concentrations during delayed ovulation in the little brown bat, *Myotis lucifugus. Biol. Reprod. Suppl., 36*, 121, (abstr).

Canpolat, S., Sandal, S., Yilmaz, B., Yasar, A., Kutlu, S., Baydas, G., et al. (2001). Effects of pinealectomy and exogenous melatonin on serum leptin levels in male rat. *Eur. J. Pharmacol., 428*, 145–148.

Carroll, J. B. (1988). Conservation programme for the Rodrigues fruit bat. In J. B. Carroll (Ed.), *Proceedings of the 5th world conference on Breeding Endangered Species in Captivity. Pteropus rodricensis* (pp. 457–475). Ohio, CT: Cincinnati Zoo and Botanical Garden.

Chanda,, Abhilasha, D., & Krishna, A. (2003). Seasonal adiposity and delayed ovulation in a vespertilionid bat, *Scotophilus heathi*: role of tumor necrosis factor-α. *Physiol. Biochem. Zool., 76*, 271–280.

Chanda,, Abhilasha, D., & Krishna, A. (2006). Hormonal induction of ovulation stimulates atresia of antral follicles in a vespertilionid bat, *Scotophilus heathi. Zoology, 109*, 208–216.

Cheke, A. S., & Dahl, J. F. (1981). The status of bats on Western Indian Ocean islands, with special reference to *Pteropus. Mammalia, 45*, 205–238.

Chen, Y. I., Payne, A. H., & Kelch, R. P. (1976). FSH stimulation of Leydig cell function in the hypophysectomized immature rat. *Proc. Soc. Exp. Biol. Med., 153*, 473–475.

Clarke, J. R., & Forsyth, I. A. (1964). Seasonal changes in the adenohypophysis of the vole (*Microtus agrestis*). *Gen. Comp. Endocrinol., 4*, 243–252.

Cohen, M., Roselle, D., & Chabner, B. (1978). Evidence for a cytoplasmic melatonin receptor. *Nature, 274*, 894–895.

Considine, R. V., Sinha, M. K., Heiman, M. L., Kriauciunas, A., Stephens, T. W., Nyce, M. R., et al. (1996). Serum immunoreactive-leptin concentrations in normal-weight and obese humans. *New Engl. J. Med., 334*, 292–295.

Conway, G. S., & Jacobs, H. S. (1993). Clinical implications of hyperinsulinemia in women. *Clin. Endocrinol. (Oxf)., 39*, 623–632.

Courrier, R. (1924). Le cycle sexual chez la female des Mammifères. Etude de la phase folliculaire. *Arch. Biol., 34*, 369–477.

Courrier, R. (1927). Etude sur le déterminisme des caractères sexuels secondaires chez quelques mammiféres à l'activité testiculaire pérodique. *Arch. Biol., 37*, 173–334.

Crichton, E. G. (2000). Sperm storage and fertilization. In E. G. Crichton, & P. H. Krutzsch (Eds.), *Reproductive Biology of Bats* (pp. 295–320). San Diego, CA: Academic Press.

Crichton, E. C., Hoyer, P. B., & Krutzsch, P. H. (1990). Cellular composition and steroidogenic capacity of the ovary of *Macrotus californicus* (Chiroptera: Phyllostomatidae) during and after delayed embryonic development. *Cell Tissue Res., 260*, 355–366.

Crichton, E. G., & Krutzsch, P. H. (1985). Reproductive biology of the female leaf-nosed bat *Macrotus californicus* in South Western United States. A morphometric analysis of the annual ovarian cycle. *Am. J. Anat., 173*, 69–87.

Crichton, E. G., & Krutzsch, P. H. (1987). Reproductive biology of the female little mastiff bat, *Mormopterus planiceps* (Chiroptera; Molossidae) in southeast Australia. *Am. J. Anat., 178*. 369–368.

Crichton, E. C., Seamark, R. F., & Krutzsch, P. H. (1989). The status of corpus luteum during pregnancy in *Miniopterus schreibersii* (Chiroptera: Vespertilionidae) with emphasis on its role in developmental delay. *Cell Tissue Res., 258*, 183–201.

Crichton, E. G., Krutzsch, P. H., & Wimsatt, W. A. (1981). Studies on prolonged spermatozoa survival in Chiroptera I. The role of uterine free fructose in the spermatozoa storage phenomenon. *Comp. Biochem. Physiol*, 387–395, 70 *A*.

Cumming, G. S., & Bernard, R. T. F. (1997). Rainfall, food abundance and timing of parturition in African bats. *Oecologia (Berl.), 111*, 309–317.

Currie, W. B., Blake, M., & Wimsatt, W. A. (1988). Fetal development and placental and maternal plasma concentrations of progesterone in the little brown bat (*Myotis lucifugus*). *J. Reprod. Fertil., 82*, 401–407.

Curlewis, J. D. (1992). Seasonal prolactin secretion and its role in seasonal reproduction: A review. *Reprod., Fert. Develop., 4*, 1–23.

Damassa, D. A., Gustafson, A. W., Kwiecinski, G. G., & Gagin, G. A. (1995). Seasonal influences on the control of plasma sex hormone-binding globulin by T4 in male little brown bats. *Am. J. Physiol., 268*, 1303–1309.

Darwin, C. (1896). *The variation of animals and plants under domestication*. New York, NY: A. Appleton and Company.

David, S. K., & Lall, S. B. (1985). Spermatogenesis in the flying fox, *Pteropus giganteus* Brunnich (Megachiroptera Mammalia). *Seventh Int. Bat. Res. Con.*, 23.

Dominic, C. J., & Krishna, A. (1989). Reproductive cycle of mammals: Chiropetera and Insectivora. In S. K. Saidapur (Ed.), *Reproductive cycle of Indian vertebrates* (pp. 311–346). New Delhi, India: Allied Publishers Ltd.

Doval, J., & Krishna, A. (1998). Ovarian androstenedione production is enhanced by insulin during the period of delayed ovulation in a vespertilionid bat, *Scotophilus heathi. J. Reprod. Fertil., 114*, 63–68.

Duggal, P. S., Van der Hoek, K. H., Milner, C. R., Ryan, N. K., Armstrong, D. T., Magoffin, D. A., et al. (2000). The *in vivo* and *in vitro* effects of exogenous leptin on ovulation in the rat. *Endocrinology, 141*, 1971–1976.

Dwyer, P. D. (1963a). *Reproduction and distribution in Miniopterus (Chiroptera). Aust. J. Sci., 25*, 435–436.

Dwyer, P. D. (1963b). The breeding biology of *Miniopterus schreibersii blepotis Temminck* (Chiroptera) in Northeastern New South Wales. *Aust. J. Zool., 11*, 219–240.

Eisentraut, M. (1937). Die wirkung niedriger Temperaturen auf die Embryonalentwicklung bie Fleder mäusen. *Biol. Zentralblatt, 57*, 59–74.

Ellis, L. C. (1972). Inhibition of rat testicular androgen synthesis *in vitro* by melatonin and serotonin. *Endocrinology, 90*, 17–27.

Entwistle, A. C. (1994). *Roost Ecology of the Brown Long-eared Bat (Plecotus auritus,* Linnaeus 1758*) in Northeast Scotland.* UK: PhD thesis, University of Aberdeen.

Erickson, G. F., Magoffin, D. A., Dyer, C. A., & Hofeditz, C. (1985). The ovarian androgen producing cells: a review of structure/function relationships. *Endocr. Rev., 6*, 371–399.

Ewing, W. G., Studier, E. H., & O'Farrell, M. J. (1970). Autumn fat deposition and gross body composition in three species of *Myotis*. *Comp. Biochem. Physiol., 47*, 597–602.

Farquhar, M. G., & Rinehart, J. F. (1954). Electron microscopic studies of the anterior pituitary gland of castrated rats. *Endocrinology, 54*, 516–541.

Fenton, M. B. (1984). Sperm competition? The case of vespertilionid and rhinolophid bats. In R. L. Smith (Ed.), *Sperm Competition and the Evolution of Animal Mating Systems* (pp. 573–587). New York, NY: Academic Press.

Fernandez, A. M., Muniz, E., Gragera, R. R., & Martinez-Rodriguez, R. (1992). Immunocytochemical localization of GnRH in the hypothalamus of the bat *Miniopterus schreibersii schreibersii. J. Hirnforschung, 33*, 195–202.

Fleming, T. H. (1971). *Artibeus jamaicensis*: delayed embryonic development in neotropical bat. *Science, 171*, 402–404.

Fleming, T. H., Hopper, E. T., & Wilson, D. E. (1972). Three central American bat communities: Structure, reproductive cycles and movement patterns. *Ecology, 53*, 555–569.

Flint, A. P. F., Renfree, M. B., & Weir, B. J. (1981). Embryonic diapause in mammals. *J. Reprod. Fertil., Suppl., 29*, 1–26.

Forbes, G. B., Brown, M. R., Welle, S. L., & Underwood, L. E. (1989). Hormonal response to overfeeding. *Am. J. Clin. Nutr., 49*, 608–611.

Frank, C. L. (1991). Adaptations for hibernation in the depots fats of a ground squirrel (*Spermophilus beldingi*). *Can. J. Zool., 69*, 2707–2711.

Fries, S. (1879). Ueber die Fortpflanzung der einheimeschen Fledermause (chiroptere). *Zool. Anz., 2*, 355–357.

Gadegone, M. M., & Sapkal, V. M. (1983). Mucins in the male accessory sex gland of pipistrellid bat *Pipistrellus dormeri* (Dobson). *Comp. Phyiol. Ecol., 8*, 219–222.

Gaisler, J. (1966). Reproduction in the lesser horse-shoe bat (*Rhinolophus hipposideros hipposideros*) Bechstein, 1800. *Bijdr. Dierkd, 36*, 45–64.

Gaytan, F., Bellido, C., Morales, C., Van Rooijen, N., & Aguilar, E. (1995). Role of testicular macrophages in the response of Leydig cells to gonadotrophins in young hypophysectomized rats. *J. Endocrinol., 147*, 463–471.

Glass, J. D. (1988). Neuroendocrine regulation of seasonal reproduction by the pineal gland and melatonin. In R. J. Reiter (Ed.), *Pineal Research Reviews, Vol. 6* (pp. 219–259). New York, NY: Alan R. Liss, Inc.

Goldman, S. D., Dirnfeld, M., Abramovici, A., & Kraiem, Z. (1993). Triiodothyronine (T3) modulates hCG-regulated progesterone secretion, camp accumulation and DNA content in cultured human luteinized granulose cells. *Mol. Cell. Endocrinol., 96*, 125–131.

Gopalakrishna, A. (1947). Studies on the embryology of Microchiroptera. Part I: Reproduction and breeding season in the South Indian vespertilionid bat, *Scotophilus wroughtoni. Proc. Indian Acad. Sci., 26 B*, 218–232.

Gopalkrishna, A. (1948). Studies on the embryology of microchiroptera. Part II. Reproduction in the male vespertilionid bats, *Scotophilus wroughtoni* (Thomas). *Proc. Indian Acad. Sci., 27 B*, 137–152.

Gopalakrishna, A. (1950). Studies on the embryology of microchiroptera. Part VI. Structure of placenta in the Indian vampire bat, *Lyroderma lyra lyra. Proc. Nat. Inst. Sci., India, 16*(B), 93–98.

Gopalakrishna, A. (1955). Observation on the breeding habits and ovarian cycle in the Indian sheath-tailed bat, *Taphozous longimanus. Proc. Nat. Inst. Sci. India, 21*(B), 29–35.

Gopalakrishna, A. (1964). Post-partum pregnancy in the Indian fruit bat, *Rousettus leschenaulti. Curr. Sci., 33*, 558–559.

Gopalakrishna, A. (1969). Unusual persistence of the corpus luteum in the Indian fruit bat, *Rousettus leschenaultia* (Desm). *Curr. Sci., 33*, 558–559.

Gopalakrishna, A., & Madhavan, A. (1971). Survival of spermatozoa in the female genital tract of the Indian vespertilionid bat, *Pipistrellus ceylonicus chrysothrix*(Wroughton). *Proc. Indian Acad. Sci., 73*(B), 43–49.

Gopalakrishna, A., & Madhavan, A. (1978). Viability of inseminated spermatozoa in the Indian vespertilionid bat, *Scotophilus heathi* (Horsefield). *Indian J. Exp. Biol., 16*, 852–854.

Gopalakrishna, A., & Murthy, K. V. R. (1960). Utero–ovarian junction in two species of bats. *Bull. Zool. Soc., College of Science, Nagpur, 3*, 19–22.

Gopalakrishna, A., & Bhatia, D. (1980). Storage of spermatozoa in the epididymis of the bat, *Hipposideros speoris* (Schneider). *Curr. Sci., 49*, 951–953.

Gopalkrishna, A., & Karim, K. B. (1971). Localized progestational endometrial reaction in the uterus of the Indian fruit bat, *Rousettus leschenaulti*, (Desmaret). *Curr. Sci., 408*, 490–491.

Gopalakrishna, A., & Karim, K. B. (1974). Secretory blebs from the seminal vesicles of the Indian fruit bat, *Rousettus leschenaulti* (Desmarest). *Curr. Sci., 43*, 383–384.

Gopalakrishna, A., Madhvan, A., Thakur, R. S., & Rajgopal, G. (1974). .The Graafian follicle in some Indian bats. *Curr. Sci., 43*, 400–401.

Green, P. K., Wilkinson, C. W., & Woods, S. C. (1992). Intraventricular corticosterone increases the rate of body weight gain in underweight adrenalectomized rats. *Endocrinology, 130*, 269–275.

Gustafson, A. W. (1976). An improved method for the collection of plasma in small mammals. *Lab. Anim. Sci., 26*, 97–98.

Gustafson, A. W. (1979). Male reproductive patterns in hibernating bats. *J. Reprod. Fertil., 56*, 317–331.

Gustafson, A. W. (1987). Changes in Leydig cell activity during the annual testicular cycle of the bat *Myotis lucifugus lucifugus*: histology and lipid histochemistry. *Am. J. Anat., 178*, 312–325.

Gustafson, A. W., & Belt, W. D. (1981). The adrenal cortex during activity and hibernation in the male little brown bat, *Myotis lucifugus lucifugus*: annual rhythm of plasma cortisol levels. *Gen. Comp. Endocrinol., 44*, 269–278.

Gustafson, A. W., & Damassa, D. A. (1984). Perinatal and postnatal patterns of plasma sex steroid-binding protein and testosterone in relation to puberty in the male little brown bat. *Endocrinology, 115*, 2347–2354.

Gustafson, A. W., & Damassa, D. A. (1985). Annual variations in plasma sex steroid-binding protein and testosterone concentrations in the adult male little brown bat: relation to the asynchronous recrudescence of the testis and accessory reproductive organs. *Biol. Reprod., 33*, 1126–1137.

Gustafson, A. W., & Shemesh, M. (1976). Changes in plasma testosterone levels during the annual reproductive cycle of the hibernating bat, *Myotis lucifugus lucifugus* with a survey of plasma testosterone levels in adult male vertebrates. *Biol. Reprod., 15*, 9–24.

Guthrie, M. J. (1933). The reproductive cycles of some cave bats. *J. Mammal., 14*, 199–216.

Guthrie, M. J., & Smith, E. W. (1940). The pregnancy urine factors as a stimulus for ovulation in the bats and augmentation of its effect. *Am. J. Anat., 144*, 66, (Abstract).

Hagino, N. (1971). Influence of hypothyroid state on ovulation in rats. *Endocrinology, 88*, 1332–1336.

Haldar, C., & Alipreeta. (2001). Reproductive phase dependent photosensitivity of gonad and pineal gland of short-nosed fruit bat, *Cynopterus sphinx*. *Zool. Sci., 18*, 843–851.

Haldar, C., & Ghosh, M. (1993). Pineal control of reproductive cycles on the Indian short-nosed fruit bat *Cynopterus sphinx*. *Bat Res. News, 34*, 21.

Haldar, C., Yadav, R., & Alipreeta. (2006). Annual reproductive synchronization in ovary and pineal gland function of female short-nosed fruit bat, *Cynopterus sphinx*. *Comp. Biochem. Physiol., 144A*, 395–400.

Happold, D. C. D., & Happold, M. (1989). Reproduction of Angola free-tailed bats (*Tadarida condylura*) and little free-tailed bats (*Tadarida pumila*) in Malawi (Central Africa) and elsewhere in Africa. *J. Reprod. Fertil., 85*, 133–149.

Happold, D. C. D., & Happold, M. (1990a). Reproductive strategies of bats from Africa. *J. Zool. Lond., 222*, 557–583.

Happold, D. C. D., & Happold, M. (1990b). The domiciles, reproduction, social organisation and sex ratios of the banana bat *Pipistrellus nanus* (Chiroptera, Vespertilionidae) in Malawi. *Z. Saugetierk., 55*, 145–160.

Hartman, C. G., & Cuyler, W. H. (1927). Is the supposed long life of the bat spermatozoa fact or fable? *Anat. Rec., 35*, 39.

Haussler, U., Moller, E., & Schmidt, U. (1981). Zur haltung und Jugendentwicklung von Molossus molossus (Chiroptera). *Z. Saugetierk., 46*, 337–351.

Heideman, P. D. (1988). The timing of reproduction in the fruit bat *Haplonycteris fischeri* (pteropodidae): geographic variation and delayed development. *J. Zool. Lond., 215*, 577–595.

Heideman, P. D. (1989). Delayed development in Fischer's pygmy fruit bat, *Haplonycteris fischeri*, in the Philippines. *J. Reprod. Fertil., 85*, 363–382.

Heideman, P. D. (1995). Synchrony and seasonality of reproduction in tropical bats. *Symp. Zool. Soc. Lond., 67*, 151–165.

Heideman, P. D. (2000). Environmental regulation of reproduction. In E. G. Crichton, & P. H. Krutzsch (Eds.), *Reproductive Biology of Bats* (pp. 469–499). San Diego, CA: Academic Press.

Heideman, P. D., & Bronson, F. H. (1990). Photoperiod, melatonin secretion, and sexul maturation in a tropical rodent. *Biol. Reprod., 43*, 745–750.

Heideman, P. D., & Bronson, F. H. (1992). A pseudo-seasonal reproductive strategy in a tropical rodent, *Peromyscus nudipes*: correlates and causes. *J. Reprod. Fertil., 95*, 57–67.

Heideman, P. D., & Bronson, F. H. (1994). An endogenous circannual rhythm of reproduction in a tropical bat, *Anoura geoffroyi*, is not entrained by photoperiod. *Biol. Reprod., 50*, 607–614.

Heideman, P. D., & Powell, K. S. (1998). Age-specific reproductive strategies and delayed embryonic development in an old world fruit bat, *Ptenochirus jagori*. *J. Mamm., 79*, 295–311.

Heideman, P. D., Bhatnagar, K. P., Hilton, F. K., & Bronson, F. H. (1996). Melatonin rhythm and pineal structure in a tropical bat, *Anoura geoffroyi*, that does not use photoperiod to regulate seasonal reproduction. *J. Pineal Res., 20*, 90–97.

Heideman, P. D., Cummings, J. A., & Heaney, L. R. (1993). Reproductive timing and early embryonic development in an old world fruit bat, *Otopteropus cartilagonodus* (Megachiroptera). *J. Mamm., 74*, 621–630.

Heideman, P. D., Deoraj, P., & Bronson, F. H. (1992). Seasonal reproduction of a tropical bat, *Anoura geoffroyi*, in relation to photoperiod. *J. Reprod. Fertil., 96*, 765–773.

Herlant, M. (1953). Etude comparative sur l'activité génitale des chiroptères. *Ann. Soc. r. zool. Belge., 84*, 87.

Herlant, M. (1954). Influence des oestrogènes chez le Muron (*Myotis myotis*) hibernant. *Bull. Acad. r. Belg. Cl. Sci., 40*, 408–415.

Herlant, M. (1956a). Correlations hypophyso-genitales chez la female de la chauve-souris *Myotis myotis* (Brokhausen). *Arch Biol., 67*, 89–180.

Herlant, M. (1956b). Les cellules basophiles du lobe anterieur de l' hypophyse chez l' Homme. *Arch Biol., 67*, 539–553.

Herlant, M. (1963). Apport de la microscopie électroni que a l'étude du lobe antérieur de l'hypophyse. In J. Benoit, & C. Daläge (Eds.), *Cytologie de l'Adenohypophyse* (pp. 73–90). Paris, France: Éditions du Centre National de la Recherche Scientifique.

Herlant, M. (1964). The cells of the adenohypophysis and their functional significance. *Int. Rev. Cytol., 17*, 299–382.

Herlant, M. (1968). Cycle sexual chez les chiroptères des regions tempérées. In R. Canivenc (Ed.), *Cycles Génitaux Saissoniers des Mamifires Sauvages* (pp. 111–126). Paris, France: Masson.

Herlant, M., & Ectors, F. (1969). Les cellules gondotropes de l'hypophyse chez le porc. *Z. Zellforsch., 101*, 212–231.

Hoffman, L. H., & Wimsatt, W. A. (1972). Histochemical and electron microscopic observations on the sperm receptacles in the garter snake oviduct. *Am. J. Anat., 134*, 71–95.

Holmes, R. L. (1963). Gonadotrophic and thyrotrophic cells of the pituitary gland of the ferret. *J. Endocrinol., 25*, 495–504.

Hood, C. S., & Smith, J. D. (1983). Histomorphology of the female reproductive tract in phyllostomid bats. *Occas. Papers Mus. Texas Tech. Univ., 86*, 1–38.

Hosken, D. J. (1997). Sperm competition in bats. *Proc. R. Soc. Lond. B. Biol. Sci., 264*, 385–392.

Hosken, D., Blackberry, M., Stewart, T., & Stucki, A. (1998). The male reproductive cycle of three species of Australian vespertilionid bat. *J. Zool. Londn., 245*, 261–270.

Hosken, D. J., O'Shea, J. E., & Blackberry, M. A. (1996). Blood plasma concentrations of progesterone, sperm storage and sperm viability and fertility in Gould's wattled bat (*Chalinolobus gouldii*). *J. Reprod Fertil., 108*, 171–177.

Hudson, J. W., & Wang, L. C. H. (1979). Hibernation, endocrinological aspects. *Ann. Rev. Physiol., 41*, 287–303.

Hudson, R. J., & White, R. G. (1985). *Bioenergetics of wild herbivores*. Boca Raton, FL: CRC Press Inc.

Hulbert, A. J., & Augee, M. L. (1982). A comparative study of thyroid function in monotreme, marsupials, and eutherians. *Physiol. Zool., 55*, 220–228.

Hunter, A. G., Barker, L. O. S., Johnson, W. L., Fahning, M. L., & Schultz, R. H. (1971a). Antigeneity toxicity, and cross reactions of male bat *(Myotis lucifugus)* reproductive organs. *J. Reprod. Fertil., 24*, 171–177.

Hunter, A. G., Johnson, W. L., Barker, L. D. S., Fahning, M. L., & Schultz, R. H. (1971b). Bat seminal vesicle protein: its characterization and physical properties. *J. Reprod. Fertil., 24*, 179–186.

Hunter, F. M., Petrie, M., Otronen, M., Birkhead, T. R., & Moller, A. P. (1993). Why do females copulate repeatedly one male? *Trends Ecol. Evol., 8*, 21–26.

Ifuta, N. B., Gevaerts, H., & Kuhn, E. R. (1988). Thyroid hormones, testosterone and estradiol 17 β in plasma of *Epomops franqueti* (Tomes, 1860) (Chiroptera) in the rain forest of the equator. *Gen. Comp. Endocrinol., 69*, 378–380.

Ishibashi, T., & Shiino, M. (1989). Subcellular localization of prolactin in the anterior pituitary cells of the female Japanese house bat, *Pipistrellus abramus. Endocrinology, 124*, 1056–1063.

Jalabert, B., & Billard, R. (1969). Etude ultrastructural du site de conservation des spermatozoides dans. l'ovaire de *Poecilia reticulata* (Poisson Teleosteen). *Ann. De Biol. Anim, Biochem. Biophys., 9*, 273–280.

Jannini, E. A., Ulisse, S., & D'Armiento, M. (1995). Thyroid hormone and male gonadal function. *Endocr. Rev., 16*, 443–459.

Jerrett, D. P. (1979). Female reproductive patterns in non-hibernating bats. *J. Repord. Fertil., 56*, 369–378.

Jolly, S. E. (1990). The biology of the common sheath-tail bat, *Taphozous georgianus* (Chiroptera: Emballonuridae) in central Queensland. *Aust. J. Zool., 38*, 65–77.

Jolly, S. E., & Blackshaw, A. W. (1987). Prolonged epididymal sperm storage and the temporal dissociation of testicular and accessory gland activity in the common sheath tailed bat, *Taphozous georginanus* of tropical Australia. *J. Reprod. Fertil., 81*, 205–211.

Jolly, S. E., & Blackshaw, A. W. (1989). Sex steroid level and Leydig cell ultra structure of in the common sheath-tailed bat, *Taphozous georginanus* of tropical Australia. *Reprod. Fretil. Dev., 1*, 47–53.

Jones, C. (1972). Comparative ecology of the three pteropodid bats in Rio Muni, West Africa. *J. Zool. Lond., 167*, 353–370.

Kappers, J. A. (1976). The mammalian pineal gland: a survey. *Acta Neurochni., 34*, 109–149.

Karim, K. B. (1973). Occurrence of a bicornuate vagina in the Indian leaf-nosed bat, *Hipposideros fulvus fulvus* (Gray). *Curr. Sci., 42*, 62–63.

Karim, K. B., & Banerjee, S. (1985). Storage of spermatozoa in the epididymis of the tropical bat, *Rhinopoma hardwickei hardwickei* (Gray). *Anat. Rec., 95*, 211, Abstract.

Karlsson, C., Lindell, K., Svensson, E., Bergh, C., Lind, P., Billig, H., et al. (1997). Expression of functional leptin receptors in the human. ovary. *J. Clin. Endocrinol. Metab., 82*, 4144–4148.

Kawakami, S., & Winters, S. J. (1999). Regulation of luteinizing hormone secretion and subunit messenger ribonucleic acid expression by gonadal steroids in perfused pituitary cells from male monkeys and rats. *Endocrinology, 140*, 3587–3593.

Kawamoto, K. (2003). Endocrine control of the reproductive activity in hibernating bats. *Zool. Sci., 20*, 1057–1069.

Kawamoto, K., Kurahashi, S., & Hayashi, T. (1998). Changes in the gonadotropin-releasing hormone (GnRH) neuronal system during the annual reproductive cycle of the horse shoe bat, *Rhinlophus ferrumequinum. Zool Sci., 15*, 779–786.

Kawamoto, K., Tanaka, S., & Hayashi, T. (2000). Secretory activity of gonadotropin and the responsiveness of gonadotrophs to gonadotropin-releasing hormone during the annual reproductive cycle of male bats, *Rhinolophus ferrumequinum*: analysis by cell immunoblot assay. *J. Expt. Zool., 287*, 213–224.

Kayser, C. (1961). *The Physiology of Natural Hibernation.* Oxford, UK: Pergamon Press.

Kayanja, F. I., & Mutere, F. A. (1975). The ovary of the insectivorous bat *Otomops martiensseni. Anat. Anz., 137*, 166–175.

Kenagy, G. J. (1987). Energy allocation for reproduction in the golden-mantled ground squirrel. *Symp. Zool. Soc. Lond., 57*, 259–274.

Khaparde, M. S. (1976). Notes on the breeding habits of the Indian sheath-tailed bat, *Taphozous melanopogon* (Temminck). *J. Bombay Nat. Hist. Soc., 73*, 321–324.

Kikuchi, N., Andoh, K., Abe, Y., Yamada, K., Mizunuma, H., & Ibuki, Y. (2001). Inhibitory action of leptin on early follicular growth differs in immature and adult female mice. *Biol. Reprod., 65*, 66–71.

Kimura, K., & Uchida, T. A. (1983). Ultrastructural observations of delayed implantation in the Japanese long-fingered bat, *Miniopterus schreibersii fulginosus. J. Reprod. Fertil., 69*, 187–193.

Kimura, K., Takeda, A., & Uchida, T. A. (1987). Changes in progesterone concentrations in the Japanese long-fingered bat, *Miniopterus schreibersii fuliginosus. J. Reprod. Fertil., 80*, 59–63.

King, J. E., Anthony, E. L. P., Gustafson, A. W., & Damassa, D. A. (1984). Luteinizing hormone releasing hormone (LH-RH) cells and their projections in the forebrain of the bat *Myotis lucifugus lucifugus. Brain Res., 298*, 289–301.

King, J. A., Steneveld, A. A., Curlewis, J. D., Rissman, E. F., & Millar, R. P. (1994). Identification of chicken GnRH II in brains of metatherian and early-evolved eutherian species of mammals. *Regulatory Peptides, 54*, 467–477.

Kitchener, D. J. (1973). Reproduction in the common sheath-tailed bat, *Taphozous georgianus* (Thomas) (Microchiroptera: Emballonuridae), in Western Australia. *Aust. J. Zool., 21*, 375–389.

Kitchener, D. J. (1975). Reproduction in female Gould's wattled bat, *Chalinolous gouldii* (Gray) (Vespertilionidae) in western Australia. *Aust. J. Zool., 23*, 701–708.

Kitchener, D. J. (1976). Further observations on the reproduction in the common sheath-tailed bat, Taphozous georgianus 1915 in western Australia, with notes on the gular pouch. In: *Rec. West. Aust. Mus., 4*. Thomas. 335–347.

Kitchener, D. J. (1980). *Taphozous hilli* sp. nov. (Chiroptera: Emballonuridae), a new sheath-tailed bat from Western Australia and Northern Territory. *Rec. West. Aust. Mus., 8*, 161–169.

Kitchener, D. J. (1983). Bats. In R. Strahan (Ed.), *The Australian museum complete book of Australian mammals* (pp. 335–347). London, UK: Angus and Robertson Publishers.

Kitchener, D. J., & Coster, P. (1981). Reproduction in female *Chalinolobus morio* (Gray) (Vespertilionidae) in South-Western Australia. *Aust. J. Zool., 29*, 305–320.

Kitchener, D. J., & Halse, S. A. (1978). Reproduction in female *Eptesicus regulus* (Thomas) (Vespertilionidae) in South Western Australia. *Aust. J. Zool., 26*, 257–267.

Kofron, C. P. (1997). Reproduction of two species of congeneric fruit bats (*Cynopterus*) in Brunei, Borneo. *J. Zool. Lond., 243*, 485–506.

Kofron, C. P., & Chapman, A. (1994). Reproduction and sexual dimorphism of the West African fruit bat, *Epomops buettikoferi*, in Liberia. *African J. Ecology, 32*, 308–316.

Kolb, A. (1950). Beitrage sur Biologie einheimischer Fledermäuse. *Zool. Fb. Abt. Syst. Okol. Geogr. Tiera, 78*, 547–573.

Koopman, K. F., & Jones, J. K., Jr. (1970). Classification of bats. In B. H. Slaughter, & D. W. Walton (Eds.), *About bats: a chiropteran symposium* (pp. 22–28). Dallas, TX: Southern Methodist University Press.

Kopelman, P. G. (1994). Investigation of obesity. *Clin. Endocrinol. (Oxf.), 41*, 703–708.

Krishna, A. (1984). Storage of spermatozoa in the female genital tract of the Indian pigmy pipistrelle bat, *Pipistrellus mimus. Wroughton. Arch. Biol., 95*, 223–229.

Krishna, A. (1985). Reproduction in Indian pigmy pipistrelle bat, *Pipistrellus mimus* Wroughton. *J. Zool. Lond., 206*, 41–51.

Krishna, A. (1997). The relationship between spermatozoa and epithelium of the female genital tract during sperm storage in the greater yellow bats *Scotophilus heathi*: the light and electron-microscopic observations. *Proc. Nat. Sci. Council. Republic of China B, 21*, 31–36.

Krishna, A. (1999). Reproductive delays in chiropterans. In K. P. Joy, A. Krishna, & C. Haldar (Eds.), *Topics in endocrinology and reproduction* (pp. 410–421). India: Narosa Publishing House.

Krishna, A., & Abhilasha. (2000a). Proliferative activity of follicles and serum steroid concentration in *Scotophilus heathi* (vespertilionid bat) during periods of delayed ovulation. *Can. J. Zool., 78*, 1301–1308.

Krishna, A., & Abhilasha. (2000b). Mechanism of delayed ovulation in a vespertilionid bat, *Scotophilus heathi*: role of gonadotrophin, insulin, and insulin like growth factor- 1. *Physiol. Biochem. Zool., 73*, 523–529.

Krishna, A., & Bahuguna, J. (2002). Relationship between corticosterone and body weight, androstendione and insulin during the period of delayed ovulation in a vespertilionid bat, *Scotophilus heathi. Acta Biol. Hung., 53*, 279–291.

Krishna, A., & Dominic, C. J. (1978). Storage of spermatozoa in the female genital tract of the Indian vespertilionid bat, *Scotophilus heathi. J. Reprod. Fertil., 54*, 319–321.

Krishna, A., & Dominic, C. J. (1981). Reproduction in an Indian vespertilionid bat, *Scotophilus heathi* Horsefield. *Arch. Biol., 92*, 247–258.

Krishna, A., & Dominic, C. J. (1982a). Reproduction in the Indian sheath-tailed bat. *Acta Theriol., 27*, 97–106.

Krishna, A., & Dominic, C. J. (1982b). Differential rates of fetal growth in two successive pregnancies in the emballonurid bat, *Taphozous longimanus* Hardwicke. *Biol. Reprod., 27*, 351–353.

Krishna, A., & Dominic, C. J. (1982c). Observation on the Graafian follicles of the vespertilionid bat, *Scotophilus heathi* horsefield. *Proc. Zool. Soc. Calcutta, 33*, 1–7.

Krishna, A., & Dominic, C. J. (1983). Reproduction in the female short-nosed fruit bat *Cynopterus sphinx* (Vahl). *Period. Biol., 85*, 23–30.

Krishna, A., & Dominic, C. J. (1984). Reproduction in the male short-nosed bat, *Cynopterus sphinx* Vahl. *Lynx, 22*, 19–26.

Krishna, A., & Dominic, C. J. (1988). Histology and histochemical observations on the corpus luteum of an Indian vespertilionid bat, *Scotophilus heathi* Horsefield. *Zool. Anz., 220*, 8–16.

Krishna, A., & Singh, K. (1997). The relationship between testicular activity, accessory sex glands, and circulating steroid concentration during the reproductive cycle in a male Indian vespertilionid bat, *Scotophilus heathi. Can. J. Zool., 75*, 1042–1050.

Krishna, A., & Singh, K. (1998). Changes in thyroid gland during reproductive cycle of the male vespertilionid bat, *Scotophilus heathi. Rev. Brasil. Biol., 58*, 489–498.

Krishna, A., & Singh, K. (1999). Asynchrony of the reproductive organs of the male vespertilionid bat, *Scotophilus heathi*: Role of gonadotrophins. *Acta Theriol., 44*, 123–131.

Krishna, A., & Singh, U. P. (1992). Morphometric changes in the ovaries of Indian vespertilionid bat, *Scotophilus heathi*, with reference to delayed ovulation. *Europ. Arch. Biol., 103*, 257–264.

Krishna, A., Singh, K., Doval, J., & Chanda, D. (1998). Changes in circulating insulin and corticosterone concentrations during different reproductive phases and their relationships to body weight and androstenedione concentration of male *Scotophilus heathi. J. Exp. Zool., 281*, 201–206.

Kronfeld-Schor, N., Richardson, C., Silvia, B. A., Kunz, T. H., & Widmaier, E. P. (2000). Dissociation of leptin secretion and adiposity during prehibernatory fattening in little brown bats. *Am. J. Physiol. Regul. Integr. Comp. Physiol., 279*, R1277–R1281.

Krutzsch, P. H. (1979). Female reproductive patterns in non-hibernating bats. *J. Reprod. Fertil., 56*, 333–344.

Krutzsch, P. H. (2000). Reproductive biology of bats. In E. G. Crichton, & P. H. Krutzsch (Eds.), *Anatomy, physiology and cyclicity of the male reproductive tract* (pp. 91–155). London, UK: Academic Press.

Krutzsch, P. H., & Crichton, E. G. (1985). Observation on the reproductive cycle of female *Molossus fortis* (Chiroptera: Molossidae) in Puerto Rico. *J. Zool. Lond., 207*, 137–150.

Krutzsch, P. H., & Crichton, E. G. (1987). The reproductive biology of the female little mastif bat, *Mormopterus planiceps* (Chiroptera: Molossidae) in southeast Australia. *Am. J. Anat., 178*, 369–386.

Krutzsch, P. H., & Hess, M. (1961). Studies on the ascorbic acid content of the adrenal of the bat (*Myotis lucifugus*). *Endocrinology, 69*, 664–666.

Krutzsch, P. H., & Crichton, E. G. (1990a). Observations on the reproductive anatomy of the male *Molossus fortis* (Chiroptera: Molossidae) with comments on the chronology of the reproductive events. *Mammalia, 54*, 287–296.

Krutzsch, P. H., Crichton, E. G., & Nagle, R. B. (1982). Studies on prolonged spermatozoa survival in Chiroptera: a morphological examination of storage and clearance of intra-uterine and caudal epididymal spermatozoa in bats, *Myotis lucifugus* and *Myotis velifer. Am. J. Anat., 165*, 421–434.

Krutzsch, P. H., Watson, R. H., & Lox, C. D. (1976). Reproductive biology of the male leaf-nosed bat, *Macrotus waterhousii*, in south-western United States. *Anat. Rec., 184*, 611–636.

Kumamoto, K., Ebara, S., Matsuura, T., & Kawata, M. (1992). Distribution of oxytocin and vasopressin neurons in the diencephalon of the Japanese horseshoe bat, *Rhinolophus ferrumequinum*: an immunohistochemical study. *Acta Anat., 144*, 80–92.

Kunz, T. H., Bicer, E., Hood, W. R., Axtell, M. J., Harrington, W. R., Silvia, B. A., et al. (1999). Plasma leptin decreases during lactation in insectivorous bats. *J. Comp. Physiol. [B], 169*, 61–66.

Kunz, T. H., Wrazen, J. A., & Burnett, C. D. (1998). Changes in body mass and body composition in pre-hibernating little brown bats (*Myotis lucifugus*). *Ecoscience, 5*, 8–17.

Kwiecinski, G. G., Damassa, D. A., & Gustafson, A. W. (1986). Control of the sex steroid-binding protein (SBP) in the male little brown bat: Relationship of plasma thyroxine levels to the induction of plasma SBP in immature males. *J. Endocrinol., 110*, 271–278.

Li, J., Stefaneanu, L., Kovacs, K., Horvath, E., & Smyth, H. (1993). Growth hormone (GH) and prolactin (PRL) gene expression and immunoreactivity in GH- and PRL-producing human pituitary adenomas. *Virchows Archive A. Pathol. Anat., 422*, 193–201.

Lindeque, M., Skinner., J. D., & Miller, R. P. (1986). Adrenal and gonadal contribution to circulating androgen in spotted Hyaenas (*Crocuta crocuta*) as revealed by LHRH, hCG and ACTH stimulation. *J. Reprod. Dev., 78*, 211–217.

Loh, H. S. F., & Gemmel, R. T. (1980). Changes in the fine structure of the testicular Leydig cells of the seasonally-breeding bat, *Myotis adversus. Cell Tissue Res., 210*, 339–347.

Lopez-Forment, W. (1981). Algunos aspectos ecologicos del murcielago *Balantiopteryx plicata plicata* Peters, 1867 (Chiroptera: Emballonuridae) in Mexico. *Ann. Inst. Biol. Univ. Nat. Autonimo Mexico, series Zoologic., 50*, 673–699.

Madhavan, A. (1981). Breeding habits and associated phenomena in some Indian bats. *J. Bombay Nat. Hist. Soc., 77*(2), 227–237.

Maffei, M., Halaas, J., Ravussin, E., Pratley, R. E., Lee, G. H., Zhang, Y., et al. (1995). Leptin levels in human and rodent: measurement of plasma leptin and ob RNA in obese and weight-reduced subjects. *Nature Medicine, 1*, 1155–1161.

Magnus, T. H., & Henderson, N. E. (1988). Thyroid hormone resistance in hibernating ground squirrels, *Spermophilus richardsoni*. *Gen. Comp. Endocrinol., 69*, 352–360.

Malpaux, B., Migaud, M., Tricoire, H., & Chemineau, P. (2001). Biology of mammalian photoperiodism and the critical role of the pineal gland and melatonin. *J. Biol. Rhythms, 16*, 336–347.

Mandal, H., Ghosh, P. K., & Biswas, N. M. (1990). Effect of dihydrotestosterone on serum concentrations of alpha 2u-globulin and on spermatogenesis in melatonin-treated rats. *J. Endocrinol., 126*, 431–435.

Marques, S. A. (1986). Activity cycle, feeding and reproduction of *Molossus ater* (Chiroptera: Molossidae) in Brazil. *Boletim do Museu paraense Emilio Goeldi Zoology, 2*, 159–179.

Marshall, A. J. (1953). The unilateral endometrial reaction in the giant fruit bat (*Pteropus giganteus* Brunnich). *J. Endocrinol., 9*, 42–44.

Marshall, A. J., & Corbet, P. S. (1959). The breeding biology of equatorial vertebrates. Reproduction of the bat *Chaerephon hindei* Thomas, at latitudes 0°26' N. *Proc. Zool. Soc. Lond., 132*, 607–616.

Marshall, F. H. A. (1942). Exteroceptive factors in sexual periodicity. *Biol. Rev., 17*, 68–89.

Martin, L., & Bernard, R. T. F. (2000). Endocrine regulation of reproduction in bats: The role of circulating gonadal hormones. In E. G. Crichton, & P. H. Krutzsch (Eds.), *Reproductive Biology of Bats* (pp. 27–64). San Diego, CA: Academic Press.

Martin, J. E., & Sattler, C. (1979). Developmental loss of the acute inhibitory effect of melatonin on the *in vitro* pituitary luteinizing hormone and follicle-stimulating hormone responses to luteinizing hormone-releasing hormone. *Endocrinology, 105*, 1007–1012.

Martin, L., Kennedy, J. H., Little, L., Luckhoff, H. C., O'Brien, G. M., Pow, C. S. T., et al. (1995). The reproductive biology of Australian flying-foxes (genus *Pteropus*). In P. A. Racey, & S. M. Swift (Eds.), *Ecology, Evolution, and Behavior of Bats* (pp. 167–184). Oxford, UK: Clarendon Press.

Maruo, T., Katayama, K., Barnea, E. R., & Mochizuki, M. (1992). A role for thyroid hormone in the induction of ovulation and corpus luteum function. *Horm. Res., 37*(1), 12–18.

Masuzaki, H., Ogawa, Y., Sagawa, N., Hosoda, K., Matsumoto, T., Mise, H., et al. (1997). Non-adipose tissue production of leptin: leptin as a novel placenta derived hormone in humans. *Nature Medicine, 3*, 1029–1033.

Matthews, L. H. (1937). The female sex cycle in the British horse-shoe bats *Rhinolophus ferrumequinum insulanus*. *Trans. Zool. Soc. Lond., 23*, 224–225.

Matthews, L. H. (1941). Notes on the genitalia and reproduction of some African bats. *Proc. Zool. Soc. Lond., 111*(B), 289–346.

McCracken, J. A., & Baird, D. T. (1969). The study of ovarian function by means of transplantation of the ovary in the ewe. In K. W. Mc Kerns (Ed.), *The Gonads* (pp. 175–209). Amsterdam, The Netherlands: North-Holland, Publishing Company.

McGuckin, M. A. (1988). *Seasonal changes in the reproductive physiology of male flying foxes (*Pteropus *spp.) in southwest Queensland*. Ph.D. dissertation. Brisbane, Australia: University of Queensland.

McGuckin, M. A., & Blackshaw, A. W. (1985). Reproductive seasonality in the male fruit bats, genus. *Pteropus. Macroderma., 1*, 59–60.

McGuckin, M. A., & Blackshaw, A. W. (1987a). The cycle of the seminiferous epithelium in the grey-headed fruit bat, *Pteropus poliocephalus*. *Aust. J. Biol. Sci., 40*, 203–210.

McGuckin, M. A., & Blackshaw, A. W. (1987b). Seasonal changes in spermatogenesis (including germ cell degeneration) and plasma testosterone concentrations in the grey-headed fruit bat, *Pteropus poliocephalus*. *Aust. J. Biol. Sci., 40*, 211–220.

McGuckin, M. A., & Blackshaw, A. W. (1988). Different responses to the same environmental cues regulate long and short day breeding seasons in two species of Pteropus (Megachiroptera) in south-east Queensland. *Proc. Int. Cong. Anim. Reprod. Artif. Insemin., 11*, 412.

McGuckin, M. A., & Blackshaw, A. W. (1991). Seasonal changes in testicular size, plasma testosterone concentration and body weight in captive flying-foxes (*Pteropus poliocephalus and P. scapulatus*). *J. Reprod. Fertil., 92*, 339–346.

McGuckin, M. A., & Blackshaw, A. W. (1992). The effects of photoperiod in the reproductive physiology of male flying foxes, *Pteropus poliocephalus*. *Reprod. Fertil. Devlop., 4*, 43–53.

McMillen, I. C., Edwards, L. J., Duffield, J., & Muhlhausler, B. S. (2006). Regulation of leptin synthesis and secretion before birth: implication for the early programming of adult obesity. *Reproduction, 131*, 415–427.

McOwat, T. P., & Andrews, P. T. (1995). The influence of climate on the growth rate of *Rhinolophus ferrumequinum* in west Wales. *Myotis, 32–33*, 69–79.

McWilliam, A. N. (1987a). The reproductive and social biology of *Coleura afra* in a seasonal environment. In M. B. Fenton, P. A. Racey, & J. M. V. Rayner (Eds.), *Recent Advances in the Study of Bats* (pp. 224–350). London, UK: Cambridge University Press.

McWilliam, A. N. (1987b). Polyoestry and postpartum oestrus in *Tadarida (Chaerephon) pumila* (Chiroptera: Molossidae) in northern Ghana, West Africa. *J. Zool. Lond., 213*, 735–739.

McWilliam, A. N. (1988). Social organization of the bat *Tadarida (Chaerephon) pumila* (Chiroptera: Molossidae) in Ghana, West Africa. *Ethology, 77*, 115–124.

Mead, R. A. (1971). Effects of light and blinding upon delayed implantation in the spotted skunk. *Biol. Reprod., 5*, 214–220.

Mead, R. A. (1993). Embryonic diapause in vertebrates. *J. Exp. Zool., 266*, 629–641.

Medway, L. (1971). Observation on social and reproductive biology of the bent-winged bat *Miniopterus australis* in northem Bomeo. *J. Zool. Linn. Soc. Lond., 165*, 261–273.

Medway, L. (1972). Reproductive cycle of the flat-headed bats *Tylonycteris pachypus* and *T. robusta* in a humid equatorial environment. *J. Zool. Linn. Soc. Lond., 51*, 33–61.

Meenakumari, K. J. (2006). *Investigation on Factors Responsible for Delayed Embryonic Development in the Indian Short-nosed Fruit Bat. Cynopterus sphinx*. Ph.D. thesis, Varanasi, India: Banaras Hindu University.

Meenakumari, K. J., & Krishna, A. (2005). Delayed embryonic development in the Indian short-nosed fruit bat, *Cynopterus sphinx*. *Zoology, 108*, 131–140.

Meenakumari, K. J., Banerjee, A., & Krishna, A. (2009). Luteal cell steroidogenesis in relation to delayed embryonic development in the Indian short-nosed fruit bat, Cynopterus sphinx. *Zoology (Jena), 112*, 151–159.

Merchant, F. W. (1974). Prolactin and luteinizing hormone cells of pregnant and lactating rats as studied by immunohistochemistry and radioimmunoassay. *Am. J. Anat., 139*, 245–268.

Mikami, S., Chiba, S., Hojo, H., Taniguchi, K., Kubokawa, K., & Ishii, S. (1988a). Immunocytochemical studies on the pituitary pars distalis of the Japanese long-fingered bat, *Miniopterus schreibersii fuliginosus*. *Cell Tissue Res., 251*, 291–299.

Mikami, S., Chiba, S., Taniguchi, K., Kubokawa, K., & Ishii, S. (1988b). Immunocytochemical localization of neuropeptides in the hypothalamus of the Japanese long-fingered bat, *Miniopterus schreibersii fuliginosus*. *Cell Tissue Res., 254*, 49–57.

Mitchell, G. C. (1965). *A Natural History Study of the Funnel-eared Bat, Natalus stramineus*. M. Sc. Thesis. Tucson: University of Arizona.

Mokkapati, S., & Dominic, C. J. (1976). Sites of production of fructose and citric acid in the accessory reproductive glands of three species of male chiropterans. *Biol. Reprod., 14*, 627–629.

Mokkapati, S., & Dominic, C. J. (1977). Accessory reproductive glands of the male Indian mongoose, *Herpestes auropunctatus* Hodgson. *J. Mamm., 58*, 85–87.

Monesi, V. (1972). Spermatogenesis and the spermatozoa. In C. R. Austin, & R. V. Short (Eds.), *Reproduction in Mammals I: Germ cells and Fertilization* (pp. 46–84). Cambridge, UK: Cambridge University Press.

Mori, T., & Uchida, T. A. (1980). Sperm storage in the reproductive tract of the female Japanese long-fingered bat, *Miniopterus schreibersii fuliginosus*. *J. Reprod. Fertil., 58*, 429–433.

Mori, T., Oh, Y. K., & Uchida, T. A. (1982). Sperm storage in the oviduct of Japanese greater horseshoe bat *Rhinolophus ferrum equinum nippon*. *J. Fac. Agr. Kyushu Univ., 27*, 47–53.

Morigaki, T., Kurohmaru, M., Kanai, Y., Mukohyama, M., Hondo, E., Yamada, J., et al. (2001). Cycle of the seminiferous epithelium in the Java fruit bat (*Pteropus vampyrus*) and the Japanese lesser horseshoe bat (*Rhinolophus cornutus*). *J. Vet. Med. Sci., 63*, 773–779.

Mrosovsky, N. (1985). Cyclical obesity in hibernators: The search for the adjustable regulator. In J. Hirsch, & B. T. Van Itallie (Eds.), *Recent Advances in Obesity, Vol. 4* (pp. 45–56). London, UK: Libbey.

Morris, D. L., & Rui, L. (2009). Recent advances in understanding leptin signaling and leptin resistance. *Am. J. Physiol. Endocrinol. Metab., 297*, E1247–E1259.

Muniz, E., Jimenez, L., Gragera, R., Fernandez, A., & Rua, C. (1991). Ultrastructural changes in the gonadotrophic and prolactin cells of *Myotis myotis* under experimental conditions. *Funct. Develop. Morphol.*, 15–18, I.

Mustonen, A. M., Nieminen, P., Asikainen, J., Saarela, S., Kukkonen, J. V., & Hyvarinen, H. (2004). Continuous melatonin treatment and fasting in the raccoon dog, Nyctereutes procyonoides—vernal body weight regulation and reproduction. *Zool. Sci., 21*(2), 163–172.

Mutere, F. A. (1965). Delayed implantation in an equatorial fruit bat. *Nature, Lond., 207*, 780.

Mutere, F. A. (1968a). The breeding biology of *Rousettus aegyptiacus* living at 0^0 22S. *Acta Trop., 25*, 97–108.

Mutere, F. A. (1968b). Breeding cycles in tropical bats in Uganda. *J. Appl. Ecol., 5*, 8–9.

Mutere, F. A. (1973). Reproduction in two species of equatorial free-tailed bats (Molossidae). *East Afr. Wildlife. J., 11*, 271–280.

Myers, P. (1977). Patterns of reproduction of four species of vespertilionid bats in Paraguay. *Univ. Calif. Pubs. Zool., 107*, 1–41.

Nakano, O. (1928). Uber die Verteilung des Glykogens bei den zyklischen Veranderungen in den Geschlechtsorganen der Fledermaus, und uber die Nahrungsaufnahme der Spermien in den weiblichen Geschlechtswege. *Folia Anat. Japan, 6*, 777–828.

Neubaum, D. M., & Wolfner, M. F. (1999). Wise, winsome, or weird? Mechanisms of sperm storage in female animals. *Curr. Top. Dev. Biol., 41*, 67–97.

Ng, T. B., & Lo, I. L. (1988). Inhibitory actions of pineal indoles on steroidogenesis in isolated rat Leydig cells. *J. Pineal. Res., 5*, 229–243.

Niklowitz, P., Khan, S., Bergman, M., Hoffmann, K., & Nieschlag, E. (1989). Differential effects of follicle-stimulating hormone and luteinizing hormone on Leydig cell function and restoration of spermatogenesis in hypophysectomized and photoinhibited Djungarian hamsters *(Phodopus sungorus)*. *Biol. Reprod., 41*, 871–880.

Nishida, S., Sato, R., Murai, I., & Nakagawa, S. (2003). Effect of pinealectomy on plasma levels of insulin and leptin and on hepatic lipids in type 2 diabetic rats. *J. Pineal Res., 35*, 251–256.

O'Brien, G. M., Curlewis, J. D., & Martin, L. (1991). Unusual reproductive photoresponsiveness of male grey headed flying foxes (*Pteropus poliocephalus*). *Proc. Austral. Soc. Reprod. Biol., 23*, 160.

O'Brien, G. M., Curlewis, J. D., & Martin, L. (1993). Effect of photoperiod on the annual cycle of testes growth in tropical mammal, the little red flying fox, *Pteropus scapulatus*. *J. Reprod. Fertil., 98*, 121–127.

O'Brien, G. M., Curlewis, J. D., & Martin, L. (1996). A heterologous assay for measuring prolactin in pituitary extracts and plasma from Australian flying foxes (genus *Pteropus*). *Gen. Comp. Endocrinol., 104*, 304–311.

O'Farrell, M. J., & Studier, E. H. (1975). Population structure and emergence activity patterns in *Myotis thysanodes* and *M. lucifugus* (Chiroptera: Vespertilionidae) in North Eastern New Mexico. *Am. Midl. Nat., 93*, 368–376.

Oelschläger, H. A., & Northcutt, R. G. (1992). Immunocytochemical localization of luteinizing hormone-releasing hormone (LHRH) in the nervus terminalis and brain of the big brown bat, *Eptesicus fuscus*. *J. Comp. Neurol., 315*, 344–363.

Oh, Y. K., Mori, T., & Uchida, T. A. (1985). Prolonged survival of the Graafian follicle and fertilization in the Japanese greater horse-shoe bat *Rhinolophus ferrumequinum nippon*. *J. Reprod. Fertil., 73*, 121–126.

Ohata, M. (1979). Electron microscope study on the bat testicular interstitial cell with special reference to the cytoplasmic crystalloid. *Arch. Histol. Jpn., 42*, 103–118.

Okia, N. O. (1974). The breeding pattern of the eastern epauletted bat, *Epomophorus anurus Heuglin*, in Uganda. *J. Reprod. Fertil., 37*, 27–31.

Oxberry, B. A. (1979). Female reproductive patterns in hibernating bats. *J. Reprod. Fertil., 56*, 359–367.

Pal, A. N. (1977). *Studies on the Male Accessory Reproductive Organs in Some Indian Bats*. PhD dissertation, Nagpur, India: Nagpur University.

Pal, A. N. (1984). Histophysiology of the male accessory glands of the Indian sheath-tailed bat, *Taphozous longimanus* (Hardwicke). *J. Curr. Biosci., 1*, 122–126.

Parker, C. R., Neaves, W. B., & Porter, J. C. (1980). Regional and subcellular localization of LHRH in the adult human brain. *Brain Res. Bull., 5*, 307–313.

Patil, D. R. (1968). *Reproduction in the Indian Leaf-nosed Bat* Hipposideros fulvus fulvus *(Gray)*. Ph.D. Thesis, India: Nagpur University.

Patil, D. R. (1974). Comparison of pituitary gland cytology in three species of leaf-nosed bats (Hipposideridae). *J. Anat., 118*, 33–51.

Pearson, O. P., & Enders, R. K. (1944). Duration of pregnancy in certain mustelids. *J. Expt. Zool., 95*, 21–35.

Pearson, O. P., Koford, M. R., & Pearson, A. K. (1952). Reproduction of the lump-nosed bat (*Corynorhinus rafinesquei)* in California. *J. Mammal., 33*, 273–320.

Peyre, A., & Herlant, M. (1963a). Correlations hypohysogenitales chez, la femelle du Minioptere, (*Miniopterus schreibersii* b.). *Gen. Comp. Endocrinol., 3*, 726–727.

Peyre, A., & Herlant, M. (1963b). Ova-implantation differee et correlations hypophyso–genitales chez la femelle du Minioptere, (*Miniopterus schreibersii* K). *Cr. Hebd. Seanc. Acad Sci. Paris D, 257*, 524–526.

Peyre, A., & Herlant, M. (1967). Ovo-implantation différée et déterminisme hormonal chez le Miniop tère, *Miniopterus schreibersii* K. (Chiroptère). *Cr. Séanc. Soc. Biol., 161*, 1779–1782.

Pickard, G. E., & Silverman, A. J. (1979). Effect of photoperiod on hypothalamic leutinizing hormone releasing hormone in the male hamster. *J. Endocrinol., 83*, 421–428.

Planel, H. G., Guilhem, A., & Soleilhavoup, J. P. (1961). Le cycle annuel du cortex surrénal d'un semihibernant *Miniopterus schreibersii. C. r. Ass. Anat., 47*, 620–633.

Pow, C. S. T., & Martin, L. (1994). The ovarian-uterine vasculature in relation to unilateral endometrial growth in flying foxes (genus Pteropus, suborder Megachiroptera, order Vhiroptera). *J. Reprod. Fertil., 101*, 247–255.

Pow, C. S. T., & Martin, L. (1995). Ovarian and uterine lymphatic drainage in Australian flying-foxes (genus *Pteropus*, suborder Megachiroptera). *Cell. Tiss. Res., 280*, 371–381.

Racey, P. A. (1969). Diagnosis of pregnancy and experimental extension of gestation in the pipistrelle bat, *Pipistrellus pipistrellus. J. Reprod. Fertil., 19*, 465–474.

Racey, P. A. (1972). Viability of bat spermatozoa after prolonged storage in the epididymis. *J. Reprod. Fertil., 28*, 309–311.

Racey, P. A. (1973). Environmental factors affecting the length of gestation in heterothermic bats. *J. Reprod. Fertil. (Suppl.), 19*, 175–189.

Racey, P. A. (1974). The reproductive cycle in male noctule bats, *Nyctalus noctula. J. Reprod. Fertil., 41*, 169–182.

Racey, P. A. (1975). The prolonged survival of spermatozoa in bats. In The Biology of the Male Gamte (J.G. Duckett and P.A. Racey, Eds). *Biol. J. Linn. Soc.* 7 (Supl. 1), 385–416.

Racey, P. A. (1976). Induction of ovulation in the pipistrelle bat, *Pipistrellus pipistrellus. J. Reprod. Fert., 46*, 481–483.

Racey, P. A. (1978). The effect of photoperiod on the initiation of spermatogenesis in pipistrelle bats, *Pipistrellus pipistrellus*. In R. J. Olembo, J. B. Castclino, & F. A. Mutere (Eds.), *Proceedings of the Fourth International Bat Research Conference* (pp. 255–258). Kenya: Kenya Literature Bureau.

Racey, P. A. (1979). The prolonged storage and survival of spermatozoa in Chiroptera. *J. Reprod. Fertil., 56*, 391–402.

Racey, P. A. (1982). Ecology of Bat reproduction. In T. H. Kunz (Ed.), *Ecology of bats* (pp. 57–104). New York and London: Plenum Press.

Racey, P. A. (1981). Enviromental factors affecting the length of gestation in mammals. In D. Gilmore, & B. Cook (Eds.), *Envir-omental factor in mammalian reproduction* (pp. 99–123). Londan: MacMillan.

Racey, P. A., & Swift, S. M. (1981). Variations in gestation length in a colony of pipistrelle bats (*Pipistrellus pipistrellus*) from year to year. *J. Reprod. Fertil., 61*, 123–129.

Racey, P. A., & Tam, H. W. (1974). Reproduction in male *Pipistrellus pipistrellus* (Mammalia: Chiroptera). *J. Zool. Lond., 172*, 101–122.

Rajalakshmi, M., & Prasad, M. R. N. (1970). Sites of formation of fructose, citric acid and sialic acid in the accessory glands of the giant fruit bat, *Pteropus giganteus giganteus* (Brünnich). *J. Endocrinol., 46*, 413–416.

Ramakrishna, P. A. (1951). Studies on the reproduction in bats. I Some aspect of reproduction in the oriental vampires, *Megaderma lyra lyra* (Geoff) and *Megaderma spasma* (Linn). *J. Mysore Univ., 11B*, 107–118.

Ramakrishna, P. A., & Rao, K. V. B. (1977). Reproductive adaptations in the Indian Rhinolophid bat, *Rhinolophus rouxi* (Temminck). *Curr. Sci., 46*, 270–271.

Ramaswami, L. S., & Anand Kumar, T. C. (1966). Effect of exogenous hormones on the reproductive structures of the female bat *Rhinolopoma* during the non breeding season. *Acta. Anat., 63*, 101–123.

Ramaswamy, K. R. (1961). Studies on the sex cycle of the Indian vampire bat, *Megaderma (Lyroderma) lyra lyra* (Geoffroy). Part I. Breeding habits. *Proc. Nat. Inst. Sci. India, Sect., 27*, 287–307.

Ransome, R. D., & McOwat, T. P. (1994). British timing and population changes in greater horseshoe bat colonies (*Rhinolophus ferrumequinum)* are synchronized by climatic temperature. *Zool. J. Linn. Soc., 112*, 337–351.

Ranta, T. (1975). Effect of dexamethasome in the rat: dose and time relations. *Endocrinology, 96*, 1566–1570.

Rasweiler, J. J., IV (1970). *The laboratory biology of the long-tongued bat. Maintenance procedures, estivation, the menstrual cycle, histophysiology of the oviduct and intramural implantation*. In: Glossophaga soricina. Ithaca, NY: Ph.D. thesis, Cornell Univ.

Rasweiler, J. J. (1972). Reproduction in the long-tongued bat, *Glossophaga sericina*. 1. Preimplantation development and histology of the oviduct. *J. Reprod. Fertil., 31*, 249–262.

Rasweiler, J. J., IV (1978). Unilateral oviductal and uterine reactions in the little bulldog bat, *Noctilio albiventris. Am. J. Anat., 150*, 269–300.

Rasweiler, J. J. (1982). The contribution of observations on early pregnancy in the little sac-winged bat, *Peropteryx kappleri*, to an understanding of the evolution of reproductive mechanisms in monovular bats. *Biol. Reprod., 27*(3), 681–702.

Rasweiler, J. J., IV (1988). Ovarian function in the captive black mastiff bat, *Molossus ater. J. Reprod. Fertil., 82*, 97–111.

Rasweiler, J. J., IV (1991). Spontaneous decidual reactions and menstruation in the black mastiff bat, *Molossus ater. Am. J. Anat., 191*, 1–22.

Rasweiler, J. J., IV, & Badwaik, N. K. (2000). Anatomy and physiology of the reproductive tract. In E. G. Crichton, & P. H. Krutzsch (Eds.), *Reproductive Biology of Bats* (pp. 157–220). London, UK: Academic Press.

Rasweiler, J. J., IV, & De Bonilla, H. (1992). Menstruation in short tailed fruit bats (*Carollia* spp.). *J. Reprod. Fertil., 95*, 231–248.

Reeder, D. M., Kosteczko, N. S., Kunz, T. H., & Widmaier, E. P. (2004). Changes in baseline and stress-induced glucocorticoid levels during the active period in free-ranging male and female little brown myotis, *Myotis lucifugus* (Chrioptera: Vespertilionidae). *Gen. Comp. Endocrinol., 136*, 260–269.

Reppert, S., Weaver, D. R., Ebisawa, T., Mahle, C. A., & Kolakowski, L. F. (1996). Cloning of a melatonin-related receptor from human pituitary. *FEBS Lett., 386*, 219–224.

Rice, D. W. (1957). Life history and ecology of *Myotis austroriparius* in Florida. *J. Mammal., 38*, 18–31.

Richardson, B. A. (1979). The anterior pituitary and.reproduction in bats. *J. Reprod. Fertil., 56*, 379–389.

Richardson, B. A. (1981). Localization of gonadotrophic hormones in the pituitary gland of the California leaf-nosed bat (*Macrotus californicus*). *Cell Tissue Res., 220*(1), 15–23.

Richardson, E. G. (1977). The biology and evolution of the reproductive cycle of *Miniopterus schreibersii* and *M. australis* (Chiroptera: Vespertilionidae). *J. Zool. Lond., 183*, 353–375.

Robinson, J. E. (1987). Photoperiodic and steroidal regulation of the luteinizing hormone pulse generator in ewe. In W. F. Crowley, Jr., & J. G. Hoffer (Eds.), *The Episodic Secretion of Hormone* (pp. 159–167). New York, NY: John Wiley.

Rollinat, R., & Trouessart, E. (1895). Sur la-reproductive des *Chauves sourris. Mem. Sac. Zool., 9*, 214–240.

Rollinat, R., & Trouessart, E. (1897). Sur la reproduction des *Chauves souris* Les Rhinolophes. *Mem. Soc. Zoo., 10*, 114–138.

Roy, V. K., & Krishna, A. (2010). Evidence of androgen-dependent sperm storage in female reproductive tract of *Scotophilus heathi. Gen. Comp. Endocrinol., 165*, 120–126.

Ruiz-Cortes, Z. T., Men, T., Palin, M. F., Downey, B. R., Lacroix, D. A., & Murphy, B. D. (2000). Porcine leptin receptor: molecular structure and expression in the ovary. *Mol. Reprod. Develop., 56*, 465–474.

Sadler, W. W., & Tyler, W. S. (1960a). Thyroidal activity in hibernating Chiroptera. I. Uptake of ^{131}I. *Acta Endocrinol., 34*, 586–596.

Sadler, W. W., & Tyler, W. S. (1960b). Thyroidal activity in hibernating chiroptera. II. Synthesis of Radio-iodinated amino acids. *Acta Endocrinol., 34*, 597–604.

Sahastrabudhe, J. (1971). Abnormality in the breeding behaviour of the Indian fruit bat, *Pteropus giganteus giganteus* (Brunnich). *J. Bombay Nat. Hist. Soc., 67*, 550–551.

Saidapur, S. K., & Patil, S. B. (1992). Kinetics of spermatogenesis in megachiropteran bat, *Rousettus leschenaulti* (Desmarset): seminiferous epithelial cycle, frequency of stages, spermatogonial renewal and germ cell degeneration. *Indian J. Exp. Biol., 30*(11), 1037–1044.

Sahu, A. (2004). Leptin signaling in the hypothalamus: emphasis on energy homeostasis and leptin resistance. *Front. Neuroendocrinol., 24*, 225–253.

Schlatt, S., De Geyter, M., Kliesch, S., Nieschlag, E., & Bergman, M. (1995). Spontaneous recrudescence of spermatogenesis in the photoinhibited male Djungariam hamster, *Phodopus sungorus. Biol. Reprod., 53*, 1169–1177.

Sempere, A. (1977). Plasma progesterone levels in the roe deer, *Capreolus capreolus. J. Reprod. Fertil., 50*, 365–366.

Shi, Z. D., & Barrell, G. K. (1992). Requirement of thyroid function for the expression of seasonal reproductive and related changes in red deer (*Cervus elaphus*) stags. *J. Reprod. Fertil., 94*, 251–259.

Silverman, A. J., Antunes, J. L., Abrams, G. M., Nilaver, G., Thau, R., Robinson, J. A., et al. (1982). The luteinizing hormone releasing hormone pathways in rhesus (*Macaca mulatta*) and pig-tailed monkeys (*Macaca nemestrina*): new observations on thick unembedded sections. *J. Comp. Neurol., 211*, 309–317.

Simoni, M., Gromoll, J., & Nieschlag, E. (1997). The follicle stimulating hormone receptor: biochemistry, molecular biology, physiology and pathophysiology. *Endocrion. Rev., 18*, 739–773.

Singh, K., & Krishna, A. (1995). Inhibitory effects of melatonin on testosterone but not on androstenedione production during winter in the vespertilionid bat, *Scotophilus heathi. J. Pineal. Res., 19*, 127–132.

Singh, U. P., & Krishna, A. (1996a). Glycogen accumulation in the ovarian follicles of Indian vespertilionid bat, *Scotophilus heathi*, during the period of delayed ovulation. *Zool. Sci., 13*, 893–897.

Singh, U. P., & Krishna, A. (1996b). Immunocytochemical studies on the pituitary pars distalis of the tropical vespertilionid bat, *Scotophilus heathi*, in reference to ovarian cycle. *Acta Anat., 155*, 104–112.

Singh, U. P., & Krishna, A. (1992). Effects of hCG and PMSG on responsiveness of ovary during period of delayed ovulation in an Indian vespertilionid bat, *Scotophilus heathi. Indian J. Exp. Biol., 30*, 670–675.

Singh, U. P., & Krishna, A. (1994a). Identification, localization and distribution of different cell types in adenohypophysis of female vespertilionid bat, *Scotophilus heathi*: a combined histochemical, immunocytochemical and electron microscopic study. *Proceedings of the Indian National Science Academy, B60*, 115–127.

Singh, U. P., & Krishna, A. (1994b). Seasonal changes in the cytochemical and ultrastructural features of the ovarian interstitial tissue in a vespertilionid bat, *Scotophilus heathi. J. Reprod Biol. Comp. Endocrinol., 6*, 33–46.

Singh, U. P., & Krishna, A. (1997). A mini review of the female chiropteran reproduction. *J. Endocrinol. Reprod., 1*(1), 1–19.

Singh, U. P., & Krishna, A. (2000). Seasonal changes in circulating testosterone and androstenedione concentration and their correlation with the anomalous reproductive pattern in the male Indian sheath-tailed bat, *Taphozous longimanus. J. Expt. Zool., 287*, 54–61.

Singh, U. P., & Krishna, A. (2002). Seasonal changes in circulating steroid concentration and their correlation with the ovarian activity in the female Indian sheath- tailed bat, *Taphozous longimanus. J. Expt. Zool., 292*, 384–392.

Singh, U. P., Krishna, A., & Bhatnagar, K. P. (2002). Seasonal changes in thyroid activity in the female sheath tailed bat, *Taphozous longimanus* (Chiroptera: emballonuridae). *Acta Biol. Hung., 53*, 267–278.

Singh, U. P., Krishna, A., & Bhatnagar, K. P. (2007). Seasonal changes in serum leptin concentrations: relationship to reproduction in the male sheath-tailed bat, *Taphozous longimanus. Acta Theriol., 52*, 129–140.

Singwi, M. S., & Lall, S. B. (1983). Spermatogenesis in non-scrotal bat, *Rhinopoma kinneari* wroughton (Microchiroptera: Mammalia). *Acta. Anat., 116*, 136–145.

Sluiter, J. W., & Bels, L. (1951). Follicular growth and spontaneous ovulation in captive bats during the hibernation period. *Proc. K. ned. Akad. Wet. C, 54*, 585–593.

Smith, E. W. (1951). Seasonal responses of follicles in the ovaries of the bat *Myotis grisescens* to pregnancy urine gonadotropin. *Endocrinology, 49*, 67–72.

Son, S. W., Yoon, M. H., Mori, T., & Uchida, T. A. (1987). Sperm-storage in the reproductive tract and prolonged survival of the Graafian follicle in the female orange whiskered bat, *Myotis formosus tsuensis* of Korea. *J. Mamm. Soc. Japan, 12*, 1–14.

Speakman, J. R., & Racey, P. A. (1986). Measurement of CO2 production by the doubly labeled water technique. *J. Appl. Physiol., 61*(3), 1200–1202.

Speakman, J. R., & Rowland, A. (1999). Preparing for inactivity: how insectivorous bats deposit a fat store for hibernation. *Proc. Nutr. Soc., 58*, 123–131.

Sinha Hikim, A. P., Bartke., A., & Russell, L. D. (1988). Morphometric studies on hamster testes in gonadally active and inactive states: light microscope findings. *Biol. Reprod., 39*, 1225–1237.

Srivastava, R. K. (2008). *Seasonal Adiposity and its Impact on Reproduction in Female Vespertilionid Bat,* Scotohilus heathi. Ph. D. Thesis. Varanasi, India: Banaras Hindu University.

Srivastava, R. K., & Krishna, A. (2006). Pathophysiology of polycystic ovary syndrome: lessons from animal studies. *Proc. Indian Natl Sci. Acad., B71*, 191–205.

Srivastava, R. K., & Krishna, A. (2007). Adiposity associated rise in leptin impairs ovarian activity during winter dormancy in vespertilionid bat, *Scotophilus heathi. Reproduction, 133*, 165–176.

Srivastava, R. K., & Krishna, A. (2008). Seasonal adiposity, correlative changes in metabolic factors and unique reproductive activity in a vespertilionid bat *Scotophilus heathi. J. Exp. Zool., 309*, 94–110.

Srivastava, R. K., & Krishna, A. (2010a). Melatonin affects steroidogenesis and delayed ovulation during winter in vespertilionid bat, *Scotophilus heathi. J. Steroid Biochem. Mole. Biol., 118*, 107–116.

Srivastava, R. K., & Krishna, A. (2010b). Melatonin modulates glucose homeostasis during winter dormancy in a vespertilionid bat, *Scotophilus heathi. Com. Biochem. Physiology, Part A.* (In Press).

Steinberger, A., & Steinberger, E. (1977). Replication patterns of Sertoli cells in maturing rat testis *in vivo* and in organ culture. *Biol. Reprod., 4*, 84–87.

Taft, L. T., & Handley, C. O. (1991). Reproduction in a captive colony. In C. O. Handley, D. E. Wilson, & A. L. Gardener (Eds.), *Demography and natural history of the common fruit bat,* Artibeus jamaicensis, *on Barro Colorado island, Panama* (pp. 942). Washington, DCWA: Smithsonian Institution Press.

Tamarkin, L., Baird, C. J., & Almeida, O. F. (1985). Melatonin: a coordinating signal for mammalian reproduction. *Science, 227*, 714–720.

Tamsitt, J. R., & Valdivieso, D. (1965a). Reproduction of the female big fruit eating bat, *Artibeus literatus palmarum* in Columbia. *Carrib. J. Sci., 5*, 157–166.

Tamsitt, J. R., & Valdivieso, D. (1965b). The male reproductive cycle of the bat, *Artibeus lituratus. Am. Midl. Nat., 73*, 150–161.

Thomas, D. W., & Marshall, A. G. (1984). Reproduction and growth in three species of West African fruit bats. *J. Zool. Lond., 202*, 265–281.

Thornhill, R., & Alcock, J. (1983). *The evolution of insect mating system.* Cambridge MA: Harvard University Press.

Tidemann, C. R. (1993). Reproduction in the bats *Vespadelus vulturnis, V. regulus* and *V. Darlingtoni* (Microchiroptera: Vespertilionidae) in coastal South-eastern Australia. *Aust. J. Zool., 41*, 21–35.

Towers, P. A., & Martin, L. (1995). Peripheral plama progesterone concentrations in pregnant and non-pregnant Greyheaded flying-foxes (*Pteropus poliocephalus*) and little red flying-foxes (*P. scapulatus*). *Reprod. Fertil. Dev., 7*, 1163–1176.

Trayhurn, P., Duncan, J. S., Hoggard, N., & Rayner, D. V. (1998). Regulation of leptin production: a dominant role for the sympathetic nervous system? *Proc. Nutr. Soc., 57*, 413–419.

Tsvetkov, T., & Takeva, T. (1988). Changes in interstitial gland in the ovary of the bat *Rinolophus eurale* Blas: arousal from hibernation. *Cryobiology, 25*, 502–507.

Turek, F. W., & Van Cauter, E. (1994). Rhythms in reproduction. In E. Knobil, & J. D. Neill (Eds.), *The physiology of Reproduction* (2nd Ed). (pp. 487–540). New York, NY: Raven Press.

Tuttle, M. D., & Stevenson, D. (1982). Growth and survival of bats. In T. H. Kunz (Ed.), *Ecology of bats* (pp. 105–150). New York, NY: Plenum Press.

Uchida, T. A., & Mori, T. (1977). Hibernation. *Metabolism Tokyo, 14*, 154–156.

Uchida, T. A., & Mori, T. (1987). Prolonged storage of spermatozoa in hibernating bats. In M. B. Fenton, P. A. Racey, & J. V. M. Rayner (Eds.), *Recent Advances in the studies of bats* (pp. 351–365). Cambridge, UK: Cambridge University Press.

Van Aarde, R. J., Van der Merwe, M., & Skinner, D. C. (1994). Progesterone concentrations and contents in the plasma, ovary, adrenal gland and placenta of the pregnant Natal clining bat *Miniopterus schreibersii natalensis. J. Zool. Lond., 232*, 457–464.

Van der Merwe, M., & Rautenbach, I. L. (1990). Reproduction in the rusty bat, *Pipistrellus rusticus* in the northern Tranvaal bushveld South Africa. *J. Reprod. Fertil., 89*, 537–542.

Van der Merwe, M. (1987). Adaptive breeding strategies in some south African bats between 220S and 280S. *Suid. Afrik. Tyds. Vir. Western, 83*, 607–609.

Van der Merwe, M., & Van Aarde., R. J. (1989). Plasma progesterone concentrations in the female Natal clining bat *(Miniopterus schreibersii natalensis). J. Reprod. Fertil., 87*, 665–669.

Van Krey, H. P., Ogasawara, F. X., & Pangborn, J. (1967). Light and electron microscopic studies of possible sperm gland emptying mechanisms. *Poultr. Sci., 46*, 69–78.

Velicky, J., & Titlbach, M. (1972). A study of the bat thyroid gland in winter and early spring. Electron microscopic observations. *Folia. Morphol., 20*, 406–415.

Velicky, J., & Titlbach, M. (1974). Electron microscopic observation in the thyroid gland of active bats. *Zool. Mikroskop. Anat. Forsch., 88*, 1069–1092.

Wade-Smith, J., Richmond, M. E., Mead, R. A., & Taylor, M. (1980). Hormonal and gestational evidence for delayed implantation in the striped skunk, *Mephitis mephitis. Gen. Comp. Endocrinol., 42*, 509–515.

Wang, M. Y., Rider, V., Heap, R. B., & Feinstein, A. (1984). Action of anti-progesterone monoclonal antibody in blocking pregnancy after post-coital administration in mice. *J. Endocrinol., 101*, 95–100.

Wang, Z., Liang, B., Racey, P. A., Wang, Y. L., & Zhang, S. Y. (2008). Sperm storage, delayed ovulation, and menstruation of the female ricketts big-footed bat (*Myotis ricketti*). *Zool. Stud., 47*, 215–221.

Webster, J. R., Moenter, S. M., Woodfil, C. J. I., & Karsh, F. J. (1991). Role of the thyroid gland in seasonal reproduction. II. Thyroxine allows a season specific suppression of gonadotropin secretion in sleep. *Endocrinology, 129*, 176–183.

Weigle, D. S., Duell, P. B., Connor, W. E., Steiner, R. A., Soules, M. R., & Kuijper, J. L. (1997). Effects of fasting, refeeding, and dietary fat restriction on plasma leptin levels. *J. Clin. Endocrinol. Met., 82*, 561–565.

Werner, S. C. (1969). The thyroid—A fundamental and clinical text. In S. C. Werner (Ed.), *Reproductive system* (pp. 656–665). Tokyo, Japan: Hakko Company Ltd.

Widmaier, E. P., & Kunz, T. H. (1993). Basal, diurnal, and stress-induced levels of glucose and glucocorticoids in captive bats. *J. Expt. Zool., 265*, 533–540.

Widmaier, E. P., Harmer, T. L., Sulak, A. M., & Kunz, T. H. (1994). Further characterization of the pituitary-adrenocortical responses to stress in Chiroptera. *J. Exp. Zool., 269*, 442–449.

Widmaier, E. P., Long, J., Cadigan, B., Gurgel, S., & Kuntz, T. H. (1997). Leptin, corticotropin-releasing hormone (CRH) and neuropeptide Y (NPY) in free-ranging pregnant bats. *Endocrine, 7*, 145–150.

Willig, M. R. (1985). Reproductive activity of female bats from Northeast Brazil. *Bat Rese. News, 26*, 17–20.

Wilson, D. E. (1973). Reproduction of neotropical bats. *Period. Biol., 75*, 215–217.

Wilson, D. E. (1979). Reproductive patterns. In R. J. Baker, J. K. Jones, Jr., & D. C. Carter (Eds.), *Biology of bats of the New World Family Phyllostomatidae* (pp. 317–378). TX: Spec. Publ. Mus. Texas, Tech Univ, Part III.

Wilson, D. E., & Findley, J. S. (1970). Reproductive cycle of a neotropical insectivorous bat, *Myotis nigricans*. *Nature, 225*, 1155.

Wimsatt, W. A. (1944). Growth of the ovarian follicle and ovulation in *Myotis lucifugus lucifugus*. *Am. J. Anat., 74*, 129–173.

Wimsatt, W. A. (1945). Notes on breeding behaviour, pregnancy and parturition in some vespertilionid bats of the eastern United States. *J. Mamm., 26*, 23–33.

Wimsatt, W. A. (1960). Some problems of reproduction in relation to hibernation in bats. *Bull. Mus. Comp. Zool. Harvard Univ., 124*, 249–267.

Wimsatt, W. A. (1969). Some interrelations of reproduction and hibernation in mammals. *Symp. Soc. Exp. Biol., 23*, 511–549.

Wimsatt, W. A. (1975). Some comparative aspects of implantation. *Biol. Reprod., 12*, 1–40.

Wimsatt, W. A. (1979). Reproductive asymmetry and unilateral pregnancy in Chiroptera. *J. Reprod. Fertil., 56*, 345–357.

Wimsatt, W. A., & Kallen, F. C. (1957). The unique maturation response of the Graafian follicles of hibernating vespertilionid bats and the question of its significance. *Anat. Rec., 129*, 115–132.

Wimsatt, W. A., & Parks, H. F. (1966). Ultrastructure of the surviving follicles of hibemation and of the ovum-follicle cell relationship in the vespertilionid bat *Myotis lucifugus*. *Symp. Zool. Soc. Lond., 15*, 419–454.

Wimsatt, W. A., & Trapido, H. (1952). Reproduction and the female reproductive cycle in the tropical American vampire bat, *Desmodus rotundus murinus*. *Am. J. Anat., 91*, 415–445.

Witkin, J. W., Paden, C. M., & Silverman, A. J. (1982). The luteinizing hormone-releasing hormone (LH-RH) sytems in rat brain. *Neuroendocrinology, 35*, 429–438.

Woitkewitsch, A. A. (1940). Dependence of seasonal periodicity in gonadal changes on the thyroid gland in *Sturnus vulgaris*. *L R Acad. Sci. URSS (Doklady), 27*, 741–745.

Wolden-Hanson, T., Mitton, D. R., McCants, R. L., Yellon, S. M., Wilkinson, C. W., Matsumoto, A. M., et al. (2000). Daily melatonin administration to middle-aged male rats suppresses body weight, intraabdominal adiposity, and plasma leptin and insulin independent of food intake and total body fat. *Endocrinology, 141*, 487–497.

Woods, S. C., Figelewiez, D. P., Lattermann, D. P., Schwartz, M. W., & Porte, D., Jr. (1990). A reassessment of the regulation of adiposity and appetite by the brain insulin system. *Int. J. Obesity, 14*, 69–76.

Wurtman, R. J., Axlrod, J., & Chu, E. W. (1963). Melatonin a pineal substance: effect on the rat ovary. *Science, 141*, 277–278.

Yellon, S. M., & Hilliker, S. (1994). The influence of acute melatonin treatment and light on the circadian melatonin rhythm in the Djungarian hamster. *J. Biol. Rhythms, 9*, 71–81.

Young, R. A. (1976). Fat, energy and mammalian survival. *Am. Zool., 16*, 699–710.

Zhang, Y., Proenca, R., Maffei, M., Barone, M., Leopold, L., & Friedman, J. M. (1994). Positional cloning of the mouse obese gene and its human homologue. *Nature, 372*, 425–432.

Chapter 13

Hormones and Reproductive Cycles in Primates

Wendy Saltzman*, Suzette D. Tardif[†] and Julienne N. Rutherford**

*University of California, Riverside, CA, USA, [†]University of Texas, San Antonio, TX, USA, **University of Illinois at Chicago, Chicago, IL, USA

SUMMARY

Primates are characterized by long lifespans, slow reproductive processes, and low fecundity. Gonadarche is a late and prolonged process in which the maturation of the hypothalamic–pituitary–gonadal (HPG) axis is followed by a period of 'adolescent infertility'. Ovarian cycles are prolonged, with spontaneous ovulation of one ovum (or several ova, in some species) followed by a spontaneous luteal phase supporting development of the uterine endometrium. Following conception, in the anthropoid primates (monkeys, apes, and humans) an invasive hemochorial placenta supports a prolonged gestation and offers the opportunity for placental endocrine signals to be transmitted directly to the maternal blood. Mothers produce dilute milk over long periods, supporting slow postnatal growth rates. Lactation induces anovulation in most primates through a process linked to the suckling stimulus. Seasonal, energetic, and social factors can all influence the course of puberty, conception, pregnancy, and, to a lesser extent, lactation. Primates experience a long reproductive life, typically with age-related reductions in female fertility and in male androgen production. As much as 25% of the female's maximal lifespan may be postreproductive, with reproductive senescence driven primarily by loss of the follicular pool.

1. OVERVIEW OF THE PRIMATES

The order Primates includes roughly 230 extant species in two suborders: the Strepsirhini, comprising the lemurs, lorises, galagos, and pottos (also referred to as prosimians, along with tarsiers); and the Haplorhini, comprising the tarsiers, Platyrrhini (New World monkeys), Cercopithecoidea (Old World monkeys), Hylobatidae (lesser apes; i.e., gibbons and siamangs), and Hominidae (humans and great apes; i.e., orangutans, gorillas, chimpanzees, and bonobos) (Bininda-Emonds et al., 2007; Hartwig, 2007) (Figure 13.1). Primates exhibit tremendous diversity in body size, ranging from the 30 to 60 g mouse lemurs (*Microcebus* spp.) (Yoder et al., 2000) to the roughly 200 kg gorillas (*Gorilla gorilla*) (Smith & Jungers, 1997), and in longevity, ranging from approximately nine years in several prosimians to roughly 60 years in great apes and over 120 years in humans (AnAge: The Animal Ageing and Longevity Database, 2009). They occupy a diversity of habitats, from tropical forests to savannas to semideserts, and exhibit a variety of lifestyles, from arboreal to terrestrial, from diurnal to nocturnal, and from largely dispersed and solitary to highly gregarious. None-the-less, primates as a group are characterized by long lifespans, delayed reproductive maturation, low fecundity, and high investment in each offspring (Zimmermann & Radespiel, 2007).

In this chapter, we summarize current perspectives on and understanding of reproductive function in human and nonhuman primates, including the hypothalamic–pituitary–gonadal (HPG) axis function across the lifespan; pregnancy and lactation; sexual behavior and its hormonal underpinnings; and seasonal, social, and energetic influences on reproduction. Descriptive data on reproductive patterns are available for several hundred primate species and subspecies studied in the wild and/or in captivity (Zimmermann & Radespiel, 2007). In contrast, experimental investigations of reproductive physiology have focused largely on humans and a small number of monkey species, especially the Old World macaques (*Macaca* spp.) and baboons (*Papio* spp.), and the New World squirrel monkeys (*Saimiri* spp.) and marmosets (*Callithrix* spp.). By necessity, we focus mainly on the best-studied taxa; however, we also attempt to highlight the diversity among primates (Table 13.1).

2. TESTICULAR FUNCTION AND ITS NEUROENDOCRINE CONTROL

As in other mammals, the two major functions of the testes—spermatogenesis and androgen production—take place in anatomically and functionally distinct testicular compartments. Spermatogenesis occurs within the seminiferous tubules, whereas androgen production occurs primarily

Hormones and Reproduction of Vertebrates, Volume 5—Mammals

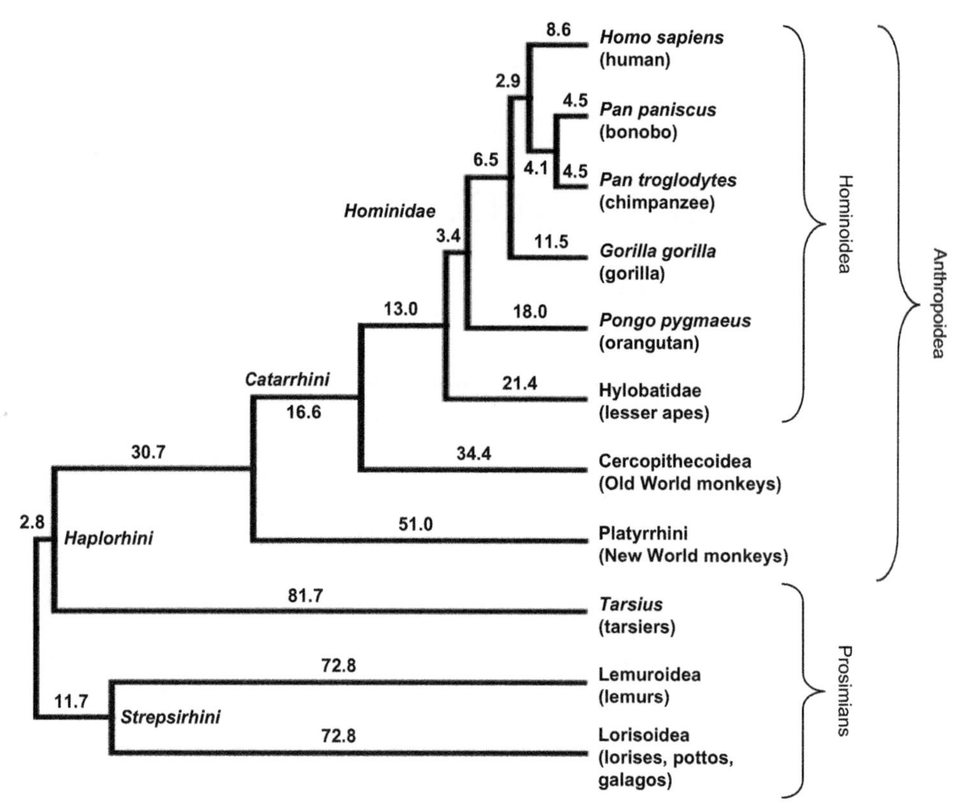

FIGURE 13.1 Cladogram showing the major extant primate taxa, with branch lengths in millions of years. *Based on Bininda-Emonds et al. (2007).*

in the Leydig cells of the testicular interstitium. Both processes are controlled by the HPG axis. Therefore, we begin by briefly reviewing this endocrine axis and its control of androgen secretion, and then discuss spermatogenesis and its hormonal control.

2.1. The Hypothalamic–pituitary–gonad (HPG) Axis and Androgen Secretion in Males

In both sexes, gametogenesis and gonadal steroid secretion are ultimately regulated by gonadotropin-releasing hormone (GnRH), a decapeptide synthesized in neuronal cell bodies in the arcuate nucleus of the medial basal hypothalamus and released into the hypothalamic–hypophysial portal blood vessels in a pulsatile manner (see Chapter 2, this volume). The cellular and molecular mechanisms responsible for the generation of GnRH pulses are not yet understood, but appear to involve endogenous oscillations within the GnRH neurons themselves (Terasawa, 2001; Zeleznik & Pohl, 2006; see also Chapter 2, this volume). Stimulatory inputs to the GnRH pulse generator in primates include kisspeptin (Kp), norepinephrine, glutamate, neuropeptide Y (NPY), and nitric oxide, while inhibitory inputs include endogenous opioids, γ-aminobutyric acid (GABA), and corticotropin-releasing hormone (CRH) (Terasawa, 2001; Zeleznik & Pohl, 2006; Plant & Ramaswamy, 2009).

At the anterior pituitary, GnRH binds to gonadotropes to stimulate synthesis and secretion of two glycoprotein hormones: luteinizing hormone (LH) and follicle-stimulating hormone (FSH). A notable exception to this pattern occurs in at least some New World monkeys (common marmoset (*Callithrix jacchus*), Bolivian squirrel monkey (*Saimiri boliviensis*), Ma's owl monkey (*Aotus nancymaee*)), in which the pituitary secretes chorionic gonadotropin (CG) instead of LH. Correspondingly, the gonads in some or all New World monkeys express a modified form of the LH receptor, in which exon 10 is not expressed and which is activated selectively by CG (Gromoll et al., 2003; Müller et al., 2004a; 2004b; Scammell, Funkhouser, Moyer, Gibson, & Willis, 2008).

The major function of LH in males is the stimulation of androgen release by the Leydig cells. Correspondingly, testosterone (T) is an important regulator of LH secretion, exerting negative feedback to reduce the frequency of LH and presumably GnRH pulses (reviewed by Tilbrook & Clarke, 2001). Testosterone-mediated feedback in primates is thought to occur primarily at the level of the brain, rather than the pituitary. The neuroanatomical substrates of this feedback, however, are not yet known. Gonadotropin-releasing hormone neurons in mammals do not appear to contain receptors for androgens, estrogens, or progesterone (P_4); thus, feedback by these steroids must be mediated by other cell populations. To date, the endogenous opiates, GABA, and Kp have all been implicated as possible mediators of androgenic feedback on GnRH in rhesus macaques (*Macaca mulatta*) (El Majdoubi, Ramaswamy, Sahu, & Plant, 2000; Shibata, Friedman, Ramaswamy, & Plant, 2007). Intracellular aromatization of T to estradiol (E_2) within the brain also has been implicated in the feedback regulation of the HPG axis in male primates, as in other mammals; however, studies have yielded inconsistent results (Tilbrook & Clarke, 2001).

In contrast to LH, the major functions of FSH in males involve development of the gonads, especially production of Leydig cells, and regulation of spermatogenesis, the latter being mediated through actions on Sertoli cells (see Section 2.2). Follicle-stimulating hormone also differs from LH in the hypothalamic and gonadal mechanisms regulating its secretion (Plant & Marshall, 2001). Whereas LH release is highly sensitive to GnRH pulse frequency, FSH release is not, so that changes in GnRH pulse frequency alter the ratio of circulating FSH to LH (Zeleznik & Pohl, 2006). The occurrence of a separate, selective hypothalamic FSH-releasing factor has been postulated but has not been confirmed, and seems unlikely to play a significant role in primates (Zeleznik & Pohl, 2006). Moreover, the inhibitory effects of T on FSH secretion, apparently mediated through aromatization to E_2, are less pronounced than those controlling LH release. Instead, the major testicular hormones controlling FSH secretion are inhibin B and activins, glycoprotein hormones produced by the Sertoli cells that inhibit and stimulate, respectively, pituitary release of FSH (McLachlan et al., 2002b).

2.2. Spermatogenesis

Spermatogenesis comprises three major processes: mitotic proliferation, which maintains the population of stem cell spermatogonia and produces differentiated spermatogonia and primary spermatocytes; meiotic division, in which each diploid primary spermatocyte gives rise to two secondary spermatocytes and, subsequently, to four haploid sperm, involving development of head and tail structures; and spermiation, or release of sperm into the lumen of the seminiferous tubules (Johnson & Everitt, 2000; Plant & Marshall, 2001; McLachlan et al., 2002b; see also Chapter 5, this volume) (Figure 13.2). In most respects, these processes are similar in all mammals; however, some noteworthy differences have been found between primates and nonprimates as well as among primate species.

Spermatogonia are classified as either type A (undifferentiated, including stem cell spermatogonia and proliferative spermatogonia) and type B (differentiated spermatogonia). In primates, but not rodents, type A spermatogonia are further divided into two morphologically distinct subtypes: A-pale, which undergo mitotic divisions to produce new type A and type B spermatogonia, and A-dark, which are thought to constitute a pool of reserves that begin to proliferate only under conditions of testicular damage (McLachlan et al., 2002b; Luetjens, Weinbauer, & Wistuba, 2005) (Figure 13.2).

Beginning at puberty, A-pale spermatogonia, occupying the basal compartment of the seminiferous tubules, undergo a limited number of mitotic divisions to yield type A as well as type B daughter cells. Subsequently, type B spermatogonia undergo one or more mitotic divisions to yield primary spermatocytes. The number of mitotic divisions at each stage differs reliably among species. The duration of the spermatogenic cycle (i.e., the time between consecutive spermatogonial divisions to produce spermatocytes in a single section of seminiferous tubule) in New World and Old World monkeys averages approximately 10 days, with longer cycles reported in chimpanzees (*Pan troglodytes*) (14.4 days) and humans (16 days) (Luetjens et al., 2005).

As in other taxa, spermatogenesis in primates is governed largely by the Sertoli cells, the only somatic cells present in the seminiferous tubules (Sofikitis et al., 2008). Functions of these cells include forming the blood–testis barrier, maintaining the cytoarchitecture of the germinal epithelium, producing nutrients that provide energy for the germ cells, regulating FSH secretion via secretion of inhibins and activins, and mediating androgenic effects on spermatogenesis through expression of androgen receptors and production of androgen-binding protein.

In rodents, each cross section of seminiferous tubule contains germ cells in only a single stage of spermatogenesis. A similar pattern is found in prosimian primates. Among great apes and humans, however, most tubular cross sections contain germ cells in multiple stages of spermatogenesis. Monkeys show intermediate patterns, but the proportion of so-called multi-stage tubules is higher in New World than Old World monkeys (Wistuba et al., 2003). The functional significance of these differences, if any, is not clear. Spermatogenic efficiency, defined as the absolute number of germ cells produced relative to the theoretical number, does not appear to vary markedly across primate species or to differ between species with single-stage vs. multi-stage tubules, although efficiency is lower in numerous primate species than in the rat (*Rattus norvegicus*) (Luetjens et al., 2005).

In rats, FSH and T play critical roles in regulating spermatogenesis, acting both separately and synergistically. Among primates, the relative roles of these two hormones are less clear but differ to some extent from those in rodents (reviewed by McLachlan et al., 2002a; 2002b; Luetjens et al., 2005; Sofikitis et al., 2008). Follicle-stimulating hormone appears to be necessary for

TABLE 13.1 Reproductive and life-history parameters for representative primate species

	Adrenarche	Age at sexual maturity—males (days)	Age at sexual maturity—females (days)	Ovarian cycle length (days)	Menstruation	Birth seasonality*	Female age at first parturition (years)
Strepsirhini							
Microcebus murinus (*gray mouse lemur*)		243[1]	243[1]	50.5[23]	Absent/covert[9]	Strong[18]	0.67[28]
Daubentonia madagascariensis (*aye-aye*)			882[1]	49.8[23]	Absent/covert[9]	Absent[18]	3.5[28]
Lemur catta (*ring-tailed lemur*)		912[1]	595[1]	39.3[23]	Absent/covert[9]	Strong[18]	2.13[28]
Eulemur fulvus (brown lemur)		548[1]	608[1]	29.5[23]	Absent/covert[9]	Strong[18]	2.41[28]
Varecia variegata (ruffed lemur)		608[1]	604[1]	30[2]	Absent/covert[9]	Strong[4]	2.0[28]
Propithecus verreauxi (Verreaux's sifaka)		912[1]	912[1]		Absent/covert[9]		3.5[28]
Otolemur crassicaudatus (thick-tailed greater galago)		639[1]	495[1]	44.0[23]	Absent/covert[9]	Strong[24]	2.17[28]
Perodicticus potto (Western potto)		547[1]	547[1]	37.9[23]	Absent/covert[9]	Weak[24]	2.03[28]
Tarsius							
Tarsius bancanus (Western tarsier)			920[1]	24.0[23]	Absent/covert[9]	Strong[19]	2.52[28]
Platyrrhini							
Callimico goeldii (Goeldi's monkey)		395[1]	365[1]	23.8[23]	Absent/covert[23]	Weak; possibly bimodal[13]	1.32[28]
Callithrix jacchus (common marmoset)	Absent[29]	382[1]	477[1]	28.6[23]	Absent/covert[9]	Weak; bimodal[13]	1.44[28]
Callithrix pygmaea (pygmy marmoset)		638[1]	684[1]	33.3[33]	Absent/covert[23]	Weak; bimodal[13]	1.88[28]
Saguinus oedpius (cotton-top tamarin)		550[1]	548[1]	22.7[23]	Absent/covert[23]	Weak[13]	1.89[28]
Leontopithecus rosalia (golden lion tamarin)		730[1]	547[1]	18.5[23]	Absent/covert[23]	Weak; sometimes bimodal[13]	2.40[28]
Saimiri sciureus (common squirrel monkey)		1,826[1]	1,003[1]	9.1[23]	Absent/covert[9]	Strong[20]	2.5[28]
Cebus apella (brown capuchin)			1,703[1]	20.0[23]	Slight[9]	Weak[20]	5.64[28]
Alouatta caraya (black howler monkey)		928[1]	1,167[1]	20.4[23]		None/weak[14]	3.71[28]
Lagothrix lagotricha (Humboldt's woolly monkey)		1,520[1]	2,555[1]	25[2]	Slight[9]	Weak[13]	6.29[28]
Ateles geoffroyi (Geoffroyi's spider monkey)		1,826[1]	1,825[1]	25.5[23]	Slight[9]	Weak[14]	6.0[28]
Brachyteles arachnoides (muriqui)			2,738[1]	21.0[23]		Weak[7]	7.5[28]
Callicebus moloch (dusky titi)			912[1]	17[12]	Absent/covert[12]	Weak[12]	3[28]
Pithecia pithecia (white-faced saki)		1,460[1]	775[1]	16-17[25]		Weak[25]	2.08[28]
Aotus trivirgatus (northern night monkey)		730[1]	821[1]	15.6[23]	Absent/covert[9]	Weak[13]	2.40[28]

Gestation length (days)	Interbirth interval (months)	Modal litter size	Age at weaning (days)	Maximum recorded lifespan in captivity (years)	% of lifespan completed at time of last birth‡	Mean body mass—wild adult females (g)	Placentation
							Superficial, epitheliochorial, villous[3]
60[28]	3[28]	2[28]	40[28]	18.2[1]		63[28]	
167[28]	20[28]	1[28]	268[28]	23.3[1]		2,531[28]	
135[28]	14[28]	1[28]	142[28]	37.3[1]	76.7% (C) (*Lemur spp.*)[17]	2,250[28]	
120[28]	24[28]	1[28]	159[28]	35.5[1]		2,228[28]	
139[28]	12[28]	2[28]	90[28]	36[1]		3,407[28]	
140[28]	12[28]	1[28]	180[28]	30.5[1]		3,285[28]	
135[28]	12[28]	2[28]	135[28]	22.7[1]		1,110[28]	
195[28]	12[28]	1[28]	150[28]	26.8[1]		836[28]	
							Superficial, epitheliochorial, villous[3]
178[28]	8[28]	1[2]	79[28]	16.3[1]		109[28]	
							Superficial, hemochorial, trabecular[3]
153[28]	9[28]	1[28]	68[28]	22.2[1]		355[28]	
148[28]	6[28]	2[28]	76[28]	22.8[1]	61.7% (C)[17]	334[28]	
137[28]	6[28]	2[28]	90[28]	18.6[1]		101[28]	
168[28]	7[28]	2[28]	50[28]	26.2[1]		404[28]	
129[28]	6[28]	2[28]	90[28]	31.6[1]	48.6% (C)[17]	579[28]	
170[28]	9[28]	1[28]	197[28]	30.2[1]	90.5% (C)[17]	681[28]	
154[28]	22[28]	1[28]	263[28]	46[1]		2,361[28]	
187[28]	11[1]	1[28]	325[28]	32.4[1]		4,606[28]	
224[28]	24[28]	1[28]	411[28]	32[1]		6,303[28]	
227[28]	37[28]	1[28]	786[28]	47.1[1]		7,480[28]	
233[28]	34[28]	1[28]	747[28]	30 (unconfirmed)[1]		8,070[28]	
164[28]	12[28]	1[28]	60[28]	26.2[1]		956[28]	
164[28]	19[28]	1[28]	122[1]	36[1]		1,580[28]	
133[28]	9[28]	1[28]	127[28]	30.1[1]		730[28]	

(Continued)

TABLE 13.1 Reproductive and life-history parameters for representative primate species—cont'd

	Adrenarche	Age at sexual maturity—males (days)	Age at sexual maturity—females (days)	Ovarian cycle length (days)	Menstruation	Birth seasonality*	Female age at first parturition (years)
Cercopithecoidea							
Semnopithecus entellus (Hanuman langur)		1,886[1]	1,162[1]	26.8[23]	Overt[9]	Absent/weak[6]	3.66[28]
Pygathrix nemaeus (red-shanked douc langur)			1,460[1]				
Nasalis larvatus (proboscis monkey)			1,460[1]			Weak[22]	4.5[28]
Colobus guereza (guereza)		2,192[1]	1,461[1]			Absent[16]	4.75[28]
Macaca mulatta (rhesus macaque)	Absent[29]	2,007[1]	1,231[1]	26.6[23]	Overt[9]	Strong[27]	3.75[28]
Macaca fascicularis (long-tailed macaque)	Absent[29]	1,544[1]	1,238[1]	29.4[23]	Overt[9]	Weak[27]	3.9[28]
Macaca arctoides (stump-tailed macaque)		2,099[1]	1,186[1]	29[2]	Slight[9]	Absent[27]	3.84[28]
Macaca fuscata (Japanese macaque)		1,369[1]	1,483[1]	28[2]	Overt[9]	Strong[27]	5.54[28]
Macaca nemestrina (pig-tailed macaque)		1,095[1]	1,125[1]		Slight[9]	Weak[27]	3.92[28]
Papio anubis (olive baboon)		973[2]		33.2[23]	Overt[9]	Absent[21]	4.5[28]
Papio hamadryas (hamadryas baboon)	Absent[29]	1,762[1]	1,514[1]	30[8]	Overt[9]	Weak[5]	6.1[28]
Mandrillus sphinx (mandrill)			1,186[1]	39.6[23]		Weak[21]	4[28]
Theropithecus gelada (gelada baboon)		2,190[1]	1,391[1]	35.5[23]	Slight[9]	Weak[5]	4[28]
Cercocebus torquatus (white-collared mangabey)			973[1]	33[2]	Slight[9]	Weak[21]	4.67[28]
Chlorocebus aethiops (vervet)		1,825[1]	1,034[1]	33[23](median)	Slight[9]	Weak[15]	4.88[28]
Cercopithecus neglectus (De Brazza's monkey)		2,555[1]	1,611[1]			Weak[15]	4.67[28]
Erythrocebus patas (patas monkey)		1,400[1]	956[1]	30.6[23]	Overt[9]	Weak[15]	3[28]
Miopithecus talapoin (Angolan talapoin)		2,008[1]	1,395[1]	36[2]	Overt[9]	Weak[15]	4.38[28]
Hominoidea							
Hylobates lar (white-handed gibbon)		1,825[1]	2,555[1]	20.2[23]	Overt[9]	Weak[30]	8.00[28]
Hylobates syndactylus (siamang)		2,190[1]	2,190[1]				7.09[28]
Pongo pygmaeus (orangutan)		2,555[1]	2,555[1]	29.6[23]	Slight[9]		10.65[28]
Gorilla gorilla (gorilla)		4,015[1]	2,829[1]	31.1[23]	Slight[9]	Absent[11]	8.1[28]
Pan paniscus (bonobo)			3,194[1]	42[31]		Weak[10]	13-15[28]

Gestation length (days)	Interbirth interval (months)	Modal litter size	Age at weaning (days)	Maximum recorded lifespan in captivity (years)	% of lifespan completed at time of last birth‡	Mean body mass—wild adult females (g)	Placentation
							Interstitial, hemochorial, villous[3]
192[28]	17[28]	1[28]	354[28]	29[1]	94.1% (FP)[17]	10,470[28]	
210[28]	20[28]	1[28]	330[1]	26[1]		8,180[28]	
166[28]	18[28]	1[28]	246[28]	25.1[1]		9,593[28]	
170[28]	20[28]	1[28]	371[28]	35[1]		8,401[28]	
165[28]	12[28]	1[28]	279[28]	40[1]	66.7% (C)[17], 73.5% (SF)[17]	6,890[28]	
164[28]	13[28]	1[28]	375[28]	39[1]		3,582[28]	
178[28]	19[28]	1[28]	393[28]	29.2[1]		8,400[28]	
173[28]	24[28]	1[28]	453[28]	38.5[1]	67.3% (FP)[17]	8,565[28]	
169[28]	14[28]	1[28]	300[28]	37.6[1]	69.2%(C)[17]	5,657[28]	
180[28]	25[28]	1[28]	592[28]	25.2[28]	92.6% (FP)[17]	13,233[28]	
170[28]	24[28]	1[28]	561[28]	37.5[1]		10,568[28]	
198[28]	17[28]	1[28]	349[28]	40[1]		12,125[28]	
170[28]	24[28]	1[28]	465[28]	36[1]		11,427[28]	
171[28]	13[28]	1[28]		46[1]		5,500[28]	
163[28]	12[28]	1[28]	262[28]	30.8[1]	66.9% (C)[17]	3,020[28]	
167[28]	12[28]	1[28]	393[28]	30.8[1]		3,816[28]	
167[28]	12[28]	1[28]	234[28]	28.3[1]		6,409[28]	
164[28]	12[28]	1[28]	188[28]	27.7[1]		1,560[28]	
							Interstitial, hemochorial, villous[3]
209[28]	30[28]	1[28]	639[28]	56[1]		5,403[28]	
232[28]	50[28]	1[28]	639[28]	43[1]		10,568[28]	
250[28]	72[28]	1[28]	1,273[28]	59[1]	68.1% (C)[17]	36,389[28]	
273[28]	47[28]	1[28]	1,061[28]	55.4[1]	51.9% (C)[17]	80,000[28]	
240[28]	48[28]	1[28]	1,080[28]	55[1]		33,200[28]	

(Continued)

TABLE 13.1 Reproductive and life-history parameters for representative primate species—cont'd

	Adrenarche	Age at sexual maturity—males (days)	Age at sexual maturity—females (days)	Ovarian cycle length (days)	Menstruation	Birth seasonality*	Female age at first parturition (years)
Pan troglodytes (chimpanzee)	Present[29]	2,920[1]	3,376[1]	37.3[23]	Overt[9]	Weak[26]	13.6[28]
Homo sapiens (human)	Present[29]	5,110[1]	4,745[1]	29.1[23]	Overt[9]	Weak[32]	14.5[28]

[1]*Data from: AnAge, The Animal Ageing and Longevity Database (http://genomics.senescence.info/species/).*
[2]*Harvey et al., 1987.*
[3]*Mossman, 1987.*
[4]*Richard, 1987.*
[5]*Stammbach, 1987.*
[6]*Struhsaker, 1987.*
[7]*Strier and Ziegler, 1994.*
[8]*Rowe, 1996.*
[9]*Strassmann, 1996.*
[10]*Furuichi* et al., *1998.*
[11]*Watts, 1998.*
[12]*Valeggia* et al., *1999.*
[13]*Di Bitetti and Janson, 2000.*
[14]*Di Fiore and Campbell, 2007.*
[15]*Enstam and Isbell, 2007.*
[16]*Fashing, 2007.*
[17]*Fedigan and Pavelka, 2007.*
[18]*Gould and Sauther, 2007.*

proliferation of A-pale spermatogonia, transition of A-pale to B spermatogonia, and spermiation. T, on the other hand, is thought to stimulate late spermatid differentiation, whereas both T and FSH may play roles in suppressing apoptosis in germ cells and in regulating meiotic divisions by spermatocytes. Other hormones that have been implicated in modulating spermatogenesis in primates include LH, insulin, inhibin, activin, follistatin, somatostatin, and estrogens (Sofikitis et al., 2008).

3. OVARIAN FUNCTION AND ITS NEUROENDOCRINE CONTROL

3.1. Overview

Ovarian cycles in primates, like those in other mammals, comprise (1) a preovulatory (follicular) phase, characterized by follicular maturation, increasing follicular secretion of estrogens, and generally low circulating levels of gonadotropins (GTHs); (2) ovulation, precipitated by midcycle surges in estrogens and GTH secretion; and (3) a postovulatory (luteal) phase, dominated by formation of a corpus luteum (or, in some species, several corpora lutea) from the ovulated follicle(s), luteal production of P_4 (and estrogens), and low pituitary GTH levels (Figure 13.3) (see also Chapter 4, this volume). In all primate species studied to date, but in contrast to numerous other mammals, follicular development, ovulation, and corpus luteum formation occur spontaneously, independent of mating-induced stimuli (Martin, 2007). Primates also tend to have extended ovarian cycles as compared to other mammals, with especially prolonged luteal phases (Johnson & Everitt, 2000). None-the-less, overall cycle length and duration of the follicular and luteal phases differ markedly among primates. Average cycle lengths tend to range from 30 to 50 days in prosimians, 16 to 30 days in New World monkeys, 24 to 35 days in Old World monkeys, 20 to 30 days in lesser apes, and 25 to 50 days in great apes, including humans (Van Horn & Eaton, 1979; Dixson, 1998; Martin, 2007; Emery Thompson, 2009; Ziegler, Strier, & Van Belle, 2009) (Table 13.1). The New World squirrel monkeys (*Saimiri* spp.), however, have a mean cycle length of seven to twelve days, with a follicular phase of only about five days (Dukelow, 1985).

As described below, menstruation occurs to some extent in all or most Old World monkeys and apes, and in several New World monkeys (Strassmann, 1996); thus, these species may be said to have a true menstrual cycle. Other species, most notably prosimians, may be considered to have an estrous cycle, as they exhibit distinct cyclical changes in sexual receptivity, with peak receptivity occurring during the periovulatory period (see Section 8). Finally, many New World monkeys exhibit neither menstruation nor strict estrous cyclicity. For consistency, therefore, we refer simply to 'ovarian cycles' in all female primates.

Dynamics of primate ovarian cycles, like many other aspects of reproductive physiology, have been characterized

Gestation length (days)	Interbirth interval (months)	Modal litter size	Age at weaning (days)	Maximum recorded lifespan in captivity (years)	% of lifespan completed at time of last birth‡	Mean body mass—wild adult females (g)	Placentation
238[28]	60[28]	1[28]	1,691[28]	59.4[1]	60.0% (C), 80.0% (FP)[17]	31,850[28]	
269[28]	36[28]	1[28]	830[28]	122.5[1]	41.7-50.0%[17]	53,733[28]	

[19]*Gursky, 2007.*
[20]*Jack, 2007.*
[21]*Jolly, 2007.*
[22]*Kirkpatrick, 2007.*
[23]*Martin, 2007.*
[24]*Nekaris and Bearder, 2007.*
[25]*Norconk, 2007.*
[26]*Stumpf, 2007.*
[27]*Thierry, 2007.*
[28]*Zimmermann and Radespiel, 2007 (if >1 value was provided for a species, the mean of all values for the species was used).*
[29]*Nguyen and Conley, 2008.*
[30]*Savini* et al., *2008.*
[31]*Emery Thompson, 2009.*
[32]*Vitzthum* et al., *2009.*
[33]*Ziegler et al., 2009.*
**Strong: highly predictable and relatively short breeding and birth periods; evidence of altered gonadal function in non-breeding season. Weak: breeding and birth period peaks are seen, but births can occur in any month. Absent: limited or no evidence of breeding or birth peaks.*
‡C: captive breeding colonies. FP: free-ranging and provisioned. SF: semi-free-ranging and provisioned

most thoroughly in macaques and women. Therefore, the following review focuses primarily on data from these species. We begin by summarizing the cyclical events occurring in the ovary. We then describe cyclical changes in the uterus and other tissues, and finally discuss the neuroendocrine control of primate ovarian cycles.

3.2. Cyclical Changes in the Ovaries

3.2.1. Folliculogenesis

As in other mammals, primate oogonial germ cells terminate mitotic division and enter their first meiotic division during prenatal development. The resulting primary oocytes become encased in primordial follicles, each consisting of a single layer of spindle-shaped granulosa cells, and abruptly suspend meiosis at the diplotene stage of the first meiotic prophase. Primordial follicles may remain in this state of suspended animation for years or even decades, until they initiate further development to the early antral stage (Johnson & Everitt, 2000).

Development of primordial follicles into early antral follicles involves growth of the oocyte, formation of a zona pellucida, and growth and proliferation of granulosa cells, followed by formation of the antral cavity and development of the thecal cell layer. This phase of follicular development, which is thought to last approximately 85 days in humans (Gougein, 1986), may occur to a small extent prepubertally (Zeleznik & Pohl, 2006). Beginning at puberty, however, several primordial follicles resume development each day, forming a continuous stream of maturing follicles. Maturation to the early antral stage appears to be independent of gonadotropic stimulation and occurs during all phases of the ovarian cycle. Granulosa cells from preantral and early antral follicles possess receptors for FSH but not LH, whereas thecal cells possess only LH receptors. These immature follicles do not secrete significant amounts of estrogens under basal conditions but can do so if stimulated with FSH for prolonged periods (Zeleznik & Pohl, 2006).

Maturation of early antral follicles to the preovulatory stage occurs exclusively during the follicular phase of the ovarian cycle, under the control of LH and FSH, and involves expansion of the antral cavity, secretion of follicular fluid into the antrum, expression of LH receptors by the granulosa cells, and follicular secretion of increasing amounts of estrogens and inhibin B. As in other mammals, estrogen production proceeds according to the 'two cell, two GTH' model: thecal cells convert C21 steroids to C19 steroids (androstenedione and T) under the influence of LH, and granulosa cells subsequently aromatize these androgens to estrogens under the influence of FSH. Steroidogenesis may be further modulated by several autocrine and paracrine factors, including estrogens, androgens, insulin-like growth factor (IGF)-II, activin, and inhibin; however, the roles of these local factors in follicular steroidogenesis are not well understood (Zeleznik & Pohl, 2006). Follicles in which development to the early antral stage does not correspond with the slight elevation in circulating GTH levels at the beginning of the follicular phase are unable to

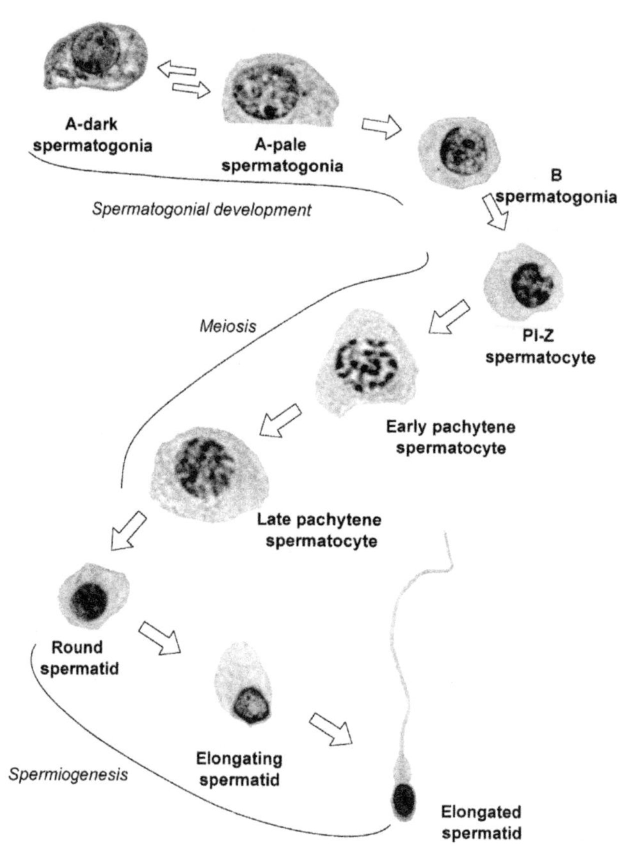

FIGURE 13.2 Spermatogenesis in the long-tailed macaque (*Macaca fascicularis*). Arrows indicate the progression of cells through spermatogonial proliferation and differentiation, meiotic division, and spermiogenesis. Dual arrows between A-dark and A-pale spermatogonia indicate the likely transdifferentiation between these cell types (see text for discussion). PI-Z indicates preleptotene-zygotene spermatocytes. *Reproduced from McLachlan et al. (2002a), with permission from the American Society of Andrology.*

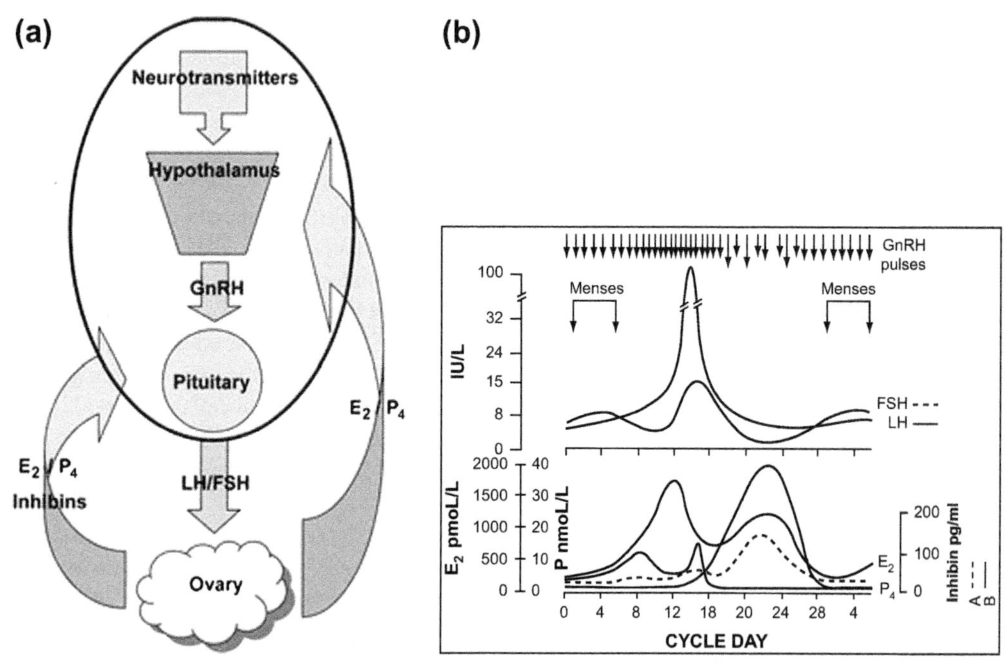

FIGURE 13.3 (a) Schematic depiction of the hypothalamic–pituitary–ovarian axis. (b) Hormonal changes across the human menstrual cycle, including relative amplitude (depicted by length of arrows) and frequency (depicted by density of arrows) of hypothalamic gonadotropin-releasing hormone GnRH release, and circulating concentrations of luteinizing hormone (LH), follicle-stimulating hormone (FSH), estradiol (E_2), progesterone (P), inhibin A, and inhibin B. *Reproduced from Randolph (2008) with permission from Wiley-Blackwell.*

mature further, and undergo atresia (Johnson & Everitt, 2000).

In most primates, only one follicle ovulates in each cycle. This 'dominant follicle' emerges during the mid-follicular phase and inhibits maturation of other follicles by secreting large amounts of estrogens and possibly inhibin B, thereby reducing FSH concentrations below the threshold level required for maturation of early antral follicles. At the same time, the dominant follicle significantly increases its own expression of FSH and LH receptors and develops a markedly denser capillary network than those supplying less mature follicles. Together, these changes ensure continuing GTH support for the dominant follicle, in spite of the declining FSH levels that lead to atresia of other follicles (Zeleznik & Pohl, 2006).

3.2.2. Ovulation

At the end of the follicular phase, sustained high concentrations of circulating estrogens from the dominant follicle exert positive feedback on the hypothalamus and pituitary (see Section 3.8) to trigger surges in secretion of GnRH, FSH, and, most dramatically, LH. Within several hours, the LH surge stimulates the primary oocyte to complete its first meiotic division. As in other mammals, this division is asymmetric, with half of the chromosomes and almost all of the cytoplasm being inherited by one daughter cell, the secondary oocyte; the other daughter cell, or first polar body, subsequently dies. The secondary oocyte immediately begins its second meiotic division but abruptly becomes arrested again at metaphase (Johnson & Everitt, 2000).

The LH surge also stimulates final preovulatory maturation and secretory activity of the dominant follicle, including vascularization of the granulosa cell layer, a large increase in the volume of follicular fluid and in follicular size, a transient rise in secretion of estrogens and androgens, and initiation of P_4 secretion. Finally, LH stimulates increased expression of collagenase, prostaglandins, vascular endothelial growth factor, matrix metalloproteinases, and their inhibitors within the follicle, leading to rupture of the follicle and ejection of the oocyte and its surrounding cluster of granulosa cells, the cumulus oophorus, out of the ovary and into the oviduct (Duffy & Stouffer, 2003; Stouffer, Xu, & Duffy, 2007).

3.2.3. Corpus luteum formation, function, and regression

Luteinization of the dominant follicle in response to the midcycle LH surge involves marked growth of the granulosa cells, proliferation of rough and smooth endoplasmic reticulum, structural modifications in the mitochondria, vascularization of the luteal cells, and increased gene expression (Zeleznik & Pohl, 2006; Stouffer et al., 2007). In humans and macaques, corpora lutea have an intrinsic lifespan of 14–16 days in nonconceptive cycles. During this time, they secrete both steroid and peptide hormones, including P_4, estrogens, relaxin, oxytocin (OXY), and inhibin A, primarily under the influence of LH. Both FSH and prolactin (PRL) have additionally been implicated as luteotropic hormones in primates, but appear to play minor roles in stimulating endocrine activity of corpora lutea in normal cycles (Zeleznik & Pohl, 2006). Interestingly, in a number of New World monkeys, atretic follicles are converted into accessory corpora lutea or interstitial glandular tissue, following luteinization of the granulosa cells or theca interna, respectively. These structures might, in turn, contribute to the extremely high circulating concentrations of estrogens and P_4 characteristic of New World primates (Dixson, 1998).

The causes of luteal regression in primates are not well understood. In contrast to many nonprimates, this process does not appear to be mediated to a significant extent by endogenous prostaglandins. Moreover, luteal regression does not appear to be determined by the decline in LH pulse frequency that occurs over the course of the luteal phase. Instead, luteal cells are thought to undergo apoptosis, associated with decreases in LH responsiveness and steroidogenic capacity (Brannian & Stouffer, 1991; Nakano, 1997).

3.3. Cyclical Changes in the Uterus

Anthropoid primates, along with several additional mammalian species (e.g., Rasweiler & De Bonilla, 1992; Zhang et al., 2007; Z. Wang, Liang, Racey, Y. Wang, & Zhang, 2008), are unique among mammals in undergoing a menstrual cycle—i.e., a cyclical pattern of changes in the nonpregnant uterus characterized by regular sloughing of the endometrium. Menstruation appears to be absent in all strepsirhines and possibly tarsiers, presumably in association with the noninvasive form of placentation used by these species (Martin, 2007) (see Section 5). Among the anthropoid primates, menstruation occurs in some New World monkeys and apparently in all catarrhines (Old World monkeys, apes, and humans) (Strassmann, 1996). The functional significance of menstruation is not yet understood. One hypothesis, that it protects females against sperm-borne pathogens (Profet, 1993), has been largely discounted (Strassmann, 1996; Martin, 2007), and an alternative hypothesis, that cyclical shedding and regrowth of the endometrium is energetically less expensive than maintaining the endometrium in a well-developed state across the entire cycle (Strassmann, 1996), has not been confirmed (Martin, 2007). Menstruation has also been suggested to be a nonadaptive consequence of the evolution of an invasive association between the embryo and an adaptive inflammatory reaction of the endometrium (Finn, 1998).

The uterine cycle in catarrhines comprises four main phases (Figure 13.4). The proliferative phase, corresponding to the mid to late follicular phase of the ovarian cycle, is characterized by increasing growth and vascularization of the endometrium under the influence of estrogens. During this time, edema and proliferation of endometrial stromal cells cause a marked increase in endometrial thickness; increases in size, number, and tortuosity of endometrial glands lead to development of a glandular network; and angiogenesis results in development of an elaborate vasculature. In addition, E_2 induces expression of both estrogen receptors (ERs) and P_4 receptors (PRs) in endometrial cells, thereby performing an obligate priming function for the subsequent secretory phase (Johnson & Everitt, 2000; Hess, Nayak, & Giudice, 2006).

The secretory phase of the uterine cycle corresponds to the luteal phase of the ovarian cycle. Under the influence of P_4 acting in concert with E_2, the endometrium undergoes additional histological and biochemical changes, including accumulation of glycogen by glandular epithelial cells; synthesis and secretion of glycoprotein-rich fluid into the glandular lumen; cell proliferation, increased capillary permeability, and edema in the stroma; and coiling of the characteristic spiral arterioles. The endometrium is receptive to implantation of a blastocyst during a narrow time window, corresponding to six to ten days after the LH surge in women (Johnson & Everitt, 2000; Hess et al., 2006).

The menstrual phase of the nonconceptive uterine cycle begins at the end of the luteal phase of the ovarian cycle, as luteal regression causes a decline in circulating P_4 and E_2 concentrations. In the absence of steroid support, lysosomal membranes in the endometrium break down, releasing lytic enzymes that degrade cellular elements and extracellular

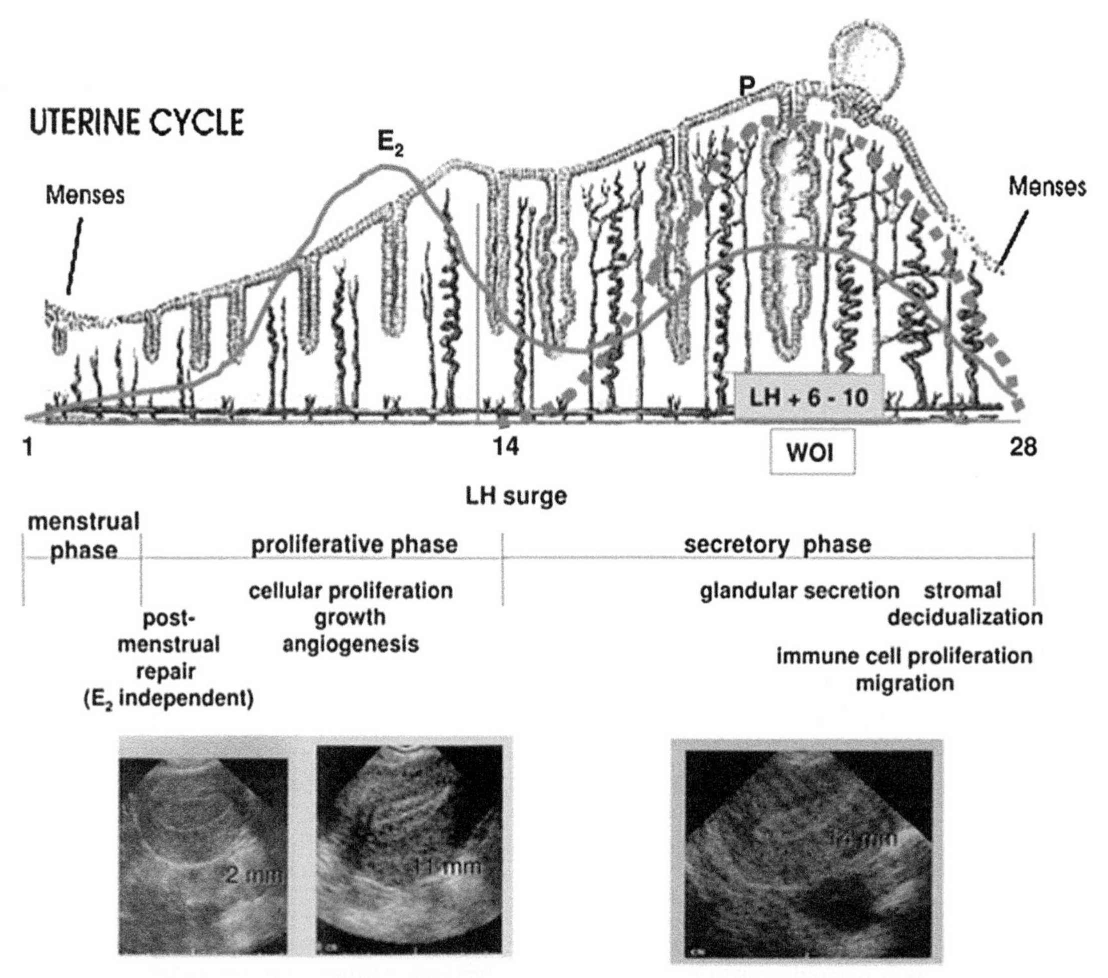

FIGURE 13.4 The menstrual cycle and its endocrine control in catarrhines, including cyclical changes in thickness, vascularization, glandular development, and secretory activity of the endometrium; phases of the uterine cycle and their major cellular events; circulating concentrations of estradiol (E_2) (solid line) and progesterone (P) (dotted line); window of implantation (WOI); and sonograms of the endometrium during the early proliferative, late proliferative, and secretory phases. *Reproduced from Giudice (2006) with permission; originally published by BioMed Central. See color plate section.*

matrix; spiral arterioles constrict, causing ischemia and vascular injury; plasminogen activators are released from the vascular endothelium, leading to production of plasmin and inhibition of clotting; and endometrial cells undergo apoptosis. Whereas the endometrium is resorbed in other mammals, in anthropoid primates the innermost two thirds of the endometrial lining and blood from the ruptured arterioles are expelled through the cervix and vagina as menstrual discharge (Lockwood, Krikun, Hausknecht, Want, & Schatz, 1997; Bergeron, 2000; Hess et al., 2006). By convention, the first day of menstruation is designated day one of a new cycle, corresponding to the beginning of the follicular phase.

Finally, menstruation is followed by a postmenstrual repair phase lasting several days in the early to mid follicular phase of the ovarian cycle. During this time, the endometrium heals and begins to regenerate its epithelial, stromal, and vascular components. These processes appear to be estrogen-independent and are thought to be initiated by a small population of endometrial stem cells (Bergeron, 2000; Hess et al., 2006).

3.4. Cyclical Changes in the Oviducts

In primates and other mammals, the oviducts undergo cyclical changes in histology, secretory function, and muscular activity under the influence of ovarian steroid hormones. During the follicular phase of the ovarian cycle, E_2 stimulates proliferation, hypertrophy and ciliation of epithelial cells, especially in the fimbrial and ampullary portions of the oviducts, as well as secretion of oviductal fluid and spontaneous muscle contractions. These changes peak during the periovulatory period and wane during the luteal phase, when the oviductal epithelial cells undergo atrophy, deciliation, and apoptosis under the influence of P_4, and secretion of oviductal fluid declines. This cyclical series of events is thought to play important roles in facilitating sperm transport, fertilization, early development of the conceptus, and implantation (Johnson & Everitt, 2000; Hess et al., 2006).

3.5. Cyclical Changes in the Cervix

The primate cervix undergoes cyclic, steroid-dependent fluctuations in muscular and secretory activity (reviewed by Johnson & Everitt, 2000; Nasir-ud-Din, Rungger-Brändle, Hussain, & Walker-Nasir, 2003; Suarez & Pacey, 2006). During the follicular phase, estrogens stimulate relaxation of the cervical muscles and increased secretion of mucus by the cervical epithelium. Cervical mucus, which contains water, glycoproteins, ions, enzymes, and immunoglobulins, plays key roles in sperm transport and defense of the female reproductive tract against microorganisms. Under the influence of estrogens, the mucus becomes profuse, highly hydrated, and highly penetrable to sperm. These characteristics peak around the time of ovulation, when cervical mucus blocks passage by microbes and abnormal sperm, while guiding normal, motile sperm through the cervix. Subsequently, during the luteal phase, P_4 increases the firmness of the cervix, decreases secretory activity of the cervical epithelium, and alters the quality and quantity of glycoproteins in the cervical mucus, causing secretion of small amounts of thick, viscous mucus that is impenetrable to sperm.

3.6. Cyclical Changes in the Vagina

Primates, like other mammals, undergo cyclical changes in the vagina under the influence of fluctuating estrogen and P_4 concentrations. During the follicular phase, estrogens stimulate increases in mitotic activity, glycogen content, thickness, and keratinization of the vaginal columnar epithelium, with these changes subsiding during the luteal phase (Farage & Maibach, 2000; Poonia et al., 2006). Cyclical changes also occur in the vaginal flora. Although the dominant microorganisms, *Lactobacillus* spp., may remain relatively constant across the menstrual cycle, levels of other microorganisms may fluctuate (Skangalis, Swenson, Mahoney, & O'Leary, 1979; Witkin, Linhares, & Giraldo, 2007). Cyclical changes in the vagina are especially pronounced in prosimians. In many of these species, as in numerous nonprimates, the vaginal orifice is imperforate throughout most of the cycle, opening—and therefore permitting intromission—only during the periovulatory period (e.g., ruffed lemur (*Varecia variegata*), Garnett's greater galago (*Otolemur garnettii*), fat-tailed dwarf lemur (*Cheirogaleus medius*)) (Van Horn & Eaton, 1979; Dixson, 1998).

3.7. Cyclical Changes in the External Genitalia and Sexual Skin

Females of many primate species undergo cyclical changes in coloration and tumescence of the external genitalia and so-called sexual skin, usually found on the rump and anogenital region (reviewed by Dixson, 1983; 1998). In numerous prosimians, the vulva becomes swollen and assumes a pink or red coloration during the periovulatory period. Such changes are greatly exaggerated in many catarrhines but are virtually nonexistent in New World monkeys.

Typically in catarrhines, estrogens stimulate tumescence—caused largely by edema—and intensification of the characteristic pink or red coloration of the sexual skin—a consequence of specialized vasculature—during the follicular phase (Dixson, 1983). Swelling and coloration peak during the periovulatory period, corresponding with the female's peak in sexual behavior. Progesterone antagonizes these effects during the luteal phase, so that detumescence occurs shortly after ovulation. Although hormonal control appears to be consistent across species, the location of sexual skin and extent of tumescence and coloration vary considerably. In geladas (*Theropithecus gelada*), e.g., pink or purple, nonedematous sexual skin is found on the circumanal, paracallosal, vulval, and clitoral regions, as well as on the lower abdomen. Additionally, white, fluid-filled vesicles develop around the edges of the sexual skin and in a figure-eight pattern on the chest during the periovulatory period (Dunbar, 1977; Dixson, 1998). Other species, including chimpanzees, bonobos (*Pan paniscus*), baboons, red colobus (*Piliocolobus badius*), and mangabeys (*Cercocebus* spp.), exhibit prominent pink swellings of the circumanal, vulval, and clitoral regions. Sex skin tumescence and coloration increase females' attractiveness to males, may facilitate intromission by males, and are thought to have evolved in response to sexual selection (Dixson, 1983; 1998).

3.8. Neuroendocrine Control of the Ovarian Cycle

The ovarian cycle in primates, as in other species, is governed by a complex interplay between the gonads, the gonadotropes in the anterior pituitary, and the GnRH pulse generator in the medial basal hypothalamus (reviewed by Johnson & Everitt, 2000; Messinis, 2006; Zeleznik & Pohl, 2006; see also Chapter 2, this volume). During the follicular phase, LH and FSH are released in low-amplitude, circhoral pulses, reflecting negative-feedback effects of estrogens on pulse amplitude but not frequency (see Figure 13.3). At midcycle, when estrogen levels exceed approximately 200 pg/ml for about 48 hours, estrogens trigger positive-feedback surges in GnRH, LH, and FSH, eliciting increases in pulse frequency and/or amplitude. Fully developed GTH surges in women require the presence of small amounts of P_4; however, no such effect is seen in rhesus macaques (Zeleznik & Pohl, 2006). Finally, the luteal phase is characterized by low-frequency, high-amplitude LH pulses, reflecting negative feedback primarily by P_4. In addition to ovarian steroids, inhibin B is secreted by the granulosa cells during the follicular phase, and inhibin A is secreted by the corpus luteum during the luteal phase, possibly exerting negative feedback specifically on FSH release; however, the precise role of inhibins in primate ovarian cycles is not yet clear (Zeleznik & Pohl, 2006; Randolph, 2008). Interestingly, FSH concentrations are elevated during the luteal phase in squirrel monkeys, suggesting that development of antral follicles may occur during this period, and possibly permitting the extremely short (~five-day) follicular phase of these species (Yeoman et al., 2000).

The sites of positive and negative feedback by estrogens have not been delineated fully in primates (reviewed by Johnson & Everitt, 2000; Messinis, 2006). A variety of experimental approaches have suggested that positive feedback by estrogens at the pituitary alone is sufficient to generate preovulatory LH surges, although GnRH plays an obligate permissive role. None-the-less, other studies have indicated that hypothalamic release of GnRH increases in response to sustained elevations of estrogens. Negative feedback by estrogens, likewise, may be mediated primarily at the level of the pituitary; however, estrogens reduce pulse amplitude (but not pulse frequency) of GnRH as well as LH, indicating that the negative feedback effects of estrogens are mediated in part at the hypothalamus (Mizuno & Terasawa, 2005; see also Chapter 2, this volume). Negative feedback effects of P_4 on GTH pulse frequency are assumed to occur at least partly within the central nervous system and are mediated, at least in part, by endogenous opioids. In addition, negative feedback effects of both estrogens and P_4 at the level of the brain appear to be mediated in part by Kp (Plant & Ramaswamy, 2009).

4. PUBERTY

In comparison to many other taxa, primates undergo an extended period of prepubertal development, as well as a prolongation of the period of reproductive maturation, known as puberty (see Table 13.1). This period is characterized by morphological, physiological, and behavioral changes driven by maturation and activation of the HPG axis (i.e., gonadarche) and, in some species, of the hypothalamus–pituitary–adrenal (HPA) axis (i.e., adrenarche). An excellent review of puberty in primates can be found in Plant and Witchel (2006).

4.1. Adrenarche

Several years prior to gonadarche, humans, chimpanzees, and possibly some other catarrhines undergo remodeling of the adrenal cortices, involving development of the zona reticularis (i.e., the innermost zone of the adrenal cortex) and increased secretion by the zona reticularis of androgens, particularly the weakly androgenic steroids dehydroepiandrosterone (DHEA) and its sulfoconjugate, DHEA-sulfate (DHEA-S). Circulating DHEA and DHEA-S levels peak during early adulthood in both males and females before declining gradually across the remaining lifespan. In other primates, the development of the zona reticularis and onset of adrenal androgen secretion occur during the prenatal and/or early postnatal period or not at all, and DHEA/DHEA-S secretion declines from high levels in infancy or remains stable throughout life (Campbell, 2006; Nguyen & Conley, 2008).

In humans and possibly chimpanzees, the pubertal development of the zona reticularis, known as adrenarche, is not associated with changes in secretion of cortisol, corticotropin (ACTH), gonadal steroids, or GTHs, and is not necessary for subsequent reproductive maturation (Auchus & Rainey, 2004; Campbell, 2006). In fact, the only known manifestations of adrenarche in humans are growth of pubic and axillary hair; development of apocrine glands in the skin, which may lead to body odor; and stimulation of sebaceous gland activity, which may cause acne (Auchus & Rainey, 2004). In addition, however, adrenarche has been hypothesized to play a role in maturation of the brain and skeletal system (Havelock, Auchus, & Rainey, 2004; Campbell, 2006), and premature (precocious) adrenarche in girls is associated with subsequent development of several clinical disorders, including polycystic ovary syndrome (PCOS) and metabolic syndrome (Auchus & Rainey, 2004; Nebesio & Eugster, 2007; Belgorosky, Baquedano, Guercio, & Rivarola, 2008).

4.2. Gonadarche

In infant primates, the pituitary and gonads secrete high levels of GTHs (i.e., LH and FSH) and steroid hormones (e.g., T, dihydrotestosterone (DHT), estrone (E_1), and E_2), respectively, for a period of weeks to months. This period of neonatal gonadal activity ends with the onset of the so-called juvenile or prepubertal hiatus, during which GTH levels drop precipitously and the gonads enter a dormant state, especially in males (reviewed by Plant & Witchel, 2006). Gonadal 'reawakening' occurs at the time of gonadarche, which begins anywhere from less than one year of age in several prosimians (e.g., fat-tailed dwarf lemur (*C. medius*) (Foerg, 1982)) to three years of age in macaques and squirrel monkeys, to seven to eleven years of age in chimpanzees, to nine to thirteen years of age in humans (Dixson, 1998; Plant & Witchel, 2006).

Gonadarche in male primates is characterized by dramatic elevations in circulating concentrations of LH and, to a lesser extent, FSH, reflecting primarily an increase in secretory pulse amplitude. These GTH increases, which are thought to reflect a concomitant amplification of pulsatile GnRH release from the hypothalamus, stimulate an increase in testicular volume (associated with growth of the seminiferous tubules, maturation of Sertoli cells, and proliferation of germ cells), development of Leydig cells, secretion of high levels of gonadal androgens, and initiation of spermatogenesis (Plant & Witchel, 2006). In humans and rhesus macaques, nocturnal elevations (known to be sleep-related in boys) in circulating LH and T levels precede diurnal elevations. The increased gonadal steroid concentrations stimulate development of species-typical secondary sexual characteristics, such as sex-specific facial and genital coloration (e.g., mandrill (*Mandrillus sphinx*)), throat sac and cheek flanges (orangutan (*Pongo pygmaeus*)), and specialized facial or body hair (e.g., hamadryas baboon (*Papio hamadryas*)), as well as the onset of sexual behavior (Dixson, 1998).

In female primates, as in males, gonadarche is triggered by a marked increase in secretion of FSH and, especially, LH by the gonadotropes, secondary to an increase in hypothalamic GnRH secretion (Watanabe & Terasawa, 1989). The surge in GTHs stimulates the initiation of cyclic ovarian activity, including the first development of Graafian (preovulatory) follicles, increased ovarian steroidogenesis, and, in catarrhines and some platyrrhines, the first menstrual period (menarche) (Plant & Witchel, 2006). Following menarche, the pituitary gonadotropes develop the capacity to exhibit GTH release via positive feedback in response to estrogens, culminating in the first ovulation. Thus, menarche precedes—by approximately a year in rhesus macaques, and by a year or more in humans and great apes—the onset of fertile ovulatory cycles, and is typically associated with a period of 'adolescent infertility,' characterized by anovulatory and irregular cycles (Bercovitch & Goy, 1990; Dixson, 1998). Across puberty, increasing E_2 concentrations stimulate uterine growth and maturation of species-typical secondary sexual characteristics, such as development of the breasts and nipples, and coloration and swelling of sexual skin (Dixson, 1998).

4.3. Neural Control of Gonadarche

Human clinical cases and studies of nonhuman primates have demonstrated that the 'reawakening' of the gonads at gonadarche is not limited by maturation of the gonads, pituitary, or hypothalamic GnRH neurons. For example, treatment of juvenile rhesus macaques with *N*-methyl-D-aspartate (NMDA), a receptor agonist of the excitatory neurotransmitter glutamate, elicits pulsatile release of GnRH from the hypothalamus, pulsatile release of LH from the pituitary, and gonadal activation, indicating that the GnRH neurons, gonadotropes, and gonads are already mature and capable of adult-like functioning prior to gonadarche (Plant, Gay, Marshall, & Arslan, 1989; Claypool, Kasuya, Saitoh, Marzban, & Terasawa, 2000). Further, developmental changes in neural or pituitary sensitivity to negative feedback by gonadal steroids (the so-called gonadostat hypothesis) do not account for the dramatic rise in GTH release during gonadarche. Instead, the proximate trigger for gonadarche involves maturation of neural inputs to the GnRH neurons, eliciting the dramatic increase in pulsatile GnRH secretion and, consequently, increases in pituitary secretion of GTHs and stimulation of gonadal endocrine and gametogenic activity.

Studies in rhesus macaques have implicated several neurotransmitters and neuropeptides in the onset of gonadarche. These include the inhibitory neurotransmitter

GABA, which plays a key role in restraining GnRH secretion during the juvenile period but exerts only modest inhibitory effects on GnRH release after the onset of puberty (Terasawa, 2005). Moreover, the developmental decrease in GABA release within the pituitary stalk-median eminence may stimulate a corresponding increase in release of the excitatory neurotransmitter glutamate, which may further elevate GnRH secretion (Terasawa, 2005). Neuropeptide Y has been implicated both in inhibiting GnRH release during the prepubertal hiatus and, paradoxically, in stimulating GnRH release during puberty and adulthood (Plant & Witchel, 2006).

In the past few years, much attention has focused on the role of the neuropeptide Kp, coded for by the *KiSS1* gene, and its receptor, GPR54 (also known as KiSS1R), in regulating gonadarche in humans and other primates (Plant & Ramaswamy 2009). In 2003, two research groups reported that members of consanguineous human families presenting with hypogonadotropic hypogonadism (i.e., impaired gonadal function secondary to GTH deficiency) and absence of puberty had homozygous mutations in the GPR54 gene (De Roux et al., 2003; Seminara et al., 2003). Since then, Kp-GPR54 signaling has been implicated compellingly in the control of hypothalamic GnRH release, pituitary GTH release, and onset of puberty in a number of mammalian species, including rhesus macaques (Roa, Aguilar, Dieguez, Pinilla, & Tena-Sempere, 2008). In macaques, both Kp and GPR54 are expressed in the arcuate nucleus of the medial basal hypothalamus (the site of the GnRH pulse generator), and expression of Kp (and of GPR54, at least in females) increases dramatically at the time of the pubertal increase in GnRH secretion (Plant, 2009). Moreover, pulsatile release of Kp and of GnRH in the median eminence is synchronized in midpubertal female rhesus macaques (Keen, Wegner, Bloom, Ghatei, & Terasawa, 2008), and treatment with exogenous Kp stimulates GnRH release in midpubertal females and pulsatile LH release in castrated juvenile males (Plant et al., 2006; Keen et al., 2008). Importantly, the stimulatory effect of Kp on LH can be blocked by simultaneous treatment with a GnRH receptor antagonist, suggesting that Kp affects the gonadotropes only indirectly, through its effects on GnRH secretion (Plant et al., 2006). Collectively, findings from humans and macaques suggest that Kp plays a critical role in triggering the pubertal increase in GnRH secretion; however, the precise nature of this role is not yet known.

4.4. Timing of Puberty

In spite of recent major advances in our understanding of the neurobiological processes governing primate puberty, the factors that determine the timing of these processes remain poorly understood (reviewed by Plant & Witchel, 2006). Clearly, the determinants of the timing of puberty are multifactorial, involving genetic, physiological, and environmental influences. Genetic factors are estimated to account for 50–80% of the variation in pubertal timing in humans, with mutations in such genes as those coding for the GnRH receptor, GPR54, and leptin leading to pathologically delayed or premature gonadarche; however, the specific genes accounting for variation in pubertal timing among clinically normal individuals remain unknown (Gajdos, Hirschhorn, & Palmert, 2009).

At the physiological level, one hypothesis is that the timing of puberty is governed by an endogenous 'pubertal clock' in the central nervous system, which initiates puberty at a specific age. This hypothesis is not widely accepted, however, in view of such findings as high variability in the age of human gonadarche, both within and among populations, as well as declines in the age of human gonadarche over recent decades (Ebling, 2005; Plant & Witchel, 2006; Euling, Selevan, Hirsch Pescovitz, & Skakkebaek, 2008).

Instead, the timing of puberty has long been thought to be governed by a putative 'somatometer' that measures some index of somatic growth. The somatometer hypothesis is supported by compelling evidence; however, the index of somatic development being monitored is not yet known. In recent years, attention has focused on a possible role of the adipocyte hormone leptin, circulating concentrations of which correlate with body fat mass. Findings in humans and rhesus macaques, as well as rodents, suggest that leptin plays a critical role in the onset of gonadarche. This role appears to be permissive, however, rather than serving as a direct trigger for puberty onset (Ebling, 2005; Plant & Witchel, 2006; Kaplowitz, 2008). Other indices of somatic development that have been implicated in determining the timing of puberty include insulin, growth hormone (GH), ghrelin, and metabolic fuels (Plant & Witchel, 2006; Kaplowitz, 2008; Tena-Sempere, 2008). Strenuous exercise, undernutrition, and chronic disease can all delay the onset of puberty, possibly acting through the putative somatometer (Plant & Witchel, 2006).

Finally, a number of environmental factors are known to modulate the timing of puberty in humans and nonhuman primates. Social influences can advance or delay puberty, as described in Section 9.2. In seasonally breeding species, aspects of pubertal maturation may be gated by seasonal cues such as photoperiod. Female rhesus macaques housed outdoors in the northern hemisphere, e.g., may undergo menarche at any time of year, but the occurrence of first ovulation is more or less restricted to the roughly three-month breeding season (autumn–winter) and is influenced by patterns of melatonin (MEL) secretion (Bercovitch & Goy, 1990; Wilson, Gordon, & Collins, 1986; Wilson & Gordon, 1989a; 1989b). Similarly, in squirrel monkeys, the onset of ovulatory cyclicity in young females and the first T surge in young males are restricted to the breeding season, presumably in response to photoperiodic cues (Coe, Chen, Lowe, Davidson, & Levine, 1981). Thus, seasonality 'imposes a quantum effect' on pubertal timing, such that gonadarche is more closely dependent on the number of breeding seasons elapsed since an individual's birth than on age *per se* (Plant & Witchel, 2006). Importantly, seasonally related cues do not necessarily govern maturation of the neural processes underlying pubertal reactivation of the GnRH neurons, but instead may play a permissive role in the expression of gonadarche, following this reactivation (Plant & Witchel, 2006).

Recently, endocrine-disrupting chemicals have gained attention as another source of variation in pubertal timing (Wang, Needham, & Barr, 2005; Cesario & Hughes, 2007; Euling et al., 2008; Schoeters, Den Hond, Dhooge, Van Larebeke, & Leijs, 2008; see also Chapter 14, this volume). In humans, pre- and/or postnatal exposure to a number of synthetic or naturally occurring chemicals can advance or delay the timing of gonadarche. For example, correlational findings suggest that puberty in girls may be advanced by *in-utero* exposure to the organochlorine dichlorodiphenyldichloroethylene (DDE) (a metabolite of the pesticide dichlorodiphenyltrichloroethane (DDT)), or by exposure to phthalates, a group of estrogenic compounds used to increase the flexibility of plastics (Cesario & Hughes, 2007). On the other hand, puberty in boys may be delayed by exposure to polychlorinated biphenyls (PCBs), industrial chemicals previously used in such products as coolants, flame retardants, and electronic components (Schoeter et al., 2008). Exposure to exogenous, naturally occurring steroids, such as phytoestrogens found in soy products and estrogens used in certain cosmetics and hair-care products, have additionally been implicated in altering the timing of gonadarche (Cesario & Hughes, 2007).

5. PREGNANCY

5.1. Overview

Numerous authors have reviewed the physiology of mammalian pregnancy (Albrecht & Pepe, 1990; Ogren & Talamantes, 1994; Solomon, 1994; Petraglia, Florio, & Simoncini, 1996; Albrecht & Pepe, 1999; see also Chapter 6, this volume). Much of this literature centers upon the best-studied primate species, i.e., humans, and is therefore relevant background to understanding primate pregnancy. In the following discussion, it can be assumed that findings for nonhuman primates are similar to those found for humans, unless otherwise stated.

Pregnancy presents unique physiological, immunological, and evolutionary challenges due to the combined presence of two or more distinct individuals (mother and fetus(es)) who are inextricably linked. The interface of this exchange between mother and fetus is the temporary organ, the placenta. This review will concentrate on the nature of the primate placenta, emphasizing recent findings on its endocrine nature.

The placenta develops from the outer cell mass, or trophectoderm, of the developing blastocyst that is in direct contact with the maternal endometrium. The outer cell mass eventually differentiates into two cell types: cytotrophoblasts and syncytiotrophoblasts. These cells form the fetal side of the boundary between mother and fetus. This boundary actively controls maternal–fetal exchange of nutrients, oxygen, and fetal wastes through alterations in passive diffusion capacity and active transport capacity. The development, maintenance, and alterations in the placenta as an exchange surface are controlled by autocrine and endocrine signals produced and received by the placenta.

The form of placentation varies greatly among mammalian taxa (Mossman, 1987; Benirschke, 2010). In the hominoid primates (apes and humans), the entire blastocyst implants into the uterine wall in a relatively invasive process that involves penetration of the maternal endometrial epithelium and invasion of the uterine vasculature (Luckett, 1974; Mossman, 1987; Lee & DeMayo, 2004). Monkeys exhibit superficial implantation, in which there is adherence to the uterine wall by the trophoblast, but without complete endometrial penetration or invasion of the deeper layers of the uterine wall (Luckett, 1974), whereas humans have a more invasive interstitial implantation with complete remodeling of uterine vessels. In all anthropoid primate placentae, however, the fetal trophoblast layer (the chorion) is in direct contact with the maternal blood supply; i.e., hemochorial placentation. It has long been proposed that this most invasive form of placentation evolved from the more shallow, epitheliochorial forms of placentation, but studies based on recent phylogenetic analysis suggest that hemochorial placentation was likely the ancestral form in mammals (Wildman et al., 2006). Hemochorial placentation offers the opportunity for endocrine signals produced by the conceptus—i.e., by the cytotrophoblast and syncytiotrophoblast cells—to be directly transmitted to the maternal bloodstream, offering a means for the placenta to affect maternal physiology in ways that may either increase or decrease fetal and placental growth (Haig, 1996; Rutherford, 2009). What follows is a description of aspects of autocrine and endocrine signaling in the primate placenta, with an emphasis on those areas that are unusual or unique to this taxonomic group. For broader characterization of the dozens of endocrine/autocrine placental processes that have been identified to date, consult Solomon (1994), Ogren and Talamantes (1994), Petraglia et al. (1996), and Petraglia, Floriom, and Vale (2005).

5.2. Steroids

The primate placenta interacts in a complex fashion with the maternal and fetal blood supplies to synthesize E_2 and P_4 and to convert cortisol to cortisone. As in other mammals, P_4 acts to alter the endometrial environment to allow implantation, including effects upon the maternal immune system. Progesterone also decreases contractility of the myometrium and inhibits lactation. Estrogens, too, prepare the uterus for implantation, but also play a critical role in the development of the endocrine/autocrine capacities of the placenta.

Albrecht and Pepe (1999) have proposed that placental estrogens are critical to the functional differentiation of the primate cytotrophoblasts into a syncytiotrophoblast. This differentiation includes an upregulation of 11β-hydroxysteroid dehydrogenase (11β-HSD) and P450 cholesterol side-chain cleavage ($P450_{scc}$); therefore, this differentiation controls the ability of the placenta to synthesize P_4 and to convert cortisol to cortisone (see Figure 13.5).

In primates, placental estrogen synthesis is dependent upon precursors supplied by the fetal adrenal gland. The primate fetal adrenal gland contains a wide inner zone, termed the fetal adrenal zone. This zone involutes rapidly after birth in humans while in rhesus monkeys it disappears more slowly (McNulty, Novy, & Walsh, 1981), but in all primates it disappears before adulthood. The fetal adrenal zone synthesizes DHEA-S, which is then used by the syncytiotrophoblast as substrate for E_2 synthesis.

At the same time, maturation of the glucocorticoid-producing zones of the fetal adrenal is controlled by fetal exposure to cortisol through the placenta. In midgestation, maternal cortisol is largely passed through the placenta as cortisol and, therefore, inhibits the fetal pituitary's production of ACTH. In late gestation, with more estrogens and, therefore, more 11β-HSD activity, maternal cortisol is converted to cortisone in the placenta, reducing fetal exposure to maternal cortisol and therefore allowing fetal ACTH production to increase. Increased fetal ACTH production leads to maturation of the fetal adrenal capacity to synthesize cortisol (Pepe & Albrecht, 1990; Mesiano & Jaffe, 1997). In this fashion, the placenta supports a timeline of fetal development that is controlled by the pace of developing placental endocrine/autocrine capacity.

5.3. Chorionic Gonadotropin (CG)

The syncytiotrophoblast synthesizes CG, a glycoprotein similar to LH synthesized by the pituitary gland. For details on the biochemistry of CG and its relation to LH and other glycoproteins, see Ogren and Talamantes (1994). Chorionic gonadotropin produced by the primate placenta functions to maintain P_4 and estrogen synthesis by the corpus luteum of the ovary. Therefore, CG is necessary for the early establishment of pregnancy in primates. In most other mammals, maintenance of corpus luteum P_4 production is accomplished through embryonic effects on prostaglandin $F_{2\alpha}$ ($PGF_{2\alpha}$), the primary signal for luteolysis (Niswender & Nett, 1994). In primates and a few other species (e.g., guinea pigs (*Cavia porcellus*)), however, CG acts trophically on the ovary, in the face of waning pituitary LH stimulation, to generate continued steroid production by the

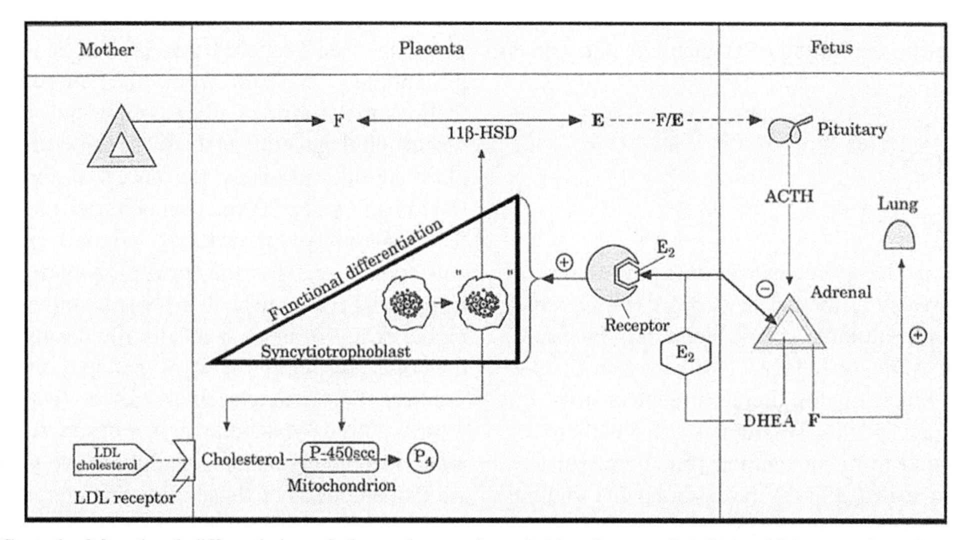

FIGURE 13.5 Control of functional differentiation of placental syncytiotrophoblast by estrodiol (E_2), which upregulates key components of the progesterone (P_4) biosynthetic pathway and the 11β-HSD (hydroxysteroid dehydrogenase)-1 and -2 system, which induces maturation of the fetal pituitary–adrenal axis. Syncytiotrophoblast is a multinucleated tissue of the placenta that produces hormones, depicted here as a drawn circle filled with smaller circles. Quotation marks signify functionally differentiated syncytiotrophoblast that is secreting hormones; e.g., placental lactogen and P_4. F, cortisol; E, cortisone; DHEA, dehydroepiandrosterone. *Reproduced from Albrecht and Pepe (1999) with permission.*

corpus luteum (Zeleznik & Benyo, 1994). Recent studies indicate that CG may also play a direct role in altering the character of the endometrium in preparation for implantation.

Numerous endocrine and autocrine factors have been found to affect CG production *in vitro*, including cortisol, CRH, triiodothyronine (T_3), thyroxine (T_4), GnRH, interleukin 6 (IL-6), and IGF-1, which increase CG secretion, and P_4 and transforming growth factor-β (TGFβ), which decrease CG secretion (summarized in Ogren & Talamantes, 1994). However, the specifics of how these interactions may function *in vivo*, separately or in concert, remain to be determined.

5.4. Chorionic Somatomammotropins (CSs)

The placentae of various mammalian taxa produce hormones that have both lactogenic and somatotropic properties. In most taxa, these hormones are derivatives of PRL (Schuler & Kessler, 1992; Soares & Linzer, 2001; Soares, 2004). In primates, however, placental lactogens are coded for by a series of genes that are part of a cluster of GH-like genes. Genetic comparisons among primates suggest that the duplication and possible selection events that led to these GH-like gene clusters occurred separately in Old World and New World primates (O. Wallis & M. Wallis, 2002; De Mendoza, Escobedo, Davila, & Saldana, 2004; Li et al., 2005). At least three of the GH-like genes are expressed in the placenta of humans (Kliman, Nestler, Sermasi, Sanger, & Strauss, 1986), rhesus macaques (Golos, Durning, Fisher, & Fowler, 1993), and baboons (Musicki, Pepe, & Albrecht, 1997). It is unknown whether the separate duplications in New World monkeys are also expressed by the placenta.

Chorionic somatomammotropin concentrations are correlated with placental weight, but not fetal weight at delivery in rhesus macaques (Novy, Aubert, Kaplan, & Grumbach, 1981). Walker, Fitzpatrick, Barrera-Saldana, Resendez-Perez, and Saunders (1991) report that CS has some direct somatotropic effects on fetal tissues, alters maternal carbohydrate and lipid metabolism, and aids in mammary cell proliferation. Reduced CS expression in human trophoblast is associated with intra-uterine growth restriction *in vivo* and with hypoxia exposure *in vitro* (Roh et al., 2005).

5.5. Corticotropin-releasing Hormone (CRH)

Corticotropin-releasing hormone was first identified as a hypothalamic neuropeptide controlling pituitary release of ACTH and hence affecting adrenal release of glucocorticoids. More recently, it has become clear that CRH (also known as corticotropin-releasing factor (CRF)) is expressed in other tissues and has different functions in those tissues. Most notably, CRH is expressed in numerous brain regions and, in primates, in the placenta. While CRH is produced in the placentae of all anthropoid primates that have been examined (Bowman et al., 2001; Smith et al., 2005), it is produced at extremely low concentrations or not at all in the placentae of other mammals (Smith et al., 2005). The pattern of circulating CRH throughout gestation differs between hominoid primates and monkeys: CRH rises continuously during the final trimester in humans and apes, peaking at term, while in monkeys it reaches its highest concentration during midpregnancy, declining thereafter (Goland, Wardlaw, Fortman, & Stark, 1992; Smith, Chan, Bowman, Harewood, & Phippard, 1993; Smith, Wickings, & Bowman, 1999; Power et al., 2006). In contrast with hypothalamic CRH, the secretion of which is inhibited by glucocorticoids, the secretion of placental CRH is enhanced by cortisol, in a positive feedforward pattern that is similar to the glucocorticoid–CRH relation in suprahypothalamic brain regions (Emanuel et al., 1994; Smith et al., 2005). This difference appears to be due to differential expression of transcription factors, coactivators, and corepressors in hypothalamic vs. placental tissue (King, Smith, & Nicholson, 2002).

Deviations from the typical CRH trajectory during gestation are often associated with preterm birth in humans, leading to the suggestion that CRH concentration may be a marker of early placentation events that set the stage for the rate of CRH change across gestation and the ultimate timing of parturition (Smith et al., 2005). Placental CRH may also affect the fetal HPA axis by stimulating fetal ACTH release. *In-vitro* evidence suggests that CRH can directly increase DHEA-S production from fetal adrenal cells, therefore providing additional substrate for placental estrogen production (Petraglia et al., 2005).

5.6. Leptin

Leptin was originally identified as a peptide, produced by adipocytes, that acts upon centers in the brain controlling satiety, energy state, and reproductive functions. Leptin is also produced by the placenta in a number of mammalian taxa; however, placental leptin production is significantly higher in primates than in other mammals. In addition, the distribution of leptin secretion to maternal vs. fetal compartments differs between rodents and primates (Henson & Castracane, 2002; Power & Tardif, 2005). In humans, 95% of placental leptin is released into the maternal circulation, suggesting that placental leptin acts mostly on maternal physiology and directly on the placenta rather than acting directly on the fetus (Hauguel-De Mouzon, Lepercq, & Catalano, 2006). Primate pregnancy can be considered a hyperleptinemic state, as leptin concentrations increase early in pregnancy and remain increased over nongravid concentrations until parturition. Local placental effects of leptin include enhancing CG production (Chardonnens et al., 1999) and enhancing synthetic processes (mitogenesis, amino-acid uptake, synthesis of extracellular matrix proteins and metalloproteinases) (Castellucci et al., 2000; N. Jansson, Greenwood, Johansson, Powell, & T. Jansson, 2003); i.e., regulation of placental growth. In the mother, it is proposed that the primary function of increased leptin concentration is to enhance mobilization of maternal fat stores. This may be of particular importance in humans, who produce fetuses with a relatively and absolutely larger amount of adipose tissue than other mammals, including other primates (Kuzawa, 1998).

6. LACTATION

Lactation is one of the defining characteristics of mammals. Milk synthesis and secretion by mammary cells into the alveolar lumen is a continuous process that requires PRL (Neville, 2001). However, PRL's influence on the process is modified by the effects of milk removal from the lumen on diffusion and transport processes. Release of OXY from the posterior pituitary causes contraction of myoepithelial cells surrounding the ducts and alveoli, forcing milk out into the nipple, where it can be accessed by the infant (Neville, 2001). If this neuroendocrine 'let-down' reflex is impaired, more milk remains in the lumen. Milk remaining in the lumen generates local factors that adjust milk secretion. In this way, the process of milk production is driven to a large extent by milk demand.

Compared to other mammalian taxa, primates produce milk that has low caloric density (i.e., high water content) and relatively low protein content (Oftedal, 1984; Milligan, Gibson, Williams, & Power, 2008). These features form part of a lactation strategy that involves frequent nursing throughout the day and night, combined with a relatively long period of exclusive milk feeding of young; i.e., weaning at a relatively late age.

Another unusual feature of lactation in most primates is the occurrence of a prolonged lactation-induced anovulatory period, termed lactational amenorrhea in primates that undergo menstrual cycles. Lactation has suppressive effects on folliculogenesis and ovulation in nonprimate mammals; however, the extent to which lactation and subsequent pregnancy are spaced apart by lactation is particularly striking in primates (McNeilly, 1994). It is well established that the suckling stimulus, rather than milk production *per se*, is the driving force behind lactational effects on the ovary. The suckling stimulus results in impaired hypothalamic GnRH release that, in turn, causes impaired pulsatile LH release from the pituitary (Weiss, Butler, Dierschke, & Knobil, 1976; McNeilly, 1994). Given this fact, it is perhaps not surprising that lactational infertility is greatly lengthened in one of the few mammalian orders providing regular and routine infant access to suckling. Primates are one of only two mammalian eutherian lineages (the other being Edentates) in which infants are routinely physically carried. In most primate species this transport is performed mostly by the mother and, as mentioned previously, frequent bouts of suckling throughout the day and night are a typical primate nursing pattern.

Early on, it was proposed that the hyperprolactinemia of lactation might suppress ovarian activity by inhibiting GnRH secretion, given the links between suckling stimulation and elevated PRL release (Freeman, Kanyicska, Lerant, & Nagy, 2000). However, the effect of PRL manipulation on time to resumption of ovulation in lactating macaques is variable (Maneckjee, Srinath, & Moudgal, 1976; Schallenberger, Richardson, & Knobil, 1981), and the correlation between circulating PRL concentration and duration of amenorrhea in lactating women also has been inconsistent (McNeilly, 1994; Tay, Glasier, & McNeilly, 1996). Further evidence that elevated PRL is not a driving force behind primate lactational anovulation is the case of the New World marmosets and tamarins. This is the only group of primates studied so far that does not display lactation-induced anovulation, with most individuals in a captive setting ovulating within 9–20 days following parturition. However, they display the same lactational hyperprolactinemia seen in other primates (McNeilly, Abbott, Lunn, Chambers, & Hearn, 1981).

Central OXY administration inhibits LH release in ovariectomized rhesus macaques (Luckhaus & Ferrin, 1989), but in marmosets OXY increases pituitary release of CG, the primary pituitary luteotropic hormone of New World monkeys (O'Byrne, Lunn, & Coen, 1990). Thus, central OXY effects may in some way mediate GnRH-induced LH/CG release from the pituitary and therefore may be tied to lactational anovulation/amenorrhea. More studies are required to define how such a mechanism might work. Opioids and dopamine also have been proposed as possible signals linking the suckling stimulus to GnRH suppression; however, the factors mediating this link remain unclear (McNeilly, 2001).

7. REPRODUCTIVE AGING

Primates, in common with many other mammals, display an inverted-U-shaped pattern relating female fertility parameters to age (e.g., Caro et al., 1995; Smucny et al., 2004). Anovulation, insufficient luteolysis, and impairment of gestational and lactational processes are all more common at the beginning and end of reproductive life (Atsalis & Margulis, 2008a). The extent to which late-life reductions in fertility are specifically due to aging neuro–endo–reproductive systems is, however, quite variable and often unclear. For example, Wright, King, Baden, and Jernvall (2008) report that aged female sifakas (*Propithecus edwardsi*), a Madagascar lemur, have decreased infant survival, but this effect is attributed to the females' aging dentition and resulting inability to support lactation. Thus, reduced fertility in old age does not, in and of itself, demonstrate impaired neuroendocrine or gonadal function.

Reproductive senescence will be used here to describe the process through which the HPG axis ages, resulting ultimately in cessation of function. vom Saal et al., (1994) provide an excellent overview of the process of reproductive senescence in laboratory rodents, and Wise (2006) provides a thoughtful perspective, comparing what is known regarding reproductive aging in rodents to that of women. Recent findings on nonhuman primate reproductive senescence, along with commentary, are found in Atsalis and Margulis (2008a). Female reproductive senescence differs among mammalian taxonomic groups. For example, in primates, the loss of the follicular pool is the primary event shaping the end of reproductive life, whereas, in rodents, striking variation is seen in the size of the follicular pool remaining at the end of reproductive life as well as at maximum lifespan (Wise, 2006).

Within primates, human females are unusual in experiencing follicular depletion relatively early in the maximal lifespan, resulting in an extended period of altered hormonal environments. These alterations stem from the declining negative feedback signals from the ovary (reduced circulating estrogens, P_4, and inhibin), resulting in elevated GTH concentrations for a time, followed by declining GTHs. These hormonal changes are believed to affect disease risks (Wise, 2006). The risk associated with bone loss due to decreasing estrogenic activity on osteoblasts is well described; however, cardiovascular effects continue to be hotly debated.

With increasing numbers of older nonhuman primates available for study, it is now clear that monkeys and apes also experience follicular depletion and associated hormonal alterations (Hodgen, Goodman, O'Connor, & Johnson, 1977; Graham, 1979; Tardif, 1985; Tardif & Ziegler, 1992; Shideler, Gee, Chen, & Lasley, 2001; Schramm, Paprocki, & Bavister, 2002; Atsalis & Margulis, 2008b; Videan, Fritz, Heward, & Murphy, 2008). However, the stage of life at which this occurs is generally later than that observed in humans (see Table 13.1). Atsalis and Margulis (2008b), in reviewing the data on monkeys and apes, conclude that 'potentially up to 25% of a female's life can be post-reproductive' (p 140). This claim is made in reference to maximal lifespan; in comparison, a human female living the maximal lifespan (now around 120 years) will spend around 58% of her life in a postreproductive state. When compared to average lifespan, as opposed to maximal lifespan, most nonhuman female primates will die at or before the point at which reproductive senescence begins. These comparisons have been controversial and will continue to be refined, given the oft-made claim that human female reproductive aging is unique and may be driven by indirect fitness advantages to postreproductive women providing resources to grandchildren; i.e., the grandmother hypothesis (Hill & Hurtado, 1991; Hawkes, 1997; Alvarez, 2000; Peccei, 2001a; 2001b).

Male primates, in common with many other male mammals, display decreases in circulating T concentrations with age (Ellison et al., 2002; Hardy & Schlegel, 2004; Tardif et al., 2008). Data from men and male rodents indicate reduced GnRH pulse amplitude with age, though LH concentrations do not decline. Old marmoset males have lower excreted T and appear to be hyper-responsive to exogenous GnRH stimulation. These findings, taken together, suggest that age-related changes in hypothalamic function may be important drivers of reduced T concentration with age.

8. SEXUAL BEHAVIOR

8.1. Description

Compared to other mammals, sexual behavior in primates is noteworthy for its flexibility—not so much with respect to the behavioral patterns used in courtship and copulation, but in the relative independence of these behaviors from hormonal control, especially in females. Here we describe the behavioral patterns commonly exhibited during primate courtship and copulation, and then discuss the role of steroid hormones in regulating these behaviors. For a comprehensive review of these and other aspects of primate sexual behavior, see Dixson (1998).

Most primate species exhibit fairly stereotyped dorsoventral mounting postures, similar to other mammals (Dixson, 1998; Campbell, 2007). Numerous variations are seen, however; e.g., the male and female may sit, stand, or hang from tree branches while mating, and may even copulate while suspended upside-down (e.g., aye-aye (*Daubentonia madagascariensis*)). Males use a variety of methods to stabilize themselves against females, including single- and double-foot clasp mounting, in which the male uses one or both feet to grip the female's ankles or legs (e.g., macaques, baboons); the leg-lock, in which a male positions his legs over the female's thighs (e.g., spider monkeys (*Ateles* spp.); brown woolly monkeys (*Lagothrix lagotricha*)); and manual grasping of the female's hips or waist (e.g., tamarins (*Saguinus* spp.); owl monkeys (*Aotus* spp.)) (Dixson, 1998). Ventro-ventral copulation is seen in a number of apes, and two species, human beings and bonobos, exhibit even more varied repertoires of copulatory positions.

Primate species differ markedly in the number and duration of intromissions prior to ejaculation (Dixson, 1998), although the most common pattern involves a single, brief intromission. For example, pygmy marmosets (*Cebuella pygmaea*) exhibit a single intromission lasting four to ten seconds and involving only several quick pelvic thrusts (Soini, 1988). In contrast, muriquis (*Brachyteles arachnoides*) perform an extended intromission, lasting an average of four minutes and consisting of a prolonged immobile period followed by five to ten pelvic thrusts and ejaculation (Milton, 1985), whereas rhesus macaques perform a series of up to twenty or more brief mounts, each lasting one to fourteen seconds, with two to fifteen pelvic thrusts per mount (Shively, Clarke, King, Schapiro, & Mitchell, 1982). Males and females often separate immediately or shortly after ejaculation, but in several species the male may remain intromitted for up to two minutes (stumptail macaque (*Macaca arctoides*) (Goldfoot et al., 1975)) or even several hours after ejaculation (thick-tailed greater galago (*Otolemur crassicaudatus*) (Eaton, Slob, & Resko, 1973)), suggesting the possibility of a genital lock.

Male primates use a variety of behaviors to initiate sexual interactions with females, including eye contact; specialized facial movements or expressions; approaches toward or following of the female; visual, olfactory, or oral investigation of the female's genitalia or sexual skin; and specific locomotor patterns (Dixson, 1998). Female sexuality, as described by Beach (1976), can be divided into three major components: attractivity (the female's attractiveness or stimulus value to a particular male), proceptivity (the female's behavioral role in initiating copulation), and receptivity (the female's willingness to copulate). Attractivity in primates may be based on both behavioral and nonbehavioral stimuli from females, including proceptive behaviors, sexual skin swellings in many Old World monkeys and apes, and olfactory cues in many prosimians and New World monkeys. Proceptivity is often a particularly striking aspect of primate sexual interactions. Females use a number of behaviors to arouse males' interest and solicit mating, including facial displays, specific body postures such as the 'sexual present' (presenting the anogenital region to the male), specialized vocalizations, and touching or even mounting the male (reviewed by Dixson, 1998). Sexual receptivity is manifest in stereotyped lordotic postures in at least some prosimian species, whereas female anthropoids do not exhibit lordosis. Consequently, receptivity in anthropoids is typically inferred from females' patterns of permitting, avoiding, refusing, or terminating mount attempts by males (Dixson, 1998).

8.2. Hormonal Influences on Sexual Behavior

As in other male mammals, sexual behavior in male primates, including men, is clearly influenced by androgens. Male sexual behavior tends to correlate with circulating androgen concentrations both across the lifespan and, in seasonal breeders, across the annual reproductive cycle (Dixson, 1998). Castration generally decreases frequencies of sexual behavior, especially intromission and ejaculation, whereas T replacement reverses these effects. Both men and monkeys, however, exhibit pronounced interindividual variation in their responses to castration: while some individuals show rapid declines in, or complete obliteration of, sexual behavior following castration, others show a more gradual cessation of sexual behavior, and still others continue to exhibit virtually normal sexual behavior even years later (Hull, Wood, & McKenna, 2006). Testicular androgens do not, therefore, appear to be essential for the expression of male sexual behavior, in at least some individuals.

The mechanisms by which androgens influence sexual behavior in male primates are not fully understood. In rodents, activational effects of androgens are mediated by intracellular aromatization of androgens to estrogens within the brain and subsequent binding to ERs. In primates, little evidence exists to support a critical role of aromatization in male sexual behavior; however, this issue has been addressed in only a small number of primate species, and therefore remains unresolved (Dixson, 1998; Wallen, 2005; Hull et al., 2006). Finally, P_4 inhibits male sexual behavior in both men and macaques (Hull et al., 2006).

Female sexual behavior in nonprimate mammals is typically exhibited only during the periovulatory period and is critically dependent on stimulation by ovarian steroid hormones. Prosimians appear to follow a similar pattern: females are sexually receptive during a limited period around ovulation – in some cases, only several hours—and ovariectomy abolishes female sexual behavior (Van Horn & Eaton, 1979; Dixson, 1998). In contrast, sexual behavior in anthropoid primates is characterized by its emancipation from strict regulation by gonadal hormones: females may engage in sexual activity at any point in the ovarian cycle, during pregnancy, or even following ovariectomy or menopause (Dixson, 1998; Campbell, 2007). Although such findings demonstrate that the expression of sexual behavior in female primates is not *dependent* upon gonadal hormones, other evidence indicates that female sexual behavior can, none-the-less, be *influenced* by them. In many New World monkeys, Old World monkeys, and apes, females show periovulatory peaks in copulatory behavior, associated with increases in proceptivity, attractivity, and, to a lesser extent, receptivity (Dixson, 1998). Moreover, ovariectomy decreases, and estrogen replacement restores, sexual behavior in a number of species. Again, proceptivity often shows a clear relationship with hormonal status in these studies, whereas effects on receptivity are much less pronounced (Dixson, 1998).

In contrast to estrogens, progestogens tend to reduce receptivity, proceptivity, and attractivity in female primates. Finally, androgens of ovarian and/or adrenal origin have been implicated in stimulating both receptivity and proceptivity in female primates, especially rhesus macaques and women; however, findings have been mixed, and the role of ovarian or adrenal androgens in female sexual behavior remains unclear (Dixson, 1998; Johnson & Everitt, 2000). A particularly controversial issue has been whether estrogens and other steroid hormones influence female sexual activity through peripheral (e.g., sexual skin, olfactory cues) and/or central (i.e., brain) actions. Current evidence suggests that estrogens act peripherally to enhance attractivity, whereas estrogens, progestogens, and androgens may all act centrally to modulate receptivity and proceptivity (Dixson, 1998).

The relationship between the ovarian cycle and sexual behavior in female primates may be influenced profoundly by the environment, especially the social environment. For example, orangutans and rhesus macaques show a pronounced peak in copulatory frequency during the periovulatory period under conditions in which females can control access to males, but not when females and males are 'forced' to interact in free-access pair tests (Wallen, 1990; Dixson, 1998). Similarly, periovulatory peaks in proceptivity and copulation are much more pronounced in rhesus macaques tested in a grouped (multi-female, single-male) situation than when males and females are tested as isolated pairs (Wallen, 1990). Such findings may reflect the use of sexual behavior for nonreproductive functions, the relative abilities of males and females to control sexual interactions under different environmental conditions, and increased agonistic interactions among females of some species (e.g., rhesus macaques) during sexual interactions with males (Wallen, 1990; Dixson, 1998). Wallen (1990) has interpreted such findings as evidence that 'hormones influence the sexual motivation required to initiate sexual activity in social circumstances requiring social effort' (p 239). Clearly, in contrast to rodents, in which ovarian steroids are essential for both sexual motivation and sexual performance (i.e., lordosis), ovarian (and possibly adrenal) steroids are not necessary for copulation in female anthropoid primates but may be one of several important factors influencing sexual desire.

9. ENVIRONMENTAL INFLUENCES ON REPRODUCTION

9.1. Reproductive Seasonality

Seasonal changes in the environment have long been known to affect reproduction in mammals. Outstanding comparative reviews of both proximate and ultimate aspects of mammalian reproductive seasonality are provided by Bronson (1989) and Malpaux (2006). For mammals in temperate regions, with strong circannual variation in both day length and temperature, the time of year at which various reproductive activities (mating, pregnancy, lactation) occur is often quite predictable, with dramatic differences in HPG function in breeding vs. nonbreeding seasons. Mammals adapted to tropical regions might be expected to have less striking reproductive seasonality, and that is the case with primates. The majority of primate species are restricted to tropical regions, macaques and humans being the best-studied exceptions. Primates exhibit an amazing array of seasonal patterning (see Table 13.1), ranging from predictable, relatively narrow breeding seasons, to seasonal, but quite variable, birth peaks, to no evidence of any seasonality. A comparison of these examples suggests that phylogeny is a poor predictor of these traits. While most prosimians are strongly seasonal and apes are generally not seasonal, the degree of seasonality in New World and Old World monkeys ranges from strong to nonexistent in a manner that is not explained by phylogeny within these groups.

Two primate genera with strongly seasonal patterns are the Central and South American squirrel monkeys (*Saimiri*) and the northernmost macaques (*Macaca*), such as rhesus macaques and Japanese macaques (*M. fuscata*). In these species, females either ovulate irregularly or fail to ovulate altogether during the nonbreeding season. Studies have suggested both central changes in GnRH pulsatility (Hendrickx & Dukelow, 1995) and direct changes in ovarian function (Hutz, Dierschke, & Wolf, 1985) as possible factors in seasonal anovulation of female rhesus macaques, but the exact mechanisms have not been fully elucidated. In squirrel monkeys, reductions in FSH secretion appear to underlie the seasonal shift from an ovulatory to an anovulatory pattern (Kuehl & Dukelow, 1975).

In both squirrel monkeys and rhesus macaques, males also display a seasonal pattern of T production, with peaks occurring as the mating season commences. Both squirrel monkey and macaque males display T-supported characteristics that arise during the breeding season. Male squirrel monkeys undergo weight gain of around 14%, caused largely by retention and deposition of water along the arms, shoulders, and back (Jack, 2007), as well as increases in testicular volume of 150% (Wiebe et al., 1988). During the nonbreeding season, male rhesus macaques display reduced LH pulsatility, reduced diurnal rhythms in pulsatility, regression of seminiferous tubules, and few spermatocytes or spermatids. In the months leading into the mating season, LH pulsatility increases, seminiferous tubular diameter increases, and spermatogenesis commences (Wickings & Nieschlag, 1980; Wickings, Marshall, & Nieschlag, 1986). Breeding males also display skin reddening, likely related to increasing T production.

The roles of specific environmental cues in generating these neuroendocrine changes in squirrel monkeys and rhesus macaques remain obscure; however, changing day length cues do not appear necessary, as the patterns are retained in controlled day length conditions (Wehrenberg & Dyrenfurth, 1983).

Circannual variation in the tropical environment is most strongly tied to rainfall that, in turn, generates seasonal variation in food availability. Where there is birth seasonality, births usually occur in the dry season; however, there are exceptions (e.g., Struhsaker & Leland, 1987). As a result of long gestation and lactation periods, primates cannot limit all reproductive investment to a single circannual period. For example, if weaning of infants is to occur during a period of relative food abundance, then other aspects of the reproductive cycle, such as mating and gestation, may occur during periods of relative food scarcity. Numerous attempts have been made to model the manner in which selective pressures may have shaped these responses, but no single model explains the wide variation seen in primate breeding schedules (Van Schaik & Brockman, 2005).

It is possible that some seasonal patterning results from alterations in physical activity or food availability. In women, e.g., high levels of physical activity are associated with suppressed ovarian function (Jasienska & Ellison, 2004). The increased day range lengths and home range sizes seen in some primate species during times of scarcity (Hemingway & Bynum, 2005) may inhibit HPG axis activity through altering energy balance or stress (See Section 9.3).

9.2. Social Influences on Reproduction

Primates exhibit a broad diversity of social systems, including a dispersed, relatively solitary lifestyle; social monogamy and nuclear family units; single-male, multi-female groups; and multi-male, multi-female societies, some of which may undergo complex fission/fusion patterns. In each of these social configurations, reproductive function of females, and in some cases males, may be modulated by salient social cues. The nature and magnitude of these socioendocrine effects, as well as the behavioral or sensory cues eliciting them, vary markedly among species, sexes, and social systems. In general, though, both males and females often exhibit enhanced activity of the HPG axis in response to interactions with or cues from unrelated, opposite-sex adults, and inhibition of HPG activity in response to same-sex adults, especially those of higher dominance status. Such effects may influence the course of reproductive maturation in adolescents, or may alter or even abolish fertility in fully mature adults. Below we describe some of the best-studied examples in each sex.

9.2.1. *Males*

9.2.1.1. Social influences on reproductive maturation in males

In several primate species, the timing and/or trajectory of reproductive maturation in males is reported to be influenced by the social environment. Most commonly, adolescent males of high dominance status, or those with high-ranking mothers, have been found to undergo earlier puberty and to have higher circulating or excreted T concentrations and larger testes, compared to lower-ranking males. Among captive rhesus macaques, e.g., both plasma T concentrations and testicular mass (either absolute or corrected for body mass) in adolescent males tend to correlate with the males' dominance status, especially at the outset of the mating season (Bercovitch, 1993; Dixson & Nevison, 1997). Moreover, adolescent males with high-ranking mothers or from high-ranking matrilines have higher plasma T levels and heavier testes, and may attain puberty earlier, than those with lower-ranking mothers (Dixson & Nevison, 1997; Mann, Akinbami, Gould, Paul, & Wallen, 1998).

Similar patterns have been described in semi-free-ranging mandrills, in which adolescent males that were relatively high-ranking for their age had higher circulating T levels, larger testes, and greater development of secondary sexual characteristics (sexual skin coloration and activity of the sternal scent gland) than lower-ranking adolescents (Setchell & Dixson, 2002). In free-ranging baboons (*Papio cynocephalus*), the onset of puberty, as determined by the age of testicular enlargement, was significantly correlated with maternal rank, with sons of high-ranking females undergoing testicular enlargement up to a year earlier than sons of low-ranking females (Alberts & Altmann, 1995). Interestingly, the age at testicular enlargement is also advanced in sons of high-ranking males if their father remains in the same social group during the son's juvenile development (Charpentier, Van Horn, Altmann, & Alberts, 2008b).

Social modulation of male reproductive development is especially pronounced in the orangutan. In both wild and captive populations, adolescent males living in proximity to fully adult males often, but not always, exhibit delayed development of secondary sexual characteristics, including cheek flanges, laryngeal sac, beard and mustache, musky odor, and large body size (Kingsley, 1982; Maggioncalda, Sapolsky, & Czekala, 1999; Setchell, 2003). Such 'developmentally arrested' adolescents are fertile and may sire offspring; however, they appear to be sexually unattractive to females and may commonly force copulations (Utami, Goossens, Bruford, De Ruiter, & Van Hooff, 2002). These males exhibit significantly reduced urinary concentrations of T, DHT, LH, cortisol, and PRL (but not FSH), as compared to adolescents undergoing development of secondary sexual characteristics (Maggioncalda et al., 1999; Maggioncalda, Czekala, & Sapolsky, 2002). The precise role of adult male orangutans in suppressing reproductive maturation in adolescents has not been tested experimentally; however, anecdotal evidence from both captive and free-living orangutans indicates that, after the removal or disappearance of a nearby, fully developed adult male, arrested adolescents may rapidly resume sexual development and attain full maturation within several months, even after a period of developmental arrest lasting 20 years or more (Maggioncalda et al., 1999; Utami, 2000, cited in Setchell, 2003).

9.2.1.2. Social influences on reproduction in adult males

Reproductive function of adult males can be influenced by both intrasexual and intersexual stimuli in a wide variety of primates. Typically, interactions with or cues from other adult males dampen activity of the HPG axis, while interactions with or cues from adult females have stimulatory effects. These two classes of socioendocrine effects may interact with each other and with responses to other environmental cues (e.g., photoperiod) in complex ways.

One of the best-studied examples is the gray mouse lemur (*Microcebus murinus*), a small, nocturnal, relatively nonsocial prosimian (reviewed in Perret, 1992). In captive, mixed-sex social groups, middle- and low-ranking males engage in very little sexual behavior and exhibit low circulating T concentrations during the annual mating period, as compared to dominant males. Moreover, circulating levels of sex hormone-binding globulin are elevated in middle- and low-ranking males, further reducing the bioavailability of T to the tissues. Circulating cortisol concentrations do not differ among high-, middle-, and low-ranking males but are elevated in all group-housed males, as compared to socially isolated males, suggesting that social housing *per se*, but not necessarily subordination, is stressful.

Circulating T levels in male gray mouse lemurs are modulated by a chemosignal found in the urine of socially dominant or isolated males (Perret, 1992). Interestingly, response to the male chemosignal varies across the annual cycle (Perret & Schilling, 1995). During the onset of photoperiodic stimulation (i.e., long days, as would occur at the outset of the breeding season), exposure to urine from an isolated or dominant male reduces T levels in other males. During continued exposure to long day lengths (i.e., later in the breeding season), exposure to male urine delays testicular regression and the decline in T levels that would normally result from photorefractoriness. Finally, during exposure to inhibitory short days (i.e., the nonbreeding season), exposure to male urine stimulates testicular recrudescence and markedly elevates plasma T levels. Under any of these photoperiodic conditions, circulating T levels can be increased in males by exposure to chemosignals from females, especially during the females' proestrus period. The neuroendocrine mechanisms mediating these stimulatory and inhibitory social effects are not fully understood but appear to involve both PRL and the endogenous opioids (Perret, 1992).

Circulating T concentrations in males are similarly influenced by social partners of both sexes in the squirrel monkey, a seasonally breeding New World primate that lives in large, multi-male, multi-female groups. In captivity, plasma T levels are higher in males housed with multiple females than in males housed with only a single female, and are higher in dominant males than in intermediate- and low-ranking males (Mendoza, Coe, Lowe, & Levine, 1979; Schiml, Mendoza, Saltzman, Lyons, & Mason, 1996). None-the-less, seasonal changes in T, as well as in circulating cortisol levels and body mass, occur in all males, regardless of social rank or access to females (Schiml et al., 1996; Schiml, Mendoza, Saltzman, Lyons, & Mason, 1999). Importantly, inter-individual differences in T levels result from, rather than cause, differences in rank: male T levels prior to group formation do not predict subsequent attainment of social status, but increase in dominant males and decrease in subordinates following both formation of all-male groups and introduction of females (Mendoza et al., 1979).

9.2.2. Females

Reproductive attempts are considerably more costly for female mammals than for males, as a consequence of the physical constraints and energetic demands imposed by pregnancy, lactation, and maternal behavior. Consequently, females are likely to undergo more intensive selection than males to initiate breeding attempts under auspicious environmental conditions, and hence to take advantage of environmental cues indicative of conditions favorable for infant survival, including cues arising from the social environment (Wasser & Barash, 1983). Not surprisingly, therefore, females across a wide range of primate taxa exhibit clear reproductive responses to social variables. As with males, social factors can modulate the timing and trajectory of reproductive maturation in young females, as well as reproductive physiology in fully mature adults.

9.2.2.1. Social influences on reproductive maturation in females

In several primate species, interactions with or cues from unrelated adult males advance puberty in young females. Female Garnett's greater galagos and Senegal lesser bush babies (*Galago senegalensis*), e.g., undergo their first vaginal estrus significantly earlier if pair-housed with an adult male than with a peer male (Izard, 1990). Similarly, cohabitation with a stepfather has been associated with earlier puberty in human girls (Ellis & Garber, 2000).

As in male primates, interactions with or cues from same-sex conspecifics tend to delay puberty in females. Among captive rhesus macaques, e.g., age at menarche is not associated with dominance status, but high- and middle-ranking females undergo their first ovulation at significantly younger ages than low-ranking females (Schwartz, Wilson, Walker, & Collins, 1985; Zehr, Van Meter, & Wallen, 2005). In free-ranging baboons (*P. cynocephalus*), menarche occurs earlier in daughters of high-ranking females than in daughters of low-ranking females (Bercovitch & Strum, 1993; Wasser, Norton, Kleindorfer, & Rhine, 2004; Charpentier, Tung, Altmann, & Alberts, 2008). Earlier menarche in baboons is also associated with a number of additional social factors, including living in a group with more maternal half-sisters or fewer adult females, and a longer period of coresidency with the father during the daughter's juvenile period (Charpentier et al., 2008a; 2008b). The mechanisms underlying such social modulation of female reproductive maturation are not known but have been suggested to involve differences in nutritional status, body mass, or psychosocial stress (Bercovitch & Strum, 1993; Wallen & Zehr, 2004; Zehr et al., 2005).

9.2.2.2. Social influences on reproduction in adult females

Social cues influence ovarian cycle dynamics in adult females of several primate species. In gray mouse lemurs, both tactile and distal cues from other females cause lengthening of the ovarian cycle, associated with an increase in luteal-phase length and a decrease in plasma P_4 concentrations (Perret, 1986). In women, axillary secretions both from other women and from men have been implicated in modulating the ovarian cycle. Axillary secretions taken from women in the late follicular phase increase LH pulse frequency, advance the timing of the preovulatory LH surge, and shorten ovarian cycle length in recipients, whereas axillary secretions collected from women during the ovulatory phase of the cycle produce the opposite effects (Stern & McClintock, 1998; Shinohara, Morofushi, Funabashi, & Kimura, 2001). Collectively, these effects might underlie the pattern of menstrual synchrony documented in many, but not all, studies of women roommates, friends, and coworkers (Weller & Weller, 1993). Luteinizing hormone pulse frequency is also increased in women exposed to axillary secretions from men (Preti, Wysocki, Barnhart, Sondheimer, & Leyden, 2003).

Perhaps the most dramatic example of social regulation of reproduction in female primates is reproductive suppression in the Callitrichidae (marmosets and tamarins). These small New World monkeys live in groups of approximately 4–15 individuals, which may include several adults of each sex as well as juveniles and infants. In most species of callitrichid, however, only a single, behaviorally dominant female breeds in each social group (Digby, Ferrari, & Saltzman, 2007). Subordinate females fail to breed, as a result of social suppression of ovulation and/or inhibition of sexual behavior, and instead serve as nonreproductive alloparents, helping to rear the infants of the dominant female. The mechanisms underlying this social control of fertility, or 'social contraception' (Abbott, 1984), differ among species (French, 1997) but have been studied most thoroughly in the common marmoset (reviewed by Abbott, Digby, & Saltzman, 2009; Saltzman, Digby, & Abbott, 2009). Most subordinate females in laboratory groups of common marmosets, and at least some in wild groups, fail to ovulate and exhibit impairments in ovarian steroidogenesis and follicular development. These deficiencies in ovarian function, which may last for periods of up to several years or more, are caused by suppressed pituitary secretion of CG, which is released by the anterior pituitary instead of LH in this species (see Section 2.1). The exact mechanism underlying CG inhibition is not yet known; however, it appears to be associated with enhanced negative-feedback sensitivity of the brain and/or pituitary to low levels of estrogen, blunted responsiveness to estrogen positive feedback, and enhanced CG inhibition by endogenous opioids. Surprisingly, however, CG suppression does not appear to be associated with altered hypothalamic secretion of GnRH and is not accompanied by manifestations of generalized stress (Abbott et al., 2009; Saltzman et al., 2009).

9.3. Energetics of Reproduction

9.3.1. Introduction

The mammalian reproductive system is adept at monitoring maternal condition and parsing the energy available for gamete production, fetal growth, and infant growth. The manner in which maternal energetic state is sensed and signals are processed to alter reproductive function has been a particularly active area of research for the past two decades. A good part of that research interest has been driven by the discoveries of a variety of endocrine and autocrine factors produced by adipose tissue, as well as identification of new hormones produced by the gastrointestinal tract, all offering a myriad of possible cues of metabolic state (Wade, Schneider, & Li, 1996; Baird, Cnattingius, Collins, & Evers, 2006; Tena-Sempere, 2007; Roa et al., 2008).

From the standpoint of tradeoffs, one would expect that early stages of reproduction, including ovulation, placentation, and fetal growth, might be more plastic in response to maternal energy stores than lactation because of the extent of investment already made late in the reproductive event. The ability of the mother to 'cut bait' early in investment and save energy stores for future investment might offer more of a selective advantage the earlier in reproduction this event occurs. Later investment, including lactation, by which time the primate mother has already invested large amounts of energy and time in the infant, should be less responsive to changes in maternal energy stores, enhancing the likelihood that the mother's investment (i.e., the offspring) survives. Therefore, certain stages of placental, fetal, and infant development may be more affected by changes in maternal status than others, and the manner in which autocrine, paracrine, and hormonal factors convey information regarding maternal energy state differs among the following points at which a female mammal might adjust investment: ovarian folliculogenesis leading to ovulation; placental transfer of resources from mother to fetus; and mammary gland transfer of resources from mother to infant. What follows is a discussion of autocrine/endocrine signals that may act at different points in the primate reproductive cascade to signal energy excess or deficit.

9.3.2. Energetics of hypothalamic–pituitary function

The cellular processes leading to maturation of gametes—both sperm and ova—are controlled by central nervous system mechanisms that respond to both negative feedback and positive feedforward from steroids and other hormones produced by the gonads. The basics of this HPG axis control system have been relatively well characterized for some time. More recent research has begun to better characterize those signals that may alter pituitary GTH release—through either altered hypothalamic GnRH pulse generation or direct effects on the pituitary—and thus alter the production and release of gametes. Recently identified signals that may indicate energetic state include leptin, produced in adipose tissue, and ghrelin, produced in the gastrointestinal tract.

As circulating leptin concentration is correlated with the amount of adipose tissue, leptin can function as a signal of stored energy availability. As such, the impact of leptin in the central nervous system is generally to reduce food intake and to enhance reproduction. Studies in rodents, monkeys, and humans reveal that leptin exposure stimulates pituitary LH release (Tena-Sempere, 2007). Normal pulsatile release of LH and FSH can be restored by leptin treatment in calorie-deprived women (Schurgin, Canavan, Koutkia, Depaoli, & Grinspoon, 2004). In rats, intrahypothalamic infusion of leptin alters GnRH pulse generation (Watanobe et al., 2002), but there is also evidence of direct effects of leptin upon pituitary gonadotropes. The interaction of leptin with GnRH neurons may be mediated by Kp (Tena-Sempere, 2007). While the overall picture of leptin as an energy signal is one in which leptin supports reproduction, leptin also has been found in some studies to perform actions that may impair GTH release (see also Section 9.3.3). In addition, studies of human obesity indicate that individuals with extremely high adiposity may become leptin-resistant. The full picture of the manner in which leptin, alone or in concert with other signals, alters HPG axis function remains, therefore, to be elucidated.

As opposed to leptin, the peptide ghrelin, produced by the gastrointestinal tract, is a potent orexigenic stimulus in the hypothalamus and has been proposed as a signal of energy insufficiency. Ghrelin, like leptin, affects reproductive aspects of HPG axis function as well as food intake. Administration of ghrelin inhibits LH pulsatility in rodents and monkeys (Vulliémoz et al., 2004). Recent studies suggest that this effect may be mediated by the HPA axis: ghrelin induces elevations in circulating glucocorticoid concentrations, and inhibition of this action abolishes ghrelin's ability to alter LH pulsatility (Vulliémoz, Xiao, Xia-Zhang, Rivier, & Ferin, 2008).

9.3.3. Energetics of gonadal function

The actions of the HPG axis on folliculogenesis and ovulation are mediated by a large number of peptide signals, many of which are produced in the ovary. Some of these signaling systems are reasonable prospects as possible metabolic cues. A primary example is the IGF system. Insulin-like growth factor-1 and IGF-2 are present in follicular fluid, as are five of the six IGF-binding proteins that control cellular access to IGFs. Locally produced IGF and local control of IGF-binding proteins are proposed to regulate the development and/or atresia of antral follicles (Giudice, 2001). However, because circulating IGF is controlled by GH, a hormone that is sensitive to energy state, the IGF system that mediates folliculogenesis also may be affected by energy state.

In contrast to the hypothalamus and pituitary, where leptin and ghrelin have largely opposing effects, the direct effect of both of these peptides on the gonads is to inhibit production of steroids (T in the testis, E_2 in the ovary). In addition, expression and immunostaining studies indicate that ghrelin is produced by the gonads, as well as by the gastrointestinal tract (Tena-Sempere, 2007). The manner in which leptin and ghrelin exert local effects on primate gonads remains to be elucidated.

9.3.4. Energetics of pregnancy

Pregnancy represents a peculiar environment in which the short- and long-term reproductive interests of the mother are played out with—or against—the short-term interests of the fetus. Variation in placental structure and function are mechanisms through which the mother may alter investment in the fetus, but they are also mechanisms through which the fetus may manipulate maternal investment (Haig, 1993; 1996; Crespi & Semeniuk, 2004). For example, increases in litter size in the common marmoset are accompanied by a dramatic expansion of the placental interface, which is the site of fetomaternal nutrient and hormonal exchange (Rutherford & Tardif, 2009). As the placenta has the same developmental legacy and genomic identity as the fetus, this functional plasticity has important implications for metabolic and hormonal signaling.

Placental growth is driven mostly by placentally derived growth factors, the most important being IGF-2, and mostly is inhibited by maternal factors (Lewis, Morlese, Sullivan, & Elder, 1993; Crossey, Pillai, & Miell, 2002). In nonprimates, deficiency of IGF-1 or IGF-2 results in a reduction in birth weight (Fowden, 2003; Gicquel & LeBouc, 2006). In cases of intrauterine growth restriction (IUGR), there is often a higher fetal : placental weight ratio, suggesting that the placenta may actually be more efficient in the transport of nutrients (Constância et al., 2002; Rutherford & Tardif, 2008), even though that increased efficiency is ultimately insufficient to sustain normal growth. Deficiency in IGF-2 leading to IUGR may be associated with upregulation of amino acid transport systems (Constância et al., 2002), indicating a specific role for placental hormonal signals in directing fetal metabolic pathways.

Two other placental hormones, GH and leptin, have been proposed to mediate a maternal insulin resistance and, therefore, a shift toward reliance on lipolysis for energy production. In humans, reduced placental GH and, therefore, reduced maternally produced IGF-1 is associated with IUGR (Lacroix, Guibourdenche, Frendo, Muller, & Evain-Brion, 2002). At the other end of the spectrum, gestational diabetes in humans is associated with markedly higher concentrations of placental leptin, and leptin pathways have been suggested as one of the possible mechanisms underlying diabetes-induced fetal macrosomia (i.e., birth weight ≥ 4000 g) (Hauguel-de Mouzon et al., 2006).

While numerous studies describe the relations among maternal condition and birth outcomes in primates, including humans (e.g., Lee, Majluf, & Gordon, 1991; Bowman & Lee, 1995; Fairbanks & McGuire, 1995; Bercovitch, Lebron, Martinez, & Kessler, 1998; Johnson, 2003), there are few primate studies in which maternal condition has been manipulated systematically. One such study was conducted in captive common marmoset monkeys exposed to a modest (25%) restriction of energy intake during either mid or late pregnancy. This restriction resulted in abortion of all mid-term pregnancies. Maternal urinary concentrations of CG, cortisol, and free E_2 were all lower in the restricted pregnancies, suggesting that the change in maternal energy availability resulted in impaired placental function. Restrictions in late pregnancy did not reliably induce pregnancy loss, though the number of preterm deliveries was higher than expected (Tardif, Power, Layne, Smucny, & Ziegler, 2004; Tardif, Ziegler, Power, & Layne, 2005). The extreme outcome in the face of a relatively modest restriction at midpregnancy suggests that placental formation and function in this very small primate is quite sensitive to maternal energy state. In contrast, a similar level of food restriction in captive baboons affected placental weight but not fetal growth (Schlabritz-Loutsevich et al., 2007), with fetal growth perhaps protected by decreased maternal activity and increased use of maternal stores.

9.3.5. *Energetics of lactation*

Given the time and energy investment represented by the primate neonate, one might propose that maternal post-parturition investment would be somewhat less sensitive to maternal condition. Neville (2001) states that, 'unlike the nutrition received by the fetuses through the placenta, the nutrition received by breastfed infants is not dependent upon the status of maternal metabolism. For most milk components, the secretory mechanisms are insulated from the regulatory mechanisms that control nutrient flux in mothers, so sufficient milk of adequate composition is available to infants even during inadequate food intake by mothers.' (p 20). Studies of primate milk are limited to a few species, with most studies being done in humans. While intra- and inter-species variation in relative fat content and, therefore, in energy density is a common finding (Power, Oftedal, & Tardif, 2002; Milligan et al., 2008), a complete failure of milk production is rare, supporting Neville's contention that milk production is more insulated from maternal insults. In lactating common marmosets that had either variable energy demand (e.g., nursing singletons vs. twins) or variable maternal energy availability (e.g., different maternal nutritive conditions or maternal energy restriction), the combination of high energy demand and low maternal stored energy availability resulted in slower growth of infants, but in no case did mothers cease investment (i.e., stop lactating) (Tardif, Power, Oftedal, Power, & Layne, 2001; Tardif & Ross, 2009).

Perhaps the largest effects of maternal condition on lactation are noted in relation to the time to weaning. Lee et al. (1991) proposed that weaning in primates was related to attainment of critical weights, so that slower-growing infants (e.g., infants of mothers with lower milk energy) would be expected to be weaned at later ages. As the mother's milk production is driven by the relative amount of emptying of the alveolar lumen, the lactating mother and the nursing infant make up a complex, behaviorally driven feedback system.

10. CONCLUSIONS AND FUTURE DIRECTIONS

Primates are morphologically generalized mammals that are distinguished by their large brains, advanced cognitive abilities, flexible behavior, sophisticated social systems, and long lives (Hartwig, 2007; Zimmermann & Radespiel, 2007). Although primate species exhibit marked diversity in morphology, ecology, life-history parameters, and social organization, they share a reproductive profile characterized by low fecundity and extensive investment in each infant, associated with delayed reproductive maturation, long gestations, small litters, large neonates, long lactational periods, and slow postnatal growth (Zimmermann & Radespiel, 2007). These trends are especially pronounced in the anthropoids (monkeys, apes, and humans), which additionally exhibit hemochorial placentation, menstrual cycles in many species, and emancipation of sexual behavior from hormonal influences, particularly in females. The evolutionary basis of these reproductive patterns is not fully understood but is thought to be associated with development of primates' characteristic large brains and cognitive sophistication (Harvey, Martin, & Clutton-Brock, 1987; Leigh, 2004; Martin, 2007; Zimmermann & Radespiel, 2007).

As reviewed above, many aspects of reproduction have been studied intensively in a small number of primates. In contrast, for most of the 637 extant species and subspecies—of which one third to one half are currently threatened with extinction (Strier, 2007; Rylands, Williamson, Hoffmann, & Mittermeier, 2008)—little or no systematic information on reproductive physiology is available. In recent years, however, the development of noninvasive methods for monitoring reproductive hormones (e.g., fecal, urinary, and salivary assays) has begun to greatly expand our knowledge of reproductive function in a wide range of primates in both captive and wild settings (Lasley & Savage, 2007). A top priority for future research on primate reproduction should be to characterize basic reproductive parameters and processes in some of the less-studied taxa, such as the tarsiers, pitheciines (New World sakis, uakaris, and titi monkeys), colobines (Old World leaf-eating monkeys), and hylobatids (gibbons and siamangs). Further, in view of the ubiquity of environmental instability, in terms of anthropogenic habitat degradation, global climate change, and, potentially, exposure to endocrine-disrupting chemicals, it is vital to deepen our understanding of environmental influences on reproductive physiology and reproductive behavior in a variety of primate taxa. Ultimately, both broadening and deepening our understanding of primate reproductive function can provide new insights into human reproduction, will illuminate the evolution of primate life histories, and may make important contributions to captive management and conservation efforts.

ABBREVIATIONS

11β-HSD 11β–hydroxysteroid dehydrogenase
ACTH Corticotropin
CG Chorionic gonadotropin
CRF Corticotropin-releasing factor
CRH Corticotropin-releasing hormone
CS Chorionic somatomammotropin
DDE Dichlorodiphenyldichloroethylene
DDT Dichlorodiphenyltrichloroethane
DHEA Dehydroepiandrosterone
DHEA-S Dehydroepiandrosterone sulfate
DHT Dihydrotestosterone
E_1 Estrone
E_2 Estradiol
ER Estrogen receptor
FSH Follicle-stimulating hormone
GABA γ-aminobutyric acid
GH Growth hormone
GnRH Gonadotropin-releasing hormone
GPR54 G-protein receptor 54
GTH Gonadotropin
HPA Hypothalamic–pituitary–adrenal
HPG Hypothalamic–pituitary–gonadal
IGF Insulin-like growth factor
IL Interleukin
IUGR Intrauterine growth restriction
Kp Kisspeptin
LH Luteinizing hormone
MEL Melatonin
NMDA *N*-methyl-D-aspartate
NPY Neuropeptide Y
OXY Oxytocin
P_4 Progesterone
$P450_{scc}$ P450 cholesterol side-chain cleavage
PCB Polycholorinated biphenyl
PCOS Polycystic ovary syndrome
$PGF_{2\alpha}$ Prostaglandin $F_{2\alpha}$
PR Progesterone receptor
PRL Prolactin
T Testosterone
T_3 Triiodothyronine
T_4 Thyroxine
TGFβ Transforming growth factor-β
TSH Thyrotropin

REFERENCES

Abbott, D. H. (1984). Behavioral and physiological suppression of fertility in subordinate marmoset monkeys. *Am. J. Primatol., 6*, 169–186.

Abbott, D. H., Digby, L. J., & Saltzman, W. (2009). Reproductive skew in female marmosets. In R. Hager, & C. B. Jones (Eds.), *Reproductive Skew in Vertebrates.* Cambridge University Press.

Alberts, S. C., & Altmann, J. (1995). Preparation and activation: determinants of age at reproductive maturity in male baboons. *Behav. Ecol. Sociobiol., 36*, 397–406.

Albrecht, E. D., & Pepe, G. J. (1990). Placental steroid-hormone biosynthesis in primate pregnancy. *Endocr. Rev., 11*, 124–150.

Albrecht, E. D., & Pepe, G. J. (1999). Central integrative role of oestrogen in modulating the communication between the placenta and fetus that results in primate fetal–placental development. *Placenta, 20*, 129–139.

Alvarez, H. P. (2000). Grandmother hypothesis and primate life histories. *Am. J. Phys. Anthropol., 113*, 435–450.

AnAge: The Animal Ageing and Longevity Database. (2009). http://genomics.senescence.info/species/; accessed 05/20/09

Atsalis, S., & Margulis, S. W. (2008a). Primate reproductive aging: from lemurs to humans. In S. Atsalis, S. W. Margulis, & P. R. Hof (Eds.), *Primate Reproductive Aging: Cross-Taxon Perspectives on Reproduction. Interdiscip. Top. Gerontol, 36* (pp. 186–194). Basel, Switzerland: Karger.

Atsalis, S., & Margulis, S. W. (2008b). Perimenopause and menopause: documenting life changes in aging female gorillas. In S. Atsalis, S. W. Margulis, & P. R. Hof (Eds.), *Primate Reproductive Aging: Cross-Taxon Perspectives on Reproduction, Interdiscip. Top. Gerontol, 36* (pp. 119–146). Basel, Switzerland: Karger.

Auchus, R. J., & Rainey, W. E. (2004). Adrenarche—physiology, biochemistry and human disease. *Clin. Endocrinol., 60*, 288–296.

Baird, D. T., Cnattingius, S., Collins, J., Evers, J. L. H., et al. (2006). Nutrition and reproduction in women. *Hum. Reprod. Update, 12*, 193–207.

Beach, F. A. (1976). Sexual attractivity, proceptivity, and receptivity in female mammals. *Horm. Behav., 7*, 105–138.

Belgorosky, A., Baquedano, M. S., Guercio, G., & Rivarola, M. A. (2008). Adrenarche: postnatal adrenal zonation and hormonal and metabolic regulation. *Horm. Res., 70*, 257–267.

Benirschke, K. (2010). Comparative Placentation. http://placentation.ucsd.edu/; accessed 06/15/10

Bercovitch, F. B. (1993). Dominance rank and reproductive maturation in male rhesus macaques (*Macaca mulatta*). *J. Reprod. Fertil., 99*, 113–120.

Bercovitch, F. B., & Goy, R. W. (1990). The socioendocrinology of reproductive development and reproductive success in macaques. In T. E. Ziegler, & F. B. Bercovitch (Eds.), *Socioendocrinology of Primate Reproduction* (pp. 59–93). New York, NY: Wiley-Liss.

Bercovitch, F. B., & Strum, S. C. (1993). Dominance rank, resource availability, and reproductive maturation in female savanna baboons. *Behav. Ecol. Sociobiol., 33*, 313–318.

Bercovitch, F. B., Lebron, M. R., Martinez, H. S., & Kessler, M. J. (1998). Primigravidity, body weight, and costs of rearing first offspring in rhesus macaques. *Am. J. Primatol., 46*, 135–144.

Bergeron, C. (2000). Morphological changes and protein secretion induced by progesterone in the endometrium during the luteal phase in preparation for nidation. *Human Reprod. 15 Suppl., 1*, 119–128.

Bininda-Emonds, O. R. P., Cardillo, M., Jones, K. E., MacPhee, R. D. E., Beck, R. M. D., Grenyer, R., et al. (2007). The delayed rise of present-day mammals. *Nature, 446*, 507–512, (Corrigendum (2008), Nature 456, 274.).

Bowman, J. E., & Lee, P. C. (1995). Growth and threshold weaning weights among captive rhesus macaques. *Am J. Phys. Anthropol., 96*, 159–175.

Bowman, M. E., Lopata, A., Jaffe, R. B., Golos, T. G., Wickings, J., & Smith, R. (2001). Corticotropin-releasing hormone-binding protein in primates. *Am. J. Primatol., 53*, 123–130.

Brannian, J. D., & Stouffer, R. L. (1991). Progesterone production by monkey luteal cell subpopulations at different stages of the menstrual cycle: changes in agonist responsiveness. *Biol. Reprod., 44*, 141–149.

Bronson, F. H. (1989). *Mammalian Reproductive Biology.* Chicago, IL: University of Chicago Press.

Campbell, B. (2006). Adrenarche and the evolution of human life history. *Am. J. Hum. Biol., 18*, 569–589.

Campbell, C. J. (2007). Primate sexuality and reproduction. In C. J. Campbell, A. Fuentes, K. C. MacKinnon, M. Panger, & S. K. Bearder (Eds.), *Primates in Perspective* (pp. 423–437). New York, NY: Oxford University Press.

Caro, T. M., Sellen, D. W., Parish, A., Frank, R., Brown, D. M., Voland, E., et al. (1995). Termination of reproduction in nonhuman and human female primates. *Int. J. Primatol., 16*, 205–220.

Castellucci, M., De Matteis, R., Meisser, A., Cancello, R., Monsurro, V., Islami, D., et al. (2000). Leptin modulates extracellular matrix molecules and metalloproteinases: possible implications for trophoblast invasion. *Mol. Human Reprod., 6*, 951–958.

Cesario, S. K., & Hughes, L. A. (2007). Precocious puberty: a comprehensive review of literature. *J. Obstet. Gynecol. Neonatal Nurs., 36*, 263–274.

Chardonnens, D., Cameo, P., Aubert, M., Pralong, R., Islami, D., Campana, A., et al. (1999). Modulation of human cytotrophoblastic leptin secretion by interleukin-1α and 17β-oestradiol and its effect on hCG secretion. *Mol. Hum. Reprod., 5*, 1077–1082.

Charpentier, M. J. E., Tung, J., Altmann, J., & Alberts, S. C. (2008a). Age at maturity in wild baboons: genetic, environmental and demographic influences. *Molec. Ecol., 17*, 2026–2040.

Charpentier, M. J. E., Van Horn, R. C., Altmann, J., & Alberts, S. C. (2008b). Paternal effects on offspring fitness in a multimale primate society. *Proc. Nat. Acad. Sci. USA, 105*, 1988–1992.

Claypool, L. E., Kasuya, E., Saitoh, Y., Marzban, F., & Terasawa, E. (2000). *N*-methyl d, l-aspartate induces the release of luteinizing hormone-releasing hormone in the prepubertal and pubertal female rhesus monkey as measured by *in vivo* push-pull perfusion in the stalk-median eminence. *Endocrinology, 141*, 219–228.

Coe, C. L., Chen, M., Lowe, E. L., Davidson, J. M., & Levine, S. (1981). Hormonal and behavioral changes at puberty in the squirrel monkey. *Horm. Behav., 15*, 36–53.

Constância, M., Hemberger, M., Hughes, J., Dean., W., Ferguson-Smith, A., Fundele, R., et al. (2002). Placental-specific IGF-II is a major modulator of placental and fetal growth. *Nature, 417*, 945–948.

Crespi, B., & Semeniuk, C. (2004). Parent–offspring conflict in the evolution of vertebrate reproductive mode. *Am. Nat., 163*, 635–653.

Crossey, P. A., Pillai, C. C., & Miell, J. P. (2002). Altered placental development and intrauterine growth restriction in IGF binding protein-1 transgenic mice. *J. Clin. Invest. 110*, 411–418.

De Mendoza, A. R., Escobedo, D. E., Davila, I. M., & Saldana, H. B. (2004). Expansion and divergence of the GH locus between spider monkey and chimpanzee. *Gene, 336*, 185–193.

De Roux, N., Genin, E., Carel, J.- C., Matsuda, F., Chaussain, J.–L., & Milgrom, E. (2003). Hypogonadotropic hypogonadism due to loss of function of the KiSS1-derived peptide receptor GPR54. *Proc. Nat. Acad. Sci. USA, 100*, 10972–10976.

Di Bitetti, M. S., & Janson, C. H. (2000). When will the stork arrive? Patterns of birth seasonality in neotropical primates. *Am. J. Primatol., 50*, 109–130.

Di Fiore, A., & Campbell, C. J. (2007). The Atelines: variation in ecology, behavior, and social organization. In C. J. Campbell, A. F. Fuentes, K. C. Mackinnon, M. Panger, & S. Bearder (Eds.), *Primates in Perspective* (pp. 155–185). New York, NY: Oxford University Press.

Digby, L. J., Ferrari, S. F., & Saltzman, W. (2007). Callitrichines: the role of competition in cooperatively breeding species. In C. J. Campbell, A. F. Fuentes, K. C. Mackinnon, M. Panger, & S. Bearder (Eds.), *Primates in Perspective* (pp. 85–106). New York, NY: Oxford University Press.

Dixson, A. F. (1983). Observations on the evolution and behavioral significance of "sexual skin" in female primates. *Adv. Stud. Behav., 13*, 63–106.

Dixson, A. F. (1998). *Primate Sexuality: Comparative Studies of the Prosimians, Monkeys, Apes, and Human Beings*. Oxford, UK: Oxford University Press.

Dixson, A. F., & Nevison, C. M. (1997). The socioendocrinology of adolescent development in male rhesus monkeys (*Macaca mulatta*). *Horm. Behav., 31*, 126–135.

Duffy, D. M., & Stouffer, R. L. (2003). Luteinizing hormone acts directly at granulosa cells to stimulate periovulatory processes. *Endocrine, 22*, 249–256.

Dukelow, W. R. (1985). Reproductive cyclicity and breeding in the squirrel monkey. In L. A. Rosenblum, & C. L. Coe (Eds.), *Handbook of Squirrel Monkey Research* (pp. 169–190). New York, NY: Plenum Press.

Dunbar, R. I. M. (1977). Age-dependent changes in sexual skin colour and associated phenomena of female gelada baboons. *J. Hum. Evol., 6*, 667–672.

Eaton, G. G., Slob, A., & Resko, J. A. (1973). Cycles of mating behaviour, oestrogen and progesterone in the thick-tailed bushbaby (*Galago crassicaudatus crassicaudatus*) under laboratory conditions. *Anim. Behav., 21*, 309–315.

Ebling, F. J. (2005). The neuroendocrine timing of puberty. *Reproduction, 129*, 675–683.

El Majdoubi, M., Ramaswamy, S., Sahu, A., & Plant, T. M. (2000). Effects of orchidectomy on levels of the mRNAs encoding gonadotropin-releasing hormone and other hypothalamic peptides in the adult male rhesus monkey (*Macaca mulatta*). *J. Neuroendocrinol., 12*, 167–176.

Ellis, B. J., & Garber, J. (2000). Psychosocial antecedents of variation in girls' pubertal timing: maternal depression, stepfather presence, and marital and family stress. *Child Dev., 71*, 485–501.

Ellison, P. T., Bribiescas, R. G., Bentley, G. R., Campbell, B. C., Lipson, S. F., Panter-Brick, C., et al. (2002). Population variation in age-related decline in male salivary testosterone. *Child Dev., 71*, 485–501.

Emanuel, R. L., Robinson, B. G., Seely, E. W., Graves, S. W., Kohane, I., Saltzman, D., et al. (1994). Corticotrophin releasing hormone levels in human plasma and amniotic fluid during gestation. *Clin. Endocrinol., 40*, 257–262.

Emery Thompson, M. (2009). The endocrinology of intersexual relationships in the apes. In P. T. Ellison, & P. B. Gray (Eds.), *Endocrinology of Social Relationships*. Cambridge, MA.: Harvard University Press.

Enstam, K. L., & Isbell, L. A. (2007). The guenons (genus *Cercopithecus*) and their allies: behavioral ecology of polyspecific associations. In C. J. Campbell, A. F. Fuentes, K. C. Mackinnon, M. Panger, & S. Bearder (Eds.), *Primates in Perspective* (pp. 252–274). New York, NY: Oxford University Press.

Euling, S. Y., Selevan, S. G., Hirsch Pescovitz, O., & Skakkebaek, N. E. (2008). Role of environmental factors in the timing of puberty. *Pediatrics, 121*, S167–S171.

Fairbanks, L. A., & McGuire, M. T. (1995). Maternal condition and the quality of maternal care in vervet monkeys. *Behaviour, 132*, 733–754.

Farage, M., & Maibach, H. (2006). Lifetime changes in the vulva and vagina. *Arch. Gynecol. Obstet., 273*, 195–202.

Fashing, P. J. (2007). African colobine monkeys: patterns of between-group interaction. In C. J. Campbell, A. F. Fuentes, K. C. Mackinnon, M. Panger, & S. Bearder (Eds.), *Primates in Perspective* (pp. 201–224). New York, NY: Oxford University Press.

Fedigan, L. M., & Pavelka, M. S. M. (2007). Reproductive cessation in female primates. In C. J. Campbell, A. Fuentes, K. C. MacKinnon, M. Panger, & S. K. Bearder (Eds.), *Primates in Perspective* (pp. 437–447). New York, NY: Oxford University Press.

Finn, C. A. (1998). Menstruation: a nonadaptive consequence of uterine evolution. *Q. Rev. Biol., 73*, 163–173.

Foerg, R. (1982). Reproduction in *Cheirogaleus medius*. *Folia Primatol., 39*, 49–62.

Fowden, A. L. (2003). The insulin-like growth factors and feto–placental growth. *Placenta, 24*, 803–812.

Freeman, M. E., Kanyicska, B., Lerant, A., & Nagy, G. (2000). Prolactin: structure, function, and regulation of secretion. *Physiol. Rev., 80*, 1523–1631.

French, J. A. (1997). Proximate regulation of singular breeding in callitrichid primates. In N. G. Solomon, & J. A. French (Eds.), *Cooperative Breeding in Mammals* (pp. 34–75). New York, NY: Cambridge University Press.

Furuichi, T., Idani, G., Ihobe, H., Kuroda, S., Kitamura, K., Mori, A., et al. (1998). Population dynamics of wild bonobos (*Pan paniscus*) at Wamba. *Int. J. Primatol., 19*, 1029–1043.

Gajdos, Z. K. Z., Hirschhorn, J. N., & Palmert, M. R. (2009). What controls the timing of puberty? An update on progress from genetic investigation. *Curr. Opin. Endocrinol. Diabetes Obes., 16*, 16–24.

Gicquel, C., & Le Bouc, Y. (2006). Hormonal regulation of fetal growth. *Horm. Res., 65*, S28–S33.

Giudice, L. C. (2001). Insulin-like growth factor family in Graafian follicle development and function. *J. Soc. Gynecol. Investig., 8*, S26–S29.

Giudice, L. C. (2006). Application of functional genomics to primate endometrium: insights into biological processes. *Reprod. Biol. Endocrinol., 4*, S4.

Goland, R. S., Wardlaw, S. L., Fortman, J. D., & Stark, R. I. (1992). Plasma corticotropin releasing factor concentrations in the baboon during pregnancy. *Endocrinol., 131*, 1782–1786.

Goldfoot, D. A., Slob, A. K., Scheffler, G., Robinson, J. A., Wiegand, S. J., & Cords, J. (1975). Multiple ejaculations during prolonged sexual tests and lack of resultant serum testosterone increases in male stumptail macaques (*M. arctoides*). *Arch. Sex. Behav., 4*, 405–420.

Golos, T. G., Durning, M., Fisher, J. M., & Fowler, P. D. (1993). Cloning of four growth hormone/chorionic somatomammotropin-related complementary deoxyribonucleic acids differentially expressed during pregnancy in the rhesus monkey placenta. *Endocrinology, 133*, 1744–1752.

Gougein, A. (1986). Dynamics of follicular growth in the human: a model from preliminary results. *Hum. Reprod., 1*, 81–87.

Gould, L., & Sauther, M. (2007). Lemuriformes. In C. J. Campbell, A. F. Fuentes, K. C. Mackinnon, M. Panger, & S. Bearder (Eds.), *Primates in Perspective* (pp. 46–72). New York, NY: Oxford University Press.

Graham, C. E. (1979). Reproductive function in aged female chimpanzees. *Am. J. Phys. Anthropol., 50*, 291–300.

Gromoll, J., Wistuba, J., Terwort, N., Godmann, M., Müller, T., & Simoni, M. (2003). A new subclass of the luteinizing hormone/chorionic gonadotropin receptor lacking exon 10 messenger RNA in the New World monkey (Platyrrhini) lineage. *Biol. Reprod., 69*, 75–80.

Gursky, S. (2007). Tarsiiformes. In C. J. Campbell, A. F. Fuentes, K. C. Mackinnon, M. Panger, & S. Bearder (Eds.), *Primates in Perspective* (pp. 73–85). New York, NY: Oxford University Press.

Haig, D. (1993). Genetic conflicts in human pregnancy. *Q. Rev. Biol., 68*, 495–532.

Haig, D. (1996). Placental hormones, genomic imprinting and maternal-fetal communication. *J. Evol. Biol., 9*, 357–380.

Hardy, M. P., & Schlegel, P. N. (2004). Testosterone production in the aging male: where does the slowdown occur? *Endocrinology, 145*, 4439–4440.

Hartwig, W. (2007). Primate evolution. In C. J. Campbell, A. Fuentes, K. C. MacKinnon, M. Panger, & S. K. Bearder (Eds.), *Primates in Perspective* (pp. 11–22). New York, NY: Oxford University Press.

Harvey, P. H., Martin, R. D., & Clutton-Brock, T. H. (1987). Life histories in comparative perspective. In B. B. Smuts, D. L. Cheney, R. M. Seyfarth, R. W. Wrangham, & T. T. Struhsaker (Eds.), *Primate Societies* (pp. 181–196). Chicago, IL: University of Chicago Press.

Hauguel-De Mouzon, S., Lepercq, J., & Catalano, P. (2006). The known and unknown of leptin in pregnancy. *Am. J. Obstet. Gynecol., 194*, 1537–1545.

Havelock, J. C., Auchus, R. J., & Rainey, W. E. (2004). The rise in adrenal androgen biosynthesis: adrenarche. *Semin. Reprod. Med., 22*, 337–347.

Hawkes, K. (1997). Hadza women's time allocation, offspring provisioning, and the evolution of long post-menopausal lifespans. *Curr. Anthropol., 38*, 551–578.

Hemingway, C. A., & Bynum, N. (2005). The influence of seasonality on primate diet and ranging. In D. K. Brockman, & C. P. Van Schaik (Eds.), *Seasonality in Primates: Studies of Living and Extinct Human and Non-human Primates* (pp. 57–104). Cambridge: Cambridge University Press.

Hendrickx, A. G., & Dukelow, W. R. (1995). Reproductive biology. In B. T. Bennett, C. R. Abee, & R. Hendickson (Eds.), *Nonhuman Primates in Biomedical Research: Biology and Management* (pp. 147–192). New York, NY: Academic Press.

Henson, M. C., & Castracane, V. D. (2002). Leptin: roles and regulation in primate pregnancy. *Semin. Reprod. Med., 20*, 113–122.

Hess, A. P., Nayak, N. R., & Giudice, L. C. (2006). Oviduct and endometrium: cyclic changes in the primate oviduct and endometrium. In J. D. Neill (Ed.) (Third Edition). *Knobil and Neill's Physiology of Reproduction, Vol. 1* (pp. 337–381). St. Louis, MO: Elsevier.

Hill, K., & Hurtado, A. M. (1991). The evolution of premature reproductive senescence and menopause in human females: an evaluation of the 'grandmother hypothesis. *Hum. Nat., 2*, 313–350.

Hodgen, G. D., Goodman, A. L., O'Connor, A., & Johnson, D. K. (1977). Menopause in rhesus monkeys: model for study of disorders in the human climacteric. *Am. J. Obstet. Gynecol., 127*, 581–584.

Hull, E. M., Wood, R. I., & McKenna, K. E. (2006). In J. D. Neill (Ed.), *Knobil and Neill's Physiology of Reproduction* (Third Edition). Neurobiology of male sexual behavior, *Vol. 2* (pp. 1729–1824). St. Louis, MO: Elsevier.

Hutz, R. J., Dierschke, D. J., & Wolf, R. C. (1985). Seasonal effects on ovarian folliculogenesis in rhesus monkeys. *Biol. Reprod., 33*, 653–659.

Izard, M. K. (1990). Social influences on the reproductive success and reproductive endocrinology of prosimian primates. In T. E. Ziegler, & F. B. Bercovitch (Eds.), *Socioendocrinology of Primate Reproduction* (pp. 159–186). New York, NY: Wiley-Liss.

Jack, K. M. (2007). The Cebines: toward an explanation of variable social structure. In C. J. Campbell, A. F. Fuentes, K. C. Mackinnon, M. Panger, & S. Bearder (Eds.), *Primates in Perspective* (pp. 107–123). New York, NY: Oxford University Press.

Jansson, N., Greenwood, S., Johansson, B., Powell, T., & Jansson, T. (2003). Leptin stimulates the activity of the system A amino acid transporter in human placental villous fragments. *J. Clin. Endocrinol. Metab., 88*, 1205–1211.

Jasienska, G., & Ellison, P. T. (2004). Energetic factors and seasonal changes in ovarian function in women from rural Poland. *Am. J. Hum. Biol., 16*, 563–580.

Johnson, M. H., & Everitt, B. J. (2000). *Essential Reproduction*. Oxford, UK: Blackwell Science.

Johnson, S. (2003). Life history and competitive environment: trajectories of growth, maturation and reproductive output among chacma baboons. *Am. J. Phys. Anthropol., 120*, 83–98.

Jolly, C. J. (2007). Baboons, mandrills, and mangabeys: Afro-Papionin socioecology in a phylogenetic perspective. In C. J. Campbell, A. F. Fuentes, K. C. Mackinnon, M. Panger, & S. Bearder (Eds.), *Primates in Perspective* (pp. 240–251). New York, NY: Oxford University Press.

Kaplowitz, P. B. (2008). Link between body fat and the timing of puberty. *Pediatrics, 121*, S208–S217.

Keen, K. L., Wegner, F. H., Bloom, S. R., Ghatei, M. A., & Terasawa, E. (2008). An increase in kisspeptin-54 release occurs with the pubertal increase in luteinizing hormone-releasing hormone-1 release in the stalk-median eminence of female rhesus monkeys *in vivo*. *Endocrinology, 149*, 4151–4157.

King, B. R., Smith, R., & Nicholson, R. C. (2002). Novel glucocorticoid and cAMP interactions on the CRH gene promoter. *Mol. Cell. Endocrinol., 194*, 19–28.

Kingsley, S. (1982). Causes of non-breeding and the development of the secondary sexual characteristics in the male orangutan: a hormonal study. In L. E. M. De Boer (Ed.), *The Orangutan: Its Biology and Conservation* (pp. 215–229). The Hague, The Netherland: Junk Publishers, Dr. W.

Kirkpatrick, R. C. (2007). The Asian colobines. In C. J. Campbell, A. F. Fuentes, K. C. Mackinnon, M. Panger, & S. Bearder (Eds.), *Primates in Perspective* (pp. 186–200). New York, NY: Oxford University Press.

Kliman, H. J., Nestler, J. E., Sermasi, E., Sanger, J. M., & Strauss, J. F. (1986). Purification, characterization, and *in vitro* differentiation of cytotrophoblasts from human term placentae. *Endocrinology, 118*, 1567–1582.

Kuehl, T. J., & Dukelow, W. R. (1975). Ovulation induction during the anovulatory season in *Saimiri sciureus*. *J. Med. Primatol., 4*, 23–31.

Kuzawa, C. W. (1998). Adipose tissue in human infancy and childhood: an evolutionary perspective. *Am. J. Phys. Anthropol. Suppl., 27*, 177–209.

Lacroix, M., -C., Guibourdenche, J., Frendo, J., -L., Muller, F., & Evain-Brion, D. (2002). Human placental growth hormone—a review. *Placenta, 23*, S87–S94.

Lasley, B. L., & Savage, A. (2007). Advances in the understanding of primate reproductive endocrinology. In C. J. Campbell, A. Fuentes, K. C. MacKinnon, M. Panger, & S. K. Bearder (Eds.), *Primates in Perspective* (pp. 356–369). New York, NY: Oxford University Press.

Lee, K. Y., & De Mayo, F. J. (2004). Animal models of implantation. *Reproduction, 128*, 679–695.

Lee, P. C., Majluf, P., & Gordon, I. J. (1991). Growth, weaning and maternal investment from a comparative perspective. *J. Zool., 225*, 99–114.

Leigh, S. R. (2004). Brain growth, life history, and cognition in primate and human evolution. *Am. J. Primatol., 62*, 139–164.

Lewis, M. P., Morlese, J. F., Sullivan, M. H., & Elder, M. G. (1993). Evidence for decidua–trophoblast interactions in early human pregnancy. *Hum. Reprod., 8*, 965–968.

Li, Y., Ye, C., Shi, P., Zou, X.- J., Xiao, R., Gong, Y.- Y., & Zhang, Y.- P. (2005). Independent origin of the growth hormone gene family in New World monkeys and Old World monkeys/hominoids. *J. Mol. Endocrinol., 35*, 399–409.

Lockwood, C. J., Krikun, G., Hausknecht, V., Want, E.-Y., & Schatz, F. (1997). Decidual cell regulation of hemostasis during implantation and menstruation. *Ann. NY Acad. Sci., 828*, 188–193.

Luckett, W. P. (1974). Comparative development and evolution of the placenta in primates. In W. P. Luckett (Ed.), *Contributions to Primatology: The Reproductive Biology of the Primates* (pp. 142–234). Basel, Switzerland: Karger.

Luckhaus, J., & Ferrin, M. (1989). Effects of intraventricular oxytocin infusion on LH and cortisol secretion in ovariectomized rhesus monkeys. *Acta Endocrinol., 120(suppl.1)*, 59.

Luetjens, C. M., Weinbauer, G. F., & Wistuba, J. (2005). Primate spermatogenesis: new insights into comparative testicular organization, spermatogenic efficiency and endocrine control. *Biol. Rev., 80*, 475–488.

Maggioncalda, A. N., Czekala, N. M., & Sapolsky, R. M. (2002). Male orangutan subadulthood: a new twist on the relationship between chronic stress and developmental arrest. *Am. J. Phys. Anthro., 118*, 25–32.

Maggioncalda, A. N., Sapolsky, R. M., & Czekala, N. M. (1999). Reproductive hormone profiles in captive male orangutans: implications for understanding developmental arrest. *Am. J. Phys. Anthro., 109*, 19–32.

Malpaux, B. (2006). Seasonal regulation of reproduction in mammals. In J. D. Neill. *Knobil and Neill's Physiology of Reproduction* (Ed.) (Third Edition)., *Vol. 2* (pp. 2231–2281). St. Louis, MO: Elsevier.

Maneckjee, R., Srinath, B. R., & Moudgal, N. R. (1976). Prolactin suppresses release of luteinising hormone during lactation in the monkey. *Nature, 262*, 507–508.

Mann, D. R., Akinbami, M. A., Gould, K. G., Paul, K., & Wallen, K. (1998). Sexual maturation in male rhesus monkeys: importance of neonatal testosterone exposure and social rank. *J. Endocrinol., 156*, 493–501.

Martin, R. D. (2007). The evolution of human reproduction: a primatological perspective. *Yearb. Phys. Anthropol., 50*, 59–84.

McLachlan, R. I., O'Donnell, L., Meachem, S. J., Stanton, P. G., De Kretser, D. M., Pratis, K., et al. (2002a). Hormonal regulation of spermatogenesis in primates and man: insights for development of the male hormonal contraceptive. *J. Androl., 23*, 149–162.

McLachlan, R. I., O'Donnell, L., Meachem, S. J., Stanton, P. G., De Kretser, D. M., Pratis, K., et al. (2002b). Identification of specific sites of hormonal regulation in spermatogenesis in rats, monkeys, and man. *Rec. Prog. Horm. Res., 57*, 149–179.

McNeilly, A. S. (1994). Suckling and the control of gonadotropin secretion. In E. Knobil, & J. D. Neill (Eds.) (2nd Ed.). *Vol. 2 The Physiology of Reproduction*, (pp. 1179–1212) New York, NY: Raven Press.

McNeilly, A. S. (2001). Lactational control of reproduction. *Reprod. Fertil. Dev., 13*, 583–590.

McNeilly, A. S., Abbott, D. H., Lunn, S. F., Chambers, P. C., & Hearn, J. P. (1981). Plasma prolactin concentrations during the ovarian cycle and lactation and their relationship to return of fertility post partum in the common marmoset (*Callithrix jacchus*). *J. Reprod. Fertil., 62*, 353–360.

McNulty, W. P., Novy, M. J., & Walsh, S. W. (1981). Fetal and postnatal development of the adrenal glands in *Macaca mulatta*. *Biol. Reprod., 25*, 1079–1089.

Mendoza, S. P., Coe, C. L., Lowe, E. L., & Levine, S. (1979). The physiological response to group formation in adult male squirrel monkeys. *Psychoneuroendocrinology, 3*, 221–229.

Mesiano, S., & Jaffe, R. B. (1997). Developmental and functional biology of the primate fetal adrenal cortex. *Endocr. Rev., 18*, 378–403.

Messinis, I. E. (2006). Ovarian feedback, mechanism of action and possible clinical implications. *Hum. Reprod. Update, 12*, 557–571.

Milligan, L. A., Gibson, S. V., Williams, L. E., & Power, M. L. (2008). The composition of milk from Bolivian squirrel monkeys (*Saimiri boliviensis boliviensis*). *Am. J. Primatol., 70*, 35–43.

Milton, K. (1985). Mating patterns of woolly spider monkeys, *Brachyteles arachnoides*: implications for female choice. *Behav. Ecol. Sociobiol., 17*, 53–59.

Mizuno, M., & Terasawa, E. (2005). Search for neural substrates mediating inhibitory effects of oestrogen on pulsatile luteinising hormone-releasing hormone release *in vivo* in ovariectomized female rhesus monkeys (*Macaca mulatta*). *J. Neuroendocrinol., 17*, 238–245.

Mossman, H. W. (1987). *Vertebrate Fetal Membranes*. New Brunswick, NJ: Rutgers University Press.

Müller, T., Gromoll, J., Simula, A. P., Norman, R., Sandhowe-Klaverkamp, R., & Simoni, M. (2004a). The carboxyterminal peptide of chorionic gonadotropin facilitates activation of the marmoset LH receptor. *Exp. Clin. Endocrinol. Diabetes, 112*, 574–579.

Müller, T., Simoni, M., Pekel, E., Luetjens, C. M., Chandolia, R., Amato, F., et al. (2004b). Chorionic gonadotrophin beta subunit mRNA but not luteinising hormone beta subunit mRNA is expressed in the pituitary of the common marmoset (*Callithrix jacchus*). *J. Mol. Endocrinol., 32*, 115–128.

Musicki, B., Pepe, G. J., & Albrecht, E. D. (1997). Functional differentiation of placental syncytiotrophoblasts during baboon pregnancy: developmental expression of chorionic somatomammotropin messenger ribonucleic acid and protein levels. *J. Clin. Endocrinol. Metab., 85*, 4105–4110.

Nakano, R. (1997). Control of the luteal function in humans. *Semin. Reprod. Endocrinol., 15*, 335–344.

Nasir-ud-Din, Hoessli, D. C., Rungger-Brändle, E., Hussain, S. A., & Walker-Nasir, E. (2003). Role of sialic acid and sulfate groups in cervical mucus physiological functions: study of *Macaca radiata* glycoproteins. *Biochim. Biophys. Acta, 1623*, 53–61.

Nebesio, T. D., & Eugster, E. A. (2007). Current concepts in normal and abnormal puberty. *Curr. Probl. Pediatr. Adolesc. Health Care, 37*, 50–72.

Nekaris, A., & Bearder, S. K. (2007). The Lorisiform primates of Asia and mainland Africa. In C. J. Campbell, A. F. Fuentes, K. C. Mackinnon, M. Panger, & S. Bearder (Eds.), *Primates in Perspective* (pp. 24–45). New York, NY: Oxford University Press.

Neville, M. C. (2001). Anatomy and physiology of lactation. *Pediatr. Clin. North Am., 48*, 13–34.

Nguyen, A. D., & Conley, A. J. (2008). Adrenal androgens in humans and nonhuman primates: production, zonation and regulation. *Endocr. Dev., 13*, 33–54.

Niswender, G. D., & Nett, T. M. (1994). Corpus luteum and its control in infraprimate species. In E. Knobil, & J. D. Neill (Eds.), *The Physiology of Reproduction* (pp. 781–816). New York, NY: Raven Press.

Norconk, M. A. (2007). Sakis, uakaris, and titi monkeys: behavioral diversity in a radiation of primate seed predators. In C. J. Campbell, A. F. Fuentes, K. C. Mackinnon, M. Panger, & S. Bearder (Eds.), *Primates in Perspective* (pp. 123–138). New York, NY: Oxford University Press.

Novy, M. J., Aubert, M. L., Kaplan, S. L., & Grumbach, M. M. (1981). Regulation of placental growth and chorionic somatomammotropin in the rhesus monkey: effects of protein deprivation, fetal anencephaly, and placental vessel ligation. *Am. J. Obstet. Gynecol., 140*, 552–562.

O'Byrne, K. T., Lunn, S. F., & Coen, C. W. (1990). Central oxytocin stimulates luteinizing hormone release in the marmoset, a primate which fails to show lactationally-induced infertility. *J. Neuroendocrinol., 2*, 419–421.

Oftedal, O. T. (1984). Milk composition, milk yield and energy output at peak lactation: a comparative review. *Symp. Zool. Soc. Lond., 51*, 33–85.

Ogren, L., & Talamantes, F. (1994). The placenta as an endocrine organ: polypeptides. In E. Knobil, & J. D. Neill (Eds.), *The Physiology of Reproduction* (2nd ed.). (pp. 875–945) New York, NY: Raven Press.

Peccei, J. S. (2001a). A critique of the grandmother hypothesis: old and new. *Am. J. Hum. Biol., 13*, 434–452.

Peccei, J. S. (2001b). Menopause: adaptation or epiphenomenon? *Evol. Anthropol., 10*, 43–57.

Pepe, G. J., & Albrecht, E. D. (1990). Regulation of the primate fetal adrenal-cortex. *Endocr. Rev., 11*, 151–176.

Perret, M. (1986). Social influences on oestrous cycle length and plasma progesterone concentrations in the female lesser mouse lemur (*Microcebus murinus*). *J. Reprod. Fertil., 77*, 303–311.

Perret, M. (1992). Environmental and social determinants of sexual function in the male lesser mouse lemur (*Microcebus murinus*). *Folia Primatol., 59*, 1–25.

Perret, M., & Schilling, A. (1995). Sexual responses to urinary chemosignals depend on photoperiod in a male primate. *Physiol. Behav., 58*, 633–639.

Petraglia, F., Florio, P., Nappi, C., & Genazzani, A. R. (1996). Peptide signaling in human placenta and membranes: autocrine, paracrine, and endocrine mechanisms. *Endocr. Rev., 17*, 156–186.

Petraglia, F., Florio, P., & Vale, W. (2005). Placental expression of neurohormones and other neuroactive molecules in human pregnancy. In M. L. Power, & J. Schulkin (Eds.), *Birth, Distress and Disease* (pp. 16–73). Cambridge, UK: Cambridge University Press.

Plant, T. M., & Marshall, G. R. (2001). The functional significance of FSH in spermatogenesis and the control of its secretion in male primates. *Endocr. Rev., 22*, 764–786.

Plant, T. M., & Ramaswamy, S. (2009). Kisspeptin and the regulation of the hypothalamic–pituitary–gonadal axis in the rhesus monkey (*Macaca mulatta*). *Peptides, 30*, 67–75.

Plant, T. M., & Witchel, S. F. (2006). Puberty in nonhuman primates and humans. In J. D. Neill (Ed.), *Knobil and Neill's Physiology of Reproduction* (Third Edition)., *Vol. 2* (pp. 2177–2230). St. Louis, MO: Elsevier.

Plant, T. M., Gay, V. L., Marshall, G. R., & Arslan, M. (1989). Puberty in monkeys is triggered by chemical stimulation of the hypothalamus. *Proc. Nat. Acad. Sci. USA, 86*, 2506–2510.

Plant, T. M., Ramaswamy, S., & Dipietro, M. J. (2006). Repetitive activation of hypothalamic G protein-coupled receptor 54 with intravenous pulses of kisspeptin in the juvenile monkey (*Macaca mulatta*) elicits a sustained train of gonadotropin-releasing hormone discharges. *Endocrinology, 147*, 1007–1013.

Poonia, B., Walter, L., Dufour, J., Harrison, R., Marx, P. A., & Veazey, R. S. (2006). Cyclic changes in the vaginal epithelium of normal rhesus macaques. *J. Endocrinol., 190*, 829–835.

Power, M. L., & Tardif, S. D. (2005). Maternal nutrition and metabolic control of pregnancy. In M. L. Power, & J. Schulkin (Eds.), *Birth, Distress and Disease* (pp. 88–113). Cambridge, UK: Cambridge University Press.

Power, M. L., Bowman, M. E., Smith, R., Ziegler, T. E., Layne, D. G., Schulkin, J., et al. (2006). The pattern of maternal serum corticotropin-releasing hormone concentration during pregnancy in the common marmoset (*Callithrix jacchus*). *Am. J. Primatol., 68*, 181–188.

Power, M. L., Oftedal, O. T., & Tardif, S. D. (2002). Does the milk of Callitrichid monkeys differ from that of larger anthropoids? *Am. J. Primatol., 56*, 117–127.

Preti, G., Wysocki, C. J., Barnhart, K. T., Sondheimer, S. J., & Leyden, J. J. (2003). Male axillary extracts contain pheromones that affect pulsatile secretion of luteinizing hormone and mood in women recipients. *Biol. Reprod., 68*, 2107–2113.

Profet, M. (1993). Menstruation as a defense against pathogens transported by sperm. *Q. Rev. Biol., 68*, 335–386.

Randolph, J. F., Jr. (2008). The endocrinology of the reproductive years. *J. Sex. Med., 5*, 2274–2281.

Rasweiler, J. J., IV, & De Bonilla, H. (1992). Menstruation in short-tailed fruit bats (*Carollia* spp.). *J. Reprod. Fertil., 95*, 231–248.

Richard, A. F. (1987). Malagasy prosimians: female dominance. In B. B. Smuts, D. L. Cheney, R. M. Seyfarth, R. W. Wrangham, & T. T. Struhsaker (Eds.), *Primate Societies* (pp. 25–33). Chicago, IL: University of Chicago Press.

Roa, J., Aguilar, E., Dieguez, C., Pinilla, L., & Tena-Sempere, M. (2008). New *frontiers* in kisspeptin/GPR54 physiology as fundamental gatekeepers of reproductive function. *Front. Neuroendocrinol., 29*, 48–69.

Roh, C. R., Budhara, V., Kim, H. S., Nelson, D. M., & Sadovsky, Y. (2005). Microarray-based identification of differentially expressed genes in hypoxic term human trophoblasts and in placental villi of pregnancies with growth restricted fetuses. *Placenta, 26*, 319–328.

Rowe, N. (1996). *The Pictorial Guide to the Living Primates*. Charlestown, RI: Pogonias Press.

Rutherford, J. N. (2009). Fetal signaling through placental structure and endocrine function: illustrations and implications from a nonhuman primate model. *Am. J. Hum. Biol., 21*, 745–753.

Rutherford, J. N., & Tardif, S. D. (2008). Placental efficiency and intrauterine resource allocation strategies in the common marmoset pregnancy. *Am. J. Phys. Anthropol., 137*, 60–68.

Rutherford, J. N., & Tardif, S. D. (2009). Developmental plasticity of the microscopic placental architecture in relation to litter size variation in the common marmoset monkey (*Callithrix jacchus*). *Placenta, 30*, 105–110.

Rylands, A. B., Williamson, E. A., Hoffmann, M., & Mittermeier, R. A. (2008). Primate surveys and conservation assessments. *Oryx, 42*, 313–314.

Saltzman, W., Digby, L. J., & Abbott, D. H. (2009). Reproductive skew in female common marmosets: what can proximate mechanisms tell us about ultimate causes? *Proc. Biol. Sci., 276*, 389–399.

Savini, T., Boesch, C., & Reichard, U. H. (2008). Home-range characteristics and the influence of seasonality on female reproduction in white-handed gibbons (*Hylobates lar*) at Khao Yai National Park, Thailand. *Am. J. Phys. Anthropol., 135*, 1–12.

Scammell, J. G., Funkhouser, J. D., Moyer, F. S., Gibson, S. V., & Willis, D. L. (2008). Molecular cloning of pituitary glycoprotein α-subunit and folicle stimulating hormone and chorionic gonadotropin β-subnits from New World squirrel monkey and owl monkey. *Gen. Comp. Endocrinol., 155*, 534–541.

Schallenberger, E., Richardson, D. W., & Knobil, E. (1981). Role of prolactin in the lactational amenorrhoea of the rhesus monkey (*Macaca mulatta*). *Biol. Reprod., 25*, 370–374.

Schiml, P. A., Mendoza, S. P., Saltzman, W., Lyons, D. M., & Mason, W. A. (1996). Seasonality in squirrel monkeys (*Saimiri sciureus*): social facilitation by females. *Physiol. Behav., 60*, 1105–1113.

Schiml, P. A., Mendoza, S. P., Saltzman, W., Lyons, D. M., & Mason, W. A. (1999). Annual physiological changes in individially housed squirrel monkeys (*Saimiri sciureus*). *Am. J. Primatol., 47*, 93–103.

Schlabritz-Loutsevitch, N. E., Dudley, C. J., Gomez, J. J., Nevill, C. H., Smith, B. K., Jenkins, S. L., et al. (2007). Metabolic adjustments to moderate maternal nutrient restriction. *Br. J. Nutr., 98*, 276–284.

Schoeters, G., Den Hond, E., Dhooge, W., Van Larebeke, N., & Leijs, M. (2008). Endocrine disruptors and abnormalities of pubertal development. *Basic Clin. Pharmacol. Toxicol., 102*, 168–175.

Schramm, R. D., Paprocki, A. M., & Bavister, B. D. (2002). Features associated with reproductive ageing in female rhesus monkeys. *Hum. Reprod., 17*, 1597–1603.

Schuler, L. A., & Kessler, M. A. (1992). Bovine placental prolactin-related hormones. *Trends Endocrinol. Metab., 3*, 334–338.

Schurgin, S., Canavan, B., Koutkia, P., Depaoli, A. M., & Grinspoon, S. (2004). Endocrine and metabolic effects of physiologic r-metHu-Leptin administration during acute caloric deprivation in normal-weight women. *J. Clin. Endocrinol. Metab., 89*, 5402–5409.

Schwartz, S. M., Wilson, M. E., Walker, M. L., & Collins, D. C. (1985). Social and growth correlates of puberty onset in female rhesus monkeys. *Nutr. Behav., 2*, 225–232.

Seminara, S. B., Messager, S., Chatzidaki, E. E., Thresher, R. R., Acierno, J. S., Jr., Shagoury, J. K., et al. (2003). The GPR54 gene as a regulator of puberty. *N. Engl. J. Med., 349*, 1614–1627.

Setchell, J. M. (2003). The evolution of alternative reproductive morphs in male primates. In C. B. Jones (Ed.), *Sexual Selection and Reproductive Competition in Primates: New Perspectives and Directions* (pp. 413–435). Norman, OK: American Society of Primatologists.

Setchell, J. M., & Dixson, A. F. (2002). Developmental variables and dominance rank in adolescent male mandrills (*Mandrillus sphinx*). *Am. J. Primatol., 56*, 9–25.

Shibata, M., Friedman, R. L., Ramaswamy, S., & Plant, T. M. (2007). Evidence that down regulation of hypothalamic *KiSS-1* expression is involved in the negative feedback action of testosterone to regulate luteinising hormone secretion in the adult male rhesus monkey (*Macaca mulatta*). *J. Neuroendocrinol., 19*, 432–438.

Shideler, S. E., Gee, N. A., Chen, J., & Lasley, B. L. (2001). Estrogen and progestogen metabolites and follicle-stimulating hormone in the aged macaque female. *Biol. Reprod., 65*, 1718–1725.

Shinohara, K., Morofushi, M., Funabashi, T., & Kimura, F. (2001). Axillary pheromones modulate pulsatile LH secretion in humans. *Neuroreport, 12*, 893–895.

Shively, C., Clarke, S., King, N., Schapiro, S., & Mitchell, G. (1982). Patterns of sexual behavior in male macaques. *Am. J. Primatol., 2*, 373–384.

Skangalis, M., Swenson, C. E., Mahoney, C. J., & O'Leary, W. M. (1979). The normal microbial flora of the baboon vagina. *J. Med. Primatol., 8*, 289–297.

Smith, R. J., & Jungers, W. L. (1997). Body mass in comparative primatology. *J. Hum. Evol., 32*, 523–559.

Smith, R., Chan, E. C., Bowman, M. E., Harewood, W. J., & Phippard, A. F. (1993). Corticotropin releasing hormone in baboon pregnancy. *J. Clin. Endocrinol. Metab., 76*, 1063–1068.

Smith, R., Mesiano, S., Nicholson, R., Clifton, V., Zakar, T., Chan, E.-C., et al. (2005). The regulation of human parturition. In M. L. Power, & J. Schulkin (Eds.), *Birth, Distress and Disease* (pp. 74–87). Cambridge, UK: Cambridge University Press.

Smith, R., Wickings, J., & Bowman, M. B. (1999). Corticotropin-releasing hormone in chimpanzee and gorilla pregnancy. *J. Clin. Endocrinol. Metab., 84*, 2820–2825.

Smucny, D. A., Abbott, D. H., Mansfield, K. G., Schultz-Darken, N. J., Yamamoto, M. E., Alencar, A. I., et al. (2004). Reproductive output, maternal age and survivorship in captive common marmoset females (*Callithrix jacchus*). *Am. J. Primatol., 64*, 107–121.

Soares, M. J. (2004). The prolactin and growth hormone families: pregnancy-specific hormones/cytokines at the maternal–fetal interface. *Reprod. Biol. Endocrinol., 2*, 51.

Soares, M. J., & Linzer, D. I. H. (2001). The rodent prolactin family and pregnancy. In N. D. Horseman (Ed.), *Prolactin* (pp. 139–167). Norwell, MA: Kluwer Academic Publishers.

Sofikitis, N., Giotitsas, N., Tsounapi, P., Baltogiannis, D., Giannakis, D., & Pardalidis, N. (2008). Hormonal regulation of spermatogenesis and spermiogenesis. *J. Steroid. Biochem. Mol. Biol., 109*, 323–330.

Soini, P. (1988). The pygmy marmoset, genus *Cebuella*. In R. A. Mittermeier, A. B. Rylands, A. F. Coimbra-Filho, & G. A. B. da Fonseca (Eds.), *Ecology and Behavior of Neotropical Primates, Vol. 2* (pp. 79–129). Washington, DC: World Wildlife Fund.

Solomon, S. (1994). The primate placenta as an endocrine organ: steroids. In E. Knobil, & J. D. Neill (Eds.), *The Physiology of Reproduction* (2nd ed.). (pp. 863–873) New York, NY: Raven Press.

Stammbach, E. (1987). Desert, forest and montane baboons: multilevel societies. In B. B. Smuts, D. L. Cheney, R. M. Seyfarth, R. W. Wrangham, & T. T. Struhsaker (Eds.), *Primate Societies* (pp. 112–120). Chicago, IL: University of Chicago Press.

Stern, K., & McClintock, M. K. (1998). Regulation of ovulation by human pheromones. *Nature, 392*, 177–179.

Stouffer, R. L., Xu, F., & Duffy, D. M. (2007). Molecular control of ovulation and luteinization in the primate follicle. *Front. Biosci., 12*, 297–307.

Strassmann, B. I. (1996). The evolution of endometrial cycles and menstruation. *Q. Rev. Biol., 71*, 181–220.

Strier, K. B. (2007). Conservation. In C. J. Campbell, A. Fuentes, K. C. MacKinnon, M. Panger, & S. K. Bearder (Eds.), *Primates in Perspective* (pp. 496–509). New York, NY: Oxford University Press.

Strier, K. B., & Ziegler, T. E. (1994). Insights into ovarian function in wild muriqui monkeys (*Brachyteles arachnoides*). *Am. J. Primatol., 32*, 31–40.

Struhsaker, T. T., & Leland, L. (1987). Colobines: infanticide by adult males. In B. B. Smuts, D. L. Cheney, R. M. Seyfarth, R. W. Wrangham, & T. T. Struhsaker (Eds.), *Primate Societies* (pp. 83–97). Chicago, IL: University of Chicago Press.

Stumpf, R. (2007). Chimpanzees and bonobos: diversity within and between species. In C. J. Campbell, A. F. Fuentes, K. C. Mackinnon, M. Panger, & S. Bearder (Eds.), *Primates in Perspective* (pp. 321–344). New York, NY: Oxford University Press.

Suarez, S. S., & Pacey, A. A. (2006). Sperm transport in the female reproductive tract. *Hum. Reprod. Update, 12*, 23–37.

Tardif, S. D. (1985). Histologic evidence for age-related differences in ovarian function in tamarins (*Saguinus* sp; Primates). *Biol. Reprod., 33*, 993–1000.

Tardif, S. D., & Ross, C. N. (2009). Integration of proximate and evolutionary explanation of reproductive strategy: the case of callitrichid primates and implications for human biology. *Am. J. Hum. Biol., 21*, 731–738.

Tardif, S. D., & Ziegler, T. E. (1992). Features of reproductive senescence in tamarins (*Saguinus* spp.), a New World primate. *J. Reprod. Fertil., 94*, 411–421.

Tardif, S. D., Araujo, A., Arruda, A., French, J. A., Sousa, M. B. C., & Yamamoto, M. E. (2008). Reproduction and aging in marmosets and tamarins. In S. Atsalis, S. W. Margulis, & P. R. Hof (Eds.), *Primate Reproductive Aging: Cross-Taxon Perspectives on Reproduction, Interdiscip. Top. Gerontol., 36* (pp. 29–48). Basel, Switzerland: Karger.

Tardif, S. D., Power, M., Oftedal, O. T., Power, R. A., & Layne, D. G. (2001). Lactation, maternal behavior and infant growth in common marmoset monkeys: effects of maternal size and litter size. *Behav. Ecol. Sociobiol., 51*, 17–25.

Tardif, S. D., Ziegler, T. E., Power, M., & Layne, D. G. (2005). Endocrine changes in full-term pregnancies and pregnancy loss due to energy restriction in the common marmoset (*Callithrix jacchus*). *J. Clin. Endocrinol. Metab., 90*, 335–339.

Tardif, S., Power, M., Layne, D., Smucny, D., & Ziegler, T. (2004). Energy restriction initiated at different gestational ages has varying effects on maternal weight gain and pregnancy outcome in common marmoset monkeys (*Callithrix jacchus*). *Br. J. Nutr., 92*, 841–849.

Tay, C. C. K., Glasier, A. F., & McNeilly, A. S. (1996). Twenty-four hour patterns of prolactin secretion during lactation ad the relationship to suckling and the resumption of fertility in breast-feeding women. *Hum. Reprod., 11*, 950–955.

Tena-Sempere, M. (2007). Roles of ghrelin and leptin in the control of reproductive function. *Neuroendocrinology, 86*, 229–241.

Tena-Sempere, M. (2008). Ghrelin as a pleotrophic modulator of gonadal function and and reproduction. *Nat. Clin. Pract. Endocrinol. Metab., 4*, 666–674.

Terasawa, E. (2001). Luteinizing hormone-releasing hormone (LHRH) neurons: mechanisms of pulsatile LHRH release. *Vitam. Horm., 63*, 91–129.

Terasawa, E. (2005). Role of GABA in the mechanism of the onset of puberty in non-human primates. *Int. Rev. Neurobiol., 71*, 113–129.

Thierry, B. (2007). The macaques: a double-layered social organization. In C. J. Campbell, A. F. Fuentes, K. C. Mackinnon, M. Panger, & S. Bearder (Eds.), *Primates in Perspective* (pp. 224–239). New York, NY: Oxford University Press.

Tilbrook, A. J., & Clarke, I. J. (2001). Negative feedback regulation of the secretion and actions of gonadotropin-releasing hormone in males. *Biol. Reprod., 64*, 735–742.

Utami, S. S. (2000). *Bimaturism in Orang-Utan males: Reproductive and Ecological Strategies.* Ph.D. dissertation. Universiteit Utrecht.

Utami, S. S., Goossens, B., Bruford, M. W., De Ruiter, J. R., & Van Hooff, J. A. R. A. M. (2002). Male bimaturism and reproductive success in Sumatran orang-utans. *Behav. Ecol., 13*, 643–652.

Valeggia, C. R., Mendoza, S. P., Fernandez-Duque, E., Mason, W. A., & Lasley, B. (1999). Reproductive biology of female titi monkeys (*Callicebus moloch*). *Am. J. Primatol., 47*, 183–195.

Van Horn, R. N., & Eaton, G. R. (1979). Reproductive physiology and behavior in prosimians. In G. A. Doyle, & R. D. Martin (Eds.), *The Study of Prosimmian Behavior* (pp. 79–122). New York, NY: Academic Press.

Van Schaik, C. P., & Brockman, D. K. (2005). Seasonality in primate ecology, reproduction and life history: an overview. In D. K. Brockman, & C. P. Van Schaik (Eds.), *Seasonality in Primates: Studies of Living and Extinct Human and Non-human Primates* (pp. 3–20). New York, NY: Cambridge University Press.

Videan, E. N., Fritz, J., Heward, C. B., & Murphy, J. (2008). Reproductive aging in chimpanzees (*Pan troglodytes*). In S. Atsalis, S. W. Margulis, & P. R. Hof (Eds.), *Primate Reproductive Aging: Cross-Taxon Perspectives on Reproduction. Interdiscip. Top. Gerontol, 36* (pp. 103–118). Basel, Switzerland: Karger.

Vitzthum, V. J., Thornburg, J., & Spielvogel, H. (2009). Seasonal modulation of reproductive effort during early pregnancy in humans. *Am. J. Hum. Biol., 21*, 548–558.

Vom Saal, F. S., Finch, C. E., & Nelson, J. F. (1994). Natural history and mechanisms of reproductive aging in humans, laboratory rodents, and other selected vertebrates. In E. Knobil, & J. D. Neill (Eds.), *The Physiology of Reproduction,* (2nd ed.). (pp. 1213–1314). NY: Raven Press.

Vulliémoz, N. R., Xiao, E. N., Xia-Zhang, L. N., Rivier, J., & Ferin, M. (2008). Astressin B, a nonselective corticotropin-releasing hormone receptor antagonist, prevents the inhibitory effect of ghrelin on luteinizing hormone pulse frequency in the ovariectomized rhesus monkey. *Endocrinology, 149*, 869–874.

Vulliémoz, N. R., Xiao, E., Xia-Zhang, L. N., Germond, M., Rivier, J., & Ferin, M. (2004). Decrease in luteinizing hormone pulse frequency during a five-hour peripheral ghrelin infusion in the ovariectomized rhesus monkey. *J. Clin. Endocrinol. Metab., 89*, 5718–5723.

Wade, G. N., Schneider, J. E., & Li, H. Y. (1996). Control of fertility by metabolic cues. *Am. J. Physiol., 33*, E1–E19.

Walker, W. H., Fitzpatrick, S. L., Barrera-Saldana, H. A., Resendez-Perez, D., & Saunders, G. F. (1991). The human placental lactogen genes: structure, function, evolution and transcription regulation. *Endocr. Rev., 12*, 316–328.

Wallen, K. (1990). Desire and ability: hormones and the regulation of female sexual behavior. *Neurosci. Biobehav. Rev., 14*, 233–241.

Wallen, K. (2005). Hormonal influences on sexually differentiated behavior in nonhuman primates. *Front. Neuroendocrinol., 26*, 7–26.

Wallen, K., & Zehr, J. L. (2004). Hormones and history: the evolution and development of primate female sexuality. *J. Sex Res., 41*, 101–112.

Wallis, O. C., & Wallis, M. (2002). Characterisation of the GH gene cluster in a new-world monkey, the marmoset (*Callithrix jacchus*). *J. Mol. Endocrinol., 29*, 89–97.

Wang, R. Y., Needham, L. L., & Barr, D. B. (2005). Effects of environmental agents on the attainment of puberty: considerations when assessing exposure to environmental chemicals in the National Children's Study. *Environ. Health Perspect., 113*, 1100–1107.

Wang, Z., Liang, B., Racey, P. A., Wang, Y.-L., & Zhang, S.-Y. (2008). Sperm storage, delayed ovulation, and menstruation of the female Rickett's big-footed bat (*Myotis ricketti*). *Zool. Stud., 47*, 215–221.

Wasser, S. K., & Barash, D. P. (1983). Reproductive suppression among female mammals: implications for biomedicine and sexual selection theory. *Q. Rev. Biol., 58m*, 513–538.

Wasser, S. K., Norton, G. W., Kleindorfer, S., & Rhine, R. J. (2004). Population trend alters the effects of maternal dominance rank on lifetime reproductive success in yellow baboons (*Papio cynocephalus*). *Behav. Ecol. Sociobiol., 56*, 338–345.

Watanabe, G., & Terasawa, E. (1989). *In vivo* release of luteinizing hormone releasing hormone increases with puberty in the female rhesus monkey. *Endocrinology, 125*, 92–99.

Watanobe, H. (2002). Leptin directly acts within the hypothalamus to stimulate gonadotropin-releasing hormone secretion *in vivo* in rats. *J. Physiol., 545*, 255–268.

Watts, D. P. (1998). Seasonality in the ecology and life histories of mountain gorillas (*Gorilla gorilla beringei*). *Int. J. Primatol., 19*, 929–948.

Wehrenberg, W. B., & Dyrenfurth, I. (1983). Photoperiod and ovulatory menstrual cycles in female macaque monkeys. *J. Reprod. Fertil., 68*, 119–122.

Weiss, G., Butler, W. R., Dierschke, D. J., & Knobil, E. (1976). Influence of suckling on gonadotropin secretion in the postpartum rhesus monkey. *Proc. Soc. Exp. Biol. Med., 153*, 330–331.

Weller, L., & Weller, A. (1993). Human menstrual synchrony: a critical assessment. *Neurosci. Biobehav. Rev., 17*, 427–439.

Wickings, E. J., & Nieschlag, E. (1980). Seasonality in endocrine and exocrine testicular function of the adult rhesus monkey (*Macaca mulatta*) maintained in a controlled laboratory environment. *Int. J. Androl., 3*, 87–104.

Wickings, E. J., Marshall, G. R., & Nieschlag, E. (1986). Endocrine regulation of male reproduction. In W. R. Dukelow, & J. Erwin (Eds.), *Comparative Primate Biology* (pp. 149–170). New York, NY: Alan R. Liss.

Wiebe, R. H., Williams, L. E., Abee, C. R., Yeoman, R. R., & Diamond, E. J. (1988). Seasonal changes in serum dehydroepiandrosterone, androstenedione and testosterone levels in the squirrel monkey (*Saimiri bolivensis bolivensis*). *Am. J. Primatol., 14*, 285–291.

Wildman, D. E., Chen, C., Erez, O., Grossman, L. I., Goodman, M., & Romero, R. (2006). Evolution of the mammalian placenta revealed by phylogenetic analysis. *Proc. Nat. Acad. Sci., 103*, 3203–3208.

Wilson, M. E., & Gordon, T. P. (1989a). Season determines timing of first ovulation in rhesus monkeys *(Macaca mulatta)* housed outdoors. *J. Reprod. Fertil., 85*, 583–591.

Wilson, M. E., & Gordon, T. P. (1989b). Short-day melatonin pattern advances puberty in seasonally breeding rhesus monkeys (*Macaca mulatta*). *J. Reprod. Fertil., 86*, 435–444.

Wilson, M. E., Gordon, T. P., & Collins, D. C. (1986). Ontogeny of luteinizing hormone secretion and first ovulation in seasonal breeding rhesus monkeys. *Endocrinology, 118*, 293–301.

Wise, P. M. (2006). Aging of the female reproductive system. In E. J. Masoro, & S. N. Austad (Eds.), *Handbook of the Biology of Aging* (pp. 570–590). London, UK: Elsevier.

Wistuba, J., Schrod, A., Greve, B., Hodges, J. K., Aslam, H., Weinbauer, G. F., et al. (2003). Organization of seminiferous epithelium in primates: relationship to spermatogenic efficiency, phylogeny, and mating system. *Biol. Reprod., 69*, 582–591.

Witkin, S. S., Linhares, A. M., & Giraldo, P. (2007). Bacterial flora of the female genital tract: function and immune regulation. *Best Pract. Res. Clin. Obstet. Gynaecol., 21*, 347–354.

Wright, P., King, S. J., Baden, A., & Jernvall, J. (2008). Aging in wild female lemurs: sustained fertility with increased infant mortality. In S. Atsalis, S. W. Margulis, & P. R. Hof (Eds.), *Primate Reproductive Aging: Cross-Taxon Perspectives on Reproduction Interdiscip. Top. Gertonol., 36* (pp. 17–28). Basel, Switzerland: Karger.

Yeoman, R. R., Wegner, F. H., Gibson, S. V., Williams, L. E., Abbott, D. H., & Abee, C. R. (2000). Midcycle and luteal elevations of follicle stimulating hormone in squirrel monkeys (*Saimiri boliviensis*) during the estrous cycle. *Am. J. Primatol., 52*, 207–211.

Yoder, A. D., Rasoloarison, R. M., Goodman, S. M., Irwin, J. A., Atsalis, S., Ravosa, M. J., et al. (2000). Remarkable species diversity in Malagasy mouse lemurs (primates, *Microcebus*). *Proc. Nat. Acad. Sci., 97*, 11325–11330.

Zehr, J. L., Van Meter, P. E., & Wallen, K. (2005). Factors regulating the timing of puberty onset in female rhesus monkeys (*Macaca mulatta*): role of prenatal androgens, social rank, and adolescent body weight. *Biol. Reprod., 72*, 1087–1094.

Zeleznik, A. J., & Benyo, D. F. (1994). Control of follicular development, corpus luteum function and the recognition of pregnancy in higher primates. In E. Knobil, & J. D. Neill (Eds.), *The Physiology of Reproduction* (2nd Edition)., *Vol. 2* (pp. 751–782). New York, NY: Raven Press.

Zeleznik, A. J., & Pohl, C. R. (2006). Control of follicular development, corpus luteum function, the maternal recognition of pregnancy, and the neuroendocrine regulation of the menstrual cycle in higher primates. In J. D. Neill (Ed.), *Knobil and Neill's Physiology of Reproduction* (Third Edition)., *Vol. 2* (pp. 2449–2510). St. Louis, MO: Elsevier.

Zhang, X., Zhu, C., Lin, H., Yang, Q., Ou, Q., Li, Y., et al. (2007). Wild fulvous fruit bats (*Rousettus leschenaulti*) exhibit human-like menstrual cycle. *Biol. Reprod., 77*, 358–364.

Ziegler, T. E., Strier, K. B., & Van Belle, S. (2009). The reproductive ecology of South American primates: ecological adaptations in ovulation and conception. In P. A. Garber, A. Estrada, J. C. Bicca-Marques, E. W. Heymann, & K. B. Strier (Eds.), *South American Primates: Comparative Perspectives in the Study of Behavior, Ecology, and Conservation* (pp. 191–210). New York, NY: Springer.

Zimmermann, E., & Radespiel, U. (2007). Primate life histories. In W. Henke, & I. Tattersall (Eds.), *Handbook of Paleoanthropology Vol. 2: Primate Evolution and Human Origins* (pp. 1163–1205). New York, NY: Springer-Verlag.

Chapter 14

Endocrine-disrupting Chemicals (EDCs) in Mammals

Katherine E. Pelch, Joseph M. Beeman, Bridget A. Niebruegge, Stacey R. Winkeler and Susan C. Nagel
University of Missouri, Columbia, MO, USA

SUMMARY

The United States presently uses more than 80 000 chemicals, a number that is estimated to grow by 2000 chemicals each year. Although generally not considered toxic at current exposure levels, many of these chemicals interfere with the endocrine system and may have profound effects on reproduction. Exposure to endocrine-disrupting chemicals (EDCs) is associated with an earlier onset of puberty, decreased fecundity and fertility, altered sexual behavior, and increased incidence of abnormalities and cancers of the reproductive tract in humans and in laboratory animals. Importantly, it appears that developing organisms may be particularly sensitive to slight variations in the hormonal milieu that occurs after exposure to some chemicals. In this chapter we review the sources and mechanisms of several well-studied EDCs and the evidence that exposure to these chemicals affects many aspects of mammalian reproduction.

1. INTRODUCTION

The United States presently uses more than 80 000 chemicals, a number that is estimated to grow by 2000 chemicals each year (National Toxicology Program (NTP), 2005). For the last half century, large amounts of these chemicals have been released into the environment and have caused harm to nature, wildlife, and humans. Several species of vertebrates have undergone distinct population declines: for example, a significant decrease in the number of bald eagles was observed following World War II due to the widespread use of the pesticide dichlorodiphenyltrichloroethane (DDT), which caused eggshell thinning (Cooke, 1973) and reduced reproductive fitness (Kozie & Anderson, 1991). Other species were reported to exhibit unusual sexual behavior: female herring gulls in Canada and the United States were observed to nest in same-sex pairings while male herring gulls displayed a noticeable disinterest in mating or parenting; behaviors that have been linked to xenobiotic exposure (Peakall & Fox, 1987). Still other vertebrates have been marked by abnormal reproductive systems and decreased fertility: after exposure to the organochlorine pesticide dicofol, male alligators have been reported to develop uncharacteristically small penises that result in vastly lower numbers of offspring (Guillette, Pickford, Crain, Rooney, & Percival, 1996).

Rachel Carson brought attention to the adverse effects of synthetic chemicals in 1962 when she wrote *Silent Spring*, a series of articles published in *The New Yorker* that depicted the declining state of the environment and human health (Carson, 1987). Following publication, the US government took heed of Carson's vivid warnings and formed the United States Environmental Protection Agency (USEPA) in 1970, which banned the use of several synthetic chemicals including DDT, kepone, and polychlorinated biphenyls (PCBs) (Markey, Coombs, Sonnenschein, & Soto, 2003). During the years after chemical regulations were put in place, gross toxic effects began to diminish and the mortality rates of chemically endangered species were significantly reduced. Despite this progress, a new object of concern manifested: the offspring of exposed adults. Although a decrease in the usage of selected chemicals resulted in a renewal of life, the offspring of previously exposed vertebrates displayed numerous developmental malformations (Colborn, vom Saal, & Soto, 1993; Guillette & Guillette, 1996). The persistence of adverse health outcomes in offspring of exposed adults presented an unexpected challenge to biologists.

Theo Colborn, an emerging ecologist, studied the declining health of vertebrates living in the Great Lakes as part of her project with the World Wildlife Fund. Along with other scientists from a number of different disciplines, she proposed the hypothesis of endocrine disruption: that chemicals in the environment were capable of acting like hormones and could alter normal physiology, reproduction, and development (Colborn, 1991; Colborn, 1992; Colborn et al., 1993; Colborn, 1994; 1995; Colborn, Dumanoski,

Hormones and Reproduction of Vertebrates, Volume 5—Mammals

& Myers, 1996). She observed that animals were affected in various ways such as reduced reproduction, thyroid problems, atypical behavior, and metabolism changes including wasting. Importantly, she noted that the observed effects were manifesting in the offspring of exposed adults in addition to the exposed adults themselves. Taking an interest in this phenomenon, Colborn organized the Wingspread Conference, a coalition of 21 scientists from across the country, in the summer of 1991. Together they discussed the evidence from wildlife populations that synthetic chemicals were causing severe health outcomes in the offspring of exposed adult animals. They linked effects observed in wildlife with the effects documented in offspring of women who took the synthetic estrogen diethylstilbestrol (DES) during pregnancy and proposed that both sets of effects were caused by hormone-like chemicals that disrupted normal endocrine function. They termed these chemicals 'endocrine disruptors' and made note of other sources of hormonally active chemicals such as plasticizers, disinfectants, and pesticides. They concluded the conference by issuing a consensus statement: 'We are certain of the following: a large number of man-made chemicals that have been released into the environment, as well as a few natural ones, have the potential to disrupt the endocrine system of animals, including humans' (Colborn, Dumanoski, & Myers 1996; pp 260–261).

Shortly after the conference, a meta-analysis was conducted to analyze human sperm production. It observed that both the quantity and quality of men's sperm decreased with the beginning of ubiquitous chemical usage (Carlsen, Giwercman, Keiding, & Skakkebaek, 1992). In addition, several reports around this time were issued that showed an increase in the cases of cryptorchidism, hypospadias, and testicular cancer (Giwercman, Carlsen, Keiding, & Skakkebaek, 1993). As soon as it became clear that human health may be at risk, the United States formed a National Academy of Science committee to assess the threat of endocrine-disrupting chemicals (EDCs). In 1996, Congress passed the Safe Drinking Water Act and the Food Quality Protection (FQP) Act, which required that the USEPA establish the Endocrine Disruptor Screening and Testing Advisory Committee (EDSTAC) to devise a plan for screening all chemicals that could potentially contaminate food and water and act as endocrine disruptors (EDSTAC, 1998).

At the same time, many in the field of toxicology were skeptical that EDCs could exert effects at low, environmentally relevant doses. However, scientific evidence had already been provided to support the potential for endocrine disruption. vom Saal and Bronson did pioneering work showing that adult physiology and behavior were affected by prior intrauterine position (IUP), specifically whether mice developed between two males or two females (vom Saal, 1989). Subsequently, it was shown that serum hormone concentrations during fetal life were altered for testosterone (T) ($2M > 1M > 0M$) and estradiol (E_2) ($0M > 1M > 2M$). Females that developed *in utero* between two females vs. two males showed earlier puberty, more regular estrus cycles, and increased receptivity and lordosis behavior, and had altered offspring sex ratios and increased reproductive longevity (vom Saal, 1981; vom Saal, Pryor, & Bronson, 1981). Males that developed *in utero* between two females vs. two males showed increased frequency of intromissions and mounting and increased infanticide behavior, and had enlarged prostates with more androgen receptors in adulthood. After experimentally manipulating E_2 within a physiologically relevant range, it was shown that a 30% increase resulted in enlarged prostates with more androgen receptors in adulthood. Interestingly, while changes within the physiological range increased prostate weight, this endpoint showed a nonmonotonic dose response where high doses decreased adult prostate weight (vom Saal et al., 1997).

In 2000, the NTP's report of the Endocrine Disruptors Low-Dose Peer Review Panel concluded that not only hormones but also synthetic chemicals can have biological consequences at environmentally relevant doses: 'low-dose effects ... were demonstrated in laboratory animals exposed to certain endocrine active agents. The effects are dependent on the compound studied and the endpoint measured ... The toxicological significance of many of these effects has not been determined' (Low-Dose Peer Review Panel, 2001; vii). The NTP report reflected the uncertainty in the toxicology community about disruption of normal physiology, development, and/or hormone homeostasis by low doses of chemicals. As the endpoints that endocrinologists and physiologists had reported often fell outside of the specific validated toxicological methods, the significance of these effects was debated. A key factor that exacerbated this skepticism was the reliance on high-dose chemical testing and the application of safety factors in the absence of low-dose testing in the toxicology community. This method assumes that the dose response is linear (monotonic); however, it has been repeatedly shown that, for hormone and EDC action, nonmonotonic dose-response curves are very common, if not typical. For example, inverted U-shaped dose-response curves have been shown for many endocrine endpoints; i.e., stimulation at low doses and inhibition of a given response at higher doses (vom Saal et al., 1997; Newbold, Jefferson, Padilla-Banks, & Haseman, 2004; Welshons, Nagel, & vom Saal, 2006).

The USEPA Endocrine Disruptor Screening and Testing Program (EDSP) was established in 1998 in response to the FQP Act and was charged with the goal of developing screening and testing methods for endocrine-active compounds. More than 10 years later, the final selection of chemicals to be tested to validate screening and testing methods was chosen (Environmental Protection Agency,

2009). Final methods for endocrine-active compounds may soon be validated.

2. ENDOCRINE-DISRUPTING CHEMICALS THAT AFFECT REPRODUCTION

There are hundreds of chemicals with reported endocrine-disrupting activity (Table 14.1) (Colborn et al., 1996). Discussed below are a few examples of EDCs with known human exposure and effects in laboratory animals and/or humans within or near the range of current human exposure.

Bisphenol A (BPA) is an estrogen receptor (ER) ligand and xenoestrogen (Nagel et al., 1997b; Nagel, Hagelbarger, & McDonnell, 2001). While it has weak agonist activity in stimulating uterine weight gain, it has relatively potent activity in stimulation of ER-dependent reporter gene activity in the uterus of ER indicator (ERIN) mice (Nagel et al., 2001). It has been widely reported to have effects at doses below the USEPA's current 'safe level' (reviewed in Welshons et al., 2006). Bisphenol A is a monomer that is polymerized to form polycarbonate plastic (hard, clear plastic). It is one of the top 50 chemicals produced worldwide, with annual production exceeding six billion pounds (Kirschner, 1996). It is released from polycarbonate products such as water bottles and baby bottles (Sun et al., 2000; Brede, Fjeldal, Skjevrak, & Herikstad, 2003), the resin lining of metal food cans (Brotons, Olea-Serrano, Villalobos, Pedraza, & Olea, 1995; Kang & Kondo, 2002; Munguia-Lopez, Peralta, Gonzalez-Leon, Vargas-Requena, & Soto-Valdez, 2002), some dental sealants (Garcia-Rodriguez et al., 1996), eye glasses, CD cases, and polycarbonate laboratory animal cages (Howdeshell et al., 2003; Hunt et al., 2003). Bisphenol A is responsible for 84% of the estrogenic activity in landfill leachate (Kawagoshi, Fujita, Kishi, & Fukunaga, 2003) and is an almost ubiquitous chemical in our environment and in humans (Vandenberg, Hauser, Marcus, Olea, & Welshons, 2007). Calafat, Wong, Ye, Reidy, and Needham (2008) recently reported that $>92\%$ of Americans have detectable levels of BPA in their urine. Importantly, children had twice the level of adults (4.5 μg/L vs. 2.5 μg/L) and women had higher levels than men. Bisphenol A has been found in human maternal and fetal serum at 0.2 to 10 ng/ml (Ikezuki, Tsutsumi, Takai, Kamei, & Taketani, 2002; Schonfelder et al., 2002a; Kuroda et al., 2003). It has been estimated that infants drinking out of polycarbonate bottles receive a daily exposure of up to 2 μg BPA/kg (Maragou, Makri, Lampi, Thomaidis, & Koupparis, 2008). Of particular concern, we and others have shown that BPA alters normal development at this dose (Nagel et al., 1997a; Steinmetz, Brown, Allen, Bigsby, & Ben-Jonathan, 1997; Steinmetz et al., 1998; vom Saal et al., 1998; Howdeshell, Hotchkiss, Thayer, Vandenbergh, & vom Saal, 1999; Gupta, 2000a; Markey, Luque, Munoz De Toro, Sonnenschein, & Soto, 2001; Takai et al., 2001; Honma et al., 2002; Palanza, Howdeshell, Parmigiani, & vom Saal, 2002; Schonfelder et al., 2002a; Hunt et al., 2003; Kubo et al., 2003; Ho, Tang, Belmonte de Frausto, & Prins, 2006; Welshons et al., 2006).

The NTP recently evaluated the potential hazard of BPA exposure for human health and found some level of concern in human fetuses, infants, and children at current exposures levels of BPA for effects on the brain, behavior, prostate gland, and mammary gland, and regarding an earlier age for puberty for females (Center for the Evaluation of Risks to Human Reproduction (CERHR), 2008). Additionally, the Endocrine Society recently recommended reducing human exposure to BPA (Diamanti-Kandarakis et al., 2009).

Genistein is a naturally occurring phytoestrogen and the predominant soy-isoflavone. The consumption of genistein has risen sharply in the form of isoflavone capsules since the Federal Drug Administration (FDA) endorsed soy isoflavones as beneficial (SERVICES, 1999) and it is under intense investigation because of the high human exposure. Adults consuming soy-based foods, particularly vegetarians, circulate 50–200 ng/ml total isoflavones, and infants fed soy-based formula circulate approximately 700 ng/ml genistein (Setchell, Zimmer-Nechemias, Cai, & Heubi, 1997). While soy-based foods have been promoted as having beneficial effects in adults, there is increasing evidence that developmental exposure during differentiation of the uterus may be detrimental (Newbold, Banks, Bullock, & Jefferson, 2001; Jefferson, Couse, Padilla-Banks, Korach, & Newbold, 2002; Jefferson, Padilla-Banks, & Newbold, 2005; Jefferson, Newbold, Padilla-Banks, & Pepling, 2006). In fact, Jefferson et al. (2005) showed that, even at their lowest-tested dose of perinatal genistein of 500 μg/kg, female mice showed impaired fertility in adulthood. This dose is an order of magnitude lower than that consumed by infants fed exclusively soy formula (Tuohy, 2003).

Ethinylestradiol (EE_2) is an orally active form of E_2 and it is a component of oral contraceptives (OCs). Millions of women use OCs and approximately 2–5% will conceive while taking them due to noncompliance (Cerel-Suhl & Yeager, 1999). Some of these women deliver infants exposed to 0.2 to 0.5 μg/kg of this potent xenoestrogen during the critical window of sexual differentiation in the first trimester of pregnancy. Importantly, EE_2 alters mouse development at levels ≤ 0.3 μg/kg (Thayer, 1999; Thayer et al., 2001). A growing concern is that humans and other animals are exposed to pharmaceuticals in water. Ethinylestradiol has been measured in waste water influent and effluent and may be a source of human exposure (Chimchirian, Suri, & Fu, 2007).

TABLE 14.1 Widespread pollutants with endocrine disrupting effects

Persistent organohalogens

Compound(s)	Hormone system affected	Mechanism if known	References
Benzenehexachloride (BHC)	Thyroid		Akhtar et al. 1996
1,2-dibromoethane	Reproductive		Brittebo et al. 1987
Chloroform	Reproductive		Brittebo et al. 1987
Dioxins and furans (in order of antiestrogenic potency: 2,3,7,8-tetrachlorodibenzo-p-dioxin > 2,3,7,8-tetrachlorodibenzofuran > 2,3,4,7,8-pentachlorodibenzofuran > 1,2,3,7,9-pentachlorodibenzofuran > 1,3,6,8-tetrachlorodibenzofuran)	Estrogen	Work as anti-estrogen through binding with Ah receptor, which then inhibits estrogen receptor binding to estrogen response elements, thereby inhibiting estrogen action	Krishnan and Safe 1993; Klinge et al. 1999
Octachlorostyrene	Thyroid		Sandan et al. 2000
PBBs	Estrogen/ Thyroid		Bahn et al. 1980; Henderson et al. 1995
PCBs (in order of antiestrogenic potency: 3,3′ - pentachlorobiphenyl > 3,3,4,4,5,5′- hexachlorobiphenyl 3,3′, 4,4- tetrachlorobiphenyl > 2,3,3′,4,4′,5′-hexa, 2,3,3′,4,4′- and 2,3,4,4′,5- pentachlorobiphenyl > Aroclors 1221, 1232. 1248, 1254, and 1260 were inactive as antiestrogens at the highest concentrations used in this study (10-6 Ni)	Estrogen/androgen/ Thyroid. Adverse outcomes in reproductive systems.	Inhibits estrogen binding to the receptor; works as anti-estrogen. Anti-androgenic via Ah receptor interaction	Korach et al. 1988; Zoeller et al. 2000; Grey et al. 1999
PCB, hydroxylated	Thyroid	Binds to thyroid hormone binding protein, but not to the thyroid hormone receptor	Cheek et al. 1999
PBDEs	Thyroid	Interferes with thyroxine (T4) binding with transthryetin	Ilonka et al. 2000
Pentachlorophenol	Thyroid	Reduces thyroid hormone possibly through a direct effect on the thyroid gland	Bear et al. 1999; Gerhard et al. 1999

Food Antioxidant

Compound	Hormone system affected	Mechanism	References
Butylated hydroxyanisole (BHA)	Estrogen	Inhibits binding to the estrogen receptor	Jobling et al. 1995

Pesticide

Compound	Hormone system affected	Mechanism	References
Acetochlor	Thyroid (decrease of thyroid hormone levels, increase in TSH)		Hurley et al. 1998
Alachlor	Thyroid (decrease of thyroid hormone levels, increase in TSH)		Wilson et al. 1996
Aldrin	Estrogen	Binds to estrogen receptors; competes with estradiol	Jorgenson 2001

TABLE 14.1 Widespread pollutants with endocrine disrupting effects—cont'd

Pesticide

Compound	Hormone system affected	Mechanism	References
Allethrin, d-trans	Estrogen		Go et al. 1999
Amitrol	Thyroid	Thyroid peroxidase inhibitors; inhibits thyroid hormone synthesis	Hurley et al. 1998
Atrazine	Neuroendo-crinepituitary (depression of LH surge), testosterone metabolism.	Inhibits ligand binding to androgen and estrogen receptors	Danzo 1997
Carbaryl	Estrogen and progesterone		Klotz et al. 1997
Chlofentezine	Thyroid	Enhances secretion of thyroid hormone	Hurley et al. 1998
Chlordane	Testosterone and progesterone		Willingham et al. 2000
Cypermethrin	Disruption of reproductive function		Moore and Waring 2001
DDT	Estrogen	DDT and related compounds act in a number of ways to disrupt endocrine function by binding with the estrogen receptor, including estrogen mimickry and antagonism, altering the pattern of synthesis or metabolism of hormones, and modifying hormone receptor levels	Soto et al. 1994; Lascombe et al. 2000; Kupfer et al. 1980; Rajapakse et al. 2001
DDT Metabolite, p,p′-DDE	Androgen	Inhibits androgen binding to the androgen receptor, androgeninduced transcriptional activity, and androgen action in developing, pubertal, and adult male rats	Kelce 1995
Dicofol (Kelthane)	Estrogen		Vinggaard et al. 1999
Dieldrin	Estrogen	Binds to estrogen receptor; competes with estradiol	Soto et al. 1994; Jorgenson 2001
Endosulfan	Estrogen		Soto et al. 1994; Soto et al. 1995
Ethylene thiourea	Thyroid	Thyroid peroxidase inhibitor	Hurley et al. 1998
Fenarimol	Estrogen	Estrogen receptor agonist	Vinggaard et al. 1999
Fenbuconazole	Thyroid	Enhances secretion of thyroid hormone	Hurley et al. 1998
Fenitrothion	Antiandrogen	Competitive androgen receptor antagonist	Tamura et al. 2001
Fenvalerate	Estrogen		Go et al. 1999
Fipronil	Thyroid	Enhances secretion of thyroid hormone	Hurley et al. 1998

(*Continued*)

TABLE 14.1 Widespread pollutants with endocrine disrupting effects—cont'd

Pesticide

Compound	Hormone system affected	Mechanism	References
Heptachlor	Thyroid		Akhtar et al. 1996; Reuber 1987
Heptachlor-epoxide	Thyroid/Reproductive	Metabolite of heptachlor	Reuber 1987
Iprodione	Inhibition of testosterone synthesis		Benhamed 1996
Karate	Thyroid	A decrease of thyroid hormone in serum; direct effect on the thyroid gland?	Akhtar et al. 1996
Kepone (Chlordecone)	Estrogen	Displays androgen and estrogen receptor-binding affinities	Waller et al. 1996; Soto et al. 1994; McLachlan (ed)
Ketoconazole	Effects on reproductive systems		Marty et al. 1999; Marty et al. 2001
Lindane (Hexachlorocyclohexane)	Estrogen/Androgen	Inhibits ligand-binding to androgen and estrogen receptors	Danzo 1997
Linuron	Androgen	Androgen receptor antagonist	Waller et al. 1996; Lambright et al. 2000; Grey et al. 1999
Malathion	Thyroid	Significant decrease of thyroid hormone in serum, with perhaps a direct effect on the thyroid gland	Akhtar et al. 1996
Mancozeb	Thyroid	Thyroid peroxidase inhibitors	Hurley et al. 1998
Maneb	Thyroid	The metabolite ethylenthiourea inhibits thyroid hormone synthesis	Toppari et al. 1995
Methomyl	Thyroid		Porter et al. 1993; Klotz et al. 1997
Methoxychlor	Estrogen	Through mechanisms other than receptor antagonism. Precise mechanism still unclear	Pickford and Morris 1999
Metribuzin	Thyroid		Porter et al. 1993
Mirex	Antiandrogenic activity; inhibits production of LH. Potentially thyroid		Chen et al. 1986; Chernoff et al. 1976
Nitrofen	Thyroid	Structural similarities to the thyroid hormones; nitrofen or its metabolite may have thyroid hormone activities	Stevens and Summer 1991
Nonachlor, trans-	Estrogen	Estrogen receptor agonist?	Willingham et al. 2000
Oxychlordane	Reproductive		Guillette et al. 1999
Pendimethalin	Thyroid	Enhances secretion of thyroid hormone	Hurley et al. 1998
Pentachloronitrobenzene	Thyroid	Enhances secretion of thyroid hormone	Hurley et al. 1998

TABLE 14.1 Widespread pollutants with endocrine disrupting effects—cont'd

Pesticide

Compound	Hormone system affected	Mechanism	References
Permethrin	Estrogenic		Go et al. 1999
Procymidone	Androgen	Androgen receptor antagonist	Ostby et al. 1999; Grey et al. 1999
Prodiamine	Thyroid	Enhances secretion of thyroid hormone	Hurley et al. 1998
Pyrimethanil	Thyroid	Enhances secretion of thyroid hormone	Hurley et al. 1998
Sumithrin	Androgen		Go et al. 1999
Tarstar	Thyroid	A decrease of thyroid hormone in serum; direct effect on the thyroid gland?	Akhtar et al. 1996
Thiazopyr	Thyroid	Enhances secretion of thyroid hormone	Hurley et al. 1998
Thiram	Neuroendo-crinepituitary (depression of LH surge), thyroid (decrease of T4, increase of TSH)		Stoker et al. 1993
Toxaphene	Estrogen/ Thyroid		Soto et al. 1994
Triadimefon	Estrogen	Estrogen receptor agonist	Vinggaard et al. 1999
Triadimenol	Estrogen	Estrogen receptor agonist	Vinggaard et al. 1999
Tributyltin	Reproductive		Horiguchi et al. 2000
Trifluralin	Reproductive/ Metabolic		Rawlings et al. 1998
Vinclozolin	Androgen	Anti-androgenic. (Competes with androgens for the androgen receptor (AR), inhibits AR-DNA binding, and alters androgendependent gene expression)	Soto et al. 1994; Soto et al. 1995; Kelce et al. 1994; Grey et al. 1999
Zineb	Thyroid	The metabolite ethylenthiourea inhibits thyroid hormone synthesis	Toppari et al. 1995
Ziram	Thyroid	Inhibits the iodide peroxidase. Structural similarities between ziram and thiram; ziram can be metabolized to thiram in the environment	Marinovich et al. 1997

Phthalate

Compound	Hormones affected	Mechanism	References
Butyl benzyl phthalate (BBP)	Estrogen	Inhibits binding to the estrogen receptor	Jobling et al. 1995
Di-n-butyl phthalate (DBP)	Estrogen; Androgen	Inhibits binding to the estrogen receptor. Anti-androgenic	Jobling et al. 1995; Harris et al. 1997; Grey et al. 1999

(*Continued*)

TABLE 14.1 Widespread pollutants with endocrine disrupting effects—cont'd

Phthalate

Compound	Hormones affected	Mechanism	References
Di-ethylhexyl phthalate (DEHP)	Estrogen; Androgen	Inhibits binding to the estrogen receptor. Anti-androgenic	Jobling et al. 1995; Moore et al. 2001; Grey e al. 1999
Diethyl Phthalate (DEP)	Estrogen		Harris et al. 1997

Other Compounds

Compound	Hormones affected	Mechanism	References
Benzophenone	Estrogen	Binds weakly to estrogen receptors, roles of its metabolite remain to be clarified	Schlumpf et al. 2001
Bisphenol A (BPA)	Estrogen	Estrogenic; binds to estrogen receptor	Fisher et al. 1999; Anderson et al. 1999; Rajapakse et al. 2001
Bisphenol F	Estrogen	Estrogenic; binds to estrogen receptor	Perez et al. 1998
Benzo(a)pyrene	Androgen	Anti-androgenic	Thomas 1990
Carbendazim	Reproductive		Gray et al. 1990
Ethane Dimethane Sulphonate	Reproductive		Gray et al. 1999
Perfluorooctane sulfonate (PFOS)	Thyroid, reproductive	Suppression of T3,T4; mechanism unknown	3M data
Nonylphenol, octylphenol	Estrogen	Estrogen receptor agonists; reduce estradiol binding to the estrogen receptor	Soto et al. 1991; Soto et al. 1995; Danzo 1997; Lascombe et al. 2000; Rajapakse et al. 2001
Resorcinol	Thyroid		Lindsay et al. 1989
Styrene dimers and trimers	Estrogen	Estrogen receptor agonists	Ohyama et al. 2001

Metals

Compound	Hormones affected	Mechanism	References
Arsenic	Glucocorticoid	Selective inhibition of DNA transcription normally stimulated by the glucocorticoid-GR complex	Kaltreider et al. 2001
Cadmium	Estrogenic	Activates estrogen receptor through an interaction with the hormone-binding domain of the receptor	Stoica et al. 2000; Johnson et al. 2003
Lead	Reproductive		Telisman et al. 2000; Hanas et al. 1999
Mercury	Reproductive/ Thyroid		Facemire et al. 1995

Reproduced with permission from http://www.ourstolenfuture.org/basics/chemlist.htm. Complete references from table are available online.

Phthalates are a group of structurally similar chemicals used in the manufacture of plastics to impart different structural characteristics. Human exposure to these chemicals is widespread from many sources, including use in plastic flooring, tubing, toys, cosmetics, perfumes, soaps, and lotions (Swan et al., 2005; Main et al., 2006; Heudorf, Mersch-Sundermann, & Angerer, 2007). Perhaps as an indicator of their ubiquitous use, phthalates have been reported at high concentrations in rain water (Peters, Beeltje, & Van Delft, 2008). Sathyanarayana, Calafat, Liu, and Swan (2008) showed that the use of baby lotion, baby powder, and baby shampoo was positively correlated with the level of three phthalates in infant urine. Phthalate esters have been shown to bind to both ERs and the androgen receptor (AR). In general, phthalates have weak estrogenic activity (Jobling, Reynolds, White, Parker, & Sumpter, 1995; Harris, Henttu, Parker, & Sumpter, 1997) and moderate antiandrogenic activity (Gray et al., 1989; Sohoni & Sumpter, 1998).

Pesticides are a group of functionally similar and dissimilar chemicals with widespread human exposure. Pesticides are often targeted to disrupt hormone signaling in insects. Given the evolutionary conservation of many physiological systems, it is not surprising that these chemicals often disrupt mammalian development and physiology. Many adverse human health trends are associated with pesticide exposure, including but not limited to prostate cancer (Forastiere et al., 1993; Mills, 1998; Alavanja et al., 2003; 2005) and poor semen quality (Swan et al., 2003b). As a result of the high human exposure and many known pesticides with hormonal activity, the US EPA focused on 70 pesticides and pesticide related chemicals to validate its EDSP methods (Environmental Protection Agency, 2009).

Persistent organic pollutants (POPs), including PCBs, polychlorinated dibenzofurans (PCDFs), polychlorinated dibenzodioxins (PCDDs), and other dioxin-like chemicals are extremely stable, lipophilic compounds that do not degrade easily and continue to contaminate the environment despite the fact that the production of some has been prohibited in the US since the 1970s. Sources of POPs have historically been from manufacturing and industrial pollution. Most human exposure today is caused by ingestion of contaminated fish, meat, or dairy products, after which the POPs are stored in fat tissue and may disrupt normal endocrine system function and hence reproductive function (Diamanti-Kandarakis et al., 2009).

3. MECHANISMS OF ENDOCRINE-DISRUPTING CHEMICAL (EDC) ACTION

The USEPA defines an EDC as 'an exogenous agent that interferes with synthesis, secretion, transport, metabolism, binding action, or elimination of natural blood-borne hormones that are present in the body and are responsible for homeostasis, reproduction, and developmental process' (Crisp et al., 1998; pp 12 and 17). Discussed below are many potential mechanisms of EDCs.

3.1. Binding to Members of the Nuclear Receptor Superfamily

Although EDCs potentially can interfere with any hormone signaling system, their effects have been described most commonly for disruption of steroid receptors and nuclear hormone signaling processes. The natural ligands for these receptors are generally small lipophilic molecules (steroids) that function by binding to receptors that reside inside cells or in some cases in the cell membrane. The intracellular receptors are part of the nuclear receptor superfamily. Endocrine-disrupting chemicals that bind to these receptors are often small lipophilic molecules that escape or partially escape metabolism in the gastrointestinal tract or are absorbed via dermal contact or inhalation. Unlike protein hormones, which are largely digested and not absorbed in a biologically active form, the absorption of lipophilic chemicals results in molecules that can passively diffuse in and out of cells and potentially bind to intracellular receptors as well as cell membrane receptors.

The nuclear receptor superfamily consists largely of ligand-activated transcription factors. In general, ligand binding induces an activating conformational change in the nuclear receptor that releases corepressor proteins and recruits coactivator proteins. Nuclear receptors dimerize, either with like receptors to form homodimers or unlike receptors to form heterodimers, and associate with regulatory regions of target genes known as hormone response elements. Coactivators possess histone acetylase activity that functions to unwind DNA from histones and facilitate the interaction with basal transcriptional machinery.

The nuclear receptor superfamily can be subdivided into two broad groups. Type I nuclear receptors are also known as steroid receptors and include ERs, progesterone (P_4) receptors (PRs), AR, glucocorticoid receptor (GR), and the mineralocorticoid receptor (MR). Type I receptors usually involve nuclear translocation from the cytosol to the nucleus (unoccupied ERs and ARs; may occur in the nucleus), where they form homodimers and bind DNA. Type II nuclear receptors are thyroid hormone receptor (TR)-like receptors, which include TRs, retinoic acid receptors (RARs), and the vitamin D receptor (VDR). Unoccupied type II nuclear receptors are found in the nucleus and bind to DNA as heterodimers, usually with the retinoid X receptor (RXR).

As a result of their broad expression and ability to directly regulate gene expression, nuclear receptors are integral regulators of development and adult physiology. Consequently, perturbation by xenobiotic chemicals (exogenous chemicals not made by the body) can have diverse and widespread effects. Highlighted below are several of the most common nuclear receptors with some examples of impacts of EDCs. More examples are given in Section 6.

3.1.1. Genomic estrogen receptors (ERs)

There are two principle ERs in mammals: ERα and ERβ. Estrogen receptor-α is expressed at low to moderate levels in all major physiological systems of the body, at moderate levels in adipose tissue and the male reproductive tract, and at high levels in the female reproductive tract, mammary gland, and pituitary (Nuclear Receptor Signaling Atlas (NURSA), 2010). Estrogen receptor-β is expressed at low to moderate levels in the central nervous system (CNS), immune system, and male reproductive tract and at high levels in the ovary and prostate (Koehler, Helguero, Haldosen, Warner, & Gustafsson, 2005).

Estradiol is the primary physiological steroidal estrogen and is synthesized by conversion from androgens by the enzyme aromatase. Aromatase is expressed primarily in the gonads, brain, and adipose tissue. Estradiol, like other steroid hormones, is carried in the blood and is associated with serum binding proteins, albumin, and sex hormone-binding globulin (SHBG) and only a small fraction is free (about 2%) and able to passively diffuse into cells. This fraction is thus considered the biologically active fraction. Xenoestrogens are chemicals made outside of the body that may bind to ERs. Many xenoestrogens have a higher bioactive fraction in human serum than E_2 (Nagel, vom Saal, & Welshons, 1998).

Estrogens, whether endogenous or xenobiotic, function within target tissues by binding to ERα and/or ERβ located within target cell nuclei. Upon ligand binding, the respective ER undergoes an activating conformational change that facilitates the interaction of ERs with comodulator proteins and regulatory regions of target genes, known as estrogen response elements. Importantly, different ER ligands induce different conformational changes in the receptor, and this is believed to regulate the interaction of ER with specific comodulatory proteins (Paige et al., 1999). Proteins that impact the ER-signaling pathway downstream of DNA binding also determine the ability of the ligand-bound ER to activate or suppress target gene transcription (McKenna, Lanz, & O'Malley, 1999). This has been most clearly demonstrated with pharmaceuticals called selective ER modulators (SERMs) such as tamoxifen and raloxifene.

Xenoestrogens can mimic (ER agonists) or antagonize (antiestrogens) the action of endogenous estrogens (e.g., E_2), or exhibit mixed and tissue-specific agonist/antagonist activity (SERMs). Xenoestrogens can act as SERMs to induce unique receptor conformations, thus allowing different xenoestrogens to bring about different responses by binding to the same receptor in the same tissue (Nagel et al., 2001; Naciff et al., 2002). Many pesticides, plastic monomers and additives, and industrial products and byproducts bind to ERs (Colborn et al., 1993). A large number of low-molecular-weight, lipophilic endocrine-active compounds in the environment can pass freely across the placenta and into the fetus, which has a limited capacity to metabolize xenobiotic chemicals. In contrast to steroidal estrogens, many xenoestrogens, such as DES, *o,p'*-DDT, PCBs, and 4-nonylphenol (NP), are more stable and remain in the body longer than natural estrogens, and tend to accumulate within the fat and tissues of animals and humans (Tapiero, Ba, & Tew, 2002).

3.1.2. Nongenomic actions of estrogens

Although the classic estrogen signaling pathway is genomic, there are other ER-mediated actions that function through other signaling pathways; e.g., interactions with activator protein-1 (AP-1) (Kushner et al., 2000; Jakacka et al., 2001) or membrane-bound ERs (Levin, 1999; Quesada et al., 2002). Thomas and Dong (2006) have characterized a seven-transmembrane receptor, termed GPR30, that binds ER ligands. Xenoestrogens also bind to this receptor. Genistein binds to GPR30 and has a binding affinity only 10-fold lower than E_2. Other chemicals with estrogenic actions, such as BPA, zearalenone, NP, kepone, *p,p'*-DDT, 2,2',5',-PCB-4-OH, and dichlorodiphenyldichloroethylene (DDE), all bind to GPR30 as well (Thomas & Dong, 2006).

Bulayeva and Watson (2004) have characterized a membrane ERα. Coumestrol, endosulfan, dieldrin, DDE, and NP induced extracellular-regulated kinases (ERKs) within minutes at nM to pM concentrations in human pituitary cancer cells (Bulayeva & Watson, 2004). Wozniak, Bulayeva, and Watson (2005) measured two membrane ER-mediated endpoints in these cells—calcium influx and prolactin (PRL) release—and found that BPA was as potent as E_2 in stimulation of these responses (Wozniak et al., 2005). This highlights the complexity of ER signaling and that xenoestrogens may be more potent *in vivo* through membrane receptors than estimated from classical tests of estrogenicity.

3.1.3. The androgen receptor (AR)

The AR is expressed at low to moderate levels throughout the body, with high expression in the reproductive tract, kidney, and adrenal gland in adulthood (NURSA, 2010). Androgen receptor expression in development is essential for differentiation of the male external and internal genitalia and reproductive behaviors and for maintenance of male secondary sexual characteristics and fertility. Androgens are synthesized in the gonads and to a lesser extent in the adrenal cortex. The principle androgens in circulation are androstenedione, dehydroepiandrosterone (DHEA), and dehydroepiandrosterone sulfate (DHEA-S); however, only T binds to AR with high affinity and it is thought that the actions of the other androgens are mediated by conversion to T (Kaufman & Vermeulen, 2005). Dihydrotestosterone (DHT) is converted in target tissues from T and binds AR with high affinity. In addition to the genomic actions of androgens, membrane actions have also been described (Shakil, Hoque, Husain, & Belsham, 2002; Papadopoulou, Papakonstanti, Kallergi, Alevizopoulos, & Stournaras, 2009).

Many EDCs are ligands for the AR, including vinclozolin, *p,p*'-DDE, *o,p*'-DDT, BPA, and butyl benzyl phthalate (BBP) (Satoh, Ohyama, Aoki, Iida, & Nagai, 2004; Xu et al., 2005). Many phthalates are antiandrogenic and this is one mechanism for their disruption of male reproductive tract development (reviewed in Howdeshell, Rider, Wilson, & Gray, 2008). Whereas most EDCs described to date have antiandrogenic actions, some have weak androgenic activity, such as NP (Sohoni & Sumpter, 1998).

3.1.4. Progesterone receptors (PRs)

Progesterone receptor is expressed in all major physiological systems throughout the body at low levels with moderate expression in the CNS and high expression in the female reproductive tract (NURSA, 2010). There are at least three functionally different PR isoforms that are transcribed from different promoters on the same gene that are capable of regulating different sets of genes (Wei & Miner, 1994; Richer et al., 2002). The primary endogenous progestin is P_4, which is synthesized in the ovary in females. Males circulate low levels of progestins, primarily as intermediates in androgen synthesis.

Several plant-derived chemicals interact with PRs. Zava, Dollbaum, and Blen (1998) found the highest consumed phytochemicals that interact with PR were from oregano, verbena, turmeric, thyme, red clover, and damiana. None of these chemicals were agonists for PR; all tended to be PR antagonists. Klotz, Ladlie, Vonier, McLachlan, and Arnold (1997) showed that DDT and its metabolites (e.g., DDE) were also antiprogestins.

3.1.5. Glucocorticoid receptor (GR)

Glucocorticoid receptor is expressed at high levels in all major physiological systems; in fact, in almost all cells of the body. It plays central roles in the immune and cardiovascular systems, the metabolism, and development (NURSA, 2010). In addition, glucocorticoids impact the reproductive system in a number of ways, including having effects on pituitary secretion of gonadotropins and fetal development (reviewed in Cooke et al., 2004). The primary glucocorticoid in humans is cortisol, which is produced in the adrenal cortex, whereas corticosterone is the primary glucocorticoid in rodents.

Chronic arsenic exposure is a known risk factor for diabetes, cancer, and cardiovascular disease. Low doses (6–120 ppb) of arsenic have been reported to stimulate GR-dependent gene expression of both an endogenous gene, tyrosine aminotransferase, and a reporter gene in rat EDR3 hepatoma cells (Bodwell, Kingsley, & Hamilton, 2004). At higher doses, arsenic inhibited these same responses up to two-fold. Thus, arsenic exhibited an inverted U dose response curve for stimulation of this gene.

3.1.6. Peroxisome proliferator-activated receptors (PPARs)

There are three peroxisome proliferator-activated receptors (PPARs): PPARα, PPARδ, and PPARγ. The PPARs are expressed at low levels throughout the body and PPARα is expressed at moderate to high levels in the heart, gastrointestinal (GI) tract, and brown adipose; PPARγ is expressed at moderate levels in the reproductive tract and CNS and at high levels in adipose tissue and the colon; PPARδ is expressed at high levels in the GI tract (NURSA, 2010). In general, PPARs play important roles in lipid and glucose metabolism and may be involved in metabolic diseases such as hyperlipidemia, insulin resistance, and coronary artery disease.

Peroxisome proliferator-activated receptors may be targets for some of the EDCs that also affect reproduction. Phthalates can activate all three PPARs (reviewed in Casals-Casas, Feige, & Desvergne, 2008; Desvergne, Feige, & Casals-Casas, 2009). Bisphenol A diglycidyl ether (BADGE) has been reported to be both a PPARγ agonist and antagonist (Bishop-Bailey, Hla, & Warner, 2000; Wright et al., 2000).

3.1.7. Pregnane X receptor (PXR)

The pregnane X receptor (PXR) is expressed at high levels in tissues involved in metabolism, particularly throughout the GI tract and the liver (NURSA, 2010). Several endogenous steroid hormones, such as P_4, and exogenous xenobiotics, such as NP, have been shown to bind to PXR and mediate gene transcription (Masuyama, Hiramatsu, Kunitomi, Kudo, & MacDonald, 2000). One set of targets is cytochrome P450 (*CYP*) 3A genes (*CYP3A*) in the mouse, rat, and human (Masuyama et al., 2000). This family of enzymes is involved in metabolism of steroid hormones and xenobiotics. The xenobiotics phthalic acid and NP stimulate PXR-mediated transcription *in vitro* in COS-7 cells and also stimulate *Cyp3a1* expression in the rat liver (Masuyama et al., 2000). Disruption of normal expression may lead to altered circulating hormone and/or xenobiotic levels.

3.1.8. The constitutive active receptor (CAR)

The constitutive active receptor (CAR) is expressed at high levels in tissues of the metabolic system, particularly in the duodenum, jejunum, kidney, and liver. An endogenous ligand is not known for CAR and is therefore classified as an orphan receptor. Similarly to PXR, CAR binds xenobiotics and regulates a number of metabolic enzymes responsible for the metabolism of endogenous lipids and steroids, and xenobiotics.

3.1.9. Thyroid hormone receptors (TRs)

There are two TRs: TRα and TRβ. Thyroid hormone receptors are expressed differentially at low to moderate levels throughout the body, with high expression in the pituitary and CNS (NURSA, 2010). Thyroid hormone receptors are essential for normal development and for normal energy homeostasis throughout life. The predominant circulating thyroid hormone is thyroxine (T_4), important for CNS development, which is converted to the higher affinity hormone triiodothyronine (T_3), in the liver and in target cells. Both T_3 and T_4 bind to both TRs. Thyroid hormones are essential for normal development and maintenance of the reproductive system, including development of the gonads, testicular Sertoli and Leydig cell number and function in men, and in ovulation and menstrual regularity in women (reviewed in Cooke et al., 2004).

Howdeshell (2002) extensively reviewed EDCs for their potential to interfere with thyroid hormone signaling and found over 150 chemicals with the ability to disrupt thyroid hormone action. Endocrine-disrupting chemicals have been found to disrupt nearly every step in the synthesis and metabolism of thyroid hormones (Howdeshell, 2002). For example, EDCs alter thyroid hormone synthesis, uptake of iodide by the thyroid gland, and binding of thyroid hormones to specific thyroid-binding proteins in the blood.

3.2. Other Mechanisms of Endocrine-disrupting Chemical (EDC) Action

3.2.1. Chemicals that alter steroid hormone synthesis

Steroid hormones are synthesized from cholesterol. One level of regulation is the expression of enzymes necessary for their synthesis. Aromatase is a key enzyme in the production of estrogens by conversion of androgen precursors.

Atrazine is the most extensively used herbicide for broadleaf and grassy weeds, and alters aromatase expression (Crain, Guillette, Rooney, & Pickford, 1997; Fan et al., 2007). The Center for Regulatory Effectiveness reports that over 76 million pounds are applied per year in the United States, primarily to cornfields, where 75% of fields are treated with atrazine. Atrazine stimulates the expression of aromatase through stimulation of steroidogenic factor-1 (SF-1) expression, a key transcription factor required for expression of aromatase (Fan et al., 2007). Atrazine levels are elevated in men with poor semen quality (Swan et al., 2003b) and are associated with altered sexual differentiation in vertebrates (Hayes et al., 2002; Hayes et al., 2003; Orton, Carr, & Handy, 2006). Whereas atrazine stimulates the production of aromatase, other xenobiotics inhibit its expression or activity, including genistein (Rice, Mason, & Whitehead, 2006), resveratrol (Eng et al., 2002), and BPA (Benachour, Moslemi, Sipahutar, & Seralini, 2007). Mono-(2-ethylhexyl) phthalate (MEHP) reduces ovarian E_2 production via the inhibition of aromatase expression. Importantly, this inhibition was within the range of current human exposure to MEHP (Davis, Weaver, Gaines, & Heindel, 1994; Reinsberg, Wegener-Toper, Van der Ven, Van der Ven, & Klingmueller, 2009).

A group of chemical fungicides known as azoles inhibit the fungal production of ergosterol, a key component of yeast and fungal cell membranes, by competitively inhibiting 14α-demethylase (Zarn, Bruschweiler, & Schlatter, 2003). These chemicals have been used clinically for treatment of yeast and fungal infections. They also are potent inhibitors of aromatase and are used in the treatment of postmenopausal breast cancer. Additionally, they inhibit aromatase and decrease fertility in laboratory animals (Shetty, H. Krishnamurthy, H.N. Krishnamurthy, Bhatnagar, & Moudgal, 1997; Shetty, H. Krishnamurthy, H.N Krishnamurthy, Bhatnagar, & Moudgal, 1998). Azoles are used agriculturally to inhibit mold and blight on crops, and it has been suggested that this use may result in altered hormone homeostasis in a broad range of species (Zarn et al., 2003).

3.2.2. Chemicals that alter steroid hormone metabolism

The nuclear receptors PXR and CAR are involved in metabolism of both endogenous hormones and xenobiotics through regulation of cytochrome P450 enzymes, specifically *CYP 2B*, *3A*, and *4A* genes. Several xenobiotics activate these receptors including NP and some phthalates. Di-n-butyl phthalate (DBP) activates both PXR and CAR *in vitro* and *in vivo* (Wyde et al., 2005). Pregnant Sprague-Dawley rats given 500 mg DBP/kg orally from gestation day 12 to 19 had upregulated *Cyp 2b1*, *Cyp 3a1*, and *Cyp 4a1* in both maternal and fetal liver on gestation day 19 (Wyde et al., 2005). In addition to their direct binding to ERs and AR, regulation of liver metabolic enzymes may be an important mechanism of phthalate endocrine disruption.

3.2.3. Small molecule enhancers of steroid receptor activity

Xenobiotics have been described recently that have the ability to enhance the action of steroid receptors without binding directly to them. For example, methoxyacetic acid (MAA) is a metabolite of ethylene glycol monomethyl ether commonly used in paints and varnishes. Occupational exposure is associated with reproductive toxicity (Correa et al., 1996). Jansen et al. (2004) showed that, in the presence of E_2, MAA enhanced the expression of estrogen-regulated gene expression but did not bind to ERs. Methoxyacetic acid also was capable of enhancing the activity of PR and other nuclear receptors. Methoxyacetic acid's enhancement was shown to be mediated by activation of p42-p44 mitogen-activated protein kinase (MAPK) and inhibition of histone deacetylases (Jansen et al., 2004).

Similarly to MAA, the antimicrobial agent triclocarban has been shown to enhance the activity of T in stimulation of AR-mediated gene expression, while it had no activity alone (Chen et al., 2008). Additionally, in castrated male mice, triclocarban plus T significantly increased male accessory reproductive organ weights, including the seminal vesicles, ventral prostate, glans penis, and Cowper's gland over either T or triclocarban alone (Chen et al., 2008).

3.2.4. Combined mechanisms of action

Endogenous hormones can send very divergent signals depending on where in the body, when, and how much is present. Xenobiotics that use endogenous hormone signaling systems can have these same divergent effects. In addition, as with chemicals such as BPA, each EDC can interact with one or many different hormone signaling systems. For example, BPA can bind to the ERs (both genomic and nongenomic), AR, TRs, aryl hydrocarbon receptor (AhR), estrogen-related receptor-γ (EERγ), and GPR30 receptors, making it difficult to predict a chemical's biological effects in the body. Finally, mixtures of EDCs with similar mechanisms of action can have additive effects (Kortenkamp, 2007). The multitude of chemicals, endpoints, and modes of action act together to increase the risk of endocrine disruption in mammals.

4. ADULT VS. DEVELOPMENTAL EFFECTS

Endocrine disruption is known to affect many different receptor systems; however, disruption of estrogen signaling was the earliest and remains the most described target for EDCs. Although many hormone systems will be discussed in this chapter, estrogen signaling will be used as a model target system for endocrine disruption.

4.1. Hormone Action in Adulthood

The effects of hormones are mostly reversible in adulthood. A well-known example of the reversible effects of steroid hormones occurs in the adult uterus. Estradiol and P_4 levels cyclically rise as a result of hormone signaling from the pituitary and ovary. In response to increasing levels of E_2, the endometrium proliferates and thickens in anticipation of implantation. If implantation does not occur, the corpus luteum involutes and E_2 and P_4 levels sharply drop. In response to the loss of P_4, maintenance of the endometrium is no longer supported and the endometrium is sloughed off during menstruation (e.g., higher primates) or resorbed (most mammals).

Similarly, EDCs tend to have reversible effects on adult mammals in the absence of continued exposure. However, humans are exposed to a low level of many xenoestrogens almost continuously. Moreover, mixtures of xenoestrogens have additive effects (Payne, Scholze, & Kortenkamp, 2001; Silva, Rajapakse, & Kortenkamp, 2002). Just as oral contraceptive pills shut down the hypothalamus–pituitary–gonad (HPG) axis and inhibit ovulation by negative feedback, sustained levels of xenoestrogens may potentially inhibit ovulation and/or lead to increased risk for hormone-dependent cancer, particularly in sensitive populations.

4.2. Hormone Action During Development

In contrast to the reversible effects seen in adulthood, the effects of hormone signaling during development are typically irreversible. In the fetus, estrogens show developmental effects in both males and females; these effects are typically permanent, in contrast to the reversible effects in adults. The human fetus first expresses ERs during the first trimester of gestation, during sexual differentiation (Taguchi, Cunha, & Robboy, 1986). A large body of work has demonstrated that the perinatal mouse, between gestation day 11 and postnatal day 7 (a period that corresponds approximately to weeks 7 to 20 in human fetal gestation), is sensitive to the permanent effects of both endogenous estrogens and xenoestrogens, prompting the term 'fragile fetus' to describe this phenomenon (Bern, 1992).

Of particular concern is xenoestrogen action during development, when estrogens can have effects that permanently program gene expression in adulthood (Colborn et al., 1993). In mice and humans, this corresponds to the *in utero* and early postnatal periods. Hormonal, nutritional, or other disruption of normal physiology during development is increasingly linked with adverse clinical symptoms in adulthood. Because many EDCs are small lipophilic molecules that can freely cross the placenta, there is concern that they may affect fetuses and permanently alter gene expression later in adults. Barker (1997) first put forth the 'fetal basis of adult disease' hypothesis, noting several studies associating low birth weight with adult disease. Later, he defined developmental plasticity as 'the phenomenon by which one genotype can give rise to a range of different physiological or morphological states in response to different environmental conditions during development. It enables the production of phenotypes that are better matched to their environment than would be possible if the same phenotype was produced in all environments' (Barker, 2003; p 733). Hence, endogenous hormones, xenobiotics, nutrition, stress and other events during development can program the fetus and increase its susceptibility to disease in adulthood.

This hypothesis has been termed 'developmental (or fetal) origins of adult disease'. In the case of EDC exposure during development, the impact of this hormonal disruption may not be expressed as clinical symptoms until adulthood. These possible long-latency birth defects from developmental exposure to xenobiotics are difficult to uncover in epidemiological studies and require the use of animal models to guide further investigation in humans (National Research Council (NRC), 1999).

One of the best examples of a long-latency, adverse outcome due to developmental xenoestrogen exposure is the case of DES daughters: women whose mothers were treated with DES during pregnancy in a mistaken attempt to promote a healthy pregnancy. It was not until after these DES daughters reached puberty that the detrimental effects were discovered. Developmental DES exposure resulted in otherwise rare vaginal adenocarcinoma and other abnormalities, including deformed uteri, immune disorders, and endometriosis (Newbold, 1995; Swan, 2000; Missmer et al., 2004). Although women no longer take DES during pregnancy, mothers and their fetuses are exposed to a wide variety of xenoestrogens that may alter normal development. Importantly, the phenotype in mice developmentally exposed to DES and BPA is strikingly similar to that seen in DES daughters (Newbold, 1995; Newbold, Jefferson, & Padilla-Banks, 2007).

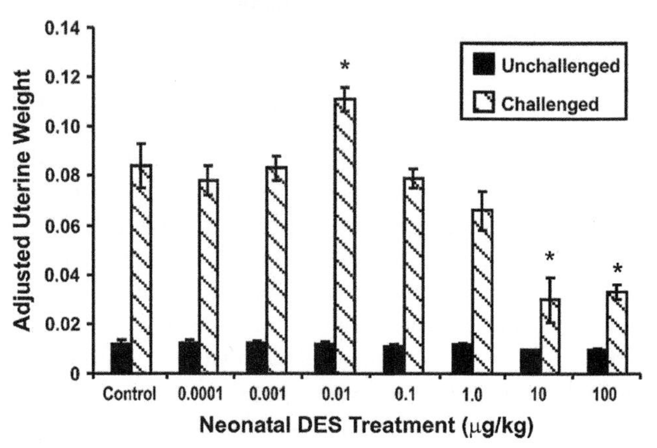

FIGURE 14.1 Uterine response in prepubescent mice (17 days old) following estrogen challenge with diethylstilbestrol (DES). Note an enhanced response in the DES 0.01 μg/kg group but a dampened response in groups exposed to higher doses of DES. Uterine weights, expressed as the mean ± standard error (SE), are adjusted for body weight. Statistical significance, indicated by an asterisk, is $P < 0.05$ vs. controls by Dunnett's test. *Reproduced from Newbold, Jefferson, Padilla-Banks, and Haseman (2004) with permission.*

4.3. Developmental Endocrine-disrupting Chemical (EDC) Exposure Programs Adult Gene Expression

Developmental xenoestrogen exposure permanently alters the estrogen responsiveness of the uterus. Upon xenoestrogen stimulation in adulthood, fetuses exposed to low doses of estrogens have significantly larger uteri than control mice, and fetuses exposed to high doses of estrogens have significantly smaller uteri (Figure 14.1) (Medlock, Branham, & Sheehan, 1992; Alworth et al., 2002; Newbold et al., 2004). This inverted-U dose response curve has also been demonstrated for developmental exposure to EE_2 (Thayer, 1999) and DES (Jefferson, Padilla-Banks, & Newbold, 2001).

There have been many reports of developmental xenoestrogen exposure reprogramming estrogen-responsive genes in the adult: ER (Medlock et al., 1992; Newbold et al., 2004), homeobox A10 (Varayoud, Ramos, Bosquiazzo, Munoz-de-Toro, & Luque, 2008), lactoferrin and epidermal growth factor in the mouse uterus (Nelson, Sakai, Eitzman, Steed, & McLachlan, 1994), and AR in the prostate (vom Saal et al., 1997). These permanent effects may result from epigenetic changes. Epigenetics refers to the heritable changes in gene function that occur in somatic or germ cells without a change in the DNA sequence (Jones & Takai, 2001). One mechanism by which this occurs is through altered DNA methylation of cytosine residues in the promoter region of genes (Jones & Takai, 2001). Hypermethylation is associated with decreased gene expression (Serman, Vlahovic, Serman, & Bulic-Jakus, 2006). For example, an increase in methylation in the gene promoter sequence can block the transcription machinery and therefore result in a heritable decrease in gene expression later in life. Likewise, hypomethylation is associated with increased gene expression.

DNA methylation is known to silence gene expression, either through blocking the machinery necessary for RNA transcription or through changing chromosome condensation. DNA methylation is under the control of three DNA methyltransferases (DNMTs). *DNMT1* is constitutively expressed and is responsible for methylating DNA after replication. In contrast, the *DNMT3* genes (*a* and *b*) appear to be developmentally regulated and can methylate both hemimethylated and unmethylated DNA equally. Proper DNA methylation is necessary for the controlled switching on and off of genes during development and for parental imprinting. When genes important for development are turned on and off incorrectly, morphological anomalies can occur. For example, developmental exposure to 10 μg

DES/kg on gestation days 9–16 resulted in a posterior shift of homeobox A gene expression and a morphologically abnormal female reproductive tract in CD-1 mice (Block, Kardana, Igarashi, & Taylor, 2000).

Many genes are differentially methylated, both hypo- and hypermethylated, by developmental xenoestrogen exposure; e.g., lactoferrin, *pS2*, and ribosomal genes, and *c-fos* (Li et al., 1997; Alworth et al., 2002; Li et al., 2003; Martin et al., 1997). Developmental DES exposure leads to uterine adenocarcinoma, and tumor development is believed to be at least partially due to changes in the DNA methylation status. Li et al. (1997) found that developmental exposure to 2 μg DES/mouse on postnatal days one to five caused a specific demethylation event in a CpG island in the promoter region of the lactoferrin gene in the mouse uterus that was dependent on adult ovarian hormones. Mature mice exposed to DES for five consecutive days (postnatal days 30–34) showed the same methylation pattern as control mice, indicating that the effect was dependent on neonatal DES exposure.

Not only have animal models been instructive for developmental effects of xenoestrogens in exposed offspring, but they have also alerted us to the possibility of transgenerational effects. Newbold and colleagues reported rare female reproductive tract abnormalities in 'DES granddaughters' in CD-1 mice (Newbold, Padilla-Banks, & Jefferson, 2006). Skinner and Crews have reported on transgenerational effects of vinclozolin and methoxychlor (Anway, Leathers, & Skinner, 2006; Anway & Skinner, 2006). In humans, DES grandsons appear to have an increased risk of hypospadias, as discussed below.

TABLE 14.2 Altered fetal estrogen and adult disease

Fetal or Maternal Condition	Fetal Estrogen	Breast Cancer	Endometriosis
DES	increased	3 [1.01-8.98][1]	1.8 [1.2-2.8][5]
Twins	increased	1.48 [1.01-2.16][2]	1.7 [1.2-2.5][6]
Smoking	decreased	0.35 [0.12-0.99][3]	0.22 [0.06-0.82][7]
Preedampsia	decreased	0.48 [0.30-0.78][4]	N/A[8]

Ratio with [95% confidence interval]

[1] *Incidence rate ratio for cancers occurring at ages >or=50 years (Palmer et al., 2006).*

[2] *Hazard ratio in meta-analysis of case-control studies for women who have a dizygotic twin (Xue and Michels, 2007).*

[3] *Adjusted (for age, nulliparity, and active smoking) relative rate for daughters born to women who smoked on average 15 cigarettes/day (Strohsnitter et al., 2005).*

[4] *Relative risk calculated from meta-analysis of case-control studies for daughters born to preeclamptic mothers (Xue and Michels, 2007).*

[5] *Rate ratio for daughters with any prenatal DES exposure (Missmer et al., 2004).*

[6] *Rate ratio for daughters who were part of a multiple gestation controlling for birth weight (Missmer et al., 2004).*

[7] *Odds ratio for daughters born to women who smoked on average 15 cigarettes/day (Buck Louis, Hediger, and Pena, 2007).*

[8] *No analysis found.*

5. LEVELS OF XENOESTROGENS RELATIVE TO ENDOGENOUS ESTROGENS *IN UTERO*

Although steroidal estrogens are high at term in human fetuses, evidence suggests that xenoestrogens may yet impact human fetal development. BPA is used as a representative estrogenic EDC for many of the points below.

(1) The human fetus appears to respond to changes in both endogenous estrogens and xenoestrogens *in utero* with altered phenotypes in adulthood, which suggests that ERs are not saturated. For example, conditions during pregnancy associated with increased exposure to estrogens, as seen in dizygotic twins and *in-utero* DES exposure, have been correlated with increased risk of breast and testicular cancer and endometriosis in adulthood (Table 14.2) (Ekbom, Trichopoulos, Adami, Hsieh, & Lan, 1992; Ekbom, Hsieh, Lipworth, Adami, & Trichopoulos, 1997; Missmer et al., 2004; Forman, Cantwell, Ronckers, & Zhang, 2005). Likewise, conditions associated with low levels of estrogens during pregnancy, such as pre-eclampsia and maternal smoking, are correlated with decreased risk of breast cancer and endometriosis in adulthood (Forman et al., 2005). See below for more detail.

(2) The concentrations of endogenous estrogens vary considerably from the first trimester of pregnancy to term, whereas xenoestrogens are likely present at the same concentration throughout pregnancy. Importantly, early fetuses may be exposed to EE_2 when their mothers are still taking oral contraceptives. These fetuses then are exposed to an extremely potent xenoestrogen at a time when endogenous estrogens are at their lowest.

(3) We, and others, have shown that BPA as well as other xenoestrogens are less tightly bound to human serum binding proteins than are endogenous estrogens, thereby increasing their available free fractions (Nagel et al., 1997b; Nagel et al., 1998).

(4) Although BPA shows low activity in stimulating uterine weight gain in the rodent, it is substantially more potent in programming the fetus and stimulating gene expression (Nagel et al., 1997b; Markey, Wadia, Rubin, Sonnenschein, & Soto, 2005).

(5) Like E_2, BPA recently has been shown to have rapid nongenomic actions. In many of these endpoints, BPA's potency is equal to or greater than E_2 (Watson, Bulayeva, Wozniak, & Finnerty, 2005; Wozniak et al., 2005; Zsarnovszky, Le, Wang, & Belcher, 2005). Since BPA circulates at ng/ml levels in maternal and fetal serum (Schonfelder et al., 2002b), BPA at current human exposure levels is theoretically sufficient to supplement endogenous estrogen levels in the fetus (Welshons et al., 2006). For

example, human fetal xenoestrogen exposure is associated with developmental programming; e.g., increased risk of breast cancer and endometriosis in adulthood (Missmer et al., 2004; Palmer et al., 2006).

(6) Kortenkamp (2007) has shown that even low levels of exogenous estrogens produce additive effects with endogenous estrogens.

(7) Neonatal exposure to xenoestrogens when endogenous estrogens are very low, for example exposure of infants to BPA from polycarbonate baby bottles, may act to permanently program postnatal development, as seen in mice (Varayoud et al., 2008).

6. ENDPOINTS

There are many endpoints reported in the literature for potential effects from exposure to EDCs. We have selected some common endpoints known to have some etiology in hormone disruption for a discussion of trends, potential mechanisms, and parallels between humans and other mammals.

6.1. Puberty

Puberty is the transition from childhood to adulthood and signifies the onset of reproductive capability. Puberty is affected by many factors including endogenous hormone levels, nutrition, race and ethnicity, geographical location, and environmental conditions, and is accompanied by several physiological and psychological changes (reviewed in Euling et al., 2008a). In females puberty is measured by the onset of breast development (thelarche), pubic hair development, and age at menarche. Male markers of puberty include genital and pubic hair development. A trend toward earlier onset of puberty, specifically menarche, has been suggested recently (Herman-Giddens et al., 1997; Freedman et al., 2002; Anderson, Dallal, & Must, 2003). Generally speaking, precocious puberty is defined as the development of secondary sexual characteristics before eight years of age in boys or girls. The identification of a population-wide trend toward earlier onset of puberty could lead to changes regarding how children are medically treated for precocious or delayed puberty (Euling, Selevan, Pescovitz, & Skakkebaek, 2008).

Endocrine-disrupting chemicals may be implicated in the trend toward earlier onset of puberty, which in turn is associated with an increased risk of adult reproductive disease (including breast and testicular cancer, polycystic ovary syndrome, and metabolic disorders), shorter adult stature, and behavioral or psychosocial problems (reviewed in Euling et al., 2008b; Golub et al., 2008).

In rodents, vaginal opening is often used as a marker for puberty, and the presence of completely cornified cells in the vaginal smear is indicative of the first ovulation (vom Saal, 1981). Small differences in endogenous hormones during fetal development as well as low doses of EDCs during this same time affect peripubertal growth and the interval between vaginal opening and first estrus (vom Saal, 1989; Howdeshell et al., 1999). For example, Howdeshell et al. (1999) found that females positioned between two females during fetal development (0M females) exposed on gestation day 11–17 to 2.4 μg BPA/kg/day were 22% larger than their 0M vehicle-treated counterparts and experienced a shorter time between vaginal opening and first estrus. Several other studies suggest that developmental exposure to BPA at doses between 25 ng/kg/day and 200 μg/kg accelerates the onset of puberty by up to four days among several strains of mice (Honma et al., 2002; Markey et al., 2003; Ryan & Vandenbergh, 2006).

Developmental or prepubertal exposure to other xenoestrogens also advances the timing of pubertal events in rodents. Synthetic xenoestrogens, EE_2 (5–10 μg/kg/day), DES (0.02–10 μg/kg/day), BPA (10 mg/kg/day), *o,p'*-DDT (10–100 mg/kg/day), BBP (182.6 μg/kg/day), zearalenone (0.5–10 mg/kg/day), 4-tert-octylphenol (200 mg/kg/day), methoxychlor (50 mg/kg/day), and NP (50 and 100 mg/kg/day) advance puberty in female rodent models by several days (Ashby et al., 1997; Laws, Carey, Ferrell, Bodman, & Cooper, 2000; Honma et al., 2002; Nikaido et al., 2004; Ryan & Vandenbergh, 2006; Rasier, Parent, Gerard, Lebrethon, & Bourguignon, 2007). Importantly, advances in puberty are not limited to synthetic EDCs. Differences in the phytoestrogen content in rodent chow have also recently been shown to affect the onset of puberty in laboratory animals. Thigpen et al. (2007) demonstrated that different mill dates of the same animal feed contain different amounts of the phytoestrogens daidzein and genistein (D&G), and this caused an advance in the time of vaginal opening in sensitive rodent strains. For example, three mill dates of Purina PMI 5K96 were found to contain 98, 223, or 431 μg D&G/g food, and, when fed to prepubertal F344 rats starting at weaning on postnatal day 19, the time of vaginal opening advanced by 2.6, 4.2, or 5.5 days, respectively. Additionally, vaginal opening was advanced in CD-1 mice fed a diet spiked with 150, 300, or 450 μg genistein/g food and in F344 rats fed the 300 and 450 μg genistein/g food diets (Thigpen et al., 2007). Each of these experiments was also performed with Sprague-Dawley rats, but no advance in vaginal opening was noted in this strain (Thigpen et al., 2007). Thus, these experiments highlight the necessity of choosing the appropriate animal species and strain for testing xenoestrogen-sensitive endpoints.

Whereas some xenoestrogens advance puberty, other EDCs delay puberty. Lindane, vinclozolin, and linuron are antiandrogenic pesticides that delay puberty in rodents (Cooper et al., 1989; Gray, Ostby, Monosson, & Kelce, 1999; Gray et al., 1999b; reviewed in Gray & Kelce, 1996). Additionally, the plasticizer DBP and the DTT metabolite *p,p'*-DDE delay preputial separation in male rats (Kelce et al., 1995; Gray et al., 1999b). Chemicals that interfere with AhR signaling, including the dioxin 2,3,7,8-tetrachlorodibenzodioxin (TCDD) and PCBs 126 and 169, also delay puberty in rodents of both sexes (Faqi & Chahoud, 1998; Faqi, Dalsenter, Merker, & Chahoud, 1998; Gray et al., 1999b). For example, vaginal opening was delayed by three days compared to controls in female Wistar rats exposed developmentally to 5 or 60 ng TCDD/kg through lactation (Faqi & Chahoud, 1998). Further, a single dose of 1 μg TCDD/kg on gestation day 15 was sufficient to delay puberty by 3.4 days in Long-Evans hooded rats (Fenton, Hamm, Birnbaum, & Youngblood, 2002).

6.1.1. Effects on puberty onset in humans

Several studies examining the timing of puberty in females have suggested that the age of onset has decreased since the mid-1900s, and an expert panel was recently convened to assess the weight of the evidence (Herman-Giddens et al., 1997; Freedman et al., 2002; Anderson et al., 2003; Euling et al., 2008b). Although most panelists agreed that the weight of evidence was strong enough to indicate a change in the onset of thelarche and/or menarche during that period of time, they disagreed with one another regarding what increment of change in puberty onset was biologically meaningful. They also concluded that there was not enough evidence to support changes in the timing of male puberty as there are only a few studies indicating a trend towards earlier onset of male puberty (Herman-Giddens, Wang, & Koch, 2001; Karpati, Rubin, Kieszak, Marcus, & Troiano, 2002; Euling et al., 2008b).

Endocrine-disrupting chemical exposure may be partially responsible for the earlier onset of puberty in females. In 1973 the Michigan food supply was accidentally contaminated with polybrominated biphenyls (PBBs) when livestock were mistakenly given the flame-retardant FireMaster in place of the nutritional supplement NutriMaster. Using serum samples collected from exposed and nonexposed women at the time of the event, Blanck et al. (2000) found an association with maternal serum PBB levels and their daughters' age of menarche. Girls exposed *in utero* and through breastfeeding to the highest levels of PBBs had an average age of menarche of 11.6 years whereas those exposed to lower levels experienced menarche at an average age of 12.2–12.6 years.

In summary, many researchers across a wide range of fields agree that there is significant evidence to indicate a population-wide trend towards earlier onset of puberty in the United States. Several EDCs already are known to advance the timing of puberty in well-established rodent models. Although there is less definitive epidemiological evidence implicating EDCs in advancing the onset of puberty in humans, there are suggestive studies. Additionally, the fact that humans are exposed to low levels of many different mixtures of EDCs, several of which are known to disrupt puberty in laboratory models, indicates that further research into this effect is warranted.

6.2. Behavior

Early behavioral studies of rats (Young, Goy, & Phoenix, 1964; Barraclough, 1966) and mice (Edwards & Burge, 1971) illustrated that neonatal androgen exposure initiated behavioral masculinization and defeminization. Later, vom Saal (1981) showed that subtle differences in fetal steroids altered female fetuses' reproductive success in adulthood. Neonatal levels of endogenous estrogens have been shown to play a major role in mammalian brain development, particularly in neural circuitry organization. For example, in the developing CNS, estrogens impact critical aspects of cellular differentiation such as neurite extension and branching, synapse formation, myelination, expression of neurotransmitters and neuropeptides, and cell death and survival (see Chapter 1, this volume). Estrogens also have been linked to sexual differentiation of the CNS: from the late stages of prenatal life until the first one to two weeks after birth, estrogens irreversibly organize male-type circuitries during brain development in both mice and rats (Arnold & Gorski, 1984). Thus, small changes in androgens and estrogens in the fetal nervous system predictably influence brain development and elicit profound effects on behavior. Endocrine-disrupting chemicals disrupt many mammalian behaviors required for reproduction, such as sexual behaviors, as well as social and nonsocial behaviors in adult life (Palanza, Gioiosa, vom Saal, & Parmigiani, 2008).

In a series of behavioral studies on the effects of prenatal and perinatal BPA exposure in CD-1 mice, 10 μg BPA/kg/day was administered to the mother orally during the final stages of gestation (prenatal exposure) or from gestation day 11 to postnatal day 7 (perinatal exposure) (Palanza et al., 2002; Laviola, Gioiosa, Adriani, & Palanza, 2005; Gioiosa, Fissore, Ghirardelli, Parmigiani, & Palanza, 2007). In both male and female offspring, developmental exposure to BPA diminished sex-specific behavioral responses to (1) novel environments, (2) exploratory behavior and level of activity in an open field test, (3) exploration and risk assessment in an elevated plus maze, and (4) sensitivity to amphetamine in a conditioned place preference test (Laviola et al., 2005; Gioiosa et al., 2007). Palanza et al. (2008) concluded that prenatal and perinatal exposure to BPA resulted in a significant elimination of sex-specific behaviors. Whereas distinct behaviors for males and females were observed in the vehicle-treated mice, a decrease in or an elimination of these behaviors was observed in BPA-treated mice.

Further, Palanza et al. (2002) noted that exposure of female mice to BPA in adulthood or during fetal life altered subsequent maternal behavior; i.e., female mice exposed to BPA during either time period displayed a considerable decrease in the amount of time they spent nursing their young.

Similar experiments with Sprague-Dawley rats have shown effects of BPA on sociosexual behavior. Dessi-Fulgheri, Porrini, and Farabollini (2002) and Farabollini, Porrini, Della Seta, Bianchi, and Dessi-Fulgheri (2002) have shown that developmental BPA exposures altered prepubertal sociosexual behavior in that animals of both sexes displayed decreased social interest (Dessi-Fulgheri et al., 2002) whereas females exhibited slightly increased sexual interest and males showed decreased sexual interest (Dessi-Fulgheri et al., 2002; Farabollini et al., 2002). These results show that developmental endocrine disruption resulted in differences in mate preference or reduced mating success and that endocrine disruption can exert effects on sexual selection that could have profound evolutionary implications for the future of exposed species.

Crews et al. (2007) examined the effects of EDCs on sociosexual behavior in Sprague-Dawley rats but from a transgenerational perspective. Vinclozolin was administered via intraperitoneal injection to pregnant female rats from gestation day 8 to 14 at a dose of 100 mg/kg/day. Vinclozolin had been shown in previous studies to demasculinize the male offspring of mothers who were exposed to both high and low doses (Gray, Ostby, & Kelce, 1994; Gray et al., 1999a). Crews et al. (2007), however, studied the F3 descendants of the pregnant mice and tested F3 females between postnatal days 90–120 for mate preference. They found that females showed a significant preference for the male descendants of vehicle-treated individuals over the vinclozolin-treated individuals (Crews et al., 2007). In a later study that employed similar methods, F3 Sprague-Dawley rats were tested for transgenerational behavioral effects of vinclozolin as they relate to anxiety. Skinner, Anway, Savenkova, Gore, and Crews (2008) observed that F3 males showed decreased anxiety-like behavior while F3 females had increased anxiety-like behavior. They hypothesized that a sex-specific effect of brain reprogramming occurred during development (Skinner et al., 2008). The findings of these transgenerational studies are particularly significant because they illustrate that the far-reaching effects of endocrine disruption two or even three generations later may still manifest significant effects on an individual's reproductive fitness and evolutionary success.

6.2.1. *Effects on human behavior*

Developmental exposure to EDCs also has been correlated with modifications of sex-specific behavior in humans (Vreugdenhil, Slijper, Mulder, & Weisglas-Kuperus, 2002). In a study conducted in the Netherlands, a cohort of 207 school-age children was assessed for prenatal PCB and dioxin levels using previously obtained umbilical cord plasma and breast milk samples and then examined for sex-specific play behavior using the Pre-School Activity Inventory. Vreugdenhil et al. (2002) found that higher prenatal PCB levels in boys correlated with decreased masculine behavior whereas in girls higher PCB levels were associated with increased masculine behavior. Conversely, higher prenatal dioxin levels were found to be linked to more feminized play in both boys and girls. They concluded that these observations indicate that prenatal exposure to EDCs may alter prenatal steroid hormone levels, which elicit profound changes in sexual differentiation of the brain and, consequently, behavior.

Although not discussed in this chapter, EDC exposure has been associated with atypical nonsocial behavior and cognitive dysfunction in humans and in animal models (Schwartz, S. Jacobson, Fein, J. Jacobson, & Price, 1983; S. Jacobson, Fein, J. Jacobson, Schwartz, & Dowler, 1985; J. Jacobson, S. Jacobson, & Humphrey, 1990; Chen, Guo, & Hsu, 1992; J. Jacobson, S. Jacobson, Padgett, Brumitt, & Billings, 1992; J. Jacobson & S. Jacobson, 1996; Winneke et al., 1998; Budtz-Jorgensen, Keiding, Grandjean, & White, 1999; Patandin et al., 1999; Darvill, Lonky, Reihman, Stewart, & Pagano, 2000; Stewart, Reihman, Lonky, Darvill, & Pagano, 2000; Grandjean et al., 2001; Walkowiak et al., 2001; Schantz, Widholm, & Rice, 2003; Stewart, Reihman, Lonky, Darvill, & Pagano, 2003; Ishido, Masuo, Kunimoto, Oka, & Morita, 2004; Ishido, Morita, Oka, & Masuo, 2005; Miyagawa, Narita, Narita, Akama, & Suzuki, 2007). Disruption of normal behavior consequently alters the natural processes of sex-specific development and may ultimately threaten the survival and reproductive success of all exposed species (Parmigiani, Palanza, & vom Saal, 1998).

6.3. Fertility

Fecundity and fertility are terms that describe an individual's potential to reproduce. Today, modern conveniences such as condoms and contraceptive methods allow the human population to control this natural, hormone-driven ability. As fertility is hormone-dependent, exposure to EDCs may decrease a population's ability to reproduce. Decreased fertility in animal models has been associated with exposure to pesticides, PCBs, phthalates, BPA, DES, and phytoestrogens; the results of selected studies are discussed below.

6.3.1. *Decreased fertility in animal models*

6.3.1.1. Pesticides

Yu, Guo, Xie, Liu, and Wang (2009) found that male mice dosed for five days with 100 and 200 mg/kg of the fungicide carbendazim had decreased testis weights, lowered sperm counts, and decreased sperm motility. In female mice, Cavieres, Jaeger, and Porter (2002) found that female mice exposed to varying levels of the commercial herbicide mixture containing 2,4-dichlorophenoxyacetic acid (2,4-D) had reductions in litter size, showing reduced fecundity.

6.3.1.2. Polychlorinated biphenyls (PCBs)

Captive and wild animal populations fed PCB-contaminated foods exhibit reproductive difficulties. Female mink fed a variety of fishes from the Great Lakes exhibited reduced litter size and decreased pup survival (Aulerich, Ringer, & Iwamoto, 1973). Females feeding on Great Lake coho salmon while mated had high reproductive failures; out of 65 females mated, 59 failed to give birth. Only 10 kits were born from the remaining six mothers, eight of which were stillborn. The remaining two kits died within 24 hours of birth (Aulerich et al., 1973). In contrast, minks fed west coast coho salmon, ocean perch, or whiting (control) did not experience such high levels of reproductive failure. Further, from 1950 to 1975, the population of the common seal in the westernmost part of the Wadden Sea collapsed, falling from 3000 to less than 500 seals living (Reijnders, 1986). The only discernable difference between the struggling seals and a thriving population of seals in a different region was their blood PCB levels. Reijnders et al. (1986) conducted a feeding trial and concluded that PCB pollution in the water was contaminating the fishes in the area, thereby altering the endocrine system of the seals. Of 12 mated seals fed contaminated fish, only four became pregnant after mating. Conversely, 10 of 12 seals fed uncontaminated fish became pregnant (Reijnders, 1986).

6.3.1.3. Phthalates

Reproduction is also disturbed by phthalates in laboratory animals. Sharpe, Fisher, Millar, Jobling, and Sumpter (1995) found that male rats born to mothers that drank 1000 μg BBP/l during gestation and for 21 days during lactation had a reduction in testicular weight and a 10–21% reduction in daily sperm production. In females, developmental exposure to 500 mg DBP/kg/day reduces fertility, as evidenced by an increased rate of spontaneous abortions (Gray & Kelce, 1996).

6.3.1.4. Bisphenol A (BPA)

Newbold et al. (2007) dosed CD-1 female mice with 10, 100, or 1000 μg BPA/kg/day on postnatal days one to five. On postnatal day 18, the reproductive tracts were examined and found to have a number of disorders including significant increases in cystic ovaries and cystic endometrial hyperplasia in the 100μ BPA/kg/day exposed group compared to controls. In all BPA-dosed groups, progressive proliferative lesions were found on the oviducts (Newbold et al., 2007). Additionally, BPA exposure to mice *in utero* and as adults has been linked to increased rates of aneuploidy in oocytes, leading to reduced fertility (Hunt et al., 2003). In males, BPA programs the reproductive tract and is associated with decreased sperm count and fertility in adulthood (vom Saal et al., 1998; Salian, Doshi, & Vanage, 2009).

6.3.1.5. Diethylstilbestrol (DES)

Diethylstilbestrol has been linked to severe effects on fertility (McLachlan, Newbold, Shah, Hogan, & Dixon, 1982). Mice exposed to DES *in utero* have decreased litter sizes and 30% fewer oocytes ovulated than controls. These mice also had numerous reproductive tract abnormalities (McLachlan et al., 1982). These findings are similar to the reproductive tract effects on humans discussed below.

6.3.1.6. Phytoestrogens

At environmentally relevant doses, developmental genistein exposure caused female mice to have abnormal estrous cycles and ovarian functions leading to subfertility or infertility (Jefferson et al., 2005). This could have several implications for infants of vegetarian mothers and infants drinking soy-based formulas. In general, phytoestrogens are linked to reduced fertility in cows and ewes following exposure to estrogenic forage, as evidenced by declining ovarian function, reduced conception rates, and increased embryonic loss (Adams, 1995).

6.3.2. Effects on fertility in humans

In males, fertility difficulties are usually caused by poor semen quality and/or reproductive tract abnormalities. Endocrine-disrupting chemicals such as pesticides, phthalates, and PCBs have been implicated. Attention has focused on male semen quality and a potential link with pesticides since Carlsen et al. (1992) reviewed 61 papers published between 1938 and 1990 about male sperm density and volume and found that male sperm density had declined by 50% worldwide over a 50-year period (Carlsen et al., 1992). Because these findings were controversial, many scientists re-evaluated aspects of the study. Swan, Elkin, and Fenster (1997) validated Carlsen's statistical analyses and found that neither changes in methods nor inherent experimental variability over time could explain the drastic decrease in sperm quality (Swan et al., 1997). Further analysis of 101 studies from 1934 to 1996 confirmed the original findings (Figure 14.2) (Swan, Elkin, & Fenster, 2000).

With such a wide range of data, time, and subjects, conclusions could not be drawn about why or how semen quality declines had been occurring or whether the pattern would continue. Yet, Swan et al. (2003a; 2003b) found that

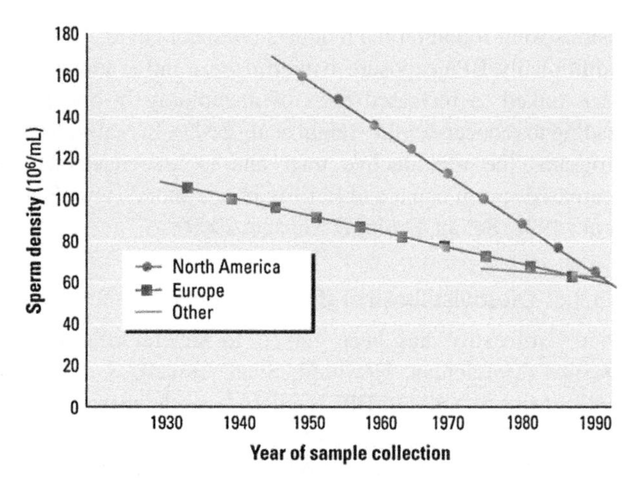

FIGURE 14.2 Interactive regression model for mean sperm density by year and geographic region, after controlling for proven fertility, abstinence time, age, specimen collection method, method of counting sperm, whether the study was included by Carlsen, Giwercman, Keiding, and Skakkebaek (1992), and interaction of region and study year. *Reproduced from Swan, Elbin and Feuster (2000) with permission.*

there were geographical differences between the sperm counts of men in the United States. More specifically, men from Missouri on average had lower sperm counts (58.7×10^6/mL) compared to men in New York, Minnesota, and California (102.9, 98.6, and 80.8×10^6/mL, respectively) (Swan et al., 2003a). They further characterized the populations of men from Missouri and Minnesota and found that, whereas pesticide levels were not associated with semen quality in Minnesota men, in Missouri men, semen quality was inversely correlated with metabolite levels of certain pesticides (Swan et al., 2003a). Men from Missouri with the highest urinary levels of the pesticides alachlor, diazinon (The particular metabolite measured was 2-isopropoxy-4-methylpyrimidinol (IMPY)), and atrazine were 30.0-, 16.7-, or 11.3-fold, respectively, more likely to have low semen quality than men with low urinary levels of the pesticides (Swan et al., 2003b). Thus, these data indicate that residence in a largely rural area with high pesticide exposures may contribute to regional differences in male fertility. Further, lifetime exposure to these EDCs may be linked to the general decline in sperm counts in the United States.

Yet, pesticides are not the only EDCs that may contribute to declining sperm quality. Phthalates may also pose a significant threat to male fertility (Diamanti-Kandarakis et al., 2009). Duty et al. (2003) studied urine and semen samples from 168 men attending Massachusetts General Hospital from 2000 to 2001 due to concerns of subfertility and found that certain phthalates were associated with sperm concentration and sperm motility. Higher levels of mono-butyl phthalate (MBP) were correlated with lower sperm motility and concentration whereas higher levels of mono-benzyl phthalate (MBzP) were correlated with lower sperm concentrations only.

Polychlorinated biphenyls also may impact semen quality. Dallinga et al. (2002) reported a possible correlation between decreasing semen quality of adult men and levels of PCBs in the blood. Further, the developmental impact of PCBs on children or unborn male fetuses is also an area of concern. Between 1978 and 1979, a rice oil contamination in Taiwan exposed more than 2000 people to high levels of PCBs, mainly PCDFs. Guo, P. Hsu, C. Hsu, and Lambert (2000) collected semen samples from 16- to 20-year-old men prenatally exposed to the rice oil contamination and control subjects to examine the effect of *prenatal* PCB exposure on adult semen quality. Prenatal PCB exposure was associated with abnormal sperm morphology, reduced motility, and reduced ability to penetrate hamster oocytes (Guo et al., 2000). Hsu et al. (2003) found that *postnatal* PCB exposure was also associated with abnormal morphology and reduced ability to penetrate a hamster oocyte. These and other studies suggest that PCB exposure during development and in adulthood could have lasting negative effects on male fertility. Developmental EDC exposure is associated with urogenital tract malformations including cryptorchidism and hypospadias that also impact fertility; these are discussed separately below.

In females, fertility problems are often more multifaceted and difficult to track. Reproductive issues in women can potentially be measured in studies of increased time-to-pregnancy, reduced fertility, miscarriage, premature ovarian failure, and reproductive tract abnormalities, many of which can be linked to EDC exposure during development or adulthood (reviewed in Diamanti-Kandarakis et al., 2009).

Just as in men, pesticide exposure and its endocrine-disrupting effects have caused similar concerns for female fertility. In the Fertility Risk Factor Study from 1997 to 2001, Greenlee, Arbuckle, and Chyou (2003) reported that women seeking treatment for infertility were 3.3 times more likely to have used fungicides and 27 times more likely to have mixed and applied herbicides two years prior to attempting to conceive (Greenlee et al., 2003). Younglai, Foster, Hughes, Trim, and Jarrell (2002) examined levels of environmental contaminants in the serum and follicular fluid of women undergoing *in-vitro* fertilization (IVF) treatment. Polychlorinated biphenyls, *p,p'*-DDE, mirex, hexachloroethane (HCE), trichlorobenzene, and heptachlor were found in over 50% of serum and follicular fluid samples. Of those contaminants, *p,p'*-DDE, a metabolite of the once widely used pesticide DDT, was most frequently detected at higher levels than other chemicals and most associated with failed IVF treatment (Younglai et al., 2002). These and other studies show that pesticides in general have a negative impact on female fertility.

Polychlorinated biphenyls also may cause problems for female reproductive health in the general population.

In many studies, the concern is mainly due to long-term PCB pollution in the US Great Lakes area (reviewed in Caserta et al., 2008). Along with paternally associated increased time to pregnancy, maternal consumption of Great Lakes fishes for three to six years was correlated with decreased fertility (Buck et al., 1999). In Germany, persistent exposure to chlorinated hydrocarbons such as PCBs, DDT metabolites, and other pesticides has been associated with recurrent pregnancy loss (Gerhard, Daniel, Link, Monga, & Runnebaum, 1998).

There is evidence that ethylene glycol ethers can act as EDCs and have reproductive toxicity. In a study on occupational exposure from 1980 to 1989, Correa et al. (1996) interviewed men and women employed by semiconductor plants in the eastern US about their reproductive health history and frequency of contact with ethylene glycol ethers. Exposure was associated with a 2.8-fold increased rate of spontaneous abortion and a 4.6-fold increased risk of subfertility in female manufacturers in the high exposure group (Correa et al., 1996). Jansen et al (2004) found that MAA, a metabolite of ethylene glycol monomethyl ether (EGME), is a hormone sensitizer and acts synergistically with steroid receptors by activating membrane-associated signaling through MAPK.

Premature ovarian failure (POF) has been linked to EDC exposure, especially during development. Any disruption to the normal programming of the ovaries during organogenesis before birth can result in POF later in life. Although this disease is rare, increasing rates within earlier age groups are concerning (reviewed in Diamanti-Kandarakis et al., 2009). In a study of DES daughters, Hatch et al. (2006) found that women exposed to DES *in utero* were more likely to experience natural menopause at an earlier age (i.e., POF) compared to unexposed women, indicating that developmental DES exposure could have negatively impacted adult ovarian reserves. Developmental EDC exposure is associated with altered reproductive tract structure, which also impacts fertility and is discussed below.

6.4. Lesions of the Female Reproductive Tract

The female reproductive tract begins developing *in utero* and is therefore susceptible to disruptions caused by altered prenatal hormonal signaling. Developmental xenoestrogen exposure alters development. *In-utero* exposure to DES alters the normal sequential expression of homeobox genes during development and results in a malformed reproductive tract (Block et al., 2000). *In-utero* exposure to DES permanently alters the estrogen-responsiveness of the uterus. Upon stimulation, females exposed to low doses of DES have significantly larger uteri than control mice, while fetuses exposed to high doses of DES have significantly smaller uteri (Newbold et al., 2004). Mice with increased estrogen responsiveness also have increased ER expression and endometrial cell proliferation in adulthood (Figure 14.1) (Newbold et al., 2004).

Newbold, Bullock, and McLachlan (1990) have extensively described the tumorigenic, carcinogenic, and mutagenic effects of DES exposure *in utero* and in subsequent generations of mice. In the F1 generation there was an increased number of animals with uterine (Newbold et al., 1990) and vaginal adenocarcinoma (McLachlan, Newbold, & Bullock, 1980; Newbold & McLachlan, 1982). In the F2 generation, low rates of uterine adenocarcinomas, uterine atypical hyperplasia, and uterine stromal cell sarcoma were seen, with rates increasing as animals reached 22–24 months of age. Ovarian lesions also were seen in older animals (Newbold et al., 1998). Additionally, Maranghi, Tassinari, Moracci, Macri, and Mantovani (2008) found ovarian lesions in mice exposed to 10 μg DES/kg/day on gestation days 9–16. Of the exposed animals, 60% had polyovular follicles, 40% had an increased presence of oocytes with condensed chromatin along the nuclear membrane, 90% had oocytes with irregular shape, and 40% showed the presence of mitoses in the ovarian follicular epithelium (Maranghi et al., 2008). The fact that increased rates of reproductive malformations can be seen in the offspring of animals exposed developmentally to DES has important implications for DES granddaughters, who have now started reaching the age when these types of lesions become prominent (discussed below).

Other xenoestrogens also alter the reproductive tracts of laboratory animals. Markey et al. (2003) found that CD-1 mice exposed to 250 ng/kg/day BPA had smaller vaginae and increased ovarian follicle volume at three months of age and had more blood-filled ovarian bursae at six months of age compared to control females. Newbold et al. (2001) found that 35% of CD-1 mice treated with 50 mg genistein/kg/day on neonatal days one to five had uterine adenocarcinoma compared to 0% of control mice. Moreover, genistein-treated mice also had cystic endometrial hyperplasia (64% of exposed; 19% of unexposed) and squamous metaplasia (64% of exposed; 0% of unexposed) (Newbold et al., 2001).

Female hypospadias has been described as the formation of a common urethral–vaginal canal accompanied by a widely cleft phallus. Miyagawa et al. (2002) dosed female ICR/Jcl mice with 0.03, 0.3, and 3 μg DES/mouse on postnatal days one to four and found hypospadias in 100% of the exposed animals. Developmental exposure to androgens is also known to induce female hypospadias. Gray and Ostby (1995) found that treatment of Long-Evans hooded rats on gestation day 8 or 15 with 1 mg TCDD/kg or Hotzman rats exposed on gestation day 15 resulted in 25%, 65%, or 100%, respectively, of females with complete or partial cleft phallus. These females also had a permanent 'thread' of tissue across the opening of the vagina

(14%, 80%, and 83%, respectively) and a 23% decrease in ovarian weight. Additionally, treatment of Long-Evans hooded rats on gestation day 8 with 1.8 mg/kg PCB 169 resulted in 57% of female offspring with mild hypospadias, 36% with a cleft phallus, and 14% with ovarian atrophy (Gray et al., 1999b).

Arctic fauna at the top of the food web are exposed to high levels of EDCs due to biomagnification because many EDCs are lipophilic and accumulate in adipose tissue. Arctic animals such as polar bears and seals have a life-cycle requiring them to accumulate large amounts of blubber or fat when food is available and then to fast for long periods of time, and cubs are exposed to large doses of these EDCs from their mothers' milk during a critical time in their development (Wiig, Derocher, Cronin, & Skaare, 1998). The presence of pseudo-hermaphroditic polar bears in Svalbard has been reported by Wiig et al. (1998), who described a pair of yearling cubs that were genetically female but had a vaginal opening with a 20 mm penis containing a baculum (determined by palpation). Additionally, Backlin, Eriksson, and Olovsson (2003) reported that 34 out of 53 (64%) Baltic gray seals autopsied between 1975 and 1997 had uterine leiomyomas, which are ovarian steroid-dependent tumors. They speculated that this finding, along with decreased reproduction rates in the 1970s, might be due to organochlorine exposure.

6.4.1. *Effects on female reproductive tract lesions in humans*

The anatomy of the female reproductive tract can develop abnormally following exposure to EDCs *in utero*. The best-known occurrence of EDC exposure resulting in female reproductive tract lesions is the case of DES daughters who were exposed *in utero*. Herbst, Ulfelder, and Poskanzer (1971) reported that the mothers of seven out of eight girls presenting with vaginal clear cell adenocarcinoma had taken DES during the first trimester of pregnancy ($p < 0.00001$ compared to control girls). Examination of cancer registries from Chicago, Illinois, and the Netherlands indicates that $\geq 60\%$ of vaginal clear cell adenocarcinoma patients born after 1947 had known exposure to DES *in utero* (Melnick, Cole, Anderson, & Herbst, 1987; Hanselaar et al., 1997). In addition to increased risk of vaginal clear cell adenocarcinoma, DES daughters exhibit several other reproductive malformations including T-shaped uteri, transverse vaginal ridges, cervical collars, hoods, etc. (Kaufman, Binder, Gray, & Adam, 1977; Rennell, 1979). Malformations, particularly severely deformed uteri, contribute to decreased fertility among DES daughters. The observations of transgenerational responses to DES reported in laboratory animals (see above) also target DES granddaughters for future studies.

Occupational exposures to EDCs in adulthood are associated with reproductive tract cancers in women. Female pesticide appliers in Iowa and North Carolina have a nearly three-fold increased incidence of ovarian cancer (Alavanja et al., 2005). Florida pesticide appliers are reported to have a 3.69-fold increased risk of cervical cancer (Fleming, Bean, Rudolph, & Hamilton, 1999). Additionally, female firefighters in Florida have a 5.24-fold increased risk of cervical cancer compared to the general Florida population (Ma, Fleming, Lee, Trapido, & Gerace, 2006).

Taken together, animal studies and the unfortunate *in-utero* exposure of DES daughters indicate that the developing reproductive tract is particularly sensitive to malformations and cancers as a result of EDC exposure.

6.5. Endometriosis

Endometriosis is a painful, chronic disease that affects millions of women worldwide and is associated with infertility (reviewed in Ozkan, Murk, & Arici, 2008). Most women seek medical attention for pelvic and/or abdominal pain, infertility, abnormal uterine bleeding, or gastrointestinal upset. Because symptoms can be diverse and noninvasive diagnostics are lacking, many women suffer for years without a diagnosis. Conversely, some women are asymptomatic, and endometriotic tissue is not detected until the time of tubal ligation (Cramer & Missmer, 2002). The costs associated with this disease are very high both financially and emotionally. Simoens, Hummelshoj, and D'Hooghe (2007) recently estimated the annual cost to be up to 22 billion dollars in the United States alone.

Importantly, endometriosis is an estrogen-dependent disease (Houston, 1984), and endometriotic tissue regresses after menopause or ovariectomy, when endogenous estrogens are low (Kitawaki et al., 2002). Exposure to estrogens represents the principal endocrine risk factor for developing the disease (Osteen, Bruner-Tran, Keller, & Eisenberg, 2002), and the primary nonsurgical treatment relies on decreasing endogenous estrogens (Rice, 2002). The etiology of endometriosis is unclear; however, it is known that only species that menstruate spontaneously develop endometriosis. Retrograde menstruation is thought to result in menstrual tissue that escapes through the fallopian tubes, where it implants in the abdomen and grows. Although 90% of women experience retrograde menstruation, only a small fraction exhibit endometriosis (Halme, Hammond, Hulka, Raj, & Talbert, 1984). Current thinking suggests that women who develop endometriosis have different endometrial tissue and/or an altered peritoneal environment that allows for the tissue to implant and grow (Sharpe-Timms, 2001).

Exposure to dioxins is correlated with increased incidence of endometriosis in humans, nonhuman primates, and rodents (Cummings & Metcalf, 1995; Cummings, Metcalf, & Birnbaum, 1996; Cummings, Hedge, & Birnbaum, 1999; Birnbaum & Cummings, 2002; Rier & Foster, 2002). The mechanism of action is unclear and is likely multifaceted as dioxins have diverse effects on the immune and endocrine systems. Importantly, dioxins also modulate ERs, displaying both estrogenic and antiestrogenic activities (Ohtake et al., 2003; Wormke et al., 2003). Developmental TCDD exposure in mice was recently reported to program the uterus and result in a phenotype similar to that seen in women with endometriosis (Nayyar, Bruner-Tran, Piestrzeniewicz-Ulanska, & Osteen, 2007).

Importantly, recent studies have linked developmental exposure to changes in endogenous or exogenous estrogens with adult endometriosis risk (Table 14.2) (Missmer et al., 2004; Buck Louis, Hediger, & Pena, 2007). Firstly, Missmer et al. (2004) reported that dizygotic twins, who circulate 60% higher estrogens during gestation than singletons, had a 70% increased risk of endometriosis. Secondly, women exposed to the xenoestrogen DES *in utero* have an 80% increased risk (Missmer et al., 2004). Further, in a subset of DES daughters with more detailed exposure data, the rate of endometriosis tripled in women exposed for more than five weeks during gestation compared to those exposed for less than five weeks. Worthy of note, this study was part of the Nurses Health Study II and was a prospective cohort study of more than 84 000 women (aged 25 to 42). This large study showed a clear increase of endometriosis in DES daughters.

Another study examined the effect of maternal smoking on the risk of endometriosis in daughters. Smoking has been reported to lower circulating estrogens by inhibiting the estrogen-synthesizing enzyme aromatase (Barbieri, McShane, & Ryan, 1986; Kitawaki et al., 1993). Daughters born to women who routinely smoked about 15 cigarettes per day had a four-fold reduced risk of endometriosis (Buck Louis et al., 2007). Changes in developmental exposure to estrogens alter the risk of endometriosis in women (Missmer et al., 2004). Xenoestrogen exposure to the fetus, throughout childhood, and/or in adulthood is a likely candidate in exacerbating this estrogen-dependent disease.

6.6. Mammary Cancer

In 2009 alone, the American Cancer Society predicted 92 370 new cases of invasive breast cancer and 40 170 deaths from breast cancer. Breast cancer is the second most fatal cancer for women and one out of eight women will be diagnosed in their lifetimes (American Cancer Society, 2009b). Despite the rise in the prevalence of breast cancer in the United States in the last 50 years, the associated mortality has declined (Ravdin et al., 2007; Kurian, Clarke, & Carlson, 2009). However, interpretation of the trends in breast cancer incidence is complicated by many factors including race, ethnicity and age, the adoption of a Western lifestyle by immigrants and residents of less developed countries, the increased use and availability of mammographic equipment, early detection and better treatment, and the change in frequency of use, type, and timing of hormone replacement therapy.

Mammary gland development begins around the fourth week of gestation in humans and the branching that occurs during fetal life is susceptible to endocrine disruption (reviewed in Pryor, Hughes, Foster, Hales, & Robaire, 2000). Growth is then arrested until puberty, when cyclical estrogens and P_4 signals once again stimulate mammary growth. During this time, terminal end buds proliferate, resulting in ductal elongation and branching. Cap cells at the end of terminal end buds are highly proliferative and are thought to play a role in carcinogenesis (Jefferson, Padilla-Banks, & Newbold, 2007). The mammary alveoli do not fully develop until pregnancy and lactation, and, when the lactational hormonal signals subside, involution occurs through apoptosis (Watson & Khaled, 2008).

Exposure to xenoestrogens increases mammary cancer susceptibility in rodents. 2,3,7,8-tetrachlorodibenzodioxin at 1 μg/kg on gestation day 15 alters mammary gland differentiation in the rat (Brown, Manzolillo, Zhang, Wang, & Lamartiniere, 1998; Lewis et al., 2001; Fenton et al., 2002; Jenkins, Rowell, Wang, & Lamartiniere, 2007), and exposed animals have nearly double the number of 7,12-dimethylbenz(a)anthracene (DMBA)-induced tumors, a shorter time to tumor detection, and a greater overall incidence of tumors compared to control rats (Brown et al., 1998). A lower dose, 2.5 ng TCDD/kg, on postnatal days 23, 25, 27, 29, and 31 also alters mammary gland development (Brown & Lamartiniere, 1995). Boylan and Calhoon (1979) reported that rats exposed to 1.2 μg DES/kg/day during the second or third week of development had more palpable DMBA-induced tumors than control rats. Further, DES induces tumors in ACI rats following *in-utero* or adult exposures (Rothschild, Boylan, Calhoon, & Vonderhaar, 1987).

Significant research has been undertaken to determine the relation of genistein to breast cancer risk. Genistein, when administered at high doses (50–500 mg/kg/day), either developmentally (Brown & Lamartiniere, 1995; Lamartiniere et al., 1995) or prepubertally (Cotroneo, Wang, Fritz, Eltoum, & Lamartiniere, 2002) suppresses mammary cancer in rats. This was also reported when Hilakivi-Clarke, Cho, Onojafe, Raygada, and Clarke (1999) administered a dose of about 1 mg/kg/day to prepubertal rats. However, when Hilakivi-Clarke et al. (1999b) administered equivalent or lower doses developmentally (~0.1, 0.5, or 1.5 mg/kg/day), rats were more susceptible to carcinogen-induced tumors. Further, Padilla-Banks, Jefferson, and Newbold (2006) reported that mice exposed on postnatal days one to five to 5 or 50 mg genistein/kg/day had hyperplastic ductal epithelia at nine months of age. Ju et al. (2001) reported that adult mice fed a genistein-supplemented diet (125, 250, 500, or 1000 μg/g diet) had a dose-dependent growth of tumors. Using ovariectomized rats as a model for postmenopausal women, Allred et al. (2004) found that that genistein can stimulate the growth of estrogen-dependent breast tumors that are already present in the body. These studies indicate that the effect of genistein on mammary cancer is sensitive to timing, route of administration, and dose (reviewed in Jefferson et al., 2007).

The effects of developmental BPA exposure at low doses (25 ng to 1 mg) have also been documented. Developmental exposure to BPA alters mammary gland development (Munoz-de-Toro et al., 2005; Moral et al., 2008), inhibits mammary duct development, increases branching (Munoz-de-Toro et al., 2005; Vandenberg et al., 2007b), and increases mammary gland density and the number of terminal end buds, a phenotype that is positively associated with mammary cancer (Markey et al., 2001). Developmental BPA exposure increased mammary sensitivity to a subcarcinogenic dose of N-nitroso-N-methylurea (NMU) (Durando et al., 2007). Further, developmental treatment with BPA has recently been found to be sufficient to induce the development of neoplastic and preneoplastic lesions in the mouse mammary gland (Murray, Maffini, Ucci, Sonnenschein, & Soto, 2007).

6.6.1. Effects on mammary cancer in humans

Similarly to endometriosis, developmental exposure to estrogens has been associated with adult breast cancer risk (Table 14.2). *In-utero* conditions resulting in increased prenatal exposure to estrogens are associated with increased breast cancer risk. For example, dizygotic twins, who circulate higher prenatal estrogens, have an increased risk of breast cancer in adulthood, while monozygotic twins, who circulate similar levels of steroidal estrogens as singletons, do not have an increased risk (Wald, Cuckle, Wu, & George, 1991; Park et al., 2008). Exposure to DES is a risk factor for developing breast cancer. Both women who took DES during their pregnancy (DES mothers) and DES daughters have an increased risk of developing breast cancer. In DES mothers, there is a 34% increased risk of death due to mammary cancer (reviewed in Giusti, Iwamoto, & Hatch, 1995; Herbst, 2000). The risk of developing breast cancer is greater in DES daughters and increases with age (Palmer et al., 2006). Palmer et al. (2006) collected data from a large US cohort of 3812 DES-exposed and 1637 unexposed women and reported a 2.05-fold increased incidence of breast cancer in DES daughters over 40 years of age and a 3.85-fold increased incidence in DES daughters over 50 years of age. Conversely, *in-utero* conditions resulting in decreased prenatal steroidal estrogens are associated with decreased breast cancer risk. Daughters born to eclamptic or pre-eclamptic mothers have a 70% decreased risk and daughters born to smokers have up to 78% reduced risk (Ekbom et al., 1997; Strohsnitter et al., 2005).

In 1976 a trichlorophenol manufacturing plant exploded near Seveso, Italy, resulting in the highest human TCDD exposures known in a residential area. Warner et al. (2002) examined the association between individual serum samples collected from 1976 to 1981 and the risk of breast cancer. Women with 10-fold higher serum concentrations of TCDD were more than twice as likely to have a breast cancer diagnosis (Warner et al., 2002). This study did not assess the risk for those who were exposed *in utero* and who may have been the most susceptible. These women are just now entering their thirties and forties and should be followed closely.

The potential benefits of a high soy diet in terms of reducing the risk of breast cancer have been debated intensely, as epidemiological data indicate that breast cancer incidence is different across the world. Asian women, for example, are at a lower risk for developing breast cancer than their European and American counterparts. Case studies by Lee et al. (1991) and Linseisen, Piller, Hermann, and Chang-Claude (2004) suggest an inverse correlation between consumption of soy products and breast cancer. Additionally, Wu et al. (2002) reported in a study of Asian-American women that the consumption of a high soy diet as an adolescent significantly decreased the risk of breast cancer compared to a high soy diet consumed only during adulthood. Importantly, associations were not found in other case-control studies of populations of Chinese, African-American, Latina, or Caucasian women (Yuan, Wang, Ross, Henderson, & Yu, 1995; Horn-Ross et al., 2001). Comparing the incidence of breast cancer among women from different races is complicated by many factors, including differences in genetics, pre- and postnatal diet, nutritional factors, and fat intake (discussed in Lagiou, Adami, & Trichopoulos, 2006).

Exposure to pesticides and organochlorine compounds is associated with breast cancer risk (Engel et al., 2005). Hoyer and colleagues analyzed serum levels of 18 organochlorines and 28 PCB congeners from women enrolled in the Copenhagen City Heart Study from 1976 to 1978 (Hoyer, Grandjean, Jorgensen, Brock, & Hartvig, 1998; Hoyer, Jorgensen, Brock, & Grandjean, 2000; Hoyer, Gerdes, Jorgensen, Rank, & Hartvig, 2002). They found dieldrin to be associated with a 2.05-fold increased breast cancer risk and a 2.61-fold increased risk of breast cancer mortality (Hoyer et al., 1998; 2000). They also suggested that mutations in the tumor suppressor gene *p53* may interact with pesticide exposure to increase breast cancer risk (Hoyer et al., 2002). Similarly, Zhang et al. (2004) found an association with specific *CYP1A1* alleles and PCB levels and an increased breast cancer risk in a US population. Additionally, data from the Agricultural Health Study in Iowa and North Carolina indicate modestly increased risk in women whose husbands applied the pesticides aldrin, carbaryl, chlordane, dieldrin, heptachlor, lindane, malathion, or 2,4,5-trichlorophenoxyacetic acid, or the fungicide captan (Engel et al., 2005). As discussed by Engel et al. (2005), there seems to be no consensus regarding the use of DDT and breast cancer risk.

A handful of reports suggest a slight increase in the incidence of breast cancer in men in the US and the UK since the 1970s (Giordano, Cohen, Buzdar, Perkins, & Hortobagyi, 2004; Stang & Thomssen, 2008; Speirs & Shaaban, 2009). A change in the incidence of male breast cancer should not be dependent on changes in estrogenic hormone replacement therapy or use of mammographies since men do not regularly seek these services. An increased incidence may be a result of an aging population, an overall increased awareness of breast cancer, and/or increased lifetime exposure to EDCs, especially estrogens.

There is strong evidence to suggest that exposure to EDCs, especially xenoestrogens, can contribute to the overall risk of breast cancer. Further, given the findings that low doses of BPA have the ability to induce neoplastic lesions and alter carcinogenic susceptibility to mice, more research into the association of BPA and other xenoestrogens with human breast cancer risk is warranted.

6.7. Testicular Dysgenesis Syndrome (TDS)

Normal sexual differentiation of the male reproductive tract is androgen-dependent. Individuals with complete androgen insensitivity syndrome have an inactivating mutation in AR, which makes the available androgen inactive. These XY individuals develop female external genitalia, lack a uterus and ovaries, but have abdominal testes, indicating that androgens are necessary for development of the male reproductive tract (except the testes) and external genitalia. It has been postulated that disorders of the male reproductive tract, including cryptorchidism, hypospadias, testicular cancer, and low sperm count, comprise a testicular dysgenesis syndrome (TDS) and may have a common etiology of deficient androgen production or action during development of the testes during fetal life (Sharpe & Skakkebaek, 2008).

Phthalates have known antiandrogenic actions, and exposure during sexual differentiation may lead to TDS (Skakkebaek, Rajpert-De Meyts, & Main, 2001). Early environmental exposure to other EDCs that can alter androgen signaling will likely increase the risk for TDS. Wilson, Blystone, Hotchkiss, Rider, and Gray (2008) reviewed different groups of EDCs with shared and unique mechanisms of action that converge on androgen signaling: antagonism of AR, inhibition of T synthesis, and mixed mechanisms of action. While the exact mechanisms may differ, the end result is a reduction in T signaling through AR. Exposure to antiandrogens during sexual differentiation produces malformations of the reproductive tract in humans and rodents. For example, perinatal exposure to the environmental antiandrogen vinclozolin during sexual differentiation resulted in reduced anogenital distance (AGD), retained nipples, cleft phallus with hypospadias, cryptorchidism, a vaginal pouch, epididymal granulomas, and small to absent sex accessory glands in male rats (Gray et al., 1999a; 1999b).

Anogenital distance is androgen-dependent, and antiandrogens reduce AGD. Numerous studies have shown that developmental exposure to EDCs, such as DBP, bis(2-ethylhexyl) phthalate (DEHP), linuron, procymidone, and DDE can reduce AGD (Gray et al., 1999a; 1999b). Importantly, Swan, et al. (2005) examined the relationship between developmental phthalate exposure and AGD in boys. Nine phthalate esters were measured during pregnancy. The ADG of 85 boys less than three years of age was measured and adjusted for age and weight. The mothers' phthalate levels during pregnancy, and presumably the boys' levels during development, of four phthalate metabolites (monoethyl phthalate (MEP), MBP, MBzP, and monoisobutyl phthalate (MiBP)) were inversely related to AGD at 2 to 36 months of age (Table 14.3). Boys with the highest prenatal exposure were 10 times more likely to have a shorter AGD (corrected for weight) than boys with the lowest prenatal exposure (Swan et al., 2005).

6.8. Testicular Cancer

Development of the testes is essential to male development in that the testes produce hormones, particularly androgens, that other parts of the male body are dependent upon for

TABLE 14.3 Mean (median) phthalate monoester metabolite levels by AGI category

Monester metabolite	AGI category [mean (median; ng/mL)]		
	Long[a] (n=17)	Intermediate[b] (n=43)	Short[c] (n=25)
MBP	13.1 (11.5)	22.2 (13.1)	38.7 (24.5)
MBzP	10.6 (6.6)	15.1 (7.7)	25.8 (16.1)
MEP	124 (47.1)	592 (112)	1076 (225)
MiBP	2.3 (1.5)	3.3 (2.1)	7.7 (4.8)

[a]*Long, AGI ≥ 75th percentile of expected AGI.*
[b]*Intermediate, 25th percentile ≤ AGI < 75th percentile of expected AGI.*
[c]*Short, AGI < 25th percentile of expected AGI.*
AGI=AGD/weight at examination.
Reproduced from Swan et al. (2005) with permission.

normal formation (Sharpe, 2001). Therefore, any alteration or disturbance to the pathway leading to normal testis development early in life can severely hinder proper development and function of other organs. Fetal germ cells exposed to an abnormal hormonal environment during development are at a higher risk of forming testicular cancer later in life (Sharpe, 2001).

Animals exposed to EDCs have an increased risk of testicular cancer. Hayes, Tarone, Casey, and Huxsoll (1990) examined testicular cancer, specifically seminoma, among US military working dogs from the Vietnam War. During their service, these dogs were potentially exposed to chemical pesticides, phenoxy herbicides, therapeutic drugs, and zoonotic infectious agents. It was shown that these animals, compared to the service dogs that remained in the US during the war, were twice as likely to have seminoma. The US military working dogs that served in Vietnam were also at higher risk for other reproductive abnormalities such as testicular degeneration, atrophy, and oligospermatogenesis. In a similar study in men, it was reported that, compared to age-matched men with nongenital cancer, service in Vietnam was associated with a two-fold increased risk of testicular cancer (Tarone et al., 1991).

Voss, Zerban, Bannasch, and Berger (2005) exposed Sprague-Dawley rats to 0, 30, 95, and 300 mg DEHP/kg/day for 159 weeks, which was meant to mimic a lifetime exposure. Rats exposed to the 300 mg/kg dose had a nearly two-fold increased incidence of testicular tumors compared to controls (Voss et al., 2005). Khan, Ball, and Hendry (1998) examined the effect of DES on the reproductive development of neonatal male hamsters. At 90 days, when hamsters were euthanized, males treated with DES had experienced cryptorchidism, reduced testicular weight, and an increased number of testicular tumors (Khan et al., 1998). Laboratory studies support the hypothesis that EDCs increase the risk of testicular cancer.

6.8.1. *Effects on testicular cancer in humans*

Over the past 50 years and more, the incidence of testicular cancer has risen at an alarming rate. In fact, testicular cancer rose 2.4% per year in Spain between 1991 and 2005 (Llanes Gonzalez, Lujan Galan, Rodriguez Garcia, Garcia Tello, & Berenguer Sanchez, 2008) and 2.6% per year in Australia between 1982 and 2004 (Baade, Carriere, & Fritschi, 2008). For European men born between 1880 and 1920, there was little risk for testicular cancer (Bergstrom et al., 1996). However, after this time, the incidence of testicular cancer reported in six European countries increased at a rate of 2.3% to 5.2% annually. Bergstrom et al. (1996) suggested that differences in the risk of testicular cancer among the countries studied were related to the extent to which each country was affected by the world war.

These results suggest that environmental differences are playing a larger role in the development of testicular cancer than genetic changes. The increasing presence of EDCs in the environment is hypothesized to be associated with the dramatic rise in incidences of testicular cancer (Carlsen, Giwercman, Keiding, & Skakkebaek, 1995). Hardell at al. (2003) investigated blood levels of EDCs in mothers of men with testicular cancer. They found that the level of PCBs, hexachlorobenzene (HCB), and chlordane combined was 3.8-fold higher in mothers who had sons with testicular cancer (Hardell et al., 2003). Similarly, DES sons who were exposed *in utero* may be at a higher risk of developing testicular cancer, but data are inconsistent or lacking (reviewed in Swan, 2000).

It has been suggested that EDCs, particularly xenoestrogens, induce testicular cancer by first acting on the development and function of the Sertoli cells (see Chapter 5, this volume) in fetal testes (reviewed in Carlsen et al., 1995). Sertoli cell development and function is controlled by T and follicle-stimulating hormone (FSH). Follicle-stimulating hormone also controls secretion of Müllerian-inhibiting substance (MIS), which is responsible for the loss of the Müllerian ducts during normal male development. Follicle-stimulating hormone is regulated by negative feedback from estrogens. Therefore, decreases in FSH due to increases of synthetic estrogens in the body may retard Sertoli cell development and release of MIS. This decrease in MIS may be the cause for maladies such as cryptorchidism and abnormal proliferation of germ cells, which later in life can lead to testicular cancer (Carlsen et al., 1995). Data from laboratory and epidemiological studies suggest that EDCs are associated with testicular dysgenesis syndrome in general and testicular cancer, specifically.

6.9. Cryptorchidism

Testicular descent from an intra-abdominal location during fetal life to an extracorporeal location is necessary for proper testicular maturation and function in many mammals including humans. Cryptorchidism occurs when one or both testes are absent from the scrotum, usually as a result of a failure to descend during fetal development, and is the most common birth defect of the male genitalia. Cryptorchidism is associated with a two- to six-fold increased risk of infertility, 20- to 46-fold increased rate of testicular cancer, and an increased risk for testicular torsion or injury against the pubic bone (reviewed in Kolon, Patel, & Huff, 2004). Developmental exposure to EDCs that are estrogenic or interfere with androgen signaling increases the risk of cryptorchidism (reviewed in Hughes & Acerini, 2008). Several studies have indicated geographical differences in the rate of cryptorchidism and also an increase in the incidence of cryptorchidism (Figure 14.3), which may

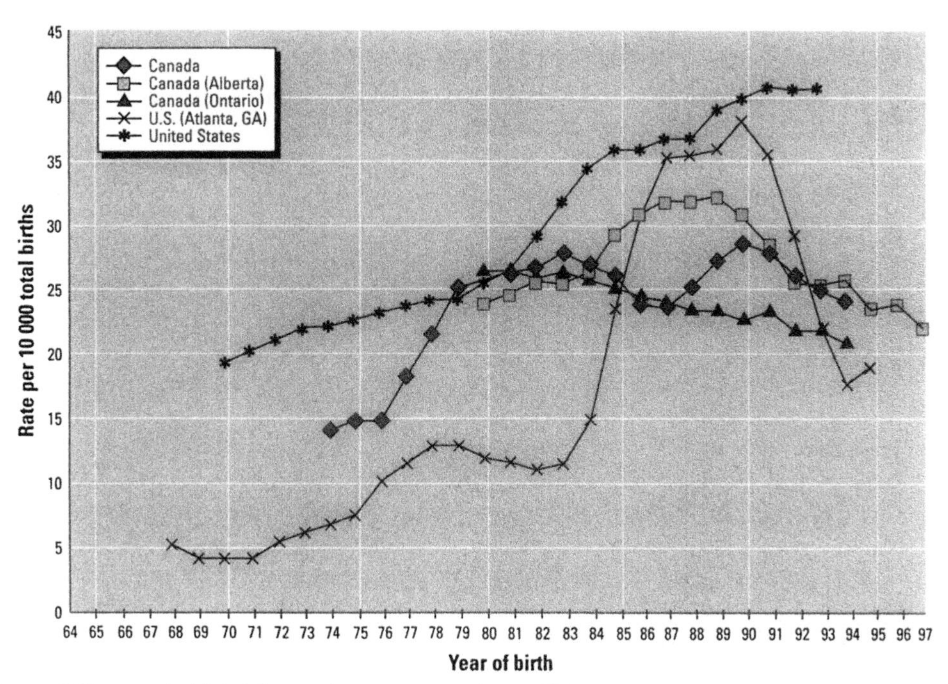

FIGURE 14.3 Cryptorchidism rates by birth defect monitoring system and year of birth (1968–1997), three-year moving averages, United States and Commonwealth groups. *Reproduced from Paulozzi (1999) with permission.*

reflect differences in EDC exposure between populations and over time (reviewed in Virtanen & Toppari, 2008).

Welsh et al. (2008) recently sought to define a critical window of exposure for the induction of cryptorchidism. They treated pregnant Wistar rats with the pure antiandrogen flutamide (100 mg/kg by oral gavage) during three different periods of fetal development: an early window (gestation days 15.5–17.5) corresponding to the period just after testicular T production begins, a middle window (gestation days 17.5–19.5) corresponding to the time just prior to morphological differentiation, and a late window (gestation days 19.5–21.5) corresponding to the period of morphological differentiation. They found that treatment with flutamide during the early window only or throughout the full window (from onset of T production until birth) induced cryptorchidism in 30% and 57% of males, respectively. The authors were surprised to find that exposure during the early window (gestation days 15.5–17.5) but not the mid or late windows resulted in an increased incidence of cryptorchidism, given that testicular descent is a late-fetal occurrence (occurring after gestation day 21.5 in the rat) (Welsh et al., 2008).

The incidence of cryptorchidism may be rising among several wildlife species, especially in polluted areas such as southern Florida. The effects of EDCs are often magnified in top predator species, such as the panther. The incidence of cryptorchidism in the Florida male panther (*Puma concolor coryi*) population appears to have risen from about 15% in 1975 to 90% in 1995 (Facemire, Gross, & Guillette, 1995; Mansfield & Land, 2002). Levels of EDCs were not determined, however, and the extent to which inbreeding contributes to the rise of cryptorchidism in this population has been debated (Mansfield & Land, 2002).

The Florida black bear (*Ursus americanus*), another top predator, may also suffer from bioaccumulation of EDCs. Dunbar, Cunningham, Wooding, and Roth (1996) assessed 20 live-trapped Florida black bears and 51 found dead from vehicular accidents, gunshots, or complications during trapping. Cryptorchidism had previously not been reported for this species (Dunbar et al., 1996). Out of 71 bears examined, 11 were found to be cryptorchid. Of the 11 cryptorchid bears, 7 may have exhibited delayed maturation, as they were less than 1.2 years of age. Unfortunately, these bears were not followed to see whether the condition corrected itself with time. Without pedigree information, biological sampling to assess EDC body burden, followup on maturation of juvenile animals, or comparisons with bears from other parts of the country, the etiology of cryptorchidism in the Florida black bears could not be clearly addressed in this study. With that said, reproductive tract abnormalities are often found in many nonmammalian species in Florida so it is possible that Florida's top mammalian predators also are affected by EDCs (reviewed in Guillette & Edwards, 2008). Further work is needed to assess the extent of EDC exposure and its relation to cryptorchidism and fertility, especially in endangered species.

6.9.1. *Effects on cryptorchidism in humans*

Bhatia and colleagues (2005) examined the association of serum *p,p'*-DDT and its primary metabolite, *p,p'*-DDE, in women enrolled in the Child Health and Development Studies with risk of genital congenital anomalies in their male children. Serum was collected from pregnant women between 1959 and 1967 and is representative of the peak use of DDT in the United States. When the data were adjusted for cholesterol, triglyceride level, and maternal race, babies born to mothers with the highest serum *p,p'*-DDT levels ($\geq$ 15 ng/mL) had nearly a two-fold increased risk of cryptorchidism when compared to those whose mothers had $<$ 15 ng DDT/mL serum (Bhatia et al., 2005).

Similarly, the concentration of several EDCs in breast milk of Finnish and Danish mothers between 1997 and 2001, as measured in a series of case reports from a prospective birth cohort, were associated with an increased risk of cryptorchidism (Damgaard et al., 2006; Main et al., 2006; 2007). Damgaard et al. (2006) measured the concentration of 21 persistent and nonpersistent EDCs with known previous worldwide use, suspicion of endocrine-disrupting activity from animal and/or *in-vitro* studies, and availability of highly sensitive analytic methods (Damgaard et al., 2006). They found that 17 out of the 21 chemicals were present at higher concentrations in the breast milk of mothers of cryptorchid boys compared to mothers of control boys, although only trans-chlordane reached significance. A combined statistical analysis for the eight most prevalent persistent EDCs (*p,p'*-DDE, *p,p'*-DDT, β-hexachlorocyclohexane (β-HCH), HCB, α-endosulfan, *cis*-heptachloroepoxide (*cis*-HE), oxychlordane, and dieldrin) revealed higher concentrations in the breast milk of case mothers. This indicates that a mixture of low concentrations of EDCs may be more important when assessing risk of cryptorchidism than the concentration of any single chemical (Damgaard et al., 2006). The concentration of the sum of seven polybrominated diphenyl ethers (PBDEs), which are commonly found in flame-retardants, was also significantly higher in the breast milk of mothers of cryptorchid boys compared to controls (Main et al., 2007). Additionally, phthalate levels in the mothers' breast milk were associated with altered levels of hormones related to Leydig cell function, which might have led to decreased androgen levels in this cohort (Main et al., 2006).

Several studies have found an association of cryptorchidism and parental involvement in agricultural occupations. Garcia-Rodriguez et al. (1996) found an increased prevalence of boys with cryptorchidism in Spanish municipalities with the highest suspected pesticide use. Paternal pesticide exposure was associated with cryptorchidism in the Netherlands (Pierik, Burdorf, Deddens, Juttmann, & Weber, 2004), whereas maternal gardening occupations were associated with an increased incidence of cryptorchidism in Denmark (Weidner, Moller, Jensen, & Skakkebaek, 1998; Andersen et al., 2008). Additionally, Garry, Schreinemachers, Harkins, and Griffith (1996) showed that pesticide appliers in rural Minnesota have a 69% increased incidence of sons born with urogenital defects.

Studies suggest that the rate of cryptorchidism is rising (Paulozzi, 1999). However, the timing of the assessment of cryptorchidism could confound the interpretation of these studies: some congenital cases are spontaneously resolved whereas other cases are not congenital and result when the testes reascend into the abdominal cavity (reviewed in Virtanen & Toppari, 2008). It is important to note that humans are exposed to low doses of several chemicals, some of which independently disrupt testicular descent in laboratory animals. Additional studies of noncaptive species are needed to further examine the contribution of EDCs to cryptorchidism and infertility in the wild.

6.10. Hypospadias

Hypospadias in males occurs when the urethra fails to open at the tip of the glans of the penis. Instead, the urethral opening can be found anywhere along the urethral groove, including the shaft and ventral side of the scrotum. Hypospadias is the second most common birth defect of the male genitalia. The etiology of hypospadias is believed to be multifactorial. Alterations in androgen receptor, homeobox genes, the steroid-converting enzymes such as 3β-hydroxysteroid dehydrogenase (3β-HSD) and CYP17A1 (17α-hydroxylase) have been implicated in the etiology of hypospadias (discussed in Baskin, Himes, & Colborn., 2001).

Several studies have demonstrated the ability of EDCs to induce hypospadias in rodent models. Vorherr, Messer, Vorherr, Jordan, and Kornfeld (1979) showed that male rat pups developed hypospadias after exposure to 15 μg DES/kg/day on gestation days 13, 16 18, and 20. Kim et al. (2004) showed that pregnant C57 mice given 50, 100, or 200 μg EE_2/kg or 100 μg DES/kg by oral gavage on gestation days 12–17 had on average 50% hypospadic male offspring compared to 0% of the control-treated animals. Vilela and colleagues (2007) demonstrated that 25, 42, and 41% of male CD-1 mice exposed developmentally to 0.17 mg/kg/day genistein, 10 mg/kg/day of the antiandrogenic fungicide vinclozolin, or the two chemicals in combination on gestation days 13–17, respectively, had hypospadias, whereas no control-treated animals had hypospadias.

Developmental exposure to high doses of the potent antiandrogenic pharmaceutical flutamide has delineated the critical window of androgen exposure necessary for proper penile formation. Welsh et al. (2008) found that 100 mg flutamide/kg during the onset of testicular T production (gestation days 15.5−17.5) and during the period just prior to morphological differentiation (gestation days 17.5−19.5) resulted in 64 and 63% of rats with hypospadias, respectively. Rats exposed during morphological differentiation (gestation days 19.5−21.5) did not have an increased incidence of hypospadias, indicating that sensitivity to EDCs and risk of hypospadias occurs earlier than previously thought. This is further supported by data from Foster and Harris (2005), who dosed pregnant Sprague-Dawley rats with 50 mg flutamide/kg on gestation days 16, 17, 18, or 19. Hypospadias was seen in 58% of pups that were dosed on gestation day 17 and < 10% of pups dosed on gestation 18, and was not seen when pups were exposed on other gestational days (Foster & Harris, 2005).

Gray et al. (1994; 1999b) performed a series of experiments with potentially antiandrogenic pesticides and toxic substances in Long-Evans hooded and Sprague-Dawley rats. They initially tested chemicals that are known AR ligands. They dosed pregnant Long-Evans hooded rats on gestation day 14 through postnatal day 3 with 100 or 200 mg/kg/day vinclozolin or 100 mg/kg/day procymidone and found that all vinclozolin-exposed rats and 40% procymidone-exposed rats had hypospadias. Interestingly, no Long-Evans hooded rats had hypospadias following exposure to 100 mg/kg/day *p,p'*-DDE through gestation day 18, whereas 7.8% of Sprague-Dawley rats had hypospadias. Linuron is a weak AR ligand, and when pregnant Sprague-Dawley rats were exposed to 40 mg linuron/kg/day on gestation days 14−18, 12.5% of the male offspring had hypospadias. They also tested the effects of two phthalates, DBP and DEHP, which previously were shown to alter reproductive development through unknown mechanisms. Developmental exposure to 500 mg DBP/kg/day on gestation day 14 though postnatal day 3 resulted in 6.2% of male offspring with hypospadias, but exposure to 750 mg DEHP/kg/day resulted in 67% of male offspring with hypospadias. Taken together, these data suggests that penile formation can be altered by EDCs of different mechanisms of action: xenoestrogens and chemicals that interfere with androgen signaling.

6.10.1. Effects on hypospadias in humans

In 1997, the CDC reported a doubling of hypospadias in the United States from 1968 to 1993 (Paulozzi, Erickson, & Jackson, 1997). European countries also reported increases in hypospadias but increases were not seen in less affluent countries (Paulozzi, 1999). Hypospadias is associated with lower birth weight, which in turn is associated with decreased androgen exposure (Baskin et al., 2001). Exposure to industrial and agricultural EDCs during development is associated with an increased risk of hypospadias, as demonstrated by Dolk et al. (1998) and Carbone et al. (2006).

Vegetarians tend to consume greater quantities of soy-based products than their omnivorous counterparts and are exposed to increased levels of phytoestrogens, chemicals in plants that mimic estrogens. A prospective cohort study in the United Kingdom that sampled 7928 males born in 1991 and 1992 found that boys with hypospadias were nearly five times more likely to be born to vegetarian mothers (North & Golding, 2000). There was also an increased incidence of hypospadias in boys born to mothers who drank soymilk or consumed soy 'meat' products, but neither was significant. In contrast, a case control study in the United Kingdom on babies born in 1997−1998 did not find any association between maternal vegetarianism and hypospadias risk (Ormond et al., 2009). However, they did find a positive association between maternal occupational exposure to hairspray and hypospadias. It is suspected that phthalates in hairspray have antiandrogen effects that contribute to the increased risk of hypospadias. Similarly, maternal progestin intake during early pregnancy to prevent fetal loss may disrupt androgen signaling and result in an increased risk of hypospadias (Carmichael et al., 2005).

Whereas a review of the data on DES sons (exposed *in utero*) does not indicate an altered risk of hypospadias, other urethral abnormalities and problems in passing urine were found (Henderson et al., 1976). Importantly, four studies suggest that effects of DES exposure can span several generations, as 'DES grandsons,' or those born to DES daughters, may have an increased incidence of hypospadias. Using a cohort of Danish women seeking treatment for infertility and receiving IVF, Klip et al. showed that DES daughters were 20.1 times more likely to have sons with hypospadias than nonexposed women (Hernandez-Diaz, 2002; Klip et al., 2002). Pons, Papiernik, Billon, Hessabi, and Duyme (2005) found a nearly five-fold increased risk of hypospadias in French boys following maternal *in-utero* DES exposure (Pons et al., 2005). Similarly, Brouwers et al. (2006; 2007) found a 4.9-fold risk of maternal *in-utero* DES exposure in Dutch boys born with hypospadias in 1987−1997 (Table 14.4). However, when a US cohort of 3916 women with confirmed DES exposure and 1746 unexposed women were asked whether any of their children had a congenital birth defect, an odds ratio for risk of having a son with hypospadias of only 1.7 was detected and was not significant (Palmer et al., 2005). The methods used in these studies are very different and may account for the striking differences in conclusions. Klip et al. (2002) used a nested case-control study and identified sons with hypospadias and controls from a cohort of women seeking treatment for infertility. Since these women were already seeking treatment for a potential effect of *in-utero* DES exposure, they may represent a subset of women who were exposed to higher doses or longer duration, or

TABLE 14.4 Prenatal exposure to 'diethylstibestrol (DES)' *in utero* and the risk of hypospadias in offspring: conditional logistic regression analyses on a 'matched' subset of cases and references[a]

	Cases (n=232) yes/no[b] (%)	Referents (n=232) yes/no[b] (%)	Conditional odds ratio (95% CI)	Adjusted odds ratio[c] (95% CI)
Mothers probably or certainly exposed to DES *in utero*	12/220 (5.2)	3/229 (1.3)	5.5(1.2-24.8)	4.9(1.1-22.3)
Fathers probably or certainly exposed to DES *in utero*	2/230 (0.9)	2/230 (0.9)	1.0(0.1-7.1)	0.9(0.1-6.7)

[a]*Analyses were performed on the subset of case parents for whom referent parents were also available.*
[b]*Parents with unknown exposure status (28 mothers and 63 fathers) were considered to be non-exposed.*
[c]*Adjusted for educational level (three levels) of the mother and father respectively.*
Reproduced from Brouwers et al. (2006) with permission of Oxford University Press.

during a critical window of fetal sexual differentiation. Additionally, the prevalence of hypospadias cases reported by Klip et al. (2002) in the unexposed group was lower than that reported for the Netherlands by the European Registration of Congenital Anomalies. Pons et al. (2005) and Brouwers et al. (2006; 2007) used a standard case-control design to recruit participants. Hypospadias cases in their study were medically confirmed but maternal *in-utero* exposure was based on recall. Finally, Palmer et al. (2005) used a retrospective cohort study to assess the prevalence of DES grandsons with hypospadias in a well-defined and closely followed US cohort of DES daughters. In this study *in-utero* DES exposure was confirmed, but hypospadias in sons were self-reported and not medically confirmed. Taken together, these data suggest that more research regarding the transgenerational effects of DES and other EDCs is necessary.

6.11. Prostate Cancer

Prostate cancer is the most common type of cancer in men and the second leading cause of cancer deaths in American men (American Cancer Society, 2009a). Prostate cancer is modulated by steroid hormones and is generally dependent on androgens for growth, so androgen ablation therapy has often been used as a treatment. Prostate cancer is also associated with endogenous estrogen exposure as men with elevated estrogens have increased prostate cancer risk (Modugno et al., 2001).

Developmental EDC exposure is implicated in altered prostate phenotypes and carcinogenesis in adulthood. Developmental exposure to xenoestrogens, such as DES and BPA, is associated with changes in adult prostates in rodents. Several investigators report enlarged prostates in mice after developmental exposure to 2, 20 or 50 μg BPA/kg body weight (Nagel et al., 1997b; Gupta, 2000a; Gupta, 2000b; Timms et al., 2005). Ho et al. (2006) reported an increase in precancerous lesions in prostates from aging male rats exposed to 10 μg BPA/kg on postnatal days one to five (reviewed in Prins, Birch, Tang, & Ho, 2007). Ogura et al. (2007) reported permanent expression of cytokeratin in mouse prostate basal cells after prenatal exposure to 20 μg/kg BPA similar to that seen with developmental DES exposure.

Unlike humans, rodents are particularly resistant to developing prostate cancer; however, a model system has been developed whereby estrogens given in adulthood induce the disease. Using this model system, 100% of males with developmental BPA treatment had precancerous prostatic lesions vs. 40% of control males. Further, developmental BPA exposure resulted in hypomethylation and persistent expression in aging males of phosphodiesterase type 4 variant 4 (*Pde4d4*), an estrogen-imprinted gene directly associated with preneoplastic prostatic lesions. Control males showed an age-related hypermethylation and thus decrease in expression of this gene (Ho et al., 2006).

6.11.1. Effects on prostate cancer in humans

In human prostate adenocarcinomas, the AR commonly undergoes mutation. The human prostate cancer cell line LNCaP contains a mutation in the AR termed AR-T877A. Importantly, BPA is an antagonist for the wild-type AR, but in LNCaP cells BPA is an agonist for AR-T877A and stimulates cell growth (Wetherill, Petre, Monk, Puga, & Knudsen, 2002). Further, BPA stimulates growth of human prostate cancer xenografts in nude mice and increases expression of the androgen-responsive gene for prostate-specific antigen (PSA) (Wetherill et al., 2005; 2006). These data suggest that BPA at current human exposure levels could stimulate growth of prostate cancers in men carrying this mutation.

Prostate cancer risk has been linked to occupational exposure. Both farmers and pesticide applicators have increased rates of prostate cancer (Alavanja et al., 2003; 2005; Van Maele-Fabry, Libotte, Willems, & Lison, 2006). In a case-control study of 405 farmers with prostate cancer and 392 age-matched controls in South Carolina, farming was associated with an 80% increased risk of prostate cancer in Caucasians but not in African-Americans (Meyer, Coker, Sanderson, & Symanski, 2007). This increase was linked to pesticide exposure. A correlation analysis in California between pesticide use in pounds per county and prostate cancer incidence found that the use of the herbicides atrazine and captan were correlated to prostate cancer in African-American men. Elevated levels of organochlorines have also been associated with increased risk of prostate cancer (Ritchie et al., 2003). Electric utility workers and capacitor manufacturing plant workers exposed to high levels of PCBs also have significantly increased risk of developing prostate cancer (Charles et al., 2003; Prince et al., 2006).

7. CONCLUSION

A recent position report published by The Endocrine Society concluded that 'Results from animal models, human clinical observations, and epidemiological studies converge to implicate EDCs as a significant concern to public health' (p 293) and highlights how much we have learned about endocrine disruption since the first Wingspread meeting in 1991 (Diamanti-Kandarakis et al., 2009). Endocrine-disrupting chemicals are ubiquitous in the environment and can act at very low doses to disrupt proper development and alter adult reproductive function. Currently, regulation of EDCs in the United States does not always follow the precautionary principle. That is, chemicals, their derivatives, and metabolites are often considered safe until proven otherwise. Typically, this means that the burden of safety falls to nonproducers of these chemicals, typically academic and government investigators. Changing policy to reflect the scientific knowledge is not an easy task and can take years or decades. As an example, consider the banning of DDT in 1972, which occurred mostly in response to a public outcry after the publication of Rachel Carson's *Silent Spring* in 1962.

It may be that other EDCs face a similar fate. In 2008 the NTP stated that there is '*some concern* for effects on the brain, behavior, and prostate gland in fetuses, infants, and children at current human exposures to BPA' (CERHR, 2008; vii and p 38). Most recently, the FDA's review of BPA safety was closely aligned with that of the NTP. On its website on January 15, 2010, the FDA announced that it now supports the 2008 NTP statement that there is reason for 'some concern' and is encouraging industry to stop producing BPA-containing products intended for newborns and infants, specifically bottles and formula containers (US Food and Drug Administration, 2010). While government regulation is slow and methodical, public pressure can have surprisingly rapid action. Growing consumer concern over the potential negative health effects of BPA has led manufacturers to market many 'BPA-free' products.

The amount of information regarding potential negative effects from EDC exposure is quite overwhelming. As summarized in this chapter, significant data exist demonstrating harmful effects of both adult and developmental EDC exposure in animal models; however, the validity of such models is sometimes called into question when considering the risk to humans. Additionally, the use of different analytical methods for the measurement of exposures, endpoints, routes, and timing of exposures, and use of different animal species and strains makes drawing conclusions about EDCs challenging. It is hoped that the USEPA's Endocrine Disruptor Screening and Testing Program will provide such validated methods. Further complicating this is the fact that we live in a 'chemical soup' and are exposed to hundreds of different chemicals daily. The implications of living in such an environment are just beginning to be studied and will be very important in terms of fully understanding endocrine disruption.

Humans have been exposed to many chemicals and pharmaceuticals later found to have negative health effects. There is a continued need for the study of long latency effects, especially from *developmental* exposures to xenobiotics. Current and past occupationally exposured adults and their offspring represent a key population for study in this regard. Hopefully, knowledge gained from epidemiological and laboratory studies regarding the risks of EDCs will be used to understand the etiology and growing incidence of several reproductive diseases to protect and improve human health.

ABBREVIATIONS

β-HCH β-hexachlorocyclohexane
2,4-D 2,4-dichlorophenoxyacetic acid
3β-HSD 3β-hydroxysteroid dehydrogenase
AGD Anogenital distance
AGI Anogenital index (AGD normalized to body weight)
AhR Aryl hydrocarbon receptor
AP-1 Activator protein-1
AR Androgen receptor
BADGE Bisphenol A diglycidyl ether
BBP Butyl benzyl phthalate
BPA Bisphenol A
CAR Constitutive active receptor
CERHR Center for the Evaluation of Risks to Human Reproduction
***cis*-HE** *Cis*-heptachloroepoxide
CNS Central nervous system
CYP Cytochrome P450 genes
D&G Daidzein and genistein
DBP Di-n-butyl phthalate
DDE Dichlorodiphenyldichloroethylene
DDT Dichlorodiphenyltrichloroethane
DEHP Bis(2-ethylhexyl) phthalate
DES Diethylstilbestrol
DHEA Dehydroepiandrosterone

DHEA-S	Dehydroepiandrosterone sulfate
DHT	Dihydrotestosterone
DMBA	7,12-dimethylbenz(a)anthracene
DNMT	DNA methyltransferase
E_2	Estradiol
EDC	Endocrine-disrupting chemical
EDSP	Endocrine Disruptor Screening and Testing program
EE_2	Ethinylestradiol
EERγ	Estrogen-related receptor-γ
EGME	Ethylene glycol monomethyl ether
ER	Estrogen receptor
ERIN	Estrogen receptor indicator
ERK	Extracellular-regulated kinase
FDA	Federal Drug Administration
FQP	Food Quality Protection (Act)
FSH	Follicle-stimulating hormone
GI	Gastrointestinal
GR	Glucocorticoid receptor
HCB	Hexachlorobenzene
HCE	Hexachloroethane
HPG	Hypothalamus–pituitary–gonad
IMPY	2-isopropoxy-4-methylpyrimidinol
IUP	Intrauterine position
IVF	*In-vitro* fertilization
MAA	Methoxyacetic acid
MAPK	Mitogen-activated protein kinase
MBP	Mono-butyl phthalate
MBzP	Mono-benzyl phthalate
MEHP	Mono-(2-ethylhexyl) phthalate
MEP	Monoethyl phthalate
MiBP	Monoisobutyl phthalate
MIS	Müllerian-inhibiting substance
MR	Mineral corticoid receptor
NMU	N-nitroso-N-methylurea
NP	4-nonylphenol
NRC	National Research Council
NTP	National Toxicology Program
NURSA	Nuclear Receptor Signaling Atlas
OC	Oral contraceptive
P_4	Progesterone
PBBs	Polybrominated biphenyls
PBDEs	Polybrominated diphenyl ethers
PCB	Polychlorinated biphenyl
PCDD	Polychlorinated dibenzodioxins
PCDF	Polychlorinated dibenzofurans
Pde4d4	Phosphodiesterase type 4 variant 4
POF	Premature ovarian failure
POP	Persistent organic pollutant
PPAR	Peroxisome proliferator-activated receptor
PR	Progesterone receptor
PRL	Prolactin
PSA	Prostate-specific antigen
PXR	Pregnane X receptor
RAR	Retinoic acid receptor
RXR	Retinoid X receptor
SE	Standard error
SERM	Selective estrogen receptor modulator
SF-1	Steroidogenic factor-1
SHBG	Sex hormone-binding globulin
T	Testosterone
T_3	Triiodothyronine
T_4	Thyroxine
TCDD	2,3,7,8-tetrachlorodibenzodioxin
TDS	Testicular dysgenesis syndrome
TR	Thyroid hormone receptor
USEPA	United States Environmental Protection Agency
VDR	Vitamin D receptor

REFERENCES

Adams, N. R. (1995). Detection of the effects of phytoestrogens on sheep and cattle. *J. Anim. Sci., 73*, 1509–1515.

Alavanja, M. C., Samanic, C., Dosemeci, M., Lubin, J., Tarone, R., Lynch, C. F., et al. (2003). Use of agricultural pesticides and prostate cancer risk in the Agricultural Health. Study cohort. *Am. J. Epidemiol., 157*, 800–814.

Alavanja, M. C., Sandler, D. P., Lynch, C. F., Knott, C., Lubin, J. H., Tarone, R., et al. (2005). Cancer incidence in the agricultural health study. *Scand. J. Work Environ. Health, 31*(Suppl. 1), 39–45, discussion 5–7.

Allred, C. D., Allred, K. F., Ju, Y. H., Clausen, L. M., Doerge, D. R., Schantz, S. L., et al. (2004). Dietary genistein results in larger MNU-induced, estrogen-dependent mammary tumors following ovariectomy of Sprague-Dawley rats. *Carcinogenesis, 25*, 211–218.

Alworth, L. C., Howdeshell, K. L., Ruhlen, R. L., Day, J. K., Lubahn, D. B., Huang, T. H., et al. (2002). Uterine responsiveness to estradiol and DNA methylation are altered by fetal exposure to diethylstilbestrol and methoxychlor in CD-1 mice: effects of low versus high doses. *Toxicol. Appl. Pharmacol., 183*, 10–22.

American Cancer Society. (2009a). *Cancer Facts & Figures 2009*. Atlanta, GA: American Cancer Society.

American Cancer Society. (2009b). *Overview: Breast Cancer. Vol. 2009*. Atlanta, GA: American Cancer Society.

Andersen, H. R., Schmidt, I. M., Grandjean, P., Jensen, T. K., Budtz-Jorgensen, E., Kjaerstad, M. B., et al. (2008). Impaired reproductive development in sons of women occupationally exposed to pesticides during pregnancy. *Environ. Health Perspect., 116*, 566–572.

Anderson, S. E., Dallal, G. E., & Must, A. (2003). Relative weight and race influence average age at menarche: results from two nationally representative surveys of US girls studied 25 years apart. *Pediatrics, 111*, 844–850.

Anway, M. D., & Skinner, M. K. (2006). Epigenetic transgenerational actions of endocrine disruptors. *Endocrinology, 147*, S43–S49.

Anway, M. D., Leathers, C., & Skinner, M. K. (2006). Endocrine disruptor vinclozolin induced epigenetic transgenerational adult-onset disease. *Endocrinology, 147*, 5515–5523.

Arnold, A. P., & Gorski, R. A. (1984). Gonadal steroid induction of structural sex differences in the central nervous system. *Annu. Rev. Neurosci., 7*, 413–442.

Ashby, J., Tinwell, H., Lefevre, P. A., Odum, J., Paton, D., Millward, S. W., et al. (1997). Normal sexual development of rats exposed to butyl benzyl phthalate from conception to weaning. *Regul. Toxicol. Pharmacol., 26*, 102–118.

Aulerich, R. J., Ringer, R. K., & Iwamoto, S. (1973). Reproductive failure and mortality in mink fed on Great Lakes fish. *J. Reprod. Fertil. Suppl., 19*, 365–376.

Baade, P., Carriere, P., & Fritschi, L. (2008). Trends in testicular germ cell cancer incidence in Australia. *Cancer Causes Control, 19*, 1043–1049.

Backlin, B. M., Eriksson, L., & Olovsson, M. (2003). Histology of uterine leiomyoma and occurrence in relation to reproductive activity in the Baltic gray seal (*Halichoerus grypus*). *Vet. Pathol., 40*, 175–180.

Barbieri, R. L., McShane, P. M., & Ryan, K. J. (1986). Constituents of cigarette smoke inhibit human granulosa cell aromatase. *Fertil. Steril., 46*, 232–236.

Barker, D. J. (1997). Maternal nutrition, fetal nutrition, and disease in later life. *Nutrition, 13*, 807–813.

Barker, D. J. (2003). The developmental origins of adult disease. *Eur. J. Epidemiol., 18*, 733–736.

Barraclough, C. A. (1966). Modifications in the CNS regulation of reproduction after exposure of prepubertal rats to steroid hormones. *Recent Prog. Horm. Res., 22*, 503–539.

Baskin, L. S., Himes, K., & Colborn, T. (2001). Hypospadias and endocrine disruption: is there a connection? *Environ. Health Perspect., 109*, 1175–1183.

Benachour, N., Moslemi, S., Sipahutar, H., & Seralini, G. E. (2007). Cytotoxic effects and aromatase inhibition by xenobiotic endocrine disrupters alone and in combination. *Toxicol. Appl. Pharmacol., 222*, 129–140.

Bergstrom, R., Adami, H. O., Mohner, M., Zatonski, W., Storm, H., Ekbom, A., et al. (1996). Increase in testicular cancer incidence in six European countries: a birth cohort phenomenon. *J. Natl. Cancer Inst., 88*, 727–733.

Bern, H. A. (1992). The fragile fetus. In T. Colborn, & C. Clement (Eds.), *Chemically-Induced Alterations in Sexual and Functional Development: The Wildlife/Human Connection, Vol. 21* (pp. 9–15). Princeton, N.J: Princeton Scientific Pub.

Bhatia, R., Shiau, R., Petreas, M., Weintraub, J. M., Farhang, L., & Eskenazi, B. (2005). Organochlorine pesticides and male genital anomalies in the child health and development studies. *Environ. Health Perspect., 113*, 220–224.

Birnbaum, L. S., & Cummings, A. M. (2002). Dioxins and endometriosis: a plausible hypothesis. *Environ. Health Perspect., 110*, 15–21.

Bishop-Bailey, D., Hla, T., & Warner, T. D. (2000). Bisphenol A diglycidyl ether (BADGE) is a PPARgamma agonist in an ECV304 cell line. *Br. J. Pharmacol., 131*, 651–654.

Blanck, H. M., Marcus, M., Tolbert, P. E., Rubin, C., Henderson, A. K., Hertzberg, V. S., et al. (2000). Age at menarche and tanner stage in girls exposed *in utero* and postnatally to polybrominated biphenyl. *Epidemiology, 11*, 641–647.

Block, K., Kardana, A., Igarashi, P., & Taylor, H. S. (2000). *In utero* diethylstilbestrol (DES) exposure alters Hox gene expression in the developing Müllerian system. *Faseb. J., 14*, 1101–1108.

Bodwell, J. E., Kingsley, L. A., & Hamilton, J. W. (2004). Arsenic at very low concentrations alters glucocorticoid receptor (GR)-mediated gene activation but not GR-mediated gene repression: complex dose-response effects are closely correlated with levels of activated GR and require a functional GR DNA binding domain. *Chem. Res. Toxicol., 17*, 1064–1076.

Boylan, E. S., & Calhoon, R. E. (1979). Mammary tumorigenesis in the rat following prenatal exposure to diethylstilbestrol and postnatal treatment with 7,12-dimethylbenz[a]anthracene. *J. Toxicol. Environ. Health, 5*, 1059–1071.

Brede, C., Fjeldal, P., Skjevrak, I., & Herikstad, H. (2003). Increased migration levels of bisphenol A from polycarbonate baby bottles after dishwashing, boiling and brushing. *Food Addit. Contam., 20*, 684–689.

Brotons, J. A., Olea-Serrano, M. F., Villalobos, M., Pedraza, V., & Olea, N. (1995). Xenoestrogens released from lacquer coatings in food cans. *Environ. Health Perspect., 103*, 608–612.

Brouwers, M. M., Feitz, W. F., Roelofs, L. A., Kiemeney, L. A., De Gier, R. P., & Roeleveld, N. (2006). Hypospadias: a transgenerational effect of diethylstilbestrol? *Hum. Reprod., 21*, 666–669.

Brouwers, M. M., Feitz, W. F., Roelofs, L. A., Kiemeney, L. A., De Gier, R. P., & Roeleveld, N. (2007). Risk factors for hypospadias. *Eur. J. Pediatr., 166*, 671–678.

Brown, N. M., & Lamartiniere, C. A. (1995). Xenoestrogens alter mammary gland differentiation and cell proliferation in the rat. *Environ. Health Perspect., 103*, 708–713.

Brown, N. M., Manzolillo, P. A., Zhang, J. X., Wang, J., & Lamartiniere, C. A. (1998). Prenatal TCDD and predisposition to mammary cancer in the rat. *Carcinogenesis, 19*, 1623–1629.

Buck, G. M., Mendola, P., Vena, J. E., Sever, L. E., Kostyniak, P., Greizerstein, H., et al. (1999). Paternal Lake Ontario fish consumption and risk of conception delay, New York State Angler Cohort. *Environ. Res., 80*, S13–S18.

Buck Louis, G. M., Hediger, M. L., & Pena, J. B. (2007). Intrauterine exposures and risk of endometriosis. *Hum. Reprod., 22*, 3232–3236.

Budtz-Jorgensen, E., Keiding, N., Grandjean, P., & White, R. F. (1999). Methylmercury neurotoxicity independent of PCB exposure. *Environ. Health Perspect., 107*, A236–A237.

Bulayeva, N. N., & Watson, C. S. (2004). Xenoestrogen-induced ERK-1 and ERK-2 activation via multiple membrane-initiated signaling pathways. *Environ. Health Perspect., 112*, 1481–1487.

Calafat, A. M., Wong, L. Y., Ye, X., Reidy, J. A., & Needham, L. L. (2008). Concentrations of the sunscreen agent benzophenone-3 in residents of the United States: National Health and Nutrition Examination Survey 2003–2004. *Environ. Health Perspect., 116*, 893–897.

Carbone, P., Giordano, F., Nori, F., Mantovani, A., Taruscio, D., Lauria, L., et al. (2006). Cryptorchidism and hypospadias in the Sicilian district of Ragusa and the use of pesticides. *Reprod. Toxicol., 22*, 8–12.

Carlsen, E., Giwercman, A., Keiding, N., & Skakkebaek, N. E. (1992). Evidence for decreasing quality of semen during past 50 years. *BMJ, 305*, 609–613.

Carlsen, E., Giwercman, A., Keiding, N., & Skakkebaek, N. E. (1995). Declining semen quality and increasing incidence of testicular cancer: is there a common cause? *Environ. Health. Perspect., 103* (Suppl. 7), 137–139.

Carmichael, S. L., Shaw, G. M., Laurent, C., Croughan, M. S., Olney, R. S., & Lammer, E. J. (2005). Maternal progestin intake and risk of hypospadias. *Arch. Pediatr. Adolesc. Med., 159*, 957–962.

Carson, R. (1987). *Silent Spring: 25th Anniversary Edition.* New York, NY: Houghton Mifflin.

Casals-Casas, C., Feige, J. N., & Desvergne, B. (2008). Interference of pollutants with PPARs: endocrine disruption meets metabolism. *Int. J. Obes. (Lond), 32*(Suppl. 6), S53–S61.

Caserta, D., Maranghi, L., Mantovani, A., Marci, R., Maranghi, F., & Moscarini, M. (2008). Impact of endocrine disruptor chemicals in gynaecology. *Hum. Reprod. Update, 14*, 59–72.

Cavieres, M. F., Jaeger, J., & Porter, W. (2002). Developmental toxicity of a commercial herbicide mixture in mice: I. Effects on embryo implantation and litter size. *Environ. Health Perspect., 110*, 1081–1085.

Cerel-Suhl, S. L., & Yeager, B. F. (1999). Update on oral contraceptive pills. *Am. Fam. Physician, 60*, 2073–2084.

Center for the Evaluation of Risks to Human Reproduction (CERHR), 2008. NTP-CERHR Monograph on the Potential Human Reproductive and Developmental Effects of Bisphenol A. NTP CERHR MON. i-III1.

Charles, L. E., Loomis, D., Shy, C. M., Newman, B., Millikan, R., Nylander-French, L. A., et al. (2003). Electromagnetic fields, polychlorinated biphenyls, and prostate cancer mortality in electric utility workers. *Am. J. Epidemiol., 157*, 683–691.

Chen, J., Ahn, K. C., Gee, N. A., Ahmed, M. I., Duleba, A. J., Zhao, L., et al. (2008). Triclocarban enhances testosterone action: a new type of endocrine disruptor? *Endocrinology, 149*, 1173–1179.

Chen, Y. C., Guo, Y. L., & Hsu, C. C. (1992). Cognitive development of children prenatally exposed to polychlorinated biphenyls (Yu-Cheng children) and their siblings. *J. Formos. Med. Assoc., 91*, 704–707.

Chimchirian, R. F., Suri, R. P., & Fu, H. (2007). Free synthetic and natural estrogen hormones in influent and effluent of three municipal wastewater treatment plants. *Water Environ. Res., 79*, 969–974.

Colborn, T. (1991). Epidemiology of Great Lakes bald eagles. *J. Toxicol. Environ. Health, 33*, 395–453.

Colborn, T. (1994). The wildlife/human connection: modernizing risk decisions. *Environ. Health Perspect., 102*(Suppl. 12), 55–59.

Colborn, T. (1995). Environmental estrogens: health implications for humans and wildlife. *Environ. Health Perspect., 103*(Suppl. 7), 135–136.

Colborn, T., Dumanoski, D., & Myers, J. P. (1996). *Our Stolen Future*. New York, NY: Plume.

Colborn, T., vom Saal, F. S., & Soto, A. M. (1993). Developmental effects of endocrine-disrupting chemicals in wildlife and humans. *Environ. Health Perspect., 101*, 378–384.

Cooke, A. S. (1973). Shell thinning in avian eggs by environmental pollutants. *Environmental Pollution, 4*, 85–152.

Cooke, P. S., Holsberger, D. R., Witorsch, R. J., Sylvester, P. W., Meredith, J. M., Treinen, K. A., et al. (2004). Thyroid hormone, glucocorticoids, and prolactin at the nexus of physiology, reproduction, and toxicology. *Toxicol. Appl. Pharmacol., 194*, 309–335.

Cooper, R. L., Chadwick, R. W., Rehnberg, G. L., Goldman, J. M., Booth, K. C., Hein, J. F., et al. (1989). Effect of lindane on hormonal control of reproductive function in the female rat. *Toxicol. Appl. Pharmacol., 99*, 384–394.

Correa, A., Gray, R. H., Cohen, R., Rothman, N., Shah, F., Seacat, H., et al. (1996). Ethylene glycol ethers and risks of spontaneous abortion and subfertility. *Am. J. Epidemiol., 143*, 707–717.

Cotroneo, M. S., Wang, J., Fritz, W. A., Eltoum, I. E., & Lamartiniere, C. A. (2002). Genistein action in the prepubertal mammary gland in a chemoprevention model. *Carcinogenesis, 23*, 1467–1474.

Crain, D. A., Guillette, L. J., Jr., Rooney, A. A., & Pickford, D. B. (1997). Alterations in steroidogenesis in alligators (*Alligator mississippiensis*) exposed naturally and experimentally to environmental contaminants. *Environ. Health Perspect., 105*, 528–533.

Cramer, D. W., & Missmer, S. A. (2002). The epidemiology of endometriosis. *Ann. NY Acad. Sci., 955*, 11–22, discussion 34-6, 396–406.

Crews, D., Gore, A. C., Hsu, T. S., Dangleben, N. L., Spinetta, M., Schallert, T., et al. (2007). Transgenerational epigenetic imprints on mate preference. *Proc. Natl. Acad. Sci. USA, 104*, 5942–5946.

Crisp, T. M., Clegg, E. D., Cooper, R. L., Wood, W. P., Anderson, D. G., Baetke, K. P., et al. (1998). Environmental endocrine disruption: an effects assessment and analysis. *Environ. Health Perspect., 106* (Suppl 1), 11–56.

Cummings, A. M., & Metcalf, J. L. (1995). Induction of endometriosis in mice: a new model sensitive to estrogen. *Reprod. Toxicol., 9*, 233–238.

Cummings, A. M., Hedge, J. M., & Birnbaum, L. S. (1999). Effect of prenatal exposure to TCDD on the promotion of endometriotic lesion growth by TCDD in adult female rats and mice. *Toxicol. Sci., 52*, 45–49.

Cummings, A. M., Metcalf, J. L., & Birnbaum, L. (1996). Promotion of endometriosis by 2,3,7,8-tetrachlorodibenzo-p-dioxin in rats and mice: time-dose dependence and species comparison. *Toxicol. Appl. Pharmacol., 138*, 131–139.

Dallinga, J. W., Moonen, E. J., Dumoulin, J. C., Evers, J. L., Geraedts, J. P., & Kleinjans, J. C. (2002). Decreased human semen quality and organochlorine compounds in blood. *Hum. Reprod., 17*, 1973–1979.

Damgaard, I. N., Skakkebaek, N. E., Toppari, J., Virtanen, H. E., Shen, H., Schramm, K. W., et al. (2006). Persistent pesticides in human breast milk and cryptorchidism. *Environ. Health Perspect., 114*, 1133–1138.

Darvill, T., Lonky, E., Reihman, J., Stewart, P., & Pagano, J. (2000). Prenatal exposure to PCBs and infant performance on the fagan test of infant intelligence. *Neurotoxicology, 21*, 1029–1038.

Davis, B. J., Weaver, R., Gaines, L. J., & Heindel, J. J. (1994). Mono-(2-ethylhexyl) phthalate suppresses estradiol production independent of FSH-cAMP stimulation in rat granulosa cells. *Toxicol. Appl. Pharmacol., 128*, 224–228.

Dessi-Fulgheri, F., Porrini, S., & Farabollini, F. (2002). Effects of perinatal exposure to bisphenol A on play behavior of female and male juvenile rats. *Environ. Health Perspect., 110*(Suppl. 3), 403–407.

Desvergne, B., Feige, J. N., & Casals-Casas, C. (2009). PPAR-mediated activity of phthalates: a link to the obesity epidemic? *Mol. Cell Endocrinol., 304*, 43–48.

Diamanti-Kandarakis, E., Bourguignon, J. P., Giudice, L. C., Hauser, R., Prins, G. S., Soto, A. M., et al. (2009). Endocrine-disrupting chemicals: an Endocrine Society scientific statement. *Endocr. Rev., 30*, 293–342.

Dolk, H., Vrijheid, M., Armstrong, B., Abramsky, L., Bianchi, F., Garne, E., et al. (1998). Risk of congenital anomalies near hazardous-waste landfill sites in Europe: the EUROHAZCON study. *Lancet., 352*, 423–427.

Dunbar, M. R., Cunningham, M. W., Wooding, J. B., & Roth, R. P. (1996). Cryptorchidism and delayed testicular descent in Florida black bears. *J. Wildl. Dis., 32*, 661–664.

Durando, M., Kass, L., Piva, J., Sonnenschein, C., Soto, A. M., Luque, E. H., et al. (2007). Prenatal bisphenol A exposure induces preneoplastic lesions in the mammary gland in Wistar rats. *Environ. Health Perspect., 115*, 80–86.

Duty, S. M., Silva, M. J., Barr, D. B., Brode, J. W., Ryan, L., Chen, Z., et al. (2003). Phthalate exposure and human semen parameters. *Epidemiology, 14*, 269–277.

Endocrine Disruptor Screening and Testing Advisory Committee (EDSTAC). (1998). Endocrine Disruptor Screening and Testing Advisory Committee (EDSTAC) Final Report. *USEPA*.

Edwards, D. A., & Burge, K. G. (1971). Early androgen treatment and male and female sexual behavior in mice. *Hormones and Behavior, 2*, 49–58.

Ekbom, A., Hsieh, C. C., Lipworth, L., Adami, H. Q., & Trichopoulos, D. (1997). Intrauterine environment and breast cancer risk in women: a population-based study. *J. Natl. Cancer Inst., 89*, 71–76.

Ekbom, A., Trichopoulos, D., Adami, H. O., Hsieh, C. C., & Lan, S. J. (1992). Evidence of prenatal influences on breast cancer risk. *Lancet., 340*, 1015–1018.

Eng, E. T., Williams, D., Mandava, U., Kirma, N., Tekmal, R. R., & Chen, S. (2002). Anti-aromatase chemicals in red wine. *Ann. NY Acad. Sci., 963*, 239–246.

Engel, L. S., Hill, D. A., Hoppin, J. A., Lubin, J. H., Lynch, C. F., Pierce, J., et al. (2005). Pesticide use and breast cancer risk among farmers' wives in the agricultural health study. *Am. J. Epidemiol., 161*, 121–135.

Enviromental Protection Agency. (2009). Final list of initial pesticide active ingredients and pesticide inert ingredients to be screened under the federal Food, Drug, and Cosmetic Act. *Federal Register, 74*, 17579–17585.

Euling, S. Y., Herman-Giddens, M. E., Lee, P. A., Selevan, S. G., Juul, A., Sorensen, T. I., et al. (2008a). Examination of US puberty-timing data from 1940 to 1994 for secular trends: panel findings. *Pediatrics, 121* (Suppl. 3), S172–S191.

Euling, S. Y., Selevan, S. G., Pescovitz, O. H., & Skakkebaek, N. E. (2008b). Role of environmental factors in the timing of puberty. *Pediatrics, 121*(Suppl 3), S167–S171.

Facemire, C. F., Gross, T. S., & Guillette, L. J., Jr. (1995). Reproductive impairment in the Florida panther: nature or nurture? *Environ. Health Perspect., 103*(Suppl. 4), 79–86.

Fan, W., Yanase, T., Morinaga, H., Gondo, S., Okabe, T., Nomura, M., et al. (2007). Atrazine-induced aromatase expression is SF-1 dependent: implications for endocrine disruption in wildlife and reproductive cancers in humans. *Environ. Health Perspect., 115*, 720–727.

Faqi, A. S., & Chahoud, I. (1998). Antiestrogenic effects of low doses of 2,3,7,8-TCDD in offspring of female rats exposed throughout pregnancy and lactation. *Bull. Environ. Contam. Toxicol., 61*, 462–469.

Faqi, A. S., Dalsenter, P. R., Merker, H. J., & Chahoud, I. (1998). Effects on developmental landmarks and reproductive capability of 3,3',4,4'-tetrachlorobiphenyl and 3,3',4,4',5-pentachlorobiphenyl in offspring of rats exposed during pregnancy. *Hum. Exp. Toxicol., 17*, 365–372.

Farabollini, F., Porrini, S., Della Seta, D., Bianchi, F., & Dessi-Fulgheri, F. (2002). Effects of perinatal exposure to bisphenol A on sociosexual behavior of female and male rats. *Environ. Health Perspect., 110*(Suppl. 3), 409–414.

Fenton, S. E., Hamm, J. T., Birnbaum, L. S., & Youngblood, G. L. (2002). Persistent abnormalities in the rat mammary gland following gestational and lactational exposure to 2,3,7,8-tetrachlorodibenzo-p-dioxin (TCDD). *Toxicol. Sci., 67*, 63–74.

Fleming, L. E., Bean, J. A., Rudolph, M., & Hamilton, K. (1999). Cancer incidence in a cohort of licensed pesticide applicators in Florida. *J. Occup. Environ. Med., 41*, 279–288.

Forastiere, F., Quercia, A., Miceli, M., Settimi, L., Terenzoni, B., Rapiti, E., et al. (1993). Cancer among farmers in central Italy. *Scand. J. Work Environ. Health, 19*, 382–389.

Forman, M. R., Cantwell, M. M., Ronckers, C., & Zhang, Y. (2005). Through the looking glass at early-life exposures and breast cancer risk. *Cancer Invest., 23*, 609–624.

Foster, P. M., & Harris, M. W. (2005). Changes in androgen-mediated reproductive development in male rat offspring following exposure to a single oral dose of flutamide at different gestational ages. *Toxicol. Sci., 85*, 1024–1032.

Freedman, D. S., Khan, L. K., Serdula, M. K., Dietz, W. H., Srinivasan, S. R., & Berenson, G. S. (2002). Relation of age at menarche to race, time period, and anthropometric dimensions: the Bogalusa Heart Study. *Pediatrics, 110*, e43.

Garcia-Rodriguez, J., Garcia-Martin, M., Nogueras-Ocana, M., De Dios Luna-del-Castillo, J., Espigares Garcia, M., Olea, N., et al. (1996). Exposure to pesticides and cryptorchidism: geographical evidence of a possible association. *Environ. Health Perspect., 104*, 1090–1095.

Garry, V. F., Schreinemachers, D., Harkins, M. E., & Griffith, J. (1996). Pesticide appliers, biocides, and birth defects in rural Minnesota. *Environ. Health Perspect., 104*, 394–399.

Gerhard, I., Daniel, V., Link, S., Monga, B., & Runnebaum, B. (1998). Chlorinated hydrocarbons in women with repeated miscarriages. *Environ. Health Perspect., 106*, 675–681.

Gioiosa, L., Fissore, E., Ghirardelli, G., Parmigiani, S., & Palanza, P. (2007). Developmental exposure to low-dose estrogenic endocrine disruptors alters sex differences in exploration and emotional responses in mice. *Horm. Behav., 52*, 307–316.

Giordano, S. H., Cohen, D. S., Buzdar, A. U., Perkins, G., & Hortobagyi, G. N. (2004). Breast carcinoma in men: a population-based study. *Cancer, 101*, 51–57.

Giusti, R. M., Iwamoto, K., & Hatch, E. E. (1995). Diethylstilbestrol revisited: a review of the long-term health effects. *Ann. Intern. Med., 122*, 778–788.

Giwercman, A., Carlsen, E., Keiding, N., & Skakkebaek, N. E. (1993). Evidence for increasing incidence of abnormalities of the human testis: a review. *Environ. Health Perspect., 101*(Suppl. 2), 65–71.

Golub, M. S., Collman, G. W., Foster, P. M., Kimmel, C. A., Rajpert-De Meyts, E., Reiter, E. O., et al. (2008). Public health implications of altered puberty timing. *Pediatrics, 121*(Suppl. 3), S218–S230.

Grandjean, P., Weihe, P., Burse, V. W., Needham, L. L., Storr-Hansen, E., Heinzow, B., et al. (2001). Neurobehavioral deficits associated with PCB in 7-year-old children prenatally exposed to seafood neurotoxicants. *Neurotoxicol. Teratol., 23*, 305–317.

Gray, L. E., Jr., & Kelce, W. R. (1996). Latent effects of pesticides and toxic substances on sexual differentiation of rodents. *Toxicol. Ind. Health, 12*, 515–531.

Gray, L. E., Jr., & Ostby, J. S. (1995). *In utero* 2,3,7,8-tetrachlorodibenzo-p-dioxin (TCDD) alters reproductive morphology and function in female rat offspring. *Toxicol. Appl. Pharmacol., 133*, 285–294.

Gray, L. E., Jr., Ostby, J., Ferrell, J., Sigmon, R., Cooper, R., Linder, R., et al. (1989). Correlation of sperm and endocrine measures with reproductive success in rodents. *Prog. Clin. Biol. Res., 302*, 193–206, discussion 206–209.

Gray, L. E., Jr., Ostby, J., Monosson, E., & Kelce, W. R. (1999a). Environmental antiandrogens: low doses of the fungicide vinclozolin alter sexual differentiation of the male rat. *Toxicol. Ind. Health, 15*, 48–64.

Gray, L. E., Jr., Ostby, J. S., & Kelce, W. R. (1994). Developmental effects of an environmental antiandrogen: the fungicide vinclozolin alters sex differentiation of the male rat. *Toxicol. Appl. Pharmacol., 129*, 46–52.

Gray, L. E., Jr., Wolf, C., Lambright, C., Mann, P., Price, M., Cooper, R. L., et al. (1999b). Administration of potentially antiandrogenic pesticides (procymidone, linuron, iprodione, chlozolinate, p, p'-DDE, and ketoconazole) and toxic substances (dibutyl- and diethylhexyl phthalate, PCB 169, and ethane dimethane sulphonate) during sexual differentiation produces diverse profiles of reproductive malformations in the male rat. *Toxicol. Ind. Health, 15*, 94–118.

Greenlee, A. R., Arbuckle, T. E., & Chyou, P. H. (2003). Risk factors for female infertility in an agricultural region. *Epidemiology, 14*, 429–436.

Guillette, L. J., Jr., & Edwards, T. M. (2008). Environmental influences on fertility: can we learn lessons from studies of wildlife? *Fertil. Steril., 89*, e21–e24.

Guillette, L. J., Jr., & Guillette, E. A. (1996). Environmental contaminants and reproductive abnormalities in wildlife: implications for public health? *Toxicol. Ind. Health, 12*, 537–550.

Guillette, L. J., Jr., Pickford, D. B., Crain, D. A., Rooney, A. A., & Percival, H. F. (1996). Reduction in penis size and plasma testosterone concentrations in juvenile alligators living in a contaminated environment. *Gen. Comp. Endocrinol., 101*, 32–42.

Guo, Y. L., Hsu, P. C., Hsu, C. C., & Lambert, G. H. (2000). Semen quality after prenatal exposure to polychlorinated biphenyls and dibenzofurans. *Lancet., 356*, 1240–1241.

Gupta, C. (2000a). Reproductive malformation of the male offspring following maternal exposure to estrogenic chemicals. *Proc. Soc. Exp. Biol. Med., 224*, 61–68.

Gupta, C. (2000b). The role of estrogen receptor, androgen receptor and growth factors in diethylstilbestrol-induced programming of prostate differentiation. *Urol. Res., 28*, 223–229.

Halme, J., Hammond, M. G., Hulka, J. F., Raj, S. G., & Talbert, L. M. (1984). Retrograde menstruation in healthy women and in patients with endometriosis. *Obstet. Gynecol., 64*, 151–154.

Hanselaar, A., Van Loosbroek, M., Schuurbiers, O., Helmerhorst, T., Bulten, J., & Bernhelm, J. (1997). Clear cell adenocarcinoma of the vagina and cervix. An update of the central Netherlands registry showing twin age incidence peaks. *Cancer, 79*, 2229–2236.

Hardell, L., Van Bavel, B., Lindstrom, G., Carlberg, M., Dreifaldt, A. C., Wijkstrom, H., et al. (2003). Increased concentrations of polychlorinated biphenyls, hexachlorobenzene, and chlordanes in mothers of men with testicular cancer. *Environ. Health Perspect., 111*, 930–934.

Harris, C. A., Henttu, P., Parker, M. G., & Sumpter, J. P. (1997). The estrogenic activity of phthalate esters *in vitro*. *Environ. Health Perspect., 105*, 802–811.

Hatch, E. E., Troisi, R., Wise, L. A., Hyer, M., Palmer, J. R., Titus-Ernstoff, L., et al. (2006). Age at natural menopause in women exposed to diethylstilbestrol *in utero*. *Am. J. Epidemiol., 164*, 682–688.

Hayes, H. M., Tarone, R. E., Casey, H. W., & Huxsoll, D. L. (1990). Excess of seminomas observed in Vietnam service U.S. military working dogs. *J. Natl. Cancer Inst., 82*, 1042–1046.

Hayes, T., Haston, K., Tsui, M., Hoang, A., Haeffele, C., & Vonk, A. (2003). Atrazine-induced hermaphroditism at 0.1 ppb in American leopard frogs (*Rana pipiens*): laboratory and field evidence. *Environ. Health Perspect., 111*, 568–575.

Hayes, T. B., Collins, A., Lee, M., Mendoza, M., Noriega, N., Stuart, A. A., et al. (2002). Hermaphroditic, demasculinized frogs after exposure to the herbicide atrazine at low ecologically relevant doses. *Proc. Natl. Acad. Sci. USA, 99*, 5476–5480.

Henderson, B. E., Benton, B., Cosgrove, M., Baptista, J., Aldrich, J., Townsend, D., et al. (1976). Urogenital tract abnormalities in sons of women treated with diethylstilbestrol. *Pediatrics, 58*, 505–507.

Herbst, A. L. (2000). Behavior of estrogen-associated female genital tract cancer and its relation to neoplasia following intrauterine exposure to diethylstilbestrol (DES). *Gynecol. Oncol., 76*, 147–156.

Herbst, A. L., Ulfelder, H., & Poskanzer, D. C. (1971). Adenocarcinoma of the vagina. Association of maternal stilbestrol therapy with tumor appearance in young women. *N. Engl. J. Med., 284*, 878–881.

Herman-Giddens, M. E., Slora, E. J., Wasserman, R. C., Bourdony, C. J., Bhapkar, M. V., Koch, G. G., et al. (1997). Secondary sexual characteristics and menses in young girls seen in office practice: a study from the Pediatric Research in Office Settings network. *Pediatrics, 99*, 505–512.

Herman-Giddens, M. E., Wang, L., & Koch, G. (2001). Secondary sexual characteristics in boys: estimates from the national health and nutrition examination survey III, 1988-1994. *Arch. Pediatr. Adolesc. Med., 155*, 1022–1028.

Hernandez-Diaz, S. (2002). Iatrogenic legacy from diethylstilbestrol exposure. *Lancet., 359*, 1081–1082.

Heudorf, U., Mersch-Sundermann, V., & Angerer, J. (2007). Phthalates: toxicology and exposure. *Int. J. Hyg. Environ. Health, 210*, 623–634.

Hilakivi-Clarke, L., Cho, E., Onojafe, I., Raygada, M., & Clarke, R. (1999a). Maternal exposure to genistein during pregnancy increases carcinogen-induced mammary tumorigenesis in female rat offspring. *Oncol. Rep., 6*, 1089–1095.

Hilakivi-Clarke, L., Onojafe, I., Raygada, M., Cho, E., Skaar, T., Russo, I., et al. (1999b). Prepubertal exposure to zearalenone or genistein reduces mammary tumorigenesis. *Br. J. Cancer., 80*, 1682–1688.

Ho, S. M., Tang, W. Y., Belmonte de Frausto, J., & Prins, G. S. (2006). Developmental exposure to estradiol and bisphenol A increases susceptibility to prostate carcinogenesis and epigenetically regulates phosphodiesterase type 4 variant 4. *Cancer Res., 66*, 5624–5632.

Honma, S., Suzuki, A., Buchanan, D. L., Katsu, Y., Watanabe, H., & Iguchi, T. (2002). Low dose effect of *in utero* exposure to bisphenol A and diethylstilbestrol on female mouse reproduction. *Reprod. Toxicol., 16*, 117–122.

Horn-Ross, P. L., John, E. M., Lee, M., Stewart, S. L., Koo, J., Sakoda, L. C., et al. (2001). Phytoestrogen consumption and breast cancer risk in a multiethnic population: the Bay Area Breast Cancer Study. *Am. J. Epidemiol., 154*, 434–441.

Houston, D. E. (1984). Evidence for the risk of pelvic endometriosis by age, race and socioeconomic status. *Epidemiol. Rev., 6*, 167–191.

Howdeshell, K. L. (2002). A model of the development of the brain as a construct of the thyroid system. *Environ. Health Perspect., 110* (Suppl. 3), 337–348.

Howdeshell, K. L., Hotchkiss, A. K., Thayer, K. A., Vandenbergh, J. G., & vom Saal, F. S. (1999). Exposure to bisphenol A advances puberty. *Nature, 401*, 763–764.

Howdeshell, K. L., Peterman, P. H., Judy, B. M., Taylor, J. A., Orazio, C. E., Ruhlen, R. L., et al. (2003). Bisphenol A is released from used polycarbonate animal cages into water at room temperature. *Environ. Health Perspect., 111*, 1180–1187.

Howdeshell, K. L., Rider, C. V., Wilson, V. S., & Gray, L. E., Jr. (2008). Mechanisms of action of phthalate esters, individually and in combination, to induce abnormal reproductive development in male laboratory rats. *Environ. Res., 108*, 168–176.

Hoyer, A. P., Gerdes, A. M., Jorgensen, T., Rank, F., & Hartvig, H. B. (2002). Organochlorines, p53 mutations in relation to breast cancer risk and survival. A Danish cohort-nested case-controls study. *Breast Cancer Res. Treat., 71*, 59–65.

Hoyer, A. P., Grandjean, P., Jorgensen, T., Brock, J. W., & Hartvig, H. B. (1998). Organochlorine exposure and risk of breast cancer. *Lancet.*, *352*, 1816–1820.

Hoyer, A. P., Jorgensen, T., Brock, J. W., & Grandjean, P. (2000). Organochlorine exposure and breast cancer survival. *J. Clin. Epidemiol.*, *53*, 323–330.

Hsu, P. C., Huang, W., Yao, W. J., Wu, M. H., Guo, Y. L., & Lambert, G. H. (2003). Sperm changes in men exposed to polychlorinated biphenyls and dibenzofurans. *Jama.*, *289*, 2943–2944.

Hughes, I. A., & Acerini, C. L. (2008). Factors controlling testis descent. *Eur. J. Endocrinol.*, *159*(Suppl. 1), S75–S82.

Hunt, P. A., Koehler, K. E., Susiarjo, M., Hodges, C. A., Ilagan, A., Voigt, R. C., et al. (2003). Bisphenol a exposure causes meiotic aneuploidy in the female mouse. *Curr. Biol.*, *13*, 546–553.

Ikezuki, Y., Tsutsumi, O., Takai, Y., Kamei, Y., & Taketani, Y. (2002). Determination of bisphenol A concentrations in human biological fluids reveals significant early prenatal exposure. *Hum. Reprod.*, *17*, 2839–2841.

Ishido, M., Masuo, Y., Kunimoto, M., Oka, S., & Morita, M. (2004). Bisphenol A causes hyperactivity in the rat concomitantly with impairment of tyrosine hydroxylase immunoreactivity. *J. Neurosci. Res.*, *76*, 423–433.

Ishido, M., Morita, M., Oka, S., & Masuo, Y. (2005). Alteration of gene expression of G protein-coupled receptors in endocrine disruptors-caused hyperactive rats. *Regul. Pept.*, *126*, 145–153.

Jacobson, J. L., & Jacobson, S. W. (1996). Dose-response in perinatal exposure to polychlorinated biphenyls (PCBs): the Michigan and North Carolina cohort studies. *Toxicol. Ind. Health.*, *12*, 435–445.

Jacobson, J. L., Jacobson, S. W., & Humphrey, H. E. (1990). Effects of *in utero* exposure to polychlorinated biphenyls and related contaminants on cognitive functioning in young children. *J. Pediatr.*, *116*, 38–45.

Jacobson, J. L., Jacobson, S. W., Padgett, R., Brumitt, G., & Billings, R. (1992). Effects of prenatal PCB exposure on cognitive processing efficiency and sustained attention. *Developmental Psychology, 28*, 297–306.

Jacobson, S. W., Fein, G. G., Jacobson, J. L., Schwartz, P. M., & Dowler, J. K. (1985). The effect of intrauterine PCB exposure on visual recognition memory. *Child. Dev.*, *56*, 853–860.

Jakacka, M., Ito, M., Weiss, J., Chien, P. Y., Gehm, B. D., & Jameson, J. L. (2001). Estrogen receptor binding to DNA is not required for its activity through the nonclassical AP1 pathway. *J. Biol. Chem.*, *276*, 13615–13621.

Jansen, M. S., Nagel, S. C., Miranda, P. J., Lobenhofer, E. K., Afshari, C. A., & McDonnell, D. P. (2004). Short-chain fatty acids enhance nuclear receptor activity through mitogen-activated protein kinase activation and histone deacetylase inhibition. *Proc. Natl. Acad. Sci. USA, 101*, 7199–7204.

Jefferson, W., Newbold, R., Padilla-Banks, E., & Pepling, M. (2006). Neonatal genistein treatment alters ovarian differentiation in the mouse: inhibition of oocyte nest breakdown and increased oocyte survival. *Biol. Reprod.*, *74*, 161–168.

Jefferson, W. N., Couse, J. F., Padilla-Banks, E., Korach, K. S., & Newbold, R. R. (2002). Neonatal exposure to genistein induces estrogen receptor (ER)alpha expression and multioocyte follicles in the maturing mouse ovary: evidence for ERbeta-mediated and non-estrogenic actions. *Biol. Reprod.*, *67*, 1285–1296.

Jefferson, W. N., Padilla-Banks, E., Newbold, R. R. (2001). Alterations in reproductive tract tissues including estrogen receptor (ER) alpha, ER beta, and lactoferrin (LF) following developmental treatment with diethylstilbestrol or genistein. *The Endocrine Society's 83rd Annual Meeting* (pp. 311–312). Denver, CO: The Endocrine Society Press.

Jefferson, W. N., Padilla-Banks, E., & Newbold, R. R. (2005). Adverse effects on female development and reproduction in CD-1 mice following neonatal exposure to the phytoestrogen genistein at environmentally relevant doses. *Biol. Reprod.*, *73*, 798–806.

Jefferson, W. N., Padilla-Banks, E., & Newbold, R. R. (2007). Disruption of the developing female reproductive system by phytoestrogens: genistein as an example. *Mol. Nutr. Food Res.*, *51*, 832–844.

Jenkins, S., Rowell, C., Wang, J., & Lamartiniere, C. A. (2007). Prenatal TCDD exposure predisposes for mammary cancer in rats. *Reprod. Toxicol.*, *23*, 391–396.

Jobling, S., Reynolds, T., White, R., Parker, M. G., & Sumpter, J. P. (1995). A variety of environmentally persistent chemicals, including some phthalate plasticizers, are weakly estrogenic. *Environ. Health Perspect.*, *103*, 582–587.

Jones, P. A., & Takai, D. (2001). The role of DNA methylation in mammalian epigenetics. *Science, 293*, 1068–1070.

Ju, Y. H., Allred, C. D., Allred, K. F., Karko, K. L., Doerge, D. R., & Helferich, W. G. (2001). Physiological concentrations of dietary genistein dose-dependently stimulate growth of estrogen-dependent human breast cancer (MCF-7) tumors implanted in athymic nude mice. *J. Nutr.*, *131*, 2957–2962.

Kang, J. H., & Kondo, F. (2002). Bisphenol A migration from cans containing coffee and caffeine. *Food Addit. Contam.*, *19*, 886–890.

Karpati, A. M., Rubin, C. H., Kieszak, S. M., Marcus, M., & Troiano, R. P. (2002). Stature and pubertal stage assessment in American boys: the 1988-1994 Third National Health and Nutrition Examination Survey. *J. Adolesc. Health, 30*, 205–212.

Kaufman, J. M., & Vermeulen, A. (2005). The decline of androgen levels in elderly men and its clinical and therapeutic implications. *Endocr. Rev.*, *26*, 833–876.

Kaufman, R. H., Binder, G. L., Gray, P. M., Jr., & Adam, E. (1977). Upper genital tract changes associated with exposure *in utero* to diethylstilbestrol. *Am. J. Obstet. Gynecol.*, *128*, 51–59.

Kawagoshi, Y., Fujita, Y., Kishi, I., & Fukunaga, I. (2003). Estrogenic chemicals and estrogenic activity in leachate from municipal waste landfill determined by yeast two-hybrid assay. *J. Environ. Monit.*, *5*, 269–274.

Kelce, W. R., Stone, C. R., Laws, S. C., Gray, L. E., Kemppainen, J. A., & Wilson, E. M. (1995). Persistent DDT metabolite p,p'-DDE is a potent androgen receptor antagonist. *Nature, 375*, 581–585.

Khan, S. A., Ball, R. B., & Hendry, W. J., 3rd (1998). Effects of neonatal administration of diethylstilbestrol in male hamsters: disruption of reproductive function in adults after apparently normal pubertal development. *Biol. Reprod.*, *58*, 137–142.

Kim, K. S., Torres, C. R., Jr., Yucel, S., Raimondo, K., Cunha, G. R., & Baskin, L. S. (2004). Induction of hypospadias in a murine model by maternal exposure to synthetic estrogens. *Environ. Res.*, *94*, 267–275.

Kirschner, E. M. (1996). Growth of top 50 chemicals slowed in 1995 from very high 1994 rate. *Chem. Eng. News. April, 8*, 16–22.

Kitawaki, J., Inoue, S., Tamura, T., Yamamoto, T., Honjo, H., Higashiyama, T., et al. (1993). Cigarette smoking during pregnancy lowers aromatase cytochrome P-450 in the human placenta. *J. Steroid. Biochem. Mol. Biol.*, *45*, 485–491.

Kitawaki, J., Kado, N., Ishihara, H., Koshiba, H., Kitaoka, Y., & Honjo, H. (2002). Endometriosis: the pathophysiology as an estrogen-dependent disease. *J. Steroid Biochem. Mol. Biol., 83*, 149–155.

Klip, H., Verloop, J., Van Gool, J. D., Koster, M. E., Burger, C. W., & Van Leeuwen, F. E. (2002). Hypospadias in sons of women exposed to diethylstilbestrol *in utero*: a cohort study. *Lancet., 359*, 1102–1107.

Klotz, D. M., Ladlie, B. L., Vonier, P. M., McLachlan, J. A., & Arnold, S. F. (1997). o,p'-DDT and its metabolites inhibit progesterone-dependent responses in yeast and human cells. *Mol. Cell Endocrinol., 129*, 63–71.

Koehler, K. F., Helguero, L. A., Haldosen, L. A., Warner, M., & Gustafsson, J. A. (2005). Reflections on the discovery and significance of estrogen receptor beta. *Endocr. Rev., 26*, 465–478.

Kolon, T. F., Patel, R. P., & Huff, D. S. (2004). Cryptorchidism: diagnosis, treatment, and long-term prognosis. *Urol. Clin. North Am., 31*, 469–480, viii-ix.

Kortenkamp, A. (2007). Ten years of mixing cocktails: a review of combination effects of endocrine-disrupting chemicals. *Environ. Health Perspect., 115*(Suppl. 1), 98–105.

Kozie, K. D., & Anderson, R. K. (1991). Productivity, diet, and environmental contaminants in bald eagles nesting near the Wisconsin shoreline of Lake Superior. *Arch. Environ. Contam. Toxicol., 20*, 41–48.

Kubo, K., Arai, O., Omura, M., Watanabe, R., Ogata, R., & Aou, S. (2003). Low dose effects of bisphenol A on sexual differentiation of the brain and behavior in rats. *Neurosci. Res., 45*, 345–356.

Kurian, A. W., Clarke, C. A., & Carlson, R. W. (2009). The decline in breast cancer incidence: real or imaginary? *Curr. Oncol. Rep., 11*, 21–28.

Kuroda, N., Kinoshita, Y., Sun, Y., Wada, M., Kishikawa, N., Nakashima, K., et al. (2003). Measurement of bisphenol A levels in human blood serum and ascitic fluid by HPLC using a fluorescent labeling reagent. *J. Pharm. Biomed. Anal., 30*, 1743–1749.

Kushner, P. J., Agard, D. A., Greene, G. L., Scanlan, T. S., Shiau, A. K., Uht, R. M., et al. (2000). Estrogen receptor pathways to AP-1. *J. Steroid. Biochem. Mol. Biol., 74*, 311–317.

Lagiou, P., Adami, H. O., & Trichopoulos, D. (2006). Early life diet and the risk for adult breast cancer. *Nutr. Cancer, 56*, 158–161.

Lamartiniere, C. A., Moore, J. B., Brown, N. M., Thompson, R., Hardin, M. J., & Barnes, S. (1995). Genistein suppresses mammary cancer in rats. *Carcinogenesis, 16*, 2833–2840.

Laviola, G., Gioiosa, L., Adriani, W., & Palanza, P. (2005). D-amphetamine-related reinforcing effects are reduced in mice exposed prenatally to estrogenic endocrine disruptors. *Brain Res. Bull., 65*, 235–240.

Laws, S. C., Carey, S. A., Ferrell, J. M., Bodman, G. J., & Cooper, R. L. (2000). Estrogenic activity of octylphenol, nonylphenol, bisphenol A and methoxychlor in rats. *Toxicol. Sci., 54*, 154–167.

Lee, H. P., Gourley, L., Duffy, S. W., Esteve, J., Lee, J., & Day, N. E. (1991). Dietary effects on breast-cancer risk in Singapore. *Lancet., 337*, 1197–1200.

Levin, E. R. (1999). Cellular Functions of the Plasma Membrane Estrogen Receptor. *Trends Endocrinol. Metab., 10*, 374–377.

Lewis, B. C., Hudgins, S., Lewis, A., Schorr, K., Sommer, R., Peterson, R. E., et al. (2001). *In-utero* and lactational treatment with 2,3,7,8-tetrachlorodibenzo-p-dioxin impairs mammary gland differentiation but does not block the response to exogenous estrogen in the postpubertal female rat. *Toxicol. Sci., 62*, 46–53.

Li, S., Hansman, R., Newbold, R., Davis, B., McLachlan, J. A., & Barrett, J. C. (2003). Neonatal diethylstilbestrol exposure induces persistent elevation of c-fos expression and hypomethylation in its exon-4 in mouse uterus. *Mol. Carcinog., 38*, 78–84.

Li, S., Washburn, K. A., Moore, R., Uno, T., Teng, C., Newbold, R. R., et al. (1997). Developmental exposure to diethylstilbestrol elicits demethylation of estrogen-responsive lactoferrin gene in mouse uterus. *Cancer Res., 57*, 4356–4359.

Linseisen, J., Piller, R., Hermann, S., & Chang-Claude, J. (2004). Dietary phytoestrogen intake and premenopausal breast cancer risk in a German case-control study. *Int. J. Cancer, 110*, 284–290.

Llanes Gonzalez, L., Lujan Galan, M., Rodriguez Garcia, N., Garcia Tello, A., & Berenguer Sanchez, A. (2008). Trends in the incidence of testicular germ cell cancer in a 300.000 inhabitants Spanish population (1991–2005). *Actas. Urol. Esp., 32*, 691–695.

Low-Dose Peer Review Panel. (2001). FINAL Report of the Endocrine Disruptors Low-Dose Peer Review. *National Toxicology Program.*

Ma, F., Fleming, L. E., Lee, D. J., Trapido, E., & Gerace, T. A. (2006). Cancer incidence in Florida professional firefighters, 1981 to 1999. *J. Occup. Environ. Med., 48*, 883–888.

Main, K. M., Kiviranta, H., Virtanen, H. E., Sundqvist, E., Tuomisto, J. T., Tuomisto, J., et al. (2007). Flame retardants in placenta and breast milk and cryptorchidism in newborn boys. *Environ. Health Perspect., 115*, 1519–1526.

Main, K. M., Mortensen, G. K., Kaleva, M. M., Boisen, K. A., Damgaard, I. N., Chellakooty, M., et al. (2006). Human breast milk contamination with phthalates and alterations of endogenous reproductive hormones in infants three months of age. *Environ. Health Perspect., 114*, 270–276.

Mansfield, K. G., & Land, E. D. (2002). Cryptorchidism in Florida panthers: prevalence, features, and influence of genetic restoration. *J. Wildl. Dis., 38*, 693–698.

Maragou, N. C., Makri, A., Lampi, E. N., Thomaidis, N. S., & Koupparis, M. A. (2008). Migration of bisphenol A from polycarbonate baby bottles under real use conditions. *Food Addit. Contam., 25*, 373–383.

Maranghi, F., Tassinari, R., Moracci, G., Macri, C., & Mantovani, A. (2008). Effects of a low oral dose of diethylstilbestrol (DES) on reproductive tract development in F1 female CD-1 mice. *Reprod. Toxicol., 26*, 146–150.

Markey, C. M., Coombs, M. A., Sonnenschein, C., & Soto, A. M. (2003). Mammalian development in a changing environment: exposure to endocrine disruptors reveals the developmental plasticity of steroid-hormone target organs. *Evol. Dev., 5*, 67–75.

Markey, C. M., Luque, E. H., Munoz De Toro, M., Sonnenschein, C., & Soto, A. M. (2001). *In utero* exposure to bisphenol A alters the development and tissue organization of the mouse mammary gland. *Biol. Reprod., 65*, 1215–1223.

Markey, C. M., Wadia, P. R., Rubin, B. S., Sonnenschein, C., & Soto, A. M. (2005). Long-term effects of fetal exposure to low doses of the xenoestrogen bisphenol-A in the female mouse genital tract. *Biol. Reprod., 72*, 1344–1351.

Martin, V., Ribieras, S., Song-Wang, X. G., Lasne, Y., Frappart, L., Rio, M. C., et al. (1997). Involvement of DNA methylation in the control of the expression of an estrogen-induced breast-cancer-associated protein (pS2) in human breast cancers. *J. Cell Biochem., 65*, 95–106.

Masuyama, H., Hiramatsu, Y., Kunitomi, M., Kudo, T., & MacDonald, P. N. (2000). Endocrine disrupting chemicals, phthalic acid and nonylphenol, activate Pregnane X receptor-mediated transcription. *Mol. Endocrinol., 14*, 421–428.

McKenna, N. J., Lanz, R. B., & O'Malley, B. W. (1999). Nuclear receptor coregulators: cellular and molecular biology. *Endocr. Rev., 20*, 321–344.

McLachlan, J. A., Newbold, R. R., & Bullock, B. C. (1980). Long-term effects on the female mouse genital tract associated with prenatal exposure to diethylstilbestrol. *Cancer Res., 40*, 3988–3999.

McLachlan, J. A., Newbold, R. R., Shah, H. C., Hogan, M. D., & Dixon, R. L. (1982). Reduced fertility in female mice exposed transplacentally to diethylstilbestrol (DES). *Fertil. Steril., 38*, 364–371.

Medlock, K. L., Branham, W. S., & Sheehan, D. M. (1992). Long-term effects of postnatal exposure to diethylstilbestrol on uterine estrogen receptor and growth. *J. Steroid. Biochem. Mol. Biol., 42*, 23–28.

Melnick, S., Cole, P., Anderson, D., & Herbst, A. (1987). Rates and risks of diethylstilbestrol-related clear-cell adenocarcinoma of the vagina and cervix. An update. *N. Engl. J. Med., 316*, 514–516.

Meyer, T. E., Coker, A. L., Sanderson, M., & Symanski, E. (2007). A case-control study of farming and prostate cancer in African-American and Caucasian men. *Occup. Environ. Med., 64*, 155–160.

Mills, P. K. (1998). Correlation analysis of pesticide use data and cancer incidence rates in California counties. *Arch. Environ. Health, 53*, 410–413.

Missmer, S. A., Hankinson, S. E., Spiegelman, D., Barbieri, R. L., Michels, K. B., & Hunter, D. J. (2004). *In utero* exposures and the incidence of endometriosis. *Fertil. Steril., 82*, 1501–1508.

Miyagawa, K., Narita, M., Narita, M., Akama, H., & Suzuki, T. (2007). Memory impairment associated with a dysfunction of the hippocampal cholinergic system induced by prenatal and neonatal exposures to bisphenol-A. *Neurosci. Lett., 418*, 236–241.

Miyagawa, S., Buchanan, D. L., Sato, T., Ohta, Y., Nishina, Y., & Iguchi, T. (2002). Characterization of diethylstilbestrol-induced hypospadias in female mice. *Anat. Rec., 266*, 43–50.

Modugno, F., Weissfeld, J. L., Trump, D. L., Zmuda, J. M., Shea, P., Cauley, J. A., et al. (2001). Allelic variants of aromatase and the androgen and estrogen receptors: toward a multigenic model of prostate cancer risk. *Clin. Cancer Res., 7*, 3092–3096.

Moral, R., Wang, R., Russo, I. H., Lamartiniere, C. A., Pereira, J., & Russo, J. (2008). Effect of prenatal exposure to the endocrine disruptor bisphenol A on mammary gland morphology and gene expression signature. *J. Endocrinol., 196*, 101–112.

Munguia-Lopez, E. M., Peralta, E., Gonzalez-Leon, A., Vargas-Requena, C., & Soto-Valdez, H. (2002). Migration of bisphenol A (BPA) from epoxy can coatings to jalapeno peppers and an acid food simulant. *J. Agric. Food Chem., 50*, 7299–7302.

Munoz-de-Toro, M., Markey, C. M., Wadia, P. R., Luque, E. H., Rubin, B. S., Sonnenschein, C., et al. (2005). Perinatal exposure to bisphenol-A alters peripubertal mammary gland development in mice. *Endocrinology, 146*, 4138–4147.

Murray, T. J., Maffini, M. V., Ucci, A. A., Sonnenschein, C., & Soto, A. M. (2007). Induction of mammary gland ductal hyperplasias and carcinoma *in situ* following fetal bisphenol A exposure. *Reprod. Toxicol., 23*, 383–390.

Naciff, J. M., Jump, M. L., Torontali, S. M., Carr, G. J., Tiesman, J. P., Overmann, G. J., et al. (2002). Gene expression profile induced by 17alpha-ethynyl estradiol, bisphenol A, and genistein in the developing female reproductive system of the rat. *Toxicol. Sci., 68*, 184–199.

Nagel, S. C., Hagelbarger, J. L., & McDonnell, D. P. (2001). Development of an ER action indicator mouse for the study of estrogens, selective ER modulators (SERMs), and Xenobiotics. *Endocrinology, 142*, 4721–4728.

Nagel, S. C., vom Saal, F. S., Thayer, K. A., Dhar, M., Boechler, M., & Welshons, W. V. (1997a). Relative binding affinity-serum modified access (RBA-SMA) assay predicts the relative *in vivo* bioactivity of the xenoestrogens bisphenol A and octylphenol. *Environ. Health Perspect., 105*, 70–76.

Nagel, S. C., vom Saal, F. S., Thayer, K. A., Dhar, M. G., Boechler, M., & Welshons, W. V. (1997b). Relative binding affinity-serum modified access (RBA-SMA) assay predicts the relative *in vivo* bioactivity of the xenoestrogens bisphenol A and octylphenol. *Environ. Health Perspect., 105*, 70–76.

Nagel, S. C., vom Saal, F. S., & Welshons, W. V. (1998). The effective free fraction of estradiol and xenoestrogens in human serum measured by whole cell uptake assays: physiology of delivery modifies estrogenic activity. *Proc. Soc. Exp. Biol. Med., 217*, 300–309.

Nayyar, T., Bruner-Tran, K. L., Piestrzeniewicz-Ulanska, D., & Osteen, K. G. (2007). Developmental exposure of mice to TCDD elicits a similar uterine phenotype in adult animals as observed in women with endometriosis. *Reprod. Toxicol., 23*, 326–336.

Nelson, K. G., Sakai, Y., Eitzman, B., Steed, T., & McLachlan, J. (1994). Exposure to diethylstilbestrol during a critical developmental period of the mouse reproductive tract leads to persistent induction of two estrogen-regulated genes. *Cell Growth Differ., 5*, 595–606.

Newbold, R. (1995). Cellular and molecular effects of developmental exposure to diethylstilbestrol: implications for other environmental estrogens. *Environ. Health Perspect., 103*, 83–87.

Newbold, R. R., & McLachlan, J. A. (1982). Vaginal adenosis and adenocarcinoma in mice exposed prenatally or neonatally to diethylstilbestrol. *Cancer Res., 42*, 2003–2011.

Newbold, R. R., Banks, E. P., Bullock, B., & Jefferson, W. N. (2001). Uterine adenocarcinoma in mice treated neonatally with genistein. *Cancer Res., 61*, 4325–4328.

Newbold, R. R., Bullock, B. C., & McLachlan, J. A. (1990). Uterine adenocarcinoma in mice following developmental treatment with estrogens: a model for hormonal carcinogenesis. *Cancer Res., 50*, 7677–7681.

Newbold, R. R., Hanson, R. B., Jefferson, W. N., Bullock, B. C., Haseman, J., & McLachlan, J. A. (1998). Increased tumors but uncompromised fertility in the female descendants of mice exposed developmentally to diethylstilbestrol. *Carcinogenesis, 19*, 1655–1663.

Newbold, R. R., Jefferson, W. N., & Padilla-Banks, E. (2007). Long-term adverse effects of neonatal exposure to bisphenol A on the murine female reproductive tract. *Reprod. Toxicol., 24*, 253–258.

Newbold, R. R., Jefferson, W. N., Padilla-Banks, E., & Haseman, J. (2004). Developmental exposure to diethylstilbestrol (DES) alters uterine response to estrogens in prepubescent mice: low versus high dose effects. *Reprod. Toxicol., 18*, 399–406.

Newbold, R. R., Padilla-Banks, E., & Jefferson, W. N. (2006). Adverse effects of the model environmental estrogen diethylstilbestrol are transmitted to subsequent generations. *Endocrinology, 147*, S11–S17.

Nikaido, Y., Yoshizawa, K., Danbara, N., Tsujita-Kyutoku, M., Yuri, T., Uehara, N., et al. (2004). Effects of maternal xenoestrogen exposure on development of the reproductive tract and mammary gland in female CD-1 mouse offspring. *Reprod. Toxicol., 18*, 803–811.

North, K., & Golding, J. (2000). A maternal vegetarian diet in pregnancy is associated with hypospadias. The ALSPAC Study Team. Avon Longitudinal Study of Pregnancy and Childhood. *BJU Int., 85*, 107–113.

National Research Council (NRC). (1999). *Hormonally Active Agents in the Environment*. Washington, DC: National Academy Press.

National Toxicology Program (NTP). (2005). About the NTP. From http://ntp.niehs.nih.gov/?objectid=7201637B-BDB7-CEBA-FS7E39896A08F1BB, last accessed 01/19/2010.

Nuclear Receptor Signaling Atlas (NURSA). (2010). *Nuclear receptor signaling atlas*. www.nursa.org.

Ogura, Y., Ishii, K., Kanda, H., Kanai, M., Arima, K., Wang, Y., et al. (2007). Bisphenol A induces permanent squamous change in mouse prostatic epithelium. *Differentiation, 75*, 745–756.

Ohtake, F., Takeyama, K., Matsumoto, T., Kitagawa, H., Yamamoto, Y., Nohara, K., et al. (2003). Modulation of oestrogen receptor signalling by association with the activated dioxin receptor. *Nature, 423*, 545–550.

Ormond, G., Nieuwenhuijsen, M. J., Nelson, P., Toledano, M. B., Iszatt, N., Geneletti, S., et al. (2009). Endocrine disruptors in the workplace, hair spray, folate supplementation, and risk of hypospadias: case-control study. *Environ. Health Perspect., 117*, 303–307.

Orton, F., Carr, J. A., & Handy, R. D. (2006). Effects of nitrate and atrazine on larval development and sexual differentiation in the northern leopard frog *Rana pipiens*. *Environ. Toxicol. Chem., 25*, 65–71.

Osteen, K. G., Bruner-Tran, K. L., Keller, N. R., & Eisenberg, E. (2002). Progesterone-mediated endometrial maturation limits matrix metalloproteinase (MMP) expression in an inflammatory-like environment: a regulatory system altered in endometriosis. *Ann. NY. Acad. Sci., 955*, 37–47, discussion 86–88, 396–406.

Our Stolen Future 2010. From http://www.ourstolenfuture.org/Basics/chemlist.htm, last accessed 02/18/10.

Ozkan, S., Murk, W., & Arici, A. (2008). Endometriosis and infertility: epidemiology and evidence-based treatments. *Ann. NY. Acad. Sci., 1127*, 92–100.

Padilla-Banks, E., Jefferson, W. N., & Newbold, R. R. (2006). Neonatal exposure to the phytoestrogen genistein alters mammary gland growth and developmental programming of hormone receptor levels. *Endocrinology, 147*, 4871–4882.

Paige, L. A., Christensen, D. J., Gron, H., Norris, J. D., Gottlin, E. B., Padilla, K. M., et al. (1999). Estrogen receptor (ER) modulators each induce distinct conformational changes in ER alpha and ER beta. *Proc. Natl. Acad. Sci. USA, 96*, 3999–4004.

Palanza, P., Gioiosa, L., vom Saal, F. S., & Parmigiani, S. (2008). Effects of developmental exposure to bisphenol A on brain and behavior in mice. *Environ. Res., 108*, 150–157.

Palanza, P. L., Howdeshell, K. L., Parmigiani, S., & vom Saal, F. S. (2002). Exposure to a low dose of bisphenol A during fetal life or in adulthood alters maternal behavior in mice. *Environ. Health Perspect., 110*(Suppl. 3), 415–422.

Palmer, J. R., Wise, L. A., Hatch, E. E., Troisi, R., Titus-Ernstoff, L., Strohsnitter, W., et al. (2006). Prenatal diethylstilbestrol exposure and risk of breast cancer. *Cancer Epidemiol. Biomarkers Prev., 15*, 1509–1514.

Palmer, J. R., Wise, L. A., Robboy, S. J., Titus-Ernstoff, L., Noller, K. L., Herbst, A. L., et al. (2005). Hypospadias in sons of women exposed to diethylstilbestrol in utero. *Epidemiology, 16*, 583–586.

Papadopoulou, N., Papakonstanti, E. A., Kallergi, G., Alevizopoulos, K., & Stournaras, C. (2009). Membrane androgen receptor activation in prostate and breast tumor cells: molecular signaling and clinical impact. *IUBMB Life, 61*, 56–61.

Park, S. K., Kang, D., McGlynn, K. A., Garcia-Closas, M., Kim, Y., Yoo, K. Y., et al. (2008). Intrauterine environments and breast cancer risk: meta-analysis and systematic review. *Breast Cancer Res., 10*, R8.

Parmigiani, S., Palanza, P., & vom Saal, F. S. (1998). Ethotoxicology: an evolutionary approach to the study of environmental endocrine-disrupting chemicals. *Toxicol. Ind. Health, 14*, 333–339.

Patandin, S., Lanting, C. I., Mulder, P. G., Boersma, E. R., Sauer, P. J., & Weisglas-Kuperus, N. (1999). Effects of environmental exposure to polychlorinated biphenyls and dioxins on cognitive abilities in Dutch children at 42 months of age. *J. Pediatr., 134*, 33–41.

Paulozzi, L. J. (1999). International trends in rates of hypospadias and cryptorchidism. *Environ. Health Perspect., 107*, 297–302.

Paulozzi, L. J., Erickson, J. D., & Jackson, R. J. (1997). Hypospadias trends in two US surveillance systems. *Pediatrics, 100*, 831–834.

Payne, J., Scholze, M., & Kortenkamp, A. (2001). Mixtures of four organochlorines enhance human breast cancer cell proliferation. *Environ. Health Perspect., 109*, 391–397.

Peakall, D. B., & Fox, G. A. (1987). Toxicological investigations of pollutant-related effects in Great Lakes gulls. *Environ. Health Perspect., 71*, 187–193.

Peters, R. J., Beeltje, H., & Van Delft, R. J. (2008). Xeno-estrogenic compounds in precipitation. *J. Environ. Monit., 10*, 760–769.

Pierik, F. H., Burdorf, A., Deddens, J. A., Juttmann, R. E., & Weber, R. F. (2004). Maternal and paternal risk factors for cryptorchidism and hypospadias: a case-control study in newborn boys. *Environ. Health Perspect., 112*, 1570–1576.

Pons, J. C., Papiernik, E., Billon, A., Hessabi, M., & Duyme, M. (2005). Hypospadias in sons of women exposed to diethylstilbestrol *in utero*. *Prenat. Diagn., 25*, 418–419.

Prince, M. M., Ruder, A. M., Hein, M. J., Waters, M. A., Whelan, E. A., Nilsen, N., et al. (2006). Mortality and exposure response among 14,458 electrical capacitor manufacturing workers exposed to polychlorinated biphenyls (PCBs). *Environ. Health Perspect., 114*, 1508–1514.

Prins, G. S., Birch, L., Tang, W. Y., & Ho, S. M. (2007). Developmental estrogen exposures predispose to prostate carcinogenesis with aging. *Reprod. Toxicol., 23*, 374–382.

Pryor, J. L., Hughes, C., Foster, W., Hales, B. F., & Robaire, B. (2000). Critical windows of exposure for children's health: the reproductive system in animals and humans. *Environ. Health Perspect., 108* (Suppl. 3), 491–503.

Quesada, I., Fuentes, E., Viso-Leon, M. C., Soria, B., Ripoll, C., & Nadal, A. (2002). Low doses of the endocrine disruptor bisphenol-A and the native hormone 17beta-estradiol rapidly activate transcription factor CREB. *Faseb. J., 16*, 1671–1673.

Rasier, G., Parent, A. S., Gerard, A., Lebrethon, M. C., & Bourguignon, J. P. (2007). Early maturation of gonadotropin-releasing hormone secretion and sexual precocity after exposure of infant female rats to estradiol or dichlorodiphenyltrichloroethane. *Biol. Reprod., 77*, 734–742.

Ravdin, P. M., Cronin, K. A., Howlader, N., Berg, C. D., Chlebowski, R. T., Feuer, E. J., Edwards, B. K., & Berry, D. A. (2007). The decrease in breast-cancer incidence in 2003 in the United States. *N. Engl. J. Med., 356*, 1670–1674.

Reijnders, P. J. (1986). Reproductive failure in common seals feeding on fish from polluted coastal waters. *Nature, 324*, 456–457.

Reinsberg, J., Wegener-Toper, P., Van der Ven, K., Van der Ven, H., & Klingmueller, D. (2009). Effect of mono-(2-ethylhexyl) phthalate on steroid production of human granulosa cells. *Toxicol. Appl. Pharmacol., 239*, 116–123.

Rennell, C. L. (1979). T-shaped uterus in diethylstilbestrol (DES) exposure. *AJR. Am. J. Roentgenol., 132*, 979–980.

Rice, S., Mason, H. D., & Whitehead, S. A. (2006). Phytoestrogens and their low dose combinations inhibit mRNA expression and activity of aromatase in human granulosa-luteal cells. *J. Steroid Biochem. Mol. Biol., 101*, 216–225.

Rice, V. M. (2002). Conventional medical therapies for endometriosis. *Ann. NY Acad. Sci., 955*, 343–352, discussion 389–393, 396–406.

Richer, J. K., Jacobsen, B. M., Manning, N. G., Abel, M. G., Wolf, D. M., & Horwitz, K. B. (2002). Differential gene regulation by the two progesterone receptor isoforms in human breast cancer cells. *J. Biol. Chem., 277*, 5209–5218.

Rier, S., & Foster, W. G. (2002). Environmental dioxins and endometriosis. *Toxicol. Sci., 70*, 161–170.

Ritchie, J. M., Vial, S. L., Fuortes, L. J., Guo, H., Reedy, V. E., & Smith, E. M. (2003). Organochlorines and risk of prostate cancer. *J. Occup. Environ. Med., 45*, 692–702.

Rothschild, T. C., Boylan, E. S., Calhoon, R. E., & Vonderhaar, B. K. (1987). Transplacental effects of diethylstilbestrol on mammary development and tumorigenesis in female ACI rats. *Cancer Res., 47*, 4508–4516.

Ryan, B. C., & Vandenbergh, J. G. (2006). Developmental exposure to environmental estrogens alters anxiety and spatial memory in female mice. *Horm. Behav., 50*, 85–93.

Salian, S., Doshi, T., & Vanage, G. (2009). Perinatal exposure of rats to Bisphenol A affects the fertility of male offspring. *Life Sci., 85*, 742–752.

Sathyanarayana, S., Calafat, A. M., Liu, F., & Swan, S. H. (2008). Maternal and infant urinary phthalate metabolite concentrations: are they related? *Environ. Res., 108*, 413–418.

Satoh, K., Ohyama, K., Aoki, N., Iida, M., & Nagai, F. (2004). Study on anti-androgenic effects of bisphenol a diglycidyl ether (BADGE), bisphenol F diglycidyl ether (BFDGE) and their derivatives using cells stably transfected with human androgen receptor, AR-EcoScreen. *Food Chem. Toxicol., 42*, 983–993.

Schantz, S. L., Widholm, J. J., & Rice, D. C. (2003). Effects of PCB exposure on neuropsychological function in children. *Environ. Health Perspect., 111*, 357–576.

Schonfelder, G., Flick, B., Mayr, E., Talsness, C., Paul, M., & Chahoud, I. (2002a). *In utero* exposure to low doses of bisphenol A lead to long-term deleterious effects in the vagina. *Neoplasia., 4*, 98–102.

Schonfelder, G., Wittfoht, W., Hopp, H., Talsness, C. E., Paul, M., & Chahoud, I. (2002b). Parent bisphenol A accumulation in the human maternal–fetal–placental unit. *Environ. Health Perspect., 110*, A703–A707.

Schwartz, P. M., Jacobson, S. W., Fein, G., Jacobson, J. L., & Price, H. A. (1983). Lake Michigan fish consumption as a source of polychlorinated biphenyls in human cord serum, maternal serum, and milk. *Am. J. Public Health, 73*, 293–296.

Serman, A., Vlahovic, M., Serman, L., & Bulic-Jakus, F. (2006). DNA methylation as a regulatory mechanism for gene expression in mammals. *Coll. Antropol., 30*, 665–671.

Services, D.O.H.A.H. (1999). Food Labeling: Health Claims; Soy Protein and Coronary Heart Disease. *Final Rule, Vol. 64*, 57699.

Setchell, K. D., Zimmer-Nechemias, L., Cai, J., & Heubi, J. E. (1997). Exposure of infants to phyto-oestrogens from soy-based infant formula. *Lancet., 350*, 23–27.

Shakil, T., Hoque, A. N., Husain, M., & Belsham, D. D. (2002). Differential regulation of gonadotropin-releasing hormone secretion and gene expression by androgen: membrane versus nuclear receptor activation. *Mol. Endocrinol., 16*, 2592–2602.

Sharpe, R. M. (2001). Hormones and testis development and the possible adverse effects of environmental chemicals. *Toxicol. Lett., 120*, 221–232.

Sharpe, R. M., & Skakkebaek, N. E. (2008). Testicular dysgenesis syndrome: mechanistic insights and potential new downstream effects. *Fertil. Steril., 89*, e33–e38.

Sharpe, R. M., Fisher, J. S., Millar, M. M., Jobling, S., & Sumpter, J. P. (1995). Gestational and lactational exposure of rats to xenoestrogens results in reduced testicular size and sperm production. *Environ. Health Perspect., 103*, 1136–1143.

Sharpe-Timms, K. L. (2001). Endometrial anomalies in women with endometriosis. *Ann. NY Acad. Sci., 943*, 131–147.

Shetty, G., Krishnamurthy, H., Krishnamurthy, H. N., Bhatnagar, S., & Moudgal, N. R. (1997). Effect of estrogen deprivation on the reproductive physiology of male and female primates. *J. Steroid Biochem. Mol. Biol., 61*, 157–166.

Shetty, G., Krishnamurthy, H., Krishnamurthy, H. N., Bhatnagar, A. S., & Moudgal, N. R. (1998). Effect of long-term treatment with aromatase inhibitor on testicular function of adult male bonnet monkeys (*M. radiata*). *Steroids, 63*, 414–420.

Silva, E., Rajapakse, N., & Kortenkamp, A. (2002). Something from "nothing"—eight weak estrogenic chemicals combined at concentrations below NOECs produce significant mixture effects. *Environ. Sci. Technol., 36*, 1751–1756.

Simoens, S., Hummelshoj, L., & D'Hooghe, T. (2007). Endometriosis: cost estimates and methodological perspective. *Hum. Reprod. Update, 13*, 395–404.

Skakkebaek, N. E., Rajpert-De Meyts, E., & Main, K. M. (2001). Testicular dysgenesis syndrome: an increasingly common developmental disorder with environmental aspects. *Hum. Reprod., 16*, 972–978.

Skinner, M. K., Anway, M. D., Savenkova, M. I., Gore, A. C., & Crews, D. (2008). Transgenerational epigenetic programming of the brain transcriptome and anxiety behavior. *PLoS One, 3*, e3745.

Sohoni, P., & Sumpter, J. P. (1998). Several environmental oestrogens are also anti-androgens. *J. Endocrinol., 158*, 327–339.

Speirs, V., & Shaaban, A. M. (2009). The rising incidence of male breast cancer. *Breast Cancer Res. Treat., 115*, 429–430.

Stang, A., & Thomssen, C. (2008). Decline in breast cancer incidence in the United States: what about male breast cancer? *Breast Cancer Res. Treat., 112*, 595–596.

Steinmetz, R., Brown, N. G., Allen, D. L., Bigsby, R. M., & Ben-Jonathan, N. (1997). The environmental estrogen bisphenol A stimulates prolactin release *in vitro* and *in vivo*. *Endocrinology, 138*, 1780–1786.

Steinmetz, R., Mitchner, N. A., Grant, A., Allen, D. L., Bigsby, R. M., & Ben-Jonathan, N. (1998). The xenoestrogen bisphenol A induces growth, differentiation, and c-fos gene expression in the female reproductive tract. *Endocrinology, 139*, 2741–2747.

Stewart, P., Reihman, J., Lonky, E., Darvill, T., & Pagano, J. (2000). Prenatal PCB exposure and neonatal behavioral assessment scale (NBAS) performance. *Neurotoxicol. Teratol., 22*, 21–29.

Stewart, P. W., Reihman, J., Lonky, E. I., Darvill, T. J., & Pagano, J. (2003). Cognitive development in preschool children prenatally exposed to PCBs and MeHg. *Neurotoxicol. Teratol., 25*, 11–22.

Strohsnitter, W. C., Noller, K. L., Titus-Ernstoff, L., Troisi, R., Hatch, E. E., Poole, C., et al. (2005). Breast cancer incidence in women prenatally exposed to maternal cigarette smoke. *Epidemiology, 16*, 342–345.

Sun, Y., Wada, M., Al-Dirbashi, O., Kuroda, N., Nakazawa, H., & Nakashima, K. (2000). High-performance liquid chromatography with peroxyoxalate chemiluminescence detection of bisphenol A migrated from polycarbonate baby bottles using 4-(4,5-diphenyl-1H-imidazol-2-yl)benzoyl chloride as a label. *J. Chromatogr. B. Biomed. Sci. Appl., 749*, 49–56.

Swan, S. H. (2000). Intrauterine exposure to diethylstilbestrol: long-term effects in humans. *Apmis., 108*, 793–804.

Swan, S. H., Brazil, C., Drobnis, E. Z., Liu, F., Kruse, R. L., Hatch, M., et al. (2003a). Geographic differences in semen quality of fertile U.S. males. *Environ. Health Perspect., 111*, 414–420.

Swan, S. H., Elkin, E. P., & Fenster, L. (1997). Have sperm densities declined? A reanalysis of global trend data. *Environ. Health Perspect., 105*, 1228–1232.

Swan, S. H., Elkin, E. P., & Fenster, L. (2000). The question of declining sperm density revisited: an analysis of 101 studies published 1934–1996. *Environ. Health Perspect., 108*, 961–966.

Swan, S. H., Kruse, R. L., Liu, F., Barr, D. B., Drobnis, E. Z., Redmon, J. B., et al. (2003b). Semen quality in relation to biomarkers of pesticide exposure. *Environ. Health Perspect., 111*, 1478–1484.

Swan, S. H., Main, K. M., Liu, F., Stewart, S. L., Kruse, R. L., Calafat, A. M., et al. (2005). Decrease in anogenital distance among male infants with prenatal phthalate exposure. *Environ. Health Perspect., 113*, 1056–1061.

Taguchi, O., Cunha, G. R., & Robboy, S. J. (1986). Expression of nuclear estrogen-binding sites within developing human fetal vagina and urogenital sinus. *Amer. J. Anat., 177*, 473–480.

Takai, Y., Tsutsumi, O., Ikezuki, Y., Kamei, Y., Osuga, Y., Yano, T., et al. (2001). Preimplantation exposure to bisphenol A advances postnatal development. *Reprod. Toxicol., 15*, 71–74.

Tapiero, H., Ba, G. N., & Tew, K. D. (2002). Estrogens and environmental estrogens. *Biomed. Pharmacother., 56*, 36–44.

Tarone, R. E., Hayes, H. M., Hoover, R. N., Rosenthal, J. F., Brown, L. M., Pottern, L. M., et al. (1991). Service in Vietnam and risk of testicular cancer. *J. Natl. Cancer Inst., 83*, 1497–1499.

Thayer, K. A. (1999). Prenatal exposure to low doses of estrogen : reproductive effects in male and female mice and implications for regulation of endocrine disrupting environmental chemicals. pp. viii.

Thayer, K. A., Ruhlen, R. L., Howdeshell, K. L., Buchanan, D. L., Cooke, P. S., Preziosi, D., et al. (2001). Altered prostate growth and daily sperm production in male mice exposed prenatally to subclinical doses of 17alpha-ethinyl oestradiol. *Hum. Reprod., 16*, 988–996.

Thigpen, J. E., Setchell, K. D., Padilla-Banks, E., Haseman, J. K., Saunders, H. E., Caviness, G. F., et al. (2007). Variations in phytoestrogen content between different mill dates of the same diet produces significant differences in the time of vaginal opening in CD-1 mice and F344 rats but not in CD Sprague-Dawley rats. *Environ. Health Perspect., 115*, 1717–1726.

Thomas, P., & Dong, J. (2006). Binding and activation of the seven-transmembrane estrogen receptor GPR30 by environmental estrogens: a potential novel mechanism of endocrine disruption. *J. Steroid. Biochem. Mol. Biol., 102*, 175–179.

Timms, B. G., Howdeshell, K. L., Barton, L., Bradley, S., Richter, C. A., & vom Saal, F. S. (2005). Estrogenic chemicals in plastic and oral contraceptives disrupt development of the fetal mouse prostate and urethra. *Proc. Natl. Acad. Sci. USA, 102*, 7014–7019.

Tuohy, P. G. (2003). Soy infant formula and phytoestrogens. *J. Paediatr. Child Health, 39*, 401–405.

US Food and Drug Administration, Update on Bisphenol A for Use in Food Contact Applications: January 2010. U.S. Food and Drug Administration, 2010. From http://www.fda.gov/NewsEvents/Public HealthFocus/ucm197739.htm, last accessed 01/29/10.

Van Maele-Fabry, G., Libotte, V., Willems, J., & Lison, D. (2006). Review and meta-analysis of risk estimates for prostate cancer in pesticide manufacturing workers. *Cancer Causes Control, 17*, 353–373.

Vandenberg, L. N., Hauser, R., Marcus, M., Olea, N., & Welshons, W. V. (2007a). Human exposure to bisphenol A (BPA). *Reprod. Toxicol., 24*, 139–177.

Vandenberg, L. N., Maffini, M. V., Wadia, P. R., Sonnenschein, C., Rubin, B. S., & Soto, A. M. (2007b). Exposure to environmentally relevant doses of the xenoestrogen bisphenol-A alters development of the fetal mouse mammary gland. *Endocrinology, 148*, 116–127.

Varayoud, J., Ramos, J. G., Bosquiazzo, V. L., Munoz-de-Toro, M., & Luque, E. H. (2008). Developmental exposure to bisphenol A impairs the uterine response to ovarian steroids in the adult. *Endocrinology, 149*, 5848–5860.

Vilela, M. L., Willingham, E., Buckley, J., Liu, B. C., Agras, K., Shiroyanagi, Y., et al. (2007). Endocrine disruptors and hypospadias: role of genistein and the fungicide vinclozolin. *Urology, 70*, 618–621.

Virtanen, H. E., & Toppari, J. (2008). Epidemiology and pathogenesis of cryptorchidism. *Hum Reprod. Update, 14*, 49–58.

vom Saal, F. S. (1981). Variation in phenotype due to random intrauterine positioning of male and female fetuses in rodents. *J. Reprod. Fertil., 62*, 633–650.

vom Saal, F. S. (1989). Sexual differentiation in litter-bearing mammals: influence of sex of adjacent fetuses in utero. *J. Anim. Sci., 67*, 1824–1840.

vom Saal, F. S., Cooke, P. S., Buchanan, D. L., Palanza, P., Thayer, K. A., Nagel, S. C., et al. (1998). A physiologically based approach to the study of bisphenol A and other estrogenic chemicals on the size of reproductive organs, daily sperm production, and behavior. *Toxicol. Ind. Health, 14*, 239–260.

vom Saal, F. S., Pryor, S., & Bronson, F. H. (1981). Effects of prior intrauterine position and housing on oestrous cycle length in adolescent mice. *J. Reprod. Fertil., 62*, 33–37.

vom Saal, F. S., Timms, B. G., Montano, M. M., Palanza, P., Thayer, K. A., Nagel, S. C., et al. (1997). Prostate enlargement in mice due to fetal exposure to low doses of estradiol or diethylstilbestrol and opposite effects at high doses. *Proc. Natl. Acad. Sci. USA, 94*, 2056–2061.

Vorherr, H., Messer, R. H., Vorherr, U. F., Jordan, S. W., & Kornfeld, M. (1979). Teratogenesis and carcinogenesis in rat offspring after transplacental and transmammary exposure to diethylstilbestrol. *Biochem. Pharmacol., 28*, 1865–1877.

Voss, C., Zerban, H., Bannasch, P., & Berger, M. R. (2005). Lifelong exposure to di-(2-ethylhexyl)-phthalate induces tumors in liver and testes of Sprague-Dawley rats. *Toxicology, 206*, 359–371.

Vreugdenhil, H. J., Slijper, F. M., Mulder, P. G., & Weisglas-Kuperus, N. (2002). Effects of perinatal exposure to PCBs and dioxins on play behavior in Dutch children at school age. *Environ. Health Perspect., 110*, A593–A598.

Wald, N., Cuckle, H., Wu, T. S., & George, L. (1991). Maternal serum unconjugated oestriol and human chorionic gonadotrophin levels in twin pregnancies: implications for screening for Down's syndrome. *Br. J. Obstet. Gynaecol., 98*, 905–908.

Walkowiak, J., Wiener, J. A., Fastabend, A., Heinzow, B., Kramer, U., Schmidt, E., et al. (2001). Environmental exposure to polychlorinated biphenyls and quality of the home environment: effects on psychodevelopment in early childhood. *Lancet, 358*, 1602–1607.

Warner, M., Eskenazi, B., Mocarelli, P., Gerthoux, P. M., Samuels, S., Needham, L., et al. (2002). Serum dioxin concentrations and breast cancer risk in the Seveso Women's Health Study. *Environ. Health Perspect., 110*, 625–628.

Watson, C. J., & Khaled, W. T. (2008). Mammary development in the embryo and adult: a journey of morphogenesis and commitment. *Development, 135*, 995–1003.

Watson, C. S., Bulayeva, N. N., Wozniak, A. L., & Finnerty, C. C. (2005). Signaling from the membrane via membrane estrogen receptor-alpha: estrogens, xenoestrogens, and phytoestrogens. *Steroids, 70*, 364–371.

Wei, L. L., & Miner, R. (1994). Evidence for the existence of a third progesterone receptor protein in human breast cancer cell line T47D. *Cancer Res., 54*, 340–343.

Weidner, I. S., Moller, H., Jensen, T. K., & Skakkebaek, N. E. (1998). Cryptorchidism and hypospadias in sons of gardeners and farmers. *Environ. Health Perspect., 106*, 793–796.

Welsh, M., Saunders, P. T., Fisken, M., Scott, H. M., Hutchison, G. R., Smith, L. B., et al. (2008). Identification in rats of a programming window for reproductive tract masculinization, disruption of which leads to hypospadias and cryptorchidism. *J. Clin. Invest., 118*, 1479–1490.

Welshons, W. V., Nagel, S. C., & vom Saal, F. S. (2006). Large effects from small exposures. III. Endocrine mechanisms mediating effects of bisphenol A at levels of human exposure. *Endocrinology, 147*, S56–S69.

Wetherill, Y. B., Fisher, N. L., Staubach, A., Danielsen, M., De Vere White, R. W., & Knudsen, K. E. (2005). Xenoestrogen action in prostate cancer: pleiotropic effects dependent on androgen receptor status. *Cancer Res, 65*, 54–65.

Wetherill, Y. B., Hess-Wilson, J. K., Comstock, C. E., Shah, S. A., Buncher, C. R., Sallans, L., et al. (2006). Bisphenol A facilitates bypass of androgen ablation therapy in prostate cancer. *Mol. Cancer Ther., 5*, 3181–3190.

Wetherill, Y. B., Petre, C. E., Monk, K. R., Puga, A., & Knudsen, K. E. (2002). The xenoestrogen bisphenol A induces inappropriate androgen receptor activation and mitogenesis in prostatic adenocarcinoma cells. *Mol. Cancer Ther., 1*, 515–524.

Wiig, O., Derocher, A. E., Cronin, M. M., & Skaare, J. U. (1998). Female pseudohermaphrodite polar bears at Svalbard. *J. Wildl. Dis., 34*, 792–796.

Wilson, V. S., Blystone, C. R., Hotchkiss, A. K., Rider, C. V., & Gray, L. E., Jr. (2008). Diverse mechanisms of anti-androgen action: impact on male rat reproductive tract development. *Int. J. Androl., 31*, 178–187.

Winneke, G., Bucholski, A., Heinzow, B., Kramer, U., Schmidt, E., Walkowiak, J., et al. (1998). Developmental neurotoxicity of polychlorinated biphenyls (PCBS): cognitive and psychomotor functions in 7-month old children. *Toxicol. Lett., 102–103*, 423–428.

Wormke, M., Stoner, M., Saville, B., Walker, K., Abdelrahim, M., Burghardt, R., et al. (2003). The aryl hydrocarbon receptor mediates degradation of estrogen receptor alpha through activation of proteasomes. *Mol. Cell Biol., 23*, 1843–1855.

Wozniak, A. L., Bulayeva, N. N., & Watson, C. S. (2005). Xenoestrogens at picomolar to nanomolar concentrations trigger membrane estrogen receptor-alpha-mediated Ca2+ fluxes and prolactin release in GH3/B6 pituitary tumor cells. *Environ. Health Perspect., 113*, 431–439.

Wright, H. M., Clish, C. B., Mikami, T., Hauser, S., Yanagi, K., Hiramatsu, R., et al. (2000). A synthetic antagonist for the peroxisome proliferator-activated receptor gamma inhibits adipocyte differentiation. *J. Biol. Chem., 275*, 1873–1877.

Wu, A. H., Wan, P., Hankin, J., Tseng, C. C., Yu, M. C., & Pike, M. C. (2002). Adolescent and adult soy intake and risk of breast cancer in Asian-Americans. *Carcinogenesis, 23*, 1491–1496.

Wyde, M. E., Kirwan, S. E., Zhang, F., Laughter, A., Hoffman, H. B., Bartolucci-Page, E., et al. (2005). Di-n-butyl phthalate activates constitutive androstane receptor and pregnane X receptor and enhances the expression of steroid-metabolizing enzymes in the liver of rat fetuses. *Toxicol. Sci., 86*, 281–290.

Xu, L. C., Sun, H., Chen, J. F., Bian, Q., Qian, J., Song, L., et al. (2005). Evaluation of androgen receptor transcriptional activities of bisphenol A, octylphenol and nonylphenol *in vitro*. *Toxicology, 216*, 197–203.

Xue, F., & Michels, K. B. (2007). Lutrauterine factors and risk of breast cancer: a systematic review and meta-analysis of current evidence. *Lancet. Oncol., 8*, 1088–1100.

Young, W. C., Goy, R. W., & Phoenix, C. H. (1964). Hormones and Sexual Behavior. *Science, 143*, 212–218.

Younglai, E. V., Foster, W. G., Hughes, E. G., Trim, K., & Jarrell, J. F. (2002). Levels of environmental contaminants in human follicular fluid, serum, and seminal plasma of couples undergoing *in-vitro* fertilization. *Arch. Environ. Contam. Toxicol., 43*, 121–126.

Yu, G., Guo, Q., Xie, L., Liu, Y., & Wang, X. (2009). Effects of subchronic exposure to carbendazim on spermatogenesis and fertility in male rats. *Toxicol. Ind. Health, 25*, 41–47.

Yuan, J. M., Wang, Q. S., Ross, R. K., Henderson, B. E., & Yu, M. C. (1995). Diet and breast cancer in Shanghai and Tianjin, China. *Br. J. Cancer, 71*, 1353–1358.

Zarn, J. A., Bruschweiler, B. J., & Schlatter, J. R. (2003). Azole fungicides affect mammalian steroidogenesis by inhibiting sterol 14 alpha-demethylase and aromatase. *Environ. Health Perspect., 111*, 255–261.

Zava, D. T., Dollbaum, C. M., & Blen, M. (1998). Estrogen and progestin bioactivity of foods, herbs, and spices. *Proc. Soc. Exp. Biol. Med., 217*, 369–378.

Zhang, Y., Wise, J. P., Holford, T. R., Xie, H., Boyle, P., Zahm, S. H., et al. (2004). Serum polychlorinated biphenyls, cytochrome P-450 1A1 polymorphisms, and risk of breast cancer in Connecticut women. *Am. J. Epidemiol., 160*, 1177–1183.

Zsarnovszky, A., Le, H. H., Wang, H. S., & Belcher, S. M. (2005). Ontogeny of rapid estrogen-mediated extracellular signal-regulated kinase signaling in the rat cerebellar cortex: potent nongenomic agonist and endocrine disrupting activity of the xenoestrogen bisphenol A. *Endocrinology, 146*, 5388–5396.

Species Index

Aconmys russatus, 222
Acrobates pygmaeus, 199
Alouatta caraya, 294
Anour goeffroyia, 269, 270, 271
Antechinus, 199, 204, 206
Antechinus flavipes, 206
Antechinus stuartii, 199, 200, 205, 206
Antechinus swainsonii, 206
Antrozous pallidus, 262, 263, 264, 270, 271, 273
Aotus, 311
Aotus nancymaee, 292
Aotus trivirgatus, 294
Aplodontia rufa, 230
Artibeus, 244
Artibeus jamaicencis, 244, 249, 263, 271, 274
Ateles, 311
Ateles geoffroyi, 294

Balantiopteryx plicata, 255
Bathyergus suillus, 217, 226
Bettongia penicillata, 197
Brachyteles arachnoides, 294, 312
Burramys parvus, 199

Calicebus moloch, 294
Callimico goeldii, 294
Callithrix, 291
Callithrix jacchus, 180, 292, 294
Callithrix pygmaea, 294
Caluromys philander, 197
Canis lupus familiaris, 90, 183
Capra aegagrus hircus, 178
Capreolus capreolus, 263
Carollia, 244
Carollia perspicillata, 249, 261, 265, 274
Cavia aperea, 229
Cavia aperea fiporcellus, 220
Cavia porcellus, 2, 144, 183, 229, 309
Cebuella pymaea, 312
Cebus apella, 294
Cercocebus, 304
Cercocebus torquatus, 296
Cercopithecus neglectus, 296
Chaerephon (Tadarida) hindei, 255
Chalinobolous morio, 264
Chalinolobous gouldi, 246, 263, 264
Cheiragaleus medius, 303, 305
Chiroderma, 244
Chlorocebus aethiops, 296
Clethrionomys glareolus, 217
Coleura afra, 244
Colobus guereza, 296
Corynorhinus rafinesquei, 65, 248, 270
Cryptomys damarensis, 223, 224, 225, 229
Cryptomys darlingi, 225
Cryptomys hottenotus natalensis, 229
Cryptomys hottentotus hottentotus, 225, 226, 229
Cryptomys hottentotus pretoriae, 226, 229
Cynopterus minutus, 255
Cynopterus sphinx, 244, 248, 249, 255, 256, 259, 260, 262, 263, 264, 270, 271, 272, 273

Dasycercus cristicauda, 203
Dasyuroides byrneii, 201
Dasyurus, 204
Dasyurus geoffroii, 198
Dasyurus hallucatus, 202, 207
Daubentonia madagascariensis, 294, 311
Demodillus auricularis, 222
Desmodus rotundus, 244, 255, 261
Dicrostonyx groenlandicus, 217
Didelphis, 204
Didelphis albiventris, 202
Didelphis virginiana, 199
Dobsonia proedatrix, 271

Eiodolon helvum, 247, 271
Elaphas maximus, 69
Eonycterus spelaea, 255
Epomops bultikoferi, 269
Epomops franqueti, 272, 273
Eptesicus, 242
Eptesicus fuscus, 250, 254, 267, 270
Eptesicus regulus, 264
Eptesicus (Vespadelus) vulturnus, 254
Equus ferus caballus, 87
Erythrocebus patasipatus, 296
Eulemur fulvus, 294

Felis catus, 65, 89
Fucomys anselli, 225, 229
Funambulus pennanti, 217

Galea musteloides, 229
Gelago senegalensis, 315
Geomyidae, 230
Georychus capensis, 229
Gerbillaris paeba, 222
Glossophaga soricina, 261, 264
Gorilla gorilla, 291, 296
Gracilinanus microtarsus, 207

Haplonycteris fischeri, 248, 249, 255, 260
Heterocephalus glaber, 223, 224, 229
Hipposideros, 242, 257
Hipposideros caffer, 255
Hipposideros fulvus, 264
Hipposideros fulvus fulvus, 243
Homo sapiens, 96, 156, 157, 158
Hydrochorus hydrochaeris, 227, 229
Hylobates lar, 296
Hylobates syndactylus, 296

Lagothrix lagotricha, 294, 311
Lama glama, 65
Lasiorhinus krefftii, 202
Lasiorhinus latifrons, 200
Lasiurus ego, 255
Leantopithecus rosalia, 294
Lemur catta, 294
Leptonycteris curasoae, 261
Loxoodonta africana, 69

Macaca, 291
Macaca arctoides, 296, 312
Macaca fascicularis, 296, 300
Macaca fuscata, 296
Macaca mulatta, 5, 293, 296
Macaca nemestrina, 296
Macropus rufogriseus, 203
Macrotis lagotis, 197
Macrotus californicus, 251, 258, 259, 262, 264, 273
Macrotus waterhousii, 273
Mandrillus sphinx, 296, 305
Marmosa incana, 207
Marmota flaviventris, 227
Marmota marmota, 227, 229
Marmota monax, 217
Megaderma lyra, 243, 255, 260
Megaderma spasma, 243
Mephitus mephitus, 263
Meriones shawi, 222
Meriones unguiculatus, 217, 227
Mesocricetus auratus, 5, 180, 217, 229
Mesocricetus unguiculatus, 229
Microcebus, 291
Microcebus murinus, 294
Micropteropus pusillus, 269
Microtus, 228
Microtus agrestis, 179, 216
Microtus californicus, 65, 229, 248, 248, 249, 263
Microtus canicaudus, 180
Microtus montanus, 220, 229
Microtus ochrogaster, 151, 176, 177, 226, 229
Microtus pennsylvanicus, 176, 217, 229
Microtus pinetorum, 177, 227, 229
Miniopterus, 242, 247, 248, 260
Miniopterus australis, 243, 255, 270
Miniopterus schreibersii, 243, 248, 249, 250, 251, 252, 253, 254, 256, 257, 258, 262, 263, 264, 265, 271, 273
Miniopterus schreibersii fulginosus, 246, 249, 263, 264

Miniopterus schreibersii natalensis, 248, 264
Miopithecus talapoin, 296
Molossus, 259
Molossus ater, 244, 261
Molossus fortis, 244
Monodelphis dimidiata, 207
Monodelphis domestica, 178
Mormopterus planiceps, 258, 264
Mus, 228
Mus musculus, 176, 215, 217
Mustela putorius furo, 65
Mustela vison, 65, 271
Myotis, 242
Myotis austroriparius, 245
Myotis daubentoni, 246
Myotis grisescens, 265
Myotis lucifugus, 244, 245, 246, 250, 251, 253, 254, 256, 257, 258, 262, 263, 264, 266, 267, 270, 272, 273, 274
Myotis myotis, 248, 249, 273
Myotis natereri, 246
Myotis nigricans, 244, 255
Myotis ricketti, 258, 261, 262
Myotis septentrionalis, 274
Myotis thysanocles, 270
Myotis velifer, 246
Mystacina tuberculata, 261

Nasalis larvatus, 296
Noctilio albiventris, 260, 261
Notomys alexis, 222
Nyctalus noctula, 245, 254, 258
Nycteris lutea, 264
Nycteris thebaica, 255
Nyctophilus geoffroyi, 258

Onychromys leucogaster, 220
Oryctolagus cuniculus, 87, 88, 89
Oryzomys palustris, 215
Otolemur crassicaudatus, 294, 312
Otolemur garnettii, 303
Otomops martiensseni, 261
Otopteropus cartilanonodus, 249, 261
Ovis aries, 177

Pan paniscus, 296, 304
Pan troglodytes, 293, 298
Papio, 291
Papio anubis, 296
Papio cyanocephalus, 314
Papio hamadryas, 296, 305
Parantechinus apicalis, 207
Perodicticus potto, 294
Peromyscus californicus, 229
Peromyscus eremicus, 226
Peromyscus leucopus, 217, 229
Peromyscus maniculatus, 144, 215, 217, 229
Peromyscus truei, 222
Petauroides volans, 204
Phascogale, 204
Phascogale calara, 200
Phascolarctos cinereus, 197
Phodopus campbelli, 218, 220, 228, 229
Phodopus sungorus, 176, 218, 229
Phyllostomus, 244
Piliocolobus badius, 304
Pipistrellus abramus, 244, 246, 253
Pipistrellus mimus, 244, 246, 255
Pipistrellus pipistrellus, 242, 246, 248, 254, 256, 258, 264, 260, 270
Pipistrellus rusticus, 243
Pithecia pithecia, 294
Plecotus auritus, 242, 273
Pongo pygmaeus, 296, 305
Procyon lotor, 65
Propithecus verreaux, 294
Pseudocheirus peregrinus, 204
Ptenochirus jagori, 249, 261
Pteropteryx kappleri, 260
Pteropus (Chalinolobous) gouldii, 264
Pteropus, 261
Pteropus geddiei, 270
Pteropus giganteus, 243, 256, 259, 260
Pteropus hypomelanus, 273, 274, 275
Pteropus poliocephalus, 253, 256, 257, 258, 264, 270, 271
Pteropus pumulus, 273, 274
Pteropus rodricensis, 270
Pteropus scapulatus, 257, 258
Pteropus vampyrus, 256, 273, 274, 275
Pygathrix nemaeus, 296

Rattus, 228
Rattus norvegicus, 5, 293
Rhabdomys pumilio, 222
Rhinolophus, 242
Rhinolophus capensis, 254, 256, 258
Rhinolophus curale, 261
Rhinolophus ferrumequinum, 242, 245, 246, 250, 251, 252, 254, 256, 262, 264, 270
Rhinolophus hipposideros, 242, 246, 270, 273
Rhinolopus cornutus, 256
Rhinolopus rouxi, 247
Rhinopoma hardwickei hardwickei, 255
Rhinopoma kinneari, 255, 256
Rousettus, 241
Rousettus aegyptiacus, 244, 255, 274, 275
Rousettus leschenaulti, 244, 256, 260, 264

Saguinus, 311
Sagulnus ordplus, 294
Saimiri, 291, 298
Saimiri boliviensis, 292
Saimiri sciureus, 151, 294
Sarcophilus harrisii, 200
Sciurus carolinensis, 230
Scotophilus heathi, 244, 245, 246, 247, 251, 252, 253, 254, 255, 256, 257, 258, 261, 262, 263, 266, 267, 268, 271, 272, 273, 274, 275, 276
Scotophilus wroughtonii, 243, 255, 268
Semnopithecus entellus, 296
Setonix brachyurus, 199
Sigmodon hispidus, 220
Sminthopsis crassicaudata, 201, 202, 204, 206
Sminthopsis macroura, 201, 203, 204
Sorex araneus, 65
Spermophilus richardsoni, 272
Spermophilus tridecemlineatus, 229, 272
Spilogale putorius latiforns, 270
Sus domestica, 84, 85, 86, 87
Sylvilagus, 65

Tadarida brasiliensis, 260
Tadarida brasiliensis mexicana, 264
Tadarida condylurua, 244
Tadarida mexicana, 273
Tadarida pumila, 244, 255
Tamiascirus hudsonicus, 230
Taphozous georgianus, 243, 254, 255, 258, 264
Taphozous hilli, 255
Taphozous longimanus, 244, 257, 258, 259, 260, 261, 263, 264, 267, 270, 272, 273
Taphozous melanopogon, 251, 255
Taphozous nudiventris, 255
Theropithecus gelada, 296, 304
Triaenops afer, 264
Trichosurus vulpecula, 200
Tylonycteris pachypus, 255
Tylonycteris robustula, 255

Varecia variegata, 294, 303
Vespdelus vulturnus, 258
Vespertilio murinus, 242
Vesperugo piccolo, 273
Vesperugo savi, 273
Vicugna pacos, 65
Vombatus ursinus, 202

Xerus inauris, 229

Zygodontomys brevicauda, 222

Subject Index

ABP, *see* Androgen-binding protein
Acetylcholine
  sexual behavior modulation, 149
ACTH, *see* Corticotropin
Aging, *see* Reproductive aging
γ-Aminobutyric acid (GABA)
  brain sexual differentiation role, 11
  gonadarche regulation in primates, 306
  neurosteroid modulation of receptors, 125
Amygdala
  pheromone responses, 176–177, 182–184
  sex differences, 7, 11
Androgen-binding protein (ABP), spermatogenesis role, 51
Androgen receptor (AR)
  endocrine-disrupting chemical binding, 337–339
  knockout mice, 145
  signaling, 4
  spermatogenesis role, 49–51
Androstenedione
  bat reproduction regulation, 257–258, 274–275
  ovarian synthesis, 62
  uterotonic mechanism, 108
Anteroventral periventricular nucleus (AVPV), sex differences, 7–8, 10, 14
AR, *see* Androgen receptor
Arcuate nucleus, sex differences, 7–8, 14
Arginine vasopressin (AVP), behavioral function, 9
Aromatase, brain sexual differentiation role, 5–6
AVP, *see* Arginine vasopressin
AVPV, *see* Anteroventral periventricular nucleus

Bats
  corticosterone effects on reproduction, 273–274
  environmental effects on reproduction
    photoperiod, 270
    rainfall, 269
    season, 268–271
  female reproductive cycle patterns
    temperate-zone bats, 242–243
    tropical bats
      aseasonal polyestry, 244
      bimodal polyestry, 244
      seasonal monoestry, 243–244
      seasonal polyestry, 244
  female reproductive tract
    asymmetry
      alternating dominance, 259
      contralateral dominance, 260
      dextral dominance, 259–260
      sinistral dominance, 260
      symmetric dominance, 260
      unilateral endometrial involvement, 260
    corpus luteum, 264–265
    menstruation, 261
    ovary
      folliculogenesis patterns, 261–262
      histology, 261
      leptin effects, 267–268
      steroidogenesis, 262–264
    ovulation induction, 265–266
  follicle-stimulating hormone function, 251–252, 257
  gonadotropin-releasing hormone seasonal dynamics, 250–251
  hibernation fat accumulation and reproduction, 266–267
  hypothalamus-pituitary-gonadal axis, 241–242
  insulin function, 275
  luteinizing hormone function, 251–253, 257
  male reproduction
    hormonal regulation overview, 253–254, 257–258
    patterns
      temperate-zone bats, 254
      tropical/subtropical bats, 254–255
    spermatogenesis regulation, 256–257
    tissue morphology changes, 256
  melatonin function, 270–272
  prolactin function, 253
  prospects for study, 275–276
  reproductive delays
    development, 248–249
    fertilization, 245–247
    implantation, 247–248
    ovulation, 244–245
  thyroid hormone function, 272–273
Bed nucleus of the stria terminalis (BNST), sex differences, 7–8
Behavior, *see* Female sexual behavior; Male sexual behavior
Bisphenol A (BPA), endocrine disruption, 331, 341, 343–344, 347
BNST, *see* Bed nucleus of the stria terminalis
BPA, *see* Bisphenol A
Brain
  functional imaging of human sexual behavior, 157
  sexual differentiation
    cell birth and death, 10–11
    cell-cell communications, 14–16
    glutamate receptors, 11–13
    history of study, 2–3
    organizational/activational hypothesis
      aromatase role, 5–6
      challenges, 8–10
      masculinization, feminization, and defeminization, 6
      sensitive periods, 5
      sexually dimorphic brain regions, 6–8
      steroid hormone action, 3–4
    overview, 1
    prospects for study, 16–17
Breast cancer, endocrine-disrupting chemical study endpoints
  humans, 352–353
  mammary cancer in animals, 351–352

Calmodulin kinase-2 (CAMK2), gonadotropin-releasing hormone pulse effects, 37
CAMK2, *see* Calmodulin kinase-2
CAR, *see* Constitutive active receptor
Cat
  estrous cycle, 89
  pregnancy, 89
Cattle, *see* Ruminants
CBG, *see* Corticosteroid-binding globulin
Cervix, cyclical changes in primates, 303
CG, *see* Chorionic gonadotropin
Chorionic gonadotropin (CG), pregnancy function, 74–75, 77–78, 80–81
Chorionic somatomammotropin (CS), pregnancy function, 77, 80–81, 309
CL, *see* Corpus luteum
Clomiphene citrate, male infertility management, 52
Constitutive active receptor (CAR), endocrine-disrupting chemical binding, 340
Corpus luteum (CL)
  bat, 264–265
  formation and regression, 73–74
  luteinizing hormone/chorionic gonadotropin receptor, 74–75
  maintenance in pregnancy, 74, 77–78
  primate formation, function, and regression, 301
  relaxin function, 76–77
Corticosteroid-binding globulin (CBG), marsupial testicular function, 206
Corticosterone, *see also* Stress
  administration studies of reproduction, 126–127

Corticosterone (*Continued*)
bat reproduction effects, 273–274
food restriction and stress response, 119
Corticotropin (ACTH), *see also* Stress
parturition regulation in marsupials, 203
Corticotropin-releasing hormone (CRH), *see also* Stress
discovery, 25
fetal signaling of labor, 102–103
food restriction and stress response, 121, 216
gestation length determination, 99–100
infection effects, 123
luteinizing hormone effects, 127–128
primate pregnancy function, 309
CRH, *see* Corticotropin-releasing hormone
CS, *see* Chorionic somatomammotropin

DDT, *see* Dichlorodiphenyltrichloroethane
Dehydroepiandrosterone sulfate (DHEAS)
fetal signaling of labor, 102–103
uterotonic mechanism, 107–108
DES, *see* Diethylstilbestrol
DHEAS, *see* Dehydroepiandrosterone sulfate
DHT, *see* Dihydrotestosterone
Dichlorodiphenyltrichloroethane (DDT), endocrine disruption, 348–349
Diethylstilbestrol (DES)
developmental effects, 342–343
fertility studies, 347, 349
hypospadias studies, 357–358
Dihydrotestosterone (DHT), male sexual behavior modulation, 141, 144
2,5-Dimethylpyrazine (DMP), female response, 178
DMP, *see* 2,5-Dimethylpyrazine
DNA methylation, endocrine-disrupting chemical exposure effects, 342–343
Dog
estrous cycle, 90
pregnancy, 90
Dopamine, sexual behavior modulation, 150

EDCs, *see* Endocrine-disrupting chemicals
Elephant, ovarian cycle, 69
Embryonic diapause, marsupials, 203
Endocrine-disrupting chemicals (EDCs), *see* specific chemicals
developmental effects of exposure, 341–342
developmental exposure and adult gene expression, 342–343
endpoints for exposure studies
breast cancer
humans, 352–353
mammary cancer in animals, 351–352
endometriosis, 350
female sexual behavior, 345–346
fertility, 346–349
human behavior, 346
hypospadias, 356–358
lesions of female reproductive tract, 349–350
prostate cancer, 358–359
puberty, 344–345
testes
cryptorchidism, 354–356
testicular cancer, 353–354
testicular dysgenesis syndrome, 353
mechanisms of action
androgen receptor binding, 337–339
combined mechanisms, 341
constitutive active receptor binding, 340
estrogen receptor binding, 339
glucocorticoid receptor binding, 339
methoxyacetic acid, 341
peroxisome proliferator-activated receptor binding, 339
pregnane X receptor binding, 339
progesterone receptor binding, 337, 339
steroid hormones
metabolism alterations, 340
synthesis alterations, 340
thyroid hormone receptor binding, 340
prospects for study, 359
regulation, 329–331
types, 331–337
xenoestrogen level comparison to *in utero* endogenous estrogens, 343–344
Endometriosis, endocrine-disrupting chemical study endpoint, 350
β-Endorphin, sexual behavior modulation, 149
ER, *see* Estrogen receptor
ERT, *see* Estrogen replacement therapy
ESR1, ESR2, *see* Estrogen receptor
Estrogen
fetal signaling of labor, 101–103
ovarian synthesis, 62
uterotonic mechanism, 104–105, 107
xenoestrogens, *see* Endocrine-disrupting chemicals
Estrogen receptor (ER)
endocrine-disrupting chemical binding, 339
insulin-like growth factor-1 crosstalk, 146
isoforms, 3–4
knockout mice, 145
orchidectomy effects on expression, 144
signaling, 4
Estrogen replacement therapy (ERT), benefits and risks, 66
Estrous cycle, *see also* Ovarian cycle
cat, 89
dog, 90
horse, 87
prospects for study, 91
rabbit, 87–88
rodents, 81, 227–229
ruminants, 82–83
swine, 84–86
Ethinylestradiol, endocrine disruption, 331, 342

FAK, *see* Focal adhesion kinase
Female germline-specific transcription factor, oogenesis role, 60
Female sexual arousal disorder (FSAD), management, 157–158
Female sexual behavior (FSB)
endocrine-disrupting chemical study endpoint, 345–346
knockout mouse studies, 154
lordosis neural circuitry, 142–144
neurotransmitter/neuropeptide modulation
acetylcholine, 149
dopamine, 150
β-endorphin, 149
gonadotropin-releasing hormone, 149
α-melanocyte-stimulating hormone, 152
norepinephrine, 148–149
overview, 147–148
oxytocin, 151–152
serotonin, 150–151
overview, 139–140
pheromone responses
attraction and investigation, 181–182
scent marking, 182
primates
description, 311–312
hormonal regulation, 312–313
sex steroid regulation, 141
steroid receptors
ligand-independent activation, 146–147
overview of function, 145
women, 157–158
FGF, *see* Fibroblast growth factor
Fibroblast growth factor (FGF)
gonadotropin-releasing hormone neuron cell fate specification and migration role, 28–30
stem spermatogonia regulation by FGF-2, 48
fMRI, *see* Functional magnetic resonance imaging
Focal adhesion kinase (FAK), brain sexual differentiation role, 14
Follicle-stimulating hormone (FSH)
bat reproduction function, 251–252, 257
folliculogenesis role, 61–62
history of study, 26
inhibin induction, 63
marsupial ovarian cycle control, 201–202
ovarian cycle regulation
follicular phase, 63–64
ovulation, 64
primate reproduction function
females, 299–301, 304
males, 292–293
Folliculogenesis, *see* Ovary
Food availability
restriction stress effects on reproduction, 119–121
rodent reproduction
developmental effects, 217
hibernation, 217
hormonal control, 216–217
overview, 215–216
FSAD, *see* Female sexual arousal disorder
FSB, *see* Female sexual behavior
FSH, *see* Follicle-stimulating hormone

Functional magnetic resonance imaging (fMRI), functional imaging of human sexual behavior, 157

GABA, *see* γ-Aminobutyric acid
GDNF, *see* Glial-derived neurotrophic factor
Genistein, endocrine disruption, 331
GFR, *see* Glomerular filtration rate
Glial-derived neurotrophic factor (GDNF), stem spermatogonia regulation, 47–48
Glomerular filtration rate (GFR), marsupials, 206
Glucocorticoid receptor, endocrine-disrupting chemical binding, 339
GnRH, *see* Gonadotropin-releasing hormone
Goat, *see* Ruminants
Gonadarche, *see* Puberty
Gonadotropin-releasing hormone (GnRH)
  bat seasonal dynamics, 250–251
  discovery, 26
  food deprivation response in rodents, 216
  gonadarche regulation in primates, 305–306
  infection effects, 124
  isoforms, 27
  marsupial ovarian cycle control, 201–202
  neurons
    anatomy, 30–31
    axonal targeting, 30
    cell fate specification, 28–29
    migration, 29–30
  pheromone responses, 177
  primate reproduction function
    females, 299–301, 304
    males, 292–293
  prospects for study, 37–38
  pulsatile release, 33–34, 36–37
  receptors
    regulation in gonadotropes, 35–36
    signaling, 32–33, 36
    structure, 34–35
    types, 34–35
  sequence conservation between species, 31–33
  sexual behavior modulation, 149
GPCR54, gonadotropin-releasing hormone pulsatile release role, 34

Hibernation
  bat fat accumulation and reproduction, 266–267
  rodents and reproduction, 217
Horse
  estrous cycle, 87
  pregnancy, 87
Hypospadia, endocrine-disrupting chemical study endpoint, 356–358

IFNT, *see* Interferon-τ
IGF-1, *see* Insulin-like growth factor-1
Insulin, bat reproduction function, 275
Insulin-like growth factor-1 (IGF-1)
  primate folliculogenesis regulation, 317
  stem spermatogonia regulation, 48
Insulin-like growth factor-1, receptor crosstalk with estrogen receptor, 146
Interferon-τ (IFNT), pregnancy maintenance in ruminants, 83–84

Kisspeptin
  food restriction and stress response, 121
  gonadarche regulation in primates, 306
  gonadotropin-releasing hormone pulsatile release role, 34
  rodent function in photoperiod regulation, 219

Labor, *see* Parturition
Lactation, primates
  energetics, 318
  regulation, 310
Leptin
  bat reproduction function, 267–268
  primate pregnancy function, 309–310
LH, *see* Luteinizing hormone
LHCGR, *see* Luteinizing hormone/chorionic gonadotropin receptor
Lordosis, *see* Female sexual behavior
Luteinizing hormone (LH)
  bat reproduction function, 251–253, 257
  food restriction and stress response, 120–121
  history of study, 26
  infection effects, 122–123
  marsupial ovarian cycle control, 201–202
  ovarian cycle regulation
    follicular phase, 64
    luteal phase, 65–66
    ovulation, 64–65
  pregnancy maintenance role, 74–75
  stress response, 126–128
Luteinizing hormone/chorionic gonadotropin receptor (LHCGR), pregnancy maintenance role, 74–75

MAA, *see* Methoxyacetic acid
Male sexual behavior (MSB)
  castration studies, 144, 152–153
  chemosensory inputs, 142
  knockout mouse studies, 153–156
  men, 156–157
  neurotransmitter/neuropeptide modulation
    dopamine, 150
    overview, 147–148
    α-melanocyte-stimulating hormone, 152
    oxytocin, 151–152
    serotonin, 150–151
  overview, 139–140
  pheromone responses
    attraction and investigation, 182–183
    copulatory behavior, 184
    vocalizations, 183–184
  primates
    description, 311–312
    hormonal regulation, 312–313
  sex steroid regulation, 141–142
  steroid receptors
    ligand-independent activation, 146–147
    overview of function, 144–145
Mammary cancer, *see* Breast cancer
Marsupials
  embryonic diapause, 203
  gestational development arrest, 203
  ovarian cycle patterns
    control, 201–202
    long gestation, delayed luteal phase, 199–201
    overview, 197–198
    short gestation, long luteal phase, 199
    short gestation, prolonged luteal phase, 199
  overview, 196
  parturition regulation, 202–203
  sexual differentiation, 196–197
  testicular function regulation
    gonadotropins, 208
    seasonal reproduction cycle followed by complete spermatogenic collapse, 204–207
    seasonal reproduction cycle with continuous spermatogenetic activity, 207–208
    seasonal reproduction cycle with testicular involution, 207
6-MBOA, 6-Methoxy-2-benzoxazolinone
α-Melanocyte-stimulating hormone (α-MSH), sexual behavior modulation, 152
MCG, *see* Midbrain central gray
Melatonin
  bat reproduction function, 270–272
  female puberty effects, 63
  rodents
    hibernation changes, 217
    photoperiodic regulation, 218
Menstrual cycle
  hormonal regulation, 77
  humans, 300
  primates, 301–303
Menstruation
  bat, 261
  primates, 301–303
Mesotocin (MST)
  parturition regulation in marsupials, 202
  testicular function in marsupials, 207
Methoxyacetic acid (MAA), endocrine disruption, 341
6-Methoxy-2-benzoxazolinone (6-MBOA), induction of reproduction readiness in rodents, 230
Midbrain central gray (MCG), lordosis neural circuitry, 143–144
Mole-rat, *see* Rodents
Monotremes
  ovarian cycle
    control, 201
    pattern, 197
  overview, 195–196
  parturition regulation, 202
  sexual differentiation, 196
  testicular function regulation, 204, 208

MSB, *see* Male sexual behavior
α-MSH, *see* α-Melanocyte-stimulating hormone
MST, *see* Mesotocin

Nasal embryonic LHRH factor (NELF), gonadotropin-releasing hormone neuron expression, 29–30
NELF, *see* Nasal embryonic LHRH factor
Nitric oxide (NO), gonadotropin-releasing hormone pulsatile release role, 33
NKCC, brain sexual differentiation role, 11
NO, *see* Nitric oxide
Norepinephrine, sexual behavior modulation, 148–149

Olfactory placode (OP), gonadotropin-releasing hormone neuron cell fate specification, 28–29
Oogenesis, *see* Ovary
OP, *see* Olfactory placode
Organizational/activational hypothesis, *see* Sexual differentiation
Ovarian cycle, *see also* Estrous cycle
  dog, 90
  elephant, 69
  follicular phase, 63–64
  luteal phase, 65–66
  marsupial patterns
    control, 201–202
    long gestation, delayed luteal phase, 199–201
    overview, 197–198
    short gestation, long luteal phase, 199
    short gestation, prolonged luteal phase, 199
  monotremes
    control, 201
    pattern, 197
  ovulation, 64–65
  pheromone effects
    facilitation, 178–179
    inhibition, 178
    synchronization, 179
  primates, 304
  rabbit, 87–88
  rat, 66–67
  sheep, 67
  women, 67–69
Ovary
  bat
    asymmetry
      alternating dominance, 259
      contralateral dominance, 260
      dextral dominance, 259–260
      sinistral dominance, 260
      symmetric dominance, 260
      unilateral endometrial involvement, 260
    folliculogenesis patterns, 261–262
    histology, 261
    ovulation
      delay, 244–245
      induction, 265–266
    steroidogenesis, 262–264
  cervical cyclical changes, 303
  development
    folliculogenesis, 60–63
    oogenesis, 60
    overview, 59–60
    puberty, 63
  lifespan and reproductive aging, 66
  primate function
    corpus luteum formation, function, and regression, 301
    folliculogenesis, 299, 301
    ovarian cycle regulation, 304
    overview, 298–299
    oviduct cyclical changes, 303
    ovulation, 301
  prospects for study, 69
  steroidogenesis, 62–63
Oxytocin
  corpus luteum formation and regression role, 73–74
  parturition regulation in marsupials, 202
  primate lactation regulation, 310
  sexual behavior modulation, 151–152, 158
  swine function, 85
  uterotonic mechanism, 109–110

Parturition
  fetal maturation synchronization, 100–101
  fetal signaling of labor
    corpus luteum-dependent species, 101–102
    corticotropin-releasing hormone, 102–103
    dehydroepiandrosterone sulfate, 102–103
    lung signals, 104
  gestation length determinants, 97–100
  marsupial regulation, 202–203
  monotreme regulation, 202
  strategies, 95–97
  uterotonic mechanism activation
    myometrium electrical conductivity, 110–111
    oxytocin, 109–110
    prostaglandins and cytokines, 108–109
    sex steroids, 104–108
PCBs, *see* Polychlorinated biphenyls
Peroxisome proliferator-activated receptors (PPARs), endocrine-disrupting chemical binding, 339
Pesticides, endocrine disruption, 337, 346–347
PET, *see* Positron emission tomography
Pheromones
  chemosensory systems, 176
  definition, 175–176
  female response
    behavior responses
      attraction and investigation, 181–182
      scent marking, 182
    ovarian cycle
      facilitation, 178–179
      inhibition, 178
      synchronization, 179
    pregnancy blockage, 179–180
    puberty
      acceleration, 176–177
      delay, 177–178
  male response
    behavior responses
      attraction and investigation, 182–183
      copulatory behavior, 184
      vocalizations, 183–184
    hormone release, 180–181
    reproductive development, 180
    sperm allocation, 180
Phosphodiesterase inhibitors, female sexual arousal disorder management, 157–158
Photoperiod, *see* Estrous cycle; Melatonin; Rodents
Phthalates, endocrine disruption, 337, 340, 347
Pig
  estrous cycle, 84–86
  pregnancy, 85–87
Pituitary
  anatomy, 27
  development, 26–27
POA, *see* Preoptic area
POF, *see* Premature ovarian failure
Polychlorinated biphenyls (PCBs), endocrine disruption, 337, 347–348
Positron emission tomography (PET), functional brain imaging of human male sexual behavior, 157
PPARs, *see* Peroxisome proliferator-activated receptors
Pregnancy
  cat, 89
  chorionic gonadotropin function, 74–75, 77–78, 80–81
  chorionic somatomammotropin hormone function, 77, 80–81
  corpus luteum
    formation and regression, 73–74
    maintenance in pregnancy, 74
  dog, 90
  gestation length determinants, 97–100
  horse, 87
  luteinizing hormone/chorionic gonadotropin receptor, 74–75
  parturition, *see* Parturition
  pheromone blockage, 179–180
  primates
    chorionic gonadotropin, 308–309
    chorionic somatomammotropin, 309
    corticotropin-releasing hormone, 309
    energetics, 317–318
    leptin, 309–310
    placentation, 307–308
    steroids, 308
  progesterone and receptors, 75–76
  prolactin function, 80–81

rabbit, 88–89
relaxin function, 76–77
rodents, 81–82
ruminants, 83–84
swine, 85–87
Pregnane X receptor (PXR), endocrine-disrupting chemical binding, 339
Premature ovarian failure (POF), diethylstilbestrol induction, 349
Preoptic area (POA)
lordosis neural circuitry, 144
pheromone responses, 176, 182
sex differences, 7–8, 10, 12–13, 15
Primates
cervical cyclical changes, 303
energetics of reproduction
gonadal function, 317
hypothalamic-pituitary function, 317
lactation, 318
overview, 316–317
pregnancy, 317–318
external cyclical changes in females, 303–304
lactation regulation, 310
ovarian function
corpus luteum formation, function, and regression, 301
folliculogenesis, 299, 301
ovarian cycle regulation, 304
overview, 298–299
oviduct cyclical changes, 303
ovulation, 301
pregnancy
chorionic gonadotropin, 308–309
chorionic somatomammotropin, 309
corticotropin-releasing hormone, 309
leptin, 309–310
placentation, 307–308
steroids, 308
prospects for study, 319
puberty
adrenarche, 304–305
gonadarche, 305–306
timing, 306–307
reproductive aging, 310–311
reproductive and life-history parameters for representative species, 294–298
seasonality of reproduction, 313–314
sexual behavior
description, 311–312
hormonal regulation, 312–313
social influences on reproduction
females
adult males, 316
maturation, 315–316
males
adult males, 315
maturation, 314–315
taxonomy, 291
testicular function
gonadotropins and androgen secretion, 292–293
spermatogenesis, 293, 298, 300
uterine cycle, 301–303
vaginal cyclical changes, 303
PRL, *see* Prolactin
Progesterone
bat circulating levels, 264
female sexual behavior modulation, 141
gestation length determination, 98
luteal phase role, 65–66
pregnancy maintenance, 75–78
primate pregnancy function, 308
stress effects, 124–12
uterotonic mechanisms, 104–108
Progesterone receptor
endocrine-disrupting chemical binding, 337, 339
isoforms, 1065
Prolactin (PRL)
bat reproduction function, 253
luteal phase role, 66
pregnancy function, 80–81
primate lactation regulation, 310
rodents, 82
Prostaglandins
brain sexual differentiation role, 12–13
parturition regulation in marsupials, 202–203
swine pregnancy function, 86
uterotonic mechanism, 108–109
Prostate cancer, endocrine-disrupting chemical study endpoint, 358–359
Prototherians, *see* Monotremes
Puberty
endocrine-disrupting chemical study endpoint, 344–345
melatonin effects in females, 63
ovary development, 63
pheromone effects in females
acceleration, 176–177
delay, 177–178
primates
adrenarche, 304–305
gonadarche, 305–306
timing, 306–307
rodent timing, 220
PXR, *see* Pregnane X receptor

Rabbit
estrous cycle, 87–88
pregnancy, 88–89
Rat, *see* Rodents
Relaxin
parturition regulation in marsupials, 202
pregnancy function, 76–77
Reproductive aging
ovary, 66
primates, 66
RFamide-related peptide (RFRP), rodent function in photoperiod regulation, 219
RFRP, *see* RFamide-related peptide
Rodents
estrous cycle, 66–67, 81, 227–229
food availability and reproduction
developmental effects, 217
hibernation, 217
hormonal control, 216–217
overview, 215–216
6-methoxy-2-benzoxazolinone induction of reproduction readiness, 230
parasite load and reproduction, 230
personality and individual differences in reproduction, 230
photoperiod and reproduction
intraspecies variation in responses, 220–221
kisspeptin function, 219
neural mechanisms, 218–219
puberty timing, 220
RFamide-related peptide, 219
seasonal breeding, 217–218
sex differences in response, 221
spontaneous recrudescence, 221
thyroid hormone, 219–220
unmasking techniques in nonphotoresponsive species, 221–222
population cycles in small rodents, 230
postpartum and lactational estrus, 228–230
pregnancy, 81–82
prospects and challenges for study, 230–231
reproductive suppression by social cues
Alpine marmot, 227
cactus mouse, 226
mole-rats
Ansell's mole-rat, 225
Cape dune mole-rat, 226
Cape mole-rat, 226
common mole-rat, 225–226
Damaraland mole-rat, 224–225
highveld mole-rat, 226
Mashona mole-rat, 225
naked mole-rat, 224
Mongolian gerbil, 227
pine vole, 227
prairie vole, 226–227
spontaneous versus induced ovulators, 227–229
temperature and reproduction, 223
water availability and reproduction, 222
Ruminants
estrous cycle, 82–83
ovarian cycle, 67
pregnancy, 83–84

Serotonin, sexual behavior modulation, 150–151
Sertoli cell, differentiated spermatogenic cell regulation, 48
Sex chromosome aneuploidy, male sexual behavior studies, 156
Sexual behavior, *see* Female sexual behavior; Male sexual behavior
Sexual differentiation
brain
cell birth and death, 10–11
cell-cell communications, 14–16

Sexual differentiation (*Continued*)
glutamate receptors, 11–13
history of study, 2–3
organizational/activational hypothesis
aromatase role, 5–6
challenges, 8–10
masculinization, feminization, and defeminization, 6
sensitive periods, 5
sexually dimorphic brain regions, 6–8
steroid hormone action, 3–4
overview, 1
prospects for study, 16–17
marsupials, 196–197
monotremes, 196
Sheep, *see* Ruminants
SNB, *see* Spinal nucleus of the bulbocavernosus
SP-A, *see* Surfactant protein-A
Sperm allocation, pheromone effects, 180
Spermatogenesis, *see also* Testes
androgen receptor expression, 49–51
bat regulation, 256–257
contraception and hormonal targeting, 52
infertility management, 52
overview, 45–46
primates, 293, 298, 300
prospects for study, 52–53
stem spermatogonia
characteristics, 46
glial-derived neurotrophic factor regulation, 47–48
replication and differentiation, 46–47
Sertoli cell function regulation, 48
testosterone role, 45–46, 49, 51–52
Spinal nucleus of the bulbocavernosus (SNB), sex differences, 7
SRY, sex determination gene, 1
Stress
challenges in reproduction studies, 117–119
food restriction effects on reproduction, 119–121
hormone modulation studies, 126–128
immune response effects on reproduction, 121–124
physical/psychological/social stress effects
females, 124–125
males, 125–126
prospects for study, 129–130
wildlife reproduction implications, 128–129
Surfactant protein-A (SP-A), fetal signaling of labor, 104
Swine, *see* Pig

Temperature, rodent reproduction effects, 223
Testes
bat
spermatogenesis regulation, 256–257
tissue morphology changes, 256
endocrine-disrupting chemical study endpoints
cryptorchidism, 354–356
testicular cancer, 353–354
testicular dysgenesis syndrome, 353
marsupial function regulation
gonadotropins, 208
seasonal reproduction cycle followed by complete spermatogenetic collapse, 204–207
seasonal reproduction cycle with continuous spermatogenetic activity, 207–208
seasonal reproduction cycle with testicular involution, 207
monotreme function regulation, 204, 208
primate function
gonadotropins and androgen secretion, 292–293
spermatogenesis, 293, 298, 300
Testis-determining factor, *see SRY*
Testosterone
bat reproduction regulation, 253–254, 257–258
human sexual behavior mediation
men, 156
women, 157–158
male hormonal contraception, 52
male sexual behavior modulation, 141–142
spermatogenesis role, 45–46, 49, 51–52
Thyroid hormone
bat reproduction function, 272–273
receptor binding by endocrine-disrupting chemicals, 340
rodent function in photoperiod regulation, 219–220

Vagina, cyclical changes in primates, 303
Ventromedial nucleus (VMN)
lordosis neural circuitry, 143
sex differences, 7, 12, 15
stress effects, 118
VMN, *see* Ventromedial nucleus
Voltage-sensitive calcium channel (VSCC), 11
Vomeronasal organ, *see* Pheromones
VSCC, *see* Voltage-sensitive calcium channel

Water availability, rodent reproduction effects, 222

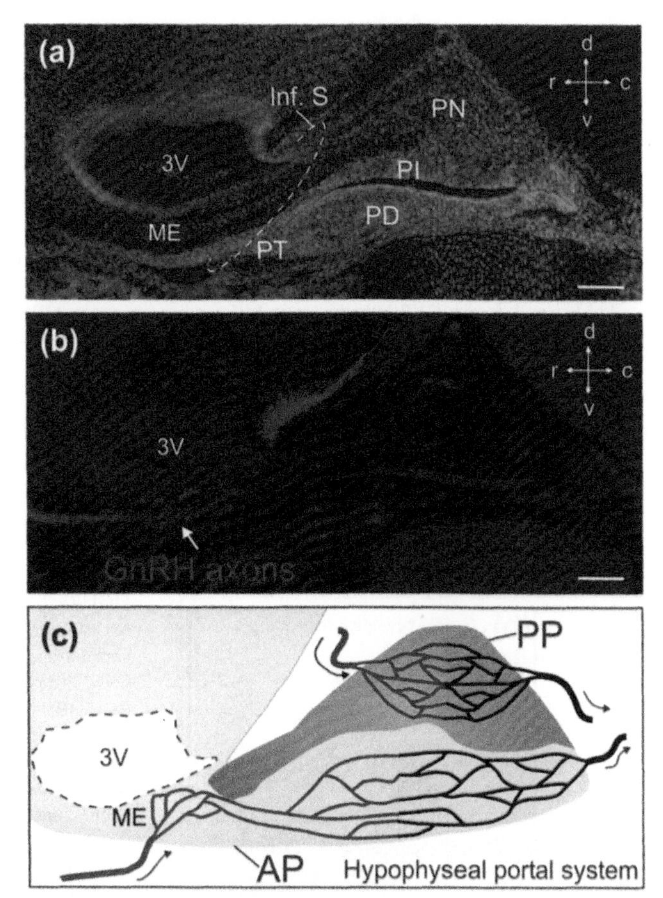

FIGURE 2.1 Parasagittal photomicrographs (a–b) and schematic drawing (c) of the postnatal day 0 mouse pituitary. (a) Anatomical organization of the pituitary as visualized by 4',6-diamidino-2-phenylindole (DAPI) nuclear stain (blue). (b) Photomicrograph depicting gonadotropin-releasing hormone (GnRH) immunofluorescent (red) axons terminating in the median eminence (arrow). (c) Schematic drawing of the hypophysial portal system in the pituitary. Note the capillary bed in the median eminence, in which GnRH peptide is released into the portal vein system in order to stimulate the release of gonadotropins from gonadotropes in the anterior pituitary (AP). Bar = 100 μm. 3V, third ventricle; c, caudal; d, dorsal; Inf. S, infundibular stalk; ME, median eminence; PD, pars distalis; PI, pars intermedialis; PN, pars nervosa; PP, posterior pituitary; PT, pars tuberalis; r, rostral; v, ventral.

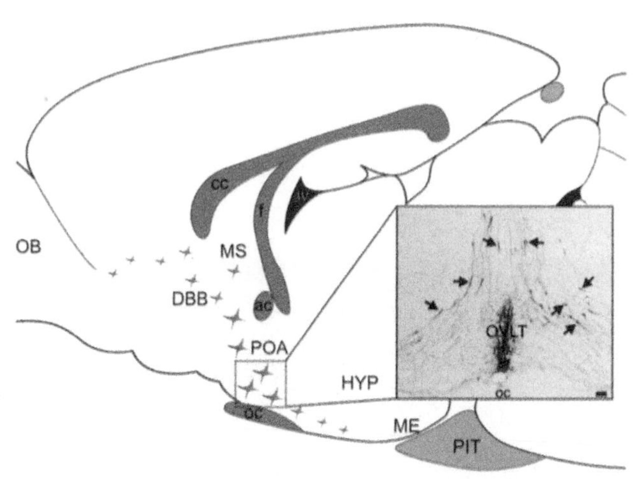

FIGURE 2.2 Schematic representation of the rostral-to-caudal distribution of gonadotropin-releasing hormone (GnRH) neurons (in red) in the postnatal rodent brain. Gonadotropin-releasing hormone neuron density is represented by the size of the red neurons. 3V, third ventricle; ac, anterior commissure; cc, corpus callosum; DBB, diagonal band of Broca; f, fornix; HYP, hypothalamus; ME, median eminence; MS, medial septum; OB, olfactory bulb; oc, optic chiasm; PIT, pituitary; POA, preoptic area. Inset is coronal photomicrograph of GnRH neurons (brown) at the level of the organum vasculosum lamina terminalis (OVLT), visualized using immunocytochemistry. Bar = 50 μm.

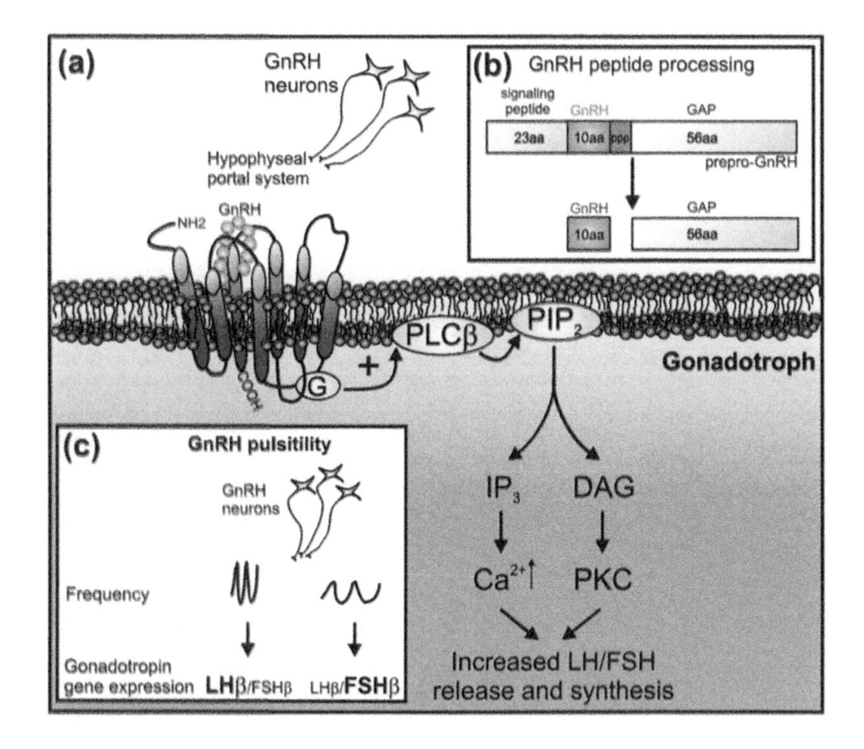

FIGURE 2.3 Schematic pathway of gonadotropin-releasing hormone (GnRH)-dependent activation of gonadotropes. (a) Upon release in the hypophysial portal system, the GnRH peptide (green) will bind the GnRH receptor (red), which is a G-protein-coupled receptor (GPCR), with seven trans-membrane alpha-helices that are connected by extracellular and intracellular loops. The activation of the GnRH receptors is conveyed by G-proteins, which signal to phospholipase C (PLCβ) to hydrolyze phosphatidylinositol 4,5 biphosphate (PIP_2) into inositol 1,4,5-triphosphate (IP_3) and diacylglycerols (DAG). This sequence of events results in the mobilization of intracellular calcium by IP_3 and the activation of PKC by DAG. (b) Schematic representation of GnRH peptide processing from the GnRH prepropeptide. aa, amino acid; GAP, GnRH-associated protein; ppp, proteolytic processing peptide. (c) Schematic representation of how GnRH pulse frequency results in the preferential expression of LHβ or FSHβ transcription. Indeed, high GnRH pulse frequency (once every 30–60 minutes) favors LHβ, whereas slow GnRH pulse frequency (once every two to three hours) results in the preferential transcription of FSHβ.

TABLE 2.1 Amino acid sequence of GnRH peptides identified to date

DEUTEROSTOMIA	1 2 3 4 5 6 7 8 9 10
Mammal	pGlu-His-Trp-Ser-Tyr-Gly-Leu-Arg-Pro-Gly-NH2
Guinea pig	pGlu-His-Trp-Ser-Tyr-Gly-Val-Arg-Pro-Gly-NH2
Chicken-I	pGlu-His-Trp-Ser-Tyr-Gly-Leu-Gln-Pro-Gly-NH2
Chicken-II	pGlu-His-Trp-Ser-His-Gly-Trp-Tyr-Pro-Gly-NH2
Salmon	pGlu-His-Trp-Ser-Tyr-Gly-Trp-Leu-Pro-Gly-NH2
Dogfish	pGlu-His-Trp-Ser-His-Gly-Trp-Leu-Pro-Gly-NH2
Catfish	pGlu-His-Trp-Ser-His-Gly-Leu-Asn-Pro-Gly-NH2
Herring	pGlu-His-Trp-Ser-His-Gly-Leu-Ser-Pro-Gly-NH2
Medaka	pGlu-His-Trp-Ser-Phe-Gly-Leu-Ser-Pro-Gly-NH2
Lamprey-I	pGlu-His-Tyr-Ser-Leu-Glu-Trp-Lys-Pro-Gly-NH2
Lamprey-III	pGlu-His-Trp-Ser-His-Asp-Trp-Lys-Pro-Gly-NH2
Rana	pGlu-His-Trp-Ser-Tyr-Gly-Leu-Trp-Pro-Gly-NH2
Sea Bream	pGlu-His-Trp-Ser-Tyr-Gly-Leu-Ser-Pro-Gly-NH2
Whitefish	pGlu-His-Trp-Ser-Tyr-Gly-Met-Asn-Pro-Gly-NH2
Tunicate-I	pGlu-His-Trp-Ser-Asp-Tyr-Phe-Lys-Pro-Gly-NH2
Tunicate-II	pGlu-His-Trp-Ser-Leu-Cys-His-Ala-Pro-Gly-NH2
Tunicate-III	pGlu-His-Trp-Ser-Tyr-Glu-Phe-Met-Pro-Gly-NH2
Tunicate-IV	pGlu-His-Trp-Ser-Asn-Gln-Leu-Thr-Pro-Gly-NH2
Tunicate-V	pGlu-His-Trp-Ser-Tyr-Glu-Tyr-Met-Pro-Gly-NH2
Tunicate-VI	pGlu-His-Trp-Ser-Lys-Gly-Tyr-Ser-Pro-Gly-NH2
Tunicate-VII	pGlu-His-Trp-Ser-Tyr-Ala-Leu-Ser-Pro-Gly-NH2
Tunicate-VIII	pGlu-His-Trp-Ser-Leu-Ala-Leu-Ser-Pro-Gly-NH2
Tunicate-IX	pGlu-His-Trp-Ser-Asn-Lys-Leu-Ala-Pro-Gly-NH2
PROTOSTOMIA	**1 2 3 4 5 6 7 8 9 10 11 12**
Octopus	pGln-Asn-Tyr-His-Phe-Ser-Asn-Gly-Trp-His-Pro-Gly-NH2
*Aplysia	pGln-Asn-Tyr-His-Phe-Ser-Asn-Gly-Trp-Tyr-Ala-NH2
*Owl Limpet	pGln-His-Tyr-His-Phe-Ser-Asn-Gly-Trp-Lys-Ser-NH2
*Annelid	pGln-Asn-Tyr-His-Phe-Ser-His-Gly-Trp-Phe-Pro-NH2

* Sequence has not yet verified by mass spectrometry.
Conserved amino acid positions are shown in red.

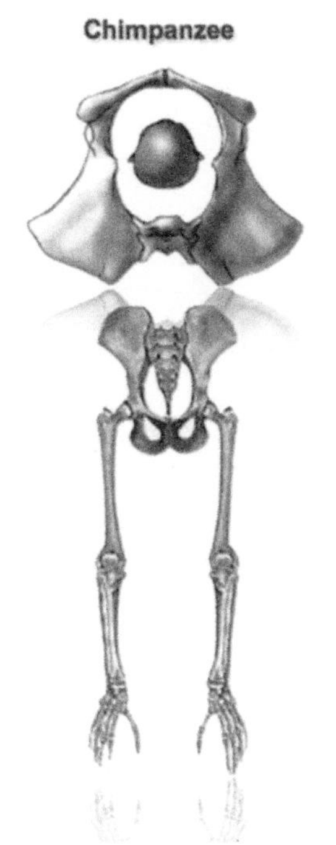

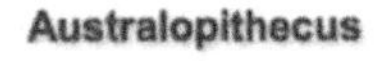

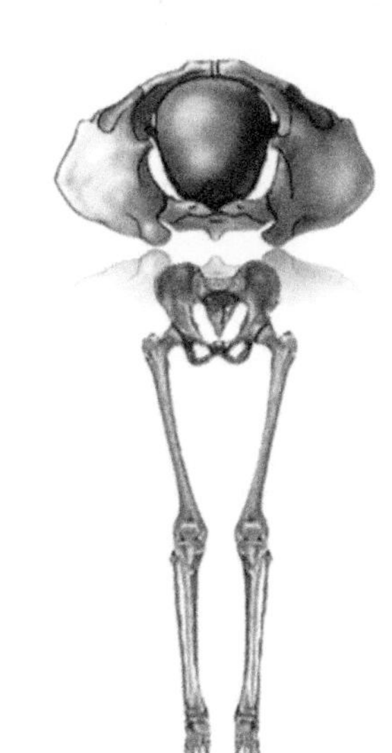

FIGURE 6.1 In the chimpanzee, as in all extant primates except *Homo sapiens*, the fetal head passes through the maternal pelvis with the face in an anterior position facing the pubic symphysis. Upright posture as present in Australopithecus was associated with a narrowing of the anterior to posterior diameter of the pelvis, forcing the fetus at delivery to rotate to traverse the birth canal. Further changes to the pelvis in *Homo sapiens* require a series of rotations to enable the large head and wide shoulders of the human fetus to negotiate the birth passage, placing a severe constraint on the size of the fetus and especially the fetal brain at the time of birth.

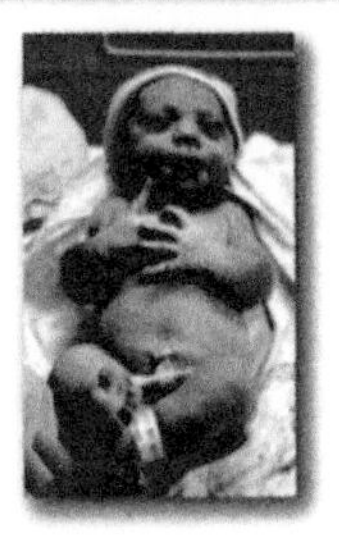

FIGURE 6.2 Various mammals adopt different strategies regarding pre- and postnatal development. Precocial species have relatively long gestational periods and deliver relatively mature, independent young; altricial species have shorter gestational periods and deliver immature, dependent young. Primates in general follow the precocial pattern, giving birth to relatively independent young after a long gestational period; however, humans have a complex pattern consisting of a long gestational period with delivery of an immature offspring that is highly dependent for an extended period of time. The human pattern seems to be a consequence of changes required for upright posture and possession of a large brain. The human brain is relatively large compared to total body size at birth and continues to grow rapidly after birth. The relatively narrow pelvis of primates has become narrowed with the adoption of upright posture (see Figure 6.1) in Australopithecus and modern humans, providing a physical limit on the size of the fetal brain that can traverse the birth canal.

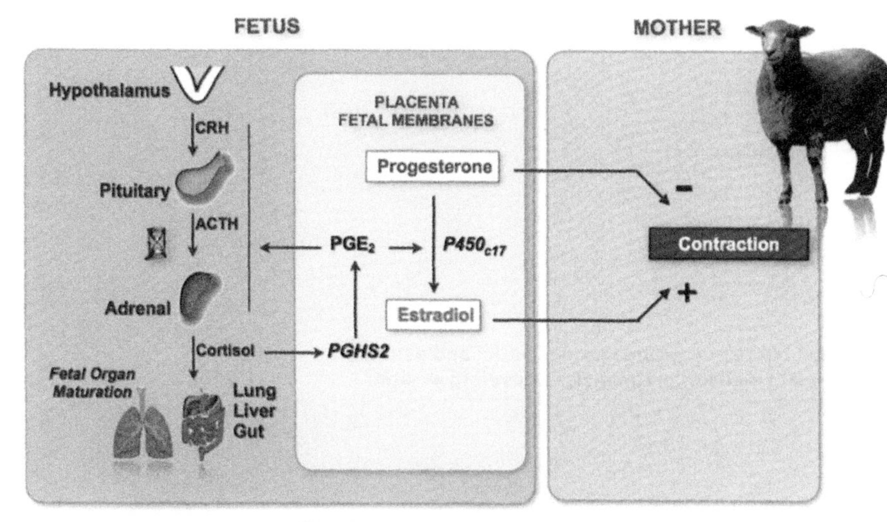

FIGURE 6.3 In the sheep, maturation of the fetal hypothalamus appears to regulate the timing of birth. ACTH, corticotropin; CRH, corticotropin-releasing hormone; $P450_{C17}$, 17-hydroxylase 17,20 lyase; PGE_2, prostaglandin E_2; PGHS2, prostaglandin synthase-2.

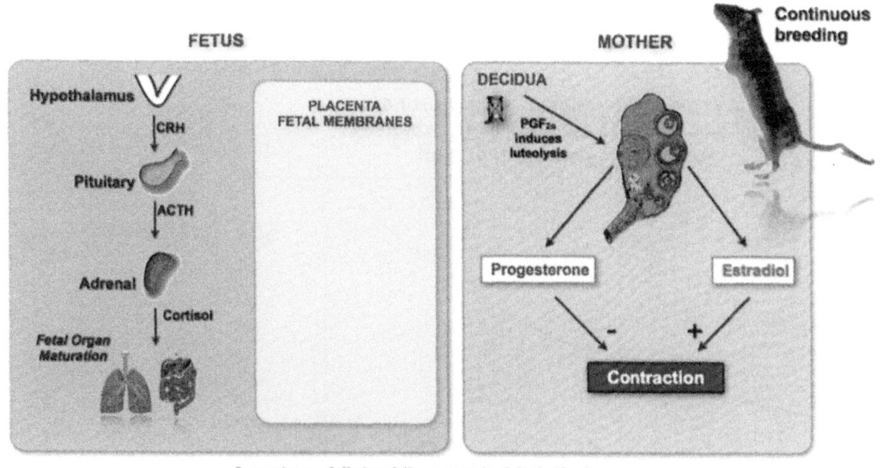

FIGURE 6.4 In rodents, the timing of birth appears to be controlled by events in the maternal tissues leading to luteolysis and a fall in progesterone (P_4) levels. ACTH, corticotropin; CRH, corticotropin-releasing hormone; PGF_2, prostaglandin F_2.

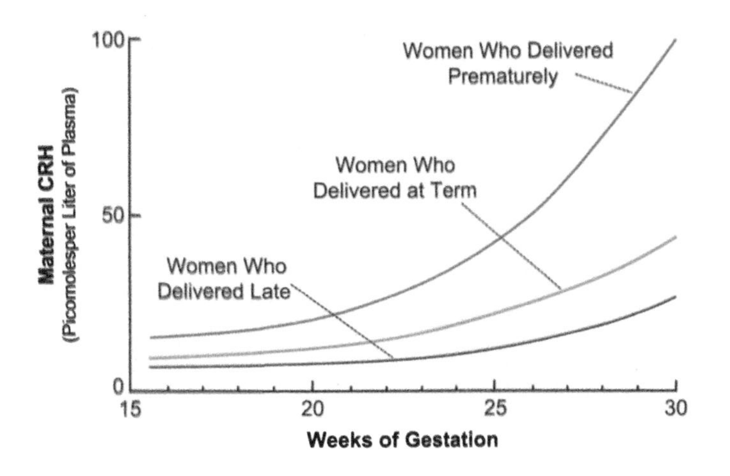

FIGURE 6.5 Placentally produced corticotropin-releasing hormone (CRH) can be measured in maternal plasma in humans. The concentration of CRH in maternal plasma follows an exponential curve that peaks at the time of delivery. In women who deliver prematurely (less than 37 completed weeks of gestation), the rate of rise of the exponential curve is greater than that observed in those that deliver at term (37–40 weeks), which is greater than that of those who deliver late (more than 40 weeks).

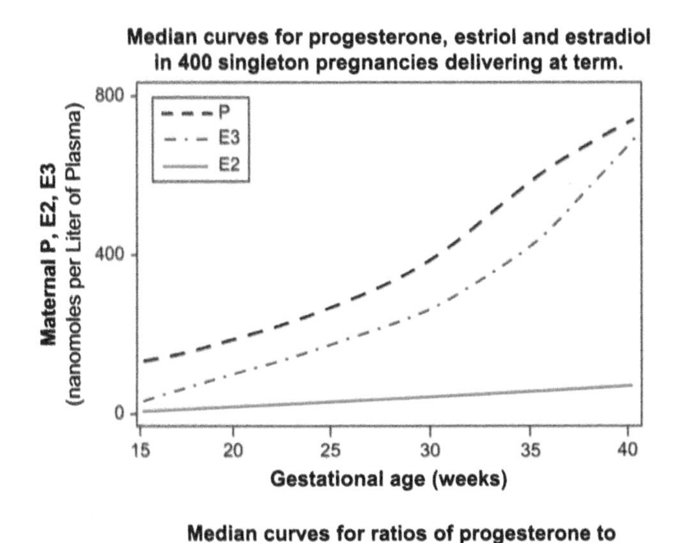

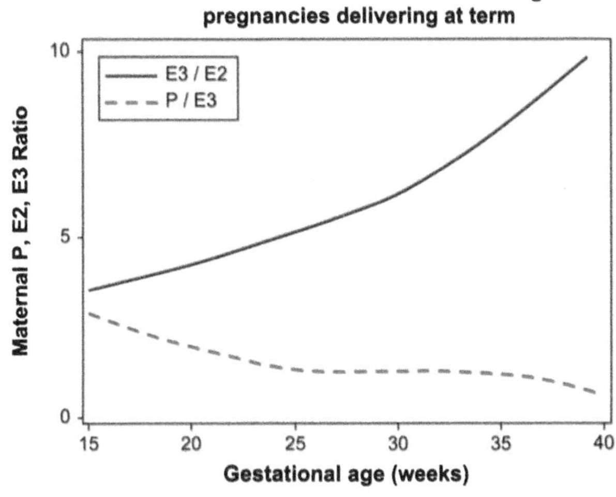

FIGURE 6.6 In human pregnancy, estriol (E_3) derived from fetal adrenal dehydroepiandrosterone sulfate (DHEA-S) increases as pregnancy advances under the drive from the exponential increase in placental corticotropin-releasing hormone (CRH) production. This leads to an increase in the ratio of E_3 to 17β-estradiol (E_2) (which is derived from maternal sources). The increase in E_3 also increases the ratio of E_3 to progesterone (P_4) in maternal plasma. Together, the changes in these ratios increase the biologically effective estrogen action at estrogen receptors and change the balance of estrogenic to progestational actions to favor estrogen and promote the onset of labor.

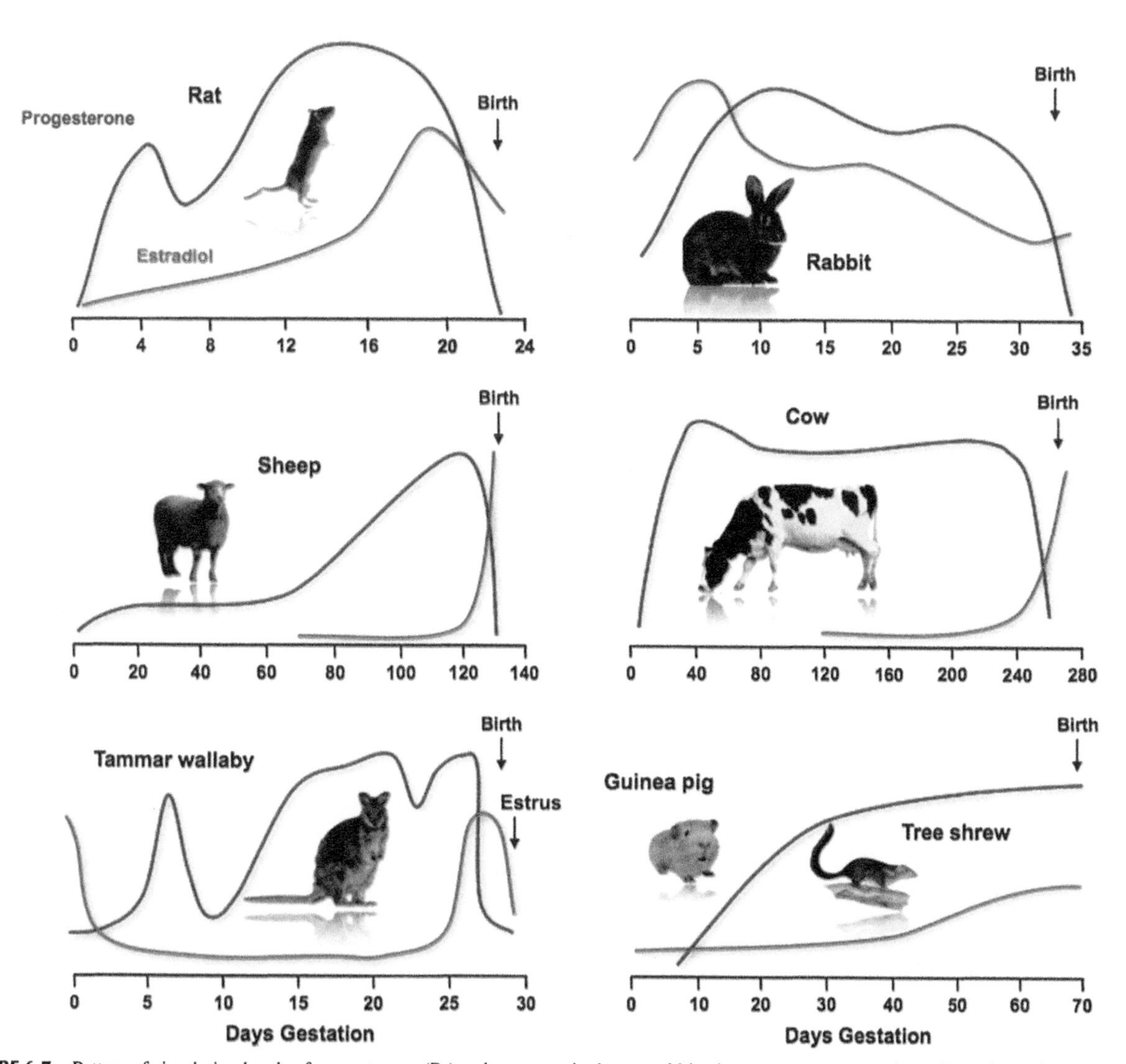

FIGURE 6.7 Pattern of circulating levels of progesterone (P_4) and estrogens in the rat, rabbit, sheep, cow, tammar, guinea pig, and tree shrew throughout gestation.

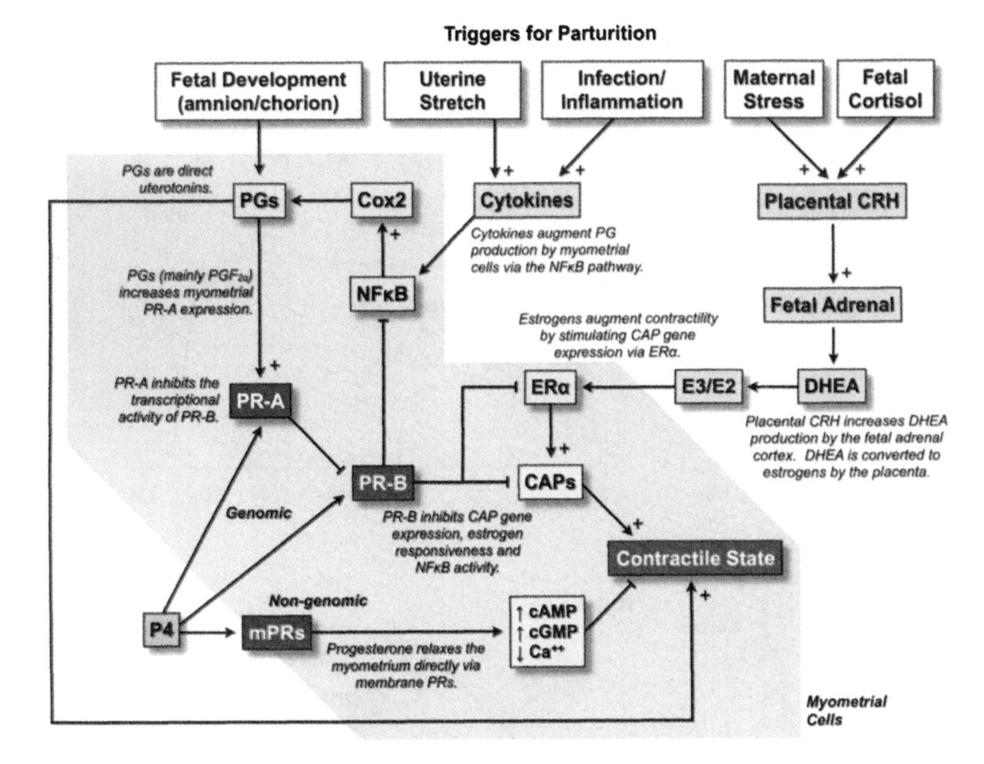

FIGURE 6.8 The onset of labor in humans. Placental production of corticotropin-releasing hormone (CRH) leads to a change in the balance of estriol (E_3) to 17β-estradiol (E_2) and of E_3 to progesterone (P_4). Contraction-associated proteins (CAPs) include connexin 43. Ca^{++}; cAMP, cyclic-3',5'-adenosine monophosphate; cGMP; CAP, contraction-associated protein; ER, estrogen receptor; IκBα; NFκB; PG, prostaglandin; PGRMC; PR, progesterone receptor.

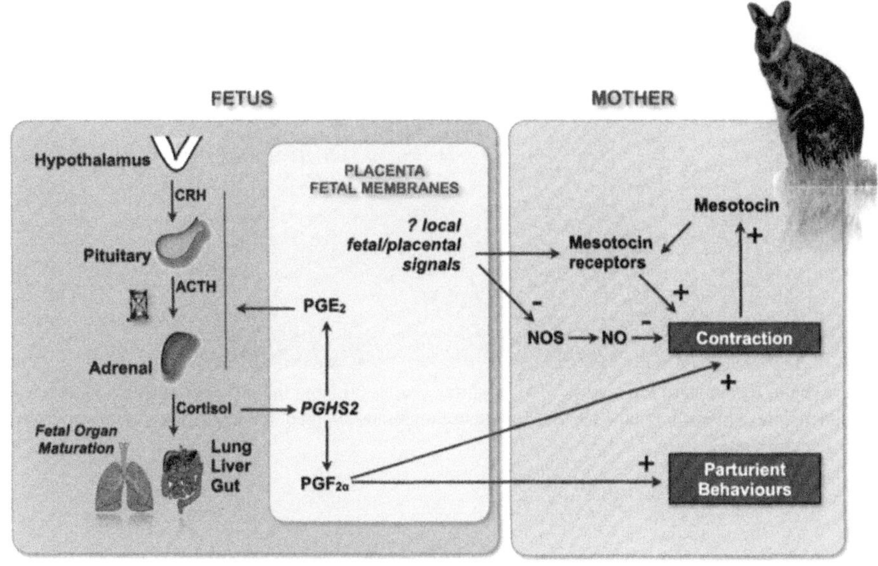

FIGURE 6.9 In the Tammar fetal production of cortisol is a dominant signal to the onset of parturition. Cortisol promotes placental transcription of the PGHS2 gene leading to increased prostaglandin production. Both PGE_2 and $PGF_{2\alpha}$ are produced within the placenta. PGE_2 feeds back to promote ACTH release from the fetal pituitary promoting the rapid rise in fetal cortisol that precedes parturition. $PGF_{2\alpha}$ is released from the placenta into the maternal circulation where it promotes parturient behaviours and myometrical contraction. Other placental factors may stimulate nitric oxide synthase (NOS) which promote myometrial relaxation during pregnancy. Mesotocin from the mother also plays a role in promoting myometrial contraction. The increase in mesotocin and $PGF_{2\alpha}$ prior to delivery occurs as a brief peak (approximately one hour).

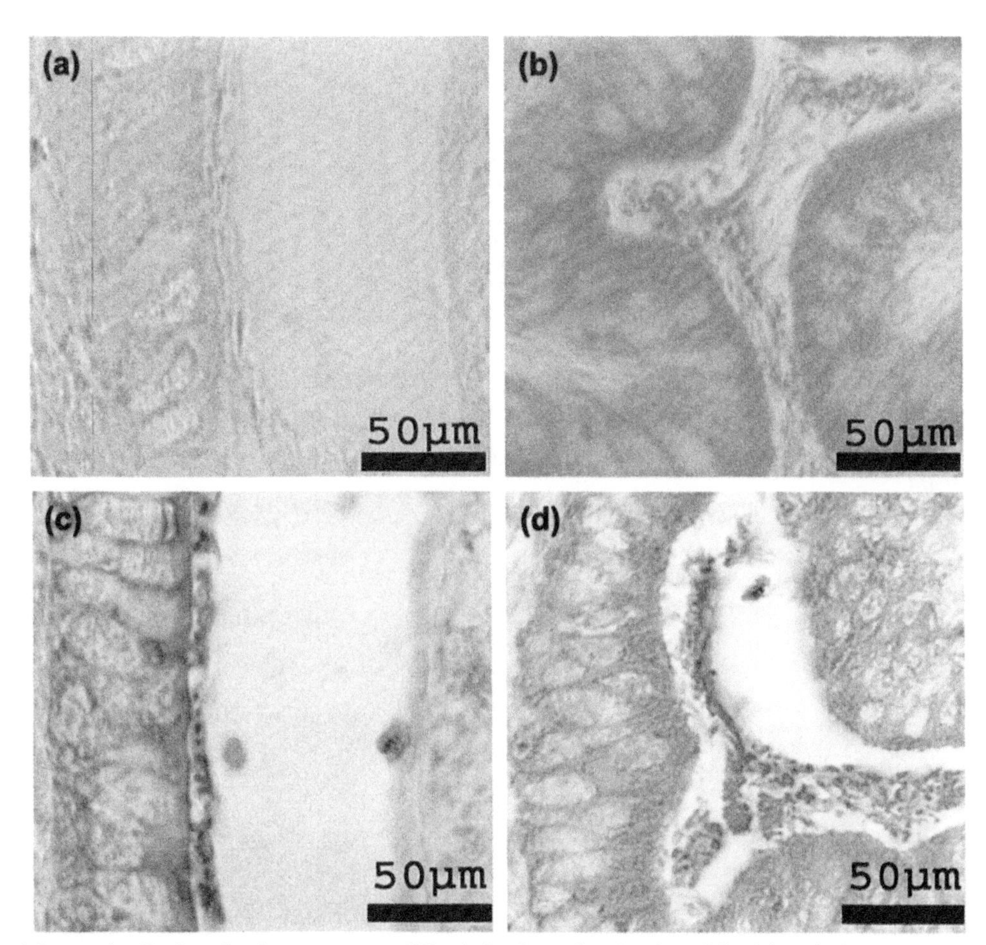

FIGURE 12.3 (a) Immunolocalization of androgen receptors (ARs) during December showing mild staining. (b) Immunolocalization of ARs during January showing strong immunostaining. (c) Immunolocalization of androgen-binding protein (ABP) during December showing mild staining in the cytoplasm of the luminal epithelial cells. (d) ABP in the epithelial lining of the uterotubal junction of *Scotophilus heathi* during the period of sperm storage.

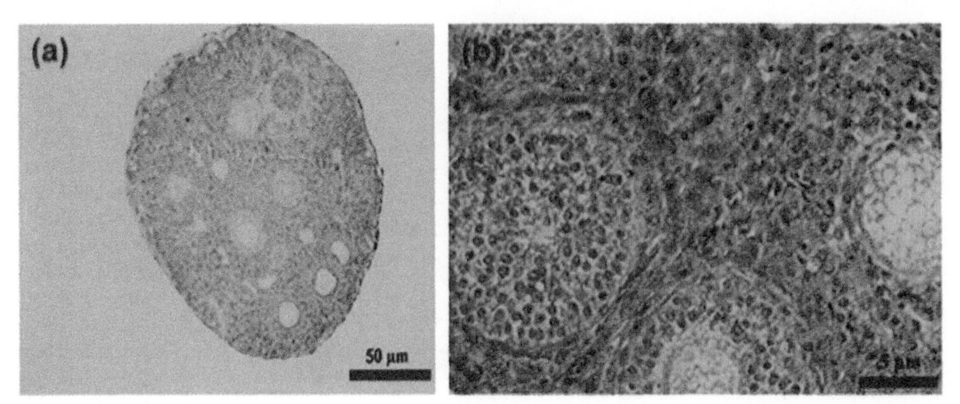

FIGURE 12.8 Immunocytochemical localization of leptin receptors in the ovaries of *Scotophilus heathi*. (a) Note the presence of intense immunostaining in the interfollicular area of the ovary. (b) Higher view demonstrating the presence of intense immunostaining of the leptin receptor, mainly in the thecal interstitial cells. Granulosa cells showed no immunostaining.

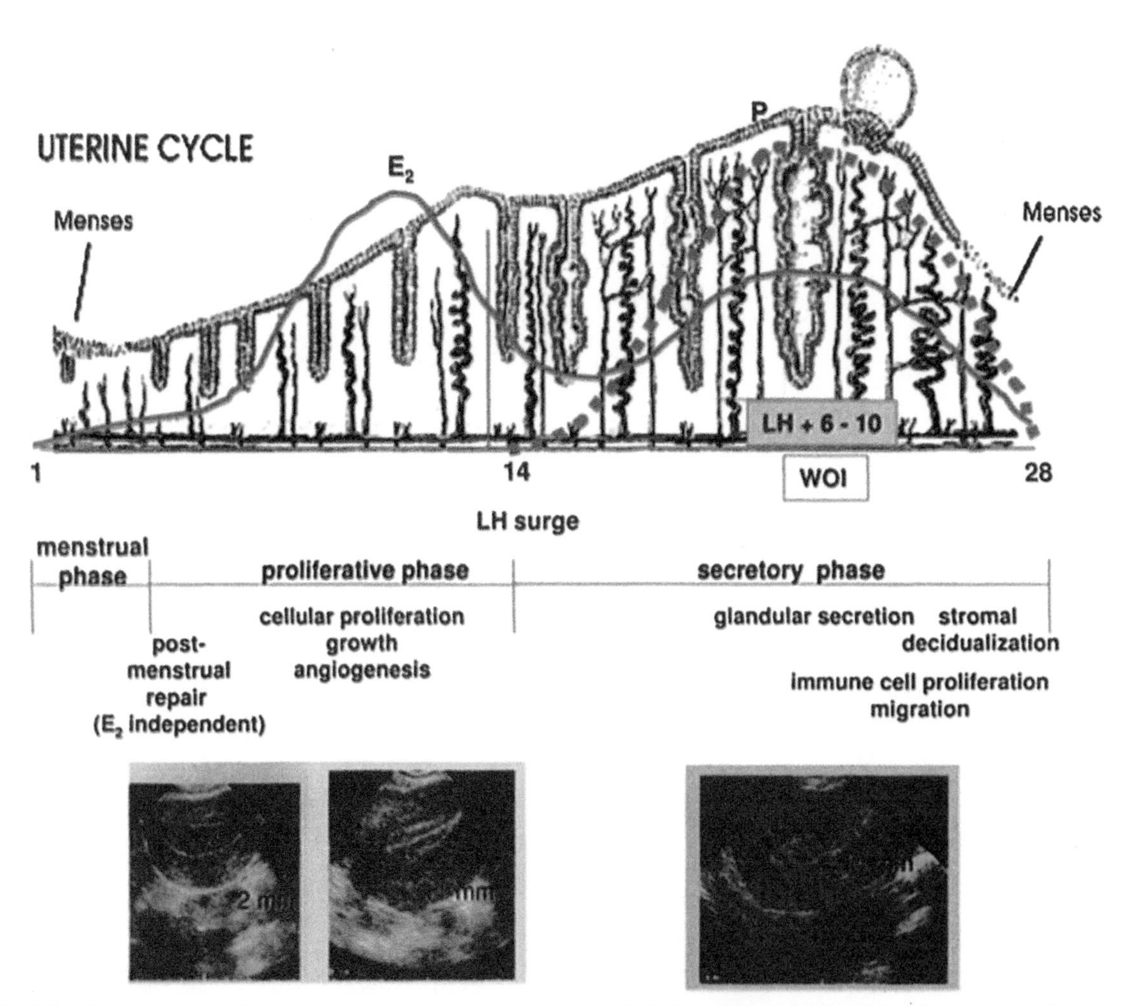

FIGURE 13.4 The menstrual cycle and its endocrine control in catarrhines, including cyclical changes in thickness, vascularization, glandular development, and secretory activity of the endometrium; phases of the uterine cycle and their major cellular events; circulating concentrations of estradiol (E_2) (solid line) and progesterone (P) (dotted line); window of implantation (WOI); and sonograms of the endometrium during the early proliferative, late proliferative, and secretory phases. *Reproduced from Giudice (2006) with permission; originally published by BioMed Central.*

Printed and bound by CPI Group (UK) Ltd, Croydon, CR0 4YY
08/05/2025
01864916-0001